STUDENT SOLUTIONS MANUAL

Applied
MATHEMATICS

for Business, Economics, Life Sciences, and Social Sciences

Sixth Edition

Raymond A. Barnett   Michael R. Ziegler

PRENTICE HALL Upper Saddle River, NJ 07458

Supplement Editor: *Audra Walsh*
Production Editor: *Kimberly Dellas*
Special Projects Manager: *Barbara A. Murray*
Production Coordinator: *Alan Fischer*
Supplement Cover Manager: *Paul Gourhan*

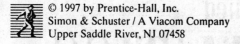
© 1997 by Prentice-Hall, Inc.
Simon & Schuster / A Viacom Company
Upper Saddle River, NJ 07458

Printed in the United States of America

10 9 8 7 6 5 4 3 2 1

ISBN 0-13-576349-5

Prentice-Hall International (UK) Limited, *London*
Prentice-Hall of Australia Pty. Limited, *Sydney*
Prentice-Hall Canada, Inc., *Toronto*
Prentice-Hall Hispanoamericana, S.A., *Mexico*
Prentice-Hall of India Private Limited, *New Delhi*
Prentice-Hall of Japan, Inc., *Tokyo*
Simon & Schuster Asia Pte. Ltd., *Singapore*
Editora Prentice-Hall do Brasil, Ltda., *Rio de Janeiro*

CONTENTS

1 A BEGINNING LIBRARY OF ELEMENTARY FUNCTIONS

EXERCISE 1-1

Things to remember:

1. A FUNCTION is a rule (process or method) that produces a correspondence between one set of elements, called a DOMAIN, and a second set of elements, called the RANGE, such that to each element in the domain there corresponds one and only one element in the range.

2. EQUATIONS AND FUNCTIONS:

   Given an equation in two variables. If there corresponds exactly one value of the dependent variable (output) to each value of the independent variable (input), then the equation defines a function. If there is more than one output for at least one input, then the equation does not define a function.

3. VERTICAL LINE TEST FOR A FUNCTION

   An equation defines a function if each vertical line in the coordinate system passes through at most one point on the graph of the equation. If any vertical line passes through two or more points on the graph of an equation, then the equation does not define a function.

4. AGREEMENT ON DOMAINS AND RANGES

   If a function is specified by an equation and the domain is not given explicitly, then assume that the domain is the set of all real number replacements of the independent variable (inputs) that produce real values for the dependent variable (outputs). The range is the set of all outputs corresponding to input values.

   In many applied problems, the domain is determined by practical considerations within the problem.

5. FUNCTION NOTATION—THE SYMBOL $f(x)$

   For any element x in the domain of the function f, the symbol $f(x)$ represents the element in the range of f corresponding to x in the domain of f. If x is an input value, then $f(x)$ is the corresponding output value. If x is an element which is not in the domain of f, then f is NOT DEFINED at x and $f(x)$ DOES NOT EXIST.

1. The table specifies a function, since for each domain value there corresponds one and only one range value.

3. The table does not specify a function, since more than one range value corresponds to a given domain value. (Range values 5, 6 correspond to domain value 3; range values 6, 7 correspond to domain value 4.)

5. This is a function.

7. The graph specifies a function; each vertical line in the plane intersects the graph in at most one point.

9. The graph does not specify a function. There are vertical lines which intersect the graph in more than one point. For example, the y-axis intersects the graph in three points.

11. The graph specifies a function.

13. $f(x) = 3x - 2$
$f(2) = 3(2) - 2 = 4$

15. $f(-1) = 3(-1) - 2$
$= -5$

17. $g(x) = x - x^2$
$g(3) = 3 - 3^2 = -6$

19. $f(0) = 3(0) - 2$
$= -2$

21. $g(-3) = -3 - (-3)^2$
$= -12$

23. $f(1) + g(2)$
$= [3(1) - 2] + (2 - 2^2) = -1$

25. $g(2) - f(2)$
$= (2 - 2^2) - [3(2) - 2]$
$= -2 - 4 = -6$

27. $g(3) \cdot f(0) = (3 - 3^2)[3(0) - 2]$
$= (-6)(-2)$
$= 12$

29. $\dfrac{g(-2)}{f(-2)} = \dfrac{-2 - (-2)^2}{3(-2) - 2} = \dfrac{-6}{-8} = \dfrac{3}{4}$

31. $y = f(-5) = 0$

33. $y = f(5) = 4$

35. $f(x) = 0$ at $x = -5, 0, 4$

37. $f(x) = -4$ at $x = 6$

39. domain: all real numbers or $(-\infty, \infty)$

41. domain: all real numbers except -4

43. $x^2 + 3x - 4 = (x + 4)(x - 1)$; domain: all real numbers except -4 and 1.

45. $x^2 + 6x + 9 = (x + 3)^2$; domain: all real numbers except -3.

47. $7 - x \geq 0$ for $x \leq 7$; domain: $x \leq 7$ or $(-\infty, 7]$

49. $7 - x > 0$ for $x < 7$; domain: $x < 7$ or $(-\infty, 7)$

51. f is not defined at the values of x where $x^2 - 9 = 0$, that is, at 3 and -3; f is defined at $x = 2$, $f(2) = \dfrac{0}{-5} = 0$.

53. $g(x) = 2x^3 - 5$

55. $G(x) = 2\sqrt{x} - x^2$

57. Function f multiplies the domain element by 2 and subtracts 3 from the result.

59. Function F multiplies the cube of the domain element by 3 and subtracts twice the square root of the domain element from the result.

61. Given $4x - 5y = 20$. Solving for y, we have:
$$-5y = -4x + 20$$
$$y = \frac{4}{5}x - 4$$

Since each input value x determines a unique output value y, the equation specifies a function. The domain is R, the set of real numbers.

63. Given $x^2 - y = 1$. Solving for y, we have:
$$-y = -x^2 + 1 \quad \text{or} \quad y = x^2 - 1$$
This equation specifies a function. The domain is R, the set of real numbers.

65. Given $x + y^2 = 10$. Solving for y, we have:
$$y^2 = 10 - x$$
$$y = \pm\sqrt{10 - x}$$
This equation does not specify a function since each value of x, $x \leq 10$, determines two values of y. For example, corresponding to $x = 1$, we have $y = 3$ and $y = -3$; corresponding to $x = 6$, we have $y = 2$ and $y = -2$.

67. Given $xy - 4y = 1$. Solving for y, we have:
$$(x - 4)y = 1 \quad \text{or} \quad y = \frac{1}{x - 4}$$
This equation specifies a function. The domain is all real numbers except $x = 4$.

69. Given $x^2 + y^2 = 25$. Solving for y, we have:
$$y^2 = 25 - x^2 \quad \text{or} \quad y = \pm\sqrt{25 - x^2}$$
Thus, the equation does not specify a function since, for $x = 0$, we have $y = \pm 5$, when $x = 4$, $y = \pm 3$, and so on.

71. Given $F(t) = 4t + 7$. Then:
$$\frac{F(3 + h) - F(3)}{h} = \frac{4(3 + h) + 7 - (4 \cdot 3 + 7)}{h}$$
$$= \frac{12 + 4h + 7 - 19}{h} = \frac{4h}{h} = 4$$

73. Given $g(w) = w^2 - 4$. Then:
$$\frac{g(1 + h) - g(1)}{h} = \frac{(1 + h)^2 - 4 - (1^2 - 4)}{h} = \frac{1 + 2h + h^2 - 4 + 3}{h}$$
$$= \frac{2h + h^2}{h} = \frac{h(2 + h)}{h} = 2 + h$$

75. Given $Q(x) = x^2 - 5x + 1$. Then:

$$\frac{Q(2 + h) - Q(2)}{h} = \frac{(2 + h)^2 - 5(2 + h) + 1 - (2^2 - 5 \cdot 2 + 1)}{h}$$

$$= \frac{4 + 4h + h^2 - 10 - 5h + 1 - (-5)}{h} = \frac{h^2 - h - 5 + 5}{h}$$

$$= \frac{h(h - 1)}{h} = h - 1$$

77. Given $f(x) = 4x - 3$. Then:

$$\frac{f(a + h) - f(a)}{h} = \frac{4(a + h) - 3 - (4a - 3)}{h}$$

$$= \frac{4a + 4h - 3 - 4a + 3}{h} = \frac{4h}{h} = 4$$

79. Given $f(x) = 4x^2 - 7x + 6$. Then:

$$\frac{f(a + h) - f(a)}{h} = \frac{4(a + h)^2 - 7(a + h) + 6 - (4a^2 - 7a - 6)}{h}$$

$$= \frac{4(a^2 + 2ah + h^2) - 7a - 7h + 6 - 4a^2 + 7a - 6}{h}$$

$$= \frac{4a^2 + 8ah + 4h^2 - 7h - 4a^2}{h} = \frac{8ah + 4h^2 - 7h}{h}$$

$$= \frac{h(8a + 4h - 7)}{h} = 8a + 4h - 7$$

81. Given $f(x) = x^3$. Then:

$$\frac{f(a + h) - f(a)}{h} = \frac{(a + h)^3 - a^3}{h} = \frac{a^3 + 3a^2h + 3ah^2 + h^3 - a^3}{h}$$

$$= \frac{h(3a^2 + 3ah + h^2)}{h} = 3a^2 + 3ah + h^2$$

83. Given $f(x) = \sqrt{x}$. Then:

$$\frac{f(a + h) - f(a)}{h} = \frac{\sqrt{a + h} - \sqrt{a}}{h}$$

$$= \frac{\sqrt{a + h} - \sqrt{a}}{h} \cdot \frac{\sqrt{a + h} + \sqrt{a}}{\sqrt{a + h} + \sqrt{a}} \quad \text{(rationalizing the numerator)}$$

$$= \frac{a + h - a}{h(\sqrt{a + h} + \sqrt{a})} = \frac{h}{h(\sqrt{a + h} + \sqrt{a})} = \frac{1}{\sqrt{a + h} + \sqrt{a}}$$

85. Given $A = \ell w = 25$.

Thus, $\ell = \dfrac{25}{w}$. Now $P = 2\ell + 2w$

$$= 2\left(\frac{25}{w}\right) + 2w = \frac{50}{w} + 2w.$$

The domain is $w > 0$.

87. Given $P = 2\ell w + 2w = 100$ or $\ell + w = 50$ and $w = 50 - \ell$.

Now $A = \ell w = \ell(50 - \ell)$ and $A = 50\ell - \ell^2$.

The domain is $0 \le \ell \le 50$. [<u>Note</u>: $\ell \le 50$ since $\ell > 50$ implies $w < 0$.]

89.

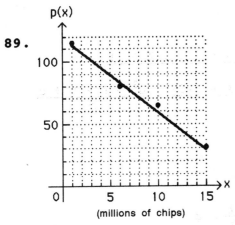

p(x)

(millions of chips)

$p(8) = 71$ dollars per chip
$p(11) = 53$ dollars per chip

91. (A) $R(x) = xp(x) = x(119 - 6x)$
Domain: $1 \le x \le 15$

(C)

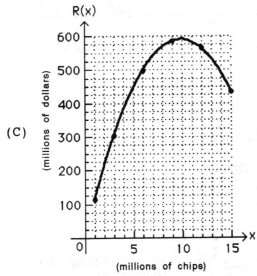

R(x)

(millions of dollars)

(millions of chips)

(B) Table 10 Revenue

x(millions)	R(x)(millions)
1	$113
3	303
6	498
9	585
12	564
15	435

93. (A) $P(x) = R(x) - C(x)$
$= x(119 - 6x) - (234 + 23x)$
$= -6x^2 + 96x - 234$ million dollars
Domain: $1 \le x \le 15$

(B) Table 12 Profit

x(millions)	P(x)(millions)
1	-$144
3	0
6	126
9	144
12	54
15	-144

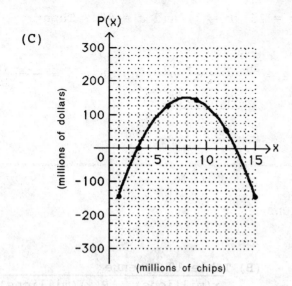

(C)

95.

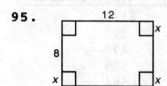

(A) $V = \text{(length)(width)(height)}$
$V(x) = (12 - 2x)(8 - 2x)x$
$\qquad = x(8 - 2x)(12 - 2x)$

(C) $V(1) = (12 - 2)(8 - 2)(1)$
$\qquad\quad = (10)(6)(1) = 60$
$V(2) = (12 - 4)(8 - 4)(2)$
$\qquad\quad = (8)(4)(2) = 64$
$V(3) = (12 - 6)(8 - 6)(3)$
$\qquad\quad = (6)(2)(3) = 36$

(B) Domain: $0 \le x \le 4$

Thus,

Volume

x	$V(x)$
1	60
2	64
3	36

(D)

V(x)

97. Given $(w + a)(v + b) = c$. Let $a = 15$, $b = 1$, and $c = 90$. Then:

$(w + 15)(v + 1) = 90$

Solving for v, we have

$v + 1 = \dfrac{90}{w + 15}$ and $v = \dfrac{90}{w + 15} - 1 = \dfrac{90 - (w + 15)}{w + 15}$, so that $v = \dfrac{75 - w}{w + 15}$.

If $w = 16$, then $v = \dfrac{75 - 16}{16 + 15} = \dfrac{59}{31} \approx 1.9032$ cm/sec.

EXERCISE 1-2

Things to remember:

<u>1.</u> LIBRARY OF ELEMENTARY FUNCTIONS

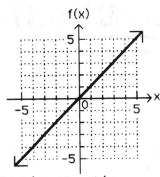

<u>Identity Function</u>

$f(x) = x$
Domain: All real numbers
Range: All real numbers
(a)

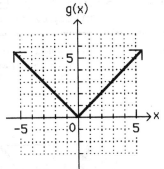

<u>Absolute Value Function</u>

$g(x) = |x|$
Domain: All real numbers
Range: $[0, \infty)$
(b)

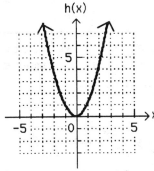

<u>Square Function</u>

$h(x) = x^2$
Domain: All real numbers
Range: $[0, \infty)$
(c)

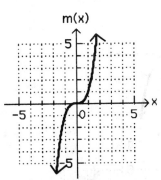

<u>Cube Function</u>

$m(x) = x^3$
Domain: All real numbers
Range: All real numbers
(d)

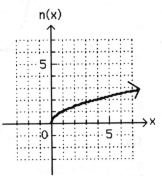

<u>Square-Root Function</u>

$n(x) = \sqrt{x}$
Domain: $[0, \infty)$
Range: $[0, \infty)$
(e)

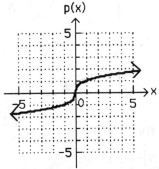

<u>Cube-Root Function</u>

$p(x) = \sqrt[3]{x}$
Domain: All real numbers
Range: All real numbers
(f)

NOTE: Letters used to designate the above functions may vary from context to context.

2. GRAPH TRANSFORMATIONS SUMMARY

Vertical Translation:

$y = f(x) + k$ $\begin{cases} k > 0 & \text{Shift graph of } y = f(x) \text{ up } k \text{ units} \\ k < 0 & \text{Shift graph of } y = f(x) \text{ down } |k| \text{ units} \end{cases}$

Horizontal Translation:

$y = f(x + h)$ $\begin{cases} h > 0 & \text{Shift graph of } y = f(x) \text{ left } h \text{ units} \\ h < 0 & \text{Shift graph of } y = f(x) \text{ right } |h| \text{ units} \end{cases}$

Reflection:

$y = -f(x)$ Reflect the graph of $y = f(x)$ in the x axis

Vertical Expansion and Contraction:

$y = Af(x)$ $\begin{cases} A > 1 & \text{Vertically expand graph of } y = f(x) \\ & \text{by multiplying each ordinate value by } A \\ \\ 0 < A < 1 & \text{Vertically contract graph of } y = f(x) \\ & \text{by multiplying each ordinate value by } A \end{cases}$

---

1. Domain: all real numbers;
Range: all real numbers

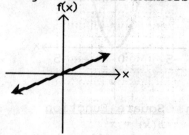

3. Domain: all real numbers;
Range: $(-\infty, 0]$

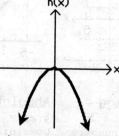

5. Domain: $[0, \infty)$;
Range: $(-\infty, 0]$

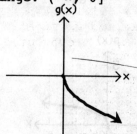

7. Domain: all real numbers;
Range: all real numbers

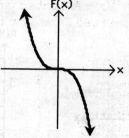

9.

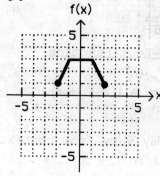

11.

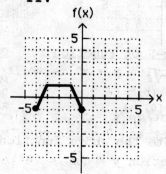

13.

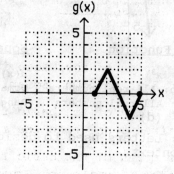

15.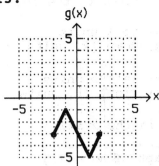

17.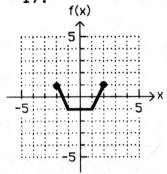

19.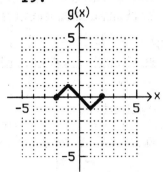

21. The graph of $g(x) = -|x + 3|$ is the graph of $y = |x|$ reflected in the x axis and shifted 3 units to the left.

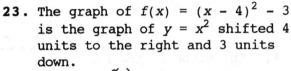

23. The graph of $f(x) = (x - 4)^2 - 3$ is the graph of $y = x^2$ shifted 4 units to the right and 3 units down.

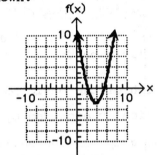

25. The graph of $f(x) = 7 - \sqrt{x}$ is the graph of $y = \sqrt{x}$ reflected in the x axis and shifted 7 units up.

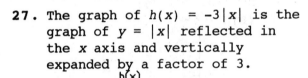

27. The graph of $h(x) = -3|x|$ is the graph of $y = |x|$ reflected in the x axis and vertically expanded by a factor of 3.

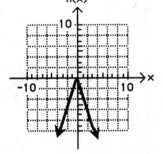

29. The graph of the basic function $y = x^2$ is shifted 2 units to the left and 3 units down. Equation: $y = (x + 2)^2 - 3$.

31. The graph of the basic function $y = x^2$ is reflected in the x axis, shifted 3 units to the right and 2 units up. Equation: $y = 2 - (x - 3)^2$.

33. The graph of the basic function $y = \sqrt{x}$ is reflected in the x axis and shifted 4 units up. Equation: $y = 4 - \sqrt{x}$.

35. The graph of the basic function $y = x^3$ is shifted 2 units to the left and 1 unit down. Equation: $y = (x + 2)^3 - 1$.

37. $g(x) = \sqrt{x - 2} - 3$ **39.** $g(x) = -|x + 3|$ **41.** $g(x) = -(x - 2)^3 - 1$

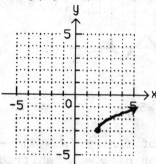 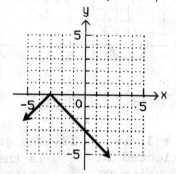 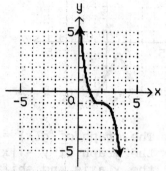

43. The graph of the basic function: $y = |x|$ is reflected in the x axis and has a vertical contraction by the factor 0.5. Equation: $y = -0.5|x|$.

45. The graph of the basic function $y = x^2$ is reflected in the x axis and is vertically expanded by the factor 2. Equation: $y = -2x^2$.

47. The graph of the basic function $y = \sqrt[3]{x}$ is reflected in the x axis and is vertically expanded by the factor 3. Equation: $y = -3\sqrt[3]{x}$.

49. (A) The graph of the basic function $y = \sqrt{x}$ is reflected in the x axis, vertically expanded by a factor of 4, and shifted up 115 units.

     (B)

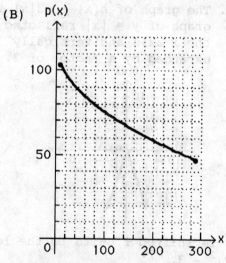

51. (A) The graph of the basic function $y = x^3$ is vertically contracted by a factor of 0.00048 and shifted right 500 units and up 60,000 units.

(B)
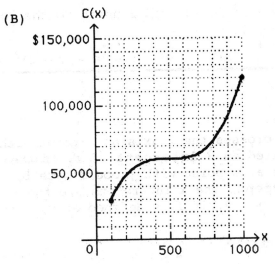

53. (A) The graph of the basic function $y = x$ is vertically expanded by a factor of 5.5 and shifted down 220 units.

(B)
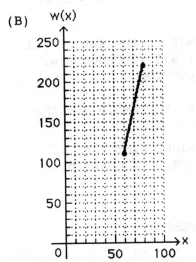

55. (A) The graph of the basic function $y = \sqrt{x}$ is vertically expanded by a factor of 7.08.

(B)
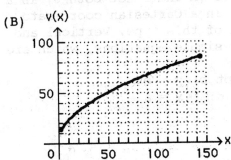

Things to remember:

<u>1</u>. INTERCEPTS

If the graph of a function *f* crosses the *x* axis at a point with *x* coordinate *a*, then *a* is called an **x intercept** of *f*. If the graph of *f* crosses the *y* axis at a point with *y* coordinate *b*, then *b* is called the **y intercept**. The *x* intercepts are the real solutions or roots of *f*(*x*) = 0; if *f* is defined at 0, then *f*(0) is the *y* intercept.

<u>2</u>. LINEAR AND CONSTANT FUNCTIONS

A function *f* is a LINEAR FUNCTION if

$$f(x) = mx + b \qquad m \neq 0$$

where *m* and *b* are real numbers. The DOMAIN is the set of all real numbers and the RANGE is the set of all real numbers. If *m* = 0, then *f* is called a CONSTANT FUNCTION

$$f(x) = b$$

which has the set of all real numbers as its DOMAIN and the constant *b* as its RANGE.

THE GRAPH OF A LINEAR FUNCTION IS A STRAIGHT LINE THAT IS NEITHER HORIZONTAL NOR VERTICAL. THE GRAPH OF A CONSTANT FUNCTION IS A HORIZONTAL STRAIGHT LINE.

<u>3</u>. GRAPH OF A LINEAR EQUATION IN TWO VARIABLES

The graph of any equation of the form

$$AX + By = C \qquad \text{Standard Form} \tag{5}$$

where *A*, *B*, and *C* are real constants (*A* and *B* not both 0) is a straight line. Every straight line in a Cartesian coordinate system is the graph of an equation of this type. Vertical and horizontal lines have particularly simple equations, which are special cases of equation (5):

Horizontal line with *y* intercept *b*: *y* = *b*
Vertical line with *x* intercept *a*:   *x* = *a*

4. SLOPE OF A LINE

If a line passes through two distinct points $P_1(x_1, y_1)$ and $P_2(x_2, y_2)$, then its slope is given by the formula

$$m = \frac{y_2 - y_1}{x_2 - x_1} \qquad x_1 \neq x_2$$

$$= \frac{\text{Vertical change (rise)}}{\text{Horizontal change (run)}}$$

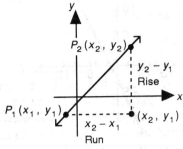

GEOMETRIC INTERPRETATION OF SLOPE

Line	Slope	Example
Rising as x moves from left to right	Positive	
Falling as x moves from left to right	Negative	
Horizontal	0	
Vertical	Not defined	

5. The equation

$$y = mx + b \qquad m = \text{slope, } b = y \text{ intercept}$$

is called the SLOPE-INTERCEPT FORM of an equation of a line.

6. An equation of the line with slope m that passes through (x_1, y_1) is:

$$y - y_1 = m(x - x_1)$$

This equation is called the POINT-SLOPE FORM of an equation of a line.

1. (d)

3. (c); The slope is 0.

5. $y = 2x - 3$

x	y
0	−3
1	−1
4	5

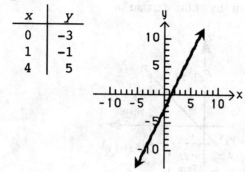

7. $2x + 3y = 12$

x	y
0	4
6	0
9	−2

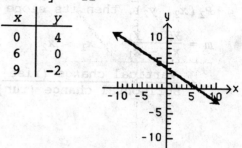

9. Slope $m = 2$

y intercept $b = -3$

11. Slope $m = -\frac{2}{3}$

y intercept $b = 2$

13. $m = -2$

$b = 4$

Using $\underline{5}$, $y = -2x + 4$.

15. $m = -\frac{3}{5}$

$b = 3$

Using $\underline{5}$, $y = -\frac{3}{5}x + 3$.

17. $y = -\frac{2}{3}x - 2$

$m = -\frac{2}{3}$, $b = -2$

x	y
0	−2
3	−4
−3	0

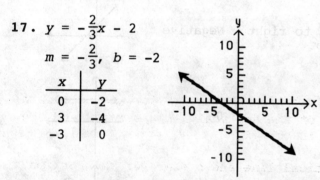

19. $3x - 2y = 10$

x	y
0	−5
10	10
−4	−11

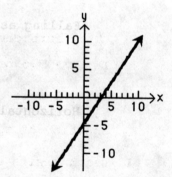

21.

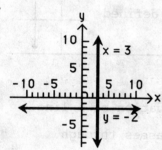

23. $3x + y = 5$

$y = -3x + 5$

$m = -3$ (using $\underline{5}$)

25. $2x + 3y = 12$

$3y = -2x + 12$

Divide both sides by 3:

$y = -\frac{2}{3}x + \frac{12}{3} = -\frac{2}{3}x + 4$

$m = -\frac{2}{3}$ (using $\underline{5}$)

27. (A)

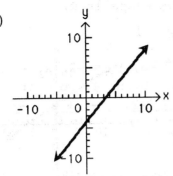

(B) x intercept--set $f(x) = 0$: $1.2x - 4.2 = 0$

$x = 3.5$

y intercept--set $x = 0$: $y = -4.2$

(C)

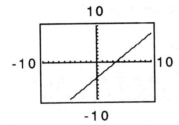

(D) x intercept: 3.5; y intercept: -4.2

(E) $x > 3.5$ or $(3.5, \infty)$

29. Using $\underline{3}$ with $a = 3$ for the vertical line and $b = -5$ for the horizontal line, we find that the equation of the vertical line is $x = 3$ and the equation of the horizontal line is $y = -5$.

31. Using $\underline{3}$ with $a = -1$ for the vertical line and $b = -3$ for the horizontal line, we find that the equation of the vertical line is $x = -1$ and the equation of the horizontal line is $y = -3$.

33. $m = -3$
For the point $(4, -1)$, $x_1 = 4$ and $y_1 = -1$. Using $\underline{6}$, we get:
$$y - (-1) = -3(x - 4)$$
$$y + 1 = -3x + 12$$
$$y = -3x + 11$$

35. $m = \dfrac{2}{3}$
For the point $(-6, -5)$, $x_1 = -6$ and $y_1 = -5$. Using $\underline{6}$, we get:
$$y - (-5) = \frac{2}{3}[x - (-6)]$$
$$y + 5 = \frac{2}{3}(x + 6)$$
$$y + 5 = \frac{2}{3}x + 4$$
$$y = \frac{2}{3}x - 1$$

37. $y - (-5) = 0(x - 3)$

$\quad y + 5 = 0 \quad$ or $\quad y = -5$

$\qquad y = 0x - 5$

39. The points are $(1, 3)$ and $(7, 5)$. Let $x_1 = 1$, $y_1 = 3$, $x_2 = 7$, and $y_2 = 5$. Using $\underline{4}$, we get:

$$m = \frac{5 - 3}{7 - 1} = \frac{2}{6} = \frac{1}{3}$$

41. Let $x_1 = -5$, $y_1 = -2$, $x_2 = 5$, and $y_2 = -4$. Using $\underline{4}$, we get:

$$m = \frac{-4 - (-2)}{5 - (-5)} = \frac{-4 + 2}{5 + 5} = \frac{-2}{10} = -\frac{1}{5}$$

43. $m = \dfrac{-3 - 7}{2 - 2} = \dfrac{-10}{0}$, the slope is not defined; the line through $(2, 7)$ and $(2, -3)$ is vertical.

45. $m = \dfrac{3 - 3}{-5 - 2} = \dfrac{0}{-7} = 0$

47. First, find the slope using $\underline{4}$:

$$m = \frac{y_2 - y_1}{x_2 - x_1} = \frac{5 - 3}{7 - 1} = \frac{2}{6} = \frac{1}{3}$$

Then, by using $\underline{6}$, $y - y_1 = m(x - x_1)$, where $m = \dfrac{1}{3}$ and $(x_1, y_1) = (1, 3)$ or $(7, 5)$, we get:

$$y - 3 = \frac{1}{3}(x - 1) \quad \text{or} \quad y - 5 = \frac{1}{3}(x - 7)$$

These two equations are equivalent. After simplifying either one of these, we obtain:

$-x + 3y = 8 \quad$ or $\quad x - 3y = -8$

49. First, find the slope using $\underline{4}$:

$$m = \frac{-4 - (-2)}{5 - (-5)} = \frac{-4 + 2}{5 + 5} = \frac{-2}{10} = -\frac{1}{5}$$

By using $\underline{6}$, and either one of these points, we obtain:

$$y - (-2) = -\frac{1}{5}[x - (-5)] \quad \text{[using } (-5, -2)\text{]}$$

$$y + 2 = -\frac{1}{5}(x + 5)$$

$$5(y + 2) = -x - 5$$

$$5y + 10 = -x - 5$$

$$x + 5y = -15$$

51. $(2, 7)$ and $(2, -3)$

Since each point has the same x coordinate, the graph of the line formed by these two points will be a *vertical line*. Then, using $\underline{3}$, with $a = 2$, we have $x = 2$ as the equation of the line.

53. $(2, 3)$ and $(-5, 3)$

Since each point has the same y coordinate, the graph of the line formed by these two points will be a *horizontal line*. Then, using $\underline{3}$, with $b = 3$, we have $y = 3$ as the equation of the line.

55. A linear function

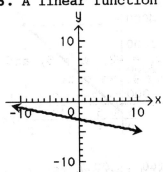

57. Not a function

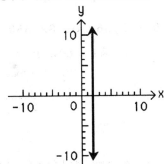

59. A constant function

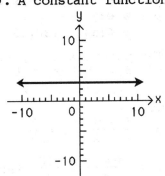

61. (A)

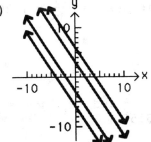

(B) Varying C produces a family of parallel lines. This is verified by observing that varying C does not change the slope of the lines but changes the intercepts.

63. $A = Prt + P$ (1)
Rate $r = 0.06$
Principal $P = 100$
Substituting in (1), we get:
$A = 6t + 100$ (2)

(A) Let $t = 5$ and $t = 20$ and substitute in (2). We get:
$A = 6(5) + 100 = \$130$
$A = 6(20) + 100 = \$220$

(B)

t	A
0	100
10	160
20	220

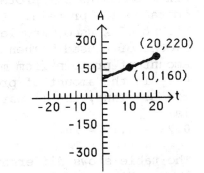

(C) Consider two points $(10, 160)$ and $(20, 220)$. Using $\underline{4}$, we have:
$$m = \frac{220 - 160}{20 - 10} = \frac{60}{10} = 6$$

65. (A) We find an equation $C(x) = mx + b$ for the line passing through $(0, 200)$ and $(20, 3800)$.
$$m = \frac{3800 - 200}{20 - 0} = \frac{3600}{20} = 180$$
Also, since $C(x) = 200$ when $x = 0$, it follows that $b = 200$.
Thus, $C(x) = 180x + 200$.

(B) The total costs at 12 boards per day are:
$$C(x) = 180(12) + 200 = 2,360 \text{ or } \$2,360$$

(C)

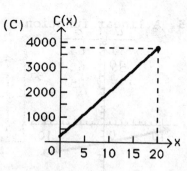

67. (A)

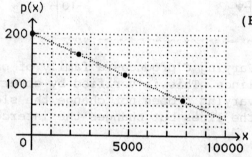

(B) slope: $m = \dfrac{160 - 200}{2,400 - 0} = \dfrac{-40}{2400} = -\dfrac{1}{60}$

y intercept: 200

equation: $p(x) = -\dfrac{1}{60}x + 200$

(C) $p(3000) = -\dfrac{1}{60}(3000) + 200 = -50 + 200 = \$150$

69. Mix A contains 20% protein. Mix B contains 10% protein. Let x be the amount of A used, and let y be the amount of B used. Then $0.2x$ is the amount of protein from mix A and $0.1y$ is the amount of protein from mix B. Thus, the linear equation is:

$0.2x + 0.1y = 20$

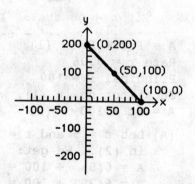

The table shows different combinations of mix A and mix B to provide 20 grams of protein.

[Note: We can get many more combinations. In fact, each point on the graph indicates a combination of mix A and mix B.]

Mix A	Mix B
x	y
100	0
0	200
50	100
10	180

71. $p = -\dfrac{1}{5}d + 70$, $30 \le d \le 175$, where d = distance in centimeters and p = pull in grams

(A) $d = 30$

$p = -\dfrac{1}{5}(30) + 70 = 64$ grams

$d = 175$

$p = -\dfrac{1}{5}(175) + 70 = 35$ grams

(B)

d	p
30	64
50	60
175	35

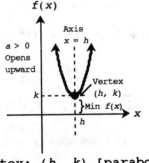

(C) Select two points $(30, 64)$ and $(50, 60)$ as (x_1, y_1) and (x_2, y_2), respectively, from part (B). Using $\underline{2}$:

$$\text{Slope } m = \frac{y_2 - y_1}{x_2 - x_1} = \frac{60 - 64}{50 - 30}$$

$$= -\frac{4}{20} = -\frac{1}{5}$$

EXERCISE 1-4

Things to remember:

$\underline{1}$. QUADRATIC FUNCTION

A function f is a QUADRATIC FUNCTION if
$$f(x) = ax^2 + bx + c \qquad a \neq 0$$
where a, b, and c are real numbers. The domain of a quadratic function is the set of all real numbers.

$\underline{2}$. PROPERTIES OF A QUADRATIC FUNCTION AND ITS GRAPH

Given a quadratic function
$$f(x) = ax^2 + bx + c \qquad a \neq 0$$
and the form obtained by completing the square
$$f(x) = a(x - h)^2 + k$$
we summarize general properties as follows:

a. The graph of f is a parabola:

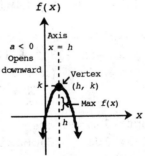

b. Vertex: (h, k) [parabola increases on one side of the vertex and decreases on the other]
c. Axis (of symmetry): $x = h$ (parallel to y axis)
d. $f(h) = k$ is the minimum if $a > 0$ and the maximum if $a < 0$
e. Domain: All real numbers
   Range: $(-\infty, k]$ if $a < 0$ or $[k, \infty)$ if $a > 0$
f. The graph of f is the graph of $g(x) = ax^2$ translated horizontally h units and vertically k units.

---

1. (a), (c), (e), (f)   **3.** (A) m   (B) g   (C) f   (D) n

5. (A) x intercepts: 1, 3; y intercept: -3   (B) Vertex: (2, 1)
   (C) Maximum: 1   (D) Range: $y \leq 1$ or $(-\infty, 1]$
   (E) Increasing interval: $x \leq 2$ or $(-\infty, 2]$
   (F) Decreasing interval: $x \geq 2$ or $[2, \infty)$

7. (A) x intercepts: -3, -1; y intercept: 3   (B) Vertex: (-2, -1)
   (C) Minimum: -1   (D) Range: $y \geq -1$ or $[-1, \infty)$
   (E) Increasing interval: $x \geq -2$ or $[-2, \infty)$
   (F) Decreasing interval: $x \leq -2$ or $(-\infty, -2]$

9. $f(x) = -(x - 2)^2 + 1 = -x^2 + 4x - 4 + 1 = -x^2 + 4x - 3 = -(x - 3)(x - 1)$
   (A) x intercepts: 1, 3; y intercepts: -3   (B) Vertex: (2, 1)
   (C) Maximum: 1   (D) Range: $y \leq 1$ or $(-\infty, 1]$

11. $M(x) = (x + 2)^2 - 1 = x^2 + 4x + 4 - 1 = x^2 + 4x + 3 = (x + 3)(x + 1)$
   (A) x intercepts: -3, -1; y intercept 3   (B) Vertex: (-2, -1)
   (C) Minimum: -1   (D) Range: $[-1, \infty)$

13. $y = -[x - (-2)]^2 + 5 = -(x + 2)^2 + 5$

15. $y = (x - 1)^2 - 3$

17. $f(x) = x^2 - 8x + 13 = x^2 - 8x + 16 - 3 = (x - 4)^2 - 3$
   (A) x intercepts: $(x - 4)^2 - 3 = 0$
$$(x - 4)^2 = 3$$
$$x - 4 = \pm\sqrt{3}$$
$$x = 4 + \sqrt{3} \approx 5.7, \ 4 - \sqrt{3} \approx 2.3$$
   y intercept: 13
   (B) Vertex: (4, -3)   (C) Minimum: -3   (D) Range: $y \geq -3$ or $[-3, \infty)$

19. $M(x) = 1 - 6x - x^2 = -(x^2 + 6x + 9) + 1 + 9 = -(x + 3)^2 + 10$
   (A) x intercepts: $-(x + 3)^2 + 10 = 0$
$$(x + 3)^2 = 10$$
$$x + 3 = \pm\sqrt{10}$$
$$x = -3 + \sqrt{10} \approx 0.2, \ -3 - \sqrt{10} = -6.2$$
   y intercept: 1
   (B) Vertex: (-3, 10)   (C) Maximum: 10   (D) Range: $y \leq 10$ or $(-\infty, 10]$

21. $G(x) = 0.5x^2 - 4x + 10 = \frac{1}{2}(x^2 - 8x + 16) + 2$
$$= \frac{1}{2}(x - 4)^2 + 2$$

   (A) x intercepts: none, since $G(x) = \frac{1}{2}(x - 4)^2 + 2 \geq 2$ for all x;
   y intercept: 10

   (B) Vertex: (4, 2)   (C) Minimum: 2   (D) Range: $y \geq 2$ or $[2, \infty)$

23. The vertex of the parabola is on the x axis.

25. $g(x) = 0.25x^2 - 1.5x - 7 = 0.25(x^2 - 6x + 9) - 2.25 - 7$
$$= 0.25(x - 3)^2 - 9.25$$

(A) x intercepts: $0.25(x - 3)^2 - 9.25 = 0$
$$(x - 3)^2 = 37$$
$$x - 3 = \pm\sqrt{37}$$
$$x = 3 + \sqrt{37} \approx 9.1, \ 3 - \sqrt{37} \approx -3.1$$

y intercept: -7

(B) Vertex: $(3, -9.25)$ (C) Minimum: -9.25

(D) Range: $y \geq -9.25$ or $[-9.25, \infty)$

27. $f(x) = -0.12x^2 + 0.96x + 1.2$
$$= -0.12(x^2 - 8x + 16) + 1.92 + 1.2$$
$$= -0.12(x - 4)^2 + 3.12$$

(A) x intercepts: $-0.12(x - 4)^2 + 3.12 = 0$
$$(x - 4)^2 = 26$$
$$x - 4 = \pm\sqrt{26}$$
$$x = 4 + \sqrt{26} \approx 9.1, \ 4 - \sqrt{26} \approx -1.1$$

y intercept: 1.2

(B) Vertex: $(4, 3.12)$ (C) Maximum: 3.12

(D) Range: $y \leq 3.12$ or $(-\infty, 3.12]$

29.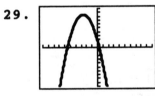

$x = -5.37, \ 0.37$

$-10 \leq x \leq 10$ xscl = 1
$-10 \leq y \leq 10$ yscl = 1

31.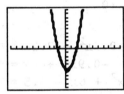

$-1.37 < x < 2.16$

$-10 \leq x \leq 10$ xscl = 1
$-10 \leq y \leq 10$ yscl = 1

33.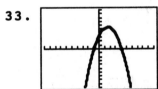

$x \leq -0.74$ or $x \geq 4.19$

$-10 \leq x \leq 10$ xscl = 1
$-10 \leq y \leq 10$ yscl = 1

35. (A)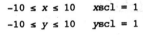

(B) $f(x) = g(x)$

$$-0.4x(x - 10) = 0.3x + 5$$
$$-0.4x^2 + 4x = 0.3x + 5$$
$$-0.4x^2 + 3.7x = 5$$
$$-0.4x^2 + 3.7x - 5 = 0$$
$$x = \frac{-3.7 \pm \sqrt{3.7^2 - 4(-0.4)(-5)}}{2(-0.4)}$$
$$x = \frac{-3.7 \pm \sqrt{5.69}}{-0.8} \approx 1.64, \ 7.61$$

(C) $f(x) > g(x)$ for $1.64 < x < 7.61$

(D) $f(x) < g(x)$ for $0 \le x < 1.64$ or $7.61 < x \le 10$

37. (A)

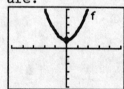

(B) $f(x) = g(x)$

$$-0.9x^2 + 7.2x = 1.2x + 5.5$$
$$-0.9x^2 + 6x = 5.5$$
$$-0.9x^2 + 6x - 5.5 = 0$$
$$x = \frac{-6 \pm \sqrt{36 - 4(-0.9)(-5.5)}}{2(-0.9)}$$
$$x = \frac{-6 \pm \sqrt{16.2}}{-1.8} \approx 1.1, \ 5.57$$

(C) $f(x) > g(x)$ for $1.10 < x < 5.57$

(D) $f(x) < g(x)$ for $0 \le x < 1.10$ or $5.57 < x \le 8$

39. $f(x) = x^2 + 1$ and $g(x) = -(x - 4)^2 - 1$ are two examples. Their graphs are:

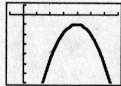

Their graphs do not intersect the x axis.

41. (A)

(B) $R(x) = x(119 - 6x) = -6x^2 + 119x$

$$= -6\left(x^2 - \frac{119}{6}x\right)$$

$$= -6\left(x^2 - 19.833x + 98.340\right) + 590.042$$

$$= -6(x - 9.917)^2 + 590.042$$

Output: 9.917 million chips, i.e., 9,917,000 chips
Maximum revenue: 590.042 million dollars, i.e. $590,042,000

(C)

(D) 9.917 million chips (9,917,000 chips)
590.042 million dollars ($590,042,000)

(E) $p(9.917) = 119 - 6(9.917) \approx \$59$

43. (A)

(B) $R(x) = C(x)$

$$x(119 - 6x) = 234 + 23x$$

$$-6x^2 + 96x = 234$$

$$x^2 - 16x = -39$$

$$x^2 - 16x + 39 = 0$$

$$(x - 13)(x - 3) = 0$$

$$x = 13, \ 3$$

Break-even at 3 million 3,000,000 and 13 million (13,000,000) chip production levels.

(C)

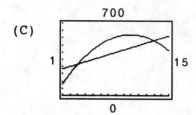

(D) Break-even at 3 million and 13 million chip production levels

(E) Loss: $1 \le x < 3$ or $13 < x \le 15$ Profit: $3 < x < 13$

45. (A) Solve: $f(x) = 1,000(0.04 - x^2) = 20$

$$40 - 1000x^2 = 20$$
$$1000x^2 = 20$$
$$x^2 = 0.02$$
$$x = 0.14 \text{ or } -0.14$$

Since we are measuring distance, we take the positive solution:
$x = 0.14$ cm

(B) $x = 0.14$ cm

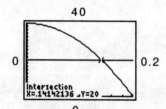

CHAPTER 1 REVIEW

1.

![graph of f(x) a downward parabola with vertex at (0,5)]

(1-1)

2. (A) Not a function; fails vertical line test   (B) A function
(C) A function   (D) Not a function; fails vertical line test   (1-1)

3. $f(x) = 2x - 1$, $g(x) = x^2 - 2x$

(A) $f(-2) + g(-1) = 2(-2) - 1 + (-1)^2 - 2(-1) = -2$

(B) $f(0) \cdot g(4) = (2 \cdot 0 - 1)(4^2 - 2 \cdot 4) = -8$

(C) $\dfrac{g(2)}{f(3)} = \dfrac{2^2 - 2 \cdot 2}{2 \cdot 3 - 1} = 0$

(D) $\dfrac{f(3)}{g(2)}$ not defined because $g(2) = 0$ (1-1)

4. (A) $y = 4$ (B) $x = 0$ (C) $y = 1$ (D) $x = -1$ or 1
    (E) $y = -2$ (F) $x = -5$ or 5
                                                         (1-1)

5. (A)   (B)

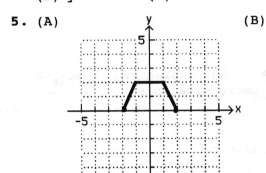 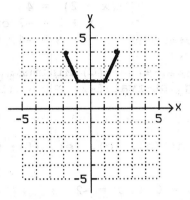

    (C)   (D)

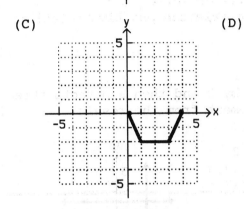 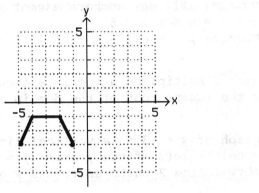

                                                                (1-2)

6. (A) (n)   (B) (g)   (C) (m); slope is zero
    (D) (f); slope is not defined
                                                           (1-3)

7. $y = -\dfrac{2}{3}x + 6$
                                                                     (1-3)

8. vertical line: $x = -6$; horizontal line: $y = 5$
                                                         (1-3)

9. x intercept: $2x = 18$, $x = 9$;
    y intercept: $-3y = 18$, $x = -6$;
    slope-intercept form: $y = \dfrac{2}{3}x - 6$; slope $= \dfrac{2}{3}$
    graph:

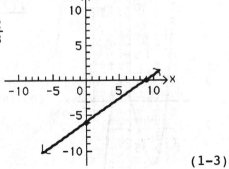

                                                                     (1-3)

10. (b), (c), (d), (f)
                                                                       (1-4)

11. (A) g (B) m (C) n (D) f
                                                          (1-2, 1-4)

12. $y = f(x) = (x + 2)^2 - 4$

    (A) x intercepts: $(x + 2)^2 - 4 = 0$
$$(x + 2)^2 = 4$$
$$x + 2 = -2 \text{ or } 2$$
$$x = -4, 0$$

        y intercept: 0

    (B) Vertex: $(-2, -4)$ (C) Minimum: -4 (D) Range: $y \geq -4$ or $[-4, \infty)$

    (E) Increasing interval $[-2, \infty)$ (F) Decreasing interval $(-\infty, -2]$

                                                                       (1-4)

13. Linear function: (a), (c), (e), (f); Constant function: (d)   (1-3)

14. (A) $x^2 - x - 6 = 0$ at $x = -2, 3$

        Domain: all real numbers except $x = -2, 3$

    (B) $5 - x > 0$ for $x < 5$

        Domain: $x < 5$ or $(-\infty, 5)$ (1-1)

15. Function g multiplies a domain element by 2 and then subtracts three times the square root of the domain element from the result.   (1-1)

16. The graph of $x = -3$ is a vertical line 3 units to the *left* of the y axis; $y = 2$ is a horizontal line 2 units *above* the x axis.

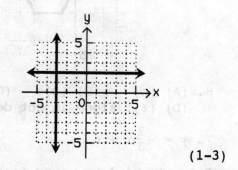

                                                                           (1-3)

17.

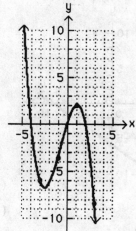

                  (1-1)

18. $f(x) = 3 - 2x$

$$\frac{f(2 + h) - f(2)}{h} = \frac{3 - 2(2 + h) - (3 - 2 \cdot 2)}{h}$$

$$= \frac{3 - 4 - 2h - 3 + 4}{h}$$

$$= \frac{-2h}{h}$$

$$= -2 \qquad (1-1)$$

19. $f(x) = x^2 - 3x + 1$

$$\frac{f(a + h) - f(a)}{h} = \frac{(a + h)^2 - 3(a + h) + 1 - (a^2 - 3a + 1)}{h}$$

$$= \frac{a^2 + 2ah + h^2 - 3a - 3h + 1 - a^2 + 3a - 1}{h}$$

$$= \frac{2ah + h^2 - 3h}{h}$$

$$= \frac{(2a + h - 3)h}{h}$$

$$= 2a + h - 3 \hspace{4cm} (1\text{-}1)$$

20. The graph of m is the graph of $y = |x|$ reflected on the x axis and shifted 4 units to the right. $\hspace{3cm} (1\text{-}2)$

21. The graph of g is the graph of $y = x^3$ vertically contracted by a factor of 0.3 and shifted up 3 units. $\hspace{5cm} (1\text{-}2)$

22. The graph of $y = x^2$ is vertically expanded by a factor of 2, reflected in the x axis and shifted to the left 3 units. Equation: $y = -2(x + 3)^2$
$$(1\text{-}2)$$

23. Equation: $f(x) = 2\sqrt{x + 3} - 1$

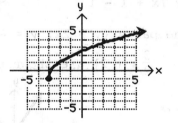

$$(1\text{-}2)$$

24. Use the point-slope form:

(A) $y - 2 = -\dfrac{2}{3}[x - (-3)]$ $\hspace{2cm}$ (B) $y - 3 = 0(x - 3)$

$\hspace{1.5cm} y - 2 = -\dfrac{2}{3}(x + 3)$ $\hspace{3.5cm} y = 3$

$\hspace{2.5cm} y = -\dfrac{2}{3}x \hspace{6cm} (1\text{-}3)$

25. (A) Slope: $\dfrac{-1 - 5}{1 - (-3)} = -\dfrac{3}{2}$ $\hspace{2cm}$ (B) Slope: $\dfrac{5 - 5}{4 - (-1)} = 0$

$\hspace{2.5cm} y - 5 = -\dfrac{3}{2}(x + 3)$ $\hspace{3cm} y - 5 = 0(x - 1)$

$\hspace{2.5cm} 3x + 2y = 1 \hspace{4cm} y = 5$

(C) Slope: $\dfrac{-2 - 7}{-2 - (-2)}$ not defined since $2 - (-2) = 0$

$\hspace{1.5cm} x = -2 \hspace{8cm} (1\text{-}3)$

26. $y = -(x - 4)^2 + 3 \hspace{7cm} (1\text{-}2, \ 1\text{-}4)$

27. $f(x) = -0.4x^2 + 3.2x - 1.2 = -0.4(x^2 - 8x + 16) + 7.6$
$$= -0.4(x - 4)^2 + 7.6$$

(A) y intercept: 1.2
  x intercepts: $-0.4(x - 4)^2 + 7.6 = 0$
$$(x - 4)^2 = 19$$
$$x = 4 + \sqrt{19} \approx 8.4, \ 4 - \sqrt{19} \approx -0.4$$

(B) Vertex: (4.0, 7.6)   (C) Maximum: 7.6

(D) Range: $x \le 7.6$ or $(-\infty, 7.6]$ (1-4)

28.

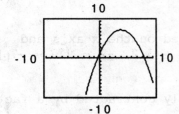

(A) y intercept: 1.2
  x intercepts: -0.4, 8.4

(B) Vertex: (4.0, 7.6)

(C) Maximum: 7.6

(D) Range: $x \le 7.6$ or $(-\infty, 7.6]$ (1-4)

29. The graph of $y = \sqrt[3]{x}$ is vertically expanded by a factor of 2, reflected in the x axis, shifted 1 unit to the left and 1 unit down.

Equation: $y = -2\sqrt[3]{x + 1} - 1$ (1-2)

30. The graphs of the pairs $\{y = 2x, \ y = -\frac{1}{2}x\}$ and

$\{y = \frac{2}{3}x + 2, \ y = -\frac{3}{2}x + 2\}$ are shown below:

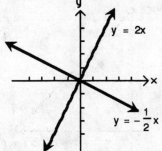 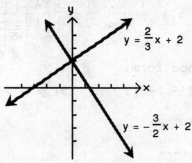

In each case, the graphs appear to be perpendicular to each other. It can be shown that two slant lines are perpendicular if and only if their slopes are negative reciprocals.   (1-3)

31. (A) $f(x) = \sqrt{x}$

$$\frac{f(x + h) - f(x)}{h} = \frac{\sqrt{x + h} - \sqrt{x}}{h} = \frac{\sqrt{x + h} - \sqrt{x}}{h} \cdot \frac{\sqrt{x + h} + \sqrt{x}}{\sqrt{x + h} + \sqrt{x}} \quad \text{rationalize the numerator}$$

$$= \frac{x + h - x}{h[\sqrt{x + h} + \sqrt{x}]} = \frac{1}{\sqrt{x + h} + \sqrt{x}}$$

(B) $f(x) = \dfrac{1}{x}$

$$\dfrac{f(x+h)-f(x)}{h} = \dfrac{\dfrac{1}{x+h} - \dfrac{1}{x}}{h} = \dfrac{\dfrac{x-(x+h)}{x(x+h)}}{h}$$

$$= \dfrac{-h}{hx(x+h)} = \dfrac{-1}{x(x+h)} \qquad (1-2)$$

32. $G(x) = 0.3x^2 + 1.2x - 6.9 = 0.3(x^2 + 4x + 4) - 8.1$
$$= 0.3(x+2)^2 - 8.1$$

(A) y intercept: -6.9
x intercepts: $0.3(x+2)^2 - 8.1 = 0$
$$(x+2)^2 = 27$$
$$x = -2 + \sqrt{27} \approx 3.2, \ -2 - \sqrt{27} \approx -7.2$$

(B) Vertex: $(-2, -8.1)$ (C) Minimum: -8.1
(D) Range: $x \geq -8.1$ or $[-8.1, \infty)$
(E) Decreasing: $(-\infty, -2]$; Increasing: $[-2, \infty)$ $(1-4)$

33.

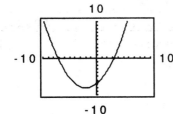

(A) y intercept: -6.9
     x intercept: $-7.2, \ 3.2$
(B) Vertex: $(-2, -8.1)$
(C) Minimum: -8.1
(D) Range: $x \geq -8.1$ or $[-8.1, \infty)$
(E) Decreasing: $(-\infty, -2]$
     Increasing: $[-2, \infty)$ $(1-4)$

34. (A) $V(0) = 12{,}000, \ V(8) = 2{,}000$
     Slope: $\dfrac{2{,}000 - 12{,}000}{8 - 0} = \dfrac{-10{,}000}{8} = -1{,}250$
     V intercept: $12{,}000$
     Equation: $V(t) = -1{,}250t + 12{,}000$

(B) $V(5) = -1{,}250(5) + 12{,}000 = \$5{,}750$

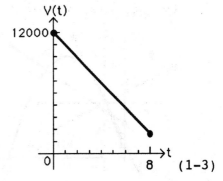

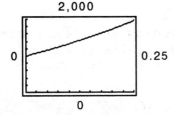

35. (A)

(B) $r = 0.1447$ or 14.7% compounded annually
     Alternative algebraic solution:
$$1000(1+r)^3 = 1500$$
$$(1+r)^3 = 1.5$$
$$1 + r = \sqrt[3]{1.5} \approx 1.1447$$
$$r = 0.1447 \qquad (1-1, \ 1-2)$$

36. (A) $R(130) = 208$, $R(50) = 80$

Slope: $\dfrac{208 - 80}{130 - 50} = \dfrac{128}{80} = 1.6$

Equation: $R - 80 = 1.6(C - 50)$ or $R = 1.6C$

(B) $R(120) = 1.6(120) = \$192$

(C) $176 = 1.6C$; $C = \$110$

(D) 1.6; The slope gives the change in retail price per unit change in the cost. (1-3)

37. (A) Let x = number of video tapes produced.
$C(x) = 84,000 + 15x$
$R(x) = 50x$

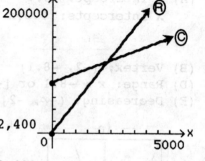

(B) $R(x) = C(x)$
$50x = 84,000 + 15x$
$35x = 84,000$
$x = 2,400$ units
$R < C$ for $0 \le x < 2,400$; $R > C$ for $x > 2,400$

(C) $R = C$ at $x = 2,400$ units
$R < C$ for $0 \le x < 2,400$; $R > C$ for $x > 2,400$ (1-3)

38. $p(x) = 50 - 1.25x$ Price-demand function
$C(x) = 160 + 10x$ Cost function
$R(x) = xp(x)$
$\quad\;\;\; = x(50 - 1.25x)$ Revenue function

(A)

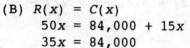

(B) $R = C$

$$x(50 - 1.25x) = 160 + 10x$$
$$-1.25x^2 + 50x = 160 + 10x$$
$$-1.25x^2 + 40x = 160$$
$$-1.25(x^2 - 32x + 256) = 160 - 320$$
$$-1.25(x - 16)^2 = -160$$
$$(x - 16)^2 = 128$$
$$x = 16 + \sqrt{128} \approx 27.314,\; 16 - \sqrt{128} \approx 4.686$$

$R = C$ at $x = 4.686$ thousand units (4,686 units) and
$x = 27.314$ thousand units (27,314 units)
$R < C$ for $1 \le x < 4.686$ or $27.314 < x \le 40$
$R > C$ for $4.686 < x < 27.314$

(C) Max Rev: $50x - 1.25x^2 = R$

$-1.25(x^2 - 40x + 400) + 500 = R$

$-1.25(x - 20)^2 + 500 = R$

Vertex at (20, 500)

Max. Rev. = 500 thousand ($500,000) occurs when <u>output</u> is 20 thousand (20,000 units)

<u>Wholesale price</u> at this output: $p(x) = 50 - 1.25x$

$p(20) = 50 - 1.25(20)$

$= \$25$ (1-3, 1-4)

39. (A) $P(x) = R(x) - C(x) = x(50 - 1.25x) - (160 + 10x)$

$= -1.25x^2 + 40x - 160$

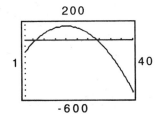

(B) $P = 0$ for $x = 4.686$ thousand uits (4,686 units) and $x = 27.314$ thousand units (27,314 units)

$P < 0$ for $1 \le x < 4.686$ or $27.314 < x \le 40$

$P > 0$ for $4.686 < x < 27.314$

(C) Maximum profit is 160 thousand dollars ($160,000), and this occurs at $x = 16$ thousand units (16,000 units). The wholesale price at this output is $p(16) = 50 - 1.25(16) = \$30$, which is $5 greater than the $25 found in 38(C).   (1-4)

40. (A) The area A enclosed by the pens is given by

$A = (2y)x$

Now, $3x + 4y = 840$

and $y = 210 - \dfrac{3}{4}x$

Thus $A(x) = 2\left(210 - \dfrac{3}{4}x\right)x$

$= 420x - \dfrac{3}{2}x^2$

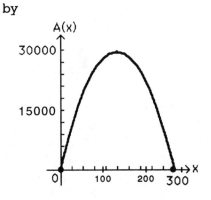

(B) Since x and y must both be nonnegative,

$210 - \dfrac{3}{4}x \ge 0$

$-\dfrac{3}{4}x \ge -210$

$x \le 280$

Domain: $0 \le x \le 280$

(C) Maximum combined area is 29,400 feet. This occurs at $x = 140$ feet, $y = 105$ feet.   (1-4)

41. (A) We are given $P(0) = 20$ and $m = 15$. Thus, $P(x) = 15x + 20$

(B) 1 PM is 5 hours after 8 AM
$$P(5) = 15(5) + 20 = 95$$

(C)
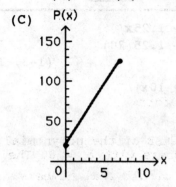

(D) Slope = 15

(1-4)

42. $\dfrac{\Delta s}{s} = k$. For $k = \dfrac{1}{30}$, $\dfrac{\Delta s}{s} = \dfrac{1}{30}$ or $\Delta s = \dfrac{1}{30}s$

(A) When $s = 30$, $\Delta s = \dfrac{1}{30}(30) = 1$ pound.

When $s = 90$, $\Delta s = \dfrac{1}{30}(90) = 3$ pounds.

(B) $\Delta s = \dfrac{1}{30}s$

Slope $m = \dfrac{1}{30}$

y intercept $b = 0$

(C) Slope $m = \dfrac{1}{30}$

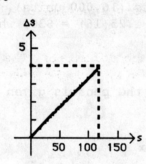

(1-4)

2 ADDITIONAL ELEMENTARY FUNCTIONS

Things to remember:

1. POLYNOMIAL FUNCTION

    A POLYNOMIAL FUNCTION is a function of the form
    $$f(x) = a_n x^n + a_{n-1} x^{n-1} + \ldots + a_1 x + a_0$$
    for n a nonnegative integer, called the DEGREE of the polynomial. The coefficients a_0, a_1, ..., a_n are real numbers with $a_n \neq 0$. The DOMAIN of a polynomial function is the set of all real numbers.

2. TURNING POINT

    A TURNING POINT on a graph is a point that separates an increasing portion from a decreasing portion, or vice versa. The graph of a polynomial function of degree $n \geq 1$ can have at most $n-1$ turning points and can cross the x axis at most n times.

3. A RATIONAL FUNCTION is any function of the form
    $$f(x) = \frac{n(x)}{d(x)} \qquad d(x) \neq 0$$
    where $n(x)$ and $d(x)$ are polynomials. The DOMAIN is the set of all real numbers such that $d(x) \neq 0$. We assume $n(x)/d(x)$ is reduced to lowest terms.

4. ASYMPTOTES OF RATIONAL FUNCTIONS

    Given the rational function
    $$f(x) = \frac{n(x)}{d(x)}$$
    where $n(x)$ and $d(x)$ are polynomials without common factors.

    (a) If a is a real number such that $d(a) = 0$, then the line $x = a$ is a VERTICAL ASYMPTOTE of the graph of $y = f(x)$.

    (b) HORIZONTAL ASYMPTOTES, if any exists, can be found by dividing each term of the numerator $n(x)$ and denominator $d(x)$ by the highest power of x that appears in the numerator and denominator.

---

1. (A) 2   (B) 1   (C) 2   (D) 0   (E) 1   (F) 1

3. (A) 5   (B) 4   (C) 5   (D) 1   (E) 1   (F) 1

5. (A) 6   (B) 5   (C) 6   (D) 0   (E) 1   (F) 1

7. (A) 3   (B) 4   (C) negative   **9.** (A) 4   (B) 5   (C) negative

11. (A) 0   (B) 1   (C) negative   **13.** (A) 5   (B) 6   (C) positive

15. $f(x) = \dfrac{x + 2}{x - 2}$

    (A) *Intercepts:*

         x intercepts: $f(x) = 0$ only if $x + 2 = 0$ or $x = -2$.
                 The x intercept is -2.

         y intercept: $f(0) = \dfrac{0 + 2}{0 - 2} = -1$
                 The y intercept is -1.

    (B) *Domain:* The denominator is 0 at $x = 2$. Thus, the domain is the set of all real numbers except 2.

    (C) *Asymptotes:*

         Vertical asymptotes: $f(x) = \dfrac{x + 2}{x - 2}$
                 The denominator is 0 at $x = 2$. Therefore, the line $x = 2$ is a vertical asymptote.

         Horizontal asymptotes: $f(x) = \dfrac{x + 2}{x - 2} = \dfrac{1 + \dfrac{2}{x}}{1 - \dfrac{2}{x}}$

                 As x increases or decreases without bound, the numerator tends to 1 and the denominator tends to 1. Therefore, the line $y = 1$ is a horizontal asymptote.

    (D)

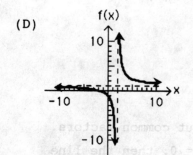

    (E)

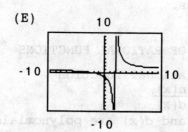

17. $f(x) = \dfrac{3x}{x + 2}$

    (A) *Intercepts:*

         x intercepts: $f(x) = 0$ only if $3x = 0$ or $x = 0$.
                 The x intercept is 0.

         y intercept: $f(0) = \dfrac{3 \cdot 0}{0 + 2} = 0$
                 The y intercept is 0.

(B) *Domain:* The denominator is 0 at $x = -2$. Thus, the domain is the set of all real numbers except -2.

(C) *Asymptotes:*

Vertical asymptotes: $f(x) = \dfrac{3x}{x + 2}$

The denominator is 0 at $x = -2$. Therefore, the line $x = -2$ is a vertical asymptote.

Horizontal asymptotes: $f(x) = \dfrac{3x}{x + 2} = \dfrac{3}{1 + \dfrac{2}{x}}$

As x increases or decreases without bound, the numerator is 3 and the denominator tends to 1. Therefore, the line $y = 3$ is a horizontal asymptote.

(D)

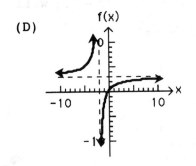

(E)

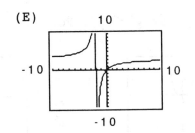

19. $f(x) = \dfrac{4 - 2x}{x - 4}$

(A) *Intercepts:*

x intercepts: $f(x) = 0$ only if $4 - 2x = 0$ or $x = 2$.
The x intercept is 2.

y intercept: $f(0) = \dfrac{4 - 2 \cdot 0}{0 - 4} = -1$
The y intercept is -1.

(B) *Domain:* The denominator is 0 at $x = 4$. Thus, the domain is the set of all real numbers except 4.

(C) *Asymptotes:*

Vertical asymptotes: $f(x) = \dfrac{4 - 2x}{x - 4}$

The denominator is 0 at $x = 4$. Therefore, the line $x = 4$ is a vertical asymptote.

Horizontal asymptotes: $f(x) = \dfrac{4 - 2x}{x - 4} = \dfrac{\dfrac{4}{x} - 2}{1 - \dfrac{4}{x}}$

As x increases or decreases without bound, the numerator tends to -2 and the denominator tends to 1. Therefore, the line $y = -2$ is a horizontal asymptote.

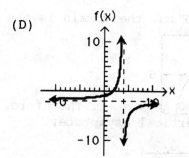

(D)

(E)

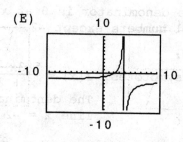

21. The graph of $f(x) = 2x^4 - 5x^2 + x + 2 = 2x^4\left(1 - \dfrac{5}{2x^2} + \dfrac{1}{2x^3} + \dfrac{1}{x^4}\right)$ will "look like" the graph of $y = 2x^4$. For large x, $f(x) \approx 2x^4$.

23. The graph of $f(x) = -x^5 + 4x^3 - 4x + 1 = -x^5\left(1 - \dfrac{4}{x^2} + \dfrac{4}{x^4} - \dfrac{1}{x^5}\right)$ will "look like" the graph of $y = -x^5$. For large x, $f(x) \approx -x^5$.

25. (A)

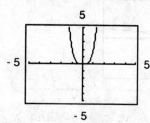

$y = 2x^4$ $y = 2x^4 - 5x^2 + x + 2$

(B)

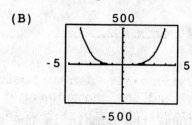

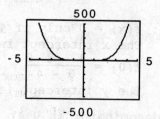

$y = 2x^4$ $y = 2x^4 - 5x^2 + x + 2$

27. (A)

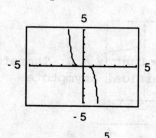

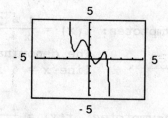

$y = -x^5$ $y = -x^5 + 4x^3 - 4x + 1$

(B)

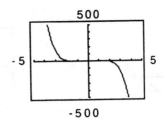

$$y = -x^5$$

$$y = -x^5 + 4x^3 - 4x + 1$$

29. $f(x) = \dfrac{2x^2}{x^2 - x - 6}$

(A) *Intercepts:*

x intercepts: $f(x) = 0$ only if $2x^2 = 0$ or $x = 0$.
The x intercept is 0.

y intercept: $f(0) = \dfrac{2 \cdot 0^2}{0^2 - 0 - 6} = 0$
The y intercept is 0.

(B) *Asymptotes:*

Vertical asymptotes: $f(x) = \dfrac{2x^2}{x^2 - x - 6} = \dfrac{2x^2}{(x - 3)(x + 2)}$

The denominator is 0 at $x = -2$ and $x = 3$.
Thus, the lines $x = -2$ and $x = 3$ are vertical asymptotes.

Horizontal asymptotes: $f(x) = \dfrac{2x^2}{x^2 - x - 6} = \dfrac{2}{1 - \dfrac{1}{x} - \dfrac{6}{x^2}}$

As x increases or decreases without bound, the numerator is 2 and the denominator tends to 1. Therefore, the line $y = 2$ is a horizontal asymptote.

(C)

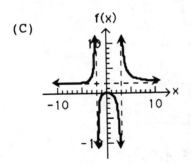

(D)

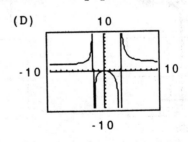

31. $f(x) = \dfrac{6 - 2x^2}{x^2 - 9}$

(A) *Intercepts:*

x intercepts: $f(x) = 0$ only if $6 - 2x^2 = 0$

$$2x^2 = 6$$
$$x^2 = 3$$
$$x = \pm\sqrt{3}$$

The x intercepts are $\pm\sqrt{3}$.

y intercept: $f(0) = \dfrac{6 - 2 \cdot 0^2}{0^2 - 9} = -\dfrac{2}{3}$

The y intercept is $-\dfrac{2}{3}$.

(B) *Asymptotes:*

Vertical asymptotes: $f(x) = \dfrac{6 - 2x^2}{x^2 - 9} = \dfrac{6 - 2x^2}{(x - 3)(x + 3)}$

The denominator is 0 at $x = -3$ and $x = 3$. Thus, the lines $x = -3$ and $x = 3$ are vertical asymptotes.

Horizontal asymptotes: $f(x) = \dfrac{6 - 2x^2}{x^2 - 9} = \dfrac{\dfrac{6}{x^2} - 2}{1 - \dfrac{9}{x^2}}$

As x increases or decreases without bound, the numerator tends to -2 and the denominator tends to 1. Therefore, the line $y = -2$ is a horizontal asymptote.

(C)

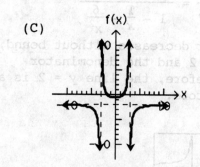

(D)

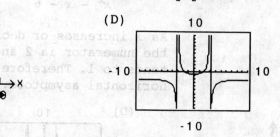

33. $f(x) = \dfrac{-4x}{x^2 + x - 6}$

(A) *Intercepts:*

x intercepts: $f(x) = 0$ only if $-4x = 0$ or $x = 0$.
The x intercept is 0.

y intercept: $f(0) = \dfrac{-4 \cdot 0}{0^2 + 0 - 6} = 0$

The y intercept is 0.

(B) *Asymptotes:*

Vertical asymptotes: $f(x) = \dfrac{-4x}{x^2 + x - 6} = \dfrac{-4x}{(x + 3)(x - 2)}$

The denominator is 0 at $x = -3$ and $x = 2$. Thus, the lines $x = -3$ and $x = 2$ are vertical asymptotes.

Horizontal asymptotes: $f(x) = \dfrac{-4x}{x^2 + x - 6} = \dfrac{-\dfrac{4}{x}}{1 + \dfrac{1}{x} - \dfrac{6}{x^2}}$

As x increases or decreases without bound, the numerator tends to 0 and the denominator tends to 1. Therefore, the line $y = 0$ (the x axis) is a horizontal asymptote.

(C)

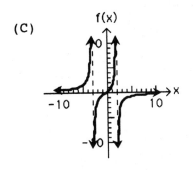

(D)

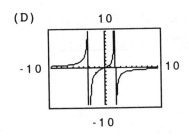

35. The graph has 1 turning point which implies degree $n = 2$.
    The x intercepts are $x = -1$ and $x = 2$.
    Thus, $f(x) = (x + 1)(x - 2) = x^2 - x - 2$.

37. The graph has 2 turning points which implies degree $n = 3$. The x intercepts are $x = -2$, $x = 0$, and $x = 2$. The direction of the graph indicates that leading coefficient is negative
    $f(x) = -(x + 2)(x)(x - 2) = 4x - x^3$.

39. (A) Since $C(x)$ is a linear function of x, it can be written in the form
    $$C(x) = mx + b$$
    Since the fixed costs are \$200, $b = 200$.
    Also, $C(20) = 3800$, so
    $$3800 = m(20) + 200$$
    $$20m = 3600$$
    $$m = 180$$
    Therefore, $C(x) = 180x + 200$

    (B) $\overline{C}(x) = \dfrac{C(x)}{x} = \dfrac{180x + 200}{x}$

(C)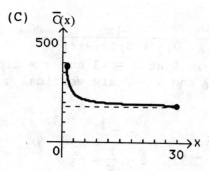

(D) $\overline{C}(x) = \dfrac{180x + 200}{x} = \dfrac{180 + \dfrac{200}{x}}{1}$

As x increases, the numerator tends to 180 and the denominator is 1. Therefore, $\overline{C}(x)$ tends to 180 or \$180 per board.

41. (A) $\overline{C}(n) = \dfrac{2500 + 175n + 25n^2}{n}$

(B)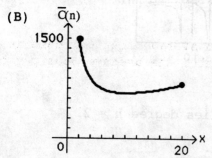

(C) Using the graph, we calculate

$C(8) = \dfrac{2500 + 175(8) + 25(8)^2}{8} = 687.50$

$C(9) = \dfrac{2500 + 175(9) + 25(9)^2}{9} = 677.78$

$C(10) = \dfrac{2500 + 175(10) + 25(10)^2}{10} = 675.00$

$C(11) = \dfrac{2500 + 175(11) + 25(11)^2}{11} = 677.27$

$C(12) = \dfrac{2500 + 175(12) + 25(12)^2}{12} = 683.33$

Thus, it appears that the average cost per year is a minimum at n = 10 years; at 10 years, the average minimum cost is \$675.00 per year.

(D) 10 years; $675.00 per year

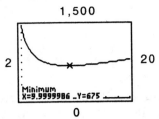

43. (A) $\overline{C}(x) = \dfrac{0.00048(x - 500)^3 + 60{,}000}{x}$

(B)

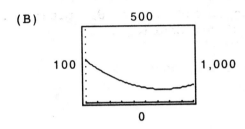

(C) The caseload which yields the minimum average cost per case is 750 cases per month. At 750 cases per month, the average cost per case is $90.

45. (A) $v(x) = \dfrac{26 + 0.06x}{x} = \dfrac{\dfrac{26}{x} + 0.06}{1}$

As x increases, the numerator tends to 0.06 and the denominator is 1. Therefore, $v(x)$ approaches 0.06 centimeters per second as x increases.

(B)

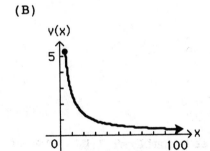

Things to remember:

<u>1</u>. EXPONENTIAL FUNCTION

The equation

$$f(x) = b^x, \ b > 0, \ b \neq 1$$

defines an EXPONENTIAL FUNCTION for each different constant b, called the BASE. The DOMAIN of f is all real numbers, and the RANGE of f is the set of positive real numbers.

<u>2</u>. BASIC PROPERTIES OF THE GRAPH OF $f(x) = b^x, \ b > 0, \ b \neq 1$

a. All graphs pass through $(0,1)$; $b^0 = 1$ for any base b.

b. All graphs are continuous curves; there are no holes or jumps.

c. The x-axis is a horizontal asymptote.

d. If $b > 1$, then b^x increases as x increases.

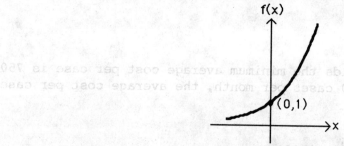

Graph of $f(x) = b^x, \ b > 1$

e. If $0 < b < 1$, then b^x decreases as x increases.

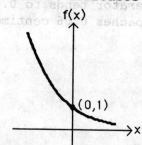

Graph of $f(x) = b^x, \ 0 < b < 1$

<u>3</u>. EXPONENTIAL FUNCTION PROPERTIES

For $a, \ b > 0, \ a \neq 1, \ b \neq 1$, and $x, \ y$ real numbers:

a. EXPONENT LAWS

(i) $a^x a^y = a^{x+y}$ (iv) $(ab)^x = a^x b^x$

(ii) $\dfrac{a^x}{a^y} = a^{x-y}$ (v) $\left(\dfrac{a}{b}\right)^x = \dfrac{a^x}{b^x}$

(iii) $(a^x)^y = a^{xy}$

b. $a^x = a^y$ if and only if $x = y$.

c. For $x \neq 0$, $a^x = b^x$ if and only if $a = b$.

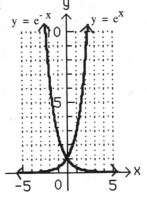

4. EXPONENTIAL FUNCTION WITH BASE $e = 2.71828...$

Exponential functions with base e and base $1/e$ are respectively defined by $y = e^x$ and $y = e^{-x}$.

    Domain: $(-\infty, \infty)$

    Range: $(0, \infty)$

5. COMPOUND INTEREST

If a principal P (present value) is invested at an annual rate r (expressed as a decimal) compounded m times per year, then the amount A (future value) in the account at the end of t years is given by:

$$A = P\left(1 + \frac{r}{m}\right)^{mt}$$

6. CONTINUOUS COMPOUND INTEREST FORMULA

If a principal P (present value) is invested at an annual rate r (expressed as a decimal) compounded continuously, then the amount A (future value) in the account at the end of t years is given by

$$A = Pe^{rt}$$

7. INTEREST FORMULAS

(a) $A = P(1 + rt)$ Simple interest

(b) $A = P\left(1 + \frac{r}{m}\right)^{mt}$ Compound interest

(c) $A = Pe^{rt}$ Continuous compound interest

---

1. $y = 5^x$, $-2 \leq x \leq 2$

x	y
-2	$\frac{1}{25}$
-1	$\frac{1}{5}$
0	1
1	5
2	25

3. $y = \left(\frac{1}{5}\right)^x = 5^{-x}$, $-2 \leq x \leq 2$

x	y
-2	25
-1	5
0	1
1	$\frac{1}{5}$
2	$\frac{1}{25}$

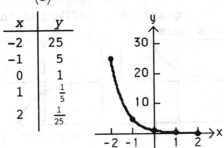

5. $f(x) = -5^x$, $-2 \le x \le 2$

x	$f(x)$
-2	$-\frac{1}{25}$
-1	$-\frac{1}{5}$
0	-1
1	-5
2	-25

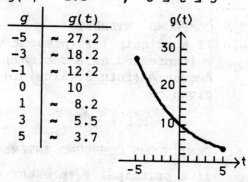

7. $y = -e^{-x}$, $-3 \le x \le 3$

x	y
-3	≈ -20
-2	≈ -7.4
-1	≈ -2.7
0	-1
1	≈ -0.4
2	≈ -0.1
3	≈ -0.05

9. $y = 100e^{0.1x}$, $-5 \le x \le 5$

x	y
-5	≈ 60
-3	≈ 74
-1	≈ 90
0	100
1	≈ 111
3	≈ 135
5	≈ 165

11. $g(t) = 10e^{-0.2t}$, $-5 \le t \le 5$

g	$g(t)$
-5	≈ 27.2
-3	≈ 18.2
-1	≈ 12.2
0	10
1	≈ 8.2
3	≈ 5.5
5	≈ 3.7

13. $(4^{3x})^{2y} = 4^{6xy}$ [see $\underline{3}$a(iii)]

15. $\dfrac{e^{x-3}}{e^{x-4}} = e^{(x-3)-(x-4)} = e^{x-3-x+4} = e$ [See $\underline{3}$a(ii)]

17. $(2e^{1.2t})^3 = 2^3 e^{3(1.2t)} = 8e^{3.6t}$ [see $\underline{3}$a (iv)]

19. $g(x) = -f(x)$; the graph of g is the graph of f reflected in the x axis.

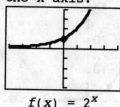

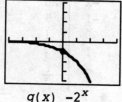

$f(x) = 2^x$ $g(x) -2^x$

21. $g(x) = f(x + 1)$; the graph of g is the graph of f shifted one unit to the left.

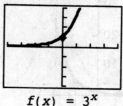

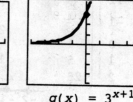

$f(x) = 3^x$ $g(x) = 3^{x+1}$

23. $g(x) = f(x) + 1$; the graph of g is the graph of f shifted one unit up.

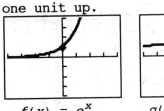

$$f(x) = e^x$$

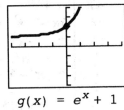

$$g(x) = e^x + 1$$

25. $g(x) = 2f(x + 2)$; the graph of g is the graph of f vertically expanded by a factor of 2 and shifted to the left 2 units.

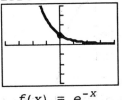

$$f(x) = e^{-x}$$

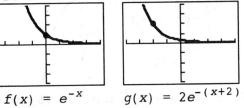

$$g(x) = 2e^{-(x+2)}$$

27. $f(t) = 2^{t/10}$, $-30 \le t \le 30$

t	$f(t)$
-30	$\frac{1}{8}$
-20	$\frac{1}{4}$
-10	$\frac{1}{2}$
0	1
10	2
20	4
30	8

29. $y = -3 + e^{1+x}$, $-4 \le x \le 2$

x	y
-4	≈ -3
-2	≈ -2.6
-1	-2
0	≈ -0.3
1	≈ 4.4
2	≈ 17.1

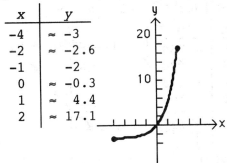

31. $y = e^{|x|}$, $-3 \le x \le 3$

x	y
-3	≈ 20.1
-1	≈ 2.7
0	1
1	≈ 2.7
3	≈ 20.1

33. $C(x) = \dfrac{e^x + e^{-x}}{2}$, $-5 \le x \le 5$

x	$C(x)$
-5	≈ 74
-3	≈ 10
0	1
3	≈ 10
5	≈ 74

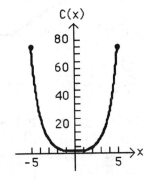

35. $y = e^{-x^2}$, $-3 \le x \le 3$

x	y
-3	0.0001
-2	0.0183
-1	0.3679
0	1
1	0.3679
2	0.0183
3	0.0001

37. The top curve is the graph of $f(x) = 2^x$, the bottom curve is the graph of $g(x) = e^x$; e^x approaches 0 more rapidly than 2^x as $x \to -\infty$.

39. The top curve is the graph of $g(x) = e^{-x}$, the bottom curve is the graph of $f(x) = 2^{-x}$; e^{-x} grows more rapidly than 2^{-x} as $x \to -\infty$.

41. $10^{2-3x} = 10^{5x-6}$ implies (see 3b)
$$2 - 3x = 5x - 6$$
$$-8x = -8$$
$$x = 1$$

43. $4^{5x-x^2} = 4^{-6}$ implies
$$5x - x^2 = -6$$
$$\text{or} \quad -x^2 + 5x + 6 = 0$$
$$x^2 - 5x - 6 = 0$$
$$(x - 6)(x + 1) = 0$$
$$x = 6, -1$$

45. $5^3 = (x + 2)^3$ implies (by property 3c)
$$5 = x + 2$$
Thus, $x = 3$.

47. $(x - 3)e^x = 0$
$$x - 3 = 0 \quad \text{(since } e^x \neq 0)$$
$$x = 3$$

49. $3xe^{-x} + x^2e^{-x} = 0$
$$e^{-x}(3x + x^2) = 0$$
$$3x + x^2 = 0 \quad \text{(since } e^{-x} \neq 0)$$
$$x(3 + x) = 0$$
$$x = 0, -3$$

51. $h(x) = x2^x$, $-5 \leq x \leq 0$

x	$h(x)$
-5	$-\frac{5}{32}$
-4	$-\frac{1}{4}$
-3	$-\frac{3}{8}$
-2	$-\frac{1}{2}$
-1	$-\frac{1}{2}$
0	0

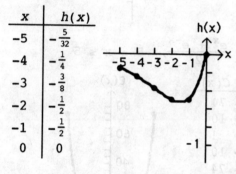

53. $N = \dfrac{100}{1 + e^{-t}}$, $0 \leq t \leq 5$

t	N
0	50
1	≈ 73.1
2	≈ 88.1
3	≈ 95.3
5	≈ 99.3

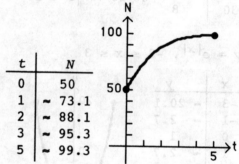

55. Using 4, $A = P\left(1 + \dfrac{r}{m}\right)^{mt}$, we have:

(A) $P = 2,500$, $r = 0.07$, $m = 4$, $t = \dfrac{3}{4}$

$$A = 2,500\left(1 + \frac{0.07}{4}\right)^{4 \cdot 3/4} = 2,500(1 + 0.0175)^3 = 2,633.56$$
Thus, $A = \$2,633.56$.

(B) $A = 2,500\left(1 + \dfrac{0.07}{4}\right)^{4 \cdot 15} = 2,500(1 + 0.0175)^{60} = 7079.54$
Thus, $A = \$7,079.54$.

57. Using 6 with $P = 7,500$ and $r = 0.0835$, we have:
$$A = 7,500e^{0.0835t}$$

(A) $A = 7,500e^{(0.0835)5.5} = 7,500e^{0.45925} \approx 11,871.65$
Thus, there will be $\$11,871.65$ in the account after 5.5 years.

(B) $A = 7,500e^{(0.0835)12} = 7,500e^{1.002} \approx 20,427.93$

Thus, there will be \$20,427.93 in the account after 12 years.

59. Using $A = P\left(1 + \dfrac{r}{m}\right)^{mt}$, we have:

$A = 15,000$, $r = 0.0975$, $m = 52$, $t = 5$

Thus, $15,000 = P\left(1 + \dfrac{0.0975}{52}\right)^{52 \cdot 5} = P(1 + 0.001875)^{260} \approx P(1.6275)$ and

$P = \dfrac{15,000}{1.6275} \approx 9,217$. Therefore, $P \approx \$9,217$.

61. Alamo Savings:

From Section 2-1, $A = P\left(1 + \dfrac{r}{m}\right)^{mt}$, where P is the principal, r is the

annual rate, and m is the number of compounding periods per year. Thus:

$A = 10,000\left(1 + \dfrac{0.0825}{4}\right)^4 = 10,000(1.020625)^4 \approx \$10,850.88$

Lamar Savings:
$A = 10,000e^{0.0805} \approx \$10,838.29$

63. In $A = Pe^{rt}$, we are given $A = 50,000$, $r = 0.1$, and $t = 5.5$. Thus:

$50,000 = Pe^{(0.1)5.5}$ or $P = \dfrac{50,000}{e^{0.55}} \approx 28,847.49$

You should be willing to pay \$28,847.49 for the note.

65. Given $N = 2(1 - e^{-0.037t})$, $0 \le t \le 50$

t	N
0	0
10	≈ 0.62
30	≈ 1.34
50	≈ 1.69

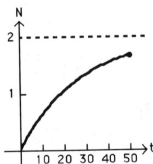

N approaches 2 as t increases without bound.

67. Given $I = I_0 e^{-0.23d}$

(A) $I = I_0 e^{-0.23(10)} = I_0 e^{-2.3} \approx I_0(0.10)$

Thus, about 10% of the surface light will reach a depth of 10 feet.

(B) $I = I_0 e^{-0.23(20)} = I_0 e^{-4.6} \approx I_0(0.010)$

Thus, about 1% of the surface light will reach a depth of 20 feet.

69. (A) Using 6 with $N_0 = 40,000$ and $r = 0.21$, we have $N = 40,000e^{0.21t}$.

(B) At the end of the year 2,000, $t = 8$ years and
$N(8) = 40,000e^{0.21(8)} = 40,000e^{1.6} \approx 215,000$

At the end of the year 2005, $t = 13$ years and
$$N(13) = 40,000e^{0.21(13)} = 40,000e^{2.73} \approx 613,000$$

(C)

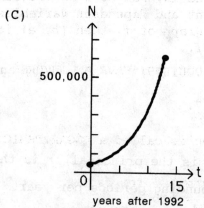

years after 1992

71. (A) Using $\underline{6}$ with $P_0 = 5.7$ and $r = 0.0114$, we have
$$P = 5.7e^{0.0114t}$$

(B) In the year 2010, $t = 15$ and
$$P = 5.7e^{0.0114(15)} = 5.7e^{0.171} \approx 6.8 \text{ billion}$$
In the year 2030, $t = 35$ and
$$P = 5.7e^{0.0114(35)} = 5.7e^{1.490} \approx 8.5 \text{ billion}$$

(C)

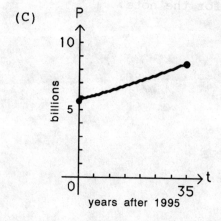

years after 1995

EXERCISE 2-3

Things to remember:

<u>1</u>. ONE-TO-ONE FUNCTIONS

A function f is said to be ONE-TO-ONE if each range value corresponds to exactly one domain value.

2. INVERSE OF A FUNCTION

If f is a one-to-one function, then the INVERSE of f is the function formed by interchanging the independent and dependent variables for f. Thus, if (a, b) is a point on the graph of f, then (b, a) is a point on the graph of the inverse of f.

Note: If f is not one-to-one, then f DOES NOT HAVE AN INVERSE.

3. LOGARITHMIC FUNCTIONS

The inverse of an exponential function is called a LOGARITHMIC FUNCTION. For $b > 0$ and $b \neq 1$,

Logarithmic form

$y = \log_b x$ is equivalent to $x = b^y$

Exponential form

The LOG TO THE BASE b OF x is the exponent to which b must be raised to obtain x. [Remember: A logarithm is an exponent.] The DOMAIN of the logarithmic function is the range of the corresponding exponential function, and the RANGE of the logarithmic function is the domain of the corresponding exponential function. Typical graphs of an exponential function and its inverse, a logarithmic function, for $b > 1$, are shown in the figure below:

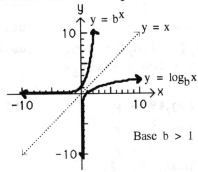

Base b > 1

4. PROPERTIES OF LOGARITHMIC FUNCTIONS

If b, M, and N are positive real numbers, $b \neq 1$, and p and x are real numbers, then:

a. $\log_b 1 = 0$

b. $\log_b b = 1$

c. $\log_b b^x = x$

d. $b^{\log_b x} = x$, $x > 0$

e. $\log_b MN = \log_b M + \log_b N$

f. $\log_b \dfrac{M}{N} = \log_b M - \log_b N$

g. $\log_b M^p = p \log_b M$

h. $\log_b M = \log_b N$ if and only if $M = N$

5. LOGARITHMIC NOTATION; LOGARITHMIC-EXPONENTIAL RELATIONSHIPS

Common logarithm $\quad \log x = \log_{10} x$

Natural logarithm $\quad \ln x = \log_e x$

$\log x = y \quad$ is equivalent to $\quad x = 10^y$

$\ln x = y \quad$ is equivalent to $\quad x = e^y$

---

1. $27 = 3^3$ (using **3**) **3.** $1 = 10^0$ **5.** $8 = 4^{3/2}$

7. $\log_7 49 = 2$ **9.** $\log_4 8 = \dfrac{3}{2}$ **11.** $\log_b A = u$

13. $\log_{10} 1 = y$ is equivalent to $10^y = 1$; $y = 0$.

15. $\log_e e = y$ is equivalent to $e^y = e$; $y = 1$.

17. $\log_{0.2} 0.2 = y$ is equivalent to $(0.2)^y = 0.2$; $y = 1$.

19. $\log_{10} 10^3 = 3$ **21.** $\log_2 2^{-3} = -3$ **23.** $\log_{10} 1,000 = \log_{10} 10^3 = 3$

(using **2a**)

25. $\log_b \dfrac{P}{Q} = \log_b P - \log_b Q$ (using **4f**) **27.** $\log_b L^5 = 5\log_b L$ (using **4g**)

29. $\log_b \dfrac{p}{qrs} = \log_b p - \log_b qrs$ $\qquad$ (using **4f**)

$\qquad = \log_b p - (\log_b q + \log_b r + \log_b s)$ $\quad$ (using **4e**)

$\qquad = \log_b p - \log_b q - \log_b r - \log_b s$

31. $\log_3 x = 2$ **33.** $\log_7 49 = y$ **35.** $\log_b 10^{-4} = -4$

$\quad x = 3^2$ (using **3**) $\quad \log_7 7^2 = y$ $\qquad\qquad 10^{-4} = b^{-4}$

$\quad x = 9$ $\qquad\qquad\quad 2 = y$ This equality implies

$\qquad\qquad\qquad$ Thus, $y = 2$. $b = 10$ (since the
exponents are the same).

37. $\log_4 x = \dfrac{1}{2}$ **39.** $\log_{1/3} 9 = y$ **41.** $\log_b 1,000 = \dfrac{3}{2}$

$\quad x = 4^{1/2}$ $\qquad 9 = \left(\dfrac{1}{3}\right)^y$ $\qquad \log_b 10^3 = \dfrac{3}{2}$

$\quad x = 2$ $\qquad 3^2 = (3^{-1})^y$ $\qquad 3\log_b 10 = \dfrac{3}{2}$

$\qquad\qquad 3^2 = 3^{-y}$ $\qquad\qquad \log_b 10 = \dfrac{1}{2}$

$\qquad$ This inequality $\qquad\qquad 10 = b^{1/2}$
$\qquad$ implies that $\qquad$ Square both sides:
$\qquad 2 = -y$ or $y = -2$. $\qquad 100 = b$, i.e., $b = 100$.

43. $\log_b \dfrac{x^5}{y^3}$

$= \log_b x^5 - \log_b y^3$

$= 5 \log_b x - 3 \log_b y$

45. $\log_b \sqrt[3]{N} = \log_b N^{1/3}$

$= \dfrac{1}{3} \log_b N$

47. $\log_b (x^2 \sqrt[3]{y}) = \log_b x^2 + \log_b y^{1/3} = 2 \log_b x + \dfrac{1}{3} \log_b y$

49. $\log_b (50 \cdot 2^{-0.2t}) = \log_b 50 + \log_b 2^{-0.2t} = \log_b 50 - 0.2t \log_b 2$

51. $\log_b P(1 + r)^t = \log_b P + \log_b (1 + r)^t = \log_b P + t \log_b (1 + r)$

53. $\log_e 100 e^{-0.01t} = \log_e 100 + \log_e e^{-0.01t}$

$\qquad = \log_e 100 - 0.01t \log_e e = \log_e 100 - 0.01t$

55. $\log_b x = \dfrac{2}{3} \log_b 8 + \dfrac{1}{2} \log_b 9 - \log_b 6 = \log_b 8^{2/3} + \log_b 9^{1/2} - \log_b 6$

$\qquad = \log_b 4 + \log_b 3 - \log_b 6 = \log_b \dfrac{4 \cdot 3}{6}$

$\log_b x = \log_b 2$

$\quad x = 2 \text{ (using } \underline{2}e)$

57. $\log_b x = \dfrac{3}{2} \log_b 4 - \dfrac{2}{3} \log_b 8 + 2 \log_b 2 = \log_b 4^{3/2} - \log_b 8^{2/3} + \log_b 2^2$

$\qquad = \log_b 8 - \log_b 4 + \log_b 4 = \log_b 8$

$\log_b x = \log_b 8$

$\quad x = 8 \text{ (using } \underline{2}e)$

59. $\log_b x + \log_b (x - 4) = \log_b 21$

$\qquad \log_b x(x - 4) = \log_b 21$

Therefore, $x(x - 4) = 21$

$\qquad x^2 - 4x - 21 = 0$

$\qquad (x - 7)(x + 3) = 0$

Thus, $x = 7$.

[Note: $x = -3$ is not a solution since $\log_b(-3)$ is not defined.]

61. $\log_{10}(x - 1) - \log_{10}(x + 1) = 1$

$\qquad \log_{10}\left(\dfrac{x - 1}{x + 1}\right) = 1$

Therefore, $\dfrac{x - 1}{x + 1} = 10^1 = 10$

$\qquad x - 1 = 10(x + 1)$

$\qquad x - 1 = 10x + 10$

$\qquad -9x = 11$

$\qquad x = -\dfrac{11}{9}$

There is *no solution*, since

$\log_{10}\left(-\dfrac{11}{9} - 1\right) = \log_{10}\left(-\dfrac{20}{9}\right)$

is not defined. Similarly,

$\log_{10}\left(-\dfrac{11}{9} + 1\right) = \log_{10}\left(-\dfrac{2}{9}\right)$

is not defined.

63. $y = \log_2(x - 2)$
$x - 2 = 2^y$
$x = 2^y + 2$

x	y
$\frac{9}{4}$	-2
$\frac{5}{2}$	-1
3	0
4	1
6	2
18	4

65. The graph of $y = \log_2(x - 2)$ is the graph of $y = \log_2 x$ shifted to the right 2 units.

67. Since logarithmic functions are defined only for positive "inputs", we must have $x + 1 > 0$ or $x > -1$; domain: $(-1, \infty)$. The range of $y = 1 + \ln(x + 1)$ is the set of all real numbers.

69. (A) 3.54743
(B) -2.16032
(C) 5.62629
(D) -3.19704

71. (A) $\log x = 1.1285$
$x = 13.4431$
(B) $\log x = -2.0497$
$x = 0.0089$
(C) $\ln x = 2.7763$
$x = 16.0595$
(D) $\ln x = -1.8879$
$x = 0.1514$

73. $10^x = 12$ (Take common logarithms of both sides)
$\log 10^x = \log 12 \approx 1.0792$
$x \approx 1.0792$ ($\log 10^x = x \log 10 = x$; $\log 10 = 1$)

75. $e^x = 4.304$ (Take natural logarithms of both sides)
$\ln e^x = \ln 4.304 \approx 1.4595$
$x \approx 1.4595$ ($\ln e^x = x \ln e = x$; $\ln e = 1$)

77. $1.03^x = 2.475$ (Take either common or natural logarithms of both sides; we use common logarithms)
$\log(1.03)^x = \log 2.475$
$x = \dfrac{\log 2.475}{\log 1.03} \approx 30.6589$

79. $1.005^{12t} = 3$ (Take either common or natural logarithms of both sides; here we'll use natural logarithms.)
$\ln 1.005^{12t} = \ln 3$
$12t = \dfrac{\ln 3}{\ln 1.005} \approx 220.2713$
$t = 18.3559$

81. $y = \ln x, \quad x > 0$

x	y
0.5	≈ -0.69
1	0
2	≈ 0.69
4	≈ 1.39
5	≈ 1.61

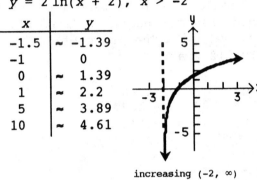

increasing $(0, \infty)$

83. $y = |\ln x|, \quad x > 0$

x	y
0.5	≈ 0.69
1	0
2	≈ 0.69
4	≈ 1.39
5	≈ 1.6

decreasing $(0, 1]$
increasing $[1, \infty)$

85. $y = 2 \ln(x + 2), \quad x > -2$

x	y
-1.5	≈ -1.39
-1	0
0	≈ 1.39
1	≈ 2.2
5	≈ 3.89
10	≈ 4.61

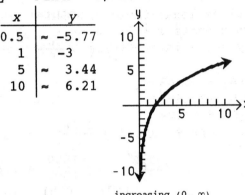

increasing $(-2, \infty)$

87. $y = 4 \ln x - 3, \quad x > 0$

x	y
0.5	≈ -5.77
1	-3
5	≈ 3.44
10	≈ 6.21

increasing $(0, \infty)$

89. The calculator interprets $\log \frac{13}{7}$ as $\frac{\log 13}{7}$ not as $\log\left(\frac{13}{7}\right)$. To find $\log\left(\frac{13}{7}\right)$, calculate $\frac{13}{7}$ and take the common logarithm of the result:

$$\log\left(\frac{13}{7}\right) = \log(1.8571...) \approx 0.2688453123$$

or calculate $\log 13 - \log 7$ to get the same result.

91. For any number b, $b > 0$, $b \neq 1$, $\log_b 1 = y$ is equivalent to $b^y = 1$ which implies $y = 0$. Thus, $\log_b 1 = 0$ for any permissible base b.

93. $\log_{10} y - \log_{10} c = 0.8x$

$\log_{10} \frac{y}{c} = 0.8x$

Therefore, $\frac{y}{c} = 10^{0.8x}$ (using $\underline{1}$)

and $y = c \cdot 10^{0.8x}$.

95.

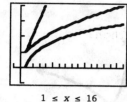

$1 \le x \le 16$

A function f is "larger than" a function g on an interval $[a, b]$ if $f(x) > g(x)$ for $a \le x \le b$. $r(x) > q(x) > p(x)$ for $1 \le x \le 16$, that is $x > \sqrt{x} > \ln x$ for $1 < x \le 16$

97. From the compound interest formula $A = P(1 + r)^t$, we have:

$2P = P(1 + .06)^t$ or $(1.06)^t = 2$

Take the natural log of both sides of this equation:

$\ln(1.06)^t = \ln 2$ [Note: The common log could have been used instead of the natural log.]

$t \ln(1.06) = \ln 2$

$$t = \frac{\ln 2}{\ln(1.06)} \approx \frac{.69315}{.05827} = 11.90 \approx 12 \text{ years}$$

99. (A) $A = P\left(1 + \dfrac{r}{m}\right)^{mt}$, $r = 0.06$, $m = 4$, $P = 1000$, $A = 1800$.

$$1800 = 1000\left(1 + \frac{0.06}{4}\right)^{4t} = 1000(1.015)^{4t}$$

$$(1.015)^{4t} = \frac{1800}{1000} = 1.8$$

$$4t \ln(1.015) = \ln(1.8)$$

$$t = \frac{\ln(1.8)}{4 \ln(1.015)} \approx 9.87$$

$1000 at 6% compounded quarterly will grow to $1800 in 9.87 years.

(B) $A = Pe^{rt}$, $r = 0.06$, $P = 1000$, $A = 1800$

$1000e^{0.06t} = 1800$

$e^{0.06t} = 1.8$

$0.06t = \ln 1.8$

$$t = \frac{\ln 1.8}{0.06} \approx 9.80$$

$1000 at 6% compounded continuously will grow to $1800 in 9.80 years.

101. $A = Pe^{rt}$, $P = 10{,}000$, $A = 20{,}000$, $t = 8$

$20{,}000 = 10{,}000e^{8r}$

$e^{8r} = 2$

$8r = \ln 2$

$$r = \frac{\ln 2}{8} \approx 0.08664$$

$10,000 invested at an annual interest rate of 8.664% compounded continuously will yield $20,000 after 8 years.

103. $I = I_0 10^{N/10}$

Take the common log of both sides of this equation. Then:
$$\log I = \log(I_0 10^{N/10}) = \log I_0 + \log 10^{N/10}$$
$$= \log I_0 + \frac{N}{10}\log 10 = \log I_0 + \frac{N}{10} \text{ (since } \log 10 = 1)$$

So, $\frac{N}{10} = \log I - \log I_0 = \log\left(\frac{I}{I_0}\right)$ and $N = 10\log\left(\frac{I}{I_0}\right)$.

105. Assuming that the world population is currently 5.8 billion and that it will grow at the rate of 1.14% compounded continuously, the population will be
$$P = 5.8e^{0.0114t}$$
after t years.

Given that there are $1.68 \times 10^{14} = 168{,}000$ billion square yards of land, we solve
$$168{,}000 = 5.8e^{0.0114t}$$
for t:
$$e^{0.0114t} \approx 28{,}966$$
$$0.0114t = \ln(28{,}966) \approx 10.2739$$
$$t \approx 901$$
It will take approximately 901 years.

CHAPTER 2 REVIEW

1. $u = e^v$
$v = \ln u$ (2-3)

2. $x = 10^y$
$y = \log x$ (2-3)

3. $\ln M = N$
$M = e^N$ (2-3)

4. $\log u = v$
$u = 10^v$ (2-3)

5. $\dfrac{5^{x+4}}{5^{4-x}} = 5^{x+4-(4-x)} = 5^{2x}$ (2-2)

6. $\left(\dfrac{e^u}{e^{-u}}\right)^u = (e^{u+u})^u = (e^{2u})^u = e^{2u^2}$ (2-2)

7. $\log_3 x = 2$
$x = 3^2$
$x = 9$ (2-3)

8. $\log_x 36 = 2$
$x^2 = 36$
$x = 6$ (2-3)

9. $\log_2 16 = x$
$2^x = 16$
$x = 4$ (2-3)

10. $10^x = 143.7$
$x = \log 143.7$
$x \approx 2.157$ (2-3)

11. $e^x = 503{,}000$
$x = \ln 503{,}000 \approx 13.128$ (2-3)

12. $\log x = 3.105$
$x = 10^{3.105} \approx 1273.503$ (2-3)

13. $\ln x = -1.147$
$x = e^{-1.147} \approx 0.318$ (2-3)

14. (A) 3   (B) 2   (C) 3   (D) 1   (E) 1   (F) 1   (2-1)

15. (A) 4   (B) 3   (C) 4   (D) 0   (E) 1   (F) 1   (2-1)

16. (A) 2   (B) 3   (C) positive   (2-1)

17. (A) 3   (B) 4   (C) negative   (2-1)

18. $f(x) = \dfrac{x + 4}{x - 2}$

(A) *Intercepts:*

     x intercepts: $f(x) = 0$ only if $x + 4 = 0$ or $x = -4$.
               The x intercept is -4.

     y intercepts: $f(0) = \dfrac{0 + 4}{0 - 2} = -2$
               The y intercept is -2.

(B) *Domain:* The denominator is 0 at $x = 2$. Thus, the domain is the set of all real numbers except 2.

(C) *Asymptotes:*

     Vertical asymptotes: $f(x) = \dfrac{x + 4}{x - 2}$

                         The denominator is 0 at $x = 2$. Therefore, the line $x = 2$ is a vertical asymptote.

     Horizontal asymptotes: $f(x) = \dfrac{x + 4}{x - 2} = \dfrac{1 + \dfrac{4}{x}}{1 - \dfrac{2}{x}}$

                         As x increases or decreases without bound, the numerator tends to 1 and the denominator tends to 1. Therefore, the line $y = 1$ is a horizontal asymptote.

(D)

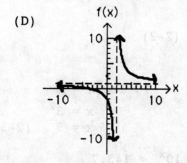

(E)

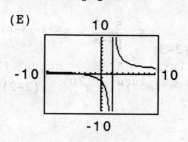

                                                         (2-1)

19. $f(x) = \dfrac{3x - 4}{2 + x}$

(A) *Intercepts:*

     x intercepts: $f(x) = 0$ only if $3x - 4 = 0$ or $x = \dfrac{4}{3}$.
                The x intercept is $\dfrac{4}{3}$.

     y intercepts: $f(0) = \dfrac{3 \cdot 0 - 4}{2 + 0} = -2$
                The y intercept is -2.

(B) *Domain:* The denominator is 0 at $x = -2$. Thus, the domain is the set of all real numbers except -2.

(C) *Asymptotes:*

Vertical asymptotes: $f(x) = \dfrac{3x - 4}{x + 2}$

The denominator is 0 at $x = -2$. Therefore, the line $x = -2$ is a vertical asymptote.

Horizontal asymptotes: $f(x) = \dfrac{3x - 4}{x + 2} = \dfrac{3 - \dfrac{4}{x}}{1 + \dfrac{2}{x}}$

As x increases or decreases without bound, the numerator tends to 3 and the denominator tends to 1. Therefore, the line $y = 3$ is a horizontal asymptote.

(D)

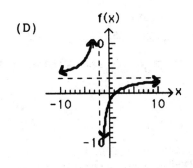

(E)

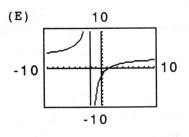

(2-1)

20. $\log(x + 5) = \log(2x - 3)$
$x + 5 = 2x - 3$
$-x = -8$
$x = 8$ (2-3)

21. $2\ln(x - 1) = \ln(x^2 - 5)$
$\ln(x - 1)^2 = \ln(x^2 - 5)$
$(x - 1)^2 = x^2 - 5$
$x^2 - 2x + 1 = x^2 - 5$
$-2x = -6$
$x = 3$ (2-3)

22. $9^{x-1} = 3^{1+x}$
$(3^2)^{x-1} = 3^{1+x}$
$3^{2x-2} = 3^{1+x}$
$2x - 2 = 1 + x$
$x = 3$ (2-2)

23. $e^{2x} = e^{x^2-3}$
$2x = x^2 - 3$
$x^2 - 2x - 3 = 0$
$(x - 3)(x + 1) = 0$
$x = 3, -1$ (2-2)

24. $2x^2 e^x = 3xe^x$
$2x^2 = 3x$ (divide both sides
$2x^2 - 3x = 0$ by e^x)
$x(2x - 3) = 0$
$x = 0, \dfrac{3}{2}$ (2-2)

25. $\log_{1/3} 9 = x$

$\left(\dfrac{1}{3}\right)^x = 9$

$\dfrac{1}{3^x} = 9$

$3^x = \dfrac{1}{9}$

$x = -2$ (2-3)

26. $\log_x 8 = -3$

$x^{-3} = 8$

$\dfrac{1}{x^3} = 8$

$x^3 = \dfrac{1}{8}$

$x = \dfrac{1}{2}$ (2-3)

27. $\log_9 x = \dfrac{3}{2}$

$9^{3/2} = x$

$x = 27$ (2-3)

28. $x = 3(e^{1.49}) \approx 13.3113$ (2-3)

29. $x = 230(10^{-0.161}) \approx 158.7552$ (2-3)

30. $\log x = -2.0144$
$x \approx 0.0097$ (2-3)

31. $\ln x = 0.3618$
$x \approx 1.4359$ (2-3)

32.

$35 = 7(3^x)$

$3^x = 5$

$\ln 3^x = \ln 5$

$x \ln 3 = \ln 5$

$x = \dfrac{\ln 5}{\ln 3} \approx 1.4650$ (2-3)

33.

$0.01 = e^{-0.05x}$

$\ln(0.01) = \ln(e^{-0.05x}) = -0.05x$

Thus, $x = \dfrac{\ln(0.01)}{-0.05} \approx 92.1034$ (2-3)

34.

$8{,}000 = 4{,}000(1.08)^x$

$(1.08)^x = 2$

$\ln(1.08)^x = \ln 2$

$x \ln 1.08 = \ln 2$

$x = \dfrac{\ln 2}{\ln 1.08} \approx 9.0065$

(2-3)

35.

$5^{2x-3} = 7.08$

$\ln(5^{2x-3}) = \ln 7.08$

$(2x - 3)\ln 5 = \ln 7.08$

$2x \ln 5 - 3 \ln 5 = \ln 7.08$

$x = \dfrac{\ln 7.08 + 3 \ln 5}{2 \ln 5}$

$= \dfrac{\ln 7.08 + \ln 5^3}{2 \ln 5}$

$= \dfrac{\ln[7.08(125)]}{2 \ln 5} \approx 2.1081$ (2-3)

36. $x = \log_2 7 = \dfrac{\log 7}{\log 2} \approx 2.8074$

or $x = \log_2 7 = \dfrac{\ln 7}{\ln 2} \approx 2.8074$

(2-3)

37. $x = \log_{0.2} 5.321 = \dfrac{\log 5.321}{\log 0.2} \approx -1.0387$

or $x = \log_{0.2} 5.321 = \dfrac{\ln 5.321}{\ln 0.2} \approx -1.0387$

(2-3)

38. The graph of $f(x) = x^4 - 4x^2 + 1 = x^4\left(1 - \dfrac{4}{x^2} + \dfrac{1}{x^4}\right)$ will "look like" the graph of $y = x^4$; for large x, $f(x) \approx x^4$.

(2-1)

39. (A) 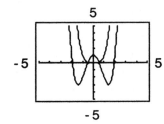 (B)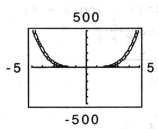

(2-1)

40. $e^x(e^{-x} + 1) - (e^x + 1)(e^{-x} - 1) = 1 + e^x - (1 - e^x + e^{-x} - 1)$
$$= 1 + e^x + e^x - e^{-x}$$
$$= 1 + 2e^x - e^{-x}$$

(2-2)

41. $(e^x - e^{-x})^2 - (e^x + e^{-x})(e^x - e^{-x})$
$$= (e^x)^2 - 2(e^x)(e^{-x}) + (e^{-x})^2 - [(e^x)^2 - (e^{-x})^2]$$
$$= e^{2x} - 2 + e^{-2x} - [e^{2x} - e^{-2x}]$$
$$= 2e^{-2x} - 2$$

(2-2)

42. $y = 2^{x-1}$, $-2 \le x \le 4$

x	y
-2	$\frac{1}{8}$
-1	$\frac{1}{4}$
0	$\frac{1}{2}$
1	1
2	2
4	8

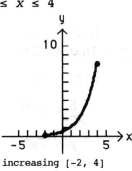

increasing [-2, 4]   (2-2)

43. $f(t) = 10e^{-0.08t}$, $t \ge 0$

t	f(t)
0	10
10	≈ 4.5
20	≈ 2
30	≈ 0.9
40	≈ 0.4

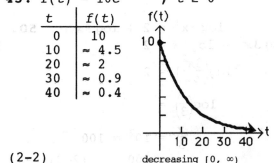

decreasing [0, ∞)   (2-2)

44. $y = \ln(x + 1)$, $-1 < x \le 10$

x	y
-0.5	≈ -0.7
0	0
4	≈ 1.6
8	≈ 2.2
10	≈ 2.4

increasing (-1, 10]   (2-2)

45. $\log 10^\pi = \pi \log 10 = \pi$ (see logarithm properties $\underline{4}$.b & g, Section 2-3)
$10^{\log\sqrt{2}} = y$ is equivalent to $\log y = \log\sqrt{2}$
which implies $y = \sqrt{2}$
Similarly, $\ln e^\pi = \pi \ln e = \pi$ (Section 2-3, $\underline{4}$.b & g) and $e^{\ln\sqrt{2}} = y$
implies $\ln y = \ln\sqrt{2}$ and $y = \sqrt{2}$.

(2-3)

46. $\log x - \log 3 = \log 4 - \log(x + 4)$

$$\log\frac{x}{3} = \log\frac{4}{x + 4}$$

$$\frac{x}{3} = \frac{4}{x + 4}$$

$$x(x + 4) = 12$$

$$x^2 + 4x - 12 = 0$$

$$(x + 6)(x - 2) = 0$$

$$x = -6, 2$$

Since $\log(-6)$ and $\log(-2)$ are not defined, -6 is not a solution. Therefore, the solution is $x = 2$. (2-3)

47. $\ln(2x - 2) - \ln(x - 1) = \ln x$

$$\ln\left(\frac{2x - 2}{x - 1}\right) = \ln x$$

$$\ln\left[\frac{2(x - 1)}{x - 1}\right] = \ln x$$

$$\ln 2 = \ln x$$

$$x = 2 \quad (2-3)$$

48. $\ln(x + 3) - \ln x = 2\ln 2$

$$\ln\left(\frac{x + 3}{x}\right) = \ln(2^2)$$

$$\frac{x + 3}{x} = 4$$

$$x + 3 = 4x$$

$$3x = 3$$

$$x = 1 \quad (2-3)$$

49. $\log 3x^2 = 2 + \log 9x$

$$\log 3x^2 - \log 9x = 2$$

$$\log\left(\frac{3x^2}{9x}\right) = 2$$

$$\log\left(\frac{x}{3}\right) = 2$$

$$\frac{x}{3} = 10^2 = 100$$

$$x = 300 \quad (2-3)$$

50. $\ln y = -5t + \ln c$

$$\ln y - \ln c = -5t$$

$$\ln\frac{y}{c} = -5t$$

$$\frac{y}{c} = e^{-5t}$$

$$y = ce^{-5t} \quad (2-3)$$

51. Let x be *any* positive real number and suppose $\log_1 x = y$. Then $1^y = x$. But, $1^y = 1$, so $x = 1$, i.e., $x = 1$ for all positive real numbers x. This is clearly impossible. (2-3)

52. $A = P\left(1 + \frac{r}{m}\right)^{mt}$.

We let $P = 5,000$, $r = 0.12$, $m = 52$, and $t = 6$. Then we have:

$$A = 5,000\left(1 + \frac{0.12}{52}\right)^{52(6)} \approx 5,000(1 + 0.0023)^{312} \approx 10,263.65$$

Thus, there will be \$10,263.65 in the account 6 years from now. (2-2)

53. $A = Pe^{rt}$. We let $P = 5,000$, $r = 0.12$, and $t = 6$. Then:

$$A = 5,000e^{(0.12)6} \approx 10,272.17$$

Thus, there will be \$10,272.17 in the account 6 years from now. (2-2)

54. The compound interest formula for money invested at 15% compounded annually is:

$A = P(1 + 0.15)^t$

To find the tripling time, we set $A = 3P$ and solve for t:

$$3P = P(1.15)^t$$
$$(1.15)^t = 3$$
$$\ln(1.15)^t = \ln 3$$
$$t \ln 1.15 = \ln 3$$
$$t = \frac{\ln 3}{\ln 1.15} \approx 7.86$$

Thus, the tripling time (to the nearest year) is 8 years. (2-2)

55. The compound interest formula for money invested at 10% compounded continuously is:

$A = Pe^{0.1t}$

To find the doubling time, we set $A = 2P$ and solve for t:

$$2P = Pe^{0.1t}$$
$$e^{0.1t} = 2$$
$$0.1t = \ln 2$$
$$t = \frac{\ln 2}{0.1} \approx 6.93 \text{ years}$$

(2-3)

56. (A) Since $C(x)$ is a linear function of x, it can be written in the form

$$C(x) = mx + b$$

Since the fixed costs are \$300, $b = 300$.

Also, $C(100) = 4300$, so

$$4300 = 100m + 300$$
$$100m = 4000$$
$$m = 40$$

Therefore,

$$C(x) = 40x + 300$$
$$\text{and} \quad \overline{C}(x) = \frac{40x + 300}{x}$$

(B)

(C) $\overline{C}(x) = \dfrac{40x + 300}{x} = \dfrac{40 + \dfrac{300}{x}}{1}$, $5 \leq x \leq 200$

As x increases, the numerator tends to 40 and the denominator is 1. Therefore, $\overline{C}(x)$ approaches 40; The line $y = 40$ is a horizontal asymptote.

(D) $\overline{C}(x)$ approaches \$40 per pair as production increases. (2-1)

57. (A) $\overline{C}(x) = \dfrac{C(x)}{x} = \dfrac{20x^3 - 360x^2 + 2,300x - 1,000}{x}$

(B)

(C) From the graph, $\overline{C}(x)$ has a minimum at $x \approx 8.667$. Thus, the minimum average cost occurs when 8.667 thousand (8,667) cases are handled per year.

$$\overline{C}(8.667) = \dfrac{20(8.667)^3 - 360(8.667)^2 + 2300(8.667) - 1000}{8.667}$$

$\approx$ \$567 per case (2-1)

58. (A) $N(0) = 1$

$N\left(\dfrac{1}{2}\right) = 2$

$N(1) = 4 = 2^2$

$N\left(\dfrac{3}{2}\right) = 8 = 2^3$

$N(2) = 16 = 2^4$

$\vdots$

Thus, we conclude that
$N(t) = 2^{2t}$ or $N = 4^t$.

(B) We need to solve:

$2^{2t} = 10^9$

$\log 2^{2t} = \log 10^9 = 9$

$2t \log 2 = 9$

$t = \dfrac{9}{2 \log 2} \approx 14.95$

Thus, the mouse will die in 15 days.

(2-2, 2-3)

59. Given $I = I_0 e^{-kd}$. When $d = 73.6$, $I = \dfrac{1}{2} I_0$. Thus, we have:

$\dfrac{1}{2} I_0 = I_0 e^{-k(73.6)}$

$e^{-k(73.6)} = \dfrac{1}{2}$

$-k(73.6) = \ln\dfrac{1}{2}$

$k = \dfrac{\ln(0.5)}{-73.6} \approx 0.00942$

Thus, $k \approx 0.00942$.

To find the depth at which 1% of the surface light remains, we set $I = 0.01 I_0$ and solve

$$0.01 I_0 = I_0 e^{-0.00942d}$$

for d:

$$0.01 = e^{-0.00942d}$$
$$-0.00942d = \ln 0.01$$
$$d = \frac{\ln 0.01}{-0.00942} \approx 488.87$$

Thus, 1% of the surface light remains at approximately 489 feet.

(2-2, 2-3)

60. Using the model $P = P_0(1 + r)^t$, we must solve $2P_0 = P_0(1 + 0.03)^t$ for t:

$$2 = (1.03)^t$$
$$\ln(1.03)^t = \ln 2$$
$$t \ln(1.03) = \ln 2$$
$$t = \frac{\ln 2}{\ln 1.03} \approx 23.4$$

Thus, at a 3% growth rate, the population will double in approximately 23.4 years.

(2-2, 2-3)

61. Using the continuous compounding model, we have:

$$2P_0 = P_0 e^{0.03t}$$
$$2 = e^{0.03t}$$
$$0.03t = \ln 2$$
$$t = \frac{\ln 2}{0.03} \approx 23.1$$

Thus, the model predicts that the population will double in approximately 23.1 years.

(2-2, 2-3)

3 MATHEMATICS OF FINANCE

Things to remember:

1. SIMPLE INTEREST

$$I = Prt$$

where P = Principal

r = Annual simple interest rate expressed as a decimal

t = Time in years

2. AMOUNT—SIMPLE INTEREST

$$A = P + Prt = P(1 + rt)$$

where P = Principal or *present value*

r = Annual simple interest rate expressed as a decimal

t = Time in years

A = Amount or *future value*

---

1. 9.5% = 0.095; 60 days = $\dfrac{60}{360} = \dfrac{1}{6}$ year

3. 0.18 = 18%; 5 months = $\dfrac{5}{12}$ year

5. P = \$500, r = 8% = 0.08, t = 6 months = $\dfrac{1}{2}$ year

$I = Prt$ (using $\underline{1}$)

$\quad = 500(0.08)\left(\dfrac{1}{2}\right) = \$20$

7. I = \$80, P = \$500, t = 2 years

$I = Prt$

$r = \dfrac{I}{Pt} = \dfrac{80}{500(2)} = 0.08$ or 8%

9. P = \$100, r = 8% = 0.08, t = 18 months = 1.5 years

$A = P(1 + rt) = 100(1 + 0.08 \cdot 1.5) = \$112$

11. A = \$1000, r = 10% = 0.1, t = 15 months = $\dfrac{15}{12}$ years

$A = P(1 + rt)$

$P = \dfrac{A}{1 + rt} = \dfrac{1000}{1 + (0.1)\left(\dfrac{15}{12}\right)} = \$888.89$

13. $I = Prt$

Divide both sides by Pt.

$$\frac{I}{Pt} = \frac{Prt}{Pt}$$

$$\frac{I}{Pt} = r \quad \text{or} \quad r = \frac{I}{Pt}$$

15. $A = P + Prt = P(1 + rt)$

Divide both sides by $(1 + rt)$.

$$\frac{A}{1 + rt} = \frac{P(1 + rt)}{1 + rt}$$

$$\frac{A}{1 + rt} = P \quad \text{or} \quad P = \frac{A}{1 + rt}$$

17. Each of the graphs is a straight line; the y intercept in each case is 1000 and their slopes are 40, 80, and 120.

19. $P = \$3000$, $r = 14\% = 0.14$, $t = 4$ months $= \frac{1}{3}$ year

$$I = Prt = 3000(0.14)\left(\frac{1}{3}\right) = \$140$$

21. $P = \$554$, $r = 20\% = 0.2$, $t = 1$ month $= \frac{1}{12}$ year

$$I = Prt = 554(0.2)\left(\frac{1}{12}\right) = \$9.23$$

23. $P = \$7250$, $r = 9\% = 0.09$, $t = 8$ months $= \frac{2}{3}$ year

$$A = 7250\left[1 + 0.09\left(\frac{2}{3}\right)\right] = 7250[1.06] = \$7685.00$$

25. $P = \$4000$, $A = \$4270$, $t = 8$ months $= \frac{2}{3}$ year

The interest on the loan is $I = A - P = \$270$. From Problem 9,

$$r = \frac{I}{Pt} = \frac{270}{4000\left(\frac{2}{3}\right)} = 0.10125. \quad \text{Thus, } r = 10.125\%.$$

27. $P = \$1000$, $I = \$30$, $t = 60$ days $= \frac{1}{6}$ year

$$r = \frac{I}{Pt} = \frac{30}{1000\left(\frac{1}{6}\right)} = 0.18. \quad \text{Thus, } r = 18\%.$$

29. $P = \$1500$. The amount of interest paid is $I = (0.5)(3)(120) = \$180$. Thus, the total amount repaid is $\$1500 + \$180 = \$1680$. To find the annual interest rate, we let $t = 120$ days $= \frac{1}{3}$ year. Then

$$r = \frac{I}{Pt} = \frac{180}{1500\left(\frac{1}{3}\right)} = 0.36. \quad \text{Thus, } r = 36\%.$$

31. $P = \$9776.94$, $A = \$10,000$, $t = 13$ weeks $= \frac{1}{4}$ year

The interest is $I = A - P = \$223.06$.

$$r = \frac{I}{Pt} = \frac{223.06}{9776.94\left(\frac{1}{4}\right)} = 0.09126. \quad \text{Thus, } r = 9.126\%.$$

33. $A = \$10,000$, $r = 12.63\% = 0.1263$, $t = 13$ weeks $= \frac{1}{4}$ year.

From Problem 15, $P = \dfrac{A}{1 + rt} = \dfrac{10,000}{1 + (0.1263)\frac{1}{4}} = \dfrac{10,000}{1.03158} = \$9693.91.$

35. Principal plus interest on the original note:

$$A = P(1 + rt) = \$5500\left[1 + 0.12\left(\frac{90}{360}\right)\right]$$

$$= \$5665$$

The third party pays $5,540 and will receive $5665 in 60 days. We want to find r given that $A = 5665$, $P = 5,540$ and $t = \dfrac{60}{360} = \dfrac{1}{6}$

$$A = P + Prt$$
$$r = \frac{A - P}{Pt}$$
$$r = \frac{5665 - 5540}{5540\left(\frac{1}{6}\right)} = 0.13538 \text{ or } 13.538\%$$

37. The principal P is the cost of the stock plus the broker's commission. The cost of the stock is $500(14.20) = \$7100$ and the commission on this is $62 + (0.003)7100 = \$83.30$. Thus, $P = \$7183.30$. The investor sells the stock for $500(16.84) = \$8420$, and the commission on this amount is $62 + (0.003)8420 = \$87.26$. Thus, the investor has $8420 - 87.26 = \$8332.74$ after selling the stock. We can now conclude that the investor has earned $8332.74 - 7183.30 = \$1149.44$.

Now, $P = \$7183.30$, $I = \$1149.44$, $t = 39$ weeks $= \dfrac{3}{4}$ year. Therefore,

$$r = \frac{I}{Pt} = \frac{1149.44}{7183.30\left(\frac{3}{4}\right)} = 0.21335 \quad \text{or} \quad r = 21.335\%.$$

39. The principal P is the cost of the stock plus the broker's commission. The cost of the stock is: $2000(23.75) = \$47,500$, and the commission on this is: $84 + 0.002(47,500) = \$179$. Thus $P = \$47,679$. The investor sells this stock for $2000(26.15) = \$52,300$, and the commission on this amount is: $134 + 0.001(52,300) = \$186.30$. Thus, the investor has $52,300 - 186.30 = \$52,113.70$ after selling the stock. We can now conclude that the investor has earned $52,113.70 - 47,679 = \$4,434.70$.

Now, $P = \$47,679$, $I = \$4434.70$, $t = 300$ days $= \dfrac{300}{360} = \dfrac{5}{6}$ year.

Therefore, $r = \dfrac{I}{Pt} = \dfrac{4434.70}{(47,679)\left(\frac{5}{6}\right)} \approx 0.11161$ or $r = 11.161\%$

Things to remember:

1. AMOUNT—COMPOUND INTEREST

$$A = P(1 + i)^n, \text{ where } i = \frac{r}{m} \text{ and}$$

r = Annual (quoted) rate

m = Number of compounding periods per year

n = Total number of compounding periods

$i = \frac{r}{m}$ = Rate per compounding period

P = Principal (present value)

A = Amount (future value) at the end of n periods

2. EFFECTIVE RATE

If principal P is invested at the (nominal) rate r, compounded m times per year, then the effecitve rate, r_e, is given by

$$r_e = \left(1 + \frac{r}{m}\right)^m - 1.$$

---

1. $P = \$100$, $i = 0.01$, $n = 12$
   Using 1,
   $$\begin{aligned} A = P(1 + i)^n &= 100(1 + 0.01)^{12} \\ &= 100(1.01)^{12} \\ &= \$112.68 \end{aligned}$$

3. $P = \$800$, $i = 0.06$, $n = 25$
   Using 1,
   $$\begin{aligned} A &= 800(1 + 0.06)^{25} \\ &= 800(1.06)^{25} \\ &= \$3433.50 \end{aligned}$$

5. $A = \$10,000$, $i = 0.03$, $n = 48$
   Using 1,
   $$A = P(1 + i)^n$$
   $$P = \frac{A}{(1 + i)^n} = \frac{10,000}{(1 + 0.03)^{48}}$$
   $$= \frac{10,000}{(1.03)^{48}} = \$2419.99$$

7. $A = \$18,000$, $i = 0.01$, $n = 90$
   Refer to Problem 5:
   $$P = \frac{A}{(1 + i)^n} = \frac{18,000}{(1 + 0.01)^{90}}$$
   $$= \frac{18,000}{(1.01)^{90}} = \$7351.04$$

9. $r = 9\%$, $m = 12$. Thus, $i = \frac{r}{m} = \frac{0.09}{12} = 0.0075$ or 0.75% per month.

11. $r = 7\%$, $m = 4$. Thus, $i = \frac{r}{m} = \frac{0.07}{4} = 0.0175$ or 1.75% per quarter.

13. $i = 0.8\%$ per month ($m = 12$). Thus, $r = i \cdot m = (0.008)12 = 0.096$ or 9.6% compounded monthly.

15. $i = 4.5\%$ per half year ($m = 2$). Thus, $r = i \cdot m = (0.045)2 = 0.09$ or 9% compounded semiannually.

17. $P = \$100$, $r = 6\% = 0.06$

    (A) $m = 1$, $i = 0.06$, $n = 4$

        $A = (1 + i)^n$
          $= 100(1 + 0.06)^4$
          $= 100(1.06)^4 = \$126.25$
        Interest $= 126.25 - 100 = \$26.25$

    (B) $m = 4$, $i = \dfrac{0.06}{4} = 0.015$

        $n = 4(4) = 16$
        $A = 100(1 + 0.015)^{16}$
          $= 100(1.015)^{16} = \$126.90$
        Interest $= 126.90 - 100 = \$26.90$

    (C) $m = 12$, $i = \dfrac{0.06}{12} = 0.005$, $n = 4(12) = 48$

        $A = 100(1 + 0.005)^{48} = 100(1.005)^{48} = \$127.05$
        Interest $= 127.05 - 100 = \$27.05$

19. $P = \$5000$, $r = 18\%$, $m = 12$

    (A) $n = 2(12) = 24$
        $i = \dfrac{0.18}{12} = 0.015$
        $A = 5000(1 + 0.015)^{24}$
          $= 5000(1.015)^{24} = \$7147.51$

    (B) $n = 4(12) = 48$
        $i = \dfrac{0.18}{12} = 0.015$
        $A = 5000(1 + 0.015)^{48}$
          $= 5000(1.015)^{48} = \$10,217.39$

21. Each of the graphs is increasing, curves upward and has y intercept 1000. The greater the interest rate, the greater the increase. The amounts at the end of 8 years are:

    At 4%: $A = 1000\left(1 + \dfrac{0.04}{12}\right)^{96} = \$1376.40$

    At 8%: $A = 1000\left(1 + \dfrac{0.08}{12}\right)^{96} = \$1892.46$

    At 12%: $A = 1000\left(1 + \dfrac{0.12}{12}\right)^{96} = \$2599.27$

23. $A = \$10,000$, $r = 8\% = 0.08$, $i = \dfrac{0.08}{2} = 0.04$

    (A) $n = 2(5) = 10$
        $A = P(1 + i)^n$
     $10,000 = P(1 + 0.04)^{10}$
          $= P(1.04)^{10}$
        $P = \dfrac{10,000}{(1.04)^{10}} = \$6755.64$

    (B) $n = 2(10) = 20$
        $P = \dfrac{A}{(1 + i)^n} = \dfrac{10,000}{(1 + 0.04)^{20}}$
           $= \dfrac{10,000}{(1.04)^{20}}$
           $= \$4563.87$

25. Use the formula for r_e in 2.

    (A) $r = 10\% = 0.1$, $m = 4$
        $r_e = \left(1 + \dfrac{0.1}{4}\right)^4 - 1 = 0.1038$
                   or 10.38%

    (B) $r = 12\% = 0.12$, $m = 12$
        $r_e = \left(1 + \dfrac{0.12}{12}\right)^{12} - 1 = 0.1268$
                   or 12.68%

27. We have $P = \$4000$, $A = \$9000$, $r = 15\% = 0.15$, $m = 12$, and $i = \dfrac{0.15}{12} = 0.0125$. Since $A = P(1 + i)^n$, we have:
$9000 = 4000(1 + 0.0125)^n$ or $(1.0125)^n = 2.25$

Method 1: Use Table II. Look down the $(1 + i)^n$ column on the page that has $i = 0.0125$. Find the value of n in this column that is closest to and greater than 2.25. In this case, $n = 66$ months or 5 years and 6 months.

Method 2: Use logarithms and a calculator.

$$\ln(1.0125)^n = \ln 2.25$$
$$n \ln 1.0125 = \ln 2.25$$
$$n = \frac{\ln 2.25}{\ln 1.0125} \approx \frac{0.8109}{0.01242} \approx 65.29$$

Thus, $n = 66$ months or 5 years and 6 months.

29. $A = 2P$, $i = 0.06$

$$A = P(1 + i)^n$$
$$2P = P(1 + 0.06)^n$$
$$(1.06)^n = 2$$
$$\ln(1.06)^n = \ln 2$$
$$n \ln(1.06) = \ln 2$$
$$n = \frac{\ln 2}{\ln 1.06} \approx \frac{0.6931}{0.0583} \approx 11.9 \approx 12$$

31. We have $A = P(1 + i)^n$. To find the doubling time, set $A = 2P$. This yields:

$$2P = P(1 + i)^n \quad \text{or} \quad (1 + i)^n = 2$$

Taking the natural logarithm of both sides, we obtain:

$$\ln(1 + i)^n = \ln 2$$
$$n \ln(1 + i) = \ln 2$$

and

$$n = \frac{\ln 2}{\ln(1 + i)}$$

(A) $r = 10\% = 0.1$, $m = 4$. Thus,

$i = \dfrac{0.1}{4} = 0.025$ and $n = \dfrac{\ln 2}{\ln(1.025)} \approx 28.07$ quarters or $7\frac{1}{4}$ years.

(B) $r = 12\% = 0.12$, $m = 4$. Thus,

$i = \dfrac{0.12}{4} = 0.03$ and $n = \dfrac{\ln 2}{\ln(1.03)} \approx 23.44$ quarters.

That is, 24 quarters or 6 years.

33. $P = \$5000$, $r = 9\% = 0.09$, $m = 4$, $i = \dfrac{0.09}{4} = 0.0225$, $n = 17(4) = 68$

Thus, $A = P(1 + i)^n$

$$= 5000(1 + 0.0225)^{68}$$
$$= 5000(1.0225)^{68}$$
$$= \$22,702.60$$

35. $P = \$110,000$, $r = 6\%$ or 0.06, $m = 1$, $i = 0.06$, $n = 10$

Thus, $A = P(1 + i)^n$

$$= 110,000(1 + 0.06)^{10}$$
$$= 110,000(1.06)^{10}$$
$$\approx \$196,993.25$$

37. $A = \$20$, $r = 7\% = 0.07$, $m = 1$, $i = 0.07$, $n = 5$

$A = P(1 + i)^n$

$P = \dfrac{A}{(1 + i)^n} = \dfrac{20}{(1.07)^5} \approx \$14.26$ per square foot per month

39. From Problem 31, the doubling time is:

$n = \dfrac{\ln 2}{\ln(1 + i)}$

Here $r = i = 0.04$. Thus,

$n = \dfrac{\ln 2}{\ln(1.04)} \approx 17.67$ or 18 years

41. The effective rate, r_e, of $r = 9\% = 0.09$ compounded monthly is:

$r_e = \left(1 + \dfrac{0.09}{12}\right)^{12} - 1 = .0938$ or 9.38%

The effective rate of 9.3% compounded annually is 9.3%. Thus, 9% compounded monthly is better than 9.3% compounded annually.

43. (A) The Declaration of Independence was signed in 1776, 222 years ago. We have $P = \$100$, $r = 0.03$, $n = 4(222) = 888$ and $i = \dfrac{0.03}{4} = 0.0075$.

Thus,
$$\begin{aligned}A &= P(1 + i)^n\\ &= 100(1 + 0.0075)^{888}\\ &= 100(761.3926)\\ &= \$76,139.26\end{aligned}$$

(B) Monthly compounding:

$n = 12(222) = 2664$; $i = \dfrac{0.03}{12} = 0.0025$

$A = 100(1 + 0.0025)^{2664}$
$\quad = \$77,409.05$

Daily compounding:

$n = 365(222) = 81,030$; $i = \dfrac{0.03}{365} = 0.000082191$

$A = 100(1 + 0.000082)^{81,030}$
$\quad = \$78,033.73$

Continuous compounding

$A = Pe^{rt}$
$\quad = 100e^{0.03(222)}$
$\quad = \$78,055.09$

(C)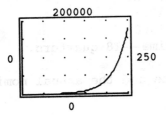

45. $P = \$7000$, $A = \$9000$, $r = 9\% = 0.09$, $m = 12$, $i = \dfrac{0.09}{12} = 0.0075$

Since $A = P(1 + i)^n$, we have:

$9000 = 7000(1 + 0.0075)^n$ or $(1.0075)^n = \dfrac{9}{7}$

Therefore, $\ln(1.0075)^n = \ln\left(\dfrac{9}{7}\right)$

$$n \ln(1.0075) = \ln\left(\dfrac{9}{7}\right)$$

$$n = \dfrac{\ln\left(\dfrac{9}{7}\right)}{\ln(1.0075)} \approx \dfrac{0.2513}{0.0075} \approx 33.6$$

Thus, it will take 34 months or 2 years and 10 months.

47. $P = \$20,000$, $r = 8\% = 0.08$, $m = 365$, $i = \dfrac{0.08}{365} \approx 0.0002192$,

$n = (365)35 = 12,775$

Since $A = P(1 + i)^n$, we have:

$$A = 20,000(1.0002192)^{12,775} \approx \$328,791.70$$

49. From Problem 31, the doubling time is:

$$n = \dfrac{\ln 2}{\ln(1 + i)}$$

(A) $r = 14\% = 0.14$, $m = 365$, $i = \dfrac{0.14}{365} \approx 0.0003836$

Thus, $n = \dfrac{\ln 2}{\ln(1.0003836)} \approx 1807.48$ days or 4.952 years.

(B) $r = 15\% = 0.15$, $m = 1$, $i = 0.15$

Thus, $n = \dfrac{\ln 2}{\ln(1.15)} \approx 4.959$ years.

51. $P = 10,000$, $r = 0.096$, $m = 4$.
The graphs of

$Y_1 = 10000\left(1 + \dfrac{0.096}{4}\right)^x$ and

$Y_2 = 15,000$ are shown in the
figure at the right.

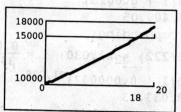

$0 \le x \le 20$ $xscl = 2$

$10,000 \le Y \le 18000$ $yscl = 1000$

intersection:
$x = 17.1$
$y = 15000$
Growth time—18 quarters.

53. The relationship between the effective rate and the annual nominal rate is

$$r_e = \left(1 + \dfrac{r}{m}\right)^m - 1$$

In this case, $r_e = 0.074$ and $m = 365$. Thus, we must solve

$$0.074 = \left(1 + \frac{r}{365}\right)^{365} - 1 \text{ for } r$$

$$\left(1 + \frac{r}{365}\right)^{365} = 1.074$$

$$1 + \frac{r}{365} = (1.074)^{1/365}$$

$$r = 365[(1.074)^{1/365} - 1] \approx 0.0714 \text{ or } r = 7.14\%$$

55. $A = \$30,000$, $r = 10\% = 0.1$, $m = 1$, $i = 0.1$, $n = 17$

From $A = P(1 + i)^n$, we have:

$$P = \frac{A}{(1 + i)^n} = \frac{30,000}{(1.1)^{17}} \approx \$5935.34$$

57. $A = \$30,000$, $P = \$6844.79$, $r = i$, $n = 17$

Using $A = P(1 + r)^n$, we have:

$$30,000 = 6844.79(1 + r)^{17}$$

$$(1 + r)^{17} = \frac{30,000}{6844.79} \approx 4.3829$$

Therefore,

$$1 + r \approx (4.3829)^{1/17} \text{ and } r \approx (4.3829)^{1/17} - 1 \approx 0.0908 \text{ or } r = 9.08\%$$

59. From <u>2</u>, $r_e = \left(1 + \frac{r}{m}\right)^m - 1.$

(A) $r = 8.28\% = 0.0828$, $m = 12$

$$r_e = \left(1 + \frac{0.0828}{12}\right)^{12} - 1 \approx 0.0860 \quad \text{or} \quad 8.60\%$$

(B) $r = 8.25\% = 0.0825$, $m = 365$

$$r_e = \left(1 + \frac{0.0825}{365}\right)^{365} - 1 \approx 0.0860 \quad \text{or} \quad 8.60\%$$

(C) $r = 8.25\% = 0.0825$, $m = 12$

$$r_e = \left(1 + \frac{0.0825}{12}\right)^{12} - 1 \approx 0.0857 \quad \text{or} \quad 8.57\%$$

61. $A = \$32,456.32$, $P = \$24,766.81$, $m = 1$, $n = 2$

$$A = P(1 + r)^n$$

$$32,456.32 = 24,766.81(1 + r)^2$$

$$(1 + r)^2 = \frac{32,456.32}{24,766.81} = 1.3105$$

Therefore, $1 + r = \sqrt{1.3105} \approx 1.1448$ and $r \approx 0.1448$ or 14.48%

63. The effective rate for 8% compounded quarterly is

$$r_e = \left(1 + \frac{0.08}{4}\right)^4 - 1 = (1.02)^4 - 1 = 0.0824 \text{ or } 8.24\%$$

To find the annual nominal rate compounded monthly which has the effective rate of 8.24%, we solve

$$0.0824 = \left(1 + \frac{r}{12}\right)^{12} - 1 \text{ for } r.$$

$$\left(1 + \frac{r}{12}\right)^{12} = 1.0824$$

$$1 + \frac{r}{12} = (1.0824)^{1/12}$$

$$\frac{r}{12} = (1.0824)^{1/12} - 1$$

$$r = 12[(1.0824)^{1/12} - 1] \approx 0.0795 \text{ or } 7.95\%$$

EXERCISE 3-3

Things to remember:

1. FUTURE VALUE OF AN ORDINARY ANNUITY

$$FV = PMT\, \frac{(1 + i)^n - 1}{i} = PMTs_{\overline{n}|i}$$

where PMT = Periodic payment
i = Rate per period
n = Number of payments (periods)
FV = Future value (amount)

(Payments are made at the end of each period.)

2. SINKING FUND PAYMENT

$$PMT = FV\, \frac{i}{(1 + i)^n - 1}$$

where PMT = Sinking fund payment
FV = Value of annuity after n payments (future value)
n = Number of payments (periods)
i = Rate per period

(Payments are made at the end of each period.)

---

1. $n = 20$, $i = 0.03$, $PMT = \$500$

$$FV = PMT\, \frac{(1 + i)^n - 1}{i}$$

$$= PMTs_{\overline{n}|i} \quad \text{(using } \underline{1})$$

$$= 500\, \frac{(1 + 0.03)^{20} - 1}{0.03} = 500s_{\overline{20}|0.03}$$

$$= 500(26.87037449) = \$13,435.19$$

3. $n = 40$, $i = 0.02$, $PMT = \$1000$

$$FV = 1000\, \frac{(1 + 0.02)^{40} - 1}{0.02}$$

$$= 1000s_{\overline{40}|0.02}$$

$$= 1000(60.40198318)$$

$$= \$60,401.98$$

5. $FV = \$3000$, $n = 20$, $i = 0.02$

$$3000 = PMT \, \frac{(1 + 0.02)^{20} - 1}{0.02}$$

$$= PMT s_{\overline{20}|0.02} \quad \text{(using } \underline{1} \text{ or } \underline{2}\text{)}$$

$$= PMT(24.29736980)$$

$$PMT = \frac{3000}{24.29736980} = \$123.47$$

7. $FV = \$5000$, $n = 15$, $i = 0.01$

$$PMT = FV \, \frac{i}{(1 + i)^n - 1}$$

$$= \frac{FV}{s_{\overline{n}|i}} \quad \text{(using } \underline{2}\text{)}$$

$$= 5000 \, \frac{0.01}{(1 + 0.01)^{15} - 1}$$

$$= \frac{5000}{16.09689554} = \$310.62$$

9. $FV = \$4000$, $i = 0.02$, $PMT = 200$, $n = ?$

$$FV = PMT \, \frac{(1 + i)^n - 1}{i}$$

$$\frac{FVi}{PMT} = (1 + i)^n - 1$$

$$(1 + i)^n = \frac{FVi}{PMT} + 1$$

$$\ln(1 + i)^n = \ln\left[\frac{FVi}{PMT} + 1\right]$$

$$n \ln(1 + i) = \ln\left[\frac{FVi}{PMT} + 1\right]$$

$$n = \frac{\ln\left[\dfrac{FVi}{PMT} + 1\right]}{\ln(1 + i)} = \frac{\ln\left[\dfrac{4000(0.02)}{200} + 1\right]}{\ln(1.02)}$$

$$= \frac{\ln(1.4)}{\ln(1.02)} \approx \frac{0.3365}{0.01980} = 16.99 \quad \text{or} \quad 17 \text{ periods}$$

11. $PMT = \$500$, $n = 10(4) = 40$, $i = \dfrac{0.08}{4} = 0.02$

$$FV = 500 \, \frac{(1 + 0.02)^{40} - 1}{0.02} = 500 s_{\overline{40}|0.02}$$

$$= 500(60.40198318) = \$30,200.99$$

Total deposits $= 500(40) = \$20,000$.
Interest $= FV - 20,000 = 30,200.99 - 20,000 = \$10,200.99$.

13. $PMT = \$300$, $i = \dfrac{0.06}{12} = 0.005$, $n = 5(12) = 60$

$$FV = 300 \, \frac{(1 + 0.005)^{60} - 1}{0.005} = 300 s_{\overline{60}|0.005} \quad \text{(using } \underline{1}\text{)}$$

$$= 300(69.77003051) = \$20,931.01$$

After five years, \$20,931.01 will be in the account.

15. $FV = \$25,000$, $i = \dfrac{0.09}{12} = 0.0075$, $n = 12(5) = 60$

$$PMT = \frac{FV}{s_{\overline{60}|0.0075}} = \frac{25,000}{75.42413693} \quad \text{(using the table)}$$

$$= \$331.46 \text{ per month}$$

17. $FV = \$100{,}000$, $i = \dfrac{0.12}{12} = 0.01$, $n = 8(12) = 96$

$$PMT = \frac{FV}{s_{\overline{96}|0.01}} = \frac{100{,}000}{159.92729236} = \$625.28 \text{ per month}$$

19. $FV = PMT\,\dfrac{(1 + i)^n - 1}{i} = 100\,\dfrac{(1 + 0.0075)^{12} - 1}{0.0075}$ (after one year)

$\qquad = 100\,\dfrac{(1.0075)^{12} - 1}{0.0075}$ $\left[\underline{\text{Note}}: PMT = \$100,\ i = \dfrac{0.09}{12} = 0.0075,\ n = 12\right]$

$\qquad = \underline{\$1250.76}$ (1)

Total deposits in one year = $12(100) = \$1200$.

Interest earned in first year = $FV - 1200 = 1250.76 - 1200 = \$50.76$.

At the end of the second year:

$FV = 100\,\dfrac{(1 + 0.0075)^{24} - 1}{0.0075}$ $\quad[\underline{\text{Note}}: n = 24]$

$\qquad = 100\,\dfrac{(1.0075)^{24} - 1}{0.0075} = \underline{\$2618.85}$ (2)

Total deposits plus interest in the second year = (2) − (1)

$\qquad\qquad\qquad\qquad\qquad\qquad\qquad\qquad = 2618.85 - 1250.76$

$\qquad\qquad\qquad\qquad\qquad\qquad\qquad\qquad = \underline{\$1368.09}$ (3)

Interest earned in the second year = (3) − 1200

$\qquad\qquad\qquad\qquad\qquad\qquad\qquad = 1368.09 - 1200$

$\qquad\qquad\qquad\qquad\qquad\qquad\qquad = \$168.09$

At the end of the third year,

$FV = 100\,\dfrac{(1 + 0.0075)^{36} - 1}{0.0075}$ $\quad[\underline{\text{Note}}: n = 36]$

$\qquad = 100\,\dfrac{(1.0075)^{36} - 1}{0.0075}$

$\qquad = \underline{\$4115.27}$ (4)

Total deposits plus interest in the third year = (4) − (2)

$\qquad\qquad\qquad\qquad\qquad\qquad\qquad\qquad = 4115.27 - 2618.85$

$\qquad\qquad\qquad\qquad\qquad\qquad\qquad\qquad = \underline{\$1496.42}$ (5)

Interest earned in the third year = (5) − 1200

$\qquad\qquad\qquad\qquad\qquad\qquad\qquad = 1496.42 - 1200$

$\qquad\qquad\qquad\qquad\qquad\qquad\qquad = \$296.42$

Thus,

Year	Interest earned
1	$ 50.76
2	$168.09
3	$296.42

21. (A) $PMT = \$2000$, $n = 8$, $i = 9\% = 0.09$

$\qquad FV = 2000\,\dfrac{(1 + 0.09)^8 - 1}{0.09} = \dfrac{2000(0.99256)}{0.09} \approx 22{,}056.95$

Thus, Jane will have \$22,056.95 in her account on her 31st birthday. On her 65th birthday, she will have:

$A = 22{,}056.95(1.09)^{34} \approx \$413{,}092$

(B) $PMT = \$2000$, $n = 34$, $i = 9\% = 0.09$

$$FV = 2000\, \frac{(1 + 0.09)^{34} - 1}{0.09} \approx \frac{2000(17.7284)}{0.09} \approx \$393,965$$

23. $FV = \$10,000$, $n = 48$, $i = \dfrac{8\%}{12} = \dfrac{0.08}{12} \approx 0.006667$

From $\underline{2}$, $PMT = \dfrac{10,000(0.006667)}{(1 + 0.006667)^{48} - 1} = \dfrac{66.67}{0.3757} \approx \$177.46$

The total of the monthly deposits for 4 years is $48 \times 177.46 = 8518.08$.
Thus, the interest earned is $10,000 - 8518.08 = \$1481.92$.

25. $PMT = \$150$, $FV = \$7000$, $i = \dfrac{8.5\%}{12} = \dfrac{0.085}{12} \approx 0.00708$. From Problem 9:

$$n = \frac{\ln\left[\dfrac{FVi}{PMT} + 1\right]}{\ln(1 + i)} = \frac{\ln\left[\dfrac{7000(0.00708)}{150} + 1\right]}{\ln(1.00708)} \approx \frac{0.28548}{0.007055} \approx 40.46$$

Thus, $n = 41$ months or 3 years and 5 months.

27. This problem was done with a graphics calculator.

Start with the equation $\dfrac{(1 + i)^n - 1}{i} - \dfrac{FV}{PMT} = 0$

where $FV = 6300$, $PMT = 1000$, $n = 5$ and $i = \dfrac{r}{1} = r$

where r is the nominal annual rate with these values, the equation is:

$$\frac{(1 + r)^5 - 1}{r} - \frac{6300}{1000} = 0$$

or $(1 + r)^5 - 1 - 6.3r = 0$

Set $y = (1 + r)^5 - 1 - 6.3r$ and use your calculator to find the zero r of the function y, where $0 < r < 1$. The result is $r = 0.1158$ or 11.58% to two decimal places.

29. Start with the equation $\dfrac{(1 + i)^n - 1}{i} - \dfrac{FV}{PMT} = 0$

where $FV = 620$, $PMT = 50$, $n = 12$ and $i = \dfrac{r}{12}$,

where r is the annual nominal rate. With these values, the equation becomes

$$\frac{(1 + i)^{12} - 1}{i} - \frac{620}{50} = 0$$

or $(1 + i)^{12} - 1 - 12.4i = 0$

Set $y = (1 + i)^{12} - 1 - 12.4i$ and use your calculator to find the zero i of the function y, where $0 < i < 1$. The result is $i = 0.005941$. Thus $r = 12(0.005941) = 0.0713$ or $r = 7.13\%$ to two decimal places.

31. Annuity: $PMT = 500$, $i = \dfrac{0.06}{4} = 0.015$

$$Y_1 = 500\, \frac{[(1 + 0.015)^{4x} - 1]}{0.015}$$

Simple interest: $P = 5000$, $r = 0.04$

$$Y_2 = 5000(1 + 0.04x)$$

The graphs of Y_1 and Y_2 are shown in the figure:

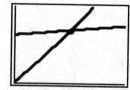

$$0 \leq x \leq 5 \qquad \text{xscl} = 1$$
$$0 \leq y \leq 8000 \qquad \text{yscl} = 1000$$

intersection:
$$x = 2.57$$
$$y = 5514$$

The annuity will be worth more after 2.57 years, or 11 quarterly payments.

EXERCISE 3-4

Things to remember:

1. PRESENT VALUE OF AN ORDINARY ANNUITY

$$PV = PMT \frac{1 - (1 + i)^{-n}}{i} = PMTa_{\overline{n}|i}$$

where PMT = Periodic payment
i = Rate per period
n = Number of periods
PV = Present value of all payments
(Payments are made at the end of each period.)

2. AMORTIZATION FORMULA

$$PMT = PV \frac{i}{1 - (1 + i)^{-n}} = PV \frac{1}{a_{\overline{n}|i}}$$

where PV = Amount of loan (present value)
i = Rate per period
n = Number of payments (periods)
PMT = Periodic payment
(Payments are made at the end of each period.)

1. $PV = 200 \dfrac{1 - (1 + 0.04)^{-30}}{0.04}$

   $= PMTa_{\overline{30}|0.04}$

   $= 200(17.29203330)$ (using the table)

   $= \$3458.41$

3. $PV = 250 \dfrac{1 - (1 + 0.025)^{-25}}{0.025}$

   $= 250a_{\overline{25}|0.025}$

   $= 250(18.42437642)$

   $= \$4606.09$

5. $PMT = 6000 \dfrac{0.01}{1 - (1 + 0.01)^{-36}}$

$\qquad = \dfrac{PV}{a_{\overline{36}|0.01}}$

$\qquad = \dfrac{6000}{30.10750504} = \$199.29$

7. $PMT = 40{,}000 \dfrac{0.0075}{1 - (1 + 0.0075)^{-96}}$

$\qquad = \dfrac{40{,}000}{a_{\overline{96}|0.0075}}$

$\qquad = \dfrac{40{,}000}{68.25843856} = \$586.01$

9. $PV = \$5000, \ i = 0.01, \ PMT = 200$

We have, $PV = PMT \dfrac{1 - (1 + i)^{-n}}{i}$

$\qquad 5000 = 200 \dfrac{1 - (1 + 0.01)^{-n}}{0.01}$

$\qquad\qquad = 20{,}000[1 - (1.01)^{-n}]$

$\qquad \dfrac{1}{4} = 1 - (1.01)^{-n}$

$\qquad (1.01)^{-n} = \dfrac{3}{4} = 0.75$

$\qquad \ln(1.01)^{-n} = \ln(0.75)$

$\qquad -n \ln(1.01) = \ln(0.75)$

$\qquad\qquad n = \dfrac{-\ln(0.75)}{\ln(1.01)} \approx 29$

11. $PMT = \$4000, \ n = 10(4) = 40$

$\qquad i = \dfrac{0.08}{4} = 0.02$

$\qquad PV = $ Present value

$\qquad = PMT \dfrac{1 - (1 + i)^{-n}}{i}$

$\qquad = PMT a_{\overline{n}|i}$

$\qquad = 4000 a_{\overline{40}|0.02}$

$\qquad = 4000(27.35547924)$

$\qquad = \$109{,}421.92$

13. This is a present value problem.

$PMT = \$350, \ n = 4(12) = 48, \ i = \dfrac{0.09}{12} = 0.0075$

Hence, $PV = PMT a_{\overline{n}|i} = 350 a_{\overline{48}|0.0075}$

$\qquad\qquad = 350(40.18478189) = \$14{,}064.67$

They should deposit \$14,064.67. The child will receive $350(48) = $ \$16,800.00.

15. (A) $PV = \$600, \ n = 18, \ i = 0.01$

Monthly payment $= PMT = PV \dfrac{i}{1 - (1 + i)^{-n}}$

$\qquad\qquad = \dfrac{PV}{a_{\overline{n}|i}} = \dfrac{600}{a_{\overline{18}|0.01}} = \dfrac{600}{16.39826858}$

$\qquad\qquad = \$36.59$ per month

The amount paid in 18 payments $= 36.59(18) = \$658.62$.
Thus, the interest paid $= 658.62 - 600 = \$58.62$.

(B) $PMT = \dfrac{600}{a_{\overline{18}|0.015}} \quad (i = 0.015)$

$\qquad = \dfrac{600}{15.67256089} = \$38.28$ per month

For 18 payments, the total amount $= 38.28(18) = \$689.04$.
Thus, the interest paid $= 689.04 - 600 = \$89.04$.

17. Amortized amount = 16,000 − (16,000)(0.25) = $12,000

Thus, $PV = \$12,000$, $n = 6(12) = 72$, $i = 0.015$

$$PMT = \text{monthly payment} = \frac{PV}{a_{\overline{n}|i}} = \frac{12,000}{a_{\overline{72}|0.015}} = \frac{12,000}{43.84466677} = \$273.69 \text{ per month}$$

The total amount paid in 72 months = 273.69(72) = $19,705.68.

Thus, the interest paid = 19,705.68 − 12,000 = $7705.68.

19. First, we compute the required quarterly payment for $PV = \$5000$, $i = 0.045$, and $n = 8$, as follows:

$$PMT = PV \frac{i}{1 - (1 + i)^{-n}} = 5000 \frac{0.045}{1 - (1 + 0.045)^{-8}} = \frac{225}{1 - (1.045)^{-8}}$$

$$= \$758.05 \text{ per quarter}$$

The amortization schedule is as follows:

Payment number	Payment	Interest	Unpaid balance reduction	Unpaid balance
0				$5000.00
1	$758.05	$225.00	$533.05	4466.95
2	758.05	201.01	557.04	3909.91
3	758.05	175.95	582.10	3327.81
4	758.05	149.75	608.30	2719.51
5	758.05	122.38	635.67	2083.84
6	758.05	93.77	664.28	1419.56
7	758.05	63.88	694.17	725.39
8	758.03	32.64	725.39	0.00
Totals	$6064.38	$1064.38	$5000.00	

21. First, we compute the required monthly payment for $PV = \$6000$, $i = \frac{12}{12(100)} = 0.01$, $n = 3(12) = 36$.

$$PMT = PV \frac{i}{1 - (1 + i)^{-n}} = 6000 \frac{0.01}{1 - (1 + 0.01)^{-36}} = \frac{60}{1 - (1.01)^{-36}}$$

$$= \$199.29$$

Now, compute the unpaid balance after 12 payments by considering 24 unpaid payments: $PMT = \$199.29$, $i = 0.01$, and $n = 24$.

$$PV = PMT \frac{1 - (1 + i)^{-n}}{i} = 199.29 \frac{1 - (1 + 0.01)^{-24}}{0.01}$$

$$= 19,929[1 - (1.01)^{-24}] = \$4233.59$$

Thus, the amount of the loan paid in 12 months is 6000 − 4233.59 = $1766.41, and the amount of total payment made during 12 months is 12(199.29) = $2391.48. The interest paid during the first 12 months (first year) is:

2391.48 − 1766.41 = $625.07

Similarly, the unpaid balance after two years can be computed by considering 12 unpaid payments: $PMT = \$199.29$, $i = 0.01$, and $n = 12$.

$$PV = 199.29 \frac{1 - (1 + 0.01)^{-12}}{0.01} = 19,929[1 - (1.01)^{-12}] = \$2243.02$$

Thus, the amount of the loan paid during 24 months is 6000 − 2243.02 = $3756.98, and the amount of the loan paid during the second year is 3756.98 − 1766.41 = $1990.57. The amount of total payment during the second year is 12(199.29) = $2391.48. The interest paid during the second year is:

2391.48 - 1990.57 = $400.91

The total amount paid in 36 months is 199.29(36) = $7174.44. Thus, the total interest paid is 7174.44 - 6000 = $1174.44 and the interest paid during the third year is 1174.44 - (625.07 + 400.91) = 1174.44 - 1025.98 = $148.46.

23. PMT = monthly payment = $525, n = 30(12) = 360, $i = \frac{0.098}{12} \approx 0.0081667$.

Thus, the present value of all payments is:

$$PV = PMT \frac{1 - (1 + i)^{-n}}{i} \approx 525 \frac{1 - (1 + 0.0081667)^{-360}}{0.0081667} \approx \$60,846.38$$

Hence, selling price = loan + down payment
$$= 60,846.38 + 25,000$$
$$= \$85,846.38$$

The total amount paid in 30 years (360 months) = 525(360) = $189,000.
The interest paid is: 189,000 - 60,846.38 = $128,153.62

25. P = $6000, n = 2(12) = 24, $i = \frac{0.035}{12} \approx 0.0029167$

The total amount owed at the end of the two years is:
$$A = P(1 + i)^n = 6000(1 + 0.0029167)^{24} = 6000(1.0029167)^{24} \approx 6434.39$$

Now, the monthly payment is:

$$PMT = PV \frac{i}{1 - (1 + i)^{-n}}$$

where n = 4(12) = 48, PV = $6434.39, $i = \frac{0.035}{12} \approx 0.0029167$. Thus,

$$PMT = 6434.39 \frac{0.0029167}{1 - (1 + 0.0029167)^{-48}} = \$143.85 \text{ per month}$$

The total amount paid in 48 payments is 143.85(48) = $6904.80. Thus, the interest paid is 6904.80 - 6000 = $904.80.

27. First, compute the monthly payment: PV = $75,000, $i = \frac{0.132}{12} = 0.011$, n = 30(12) = 360.

$$\text{Monthly payment} = PV \frac{i}{1 - (1 + i)^{-n}} = 75,000 \frac{0.011}{1 - (1 + 0.011)^{-360}}$$

$$= 75,000 \frac{0.011}{1 - (1.011)^{-360}} = \$841.39$$

(A) Now, to compute the balance after 10 years (with balance of loan to be paid in 20 years), use PMT = $841.39, i = 0.011, and n = 20(12) = 240.

$$\text{Balance after 10 years} = PMT \frac{1 - (1 + i)^{-n}}{i}$$

$$= 841.39 \frac{1 - (1 + 0.011)^{-240}}{0.011}$$

$$= 841.39 \frac{1 - (1.011)^{-240}}{0.011} = \$70,952.33$$

(B) Similarly, the balance of the loan after 20 years (with remainder of loan to be paid in 10 years) is:

$$841.39 \, \frac{1 - (1 + 0.011)^{-120}}{0.011} \quad [\underline{\text{Note}}: n = 12(10) = 120]$$

$$= 841.39 \, \frac{1 - (1.011)^{-120}}{0.011} = \$55,909.02$$

(C) The balance of the loan after 25 years (with remainder of loan to be paid in 5 years) is:

$$841.39 \, \frac{1 - (1 + 0.011)^{-60}}{0.011} \quad [\underline{\text{Note}}: n = 12(5) = 60]$$

$$= 841.39 \, \frac{1 - (1.011)^{-60}}{0.011} = \$36,813.32$$

29. (A) $PV = \$30,000$, $i = \dfrac{0.15}{12} = 0.0125$, $n = 20(12) = 240$.

Monthly payment $PMT = PV \, \dfrac{i}{1 - (1 + i)^{-n}}$

$$= 30,000 \, \frac{0.0125}{1 - (1 + 0.0125)^{-240}}$$

$$= 30,000 \, \frac{0.0125}{1 - (1.0125)^{-240}} = \$395.04$$

The total amount paid in 240 payments is:
395.04(240) = \$94,809.60
Thus, the interest paid is:
\$94,809.60 − \$30,000 = \$64,809.60

(B) New payment = PMT = \$395.04 + \$100.00 = \$495.04. $PV = \$30,000$, $i = 0.0125$.

$$PMT = PV \, \frac{i}{1 - (1 + i)^{-n}}$$

$$495.04 = 30,000 \, \frac{0.0125}{1 - (1 + 0.0125)^{-n}} = \frac{375}{1 - (1.0125)^{-n}}$$

Therefore,

$$1 - (1.0125)^{-n} = \frac{375}{495.04} = 0.7575$$

$$(1.0125)^{-n} = 1 - 0.7575 = 0.2425$$

$$\ln(1.0125)^{-n} = \ln(0.2425)$$

$$-n \ln(1.0125) = \ln(0.2425)$$

$$= \frac{-\ln(0.2425)}{\ln(1.0125)} \approx 114.047 \approx 114 \text{ months or 9.5 years}$$

The total amount paid in 114 payments of \$495.04 is:
495.04(114) = \$56,434.56
Thus, the interest paid is:
\$56,434.56 − \$30,000 = \$26,434.56
The savings on interest is:
\$64,809.60 − \$26,434.56 = \$38,375.04

31. $PV = (\$79,000)(0.80) = \$63,200$, $i = \dfrac{0.12}{12} = 0.01$, $n = 12(30) = 360$.

Monthly payment $PMT = PV \dfrac{i}{1 - (1 + i)^{-n}} = 63,200 \dfrac{0.01}{1 - (1 + 0.01)^{-360}}$

$$= \dfrac{632}{1 - (1.01)^{-360}} = \$650.08$$

Next, we find the present value of a \$650.08 per month, 18-year annuity.
$PMT = \$650.08$, $i = 0.01$, and $n = 12(18) = 216$.

$$PV = PMT \dfrac{1 - (1 + i)^{-n}}{i} = 650.08 \dfrac{1 - (1.01)^{-216}}{0.01}$$

$$= \dfrac{650.08(0.8834309)}{0.01} = \$57,430.08$$

Finally,
Equity = (current market value) − (unpaid loan balance)
  = \$100,000 − \$57,430.08 = \$42,569.92
The couple can borrow (\$42,569.92)(0.70) = \$29,799.

Problems 33 thru 37 start from the equation

$$(*) \quad \dfrac{1 - (1 + i)^{-n}}{i} - \dfrac{PV}{PMT} = 0$$

A graphics calculator was used to solve these problems.

33. The graphs are decreasing, curve downward, and have x intercept 30. The unpaid balances are always in the ratio 2:3:4. The monthly payments and total interest in each case are:

(use $PMT = PV \dfrac{i}{1 - (1 + i)^{-n}}$ where $i = \dfrac{0.09}{12} = 0.0075$, $n = 360$)

$\underline{\$50,000 \text{ mortgage}}$

$PMT = 50,000 \dfrac{0.0075}{1 - (1.0075)^{-360}}$

  $= 50,000(0.0080462)$

  $= \$402.31$ per month

Total interest paid = $360(402.31) - 50,000 = \$94,831.60$

$\underline{\$75,000 \text{ mortgage}}$

$PMT = 75,000 \dfrac{0.0075}{1 - (1.0075)^{-360}}$

  $= 75,000(0.0080462)$

  $= \$603.47$ per month

Total interest paid = $360(603.47) - 75,000 = \$142,249.20$

$\underline{\$100,000 \text{ mortgage}}$

$PMT = 100,000 \dfrac{0.0075}{1 - (1.0075)^{-360}}$

  $= 100,000(0.0080462)$

  $= \$804.62$ per month

Total interest paid = $360(804.62) - 100,000 = \$189,663.20$

35. $PV = 1000$, $PMT = 90$, $n = 12$, $i = \dfrac{r}{12}$ where r is the annual nominal rate. With these values, the equation (*) becomes

$$\frac{1 - (1 + i)^{-12}}{i} - \frac{1000}{90} = 0$$

or $1 - (1 + i)^{-12} - 11.11i = 0$

Put $y = 1 - (1 + i)^{-12} - 11.11i$ and use your calculator to find the zero i of y, where $0 < i < 1$. The result is $i \approx 0.01204$ and $r = 12(0.01204) = 0.14448$. Thus, $r = 14.45\%$ (two decimal places).

37. $PV = 90{,}000$, $PMT = 1200$, $n = 12(10) = 120$, $i = \dfrac{r}{12}$ where r is the annual nominal rate. With these values, the equation (*) becomes

$$\frac{1 - (1 + i)^{-120}}{i} - \frac{90{,}000}{1200} = 0$$

or $1 - (1 + i)^{-120} - 75i = 0$

This equation can be written

$$(1 + i)^{120} - (1 - 75i)^{-1} = 0$$

Put $y = (1 + i)^{120} - (1 - 75i)^{-1}$ and use your calculator to find the zero i of y where $0 < i < 1$. The result is $i \approx 0.00851$ and $r = 12(0.00851) = 0.10212$. Thus $r = 10.21\%$ (two decimal places).

CHAPTER 3 REVIEW

1. $A = 100\left(1 + 0.09 \cdot \dfrac{1}{2}\right)$

$= 100(1.045) = \$104.50$ (3-1)

2. $808 = P\left(1 + 0.12 \cdot \dfrac{1}{12}\right)$

$P = \dfrac{808}{1.01} = \$800$ (3-1)

3. $212 = 200(1 + 0.08 \cdot t)$

$1 + 0.08t = \dfrac{212}{200}$

$0.08t = \dfrac{212}{200} - 1 = \dfrac{12}{200}$

$t = \dfrac{0.06}{0.08} = 0.75$ yr. or 9 mos.

(3-1)

4. $4120 = 4000\left(1 + r \cdot \dfrac{1}{2}\right)$

$1 + \dfrac{r}{2} = \dfrac{4120}{4000}$

$\dfrac{r}{2} = \dfrac{4120}{4000} - 1 = \dfrac{120}{4000} = 0.03$

$r = 0.06$ or   6%

(3-1)

5. $A = 1200(1 + 0.005)^{30}$

$= 1200(1.005)^{30} = \$1393.68$

(3-2)

6. $P = \dfrac{5000}{(1 + 0.0075)^{60}} = \dfrac{5000}{(1.0075)^{60}}$

$= \$3193.50$

(3-2)

7. $FV = 1000s_{\overline{60}|0.005}$

$= 1000 \cdot 69.77003051$

$= \$69{,}770.03$ (3-3)

8. $PMT = \dfrac{FV}{s_{\overline{n}|i}} = \dfrac{8000}{s_{\overline{48}|0.015}}$

$= \dfrac{8000}{69.56321929} = \$115.00$ (3-3)

9. $PV = PMT a_{\overline{n}|i} = 2500 a_{\overline{16}|0.02}$

$\quad = 2500 \cdot 13.57770931$

$\quad = \$33,944.27 \qquad (3\text{-}4)$

10. $PMT = \dfrac{PV}{a_{\overline{n}|i}} = \dfrac{8000}{a_{\overline{60}|0.0075}}$

$\quad = \dfrac{8000}{48.17337352} = \$166.07 \qquad (3\text{-}4)$

11. (A)
$$2500 = 1000(1.06)^n$$
$$(1.06)^n = 2.5$$
$$\ln(1.06)^n = \ln 2.5$$
$$n \ln 1.06 = \ln 2.5$$
$$n = \frac{\ln 2.5}{\ln 1.06} = 15.73 \approx 16$$

(B) We find the intersection of
$$Y_1 = 1000(1.06)^{\wedge}X \quad \text{and} \quad Y_2 = 2500$$
The graphs are shown in the figure:

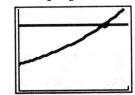

$\quad 0 \le x \le 20 \qquad \text{xscl} = 2$

$\quad 0 \le y \le 3000 \qquad \text{yscl} = 500$

intersection:

$\quad x = 15.73$

$\quad y = 2500 \qquad\qquad (3\text{-}2)$

12. (A)
$$5000 = 100 \frac{(1.01)^n - 1}{0.01}$$
$$5000 = 10{,}000[(1.01)^n - 1]$$
$$0.5 = (1.01)^n - 1$$
$$(1.01)^n = 1.5$$
$$n \ln(1.01) = \ln(1.5)$$
$$n = \frac{\ln(1.5)}{\ln(1.01)} = 40.75 \approx 41$$

(B) We find the intersection of
$$Y_1 = 100 \frac{(1.01)^X - 1}{0.01} = 10{,}000[(1.01)^{\wedge}X - 1] \quad \text{and} \quad Y_2 = 5000$$
The graphs are shown in the figure:

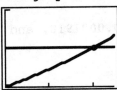

$\quad 0 \le x \le 50 \qquad \text{xscl} = 10$

$\quad 0 \le y \le 10{,}000 \qquad \text{yscl} = 1000$

intersection:

$\quad x = 40.75$

$\quad y = 5000 \qquad\qquad (3\text{-}3)$

13. $P = \$3000$, $r = 0.14$, $t = \dfrac{10}{12}$

$A = 3000\left(1 + 0.14 \cdot \dfrac{10}{12}\right)$ [using $A = P(1 + rt)$]

$\quad = \$3350$

Interest $= 3350 - 3000 = \$350 \qquad\qquad (3\text{-}1)$

14. $P = \$6000$, $r = 9\% = 0.09$, $m = 12$, $i = \dfrac{0.09}{12} = 0.0075$, $n = 12(17) = 204$

$A = P(1 + i)^n = 6000(1 + 0.0075)^{204} = 6000(1.0075)^{204} \approx \$27,551.32$ (3-2)

15. $A = \$25,000$, $r = 10\% = 0.10$, $m = 2$, $i = \dfrac{0.10}{2} = 0.05$, $n = 2(10) = 20$

$P = \dfrac{A}{(1 + i)^n} = \dfrac{25,000}{(1 + 0.05)^{20}} = \dfrac{25,000}{(1.05)^{20}} \approx \$9422.24$ (3-2)

16. (A) The present value of an annuity which provides for quarterly withdrawals of \$5000 for 10 years at 12% interest compounded quarterly is given by:

$PV = PMT\,\dfrac{1 - (1 + i)^{-n}}{i}$ with $PMT = \$5000$, $i = \dfrac{0.12}{4} = 0.03$,

and $n = 10(4) = 40$

$= 5000\,\dfrac{1 - (1 + 0.03)^{-40}}{0.03}$

$= 166,666.67[1 - (1.03)^{-40}] = \$115,573.86$

This amount will have to be in the account when he retires.

(B) To determine the quarterly deposit to accumulate the amount in part (A), we use the formula:

$PMT = FV\,\dfrac{i}{(1 + i)^n - 1}$ where $FV = \$115,573.86$, $i = 0.03$,

and $n = 4(20) = 80$

$= 115,573.86\,\dfrac{0.03}{(1 + 0.03)^{80} - 1}$

$= \dfrac{3467.22}{(1.03)^{80} - 1} = \$359.64$ quarterly payment

(C) The amount collected during the 10-year period is:
(\$5000)40 = \$200,000
The amount deposited during the 20-year period is:
(\$359.64)80 = \$28,771.20
Thus, the interest earned during the 30-year period is:
\$200,000 - \$28,771.20 = \$171,228.80 (3-4)

17. $P = \$10,000$, $r = 7\% = 0.07$, $m = 365$, $i = \dfrac{0.07}{365} = 0.0001918$, and

$n = 40(365) = 14,600$

$A = P(1 + i)^n = 10,000(1 + 0.0001918)^{14,600}$

$= 10,000(1.0001918)^{14,600}$

$= \$164,402$ (3-2)

18. The effective rate for 9% compounded quarterly is:

$r_e = \left(1 + \dfrac{r}{m}\right)^m - 1$, $r = 0.09$, $m = 4$

$= \left(1 + \dfrac{0.09}{4}\right)^4 - 1 = (1.0225)^4 - 1 \approx 0.0931$ or 9.31%

The effective rate for 9.25% compounded annually is 9.25%. Thus, 9% compounded quarterly is the better investment. (3-2)

19. $PMT = \$200$, $r = 9\% = 0.09$, $m = 12$, $i = \dfrac{0.09}{12} = 0.0075$, $n = 12(8) = 96$

$FV = PMT\ \dfrac{(1 + i)^n - 1}{i}$

$= 200\ \dfrac{(1 + 0.0075)^{96} - 1}{0.0075} = 200\ \dfrac{(1.0075)^{96} - 1}{0.0075} \approx \$27{,}971.23$

The total amount invested with 96 payments of $200 is: $96(200) = \$19{,}200$
Thus, the interest earned with this annuity is:
$I = \$27{,}971.23 - \$19{,}200 = \$8771.23$ $\hspace{2em}$ (3-3)

20. $P = \$635$, $r = 22\% = 0.22$, $t = \dfrac{1}{12}$, $I = Prt = 635(0.22)\dfrac{1}{12} = \$11.64$ $\hspace{1em}$ (3-1)

21. $P = \$8000$, $r = 5\% = 0.05$, $m = 1$, $i = \dfrac{0.05}{1} = 0.05$, $n = 5$

$A = P(1 + i)^n = 8000(1 + 0.05)^5 = 8000(1.05)^5 \approx \$10{,}210.25$ $\hspace{1em}$ (3-2)

22. $A = \$8000$, $r = 5\% = 0.05$, $m = 1$, $i = \dfrac{0.05}{1} = 0.05$, $n = 5$

$P = \dfrac{A}{(1 + i)^n} = \dfrac{8000}{(1 + 0.05)^5} = \dfrac{8000}{(1.05)^5} \approx \$6268.21$ $\hspace{1em}$ (3-2)

23. The interest paid was $\$2812.50 - \$2500 = \$312.50$. $\hspace{0.5em}$ $P = \$2500$, $t = \dfrac{10}{12} = \dfrac{5}{6}$

Solving $I = Prt$ for r, we have:

$r = \dfrac{I}{Pt} = \dfrac{312.50}{2500\left(\dfrac{5}{6}\right)} = 0.15$ $\hspace{0.5em}$ or $\hspace{0.5em}$ 15% $\hspace{3em}$ (3-1)

24. The present value, PV, of an annuity of $200 per month for 48 months at 14% interest compounded monthly is given by:

$PV = PMT\ \dfrac{1 - (1 + i)^{-n}}{i}$ $\hspace{0.5em}$ where $PMT = \$200$, $i = \dfrac{0.14}{12} = 0.0116667$
$\hspace{9em}$ and $n = 48$

$= 200\ \dfrac{1 - (1 + 0.0116667)^{-48}}{0.0116667}$

$= 17{,}142.857[1 - (1.0116667)^{-48}] = \$7318.91$

With the $3000 down payment, the selling price of the car is $10,318.91.
The total amount paid is: $\$3000 + 48(\$200) = \$12{,}600$
Thus, the interest paid is: $I = \$12{,}600 - \$10{,}318.91 = \$2281.09$ $\hspace{1em}$ (3-4)

25. $P = \$2500$, $r = 9\% = 0.09$, $m = 4$, $i = \dfrac{0.09}{4} = 0.0225$, $A = \$3000$

$A = P(1 + i)^n$
$3000 = 2500(1 + 0.0225)^n$
$(1.0225)^n = \dfrac{3000}{2500} = 1.2$
$\ln(1.0225)^n = \ln 1.2$
$n \ln 1.0225 = \ln 1.2$
$n = \dfrac{\ln 1.2}{\ln 1.0225} \approx 8.19$

Thus, it will take 9 quarters, or 2 years and 3 months. $\hspace{3em}$ (3-2)

26. (A) $r = 12\% = 0.12$, $m = 12$, $i = \dfrac{0.12}{12} = 0.01$

If we invest P dollars, then we want to know how long it will take to have $2P$ dollars:

$$A = P(1 + i)^n$$
$$2P = P(1 + 0.01)^n$$
$$(1.01)^n = 2$$
$$\ln(1.01)^n = \ln 2$$
$$n \ln 1.01 = \ln 2$$
$$n = \frac{\ln 2}{\ln 1.01} \approx 69.66$$

Thus, it will take 70 months, or 5 years and 10 months, for an investment to double at 12% interest compounded monthly.

(B) $r = 18\% = 0.18$, $m = 12$, $i = \dfrac{0.18}{12} = 0.015$

$$2P = P(1 + 0.015)^n$$
$$(1.015)^n = 2$$
$$\ln(1.015)^n = \ln 2$$
$$n = \frac{\ln 2}{\ln 1.015} \approx 46.56$$

Thus, it will take 47 months, or 3 years and 11 months, for an investment to double at 18% compounded monthly. (3-2)

27. (A) $PMT = \$2000$, $m = 1$, $r = i = 7\% = 0.07$, $n = 45$

$$FV = PMT \frac{(1 + i)^n - 1}{i}$$

$$= 2000 \frac{(1 + 0.07)^{45} - 1}{0.07} = 2000 \frac{(1.07)^{45} - 1}{0.07} \approx \$571,499$$

(B) $PMT = \$2000$, $m = 1$, $r = i = 11\% = 0.11$, $n = 45$

$$FV = PMT \frac{(1 + i)^n - 1}{i}$$

$$= 2000 \frac{(1 + 0.11)^{45} - 1}{0.11} = 2000 \frac{(1.11)^{45} - 1}{0.11} \approx \$1,973,277$$ (3-3)

28. $A = \$17,388.17$, $P = \$12,903.28$, $m = 1$, $r = i$, $n = 3$

$$A = P(1 + i)^n$$
$$17,388.17 = 12,903.28(1 + i)^3$$
$$(1 + i)^3 = \frac{17,388.17}{12,903.28} \approx 1.3475775$$
$$3 \ln(1 + i) = \ln(1.3475775)$$
$$\ln(1 + i) \approx \frac{0.2983085}{3} \approx 0.0994362$$
$$1 + i = e^{0.0994362} \approx 1.1045$$
$$i = 0.1045 \text{ or } 10.45\%$$ (3-2)

29. $P = \$1500$, $I = \$100$,

$t = \dfrac{120}{360} = \dfrac{1}{3}$ year

From Problem 23,

$r = \dfrac{I}{Pt} = \dfrac{100}{1500\left(\frac{1}{3}\right)} = 0.20$ or 20% (3-1)

30. $PMT = \$1500$, $r = 8\% = 0.08$, $m = 4$, $i = \dfrac{0.08}{4} = 0.02$, $n = 2(4) = 8$

We want to find the present value, PV, of this annuity.

$PV = PMT\,\dfrac{1 - (1 + i)^{-n}}{i}$

$= 1500\,\dfrac{1 - (1 + 0.02)^{-8}}{0.02} = 1500\,\dfrac{1 - (1.02)^{-8}}{0.02} = \$10{,}988.22$

The student will receive $8(\$1500) = \$12{,}000$.   (3-4)

31. The amount of the loan is $\$3000\left(\dfrac{2}{3}\right) = \$2000$. The monthly interest rate

is $i = 1.5\% = 0.015$ and $n = 2(12) = 24$.

$PMT = PV\,\dfrac{i}{1 - (1 + i)^{-n}} = 2000\,\dfrac{0.015}{1 - (1 + 0.015)^{-24}} = \dfrac{30}{1 - (1.015)^{-24}}$

$= \$99.85$ per month

The amount paid in 24 payments is

$99.85(24) = \$2396.40$. So, the interest paid is $2396.40 - 2000 = \$396.40$.   (3-4)

32. $FV = \$50{,}000$, $r = 9\% = 0.09$, $m = 12$, $i = \dfrac{0.09}{12} = 0.0075$, $n = 12(6) = 72$

$PMT = FV\,\dfrac{i}{(1 + i)^n - 1} = \dfrac{FV}{s_{\overline{n}|i}} = \dfrac{50{,}000}{s_{\overline{72}|0.0075}}$

$= \dfrac{50{,}000}{95.007028}$ (from Table II)

$= \$526.28$ per month   (3-3)

33. To determine how long it will take money to double, we need to solve the equation $2P = P(1 + i)^n$ for n. From this equation, we obtain:

$(1 + i)^n = 2$

$\ln(1 + i)^n = \ln 2$

$n \ln(1 + i) = \ln 2$

$n = \dfrac{\ln 2}{\ln(1 + i)}$

(A) $i = \dfrac{0.10}{365} = 0.000274$

Thus, $n = \dfrac{\ln 2}{\ln(1.000274)} \approx 2530.08$ days or 6.93 years.

(B) $i = 0.10$

Thus, $n = \dfrac{\ln 2}{\ln(1.1)} \approx 7.27$ years.   (3-2)

34. First, we must calculate the future value of $8000 at 5.5% interest compounded monthly for 2.5 years.

$A = P(1 + i)^n$ where $P = \$8000$, $i = \dfrac{0.055}{12}$ and $n = 30$

$$= 8000\left(1 + \frac{0.055}{12}\right)^{30} = \$9176.33$$

Now, we calculate the monthly payment to amortize this debt at 5.5% interest compounded monthly over 5 years.

$PMT = PV\dfrac{i}{1 + (1 + i)^{-n}}$ where $PV = \$9176.33$, $i = \dfrac{0.055}{12} \approx 0.0045833$ and $n = 12(5) = 60$

$$= 9176.33\frac{0.0045833}{1 - (1 + 0.0045833)^{-60}} = \frac{42.058179}{1 - (1.0045833)^{-60}} \approx \$175.28$$

The total amount paid on the loan is:
$175.28(60) = \$10,516.80$
Thus, the interest paid is:
$I = \$10,516.80 - \$8000 = \$2516.80$

(3-4)

35. Use $FV = PMT\dfrac{(1 + i)^n - 1}{i}$ where $PMT = 1200$ and $i = \dfrac{0.06}{12} = 0.005$. The graphs of

$$Y_1 = 1200\frac{(1.005)^x - 1}{0.005} = 240,000[(1.005)^x - 1]$$
$$Y_2 = 100,000$$

are shown in the figure:

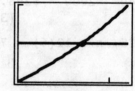

$0 \le x \le 120$ xscl = 12
$0 \le y \le 200,000$ yscl = 10,000

intersection:
$x \approx 70$
$y = 100,000$

The fund will be worth $100,000 after 70 payments, that is, after 5 years, 10 months.

(3-3)

36. We first find the monthly payment:
$PV = \$50,000$
$PMT = PV\dfrac{i}{1 - (1 + i)^{-n}}$ $i = \dfrac{0.09}{12} = 0.0075$
$n = 12(20) = 240$

$$= 50,000\frac{0.0075}{1 - (1.0075)^{-240}}$$
$$= \$449.86 \text{ per month}$$

The present value of the $449.86 per month, 20 year annuity at 9%, after x years, is given by

$$y = 449.86\frac{1 - (1.0075)^{-12(20-x)}}{0.0075}$$

$$= 59,981.33\left[1 - (1.0075)^{-12(20-x)}\right]$$

The graphs of

$$Y_1 = 59,981.33\left[1 - (1.0075)^{-12(20-x)}\right]$$
$$Y_2 = 10,000$$

are shown in the figure:

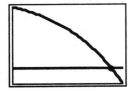

$$0 \le x \le 20 \qquad \text{xscl} = 2$$
$$0 \le y \le 50,000 \qquad \text{yscl} = 10,000$$

intersection:
$$x \approx 18$$
$$y = 10,000$$

The unpaid balance will be below $10,000 after 18 years. (3-4)

37. $P = \$100$, $I = \$0.08$, $t = \dfrac{1}{360}$

From Problem 23, $r = \dfrac{I}{Pt} = \dfrac{0.08}{100\left(\dfrac{1}{360}\right)} = 0.288$ or 28.8% (3-1)

38. $PV = \$1000$, $i = 0.025$, $n = 4$
The quarterly payment is:
$$PMT = PV\,\frac{i}{1 - (1 + i)^{-n}}$$
$$= 1000\,\frac{0.025}{1 - (1 + 0.025)^{-4}} = \frac{25}{1 - (1.025)^{-4}} \approx \$265.82$$

Payment number	Payment	Interest	Unpaid balance reduction	Unpaid balance
0				$1000.00
1	$265.82	$25.00	$240.82	759.18
2	265.82	18.98	246.84	512.34
3	265.82	12.81	253.01	259.33
4	265.81	6.48	259.33	0.00
Totals	$1063.27	$63.27	$1000.00	

(3-4)

39. $PMT = \$200$, $FV = \$2500$, $i = \dfrac{0.0798}{12} = 0.00665$

$$FV = PMT\,\frac{(1 + i)^n - 1}{i}$$

$$2500 = 200\,\frac{(1 + 0.00665)^n - 1}{0.00665} = 30,075.188[(1.00665)^n - 1]$$

$$(1.00665)^n - 1 = \frac{2500}{30,075.188} \approx 0.083125$$

$$(1.0065)^n = 1.083125$$

$$n \ln 1.0065 = \ln 1.083125$$

$$n = \frac{\ln 1.083125}{\ln 1.0065} \approx 12.32 \text{ months}$$

Thus, it will take 13 months, or 1 year and 1 month. (3-2)

40. $FV = \$850,000$, $r = 8.76\% = 0.0876$, $m = 2$, $i = \dfrac{0.0876}{2} = 0.0438$,

$n = 2(6) = 12$

$PMT = FV \dfrac{i}{(1 + i)^n - 1}$

$ = 850{,}000 \dfrac{0.0438}{(1 + 0.0438)^{12} - 1} = \dfrac{37{,}230}{(1.0438)^{12} - 1} \approx \$55{,}347.48$

The total amount invested is:

$12(55{,}347.48) = \$664{,}169.76$

Thus, the interest earned with this annuity is:

$I = \$850{,}000 - \$664{,}169.76 = \$185{,}830.24$ (3-3)

41. The effective rate for Security S & L is:

$r_e = \left(1 + \dfrac{r}{m}\right)^m - 1$ where $r = 9.38\% = 0.0938$ and $m = 12$

$ = \left(1 + \dfrac{0.0938}{12}\right)^{12} - 1 \approx 0.09794$ or 9.794%

The effective rate for West Lake S & L is:

$r_e = \left(1 + \dfrac{r}{m}\right)^m - 1$ where $r = 9.35\% = 0.0935$ and $m = 365$

$ = \left(1 + \dfrac{0.0935}{365}\right)^{365} - 1 \approx 0.09799$ or 9.8%

Thus, West Lake S & L is a better investment. (3-2)

42. $A = \$5000$, $P = \$4899.08$, $t = \dfrac{13}{52} = 0.25$

The interest earned is $I = \$5000.00 - \$4899.08 = \$100.92$. Thus:

$r = \dfrac{I}{Pt} = \dfrac{100.92}{(4899.08)(0.25)} \approx 0.0824$ or 8.24% (3-1)

43. Using the sinking fund formula

$PMT = FV \dfrac{i}{(1 + i)^n - 1}$

with $PMT = \$200$, $FV = \$10,000$, and $i = \dfrac{0.09}{12} = 0.0075$, we have:

$200 = 10{,}000 \dfrac{0.0075}{(1 + 0.0075)^n - 1} = \dfrac{75}{(1.0075)^n - 1}$

Therefore,

$(1.0075)^n - 1 = \dfrac{75}{200} = 0.375$

$ (1.0075)^n = 0.375 + 1 = 1.375$

$\ln(1.0075)^n = \ln 1.375$

$ n = \dfrac{\ln 1.375}{1.0075} \approx 42.62$

The couple will have to make 43 deposits. (3-3)

44. $PV = \$80,000$, $i = \dfrac{0.15}{12} = 0.0125$, $n = 8(12) = 96$

(A) $PMT = PV \dfrac{i}{1 - (1 + i)^{-n}}$

$= 80,000 \dfrac{0.0125}{1 - (1 + 0.0125)^{-96}} = \dfrac{1000}{1 - (1.0125)^{-96}}$

$= \$1435.63$ monthly payment

(B) Now use $PMT = \$1435.63$, $i = 0.0125$, and $n = 96 - 12 = 84$ to calculate the unpaid balance.

$PV = PMT \dfrac{1 - (1 + i)^{-n}}{i}$

$= 1435.63 \dfrac{1 - (1 + 0.0125)^{-84}}{0.0125} = 114,850.40[1 - (1.0125)^{-84}]$

$= \$74,397.48$ unpaid balance after the first year

(C) Amount of loan paid during the first year:
$\$80,000 - \$74,397.48 = \$5602.52$
Amount of payments during the first year:
$12(\$1435.63) = \$17,227.56$
Thus, the interest paid during the first year is:
$\$17,227.56 - \$5602.52 = \$11,625.04$ (3–4)

45. <u>Certificate of Deposit</u>: $\$10,000$ at 8.75% compounded monthly for 288 months

$A = P(1 + i)^n$ $\qquad P = 10,000$

$= 10,000(1.00729)^{288}$ $\qquad i = \dfrac{0.875}{12} = 0.00729$

$= \$81,041.86$ $\qquad n = 288$

<u>Reduce the principal</u>:
Step 1: Find the monthly payment.

$PMT = PV \dfrac{i}{1 - (1 + i)^{-n}}$ $\qquad PV = \$60,000$

$= 60,000 \dfrac{0.00854}{1 - (1.00854)^{-360}}$ $\qquad i = \dfrac{0.1025}{12} = 0.00854$

$= 60,000(0.008961)$ $\qquad n = 12(30) = 360$

$= \$537.66$ per month

Step 2: Find the unpaid balance after 72 payments, that is, find the present value of a $537.66 per month annuity at 10.25% for 288 payments.

$PV = PMT \dfrac{1 - (1 + i)^{-n}}{i}$ $\qquad PMT = \$537.66$

$= 537.66(106.9659599)$ $\qquad i = \dfrac{0.1025}{12} = 0.00854167$

$= \$57,511.32$ $\qquad n = 288$

Step 3: Reduce the principal by \$10,000 and determine the length of time required to pay off the loan.

$$47{,}511.32 = 537.66 \frac{1 - (1.00854167)^{-n}}{0.00854167}$$

$$1 - (1.00854167)^{-n} = 0.7548005$$

$$(1.00854167)^{-n} = 0.2451995$$

$$n = -\frac{\ln(0.2451995)}{\ln(1.00854167)} \approx 165.3$$

The loan will be paid off after 166 more payments. Thus, by reducing the principal after 72 payments, the entire loan will be paid after 72 + 166 = 238 payments.

Step 4: Calculate the future value of a \$537.66 per month annuity at 8.75% for 360 − 238 = 122 months.

$$FV = PMT \frac{(1 + i)^n - 1}{i} \qquad PMT = \$537.66$$

$$= 537.66 \frac{(1.00729167)^{122} - 1}{0.00729167} \qquad i = \frac{0.0875}{12} = 0.00729167$$

$$= 537.66(195.6030015) \qquad n = 122$$

$$= 105{,}155.54$$

Conclusion: Use the \$10,000 to reduce the principal and invest the monthly payment. (3-2, 3-3, 3-4)

46. We find the monthly payment and the total interest for each of the options. The monthly payment is given by:

$$PMT = PV \frac{i}{1 - (1 + i)^{-n}} \qquad \begin{aligned} PV &= \$75{,}000 \\ n &= 12(30) = 360 \end{aligned}$$

<u>12% mortgage</u>: $i = \dfrac{0.12}{12} = 0.01$

$$PMT = 75{,}000 \frac{0.01}{1 - (1.01)^{-360}}$$

$$= 75{,}000(0.010286)$$

$$= 771.46 \text{ per month}$$

Total interest paid = 360(771.46) − 75,000

$$= \$202{,}725.60$$

<u>11.25% mortgage</u>: $i = \dfrac{0.1125}{12} = 0.009375$

$$PMT = 75{,}000 \frac{0.009375}{1 - (1.009375)^{-360}}$$

$$= 75{,}000(0.0097126)$$

$$= \$728.45 \text{ per month}$$

Total interest paid = 360(728.47) − 75,000

$$= \$187{,}242.00$$

The lower rate would save over \$15,000 in interest. (3-4)

47. $A = \$5000$, $r = i = 9.5\% = 0.095$, $n = 5$

$$P = \frac{A}{(1 + i)^n} = \frac{5000}{(1 + 0.095)^5} = \frac{5000}{(1.095)^5} \approx \$3176.14 \tag{3-2}$$

48. $P = \$4476.20$, $A = \$10,000$, $m = 1$, $r = i$, $n = 10$

$$A = P(1 + i)^n$$

$$10,000 = 4476.20(1 + i)^{10}$$

$$(1 + i)^{10} = \frac{10,000}{4476.20} \approx 2.23404$$

$$10 \ln(1 + i) = \ln(2.23404)$$

$$\ln(1 + i) = \frac{\ln(2.23404)}{10} \approx 0.0803811$$

$$1 + i = e^{0.0803811} \approx 1.0837$$

$$i = 0.0837 \text{ or } 8.37\% \tag{3-2}$$

49. $A = \$5000$, $r = 10.76\% = 0.1076$, $t = \frac{26}{52} = 0.5$

$$P = \frac{A}{1 + rt} = \frac{5000}{1 + (0.1076)(0.5)} = \$4744.73 \tag{3-1}$$

50. We first compute the monthly payment using $PV = \$10,000$, $i = \frac{0.12}{12} = 0.01$, and $n = 5(12) = 60$.

$$PMT = PV \frac{i}{1 - (1 + i)^{-n}}$$

$$= 10,000 \frac{0.01}{1 - (1 + 0.01)^{-60}} = \frac{100}{1 - (1.01)^{-60}} = \$222.44 \text{ per month}$$

Now, we calculate the unpaid balance after 24 payments by using $PMT = \$222.44$, $i = 0.01$, and $n = 60 - 24 = 36$.

$$PV = PMT \frac{1 - (1 + i)^{-n}}{i}$$

$$= 222.44 \frac{1 - (1 + 0.01)^{-36}}{0.01} = 22,244[1 - (1.01)^{-36}] = \$6697.11$$

Thus, the unpaid balance after 2 years is $6697.11. $\tag{3-4}$

51. $r = 9\% = 0.09$, $m = 12$

$$r_e = \left(1 + \frac{r}{m}\right)^m - 1 = \left(1 + \frac{0.09}{12}\right)^{12} - 1$$

$$= (1.0075)^{12} - 1 \approx 0.0938 \text{ or } 9.38\% \tag{3-2}$$

52. (A) We first calculate the future value of an annuity of $2000 at 8% compounded annually for 9 years.

$$FV = PMT \frac{(1 + i)^n - 1}{i} \text{ where } PMT = \$2000, i = 0.08, \text{ and } n = 9$$

$$= 2000 \frac{(1 + 0.08)^9 - 1}{0.08} = 25,000[(1.08)^9 - 1] \approx \$24,975.12$$

Now, we calculate the future value of this amount at 8% compounded annually for 36 years.

$A = P(1 + i)^n$, where $P = \$24,975.12$, $i = 0.08$, and $n = 36$

$= 24,975.12(1 + 0.08)^{36} = 24,975.12(1.08)^{36} \approx \$398,807$

(B) This is the future value of a $2000 annuity at 8% compounded annually for 35 years.

$FV = PMT \dfrac{(1 + i)^n - 1}{i}$ where $PMT = \$2000$, $i = 0.08$, and $n = 36$

$= 2000 \dfrac{(1 + 0.08)^{36} - 1}{0.08} = 25,000[(1.08)^{36} - 1] \approx \$374,204$ (3-3)

53. The amount of the loan is ($100,000)(0.8) = $80,000 and

$PMT = PV \dfrac{i}{1 - (1 + i)^{-n}}.$

(A) First, let $i = \dfrac{0.1075}{12} = 0.0089583$, $n = 12(30) = 360$. Then,

$PMT = 80,000 \dfrac{0.0089583}{1 - (1 + 0.0089583)^{-360}} = \dfrac{716.66667}{0.9596687}$

$\approx \$746.79$ monthly payment for 30 years.

Next, let $i = \dfrac{0.1075}{12} = 0.0089583$, $n = 12(15) = 180$. Then

$PMT = 80,000 \dfrac{0.0089583}{1 - (1 + 0.0089593)^{-180}} = \dfrac{716.66667}{0.7991735}$

$\approx \$896.76$ monthly payment for 15 years.

(B) To find the unpaid balance after 10 years, we use

$PV = PMT \dfrac{1 - (1 + i)^{-n}}{i}$

First, for the 30-year mortgage:

$PMT = \$746.79$, $i = \dfrac{0.1075}{12} = 0.0089583$, $n = 12(20) = 240$

$PV = 746.79 \dfrac{1 - (1 + 0.0089583)^{-240}}{0.0089583} = 83,362.915[1 - (1.0089583)^{-240}]$

$= \$73,558.78$ unpaid balance for the 30-year mortgage

Next, for the 15-year mortgage:

$PMT = \$896.76$, $i = 0.0089583$, $n = 5(12) = 60$

$PV = 896.76 \dfrac{1 - (1.0089583)^{-60}}{0.0089583} = 100,103.81[1 - (1.0089583)^{-60}]$

$\approx \$41,482.19$ unpaid balance for the 15-year mortgage.   (3-4)

54. The amount of the mortgage is:
($83,000)(0.8) = $66,400

The monthly payment is given by:

$$PMT = PV \frac{i}{1 - (1 + i)^{-n}} \quad \text{where } PV = \$66,400, \; i = \frac{0.1125}{12} = 0.009375,$$
$$\text{and } n = 12(30) = 360$$

$$= 66,400 \frac{0.009375}{1 - (1 + 0.009375)^{-360}}$$

$$= \frac{622.50}{1 - (1.009375)^{-360}} \approx \$644.92$$

Next, we find the present value of a $644.92 per month, 22-year annuity:

$$PV = PMT \frac{1 - (1 + i)^{-n}}{i} \quad \text{where } PMT = \$644.92, \; i = 0.009375,$$
$$\text{and } n = 12(22) = 264$$

$$= 644.92 \frac{1 - (1 + 0.009375)^{-264}}{0.009375}$$

$$= 68,791.467[1 - (1.009375)^{-264}] = \$62,934.63$$

Finally,

Equity = (current market value) − (unpaid loan balance)
= $95,000 − $62,934.63 = $32,065.37

The family can borrow up to ($32,065.37)(0.60) = $19,239.   (3-4)

4 SYSTEMS OF LINEAR EQUATIONS; MATRICES

EXERCISE 4-1

Things to remember:

1. **SYSTEMS OF TWO EQUATIONS IN TWO VARIABLES**

Given the LINEAR SYSTEM
$$ax + by = h$$
$$cx + dy = k$$
where a, b, c, d, h, and k are real constants, a pair of numbers $x = x_0$ and $y = y_0$ [also written as an ordered pair (x_0, y_0)] is a SOLUTION to this system if each equation is satisfied by the pair. The set of all such ordered pairs is called the SOLUTION SET for the system. To SOLVE a system is to find its solution set.

2. **SYSTEMS OF LINEAR EQUATIONS: BASIC TERMS**

A system of linear equations is CONSISTENT if it has one or more solutions and INCONSISTENT if no solutions exist. Furthermore, a consistent system is said to be INDEPENDENT if it has exactly one solution (often referred to as the UNIQUE SOLUTION) and DEPENDENT if it has more than one solution.

3. The system of two linear equations in two variables
$$ax + by = h$$
$$cx + dy = k$$
can be solved by:
(a) graphing;
(b) substitution;
(c) elimination by addition.

4. **POSSIBLE SOLUTIONS TO A LINEAR SYSTEM**
The linear system
$$ax + by = h$$
$$cx + dy = k$$
must have:
(a) exactly one solution (consistent and independent); or
(b) no solution (inconsistent); or
(c) infinitely many solutions (consistent and dependent).

<u>5.</u> Two systems of linear equations are EQUIVALENT if they have exactly the same solution set. A system of linear equations is transformed into an equivalent system if:

(a) two equations are interchanged;

(b) an equation is multiplied by a nonzero constant;

(c) a constant multiple of one equation is added to another equation.

---

1. (B); no solution

3. (A); $x = -3$, $y = 1$

5. $x + y = 5$
$x - y = 1$
Point of intersection: (3, 2)
Solution: $x = 3$; $y = 2$

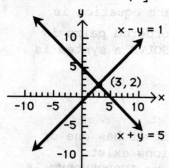

7. $3x - y = 2$
$x + 2y = 10$
Point of intersection: (2, 4)
Solution: $x = 2$; $y = 4$

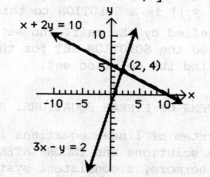

9. $m + 2n = 4$
$2m + 4n = -8$
Since the graphs of the given equations are parallel lines, there is no solution.

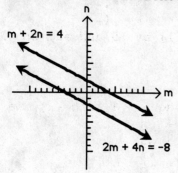

11. $y = 2x - 3$ (1)
$x + 2y = 14$ (2)
By substituting y from (1) into (2), we get:
$x + 2(2x - 3) = 14$
$x + 4x - 6 = 14$
$5x = 20$
$x = 4$
Now, substituting $x = 4$ into (1), we have:
$y = 2(4) - 3$
$y = 5$

Solution: $x = 4$
$y = 5$

13. $2x + y = 6$ (1)
    $x - y = -3$ (2)

Solve (2) for y to obtain the system:

$2x + y = 6$ (3)
    $y = x + 3$ (4)

Substitute y from (4) into (3):

$2x + x + 3 = 6$
       $3x = 3$
        $x = 1$

Now, substituting $x = 1$ into (4), we get:

$y = 1 + 3$
$y = 4$

Solution: $x = 1$
         $y = 4$

15. $3u - 2v = 12$ (1)
    $7u + 2v = 8$ (2)

Add (1) and (2):

$10u = 20$
   $u = 2$

Substituting $u = 2$ into (2), we get:

$7(2) + 2v = 8$
       $2v = -6$
        $v = -3$

Solution: $u = 2$
         $v = -3$

17. $2m - n = 10$ (1)
    $m - 2n = -4$ (2)

Multiply (1) by -2 and add to (2) to obtain:

$-3m = -24$
   $m = 8$

Substituting $m = 8$ into (2), we get:

$8 - 2n = -4$
    $-2n = -12$
      $n = 6$

Solution: $m = 8$
         $n = 6$

19. $9x - 3y = 24$ (1)
    $11x + 2y = 1$ (2)

Solve (1) for y to obtain:

$y = 3x - 8$ (3)

and substitute into (2):

$11x + 2(3x - 8) = 1$
   $11x + 6x - 16 = 1$
          $17x = 17$
            $x = 1$

Now, substitute $x = 1$ into (3):

$y = 3(1) - 8$
$y = -5$

Solution: $x = 1$
         $y = -5$

21. $2x - 3y = -2$ (1)
    $-4x + 6y = 7$ (2)

Multiply (1) by 2 and add to (2) to get:

$0 = 3$

This implies that the system is inconsistent, and thus there is no solution.

23. $3x + 8y = 4$ (1)
    $15x + 10y = -10$ (2)

Multiply (1) by -5 and add to (2) to get:

$-30y = -30$
   $y = 1$

Substituting $y = 1$ into (1), we get:

$3x + 8(1) = 4$
       $3x = -4$
        $x = -\dfrac{4}{3}$

Solution: $x = -\dfrac{4}{3}$; $y = 1$

25. $-6x + 10y = -30$ (1)
     $3x - 5y = 15$ (2)

Multiply (2) by 2 and add to (1). This yields:

$0 = 0$

which implies that (1) and (2) are equivalent equations and there are infinitely many solutions. Geometrically, the two lines are coincident. The system is dependent.

27. $y = 0.07x$ (1)
    $y = 80 + 0.05x$ (2)

Substitute y from (1) into (2):

$0.07x = 80 + 0.05x$

$0.02x = 80$

$x = \dfrac{80}{0.02}$

$x = 4000$

Next, by substituting $x = 4000$ into (1), we get:

$y = 0.07(4000) = 280$

Solution: $x = 4000$; $y = 280$

29. $x + y = 1$ (1)
   $0.3x - 0.4y = 0$ (2)

Multiply equation (2) by 10 to remove the decimals

    $x + y = 1$ (1)
  $3x - 4y = 0$ (3)

Multiply (1) by 4 and add to (2) to get

$7x = 4$

$x = \dfrac{4}{7}$

Now substitute $x = \dfrac{4}{7}$ in (1):

$\dfrac{4}{7} + y = 1$

     $y = 1 - \dfrac{4}{7} = \dfrac{3}{7}$

Solution: $x = \dfrac{4}{7}$, $y = \dfrac{3}{7}$

31. $0.2x - 0.5y = 0.07$ (1)
   $0.8x - 0.3y = 0.79$ (2)

Clear the decimals from (1) and (2) by multiplying each equation by 100.

$20x - 50y = 7$ (3)
$80x - 30y = 79$ (4)

Multiply (3) by -4 and add to (4) to get:

$170y = 51$

    $y = \dfrac{51}{170}$

    $y = 0.3$

Now, substitute $y = 0.3$ into (1):

$0.2x - 0.5(0.3) = 0.07$

    $0.2x - 0.15 = 0.07$

        $0.2x = 0.22$

          $x = 1.1$

Solution: $x = 1.1$; $y = 0.3$

33. $\dfrac{2}{5}x + \dfrac{3}{2}y = 2$ (1)

   $\dfrac{7}{3}x - \dfrac{5}{4}y = -5$ (2)

Multiply (1) by 10 and (2) by 12 to remove the fractions

  $4x + 15y = 20$ (3)
$28x - 15y = -60$ (4)

System (3), (4) is equivalent to system (1), (2). Now add equations (3) and (4) to get

$32x = -40$

   $x = -\dfrac{40}{32} = -\dfrac{5}{4}$

Now substitute $x = -\frac{5}{4}$ into either (1), (2), (3), or (4) — (3) is probably the easiest

$$4\left(-\frac{5}{4}\right) + 15y = 20$$
$$-5 + 15y = 20$$
$$15y = 25$$
$$y = \frac{25}{15} = \frac{5}{3}$$

Solution: $x = -\frac{5}{4}$, $y = \frac{5}{3}$

35. First solve each equation for y:
$$y = \frac{3}{2}x - \frac{5}{2}$$
$$y = -\frac{4}{3}x + \frac{13}{3}$$

The graphs of the equations are:

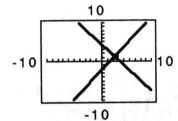

intersection: $x = 2.41$
$\qquad\qquad y = 1.12$

$(2.41, 1.12)$

37. Multiply each equation by 10 and then solve for y:
$$y = \frac{24}{35}x + \frac{1}{35}$$
$$y = \frac{17}{26}x - \frac{1}{13}$$

The graphs of these equations are almost indistinguishable:

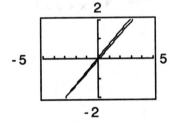

intersection: $x = -3.31$
$\qquad\qquad y = -2.24$

$(-3.31, -2.24)$

39. $x - 2y = -6$ $\qquad (L_1)$
$2x + y = 8$ $\qquad (L_2)$
$x + 2y = -2$ $\qquad (L_3)$

(A) L_1 and L_2 intersect:
$\qquad x - 2y = -6$ $\qquad (1)$
$\qquad 2x + y = 8$ $\qquad (2)$

Multiply (2) by 2 and add to (1):
$\qquad 5x = 10$
$\qquad x = 2$

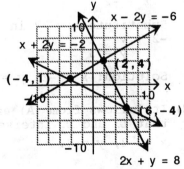

Substitute $x = 2$ in (1) to get
$$2 - 2y = -6$$
$$-2y = -8$$
$$y = 4$$
Solution: $x = 2$, $y = 4$

(B) L_1 and L_3 intersect:
$$x - 2y = -6 \qquad (3)$$
$$x + 2y = -2 \qquad (4)$$
Add (3) and (4):
$$2x = -8$$
$$x = -4$$
Substitute $x = -4$ in (3) to get
$$-4 - 2y = -6$$
$$-2y = -2$$
$$y = 1$$
Solution: $x = -4$, $y = 1$

(C) L_2 and L_3 intersect:
$$2x + y = 8 \qquad (5)$$
$$x + 2y = -2 \qquad (6)$$
Multiply (6) by -2 and add to (5)
$$-3y = 12$$
$$y = -4$$
Substitute $y = -4$ in (5) to get
$$2x - 4 = 8$$
$$2x = 12$$
$$x = 6$$
Solution: $x = 6$, $y = -4$

41. $x + y = 1 \qquad (L_1)$
$x - 2y = -8 \qquad (L_2)$
$3x + y = -3 \qquad (L_3)$

(A) L_1 and L_2 intersect
$$x + y = 1 \qquad (1)$$
$$x - 2y = -8 \qquad (2)$$
Subtract (2) from (1):
$$3y = 9$$
$$y = 3$$
Substitute $y = 3$ in (1) to get
$$x + 3 = 1$$
$$x = -2$$
Solution: $x = -2$, $y = 3$

(B) L_1 and L_3 intersect:
$$x + y = 1 \qquad (3)$$
$$3x + y = -3 \qquad (4)$$
Subtract (4) from (3):
$$-2x = 4$$
$$x = -2$$
Substitute $x = -2$ in (3) to get
$$-2 + y = 1$$
$$y = 3$$
Solution: $x = -2$, $y = 3$

(C) It follows from (A) and (B),
that L_2 and L_3 intersect at
$x = -2$, $y = 3$.

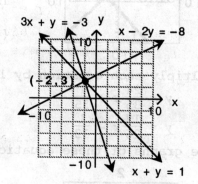

43. $4x - 3y = -24$ $\quad(L_1)$
$2x + 3y = 12$ $\quad(L_2)$
$8x - 6y = 24$ $\quad(L_3)$

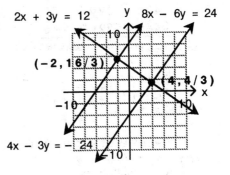

(A) L_1 and L_2 intersect:
$4x - 3y = -24$ $\quad(1)$
$2x + 3y = 12$ $\quad(2)$

Add (1) and (2):
$6x = -12$
$x = -2$

Substitute $x = -2$ in (2) to get
$2(-2) + 3y = 12$
$3y = 16$
$y = \dfrac{16}{3}$

Solution: $x = -2$, $y = \dfrac{16}{3}$

(B) In slope-intercept form, (L_1) and (L_3) have equations

$y = \dfrac{4}{3}x + 8$ $\quad(L_1)$

$y = \dfrac{4}{3}x - 4$ $\quad(L_2)$

Thus, (L_1) and (L_3) have the same slope and different y-intercepts; (L_1) and (L_3) are parallel; they do not intersect.

(C) L_2 and L_3 intersect:
$2x + 3y = 12$ $\quad(3)$
$8x - 6y = 24$ $\quad(4)$

Multiply (3) by 2 and add to (4):
$12x = 48$
$x = 4$

Substitute $x = 4$ in (3) to get
$2(4) + 3y = 12$
$3y = 4$
$y = \dfrac{4}{3}$

Solution: $x = 4$, $y = \dfrac{4}{3}$

45. (A) $\quad 5x + 4y = 4$ Multiply the top equation by 9 and the bottom
$\quad 11x + 9y = 4$ equation by -4.

$\quad 45x + 36y = 36$ Add the equations.
$\quad \underline{-44x - 36y = -16}$
$\quad\quad\ x \quad\quad = 20$

$5(20) + 4y = 4$ Substitute $x = 20$ in the first equation.
$4y = -96$
$y = -24$

Solution: $(20, -24)$

(B) $5x + 4y = 4$ Multiply the top equation by 8 and the bottom
      $11x + 8y = 4$ equation by -4.

      $40x + 32y = 32$ Add the equations.
      $\underline{-44x - 32y = -16}$
      $-4x \quad\quad = 16$
      $\quad\quad x = -4$

      $5(-4) + 4y = 4$ Substitute $x = 4$ in the first equation.
      $\quad\quad 4y = 24$
      $\quad\quad\, y = 6$

      Solution: $(-4, 6)$

(C) $5x + 4y = 4$ Multiply the top equation by 8 and the bottom
      $10x + 8y = 4$ equation by -4.

      $40x + 32y = 32$ Add the equations.
      $\underline{-40x - 32y = -16}$
      $\quad\quad\quad\quad 0 = -16$ This system has no solutions.

47. (A) $p = 0.7q + 3$ (1)
        $p = -1.7q + 15$ (2)
        Solve the above system for
        equilibrium price p and the
        equilibrium quantity q.
        $0.7q + 3 = -1.7q + 15$
        $0.7q + 1.7q = 15 - 3$
        $2.4q = 12$
        $q = \dfrac{12}{2.4}$
        $q = 5$ (5 hundreds
                  or 500)
        Equilibrium quantity $q = 5$.
        Substitute $q = 5$ in (1):
        $p = 0.7(5) + 3$
        $p = 3.5 + 3$
        $p = 6.50$
        Equilibrium price $p = \$6.50$

(B)
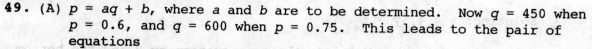

49. (A) $p = aq + b$, where a and b are to be determined. Now $q = 450$ when
        $p = 0.6$, and $q = 600$ when $p = 0.75$. This leads to the pair of
        equations
        $450a + b = 0.6$ (1)
        $600a + b = 0.75$ (2)

        Subtracting (1) from (2), we have
        $150a = 0.15$
        $a = 0.001$

        Substituting $a = 0.001$ in (2), we get
        $600(0.001) + b = 0.75$
        $b = 0.15$

        Thus $p = 0.001q + 0.15$ <u>Supply equation</u>

(B) $p = aq + b$; $q = 570$ when $p = 0.6$ and $q = 495$ when $p = 0.75$. This leads to the pair of equations

$570a + b = 0.6$ (3)
$495a + b = 0.75$ (4)

Subtracting (3) from (4), we have
$-75a = 0.15$
   $a = -0.002$
Substituting $a = -0.002$ in (3), we get
$570(-0.002) + b = 0.6$
           $b = 1.74$

Thus $p = -0.002q + 1.74$ <u>Demand equation</u>

(C) Equilibrium occurs when supply equals demand. Equating the supply and demand equations, yields
$0.001q + 0.15 = -0.002q + 1.74$
      $0.003q = 1.59$
         $q = 530$ <u>Equilibrium quantity</u>

Substituting $q = 530$ into the supply equation (or into the demand equation), we get
  $p = 0.001(530) + 0.15$
  $p = 0.68$ or $0.68   <u>Equilibrium price</u>

(D)

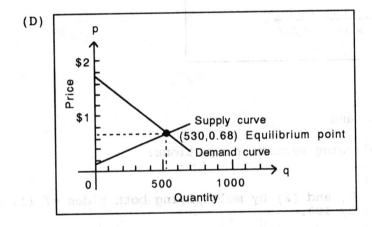

51. (A) The company breaks even when:
             Cost = Revenue
$48,000 + 1400x = 1800x$
       $48,000 = 1800x - 1400x$
or $400x = 48,000$
           $x = \dfrac{48,000}{400}$
           $x = 120$
Thus, 120 units must be manufactured and sold to break even.
Cost $= 48,000 + 1400(120)$
     $= 216,000 =$ Revenue

(B)

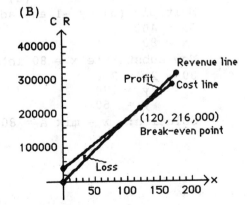

53. Let x = number of tapes marketed per month

  (A) Revenue: $R = 19.95x$
  Cost: $C = 7.45x + 24,000$
  At the break-even point Revenue = Cost, that is
  $19.95x = 7.45x + 24,000$
  $12.50x = 24,000$
  $\quad\quad x = 1920$
  Thus, 1920 tapes must be sold per month to break even.
  Cost = Revenue = \$38,304 at the break-even point.

  (B)

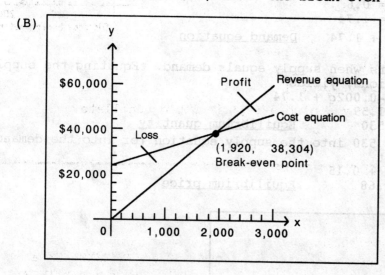

55. Let x = amount of mix A, and
  $\quad\quad y$ = amount of mix B.
  We want to solve the following system of equations:
  $0.1x + 0.2y = 20 \quad (1)$
  $0.06x + 0.02y = 6 \quad (2)$
  Clear the decimals from (1) and (2) by multiplying both sides of (1) by 10 and both sides of (2) by 100.
  $\quad x + 2y = 200 \quad\quad (3)$
  $6x + 2y = 600 \quad\quad (4)$
  Multiply (3) by −1 and add to (4):
  $5x = 400$
  $\quad x = 80$

  Now substitute $x = 80$ into (3):
  $80 + 2y = 200$
  $\quad\quad 2y = 120$
  $\quad\quad\; y = 60$

  Solution: x = mix A = 80 grams; y = mix B = 60 grams

57. $p = -\frac{1}{5}d + 70$ [Approach equation]

$p = -\frac{4}{3}d + 230$ [Avoidance equation]

(A) The figure shows the graphs of the two equations.

(B) Setting the two equations equal to each other, we have

$$-\frac{1}{5}d + 70 = -\frac{4}{3}d + 230$$

$$-\frac{1}{5}d + \frac{4}{3}d = 230 - 70$$

$$\frac{17}{15}d = 160$$

$$d = 141 \text{ cm (approx.)}$$

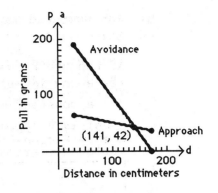

(C) The rat would be very confused (!); it would vacillate.

EXERCISE 4-2

Things to remember:

1. **MATRICES**

A MATRIX is a rectangular array of numbers written within brackets. Each number in a matrix is called an ELEMENT. If a matrix has m rows and n columns, it is called an $m \times n$ MATRIX; $m \times n$ is the SIZE; m and n are the DIMENSIONS. A matrix with n rows and n columns is a SQUARE MATRIX OF ORDER n. A matrix with only one column is a COLUMN MATRIX; a matrix with only one row is a ROW MATRIX. The element in the ith row and jth column of a matrix A is denoted a_{ij}. The PRINCIPAL DIAGONAL of a matrix A consists of the elements a_{11}, a_{22}, a_{33}, ...

2. A system of linear equations is transformed into an equivalent system if:

(a) two equations are interchanged;
(b) an equation is multiplied by a nonzero constant;
(c) a constant multiple of one equation is added to another equation.

3. Associated with the linear system

$$\begin{aligned} a_1 x_1 + b_1 x_2 &= k_1 \\ a_2 x_1 + b_2 x_2 &= k_2 \end{aligned} \qquad \text{(I)}$$

is the AUGMENTED MATRIX of the system

$$\left[\begin{array}{cc|c} a_1 & b_1 & k_1 \\ a_2 & b_2 & k_2 \end{array} \right]. \qquad \text{(II)}$$

<u>4</u>. An augmented matrix is transformed into a row-equivalent matrix if:

(a) two rows are interchanged ($R_i \leftrightarrow R_j$);

(b) a row is multiplied by a nonzero constant ($kR_i \rightarrow R_i$);

(c) a constant multiple of one row is added to another row ($kR_j + R_i \rightarrow R_i$).

(Note: The arrow $\rightarrow$ means "replaces.")

<u>5</u>. Given the system of linear equations (I) and its associated augmented matrix (II). If (II) is row equivalent to a matrix of the form:

(1) $\begin{bmatrix} 1 & 0 & m \\ 0 & 1 & n \end{bmatrix}$, then (I) has a unique solution; (consistent and independent);

(2) $\begin{bmatrix} 1 & m & n \\ 0 & 0 & 0 \end{bmatrix}$, then (I) has infinitely many solutions (consistent and dependent);

(3) $\begin{bmatrix} 1 & m & n \\ 0 & 0 & p \end{bmatrix}$, $p \neq 0$, then (I) has no solution (inconsistent).

_____

1. A is 2×3; C is 1×3 **3.** C **5.** B

7. $a_{12} = -4$, $a_{23} = -5$ **9.** -1, 8, 0

11. Interchange row 1 and row 2.

$\begin{bmatrix} 4 & -6 & -8 \\ 1 & -3 & 2 \end{bmatrix}$

13. Multiply row 1 by -4.

$\begin{bmatrix} -4 & 12 & -8 \\ 4 & -6 & -8 \end{bmatrix}$

15. Multiply row 2 by 2.

$\begin{bmatrix} 1 & -3 & 2 \\ 8 & -12 & -16 \end{bmatrix}$

17. Replace row 2 by the sum of row 2 and -4 times row 1.

$\begin{bmatrix} 1 & -3 & 2 \\ 0 & 6 & -16 \end{bmatrix}$

19. Replace row 2 by the sum of row 2 and -2 times row 1.

$\begin{bmatrix} 1 & -3 & 2 \\ 2 & 0 & -12 \end{bmatrix}$

21. Replace row 2 by the sum of row 2 and -1 times row 1.

$\begin{bmatrix} 1 & -3 & 2 \\ 3 & -3 & -10 \end{bmatrix}$

23. System $\qquad$ Augmented matrix $\qquad$ Graphs:

$$x_1 + x_2 = 5$$
$$x_1 - x_2 = 1$$

$$\begin{bmatrix} 1 & 1 & | & 5 \\ 1 & -1 & | & 1 \end{bmatrix}$$

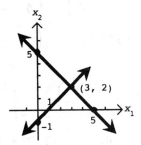

$$\begin{bmatrix} 1 & 1 & | & 5 \\ 1 & -1 & | & 1 \end{bmatrix} \quad (-1)R_1 + R_2 \rightarrow R_2 \quad \begin{bmatrix} 1 & 1 & | & 5 \\ 0 & -2 & | & -4 \end{bmatrix}$$

$$x_1 + x_2 = 5$$
$$-2x_2 = -4$$

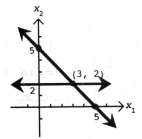

$$\begin{bmatrix} 1 & 1 & | & 5 \\ 0 & -2 & | & -4 \end{bmatrix} \quad -\frac{1}{2}R_2 \rightarrow R_2 \quad \begin{bmatrix} 1 & 1 & | & 5 \\ 0 & 1 & | & 2 \end{bmatrix}$$

$$x_1 + x_2 = 5$$
$$x_2 = 2$$

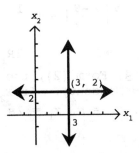

$$\begin{bmatrix} 1 & 1 & | & 5 \\ 0 & 1 & | & 2 \end{bmatrix} \quad (-1)R_2 + R_1 \rightarrow R_1 \quad \begin{bmatrix} 1 & 0 & | & 3 \\ 0 & 1 & | & 2 \end{bmatrix}$$

$$x_1 \quad = 3$$
$$x_2 = 2$$

Solution: $x_1 = 3$, $x_2 = 2$. Each pair of lines has the same intersection point.

25. $\begin{bmatrix} 1 & -2 & | & 1 \\ 2 & -1 & | & 5 \end{bmatrix} \sim \begin{bmatrix} 1 & -2 & | & 1 \\ 0 & 3 & | & 3 \end{bmatrix} \sim \begin{bmatrix} 1 & -2 & | & 1 \\ 0 & 1 & | & 1 \end{bmatrix} \sim \begin{bmatrix} 1 & 0 & | & 3 \\ 0 & 1 & | & 1 \end{bmatrix}$ $\quad$ Thus, $x_1 = 3$ and $x_2 = 1$.

$$(-2)R_1 + R_2 \rightarrow R_2 \qquad \frac{1}{3}R_2 \rightarrow R_2 \qquad 2R_2 + R_1 \rightarrow R_1$$

27. $\begin{bmatrix} 1 & -4 & | & -2 \\ -2 & 1 & | & -3 \end{bmatrix} \sim \begin{bmatrix} 1 & -4 & | & -2 \\ 0 & -7 & | & -7 \end{bmatrix} \sim \begin{bmatrix} 1 & -4 & | & -2 \\ 0 & 1 & | & 1 \end{bmatrix} \sim \begin{bmatrix} 1 & 0 & | & 2 \\ 0 & 1 & | & 1 \end{bmatrix}$ Thus, $x_1 = 2$ and $x_2 = 1$.

$\qquad 2R_1 + R_2 \to R_2 \qquad\quad \left(-\frac{1}{7}\right)R_2 \to R_2 \qquad\quad 4R_2 + R_1 \to R_1$

29. $\begin{bmatrix} 3 & -1 & | & 2 \\ 1 & 2 & | & 10 \end{bmatrix} \sim \begin{bmatrix} 1 & 2 & | & 10 \\ 3 & -1 & | & 2 \end{bmatrix} \sim \begin{bmatrix} 1 & 2 & | & 10 \\ 0 & -7 & | & -28 \end{bmatrix} \sim \begin{bmatrix} 1 & 2 & | & 10 \\ 0 & 1 & | & 4 \end{bmatrix} \sim \begin{bmatrix} 1 & 0 & | & 2 \\ 0 & 1 & | & 4 \end{bmatrix}$

$\qquad\quad R_1 \leftrightarrow R_2 \qquad (-3)R_1 + R_2 \to R_2 \qquad \left(-\frac{1}{7}\right)R_2 \to R_2 \qquad (-2)R_2 + R_1 \to R_1$

Thus, $x_1 = 2$ and $x_2 = 4$.

31. $\begin{bmatrix} 1 & 2 & | & 4 \\ 2 & 4 & | & -8 \end{bmatrix} \sim \begin{bmatrix} 1 & 2 & | & 4 \\ 0 & 0 & | & -16 \end{bmatrix}$ From **4**, Form (3), the system is inconsistent; there is no solution.

$\quad (-2)R_1 + R_2 \to R_2$

33. $\begin{bmatrix} 2 & 1 & | & 6 \\ 1 & -1 & | & -3 \end{bmatrix} \sim \begin{bmatrix} 1 & -1 & | & -3 \\ 2 & 1 & | & 6 \end{bmatrix} \sim \begin{bmatrix} 1 & -1 & | & -3 \\ 0 & 3 & | & 12 \end{bmatrix} \sim \begin{bmatrix} 1 & -1 & | & -3 \\ 0 & 1 & | & 4 \end{bmatrix} \sim \begin{bmatrix} 1 & 0 & | & 1 \\ 0 & 1 & | & 4 \end{bmatrix}$

$\qquad\quad R_1 \leftrightarrow R_2 \qquad (-2)R_1 + R_2 \to R_2 \qquad \frac{1}{3}R_2 \to R_2 \qquad R_2 + R_1 \to R_1$

Thus, $x_1 = 1$,and $x_2 = 4$.

35. $\begin{bmatrix} 3 & -6 & | & -9 \\ -2 & 4 & | & 6 \end{bmatrix} \sim \begin{bmatrix} 1 & -2 & | & -3 \\ -2 & 4 & | & 6 \end{bmatrix} \sim \begin{bmatrix} 1 & -2 & | & -3 \\ 0 & 0 & | & 0 \end{bmatrix}$

$\quad \frac{1}{3}R_1 \to R_1 \qquad\qquad 2R_1 + R_2 \to R_2$

From **4**, Form (2), the system has infinitely many solutions (consistent and dependent). If $x_2 = s$, then $x_1 - 2s = -3$ or $x_1 = 2s - 3$.
Thus, $x_2 = s$, $x_1 = 2s - 3$, for any real number s, are the solutions.

37. $\begin{bmatrix} 4 & -2 & | & 2 \\ -6 & 3 & | & -3 \end{bmatrix} \sim \begin{bmatrix} 1 & -\frac{1}{2} & | & \frac{1}{2} \\ -6 & 3 & | & -3 \end{bmatrix} \sim \begin{bmatrix} 1 & -\frac{1}{2} & | & \frac{1}{2} \\ 0 & 0 & | & 0 \end{bmatrix}$

$\quad \frac{1}{4}R_1 \to R_1 \qquad\qquad 6R_1 + R_2 \to R_2$

Thus, the system has infinitely many solutions (consistent and dependent). Let $x_2 = s$. Then

$x_1 - \frac{1}{2}s = \frac{1}{2}$ or $x_1 = \frac{1}{2}s + \frac{1}{2}$.

The set of solutions is $x_2 = s$, $x_1 = \frac{1}{2}s + \frac{1}{2}$ for any real number s.

39. $\begin{bmatrix} 2 & 1 & | & 1 \\ 4 & -1 & | & -7 \end{bmatrix} \sim \begin{bmatrix} 1 & \frac{1}{2} & | & \frac{1}{2} \\ 4 & -1 & | & -7 \end{bmatrix} \sim \begin{bmatrix} 1 & \frac{1}{2} & | & \frac{1}{2} \\ 0 & -3 & | & -9 \end{bmatrix} \sim \begin{bmatrix} 1 & \frac{1}{2} & | & \frac{1}{2} \\ 0 & 1 & | & 3 \end{bmatrix} \sim \begin{bmatrix} 1 & 0 & | & -1 \\ 0 & 1 & | & 3 \end{bmatrix}$

$\quad\quad \frac{1}{2}R_1 \to R_1 \quad\quad (-4)R_1 + R_2 \to R_2 \quad \left(-\frac{1}{3}\right)R_2 \to R_2 \quad \left(-\frac{1}{2}\right)R_2 + R_1 \to R_1$

Thus, $x_1 = -1$ and $x_2 = 3$.

41. $\begin{bmatrix} 4 & -6 & | & 8 \\ -6 & 9 & | & -10 \end{bmatrix} \sim \begin{bmatrix} 1 & -\frac{3}{2} & | & 2 \\ -6 & 9 & | & -10 \end{bmatrix} \sim \begin{bmatrix} 1 & -\frac{3}{2} & | & 2 \\ 0 & 0 & | & 2 \end{bmatrix}$

$\quad\quad\quad \frac{1}{4}R_1 \to R_1 \quad\quad\quad 6R_1 + R_2 \to R_2$

The second row of the final augmented matrix corresponds to the equation
$0x_1 + 0x_2 = 2$
which has no solution. Thus, the system has no solution; it is inconsistent.

43. $\begin{bmatrix} -4 & 6 & | & -8 \\ 6 & -9 & | & 12 \end{bmatrix} \sim \begin{bmatrix} 1 & -\frac{3}{2} & | & 2 \\ 6 & -9 & | & 12 \end{bmatrix} \sim \begin{bmatrix} 1 & -\frac{3}{2} & | & 2 \\ 0 & 0 & | & 0 \end{bmatrix}$

$\quad \left(-\frac{1}{4}\right)R_1 \to R_1 \quad\quad (-6)R_1 + R_2 \to R_2$

The system has infinitely many solutions (consistent and dependent).
If $x_2 = t$, then

$x_1 - \frac{3}{2}t = 2 \quad$ or $\quad x_1 = \frac{3}{2}t + 2$

Thus, the set of solutions is

$x_2 = t, \; x_1 = \frac{3}{2}t + 2$

for any real number t.

45. $\begin{bmatrix} 3 & -1 & | & 7 \\ 2 & 3 & | & 1 \end{bmatrix} \sim \begin{bmatrix} 1 & -\frac{1}{3} & | & \frac{7}{3} \\ 2 & 3 & | & 1 \end{bmatrix} \sim \begin{bmatrix} 1 & -\frac{1}{3} & | & \frac{7}{3} \\ 0 & \frac{11}{3} & | & -\frac{11}{3} \end{bmatrix} \sim \begin{bmatrix} 1 & -\frac{1}{3} & | & \frac{7}{3} \\ 0 & 1 & | & -1 \end{bmatrix} \sim \begin{bmatrix} 1 & 0 & | & 2 \\ 0 & 1 & | & -1 \end{bmatrix}$

$\quad\quad \frac{1}{3}R_1 \to R_1 \quad (-2)R_1 + R_2 \to R_2 \quad\quad \frac{3}{11}R_2 \to R_2 \quad\quad \frac{1}{3}R_2 + R_1 \to R_1 \quad$ Thus, $x_1 = 2$ and $x_2 = -1$.

47. $\begin{bmatrix} 3 & 2 & | & 4 \\ 2 & -1 & | & 5 \end{bmatrix} \sim \begin{bmatrix} 1 & \frac{2}{3} & | & \frac{4}{3} \\ 2 & -1 & | & 5 \end{bmatrix} \sim \begin{bmatrix} 1 & \frac{2}{3} & | & \frac{4}{3} \\ 0 & -\frac{7}{3} & | & \frac{7}{3} \end{bmatrix} \sim \begin{bmatrix} 1 & \frac{2}{3} & | & \frac{4}{3} \\ 0 & 1 & | & -1 \end{bmatrix} \sim \begin{bmatrix} 1 & 0 & | & 2 \\ 0 & 1 & | & -1 \end{bmatrix}$

$\quad\quad \frac{1}{3}R_1 \to R_1 \quad (-2)R_1 + R_2 \to R_2 \quad \left(-\frac{3}{7}\right)R_2 \to R_2 \quad \left(-\frac{2}{3}\right)R_2 + R_1 \to R_1 \quad$ Thus, $x_1 = 2$ and $x_2 = -1$.

49. $\begin{bmatrix} 0.2 & -0.5 & | & 0.07 \\ 0.8 & -0.3 & | & 0.79 \end{bmatrix} \sim \begin{bmatrix} 1 & -2.5 & | & 0.35 \\ 0.8 & -0.3 & | & 0.79 \end{bmatrix} \sim \begin{bmatrix} 1 & -2.5 & | & 0.35 \\ 0 & 1.7 & | & 0.51 \end{bmatrix}$

$\quad\quad \frac{1}{0.2}R_1 \to R_1 \quad\quad (-0.8)R_1 + R_2 \to R_2 \quad\quad \frac{1}{1.7}R_2 \to R_2$

$$\sim \begin{bmatrix} 1 & -2.5 & \bigm| & 0.35 \\ 0 & 1 & \bigm| & 0.3 \end{bmatrix} \sim \begin{bmatrix} 1 & 0 & \bigm| & 1.1 \\ 0 & 1 & \bigm| & 0.3 \end{bmatrix} \quad \text{Thus, } x_1 = 1.1 \text{ and } x_2 = 0.3.$$

$$2.5R_2 + R_1 \rightarrow R_1$$

51. $0.8x_1 + 2.88x_2 = 4$
$1.25x_1 + 4.34x_2 = 5$

$$\begin{bmatrix} 0.8 & 2.88 & \bigm| & 4 \\ 1.25 & 4.34 & \bigm| & 5 \end{bmatrix} (1.25)R_1 \rightarrow R_1$$

$$\begin{bmatrix} 1 & 3.6 & \bigm| & 5 \\ 1.25 & 4.34 & \bigm| & 5 \end{bmatrix} (-1.25)R_1 + R_2 \rightarrow R_2$$

$$\begin{bmatrix} 1 & 3.6 & \bigm| & 5 \\ 0 & -0.16 & \bigm| & -1.25 \end{bmatrix} (-6.25)R_2 \rightarrow R_2$$

$$\begin{bmatrix} 1 & 3.6 & \bigm| & 5 \\ 0 & 1 & \bigm| & 7.8125 \end{bmatrix} (-3.6)R_2 + R_1 \rightarrow R_1$$

$$\begin{bmatrix} 1 & 0 & \bigm| & -23.125 \\ 0 & 1 & \bigm| & 7.8125 \end{bmatrix}$$

Solution: $x_1 = -23.125$, $x_2 = 7.8125$

53. $4.8x_1 - 40.32x_2 = 295.2$
$-3.75x_1 + 28.7x_2 = -211.2$

$$\begin{bmatrix} 4.8 & -40.32 & \bigm| & 295.2 \\ -3.75 & 28.7 & \bigm| & -211.2 \end{bmatrix} (0.20833)R_1 \rightarrow R_1$$

$$\begin{bmatrix} 1 & -8.4 & \bigm| & 61.5 \\ -3.75 & 28.7 & \bigm| & -211.2 \end{bmatrix} (3.75)R_1 + R_2 \rightarrow R_2$$

$$\begin{bmatrix} 1 & -8.4 & \bigm| & 61.5 \\ 0 & -2.8 & \bigm| & 19.425 \end{bmatrix} (-0.35714)R_2 \rightarrow R_2$$

$$\begin{bmatrix} 1 & -8.4 & \bigm| & 61.5 \\ 0 & 1 & \bigm| & -6.9375 \end{bmatrix} (8.4)R_2 + R_1 \rightarrow R_1$$

$$\begin{bmatrix} 1 & 0 & \bigm| & 3.225 \\ 0 & 1 & \bigm| & -6.9375 \end{bmatrix}$$

Solution: $x_1 = 3.225$, $x_2 = -6.9375$

EXERCISE 4-3

Things to remember:

1. A matrix is in REDUCED FORM if

   (a) each row consisting entirely of zeros is below any row
   having at least one nonzero element;

(b) the left-most nonzero element in each row is 1;

(c) all other elements in the column containing the left-most 1 of a given row are zeros;

(d) the left-most 1 in any row is to the right of the left-most 1 in any row above.

2. GAUSS-JORDAN ELIMINATION

Step 1. Choose the leftmost nonzero column and use appropriate row operations to get a 1 at the top.

Step 2. Use multiples of the row containing the 1 from step 1 to get zeros in all remaining places in the column containing this 1.

Step 3. Repeat step 1 with the SUBMATRIX formed by (mentally) deleting the row used in step 2 and all rows above this row.

Step 4. Repeat step 2 with the ENTIRE MATRIX, including the mentally deleted rows. Continue this process until it is impossible to go further.

[*Note:* If at any point in this process we obtain a row with all zeros to the left of the vertical line and a nonzero number to the right, we can stop, since we will have a contradiction: $0 = n$, $n \neq 0$. We can then conclude that the system has no solution.]

---

1. $\begin{bmatrix} 1 & 0 & | & 2 \\ 0 & 1 & | & -1 \end{bmatrix}$

Is in reduced form. Use $\underline{1}$.

3. $\begin{bmatrix} 1 & 0 & 2 & | & 3 \\ 0 & 0 & 0 & | & 0 \\ 0 & 1 & -1 & | & 4 \end{bmatrix}$

Is not in reduced form. Condition (a) has been violated. The second row should be at the bottom.

5. $\begin{bmatrix} 0 & 1 & 0 & | & 2 \\ 0 & 0 & 3 & | & -1 \\ 0 & 0 & 0 & | & 0 \end{bmatrix}$

Is not in reduced form. Condition (b) has been violated. The left-most nonzero element in the second row should be 1, not 3.

7. $\begin{bmatrix} 1 & 2 & 0 & 3 & | & 2 \\ 0 & 0 & 1 & -1 & | & 0 \end{bmatrix}$

Is in reduced form.

9. $x_1 \qquad = -2$
$\qquad x_2 \qquad = 3$
$\qquad\qquad x_3 = 0$

11. $x_1 \qquad\quad - 2x_3 = 3 \quad (1)$
$\qquad x_2 + x_3 = -5 \quad (2)$

Let $x_3 = t$. From (2), $x_2 = -5 - t$. From (1), $x_1 = 3 + 2t$. Thus, the solution is

$x_1 = 2t + 3$
$x_2 = -t - 5$
$x_3 = t$

t any real number.

13. $x_1 \quad\ = 0$
$\quad\quad x_2 = 0$
$\quad\quad\ 0 = 1$

Inconsistent; no solution.

15. $x_1 - 2x_2 \quad\ - 3x_4 = -5$
$\quad\quad\quad\quad x_3 + 3x_4 = \ 2$

Let $x_2 = s$ and $x_4 = t$. Then

$x_1 = 2s + 3t - 5$
$x_2 = s$
$x_3 = -3t + 2$
$x_4 = t$

s and t any real numbers.

17. $\begin{bmatrix} 1 & 2 & | & -1 \\ 0 & 1 & | & 3 \end{bmatrix} \sim \begin{bmatrix} 1 & 0 & | & -7 \\ 0 & 1 & | & 3 \end{bmatrix}$

$\quad (-2)R_2 + R_1 \to R_1$

19. $\begin{bmatrix} 1 & 0 & -3 & | & 1 \\ 0 & 1 & 2 & | & 0 \\ 0 & 0 & 3 & | & -6 \end{bmatrix} \sim \begin{bmatrix} 1 & 0 & -3 & | & 1 \\ 0 & 1 & 2 & | & 0 \\ 0 & 0 & 1 & | & -2 \end{bmatrix} \sim \begin{bmatrix} 1 & 0 & 0 & | & -5 \\ 0 & 1 & 0 & | & 4 \\ 0 & 0 & 1 & | & -2 \end{bmatrix}$

$\quad\quad \frac{1}{3}R_3 \to R_3 \quad\quad 3R_3 + R_1 \to R_1$

$\quad\quad\quad\quad\quad\quad\quad (-2)R_3 + R_2 \to R_2$

21. $\begin{bmatrix} 1 & 2 & -2 & | & -1 \\ 0 & 3 & -6 & | & 1 \\ 0 & -1 & 2 & | & -\frac{1}{3} \end{bmatrix} \sim \begin{bmatrix} 1 & 2 & -2 & | & -1 \\ 0 & 1 & -2 & | & \frac{1}{3} \\ 0 & -1 & 2 & | & -\frac{1}{3} \end{bmatrix} \sim \begin{bmatrix} 1 & 2 & -2 & | & -1 \\ 0 & 1 & -2 & | & \frac{1}{3} \\ 0 & 0 & 0 & | & 0 \end{bmatrix} \sim \begin{bmatrix} 1 & 0 & 2 & | & -\frac{5}{3} \\ 0 & 1 & -2 & | & \frac{1}{3} \\ 0 & 0 & 0 & | & 0 \end{bmatrix}$

$\quad \frac{1}{3}R_2 \to R_2 \quad\quad\quad R_2 + R_3 \to R_3 \quad\quad (-2)R_2 + R_1 \to R_1$

23. The corresponding augmented matrix is:

$\begin{bmatrix} 2 & 4 & -10 & | & -2 \\ 3 & 9 & -21 & | & 0 \\ 1 & 5 & -12 & | & 1 \end{bmatrix} \sim \begin{bmatrix} 1 & 2 & -5 & | & -1 \\ 3 & 9 & -21 & | & 0 \\ 1 & 5 & -12 & | & 1 \end{bmatrix} \sim \begin{bmatrix} 1 & 2 & -5 & | & -1 \\ 0 & 3 & -6 & | & 3 \\ 0 & 3 & -7 & | & 2 \end{bmatrix} \sim \begin{bmatrix} 1 & 2 & -5 & | & -1 \\ 0 & 1 & -2 & | & 1 \\ 0 & 3 & -7 & | & 2 \end{bmatrix}$

$\quad \frac{1}{2}R_1 \to R_1 \quad\quad (-3)R_1 + R_2 \to R_2 \quad\quad \frac{1}{3}R_2 \to R_2 \quad\quad (-3)R_2 + R_3 \to R_3$

$\quad\quad\quad\quad\quad (-1)R_1 + R_3 + \to R_3 \quad\quad\quad\quad\quad\quad\quad (-2)R_2 + R_1 \to R_1$

$\sim \begin{bmatrix} 1 & 0 & -1 & | & -3 \\ 0 & 1 & -2 & | & 1 \\ 0 & 0 & -1 & | & -1 \end{bmatrix} \sim \begin{bmatrix} 1 & 0 & -1 & | & -3 \\ 0 & 1 & -2 & | & 1 \\ 0 & 0 & 1 & | & 1 \end{bmatrix} \sim \begin{bmatrix} 1 & 0 & 0 & | & -2 \\ 0 & 1 & 0 & | & 3 \\ 0 & 0 & 1 & | & 1 \end{bmatrix}$

$\quad (-1)R_3 \to R_3 \quad\quad 2R_3 + R_2 + \to R_2$

$\quad\quad\quad\quad\quad\quad R_3 + R_1 \to R_1$

Thus, $x_1 = -2; \ x_2 = 3; \ x_3 = 1.$

25. The corresponding augmented matrix is:

$$\begin{bmatrix} 3 & 8 & -1 & | & -18 \\ 2 & 1 & 5 & | & 8 \\ 2 & 4 & 2 & | & -4 \end{bmatrix} \sim \begin{bmatrix} 2 & 4 & 2 & | & -4 \\ 2 & 1 & 5 & | & 8 \\ 3 & 8 & -1 & | & -18 \end{bmatrix} \sim \begin{bmatrix} 1 & 2 & 1 & | & -2 \\ 2 & 1 & 5 & | & 8 \\ 3 & 8 & -1 & | & -18 \end{bmatrix}$$

$$R_1 \leftrightarrow R_3 \qquad\qquad \frac{1}{2}R_1 \to R_1 \qquad\qquad (-2)R_1 + R_2 \to R_2$$
$$(-3)R_1 + R_3 \to R_3$$

$$\sim \begin{bmatrix} 1 & 2 & 1 & | & -2 \\ 0 & -3 & 3 & | & 12 \\ 0 & 2 & -4 & | & -12 \end{bmatrix} \sim \begin{bmatrix} 1 & 2 & 1 & | & -2 \\ 0 & 1 & -1 & | & -4 \\ 0 & 2 & -4 & | & -12 \end{bmatrix} \sim \begin{bmatrix} 1 & 0 & 3 & | & 6 \\ 0 & 1 & -1 & | & -4 \\ 0 & 0 & -2 & | & -4 \end{bmatrix}$$

$$\left(-\frac{1}{3}\right)R_2 \to R_2 \qquad\qquad (-2)R_2 + R_3 \to R_3 \qquad\qquad \left(-\frac{1}{2}\right)R_3 \to R_3$$
$$(-2)R_2 + R_1 \to R_1$$

$$\sim \begin{bmatrix} 1 & 0 & 3 & | & 6 \\ 0 & 1 & -1 & | & -4 \\ 0 & 0 & 1 & | & 2 \end{bmatrix} \sim \begin{bmatrix} 1 & 0 & 0 & | & 0 \\ 0 & 1 & 0 & | & -2 \\ 0 & 0 & 1 & | & 2 \end{bmatrix} \qquad \text{Thus, } \begin{array}{l} x_1 = 0 \\ x_2 = -2 \\ x_3 = 2. \end{array}$$

$$(-3)R_3 + R_1 \to R_1$$
$$R_3 + R_2 \to R_2$$

27. $\begin{bmatrix} 2 & -1 & -3 & | & 8 \\ 1 & -2 & 0 & | & 7 \end{bmatrix} \sim \begin{bmatrix} 1 & -2 & 0 & | & 7 \\ 2 & -1 & -3 & | & 8 \end{bmatrix} \sim \begin{bmatrix} 1 & -2 & 0 & | & 7 \\ 0 & 3 & -3 & | & -6 \end{bmatrix} \sim \begin{bmatrix} 1 & -2 & 0 & | & 7 \\ 0 & 1 & -1 & | & -2 \end{bmatrix}$

$$R_1 \leftrightarrow R_2 \qquad (-2)R_1 + R_2 \to R_2 \qquad \frac{1}{3}R_2 \to R_2 \qquad 2R_2 + R_1 \to R_1$$

$$\sim \begin{bmatrix} 1 & 0 & -2 & | & 3 \\ 0 & 1 & -1 & | & -2 \end{bmatrix} \qquad \text{Thus, } \begin{array}{l} x_1 \quad - 2x_3 = 3 \quad (1) \\ x_2 - x_3 = -2 \quad (2) \end{array}$$

Let $x_3 = t$, where t is any real number. Then:
$x_1 = 2t + 3$
$x_2 = t - 2$
$x_3 = t$

29. $\begin{bmatrix} 2 & -1 & | & 0 \\ 3 & 2 & | & 7 \\ 1 & -1 & | & -1 \end{bmatrix} \sim \begin{bmatrix} 1 & -1 & | & -1 \\ 3 & 2 & | & 7 \\ 2 & -1 & | & 0 \end{bmatrix} \sim \begin{bmatrix} 1 & -1 & | & -1 \\ 0 & 5 & | & 10 \\ 0 & 1 & | & 2 \end{bmatrix} \sim \begin{bmatrix} 1 & -1 & | & -1 \\ 0 & 1 & | & 2 \\ 0 & 5 & | & 10 \end{bmatrix}$

$$R_1 \leftrightarrow R_3 \qquad (-3)R_1 + R_2 \to R_2 \qquad R_2 \leftrightarrow R_3 \qquad R_2 + R_1 \to R_1$$
$$(-2)R_1 + R_3 \to R_3 \qquad\qquad\qquad (-5)R_2 + R_3 \to R_3$$

$$\sim \begin{bmatrix} 1 & 0 & | & 1 \\ 0 & 1 & | & 2 \\ 0 & 0 & | & 0 \end{bmatrix} \qquad \text{Thus, } \begin{array}{l} x_1 = 1 \\ x_2 = 2 \end{array}$$

31. $\begin{bmatrix} 3 & -4 & -1 & | & 1 \\ 2 & -3 & 1 & | & 1 \\ 1 & -2 & 3 & | & 2 \end{bmatrix} \sim \begin{bmatrix} 1 & -2 & 3 & | & 2 \\ 2 & -3 & 1 & | & 1 \\ 3 & -4 & -1 & | & 1 \end{bmatrix} \sim \begin{bmatrix} 1 & -2 & 3 & | & 2 \\ 0 & 1 & -5 & | & -3 \\ 0 & 2 & -10 & | & -5 \end{bmatrix} \sim \begin{bmatrix} 1 & -2 & 3 & | & 2 \\ 0 & 1 & -5 & | & -3 \\ 0 & 0 & 0 & | & 1 \end{bmatrix}$

$\qquad R_1 \leftrightarrow R_3 \qquad\qquad (-2)R_1 + R_2 \to R_2 \qquad (-2)R_2 + R_3 \to R_3$

$\qquad\qquad\qquad\qquad\quad (-3)R_1 + R_3 \to R_3$

From the last row, we conclude that there is no solution; the system is inconsistent.

33. $\begin{bmatrix} 3 & -2 & 1 & | & -7 \\ 2 & 1 & -4 & | & 0 \\ 1 & 1 & -3 & | & 1 \end{bmatrix} \sim \begin{bmatrix} 1 & 1 & -3 & | & 1 \\ 2 & 1 & -4 & | & 0 \\ 3 & -2 & 1 & | & -7 \end{bmatrix} \sim \begin{bmatrix} 1 & 1 & -3 & | & 1 \\ 0 & -1 & 2 & | & -2 \\ 0 & -5 & 10 & | & -10 \end{bmatrix}$

$\qquad R_1 \leftrightarrow R_3 \qquad\qquad\quad (-2)R_1 + R_2 \to R_2 \qquad\qquad (-1)R_2 \to R_2$

$\qquad\qquad\qquad\qquad\qquad\quad (-3)R_1 + R_3 \to R_3$

$\begin{bmatrix} 1 & 1 & -3 & | & 1 \\ 0 & 1 & -2 & | & 2 \\ 0 & -5 & 10 & | & -10 \end{bmatrix} \sim \begin{bmatrix} 1 & 0 & -1 & | & -1 \\ 0 & 1 & -2 & | & 2 \\ 0 & 0 & 0 & | & 0 \end{bmatrix}$

$(-1)R_2 + R_1 \to R_1$

$5R_2 + R_3 \to R_3$

From this matrix, $x_1 - x_3 = -1$ and $x_2 - 2x_3 = 2$. Let $x_3 = t$ be any real number, then $x_1 = t - 1$, $x_2 = 2t + 2$, and $x_3 = t$.

35. $\begin{bmatrix} 2 & 4 & -2 & | & 2 \\ -3 & -6 & 3 & | & -3 \end{bmatrix} \sim \begin{bmatrix} 1 & 2 & -1 & | & 1 \\ -3 & -6 & 3 & | & -3 \end{bmatrix} \sim \begin{bmatrix} 1 & 2 & -1 & | & 1 \\ 0 & 0 & 0 & | & 0 \end{bmatrix}$

$\qquad \frac{1}{2}R_1 \to R_1 \qquad\qquad 3R_1 + R_2 \to R_2$

From this matrix, $x_1 + 2x_2 - x_3 = 1$. Let $x_2 = s$ and $x_3 = t$. Then $x_1 = -2s + t + 1$, $x_2 = s$, and $x_3 = t$, s and t any real numbers.

37. $\begin{bmatrix} 4 & -1 & 2 & | & 3 \\ -4 & 1 & -3 & | & -10 \\ 8 & -2 & 9 & | & -1 \end{bmatrix} \; \frac{1}{4}R_1 \to R_1 \sim \begin{bmatrix} 1 & -\frac{1}{4} & \frac{1}{2} & | & \frac{3}{4} \\ -4 & 1 & -3 & | & -10 \\ 8 & -2 & 9 & | & -1 \end{bmatrix} \; \begin{array}{l} 4R_1 + R_2 \to R_2 \\ (-8)R_1 + R_3 \to R_3 \end{array}$

$\sim \begin{bmatrix} 1 & -\frac{1}{4} & \frac{1}{2} & | & \frac{3}{4} \\ 0 & 0 & -1 & | & -7 \\ 0 & 0 & 5 & | & -7 \end{bmatrix} \; (-1)R_2 \to R_2 \sim \begin{bmatrix} 1 & -\frac{1}{4} & \frac{1}{2} & | & \frac{3}{4} \\ 0 & 0 & 1 & | & 7 \\ 0 & 0 & 5 & | & -7 \end{bmatrix} \; (-5)R_2 + R_3 \to R_3$

$\sim \begin{bmatrix} 1 & -\frac{1}{4} & \frac{1}{2} & | & \frac{3}{4} \\ 0 & 0 & 1 & | & 7 \\ 0 & 0 & 0 & | & -42 \end{bmatrix}$

No solution.

39. $\begin{bmatrix} 2 & -5 & -3 & | & 7 \\ -4 & 10 & 2 & | & 6 \\ 6 & -15 & -1 & | & -19 \end{bmatrix}$ $\frac{1}{2}R_1 \rightarrow R_1$ $\sim$ $\begin{bmatrix} 1 & -\frac{5}{2} & -\frac{3}{2} & | & \frac{7}{2} \\ -4 & 10 & 2 & | & 6 \\ 6 & -15 & -1 & | & -19 \end{bmatrix}$ $\begin{array}{l} 4R_1 + R_2 \rightarrow R_2 \\ (-6)R_1 + R_3 \rightarrow R_3 \end{array}$

$\sim$ $\begin{bmatrix} 1 & -\frac{5}{2} & -\frac{3}{2} & | & \frac{7}{2} \\ 0 & 0 & -4 & | & 20 \\ 0 & 0 & 8 & | & -40 \end{bmatrix}$ $\left(-\frac{1}{4}\right)R_2 \rightarrow R_2$ $\sim$ $\begin{bmatrix} 1 & -\frac{5}{2} & -\frac{3}{2} & | & \frac{7}{2} \\ 0 & 0 & 1 & | & -5 \\ 0 & 0 & 8 & | & -40 \end{bmatrix}$ $(-8)R_2 + R_3 \rightarrow R_3$

$\sim$ $\begin{bmatrix} 1 & -\frac{5}{2} & -\frac{3}{2} & | & \frac{7}{2} \\ 0 & 0 & 1 & | & -5 \\ 0 & 0 & 0 & | & 0 \end{bmatrix}$ $\frac{3}{2}R_2 + R_1 \rightarrow R_1$ $\sim$ $\begin{bmatrix} 1 & -\frac{5}{2} & 0 & | & -4 \\ 0 & 0 & 1 & | & -5 \\ 0 & 0 & 0 & | & 0 \end{bmatrix}$

The system of equations is:

$$x_1 - \frac{5}{2}x_2 \qquad = -4$$
$$x_3 = -5$$

and

$$x_1 = 2.5x_2 - 4$$
$$x_3 = -5$$

Let $x_2 = t$. Then for any real number t,

$$x_1 = 2.5t - 4$$
$$x_2 = t$$
$$x_3 = -5$$

is a solution.

41. $\begin{bmatrix} 5 & -3 & 2 & | & 13 \\ 2 & -1 & -3 & | & 1 \\ 4 & -2 & 4 & | & 12 \end{bmatrix}$ $(-1)R_3 + R_1 \rightarrow R_1$ (to simplify the arithmetic)

$\sim$ $\begin{bmatrix} 1 & -1 & -2 & | & 1 \\ 2 & -1 & -3 & | & 1 \\ 4 & -2 & 4 & | & 12 \end{bmatrix}$ $\begin{array}{l} (-2)R_1 + R_2 \rightarrow R_2 \\ (-4)R_1 + R_3 \rightarrow R_3 \end{array}$ $\sim$ $\begin{bmatrix} 1 & -1 & -2 & | & 1 \\ 0 & 1 & 1 & | & -1 \\ 0 & 2 & 12 & | & 8 \end{bmatrix}$ $\begin{array}{l} R_2 + R_1 \rightarrow R_1 \\ (-2)R_2 + R_3 \rightarrow R_3 \end{array}$

$\sim$ $\begin{bmatrix} 1 & 0 & -1 & | & 0 \\ 0 & 1 & 1 & | & -1 \\ 0 & 0 & 10 & | & 10 \end{bmatrix}$ $\frac{1}{10}R_3 \rightarrow R_3$ $\sim$ $\begin{bmatrix} 1 & 0 & -1 & | & 0 \\ 0 & 1 & 1 & | & -1 \\ 0 & 0 & 1 & | & 1 \end{bmatrix}$ $\begin{array}{l} R_3 + R_1 \rightarrow R_1 \\ (-1)R_3 + R_2 \rightarrow R_2 \end{array}$

$\sim$ $\begin{bmatrix} 1 & 0 & 0 & | & 1 \\ 0 & 1 & 0 & | & -2 \\ 0 & 0 & 1 & | & 1 \end{bmatrix}$

The system of equations is:

$$x_1 \qquad\qquad = 1$$
$$x_2 \qquad = -2$$
$$x_3 = 1$$

Solution: $x_1 = 1$, $x_2 = -2$, $x_3 = 1$

43. (A) The system is dependent with two parameters and an infinite number of solutions.

(B) The system is dependent with one parameter and an infinite number of solutions.

(C) The system is independent with a unique solution.

(D) Impossible.

45.
$$\begin{bmatrix} 1 & 2 & -4 & -1 & | & 7 \\ 2 & 5 & -9 & -4 & | & 16 \\ 1 & 5 & -7 & -7 & | & 13 \end{bmatrix} \sim \begin{bmatrix} 1 & 2 & -4 & -1 & | & 7 \\ 0 & 1 & -1 & -2 & | & 2 \\ 0 & 3 & -3 & -6 & | & 6 \end{bmatrix} \sim \begin{bmatrix} 1 & 0 & -2 & 3 & | & 3 \\ 0 & 1 & -1 & -2 & | & 2 \\ 0 & 0 & 0 & 0 & | & 0 \end{bmatrix}$$

$(-2)R_1 + R_2 \to R_2$ $\qquad (-2)R_2 + R_1 \to R_1$

$(-1)R_1 + R_3 \to R_3$ $\qquad (-3)R_2 + R_3 \to R_3$

Thus, $x_1 - 2x_3 + 3x_4 = 3$ and $x_2 - x_3 - 2x_4 = 2$. Let $x_3 = s$ and $x_4 = t$. Then $x_1 = 2s - 3t + 3$, $x_2 = s + 2t + 2$, $x_3 = s$, $x_4 = t$, where s, t are any real numbers.

47.
$$\begin{bmatrix} 1 & -1 & 3 & -2 & | & 1 \\ -2 & 4 & -3 & 1 & | & 0.5 \\ 3 & -1 & 10 & -4 & | & 2.9 \\ 4 & -3 & 8 & -2 & | & 0.6 \end{bmatrix} \begin{array}{l} 2R_1 + R_2 \to R_2 \\ (-3)R_1 + R_3 \to R_3 \\ (-4)R_1 + R_4 \to R_4 \end{array} \sim \begin{bmatrix} 1 & -1 & 3 & -2 & | & 1 \\ 0 & 2 & 3 & -3 & | & 2.5 \\ 0 & 2 & 1 & 2 & | & -0.1 \\ 0 & 1 & -4 & 6 & | & -3.4 \end{bmatrix} R_2 \leftrightarrow R_4$$

$$\sim \begin{bmatrix} 1 & -1 & 3 & -2 & | & 1 \\ 0 & 1 & -4 & 6 & | & -3.4 \\ 0 & 2 & 1 & 2 & | & -0.1 \\ 0 & 2 & 3 & -3 & | & 2.5 \end{bmatrix} \begin{array}{l} R_2 + R_1 \to R_1 \\ (-2)R_2 + R_3 \to R_3 \\ (-2)R_2 + R_4 \to R_4 \end{array}$$

$$\sim \begin{bmatrix} 1 & 0 & -1 & 4 & | & -2.4 \\ 0 & 1 & -4 & 6 & | & -3.4 \\ 0 & 0 & 9 & -10 & | & 6.7 \\ 0 & 0 & 11 & -15 & | & 9.3 \end{bmatrix} \quad (-1)R_4 + R_3 \to R_3 \text{ (to simplify arithmetic)}$$

$$\sim \begin{bmatrix} 1 & 0 & -1 & 4 & | & -2.4 \\ 0 & 1 & -4 & 6 & | & -3.4 \\ 0 & 0 & -2 & 5 & | & -2.6 \\ 0 & 0 & 11 & -15 & | & 9.3 \end{bmatrix} \quad \left(-\frac{1}{2}\right) R_3 \to R_3$$

$$\sim \begin{bmatrix} 1 & 0 & -1 & 4 & | & -2.4 \\ 0 & 1 & -4 & 6 & | & -3.4 \\ 0 & 0 & 1 & -2.5 & | & 1.3 \\ 0 & 0 & 11 & -15 & | & 9.3 \end{bmatrix} \begin{array}{l} R_3 + R_1 \to R_1 \\ 4R_3 + R_2 \to R_2 \\ (-11)R_3 + R_4 \to R_4 \end{array}$$

$$\sim \begin{bmatrix} 1 & 0 & 0 & 1.5 & | & -1.1 \\ 0 & 1 & 0 & -4 & | & 1.8 \\ 0 & 0 & 1 & -2.5 & | & 1.3 \\ 0 & 0 & 0 & 12.5 & | & -5 \end{bmatrix} \frac{1}{12.5} R_4 \to R_4$$

$$\sim \begin{bmatrix} 1 & 0 & 0 & 1.5 & | & -1.1 \\ 0 & 1 & 0 & -4 & | & 1.8 \\ 0 & 0 & 1 & -2.5 & | & 1.3 \\ 0 & 0 & 0 & 1 & | & -0.4 \end{bmatrix} \begin{matrix} (-1.5)R_4 + R_1 \rightarrow R_1 \\ 4R_4 + R_2 \rightarrow R_2 \\ 2.5R_4 + R_3 \rightarrow R_3 \end{matrix}$$

$$\sim \begin{bmatrix} 1 & 0 & 0 & 0 & | & -0.5 \\ 0 & 1 & 0 & 0 & | & 0.2 \\ 0 & 0 & 1 & 0 & | & 0.3 \\ 0 & 0 & 0 & 1 & | & -0.4 \end{bmatrix}$$

The system of equations is:

$$
\begin{aligned}
x_1 &= -0.5 \\
x_2 &= 0.2 \\
x_3 &= 0.3 \\
x_4 &= -0.4
\end{aligned}
$$

Solution: $x_1 = -0.5$, $x_2 = 0.2$, $x_3 = 0.3$, $x_2 = -0.4$

49. $\begin{bmatrix} 1 & -2 & 1 & 1 & 2 & | & 2 \\ -2 & 4 & 2 & 2 & -2 & | & 0 \\ 3 & -6 & 1 & 1 & 5 & | & 4 \\ -1 & 2 & 3 & 1 & 1 & | & 3 \end{bmatrix} \begin{matrix} 2R_1 + R_2 \rightarrow R_2 \\ (-3)R_1 + R_3 \rightarrow R_3 \\ R_1 + R_4 \rightarrow R_4 \end{matrix}$

$$\sim \begin{bmatrix} 1 & -2 & 1 & 1 & 2 & | & 2 \\ 0 & 0 & 4 & 4 & 2 & | & 4 \\ 0 & 0 & 2 & -2 & -1 & | & -2 \\ 0 & 0 & 4 & 2 & 3 & | & 5 \end{bmatrix} \frac{1}{4}R_2 \rightarrow R_2$$

$$\sim \begin{bmatrix} 1 & -2 & 1 & 1 & 2 & | & 2 \\ 0 & 0 & 1 & 1 & \frac{1}{2} & | & 1 \\ 0 & 0 & -2 & -2 & -1 & | & -2 \\ 0 & 0 & 4 & 2 & 3 & | & 5 \end{bmatrix} \begin{matrix} (-1)R_2 + R_1 \rightarrow R_1 \\ 2R_2 + R_3 \rightarrow R_3 \\ (-4)R_2 + R_4 \rightarrow R_4 \end{matrix}$$

$$\sim \begin{bmatrix} 1 & -2 & 1 & 1 & \frac{3}{2} & | & 1 \\ 0 & 0 & 1 & 1 & \frac{1}{2} & | & 1 \\ 0 & 0 & 0 & 0 & 0 & | & 0 \\ 0 & 0 & 0 & -2 & 1 & | & 1 \end{bmatrix} R_3 \leftrightarrow R_4 \sim \begin{bmatrix} 1 & -2 & 0 & 0 & \frac{3}{2} & | & 1 \\ 0 & 0 & 1 & 1 & \frac{1}{2} & | & 1 \\ 0 & 0 & 0 & -2 & 1 & | & 1 \\ 0 & 0 & 0 & 0 & 0 & | & 0 \end{bmatrix} \left(-\frac{1}{2}\right)R_3 \rightarrow R_3$$

$$\sim \begin{bmatrix} 1 & -2 & 0 & 0 & \frac{3}{2} & | & 1 \\ 0 & 0 & 1 & 1 & \frac{1}{2} & | & 1 \\ 0 & 0 & 0 & 1 & -\frac{1}{2} & | & -\frac{1}{2} \\ 0 & 0 & 0 & 0 & 0 & | & 0 \end{bmatrix} (-1)R_3 + R_2 \rightarrow R_2 \sim \begin{bmatrix} 1 & -2 & 0 & 0 & \frac{3}{2} & | & 1 \\ 0 & 0 & 1 & 0 & 1 & | & \frac{3}{2} \\ 0 & 0 & 0 & 1 & -\frac{1}{2} & | & -\frac{1}{2} \\ 0 & 0 & 0 & 0 & 0 & | & 0 \end{bmatrix}$$

The system of equations is:

$$x_1 - 2x_2 \qquad\qquad + \frac{3}{2}x_5 = 1$$
$$x_3 \qquad + x_5 = \frac{3}{2}$$
$$x_4 - \frac{1}{2}x_5 = -\frac{1}{2}$$

or $\quad x_1 = 2x_2 - \frac{3}{2}x_5 + 1$

$$x_3 = -x_5 + \frac{3}{2}$$
$$x_4 = \frac{1}{2}x_5 - \frac{1}{2}$$

Let $x_2 = s$ and $x_5 = t$. Then $x_1 = 2s - \frac{3}{2}t + 1$, $x_2 = s$, $x_3 = -t + \frac{3}{2}$, $x_4 = \frac{1}{2}t - \frac{1}{2}$, $x_5 = t$ for any real numbers s and t.

51. Let x_1 = number of one-person boats,
$\qquad x_2$ = number of two-person boats,
and x_3 = number of four-person boats.

We have the following system of linear equations:
$$0.5x_1 + \qquad x_2 + 1.5x_3 = 380$$
$$0.6x_1 + 0.9x_2 + 1.2x_3 = 330$$
$$0.2x_1 + 0.3x_2 + 0.5x_3 = 120$$

$$\begin{bmatrix} 0.5 & 1 & 1.5 & | & 380 \\ 0.6 & 0.9 & 1.2 & | & 330 \\ 0.2 & 0.3 & 0.5 & | & 120 \end{bmatrix} \sim \begin{bmatrix} 1 & 2 & 3 & | & 760 \\ 0.6 & 0.9 & 1.2 & | & 330 \\ 0.2 & 0.3 & 0.5 & | & 120 \end{bmatrix} \sim \begin{bmatrix} 1 & 2 & 3 & | & 760 \\ 0 & -0.3 & -0.6 & | & -126 \\ 0 & -0.1 & -0.1 & | & -32 \end{bmatrix}$$

$$2R_1 \to R_1 \qquad\qquad (-0.6)R_1 + R_2 \to R_2 \qquad\qquad \left(-\frac{1}{0.3}\right)R_2 \to R_2$$
$$(-0.2)R_1 + R_3 \to R_3$$

$$\sim \begin{bmatrix} 1 & 2 & 3 & | & 760 \\ 0 & 1 & 2 & | & 420 \\ 0 & -0.1 & -0.1 & | & -32 \end{bmatrix} \sim \begin{bmatrix} 1 & 0 & -1 & | & -80 \\ 0 & 1 & 2 & | & 420 \\ 0 & 0 & 0.1 & | & 10 \end{bmatrix} \sim \begin{bmatrix} 1 & 0 & -1 & | & -80 \\ 0 & 1 & 2 & | & 420 \\ 0 & 0 & 1 & | & 100 \end{bmatrix}$$

$$(0.1)R_2 + R_3 \to R_3 \qquad\qquad 10R_3 \to R_3 \qquad\qquad R_3 + R_1 \to R_1$$
$$(-2)R_2 + R_1 \to R_1 \qquad\qquad\qquad\qquad (-2)R_3 + R_2 \to R_2$$

$$\sim \begin{bmatrix} 1 & 0 & 0 & | & 20 \\ 0 & 1 & 0 & | & 220 \\ 0 & 0 & 1 & | & 100 \end{bmatrix}$$

Thus, $x_1 = 20$, $x_2 = 220$, and $x_3 = 100$, or 20 one-person boats, 220 two-person boats, and 100 four-person boats.

53. Referring to Problem 51, we now have the following system of equations to solve:

$$0.5x_1 + x_2 + 1.5x_3 = 380$$
$$0.6x_1 + 0.9x_2 + 1.2x_3 = 330$$

$$\begin{bmatrix} 0.5 & 1 & 1.5 & | & 380 \\ 0.6 & 0.9 & 1.2 & | & 330 \end{bmatrix} \sim \begin{bmatrix} 1 & 2 & 3 & | & 760 \\ 0.6 & 0.9 & 1.2 & | & 330 \end{bmatrix} \sim \begin{bmatrix} 1 & 2 & 3 & | & 760 \\ 0 & -0.3 & -0.6 & | & -126 \end{bmatrix}$$

$$2R_1 \to R_1 \qquad\qquad (-0.6)R_1 + R_2 \to R_2 \qquad\qquad \left(-\frac{1}{0.3}\right)R_2 \to R_2$$

$$\sim \begin{bmatrix} 1 & 2 & 3 & | & 760 \\ 0 & 1 & 2 & | & 420 \end{bmatrix} \sim \begin{bmatrix} 1 & 0 & -1 & | & -80 \\ 0 & 1 & 2 & | & 420 \end{bmatrix}$$

$$(-2)R_2 + R_1 \to R_1$$

Thus, $x_1 - x_3 = -80 \qquad$ (1)
$$x_2 + 2x_3 = 420 \qquad (2)$$

Let $x_3 = t \qquad$ (t any real number)

Then, $x_2 = 420 - 2t$ [from (2)]
$x_1 = t - 80$ [from (1)]

In order to keep x_1 and x_2 positive, $t \le 210$ and $t \ge 80$.

Thus, $x_1 = t - 80$ (one-person boats)
$x_2 = 420 - 2t$ (two-person boats)
$x_3 = t$ (four-person boats)

where $80 \le t \le 210$ and t is an integer.

55. Again referring to Problem 51, we have have the following system:

$$0.5x_1 + x_2 = 380$$
$$0.6x_1 + 0.9x_2 = 330$$
$$0.2x_1 + 0.3x_2 = 120$$

$$\begin{bmatrix} 0.5 & 1 & | & 380 \\ 0.6 & 0.9 & | & 330 \\ 0.2 & 0.3 & | & 120 \end{bmatrix} \sim \begin{bmatrix} 1 & 2 & | & 760 \\ 0.6 & 0.9 & | & 330 \\ 0.2 & 0.3 & | & 120 \end{bmatrix} \sim \begin{bmatrix} 1 & 2 & | & 760 \\ 0 & -0.3 & | & -126 \\ 0 & -0.1 & | & -32 \end{bmatrix} \sim \begin{bmatrix} 1 & 2 & | & 760 \\ 0 & 1 & | & 420 \\ 0 & -0.1 & | & -32 \end{bmatrix}$$

$$2R_1 \to R_1 \qquad (-0.6R_1) + R_2 \to R_2 \qquad \left(-\frac{1}{0.3}\right)R_2 \to R_2 \qquad 0.1R_2 + R_3 \to R_3$$

$$(-0.2R_1) + R_3 \to R_3$$

$$\sim \begin{bmatrix} 1 & 2 & | & 760 \\ 0 & 1 & | & 420 \\ 0 & 0 & | & 10 \end{bmatrix}$$ From this matrix, we conclude that there is no solution; there is no production schedule that will use all the labor-hours in all departments.

57. Let x_1 = number of 6000 gallon tank cars
x_2 = number of 8000 gallon tank cars
x_3 = number of 18,000 gallon tank cars

Then
$$x_1 + x_2 + x_3 = 24$$
and $6000x_1 + 8000x_2 + 18,000x_3 = 250,000$

Dividing the second equation by 2000, we get the system
$$x_1 + x_2 + x_3 = 24$$
$$3x_1 + 4x_2 + 9x_3 = 125$$

The augmented matrix corresponding to this system is:

$$\begin{bmatrix} 1 & 1 & 1 & | & 24 \\ 3 & 4 & 9 & | & 125 \end{bmatrix} \sim \begin{bmatrix} 1 & 1 & 1 & | & 24 \\ 0 & 1 & 6 & | & 53 \end{bmatrix} \sim \begin{bmatrix} 1 & 0 & -5 & | & -29 \\ 0 & 1 & 6 & | & 53 \end{bmatrix}$$

$(-3)R_1 + R_2 \to R_2$ $(-1)R_2 + R_1 \to R_1$

Thus $x_1 \qquad - 5x_3 = -29$
$\qquad \quad x_2 + 6x_3 = 53$

Let $x_3 = t$. Then $x_1 = 5t - 29$ and $x_2 = 53 - 6t$

Thus, $(5t - 29)$ 6000-gallon tank cars, $(53 - 6t)$ 8000-gallon tank cars and (t) 18000-gallon tank cars should be purchased. Also, since t, $5t - 29$ and $53 - 6t$ must each be non-negative integers, it follows that $t = 6, 7$ or 8.

59. Let x_1 = federal income tax
x_2 = state income tax
x_3 = local income tax
Then
$$x_1 = 0.50[7,650,000 - (x_2 + x_3)]$$
$$x_2 = 0.20[7,650,000 - (x_1 + x_3)]$$
$$x_3 = 0.10[7,650,000 - (x_1 + x_2)]$$
and
$$x_1 + 0.5x_2 + 0.5x_3 = 3,825,000$$
$$0.2x_1 + x_2 + 0.2x_3 = 1,530,000$$
$$0.1x_1 + 0.1x_2 + x_3 = 765,000$$

The corresponding augmented matrix is:

$$\begin{bmatrix} 1 & 0.5 & 0.5 & | & 3,825,000 \\ 0.2 & 1 & 0.2 & | & 1,530,000 \\ 0.1 & 0.1 & 1 & | & 765,000 \end{bmatrix} \quad \begin{array}{l} (-0.2)R_1 + R_2 \to R_2 \\ \\ (-0.1)R_1 + R_3 \to R_3 \end{array}$$

$$\sim \begin{bmatrix} 1 & 0.5 & 0.5 & | & 3,825,000 \\ 0 & 0.9 & 0.1 & | & 765,000 \\ 0 & 0.05 & 0.95 & | & 382,500 \end{bmatrix} \quad 20R_3 \to R_3 \text{ (simplify arithmetic)}$$

$$\sim \begin{bmatrix} 1 & 0.5 & 0.5 & | & 3,825,000 \\ 0 & 0.9 & 0.1 & | & 765,000 \\ 0 & 1 & 19 & | & 7,650,000 \end{bmatrix} \quad R_2 \leftrightarrow R_3$$

$$\sim \begin{bmatrix} 1 & 0.5 & 0.5 & | & 3,825,000 \\ 0 & 1 & 19 & | & 7,650,000 \\ 0 & 0.9 & 0.1 & | & 765,000 \end{bmatrix} \begin{matrix} (-0.5)R_2 + R_1 \to R_1 \\ (-0.9)R_2 + R_3 \to R_3 \end{matrix}$$

$$\sim \begin{bmatrix} 1 & 0 & -9 & | & 0 \\ 0 & 1 & 19 & | & 7,650,000 \\ 0 & 0 & -17 & | & -6,120,000 \end{bmatrix} \left(-\frac{1}{17}\right)R_3 \to R_3$$

$$\sim \begin{bmatrix} 1 & 0 & -9 & | & 0 \\ 0 & 1 & 19 & | & 7,650,000 \\ 0 & 0 & 1 & | & 360,000 \end{bmatrix} \begin{matrix} 9R_3 + R_1 \to R_1 \\ (-19)R_3 + R_2 \to R_2 \end{matrix}$$

$$\sim \begin{bmatrix} 1 & 0 & 0 & | & 3,240,000 \\ 0 & 1 & 0 & | & 810,000 \\ 0 & 0 & 1 & | & 360,000 \end{bmatrix}$$

Thus, $x_1 = \$3,240,000$, $x_2 = \$810,000$, $x_3 = \$360,000$. The total tax liability is $x_1 + x_2 + x_3 = \$4,410,000$ which is 57.65% of the taxable income $\left(\frac{4,410,000}{7,650,000} = 0.5765 \text{ or } 57.65\%\right)$.

61. Let x_1 = number of ounces of food A,
x_2 = number of ounces of food B,
and x_3 = number of ounces of food C.

We have the following system of equations to solve:
$30x_1 + 10x_2 + 20x_3 = 340$
$10x_1 + 10x_2 + 20x_3 = 180$
$10x_1 + 30x_2 + 20x_3 = 220$

$$\begin{bmatrix} 30 & 10 & 20 & | & 340 \\ 10 & 10 & 20 & | & 180 \\ 10 & 30 & 20 & | & 220 \end{bmatrix} \sim \begin{bmatrix} 10 & 10 & 20 & | & 180 \\ 30 & 10 & 20 & | & 340 \\ 10 & 30 & 20 & | & 220 \end{bmatrix} \sim \begin{bmatrix} 1 & 1 & 2 & | & 18 \\ 3 & 1 & 2 & | & 34 \\ 1 & 3 & 2 & | & 22 \end{bmatrix}$$

$$\begin{matrix} R_1 \leftrightarrow R_2 & & \frac{1}{10}R_1 \to R_1 & & (-3)R_1 + R_2 \to R_2 \\ & & \frac{1}{10}R_2 \to R_2 & & (-1)R_1 + R_3 \to R_3 \\ & & \frac{1}{10}R_3 \to R_3 & & \end{matrix}$$

$$\sim \begin{bmatrix} 1 & 1 & 2 & | & 18 \\ 0 & -2 & -4 & | & -20 \\ 0 & 2 & 0 & | & 4 \end{bmatrix} \sim \begin{bmatrix} 1 & 1 & 2 & | & 18 \\ 0 & 1 & 2 & | & 10 \\ 0 & 2 & 0 & | & 4 \end{bmatrix} \sim \begin{bmatrix} 1 & 0 & 0 & | & 8 \\ 0 & 1 & 2 & | & 10 \\ 0 & 0 & -4 & | & -16 \end{bmatrix}$$

$$\begin{matrix} -\frac{1}{2}R_2 \to R_2 & & (-1)R_2 + R_1 \to R_1 & & -\frac{1}{4}R_3 \to R_3 \\ & & (-2)R_2 + R_3 \to R_3 & & \end{matrix}$$

$$\sim \begin{bmatrix} 1 & 0 & 0 & | & 8 \\ 0 & 1 & 2 & | & 10 \\ 0 & 0 & 1 & | & 4 \end{bmatrix} \sim \begin{bmatrix} 1 & 0 & 0 & | & 8 \\ 0 & 1 & 0 & | & 2 \\ 0 & 0 & 1 & | & 4 \end{bmatrix}$$

$$(-2)R_3 + R_2 \rightarrow R_2$$

Thus, $x_1 = 8$, $x_2 = 2$, and $x_3 = 4$ or 8 ounces of food A, 2 ounces of food B, and 4 ounces of food C.

63. Referring to Problem 61, we have:

$$30x_1 + 10x_2 = 340$$
$$10x_1 + 10x_2 = 180$$
$$10x_1 + 30x_2 = 220$$

$$\begin{bmatrix} 30 & 10 & | & 340 \\ 10 & 10 & | & 180 \\ 10 & 30 & | & 220 \end{bmatrix} \sim \begin{bmatrix} 3 & 1 & | & 34 \\ 1 & 1 & | & 18 \\ 1 & 3 & | & 22 \end{bmatrix} \sim \begin{bmatrix} 1 & 1 & | & 18 \\ 3 & 1 & | & 34 \\ 1 & 3 & | & 22 \end{bmatrix} \sim \begin{bmatrix} 1 & 1 & | & 18 \\ 0 & -2 & | & -20 \\ 0 & 2 & | & 4 \end{bmatrix} \sim \begin{bmatrix} 1 & 1 & | & 18 \\ 0 & 1 & | & 10 \\ 0 & 2 & | & 4 \end{bmatrix}$$

$$\frac{1}{10}R_1 \rightarrow R_1 \qquad R_1 \leftrightarrow R_2 \qquad (-3)R_1 + R_2 \rightarrow R_2 \qquad \left(-\frac{1}{2}\right)R_2 \rightarrow R_2 \qquad (-2)R_2 + R_3 \rightarrow R_3$$

$$\frac{1}{10}R_2 \rightarrow R_2 \qquad \qquad (-1)R_1 + R_3 \rightarrow R_3$$

$$\frac{1}{10}R_3 \rightarrow R_3$$

$$\sim \begin{bmatrix} 1 & 1 & | & 18 \\ 0 & 1 & | & 10 \\ 0 & 0 & | & -16 \end{bmatrix}$$ From this matrix, we conclude that there is no solution.

65. Referring to Problem 61, we have the following system of equations to solve:

$$30x_1 + 10x_2 + 20x_3 = 340$$
$$10x_1 + 10x_2 + 20x_3 = 180$$

$$\begin{bmatrix} 30 & 10 & 20 & | & 340 \\ 10 & 10 & 20 & | & 180 \end{bmatrix} \sim \begin{bmatrix} 10 & 10 & 20 & | & 180 \\ 30 & 10 & 20 & | & 340 \end{bmatrix} \sim \begin{bmatrix} 1 & 1 & 2 & | & 18 \\ 3 & 1 & 2 & | & 34 \end{bmatrix} \sim \begin{bmatrix} 1 & 1 & 2 & | & 18 \\ 0 & -2 & -4 & | & -20 \end{bmatrix}$$

$$R_1 \leftrightarrow R_2 \qquad \qquad \frac{1}{10}R_1 \rightarrow R_1 \qquad (-3)R_1 + R_2 \rightarrow R_2 \qquad \left(-\frac{1}{2}\right)R_2 \rightarrow R_2$$

$$\frac{1}{10}R_2 \rightarrow R_2$$

$$\sim \begin{bmatrix} 1 & 1 & 2 & | & 18 \\ 0 & 1 & 2 & | & 10 \end{bmatrix} \sim \begin{bmatrix} 1 & 0 & 0 & | & 8 \\ 0 & 1 & 2 & | & 10 \end{bmatrix}$$ Thus, $x_1 \qquad = 8$
$\qquad\qquad x_2 + 2x_3 = 10$

$$(-1)R_2 + R_1 \rightarrow R_1$$

Let $x_3 = t$ (t any real number). Then, $x_2 = 10 - 2t$, $0 \le t \le 5$, for x_2 to be positive.

The solution is: $x_1 = 8$ ounces of food A; $x_2 = 10 - 2t$ ounces of food B; $x_3 = t$ ounces of food C, $0 \le t \le 5$.

67. Let x_1 = number of barrels of mix A,

x_2 = number of barrels of mix B,

x_3 = number of barrels of mix C,

and x_4 = number of barrels of mix D,

Then,

$$30x_1 + 30x_2 + 30x_3 + 60x_4 = 900 \quad (1)$$
$$50x_1 + 75x_2 + 25x_3 + 25x_4 = 750 \quad (2)$$
$$30x_1 + 20x_2 + 20x_3 + 50x_4 = 700 \quad (3)$$

Divide each side of equation (1) by 30, each side of equation (2) by 25, and each side of equation (3) by 10. This yields the system of linear equations:

$$x_1 + x_2 + x_3 + 2x_4 = 30$$
$$2x_1 + 3x_2 + x_3 + x_4 = 30$$
$$3x_1 + 2x_2 + 2x_3 + 5x_4 = 70$$

$$
\begin{bmatrix}
1 & 1 & 1 & 2 & | & 30 \\
2 & 3 & 1 & 1 & | & 30 \\
3 & 2 & 2 & 5 & | & 70
\end{bmatrix}
\sim
\begin{bmatrix}
1 & 1 & 1 & 2 & | & 30 \\
0 & 1 & -1 & -3 & | & -30 \\
0 & -1 & -1 & -1 & | & -20
\end{bmatrix}
\sim
\begin{bmatrix}
1 & 0 & 2 & 5 & | & 60 \\
0 & 1 & -1 & -3 & | & -30 \\
0 & 0 & -2 & -4 & | & -50
\end{bmatrix}
$$

$(-2)R_1 + R_2 \to R_2$ $\qquad$ $R_3 + R_2 \to R_3$ $\qquad$ $\left(-\dfrac{1}{2}\right)R_3 \to R_3$

$(-3)R_1 + R_3 \to R_3$ $\qquad$ $(-1)R_2 + R_1 \to R_1$

$$
\sim
\begin{bmatrix}
1 & 0 & 2 & 5 & | & 60 \\
0 & 1 & -1 & -3 & | & -30 \\
0 & 0 & 1 & 2 & | & 25
\end{bmatrix}
\sim
\begin{bmatrix}
1 & 0 & 0 & 1 & | & 10 \\
0 & 1 & 0 & -1 & | & -5 \\
0 & 0 & 1 & 2 & | & 25
\end{bmatrix}
$$

$R_3 + R_2 \to R_2$

$(-2)R_3 + R_1 \to R_1$

Thus, $\quad x_1 \qquad + x_4 = 10$

$\qquad\qquad x_2 \quad - x_4 = -5$

$\qquad\qquad\qquad x_3 + 2x_4 = 25$

Let $x_4 = t$ = number of barrels of mix D. Then $x_1 = 10 - t$ = number of barrels of mix A, $x_2 = t - 5$ = number of barrels of mix B, and $x_3 = 25 - 2t$ = number of barrels of mix C. Since the number of barrels of each mix must be nonnegative, $5 \le t \le 10$. Also, t is an integer.

69. Let x_1 = number of hours for Company A,

and x_2 = number of hours for Company B.

Then, $\quad 30x_1 + 20x_2 = 600$

$\qquad\qquad 10x_1 + 20x_2 = 400$

Divide each side of each equation by 10. This yields the system of linear equations:

$$3x_1 + 2x_2 = 60$$
$$x_1 + 2x_2 = 40$$

$$\begin{bmatrix} 3 & 2 & | & 60 \\ 1 & 2 & | & 40 \end{bmatrix} \sim \begin{bmatrix} 1 & 2 & | & 40 \\ 3 & 2 & | & 60 \end{bmatrix} \sim \begin{bmatrix} 1 & 2 & | & 40 \\ 0 & -4 & | & -60 \end{bmatrix} \sim \begin{bmatrix} 1 & 2 & | & 40 \\ 0 & 1 & | & 15 \end{bmatrix} \sim \begin{bmatrix} 1 & 0 & | & 10 \\ 0 & 1 & | & 15 \end{bmatrix}$$

$R_1 \leftrightarrow R_2 \qquad (-3)R_1 + R_2 \rightarrow R_2 \qquad \left(-\dfrac{1}{4}\right) R_2 \rightarrow R_2 \qquad (-2)R_2 + R_1 \rightarrow R_1$

Thus, $x_1 = 10$ and $x_2 = 15$, or 10 hours for Company A and 15 hours for Company B.

71. (A) 6th and Washington Ave.: $x_1 + x_2 = 1200$
    6th and Lincoln Ave.: $\quad x_2 + x_3 = 1000$
    5th and Lincoln Ave.: $\quad x_3 + x_4 = 1300$

(B) The system of equations is:

$$\begin{aligned} x_1 \qquad\quad + x_4 &= 1500 \\ x_1 + x_2 \qquad\quad &= 1200 \\ x_2 + x_3 \quad\;\; &= 1000 \\ x_3 + x_4 &= 1300 \end{aligned}$$

$$\begin{bmatrix} 1 & 0 & 0 & 1 & | & 1500 \\ 1 & 1 & 0 & 0 & | & 1200 \\ 0 & 1 & 1 & 0 & | & 1000 \\ 0 & 0 & 1 & 1 & | & 1300 \end{bmatrix} \sim \begin{bmatrix} 1 & 0 & 0 & 1 & | & 1500 \\ 0 & 1 & 0 & -1 & | & -300 \\ 0 & 1 & 1 & 0 & | & 1000 \\ 0 & 0 & 1 & 1 & | & 1300 \end{bmatrix} \sim \begin{bmatrix} 1 & 0 & 0 & 1 & | & 1500 \\ 0 & 1 & 0 & -1 & | & -300 \\ 0 & 0 & 1 & 1 & | & 1300 \\ 0 & 0 & 1 & 1 & | & 1300 \end{bmatrix}$$

$\qquad\qquad (-1)R_1 + R_2 \rightarrow R_2 \qquad\qquad (-1)R_2 + R_3 \rightarrow R_3 \qquad\qquad (-1)R_3 + R_4 \rightarrow R_4$

$$\sim \begin{bmatrix} 1 & 0 & 0 & 1 & | & 1500 \\ 0 & 1 & 0 & -1 & | & -300 \\ 0 & 0 & 1 & 1 & | & 1300 \\ 0 & 0 & 0 & 0 & | & 0 \end{bmatrix}$$

Thus
$$\begin{aligned} x_1 \qquad\quad + x_4 &= 1500 \\ x_2 \quad\; - x_4 &= -300 \\ x_3 + x_4 &= 1300 \end{aligned}$$

Let $x_4 = t$. Then $x_1 = 1500 - t$, $x_2 = t - 300$ and $x_3 = 1300 - t$. Since x_1, x_2, x_3, and x_4 must be nonnegative integers, we have $300 \le t \le 1300$.

(C) The flow from Washington Ave. to Lincoln Ave. on 5th Street is given by $x_4 = t$. As shown in part (B), $300 \le t \le 1300$, that is, the maximum number of vehicles is 1300 and the minimum number is 300.

(D) If $x_4 = t = 1000$, then
    Washington Ave.: $x_1 = 1500 - 1000 = 500$
    6th St.: $x_2 = 1000 - 300 = 700$
    Lincoln Ave.: $x_3 = 1300 - 1000 = 300$

EXERCISE 4-4

Things to remember:

1. A matrix with m rows and n columns is said to have SIZE or DIMENSION $m \times n$. If a matrix has the same number of rows and columns, then it is called a SQUARE MATRIX. A matrix with only one column is a COLUMN MATRIX, and a matrix with only one row is a ROW MATRIX.

2. Two matrices are EQUAL if they have the same dimension and their corresponding elements are equal.

3. The SUM of two matrices of the same dimension, $m \times n$, is an $m \times n$ matrix whose elements are the sum of the corresponding elements of the two given matrices. Addition is not defined for matrices with different dimensions. Matrix addition is commutative: $A + B = B + A$, and assocative: $(A + B) + C = A + (B + C)$.

4. A matrix with all elements equal to zero is called a ZERO MATRIX.

5. The NEGATIVE OF A MATRIX M, denoted by $-M$, is the matrix whose elements are the negatives of the elements of M.

6. If A and B are matrices of the same dimension, then subtraction is defined by $A - B = A + (-B)$. Thus, to subtract B from A, simply subtract corresponding elements.

7. If M is a matrix and k is a number, then kM is the matrix formed by multiplying each element of M by k.

8. PRODUCT OF A ROW MATRIX AND A COLUMN MATRIX

   The product of a $1 \times n$ row matrix and an $n \times 1$ column matrix is the 1×1 matrix given by

   $$\underset{1 \times n}{[a_1 \quad a_2 \quad \cdots \quad a_n]} \overset{n \times 1}{\begin{bmatrix} b_1 \\ b_2 \\ \vdots \\ b_n \end{bmatrix}} = [a_1 b_1 + a_2 b_2 + \cdots + a_n b_n]$$

   Note that the number of elements in the row matrix and the number of elements in the column matrix must be the same for the product to be defined.

9. Let A be an $m \times p$ matrix and B be a $p \times n$ matrix. The MATRIX PRODUCT of A and B, denoted AB, is the $m \times n$ matrix whose element in the ith row and the jth column is the real number obtained from the product of the ith row of A and the jth column of B. If the number of columns in A does not equal the number of rows in B, then the matrix product AB is not defined.

NOTE: Matrix multiplication is *not* commutative. That is AB does not always equal BA, even when both multiplications are defined.

---

1. $\begin{bmatrix} 2 & -1 \\ 3 & 0 \end{bmatrix} + \begin{bmatrix} -3 & 1 \\ 2 & -3 \end{bmatrix} = \begin{bmatrix} 2 + (-3) & -1 + 1 \\ 3 + 2 & 0 + (-3) \end{bmatrix} = \begin{bmatrix} -1 & 0 \\ 5 & -3 \end{bmatrix}$

3. $\begin{bmatrix} 4 & -1 & 0 \\ 2 & 1 & 3 \end{bmatrix} + \begin{bmatrix} -2 & 1 & 3 \\ 5 & 6 & -8 \end{bmatrix} = \begin{bmatrix} 4 + (-2) & -1 + 1 & 0 + 3 \\ 2 + 5 & 1 + 6 & 3 + (-8) \end{bmatrix} = \begin{bmatrix} 2 & 0 & 3 \\ 7 & 7 & -5 \end{bmatrix}$

5. Addition not defined; the matrices have different dimensions.

7. $\begin{bmatrix} 4 & -5 \\ 1 & 0 \\ 1 & -3 \end{bmatrix} - \begin{bmatrix} -1 & 2 \\ 6 & -2 \\ 1 & -7 \end{bmatrix} = \begin{bmatrix} 4 - (-1) & -5 - 2 \\ 1 - 6 & 0 - (-2) \\ 1 - 1 & -3 - (-7) \end{bmatrix} = \begin{bmatrix} 5 & -7 \\ -5 & 2 \\ 0 & 4 \end{bmatrix}$

9. $5 \begin{bmatrix} 1 & -2 & 0 & 4 \\ -3 & 2 & -1 & 6 \end{bmatrix} = \begin{bmatrix} 5(1) & 5(-2) & 5(0) & 5(4) \\ 5(-3) & 5(2) & 5(-1) & 5(6) \end{bmatrix} = \begin{bmatrix} 5 & -10 & 0 & 20 \\ -15 & 10 & -5 & 30 \end{bmatrix}$

11. $\begin{bmatrix} 2 & 4 \end{bmatrix} \begin{bmatrix} 3 \\ 1 \end{bmatrix} = [2 \cdot 3 + 4 \cdot 1]$ (using $\underline{8}$)

$= [10]$

13. $\begin{bmatrix} 3 & 4 \\ -1 & -2 \end{bmatrix} \begin{bmatrix} -1 \\ 2 \end{bmatrix} = \begin{bmatrix} \begin{bmatrix} 3 & 4 \end{bmatrix} \begin{bmatrix} -1 \\ 2 \end{bmatrix} \\ \begin{bmatrix} -1 & -2 \end{bmatrix} \begin{bmatrix} -1 \\ 2 \end{bmatrix} \end{bmatrix} = \begin{bmatrix} -3 + 8 \\ 1 - 4 \end{bmatrix} = \begin{bmatrix} 5 \\ -3 \end{bmatrix}$

15. $\begin{bmatrix} 2 & -3 \\ 1 & 2 \end{bmatrix} \begin{bmatrix} 1 & -1 \\ 0 & -2 \end{bmatrix} = \begin{bmatrix} \begin{bmatrix} 2 & -3 \end{bmatrix} \begin{bmatrix} 1 \\ 0 \end{bmatrix} & \begin{bmatrix} 2 & -3 \end{bmatrix} \begin{bmatrix} -1 \\ -2 \end{bmatrix} \\ \begin{bmatrix} 1 & 2 \end{bmatrix} \begin{bmatrix} 1 \\ 0 \end{bmatrix} & \begin{bmatrix} 1 & 2 \end{bmatrix} \begin{bmatrix} -1 \\ -2 \end{bmatrix} \end{bmatrix} = \begin{bmatrix} 2 + 0 & -2 + 6 \\ 1 + 0 & -1 - 4 \end{bmatrix} = \begin{bmatrix} 2 & 4 \\ 1 & -5 \end{bmatrix}$

17. $\begin{bmatrix} 1 & -1 \\ 0 & -2 \end{bmatrix} \begin{bmatrix} 2 & -3 \\ 1 & 2 \end{bmatrix} = \begin{bmatrix} \begin{bmatrix} 1 & -1 \end{bmatrix} \begin{bmatrix} 2 \\ 1 \end{bmatrix} & \begin{bmatrix} 1 & -1 \end{bmatrix} \begin{bmatrix} -3 \\ 2 \end{bmatrix} \\ \begin{bmatrix} 0 & -2 \end{bmatrix} \begin{bmatrix} 2 \\ 1 \end{bmatrix} & \begin{bmatrix} 0 & -2 \end{bmatrix} \begin{bmatrix} -3 \\ 2 \end{bmatrix} \end{bmatrix} = \begin{bmatrix} 2 - 1 & -3 - 2 \\ 0 - 2 & 0 - 4 \end{bmatrix} = \begin{bmatrix} 1 & -5 \\ -2 & -4 \end{bmatrix}$

19. $\begin{bmatrix} 5 & -2 \end{bmatrix} \begin{bmatrix} -3 \\ -4 \end{bmatrix} = [-15 + 8] = [-7]$

21. $\begin{bmatrix} -3 \\ -4 \end{bmatrix} \begin{bmatrix} 5 & -2 \end{bmatrix} = \begin{bmatrix} (-3)(5) & (-3)(-2) \\ (-4)(5) & (-4)(-2) \end{bmatrix} = \begin{bmatrix} -15 & 6 \\ -20 & 8 \end{bmatrix}$

23. $\begin{bmatrix} 3 & -2 & -4 \end{bmatrix} \begin{bmatrix} 1 \\ 2 \\ -3 \end{bmatrix} = [(3 - 4 + 12)] = [11]$

25. $\begin{bmatrix} 1 \\ 2 \\ -3 \end{bmatrix} [3 \quad -2 \quad -4] = \begin{bmatrix} (1)(3) & (1)(-2) & (1)(-4) \\ (2)(3) & (2)(-2) & (2)(-4) \\ (-3)(3) & (-3)(-2) & (-3)(-4) \end{bmatrix} = \begin{bmatrix} 3 & -2 & -4 \\ 6 & -4 & -8 \\ -9 & 6 & 12 \end{bmatrix}$

27. $AC = \begin{bmatrix} 2 & -1 & 3 \\ 0 & 4 & -2 \end{bmatrix} \begin{bmatrix} -1 & 0 & 2 \\ 4 & -3 & 1 \\ -2 & 3 & 5 \end{bmatrix} = \begin{bmatrix} -12 & 12 & 18 \\ 20 & -18 & -6 \end{bmatrix}$

29. AB is not defined; the number of columns of A (3) does not equal the number of rows of B (2).

31. $B^2 = BB = \begin{bmatrix} -3 & 1 \\ 2 & 5 \end{bmatrix} \begin{bmatrix} -3 & 1 \\ 2 & 5 \end{bmatrix} = \begin{bmatrix} 11 & 2 \\ 4 & 27 \end{bmatrix}$

33. $B + AD = \begin{bmatrix} -3 & 1 \\ 2 & 5 \end{bmatrix} + \begin{bmatrix} 2 & -1 & 3 \\ 0 & 4 & -2 \end{bmatrix} \begin{bmatrix} 3 & -2 \\ 0 & -1 \\ 1 & 2 \end{bmatrix}$

$= \begin{bmatrix} -3 & 1 \\ 2 & 5 \end{bmatrix} + \begin{bmatrix} 9 & 3 \\ -2 & -8 \end{bmatrix} = \begin{bmatrix} 6 & 4 \\ 0 & -3 \end{bmatrix}$

35. $0.1DB = 0.1 \begin{bmatrix} 3 & -2 \\ 0 & -1 \\ 1 & 2 \end{bmatrix} \begin{bmatrix} -3 & 1 \\ 2 & 5 \end{bmatrix}$

$= 0.1 \begin{bmatrix} -13 & -7 \\ -2 & -5 \\ 1 & 11 \end{bmatrix} = \begin{bmatrix} -1.3 & -0.7 \\ -0.2 & -0.5 \\ 0.1 & 1.1 \end{bmatrix}$

37. $3BA + 4AC = 3 \begin{bmatrix} -3 & 1 \\ 2 & 5 \end{bmatrix} \begin{bmatrix} 2 & -1 & 3 \\ 0 & 4 & -2 \end{bmatrix} + 4 \begin{bmatrix} 2 & -1 & 3 \\ 0 & 4 & -2 \end{bmatrix} \begin{bmatrix} -1 & 0 & 2 \\ 4 & -3 & 1 \\ -2 & 3 & 5 \end{bmatrix}$

$= 3 \begin{bmatrix} -6 & 7 & -11 \\ 4 & 18 & -4 \end{bmatrix} + 4 \begin{bmatrix} -12 & 12 & 18 \\ 20 & -18 & -6 \end{bmatrix}$

$= \begin{bmatrix} -18 & 21 & -33 \\ 12 & 54 & -12 \end{bmatrix} + \begin{bmatrix} -48 & 48 & 72 \\ 80 & -72 & -24 \end{bmatrix}$

$= \begin{bmatrix} -66 & 69 & 39 \\ 92 & -18 & -36 \end{bmatrix}$

39. $(-2)BA + 6CD$ is not defined; $-2BA$ is 2×3, $6CD$ is 3×2

41. $ACD = A(CD) = \begin{bmatrix} 2 & -1 & 3 \\ 0 & 4 & -2 \end{bmatrix} \left(\begin{bmatrix} -1 & 0 & 2 \\ 4 & -3 & 1 \\ -2 & 3 & 5 \end{bmatrix} \begin{bmatrix} 3 & -2 \\ 0 & -1 \\ 1 & 2 \end{bmatrix} \right)$

$= \begin{bmatrix} 2 & -1 & 3 \\ 0 & 4 & -2 \end{bmatrix} \begin{bmatrix} -1 & 6 \\ 13 & -3 \\ -1 & 11 \end{bmatrix}$

$= \begin{bmatrix} -18 & 48 \\ 54 & -34 \end{bmatrix}$

43. $DBA = D(BA) = \begin{bmatrix} 3 & -2 \\ 0 & -1 \\ 1 & 2 \end{bmatrix} \left(\begin{bmatrix} -3 & 1 \\ 2 & 5 \end{bmatrix} \begin{bmatrix} 2 & -1 & 3 \\ 0 & 4 & -2 \end{bmatrix} \right)$

$= \begin{bmatrix} 3 & -2 \\ 0 & -1 \\ 1 & 2 \end{bmatrix} \begin{bmatrix} -6 & 7 & -11 \\ 4 & 18 & -4 \end{bmatrix}$

$= \begin{bmatrix} -26 & -15 & -25 \\ -4 & -18 & 4 \\ 2 & 43 & -19 \end{bmatrix}$

45. $\begin{bmatrix} a & b \\ c & d \end{bmatrix} + \begin{bmatrix} 2 & -3 \\ 0 & 1 \end{bmatrix} = \begin{bmatrix} a+2 & b-3 \\ c+0 & d+1 \end{bmatrix} = \begin{bmatrix} 1 & -2 \\ 3 & -4 \end{bmatrix}$ Thus, $a+2 = 1,\ a = -1$
$\qquad b-3 = -2,\ b = 1$
$\qquad c+0 = 3,\ c = 3$
$\qquad d+1 = -4,\ d = -5$

47. $\begin{bmatrix} 2x & 4 \\ -3 & 5x \end{bmatrix} + \begin{bmatrix} 3y & -2 \\ -2 & -y \end{bmatrix} = \begin{bmatrix} -5 & 2 \\ -5 & 13 \end{bmatrix}$

$\begin{bmatrix} 2x+3y & 4-2 \\ -3-2 & 5x-y \end{bmatrix} = \begin{bmatrix} -5 & 2 \\ -5 & 13 \end{bmatrix}$

$\begin{bmatrix} 2x+3y & 2 \\ -5 & 5x-y \end{bmatrix} = \begin{bmatrix} -5 & 2 \\ -5 & 13 \end{bmatrix}$ Thus, $2x + 3y = -5$ (1)
$\qquad\qquad 5x - y = 13$ (2)
Solve the above system for $x,\ y$.

From (2), $y = 5x - 13$. Substitute $y = 5x - 13$ in (1):

$2x + 3(5x - 13) = -5$
$\quad 2x + 15x - 39 = -5$
$\qquad\qquad 17x = -5 + 39$
$\qquad\qquad 17x = 34$
$\qquad\qquad\ x = 2$

Substitute $x = 2$ in (1):

$2(2) + 3y = -5$
$\quad 4 + 3y = -5$
$\qquad 3y = -9$
$\qquad\ y = -3$

Thus, the solution is $x = 2$ and $y = -3$.

49. $\begin{bmatrix} x & -1 \\ 1 & 0 \end{bmatrix}\begin{bmatrix} 2 & 1 \\ 4 & 1 \end{bmatrix} = \begin{bmatrix} y & y \\ 2 & 1 \end{bmatrix}$

$\begin{bmatrix} 2x - 4 & x - 1 \\ 2 & 1 \end{bmatrix} = \begin{bmatrix} y & y \\ 2 & 1 \end{bmatrix}$

implies $2x - 4 = y$

$\qquad\qquad x - 1 = y$

Thus, $2x - 4 = x - 1$

and $\qquad\qquad x = 3, \ y = 2$

51. $\begin{bmatrix} 1 & -2 \\ 2 & -3 \end{bmatrix}\begin{bmatrix} a & b \\ c & d \end{bmatrix} = \begin{bmatrix} 1 & 0 \\ 3 & 2 \end{bmatrix}$

$\begin{bmatrix} a - 2c & b - 2d \\ 2a - 3c & 2b - 3d \end{bmatrix} = \begin{bmatrix} 1 & 0 \\ 3 & 2 \end{bmatrix}$

implies $\quad a - 2c = 1 \qquad\qquad b - 2d = 0$

$\qquad\qquad 2a - 3c = 3 \qquad\qquad 2b - 3d = 2$

The augmented matrix for the first system is:

$\begin{bmatrix} 1 & -2 & | & 1 \\ 2 & -3 & | & 3 \end{bmatrix} (-2)R_1 + R_2 \rightarrow R_2 \sim \begin{bmatrix} 1 & -2 & | & 1 \\ 0 & 1 & | & 1 \end{bmatrix} 2R_2 + R_1 \rightarrow R_1$

$\sim \begin{bmatrix} 1 & 0 & | & 3 \\ 0 & 1 & | & 1 \end{bmatrix}$ Thus, $a = 3, \ c = 1$.

For the second system, substitute $b = 2d$ from the first equation into the second equation:

$\qquad\qquad 2(2d) - 3d = 2$

$\qquad\qquad\qquad\qquad d = 2$

$\qquad\qquad\qquad\qquad b = 4$

Solution: $a = 3, \ b = 4, \ c = 1, \ d = 2$

53. Let $A = \begin{bmatrix} a_1 & 0 \\ 0 & a_2 \end{bmatrix}$ and $B = \begin{bmatrix} b_1 & 0 \\ 0 & b_2 \end{bmatrix}$

(A) Always true:

$\qquad A + B = \begin{bmatrix} a_1 & 0 \\ 0 & a_2 \end{bmatrix} + \begin{bmatrix} b_1 & 0 \\ 0 & b_2 \end{bmatrix} = \begin{bmatrix} a_1 + b_1 & 0 \\ 0 & a_2 + b_2 \end{bmatrix}$

(B) Always true: matrix addition is commutative, $A + B = B + A$ for *any* pair of matrices of the same size.

(C) Always true:

$\qquad AB = \begin{bmatrix} a_1 & 0 \\ 0 & a_2 \end{bmatrix}\begin{bmatrix} b_1 & 0 \\ 0 & b_2 \end{bmatrix} = \begin{bmatrix} a_1 b_1 & 0 \\ 0 & a_2 b_2 \end{bmatrix}$

(D) Always true:

$\qquad BA = \begin{bmatrix} b_1 & 0 \\ 0 & b_2 \end{bmatrix}\begin{bmatrix} a_1 & 0 \\ 0 & a_2 \end{bmatrix} = \begin{bmatrix} b_1 a_1 & 0 \\ 0 & b_2 a_2 \end{bmatrix} = \begin{bmatrix} a_1 b_1 & 0 \\ 0 & a_2 b_2 \end{bmatrix} = AB$

55. $A + B = \begin{bmatrix} \$30 & \$25 \\ \$60 & \$80 \end{bmatrix} + \begin{bmatrix} \$36 & \$27 \\ \$54 & \$74 \end{bmatrix} = \begin{bmatrix} \$66 & \$52 \\ \$114 & \$154 \end{bmatrix}$

$\frac{1}{2}(A + B) = \frac{1}{2}\begin{bmatrix} \$66 & \$52 \\ \$114 & \$154 \end{bmatrix} = \begin{bmatrix} \$33 & \$26 \\ \$57 & \$77 \end{bmatrix} \begin{matrix} \text{Materials} \\ \text{Labor} \end{matrix}$

with column labels Guitar, Banjo above.

57. The dealer is increasing the retail prices by 10%. Thus, the new retail price matrix M' (to the nearest dollar) is given by

$$M' = M + 0.1M = 1.1M = 1.1 \begin{bmatrix} 10900 & 683 & 253 & 195 \\ 13000 & 738 & 382 & 206 \\ 16300 & 867 & 537 & 225 \end{bmatrix} = \begin{bmatrix} 11990 & 751 & 278 & 214 \\ 14300 & 812 & 420 & 227 \\ 17930 & 954 & 591 & 248 \end{bmatrix}$$

The new dealer invoice matrix N' is given by

$$N' = N + 0.15N = 1.15N = 1.15 \begin{bmatrix} 9400 & 582 & 195 & 160 \\ 11500 & 621 & 295 & 171 \\ 14100 & 737 & 420 & 184 \end{bmatrix}$$

$$= \begin{bmatrix} 10810 & 669 & 224 & 184 \\ 13225 & 714 & 339 & 197 \\ 16215 & 848 & 483 & 212 \end{bmatrix}$$

The new markup is:

$$M' - N' = \begin{bmatrix} 11990 & 751 & 278 & 214 \\ 14300 & 812 & 420 & 227 \\ 17930 & 954 & 591 & 248 \end{bmatrix} - \begin{bmatrix} 10810 & 669 & 224 & 184 \\ 13225 & 714 & 339 & 197 \\ 16215 & 848 & 483 & 212 \end{bmatrix}$$

$$= \begin{matrix} \text{Model A} \\ \text{Model B} \\ \text{Model C} \end{matrix} \begin{bmatrix} \$1180 & \$82 & \$54 & \$30 \\ \$1075 & \$98 & \$81 & \$30 \\ \$1715 & \$106 & \$108 & \$36 \end{bmatrix}$$

with column labels Basic Car, Air/Cond., AM/FM, Cruise above.

59. (A) $[0.6 \quad 0.6 \quad 0.2] \begin{bmatrix} 8 \\ 10 \\ 5 \end{bmatrix} = [4.8 + 6.0 + 1.0] = [11.8]$

Thus, the labor cost per boat for one-person boats at plant I is $11.80.

(B) $[1.5 \quad 1.2 \quad 0.4] \begin{bmatrix} 9 \\ 12 \\ 6 \end{bmatrix} = [13.5 + 14.4 + 2.4] = [30.3]$

Thus, the labor cost per boat for four-person boats at plant II is $30.30.

(C) MN gives the labor costs per boat at each plant; NM is not defined.

$$(D) \quad MN = \begin{bmatrix} 0.6 & 0.6 & 0.2 \\ 1.0 & 0.9 & 0.3 \\ 1.5 & 1.2 & 0.4 \end{bmatrix} \begin{bmatrix} 8 & 9 \\ 10 & 12 \\ 5 & 6 \end{bmatrix}$$

$$= \begin{bmatrix} \$11.80 & \$13.80 \\ \$18.50 & \$21.60 \\ \$26.00 & \$30.30 \end{bmatrix} \begin{matrix} \text{One-person boat} \\ \text{Two-person boat} \\ \text{Four-person boat} \end{matrix}$$

Plant I   Plant II

This matrix represents the labor cost per boat for each kind of boat at each plant. For example, $21.60 represents labor costs per boat for two-person boats at plant II.

61. $A = \begin{bmatrix} 0 & 1 & 0 & 1 & 0 \\ 0 & 0 & 1 & 0 & 0 \\ 1 & 0 & 0 & 0 & 1 \\ 0 & 0 & 1 & 0 & 0 \\ 0 & 0 & 0 & 1 & 0 \end{bmatrix}$

$$(A) \quad A^2 = \begin{bmatrix} 0 & 1 & 0 & 1 & 0 \\ 0 & 0 & 1 & 0 & 0 \\ 1 & 0 & 0 & 0 & 1 \\ 0 & 0 & 1 & 0 & 0 \\ 0 & 0 & 0 & 1 & 0 \end{bmatrix} \begin{bmatrix} 0 & 1 & 0 & 1 & 0 \\ 0 & 0 & 1 & 0 & 0 \\ 1 & 0 & 0 & 0 & 1 \\ 0 & 0 & 1 & 0 & 0 \\ 0 & 0 & 0 & 1 & 0 \end{bmatrix} = \begin{bmatrix} 0 & 0 & 2 & 0 & 0 \\ 1 & 0 & 0 & 0 & 1 \\ 0 & 1 & 0 & 2 & 0 \\ 1 & 0 & 0 & 0 & 1 \\ 0 & 0 & 1 & 0 & 0 \end{bmatrix}$$

The 1 in row two, column one indicates that there is one way to travel from Baltimore to Atlanta with one intermediate connection, namely Baltimore-to-Chicago-to-Atlanta.

The 2 in row one, column three indicates that there are two ways to travel from Atlanta to Chicago with one intermediate connection, namely Atlanta-to-Baltimore-to-Chicago, and Atlanta-to-Denver-to-Chicago.

In general, the element b_{ij}, $i \neq j$, in A^2 indicates the number of different ways to travel from the ith city to the jth city with one intermediate connection.

$$(B) \quad A^3 = AA^2 = \begin{bmatrix} 0 & 1 & 0 & 1 & 0 \\ 0 & 0 & 1 & 0 & 0 \\ 1 & 0 & 0 & 0 & 1 \\ 0 & 0 & 1 & 0 & 0 \\ 0 & 0 & 0 & 1 & 0 \end{bmatrix} \begin{bmatrix} 0 & 0 & 2 & 0 & 0 \\ 1 & 0 & 0 & 0 & 1 \\ 0 & 1 & 0 & 2 & 0 \\ 1 & 0 & 0 & 0 & 1 \\ 0 & 0 & 1 & 0 & 0 \end{bmatrix} = \begin{bmatrix} 2 & 0 & 0 & 0 & 2 \\ 0 & 1 & 0 & 2 & 0 \\ 0 & 0 & 3 & 0 & 0 \\ 0 & 1 & 0 & 2 & 0 \\ 1 & 0 & 0 & 0 & 1 \end{bmatrix}$$

The 1 in row four, column 2 indicates that there is one way to travel from Denver to Baltimore with two intermediate connections.

The 2 in row one, column five indicates that there are two ways to travel from Atlanta to El Paso with two intermediate connections.

In general, the element c_{ij}, $i \neq j$, in A^3 indicates the number of ways to travel from the ith city to the jth city with two intermediate connections.

(C) From parts (A) and (B)

$$A + A^2 = \begin{bmatrix} 0 & 1 & 0 & 1 & 0 \\ 0 & 0 & 1 & 0 & 0 \\ 1 & 0 & 0 & 0 & 1 \\ 0 & 0 & 1 & 0 & 0 \\ 0 & 0 & 0 & 1 & 0 \end{bmatrix} + \begin{bmatrix} 0 & 0 & 2 & 0 & 0 \\ 1 & 0 & 0 & 0 & 1 \\ 0 & 1 & 0 & 2 & 0 \\ 1 & 0 & 0 & 0 & 1 \\ 0 & 0 & 1 & 0 & 0 \end{bmatrix} = \begin{bmatrix} 0 & 1 & 2 & 1 & 0 \\ 1 & 0 & 1 & 0 & 1 \\ 1 & 1 & 0 & 2 & 1 \\ 1 & 0 & 1 & 0 & 1 \\ 0 & 0 & 1 & 1 & 0 \end{bmatrix}$$

$$A + A^2 + A^3 = \begin{bmatrix} 0 & 1 & 2 & 1 & 0 \\ 1 & 0 & 1 & 0 & 1 \\ 1 & 1 & 0 & 2 & 1 \\ 1 & 0 & 1 & 0 & 1 \\ 0 & 0 & 1 & 1 & 0 \end{bmatrix} + \begin{bmatrix} 2 & 0 & 0 & 0 & 2 \\ 0 & 1 & 0 & 2 & 0 \\ 0 & 0 & 3 & 0 & 0 \\ 0 & 1 & 0 & 2 & 0 \\ 1 & 0 & 0 & 0 & 1 \end{bmatrix} = \begin{bmatrix} 2 & 1 & 2 & 1 & 2 \\ 1 & 1 & 1 & 2 & 1 \\ 1 & 1 & 3 & 2 & 1 \\ 1 & 1 & 1 & 2 & 1 \\ 1 & 0 & 1 & 1 & 1 \end{bmatrix}$$

$$A^4 = AA^3 = \begin{bmatrix} 0 & 1 & 0 & 1 & 0 \\ 0 & 0 & 1 & 0 & 0 \\ 1 & 0 & 0 & 0 & 1 \\ 0 & 0 & 1 & 0 & 0 \\ 0 & 0 & 0 & 1 & 0 \end{bmatrix} \begin{bmatrix} 2 & 0 & 0 & 0 & 2 \\ 0 & 1 & 0 & 2 & 0 \\ 0 & 0 & 3 & 0 & 0 \\ 0 & 1 & 0 & 2 & 0 \\ 1 & 0 & 0 & 0 & 1 \end{bmatrix} = \begin{bmatrix} 0 & 2 & 0 & 4 & 0 \\ 0 & 0 & 3 & 0 & 0 \\ 3 & 0 & 0 & 0 & 3 \\ 0 & 0 & 3 & 0 & 0 \\ 0 & 1 & 0 & 2 & 0 \end{bmatrix}$$

$$A + A^2 + A^3 + A^4 = \begin{bmatrix} 2 & 1 & 2 & 1 & 2 \\ 1 & 1 & 1 & 2 & 1 \\ 1 & 1 & 3 & 2 & 1 \\ 1 & 1 & 1 & 2 & 1 \\ 1 & 0 & 1 & 1 & 1 \end{bmatrix} + \begin{bmatrix} 0 & 2 & 0 & 4 & 0 \\ 0 & 0 & 3 & 0 & 0 \\ 3 & 0 & 0 & 0 & 3 \\ 0 & 0 & 3 & 0 & 0 \\ 0 & 1 & 0 & 2 & 0 \end{bmatrix}$$

$$= \begin{bmatrix} 2 & 3 & 2 & 5 & 2 \\ 1 & 1 & 4 & 2 & 1 \\ 4 & 1 & 3 & 2 & 4 \\ 1 & 1 & 4 & 2 & 1 \\ 1 & 1 & 1 & 3 & 1 \end{bmatrix}$$

It is possible to travel from any origin to any destination with at most 3 intermediate connections.

63. (A) $[4 \quad 2]\begin{bmatrix} 15 \\ 5 \end{bmatrix} = 70$

There are 70 g of protein in Mix X.

(B) $[3 \quad 1]\begin{bmatrix} 5 \\ 15 \end{bmatrix} = 30$

There are 30 g of fat in Mix Z.

(C) *MN* gives the amount (in grams) of protein, carbohydrates and fat in 20 ounces of each mix. The product *NM* is not defined.

$$(D) \quad MN = \begin{bmatrix} 4 & 2 \\ 20 & 16 \\ 3 & 1 \end{bmatrix} \begin{bmatrix} 15 & 10 & 5 \\ 5 & 10 & 15 \end{bmatrix} = \begin{array}{ccc} \text{Mix X} & \text{Mix Y} & \text{Mix Z} \end{array}$$

$$\begin{bmatrix} 70 & 60 & 50 \\ 380 & 360 & 340 \\ 50 & 40 & 30 \end{bmatrix} \begin{array}{l} \text{Protein} \\ \text{Carbohydrate} \\ \text{Fat} \end{array}$$

65. (A) $[1000 \quad 500 \quad 5000] \begin{bmatrix} 0.40 \\ 1.00 \\ 0.35 \end{bmatrix} = 2{,}650$

Total amount spent in Berkeley = $2,650.

(B) $[2000 \quad 800 \quad 8000] \begin{bmatrix} 0.40 \\ 1.00 \\ 0.35 \end{bmatrix} = 4{,}400$

Total amount spent in Oakland = $4,400.

(C) NM gives the total cost per town.

(D) $NM = \begin{bmatrix} 1000 & 500 & 5000 \\ 2000 & 800 & 8000 \end{bmatrix} \begin{bmatrix} 0.40 \\ 1.00 \\ 0.35 \end{bmatrix} = \begin{array}{c} \text{Cost/Town} \end{array} \begin{bmatrix} 2{,}650 \\ 4{,}400 \end{bmatrix} \begin{array}{l} \text{Berkeley} \\ \text{Oakland} \end{array}$

(E) $[1 \quad 1]N = [1 \quad 1] \begin{bmatrix} 1000 & 500 & 5000 \\ 2000 & 800 & 8000 \end{bmatrix}$

$$= \begin{array}{ccc} \text{Telephone} & \text{House} & \text{Letters} \\ \text{Calls} & \text{Calls} & \\ [3{,}000 & 1300 & 13{,}000] \end{array}$$

(F) $N \begin{bmatrix} 1 \\ 1 \\ 1 \end{bmatrix} = \begin{bmatrix} 1{,}000 & 500 & 5{,}000 \\ 2{,}000 & 800 & 8{,}000 \end{bmatrix} \begin{bmatrix} 1 \\ 1 \\ 1 \end{bmatrix} = \begin{array}{c} \text{Total contacts} \end{array} \begin{bmatrix} 6{,}500 \\ 10{,}800 \end{bmatrix} \begin{array}{l} \text{Berkeley} \\ \text{Oakland} \end{array}$

EXERCISE 4-5

Things to remember:

1. The IDENTITY element for multiplication for the set of square
    matrices of order n (dimension $n \times n$) is the square matrix I of
    order n which has 1's on the principal diagonal (upper left
    corner to lower right corner) and 0's elsewhere. The identity
    matrices of order 2 and 3, respectively, are

    $$I = \begin{bmatrix} 1 & 0 \\ 0 & 1 \end{bmatrix} \quad \text{and} \quad I = \begin{bmatrix} 1 & 0 & 0 \\ 0 & 1 & 0 \\ 0 & 0 & 1 \end{bmatrix}.$$

2. If M is any square matrix of order n and I is the identity
    matrix of order n, then

    $$IM = MI = M.$$

3. INVERSE OF A SQUARE MATRIX

Let M be a square matrix of order n and I be the identity matrix of order n. If there exists a matrix M^{-1} such that

$$MM^{-1} = M^{-1}M = I$$

then M^{-1} is called the MULTIPLICATIVE INVERSE OF M or, more simply, the INVERSE OF M. M^{-1} is read "M inverse."

4.

If the augmented matrix $[M \mid I]$ is transformed by row operations into $[I \mid B]$, then the resulting matrix B is M^{-1}. However, if all zeros are obtained in one or more rows to the left of the vertical line during the row transformation procedure, then M^{-1} does not exist.

---

1. $\begin{bmatrix} 1 & 0 \\ 0 & 1 \end{bmatrix}\begin{bmatrix} 2 & -3 \\ 4 & 5 \end{bmatrix} = \begin{bmatrix} 1\cdot2 + 0\cdot4 & 1(-3) + 0\cdot5 \\ 0\cdot2 + 1\cdot4 & 0(-3) + 1\cdot5 \end{bmatrix} = \begin{bmatrix} 2 & -3 \\ 4 & 5 \end{bmatrix}$

3. $\begin{bmatrix} 2 & -3 \\ 4 & 5 \end{bmatrix}\begin{bmatrix} 1 & 0 \\ 0 & 1 \end{bmatrix} = \begin{bmatrix} 2\cdot1 + (-3)0 & 2\cdot0 + (-3)1 \\ 4\cdot1 + 5\cdot0 & 4\cdot0 + 5\cdot1 \end{bmatrix} = \begin{bmatrix} 2 & -3 \\ 4 & 5 \end{bmatrix}$

5. $\begin{bmatrix} 1 & 0 & 0 \\ 0 & 1 & 0 \\ 0 & 0 & 1 \end{bmatrix}\begin{bmatrix} -2 & 1 & 3 \\ 2 & 4 & -2 \\ 5 & 1 & 0 \end{bmatrix}$

$= \begin{bmatrix} 1(-2) + 0\cdot2 + 0\cdot5 & 1\cdot1 + 0\cdot4 + 0\cdot1 & 1\cdot3 + 0(-2) + 0\cdot0 \\ 0(-2) + 1\cdot2 + 0\cdot5 & 0\cdot1 + 1\cdot4 + 0\cdot1 & 0\cdot3 + 1(-2) + 0\cdot0 \\ 0(-2) + 0\cdot2 + 1\cdot5 & 0\cdot1 + 0\cdot4 + 1\cdot1 & 0\cdot3 + 0(-2) + 1\cdot0 \end{bmatrix} = \begin{bmatrix} -2 & 1 & 3 \\ 2 & 4 & -2 \\ 5 & 1 & 0 \end{bmatrix}$

7. $\begin{bmatrix} -2 & 1 & 3 \\ 2 & 4 & -2 \\ 5 & 1 & 0 \end{bmatrix}\begin{bmatrix} 1 & 0 & 0 \\ 0 & 1 & 0 \\ 0 & 0 & 1 \end{bmatrix}$

$= \begin{bmatrix} (-2)\cdot1 + 1\cdot0 + 3\cdot0 & (-2)0 + 1\cdot1 + 3\cdot0 & (-2)0 + 1\cdot0 + 3\cdot1 \\ 2\cdot1 + 4\cdot0 + (-2)0 & 2\cdot0 + 4\cdot1 + (-2)0 & 2\cdot0 + 4\cdot0 + (-2)1 \\ 5\cdot1 + 1\cdot0 + 0\cdot0 & 5\cdot0 + 1\cdot1 + 0\cdot0 & 5\cdot0 + 1\cdot0 + 0\cdot1 \end{bmatrix}$

$= \begin{bmatrix} -2 & 1 & 3 \\ 2 & 4 & -2 \\ 5 & 1 & 0 \end{bmatrix}$

9. $\begin{bmatrix} 3 & -4 \\ -2 & 3 \end{bmatrix}\begin{bmatrix} 3 & 4 \\ 2 & 3 \end{bmatrix} = \begin{bmatrix} 3\cdot3 + (-4)2 & 3\cdot4 + (-4)3 \\ (-2)3 + 3\cdot2 & (-2)4 + 3\cdot3 \end{bmatrix} = \begin{bmatrix} 1 & 0 \\ 0 & 1 \end{bmatrix}$

11. $\begin{bmatrix} -5 & 2 \\ -8 & 3 \end{bmatrix}\begin{bmatrix} 3 & -2 \\ 8 & -5 \end{bmatrix} = \begin{bmatrix} (-5)3 + 2\cdot8 & (-5)(-2) + 2(-5) \\ (-8)3 + 3\cdot8 & (-8)(-2) + 3(-5) \end{bmatrix} = \begin{bmatrix} 1 & 0 \\ 0 & 1 \end{bmatrix}$

13. $\begin{bmatrix} 1 & -1 & 1 \\ 0 & 2 & -1 \\ 2 & 3 & 0 \end{bmatrix} \begin{bmatrix} 3 & 3 & -1 \\ -2 & -2 & 1 \\ -4 & -5 & 2 \end{bmatrix}$

$= \begin{bmatrix} 1 \cdot 3 + (-1)(-2) + 1(-4) & 1 \cdot 3 + (-1)(-2) + 1(-5) & 1(-1) + (-1)1 + 1 \cdot 2 \\ 0 \cdot 3 + 2(-2) + (-1)(-4) & 0 \cdot 3 + 2(-2) + (-1)(-5) & 0(-1) + 2 \cdot 1 + (-1)2 \\ 2 \cdot 3 + 3(-2) + 0(-4) & 2 \cdot 3 + 3(-2) + 0(-5) & 2(-1) + 3 \cdot 1 + 0 \cdot 2 \end{bmatrix}$

$= \begin{bmatrix} 1 & 0 & 0 \\ 0 & 1 & 0 \\ 0 & 0 & 1 \end{bmatrix}$

15. $\left[\begin{array}{cc|cc} -1 & 0 & 1 & 0 \\ -3 & 1 & 0 & 1 \end{array}\right] \sim \left[\begin{array}{cc|cc} 1 & 0 & -1 & 0 \\ -3 & 1 & 0 & 1 \end{array}\right] \sim \left[\begin{array}{cc|cc} 1 & 0 & -1 & 0 \\ 0 & 1 & -3 & 1 \end{array}\right]$

$\qquad (-1)R_1 \rightarrow R_1 \qquad\qquad 3R_1 + R_2 \rightarrow R_2$

Thus, $M^{-1} = \begin{bmatrix} -1 & 0 \\ -3 & 1 \end{bmatrix}$

Check:

$M \cdot M^{-1} = \begin{bmatrix} -1 & 0 \\ -3 & 1 \end{bmatrix} \begin{bmatrix} -1 & 0 \\ -3 & 1 \end{bmatrix} = \begin{bmatrix} (-1)(-1) + 0(-3) & (-1)0 + 0 \cdot 1 \\ (-3)(-1) + 1(-3) & (-3)0 + 1 \cdot 1 \end{bmatrix} = \begin{bmatrix} 1 & 0 \\ 0 & 1 \end{bmatrix}$

17. $\left[\begin{array}{cc|cc} 1 & 2 & 1 & 0 \\ 1 & 3 & 0 & 1 \end{array}\right] \sim \left[\begin{array}{cc|cc} 1 & 2 & 1 & 0 \\ 0 & 1 & -1 & 1 \end{array}\right] \sim \left[\begin{array}{cc|cc} 1 & 0 & 3 & -2 \\ 0 & 1 & -1 & 1 \end{array}\right]$

$\quad (-1)R_1 + R_2 \rightarrow R_2 \qquad (-2)R_2 + R_1 \rightarrow R_1$

Thus, $M^{-1} = \begin{bmatrix} 3 & -2 \\ -1 & 1 \end{bmatrix}.$

Check:

$M \cdot M^{-1} = \begin{bmatrix} 1 & 2 \\ 1 & 3 \end{bmatrix} \begin{bmatrix} 3 & -2 \\ -1 & 1 \end{bmatrix} = \begin{bmatrix} 1 \cdot 3 + 2(-1) & 1(-2) + 2 \cdot 1 \\ 1 \cdot 3 + 3(-1) & 1(-2) + 3 \cdot 1 \end{bmatrix} = \begin{bmatrix} 1 & 0 \\ 0 & 1 \end{bmatrix}$

19. $\left[\begin{array}{cc|cc} 1 & 3 & 1 & 0 \\ 2 & 7 & 0 & 1 \end{array}\right] \sim \left[\begin{array}{cc|cc} 1 & 3 & 1 & 0 \\ 0 & 1 & -2 & 1 \end{array}\right] \sim \left[\begin{array}{cc|cc} 1 & 0 & 7 & -3 \\ 0 & 1 & -2 & 1 \end{array}\right]$

$\quad (-2)R_1 + R_2 \rightarrow R_2 \qquad (-3)R_2 + R_1 \rightarrow R_1$

Thus, $M^{-1} = \begin{bmatrix} 7 & -3 \\ -2 & 1 \end{bmatrix}.$

Check:

$\begin{bmatrix} 1 & 3 \\ 2 & 7 \end{bmatrix} \begin{bmatrix} 7 & -3 \\ -2 & 1 \end{bmatrix} = \begin{bmatrix} 1 \cdot 7 + 3(-2) & 1(-3) + 3 \cdot 1 \\ 2 \cdot 7 + 7(-2) & 2(-3) + 7 \cdot 1 \end{bmatrix} = \begin{bmatrix} 1 & 0 \\ 0 & 1 \end{bmatrix}$

21. $\begin{bmatrix} 1 & -3 & 0 & | & 1 & 0 & 0 \\ 0 & 1 & 1 & | & 0 & 1 & 0 \\ 2 & -1 & 4 & | & 0 & 0 & 1 \end{bmatrix}$ $(-2)R_1 + R_3 \rightarrow R_3$

$\sim \begin{bmatrix} 1 & -3 & 0 & | & 1 & 0 & 0 \\ 0 & 1 & 1 & | & 0 & 1 & 0 \\ 0 & 5 & 4 & | & -2 & 0 & 1 \end{bmatrix}$ $\begin{array}{l} 3R_2 + R_1 \rightarrow R_1 \\ (-5)R_2 + R_3 \rightarrow R_3 \end{array}$

$\sim \begin{bmatrix} 1 & 0 & 3 & | & 1 & 3 & 0 \\ 0 & 1 & 1 & | & 0 & 1 & 0 \\ 0 & 0 & -1 & | & -2 & -5 & 1 \end{bmatrix}$ $(-1)R_3 \rightarrow R_3$

$\sim \begin{bmatrix} 1 & 0 & 3 & | & 1 & 3 & 0 \\ 0 & 1 & 1 & | & 0 & 1 & 0 \\ 0 & 0 & 1 & | & 2 & 5 & -1 \end{bmatrix}$ $\begin{array}{l} (-3)R_3 + R_1 \rightarrow R_1 \\ (-1)R_3 + R_2 \rightarrow R_2 \end{array}$

$\sim \begin{bmatrix} 1 & 0 & 0 & | & -5 & -12 & 3 \\ 0 & 1 & 0 & | & -2 & -4 & 1 \\ 0 & 0 & 1 & | & 2 & 5 & -1 \end{bmatrix}$ $M^{-1} = \begin{bmatrix} -5 & -12 & 3 \\ -2 & -4 & 1 \\ 2 & 5 & -1 \end{bmatrix}$

$M \cdot M^{-1} = \begin{bmatrix} 1 & -3 & 0 \\ 0 & 1 & 1 \\ 2 & -1 & 4 \end{bmatrix} \begin{bmatrix} -5 & -12 & 3 \\ -2 & -4 & 1 \\ 2 & 5 & -1 \end{bmatrix} = \begin{bmatrix} 1 & 0 & 0 \\ 0 & 1 & 0 \\ 0 & 0 & 1 \end{bmatrix}$

23. $\begin{bmatrix} 1 & 1 & 0 & | & 1 & 0 & 0 \\ 2 & 3 & -1 & | & 0 & 1 & 0 \\ 1 & 0 & 2 & | & 0 & 0 & 1 \end{bmatrix}$ $\begin{array}{l} (-2)R_1 + R_2 \rightarrow R_2 \\ (-1)R_1 + R_3 \rightarrow R_3 \end{array}$

$\sim \begin{bmatrix} 1 & 1 & 0 & | & 1 & 0 & 0 \\ 0 & 1 & -1 & | & -2 & 1 & 0 \\ 0 & -1 & 2 & | & -1 & 0 & 1 \end{bmatrix}$ $\begin{array}{l} (-1)R_2 + R_1 \rightarrow R_1 \\ R_2 + R_3 \rightarrow R_3 \end{array}$

$\sim \begin{bmatrix} 1 & 0 & 1 & | & 3 & -1 & 0 \\ 0 & 1 & -1 & | & -2 & 1 & 0 \\ 0 & 0 & 1 & | & -3 & 1 & 1 \end{bmatrix}$ $\begin{array}{l} (-1)R_3 + R_1 \rightarrow R_1 \\ R_3 + R_2 \rightarrow R_2 \end{array}$

$\sim \begin{bmatrix} 1 & 0 & 0 & | & 6 & -2 & -1 \\ 0 & 1 & 0 & | & -5 & 2 & 1 \\ 0 & 0 & 1 & | & -3 & 1 & 1 \end{bmatrix}$ $M^{-1} = \begin{bmatrix} 6 & -2 & -1 \\ -5 & 2 & 1 \\ -3 & 1 & 1 \end{bmatrix}$

$M \cdot M^{-1} = \begin{bmatrix} 1 & 1 & 0 \\ 2 & 3 & -1 \\ 1 & 0 & 2 \end{bmatrix} \begin{bmatrix} 6 & -2 & -1 \\ -5 & 2 & 1 \\ -3 & 1 & 1 \end{bmatrix} = \begin{bmatrix} 1 & 0 & 0 \\ 0 & 1 & 0 \\ 0 & 0 & 1 \end{bmatrix}$

25. $\begin{bmatrix} 4 & 3 & | & 1 & 0 \\ -3 & -2 & | & 0 & 1 \end{bmatrix} R_2 + R_1 \rightarrow R_1$

$\sim \begin{bmatrix} 1 & 1 & | & 1 & 1 \\ -3 & -2 & | & 0 & 1 \end{bmatrix} 3R_1 + R_2 \rightarrow R_2$

$\sim \begin{bmatrix} 1 & 1 & | & 1 & 1 \\ 0 & 1 & | & 3 & 4 \end{bmatrix} (-1)R_2 + R_1 \rightarrow R_1$

$\sim \begin{bmatrix} 1 & 0 & | & -2 & -3 \\ 0 & 1 & | & 3 & 4 \end{bmatrix} (-1)R_2 + R_1 \rightarrow R_1$

$M^{-1} = \begin{bmatrix} -2 & -3 \\ 3 & 4 \end{bmatrix}$

27. $\begin{bmatrix} 2 & 6 & | & 1 & 0 \\ 3 & 9 & | & 0 & 1 \end{bmatrix} \frac{1}{2}R_1 \rightarrow R_1$

$\sim \begin{bmatrix} 1 & 3 & | & \frac{1}{2} & 0 \\ 3 & 9 & | & 0 & 1 \end{bmatrix} (-3)R_1 + R_2 \rightarrow R_2$

$\sim \begin{bmatrix} 1 & 3 & | & \frac{1}{2} & 0 \\ 0 & 0 & | & -\frac{3}{2} & 1 \end{bmatrix}$ The inverse does not exist.

29. $\begin{bmatrix} 2 & 1 & | & 1 & 0 \\ 4 & 3 & | & 0 & 1 \end{bmatrix} \frac{1}{2}R_1 \rightarrow R_1$

$\sim \begin{bmatrix} 1 & \frac{1}{2} & | & \frac{1}{2} & 0 \\ 4 & 3 & | & 0 & 1 \end{bmatrix} (-4)R_1 + R_2 \rightarrow R_2$

$\sim \begin{bmatrix} 1 & \frac{1}{2} & | & \frac{1}{2} & 0 \\ 0 & 1 & | & -2 & 1 \end{bmatrix} \left(-\frac{1}{2}\right)R_2 + R_1 \rightarrow R_1$

$\sim \begin{bmatrix} 1 & 0 & | & \frac{3}{2} & -\frac{1}{2} \\ 0 & 1 & | & -2 & 1 \end{bmatrix}$

$M^{-1} = \begin{bmatrix} \frac{3}{2} & -\frac{1}{2} \\ -2 & 1 \end{bmatrix} = \begin{bmatrix} 1.5 & -0.5 \\ -2 & 1 \end{bmatrix}$

31. $\begin{bmatrix} -5 & -2 & -2 & | & 1 & 0 & 0 \\ 2 & 1 & 0 & | & 0 & 1 & 0 \\ 1 & 0 & 1 & | & 0 & 0 & 1 \end{bmatrix} \sim \begin{bmatrix} 1 & 0 & 1 & | & 0 & 0 & 1 \\ 2 & 1 & 0 & | & 0 & 1 & 0 \\ -5 & -2 & -2 & | & 1 & 0 & 0 \end{bmatrix}$

$\qquad\qquad R_1 \leftrightarrow R_3 \qquad\qquad\qquad\qquad (-2)R_1 + R_2 \rightarrow R_2$

$\qquad\qquad\qquad\qquad\qquad\qquad\qquad\qquad 5R_1 + R_3 \rightarrow R_3$

$$\sim \begin{bmatrix} 1 & 0 & 1 & | & 0 & 0 & 1 \\ 0 & 1 & -2 & | & 0 & 1 & -2 \\ 0 & -2 & 3 & | & 1 & 0 & 5 \end{bmatrix} \sim \begin{bmatrix} 1 & 0 & 1 & | & 0 & 0 & 1 \\ 0 & 1 & -2 & | & 0 & 1 & -2 \\ 0 & 0 & -1 & | & 1 & 2 & 1 \end{bmatrix}$$

$$\qquad\qquad 2R_2 + R_3 \rightarrow R_3 \qquad\qquad\qquad\qquad (-1)R_3 \rightarrow R_3$$

$$\sim \begin{bmatrix} 1 & 0 & 1 & | & 0 & 0 & 1 \\ 0 & 1 & -2 & | & 0 & 1 & -2 \\ 0 & 0 & 1 & | & -1 & -2 & -1 \end{bmatrix} \sim \begin{bmatrix} 1 & 0 & 0 & | & 1 & 2 & 2 \\ 0 & 1 & 0 & | & -2 & -3 & -4 \\ 0 & 0 & 1 & | & -1 & -2 & -1 \end{bmatrix}$$

$$\qquad\qquad 2R_3 + R_2 \rightarrow R_2$$
$$\qquad\qquad (-1)R_3 + R_1 \rightarrow R_1$$

Thus, the inverse is $\begin{bmatrix} 1 & 2 & 2 \\ -2 & -3 & -4 \\ -1 & -2 & -1 \end{bmatrix}$.

33. $\begin{bmatrix} 2 & 1 & 1 & | & 1 & 0 & 0 \\ 1 & 1 & 0 & | & 0 & 1 & 0 \\ -1 & -1 & 0 & | & 0 & 0 & 1 \end{bmatrix} \sim \begin{bmatrix} 1 & 1 & 0 & | & 0 & 1 & 0 \\ 2 & 1 & 1 & | & 1 & 0 & 0 \\ -1 & -1 & 0 & | & 0 & 0 & 1 \end{bmatrix} \sim \begin{bmatrix} 1 & 1 & 0 & | & 0 & 1 & 0 \\ 0 & -1 & 1 & | & 1 & -2 & 0 \\ 0 & 0 & 0 & | & 0 & 1 & 1 \end{bmatrix}$

$$\qquad\quad R_1 \leftrightarrow R_2 \qquad\qquad\qquad (-2)R_1 + R_2 \rightarrow R_2$$
$$\qquad\qquad\qquad\qquad\qquad\qquad R_1 + R_3 \rightarrow R_3$$

From this matrix, we conclude that the inverse does not exist.

35. $\begin{bmatrix} -1 & -2 & 2 & | & 1 & 0 & 0 \\ 4 & 3 & 0 & | & 0 & 1 & 0 \\ 4 & 0 & 4 & | & 0 & 0 & 1 \end{bmatrix} (-1)R_1 \rightarrow R_1$

$$\sim \begin{bmatrix} 1 & 2 & -2 & | & -1 & 0 & 0 \\ 4 & 3 & 0 & | & 0 & 1 & 0 \\ 4 & 0 & 4 & | & 0 & 0 & 1 \end{bmatrix} \begin{array}{l} (-4)R_1 + R_2 \rightarrow R_2 \\ (-4)R_1 + R_3 \rightarrow R_3 \end{array}$$

$$\sim \begin{bmatrix} 1 & 2 & -2 & | & -1 & 0 & 0 \\ 0 & -5 & 8 & | & 4 & 1 & 0 \\ 0 & -8 & 12 & | & 4 & 0 & 1 \end{bmatrix} \left(-\frac{1}{5}\right)R_2 \rightarrow R_2$$

$$\sim \begin{bmatrix} 1 & 2 & -2 & | & -1 & 0 & 0 \\ 0 & 1 & -\frac{8}{5} & | & -\frac{4}{5} & -\frac{1}{5} & 0 \\ 0 & -8 & 12 & | & 4 & 0 & 1 \end{bmatrix} \begin{array}{l} (-2)R_2 + R_1 \rightarrow R_1 \\ 8R_2 + R_3 \rightarrow R_3 \end{array}$$

$$\sim \begin{bmatrix} 1 & 0 & \frac{6}{5} & | & \frac{3}{5} & \frac{2}{5} & 0 \\ 0 & 1 & -\frac{8}{5} & | & -\frac{4}{5} & -\frac{1}{5} & 0 \\ 0 & 0 & -\frac{4}{5} & | & -\frac{12}{5} & -\frac{8}{5} & 1 \end{bmatrix} \left(-\frac{5}{4}\right)R_3 \rightarrow R_3$$

$$\sim \begin{bmatrix} 1 & 0 & \frac{6}{5} & \bigg| & \frac{3}{5} & \frac{2}{5} & 0 \\ 0 & 1 & -\frac{8}{5} & \bigg| & -\frac{4}{5} & -\frac{1}{5} & 0 \\ 0 & 0 & 1 & \bigg| & 3 & 2 & -\frac{5}{4} \end{bmatrix} \begin{array}{l} \left(-\dfrac{6}{5}\right)R_3 + R_1 \rightarrow R_1 \\[2mm] \left(\dfrac{8}{5}\right)R_3 + R_2 \rightarrow R_2 \end{array}$$

$$\sim \begin{bmatrix} 1 & 0 & 0 & \bigg| & -3 & -2 & \frac{3}{2} \\ 0 & 1 & 0 & \bigg| & 4 & 3 & -2 \\ 0 & 0 & 1 & \bigg| & 3 & 2 & -\frac{5}{4} \end{bmatrix}$$

$$M^{-1} = \begin{bmatrix} -3 & -2 & \frac{3}{2} \\ 4 & 3 & -2 \\ 3 & 2 & -\frac{5}{4} \end{bmatrix} = \begin{bmatrix} -3 & -2 & 1.5 \\ 4 & 3 & -2 \\ 3 & 2 & -1.25 \end{bmatrix}$$

37. $\begin{bmatrix} 2 & -1 & -2 & \bigg| & 1 & 0 & 0 \\ -4 & 2 & 8 & \bigg| & 0 & 1 & 0 \\ 6 & -2 & -1 & \bigg| & 0 & 0 & 1 \end{bmatrix} \dfrac{1}{2}R_1 \rightarrow R_1$

$$\sim \begin{bmatrix} 1 & -\frac{1}{2} & -1 & \bigg| & \frac{1}{2} & 0 & 0 \\ -4 & 2 & 8 & \bigg| & 0 & 1 & 0 \\ 6 & -2 & -1 & \bigg| & 0 & 0 & 1 \end{bmatrix} \begin{array}{l} 4R_1 + R_2 \rightarrow R_2 \\[2mm] (-6)R_1 + R_3 \rightarrow R_3 \end{array}$$

$$\sim \begin{bmatrix} 1 & -\frac{1}{2} & -1 & \bigg| & \frac{1}{2} & 0 & 0 \\ 0 & 0 & 4 & \bigg| & 2 & 1 & 0 \\ 0 & 1 & 5 & \bigg| & -3 & 0 & 1 \end{bmatrix} R_2 \leftrightarrow R_3$$

$$\sim \begin{bmatrix} 1 & -\frac{1}{2} & -1 & \bigg| & \frac{1}{2} & 0 & 0 \\ 0 & 1 & 5 & \bigg| & -3 & 0 & 1 \\ 0 & 0 & 4 & \bigg| & 2 & 1 & 0 \end{bmatrix} \begin{array}{l} \dfrac{1}{2}R_2 + R_1 \rightarrow R_1 \\[2mm] \dfrac{1}{4}R_3 \rightarrow R_3 \end{array}$$

$$\sim \begin{bmatrix} 1 & 0 & \frac{3}{2} & \bigg| & -1 & 0 & \frac{1}{2} \\ 0 & 1 & 5 & \bigg| & -3 & 0 & 1 \\ 0 & 0 & 1 & \bigg| & \frac{1}{2} & \frac{1}{4} & 0 \end{bmatrix} \begin{array}{l} \left(-\dfrac{3}{2}\right)R_3 + R_1 \rightarrow R_1 \\[2mm] (-5)R_3 + R_2 \rightarrow R_2 \end{array}$$

$$\sim \begin{bmatrix} 1 & 0 & 0 & \bigg| & -\frac{7}{4} & -\frac{3}{8} & \frac{1}{2} \\ 0 & 1 & 0 & \bigg| & -\frac{11}{2} & -\frac{5}{4} & 1 \\ 0 & 0 & 1 & \bigg| & \frac{1}{2} & \frac{1}{4} & 0 \end{bmatrix}$$

$$M^{-1} = \begin{bmatrix} -\frac{7}{4} & -\frac{3}{8} & \frac{1}{2} \\ -\frac{11}{2} & -\frac{5}{4} & 1 \\ \frac{1}{2} & \frac{1}{4} & 0 \end{bmatrix} = \begin{bmatrix} -1.75 & -0.375 & 0.5 \\ -5.5 & -1.25 & 1 \\ 0.5 & 0.25 & 0 \end{bmatrix}$$

39. $A = \begin{bmatrix} 4 & 3 \\ 3 & 2 \end{bmatrix}$; $\begin{bmatrix} 4 & 3 & | & 1 & 0 \\ 3 & 2 & | & 0 & 1 \end{bmatrix}$ $(-1)R_2 + R_1 \rightarrow R_1$

$\sim \begin{bmatrix} 1 & 1 & | & 1 & -1 \\ 3 & 2 & | & 0 & 1 \end{bmatrix}$ $(-3)R_1 + R_2 \rightarrow R_2$ $\sim \begin{bmatrix} 1 & 1 & | & 1 & -1 \\ 0 & -1 & | & -3 & 4 \end{bmatrix}$ $(-1)R_2 \rightarrow R_2$

$\sim \begin{bmatrix} 1 & 1 & | & 1 & -1 \\ 0 & 1 & | & 3 & -4 \end{bmatrix}$ $(-1)R_2 + R_1 \rightarrow R_1$ $\sim \begin{bmatrix} 1 & 0 & | & -2 & 3 \\ 0 & 1 & | & 3 & -4 \end{bmatrix}$

$A^{-1} = \begin{bmatrix} -2 & 3 \\ 3 & -4 \end{bmatrix}$; $\begin{bmatrix} -2 & 3 & | & 1 & 0 \\ 3 & -4 & | & 0 & 1 \end{bmatrix}$ $R_2 + R_1 \rightarrow R_1$

$\sim \begin{bmatrix} 1 & -1 & | & 1 & 1 \\ 3 & -4 & | & 0 & 1 \end{bmatrix}$ $(-3)R_1 + R_2 \rightarrow R_2$ $\sim \begin{bmatrix} 1 & -1 & | & 1 & 1 \\ 0 & -1 & | & -3 & -2 \end{bmatrix}$ $(-1)R_2 \rightarrow R_2$

$\sim \begin{bmatrix} 1 & -1 & | & 1 & 1 \\ 0 & 1 & | & 3 & 2 \end{bmatrix}$ $R_2 + R_1 \rightarrow R_1$ $\sim \begin{bmatrix} 1 & 0 & | & 4 & 3 \\ 0 & 1 & | & 3 & 2 \end{bmatrix}$ Thus, $(A^{-1})^{-1} = \begin{bmatrix} 4 & 3 \\ 3 & 2 \end{bmatrix} = A$

41. $\begin{bmatrix} a & 0 & | & 1 & 0 \\ 0 & d & | & 0 & 1 \end{bmatrix}$ $\begin{array}{l} \frac{1}{a}R_1 \rightarrow R_1, \text{ provided } a \neq 0 \\ \frac{1}{d}R_2 \rightarrow R_2, \text{ provided } d \neq 0 \end{array}$

$\begin{bmatrix} 1 & 0 & | & \frac{1}{a} & 0 \\ 0 & 1 & | & 0 & \frac{1}{d} \end{bmatrix}$

M^{-1} exists and equals $\begin{bmatrix} \frac{1}{a} & 0 \\ 0 & \frac{1}{d} \end{bmatrix}$ if and only if $a \neq 0$, $d \neq 0$. In general,

the inverse of a diagonal matrix exists if and only if each of the diagonal elements is non-zero.

43. $A = \begin{bmatrix} 6 & 2 & 0 & 4 \\ 5 & 3 & 2 & 1 \\ 0 & -1 & 1 & -2 \\ 2 & -3 & 1 & 0 \end{bmatrix}$; $A^{-1} = \begin{bmatrix} 0.5 & -0.3 & 0.85 & -0.25 \\ 0 & 0.1 & 0.05 & -0.25 \\ -1 & 0.9 & -1.55 & 0.75 \\ -0.5 & 0.4 & -1.3 & 0.5 \end{bmatrix}$

45. $A = \begin{bmatrix} 3 & 2 & 3 & 4 & 4 \\ 5 & 4 & 3 & 2 & 1 \\ -1 & -1 & 2 & -2 & 3 \\ 3 & -3 & 1 & 0 & 1 \\ 1 & 1 & 2 & 0 & 2 \end{bmatrix}$;

$A^{-1} = \begin{bmatrix} 1.75 & 5.25 & 8.75 & -1 & -18.75 \\ 1.25 & 3.75 & 6.25 & -1 & -13.25 \\ -4.75 & -13.25 & -22.75 & 3 & 48.75 \\ -1.375 & -4.625 & -7.875 & 1 & 16.375 \\ 3.25 & 8.75 & 15.25 & -2 & -32.25 \end{bmatrix}$

47. $A = \begin{bmatrix} 1 & 2 \\ 1 & 3 \end{bmatrix}$

Assign the numbers 1—26 to the letters of the alphabet, in order, and let 27 correspond to a blank space. Then the message "THE SUN ALSO RISES" corresponds to the sequence

20 8 5 27 19 21 14 27 1 12 19 15 27 18 9 19 5 19

To encode this message, divide the numbers into groups of two and use the groups as columns of a matrix B with two rows

$$B = \begin{bmatrix} 20 & 5 & 19 & 14 & 1 & 19 & 27 & 9 & 5 \\ 8 & 27 & 21 & 27 & 12 & 15 & 18 & 19 & 19 \end{bmatrix}$$

Now

$$AB = \begin{bmatrix} 1 & 2 \\ 1 & 3 \end{bmatrix} \begin{bmatrix} 20 & 5 & 19 & 14 & 1 & 19 & 27 & 9 & 5 \\ 8 & 27 & 21 & 27 & 12 & 15 & 18 & 19 & 19 \end{bmatrix}$$

$$= \begin{bmatrix} 36 & 59 & 61 & 68 & 25 & 49 & 63 & 47 & 43 \\ 44 & 86 & 82 & 95 & 37 & 64 & 81 & 66 & 62 \end{bmatrix}$$

The coded message is:
36 44 59 86 61 82 68 95 25 37 49 64 63 81 47 66 43 62.

49. First, we must find the inverse of $A = \begin{bmatrix} 1 & 2 \\ 1 & 3 \end{bmatrix}$

$$\begin{bmatrix} 1 & 2 & | & 1 & 0 \\ 1 & 3 & | & 0 & 1 \end{bmatrix} \sim \begin{bmatrix} 1 & 2 & | & 1 & 0 \\ 0 & 1 & | & -1 & 1 \end{bmatrix} \sim \begin{bmatrix} 1 & 0 & | & 3 & -2 \\ 0 & 1 & | & -1 & 1 \end{bmatrix}$$

$(-1)R_1 + R_2 \rightarrow R_2 \qquad (-2)R_2 + R_1 \rightarrow R_1$

Thus, $A^{-1} = \begin{bmatrix} 3 & -2 \\ -1 & 1 \end{bmatrix}$

Now $\begin{bmatrix} 3 & -2 \\ -1 & 1 \end{bmatrix} \begin{bmatrix} 37 & 24 & 73 & 49 & 62 & 36 & 59 & 41 & 22 \\ 52 & 29 & 96 & 69 & 89 & 44 & 86 & 50 & 26 \end{bmatrix}$

$$= \begin{bmatrix} 7 & 14 & 27 & 9 & 8 & 20 & 5 & 23 & 14 \\ 15 & 5 & 23 & 20 & 27 & 8 & 27 & 9 & 4 \end{bmatrix}$$

Thus, the decoded message is
7 15 14 5 27 23 9 20 8 27 20 8 5 27 23 9 14 4

which corresponds to

GONE WITH THE WIND

51. "THE BEST YEARS OF OUR LIVES" corresponds to the sequence
20 8 5 27 2 5 19 20 27 25 5 1 18 19 27 15 6 27 15
21 18 27 12 9 22 5 19

We divide the numbers in the sequence into groups of 5 and use these groups as the columns of a matrix with 5 rows, adding 3 blanks at the end to make the columns come out even. Then we multiply this matrix on the left by the given matrix B.

$$\begin{bmatrix} 1 & 0 & 1 & 0 & 1 \\ 0 & 1 & 1 & 0 & 3 \\ 2 & 1 & 1 & 1 & 1 \\ 0 & 0 & 1 & 0 & 2 \\ 1 & 1 & 1 & 2 & 1 \end{bmatrix} \begin{bmatrix} 20 & 5 & 5 & 15 & 18 & 5 \\ 8 & 19 & 1 & 6 & 27 & 19 \\ 5 & 20 & 18 & 27 & 12 & 27 \\ 27 & 27 & 19 & 15 & 9 & 27 \\ 2 & 25 & 27 & 21 & 22 & 27 \end{bmatrix}$$

$$= \begin{bmatrix} 27 & 50 & 50 & 63 & 52 & 59 \\ 19 & 114 & 100 & 96 & 105 & 127 \\ 82 & 101 & 75 & 99 & 106 & 110 \\ 9 & 70 & 72 & 69 & 56 & 81 \\ 89 & 123 & 89 & 99 & 97 & 132 \end{bmatrix}$$

The encoded message is:

27 19 82 9 89 50 114 101 70 123 50 100 75 72 89 63 96 99 69 99 52 105 106 56 97 59 127 110 81 132.

53. First, we must find the inverse of B:

$$B^{-1} = \begin{bmatrix} -2 & -1 & 2 & 2 & -1 \\ 3 & 2 & -2 & -4 & 1 \\ 6 & 2 & -4 & -5 & 2 \\ -2 & -1 & 1 & 2 & -7 \\ 3 & -1 & 2 & 3 & -1 \end{bmatrix}$$

Now B^{-1} $\begin{bmatrix} 32 & 24 & 54 & 56 & 48 & 62 \\ 34 & 21 & 71 & 92 & 66 & 135 \\ 87 & 67 & 112 & 109 & 98 & 124 \\ 19 & 11 & 43 & 55 & 41 & 81 \\ 94 & 69 & 112 & 109 & 89 & 143 \end{bmatrix} = \begin{bmatrix} 20 & 18 & 19 & 15 & 27 & 8 \\ 8 & 5 & 20 & 23 & 5 & 27 \\ 5 & 1 & 27 & 27 & 1 & 27 \\ 27 & 20 & 19 & 15 & 18 & 27 \\ 7 & 5 & 8 & 14 & 20 & 27 \end{bmatrix}$

Thus, the decoded message is

20 8 5 27 7 18 5 1 20 5 19 20 27 19 8 15 23 27 15 14 27 5 1 18 20 8

which corresponds to

THE GREATEST SHOW ON EARTH

EXERCISE 4-6

Things to remember:

1. BASIC PROPERTIES OF MATRICES

   Assuming all products and sums are defined for the indicated matrices A, B, C, I, and O, then

   ADDITION PROPERTIES

ASSOCIATIVE:	$(A + B) + C = A + (B + C)$
COMMUTATIVE:	$A + B = B + A$
ADDITIVE IDENTITY:	$A + 0 = 0 + A = A$
ADDITIVE INVERSE:	$A + (-A) = (-A) + A = 0$

MULTIPLICATION PROPERTIES

ASSOCIATIVE PROPERTY:	$A(BC) = (AB)C$
MULTIPLICATIVE IDENTITY:	$AI = IA = A$
MULTIPLICATIVE INVERSE:	If A is a square matrix and A^{-1} exists, then $AA^{-1} = A^{-1}A = I$.

COMBINED PROPERTIES

LEFT DISTRIBUTIVE:	$A(B + C) = AB + AC$
RIGHT DISTRIBUTIVE:	$(B + C)A = BA + CA$

EQUALITY

ADDITION:	If $A = B$ then $A + C = B + C$.
LEFT MULTIPLICATION:	If $A = B$, then $CA = CB$.
RIGHT MULTIPLICATION:	If $A = B$, then $AC = BC$.

2. USING INVERSE METHODS TO SOLVE SYSTEMS OF EQUATIONS

If the number of equations in a system equals the number of variables and the coefficient matrix has an inverse, then the system will always have a unique solution that can be found by using the inverse of the coefficient matrix to solve the corresponding matrix equation.

Matrix Equation	Solution
$AX = B$	$X = A^{-1}B$

1. $\begin{bmatrix} 3 & 1 \\ 2 & -1 \end{bmatrix} \begin{bmatrix} x_1 \\ x_2 \end{bmatrix} = \begin{bmatrix} 5 \\ -4 \end{bmatrix}$

$\begin{bmatrix} 3x_1 + x_2 \\ 2x_1 - x_2 \end{bmatrix} = \begin{bmatrix} 5 \\ -4 \end{bmatrix}$

Thus, $3x_1 + x_2 = 5$
$2x_1 - x_2 = -4$

3. $\begin{bmatrix} -3 & 1 & 0 \\ 2 & 0 & 1 \\ -1 & 3 & -2 \end{bmatrix} \begin{bmatrix} x_1 \\ x_2 \\ x_3 \end{bmatrix} = \begin{bmatrix} 3 \\ -4 \\ 2 \end{bmatrix}$

$\begin{bmatrix} -3x_1 + x_2 \\ 2x_1 + x_3 \\ -x_1 + 3x_2 - 2x_3 \end{bmatrix} = \begin{bmatrix} 3 \\ -4 \\ 2 \end{bmatrix}$

Thus, $-3x_1 + x_2 = 3$
$2x_1 + x_3 = -4$
$-x_1 + 3x_2 - 2x_3 = 2$

5. $3x_1 - 4x_2 = 1$
$2x_1 + x_2 = 5$

$\begin{bmatrix} 3x_1 - 4x_2 \\ 2x_1 + x_2 \end{bmatrix} = \begin{bmatrix} 1 \\ 5 \end{bmatrix}$ and $\begin{bmatrix} 3 & -4 \\ 2 & 1 \end{bmatrix} \begin{bmatrix} x_1 \\ x_2 \end{bmatrix} = \begin{bmatrix} 1 \\ 5 \end{bmatrix}$

7.
$$x_1 - 3x_2 + 2x_3 = -3$$
$$-2x_1 + 3x_2 \qquad = 1$$
$$x_1 + x_2 + 4x_3 = -2$$

$$\begin{bmatrix} x_1 - 3x_2 + 2x_3 \\ -2x_1 + 3x_2 \\ x_1 + x_2 + 4x_3 \end{bmatrix} = \begin{bmatrix} -3 \\ 1 \\ -2 \end{bmatrix} \quad \text{and} \quad \begin{bmatrix} 1 & -3 & 2 \\ -2 & 3 & 0 \\ 1 & 1 & 4 \end{bmatrix} \begin{bmatrix} x_1 \\ x_2 \\ x_3 \end{bmatrix} = \begin{bmatrix} -3 \\ 1 \\ -2 \end{bmatrix}$$

9. $\begin{bmatrix} x_1 \\ x_2 \end{bmatrix} = \begin{bmatrix} 3 & -2 \\ 1 & 4 \end{bmatrix} \begin{bmatrix} -2 \\ 1 \end{bmatrix} = \begin{bmatrix} 3(-2) + (-2)1 \\ 1(-2) + 4 \cdot 1 \end{bmatrix} = \begin{bmatrix} -8 \\ 2 \end{bmatrix}$ Thus, $x_1 = -8$ and $x_2 = 2$

11. $\begin{bmatrix} x_1 \\ x_2 \end{bmatrix} = \begin{bmatrix} -2 & 3 \\ 2 & -1 \end{bmatrix} \begin{bmatrix} 3 \\ 2 \end{bmatrix} = \begin{bmatrix} (-2)3 + 3 \cdot 2 \\ 2 \cdot 3 + (-1)2 \end{bmatrix} = \begin{bmatrix} 0 \\ 4 \end{bmatrix}$ Thus, $x_1 = 0$ and $x_2 = 4$

13. The matrix equation for the given system is:
$$\begin{bmatrix} 1 & 2 \\ 1 & 3 \end{bmatrix} \begin{bmatrix} x_1 \\ x_2 \end{bmatrix} = \begin{bmatrix} k_1 \\ k_2 \end{bmatrix}$$

From Exercise 4-5, Problem 17, $\begin{bmatrix} 1 & 2 \\ 1 & 3 \end{bmatrix}^{-1} = \begin{bmatrix} 3 & -2 \\ -1 & 1 \end{bmatrix}$

Thus, $\begin{bmatrix} x_1 \\ x_2 \end{bmatrix} = \begin{bmatrix} 3 & -2 \\ -1 & 1 \end{bmatrix} \begin{bmatrix} k_1 \\ k_2 \end{bmatrix}$

(A) $\begin{bmatrix} x_1 \\ x_2 \end{bmatrix} = \begin{bmatrix} 3 & -2 \\ -1 & 1 \end{bmatrix} \begin{bmatrix} 1 \\ 3 \end{bmatrix} = \begin{bmatrix} -3 \\ 2 \end{bmatrix}$ Thus, $x_1 = -3$ and $x_2 = 2$

(B) $\begin{bmatrix} x_1 \\ x_2 \end{bmatrix} = \begin{bmatrix} 3 & -2 \\ -1 & 1 \end{bmatrix} \begin{bmatrix} 3 \\ 5 \end{bmatrix} = \begin{bmatrix} -1 \\ 2 \end{bmatrix}$ Thus, $x_1 = -1$ and $x_2 = 2$

(C) $\begin{bmatrix} x_1 \\ x_2 \end{bmatrix} = \begin{bmatrix} 3 & -2 \\ -1 & 1 \end{bmatrix} \begin{bmatrix} -2 \\ 1 \end{bmatrix} = \begin{bmatrix} -8 \\ 3 \end{bmatrix}$ Thus, $x_1 = -8$ and $x_2 = 3$

15. The matrix equation for the given system is:
$$\begin{bmatrix} 1 & 3 \\ 2 & 7 \end{bmatrix} \begin{bmatrix} x_1 \\ x_2 \end{bmatrix} = \begin{bmatrix} k_1 \\ k_2 \end{bmatrix}$$

From Exercise 4-5, Problem 19, $\begin{bmatrix} 1 & 3 \\ 2 & 7 \end{bmatrix}^{-1} = \begin{bmatrix} 7 & -3 \\ -2 & 1 \end{bmatrix}$

Thus, $\begin{bmatrix} x_1 \\ x_2 \end{bmatrix} = \begin{bmatrix} 7 & -3 \\ -2 & 1 \end{bmatrix} \begin{bmatrix} k_1 \\ k_2 \end{bmatrix}$

(A) $\begin{bmatrix} x_1 \\ x_2 \end{bmatrix} = \begin{bmatrix} 7 & -3 \\ -2 & 1 \end{bmatrix} \begin{bmatrix} 2 \\ -1 \end{bmatrix} = \begin{bmatrix} 17 \\ -5 \end{bmatrix}$ Thus, $x_1 = 17$ and $x_2 = -5$

(B) $\begin{bmatrix} x_1 \\ x_2 \end{bmatrix} = \begin{bmatrix} 7 & -3 \\ -2 & 1 \end{bmatrix} \begin{bmatrix} 1 \\ 0 \end{bmatrix} = \begin{bmatrix} 7 \\ -2 \end{bmatrix}$ Thus, $x_1 = 7$ and $x_2 = -2$

(C) $\begin{bmatrix} x_1 \\ x_2 \end{bmatrix} = \begin{bmatrix} 7 & -3 \\ -2 & 1 \end{bmatrix} \begin{bmatrix} 3 \\ -1 \end{bmatrix} = \begin{bmatrix} 24 \\ -7 \end{bmatrix}$ Thus, $x_1 = 24$ and $x_2 = -7$

17. The matrix equation for the given system is:

$$\begin{bmatrix} 1 & -3 & 0 \\ 0 & 1 & 1 \\ 2 & -1 & 4 \end{bmatrix} \begin{bmatrix} x_1 \\ x_2 \\ x_3 \end{bmatrix} = \begin{bmatrix} k_1 \\ k_2 \\ k_3 \end{bmatrix}$$

From Exercise 4-5, Problem 21, $\begin{bmatrix} 1 & -3 & 0 \\ 0 & 1 & 1 \\ 2 & -1 & 4 \end{bmatrix}^{-1} = \begin{bmatrix} -5 & -12 & 3 \\ -2 & -4 & 1 \\ 2 & 5 & -1 \end{bmatrix}$

Thus,

$$\begin{bmatrix} x_1 \\ x_2 \\ x_3 \end{bmatrix} = \begin{bmatrix} -5 & -12 & 3 \\ -2 & -4 & 1 \\ 2 & 5 & -1 \end{bmatrix} \begin{bmatrix} k_1 \\ k_2 \\ k_3 \end{bmatrix}$$

(A) $\begin{bmatrix} x_1 \\ x_2 \\ x_3 \end{bmatrix} = \begin{bmatrix} -5 & -12 & 3 \\ -2 & -4 & 1 \\ 2 & 5 & -1 \end{bmatrix} \begin{bmatrix} 1 \\ 0 \\ 2 \end{bmatrix} = \begin{bmatrix} 1 \\ 0 \\ 0 \end{bmatrix}$; $x_1 = 1$, $x_2 = 0$, $x_3 = 0$

(B) $\begin{bmatrix} x_1 \\ x_2 \\ x_3 \end{bmatrix} = \begin{bmatrix} -5 & -12 & 3 \\ -2 & -4 & 1 \\ 2 & 5 & -1 \end{bmatrix} \begin{bmatrix} -1 \\ 1 \\ 0 \end{bmatrix} = \begin{bmatrix} -7 \\ -2 \\ 3 \end{bmatrix}$; $x_1 = -7$, $x_2 = -2$, $x_3 = 3$

(C) $\begin{bmatrix} x_1 \\ x_2 \\ x_3 \end{bmatrix} = \begin{bmatrix} -5 & -12 & 3 \\ -2 & -4 & 1 \\ 2 & 5 & -1 \end{bmatrix} \begin{bmatrix} 2 \\ -2 \\ 1 \end{bmatrix} = \begin{bmatrix} 17 \\ 5 \\ -7 \end{bmatrix}$; $x_1 = 17$, $x_2 = 5$, $x_3 = -7$

19. The matrix equation for the given system is:

$$\begin{bmatrix} 1 & 1 & 0 \\ 2 & 3 & -1 \\ 1 & 0 & 2 \end{bmatrix} \begin{bmatrix} x_1 \\ x_2 \\ x_3 \end{bmatrix} = \begin{bmatrix} k_1 \\ k_2 \\ k_3 \end{bmatrix}$$

From Exercise 4-5, Problem 23, $\begin{bmatrix} 1 & 1 & 0 \\ 2 & 3 & -1 \\ 1 & 0 & 2 \end{bmatrix}^{-1} = \begin{bmatrix} 6 & -2 & -1 \\ -5 & 2 & 1 \\ -3 & 1 & 1 \end{bmatrix}$

Thus,

$$\begin{bmatrix} x_1 \\ x_2 \\ x_3 \end{bmatrix} = \begin{bmatrix} 6 & -2 & -1 \\ -5 & 2 & 1 \\ -3 & 1 & 1 \end{bmatrix} \begin{bmatrix} k_1 \\ k_2 \\ k_3 \end{bmatrix}$$

(A) $\begin{bmatrix} x_1 \\ x_2 \\ x_3 \end{bmatrix} = \begin{bmatrix} 6 & -2 & -1 \\ -5 & 2 & 1 \\ -3 & 1 & 1 \end{bmatrix} \begin{bmatrix} 2 \\ 0 \\ 4 \end{bmatrix} = \begin{bmatrix} 8 \\ -6 \\ -2 \end{bmatrix}$; $x_1 = 8$, $x_2 = -6$, $x_3 = -2$

(B) $\begin{bmatrix} x_1 \\ x_2 \\ x_3 \end{bmatrix} = \begin{bmatrix} 6 & -2 & -1 \\ -5 & 2 & 1 \\ -3 & 1 & 1 \end{bmatrix} \begin{bmatrix} 0 \\ 4 \\ -2 \end{bmatrix} = \begin{bmatrix} -6 \\ 6 \\ 2 \end{bmatrix}$; $x_1 = -6$, $x_2 = 6$, $x_3 = 2$

(C) $\begin{bmatrix} x_1 \\ x_2 \\ x_3 \end{bmatrix} = \begin{bmatrix} 6 & -2 & -1 \\ -5 & 2 & 1 \\ -3 & 1 & 1 \end{bmatrix} \begin{bmatrix} 4 \\ 2 \\ 0 \end{bmatrix} = \begin{bmatrix} 20 \\ -16 \\ -10 \end{bmatrix}$; $x_1 = 20$, $x_2 = -16$, $x_3 = -10$

21. $AX - BX = C$
$(A - B)X = C$
$X = (A - B)^{-1}C$

23. $AX + X = C$
$(A + I)X = C$, where I is the identity matrix of
order n
$X = (A + I)^{-1}C$

25. $AX - C = D - BX$
$AX + BX = C + D$
$(A + B)X = C + D$
$X = (A + B)^{-1}(C + D)$

27. The matrix equation for the given system is:

$$\begin{bmatrix} 1 & 2.001 \\ 1 & 2 \end{bmatrix} \begin{bmatrix} x_1 \\ x_2 \end{bmatrix} = \begin{bmatrix} k_1 \\ k_2 \end{bmatrix}$$

First we compute the inverse of $\begin{bmatrix} 1 & 2.001 \\ 1 & 2 \end{bmatrix}$

$$\begin{bmatrix} 1 & 2.001 & | & 1 & 0 \\ 1 & 2 & | & 0 & 1 \end{bmatrix} (-1)R_1 + R_2 \rightarrow R_2$$

$$\sim \begin{bmatrix} 1 & 2.001 & | & 1 & 0 \\ 0 & -0.001 & | & -1 & 1 \end{bmatrix} (-1000)R_2 \rightarrow R_2$$

$$\sim \begin{bmatrix} 1 & 2.001 & | & 1 & 0 \\ 0 & 1 & | & 1000 & -1000 \end{bmatrix} (-2.001)R_2 + R_1 \rightarrow R_1$$

$$\begin{bmatrix} 1 & 0 & | & -2000 & 2001 \\ 0 & 1 & | & 1000 & -1000 \end{bmatrix}$$

Thus, $\begin{bmatrix} 1 & 2.001 \\ 1 & 2 \end{bmatrix}^{-1} = \begin{bmatrix} -2000 & 2001 \\ 1000 & -1000 \end{bmatrix}$ and $\begin{bmatrix} x_1 \\ x_2 \end{bmatrix} = \begin{bmatrix} -2000 & 2001 \\ 1000 & -1000 \end{bmatrix} \begin{bmatrix} k_1 \\ k_2 \end{bmatrix}$

(A) $\begin{bmatrix} x_1 \\ x_2 \end{bmatrix} = \begin{bmatrix} -2000 & 2001 \\ 1000 & -1000 \end{bmatrix} \begin{bmatrix} 1 \\ 1 \end{bmatrix} = \begin{bmatrix} 1 \\ 0 \end{bmatrix}$; $x_1 = 1$, $x_2 = 0$

(B) $\begin{bmatrix} x_1 \\ x_2 \end{bmatrix} = \begin{bmatrix} -2000 & 2001 \\ 1000 & -1000 \end{bmatrix} \begin{bmatrix} 1 \\ 0 \end{bmatrix} = \begin{bmatrix} -2000 \\ 1000 \end{bmatrix}$; $x_1 = -2,000$, $x_2 = 1,000$

(C) $\begin{bmatrix} x_1 \\ x_2 \end{bmatrix} = \begin{bmatrix} -2000 & 2001 \\ 1000 & -1000 \end{bmatrix} \begin{bmatrix} 0 \\ 1 \end{bmatrix} = \begin{bmatrix} 2001 \\ -1000 \end{bmatrix}$; $x_1 = 2,001$, $x_2 = -1,000$

29. The matrix equation for the given system is:

$$\begin{bmatrix} 1 & 8 & 7 \\ 6 & 6 & 8 \\ 3 & 4 & 6 \end{bmatrix} \begin{bmatrix} x_1 \\ x_2 \\ x_3 \end{bmatrix} = \begin{bmatrix} 135 \\ 155 \\ 75 \end{bmatrix}$$

Thus, $\begin{bmatrix} x_1 \\ x_2 \\ x_3 \end{bmatrix} = \begin{bmatrix} 1 & 8 & 7 \\ 6 & 6 & 8 \\ 3 & 4 & 6 \end{bmatrix}^{-1} \begin{bmatrix} 135 \\ 155 \\ 75 \end{bmatrix} = \begin{bmatrix} -0.08 & 0.4 & -0.44 \\ 0.24 & 0.3 & -0.68 \\ -0.12 & -0.4 & 0.84 \end{bmatrix} \begin{bmatrix} 135 \\ 155 \\ 75 \end{bmatrix}$

$$= \begin{bmatrix} 18.2 \\ 27.9 \\ -15.2 \end{bmatrix} \quad \text{and} \quad \begin{array}{l} x_1 = 18.2 \\ x_2 = 27.9 \\ x_3 = -15.2 \end{array}$$

31. The matrix equation for the given system is:

$$\begin{bmatrix} 6 & 9 & 7 & 5 \\ 6 & 4 & 7 & 3 \\ 4 & 5 & 3 & 2 \\ 4 & 3 & 8 & 2 \end{bmatrix} \begin{bmatrix} x_1 \\ x_2 \\ x_3 \\ x_4 \end{bmatrix} = \begin{bmatrix} 250 \\ 195 \\ 145 \\ 125 \end{bmatrix}$$

Thus

$$
\begin{bmatrix} x_1 \\ x_2 \\ x_3 \\ x_4 \end{bmatrix} = \begin{bmatrix} 6 & 9 & 7 & 5 \\ 6 & 4 & 7 & 3 \\ 4 & 5 & 3 & 2 \\ 4 & 3 & 8 & 2 \end{bmatrix}^{-1} \begin{bmatrix} 250 \\ 195 \\ 145 \\ 125 \end{bmatrix} = \begin{bmatrix} -0.25 & 0.37 & 0.28 & -0.21 \\ 0 & -0.4 & 0.4 & 0.2 \\ 0 & -0.16 & -0.04 & 0.28 \\ 0.5 & 0.5 & -1 & -0.5 \end{bmatrix} \begin{bmatrix} 250 \\ 195 \\ 145 \\ 125 \end{bmatrix}
$$

$$
= \begin{bmatrix} 24 \\ 5 \\ -2 \\ 15 \end{bmatrix} \quad \text{and} \quad
\begin{aligned}
x_1 &= 24 \\
x_2 &= 5 \\
x_3 &= -2 \\
x_4 &= 15
\end{aligned}
$$

33. Let x_1 = number of \$4 tickets sold
and x_2 = number of \$8 tickets sold.

For the first return of \$56,000 we have the following system to solve:
$$x_1 + x_2 = 10,000$$
$$4x_1 + 8x_2 = 56,000$$

The corresponding matrix equation is: $\begin{bmatrix} 1 & 1 \\ 4 & 8 \end{bmatrix} \begin{bmatrix} x_1 \\ x_2 \end{bmatrix} = \begin{bmatrix} 10,000 \\ 56,000 \end{bmatrix}$.

First, we compute the inverse of $\begin{bmatrix} 1 & 1 \\ 4 & 8 \end{bmatrix}$.

$$
\left[\begin{array}{cc|cc} 1 & 1 & 1 & 0 \\ 4 & 8 & 0 & 1 \end{array} \right] \sim
\left[\begin{array}{cc|cc} 1 & 1 & 1 & 0 \\ 0 & 4 & -4 & 1 \end{array} \right] \sim
\left[\begin{array}{cc|cc} 1 & 1 & 1 & 0 \\ 0 & 1 & -1 & \frac{1}{4} \end{array} \right] \sim
\left[\begin{array}{cc|cc} 1 & 0 & 2 & -\frac{1}{4} \\ 0 & 1 & -1 & \frac{1}{4} \end{array} \right]
$$

$(-4)R_1 + R_2 \to R_2 \qquad \frac{1}{4} R_2 \to R_2 \qquad (-1)R_2 + R_1 \to R_1$

Thus, $\begin{bmatrix} 1 & 1 \\ 4 & 8 \end{bmatrix}^{-1} = \begin{bmatrix} 2 & -\frac{1}{4} \\ -1 & \frac{1}{4} \end{bmatrix}$ and $\begin{bmatrix} x_1 \\ x_2 \end{bmatrix} = \begin{bmatrix} 2 & -\frac{1}{4} \\ -1 & \frac{1}{4} \end{bmatrix} \begin{bmatrix} 10,000 \\ 56,000 \end{bmatrix}$

$$
= \begin{bmatrix} 20,000 - 14,000 \\ -10,000 + 14,000 \end{bmatrix} = \begin{bmatrix} 6000 \\ 4000 \end{bmatrix}.
$$

So, for Concert 1, $x_1 = 6000$ \$4 tickets
$x_2 = 4000$ \$8 tickets.

For a return of \$60,000:

$$
\begin{bmatrix} x_1 \\ x_2 \end{bmatrix} = \begin{bmatrix} 2 & -\frac{1}{4} \\ -1 & \frac{1}{4} \end{bmatrix} \begin{bmatrix} 10,000 \\ 60,000 \end{bmatrix} = \begin{bmatrix} 20,000 - 15,000 \\ -10,000 + 15,000 \end{bmatrix} = \begin{bmatrix} 5000 \\ 5000 \end{bmatrix}.
$$

For Concert 2, $x_1 = 5000$ \$4 tickets
$x_2 = 5000$ \$8 tickets.

Finally, for a return of $68,000:

$$\begin{bmatrix} x_1 \\ x_2 \end{bmatrix} = \begin{bmatrix} 2 & -\frac{1}{4} \\ -1 & \frac{1}{4} \end{bmatrix} \begin{bmatrix} 10,000 \\ 68,000 \end{bmatrix} = \begin{bmatrix} 20,000 - 17,000 \\ -10,000 + 17,000 \end{bmatrix} = \begin{bmatrix} 3000 \\ 7000 \end{bmatrix}.$$

Thus, for Concert 3, x_1 = 3000 $4 tickets
x_2 = 7000 $8 tickets.

35. Let x_1 = number of hours at Plant A
and x_2 = number of hours at Plant B

Then $10x_1 + 8x_2 = k_1$ (number of car frames)
       $5x_1 + 8x_2 = k_2$ (number of truck frames)

The corresponding matrix equation is:

$$\begin{bmatrix} 10 & 8 \\ 5 & 8 \end{bmatrix} \begin{bmatrix} x_1 \\ x_2 \end{bmatrix} = \begin{bmatrix} k_1 \\ k_2 \end{bmatrix}$$

First we compute the inverse of $\begin{bmatrix} 10 & 8 \\ 5 & 8 \end{bmatrix}$

$$\left[\begin{array}{cc|cc} 10 & 8 & 1 & 0 \\ 5 & 8 & 0 & 1 \end{array}\right] \sim \left[\begin{array}{cc|cc} 1 & \frac{4}{5} & \frac{1}{10} & 0 \\ 5 & 8 & 0 & 1 \end{array}\right] \sim \left[\begin{array}{cc|cc} 1 & \frac{4}{5} & \frac{1}{10} & 0 \\ 0 & 4 & -\frac{1}{2} & 1 \end{array}\right]$$

$\frac{1}{10}R_1 \rightarrow R_1$ $(-5)R_1 + R_2 \rightarrow R_2$ $\frac{1}{4}R_2 \rightarrow R_2$

$$\sim \left[\begin{array}{cc|cc} 1 & \frac{4}{5} & \frac{1}{10} & 0 \\ 0 & 1 & -\frac{1}{8} & \frac{1}{4} \end{array}\right] \sim \left[\begin{array}{cc|cc} 1 & 0 & \frac{1}{5} & -\frac{1}{5} \\ 0 & 1 & -\frac{1}{8} & \frac{1}{4} \end{array}\right]$$

$\left(-\frac{4}{5}\right)R_2 + R_1 \rightarrow R_1$

Thus $\begin{bmatrix} 10 & 8 \\ 5 & 8 \end{bmatrix}^{-1} = \begin{bmatrix} \frac{1}{5} & -\frac{1}{5} \\ -\frac{1}{8} & \frac{1}{4} \end{bmatrix}$ and $\begin{bmatrix} x_1 \\ x_2 \end{bmatrix} = \begin{bmatrix} \frac{1}{5} & -\frac{1}{5} \\ -\frac{1}{8} & \frac{1}{4} \end{bmatrix} \begin{bmatrix} k_1 \\ k_2 \end{bmatrix}$

Now, for order 1:

$$\begin{bmatrix} x_1 \\ x_2 \end{bmatrix} = \begin{bmatrix} \frac{1}{5} & -\frac{1}{5} \\ -\frac{1}{8} & \frac{1}{4} \end{bmatrix} \begin{bmatrix} 3000 \\ 1600 \end{bmatrix} = \begin{bmatrix} 280 \\ 25 \end{bmatrix}$$ and $\begin{array}{l} x_1 = 280 \text{ hours at Plant A} \\ x_2 = 25 \text{ hours at Plant B} \end{array}$

For order 2:

$$\begin{bmatrix} x_1 \\ x_2 \end{bmatrix} = \begin{bmatrix} \frac{1}{5} & -\frac{1}{5} \\ -\frac{1}{8} & \frac{1}{4} \end{bmatrix} \begin{bmatrix} 2800 \\ 2000 \end{bmatrix} = \begin{bmatrix} 160 \\ 150 \end{bmatrix}$$ and $\begin{array}{l} x_1 = 160 \text{ hours at Plant A} \\ x_2 = 150 \text{ hours at Plant B} \end{array}$

For order 3:

$$\begin{bmatrix} x_1 \\ x_2 \end{bmatrix} = \begin{bmatrix} \frac{1}{5} & -\frac{1}{5} \\ -\frac{1}{8} & \frac{1}{4} \end{bmatrix} \begin{bmatrix} 2600 \\ 2200 \end{bmatrix} = \begin{bmatrix} 80 \\ 225 \end{bmatrix} \quad \text{and} \quad \begin{array}{l} x_1 = 80 \text{ hours at Plant A} \\ x_2 = 225 \text{ hours at Plant B} \end{array}$$

37. Let x_1 = number of ounces of mix A

and x_2 = number of ounces of mix B.

For Diet 1, we have the following system to solve:

$0.2x_1 + 0.1x_2 = 20$

$0.02x_1 + 0.06x_2 = 6$

or

$$\begin{array}{ccc} \text{Diet 1} & \text{Diet 2} & \text{Diet 3} \\ 2x_1 + x_2 = 200 & = 100 & = 100 \\ 2x_1 + 6x_2 = 600 & = 400 & = 600 \end{array}$$

First, compute the inverse matrix of $\begin{bmatrix} 2 & 1 \\ 2 & 6 \end{bmatrix}$.

$$\begin{bmatrix} 2 & 1 & | & 1 & 0 \\ 2 & 6 & | & 0 & 1 \end{bmatrix} \sim \begin{bmatrix} 1 & \frac{1}{2} & | & \frac{1}{2} & 0 \\ 2 & 6 & | & 0 & 1 \end{bmatrix} \sim \begin{bmatrix} 1 & \frac{1}{2} & | & \frac{1}{2} & 0 \\ 0 & 5 & | & -1 & 1 \end{bmatrix} \sim \begin{bmatrix} 1 & \frac{1}{2} & | & \frac{1}{2} & 0 \\ 0 & 1 & | & -\frac{1}{5} & \frac{1}{5} \end{bmatrix}$$

$$\frac{1}{2}R_1 \rightarrow R_1 \qquad (-2)R_1 + R_2 \rightarrow R_2 \qquad \frac{1}{5}R_2 \rightarrow R_2 \qquad \left(-\frac{1}{2}\right)R_2 + R_1 \rightarrow R_1$$

$$\sim \begin{bmatrix} 1 & 0 & | & \frac{6}{10} & -\frac{1}{10} \\ 0 & 1 & | & -\frac{1}{5} & \frac{1}{5} \end{bmatrix} \quad \text{Thus,} \quad \begin{bmatrix} 2 & 1 \\ 2 & 6 \end{bmatrix}^{-1} = \begin{bmatrix} \frac{3}{5} & -\frac{1}{10} \\ -\frac{1}{5} & \frac{1}{5} \end{bmatrix}$$

$$\text{and} \quad \begin{bmatrix} x_1 \\ x_2 \end{bmatrix} = \begin{bmatrix} \frac{3}{5} & -\frac{1}{10} \\ -\frac{1}{5} & \frac{1}{5} \end{bmatrix} \begin{bmatrix} 200 \\ 600 \end{bmatrix} = \begin{bmatrix} 120 - 60 \\ -40 + 120 \end{bmatrix} = \begin{bmatrix} 60 \\ 80 \end{bmatrix}.$$

So, for Diet 1, $x_1 = 60$ ounces of mix A

$x_2 = 80$ ounces of mix B.

For Diet 2, the solution is:

$$\begin{bmatrix} x_1 \\ x_2 \end{bmatrix} = \begin{bmatrix} \frac{3}{5} & -\frac{1}{10} \\ -\frac{1}{5} & \frac{1}{5} \end{bmatrix} \begin{bmatrix} 100 \\ 400 \end{bmatrix} = \begin{bmatrix} 60 - 40 \\ -20 + 80 \end{bmatrix} = \begin{bmatrix} 20 \\ 60 \end{bmatrix}$$

So for Diet 2, $x_1 = 20$ ounces of mix A

$x_2 = 60$ ounces of mix B.

For Diet 3, we have:

$$\begin{bmatrix} x_1 \\ x_2 \end{bmatrix} = \begin{bmatrix} \frac{3}{5} & -\frac{1}{10} \\ -\frac{1}{5} & \frac{1}{5} \end{bmatrix} \begin{bmatrix} 100 \\ 600 \end{bmatrix} = \begin{bmatrix} 60 - 60 \\ -20 + 120 \end{bmatrix} = \begin{bmatrix} 0 \\ 100 \end{bmatrix}$$

Thus, for Diet 3, $x_1 = 0$ ounces of mix A

$x_2 = 100$ ounces of mix B.

39. Let x_1 = President's bonus
x_2 = Executive Vice President's bonus
x_3 = Associate Vice President's bonus
x_4 = Assistant Vice President's bonus

Then

$x_1 = 0.03(2,000,000 - x_2 - x_3 - x_4)$
$x_2 = 0.025(2,000,000 - x_1 - x_3 - x_4)$
$x_3 = 0.02(2,000,000 - x_1 - x_2 - x_4)$
$x_4 = 0.015(2,000,000 - x_1 - x_2 - x_3)$

or

$$
\begin{aligned}
x_1 + 0.03x_2 + 0.03x_3 + 0.03x_4 &= 60,000 \\
0.025x_1 + x_2 + 0.025x_3 + 0.025x_4 &= 50,000 \\
0.02x_1 + 0.02x_2 + x_3 + 0.02x_4 &= 40,000 \\
0.015x_1 + 0.015x_2 + 0.015x_3 + x_4 &= 30,000
\end{aligned}
$$

and

$$
\begin{bmatrix}
1 & 0.03 & 0.03 & 0.03 \\
0.025 & 1 & 0.025 & 0.025 \\
0.02 & 0.02 & 1 & 0.02 \\
0.015 & 0.015 & 0.015 & 1
\end{bmatrix}
\begin{bmatrix}
x_1 \\ x_2 \\ x_3 \\ x_4
\end{bmatrix}
=
\begin{bmatrix}
60,000 \\ 50,000 \\ 40,000 \\ 30,000
\end{bmatrix}
$$

Thus

$$
\begin{bmatrix}
x_1 \\ x_2 \\ x_3 \\ x_4
\end{bmatrix}
=
\begin{bmatrix}
1 & 0.03 & 0.03 & 0.03 \\
0.025 & 1 & 0.025 & 0.025 \\
0.02 & 0.02 & 1 & 0.02 \\
0.015 & 0.015 & 0.015 & 1
\end{bmatrix}^{-1}
\begin{bmatrix}
60,000 \\ 50,000 \\ 40,000 \\ 30,000
\end{bmatrix}
\approx
\begin{bmatrix}
56,600 \\ 47,000 \\ 37,400 \\ 27,900
\end{bmatrix}
$$

or $x_1 = \$56,600$, $x_2 = \$47,000$, $x_3 = \$37,400$, $x_4 = \$27,900$ to the nearest hundred dollars.

41. Let x_1 = Taxable income of company A
x_2 = Taxable income of company B
x_3 = Taxable income of company C
x_4 = Taxable income of company D

The taxable income of each company is given by the system of equations:
$x_1 = 0.82(3.2) + 0.08x_2 + 0.03x_3 + 0.07x_4$
$x_2 = 0.12x_1 + 0.64(2.6) + 0.11x_3 + 0.13x_4$
$x_3 = 0.11x_1 + 0.09x_2 + 0.72(3.8) + 0.08x_4$
$x_4 = 0.06x_1 + 0.02x_2 + 0.14x_3 + 0.78(4.4)$

which is the same as:
$x_1 - 0.08x_2 - 0.03x_3 - 0.07x_4 = 2.624$
$-0.12x_1 + x_2 - 0.11x_3 - 0.13x_4 = 1.664$
$-0.11x_1 - 0.09x_2 + x_3 - 0.08x_4 = 2.736$
$-0.06x_1 - 0.02x_2 - 0.14x_3 + x_4 = 3.432$

Written in matrix form, we have

$$\begin{bmatrix} 1 & -0.08 & -0.03 & -0.07 \\ -0.12 & 1 & -0.11 & -0.13 \\ -0.11 & -0.09 & 1 & -0.08 \\ -0.06 & -0.02 & -0.14 & 1 \end{bmatrix} \begin{bmatrix} x_1 \\ x_2 \\ x_3 \\ x_4 \end{bmatrix} = \begin{bmatrix} 2.624 \\ 1.664 \\ 2.736 \\ 3.432 \end{bmatrix}$$

and

$$\begin{bmatrix} x_1 \\ x_2 \\ x_3 \\ x_4 \end{bmatrix} = \begin{bmatrix} 1 & -0.08 & -0.03 & -0.07 \\ -0.12 & 1 & -0.11 & -0.13 \\ -0.11 & -0.09 & 1 & -0.08 \\ -0.06 & -0.02 & -0.14 & 1 \end{bmatrix}^{-1} \begin{bmatrix} 2.624 \\ 1.664 \\ 2.736 \\ 3.432 \end{bmatrix}$$

$$\sim \begin{bmatrix} 1.022 & 0.088 & 0.053 & 0.087 \\ 0.148 & 1.027 & 0.139 & 0.155 \\ 0.132 & 0.105 & 1.030 & 0.105 \\ 0.083 & 0.040 & 0.150 & 1.023 \end{bmatrix} \begin{bmatrix} 2.624 \\ 1.664 \\ 2.736 \\ 3.432 \end{bmatrix} \sim \begin{bmatrix} 3.270 \\ 3.011 \\ 3.703 \\ 4.207 \end{bmatrix}$$

Thus the taxable incomes are:

Company A: \$3,270,000; Company B: \$3,011,000;
Company C: \$3,703,000; Company D: \$4,207,000.

(Taxable income) − (net income) = [3.270 + 3.011 + 3.703 + 4.207] − [3.2 + 2.6 + 3.8 + 4.4] = 0.191 or \$191,000

43. Let F = total cost for the Freezer Department
R = total cost for the Refrigerator Department
A = total cost for the Accounting Department
M = total cost for the Maintenance Department

(A) $F = 260,000 + 0.38A + 0.42M$
$R = 190,000 + 0.34A + 0.28M$
$A = 55,000 + 0.16A + 0.13M$
$M = 95,000 + 0.12A + 0.17M$

This system of equations can be written as
$$F - 0.38A - 0.42M = 260,000$$
$$R - 0.34A - 0.28M = 190,000$$
$$0.84A - 0.13M = 55,000$$
$$-0.12A + 0.83M = 95,000$$

The matrix equation for this system is:

$$\begin{bmatrix} 1 & 0 & -0.38 & -0.42 \\ 0 & 1 & -0.34 & -0.28 \\ 0 & 0 & 0.84 & -0.13 \\ 0 & 0 & -0.12 & 0.83 \end{bmatrix} \begin{bmatrix} F \\ R \\ A \\ M \end{bmatrix} = \begin{bmatrix} 260,000 \\ 190,000 \\ 55,000 \\ 95,000 \end{bmatrix}$$

and

$$\begin{bmatrix} F \\ R \\ A \\ M \end{bmatrix} = \begin{bmatrix} 1 & 0 & -0.38 & -0.42 \\ 0 & 1 & -0.34 & -0.28 \\ 0 & 0 & 0.84 & -0.13 \\ 0 & 0 & -0.12 & 0.83 \end{bmatrix}^{-1} \begin{bmatrix} 260,000 \\ 190,000 \\ 55,000 \\ 95,000 \end{bmatrix}$$

$$\approx \begin{bmatrix} 1 & 0 & 0.537 & 0.590 \\ 0 & 1 & 0.463 & 0.410 \\ 0 & 0 & 1.218 & 0.190 \\ 0 & 0 & 0.176 & 1.232 \end{bmatrix} \begin{bmatrix} 260,000 \\ 190,000 \\ 55,000 \\ 95,000 \end{bmatrix} = \begin{bmatrix} 345,575 \\ 254,424 \\ 85,093 \\ 126,760 \end{bmatrix}$$

$$\approx \begin{bmatrix} 346,000 \\ 254,000 \\ 85,000 \\ 127,000 \end{bmatrix} \quad \text{(to the nearest thousand)}$$

The total cost of each department is:
Freezers - $346,000, Refrigerators - $254,000,
Accounting - $85,000, Maintenance - $127,000

(B) Sum of direct costs:
$260,000 + $190,000 + $55,000 + $95,000 = $600,000
total costs of production departments:
$346,000 + $254,000 = $600,000

The direct costs of the service departments are distributed among the production departments.

EXERCISE 4-7

Things to remember:

1. Given two industries C_1 and C_2, with

$$M = \begin{array}{c} \\ C_1 \\ C_2 \end{array}\begin{array}{cc} C_1 & C_2 \\ \begin{bmatrix} a_{11} & a_{12} \\ a_{21} & a_{22} \end{bmatrix} \end{array}, \quad X = \begin{bmatrix} x_1 \\ x_2 \end{bmatrix}, \quad D = \begin{bmatrix} d_1 \\ d_2 \end{bmatrix},$$

$\quad\quad$ Technology $\quad\quad$ Output $\quad$ Final Demand
$\quad\quad\quad$ Matrix $\quad\quad\quad$ Matrix $\quad\quad$ Matrix

where a_{ij} is the input required from C_i to produce a dollar's worth of output for C_j. The solution to the input-output matrix equation $X = MX + D$ is

$$X = (I - M)^{-1}D,$$

where I is the identity matrix, assuming $I - M$ has an inverse.

1. 40¢ from A and 20¢ from E are required to produce a dollar's worth of output for A.

3. $I - M = \begin{bmatrix} 1 & 0 \\ 0 & 1 \end{bmatrix} - \begin{bmatrix} 0.4 & 0.2 \\ 0.2 & 0.1 \end{bmatrix} = \begin{bmatrix} 0.6 & -0.2 \\ -0.2 & 0.9 \end{bmatrix}$

Converting the decimals to fractions to calculate the inverse, we have:

$\left[\begin{array}{cc|cc} \frac{3}{5} & -\frac{1}{5} & 1 & 0 \\ -\frac{1}{5} & \frac{9}{10} & 0 & 1 \end{array}\right] \sim \left[\begin{array}{cc|cc} 1 & -\frac{1}{3} & \frac{5}{3} & 0 \\ -\frac{1}{5} & \frac{9}{10} & 0 & 1 \end{array}\right] \sim \left[\begin{array}{cc|cc} 1 & -\frac{1}{3} & \frac{5}{3} & 0 \\ 0 & \frac{5}{6} & \frac{1}{3} & 1 \end{array}\right] \sim \left[\begin{array}{cc|cc} 1 & -\frac{1}{3} & \frac{5}{3} & 0 \\ 0 & 1 & \frac{2}{5} & \frac{6}{5} \end{array}\right]$

$\frac{5}{3}R_1 \rightarrow R_1$ $\qquad \frac{1}{5}R_1 + R_2 \rightarrow R_2$ $\qquad \frac{6}{5}R_2 \rightarrow R_2$ $\qquad \frac{1}{3}R_2 + R_1 \rightarrow R_1$

$\sim \left[\begin{array}{cc|cc} 1 & 0 & \frac{9}{5} & \frac{2}{5} \\ 0 & 1 & \frac{2}{5} & \frac{6}{5} \end{array}\right]$ Thus, $I - M = \begin{bmatrix} 0.6 & -0.2 \\ -0.2 & 0.9 \end{bmatrix}$ and $(I - M)^{-1} = \begin{bmatrix} 1.8 & 0.4 \\ 0.4 & 1.2 \end{bmatrix}$

5. $X = (I - M)^{-1}D_2 = \begin{bmatrix} 1.8 & 0.4 \\ 0.4 & 1.2 \end{bmatrix}\begin{bmatrix} 8 \\ 5 \end{bmatrix}$ Thus, $\begin{bmatrix} x_1 \\ x_2 \end{bmatrix} = \begin{bmatrix} 16.4 \\ 9.2 \end{bmatrix}$ and $x_1 = 16.4$, $x_2 = 9.2$.

7. 20¢ from A, 10¢ from B, and 10¢ from E are required to produce a dollar's worth of output for B.

9. $\begin{bmatrix} 1 & 0 & 0 \\ 0 & 1 & 0 \\ 0 & 0 & 1 \end{bmatrix} - \begin{bmatrix} 0.3 & 0.2 & 0.2 \\ 0.1 & 0.1 & 0.1 \\ 0.2 & 0.1 & 0.1 \end{bmatrix} = \begin{bmatrix} 0.7 & -0.2 & -0.2 \\ -0.1 & 0.9 & -0.1 \\ -0.2 & -0.1 & 0.9 \end{bmatrix}$

11. $X = (I - M)^{-1}D_1$

Therefore, $\begin{bmatrix} x_1 \\ x_2 \\ x_3 \end{bmatrix} = \begin{bmatrix} 1.6 & 0.4 & 0.4 \\ 0.22 & 1.18 & 0.18 \\ 0.38 & 0.22 & 1.22 \end{bmatrix}\begin{bmatrix} 5 \\ 10 \\ 15 \end{bmatrix} = \begin{bmatrix} (1.6)5 + (0.4)10 + (0.4)15 \\ (0.22)5 + (1.18)10 + (0.18)15 \\ (0.38)5 + (0.22)10 + (1.22)15 \end{bmatrix}$

$= \begin{bmatrix} 8 + 4 + 6 \\ 1.1 + 11.8 + 2.7 \\ 1.9 + 2.2 + 18.3 \end{bmatrix} = \begin{bmatrix} 18 \\ 15.6 \\ 22.4 \end{bmatrix}$

Thus, agriculture, \$18 billion; building, \$15.6 billion; and energy, \$22.4 billion.

13. $I - M = \begin{bmatrix} 1 & 0 \\ 0 & 1 \end{bmatrix} - \begin{bmatrix} 0.2 & 0.2 \\ 0.3 & 0.3 \end{bmatrix} = \begin{bmatrix} 0.8 & -0.2 \\ -0.3 & 0.7 \end{bmatrix} = \begin{bmatrix} \frac{4}{5} & -\frac{1}{5} \\ -\frac{3}{10} & \frac{7}{10} \end{bmatrix}$,

converting the decimals to fractions.

$\left[\begin{array}{cc|cc} \frac{4}{5} & -\frac{1}{5} & 1 & 0 \\ -\frac{3}{10} & \frac{7}{10} & 0 & 1 \end{array}\right] \sim \left[\begin{array}{cc|cc} 1 & -\frac{1}{4} & \frac{5}{4} & 0 \\ -\frac{3}{10} & \frac{7}{10} & 0 & 1 \end{array}\right] \sim \left[\begin{array}{cc|cc} 1 & -\frac{1}{4} & \frac{5}{4} & 0 \\ 0 & \frac{5}{8} & \frac{3}{8} & 1 \end{array}\right] \sim \left[\begin{array}{cc|cc} 1 & -\frac{1}{4} & \frac{5}{4} & 0 \\ 0 & 1 & \frac{3}{5} & \frac{8}{5} \end{array}\right]$

$\frac{5}{4}R_1 \rightarrow R_1$ $\qquad \frac{3}{10}R_1 + R_2 \rightarrow R_2$ $\qquad \frac{8}{5}R_2 \rightarrow R_2$ $\qquad \frac{1}{4}R_2 + R_1 \rightarrow R_1$

$$\sim \begin{bmatrix} 1 & 0 & | & \frac{7}{5} & \frac{2}{5} \\ 0 & 1 & | & \frac{3}{5} & \frac{8}{5} \end{bmatrix} \quad \text{Thus, } (I - M)^{-1} = \begin{bmatrix} 1.4 & 0.4 \\ 0.6 & 1.6 \end{bmatrix}.$$

Now, $X = (I - M)^{-1}D = \begin{bmatrix} 1.4 & 0.4 \\ 0.6 & 1.6 \end{bmatrix} \begin{bmatrix} 10 \\ 25 \end{bmatrix} = \begin{bmatrix} 24 \\ 46 \end{bmatrix}.$

15. $I - M = \begin{bmatrix} 1 & 0 & 0 \\ 0 & 1 & 0 \\ 0 & 0 & 1 \end{bmatrix} - \begin{bmatrix} 0.3 & 0.1 & 0.3 \\ 0.2 & 0.1 & 0.2 \\ 0.1 & 0.1 & 0.1 \end{bmatrix} = \begin{bmatrix} 0.7 & -0.1 & -0.3 \\ -0.2 & 0.9 & -0.2 \\ -0.1 & -0.1 & 0.9 \end{bmatrix}$

$$\begin{bmatrix} 0.7 & -0.1 & -0.3 & | & 1 & 0 & 0 \\ -0.2 & 0.9 & -0.2 & | & 0 & 1 & 0 \\ -0.1 & -0.1 & 0.9 & | & 0 & 0 & 1 \end{bmatrix} \sim \begin{bmatrix} 7 & -1 & -3 & | & 10 & 0 & 0 \\ -2 & 9 & -2 & | & 0 & 10 & 0 \\ 1 & 1 & -9 & | & 0 & 0 & -10 \end{bmatrix}$$

$$\begin{array}{c} 10R_1 \to R_1 \\ 10R_2 \to R_2 \\ -10R_3 \to R_3 \end{array} \qquad\qquad R_1 \leftrightarrow R_3$$

$$\sim \begin{bmatrix} 1 & 1 & -9 & | & 0 & 0 & -10 \\ -2 & 9 & -2 & | & 0 & 10 & 0 \\ 7 & -1 & -3 & | & 10 & 0 & 0 \end{bmatrix} \sim \begin{bmatrix} 1 & 1 & -9 & | & 0 & 0 & -10 \\ 0 & 11 & -20 & | & 0 & 10 & -20 \\ 0 & -8 & 60 & | & 10 & 0 & 70 \end{bmatrix}$$

$$\begin{array}{c} 2R_1 + R_2 \to R_2 \\ (-7)R_1 + R_3 \to R_3 \end{array} \qquad\qquad \frac{1}{11}R_2 \to R_2$$

$$\sim \begin{bmatrix} 1 & 1 & -9 & | & 0 & 0 & -10 \\ 0 & 1 & -1.82 & | & 0 & 0.91 & -1.82 \\ 0 & -8 & 60 & | & 10 & 0 & 70 \end{bmatrix} \sim \begin{bmatrix} 1 & 0 & -7.18 & | & 0 & -0.91 & -8.18 \\ 0 & 1 & -1.82 & | & 0 & 0.91 & -1.82 \\ 0 & 0 & 45.44 & | & 10 & 7.28 & 55.44 \end{bmatrix}$$

$$8R_2 + R_3 \to R_3 \qquad\qquad \frac{1}{45.44}R_3 \to R_3$$

$$\sim \begin{bmatrix} 1 & 0 & -7.18 & | & 0 & -0.91 & -8.18 \\ 0 & 1 & -1.82 & | & 0 & 0.91 & -1.82 \\ 0 & 0 & 1 & | & 0.22 & 0.16 & 1.22 \end{bmatrix} \sim \begin{bmatrix} 1 & 0 & 0 & | & 1.58 & 0.24 & 0.58 \\ 0 & 1 & 0 & | & 0.4 & 1.2 & 0.4 \\ 0 & 0 & 1 & | & 0.22 & 0.16 & 1.22 \end{bmatrix}$$

$$\begin{array}{c} 1.82R_3 + R_2 \to R_2 \\ 7.18R_3 + R_1 \to R_1 \end{array}$$

Thus, $(I - M)^{-1} = \begin{bmatrix} 1.58 & 0.24 & 0.58 \\ 0.4 & 1.2 & 0.4 \\ 0.22 & 0.16 & 1.22 \end{bmatrix}$,

and $X = (I - M)^{-1}D = \begin{bmatrix} 1.58 & 0.24 & 0.58 \\ 0.4 & 1.2 & 0.4 \\ 0.22 & 0.16 & 1.22 \end{bmatrix} \begin{bmatrix} 20 \\ 5 \\ 10 \end{bmatrix}$

$$= \begin{bmatrix} (1.58)20 + (0.24)5 + (0.58)10 \\ (0.4)20 + (1.2)5 + (0.4)10 \\ (0.22)20 + (0.16)5 + (1.22)10 \end{bmatrix} = \begin{bmatrix} 38.6 \\ 18 \\ 17.4 \end{bmatrix}.$$

17. **(A)** The technology matrix $M = \begin{bmatrix} 0.3 & 0.25 \\ 0.1 & 0.25 \end{bmatrix}$ and the final demand matrix

$D = \begin{bmatrix} 40 \\ 40 \end{bmatrix}$. The input-output matrix equation is $X = MX + D$ or

$X = \begin{bmatrix} 0.3 & 0.25 \\ 0.1 & 0.25 \end{bmatrix} X + \begin{bmatrix} 40 \\ 40 \end{bmatrix}$, where $X = \begin{bmatrix} x_1 \\ x_2 \end{bmatrix}$.

The solution is $X = (I - M)^{-1}D$, provided $I - M$ has an inverse. Now,

$I - M = \begin{bmatrix} 1 & 0 \\ 0 & 1 \end{bmatrix} - \begin{bmatrix} 0.3 & 0.25 \\ 0.1 & 0.25 \end{bmatrix} = \begin{bmatrix} 0.7 & -0.25 \\ -0.1 & 0.75 \end{bmatrix}$

$(I - M)^{-1}$: $\begin{bmatrix} 0.7 & -0.25 & | & 1 & 0 \\ -0.1 & 0.75 & | & 0 & 1 \end{bmatrix}$ $-10R_2 \rightarrow R_2$

$\sim \begin{bmatrix} 0.7 & -0.25 & | & 1 & 0 \\ 1 & -7.5 & | & 0 & -10 \end{bmatrix}$ $R_1 \leftrightarrow R_2$

$\sim \begin{bmatrix} 1 & -7.5 & | & 0 & -10 \\ 0.7 & -0.25 & | & 1 & 0 \end{bmatrix}$ $(-0.7)R_1 + R_2 \rightarrow R_2$

$\sim \begin{bmatrix} 1 & -7.5 & | & 0 & -10 \\ 0 & 5 & | & 1 & 7 \end{bmatrix}$ $(0.2)R_2 \rightarrow R_2$

$\sim \begin{bmatrix} 1 & -7.5 & | & 0 & -10 \\ 0 & 1 & | & 0.2 & 1.4 \end{bmatrix}$ $(7.5)R_2 + R_1 \rightarrow R_1$

$\sim \begin{bmatrix} 1 & 0 & | & 1.5 & 0.5 \\ 0 & 1 & | & 0.2 & 1.4 \end{bmatrix}$

Thus, $(I - M)^{-1} = \begin{bmatrix} 1.5 & 0.5 \\ 0.2 & 1.4 \end{bmatrix}$ and $X = \begin{bmatrix} 1.5 & 0.5 \\ 0.2 & 1.4 \end{bmatrix} \begin{bmatrix} 40 \\ 40 \end{bmatrix} = \begin{bmatrix} 80 \\ 64 \end{bmatrix}$

Thus, the output for each sector is:
Agriculture: $80 million; Manufacturing: $64 million

(B) If the agricultural output is increased by $20 million and the
manufacturing output remains at $64 million, then the final demand
D is given by

$D = (I - M)X = \begin{bmatrix} 0.7 & -0.25 \\ -0.1 & 0.75 \end{bmatrix} \begin{bmatrix} 100 \\ 64 \end{bmatrix} = \begin{bmatrix} 54 \\ 38 \end{bmatrix}$

The final demand for agriculture increases to $54 million and the
final demand for manufacturing decreases to $38 million.

19. Let x_1 = total output of energy
x_2 = total output of mining

Then the final demand matrix $D = \begin{pmatrix} 0.4x_1 \\ 0.4x_2 \end{pmatrix}$ and the input-output matrix
equation is:

$$\begin{bmatrix} x_1 \\ x_2 \end{bmatrix} = \begin{bmatrix} 0.2 & 0.3 \\ 0.4 & 0.3 \end{bmatrix} \begin{bmatrix} x_1 \\ x_2 \end{bmatrix} + \begin{bmatrix} 0.4x_1 \\ 0.4x_2 \end{bmatrix}$$

This yields the dependent system of equations

$$0.6x_1 = 0.2x_1 + 0.3x_2$$
$$0.6x_2 = 0.4x_1 + 0.3x_2$$

which is equivalent to

$$0.4x_1 - 0.3x_2 = 0$$

or $\qquad x_1 = \dfrac{3}{4}x_2$

Thus, the total ouput of the energy sector should be 75% of the total output of the mining sector.

21. Each element of a technology matrix represents the input needed from C_i to produce \$1 dollar's worth of output for C_j. Hence, each element must be a number between 0 and 1, inclusive.

23. The technology matrix $M = \begin{bmatrix} 0.1 & 0.2 \\ 0.2 & 0.4 \end{bmatrix}$ and the final demand matrix $D = \begin{bmatrix} 20 \\ 10 \end{bmatrix}$.

The input-output matrix equation is $X = MX + D$ or

$$X = \begin{bmatrix} 0.1 & 0.2 \\ 0.2 & 0.4 \end{bmatrix} X + \begin{bmatrix} 20 \\ 10 \end{bmatrix} \text{ where } X = \begin{bmatrix} x_1 \\ x_2 \end{bmatrix}.$$

The solution is $X = (I - M)^{-1}D$, provided $(I - M)$ has an inverse. Now,

$$I - M = \begin{bmatrix} 1 & 0 \\ 0 & 1 \end{bmatrix} - \begin{bmatrix} 0.1 & 0.2 \\ 0.2 & 0.4 \end{bmatrix} = \begin{bmatrix} 0.9 & -0.2 \\ -0.2 & 0.6 \end{bmatrix} = \begin{bmatrix} \frac{9}{10} & -\frac{1}{5} \\ -\frac{1}{5} & \frac{3}{5} \end{bmatrix}$$

$$\begin{bmatrix} \frac{9}{10} & -\frac{1}{5} & 1 & 0 \\ -\frac{1}{5} & \frac{3}{5} & 0 & 1 \end{bmatrix} \sim \begin{bmatrix} 1 & -\frac{2}{9} & \frac{10}{9} & 0 \\ -\frac{1}{5} & \frac{3}{5} & 0 & 1 \end{bmatrix} \sim \begin{bmatrix} 1 & -\frac{2}{9} & \frac{10}{9} & 0 \\ 0 & \frac{5}{9} & \frac{2}{9} & 1 \end{bmatrix} \sim \begin{bmatrix} 1 & -\frac{2}{9} & \frac{10}{9} & 0 \\ 0 & 1 & \frac{2}{5} & \frac{9}{5} \end{bmatrix}$$

$$\frac{10}{9}R_1 \to R_1 \qquad \frac{1}{5}R_1 + R_2 \to R_2 \qquad \frac{9}{5}R_2 \to R_2 \qquad \frac{2}{9}R_2 + R_1 \to R_1$$

$$\sim \begin{bmatrix} 1 & 0 & \frac{6}{5} & \frac{2}{5} \\ 0 & 1 & \frac{2}{5} & \frac{9}{5} \end{bmatrix} \text{ Thus, } (I - M)^{-1} = \begin{bmatrix} \frac{6}{5} & \frac{2}{5} \\ \frac{2}{5} & \frac{9}{5} \end{bmatrix} = \begin{bmatrix} 1.2 & 0.4 \\ 0.4 & 1.8 \end{bmatrix}, \text{ and}$$

$$X = \begin{bmatrix} 1.2 & 0.4 \\ 0.4 & 1.8 \end{bmatrix} \begin{bmatrix} 20 \\ 10 \end{bmatrix} = \begin{bmatrix} 28 \\ 26 \end{bmatrix}.$$

Therefore, the output for each sector is: coal, \$28 billion; steel, \$26 billion.

25. The technology matrix $M = \begin{bmatrix} 0.20 & 0.40 \\ 0.15 & 0.30 \end{bmatrix} = \begin{bmatrix} \frac{1}{5} & \frac{2}{5} \\ \frac{3}{20} & \frac{3}{10} \end{bmatrix}$ and the final demand

matrix $D = \begin{bmatrix} 60 \\ 80 \end{bmatrix}$. The input-output matrix equation is $X = MX + D$ or

$$X = \begin{bmatrix} \frac{1}{5} & \frac{2}{5} \\ \frac{3}{20} & \frac{3}{10} \end{bmatrix} X + \begin{bmatrix} 60 \\ 80 \end{bmatrix} \text{ where } X = \begin{bmatrix} x_1 \\ x_2 \end{bmatrix}. \text{ The solution is } X = (I - M)^{-1}D,$$

provided $(I - M)$ has an inverse. Now

$$I - M = \begin{bmatrix} 1 & 0 \\ 0 & 1 \end{bmatrix} - \begin{bmatrix} \frac{1}{5} & \frac{2}{5} \\ \frac{3}{20} & \frac{3}{10} \end{bmatrix} = \begin{bmatrix} \frac{4}{5} & -\frac{2}{5} \\ -\frac{3}{20} & \frac{7}{10} \end{bmatrix}.$$

$$\begin{bmatrix} \frac{4}{5} & -\frac{2}{5} & \Big| & 1 & 0 \\ -\frac{3}{20} & \frac{7}{10} & \Big| & 0 & 1 \end{bmatrix} \sim \begin{bmatrix} 1 & -\frac{1}{2} & \Big| & \frac{5}{4} & 0 \\ -\frac{3}{20} & \frac{7}{10} & \Big| & 0 & 1 \end{bmatrix} \sim \begin{bmatrix} 1 & -\frac{1}{2} & \Big| & \frac{5}{4} & 0 \\ 0 & \frac{5}{8} & \Big| & \frac{3}{16} & 1 \end{bmatrix}$$

$$\frac{5}{4} R_1 \rightarrow R_1 \qquad\qquad \frac{3}{20} R_1 + R_2 \rightarrow R_2 \qquad\qquad \frac{8}{5} R_2 \rightarrow R_2$$

$$\sim \begin{bmatrix} 1 & -\frac{1}{2} & \Big| & \frac{5}{4} & 0 \\ 0 & 1 & \Big| & \frac{3}{10} & \frac{8}{5} \end{bmatrix} \sim \begin{bmatrix} 1 & 0 & \Big| & \frac{7}{5} & \frac{4}{5} \\ 0 & 1 & \Big| & \frac{3}{10} & \frac{8}{5} \end{bmatrix}$$

$$\frac{1}{2} R_2 + R_1 \rightarrow R_1$$

Thus $(I - M)^{-1} = \begin{bmatrix} \frac{7}{5} & \frac{4}{5} \\ \frac{3}{10} & \frac{8}{5} \end{bmatrix}$ and $X = \begin{bmatrix} \frac{7}{5} & \frac{4}{5} \\ \frac{3}{10} & \frac{8}{5} \end{bmatrix} \begin{bmatrix} 60 \\ 80 \end{bmatrix} = \begin{bmatrix} 148 \\ 146 \end{bmatrix}$

Therefore, the output for each sector is:
 Agriculture—$148 million; Tourism—$146 million

27. The technology matrix $M = \begin{bmatrix} 0.2 & 0.4 & 0.3 \\ 0.2 & 0.1 & 0.1 \\ 0.2 & 0.1 & 0.1 \end{bmatrix}$ and the final demand matrix

$D = \begin{bmatrix} 10 \\ 15 \\ 20 \end{bmatrix}.$

The input-output matrix equation is $X = MX + D$ or

$$X = \begin{bmatrix} 0.2 & 0.4 & 0.3 \\ 0.2 & 0.1 & 0.1 \\ 0.2 & 0.1 & 0.1 \end{bmatrix} X + \begin{bmatrix} 10 \\ 15 \\ 20 \end{bmatrix}.$$

The solution is $X = (I - M)^{-1}D$, provided $I - M$ has an inverse. Now,

$$I - M = \begin{bmatrix} 1 & 0 & 0 \\ 0 & 1 & 0 \\ 0 & 0 & 1 \end{bmatrix} - \begin{bmatrix} 0.2 & 0.4 & 0.3 \\ 0.2 & 0.1 & 0.1 \\ 0.2 & 0.1 & 0.1 \end{bmatrix} = \begin{bmatrix} 0.8 & -0.4 & -0.3 \\ -0.2 & 0.9 & -0.1 \\ -0.2 & -0.1 & 0.9 \end{bmatrix}.$$

$$\begin{bmatrix} 0.8 & -0.4 & -0.3 & \Big| & 1 & 0 & 0 \\ -0.2 & 0.9 & -0.1 & \Big| & 0 & 1 & 0 \\ -0.2 & -0.1 & 0.9 & \Big| & 0 & 0 & 1 \end{bmatrix} \sim \begin{bmatrix} 8 & -4 & -3 & \Big| & 10 & 0 & 0 \\ -2 & 9 & -1 & \Big| & 0 & 10 & 0 \\ -2 & -1 & 9 & \Big| & 0 & 0 & 10 \end{bmatrix}$$

$$\begin{aligned} 10R_1 &\rightarrow R_1 \\ 10R_2 &\rightarrow R_2 \\ 10R_3 &\rightarrow R_3 \end{aligned} \qquad\qquad \left(-\frac{1}{2}\right) R_2 \rightarrow R_2$$

$$\sim \begin{bmatrix} 8 & -4 & -3 & | & 10 & 0 & 0 \\ 1 & -\frac{9}{2} & \frac{1}{2} & | & 0 & -5 & 0 \\ -2 & -1 & 9 & | & 0 & 0 & 10 \end{bmatrix} \sim \begin{bmatrix} 1 & -\frac{9}{2} & \frac{1}{2} & | & 0 & -5 & 0 \\ 8 & -4 & -3 & | & 10 & 0 & 0 \\ -2 & -1 & 9 & | & 0 & 0 & 10 \end{bmatrix}$$

$$R_1 \leftrightarrow R_2 \qquad\qquad\qquad (-8)R_1 + R_2 \rightarrow R_2$$
$$2R_1 + R_3 \rightarrow R_3$$

$$\begin{bmatrix} 1 & -\frac{9}{2} & \frac{1}{2} & | & 0 & -5 & 0 \\ 0 & 32 & -7 & | & 10 & 40 & 0 \\ 0 & -10 & 10 & | & 0 & -10 & 10 \end{bmatrix} \sim \begin{bmatrix} 1 & -\frac{9}{2} & \frac{1}{2} & | & 0 & -5 & 0 \\ 0 & 32 & -7 & | & 10 & 40 & 0 \\ 0 & 1 & -1 & | & 0 & 1 & -1 \end{bmatrix}$$

$$\left(-\frac{1}{10}\right) R_3 \rightarrow R_3 \qquad\qquad\qquad R_2 \leftrightarrow R_3$$

$$\begin{bmatrix} 1 & -\frac{9}{2} & \frac{1}{2} & | & 0 & -5 & 0 \\ 0 & 1 & -1 & | & 0 & 1 & -1 \\ 0 & 32 & -7 & | & 10 & 40 & 0 \end{bmatrix} \sim \begin{bmatrix} 1 & 0 & -4 & | & 0 & -\frac{1}{2} & -\frac{9}{2} \\ 0 & 1 & -1 & | & 0 & 1 & -1 \\ 0 & 0 & 25 & | & 10 & 8 & 32 \end{bmatrix}$$

$$(-32)R_2 + R_3 \rightarrow R_3 \qquad\qquad \frac{1}{25}R_3 \rightarrow R_3$$
$$\frac{9}{2}R_2 + R_1 \rightarrow R_1$$

$$\sim \begin{bmatrix} 1 & 0 & -4 & | & 0 & -\frac{1}{2} & -\frac{9}{2} \\ 0 & 1 & -1 & | & 0 & 1 & -1 \\ 0 & 0 & 1 & | & 0.4 & 0.32 & 1.28 \end{bmatrix} \sim \begin{bmatrix} 1 & 0 & 0 & | & 1.6 & 0.78 & 0.62 \\ 0 & 1 & 0 & | & 0.4 & 1.32 & 0.28 \\ 0 & 0 & 1 & | & 0.4 & 0.32 & 1.28 \end{bmatrix}$$

$$R_3 + R_2 \rightarrow R_2$$
$$4R_3 + R_1 \rightarrow R_1$$

Thus, $(I - M)^{-1} = \begin{bmatrix} 1.6 & 0.78 & 0.62 \\ 0.4 & 1.32 & 0.28 \\ 0.4 & 0.32 & 1.28 \end{bmatrix}$, and

$$X = (I - M)^{-1}D = \begin{bmatrix} 1.6 & 0.78 & 0.62 \\ 0.4 & 1.32 & 0.28 \\ 0.4 & 0.32 & 1.28 \end{bmatrix} \begin{bmatrix} 10 \\ 15 \\ 20 \end{bmatrix}$$

$$= \begin{bmatrix} (1.6)10 + (0.78)15 + (0.62)20 \\ (0.4)10 + (1.32)15 + (0.28)20 \\ (0.4)10 + (0.32)15 + (1.28)20 \end{bmatrix} = \begin{bmatrix} 40.1 \\ 29.4 \\ 34.4 \end{bmatrix}.$$

Therefore, agriculture, \$40.1 billion; manufacturing, \$29.4 billion; and energy, \$34.4 billion.

29. The technology matrix is $M = \begin{bmatrix} 0.05 & 0.17 & 0.23 & 0.09 \\ 0.07 & 0.12 & 0.15 & 0.19 \\ 0.25 & 0.08 & 0.03 & 0.32 \\ 0.11 & 0.19 & 0.28 & 0.16 \end{bmatrix}.$

The input-output matrix equation is $X = MX + D$

where $X = \begin{bmatrix} A \\ E \\ L \\ M \end{bmatrix}$ and D is the final demand matrix. Thus, $X = (I - M)^{-1}D$, where $I - M = \begin{bmatrix} 0.95 & -0.17 & -0.23 & -0.09 \\ -0.07 & 0.88 & -0.15 & -0.19 \\ -0.25 & -0.08 & 0.97 & -0.32 \\ -0.11 & -0.19 & -0.28 & 0.84 \end{bmatrix}$

Now, $X = \begin{bmatrix} 1.25 & 0.37 & 0.47 & 0.40 \\ 0.26 & 1.33 & 0.41 & 0.48 \\ 0.47 & 0.36 & 1.39 & 0.66 \\ 0.38 & 0.47 & 0.62 & 1.57 \end{bmatrix} D$

Year 1: $D = \begin{bmatrix} 23 \\ 41 \\ 18 \\ 31 \end{bmatrix}$ and $(I - M)^{-1}D \sim \begin{bmatrix} 65 \\ 83 \\ 71 \\ 88 \end{bmatrix}$

Agriculture: \$65 billion; Energy: \$83 billion; Labor: \$71 billion; Manufacturing: \$88 billion

Year 2: $D = \begin{bmatrix} 32 \\ 48 \\ 21 \\ 33 \end{bmatrix}$ and $(I - M)^{-1}D \approx \begin{bmatrix} 81 \\ 97 \\ 83 \\ 99 \end{bmatrix}$

Agriculture: \$81 billion; Energy: \$97 billion; Labor: \$83 billion; Manufacturing: \$99 billion

Year 3: $D = \begin{bmatrix} 55 \\ 62 \\ 25 \\ 35 \end{bmatrix}$ and $(I - M)^{-1}D \sim \begin{bmatrix} 117 \\ 124 \\ 106 \\ 120 \end{bmatrix}$

Agriculture: \$117 billion; Energy: \$124 billion; Labor: \$106 billion; Manufacturing: \$120 billion

CHAPTER 4 REVIEW

1. $y = 2x - 4$ (1)
$y = \frac{1}{2}x + 2$ (2)

The point of intersection is the solution. This is $x = 4, y = 4$.

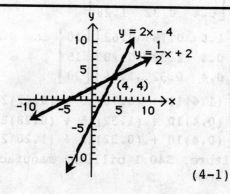

(4-1)

2. Substitute equation (1) into (2):

$2x - 4 = \frac{1}{2}x + 2$

$\frac{3}{2}x = 6$

$x = 4$

Substitute $x = 4$ into (1):

$y = 2 \cdot 4 - 4 = 4$

Solution:
$x = 4, y = 4$ (4-1)

3. $A + B = \begin{bmatrix} 1 + 2 & 2 + 1 \\ 3 + 1 & 1 + 1 \end{bmatrix} = \begin{bmatrix} 3 & 3 \\ 4 & 2 \end{bmatrix}$

(4-4)

4. $B + D = \begin{bmatrix} 2 & 1 \\ 1 & 1 \end{bmatrix} + \begin{bmatrix} 1 \\ 2 \end{bmatrix}$

The matrices B and D cannot be added because their dimensions are different.   (4-4)

5. $A - 2B = \begin{bmatrix} 1 & 2 \\ 3 & 1 \end{bmatrix} - 2\begin{bmatrix} 2 & 1 \\ 1 & 1 \end{bmatrix} = \begin{bmatrix} 1 & 2 \\ 3 & 1 \end{bmatrix} + \begin{bmatrix} -4 & -2 \\ -2 & -2 \end{bmatrix} = \begin{bmatrix} -3 & 0 \\ 1 & -1 \end{bmatrix}$ (4-4)

6. $AB = \begin{bmatrix} 1 & 2 \\ 3 & 1 \end{bmatrix}\begin{bmatrix} 2 & 1 \\ 1 & 1 \end{bmatrix} = \begin{bmatrix} [1 \quad 2]\begin{bmatrix}2\\1\end{bmatrix} & [1 \quad 2]\begin{bmatrix}1\\1\end{bmatrix} \\ [3 \quad 1]\begin{bmatrix}2\\1\end{bmatrix} & [3 \quad 1]\begin{bmatrix}1\\1\end{bmatrix} \end{bmatrix} = \begin{bmatrix} 4 & 3 \\ 7 & 4 \end{bmatrix}$ (4-4)

7. AC is *not defined* because the dimension of A is 2×2 and the dimension of C is 1×2. So, the number of columns in A is not equal to the number of rows in C. (4-4)

8. $AD = \begin{bmatrix} 1 & 2 \\ 3 & 1 \end{bmatrix}\begin{bmatrix} 1 \\ 2 \end{bmatrix} = \begin{bmatrix} [1 \quad 2]\begin{bmatrix}1\\2\end{bmatrix} \\ [3 \quad 1]\begin{bmatrix}1\\2\end{bmatrix} \end{bmatrix} = \begin{bmatrix} 5 \\ 5 \end{bmatrix}$ (4-4)

9. $DC = \begin{bmatrix} 1 \\ 2 \end{bmatrix}[2 \quad 3] = \begin{bmatrix} (1)\cdot(2) & (1)\cdot(3) \\ (2)\cdot(2) & (2)\cdot(3) \end{bmatrix} = \begin{bmatrix} 2 & 3 \\ 4 & 6 \end{bmatrix}$ (4-4)

10. $CD = [2 \quad 3]\begin{bmatrix} 1 \\ 2 \end{bmatrix} = [2 + 6] = [8]$ (4-4)

11. $C + D = [2 \quad 3] + \begin{bmatrix} 1 \\ 2 \end{bmatrix}$

Not defined because the dimensions of C and D are different. (4-4)

12. $\begin{bmatrix} 4 & 3 & | & 1 & 0 \\ 3 & 2 & | & 0 & 1 \end{bmatrix}$ $(-1)R_2 + R_1 \rightarrow R_1$

$\sim \begin{bmatrix} 1 & 1 & | & 1 & -1 \\ 3 & 2 & | & 0 & 1 \end{bmatrix}$ $(-3)R_1 + R_2 \rightarrow R_2$

$\sim \begin{bmatrix} 1 & 1 & | & 1 & -1 \\ 0 & -1 & | & -3 & 4 \end{bmatrix}$ $(-1)R_2 \rightarrow R_2$

$\sim \begin{bmatrix} 1 & 1 & | & 1 & -1 \\ 0 & 1 & | & 3 & -4 \end{bmatrix}$ $(-1)R_2 + R_1 \rightarrow R_1$

$\begin{bmatrix} 1 & 0 & | & -2 & 3 \\ 0 & 1 & | & 3 & -4 \end{bmatrix}$

Thus, $A^{-1} = \begin{bmatrix} -2 & 3 \\ 3 & -4 \end{bmatrix}$ and $A^{-1}A = \begin{bmatrix} -2 & 3 \\ 3 & -4 \end{bmatrix}\begin{bmatrix} 4 & 3 \\ 3 & 2 \end{bmatrix} = \begin{bmatrix} 1 & 0 \\ 0 & 1 \end{bmatrix}$ (4-5)

13. (1) $4x_1 + 3x_2 = 3$ Multiply (1) by 2 and (2) by -3.
   (2) $3x_1 + 2x_2 = 5$

$$8x_1 + 6x_2 = 6$$
$$-9x_1 - 6x_2 = -15$$

Add the two equations.

$$-x_1 = -9$$
$$x_1 = 9$$

Substitute $x_1 = 9$ into either (1) or (2); we choose (2).

$$3(9) + 2x_2 = 5$$
$$27 + 2x_2 = 5$$
$$2x_2 = -22$$
$$x_2 = -11$$

Solution: $x_1 = 9$, $x_2 = -11$ (4-1)

14. The augmented matrix of the system is:

$$\begin{bmatrix} 4 & 3 & | & 3 \\ 3 & 2 & | & 5 \end{bmatrix} \quad (-1)R_2 + R_1 \rightarrow R_1$$

$$\sim \begin{bmatrix} 1 & 1 & | & -2 \\ 3 & 2 & | & 5 \end{bmatrix} \quad (-3)R_1 + R_2 \rightarrow R_2 \sim \begin{bmatrix} 1 & 1 & | & -2 \\ 0 & -1 & | & 11 \end{bmatrix} \quad (-1)R_2 \rightarrow R_2$$

$$\sim \begin{bmatrix} 1 & 1 & | & -2 \\ 0 & 1 & | & -11 \end{bmatrix} \quad (-1)R_2 + R_1 \rightarrow R_1$$

$$\sim \begin{bmatrix} 1 & 0 & | & 9 \\ 0 & 1 & | & -11 \end{bmatrix} \quad \begin{array}{l} \text{The system of equations is:} \\ \quad x_1 = 9 \\ \quad\quad x_2 = -11 \end{array}$$

Solution: $x_1 = 9$, $x_2 = -11$ (4-2)

15. The system of equations in matrix form is:

$$\begin{bmatrix} 4 & 3 \\ 3 & 2 \end{bmatrix} \begin{bmatrix} x_1 \\ x_2 \end{bmatrix} = \begin{bmatrix} 3 \\ 5 \end{bmatrix}$$

Thus, $\begin{bmatrix} x_1 \\ x_2 \end{bmatrix} = \begin{bmatrix} 4 & 3 \\ 3 & 2 \end{bmatrix}^{-1} \begin{bmatrix} 3 \\ 5 \end{bmatrix} = \begin{bmatrix} -2 & 3 \\ 3 & -4 \end{bmatrix} \begin{bmatrix} 3 \\ 5 \end{bmatrix}$ (by Problem 12) $= \begin{bmatrix} 9 \\ -11 \end{bmatrix}$

Solution: $x_1 = 9$, $x_2 = -11$.

Replacing the constants 3, 5 by 7, 10, respectively:

$$\begin{bmatrix} x_1 \\ x_2 \end{bmatrix} = \begin{bmatrix} -2 & 3 \\ 3 & -4 \end{bmatrix} \begin{bmatrix} 7 \\ 10 \end{bmatrix} = \begin{bmatrix} 16 \\ -19 \end{bmatrix}$$

Solution: $x_1 = 16$, $x_2 = -19$

Replacing the constants 3, 5 by 4, 2, respectively:

$$\begin{bmatrix} x_1 \\ x_2 \end{bmatrix} = \begin{bmatrix} -2 & 3 \\ 3 & -4 \end{bmatrix} \begin{bmatrix} 4 \\ 2 \end{bmatrix} = \begin{bmatrix} -2 \\ 4 \end{bmatrix}$$

Solution: $x_1 = -2$, $x_2 = 4$ (4-6)

16. $A + D = \begin{bmatrix} 2 & -2 \\ 1 & 0 \\ 3 & 2 \end{bmatrix} + \begin{bmatrix} 3 & -2 & 1 \\ -1 & 1 & 2 \end{bmatrix}$ Not defined, because the dimensions of A and D are different.   (4-4)

17. $E + DA = \begin{bmatrix} 3 & -4 \\ -1 & 0 \end{bmatrix} + \begin{bmatrix} 3 & -2 & 1 \\ -1 & 1 & 2 \end{bmatrix} \begin{bmatrix} 2 & -2 \\ 1 & 0 \\ 3 & 2 \end{bmatrix} = \begin{bmatrix} 3 & -4 \\ -1 & 0 \end{bmatrix} + \begin{bmatrix} 7 & -4 \\ 5 & 6 \end{bmatrix} = \begin{bmatrix} 10 & -8 \\ 4 & 6 \end{bmatrix}$

(4-4)

18. From Problem 17, $DA = \begin{bmatrix} 7 & -4 \\ 5 & 6 \end{bmatrix}$.   Thus,

$DA - 3E = \begin{bmatrix} 7 & -4 \\ 5 & 6 \end{bmatrix} - 3\begin{bmatrix} 3 & -4 \\ -1 & 0 \end{bmatrix} = \begin{bmatrix} 7 & -4 \\ 5 & 6 \end{bmatrix} + \begin{bmatrix} -9 & 12 \\ 3 & 0 \end{bmatrix} = \begin{bmatrix} -2 & 8 \\ 8 & 6 \end{bmatrix}$ (4-4)

19. $BC = \begin{bmatrix} -1 \\ 2 \\ 3 \end{bmatrix} [2 \quad 1 \quad 3] = \begin{bmatrix} -2 & -1 & -3 \\ 4 & 2 & 6 \\ 6 & 3 & 9 \end{bmatrix}$ (4-4)

20. $CB = [2 \quad 1 \quad 3] \begin{bmatrix} -1 \\ 2 \\ 3 \end{bmatrix} = [-2 + 2 + 9] = [9]$ (a 1×1 matrix)   (4-4)

21. $AD - BC$

$AD = \begin{bmatrix} 2 & -2 \\ 1 & 0 \\ 3 & 2 \end{bmatrix} \begin{bmatrix} 3 & -2 & 1 \\ -1 & 1 & 2 \end{bmatrix} = \begin{bmatrix} [2 \;\; -2]\begin{bmatrix}3\\-1\end{bmatrix} & [2 \;\; -2]\begin{bmatrix}-2\\1\end{bmatrix} & [2 \;\; -2]\begin{bmatrix}1\\2\end{bmatrix} \\ [1 \;\; 0]\begin{bmatrix}3\\-1\end{bmatrix} & [1 \;\; 0]\begin{bmatrix}-2\\1\end{bmatrix} & [1 \;\; 0]\begin{bmatrix}1\\2\end{bmatrix} \\ [3 \;\; 2]\begin{bmatrix}3\\-1\end{bmatrix} & [3 \;\; 2]\begin{bmatrix}-2\\1\end{bmatrix} & [3 \;\; 2]\begin{bmatrix}1\\2\end{bmatrix} \end{bmatrix}$

$= \begin{bmatrix} 8 & -6 & -2 \\ 3 & -2 & 1 \\ 7 & -4 & 7 \end{bmatrix}$

$BC = \begin{bmatrix} -1 \\ 2 \\ 3 \end{bmatrix} [2 \quad 1 \quad 3] = \begin{bmatrix} -2 & -1 & -3 \\ 4 & 2 & 6 \\ 6 & 3 & 9 \end{bmatrix}$

$AD - BC = \begin{bmatrix} 8 & -6 & -2 \\ 3 & -2 & 1 \\ 7 & -4 & 7 \end{bmatrix} - \begin{bmatrix} -2 & -1 & -3 \\ 4 & 2 & 6 \\ 6 & 3 & 9 \end{bmatrix} = \begin{bmatrix} 8-(-2) & -6-(-1) & -2-(-3) \\ 3-4 & -2-2 & 1-6 \\ 7-6 & -4-3 & 7-9 \end{bmatrix} = \begin{bmatrix} 10 & -5 & 1 \\ -1 & -4 & -5 \\ 1 & -7 & -2 \end{bmatrix}$

(4-4)

22. $\begin{bmatrix} 1 & 2 & 3 & | & 1 & 0 & 0 \\ 2 & 3 & 4 & | & 0 & 1 & 0 \\ 1 & 2 & 1 & | & 0 & 0 & 1 \end{bmatrix} \sim \begin{bmatrix} 1 & 2 & 3 & | & 1 & 0 & 0 \\ 0 & -1 & -2 & | & -2 & 1 & 0 \\ 0 & 0 & -2 & | & -1 & 0 & 1 \end{bmatrix} \sim \begin{bmatrix} 1 & 2 & 3 & | & 1 & 0 & 0 \\ 0 & 1 & 2 & | & 2 & -1 & 0 \\ 0 & 0 & -2 & | & -1 & 0 & 1 \end{bmatrix}$

$(-2)R_1 + R_2 \rightarrow R_2$ $\qquad$ $(-1)R_1 \rightarrow R_2$ $\qquad$ $(-2)R_2 + R_1 \rightarrow R_1$

$(-1)R_1 + R_3 \rightarrow R_3$ $\qquad\qquad\qquad\qquad\qquad\qquad\qquad$ $\left(-\dfrac{1}{2}\right)R_3 \rightarrow R_3$

$$\sim \begin{bmatrix} 1 & 0 & -1 \\ 0 & 1 & 2 \\ 0 & 0 & 1 \end{bmatrix} \left.\begin{matrix} -3 & 2 & 0 \\ 2 & -1 & 0 \\ \frac{1}{2} & 0 & -\frac{1}{2} \end{matrix}\right] \sim \begin{bmatrix} 1 & 0 & 0 \\ 0 & 1 & 0 \\ 0 & 0 & 1 \end{bmatrix} \left.\begin{matrix} -\frac{5}{2} & 2 & -\frac{1}{2} \\ 1 & -1 & 1 \\ \frac{1}{2} & 0 & -\frac{1}{2} \end{matrix}\right]; A^{-1} = \begin{bmatrix} -\frac{5}{2} & 2 & -\frac{1}{2} \\ 1 & -1 & 1 \\ \frac{1}{2} & 0 & -\frac{1}{2} \end{bmatrix}$$

$$R_3 + R_1 \to R_1$$
$$(-2)R_3 + R_2 \to R_2$$

Check:

$$A^{-1}A = \begin{bmatrix} -\frac{5}{2} & 2 & -\frac{1}{2} \\ 1 & -1 & 1 \\ \frac{1}{2} & 0 & -\frac{1}{2} \end{bmatrix} \begin{bmatrix} 1 & 2 & 3 \\ 2 & 3 & 4 \\ 1 & 2 & 1 \end{bmatrix} = \begin{bmatrix} -\frac{5}{2} + 4 - \frac{1}{2} & -5 + 6 - 1 & -\frac{15}{2} + 8 - \frac{1}{2} \\ 1 - 2 + 1 & 2 - 3 + 2 & 3 - 4 + 1 \\ \frac{1}{2} + 0 - \frac{1}{2} & 1 + 0 - 1 & \frac{3}{2} + 0 - \frac{1}{2} \end{bmatrix}$$

$$= \begin{bmatrix} 1 & 0 & 0 \\ 0 & 1 & 0 \\ 0 & 0 & 1 \end{bmatrix} \tag{4-5}$$

23. (A) The augmented matrix corresponding to the given system is:

$$\begin{bmatrix} 1 & 2 & 3 \\ 2 & 3 & 4 \\ 1 & 2 & 1 \end{bmatrix}\left.\begin{matrix} 1 \\ 3 \\ 3 \end{matrix}\right] \sim \begin{bmatrix} 1 & 2 & 3 \\ 0 & -1 & -2 \\ 0 & 0 & -2 \end{bmatrix}\left.\begin{matrix} 1 \\ 1 \\ 2 \end{matrix}\right] \sim \begin{bmatrix} 1 & 2 & 3 \\ 0 & 1 & 2 \\ 0 & 0 & -2 \end{bmatrix}\left.\begin{matrix} 1 \\ -1 \\ 2 \end{matrix}\right] \sim \begin{bmatrix} 1 & 0 & -1 \\ 0 & 1 & 2 \\ 0 & 0 & -2 \end{bmatrix}\left.\begin{matrix} 3 \\ -1 \\ 2 \end{matrix}\right]$$

$$(-2)R_1 + R_2 \to R_2 \qquad (-1)R_2 \to R_2 \qquad (-2)R_2 + R_1 \to R_1 \qquad \left(-\frac{1}{2}\right)R_3 \to R_3$$
$$(-1)R_1 + R_3 \to R_3$$

$$\sim \begin{bmatrix} 1 & 0 & -1 \\ 0 & 1 & 2 \\ 0 & 0 & 1 \end{bmatrix}\left.\begin{matrix} 3 \\ -1 \\ -1 \end{matrix}\right] \sim \begin{bmatrix} 1 & 0 & 0 \\ 0 & 1 & 0 \\ 0 & 0 & 1 \end{bmatrix}\left.\begin{matrix} 2 \\ 1 \\ -1 \end{matrix}\right]$$ Thus, the solution is: $x_1 = 2$
$$x_2 = 1$$
$$x_3 = -1.$$

$$(-2)R_3 + R_2 \to R_2$$
$$R_3 + R_1 \to R_1$$

$$\tag{4-3}$$

(B) The augmented matrix corresponding to the given system is:

$$\begin{bmatrix} 1 & 2 & -1 \\ 2 & 3 & 1 \\ 3 & 5 & 0 \end{bmatrix}\left.\begin{matrix} 2 \\ -3 \\ -1 \end{matrix}\right] \sim \begin{bmatrix} 1 & 2 & -1 \\ 0 & -1 & 3 \\ 0 & -1 & 3 \end{bmatrix}\left.\begin{matrix} 2 \\ -7 \\ -7 \end{matrix}\right] \sim \begin{bmatrix} 1 & 2 & -1 \\ 0 & 1 & -3 \\ 0 & -1 & 3 \end{bmatrix}\left.\begin{matrix} 2 \\ 7 \\ -7 \end{matrix}\right] \sim \begin{bmatrix} 1 & 0 & 5 \\ 0 & 1 & -3 \\ 0 & 0 & 0 \end{bmatrix}\left.\begin{matrix} -12 \\ 7 \\ 0 \end{matrix}\right]$$

$$(-2)R_1 + R_2 \to R_2 \qquad (-1)R_2 \to R_2 \qquad R_2 + R_3 \to R_3$$
$$(-3)R_1 + R_3 \to R_3 \qquad\qquad\qquad (-2)R_2 + R_1 \to R_1$$

Thus, $x_1 \quad + 5x_3 = -12$ (1)
$$x_2 - 3x_3 = 7 \quad (2)$$

Let $x_3 = t$ (t any real number). Then, from (1),
$x_1 = -5t - 12$
and, from (2),
$x_2 = 3t + 7$.
Thus, the solution is $x_1 = -5t - 12$, $x_2 = 3t + 7$, $x_3 = t$. $\qquad$ (4-3)

24. (A) The matrix equation for the given system is:

$$\begin{bmatrix} 1 & 2 & 3 \\ 2 & 3 & 4 \\ 1 & 2 & 1 \end{bmatrix} \begin{bmatrix} x_1 \\ x_2 \\ x_3 \end{bmatrix} = \begin{bmatrix} 1 \\ 3 \\ 3 \end{bmatrix}$$

The inverse matrix of the coefficient matrix of the system, from Problem 22, is:

$$\begin{bmatrix} -\frac{5}{2} & 2 & -\frac{1}{2} \\ 1 & -1 & 1 \\ \frac{1}{2} & 0 & -\frac{1}{2} \end{bmatrix} \quad \text{Thus,}$$

$$\begin{bmatrix} x_1 \\ x_2 \\ x_3 \end{bmatrix} = \begin{bmatrix} -\frac{5}{2} & 2 & -\frac{1}{2} \\ 1 & -1 & 1 \\ \frac{1}{2} & 0 & -\frac{1}{2} \end{bmatrix} \begin{bmatrix} 1 \\ 3 \\ 3 \end{bmatrix} = \begin{bmatrix} \frac{-5+12-3}{2} \\ 1-3+3 \\ \frac{1+0-3}{2} \end{bmatrix} = \begin{bmatrix} 2 \\ 1 \\ -1 \end{bmatrix} \qquad \text{Solution: } x_1 = 2, \\ x_2 = 1, \\ x_3 = -1.$$

(B) $$\begin{bmatrix} x_1 \\ x_2 \\ x_3 \end{bmatrix} = \begin{bmatrix} -\frac{5}{2} & 2 & -\frac{1}{2} \\ 1 & -1 & 1 \\ \frac{1}{2} & 0 & -\frac{1}{2} \end{bmatrix} \begin{bmatrix} 0 \\ 0 \\ -2 \end{bmatrix} = \begin{bmatrix} 1 \\ -2 \\ 1 \end{bmatrix} \qquad \text{Solution: } x_1 = 1, \\ x_2 = -2, \\ x_3 = 1.$$

(C) $$\begin{bmatrix} x_1 \\ x_2 \\ x_3 \end{bmatrix} = \begin{bmatrix} -\frac{5}{2} & 2 & -\frac{1}{2} \\ 1 & -1 & 1 \\ \frac{1}{2} & 0 & -\frac{1}{2} \end{bmatrix} \begin{bmatrix} -3 \\ -4 \\ 1 \end{bmatrix} = \begin{bmatrix} -1 \\ 2 \\ -2 \end{bmatrix} \qquad \text{Solution: } x_1 = -1, \\ x_2 = 2, \\ x_3 = -2.$$ (4-6)

25. $M = \begin{bmatrix} 0.2 & 0.15 \\ 0.4 & 0.3 \end{bmatrix}$, $D = \begin{bmatrix} 30 \\ 20 \end{bmatrix}$

$$I - M = \begin{bmatrix} 1 & 0 \\ 0 & 1 \end{bmatrix} - \begin{bmatrix} 0.2 & 0.15 \\ 0.4 & 0.3 \end{bmatrix} = \begin{bmatrix} 0.8 & -0.15 \\ -0.4 & 0.7 \end{bmatrix} = \begin{bmatrix} \frac{4}{5} & -\frac{3}{20} \\ -\frac{2}{5} & \frac{7}{10} \end{bmatrix}$$

Now,

$$\left[\begin{array}{cc|cc} \frac{4}{5} & -\frac{3}{20} & 1 & 0 \\ -\frac{2}{5} & \frac{7}{10} & 0 & 1 \end{array}\right] \sim \left[\begin{array}{cc|cc} 1 & -\frac{3}{16} & \frac{5}{4} & 0 \\ -\frac{2}{5} & \frac{7}{10} & 0 & 1 \end{array}\right] \sim \left[\begin{array}{cc|cc} 1 & -\frac{3}{16} & \frac{5}{4} & 0 \\ 0 & \frac{5}{8} & \frac{1}{2} & 1 \end{array}\right] \sim \left[\begin{array}{cc|cc} 1 & -\frac{3}{16} & \frac{5}{4} & 0 \\ 0 & 1 & \frac{4}{5} & \frac{8}{5} \end{array}\right]$$

$$\frac{5}{4}R_1 \to R_1 \qquad \frac{2}{5}R_1 + R_2 \to R_2 \qquad \frac{8}{5}R_2 \to R_2 \qquad \frac{3}{16}R_2 + R_1 \to R_1$$

$$\sim \left[\begin{array}{cc|cc} 1 & 0 & \frac{7}{5} & \frac{3}{10} \\ 0 & 1 & \frac{4}{5} & \frac{8}{5} \end{array}\right] \quad \text{Thus, } (I - M)^{-1} = \begin{bmatrix} \frac{7}{5} & \frac{3}{10} \\ \frac{4}{5} & \frac{8}{5} \end{bmatrix} = \begin{bmatrix} 1.4 & 0.3 \\ 0.8 & 1.6 \end{bmatrix}$$

The output matrix $X = (I - M)^{-1}D = \begin{bmatrix} 1.4 & 0.3 \\ 0.8 & 1.6 \end{bmatrix} \begin{bmatrix} 30 \\ 20 \end{bmatrix} = \begin{bmatrix} 48 \\ 56 \end{bmatrix}$ (4-7)

26. The graphs of the two equations are:

$$x \approx 3.46, \quad y \approx 1.69 \qquad (4\text{-}1)$$

27.
$$\left[\begin{array}{ccc|ccc} 4 & 5 & 6 & 1 & 0 & 0 \\ 4 & 5 & -4 & 0 & 1 & 0 \\ 1 & -1 & 1 & 0 & 0 & 1 \end{array}\right] R_1 \leftrightarrow R_3$$

$$\sim \left[\begin{array}{ccc|ccc} 1 & 1 & 1 & 0 & 0 & 1 \\ 4 & 5 & -4 & 0 & 1 & 0 \\ 4 & 5 & 6 & 1 & 0 & 0 \end{array}\right] \begin{array}{l}(-4)R_1 + R_2 \rightarrow R_2 \\ (-4)R_1 + R_3 \rightarrow R_3\end{array}$$

$$\sim \left[\begin{array}{ccc|ccc} 1 & 1 & 1 & 0 & 0 & 1 \\ 0 & 1 & -8 & 0 & 1 & -4 \\ 0 & 1 & 2 & 1 & 0 & -4 \end{array}\right] \begin{array}{l}(-1)R_2 + R_1 \rightarrow R_1 \\ (-1)R_2 + R_3 \rightarrow R_3\end{array}$$

$$\sim \left[\begin{array}{ccc|ccc} 1 & 0 & 9 & 0 & -1 & 5 \\ 0 & 1 & -8 & 0 & 1 & -4 \\ 0 & 0 & 10 & 1 & -1 & 0 \end{array}\right] \tfrac{1}{10}R_3 \rightarrow R_3$$

$$\sim \left[\begin{array}{ccc|ccc} 1 & 0 & 9 & 0 & -1 & 5 \\ 0 & 1 & -8 & 0 & 1 & -4 \\ 0 & 0 & 1 & \frac{1}{10} & -\frac{1}{10} & 0 \end{array}\right] \begin{array}{l}(-9)R_3 + R_1 \rightarrow R_1 \\ 8R_3 + R_2 \rightarrow R_2\end{array}$$

$$\sim \left[\begin{array}{ccc|ccc} 1 & 0 & 0 & -\frac{9}{10} & -\frac{1}{10} & 5 \\ 0 & 1 & 0 & \frac{8}{10} & \frac{2}{10} & -4 \\ 0 & 0 & 1 & \frac{1}{10} & -\frac{1}{10} & 0 \end{array}\right]$$

Thus, $A^{-1} = \begin{bmatrix} -0.9 & -0.1 & 5 \\ 0.8 & 0.2 & -4 \\ 0.1 & -0.1 & 0 \end{bmatrix}$;

$$A^{-1}A = \begin{bmatrix} -0.9 & -0.1 & 5 \\ 0.8 & 0.2 & -4 \\ 0.1 & -0.1 & 0 \end{bmatrix}\begin{bmatrix} 4 & 5 & 6 \\ 4 & 5 & -4 \\ 1 & 1 & 1 \end{bmatrix} = \begin{bmatrix} 1 & 0 & 0 \\ 0 & 1 & 0 \\ 0 & 0 & 1 \end{bmatrix} \qquad (4\text{-}5)$$

28. The given system is equivalent to:

$$4x_1 + 5x_2 + 6x_3 = 36,000$$
$$4x_1 + 5x_2 - 4x_3 = 12,000$$
$$x_1 + x_2 + x_3 = 7,000$$

In matrix form, this system is:

$$\begin{bmatrix} 4 & 5 & 6 \\ 4 & 5 & -4 \\ 1 & 1 & 1 \end{bmatrix} \begin{bmatrix} x_1 \\ x_2 \\ x_3 \end{bmatrix} = \begin{bmatrix} 36,000 \\ 12,000 \\ 7,000 \end{bmatrix}$$

Thus,

$$\begin{bmatrix} x_1 \\ x_2 \\ x_3 \end{bmatrix} = \begin{bmatrix} 4 & 5 & 6 \\ 4 & 5 & -4 \\ 1 & 1 & 1 \end{bmatrix}^{-1} \begin{bmatrix} 36,000 \\ 12,000 \\ 7,000 \end{bmatrix} = \begin{bmatrix} -0.9 & -0.1 & 5 \\ 0.8 & 0.2 & -4 \\ 0.1 & -0.1 & 0 \end{bmatrix} \begin{bmatrix} 36,000 \\ 12,000 \\ 7,000 \end{bmatrix} = \begin{bmatrix} 1,400 \\ 3,200 \\ 2,400 \end{bmatrix}$$

Solution: $x_1 = 1,400$, $x_2 = 3,200$, $x_3 = 2,400$　　　　　(4-6)

29. First, multiply the first two equations of the system by 100. Then the augmented matrix of the resulting system is:

$$\begin{bmatrix} 4 & 5 & 6 & | & 36,000 \\ 4 & 5 & -4 & | & 12,000 \\ 1 & 1 & 1 & | & 7,000 \end{bmatrix} \quad R_1 \leftrightarrow R_3$$

$$\sim \begin{bmatrix} 1 & 1 & 1 & | & 7,000 \\ 4 & 5 & -4 & | & 12,000 \\ 4 & 5 & 6 & | & 36,000 \end{bmatrix} \quad \begin{matrix} (-4)R_1 + R_2 \rightarrow R_2 \\ (-4)R_1 + R_3 \rightarrow R_3 \end{matrix}$$

$$\sim \begin{bmatrix} 1 & 1 & 1 & | & 7,000 \\ 0 & 1 & -8 & | & -16,000 \\ 0 & 1 & 2 & | & 8,000 \end{bmatrix} \quad \begin{matrix} (-1)R_2 + R_1 \rightarrow R_1 \\ (-1)R_2 + R_3 \rightarrow R_3 \end{matrix}$$

$$\sim \begin{bmatrix} 1 & 0 & 9 & | & 23,000 \\ 0 & 1 & -8 & | & -16,000 \\ 0 & 0 & 10 & | & 24,000 \end{bmatrix} \quad \frac{1}{10}R_3 \rightarrow R_3$$

$$\sim \begin{bmatrix} 1 & 0 & 9 & | & 23,000 \\ 0 & 1 & -8 & | & -16,000 \\ 0 & 0 & 1 & | & 2,400 \end{bmatrix} \quad \begin{matrix} (-9)R_3 + R_1 \rightarrow R_1 \\ 8R_3 + R_2 \rightarrow R_2 \end{matrix}$$

$$\begin{bmatrix} 1 & 0 & 0 & | & 1,400 \\ 0 & 1 & 0 & | & 3,200 \\ 0 & 0 & 1 & | & 2,400 \end{bmatrix} \quad \begin{matrix} x_1 = 1,400 \\ x_2 = 3,200 \\ x_3 = 2,400 \end{matrix}$$

Solution: $x_1 = 1,400$, $x_2 = 3,200$, $x_3 = 2,400$　　　　　(4-3)

30. $M = \begin{bmatrix} 0.2 & 0 & 0.4 \\ 0.1 & 0.3 & 0.1 \\ 0 & 0.4 & 0.2 \end{bmatrix} = \begin{bmatrix} \frac{1}{5} & 0 & \frac{2}{5} \\ \frac{1}{10} & \frac{3}{10} & \frac{1}{10} \\ 0 & \frac{2}{5} & \frac{1}{5} \end{bmatrix}$ and $D = \begin{bmatrix} 40 \\ 20 \\ 30 \end{bmatrix}$

$$I - M = \begin{bmatrix} 1 & 0 & 0 \\ 0 & 1 & 0 \\ 0 & 0 & 1 \end{bmatrix} - \begin{bmatrix} \frac{1}{5} & 0 & \frac{2}{5} \\ \frac{1}{10} & \frac{3}{10} & \frac{1}{10} \\ 0 & \frac{2}{5} & \frac{1}{5} \end{bmatrix} = \begin{bmatrix} \frac{4}{5} & 0 & -\frac{2}{5} \\ -\frac{1}{10} & \frac{7}{10} & -\frac{1}{10} \\ 0 & -\frac{2}{5} & \frac{4}{5} \end{bmatrix} .$$

$$\left[\begin{array}{ccc|ccc} \frac{4}{5} & 0 & -\frac{2}{5} & 1 & 0 & 0 \\ -\frac{1}{10} & \frac{7}{10} & -\frac{1}{10} & 0 & 1 & 0 \\ 0 & -\frac{2}{5} & \frac{4}{5} & 0 & 0 & 1 \end{array} \right] \sim \left[\begin{array}{ccc|ccc} 1 & 0 & -\frac{1}{2} & \frac{5}{4} & 0 & 0 \\ -\frac{1}{10} & \frac{7}{10} & -\frac{1}{10} & 0 & 1 & 0 \\ 0 & -\frac{2}{5} & \frac{4}{5} & 0 & 0 & 1 \end{array} \right]$$

$$\frac{5}{4} R_1 \rightarrow R_1 \qquad\qquad\qquad \frac{1}{10} R_1 + R_2 \rightarrow R_2$$

$$\sim \left[\begin{array}{ccc|ccc} 1 & 0 & -\frac{1}{2} & \frac{5}{4} & 0 & 0 \\ 0 & \frac{7}{10} & -\frac{3}{20} & \frac{1}{8} & 1 & 0 \\ 0 & -\frac{2}{5} & \frac{4}{5} & 0 & 0 & 1 \end{array} \right] \sim \left[\begin{array}{ccc|ccc} 1 & 0 & -\frac{1}{2} & \frac{5}{4} & 0 & 0 \\ 0 & -\frac{2}{5} & \frac{4}{5} & 0 & 0 & 1 \\ 0 & \frac{7}{10} & -\frac{3}{20} & \frac{1}{8} & 1 & 0 \end{array} \right]$$

$$R_2 \leftrightarrow R_3 \qquad\qquad\qquad \left(-\frac{5}{2} \right) R_2 \rightarrow R_2$$

$$\sim \left[\begin{array}{ccc|ccc} 1 & 0 & -\frac{1}{2} & \frac{5}{4} & 0 & 0 \\ 0 & 1 & -2 & 0 & 0 & -\frac{5}{2} \\ 0 & \frac{7}{10} & -\frac{3}{20} & \frac{1}{8} & 1 & 0 \end{array} \right] \sim \left[\begin{array}{ccc|ccc} 1 & 0 & -\frac{1}{2} & \frac{5}{4} & 0 & 0 \\ 0 & 1 & -2 & 0 & 0 & -\frac{5}{2} \\ 0 & 0 & \frac{5}{4} & \frac{1}{8} & 1 & \frac{7}{4} \end{array} \right]$$

$$\left(-\frac{7}{10} \right) R_2 + R_3 \rightarrow R_3 \qquad\qquad \frac{4}{5} R_3 \rightarrow R_3$$

$$\sim \left[\begin{array}{ccc|ccc} 1 & 0 & -\frac{1}{2} & \frac{5}{4} & 0 & 0 \\ 0 & 1 & -2 & 0 & 0 & -\frac{5}{2} \\ 0 & 0 & 1 & \frac{1}{10} & \frac{4}{5} & \frac{7}{5} \end{array} \right] \sim \left[\begin{array}{ccc|ccc} 1 & 0 & 0 & \frac{13}{10} & \frac{2}{5} & \frac{7}{10} \\ 0 & 1 & 0 & \frac{1}{5} & \frac{8}{5} & \frac{3}{10} \\ 0 & 0 & 1 & \frac{1}{10} & \frac{4}{5} & \frac{7}{5} \end{array} \right]$$

$$2R_3 + R_2 \rightarrow R_2, \ \ \frac{1}{2} R_3 + R_1 \rightarrow R_1$$

Thus $(I - M)^{-1} = \begin{bmatrix} \frac{13}{10} & \frac{2}{5} & \frac{7}{10} \\ \frac{1}{5} & \frac{8}{5} & \frac{3}{10} \\ \frac{1}{10} & \frac{4}{5} & \frac{7}{5} \end{bmatrix} = \begin{bmatrix} 1.3 & 0.4 & 0.7 \\ 0.2 & 1.6 & 0.3 \\ 0.1 & 0.8 & 1.4 \end{bmatrix}$ and

$$X = (I - M)^{-1}D = \begin{bmatrix} 1.3 & 0.4 & 0.7 \\ 0.2 & 1.6 & 0.3 \\ 0.1 & 0.8 & 1.4 \end{bmatrix} \begin{bmatrix} 40 \\ 20 \\ 30 \end{bmatrix} = \begin{bmatrix} 81 \\ 49 \\ 62 \end{bmatrix} \qquad (4-7)$$

31. (A) The system has a unique solution.

    (B) The system either has no solutions or infinitely many solutions.

    $\qquad\qquad\qquad\qquad\qquad\qquad\qquad\qquad\qquad\qquad\qquad\qquad\qquad$ (4-6)

32. (A) The system has a unique solution.

    (B) The system has **no** solutions.

    (C) The system has infinitely many solutions. $\qquad\qquad\qquad\qquad$ (4-3)

33. The third step in (A) is incorrect:
    $$X - MX = (I - M)X \quad \textbf{not} \quad X(I - M)$$
    Each step in (B) is correct.

    $\qquad\qquad\qquad\qquad\qquad\qquad\qquad\qquad\qquad\qquad\qquad\qquad\qquad$ (4-6)

34. Let x = the number of machines produced.

    (A) $C(x) = 243,000 + 22.45x$
        $R(x) = 59.95x$

    (B) Set $C(x) = R(x)$:
    $$\begin{aligned} 59.95x &= 243,000 + 22.45x \\ 37.5x &= 243,000 \\ x &= 6,480 \end{aligned}$$

    If 6,480 machines are produced, $C = R = \$388,476$;
    break-even point (6,480, 388,476).

    (C) A profit occurs if $x > 6,480$; a loss occurs if $x < 6,480$

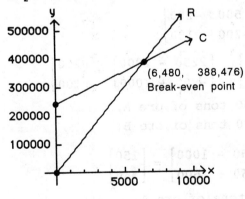

$\qquad\qquad\qquad\qquad\qquad\qquad\qquad\qquad\qquad\qquad\qquad\qquad\qquad\qquad$ (4-1)

35. Let x_1 = number of tons of ore A
and x_2 = number of tons of ore B.

Then, we have the following system of equations:

$0.01x_1 + 0.02x_2 = 4.5$
$0.02x_1 + 0.05x_2 = 10$

Multiply each equation by 100. This yields

$x_1 + 2x_2 = 450$
$2x_1 + 5x_2 = 1000$

The augmented matrix corresponding to this system is:

$$\begin{bmatrix} 1 & 2 & | & 450 \\ 2 & 5 & | & 1000 \end{bmatrix} \sim \begin{bmatrix} 1 & 2 & | & 450 \\ 0 & 1 & | & 100 \end{bmatrix} \sim \begin{bmatrix} 1 & 0 & | & 250 \\ 0 & 1 & | & 100 \end{bmatrix}$$

$(-2)R_1 + R_2 \rightarrow R_2 \quad (-2)R_2 + R_1 \rightarrow R_1$

Thus, the solution is: x_1 = 250 tons of ore A, x_2 = 100 tons of ore B.

(4-3)

36. (A) The matrix equation for Problem 30 is:

$$\begin{bmatrix} 0.01 & 0.02 \\ 0.02 & 0.05 \end{bmatrix} \begin{bmatrix} x_1 \\ x_2 \end{bmatrix} = \begin{bmatrix} 4.5 \\ 10 \end{bmatrix}$$

First, compute the inverse of $\begin{bmatrix} 0.01 & 0.02 \\ 0.02 & 0.05 \end{bmatrix}$;

$$\begin{bmatrix} 0.01 & 0.02 & | & 1 & 0 \\ 0.02 & 0.05 & | & 0 & 1 \end{bmatrix} \sim \begin{bmatrix} 1 & 2 & | & 100 & 0 \\ 0.02 & 0.05 & | & 0 & 1 \end{bmatrix} \sim \begin{bmatrix} 1 & 2 & | & 100 & 0 \\ 0 & 0.01 & | & -2 & 1 \end{bmatrix}$$

$100R_1 \rightarrow R_1 \qquad\quad (-0.02)R_1 + R_2 \rightarrow R_2 \qquad\quad 100R_2 \rightarrow R_2$

$$\sim \begin{bmatrix} 1 & 2 & | & 100 & 0 \\ 0 & 1 & | & -200 & 100 \end{bmatrix} \sim \begin{bmatrix} 1 & 0 & | & 500 & -200 \\ 0 & 1 & | & -200 & 100 \end{bmatrix}$$

$(-2)R_2 + R_1 \rightarrow R_1$

Thus, the inverse matrix is $\begin{bmatrix} 500 & -200 \\ -200 & 100 \end{bmatrix}$.

Hence, $\begin{bmatrix} x_1 \\ x_2 \end{bmatrix} = \begin{bmatrix} 500 & -200 \\ -200 & 100 \end{bmatrix} \begin{bmatrix} 4.5 \\ 10 \end{bmatrix} = \begin{bmatrix} 2250 - 2000 \\ -900 + 1000 \end{bmatrix} = \begin{bmatrix} 250 \\ 100 \end{bmatrix}$

Again the solution is: x_1 = 250 tons of ore A.
x_2 = 100 tons of ore B.

(B) $\begin{bmatrix} x_1 \\ x_2 \end{bmatrix} = \begin{bmatrix} 500 & -200 \\ -200 & 100 \end{bmatrix} \begin{bmatrix} 2.3 \\ 5 \end{bmatrix} = \begin{bmatrix} 1150 - 1000 \\ -460 + 500 \end{bmatrix} = \begin{bmatrix} 150 \\ 40 \end{bmatrix}$

Now the solution is: x_1 = 150 tons of ore A.
x_2 = 40 tons of ore B.

(4-6)

37. Let x_1 = number of model A trucks
x_2 = number of model B trucks
x_3 = number of model C trucks

Then $x_1 + x_2 + x_3 = 12$
and $18{,}000x_1 + 22{,}000x_2 + 30{,}000x_3 = 300{,}000$
or $x_1 + x_2 + x_3 = 12$
$9x_1 + 11x_2 + 15x_3 = 150$

The augmented matrix corresponding to this system is $\begin{bmatrix} 1 & 1 & 1 & | & 12 \\ 9 & 11 & 15 & | & 150 \end{bmatrix}$

Now

$$\begin{bmatrix} 1 & 1 & 1 & | & 12 \\ 9 & 11 & 15 & | & 150 \end{bmatrix} \sim \begin{bmatrix} 1 & 1 & 1 & | & 12 \\ 0 & 2 & 6 & | & 42 \end{bmatrix} \sim \begin{bmatrix} 1 & 1 & 1 & | & 12 \\ 0 & 1 & 3 & | & 21 \end{bmatrix} \sim \begin{bmatrix} 1 & 0 & -2 & | & -9 \\ 0 & 1 & 3 & | & 21 \end{bmatrix}$$

$(-9)R_1 + R_2 \rightarrow R_2 \qquad \frac{1}{2}R_2 \rightarrow R_2 \qquad (-1)R_2 + R_1 \rightarrow R_1$

The corresponding system of equations is

$\begin{aligned} x_1 \qquad - 2x_3 &= -9 \\ x_2 + 3x_3 &= 21 \end{aligned}$ and the solutions are $\begin{aligned} x_1 &= 2t - 9 \\ x_2 &= 21 - 3t \\ x_3 &= t \end{aligned}$

Now, since x_1, x_2, and x_3 are nonnegative integers, we must have

$\frac{9}{2} \le t \le 7$ or $t = 5$, 6, or 7.

For $t = 5$: 1 model A truck, 6 model B trucks, 5 model C trucks
$t = 6$: 3 model A trucks, 3 model B trucks, 6 model C trucks
$t = 7$: 5 model A trucks, 0 model B trucks, 7 model C trucks   (4-3)

38. (A) The elements of MN give the cost of materials for each alloy from each supplier. The product NM is also defined, but does not have an interpretation in the context of this problem.

(B) $MN = \begin{bmatrix} 4{,}800 & 600 & 300 \\ 6{,}000 & 1{,}400 & 700 \end{bmatrix} \begin{bmatrix} 0.75 & 0.70 \\ 6.50 & 6.70 \\ 0.40 & 0.50 \end{bmatrix}$

$= \begin{bmatrix} \$ 7{,}620 & \$ 7{,}530 \\ \$13{,}880 & \$13{,}930 \end{bmatrix} \begin{matrix} \text{Alloy 1} \\ \text{Alloy 2} \end{matrix}$

$\qquad$ Supplier A $\quad$ Supplier B

(C) The total costs of materials from Supplier A is:
$\$7{,}620 + \$13{,}880 = \$21{,}500$

The total costs of materials from Supplier B is:
$\$7{,}530 + \$13{,}930 = \$21{,}460$

These values can be obtained from the matrix product

$[1 \quad 1] \begin{bmatrix} 7{,}620 & 7{,}530 \\ 13{,}880 & 13{,}930 \end{bmatrix}$

Supplier B will provide the materials at lower cost.   (4-4)

39. (A) $[0.25 \quad 0.20 \quad 0.05] \begin{bmatrix} 15 \\ 12 \\ 4 \end{bmatrix} = 6.35$

The labor cost for one model B calculator at the California plant is $6.35.

(B) The elements of MN give the total labor costs for each calculator at each plant. The product NM is also defined, but does not have an interpretation in the context of this problem.

(C) $MN = \begin{bmatrix} 0.15 & 0.10 & 0.05 \\ 0.25 & 0.20 & 0.05 \end{bmatrix} \begin{bmatrix} 15 & 12 \\ 12 & 10 \\ 4 & 4 \end{bmatrix} = \begin{matrix} \text{Calif.} \quad \text{Texas} \\ \begin{bmatrix} \$3.65 & \$3.00 \\ \$6.35 & \$5.20 \end{bmatrix} \begin{matrix} \text{Model A} \\ \text{Model B} \end{matrix} \end{matrix}$ (4-4)

40. Let x_1 = amount invested at 5%
and x_2 = amount invested at 10%.

Then, $\quad x_1 + \quad x_2 = 5000$
$\quad 0.05x_1 + 0.1x_2 = \quad 400$

The augmented matrix for the system given above is:

$\begin{bmatrix} 1 & 1 & | & 5000 \\ 0.05 & 0.1 & | & 400 \end{bmatrix} \sim \begin{bmatrix} 1 & 1 & | & 5000 \\ 0 & 0.05 & | & 150 \end{bmatrix} \sim \begin{bmatrix} 1 & 1 & | & 5000 \\ 0 & 1 & | & 3000 \end{bmatrix} \sim \begin{bmatrix} 1 & 0 & | & 2000 \\ 0 & 1 & | & 3000 \end{bmatrix}$

$(-0.05)R_1 + R_2 \rightarrow R_2 \qquad \dfrac{1}{0.05} R_2 \rightarrow R_2 \qquad (-1)R_2 + R_1 \rightarrow R_1$

Hence, $x_1 = \$2000$ at 5%, $x_2 = \$3000$ at 10%. (4-3)

41. The matrix equation corresponding to the system in Problem 35 is:

$\begin{bmatrix} 1 & 1 \\ 0.05 & 0.1 \end{bmatrix} \begin{bmatrix} x_1 \\ x_2 \end{bmatrix} = \begin{bmatrix} 5000 \\ 400 \end{bmatrix}$

Now we compute the inverse matrix of $\begin{bmatrix} 1 & 1 \\ 0.05 & 0.1 \end{bmatrix}$.

$\begin{bmatrix} 1 & 1 & | & 1 & 0 \\ 0.05 & 0.1 & | & 0 & 1 \end{bmatrix} \sim \begin{bmatrix} 1 & 1 & | & 1 & 0 \\ 0 & 0.05 & | & -0.05 & 1 \end{bmatrix} \sim \begin{bmatrix} 1 & 1 & | & 1 & 0 \\ 0 & 1 & | & -1 & 20 \end{bmatrix} \sim \begin{bmatrix} 1 & 0 & | & 2 & -20 \\ 0 & 1 & | & -1 & 20 \end{bmatrix}$

$(-0.05)R_1 + R_2 \rightarrow R_2 \qquad \dfrac{1}{0.05} R_2 \rightarrow R_2 \qquad (-1)R_2 + R_1 \rightarrow R_1$

Thus, the inverse of the coefficient matrix is $\begin{bmatrix} 2 & -20 \\ -1 & 20 \end{bmatrix}$, and

$\begin{bmatrix} x_1 \\ x_2 \end{bmatrix} = \begin{bmatrix} 2 & -20 \\ -1 & 20 \end{bmatrix} \begin{bmatrix} 5000 \\ 400 \end{bmatrix} = \begin{bmatrix} 10,000 - 8,000 \\ -5,000 + 8,000 \end{bmatrix} = \begin{bmatrix} 2000 \\ 3000 \end{bmatrix}.$ So, $x_1 = \$2000$ at 5%, $x_2 = \$3000$ at 10%.

(4-6)

42. Let x_1 = number of \$8 tickets
x_2 = number of \$12 tickets
x_3 = number of \$20 tickets

Since the number of \$8 tickets must equal the number of \$20 tickets, we have
$$x_1 = x_3 \quad \text{or} \quad x_1 - x_3 = 0$$
Also, since all seats are sold
$$x_1 + x_2 + x_3 = 25,000$$
Finally, the return is
$$8x_1 + 12x_2 + 20x_3 = R \quad \text{(where } R \text{ is the return required)}.$$
Thus, the system of equations is:

$$\begin{array}{rcrcrcl} x_1 & & & - & x_3 & = & 0 \\ x_1 & + & x_2 & + & x_3 & = & 25,000 \\ 8x_1 & + & 12x_2 & + & 20x_3 & = & R \end{array} \quad \text{or} \quad \begin{bmatrix} 1 & 0 & -1 \\ 1 & 1 & 1 \\ 8 & 12 & 20 \end{bmatrix} \begin{bmatrix} x_1 \\ x_2 \\ x_3 \end{bmatrix} = \begin{bmatrix} 0 \\ 25,000 \\ R \end{bmatrix}$$

First, we compute the inverse of the coefficient matrix

$$\left[\begin{array}{ccc|ccc} 1 & 0 & -1 & 1 & 0 & 0 \\ 1 & 1 & 1 & 0 & 1 & 0 \\ 8 & 12 & 20 & 0 & 0 & 1 \end{array}\right] \sim \left[\begin{array}{ccc|ccc} 1 & 0 & -1 & 1 & 0 & 0 \\ 0 & 1 & 2 & -1 & 1 & 0 \\ 0 & 12 & 28 & -8 & 0 & 1 \end{array}\right]$$

$$\begin{array}{l} (-1)R_1 + R_2 \rightarrow R_2 \\ (-8)R_1 + R_3 \rightarrow R_3 \end{array} \qquad (-12)R_2 + R_3 \rightarrow R_3$$

$$\sim \left[\begin{array}{ccc|ccc} 1 & 0 & -1 & 1 & 0 & 0 \\ 0 & 1 & 2 & -1 & 1 & 0 \\ 0 & 0 & 4 & 4 & -12 & 1 \end{array}\right] \sim \left[\begin{array}{ccc|ccc} 1 & 0 & -1 & 1 & 0 & 0 \\ 0 & 1 & 2 & -1 & 1 & 0 \\ 0 & 0 & 1 & 1 & -3 & \frac{1}{4} \end{array}\right]$$

$$\frac{1}{4}R_3 \rightarrow R_3 \qquad\qquad \begin{array}{l} (-2)R_3 + R_2 \rightarrow R_2 \\ R_3 + R_1 \rightarrow R_1 \end{array}$$

$$\sim \left[\begin{array}{ccc|ccc} 1 & 0 & 0 & 2 & -3 & \frac{1}{4} \\ 0 & 1 & 0 & -3 & 7 & -\frac{1}{2} \\ 0 & 0 & 1 & 1 & -3 & \frac{1}{4} \end{array}\right]. \quad \text{Thus, the inverse is} \quad \begin{bmatrix} 2 & -3 & \frac{1}{4} \\ -3 & 7 & -\frac{1}{2} \\ 1 & -3 & \frac{1}{4} \end{bmatrix}$$

Concert 1:

$$\begin{array}{rcrcrcl} x_1 & & & - & x_3 & = & 0 \\ x_1 & + & x_2 & + & x_3 & = & 25,000 \\ 8x_1 & + & 12x_2 & + & 20x_3 & = & 320,000 \end{array} \quad \text{or} \quad \begin{bmatrix} 1 & 0 & -1 \\ 1 & 1 & 1 \\ 8 & 12 & 20 \end{bmatrix} \begin{bmatrix} x_1 \\ x_2 \\ x_3 \end{bmatrix} = \begin{bmatrix} 0 \\ 25,000 \\ 320,000 \end{bmatrix}$$

$$\text{Thus} \quad \begin{bmatrix} x_1 \\ x_2 \\ x_3 \end{bmatrix} = \begin{bmatrix} 2 & -3 & \frac{1}{4} \\ -3 & 7 & -\frac{1}{2} \\ 1 & -3 & \frac{1}{4} \end{bmatrix} \begin{bmatrix} 0 \\ 25,000 \\ 320,000 \end{bmatrix} = \begin{bmatrix} 5,000 \\ 15,000 \\ 5,000 \end{bmatrix}$$

and x_1 = 5,000 $8 tickets
    x_2 = 15,000 $12 tickets
    x_3 = 5,000 $20 tickets

Concert 2:

$$\begin{aligned} x_1 \qquad\quad - \quad x_3 &= 0 \\ x_1 + \quad x_2 + \quad x_3 &= 25{,}000 \\ 8x_1 + 12x_2 + 20x_3 &= 330{,}000 \end{aligned} \quad \text{or} \quad \begin{bmatrix} 1 & 0 & -1 \\ 1 & 1 & 1 \\ 8 & 12 & 20 \end{bmatrix} \begin{bmatrix} x_1 \\ x_2 \\ x_3 \end{bmatrix} = \begin{bmatrix} 0 \\ 25{,}000 \\ 330{,}000 \end{bmatrix}$$

Thus $\begin{bmatrix} x_1 \\ x_2 \\ x_3 \end{bmatrix} = \begin{bmatrix} 2 & -3 & \frac{1}{4} \\ -3 & 7 & -\frac{1}{2} \\ 1 & -3 & \frac{1}{4} \end{bmatrix} \begin{bmatrix} 0 \\ 25{,}000 \\ 330{,}000 \end{bmatrix} = \begin{bmatrix} 7{,}500 \\ 10{,}000 \\ 7{,}500 \end{bmatrix}$

and x_1 = 7,500 $8 tickets
    x_2 = 10,000 $12 tickets
    x_3 = 7,500 $20 tickets

Concert 3:

$$\begin{aligned} x_1 \qquad\quad - \quad x_3 &= 0 \\ x_1 + \quad x_2 + \quad x_3 &= 25{,}000 \\ 8x_1 + 12x_2 + 20x_3 &= 340{,}000 \end{aligned} \quad \text{or} \quad \begin{bmatrix} 1 & 0 & -1 \\ 1 & 1 & 1 \\ 8 & 12 & 20 \end{bmatrix} \begin{bmatrix} x_1 \\ x_2 \\ x_3 \end{bmatrix} = \begin{bmatrix} 0 \\ 25{,}000 \\ 340{,}000 \end{bmatrix}$$

Thus $\begin{bmatrix} x_1 \\ x_2 \\ x_3 \end{bmatrix} = \begin{bmatrix} 2 & -3 & \frac{1}{4} \\ -3 & 7 & -\frac{1}{2} \\ 1 & -3 & \frac{1}{4} \end{bmatrix} \begin{bmatrix} 0 \\ 25{,}000 \\ 340{,}000 \end{bmatrix} = \begin{bmatrix} 10{,}000 \\ 5{,}000 \\ 10{,}000 \end{bmatrix}$

and x_1 = 10,000 $8 tickets
    x_2 = 5,000 $12 tickets
    x_3 = 10,000 $20 tickets

(4-6)

43. The technology matrix is

$$M = \begin{array}{c} \\ \text{Agriculture} \\ \text{Fabrication} \end{array} \begin{array}{cc} \text{Agriculture} & \text{Fabrication} \\ \begin{bmatrix} 0.30 & 0.10 \\ 0.20 & 0.40 \end{bmatrix} \end{array}$$

Now $I - M = \begin{bmatrix} 1 & 0 \\ 0 & 1 \end{bmatrix} - \begin{bmatrix} 0.30 & 0.10 \\ 0.20 & 0.40 \end{bmatrix} = \begin{bmatrix} 0.70 & -0.10 \\ -0.20 & 0.60 \end{bmatrix} = \begin{bmatrix} \frac{7}{10} & -\frac{1}{10} \\ -\frac{1}{5} & \frac{3}{5} \end{bmatrix}$

Next, we calculate the inverse of $I - M$

$$\left[\begin{array}{cc|cc} \frac{7}{10} & -\frac{1}{10} & 1 & 0 \\ -\frac{1}{5} & \frac{3}{5} & 0 & 1 \end{array} \right] \sim \left[\begin{array}{cc|cc} 1 & -3 & 0 & -5 \\ \frac{7}{10} & -\frac{1}{10} & 1 & 0 \end{array} \right] \sim \left[\begin{array}{cc|cc} 1 & -3 & 0 & -5 \\ 0 & 2 & 1 & \frac{7}{2} \end{array} \right] \sim \left[\begin{array}{cc|cc} 1 & -3 & 0 & -5 \\ 0 & 1 & \frac{1}{2} & \frac{7}{4} \end{array} \right]$$

$$\begin{array}{cccc} -5R_2 \to R_2 & \left(-\frac{7}{10}\right)R_1 + R_2 \to R_2 & \frac{1}{2}R_2 \to R_2 & 3R_2 + R_1 \to R_1 \\ R_1 \leftrightarrow R_2 & & & \end{array}$$

$$\sim \begin{bmatrix} 1 & 0 & \frac{3}{2} & \frac{1}{4} \\ 0 & 1 & \frac{1}{2} & \frac{7}{4} \end{bmatrix} \quad \text{Thus,} \quad (I - M)^{-1} = \begin{bmatrix} \frac{3}{2} & \frac{1}{4} \\ \frac{1}{2} & \frac{7}{4} \end{bmatrix}.$$

Let x_1 = output for agriculture and x_2 = output for fabrication.

Then the output $X = \begin{bmatrix} x_1 \\ x_2 \end{bmatrix}$ needed to satisfy a final demand $D = \begin{bmatrix} d_1 \\ d_2 \end{bmatrix}$ for agriculture and fabrication is given by $X = (I - M)^{-1}D$.

(A) Let $D = \begin{bmatrix} 50 \\ 20 \end{bmatrix}$. Then $X = \begin{bmatrix} \frac{3}{2} & \frac{1}{4} \\ \frac{1}{2} & \frac{7}{4} \end{bmatrix} \begin{bmatrix} 50 \\ 20 \end{bmatrix} = \begin{bmatrix} 75 + 5 \\ 25 + 35 \end{bmatrix} = \begin{bmatrix} 80 \\ 60 \end{bmatrix}$

Thus, the total output for agriculture is \$80 billion; the total output for fabrication is \$60 billion.

(B) Let $D = \begin{bmatrix} 80 \\ 60 \end{bmatrix}$. Then $X = \begin{bmatrix} \frac{3}{2} & \frac{1}{4} \\ \frac{1}{2} & \frac{7}{4} \end{bmatrix} \begin{bmatrix} 80 \\ 60 \end{bmatrix} = \begin{bmatrix} 120 + 15 \\ 40 + 105 \end{bmatrix} = \begin{bmatrix} 135 \\ 145 \end{bmatrix}$

Thus, the total output for agriculture is \$135 billion; the total output for fabrication is \$145 billion. **(4-7)**

44. First we find the inverse of $B = \begin{bmatrix} 1 & 1 & 0 \\ 1 & 0 & 1 \\ 1 & 1 & 1 \end{bmatrix}$.

$$\begin{bmatrix} 1 & 1 & 0 & | & 1 & 0 & 0 \\ 1 & 0 & 1 & | & 0 & 1 & 0 \\ 1 & 1 & 1 & | & 0 & 0 & 1 \end{bmatrix} \sim \begin{bmatrix} 1 & 1 & 0 & | & 1 & 0 & 0 \\ 0 & -1 & 1 & | & -1 & 1 & 0 \\ 0 & 0 & 1 & | & -1 & 0 & 1 \end{bmatrix}$$

$(-1)R_1 + R_2 \rightarrow R_2 \qquad\qquad (-1)R_2 \rightarrow R_2$

$(-1)R_1 + R_3 \rightarrow R_3$

$$\sim \begin{bmatrix} 1 & 1 & 0 & | & 1 & 0 & 0 \\ 0 & 1 & -1 & | & 1 & -1 & 0 \\ 0 & 0 & 1 & | & -1 & 0 & 1 \end{bmatrix} \sim \begin{bmatrix} 1 & 0 & 1 & | & 0 & 1 & 0 \\ 0 & 1 & -1 & | & 1 & -1 & 0 \\ 0 & 0 & 1 & | & -1 & 0 & 1 \end{bmatrix}$$

$(-1)R_2 + R_1 \rightarrow R_1 \qquad\qquad (-1)R_3 + R_1 \rightarrow R_1$

$R_3 + R_2 \rightarrow R_2$

$$\sim \begin{bmatrix} 1 & 0 & 0 & | & 1 & 1 & -1 \\ 0 & 1 & 0 & | & 0 & -1 & 1 \\ 0 & 0 & 1 & | & -1 & 0 & 1 \end{bmatrix} \quad \text{Thus,} \quad B^{-1} = \begin{bmatrix} 1 & 1 & -1 \\ 0 & -1 & 1 \\ -1 & 0 & 1 \end{bmatrix}$$

Now, $\begin{bmatrix} 1 & 1 & -1 \\ 0 & -1 & 1 \\ -1 & 0 & 1 \end{bmatrix} \begin{bmatrix} 25 & 24 & 21 & 41 & 21 & 52 \\ 8 & 25 & 41 & 30 & 32 & 52 \\ 26 & 33 & 48 & 50 & 41 & 79 \end{bmatrix} = \begin{bmatrix} 7 & 16 & 14 & 21 & 12 & 25 \\ 18 & 8 & 7 & 20 & 9 & 27 \\ 1 & 9 & 27 & 9 & 20 & 27 \end{bmatrix}$

Thus, the decoded message is

7 18 1 16 8 9 14 7 27 21 20 9 12 9 20 25 27 27

which corresponds to GRAPHING UTILITY.   (4–5)

45. (A) 1st & Elm: $x_1 + x_4 = 1300$
       2nd & Elm: $x_1 - x_2 = 400$
       2nd & Oak: $x_2 + x_3 = 700$
       1st & Oak: $x_3 - x_4 = -200$

(B) The augmented matrix for the system in part A is:

$$\begin{bmatrix} 1 & 0 & 0 & 1 & | & 1300 \\ 1 & -1 & 0 & 0 & | & 400 \\ 0 & 1 & 1 & 0 & | & 700 \\ 0 & 0 & 1 & -1 & | & -200 \end{bmatrix} \sim \begin{bmatrix} 1 & 0 & 0 & 1 & | & 1300 \\ 0 & -1 & 0 & -1 & | & -900 \\ 0 & 1 & 1 & 0 & | & 700 \\ 0 & 0 & 1 & -1 & | & -200 \end{bmatrix}$$

$$(-1)R_1 + R_2 \rightarrow R_2 \qquad\qquad (-1)R_2 \rightarrow R_2$$

$$\sim \begin{bmatrix} 1 & 0 & 0 & 1 & | & 1300 \\ 0 & 1 & 0 & 1 & | & 900 \\ 0 & 1 & 1 & 0 & | & 700 \\ 0 & 0 & 1 & -1 & | & -200 \end{bmatrix} \sim \begin{bmatrix} 1 & 0 & 0 & 1 & | & 1300 \\ 0 & 1 & 0 & 1 & | & 900 \\ 0 & 0 & 1 & -1 & | & -200 \\ 0 & 0 & 1 & -1 & | & -200 \end{bmatrix}$$

$$(-1)R_2 + R_3 \rightarrow R_3 \qquad\qquad (-1)R_3 + R_4 \rightarrow R_4$$

$$\sim \begin{bmatrix} 1 & 0 & 0 & 1 & | & 1300 \\ 0 & 1 & 0 & 1 & | & 900 \\ 0 & 0 & 1 & -1 & | & -200 \\ 0 & 0 & 0 & 0 & | & 0 \end{bmatrix}$$

The corresponding system of equations is

$$x_1 \quad + x_4 = 1300$$
$$x_2 + x_4 = 900$$
$$x_3 - x_4 = -1200$$

Let $x_4 = t$. Then, $x_1 = 1300 - t$, $x_2 = 900 - t$, $x_3 = t - 200$, $x_4 = t$ where $200 \leq t \leq 900$. *($t \geq 200$ so that x_3 is non–negative; $t \leq 900$ so that x_2 is non–negative)

(C) maximum: 900
    minimum: 200

(D) Elm St.: $x_1 = 800$; 2nd St.: $x_2 = 400$; Oak St.: $x_3 = 300$.   (4–3)

5 LINEAR INEQUALITIES AND LINEAR PROGRAMMING

EXERCISE 5-1

Things to remember:

1. The graph of the linear inequality

   $$Ax + By < C \quad \text{or} \quad Ax + By > C$$

   with $B \neq 0$ is either the upper half-plane or the lower half-plane (but not both) determined by the line $Ax + By = C$.
   If $B = 0$, the graph of

   $$Ax < C \quad \text{or} \quad Ax > C$$

   is either the right half-plane or the left half-plane (but not both) determined by the vertical line $Ax = C$.

2. For strict inequalities ("<" or ">"), the line is not included in the graph. For weak inequalities ("≤" or "≥"), the line is included in the graph.

3. PROCEDURE FOR GRAPHING LINEAR INEQUALITIES

   (a) First graph $Ax + By = C$ as a broken line if equality is not included in the original statement or as a solid line if equality is included.

   (b) Choose a test point anywhere in the plane not on the line [the origin $(0, 0)$ often requires the least computation] and substitute the coordinates into the inequality.

   (c) The graph of the original inequality includes the half-plane containing the test point if the inequality is satisfied by that point or the half-plane not containing the test point if the inequality is not satisfied by that point.

4. To solve a system of linear inequalities graphically, graph each inequality in the system and then take the intersection of all the graphs. The resulting graph is called the SOLUTION REGION, or FEASIBLE REGION.

5. A CORNER POINT of a solution region is a point in the solution region which is the intersection of two boundary lines.

6. The solution region of a system of linear inequalities is BOUNDED if it can be enclosed within a circle; if it cannot be enclosed within a circle, then it is UNBOUNDED.

1. $y \le x - 1$

Graph $y = x - 1$ as a solid line.

Test point $(0, 0)$:

$0 \le 0 - 1$

$0 \le -1$

The inequality is false. Thus, the graph is below the line $y = x - 1$, including the line.

x	y
0	-1
1	0

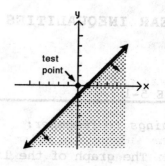

3. $3x - 2y > 6$

Graph $3x - 2y = 6$ as a broken line.

Test point $(0, 0)$:

$3 \cdot 0 - 2 \cdot 0 > 6$

$0 > 6$

The inequality is false. Thus, the graph is below the line $3x - 2y = 6$, not including the line.

x	y
0	-3
2	0

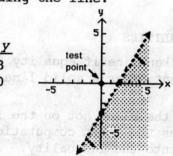

5. $x \ge -4$

Graph $x = -4$ [the vertical line through $(-4, 0)$] as a solid line.

Test point $(0, 0)$:

$0 \ge -4$

The inequality is true. Thus, the graph is to the right of the line $x = -4$, including the line.

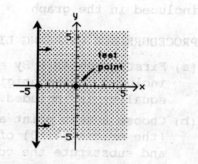

7. $-4 \le y < 4$

Graph $y = -4$ as a solid line and $y = 4$ as a broken line [horizontal lines through $(0, -4)$ and $(0, 4)$, respectively].

Test point $(0, 0)$:

$-4 \le 0$ and $0 < 4$

i.e., $-4 \le 0 < 4$

Both inequalities are true. Thus, the graph is between the lines $y = -4$ and $y = 4$, including the line $y = -4$ but not including the line $y = 4$.

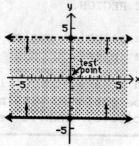

9. $6x + 4y \geq 24$

Graph the line $6x + 4y = 24$ as a solid line.

Test point $(0, 0)$:

$6 \cdot 0 + 4 \cdot 0 \geq 24$

$\qquad 0 \geq 24$

The inequality is false. Thus, the graph is the region above the line, including the line.

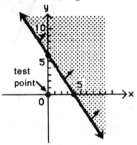

11. $5x \leq -2y$ or $5x + 2y \leq 0$

Graph the line $5x + 2y = 0$ as a solid line. Since the line **passes** through the origin $(0, 0)$, we use $(1, 0)$ as a test point:

$5 \cdot 1 + 2 \cdot 0 \leq 0$

$\qquad 5 \leq 0$

This inequality is false. Thus, **the** graph is below the line $5x + 2y = 0$, including the line.

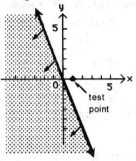

13. The graph of $x + 2y \leq 8$ is the region below the line $x + 2y = 8$ [e.g., $(0, 0)$ satisfies the inequality]. The graph of $3x - 2y \geq 0$ is the region below the line $3x - 2y = 0$ [e.g., $(1, 0)$ satisfies the inequality]. The intersection of these two regions is region IV.

15. The graph of $x + 2y \geq 8$ is the region above the line $x + 2y = 8$ [e.g., $(0, 0)$ does not satisfy the inequality]. The graph of $3x - 2y \geq 0$ is the region below the line $3x - 2y = 0$ [e.g., $(1, 0)$ satisfies the inequality]. The intersection of these two regions is region I.

17. The graphs of the inequalities $3x + y \geq 6$ and $x \leq 4$ are:

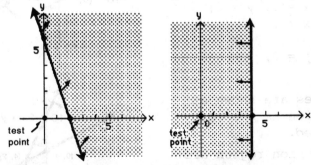

The intersection of these regions (drawn on the same coordinate plane) is shown in the graph at the right.

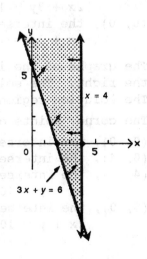

19. The graphs of the inequalities $x - 2y \leq 12$ and $2x + y \geq 4$ are:

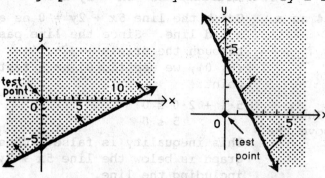

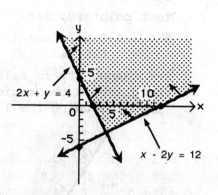

The intersection of these regions (drawn on the same coordinate plane) is shown in the graph at the right.

21. The graph of $x + 3y \leq 18$ is the region below the line $x + 3y = 18$ and the graph of $2x + y \geq 16$ is the region above the line $2x + y = 16$. The graph of $x \geq 0$, $y \geq 0$ is the first quadrant. The intersection of these regions is region IV. The corner points are $(8, 0)$, $(18, 0)$, and $(6, 4)$.

23. The graph of $x + 3y \geq 18$ is the region above the line $x + 3y = 18$ and the graph of $2x + y \geq 16$ is the region above the line $2x + y = 16$. The graph of $x \geq 0$, $y \geq 0$ is the first quadrant. The intersection of these regions is region I. The corner points are $(0, 16)$, $(6, 4)$, and $(18, 0)$.

25. The graphs of the inequalities are shown at the right. The solution region is indicated by the shaded region. The solution region is *bounded*.

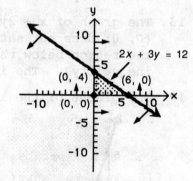

The corner points of the solution region are:

$(0, 0)$, the intersection of $x = 0$, $y = 0$;
$(0, 4)$, the intersection of $x = 0$,
$\qquad 2x + 3y = 12$;
$(6, 0)$, the intersection of $y = 0$,
$\qquad 2x + 3y = 12$.

27. The graphs of the inequalities are shown at the right. The solution region is shaded. The solution region is *bounded*.

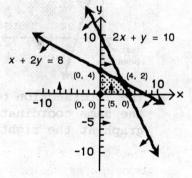

The corner points of the solution region are:

$(0, 0)$, the intersection of $x = 0$, $y = 0$;
$(0, 4)$, the intersection of $x = 0$, $x + 2y = 8$;
$(4, 2)$, the intersection of $x + 2y = 8$,
$\qquad 2x + y = 10$;
$(5, 0)$, the intersection of $y = 0$,
$\qquad 2x + y = 10$.

29. The graphs of the inequalities are shown at the right. The solution region is shaded. The solution region is *unbounded*.

The corner points of the solution region are:

(0, 10), the intersection of $x = 0$,
  $2x + y = 10$;
(4, 2), the intersection of $x + 2y = 8$,
  $2x + y = 10$;
(8, 0), the intersection of $y = 0$, $x + 2y = 8$.

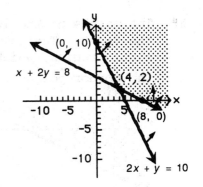

31. The graphs of the inequalities are shown at the right. The solution is indicated by the shaded region. The solution region is *bounded*.

The corner points of the solution region are:

(0, 0), the intersection of $x = 0$, $y = 0$,
(0, 6), the intersection of $x = 0$,
  $x + 2y = 12$;
(2, 5), the intersection of $x + 2y = 12$,
  $x + y = 7$;
(3, 4), the intersection of $x + y = 7$,
  $2x + y = 10$;
(5, 0), the intersection of $y = 0$,
  $2x + y = 10$.

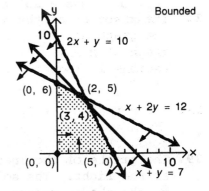

Note that the point of intersection of the lines $2x + y = 10$, $x + 2y = 12$ is not a corner point because it is not in the solution region.

33. The graphs of the inequalities are shown at the right. The solution is indicated by the shaded region, which is *unbounded*.

The corner points are:

(0, 16), the intersection of $x = 0$,
  $2x + y = 16$;
(4, 8), the intersection of $2x + y = 16$,
  $x + y = 12$;
(10, 2), the intersection of $x + y = 12$,
  $x + 2y = 14$;
(14, 0), the intersection of $y = 0$,
  $x + 2y = 14$.

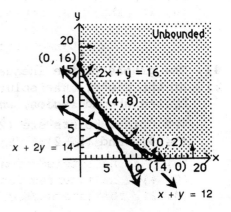

The intersection of $x + 2y = 14$, $2x + y = 16$ is not a corner point because it is not in the solution region.

35. The graphs of the inequalities are shown at the right. The solution is indicated by the shaded region, which is *bounded*.

The corner points are (8, 6), (4, 7), and (9, 3).

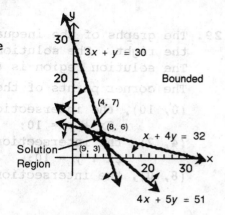

37. The graphs of the inequalities are shown at the right. The system of inequalities does not have a solution because the intersection of the graphs is empty.

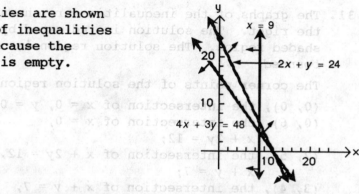

39. The graphs of the inequalities are shown at the right. The solution is indicated by the shaded region, which is *unbounded*.

The corner points are (0, 0), (4, 4), and (8, 12).

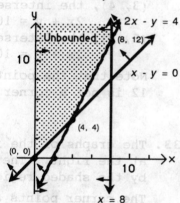

41. The graphs of the inequalities are shown at the right. The solution is indicated by the shaded region, which is *bounded*.

The corner points are (2, 1), (3, 6), (5, 4), and (5, 2).

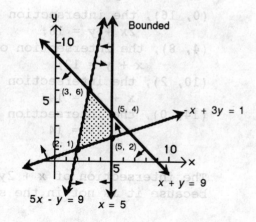

43. The graphs of the inequalities are shown at the right. The solution is indicated by the shaded region, which is *bounded*. The corner points are (1.27, 5.36), (2.14, 6.52), and (5.91, 1.88).

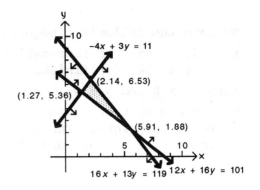

45. (A)
$$3x + 4y = 36$$
$$\underline{3x + 2y = 30} \quad \text{subtract}$$
$$\begin{aligned} 2y &= 6 \\ y &= 3 \\ x &= 8 \end{aligned}$$
intersection point: (8, 3)

$$3x + 4y = 36$$
$$\underline{\qquad y = 0}$$
$$\begin{aligned} 3x &= 36 \\ x &= 12 \end{aligned}$$
intersection point: (12, 0)

$$3x + 2y = 30$$
$$\underline{\qquad y = 0}$$
$$\begin{aligned} 3x &= 30 \\ x &= 10 \end{aligned}$$
intersection point: (10, 0)

$$3x + 4y = 36$$
$$\underline{\quad x = 0}$$
$$\begin{aligned} 4y &= 36 \\ y &= 9 \end{aligned}$$
intersection point: (0, 9)

$$3x + 2y = 30$$
$$\underline{\quad x = 0}$$
$$\begin{aligned} 2y &= 30 \\ y &= 15 \end{aligned}$$
intersection point: (0, 15)

$$x = 0$$
$$y = 0$$
intersection point: (0, 0)

(B) The corner points are: (8, 3), (0, 9), (10, 0), (0, 0); (0, 15) does not satisfy $3x + 4y \leq 36$, (12, 0) does not satisfy $3x + 2y \leq 30$.

47. Let x = the number of trick skis and y = the number of slalom skis produced per day. The information is summarized in the following table.

	Hours per ski		
	Trick ski	Slalom ski	Maximum labor-hours per day available
Fabrication	6 hrs	4 hrs	108 hrs
Finishing	1 hr	1 hr	24 hrs

We have the following inequalities:

$6x + 4y \leq 108$ for fabrication
$x + y \leq 24$ for finishing

Also, $x \geq 0$ and $y \geq 0$.

The graphs of these inequalities are shown at the right. The shaded region indicates the set of feasible solutions.

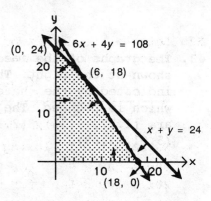

49. (A) If x is the number of trick skis and y is the number of slalom skis per day, then the profit per day is given by
$$P(x, y) = 50x + 60y$$
All the production schedules in the feasible region that lie on the graph of the line
$$50x + 60y = 1,100$$
will provide a profit of \$1,100.

(B) There are many possible choices. For example, producing 5 trick skis and 15 slalom skis per day will produce a profit of
$$P(5, 15) = 50(5) + 60(15) = 1,150$$
All the production schedules in the feasible region that lie on the graph of
$$50x + 60y = 1,150$$
will provide a profit of \$1,150.

51. Let x = the number of cubic yards of mix A and y = the number of cubic yards of mix B. The information is summarized in the following table:

| | Amount of substance per cubic yard | | Minimum monthly requirement |
	Mix A	Mix B	
Phosphoric acid	20 lbs	10 lbs	460 lbs
Nitrogen	30 lbs	30 lbs	960 lbs
Potash	5 lbs	10 lbs	220 lbs

We have the following inequalities:

$20x + 10y \geq 460$
$30x + 30y \geq 960$
$5x + 10y \geq 220$

Also, $x \geq 0$ and $y \geq 0$.

The graphs of these inequalities are shown at the right. The shaded region indicates the set of feasible solutions.

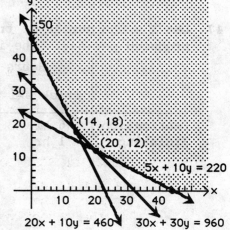

53. Let x = the number of mice used and y = the number of rats used. The information is summarized in the following table.

	Mice	Rats	Maximum time available per day
Box A	10 min	20 min	800 min
Box B	20 min	10 min	640 min

We have the following inequalities:

$10x + 20y \leq 800$ for box A
$20x + 10y \leq 640$ for box B

Also, $x \geq 0$ and $y \geq 0$.

The graphs of these inequalities are shown at the right. The shaded region indicates the set of feasible solutions.

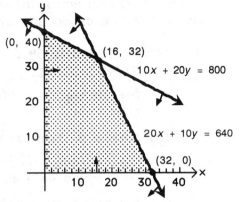

EXERCISE 5-2

Things to remember:

1. A LINEAR PROGRAMMING PROBLEM is a problem concerned with finding the maximum or minimum value of a linear OBJECTIVE FUNCTION of the form

   $$z = c_1 x_1 + c_2 x_2 + \cdots + c_n x_n,$$

   where the DECISION VARIABLES x_1, x_2, ..., x_n are subject to PROBLEM CONSTRAINTS in the form of linear inequalities and equations. In addition, the decision variables must satisfy the NONNEGATIVE CONSTRAINTS $x_i \geq 0$, for i = 1, 2, ..., n. The set of points satisfying both the problem constraints and the nonnegative constraints is called the FEASIBLE REGION for the problem. Any point in the feasible region that produces the optimal value of the objective function over the feasible region is called an OPTIMAL SOLUTION.

2. FUNDAMENTAL THEOREM OF LINEAR PROGRAMMING

   If the optimal value of the objective function in a linear programming problem exists, then that value must occur at one (or more) of the corner points of the feasible region.

3. EXISTENCE OF SOLUTIONS

   (A) If the feasible region for a linear programming problem is bounded, then both the maximum value and the minimum value of the objective function always exist.

(B) If the feasible region is unbounded, and the coefficients of the objective function are positive, then the minimum value of the objective function exists, but the maximum value does not.

(C) If the feasible region is empty (that is, there are no points that satisfy all the constraints), then both the maximum value and the minimum value of the objective function do not exist.

<u>4</u>. GEOMETRIC SOLUTION OF A LINEAR PROGRAMMING PROBLEM WITH TWO DECISION VARIABLES.

(1) For an applied problem, summarize relevant material in table form.

(2) Form a mathematical model for the problem:

   (a) Introduce decision variables and write a linear objective function.

   (b) Write problem constraints using linear inequalities and/or equations.

   (c) Write nonnegative constraints.

(3) Graph the feasible region. Then, if according to <u>3</u> an optimal solution exists, find the coordinates of each corner point.

(4) Make a table listing the value of the objective function at each corner point.

(5) Determine the optimal solution(s) from the table in Step (4).

(6) For an applied problem, interpret the optimal solution(s) in terms of the original problem.

---

1. Steps (1)—(3) in <u>4</u> do not apply. Thus, we begin with Step (4).

<u>Step (4)</u>: Evaluate the objective function at each corner point.

Corner Point	$z = x + y$
(0, 0)	0
(0, 12)	12
(7, 9)	16
(10, 0)	10

<u>Step (5)</u>: Determine the optimal solution from Step (4). The maximum value of z is 16 at (7, 9).

3. Steps (1)—(3) in <u>4</u> do not apply. Thus, we begin with Step (4).

<u>Step (4)</u>: Evaluate the objective function at each corner point.

Corner Point	$z = 3x + 7y$
(0, 0)	0
(0, 12)	84
(7, 9)	84
(10, 0)	30

Step (5): Determine the optimal solution from Step (4).
The maximum value of z is 84 at (0, 12) and (7, 9). This is
a multiple optimal solution.

5. Steps (1)—(3) in 4 do not apply. Thus, we begin with Step (4).

Step (4): Evaluate the objective function at each corner point.

Corner Point	$z = 7x + 4y$
(0, 12)	48
(0, 8)	32
(4, 3)	40
(12, 0)	84

Step (5): Determine the optimal solution from Step (4).
The minimum value of z is 32 at (0, 8).

7. Steps (1)—(3) in 4 do not apply. Thus, we begin with Step (4).

Step (4): Evaluate the objective function at each corner point.

Corner Point	$z = 3x + 8y$
(0, 12)	96
(0, 8)	64
(4, 3)	36
(12, 0)	36

Step (5): Determine the optimal solution from Step (4).
The minimum value of z is 36 at (4, 3) and (12, 0). This is
a multiple optimal solution.

9. Step (3): Graph the feasible region
and find the corner points.

The feasible region S is the
solution set of the given
inequalities. This region is
indicated by the shading in the
graph at the right.

The corner points are (0, 0),
(0, 4), (4, 2), and (5, 0).

Since S is bounded, it follows from
3(a) that P has a maximum value.

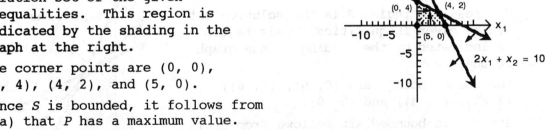

Step (4): Evaluate the objective function at each corner point.

The value of P at each corner point is given in the following
table.

Corner Point	$P = 5x_1 + 5x_2$
(0, 0)	$P = 5(0) + 5(0) = 0$
(0, 4)	$P = 5(0) + 5(4) = 20$
(4, 2)	$P = 5(4) + 5(2) = 30$
(5, 0)	$P = 5(5) + 5(0) = 25$

Step (5): Determine the optimal solution.

The maximum value of P is 30 at $x_1 = 4$, $x_2 = 2$.

11. Step (3): Graph the feasible region and find the corner points.

The feasible region S is the solution set of the given inequalities. This region is indicated by the shading in the graph at the right.

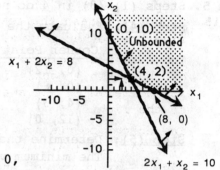

The corner points are (0, 10), (4, 2), and (8, 0).

Since S is unbounded and $a = 2 > 0$, $b = 3 > 0$, it follows from 3(b) that P has a minimum value but not a maximum value.

Step (4): Evaluate the objective function at each corner point.

The value of P at each corner point is given in the following table:

Corner Point	$z = 2x_1 + 3x_2$
(0, 10)	$z = 2(0) + 3(10) = 30$
(4, 2)	$z = 2(4) + 3(2) = 14$
(8, 0)	$z = 2(8) + 3(0) = 16$

Step (5): Determine the optimal solutions.

The minimum occurs at $x_1 = 4$, $x_2 = 2$, and the minimum value is $z = 14$; z does not have a maximum value.

13. Step (3): Graph the feasible region and find the corner points.

The feasible region S is the solution set of the given inequalities. This region is indicated by the shading in the graph at the right.

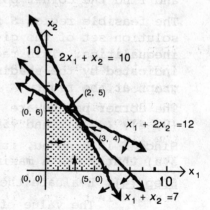

The corner points are (0, 0), (0, 6), (2, 5), (3, 4), and (5, 0).

Since S is bounded, it follows from 3(a) that P has a maximum value.

Step (4): Evaluate the objective function at each corner point.

The value of P at each corner point is:

Corner Point	$P = 30x_1 + 40x_2$
(0, 0)	$P = 30(0) + 40(0) = \quad 0$
(0, 6)	$P = 30(0) + 40(6) = 240$
(2, 5)	$P = 30(2) + 40(5) = 260$
(3, 4)	$P = 30(3) + 40(4) = 250$
(5, 0)	$P = 30(5) + 40(0) = 150$

Step (5): Determine the optimal solution.

The maximum occurs at $x_1 = 2$, $x_2 = 5$, and the maximum value is $P = 260$.

15. Step (3): Graph the feasible region and find the corner points.

The feasible region S is the solution set of the given inequalities. This region is indicated by the shading in the graph at the right.

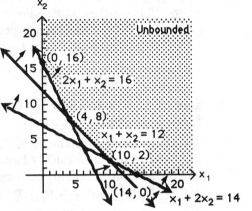

The corner points are (0, 16), (4, 8), (10, 2), and (14, 0).

Since S is unbounded and $a = 10 > 0$, $b = 30 > 0$, it follows from 3(b) that z has a minimum value but not a maximum value.

Step (4): Evaluate the objective function at each corner point.

The value of z at each corner point is:

Corner Point	$z = 10x_1 + 30x_2$
(0, 16)	$z = 10(0) + 30(16) = 480$
(4, 8)	$z = 10(4) + 30(8) = 280$
(10, 2)	$z = 10(10) + 30(2) = 160$
(14, 0)	$z = 10(14) + 30(0) = 140$

Step (5): Determine the optimal solution.

The minimum occurs at $x_1 = 14$, $x_2 = 0$, and the minimum value is $z = 140$; z does not have a maximum value.

17. Step (3): Graph the feasible region and find the corner points.

The feasible region S is the solution set of the given inequalities, and is indicated by the shading in the graph at the right.

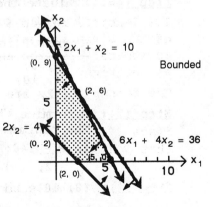

The corner points are (0, 2), (0, 9), (2, 6), (5, 0), and (2, 0).

Since S is bounded, it follows from 3(a) that P has a maximum value and a minimum value.

Step (4): Evaluate the objective function at each corner point.

The value of P at each corner point is given in the following table:

Corner Point	$P = 30x_1 + 10x_2$
(0, 2)	$P = 30(0) + 10(2) = 20$
(0, 9)	$P = 30(0) + 10(9) = 90$
(2, 6)	$P = 30(2) + 10(6) = 120$
(5, 0)	$P = 30(5) + 10(0) = 150$
(2, 0)	$P = 30(2) + 10(0) = 60$

Step (5): Determine the optimal solutions.

The maximum occurs at $x_1 = 5$, $x_2 = 0$, and the maximum value is $P = 150$; the minimum occurs at $x_1 = 0$, $x_2 = 2$, and the minimum value is $P = 20$.

19. Step (3): Graph the feasible region and find the corner points.

The feasible region S is the solution set of the given inequalities. As indicated, the feasible region is empty. Thus, by 3(c), there are no optimal solutions.

21. Step (3): Graph the feasible region and find the corner points.

The feasible region S is the solution set of the given inequalities, and is indicated by the shading in the graph at the right.

The corner points are (3, 8), (8, 10), and (12, 2).

Since S is bounded, it follows from 3(a) that P has a maximum value and a minimum value.

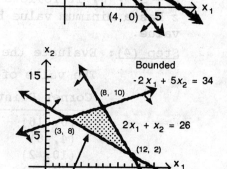

Step (4): Evaluate the objective function at each corner point.

The value of P at each corner point is:

Corner Point	$P = 20x_1 + 10x_2$
(3, 8)	$P = 20(3) + 10(8) = 140$
(8, 10)	$P = 20(8) + 10(10) = 260$
(12, 2)	$P = 20(12) + 10(2) = 260$

Step (5): Determine the optimal solutions.

The minimum occurs at $x_1 = 3$, $x_2 = 8$, and the minimum value is $P = 140$; the maximum occurs at $x_1 = 8$, $x_2 = 10$, at $x_1 = 12$, $x_2 = 2$, and at any point along the line segment joining (8, 10) and (12, 2). The maximum value is $P = 260$.

23. Step (3): Graph the feasible region and find the corner points.

The feasible region S is the set of solutions of the given inequalities, and is indicated by the shading in the graph at the right.

The corner points are $(0, 0)$, $(0, 800)$, $(400, 600)$, $(600, 450)$, and $(900, 0)$. Since S is bounded, it follows from 3(a) that P has a maximum value.

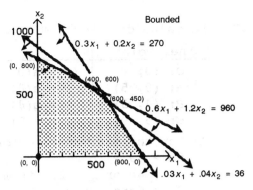

Step (4): Evaluate the objective function at each corner point.

The value of P at each corner point is:

Corner Point	$P = 20x_1 + 30x_2$
$(0, 0)$	$P = 20(0) + 30(0) = 0$
$(0, 800)$	$P = 20(0) + 30(800) = 24,000$
$(400, 600)$	$P = 20(400) + 30(600) = 26,000$
$(600, 450)$	$P = 20(600) + 30(450) = 25,500$
$(900, 0)$	$P = 20(900) + 30(0) = 18,000$

Step (5): Determine the optimal solution.

The maximum occurs at $x_1 = 400$, $x_2 = 600$, and the maximum value is $P = 26,000$.

25. ℓ_1: $275x_1 + 322x_2 = 3,381$
ℓ_2: $350x_1 + 340x_2 = 3,762$
ℓ_3: $425x_1 + 306x_2 = 4,114$.

Step (3): Graph the feasible region and find the corner points.

The feasible region S is the solution set of the given inequalities, and is indicated by the shading in the graph at the right.

The corner points are $(0, 0)$, $(0, 10.5)$, $(3.22, 7.75)$, $(6.62, 4.25)$, $(9.68, 0)$.

Step (4): Evaluate the objective function at each corner point.

The value of P at each corner point is

Corner Point	$P = 525x_1 + 478x_2$
$(0, 0)$	$P = 525(0) + 478(0) = 0$
$(0, 10.5)$	$P = 525(0) + 478(10.5) = 5,019$
$(3.22, 7.75)$	$P = 525(3.22) + 478(7.75) = 5,395$
$(6.62, 4.25)$	$P = 525(6.62) + 478(4.25) = 5,507$
$(9.68, 0)$	$P = 525(9.68) + 478(0) = 5,082$

Step (5): Determine the optimal solution.

The maximum occurs at $x_1 = 6.62$, $x_2 = 4.25$, and the maximum value is $P = 5,507$.

27. The value of $P = ax_1 + bx_2$, $a > 0$, $b > 0$, at each corner point is:

Corner Point	P
O: (0, 0)	$P = a(0) + b(0) = 0$
A: (0, 5)	$P = a(0) + b(5) = 5b$
B: (4, 3)	$P = a(4) + b(3) = 4a + 3b$
C: (5, 0)	$P = a(5) + b(0) = 5a$

(A) For the maximum value of P to occur at A only, we must have $5b > 4a + 3b$ and $5b > 5a$. Solving the first inequality, we get $2b > 4a$ or $b > 2a$; from the second inequality, we get $b > a$. Therefore, we must have $b > 2a$ or $2a < b$ in order for P to have its maximum value at A only.

(B) For the maximum value of P to occur at B only, we must have $4a + 3b > 5b$ and $4a + 3b > 5a$. Solving this pair of inequalities, we get $4a > 2b$ and $3b > a$, which is the same as $\frac{a}{3} < b < 2a$.

(C) For the maximum value of P to occur at C only, we must have $5a > 4a + 3b$ and $5a > 5b$. This pair of inequalities implies that $a > 3b$ or $b < \frac{a}{3}$.

(D) For the maximum value of P to occur at both A and B, we must have $5b = 4a + 3b$ or $b = 2a$.

(E) For the maximum value of P to occur at both B and C, we must have $4a + 3b = 5a$ or $b = \frac{a}{3}$.

29. (A) <u>Step (1)</u>: Has been done.

<u>Step (2)</u>: Form a mathematical model for the problem.

Let x_1 = the number of trick skis
and x_2 = the number of slalom skis produced per day.

The mathematical model for this problem is:
$$\text{Maximize } P = 40x_1 + 30x_2$$
$$\text{Subject to: } 6x_1 + 4x_2 \leq 108$$
$$x_1 + x_2 \leq 24$$
$$x_1 \geq 0, \ x_2 \geq 0$$

<u>Step (3)</u>: Graph the feasible region and find the corner points.

The feasible region S is the solution set of the given system of inequalities, and is indicated by the shading in the graph below.

The corner points are (0, 0), (0, 24), (6, 18), and (18, 0).

Since S is bounded, P has a maximum value by <u>3</u>(a).

Step (4): Evaluate the objective function at each corner point. The value of P at each corner point is:

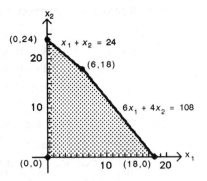

Corner Point	$P = 40x_1 + 30x_2$
(0, 0)	$P = 40(0) + 30(0) = 0$
(0, 24)	$P = 40(0) + 30(24) = 720$
(6, 18)	$P = 40(6) + 30(18) = 780$
(18, 0)	$P = 40(18) + 30(0) = 720$

Step (5): Determine the optimal solution. The maximum occurs when $x_1 = 6$ (trick skis) and $x_2 = 18$ (slalom skis) are produced. The maximum profit is $P = \$780$.

(B)
Corner Point	$P = 40x_1 + 25x_2$
(0, 0)	$P = 40(0) + 25(0) = 0$
(0, 24)	$P = 40(0) + 25(24) = 600$
(6, 18)	$P = 40(6) + 25(18) = 690$
(18, 0)	$P = 40(18) + 25(0) = 720$

The maximum profit decreases to $720 when 18 trick skis and no slalom skis are produced.

(C)
Corner Point	$P = 40x_1 + 45x_2$
(0, 0)	$P = 40(0) + 45(0) = 0$
(0, 24)	$P = 40(0) + 45(24) = 1080$
(6, 18)	$P = 40(6) + 45(18) = 1050$
(18, 0)	$P = 40(18) + 45(0) = 720$

The maximum profit increases to $1,080 when no trick skis and 24 slalom skis are produced.

31. (A) Step (1): Summarize relevant material in table form.

	Plant A	Plant B	Amount required
Tables	20	25	200
Chairs	60	50	500
Cost per day	$1000	$900	

Step (2): Form a mathematical model for the problem.

Let x_1 = the number of days to operate Plant A and x_2 = the number of days to operate Plant B.

The mathematical model for this problem is:

Minimize $C = 1000x_1 + 900x_2$
Subject to: $20x_1 + 25x_2 \geq 200$
$60x_1 + 50x_2 \geq 500$
$x_1 \geq 0, \; x_2 \geq 0$

<u>Step (3)</u>: Graph the feasible region and find the corner points.

The feasible region S is the solution set of the system of inequalities, and is indicated by the shading in the graph shown below.

The corner points are $(0, 10)$, $(5, 4)$, and $(10, 0)$.

Since S is unbounded and $a = 1000 > 0$, $b = 900 > 0$, C has a minimum value by <u>3</u>(b).

<u>Step (4)</u>: Evaluate the objective function at each corner point.

The value of C at each corner point is:

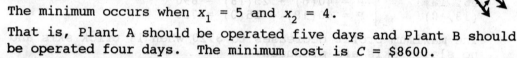

Corner Point	$C = 1000x_1 + 900x_2$
$(0, 10)$	$C = 1000(0) + 900(10) = 9,000$
$(5, 4)$	$C = 1000(5) + 900(4) = 8,600$
$(10, 0)$	$C = 1000(10) + 900(0) = 10,000$

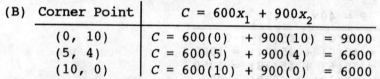

<u>Step (5)</u>: Determine the optimal solution.

The minimum occurs when $x_1 = 5$ and $x_2 = 4$.

That is, Plant A should be operated five days and Plant B should be operated four days. The minimum cost is $C = \$8600$.

(B)

Corner Point	$C = 600x_1 + 900x_2$
$(0, 10)$	$C = 600(0) + 900(10) = 9000$
$(5, 4)$	$C = 600(5) + 900(4) = 6600$
$(10, 0)$	$C = 600(10) + 900(0) = 6000$

The minimum cost decreases to $6,000 per day when Plant A is operated 10 days and Plant B is operated 0 days.

(C)

Corner Point	$C = 1000x_1 + 800x_2$
$(0, 10)$	$C = 600(0) + 800(10) = 8,000$
$(5, 4)$	$C = 1000(5) + 800(4) = 8,200$
$(10, 0)$	$C = 1000(10) + 800(0) = 10,000$

The minimum cost decreases to $8,000 per day when Plant A is operated 0 days and Plant B is operated 10 days.

33. (A) <u>Step (1)</u>: Summarize relevant material.

	Buses	Vans	Number to accommodate
Students	40	8	400
Chaperones	3	1	36
Rental cost	$1200 per bus	$100 per van	

<u>Step (2)</u>: Form a mathematical model for the problem.

Let x_1 = the number of buses

and x_2 = the number of vans.

The mathematical model for this problem is:

Minimize $C = 1200x_1 + 100x_2$

Subject to: $40x_1 + 8x_2 \geq 400$

$3x_1 + x_2 \leq 36$

$x_1 \geq 0, \; x_2 \geq 0$

<u>Step (3)</u>: Graph the feasible region and find the corner points.

The feasible region S is the solution set of the system of inequalities, and is indicated by the shading in the graph at the right.

The corner points are $(10, 0)$, $(7, 15)$, and $(12, 0)$.

Since S is bounded, C has a minimum value by <u>3</u>(a).

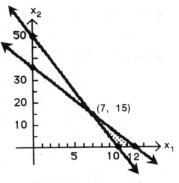

<u>Step (4)</u>: Evaluate the objective function at each corner point.

The value of C at each corner point is:

Corner Point	$C = 1200x_1 + 100x_2$
$(10, 0)$	$C = 1200(10) + 100(0) = 12{,}000$
$(7, 15)$	$C = 1200(7) + 100(15) = 9{,}900$
$(12, 0)$	$C = 1200(12) + 100(0) = 14{,}400$

<u>Step (5)</u>: Determine the optimal solution.

The minimum occurs when $x_1 = 7$ and $x_2 = 15$. That is, the officers should rent 7 buses and 15 vans at the minimum cost of $9900.

35. (A) <u>Step (1)</u>: Summarize relevant material.

	Grams per Gallon Emitted by:		Maximum allowed
	Old process	New Process	
Sulfur dioxide	20	5	16,000
Particulate	40	20	30,000

<u>Step (2)</u>: Form a mathematical model for the problem.

Let x_1 = the number of gallons produced by the old process and x_2 = the number of gallons produced by the new process.

The mathematical model for this problem is:

Maximize $P = 60x_1 + 20x_2$

Subject to: $20x_1 + 5x_2 \leq 16{,}000$

$40x_1 + 20x_2 \leq 30{,}000$

$x_1 \geq 0, \; x_2 \geq 0$

Step (3): Graph the feasible region and find the corner points.

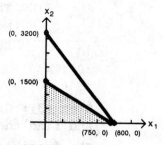

The feasible region S is the solution set of the given inequalities, and is indicated by the shading in the graph at the right.

The corner points are $(0, 0)$, $(0, 1500)$, and $(750, 0)$.

Since S is bounded, P has a maximum value by 3(a).

Step (4): Evaluate the objective function at each corner point.

The value of P at each corner point is:

Corner Point	$P = 60x_1 + 20x_2$
$(0, 0)$	$P = 60(0) + 20(0) = 0$
$(0, 1500)$	$P = 60(0) + 20(1500) = 30,000$
$(750, 0)$	$P = 60(750) + 20(0) = 45,000$

The maximum profit is \$450 when 750 gallons are produced using the old process exclusively.

(B) The mathematical model for this problem is:
Maximize $P = 60x_1 + 20x_2$
Subject to: $20x_1 + 5x_2 \le 11,500$
$\qquad\qquad 40x_1 + 20x_2 \le 30,000$

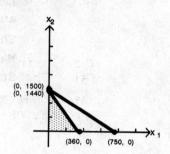

The feasible region S for this problem is indicated by the shading in the graph at the right.

The corner points are $(0, 0)$, $(0, 1500)$, $(400, 700)$, and $(575, 0)$.

The value P at each corner point is:

Corner Point	$P = 60x_1 + 20x_2$
$(0, 0)$	$P = 60(0) + 20(0) = 0$
$(0, 1500)$	$P = 60(0) + 20(1,500) = 30,000$
$(400, 700)$	$P = 60(400) + 20(700) = 38,000$
$(575, 0)$	$P = 60(575) + 20(0) = 34,500$

The maximum profit is \$380 when 400 gallons are produced using the old process and 700 gallons using the new process.

(C) The mathematical model for this problem is:
Maximize $P = 60x_1 + 20x_2$
Subject to: $20x_1 + 5x_2 \le 7,200$
$\qquad\qquad 40x_1 + 20x_2 \le 30,000$

The feasible region S for this problem is indicated by the shading in the graph at the right.

The corner points are $(0, 0)$, $(0, 1440)$, and $(360, 0)$.

The value of P at each corner point is:

Corner Point	$P = 60x_1 + 20x_2$
(0, 0)	$P = 60(0) + 20(0) = 0$
(0, 1440)	$P = 60(0) + 20(1,440) = 28,800$
(360, 0)	$P = 60(360) + 20(0) = 21,600$

The maximum profit is $288 when 1,440 gallons are produced by the new process exclusively.

37. Let x_1 = the number of bags of Brand A
and x_2 = the number of bags of Brand B.

(A) The mathematical model for this problem is:

Maximize $N = 8x_1 + 3x_2$
Subject to: $4x_1 + 4x_2 \geq 1000$
$2x_1 + x_2 \leq 400$
$x_1 \geq 0,\ x_2 \geq 0$

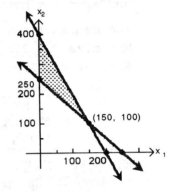

The feasible region S is the solution set of the system of inequalities, and is indicated by the shading in the graph at the right.
The corner points are (0, 250), (0, 400), and (150, 100).
Since S is bounded, N has a maximum value by <u>3</u>(a).

The value of N at each corner point is given in the table below:

Corner Point	$N = 8x_1 + 3x_2$
(0, 250)	$N = 8(0) + 3(250) = 750$
(150, 100)	$N = 8(150) + 3(100) = 1500$
(0, 400)	$N = 8(0) + 3(400) = 1200$

Thus, the maximum occurs when $x_1 = 150$ and $x_2 = 100$. That is, the grower should use 150 bags of Brand A and 100 bags of Brand B. The maximum number of pounds of nitrogen is 1500.

(B) The mathematical model for this problem is:

Minimize $N = 8x_1 + 3x_2$
Subject to: $4x_1 + 4x_2 \geq 1000$
$2x_1 + x_2 \leq 400$
$x_1 \geq 0,\ x_2 \geq 0$

The feasible region S and the corner points are the same as in part (A). Thus, the minimum occurs when $x_1 = 0$ and $x_2 = 250$. That is, the grower should use 0 bags of Brand A and 250 bags of Brand B. The minimum number of pounds of nitrogen is 750.

39.

	Amount per Cubic Yard (in pounds)		Minimum monthly requirement
	Mix A	Mix B	
Phosphoric acid	20	10	460
Nitrogen	30	30	960
Potash	5	10	220
Cost/cubic yd.	$30	$35	

Let x_1 = the number of cubic yards of mix A
and x_2 = the number of cubic yards of mix B.

The mathematical model for this problem is:

Minimize $C = 30x_1 + 35x_2$

Subject to: $20x_1 + 10x_2 \geq 460$
$30x_1 + 30x_2 \geq 960$
$5x_1 + 10x_2 \geq 220$
$x_1 \geq 0, \ x_2 \geq 0$

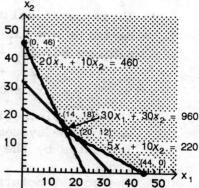

The feasible region S is the solution set of the given inequalities and is indicated by the shading in the graph at the right.

The corner points are $(0, 46)$, $(14, 18)$, $(20, 12)$, and $(44, 0)$.

Since S is unbounded and $a = 30 > 0$, $b = 35 > 0$, C has a minimum value by $\underline{3}$(b).

The value of C at each corner point is:

Corner Point	$C = 30x_1 + 35x_2$
$(0, 46)$	$C = 30(0) + 35(46) = 1610$
$(14, 18)$	$C = 30(14) + 35(18) = 1050$
$(20, 12)$	$C = 30(20) + 35(12) = 1020$
$(44, 0)$	$C = 30(44) + 35(0) = 1320$

Thus, the minimum occurs when the amount of mix A used is 20 cubic yards and the amount of mix B used is 12 cubic yards. The minimum cost is $C = \$1020$.

41. Let x_1 = the number of mice used
and x_2 = the number of rats used.

The mathematical model for this problem is:

Maximize $P = x_1 + x_2$

Subject to: $10x_1 + 20x_2 \leq 800$
$20x_1 + 10x_2 \leq 640$
$x_1 \geq 0, \ x_2 \geq 0$

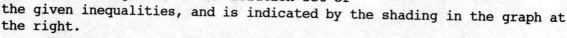

The feasible region S is the solution set of the given inequalities, and is indicated by the shading in the graph at the right.

The corner points are $(0, 0)$, $(0, 40)$, $(16, 32)$, and $(32, 0)$.

Since S is bounded, P has a maximum value by <u>3</u>(a).

The value of P at each corner point is:

Corner Point	$P = x_1 + x_2$
(0, 0)	$P = 0 + 0 = 0$
(0, 40)	$P = 0 + 40 = 40$
(16, 32)	$P = 16 + 32 = 48$
(32, 0)	$P = 32 + 0 = 32$

Thus, the maximum occurs when the number of mice used is 16 and the number of rats used is 32. The maximum number of mice and rats that can be used is 48.

EXERCISE 5-3

Things to remember:

<u>1</u>. STANDARD MAXIMIZATION PROBLEM IN STANDARD FORM

A linear programming problem is said to be a STANDARD MAXIMIZATION PROBLEM IN STANDARD FORM if its mathematical model is of the form:

Maximize $P = c_1 x_1 + c_2 x_2 + \cdots + c_n x_n$

Subject to problem constraints of the form:

$a_1 x_1 + a_2 x_2 + \cdots + a_n x_n \le b, \quad b \ge 0$

with nonnegative constraints:

$x_1, \; x_2, \; \ldots, \; x_n \ge 0.$

[<u>Note</u>: The coefficients of the objective function can be any real numbers.]

<u>2</u>. SLACK VARIABLES

Given a linear programming problem. SLACK VARIABLES are nonnegative quantities that are introduced to convert problem constraint inequalities into equations.

<u>3</u>. BASIC VARIABLES AND NONBASIC VARIABLES; BASIC SOLUTIONS AND BASIC FEASIBLE SOLUTIONS

Given a system of linear equations associated with a linear programming problem. (Such a system will always have more variables than equations.)

The variables are divided into two (mutually exclusive) groups, called BASIC VARIABLES and NONBASIC VARIABLES, as follows: Basic variables are selected arbitrarily with the one restriction that there be as many basic variables as there are equations. The remaining variables are called nonbasic variables.

A solution found by setting the nonbasic variables equal to zero and solving for the basic variables is called a BASIC SOLUTION. If a basic solution has no negative values, it is a BASIC FEASIBLE SOLUTION.

4. FUNDAMENTAL THEOREM OF LINEAR PROGRAMMING

If the optimal value of the objective function in a linear programming problem exists, then that value must occur at one (or more) of the basic feasible solutions.

---

1. (A) Since there are 2 problem constraints, 2 slack variables are introduced.

(B) Since there are two equations (from the two problem constraints) and three decision variables, there are two basic variables and three nonbasic variables.

(C) There will be two linear equations and two variables.

3. (A) There are 5 constraint equations; the number of equations is the same as the number of slack variables.

(B) There are 4 decision variables since there are 9 variables altogether, and 5 of them are slack variables.

(C) There are 5 basic variables and 4 nonbasic variables; the number of basic variables equals the number of equations.

(D) Five linear equations with 5 variables.

5.

	Nonbasic	Basic	Feasible?
(A)	x_1, x_2	s_1, s_2	Yes, all values are nonnegative.
(B)	x_1, s_1	x_2, s_2	Yes, all values are nonnegative.
(C)	x_1, s_2	x_2, s_1	No, $s_1 = -12 < 0$.
(D)	x_2, s_1	x_1, s_2	No, $s_2 = -12 < 0$.
(E)	x_2, s_2	x_1, s_1	Yes, all values are nonnegative.
(F)	s_1, s_2	x_1, x_2	Yes, all values are nonnegative.

7.

	x_1	x_2	s_1	s_2	Feasible?
(A)	0	0	50	40	Yes, all values are nonnegative.
(B)	0	50	0	-60	No, $s_2 = -60 < 0$.
(C)	0	20	30	0	Yes, all values are nonnegative.
(D)	25	0	0	15	Yes, all values are nonnegative.
(E)	40	0	-30	0	No, $s_1 = -30 < 0$.
(F)	20	10	0	0	Yes, all values are nonnegative.

9.

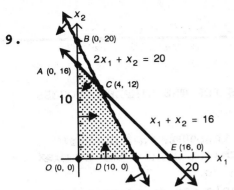

Introduce slack variables s_1 and s_2 to obtain the system of equations:

$$x_1 + x_2 + s_1 \qquad = 16$$
$$2x_1 + x_2 \qquad + s_2 = 20$$

x_1	x_2	s_1	s_2	Intersection Point	Feasible?
0	0	16	20	O	Yes
0	16	0	4	A	Yes
0	20	-4	0	B	No, $s_1 = -4 < 0$
16	0	0	-12	E	No, $s_2 = -12 < 0$
10	0	6	0	D	Yes
4	12	0	0	C	Yes

11.

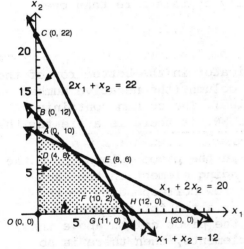

Introduce slack variables s_1, s_2, and s_3 to obtain the system of equations:

$$2x_1 + x_2 + s_1 \qquad = 22$$
$$x_1 + x_2 \qquad + s_2 \qquad = 12$$
$$x_1 + 2x_2 \qquad + s_3 = 20$$

x_1	x_2	s_1	s_2	s_3	Intersection Point	Feasible?
0	0	22	12	20	O	Yes
0	22	0	-10	-24	C	No
0	12	10	0	-4	B	No
0	10	12	2	0	A	Yes
11	0	0	1	9	G	Yes
12	0	-2	0	8	H	No
20	0	-18	-8	0	I	No
10	2	0	0	6	F	Yes
8	6	0	-2	0	E	No
4	8	6	0	0	D	Yes

Things to remember:

1. **SELECTING BASIC AND NONBASIC VARIABLES FOR THE SIMPLEX PROCESS**

   Given a simplex tableau.

   (a) Determine the number of basic and the number of nonbasic variables. These numbers do not change during the simplex process.

   (b) SELECTING BASIC VARIABLES: A variable can be selected as a basic variable only if it corresponds to a column in the tableau that has exactly one nonzero element (usually 1) and the nonzero element in the column is not in the same row as the nonzero element in the column of another basic variable. (This procedure always selects P as a basic variable, since the P column never changes during the simplex process.)

   (c) SELECTING NONBASIC VARIABLES: After the basic variables are selected in Step (b), the remaining variables are selected as the nonbasic variables. (The tableau columns under the nonbasic variables will usually contain more than one nonzero element.)

2. **SELECTING THE PIVOT ELEMENT**

   (a) Locate the most negative indicator in the bottom row of the tableau to the left of the P column (the negative number with the largest absolute value). The column containing this element is the PIVOT COLUMN. If there is a tie for the most negative, choose either.

   (b) Divide each POSITIVE element in the pivot column above the dashed line into the corresponding element in the last column. The PIVOT ROW is the row corresponding to the smallest quotient. If there is a tie for the smallest quotient, choose either. If the pivot column above the dashed line has no positive elements, then there is no solution and we stop.

   (c) The PIVOT (or PIVOT ELEMENT) is the element in the intersection of the pivot column and pivot row. [Note: The pivot element is always positive and is never in the bottom row.]

   [Remember: The entering variable is at the top of the pivot column and the exiting variable is at the left of the pivot row.]

<u>3</u>. PERFORMING THE PIVOT OPERATION

A PIVOT OPERATION or PIVOTING consists of performing row operations as follows:

(a) Multiply the pivot row by the reciprocal of the pivot element to transform the pivot element into a 1. (If the pivot element is already a 1, omit this step.)

(b) Add multiples of the pivot row to other rows in the tableau to transform all other nonzero elements in the pivot column into 0's.

[<u>Note</u>: Rows are not to be interchanged while performing a pivot operation. The only way the (positive) pivot element can be transformed into 1 (if it is not a 1 already) is for the pivot row to be multiplied by the reciprocal of the pivot element.]

<u>4</u>. SIMPLEX ALGORITHM FOR STANDARD MAXIMIZATION PROBLEMS

Problem constraints are of the ≤ form with nonnegative constants on the right hand side. The coefficients of the objective function can be any real numbers.

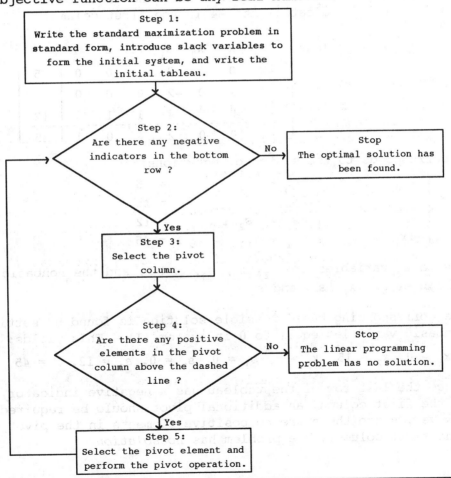

1. Given the simplex tableau:

$$\begin{array}{ccccc} x_1 & x_2 & s_1 & s_2 & P \\ \left[\begin{array}{ccccc|c} 2 & 1 & 0 & 3 & 0 & 12 \\ 3 & 0 & 1 & -2 & 0 & 15 \\ \hline -4 & 0 & 0 & 4 & 1 & 20 \end{array}\right] \end{array}$$

which corresponds to the system of equations:

$$(I) \begin{cases} 2x_1 + x_2 \qquad\quad + 3s_2 \qquad = 12 \\ 3x_1 \qquad\quad + s_1 - 2s_2 \qquad = 15 \\ -4x_1 \qquad\qquad\quad + 4s_2 + P = 20 \end{cases}$$

(A) The basic variables are x_2, s_1, and P, and the nonbasic variables are x_1 and s_2.

(B) The corresponding basic feasible solution is found by setting the nonbasic variables equal to 0 in system (I). This yields:

$$x_1 = 0, \ x_2 = 12, \ s_1 = 15, \ s_2 = 0, \ P = 20$$

(C) An additional pivot is required, since the last row of the tableau has a negative indicator, the -4 in the first column.

3. Given the simplex tableau:

$$\begin{array}{ccccccc} x_1 & x_2 & x_3 & s_1 & s_2 & s_3 & P \\ \left[\begin{array}{ccccccc|c} -2 & 0 & 1 & 3 & 1 & 0 & 0 & 5 \\ 0 & 1 & 0 & -2 & 0 & 0 & 0 & 15 \\ -1 & 0 & 0 & 4 & 1 & 1 & 0 & 12 \\ \hline -4 & 0 & 0 & 2 & 4 & 0 & 1 & 45 \end{array}\right] \end{array}$$

which corresponds to the system of equations:

$$(I) \begin{cases} -2x_1 \qquad + x_3 + 3s_1 + \ s_2 \qquad\qquad = 5 \\ \qquad x_2 \qquad\quad - 2s_1 \qquad\qquad\qquad = 15 \\ -x_1 \qquad\quad + 4s_1 + \ s_2 + s_3 \qquad = 12 \\ -4x_1 \qquad\quad + 2s_1 + 4s_2 \qquad + P = 45 \end{cases}$$

(A) The basic variables are x_2, x_3, s_3, and P, and the nonbasic variables are x_1, s_1, and s_2.

(B) The corresponding basic feasible solution is found by setting the nonbasic variables equal to 0 in system (I). This yields:

$$x_1 = 0, \ x_2 = 15, \ x_3 = 5, \ s_1 = 0, \ s_2 = 0, \ s_3 = 12, \ P = 45$$

(C) Since the last row of the tableau has a negative indicator, the -4 in the first column, an additional pivot should be required. However, since there are no positive elements in the pivot column (the first column), the problem has *no solution*.

5. Given the simplex tableau:

$$\begin{array}{ccccc} x_1 & x_2 & s_1 & s_2 & P \end{array}$$

$$\left[\begin{array}{ccccc|c} 1 & 4 & 1 & 0 & 0 & 4 \\ 3 & 5 & 0 & 1 & 0 & 24 \\ \hline -8 & -5 & 0 & 0 & 1 & 0 \end{array}\right]$$

The most negative indicator is -8 in the first column. Thus, the first column is the pivot column. Now, $\frac{4}{1} = 4$ and $\frac{24}{3} = 8$. Thus, the first row is the pivot row and the pivot element is the element in the first row, first column. These are indicated in the following tableau.

Enter

$$\begin{array}{ccccc} & x_1 & x_2 & s_1 & s_2 & P \end{array}$$

Exit s_1
$$\left[\begin{array}{ccccc|c} \textcircled{1} & 4 & 1 & 0 & 0 & 4 \\ 3 & 5 & 0 & 1 & 0 & 24 \\ \hline -8 & -5 & 0 & 0 & 1 & 0 \end{array}\right] \begin{array}{l} \frac{4}{1} = 4 \text{ (minimum)} \\ \frac{24}{3} = 8 \end{array}$$

with rows labeled s_2 and P.

$$\left[\begin{array}{ccccc|c} \textcircled{1} & 4 & 1 & 0 & 0 & 4 \\ 3 & 5 & 0 & 1 & 0 & 24 \\ \hline -8 & -5 & 0 & 0 & 1 & 0 \end{array}\right] \sim \left[\begin{array}{ccccc|c} 1 & 4 & 1 & 0 & 0 & 4 \\ 0 & -7 & -3 & 1 & 0 & 12 \\ \hline 0 & 27 & 8 & 0 & 1 & 32 \end{array}\right]$$

$$(-3)R_1 + R_2 \rightarrow R_2$$
$$8R_1 + R_3 \rightarrow R_3$$

7. Given the simplex tableau:

$$\begin{array}{cccccc} x_1 & x_2 & s_1 & s_2 & s_3 & P \end{array}$$

$$\left[\begin{array}{cccccc|c} 2 & 1 & 1 & 0 & 0 & 0 & 4 \\ 3 & 0 & 1 & 1 & 0 & 0 & 8 \\ 0 & 0 & 2 & 0 & 1 & 0 & 2 \\ \hline -4 & 0 & -3 & 0 & 0 & 1 & 5 \end{array}\right]$$

The most negative indicator is -4. Thus, the first column is the pivot column. Now, $\frac{4}{2} = 2$, $\frac{8}{3} = 2\frac{2}{3}$. Thus, the first row is the pivot row, and the pivot element is the element in the first row, first column. These are indicated in the tableau.

Enter

$$\begin{array}{cccccc} & x_1 & x_2 & s_1 & s_2 & s_3 & P \end{array}$$

Exit x_2
$$\left[\begin{array}{cccccc|c} \textcircled{2} & 1 & 1 & 0 & 0 & 0 & 4 \\ 3 & 0 & 1 & 1 & 0 & 0 & 8 \\ 0 & 0 & 2 & 0 & 1 & 0 & 2 \\ \hline -4 & 0 & -3 & 0 & 0 & 1 & 5 \end{array}\right] \begin{array}{l} \frac{4}{2} = 2 \text{ (minimum)} \\ \frac{8}{3} = 2\frac{2}{3} \end{array}$$

with rows labeled s_2, s_3, and P.

$$\begin{bmatrix} \begin{array}{cccccc|c} ② & 1 & 1 & 0 & 0 & 0 & 4 \\ 3 & 0 & 1 & 1 & 0 & 0 & 8 \\ 0 & 0 & 2 & 0 & 1 & 0 & 2 \\ \hline -4 & 0 & -3 & 0 & 0 & 1 & 5 \end{array} \end{bmatrix} \sim \begin{array}{cccccc} x_1 & x_2 & s_1 & s_2 & s_3 & P \end{array}$$

$$\begin{bmatrix} \begin{array}{cccccc|c} ① & \frac{1}{2} & \frac{1}{2} & 0 & 0 & 0 & 2 \\ 3 & 0 & 1 & 1 & 0 & 0 & 8 \\ 0 & 0 & 2 & 0 & 1 & 0 & 2 \\ \hline -4 & 0 & -3 & 0 & 0 & 1 & 5 \end{array} \end{bmatrix}$$

$$\tfrac{1}{2}R_1 \to R_1 \qquad\qquad (-3)R_1 + R_2 \to R_2, \quad 4R_1 + R_4 \to R_4$$

$$\sim \begin{bmatrix} \begin{array}{cccccc|c} 1 & \frac{1}{2} & \frac{1}{2} & 0 & 0 & 0 & 2 \\ 0 & -\frac{3}{2} & -\frac{1}{2} & 1 & 0 & 0 & 2 \\ 0 & 0 & 2 & 0 & 1 & 0 & 2 \\ \hline 0 & 2 & -1 & 0 & 0 & 1 & 13 \end{array} \end{bmatrix}$$

9. (A) Introduce slack variables s_1 and s_2 to obtain:

Maximize $P = 15x_1 + 10x_2$
Subject to: $2x_1 + x_2 + s_1 \qquad\quad = 10$
$\qquad\qquad x_1 + 3x_2 \qquad + s_2 = 10$
$\qquad\qquad x_1, \ x_2, \ s_1, \ s_2 \ge 0$

This system can be written in initial form:

$$2x_1 + \ x_2 + s_1 \qquad\qquad\quad = 10$$
$$x_1 + 3x_2 \qquad + s_2 \qquad = 10$$
$$-15x_1 - 10x_2 \qquad\qquad + P = 0$$
$$x_1, \ x_2, \ s_1, \ s_2 \ge 0$$

(B) The simplex tableau for this problem is:

$$\begin{array}{c} \text{Enter} \\ \begin{array}{c} \\ \text{Exit } s_1 \\ s_2 \\ P \end{array} \begin{array}{ccccc} x_1 & x_2 & s_1 & s_2 & P \end{array} \end{array}$$

$$\begin{bmatrix} \begin{array}{ccccc|c} ② & 1 & 1 & 0 & 0 & 10 \\ 1 & 3 & 0 & 1 & 0 & 10 \\ \hline -15 & -10 & 0 & 0 & 1 & 0 \end{array} \end{bmatrix} \quad \begin{array}{l} \frac{10}{2} = 5 \ (\text{minimum}) \\[4pt] \frac{10}{1} = 10 \end{array}$$

Column 1 is the pivot column (-15 is the most negative indicator). Row 1 is the pivot row (5 is the smallest positive quotient). Thus, the pivot element is the circled 2.

(C) We use the simplex method as outlined above. The pivot elements are circled.

$$\begin{array}{c} \begin{array}{c} s_1 \\ s_2 \\ P \end{array} \begin{array}{ccccc} x_1 & x_2 & s_1 & s_2 & P \end{array} \end{array}$$

$$\begin{bmatrix} \begin{array}{ccccc|c} ② & 1 & 1 & 0 & 0 & 10 \\ 1 & 3 & 0 & 1 & 0 & 10 \\ \hline -15 & -10 & 0 & 0 & 1 & 0 \end{array} \end{bmatrix} \sim \begin{bmatrix} \begin{array}{ccccc|c} ① & \frac{1}{2} & \frac{1}{2} & 0 & 0 & 5 \\ 1 & 3 & 0 & 1 & 0 & 10 \\ \hline -15 & -10 & 0 & 0 & 1 & 0 \end{array} \end{bmatrix} \sim$$

$$\tfrac{1}{2}R_1 \to R_1 \qquad\qquad (-1)R_1 + R_2 \to R_2$$
$$15R_1 + R_3 \to R_3$$

$$\sim \begin{bmatrix} 1 & \frac{1}{2} & \frac{1}{2} & 0 & 0 & | & 5 \\ 0 & \boxed{\frac{5}{2}} & -\frac{1}{2} & 1 & 0 & | & 5 \\ \hdashline 0 & -\frac{5}{2} & \frac{15}{2} & 0 & 1 & | & 75 \end{bmatrix} \sim \begin{bmatrix} 1 & \frac{1}{2} & \frac{1}{2} & 0 & 0 & | & 5 \\ 0 & \boxed{1} & -\frac{1}{5} & \frac{2}{5} & 0 & | & 2 \\ \hdashline 0 & -\frac{5}{2} & \frac{15}{2} & 0 & 1 & | & 75 \end{bmatrix}$$

$$\frac{2}{5} R_2 \rightarrow R_2 \qquad\qquad \left(-\frac{1}{2}\right) R_2 + R_1 \rightarrow R_1$$

$$\left(\frac{5}{2}\right) R_2 + R_3 \rightarrow R_3$$

$$\sim \begin{array}{c} \\ x_1 \\ x_2 \\ P \end{array} \begin{array}{cc} \begin{array}{ccccc} x_1 & x_2 & s_1 & s_2 & P \end{array} \\ \begin{bmatrix} 1 & 0 & \frac{3}{5} & -\frac{1}{5} & 0 & | & 4 \\ 0 & 1 & -\frac{1}{5} & \frac{2}{5} & 0 & | & 2 \\ \hdashline 0 & 0 & 7 & 1 & 1 & | & 80 \end{bmatrix} \end{array}$$

All elements in the last row are nonnegative. Thus, max $P = 80$ at $x_1 = 4$, $x_2 = 2$, $s_1 = 0$, $s_2 = 0$.

11. (A) Introduce slack variables s_1 and s_2 to obtain:

$$\begin{aligned}
\text{Maximize } P &= 30x_1 + x_2 \\
\text{Subject to: } 2x_1 &+ x_2 + s_1 && = 10 \\
x_1 &+ 3x_2 && + s_2 = 10 \\
x_1, x_2, &\ s_1, s_2 \geq 0
\end{aligned}$$

This system can be written in the initial form:

$$\begin{aligned}
2x_1 + x_2 + s_1 &&&= 10 \\
x_1 + 3x_2 &&+ s_2 &= 10 \\
-30x_1 - x_2 &&+ P &= 0
\end{aligned}$$

(B) The simplex tableau for this problem is:

$$\begin{array}{c} \\ \text{Exit } s_1 \\ s_2 \\ P \end{array} \begin{array}{c} \text{Enter} \\ \begin{array}{ccccc} x_1 & x_2 & s_1 & s_2 & P \end{array} \\ \begin{bmatrix} \boxed{2} & 1 & 1 & 0 & 0 & | & 10 \\ 1 & 3 & 0 & 1 & 0 & | & 10 \\ \hdashline -30 & -1 & 0 & 0 & 1 & | & 0 \end{bmatrix} \end{array} \quad \begin{array}{l} \frac{10}{2} = 5 \text{ (minimum)} \\ \frac{10}{1} = 10 \end{array}$$

↑
pivot
column

(C)

$$
\begin{array}{c}
\begin{array}{ccccc} x_1 & x_2 & s_1 & s_2 & P \end{array}\\
\begin{array}{c} s_1 \\ s_2 \\ P \end{array}
\left[\begin{array}{ccccc|c}
② & 1 & 1 & 0 & 0 & 10 \\
1 & 3 & 0 & 1 & 0 & 10 \\
\hline
-30 & -1 & 0 & 0 & 1 & 0
\end{array}\right]
\end{array}
\sim
\left[\begin{array}{ccccc|c}
① & \frac{1}{2} & \frac{1}{2} & 0 & 0 & 5 \\
1 & 3 & 0 & 1 & 0 & 10 \\
\hline
-30 & -1 & 0 & 0 & 1 & 0
\end{array}\right]
$$

$$\tfrac{1}{2}R_1 \to R_1 \qquad\qquad\qquad (-1)R_1 + R_2 \to R_2$$
$$30R_1 + R_3 \to R_3$$

$$
\begin{array}{c}
\begin{array}{ccccc} x_1 & x_2 & s_1 & s_2 & P \end{array}\\
\begin{array}{c} x_1 \\ s_2 \\ P \end{array}
\sim
\left[\begin{array}{ccccc|c}
1 & \frac{1}{2} & \frac{1}{2} & 0 & 0 & 5 \\
0 & \frac{5}{2} & -\frac{1}{2} & 1 & 0 & 5 \\
\hline
0 & 14 & 15 & 0 & 1 & 150
\end{array}\right]
\end{array}
$$

All the elements in the last row are nonnegative. Thus, max $P = 150$ at $x_1 = 5$, $x_2 = 0$, $s_1 = 0$, $s_2 = 5$.

13. The simplex tableau for this problem is:

$$
\begin{array}{c}
\text{Enter}\\
\begin{array}{cccccc} x_1 & x_2 & s_1 & s_2 & s_3 & P \end{array}\\
\begin{array}{c} s_1 \\ s_2 \\ s_3 \\ {} \end{array}
\left[\begin{array}{cccccc|c}
2 & 1 & 1 & 0 & 0 & 0 & 10 \\
1 & 1 & 0 & 1 & 0 & 0 & 7 \\
1 & ② & 0 & 0 & 1 & 0 & 12 \\
\hline
-30 & -40 & 0 & 0 & 0 & 1 & 0
\end{array}\right]
\begin{array}{l}
10 \\
7 \\
\frac{12}{2} = 6 \ \text{(minimum)}
\end{array}
\end{array}
$$

pivot → s_3 row

Exit

↑ pivot column $\tfrac{1}{2}R_3 \to R_3$

[**Note:** The pivot elements have been circled.]

$$
\sim
\left[\begin{array}{cccccc|c}
2 & 1 & 1 & 0 & 0 & 0 & 10 \\
1 & 1 & 0 & 1 & 0 & 0 & 7 \\
\frac{1}{2} & ① & 0 & 0 & \frac{1}{2} & 0 & 6 \\
\hline
-30 & -40 & 0 & 0 & 0 & 1 & 0
\end{array}\right]
$$

$$(-1)R_3 + R_1 \to R_1, \ (-1)R_3 + R_2 \to R_2, \text{ and } 40R_3 + R_4 \to R_4$$

$$
\sim
\left[\begin{array}{cccccc|c}
\frac{3}{2} & 0 & 1 & 0 & -\frac{1}{2} & 0 & 4 \\
① & 0 & 0 & 1 & -\frac{1}{2} & 0 & 1 \\
\frac{1}{2} & 1 & 0 & 0 & \frac{1}{2} & 0 & 6 \\
\hline
-10 & 0 & 0 & 0 & 20 & 1 & 240
\end{array}\right]
\begin{array}{l}
\frac{4}{3/2} = \frac{8}{3} \\
\frac{1}{1/2} = 2 \ \text{(minimum)} \\
\frac{6}{1/2} = 12
\end{array}
$$

pivot → row

↑ pivot column $2R_2 \to R_2$

$$\sim \begin{bmatrix} \frac{3}{2} & 0 & 1 & 0 & -\frac{1}{2} & 0 & | & 4 \\ \boxed{1} & 0 & 0 & 2 & -1 & 0 & | & 2 \\ \frac{1}{2} & 1 & 0 & 0 & \frac{1}{2} & 0 & | & 6 \\ \hline -10 & 0 & 0 & 0 & 20 & 1 & | & 240 \end{bmatrix}$$

$$\left(-\frac{3}{2}\right)R_2 + R_1 \to R_1, \quad \left(-\frac{1}{2}\right)R_2 + R_3 \to R_3, \text{ and } 10R_2 + R_4 \to R_4$$

$$\sim \begin{array}{c} s_1 \\ x_1 \\ x_2 \\ \\ \end{array} \begin{bmatrix} \begin{array}{cccccc} x_1 & x_2 & s_1 & s_2 & s_3 & P \end{array} \\ \begin{array}{cccccc} 0 & 0 & 1 & -3 & 1 & 0 \end{array} & | & 1 \\ \begin{array}{cccccc} 1 & 0 & 0 & 2 & -1 & 0 \end{array} & | & 2 \\ \begin{array}{cccccc} 0 & 1 & 0 & -1 & 1 & 0 \end{array} & | & 5 \\ \hline \begin{array}{cccccc} 0 & 0 & 0 & 20 & 10 & 1 \end{array} & | & 260 \end{bmatrix}$$

Optimal solution: max $P = 260$ at $x_1 = 2$, $x_2 = 5$, $s_1 = 1$, $s_2 = 0$, $s_3 = 0$.

15. The simplex tableau for this problem is:

Enter

Exit $\quad x_1 \quad x_2 \quad s_1 \quad s_2 \quad s_3 \quad P$

$$\begin{array}{c} \text{pivot} \to s_1 \\ \text{row} \\ s_2 \\ s_3 \\ P \end{array} \begin{bmatrix} -2 & \boxed{1} & 1 & 0 & 0 & 0 & | & 2 \\ -1 & 1 & 0 & 1 & 0 & 0 & | & 5 \\ 0 & 1 & 0 & 0 & 1 & 0 & | & 6 \\ \hline -2 & -3 & 0 & 0 & 0 & 1 & | & 0 \end{bmatrix} \begin{array}{l} \frac{2}{1} = 2 \text{ (minimum)} \\ \frac{5}{1} = 5 \\ \frac{6}{1} = 6 \\ \\ \end{array}$$

pivot column

$$(-1)R_1 + R_2 \to R_2, \quad (-1)R_1 + R_3 \to R_3, \text{ and } 3R_1 + R_4 \to R_4$$

$$\sim \begin{bmatrix} -2 & 1 & 1 & 0 & 0 & 0 & | & 2 \\ 1 & 0 & -1 & 1 & 0 & 0 & | & 3 \\ \boxed{2} & 0 & -1 & 0 & 1 & 0 & | & 4 \\ \hline -8 & 0 & 3 & 0 & 0 & 1 & | & 6 \end{bmatrix} \begin{array}{l} \\ \frac{3}{1} = 3 \\ \frac{4}{2} = 2 \text{ (minimum)} \\ \\ \end{array}$$

pivot row

pivot column

$$\frac{1}{2}R_3 \to R_3$$

$$\sim \begin{bmatrix} -2 & 1 & 1 & 0 & 0 & 0 & | & 2 \\ 1 & 0 & -1 & 1 & 0 & 0 & | & 3 \\ \boxed{1} & 0 & -\frac{1}{2} & 0 & \frac{1}{2} & 0 & | & 2 \\ \hline -8 & 0 & 3 & 0 & 0 & 1 & | & 6 \end{bmatrix} \sim \begin{bmatrix} \begin{array}{cccccc} x_1 & x_2 & s_1 & s_2 & s_3 & P \end{array} \\ \begin{array}{cccccc} 0 & 1 & 0 & 0 & 1 & 0 \end{array} & | & 6 \\ \begin{array}{cccccc} 0 & 0 & -\frac{1}{2} & 1 & -\frac{1}{2} & 0 \end{array} & | & 1 \\ \begin{array}{cccccc} 1 & 0 & -\frac{1}{2} & 0 & \frac{1}{2} & 0 \end{array} & | & 2 \\ \hline \begin{array}{cccccc} 0 & 0 & -1 & 0 & 4 & 1 \end{array} & | & 22 \end{bmatrix}$$

pivot column

$$2R_3 + R_1 \to R_1, \quad (-1)R_3 + R_2 \to R_2,$$
$$\text{and } 8R_3 + R_4 \to R_4$$

Since there are no positive elements in the pivot column (above the dashed line), we conclude that there is no solution.

17. The simplex tableau for this problem is:

$$
\begin{array}{c}
\text{pivot} \to s_1 \\
\text{row} \quad s_2 \\
s_3 \\
P
\end{array}
\begin{array}{c}
\quad x_1 \quad x_2 \quad s_1 \quad s_2 \quad s_3 \quad P \\
\left[
\begin{array}{cccccc|c}
-1 & \boxed{1} & 1 & 0 & 0 & 0 & 2 \\
-1 & 3 & 0 & 1 & 0 & 0 & 12 \\
1 & -4 & 0 & 0 & 1 & 0 & 4 \\
\hdashline
1 & -2 & 0 & 0 & 0 & 1 & 0
\end{array}
\right]
\end{array}
\quad
\begin{array}{l}
\frac{2}{1} = 2 \text{ (minimum)} \\[6pt]
\frac{12}{3} = 4
\end{array}
$$

pivot column

$(-3)R_1 + R_2 \to R_2$, $4R_1 + R_3 \to R_3$, and $2R_1 + R_4 \to R_4$

$$
\begin{array}{c}
\text{pivot} \to \\
\text{row}
\end{array}
\sim
\left[
\begin{array}{cccccc|c}
-1 & 1 & 1 & 0 & 0 & 0 & 2 \\
\boxed{2} & 0 & -3 & 1 & 0 & 0 & 6 \\
-3 & 0 & 4 & 0 & 1 & 0 & 12 \\
\hdashline
-1 & 0 & 2 & 0 & 0 & 1 & 4
\end{array}
\right]
\quad
\begin{array}{l}
\frac{6}{2} = 3 \leftarrow \text{pivot row} \\[6pt]
\text{[\underline{Note}: We only use the} \\
\textit{positive} \text{ elements above} \\
\text{the dashed line in the} \\
\text{pivot column.]}
\end{array}
$$

pivot column $\frac{1}{2}R_2 \to R_2$

$$
\sim
\left[
\begin{array}{cccccc|c}
-1 & 1 & 1 & 0 & 0 & 0 & 2 \\
\boxed{1} & 0 & -\frac{3}{2} & \frac{1}{2} & 0 & 0 & 3 \\
-3 & 0 & 4 & 0 & 1 & 0 & 12 \\
\hdashline
-1 & 0 & 2 & 0 & 0 & 1 & 4
\end{array}
\right]
\sim
\begin{array}{c}
x_2 \\
x_1 \\
s_3 \\
\;
\end{array}
\begin{array}{c}
x_1 \quad x_2 \quad s_1 \quad s_2 \quad s_3 \quad P \\
\left[
\begin{array}{cccccc|c}
0 & 1 & -\frac{1}{2} & \frac{1}{2} & 0 & 0 & 5 \\
1 & 0 & -\frac{3}{2} & \frac{1}{2} & 0 & 0 & 3 \\
0 & 0 & -\frac{1}{2} & \frac{3}{2} & 1 & 0 & 21 \\
\hdashline
0 & 0 & \frac{1}{2} & \frac{1}{2} & 0 & 1 & 7
\end{array}
\right]
\end{array}
$$

$R_2 + R_1 \to R_1$, $3R_2 + R_3 \to R_3$,
and $R_2 + R_4 \to R_4$

Optimal solution: max $P = 7$ at $x_1 = 3$,
$x_2 = 5$, $s_1 = 0$, $s_2 = 0$, $s_3 = 21$.

19. The simplex tableau for this problem is:

$$
\begin{array}{c}
\text{pivot} \to s_1 \\
\text{row} \quad s_2 \\
P
\end{array}
\begin{array}{c}
\; x_1 \quad x_2 \quad x_3 \quad s_1 \quad s_2 \quad P \\
\left[
\begin{array}{cccccc|c}
\boxed{1} & 1 & -1 & 1 & 0 & 0 & 10 \\
2 & 4 & 3 & 0 & 1 & 0 & 30 \\
\hdashline
-5 & -2 & 1 & 0 & 0 & 1 & 0
\end{array}
\right]
\end{array}
\quad
\begin{array}{l}
\frac{10}{1} = 10 \text{ (minimum)} \\[6pt]
\frac{30}{2} = 15
\end{array}
$$

pivot column $(-2)R_1 + R_2 \to R_2$, $5R_1 + R_3 \to R_3$

$$
\sim
\left[
\begin{array}{cccccc|c}
1 & 1 & -1 & 1 & 0 & 0 & 10 \\
0 & 2 & \boxed{5} & -2 & 1 & 0 & 10 \\
\hdashline
0 & 3 & -4 & 5 & 0 & 1 & 50
\end{array}
\right]
\sim
\left[
\begin{array}{cccccc|c}
1 & 1 & -1 & 1 & 0 & 0 & 10 \\
0 & \frac{2}{5} & \boxed{1} & -\frac{2}{5} & \frac{1}{5} & 0 & 2 \\
\hdashline
0 & 3 & -4 & 5 & 0 & 1 & 50
\end{array}
\right]
$$

$\frac{1}{5}R_2 \to R_2$ 　　　　　　$R_2 + R_1 \to R_1$, $4R_2 + R_3 \to R_3$

$$\begin{array}{c}
\quad\quad x_1 \quad x_2 \quad x_3 \quad s_1 \quad s_2 \quad P \\
\begin{array}{c} x_1 \\ \\ x_3 \\ \\ P \end{array}
\left[\begin{array}{ccccccc}
1 & \frac{7}{5} & 0 & \frac{3}{5} & \frac{1}{5} & 0 & 12 \\
0 & \frac{2}{5} & 1 & -\frac{2}{5} & \frac{1}{5} & 0 & 2 \\
\hdashline
0 & \frac{23}{5} & 0 & \frac{17}{5} & \frac{4}{5} & 1 & 58
\end{array}\right]
\end{array}$$

Optimal solution: max $P = 58$ at $x_1 = 12$, $x_2 = 0$, $x_3 = 2$, $s_1 = 0$, $s_2 = 0$.

21. The simplex tableau for this problem is:

$$\begin{array}{c}
\quad\quad\quad\quad x_1 \quad x_2 \quad x_3 \quad s_1 \quad s_2 \quad P \\
\begin{array}{c} s_1 \\ \\ \text{pivot} \to \ s_2 \\ \text{row} \\ P \end{array}
\left[\begin{array}{cccccc|c}
1 & 0 & 1 & 1 & 0 & 0 & 4 \\
0 & 1 & ① & 0 & 1 & 0 & 3 \\
\hline
-2 & -3 & -4 & 0 & 0 & 1 & 0
\end{array}\right]
\begin{array}{l} \frac{4}{1} = 4 \\ \\ \frac{3}{1} = 3 \text{ (minimum)} \end{array}
\end{array}$$

pivot column $\quad (-1)R_2 + R_1 \to R_1, \quad 4R_2 + R_3 \to R_3$

$$\sim \left[\begin{array}{cccccc|c}
① & -1 & 0 & 1 & -1 & 0 & 1 \\
0 & 1 & 1 & 0 & 1 & 0 & 3 \\
\hline
-2 & 1 & 0 & 0 & 4 & 1 & 12
\end{array}\right]
\sim \left[\begin{array}{cccccc|c}
1 & -1 & 0 & 1 & -1 & 0 & 1 \\
0 & ① & 1 & 0 & 1 & 0 & 3 \\
\hline
0 & -1 & 0 & 2 & 2 & 1 & 14
\end{array}\right]$$

$2R_1 + R_3 \to R_3$ $\qquad\qquad\qquad R_2 + R_1 \to R_1$ and $R_2 + R_3 \to R_3$

$$\begin{array}{c}
x_1 \quad x_2 \quad x_3 \quad s_1 \quad s_2 \quad P \\
\sim \left[\begin{array}{cccccc|c}
1 & 0 & 1 & 1 & 0 & 0 & 4 \\
0 & 1 & 1 & 0 & 1 & 0 & 3 \\
\hdashline
0 & 0 & 1 & 2 & 3 & 1 & 17
\end{array}\right]
\end{array}$$

Optimal solution: max $P = 17$ at $x_1 = 4$, $x_2 = 3$, $x_3 = 0$, $s_1 = 0$, $s_2 = 0$.

23. The simplex tableau for this problem is:

$$\begin{array}{c}
\quad\quad\quad\quad x_1 \quad x_2 \quad x_3 \quad s_1 \quad s_2 \quad s_3 \quad P \\
\begin{array}{c} s_1 \\ \\ \text{pivot} \to \ s_2 \\ \text{row} \\ s_3 \\ \\ \end{array}
\left[\begin{array}{ccccccc|c}
3 & 2 & 5 & 1 & 0 & 0 & 0 & 23 \\
② & 1 & 1 & 0 & 1 & 0 & 0 & 8 \\
1 & 1 & 2 & 0 & 0 & 1 & 0 & 7 \\
\hline
-4 & -3 & -2 & 0 & 0 & 0 & 1 & 0
\end{array}\right]
\begin{array}{l} \frac{23}{3} = 7\frac{2}{3} \\ \\ \frac{8}{2} = 4 \text{ (minimum)} \\ \\ \frac{7}{1} = 7 \end{array}
\end{array}$$

pivot column $\quad \frac{1}{2}R_2 \to R_2$

$$\sim \left[\begin{array}{ccccccc|c}
3 & 2 & 5 & 1 & 0 & 0 & 0 & 23 \\
① & \frac{1}{2} & \frac{1}{2} & 0 & \frac{1}{2} & 0 & 0 & 4 \\
1 & 1 & 2 & 0 & 0 & 1 & 0 & 7 \\
\hline
-4 & -3 & -2 & 0 & 0 & 0 & 1 & 0
\end{array}\right]
\sim \left[\begin{array}{ccccccc|c}
0 & \frac{1}{2} & \frac{7}{2} & 1 & -\frac{3}{2} & 0 & 0 & 11 \\
1 & \frac{1}{2} & \frac{1}{2} & 0 & \frac{1}{2} & 0 & 0 & 4 \\
0 & ⑫ & \frac{3}{2} & 0 & -\frac{1}{2} & 1 & 0 & 3 \\
\hline
0 & -1 & 0 & 0 & 2 & 0 & 1 & 16
\end{array}\right]$$

$(-3)R_2 + R_1 \to R_1$, $(-1)R_2 + R_3 \to R_3$, and $\qquad 2R_3 \to R_3$
$4R_2 + R_4 \to R_4$

$$\sim \begin{bmatrix} 0 & \frac{1}{2} & \frac{7}{2} & 1 & -\frac{3}{2} & 0 & 0 & 11 \\ 1 & \frac{1}{2} & \frac{1}{2} & 0 & \frac{1}{2} & 0 & 0 & 4 \\ 0 & \textcircled{1} & 3 & 0 & -1 & 2 & 0 & 6 \\ \hdashline 0 & -1 & 0 & 0 & 2 & 0 & 1 & 16 \end{bmatrix}$$

	x_1	x_2	x_3	s_1	s_2	s_3	P	
s_1	0	0	2	1	-1	-1	0	8
x_1	1	0	-1	0	1	-1	0	1
x_2	0	1	3	0	-1	2	0	6
P	0	0	3	0	1	2	1	22

$\left(-\frac{1}{2}\right)R_3 + R_1 \rightarrow R_1,\ \left(-\frac{1}{2}\right)R_3 + R_2 \rightarrow R_2$, and

$R_3 + R_4 \rightarrow R_4$

Optimal solution: max $P = 22$ at $x_1 = 1$, $x_2 = 6$, $x_3 = 0$, $s_1 = 8$, $s_2 = 0$, $s_3 = 0$.

25. Multiply the first problem constraint by $\frac{10}{6}$, the second by 100, and the third by 10 to clear the fractions. Then, the simplex tableau for this problem is:

	x_1	x_2	s_1	s_2	s_3	P	
s_1	1	②	1	0	0	0	1,600
s_2	3	4	0	1	0	0	3,600
s_3	3	2	0	0	1	0	2,700
P	-20	-30	0	0	0	1	0

$\dfrac{1,600}{2} = 800$

$\dfrac{3,600}{4} = 900$

$\dfrac{2,700}{2} = 1,350$

$\frac{1}{2}R_1 \rightarrow R_1$

$$\sim \begin{bmatrix} \frac{1}{2} & \textcircled{1} & \frac{1}{2} & 0 & 0 & 0 & 800 \\ 3 & 4 & 0 & 1 & 0 & 0 & 3,600 \\ 3 & 2 & 0 & 0 & 1 & 0 & 2,700 \\ \hdashline -20 & -30 & 0 & 0 & 0 & 1 & 0 \end{bmatrix}$$

$(-4)R_1 + R_2 \rightarrow R_2,\ (-2)R_1 + R_3 \rightarrow R_3$, and $30R_1 + R_4 \rightarrow R_4$

$$\sim \begin{bmatrix} \frac{1}{2} & 1 & \frac{1}{2} & 0 & 0 & 0 & 800 \\ \textcircled{1} & 0 & -2 & 1 & 0 & 0 & 400 \\ 2 & 0 & -1 & 0 & 1 & 0 & 1,100 \\ \hdashline -5 & 0 & 15 & 0 & 0 & 1 & 24,000 \end{bmatrix}$$

$\dfrac{800}{1/2} = 1,600$

$\dfrac{400}{1} = 400$

$\dfrac{1,100}{2} = 550$

$\left(-\frac{1}{2}\right)R_2 + R_1 \rightarrow R_1,\ (-2)R_2 + R_3 \rightarrow R_3$, and $5R_2 + R_4 \rightarrow R_4$

$$\sim \quad \begin{array}{c} \\ x_2 \\ x_1 \\ s_3 \\ \hline P \end{array} \begin{bmatrix} \begin{array}{cccccc|c} x_1 & x_2 & s_1 & s_2 & s_3 & P & \\ 0 & 1 & \frac{3}{2} & -\frac{1}{2} & 0 & 0 & 600 \\ 1 & 0 & -2 & 1 & 0 & 0 & 400 \\ 0 & 0 & 3 & -2 & 1 & 0 & 300 \\ \hline 0 & 0 & 5 & 5 & 0 & 1 & 26,000 \end{array} \end{bmatrix}$$

Optimal solution: max $P = 26{,}000$ at $x_1 = 400$, $x_2 = 600$, $s_1 = 0$, $s_2 = 0$, $s_3 = 300$.

27. The simplex tableau for this problem is:

$$\begin{array}{c} \\ s_1 \\ s_2 \\ s_3 \\ \hline P \end{array} \begin{bmatrix} \begin{array}{ccccccc|c} x_1 & x_2 & x_3 & s_1 & s_2 & s_3 & P & \\ 2 & 2 & \circled{8} & 1 & 0 & 0 & 0 & 600 \\ 1 & 3 & 2 & 0 & 1 & 0 & 0 & 600 \\ 3 & 2 & 1 & 0 & 0 & 1 & 0 & 400 \\ \hline -1 & -2 & -3 & 0 & 0 & 0 & 1 & 0 \end{array} \end{bmatrix} \begin{array}{l} \frac{600}{8} = 75 \\ \\ \frac{600}{2} = 300 \\ \\ \frac{400}{1} = 400 \end{array}$$

$$\frac{1}{8} R_1 \to R_1$$

$$\sim \begin{bmatrix} \begin{array}{ccccccc|c} \frac{1}{4} & \frac{1}{4} & \circled{1} & \frac{1}{8} & 0 & 0 & 0 & 75 \\ 1 & 3 & 2 & 0 & 1 & 0 & 0 & 600 \\ 3 & 2 & 1 & 0 & 0 & 1 & 0 & 400 \\ \hline -1 & -2 & -3 & 0 & 0 & 0 & 1 & 0 \end{array} \end{bmatrix}$$

$$(-2) R_1 + R_2 \to R_2, \quad (-1) R_1 + R_3 \to R_3, \text{ and } 3 R_1 + R_4 \to R_4$$

$$\sim \begin{bmatrix} \begin{array}{ccccccc|c} \frac{1}{4} & \frac{1}{4} & 1 & \frac{1}{8} & 0 & 0 & 0 & 75 \\ \frac{1}{2} & \circled{\frac{5}{2}} & 0 & -\frac{1}{4} & 1 & 0 & 0 & 450 \\ \frac{11}{4} & \frac{7}{4} & 0 & -\frac{1}{8} & 0 & 1 & 0 & 325 \\ \hline -\frac{1}{4} & -\frac{5}{4} & 0 & \frac{3}{8} & 0 & 0 & 1 & 225 \end{array} \end{bmatrix} \begin{array}{l} \frac{75}{1/4} = 300 \\ \\ \frac{450}{5/2} = 180 \\ \\ \frac{325}{7/4} = 185.71 \end{array}$$

$$\frac{2}{5} R_2 \to R_2$$

$$\sim \begin{bmatrix} \begin{array}{ccccccc|c} \frac{1}{4} & \frac{1}{4} & 1 & \frac{1}{8} & 0 & 0 & 0 & 75 \\ \frac{1}{5} & \circled{1} & 0 & -\frac{1}{10} & \frac{2}{5} & 0 & 0 & 180 \\ \frac{11}{4} & \frac{7}{4} & 0 & -\frac{1}{8} & 0 & 1 & 0 & 325 \\ \hline -\frac{1}{4} & -\frac{5}{4} & 0 & \frac{3}{8} & 0 & 0 & 1 & 225 \end{array} \end{bmatrix}$$

$$\sim \quad \begin{array}{c} \\ x_3 \\ x_2 \\ s_3 \\ \hline P \end{array} \begin{bmatrix} \begin{array}{ccccccc|c} x_1 & x_2 & x_3 & s_1 & s_2 & s_3 & P & \\ \frac{1}{5} & 0 & 1 & \frac{3}{20} & -\frac{1}{10} & 0 & 0 & 30 \\ \frac{1}{5} & 1 & 0 & -\frac{1}{10} & \frac{2}{5} & 0 & 0 & 180 \\ \frac{12}{5} & 0 & 0 & \frac{1}{20} & -\frac{7}{10} & 1 & 0 & 10 \\ \hline 0 & 0 & 0 & \frac{1}{4} & \frac{1}{2} & 0 & 1 & 450 \end{array} \end{bmatrix}$$

$$\left(-\frac{1}{4}\right) R_2 + R_1 \to R_1, \quad \left(-\frac{7}{4}\right) R_2 + R_3 \to R_3,$$
and $\frac{5}{4} R_2 + R_4 \to R_4$

Optimal solution: max $P = 450$ at $x_1 = 0$, $x_2 = 180$, $x_3 = 30$, $s_1 = 0$, $s_2 = 0$, $s_3 = 10$.

29. The simplex tableau for this problem is:

$$
\begin{array}{c}
\\
s_1 \\
s_2 \\
s_3 \\
s_4 \\
P
\end{array}
\begin{array}{c}
\begin{array}{ccccccc}
x_1 & x_2 & s_1 & s_2 & s_3 & s_4 & P
\end{array} \\
\left[
\begin{array}{ccccccc|c}
1 & 2 & 1 & 0 & 0 & 0 & 0 & 40 \\
1 & 3 & 0 & 1 & 0 & 0 & 0 & 48 \\
1 & 4 & 0 & 0 & 1 & 0 & 0 & 60 \\
0 & ① & 0 & 0 & 0 & 1 & 0 & 14 \\
\hdashline
-2 & -5 & 0 & 0 & 0 & 0 & 1 & 0
\end{array}
\right]
\end{array}
\begin{array}{l}
\frac{40}{2} = 20 \\[4pt]
\frac{48}{3} = 16 \\[4pt]
\frac{60}{4} = 15 \\[4pt]
\frac{14}{1} = 14 \\[4pt]
\;
\end{array}
$$

$(-2)R_4 + R_1 \rightarrow R_1$, $(-3)R_4 + R_2 \rightarrow R_2$, $(-4)R_4 + R_3 \rightarrow R_3$, and $5R_4 + R_5 \rightarrow R_5$

$$
\sim
\left[
\begin{array}{ccccccc|c}
1 & 0 & 1 & 0 & 0 & -2 & 0 & 12 \\
1 & 0 & 0 & 1 & 0 & -3 & 0 & 6 \\
① & 0 & 0 & 0 & 1 & -4 & 0 & 4 \\
0 & 1 & 0 & 0 & 0 & 1 & 0 & 14 \\
\hdashline
-2 & 0 & 0 & 0 & 0 & 5 & 1 & 70
\end{array}
\right]
\begin{array}{l}
\frac{12}{1} = 12 \\[4pt]
\frac{6}{1} = 6 \\[4pt]
\frac{4}{1} = 4 \\[4pt]
\;
\end{array}
$$

$(-1)R_3 + R_1 \rightarrow R_1$, $(-1)R_3 + R_2 \rightarrow R_2$, and $2R_3 + R_5 \rightarrow R_5$

$$
\sim
\left[
\begin{array}{ccccccc|c}
0 & 0 & 1 & 0 & -1 & 2 & 0 & 8 \\
0 & 0 & 0 & 1 & -1 & ① & 0 & 2 \\
1 & 0 & 0 & 0 & 1 & -4 & 0 & 4 \\
0 & 1 & 0 & 0 & 0 & 1 & 0 & 14 \\
\hdashline
0 & 0 & 0 & 0 & 2 & -3 & 1 & 78
\end{array}
\right]
\begin{array}{l}
\frac{8}{2} = 4 \\[4pt]
\frac{2}{1} = 2 \\[4pt]
\; \\[4pt]
\frac{14}{1} = 14 \\[4pt]
\;
\end{array}
$$

$(-2)R_2 + R_1 \rightarrow R_1$, $4R_2 + R_3 \rightarrow R_3$, $(-1)R_2 + R_4 \rightarrow R_4$, and $3R_2 + R_5 \rightarrow R_5$

$$
\sim
\left[
\begin{array}{ccccccc|c}
0 & 0 & 1 & -2 & ① & 0 & 0 & 4 \\
0 & 0 & 0 & 1 & -1 & 1 & 0 & 2 \\
1 & 0 & 0 & 4 & -3 & 0 & 0 & 12 \\
0 & 1 & 0 & -1 & 1 & 0 & 0 & 12 \\
\hdashline
0 & 0 & 0 & 3 & -1 & 0 & 1 & 84
\end{array}
\right]
\begin{array}{l}
\frac{4}{1} = 4 \\[4pt]
\; \\[4pt]
\; \\[4pt]
\frac{12}{1} = 12 \\[4pt]
\;
\end{array}
$$

$R_1 + R_2 \rightarrow R_2$, $3R_1 + R_3 \rightarrow R_3$, $(-1)R_1 + R_4 \rightarrow R_4$, and $R_1 + R_5 \rightarrow R_5$

	x_1	x_2	s_1	s_2	s_3	s_4	P	
s_3	0	0	1	-2	1	0	0	4
s_4	0	0	1	-1	0	1	0	6
~ x_1	1	0	3	-2	0	0	0	24
x_2	0	1	-1	1	0	0	0	8
P	0	0	1	1	0	0	1	88

Optimal solution: max $P = 88$ at $x_1 = 24$, $x_2 = 8$, $s_1 = 0$, $s_2 = 0$, $s_3 = 4$, $s_4 = 6$.

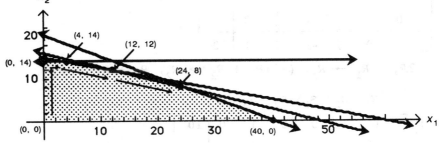

31. The simplex tableau for this problem is:

	x_1	x_2	s_1	s_2	s_3	P	
s_1	2	1	1	0	0	0	16
s_2	1	0	0	1	0	0	6
s_3	0	1	0	0	1	0	10
P	-1	-1	0	0	0	1	0

(A) Solution using the first column as the pivot column

	x_1	x_2	s_1	s_2	s_3	P		
	2	1	1	0	0	0	16	$\dfrac{16}{2} = 8$
	①	0	0	1	0	0	6	$\dfrac{6}{1} = 6$
	0	1	0	0	1	0	10	
	-1	-1	0	0	0	1	0	

$$(-2)R_2 + R_1 \rightarrow R_1, \quad R_2 + R_4 \rightarrow R_4$$

	x_1	x_2	s_1	s_2	s_3	P		
~	0	①	1	-2	0	0	4	$\dfrac{4}{1} = 4$
	1	0	0	1	0	0	6	
	0	1	0	0	1	0	10	$\dfrac{10}{1} = 10$
	0	-1	0	1	0	1	6	

$$(-1)R_1 + R_3 \rightarrow R_3, \quad R_1 + R_4 \rightarrow R_4$$

$$\sim \begin{bmatrix} 0 & 1 & 1 & -2 & 0 & 0 & | & 4 \\ 1 & 0 & 0 & 1 & 0 & 0 & | & 6 \\ 0 & 0 & -1 & ② & 1 & 0 & | & 6 \\ \text{-----} & & & & & & & \\ 0 & 0 & 1 & -1 & 0 & 1 & | & 10 \end{bmatrix} \quad \begin{array}{l} \frac{6}{1} = 6 \\ \\ \frac{6}{2} = 3 \end{array}$$

$$\tfrac{1}{2}R_3 \rightarrow R_3$$

$$\sim \begin{bmatrix} 0 & 1 & 1 & -2 & 0 & 0 & | & 4 \\ 1 & 0 & 0 & 1 & 0 & 0 & | & 6 \\ 0 & 0 & -\tfrac{1}{2} & 1 & \tfrac{1}{2} & 0 & | & 3 \\ \text{-----} & & & & & & & \\ 0 & 0 & 1 & -1 & 0 & 1 & | & 10 \end{bmatrix}$$

$$2R_3 + R_1 \rightarrow R_1, \quad (-1)R_3 + R_2 \rightarrow R_2, \quad R_3 + R_4 \rightarrow R_4$$

$$\begin{array}{c} \\ x_2 \\ x_1 \\ \sim \quad s_2 \\ \\ P \end{array} \begin{array}{cccccc} x_1 & x_2 & s_1 & s_2 & s_3 & P \\ \left[\begin{array}{cccccc} 0 & 1 & 0 & 0 & 1 & 0 \\ 1 & 0 & \tfrac{1}{2} & 0 & -\tfrac{1}{2} & 0 \\ 0 & 0 & -\tfrac{1}{2} & 1 & \tfrac{1}{2} & 0 \\ \text{-----} & & & & & \\ 0 & 0 & \tfrac{1}{2} & 0 & \tfrac{1}{2} & 1 \end{array} \right. & \left. \begin{array}{c} | & 10 \\ | & 3 \\ | & 3 \\ \\ | & 13 \end{array} \right] \end{array}$$

Optimal solution: max $P = 13$
at $x_1 = 3$, $x_2 = 10$, $s_1 = 0$,
$s_2 = 3$, $s_3 = 0$

(B) Solution using the second column as the pivot column

$$\begin{array}{c} \\ s_1 \\ s_2 \\ s_3 \\ P \end{array} \begin{array}{cccccc} x_1 & x_2 & s_1 & s_2 & s_3 & P \\ \left[\begin{array}{cccccc} 2 & 1 & 1 & 0 & 0 & 0 \\ 1 & 0 & 0 & 1 & 0 & 0 \\ 0 & ① & 0 & 0 & 1 & 0 \\ \text{-----} & & & & & \\ -1 & -1 & 0 & 0 & 0 & 1 \end{array} \right. & \left. \begin{array}{c} | & 16 \\ | & 6 \\ | & 10 \\ \\ | & 0 \end{array} \right] \end{array} \quad \begin{array}{l} \frac{16}{1} = 16 \\ \\ \frac{10}{1} = 10 \end{array}$$

$$(-1)R_3 + R_1 \rightarrow R_1, \quad R_3 + R_4 \rightarrow R_4$$

$$\sim \begin{bmatrix} ② & 0 & 1 & 0 & -1 & 0 & | & 6 \\ 1 & 0 & 0 & 1 & 0 & 0 & | & 6 \\ 0 & 1 & 0 & 0 & 1 & 0 & | & 10 \\ \text{-----} & & & & & & & \\ -1 & 0 & 0 & 0 & 1 & 1 & | & 10 \end{bmatrix} \quad \begin{array}{l} \frac{6}{2} = 3 \\ \\ \frac{6}{1} = 6 \end{array}$$

$$\tfrac{1}{2}R_1 \rightarrow R_1$$

$$\sim \begin{bmatrix} 1 & 0 & \frac{1}{2} & 0 & -\frac{1}{2} & 0 & 3 \\ 1 & 0 & 0 & 1 & 0 & 0 & 6 \\ 0 & 1 & 0 & 0 & 1 & 0 & 10 \\ \hline -1 & 0 & 0 & 0 & 1 & 1 & 10 \end{bmatrix}$$

$$(-1)R_1 + R_2 \rightarrow R_2, \quad R_1 + R_4 \rightarrow R_4$$

$$
\sim
\begin{array}{c}
 \\ x_1 \\ s_2 \\ x_2 \\ P
\end{array}
\begin{array}{c}
\begin{array}{cccccc} x_1 & x_2 & s_1 & s_2 & s_3 & P \end{array} \\
\left[\begin{array}{cccccc|c}
1 & 0 & \frac{1}{2} & 0 & -\frac{1}{2} & 0 & 3 \\
0 & 0 & -\frac{1}{2} & 1 & \frac{1}{2} & 0 & 3 \\
0 & ① & 0 & 0 & 1 & 0 & 10 \\
\hline
0 & 0 & \frac{1}{2} & 0 & \frac{1}{2} & 1 & 13
\end{array} \right]
\end{array}
$$

Optimal solution: max $P = 13$ at $x_1 = 3$, $x_2 = 10$, $s_1 = 0$, $s_2 = 3$, $s_3 = 0$

Choosing either solution produces the *same* optimal solution.

33. The simplex tableau for this problem is:

$$
\begin{array}{c}
 \\ s_1 \\ s_2 \\ P
\end{array}
\begin{array}{c}
\begin{array}{cccccc} x_1 & x_2 & x_3 & s_1 & s_2 & P \end{array} \\
\left[\begin{array}{cccccc|c}
1 & 1 & 2 & 1 & 0 & 0 & 20 \\
2 & 1 & 4 & 0 & 1 & 0 & 32 \\
\hline
-3 & -3 & -2 & 0 & 0 & 1 & 0
\end{array} \right]
\end{array}
$$

(A) Solution using the first column as the pivot column

$$
\begin{array}{c}
\begin{array}{cccccc} x_1 & x_2 & x_3 & s_1 & s_2 & P \end{array} \\
\left[\begin{array}{cccccc|c}
1 & 1 & 2 & 1 & 0 & 0 & 20 \\
② & 1 & 4 & 0 & 1 & 0 & 32 \\
\hline
-3 & -3 & -2 & 0 & 0 & 1 & 0
\end{array} \right]
\end{array}
\begin{array}{l}
\frac{20}{1} = 20 \\[4pt]
\frac{32}{2} = 16
\end{array}
$$

$$\tfrac{1}{2} R_2 \rightarrow R_2$$

$$
\sim
\left[\begin{array}{cccccc|c}
1 & 1 & 2 & 1 & 0 & 0 & 20 \\
1 & \frac{1}{2} & 2 & 0 & \frac{1}{2} & 0 & 16 \\
\hline
-3 & -3 & -2 & 0 & 0 & 1 & 0
\end{array} \right]
$$

$$(-1)R_2 + R_1 \rightarrow R_1, \quad 3R_2 + R_3 \rightarrow R_3$$

$$
\sim
\left[\begin{array}{cccccc|c}
0 & \frac{1}{2} & 0 & 1 & -\frac{1}{2} & 0 & 4 \\
1 & \frac{1}{2} & 2 & 0 & \frac{1}{2} & 0 & 16 \\
\hline
0 & -\frac{3}{2} & 4 & 0 & \frac{3}{2} & 1 & 48
\end{array} \right]
\begin{array}{l}
\frac{4}{1/2} = 8 \\[4pt]
\frac{16}{1/2} = 32
\end{array}
$$

$$2R_1 \rightarrow R_1$$

$$\sim \begin{bmatrix} 0 & 1 & 0 & 2 & -1 & 0 & | & 8 \\ 1 & \frac{1}{2} & 2 & 0 & \frac{1}{2} & 0 & | & 16 \\ \hline 0 & -\frac{3}{2} & 4 & 0 & \frac{3}{2} & 1 & | & 48 \end{bmatrix}$$

$$\left(-\frac{1}{2}\right)R_1 + R_2 \rightarrow R_2, \quad \frac{3}{2}R_1 + R_3 \rightarrow R_3$$

$$
\begin{array}{c}
\\
x_2 \\
\sim \quad x_1 \\
P
\end{array}
\begin{array}{cccccc}
x_1 & x_2 & x_3 & s_1 & s_2 & P \\
\end{array}
$$

$$
\begin{array}{c}
x_2 \\
x_1 \\
P
\end{array}
\begin{bmatrix}
0 & 1 & 0 & 2 & -1 & 0 & | & 8 \\
1 & 0 & 2 & -1 & 1 & 0 & | & 12 \\
\hline
0 & 0 & 4 & 3 & 0 & 1 & | & 60
\end{bmatrix}
$$

Optimal solution: max $P = 60$ at $x_1 = 12$, $x_2 = 8$, $x_3 = 0$, $s_1 = 0$, $s_2 = 0$

(B) Solution using the second column as the pivot column

$$
\begin{array}{c}
s_1 \\
s_2 \\
P
\end{array}
\begin{array}{cccccc}
x_1 & x_2 & x_3 & s_1 & s_2 & P \\
\end{array}
$$

$$
\begin{array}{c}
s_1 \\
s_2 \\
P
\end{array}
\begin{bmatrix}
1 & \textcircled{1} & 2 & 1 & 0 & 0 & | & 20 \\
2 & 1 & 4 & 0 & 1 & 0 & | & 32 \\
\hline
-3 & -3 & -2 & 0 & 0 & 1 & | & 0
\end{bmatrix}
\quad
\begin{array}{l}
\frac{20}{1} = 20 \\[6pt]
\frac{32}{1} = 32
\end{array}
$$

$$(-1)R_1 + R_2 \rightarrow R_2, \quad 3R_1 + R_3 \rightarrow R_3$$

$$
\begin{array}{c}
x_2 \\
\sim \quad s_2 \\
P
\end{array}
\begin{array}{cccccc}
x_1 & x_2 & x_3 & s_1 & s_2 & P \\
\end{array}
$$

$$
\begin{array}{c}
x_2 \\
s_2 \\
P
\end{array}
\begin{bmatrix}
1 & 1 & 2 & 1 & 0 & 0 & | & 20 \\
1 & 0 & 2 & -1 & 1 & 0 & | & 12 \\
\hline
0 & 0 & 4 & 3 & 0 & 1 & | & 60
\end{bmatrix}
$$

Optimal solution: max $P = 60$ at $x_1 = 0$, $x_2 = 20$, $x_3 = 0$, $s_1 = 0$, $s_2 = 12$

The maximum value of P is 60. Since the optimal solution is obtained at two corner points, $(12, 8, 0)$ and $(0, 20, 0)$, every point on the line segment connecting these points is also an optimal solution.

35. Let $x_1 =$ the number of A components
$x_2 =$ the number of B components
$x_3 =$ the number of C components

The mathematical model for this problem is:
Maximize $P = 7x_1 + 8x_2 + 10x_3$
Subject to
$$
\begin{aligned}
2x_1 + 3x_2 + 2x_3 &\le 1000 \\
x_1 + x_2 + 2x_3 &\le 800 \\
x_1, \ x_2, \ x_3 &\ge 0
\end{aligned}
$$

We introduce slack variables s_1, s_2 to obtain the equivalent form:
$$
\begin{aligned}
2x_1 + 3x_2 + 2x_3 + s_1 \qquad\qquad &= 1000 \\
x_1 + x_2 + 2x_3 \qquad + s_2 \quad &= 800 \\
-7x_1 - 8x_2 - 10x_3 \qquad\qquad + P &= 0
\end{aligned}
$$

The simplex tableau for this problem is:

$$
\begin{array}{c}
\;\;\begin{array}{cccccc} x_1 & x_2 & x_3 & s_1 & s_2 & P \end{array}\\
\begin{array}{c} s_1 \\ s_2 \\ P \end{array}
\left[\begin{array}{cccccc|c}
2 & 3 & 2 & 1 & 0 & 0 & 1000 \\
1 & 1 & ② & 0 & 1 & 0 & 800 \\
\hline
-7 & -8 & -10 & 0 & 0 & 1 & 0
\end{array}\right]
\begin{array}{l} \dfrac{1000}{2}=500 \\[6pt] \dfrac{800}{2}=400 \end{array}
\end{array}
$$

$$\tfrac{1}{2}R_2 \rightarrow R_2$$

$$
\sim
\left[\begin{array}{cccccc|c}
2 & 3 & 2 & 1 & 0 & 0 & 1000 \\
\frac{1}{2} & \frac{1}{2} & ① & 0 & \frac{1}{2} & 0 & 400 \\
\hline
-7 & -8 & -10 & 0 & 0 & 1 & 0
\end{array}\right]
$$

$$(-2)R_2 + R_1 \rightarrow R_1, \quad 10R_2 + R_3 \rightarrow R_3$$

$$
\sim
\left[\begin{array}{cccccc|c}
1 & ② & 0 & 1 & -1 & 0 & 200 \\
\frac{1}{2} & \frac{1}{2} & 1 & 0 & \frac{1}{2} & 0 & 400 \\
\hline
-2 & -3 & 0 & 0 & 5 & 1 & 4000
\end{array}\right]
\begin{array}{l} \dfrac{200}{2}=100 \\[6pt] \dfrac{400}{1/2}=800 \end{array}
$$

$$\tfrac{1}{2}R_1 \rightarrow R_1$$

$$
\sim
\left[\begin{array}{cccccc|c}
\frac{1}{2} & ① & 0 & \frac{1}{2} & -\frac{1}{2} & 0 & 100 \\
\frac{1}{2} & \frac{1}{2} & 1 & 0 & \frac{1}{2} & 0 & 400 \\
\hline
-2 & -3 & 0 & 0 & 5 & 1 & 4000
\end{array}\right]
$$

$$\left(-\tfrac{1}{2}\right)R_1 + R_2 \rightarrow R_2, \quad 3R_1 + R_3 \rightarrow R_3$$

$$
\sim
\left[\begin{array}{cccccc|c}
\boxed{\tfrac{1}{2}} & 1 & 0 & \frac{1}{2} & -\frac{1}{2} & 0 & 100 \\
\frac{1}{4} & 0 & 1 & -\frac{1}{4} & \frac{3}{4} & 0 & 350 \\
\hline
-\frac{1}{2} & 0 & 0 & \frac{3}{2} & \frac{7}{2} & 1 & 4300
\end{array}\right]
\begin{array}{l} \dfrac{100}{1/2}=200 \\[6pt] \dfrac{350}{1/4}=1400 \end{array}
$$

$$2R_1 \rightarrow R_1$$

$$
\sim
\left[\begin{array}{cccccc|c}
1 & 2 & 0 & 1 & -1 & 0 & 200 \\
\frac{1}{4} & 0 & 1 & -\frac{1}{4} & \frac{3}{4} & 0 & 350 \\
\hline
-\frac{1}{2} & 0 & 0 & \frac{3}{2} & \frac{7}{2} & 1 & 4300
\end{array}\right]
$$

$$\left(-\tfrac{1}{4}\right)R_1 + R_2 \rightarrow R_2, \quad \tfrac{1}{2}R_1 + R_3 \rightarrow R_3$$

$$\begin{array}{c} \\ x_1 \\ \sim \quad x_3 \\ P \end{array} \begin{array}{cccccc} x_1 & x_2 & x_3 & s_1 & s_2 & P \\ \left[\begin{array}{cccccc|c} 1 & 2 & 0 & 1 & -1 & 0 & 200 \\ 0 & -\frac{1}{2} & 1 & -\frac{1}{2} & 1 & 0 & 300 \\ \hline 0 & 1 & 0 & 2 & 3 & 1 & 4400 \end{array}\right] \end{array}$$

Optimal solution: the maximum profit is \$4400 when 200 A components, 0 B components and 300 C components are manufactured.

37. Let x_1 = the amount invested in government bonds,
x_2 = the amount invested in mutual funds,
and x_3 = the amount invested in money market funds.

The mathematical model for this problem is:

Maximize $P = .08x_1 + .13x_2 + .15x_3$
Subject to: $x_1 + x_2 + x_3 \le 100{,}000$
$$x_2 + x_3 \le x_1$$
$$x_1, \ x_2, \ x_3 \ge 0$$

We introduce slack variables s_1 and s_2 to obtain the equivalent form:

$$\begin{array}{rrrrrl}
x_1 + & x_2 + & x_3 + s_1 & & = & 100{,}000 \\
-x_1 + & x_2 + & x_3 & + s_2 & = & 0 \\
-.08x_1 - & .13x_2 - & .15x_3 & + P & = & 0
\end{array}$$

The simplex tableau for this problem is:

$$\begin{array}{c} \\ s_1 \\ s_2 \\ P \end{array} \begin{array}{cccccc} x_1 & x_2 & x_3 & s_1 & s_2 & P \\ \left[\begin{array}{cccccc|c} 1 & 1 & 1 & 1 & 0 & 0 & 100{,}000 \\ -1 & 1 & \boxed{1} & 0 & 1 & 0 & 0 \\ \hline -.08 & -.13 & -.15 & 0 & 0 & 1 & 0 \end{array}\right] \end{array} \quad \dfrac{100{,}000}{1} = 100{,}000$$

$(-1)R_2 + R_1 \to R_1$ and $.15R_2 + R_3 \to R_3$

$$\sim \left[\begin{array}{cccccc|c} \boxed{2} & 0 & 0 & 1 & -1 & 0 & 100{,}000 \\ -1 & 1 & 1 & 0 & 1 & 0 & 0 \\ \hline -.23 & .02 & 0 & 0 & .15 & 1 & 0 \end{array}\right] \sim \left[\begin{array}{cccccc|c} \boxed{1} & 0 & 0 & \frac{1}{2} & -\frac{1}{2} & 0 & 50{,}000 \\ -1 & 1 & 1 & 0 & 1 & 0 & 0 \\ \hline -.23 & .02 & 0 & 0 & .15 & 1 & 0 \end{array}\right]$$

$\dfrac{1}{2}R_1 \to R_1$

$R_1 + R_2 \to R_2$ and $.23R_1 + R_3 \to R_3$

$$\begin{array}{c} \\ x_1 \\ \sim \quad x_2 \\ P \end{array} \begin{array}{cccccc} x_1 & x_2 & x_3 & s_1 & s_2 & P \\ \left[\begin{array}{cccccc|c} 1 & 0 & 0 & \frac{1}{2} & -\frac{1}{2} & 0 & 50{,}000 \\ 0 & 1 & 1 & \frac{1}{2} & \frac{1}{2} & 0 & 50{,}000 \\ \hline 0 & .02 & 0 & .115 & .035 & 1 & 11{,}500 \end{array}\right] \end{array}$$

Optimal solution: the maximum return is \$11,500 when $x_1 =$ \$50,000 is invested in government bonds, $x_2 =$ \$0 is invested in mutual funds, and $x_3 =$ \$50,000 is invested in money market funds.

39. Let x_1 = the number of daytime ads,

x_2 = the number of prime-time ads,

and x_3 = the number of late-night ads.

The mathematical model for this problem is:

Maximize $P = 14,000x_1 + 24,000x_2 + 18,000x_3$

Subject to: $1000x_1 + 2000x_2 + 1500x_3 \le 20,000$

$$x_1 + x_2 + x_3 \le 15$$

$$x_1, \ x_2, \ x_3 \ge 0$$

We introduce slack variables to obtain the following initial form:

$$1000x_1 + 2000x_2 + 1500x_3 + s_1 = 20,000$$

$$x_1 + x_2 + x_3 + s_2 = 15$$

$$-14,000x_1 - 24,000x_2 - 18,000x_3 + P = 0$$

The simplex tableau for this problem is:

	x_1	x_2	x_3	s_1	s_2	P		
s_1	1000	(2000)	1500	1	0	0	20,000	$\dfrac{20,000}{2000} = 10$
s_2	1	1	1	0	1	0	15	$\dfrac{15}{1} = 15$
P	$-14,000$	$-24,000$	$-18,000$	0	0	1	0	

$$\frac{1}{2000}R_1 \rightarrow R_1$$

$$\sim \begin{bmatrix} \frac{1}{2} & (1) & \frac{3}{4} & \frac{1}{2000} & 0 & 0 & | & 10 \\ 1 & 1 & 1 & 0 & 1 & 0 & | & 15 \\ \hline -14,000 & -24,000 & -18,000 & 0 & 0 & 1 & | & 0 \end{bmatrix}$$

$$(-1)R_1 + R_2 \rightarrow R_2, \quad 24,000R_1 + R_3 \rightarrow R_3$$

$$\sim \begin{bmatrix} \frac{1}{2} & 1 & \frac{3}{4} & \frac{1}{2000} & 0 & 0 & | & 10 \\ \left(\frac{1}{2}\right) & 0 & \frac{1}{4} & -\frac{1}{2000} & 1 & 0 & | & 5 \\ \hline -2000 & 0 & 0 & 12 & 0 & 1 & | & 240,000 \end{bmatrix}$$

$$2R_2 \rightarrow R_2$$

$$\sim \begin{bmatrix} \frac{1}{2} & 1 & \frac{3}{4} & \frac{1}{2000} & 0 & 0 & | & 10 \\ (1) & 0 & \frac{1}{2} & -\frac{1}{1000} & 2 & 0 & | & 10 \\ \hline -2000 & 0 & 0 & 12 & 0 & 1 & | & 240,000 \end{bmatrix}$$

$$\left(-\frac{1}{2}\right)R_2 + R_1 \rightarrow R_1, \quad 2000R_2 + R_3 \rightarrow R_3$$

$$\sim \begin{array}{c} x_2 \\ x_1 \\ P \end{array} \left[\begin{array}{cccccc|c} x_1 & x_2 & x_3 & s_1 & s_2 & P & \\ 0 & 1 & \frac{1}{2} & \frac{1}{1000} & -1 & 0 & 5 \\ 1 & 0 & \frac{1}{2} & -\frac{1}{1000} & 2 & 0 & 10 \\ \hline 0 & 0 & 1000 & 10 & 4000 & 1 & 260,000 \end{array} \right]$$

Optimal solution: maximum number of potential customers is 260,000 when $x_1 = 10$ daytime ads, $x_2 = 5$ prime-time ads, and $x_3 = 0$ late-night ads are placed.

41. Let x_1 = the number of colonial houses,
$\quad\quad x_2$ = the number of split-level houses,
and x_3 = the number of ranch-style houses.

(A) The mathematical model for this problem is:

Maximize $P = 20,000x_1 + 18,000x_2 + 24,000x_3$

Subject to:
$$\frac{1}{2}x_1 + \frac{1}{2}x_2 + x_3 \leq 30$$
$$60,000x_1 + 60,000x_2 + 80,000x_3 \leq 3,200,000$$
$$4,000x_1 + 3,000x_2 + 4,000x_3 \leq 180,000$$
$$x_1, x_2, x_3 \geq 0$$

We simplify the inequalities and then introduce slack variables to obtain the initial form:

$$\frac{1}{2}x_1 + \frac{1}{2}x_2 + x_3 + s_1 = 30$$
$$6x_1 + 6x_2 + 8x_3 + s_2 = 320$$
$$4x_1 + 3x_2 + 4x_3 + s_3 = 180$$
$$-20,000x_1 - 18,000x_2 - 24,000x_3 + P = 0$$

[Note: This simplification will change the interpretation of the slack variables.]

The simplex tableau for this problem is:

$$\begin{array}{c} s_1 \\ s_2 \\ s_3 \\ P \end{array} \left[\begin{array}{ccccccc|c} x_1 & x_2 & x_3 & s_1 & s_2 & s_3 & P & \\ \frac{1}{2} & \frac{1}{2} & \textcircled{1} & 1 & 0 & 0 & 0 & 30 \\ 6 & 6 & 8 & 0 & 1 & 0 & 0 & 320 \\ 4 & 3 & 4 & 0 & 0 & 1 & 0 & 180 \\ \hline -20,000 & -18,000 & -24,000 & 0 & 0 & 0 & 1 & 0 \end{array} \right] \begin{array}{l} \frac{30}{1} = 30 \\ \frac{320}{8} = 40 \\ \frac{180}{4} = 45 \\ \end{array}$$

$(-8)R_1 + R_2 \rightarrow R_2, \quad (-4)R_1 + R_3 \rightarrow R_3, \quad 24,000R_1 + R_4 \rightarrow R_4$

$$\sim \left[\begin{array}{ccccccc|c} \frac{1}{2} & \frac{1}{2} & 1 & 1 & 0 & 0 & 0 & 30 \\ 2 & 2 & 0 & -8 & 1 & 0 & 0 & 80 \\ \textcircled{2} & 1 & 0 & -4 & 0 & 1 & 0 & 60 \\ \hline -8000 & -6000 & 0 & 24,000 & 0 & 0 & 1 & 720,000 \end{array} \right]$$

$\frac{1}{2}R_3 \rightarrow R_3$

$$\sim \begin{bmatrix} \frac{1}{2} & \frac{1}{2} & 1 & 1 & 0 & 0 & 0 & | & 30 \\ 2 & 2 & 0 & -8 & 1 & 0 & 0 & | & 80 \\ \boxed{1} & \frac{1}{2} & 0 & -2 & 0 & \frac{1}{2} & 0 & | & 30 \\ \hline -8000 & -6000 & 0 & 24{,}000 & 0 & 0 & 1 & | & 720{,}000 \end{bmatrix}$$

$$\left(-\frac{1}{2}\right)R_3 + R_1 \rightarrow R_1, \quad (-2)R_3 + R_2 \rightarrow R_2, \quad 8000R_3 + R_4 \rightarrow R_4$$

$$\sim \begin{bmatrix} 0 & \frac{1}{4} & 1 & 2 & 0 & -\frac{1}{4} & 0 & | & 15 \\ 0 & \boxed{1} & 0 & -4 & 1 & -1 & 0 & | & 20 \\ 1 & \frac{1}{2} & 0 & -2 & 0 & \frac{1}{2} & 0 & | & 30 \\ \hline 0 & -2000 & 0 & 8000 & 0 & 4000 & 1 & | & 960{,}000 \end{bmatrix}$$

$$\left(-\frac{1}{4}\right)R_2 + R_1 \rightarrow R_1, \quad \left(-\frac{1}{2}\right)R_2 + R_3 \rightarrow R_3, \quad 2000R_2 + R_4 \rightarrow R_4$$

$$\sim \begin{array}{c} \\ x_3 \\ x_2 \\ x_1 \\ P \end{array} \begin{array}{cccccccc} x_1 & x_2 & x_3 & s_1 & s_2 & s_3 & P & \\ \left[\, 0\right. & 0 & 1 & 3 & -\frac{1}{4} & 0 & 0 & |\quad 10 \\ 0 & 1 & 0 & -4 & 1 & -1 & 0 & |\quad 20 \\ 1 & 0 & 0 & 0 & -\frac{1}{2} & 1 & 0 & |\quad 20 \\ \hline 0 & 0 & 0 & 0 & 2000 & 2000 & 1 & |\quad 1{,}000{,}000 \left.\right] \end{array}$$

Optimal solution: maximum profit is \$1,000,000 when $x_1 = 20$ colonial houses, $x_2 = 20$ split-level houses, and $x_3 = 10$ ranch-style houses are built.

(B) The mathematical model for this problem is:

Maximize $P = 17{,}000x_1 + 18{,}000x_2 + 24{,}000x_3$

Subject to:
$$\frac{1}{2}x_1 + \frac{1}{2}x_2 + x_3 \le 30$$
$$60{,}000x_1 + 60{,}000x_2 + 80{,}000x_3 \le 3{,}200{,}000$$
$$4{,}000x_1 + 3{,}000x_2 + 4{,}000x_3 \le 180{,}000$$

Following the solution in part (A), we obtain the simplex tableau:

$$\begin{array}{c} \\ s_1 \\ s_2 \\ s_3 \\ P \end{array} \begin{array}{ccccccc} x_1 & x_2 & x_3 & s_1 & s_2 & s_3 & P \\ \left[\, \frac{1}{2}\right. & \frac{1}{2} & \boxed{1} & 1 & 0 & 0 & 0 & |\; 30 \\ 6 & 6 & 8 & 0 & 1 & 0 & 0 & |\; 320 \\ 4 & 3 & 4 & 0 & 0 & 1 & 0 & |\; 180 \\ \hline -17{,}000 & -18{,}000 & -24{,}000 & 0 & 0 & 0 & 1 & |\; 0 \left.\right] \end{array} \begin{array}{l} \frac{30}{1} = 30 \\ \frac{320}{8} = 40 \\ \frac{180}{4} = 45 \end{array}$$

$$(-8)R_1 + R_2 \rightarrow R_2, \quad (-4)R_1 + R_3 \rightarrow R_3, \quad 24{,}000R_1 + R_4 \rightarrow R_4$$

$$\sim \begin{bmatrix} \frac{1}{2} & \frac{1}{2} & 1 & 1 & 0 & 0 & 0 & 30 \\ 2 & ② & 0 & -8 & 1 & 0 & 0 & 80 \\ 2 & 1 & 0 & -4 & 0 & 1 & 0 & 60 \\ \hline -5000 & -6000 & 0 & 24{,}000 & 0 & 0 & 1 & 720{,}000 \end{bmatrix} \begin{matrix} \frac{30}{1/2}=60 \\ \frac{80}{2}=40 \\ \frac{60}{1}=60 \\ \\ \end{matrix}$$

$$\tfrac{1}{2}R_2 \to R_2$$

$$\sim \begin{bmatrix} \frac{1}{2} & \frac{1}{2} & 1 & 1 & 0 & 0 & 0 & 30 \\ 1 & ① & 0 & -4 & \frac{1}{2} & 0 & 0 & 40 \\ 2 & 1 & 0 & -4 & 0 & 1 & 0 & 60 \\ \hline -5000 & -6000 & 0 & 24{,}000 & 0 & 0 & 1 & 720{,}000 \end{bmatrix}$$

$$\left(-\tfrac{1}{2}\right)R_2 + R_1 \to R_1, \quad (-1)R_2 + R_3 \to R_3, \quad 6{,}000R_2 + R_4 \to R_4$$

	x_1	x_2	x_3	s_1	s_2	s_3	P	
x_3	0	0	1	3	$-\frac{1}{4}$	0	0	10
x_2	1	1	0	-4	$\frac{1}{2}$	0	0	40
s_3	1	0	0	0	$-\frac{1}{2}$	1	0	20
P	1000	0	0	0	3000	0	1	960,000

Optimal solution: maximum profit is \$960,000 when $x_1 = 0$ colonial houses, $x_2 = 40$ split level houses and $x_3 = 10$ ranch houses are built. In this case, $s_3 = 20$ (thousand) labor hours are not used.

(C) The mathematical model for this problem is:
Maximize $P = 25{,}000x_1 + 18{,}000x_2 + 24{,}000x_3$

Subject to:
$$\tfrac{1}{2}x_1 + \tfrac{1}{2}x_2 + x_3 \le 30$$
$$60{,}000x_1 + 60{,}000x_2 + 80{,}000x_3 \le 3{,}200{,}000$$
$$4{,}000x_1 + 3{,}000x_2 + 4{,}000x_3 \le 180{,}000$$

Following the solutions in parts (A) and (B), we obtain the simplex tableau:

	x_1	x_2	x_3	s_1	s_2	s_3	P		
s_1	$\frac{1}{2}$	$\frac{1}{2}$	1	1	0	0	0	30	$\frac{30}{1/2}=60$
s_2	6	6	8	0	1	0	0	320	$\frac{320}{6}=53.33$
s_3	④	3	4	0	0	1	0	180	$\frac{180}{4}=45$
P	-25,000	-18,000	-24,000	0	0	0	1	0	

$$\tfrac{1}{4}R_3 \to R_3$$

$$\sim \begin{bmatrix} \frac{1}{2} & \frac{1}{2} & 1 & 1 & 0 & 0 & 0 & 30 \\ 6 & 6 & 8 & 0 & 1 & 0 & 0 & 320 \\ \textcircled{1} & \frac{3}{4} & 1 & 0 & 0 & \frac{1}{4} & 0 & 45 \\ \hdashline -25{,}000 & -18{,}000 & -24{,}000 & 0 & 0 & 0 & 1 & 0 \end{bmatrix}$$

$$\left(-\frac{1}{2}\right) R_3 + R_1 \to R_1, \quad -6R_3 + R_2 \to R_2, \quad 25{,}000 R_3 + R_4 \to R_4$$

	x_1	x_2	x_3	s_1	s_2	s_3	P	
s_1	0	$\frac{1}{8}$	$\frac{1}{2}$	1	0	$-\frac{1}{8}$	0	7.5
s_2	0	$\frac{3}{2}$	2	0	1	$\frac{3}{2}$	0	50
x_1	1	$\frac{3}{4}$	1	0	0	$\frac{1}{4}$	0	45
P	0	750	1000	0	0	6250	1	1,125,000

(The whole is wrapped with a $\sim$ at the left.)

Optimal solution: maximum profit is $1,125,000 when x_1 = 45 colonial houses, x_2 = 0 split level houses and x_3 = 0 ranch houses are built. In this case, s_1 = 7.5 acres of land, and s_2 = 50(10,000) = $500,000 of capital are not used.

43. Let x_1 = the number of boxes of Assortment I,
   x_2 = the number of boxes of Assortment II,
and x_3 = the number of boxes of Assortment III.

(A) The profit per box of Assortment I is:
   9.40 − [4(0.20) + 4(0.25) + 12(0.30)] = $4.00
The profit per box of Assortment II is:
   7.60 − [12(0.20) + 4(0.25) + 4(0.30)] = $3.00
The profit per box of Assortment III is:
   11.00 − [8(0.20) + 8(0.25) + 8(0.30)] = $5.00
The mathematical model for this problem is:
Maximize $P = 4x_1 + 3x_2 + 5x_3$
Subject to: $4x_1 + 12x_2 + 8x_3 \le 4800$
$\qquad\qquad 4x_1 + 4x_2 + 8x_3 \le 4000$
$\qquad\quad 12x_1 + 4x_2 + 8x_3 \le 5600$
$\qquad\qquad\qquad x_1, \ x_2, \ x_3 \ge 0$
We introduce slack variables to obtain the initial form:
$4x_1 + 12x_2 + 8x_3 + s_1 \qquad\qquad\quad = 4800$
$4x_1 + 4x_2 + 8x_3 \qquad + s_2 \qquad\quad = 4000$
$12x_1 + 4x_2 + 8x_3 \qquad\qquad + s_3 \quad = 5600$
$-4x_1 - 3x_2 - 5x_3 \qquad\qquad\qquad + P = 0$

$$\begin{array}{c} \\ s_1 \\ s_2 \\ s_3 \\ P \end{array}
\begin{array}{ccccccc} x_1 & x_2 & x_3 & s_1 & s_2 & s_3 & P \end{array}
\left[\begin{array}{ccccccc|c}
4 & 12 & 8 & 1 & 0 & 0 & 0 & 4800 \\
4 & 4 & \boxed{8} & 0 & 1 & 0 & 0 & 4000 \\
12 & 4 & 8 & 0 & 0 & 1 & 0 & 5600 \\
\hdashline
-4 & -3 & -5 & 0 & 0 & 0 & 1 & 0
\end{array}\right]
\begin{array}{l}
\frac{4800}{8} = 600 \\[4pt]
\frac{4000}{8} = 500 \\[4pt]
\frac{5600}{8} = 700
\end{array}$$

$$\frac{1}{8}R_2 \rightarrow R_2$$

$$\sim \left[\begin{array}{ccccccc|c}
4 & 12 & 8 & 1 & 0 & 0 & 0 & 4800 \\
\frac{1}{2} & \frac{1}{2} & \textcircled{1} & 0 & \frac{1}{8} & 0 & 0 & 500 \\
12 & 4 & 8 & 0 & 0 & 1 & 0 & 5600 \\
\hdashline
-4 & -3 & -5 & 0 & 0 & 0 & 1 & 0
\end{array}\right]
\sim \left[\begin{array}{ccccccc|c}
0 & 8 & 0 & 1 & -1 & 0 & 0 & 800 \\
\frac{1}{2} & \frac{1}{2} & 1 & 0 & \frac{1}{8} & 0 & 0 & 500 \\
\textcircled{8} & 0 & 0 & 0 & -1 & 1 & 0 & 1600 \\
\hdashline
-\frac{3}{2} & -\frac{1}{2} & 0 & 0 & \frac{5}{8} & 0 & 1 & 2500
\end{array}\right]$$

$$(-8)R_2 + R_1 \rightarrow R_1, \quad (-8)R_2 + R_3 \rightarrow R_3, \qquad \frac{1}{8}R_3 \rightarrow R_3$$
$$5R_2 + R_4 \rightarrow R_4$$

$$\sim \left[\begin{array}{ccccccc|c}
0 & 8 & 0 & 1 & -1 & 0 & 0 & 800 \\
\frac{1}{2} & \frac{1}{2} & 1 & 0 & \frac{1}{8} & 0 & 0 & 500 \\
\textcircled{1} & 0 & 0 & 0 & -\frac{1}{8} & \frac{1}{8} & 0 & 200 \\
\hdashline
-\frac{3}{2} & -\frac{1}{2} & 0 & 0 & \frac{5}{8} & 0 & 1 & 2500
\end{array}\right]
\sim \left[\begin{array}{ccccccc|c}
0 & \textcircled{8} & 0 & 1 & -1 & 0 & 0 & 800 \\
0 & \frac{1}{2} & 1 & 0 & \frac{3}{16} & -\frac{1}{16} & 0 & 400 \\
1 & 0 & 0 & 0 & -\frac{1}{8} & \frac{1}{8} & 0 & 200 \\
\hdashline
0 & -\frac{1}{2} & 0 & 0 & \frac{7}{16} & \frac{3}{16} & 1 & 2800
\end{array}\right]$$

$$\left(-\frac{1}{2}\right)R_3 + R_2 \rightarrow R_2, \quad \frac{3}{2}R_3 + R_4 \rightarrow R_4 \qquad \frac{1}{8}R_1 \rightarrow R_1$$

$$\sim \left[\begin{array}{ccccccc|c}
0 & \textcircled{1} & 0 & \frac{1}{8} & -\frac{1}{8} & 0 & 0 & 100 \\
0 & \frac{1}{2} & 1 & 0 & \frac{3}{16} & -\frac{1}{16} & 0 & 400 \\
1 & 0 & 0 & 0 & -\frac{1}{8} & \frac{1}{8} & 0 & 200 \\
\hdashline
0 & -\frac{1}{2} & 0 & 0 & \frac{7}{16} & \frac{3}{16} & 1 & 2800
\end{array}\right]$$

$$\begin{array}{ccccccc} x_1 & x_2 & x_3 & s_1 & s_2 & s_3 & P \end{array}$$
$$\sim \left[\begin{array}{ccccccc|c}
0 & 1 & 0 & \frac{1}{8} & -\frac{1}{8} & 0 & 0 & 100 \\
0 & 0 & 1 & -\frac{1}{16} & \frac{1}{4} & -\frac{1}{16} & 0 & 350 \\
1 & 0 & 0 & 0 & -\frac{1}{8} & \frac{1}{8} & 0 & 200 \\
\hdashline
0 & 0 & 0 & \frac{1}{16} & \frac{3}{8} & \frac{3}{16} & 1 & 2850
\end{array}\right]$$

$$\left(-\frac{1}{2}\right)R_1 + R_2 \rightarrow R_2, \quad \frac{1}{2}R_1 + R_4 \rightarrow R_4$$

Optimal solution: maximum profit is \$2850 when 200 boxes of Assortment I, 100 boxes of Assortment II, and 350 boxes of Assortment III are made.

(B) The mathematical model for this problem is:

Maximize $P = 4x_1 + 3x_2 + 5x_3$

Subject to:
$$4x_1 + 12x_2 + 8x_3 \leq 4800$$
$$4x_1 + 4x_2 + 8x_3 \leq 5000$$
$$12x_1 + 4x_2 + 8x_3 \leq 5600$$

Following the solution in part (A), we obtain the simplex tableau:

$$
\begin{array}{c}
\\ s_1 \\ s_2 \\ s_3 \\ \\ P
\end{array}
\begin{array}{c}
\begin{array}{ccccccc}
x_1 & x_2 & x_3 & s_1 & s_2 & s_3 & P
\end{array}\\
\left[
\begin{array}{ccccccc|c}
4 & 12 & ⑧ & 1 & 0 & 0 & 0 & 4800 \\
4 & 4 & 8 & 0 & 1 & 0 & 0 & 5000 \\
12 & 4 & 8 & 0 & 0 & 1 & 0 & 5600 \\
\hdashline
-4 & -3 & -5 & 0 & 0 & 0 & 1 & 0
\end{array}
\right]
\end{array}
\begin{array}{l}
\dfrac{4800}{8} = 600 \\[4pt]
\dfrac{5000}{8} = 625 \\[4pt]
\dfrac{5600}{8} = 700
\end{array}
$$

$\dfrac{1}{8} R_1 \rightarrow R_1$

$$
\sim
\left[
\begin{array}{ccccccc|c}
\frac{1}{2} & \frac{3}{2} & ① & \frac{1}{8} & 0 & 0 & 0 & 600 \\
4 & 4 & 8 & 0 & 1 & 0 & 0 & 5000 \\
12 & 4 & 8 & 0 & 0 & 1 & 0 & 5600 \\
\hdashline
-4 & -3 & -5 & 0 & 0 & 0 & 1 & 0
\end{array}
\right]
$$

$(-8)R_1 + R_2 \rightarrow R_2, \quad (-8)R_1 + R_3 \rightarrow R_3, \quad 5R_1 + R_4 \rightarrow R_4$

$$
\sim
\left[
\begin{array}{ccccccc|c}
\frac{1}{2} & \frac{3}{2} & 1 & \frac{1}{8} & 0 & 0 & 0 & 600 \\
0 & -8 & 0 & -1 & 1 & 0 & 0 & 200 \\
⑧ & -8 & 0 & -1 & 0 & 1 & 0 & 800 \\
\hdashline
-\frac{3}{2} & \frac{9}{2} & 0 & \frac{5}{8} & 0 & 0 & 1 & 3000
\end{array}
\right]
\begin{array}{l}
\dfrac{600}{1/2} = 1200 \\[10pt]
\dfrac{800}{8} = 100
\end{array}
$$

$\dfrac{1}{8} R_3 \rightarrow R_3$

$$
\sim
\left[
\begin{array}{ccccccc|c}
\frac{1}{2} & \frac{3}{2} & 1 & \frac{1}{8} & 0 & 0 & 0 & 600 \\
0 & -8 & 0 & -1 & 1 & 0 & 0 & 200 \\
① & -1 & 0 & -\frac{1}{8} & 0 & \frac{1}{8} & 0 & 100 \\
\hdashline
-\frac{3}{2} & \frac{9}{2} & 0 & \frac{5}{8} & 0 & 0 & 1 & 3000
\end{array}
\right]
$$

$\left(-\dfrac{1}{2}\right) R_3 + R_1 \rightarrow R_1, \quad \left(\dfrac{3}{2}\right) R_3 + R_4 \rightarrow R_4$

$$
\begin{array}{c}
\\ x_3 \\ s_2 \\ x_1 \\ \\ P
\end{array}
\begin{array}{c}
\begin{array}{ccccccc}
x_1 & x_2 & x_3 & s_1 & s_2 & s_3 & P
\end{array}\\
\sim
\left[
\begin{array}{ccccccc|c}
0 & 2 & 1 & \frac{3}{16} & 0 & -\frac{1}{16} & 0 & 550 \\
0 & -8 & 0 & -1 & 1 & 0 & 0 & 200 \\
1 & -1 & 0 & -\frac{1}{8} & 0 & \frac{1}{8} & 0 & 100 \\
\hdashline
0 & 3 & 0 & \frac{7}{16} & 0 & \frac{3}{16} & 1 & 3150
\end{array}
\right]
\end{array}
$$

Optimal solution: maximum profit is \$3,150 when $x_1 = 100$ boxes of
assortment I, $x_2 = 0$ boxes of assortment II, and $x_3 = 550$ boxes of
assortment III are produced. In this case, $s_2 = 200$ fruit-filled
candies are not used.

(C) The mathematical model for this problem is:

Maximize $P = 4x_1 + 3x_2 + 5x_3$

Subject to:
$$4x_1 + 12x_2 + 8x_3 \le 6000$$
$$4x_1 + 4x_2 + 8x_3 \le 6000$$
$$12x_1 + 4x_2 + 8x_3 \le 5600$$

Following the solutions in parts (A) and (B), we obtain the simplex tableau:

$$
\begin{array}{c}
\\ s_1 \\ s_2 \\ s_3 \\ P
\end{array}
\begin{array}{c}
x_1 \quad x_2 \quad x_3 \quad s_1 \quad s_2 \quad s_3 \quad P \\
\left[\begin{array}{ccccccc|c}
4 & 12 & 8 & 1 & 0 & 0 & 0 & 6000 \\
4 & 4 & 8 & 0 & 1 & 0 & 0 & 6000 \\
12 & 4 & \boxed{8} & 0 & 0 & 1 & 0 & 5600 \\
\hline
-4 & -3 & -5 & 0 & 0 & 0 & 1 & 0
\end{array}\right]
\end{array}
\quad
\begin{array}{l}
\frac{6000}{8} = 750 \\[4pt]
\frac{6000}{8} = 750 \\[4pt]
\frac{5600}{8} = 700 \\[4pt]

\end{array}
$$

$$\tfrac{1}{8}R_3 \rightarrow R_3$$

$$
\sim
\left[\begin{array}{ccccccc|c}
4 & 12 & 8 & 1 & 0 & 0 & 0 & 6000 \\
4 & 4 & 8 & 0 & 1 & 0 & 0 & 6000 \\
\frac{3}{2} & \frac{1}{2} & 1 & 0 & 0 & \frac{1}{8} & 0 & 700 \\
\hline
-4 & -3 & -5 & 0 & 0 & 0 & 1 & 0
\end{array}\right]
$$

$$(-8)R_3 + R_1 \rightarrow R_1, \quad (-8)R_3 + R_2 \rightarrow R_2, \quad 5R_3 + R_4 \rightarrow R_4$$

$$
\sim
\left[\begin{array}{ccccccc|c}
-8 & \boxed{8} & 0 & 1 & 0 & -1 & 0 & 400 \\
-8 & 0 & 0 & 0 & 1 & -1 & 0 & 400 \\
\frac{3}{2} & \frac{1}{2} & 1 & 0 & 0 & \frac{1}{8} & 0 & 700 \\
\hline
\frac{7}{2} & -\frac{1}{2} & 0 & 0 & 0 & \frac{5}{8} & 1 & 3500
\end{array}\right]
\quad
\begin{array}{l}
\frac{400}{8} = 50 \\[10pt]
\frac{700}{1/2} = 1400
\end{array}
$$

$$\tfrac{1}{8}R_1 \rightarrow R_1$$

$$
\sim
\left[\begin{array}{ccccccc|c}
-1 & \boxed{1} & 0 & \frac{1}{8} & 0 & -\frac{1}{8} & 0 & 50 \\
-8 & 0 & 0 & 0 & 1 & -1 & 0 & 400 \\
\frac{3}{2} & \frac{1}{2} & 1 & 0 & 0 & \frac{1}{8} & 0 & 700 \\
\hline
\frac{7}{2} & -\frac{1}{2} & 0 & 0 & 0 & \frac{5}{8} & 1 & 3500
\end{array}\right]
$$

$$\left(-\tfrac{1}{2}\right)R_1 + R_3 \rightarrow R_3, \quad \tfrac{1}{2}R_1 + R_4 \rightarrow R_4$$

$$
\begin{array}{c}
\quad\ \ \begin{array}{ccccccc} x_1 & x_2 & x_3 & s_1 & s_2 & s_3 & P \end{array}\\[2pt]
\sim\ \begin{array}{c} x_2 \\ s_2 \\ x_3 \\ \hline P \end{array}
\left[\begin{array}{ccccccc|c}
-1 & 1 & 0 & \tfrac{1}{8} & 0 & -\tfrac{1}{8} & 0 & 50 \\
-8 & 0 & 0 & 0 & 1 & -1 & 0 & 400 \\
2 & 0 & 1 & -\tfrac{1}{16} & 0 & \tfrac{3}{16} & 0 & 675 \\
\hline
3 & 0 & 0 & \tfrac{1}{16} & 0 & \tfrac{9}{16} & 1 & 3525
\end{array}\right]
\end{array}
$$

Optimal solution: maximum profit is $3,525 when $x_1 = 0$ boxes of assortment I, $x_2 = 50$ boxes of assortment II, and $x_3 = 675$ boxes of assortment III are produced. In this case, $s_2 = 400$ fruit-filled candies are not used.

45. Let x_1 = the number of grams of food A,
$\quad\ \ x_2$ = the number of grams of food B,
and x_3 = the number of grams of food C.

The mathematical model for this problem is:
$\quad$ Maximize $P = 3x_1 + 4x_2 + 5x_3$
$\quad$ Subject to: $\ x_1 + 3x_2 + 2x_3 \le 30$
$$2x_1 + x_2 + 2x_3 \le 24$$
$$x_1,\ x_2,\ x_3 \ge 0$$

We introduce slack variables s_1 and s_2 to obtain the initial form:
$$x_1 + 3x_2 + 2x_3 + s_1 \qquad\quad = 30$$
$$2x_1 + x_2 + 2x_3 \qquad + s_2 \quad\ = 24$$
$$-3x_1 - 4x_2 - 5x_3 \qquad\qquad + P = 0$$

The simplex tableau for this problem is:

$$
\begin{array}{c}
\quad\ \begin{array}{cccccc} x_1 & x_2 & x_3 & s_1 & s_2 & P \end{array}\\[2pt]
\begin{array}{c} s_1 \\ s_2 \\ \hline P \end{array}
\left[\begin{array}{cccccc|c}
1 & 3 & 2 & 1 & 0 & 0 & 30 \\
2 & 1 & \textcircled{2} & 0 & 1 & 0 & 24 \\
\hline
-3 & -4 & -5 & 0 & 0 & 1 & 0
\end{array}\right]
\begin{array}{l} \tfrac{30}{2} = 15 \\[6pt] \tfrac{24}{2} = 12 \end{array}
\end{array}
$$

$$\tfrac{1}{2}R_2 \to R_2$$

$$
\sim
\left[\begin{array}{cccccc|c}
1 & 3 & 2 & 1 & 0 & 0 & 30 \\
1 & \tfrac{1}{2} & \textcircled{1} & 0 & \tfrac{1}{2} & 0 & 12 \\
\hline
-3 & -4 & -5 & 0 & 0 & 1 & 0
\end{array}\right]
\sim
\left[\begin{array}{cccccc|c}
-1 & \textcircled{2} & 0 & 1 & -1 & 0 & 6 \\
1 & \tfrac{1}{2} & 1 & 0 & \tfrac{1}{2} & 0 & 12 \\
\hline
2 & -\tfrac{3}{2} & 0 & 0 & \tfrac{5}{2} & 1 & 60
\end{array}\right]
\begin{array}{l} \tfrac{6}{2} = 3 \\[6pt] \tfrac{12}{1/2} = 24 \end{array}
$$

$$(-2)R_2 + R_1 \to R_1,\ 5R_2 + R_3 \to R_3 \qquad\qquad \tfrac{1}{2}R_1 \to R_1$$

$$\sim \begin{bmatrix} -\frac{1}{2} & \boxed{1} & 0 & \frac{1}{2} & -\frac{1}{2} & 0 & 3 \\ 1 & \frac{1}{2} & 1 & 0 & -\frac{1}{2} & 0 & 12 \\ \hdashline 2 & -\frac{3}{2} & 0 & 0 & \frac{5}{2} & 1 & 60 \end{bmatrix}$$

$$\sim \begin{array}{c} \\ x_2 \\ x_3 \\ P \end{array} \begin{array}{cccccc} x_1 & x_2 & x_3 & s_1 & s_2 & P \\ \end{array}$$

$$\begin{array}{c} x_2 \\ x_3 \\ P \end{array} \begin{bmatrix} -\frac{1}{2} & 1 & 0 & \frac{1}{2} & -\frac{1}{2} & 0 & 3 \\ \frac{5}{4} & 0 & 1 & -\frac{1}{4} & \frac{3}{4} & 0 & \frac{21}{2} \\ \hdashline \frac{5}{4} & 0 & 0 & \frac{3}{4} & \frac{7}{4} & 1 & \frac{129}{2} \end{bmatrix}$$

$$\left(-\frac{1}{2}\right)R_1 + R_2 \to R_2, \quad \frac{3}{2}R_1 + R_3 \to R_3$$

Optimal solution: the maximum amount of protein is 64.5 units when $x_1 = 0$ grams of food A, $x_2 = 3$ grams of food B and $x_3 = 10.5$ grams of food C are used.

47. Let x_1 = the number of undergraduate students,
x_2 = the number of graduate students,
and x_3 = the number of faculty members.

The mathematical model for this problem is:
Maximize $P = 18x_1 + 25x_2 + 30x_3$
Subject to:
$$\begin{aligned} x_1 + x_2 + x_3 &\leq 20 \\ 100x_1 + 150x_2 + 200x_3 &\leq 3200 \\ x_1,\ x_2,\ x_3 &\geq 0 \end{aligned}$$

Divide the second inequality by 50 to simplify the arithmetic. Then introduce slack variables s_1 and s_2 to obtain the initial form.

$$\begin{aligned} x_1 + x_2 + x_3 + s_1 &= 20 \\ 2x_1 + 3x_2 + 4x_3 + s_2 &= 64 \\ -18x_1 - 25x_2 - 30x_3 + P &= 0 \end{aligned}$$

The simplex tableau for this problem is:

$$\begin{array}{c} \\ s_1 \\ s_2 \\ P \end{array} \begin{array}{cccccc} x_1 & x_2 & x_3 & s_1 & s_2 & P \\ \end{array}$$

$$\begin{array}{c} s_1 \\ s_2 \\ P \end{array} \begin{bmatrix} 1 & 1 & 1 & 1 & 0 & 0 & 20 \\ 2 & 3 & \boxed{4} & 0 & 1 & 0 & 64 \\ \hdashline -18 & -25 & -30 & 0 & 0 & 1 & 0 \end{bmatrix} \begin{array}{l} \frac{20}{1} = 20 \\ \frac{64}{4} = 16 \end{array}$$

$$\frac{1}{4}R_2 \to R_2$$

$$\sim \begin{bmatrix} 1 & 1 & 1 & 1 & 0 & 0 & 20 \\ \frac{1}{2} & \frac{3}{4} & \boxed{1} & 0 & \frac{1}{4} & 0 & 16 \\ \hdashline -18 & -25 & -30 & 0 & 0 & 1 & 0 \end{bmatrix}$$

$$\sim \begin{bmatrix} \boxed{\frac{1}{2}} & \frac{1}{4} & 0 & 1 & -\frac{1}{4} & 0 & 4 \\ \frac{1}{2} & \frac{3}{4} & 1 & 0 & \frac{1}{4} & 0 & 16 \\ \hdashline -3 & -\frac{5}{2} & 0 & 0 & \frac{15}{2} & 1 & 480 \end{bmatrix} \begin{array}{l} \frac{4}{1/2} = 8 \\ \frac{16}{1/2} = 32 \end{array}$$

$$(-1)R_2 + R_1 \to R_1, \quad 30R_2 + R_3 \to R_3 \qquad 2R_1 \to R_1$$

$$\sim \begin{bmatrix} \boxed{1} & \frac{1}{2} & 0 & 2 & -\frac{1}{2} & 0 & 8 \\ \frac{1}{2} & \frac{3}{4} & 1 & 0 & \frac{1}{4} & 0 & 16 \\ \hdashline -3 & -\frac{5}{2} & 0 & 0 & \frac{15}{2} & 1 & 480 \end{bmatrix}$$

$$\sim \begin{bmatrix} 1 & \boxed{\frac{1}{2}} & 0 & 2 & -\frac{1}{2} & 0 & 8 \\ 0 & \frac{1}{2} & 1 & -1 & \frac{1}{2} & 0 & 12 \\ \hdashline 0 & -1 & 0 & 6 & 6 & 1 & 504 \end{bmatrix} \begin{array}{l} \frac{8}{1/2} = 16 \\ \frac{12}{1/2} = 24 \end{array}$$

$$\left(-\frac{1}{2}\right)R_1 + R_2 \to R_2, \quad 3R_1 + R_3 \to R_3 \qquad 2R_1 \to R_1$$

$$\sim \begin{bmatrix} 2 & 1 & 0 & 4 & -1 & 0 & | & 16 \\ 0 & \frac{1}{2} & 1 & -1 & \frac{1}{2} & 0 & | & 12 \\ \hdashline 0 & -1 & 0 & 6 & 6 & 1 & | & 504 \end{bmatrix} \sim \begin{array}{c} x_2 \\ x_3 \\ \\ P \end{array} \begin{bmatrix} x_1 & x_2 & x_3 & s_1 & s_2 & P & \\ 2 & 1 & 0 & 4 & -1 & 0 & | & 16 \\ -1 & 0 & 1 & -3 & 1 & 0 & | & 4 \\ \hdashline 2 & 0 & 0 & 10 & 5 & 1 & | & 520 \end{bmatrix}$$

$$\left(-\frac{1}{2}\right)R_1 + R_2 \rightarrow R_2, \quad R_1 + R_3 \rightarrow R_3$$

Optimal solution: the maximum number of interviews is 520 when $x_1 = 0$ undergraduates, $x_2 = 16$ graduate students, and $x_3 = 4$ faculty members are hired.

EXERCISE 5-5

Things to remember:

1. Given a matrix A. The transpose of A, denoted A^T, is the matrix formed by interchanging the rows and corresponding columns of A (first row with first columnn, second row with second column, and so on.)

2. FORMATION OF THE DUAL PROBLEM

    Given a minimization problem with $\geq$ problem constraints:

    Step 1. Use the coefficients and constants in the problem constraints and the objective function to form a matrix A with the coefficients of the objective function in the last row.

    Step 2. Interchange the rows and columns of matrix A to form the matrix A^T, the transpose of A.

    Step 3. Use the rows of A^T to form a maximization problem with $\leq$ problem constraints.

3. THE FUNDAMENTAL PRINCIPLE OF DUALITY

    A minimization problem has a solution if and only if its dual problem has a solution. If a solution exists, then the optimal value of the minimization problem is the same as the optimal value of the dual problem.

4. SOLUTION OF A MINIMIZATION PROBLEM

    Given a minimization problem with nonnegative coefficients in the objective function:

    (i) Write all problem constraints as $\geq$ inequalities. (This may introduce negative numbers on the right side of the problem constraints.)

(ii) Form the dual problem.

(iii) Write the initial system of the dual problem, using the variables from the minimization problem as the slack variables.

(iv) Use the simplex method to solve the dual problem.

(v) Read the solution of the minimization problem from the bottom row of the final simplex tableau in Step (iv).

[Note: If the dual problem has no solution, then the minimization problem has no solution.]

---

1. (A) Given the minimization problem:

Minimize $C = 8x_1 + 9x_2$

Subject to:
$$x_1 + 3x_2 \geq 4$$
$$2x_1 + x_2 \geq 5$$
$$x_1, x_2 \geq 0$$

The matrix A^T corresponding to this problem is: $A = \begin{bmatrix} 1 & 3 & | & 4 \\ 2 & 1 & | & 5 \\ 8 & 9 & | & 1 \end{bmatrix}$

The matrix A^T corresponding to the dual problem has the rows of A as its columns. Thus:

$$A^T = \begin{bmatrix} 1 & 2 & | & 8 \\ 3 & 1 & | & 9 \\ 4 & 5 & | & 1 \end{bmatrix}$$

The dual problem is: Maximize $P = 4y_1 + 5y_2$

Subject to:
$$y_1 + 2y_2 \leq 8$$
$$3y_1 + y_2 \leq 9$$
$$y_1, y_2 \geq 0$$

(B) Letting x_1 and x_2 be slack variables, the initial system for the dual problem is:

$$y_1 + 2y_2 + x_1 \qquad = 8$$
$$3y_1 + y_2 \qquad + x_2 \qquad = 9$$
$$-4y_1 - 5y_2 \qquad \qquad + P = 0$$

(C) The simplex tableau for this problem is:

	y_1	y_2	x_1	x_2	P	
x_1	1	2	1	0	0	8
x_2	3	1	0	1	0	9
P	-4	-5	0	0	1	0

3. From the final simplex tableau,

$$
\begin{array}{c}
\quad\; y_1 \;\; y_2 \;\; x_1 \;\; x_2 \;\; P \\
\begin{array}{c} y_2 \\ y_1 \\ P \end{array}
\left[
\begin{array}{ccccc|c}
0 & 1 & 5 & -2 & 0 & 5 \\
1 & 0 & -7 & 3 & 0 & 3 \\
\hline
0 & 0 & 1 & 2 & 1 & 121
\end{array}
\right]
\end{array}
$$

(A) the optimal solution of the dual problem is:
   maximum value of $P = 121$ at $y_1 = 3$ and $y_2 = 5$;

(B) the optimal solution of the minimization problem is:
   minimum value of $C = 121$ at $x_1 = 1$, $x_2 = 2$.

5. (A) The matrix corresponding to the given problem is: $A = \begin{bmatrix} 4 & 1 & | & 13 \\ 3 & 1 & | & 12 \\ 9 & 2 & | & 1 \end{bmatrix}$

The matrix A^T corresponding to the dual problem has the rows of A as its columns, that is:

$$
A^T = \begin{bmatrix} 4 & 3 & | & 9 \\ 1 & 1 & | & 2 \\ 13 & 12 & | & 1 \end{bmatrix}
$$

Thus, the dual problem is: Maximize $P = 13y_1 + 12y_2$
$$\text{Subject to: } 4y_1 + 3y_2 \le 9$$
$$y_1 + y_2 \le 2$$
$$y_1,\ y_2 \ge 0$$

(B) We introduce slack variables x_1 and x_2 to obtain the initial system for the dual problem:
$$
\begin{aligned}
4y_1 + 3y_2 + x_1 \qquad\quad &= 9 \\
y_1 + y_2 \qquad + x_2 \quad &= 2 \\
-13y_1 - 12y_2 \qquad\qquad + P &= 0
\end{aligned}
$$

The simplex tableau for this problem is:

$$
\begin{array}{c}
\quad\; y_1 \;\;\; y_2 \;\; x_1 \;\; x_2 \;\; P \\
\begin{array}{c} x_1 \\ x_2 \\ P \end{array}
\left[
\begin{array}{ccccc|c}
4 & 3 & 1 & 0 & 0 & 9 \\
① & 1 & 0 & 1 & 0 & 2 \\
\hline
-13 & -12 & 0 & 0 & 1 & 0
\end{array}
\right]
\begin{array}{l} \tfrac{9}{4} = 2.25 \\ \tfrac{2}{1} = 2 \end{array}
\end{array}
\quad\sim\quad
\begin{array}{c}
\quad\; y_1 \;\; y_2 \;\; x_1 \;\; x_2 \;\; P \\
\begin{array}{c} x_1 \\ y_1 \\ P \end{array}
\left[
\begin{array}{ccccc|c}
0 & -1 & 1 & -4 & 0 & 1 \\
1 & 1 & 0 & 1 & 0 & 2 \\
\hline
0 & 1 & 0 & 13 & 1 & 26
\end{array}
\right]
\end{array}
$$

$(-4)R_2 + R_1 \to R_1$ and $13R_2 + R_3 \to R_3$

Optimal solution: min $C = 26$ at $x_1 = 0$, $x_2 = 13$.

7. (A) The matrix corresponding to the given problem is: $A = \begin{bmatrix} 2 & 3 & | & 15 \\ 1 & 2 & | & 8 \\ 7 & 12 & | & 1 \end{bmatrix}$

The matrix A^T corresponding to the dual problem has the rows of A as its columns, that is:

$$A^T = \begin{bmatrix} 2 & 1 & | & 7 \\ 3 & 2 & | & 12 \\ \hline 15 & 8 & | & 1 \end{bmatrix}$$

Thus, the dual problem is: Maximize $P = 15y_1 + 8y_2$

Subject to: $2y_1 + y_2 \leq 7$
$3y_1 + 2y_2 \leq 12$
$y_1, y_2 \geq 0$

(B) We introduce slack variables x_1 and x_2 to obtain the initial system for the dual problem:

$$2y_1 + y_2 + x_1 \qquad = 7$$
$$3y_1 + 2y_2 \qquad + x_2 \qquad = 12$$
$$-15y_1 - 8y_2 \qquad\qquad + P = 0$$

The simplex tableau for this problem is:

$$\begin{array}{c} \\ x_1 \\ x_2 \\ P \end{array}
\begin{array}{c} y_1 \quad y_2 \quad x_1 \quad x_2 \quad P \\ \left[\begin{array}{ccccc|c} ② & 1 & 1 & 0 & 0 & 7 \\ 3 & 2 & 0 & 1 & 0 & 12 \\ \hline -15 & -8 & 0 & 0 & 1 & 0 \end{array}\right] \end{array}
\begin{array}{c} \frac{7}{2} = 3.5 \\ \frac{12}{3} = 4 \\ {} \end{array}
\sim
\left[\begin{array}{ccccc|c} ① & \frac{1}{2} & \frac{1}{2} & 0 & 0 & \frac{7}{2} \\ 3 & 2 & 0 & 1 & 0 & 12 \\ \hline -15 & -8 & 0 & 0 & 1 & 0 \end{array}\right]$$

$$\frac{1}{2}R_1 \to R_1 \qquad\qquad\qquad (-3)R_1 + R_2 \to R_2 \text{ and } 15R_1 + R_3 \to R_3$$

$$\sim \left[\begin{array}{ccccc|c} 1 & \frac{1}{2} & \frac{1}{2} & 0 & 0 & \frac{7}{2} \\ 0 & ⑫ & -\frac{3}{2} & 1 & 0 & \frac{3}{2} \\ \hline 0 & -\frac{1}{2} & \frac{15}{2} & 0 & 1 & \frac{105}{2} \end{array}\right]
\begin{array}{c} \frac{7/2}{1/2} = 7 \\ \frac{3/2}{1/2} = 3 \\ {} \end{array}
\sim
\left[\begin{array}{ccccc|c} 1 & \frac{1}{2} & \frac{1}{2} & 0 & 0 & \frac{7}{2} \\ 0 & ① & -3 & 2 & 0 & 3 \\ \hline 0 & -\frac{1}{2} & \frac{15}{2} & 0 & 1 & \frac{105}{2} \end{array}\right]$$

$$2R_2 \to R_2 \qquad\qquad\qquad \left(-\frac{1}{2}\right)R_2 + R_1 \to R_1 \text{ and } \frac{1}{2}R_2 + R_3 \to R_3$$

$$\begin{array}{c} \\ y_1 \\ \sim \ y_2 \\ P \end{array}
\begin{array}{c} y_1 \quad y_2 \quad x_1 \quad x_2 \quad P \\ \left[\begin{array}{ccccc|c} 1 & 0 & 2 & -1 & 0 & 2 \\ 0 & 1 & -3 & 2 & 0 & 3 \\ \hline 0 & 0 & 6 & 1 & 1 & 54 \end{array}\right] \end{array}$$

Optimal solution: min $C = 54$ at $x_1 = 6$, $x_2 = 1$.

9. (A) The matrices corresponding to the given problem and to the dual problem are:

$$A = \begin{bmatrix} 2 & 1 & | & 8 \\ -2 & 3 & | & 4 \\ \hline 11 & 4 & | & 1 \end{bmatrix} \quad \text{and} \quad A^T = \begin{bmatrix} 2 & -2 & | & 11 \\ 1 & 3 & | & 4 \\ \hline 8 & 4 & | & 1 \end{bmatrix}$$

respectively.

Thus, the dual problem is:

Maximize $P = 8y_1 + 4y_2$

Subject to: $2y_1 - 2y_2 \leq 11$

$y_1 + 3y_2 \leq 4$

$y_1, y_2 \geq 0$

(B) We introduce slack variables x_1 and x_2 to obtain the initial system for the dual problem:

$2y_1 - 2y_2 + x_1 \quad\quad = 11$

$y_1 + 3y_2 \quad + x_2 \quad = 4$

$-8y_1 - 4y_2 \quad\quad + P = 0$

The simplex tableau for this problem is:

$$
\begin{array}{c}
\begin{array}{ccccc}
y_1 & y_2 & x_1 & x_2 & P
\end{array} \\
\begin{array}{c} x_1 \\ \sim x_2 \\ P \end{array}
\left[
\begin{array}{ccccc|c}
2 & -2 & 1 & 0 & 0 & 11 \\
① & 3 & 0 & 1 & 0 & 4 \\
\hline
-8 & -4 & 0 & 0 & 1 & 0
\end{array}
\right]
\begin{array}{l}
\frac{11}{2} = 5.5 \\
\frac{4}{1} = 4 \\
\end{array}
\end{array}
$$

$$
\sim
\begin{array}{c}
\begin{array}{ccccc}
y_1 & y_2 & x_1 & x_2 & P
\end{array} \\
\begin{array}{c} x_1 \\ y_1 \\ P \end{array}
\left[
\begin{array}{ccccc|c}
0 & -8 & 1 & -2 & 0 & 3 \\
1 & 3 & 0 & 1 & 0 & 4 \\
\hline
0 & 20 & 0 & 8 & 1 & 32
\end{array}
\right]
\end{array}
$$

$(-2)R_2 + R_1 \to R_1$ and $8R_2 + R_3 \to R_3$

Optimal solution: min $C = 32$ at $x_1 = 0$, $x_2 = 8$.

11. (A) The matrices corresponding to the given problem and the dual problem are:

$$
A = \left[
\begin{array}{cc|c}
-3 & 1 & 6 \\
1 & -2 & 4 \\
7 & 9 & 1
\end{array}
\right]
\quad\text{and}\quad
A^T = \left[
\begin{array}{cc|c}
-3 & 1 & 7 \\
1 & -2 & 9 \\
6 & 4 & 1
\end{array}
\right]
$$

respectively. Thus, the dual problem is:

Maximize $P = 6y_1 + 4y_2$

Subject to: $-3y_1 + y_2 \leq 7$

$y_1 - 2y_2 \leq 9$

$y_1, y_2 \geq 0$

(B) We introduce slack variables x_1 and x_2 to obtain the initial system for the dual problem:

$-3y_1 + y_2 + x_1 \quad\quad = 7$

$y_1 - 2y_2 \quad + x_2 \quad = 9$

$-6y_1 - 4y_2 \quad\quad + P = 0$

The simplex tableau for this problem is:

$$
\begin{array}{c}
\quad\quad \begin{array}{ccccc} y_1 & y_2 & x_1 & x_2 & P \end{array} \\
\begin{array}{c} x_1 \\ x_2 \\ P \end{array}
\left[
\begin{array}{ccccc|c}
-3 & 1 & 1 & 0 & 0 & 7 \\
\boxed{1} & -2 & 0 & 1 & 0 & 9 \\
\hline
-6 & -4 & 0 & 0 & 1 & 0
\end{array}
\right]
\end{array}
\sim
\begin{array}{c}
\quad\quad \begin{array}{ccccc} y_1 & y_2 & x_1 & x_2 & P \end{array} \\
\left[
\begin{array}{ccccc|c}
0 & -5 & 1 & 3 & 0 & 34 \\
1 & -2 & 0 & 1 & 0 & 9 \\
\hline
0 & -16 & 0 & 6 & 1 & 54
\end{array}
\right]
\end{array}
$$

$3R_2 + R_1 \to R_1$ and $6R_2 + R_3 \to R_3$ The negative elements in the second column above the dashed line indicate that the problem does not have a solution.

13. The matrices corresponding to the given problem and the dual problem are:

$$
A = \left[
\begin{array}{cc|c}
2 & 1 & 8 \\
1 & 2 & 8 \\
\hline
3 & 9 & 1
\end{array}
\right]
\quad \text{and} \quad
A^T = \left[
\begin{array}{cc|c}
2 & 1 & 3 \\
1 & 2 & 9 \\
\hline
8 & 8 & 1
\end{array}
\right]
$$

respectively. Thus, the dual problem is: Maximize $P = 8y_1 + 8y_2$

$$
\begin{aligned}
\text{Subject to: } 2y_1 + \; y_2 &\le 3 \\
y_1 + 2y_2 &\le 9 \\
y_1, \; y_2 &\ge 0
\end{aligned}
$$

We introduce slack variables x_1 and x_2 to obtain the initial system:

$$
\begin{aligned}
2y_1 + \; y_2 + x_1 \quad\quad\quad\quad &= 3 \\
y_1 + 2y_2 \quad\quad + x_2 \quad\quad &= 9 \\
-8y_1 - 8y_2 \quad\quad\quad\quad + P &= 0
\end{aligned}
$$

The simplex tableau for this problem is:

$$
\begin{array}{c}
\quad\quad \begin{array}{ccccc} y_1 & y_2 & x_1 & x_2 & P \end{array} \\
\begin{array}{c} x_1 \\ x_2 \\ P \end{array}
\left[
\begin{array}{ccccc|c}
2 & \boxed{1} & 1 & 0 & 0 & 3 \\
1 & 2 & 0 & 1 & 0 & 9 \\
\hline
-8 & -8 & 0 & 0 & 1 & 0
\end{array}
\right]
\begin{array}{l} \frac{3}{1} = 3 \\ \frac{9}{2} = 4.5 \end{array}
\end{array}
\sim
\begin{array}{c}
\quad\quad \begin{array}{ccccc} y_1 & y_2 & x_1 & x_2 & P \end{array} \\
\begin{array}{c} y_2 \\ x_2 \\ P \end{array}
\left[
\begin{array}{ccccc|c}
2 & 1 & 1 & 0 & 0 & 3 \\
-3 & 0 & -2 & 1 & 0 & 3 \\
\hline
8 & 0 & 8 & 0 & 1 & 24
\end{array}
\right]
\end{array}
$$

$(-2)R_1 + R_2 \to R_2$ and $8R_1 + R_3 \to R_3$ Optimal solution: min $C = 24$ at $x_1 = 8$, $x_2 = 0$.

[Note: We could use either column 1 or column 2 as the pivot column. Column 2 involves slightly simpler calculations.]

15. The matrices corresponding to the given problem and the dual problem are:

$$A = \begin{bmatrix} 1 & 1 & | & 4 \\ 1 & -2 & | & -8 \\ -2 & 1 & | & -8 \\ \hline 7 & 5 & | & 1 \end{bmatrix} \quad \text{and} \quad A^T = \begin{bmatrix} 1 & 1 & -2 & | & 7 \\ 1 & -2 & 1 & | & 5 \\ \hline 4 & -8 & -8 & | & 1 \end{bmatrix} \quad \text{respectively.}$$

Thus, the dual problem is: Maximize $P = 4y_1 - 8y_2 - 8y_3$

Subject to: $y_1 + y_2 - 2y_3 \le 7$
$y_1 - 2y_2 + y_3 \le 5$
$y_1, y_2, y_3 \ge 0$

We introduce slack variables x_1 and x_2 to obtain the initial system:

$$\begin{aligned} y_1 + y_2 - 2y_3 + x_1 \quad\quad &= 7 \\ y_1 - 2y_2 + y_3 \quad\quad + x_2 \quad &= 5 \\ -4y_1 + 8y_2 + 8y_3 \quad\quad + P &= 0 \end{aligned}$$

The simplex tableau for this problem is:

$$\begin{array}{c} \begin{array}{cccccc} y_1 & y_2 & y_3 & x_1 & x_2 & P \end{array} \\ \begin{array}{c} x_1 \\ x_2 \\ P \end{array} \left[\begin{array}{cccccc|c} 1 & 1 & -2 & 1 & 0 & 0 & 7 \\ \textcircled{1} & -2 & 1 & 0 & 1 & 0 & 5 \\ \hline -4 & 8 & 8 & 0 & 0 & 1 & 0 \end{array}\right] \end{array} \begin{array}{c} \frac{7}{1} = 7 \\ \frac{5}{1} = 5 \\ \\ \end{array}$$

$$\begin{array}{c} \begin{array}{cccccc} y_1 & y_2 & y_3 & x_1 & x_2 & P \end{array} \\ \begin{array}{c} x_1 \\ \sim y_1 \\ P \end{array} \left[\begin{array}{cccccc|c} 0 & 3 & -3 & 1 & -1 & 0 & 2 \\ 1 & -2 & 1 & 0 & 1 & 0 & 5 \\ \hline 0 & 0 & 12 & 0 & 4 & 1 & 20 \end{array}\right] \end{array}$$

$(-1)R_2 + R_1 \to R_1$ and $4R_2 + R_3 \to R_3$

Optimal solution: min $C = 20$ at $x_1 = 0$, $x_2 = 4$.

17. The matrices corresponding to the given problem and the dual problem are:

$$A = \begin{bmatrix} 2 & 1 & | & 16 \\ 1 & 1 & | & 12 \\ 1 & 2 & | & 14 \\ \hline 10 & 30 & | & 1 \end{bmatrix} \quad \text{and} \quad A^T = \begin{bmatrix} 2 & 1 & 1 & | & 10 \\ 1 & 1 & 2 & | & 30 \\ \hline 16 & 12 & 14 & | & 1 \end{bmatrix} \quad \text{respectively.}$$

Thus, the dual problem is: Maximize $P = 16y_1 + 12y_2 + 14y_3$

Subject to: $2y_1 + y_2 + y_3 \le 10$
$y_1 + y_2 + 2y_3 \le 30$
$y_1, y_2, y_3 \ge 0$

We introduce slack variables x_1 and x_2 to obtain the initial system:

$$\begin{aligned} 2y_1 + y_2 + y_3 + x_1 \quad\quad &= 10 \\ y_1 + y_2 + 2y_3 \quad\quad + x_2 \quad &= 30 \\ -16y_1 - 12y_2 - 14y_3 \quad\quad + P &= 0 \end{aligned}$$

The simplex tableau for this problem is:

$$
\begin{array}{c}
 \\ x_1 \\ x_2 \\ P
\end{array}
\begin{array}{c}
\begin{array}{cccccc} y_1 & y_2 & y_3 & x_1 & x_2 & P \end{array} \\
\left[\begin{array}{cccccc|c}
②\ & 1 & 1 & 1 & 0 & 0 & 10 \\
1 & 1 & 2 & 0 & 1 & 0 & 30 \\
\hdashline
-16 & -12 & -14 & 0 & 0 & 1 & 0
\end{array}\right]
\end{array}
\begin{array}{l}
\frac{10}{2} = 5 \\[4pt]
\frac{30}{1} = 30
\end{array}
\sim
\left[\begin{array}{cccccc|c}
①\ & \frac{1}{2} & \frac{1}{2} & \frac{1}{2} & 0 & 0 & 5 \\
1 & 1 & 2 & 0 & 1 & 0 & 30 \\
\hdashline
-16 & -12 & -14 & 0 & 0 & 1 & 0
\end{array}\right]
$$

$$\tfrac{1}{2}R_1 \to R_1 \qquad\qquad\qquad\qquad\qquad (-1)R_1 + R_2 \to R_2 \text{ and } 16R_1 + R_3 \to R_3$$

$$
\sim
\left[\begin{array}{cccccc|c}
1 & \frac{1}{2} & ⨍\tfrac{1}{2} & \frac{1}{2} & 0 & 0 & 5 \\
0 & \frac{1}{2} & \frac{3}{2} & -\frac{1}{2} & 1 & 0 & 25 \\
\hdashline
0 & -4 & -6 & 8 & 0 & 1 & 80
\end{array}\right]
\begin{array}{l}
\frac{5}{1/2} = 10 \\[4pt]
\frac{25}{3/2} = 16.66
\end{array}
\sim
\left[\begin{array}{cccccc|c}
2 & 1 & ①\ & 1 & 0 & 0 & 10 \\
0 & \frac{1}{2} & \frac{3}{2} & -\frac{1}{2} & 1 & 0 & 25 \\
\hdashline
0 & -4 & -6 & 8 & 0 & 1 & 80
\end{array}\right]
$$

$$2R_1 \to R_1 \qquad\qquad\qquad\qquad\qquad \left(-\tfrac{3}{2}\right)R_1 + R_2 \to R_2 \text{ and } 6R_1 + R_3 \to R_3$$

$$
\begin{array}{c}
 \\ y_3 \\ x_2 \\ P
\end{array}
\begin{array}{c}
\begin{array}{cccccc} y_1 & y_2 & y_3 & x_1 & x_2 & P \end{array} \\
\sim
\left[\begin{array}{cccccc|c}
2 & 1 & 1 & 1 & 0 & 0 & 10 \\
-3 & -1 & 0 & -2 & 1 & 0 & 10 \\
\hdashline
12 & 2 & 0 & 14 & 0 & 1 & 140
\end{array}\right]
\end{array}
$$

Optimal solution: $\min C = 140$ at $x_1 = 14$, $x_2 = 0$.

19. The matrices corresponding to the given problem and the dual problem are:

$$
A = \begin{bmatrix}
1 & 0 & 4 \\
1 & 1 & 8 \\
1 & 2 & 10 \\
5 & 7 & 1
\end{bmatrix}
\quad \text{and} \quad
A^T = \begin{bmatrix}
1 & 1 & 1 & 5 \\
0 & 1 & 2 & 7 \\
4 & 8 & 10 & 1
\end{bmatrix}
\quad \text{respectively.}
$$

Thus, the dual problem is: Maximize $P = 4y_1 + 8y_2 + 10y_3$
Subject to: $y_1 + y_2 + y_3 \le 5$
$y_2 + 2y_3 \le 7$
$y_1,\ y_2,\ y_3 \ge 0$

We introduce slack variables x_1 and x_2 to obtain the initial system:

$$
\begin{array}{rcl}
y_1 + y_2 + y_3 + x_1 & = & 5 \\
y_2 + 2y_3 + x_2 & = & 7 \\
-4y_1 - 8y_2 - 10y_3 + P & = & 0
\end{array}
$$

The simplex tableau for this problem is:

$$
\begin{array}{c}
 \\ x_1 \\ x_2 \\ P
\end{array}
\begin{array}{c}
\begin{array}{cccccc} y_1 & y_2 & y_3 & x_1 & x_2 & P \end{array} \\
\left[\begin{array}{cccccc|c}
1 & 1 & 1 & 1 & 0 & 0 & 5 \\
0 & 1 & ②\ & 0 & 1 & 0 & 7 \\
\hdashline
-4 & -8 & -10 & 0 & 0 & 1 & 0
\end{array}\right]
\end{array}
\begin{array}{l}
\frac{5}{1} = 5 \\[4pt]
\frac{7}{2} = 3.5
\end{array}
\sim
\left[\begin{array}{cccccc|c}
1 & 1 & 1 & 1 & 0 & 0 & 5 \\
0 & \frac{1}{2} & ①\ & 0 & \frac{1}{2} & 0 & \frac{7}{2} \\
\hdashline
-4 & -8 & -10 & 0 & 0 & 1 & 0
\end{array}\right]
$$

$$\tfrac{1}{2}R_2 \to R_2 \qquad\qquad\qquad\qquad (-1)R_2 + R_1 \to R_1 \text{ and } 10R_2 + R_3 \to R_3$$

$$\sim \begin{bmatrix} \boxed{1} & \frac{1}{2} & 0 & 1 & -\frac{1}{2} & 0 & | & \frac{3}{2} \\ 0 & \frac{1}{2} & 1 & 0 & \frac{1}{2} & 0 & | & \frac{7}{2} \\ \hdashline -4 & -3 & 0 & 0 & 5 & 1 & | & 35 \end{bmatrix} \qquad \sim \begin{bmatrix} 1 & \boxed{\frac{1}{2}} & 0 & 1 & -\frac{1}{2} & 0 & | & \frac{3}{2} \\ 0 & \frac{1}{2} & 1 & 0 & \frac{1}{2} & 0 & | & \frac{7}{2} \\ \hdashline 0 & -1 & 0 & 4 & 3 & 1 & | & 41 \end{bmatrix} \begin{matrix} \frac{3/2}{1/2} = 3 \\ \frac{7/2}{1/2} = 7 \\ \\ \end{matrix}$$

$$4R_1 + R_3 \to R_3 \qquad\qquad\qquad 2R_1 \to R_1$$

$$\qquad\qquad\qquad\qquad\qquad\qquad\qquad\qquad Y_1 \quad Y_2 \quad Y_3 \quad X_1 \quad X_2 \quad P$$

$$\sim \begin{bmatrix} 2 & \boxed{1} & 0 & 2 & -1 & 0 & | & 3 \\ 0 & \frac{1}{2} & 1 & 0 & \frac{1}{2} & 0 & | & \frac{7}{2} \\ \hdashline 0 & -1 & 0 & 4 & 3 & 1 & | & 41 \end{bmatrix} \qquad \sim \begin{matrix} Y_2 \\ Y_3 \\ P \end{matrix} \begin{bmatrix} 2 & 1 & 0 & 2 & -1 & 0 & | & 3 \\ -1 & 0 & 1 & -1 & 1 & 0 & | & 2 \\ \hdashline 2 & 0 & 0 & 6 & 2 & 1 & | & 44 \end{bmatrix}$$

$$\left(-\frac{1}{2}\right)R_1 + R_2 \to R_2 \text{ and } R_3 + R_1 \to R_3 \quad \text{Optimal solution: min } C = 44 \text{ at } x_1 = 6,$$
$$x_2 = 2.$$

21. The matrices corresponding to the given problem and the dual problem are:

$$A = \begin{bmatrix} 1 & 1 & 2 & | & 7 \\ 2 & 1 & 1 & | & 4 \\ \hline 10 & 7 & 12 & | & 1 \end{bmatrix} \qquad \text{and} \qquad A^T = \begin{bmatrix} 1 & 2 & | & 10 \\ 1 & 1 & | & 7 \\ 2 & 1 & | & 12 \\ \hline 7 & 4 & | & 1 \end{bmatrix} \qquad \text{respectively.}$$

Thus, the dual problem is: Maximize $P = 7y_1 + 4y_2$

$$\begin{aligned} \text{Subject to:} \quad y_1 + 2y_2 &\leq 10 \\ y_1 + y_2 &\leq 7 \\ 2y_1 + y_2 &\leq 12 \\ y_1, \ y_2 &\geq 0 \end{aligned}$$

We introduce slack variables x_1, x_2, and x_3 to obtain the initial system:

$$\begin{aligned} y_1 + 2y_2 + x_1 &&&= 10 \\ y_1 + y_2 &+ x_2 &&= 7 \\ 2y_1 + y_2 && + x_3 &= 12 \\ -7y_1 - 4y_2 &&&+ P = 0 \end{aligned}$$

The simplex tableau for this problem is:

$$\begin{matrix} & Y_1 & Y_2 & X_1 & X_2 & X_3 & P & \\ x_1 \\ x_2 \\ \sim \ x_3 \\ P \end{matrix} \begin{bmatrix} 1 & 2 & 1 & 0 & 0 & 0 & | & 10 \\ 1 & 1 & 0 & 1 & 0 & 0 & | & 7 \\ \boxed{2} & 1 & 0 & 0 & 1 & 0 & | & 12 \\ \hdashline -7 & -4 & 0 & 0 & 0 & 1 & | & 0 \end{bmatrix} \begin{matrix} \frac{10}{1} = 10 \\ \frac{7}{1} = 7 \\ \frac{12}{2} = 6 \\ \\ \end{matrix} \quad \sim \begin{bmatrix} 1 & 2 & 1 & 0 & 0 & 0 & | & 10 \\ 1 & 1 & 0 & 1 & 0 & 0 & | & 7 \\ \boxed{1} & \frac{1}{2} & 0 & 0 & \frac{1}{2} & 0 & | & 6 \\ \hdashline -7 & -4 & 0 & 0 & 0 & 1 & | & 0 \end{bmatrix}$$

$$\frac{1}{2}R_3 \to R_3 \qquad\qquad\qquad\qquad (-1)R_3 + R_1 \to R_1 , \ (-1)R_3 + R_2 \to R_2 ,$$
$$\text{and } 7R_3 + R_4 \to R_4$$

$$\sim \begin{bmatrix} 0 & \frac{3}{2} & 1 & 0 & -\frac{1}{2} & 0 & 4 \\ 0 & \boxed{\frac{1}{2}} & 0 & 1 & -\frac{1}{2} & 0 & 1 \\ 1 & \frac{1}{2} & 0 & 0 & \frac{1}{2} & 0 & 6 \\ \hdashline 0 & -\frac{1}{2} & 0 & 0 & \frac{7}{2} & 1 & 42 \end{bmatrix} \begin{matrix} \frac{4}{3/2} = \frac{8}{3} \\ \frac{1}{1/2} = 2 \\ \frac{6}{1/2} = 12 \end{matrix} \sim \begin{bmatrix} 0 & \frac{3}{2} & 1 & 0 & -\frac{1}{2} & 0 & 4 \\ 0 & \boxed{1} & 0 & 2 & -1 & 0 & 2 \\ 1 & \frac{1}{2} & 0 & 0 & \frac{1}{2} & 0 & 6 \\ \hdashline 0 & -\frac{1}{2} & 0 & 0 & \frac{7}{2} & 1 & 42 \end{bmatrix}$$

$$2R_2 \to R_2 \qquad\qquad\qquad \left(-\frac{3}{2}\right)R_2 + R_1 \to R_1, \ \left(-\frac{1}{2}\right)R_2 + R_3 \to R_3,$$
$$\text{and } \frac{1}{2}R_2 + R_4 \to R_4$$

$$\sim \begin{array}{c} \\ x_1 \\ y_2 \\ y_1 \\ P \end{array} \begin{array}{cccccc} y_1 & y_2 & x_1 & x_2 & x_3 & P \\ \end{array} \begin{bmatrix} 0 & 0 & 1 & -3 & 1 & 0 & 1 \\ 0 & 1 & 0 & 2 & -1 & 0 & 2 \\ 1 & 0 & 0 & -1 & 1 & 0 & 5 \\ \hdashline 0 & 0 & 0 & 1 & 3 & 1 & 43 \end{bmatrix}$$

Optimal solution: min $C = 43$ at $x_1 = 0$, $x_2 = 1$, $x_3 = 3$.

23. The matrices corresponding to the given problem and the dual problem are:

$$A = \begin{bmatrix} 1 & -4 & 1 & 6 \\ -1 & 1 & -2 & 4 \\ \hline 5 & 2 & 2 & 1 \end{bmatrix} \quad \text{and} \quad A^T = \begin{bmatrix} 1 & -1 & 5 \\ -4 & 1 & 2 \\ 1 & -2 & 2 \\ \hline 6 & 4 & 1 \end{bmatrix} \quad \text{respectively.}$$

Thus, the dual problem is: Maximize $P = 6y_1 + 4y_2$

$$\begin{aligned} \text{Subject to:} \quad y_1 - \ y_2 &\le 5 \\ -4y_1 + \ y_2 &\le 2 \\ y_1 - 2y_2 &\le 2 \\ y_1, \ y_2 &\ge 0 \end{aligned}$$

We introduce slack variables x_1, x_2, and x_3 to obtain the initial system:

$$\begin{aligned} y_1 - \ y_2 + x_1 &= 5 \\ -4y_1 + \ y_2 + x_2 &= 2 \\ y_1 - 2y_2 + x_3 &= 2 \\ -6y_1 - 4y_2 + P &= 0 \end{aligned}$$

The simplex tableau for this problem is:

$$\begin{array}{c} \\ x_1 \\ x_2 \\ x_3 \\ P \end{array} \begin{array}{cccccc} y_1 & y_2 & x_1 & x_2 & x_3 & P \\ \end{array} \begin{bmatrix} 1 & -1 & 1 & 0 & 0 & 0 & 5 \\ -4 & 1 & 0 & 1 & 0 & 0 & 2 \\ \boxed{1} & -2 & 0 & 0 & 1 & 0 & 2 \\ \hdashline -6 & -4 & 0 & 0 & 0 & 1 & 0 \end{bmatrix} \begin{matrix} \frac{5}{1} = 5 \\ \\ \frac{2}{1} = 2 \\ \end{matrix} \sim \begin{bmatrix} 0 & \boxed{1} & 1 & 0 & -1 & 0 & 3 \\ 0 & -7 & 0 & 1 & 4 & 0 & 10 \\ 1 & -2 & 0 & 0 & 1 & 0 & 2 \\ \hdashline 0 & -16 & 0 & 0 & 6 & 1 & 12 \end{bmatrix}$$

$$(-1)R_3 + R_1 \to R_1, \ 4R_3 + R_2 \to R_2, \qquad\qquad 7R_1 + R_2 \to R_2, \ 2R_1 + R_3 \to R_3,$$
$$\text{and } 6R_3 + R_4 \to R_4 \qquad\qquad\qquad\qquad \text{and } 16R_1 + R_4 \to R_4$$

$$
\begin{array}{c}
\begin{array}{cccccc}
\ y_1 & y_2 & x_1 & x_2 & x_3 & P
\end{array}\\
\begin{array}{c}
y_2\\ x_2\\ \sim\ y_1\\ \\ P
\end{array}
\left[
\begin{array}{cccccc|c}
0 & 1 & 1 & 0 & -1 & 0 & 3\\
0 & 0 & 7 & 1 & -3 & 0 & 31\\
1 & 0 & 2 & 0 & -1 & 0 & 8\\
\hdashline
0 & 0 & 16 & 0 & -10 & 1 & 60
\end{array}
\right]
\end{array}
$$

Since all the entries above the dashed line in the pivot column, the x_3 column, are negative, the problem does not have an optimal solution.

25. The dual problem has 2 variables and 4 problem constraints.

27. The original problem must have two problem constraints, and any number of variables.

29. No. The dual problem will not be a standard maximization problem (one of the elements in the last column will be negative.)

31. Yes. Multiply both sides of the inequality by -1.

33. The matrices corresponding to the given problem and the dual problem are:

$$
A = \left[
\begin{array}{ccc|c}
3 & 2 & 2 & 16\\
4 & 3 & 1 & 14\\
5 & 3 & 1 & 12\\
\hline
16 & 8 & 4 & 1
\end{array}
\right]
\quad\text{and}\quad
A^T = \left[
\begin{array}{ccc|c}
3 & 4 & 5 & 16\\
2 & 3 & 3 & 8\\
2 & 1 & 1 & 4\\
\hline
16 & 14 & 12 & 1
\end{array}
\right]
\quad\text{respectively.}
$$

Thus, the dual problem is: Maximize $P = 16y_1 + 14y_2 + 12y_3$

Subject to:
$$
\begin{aligned}
3y_1 + 4y_2 + 5y_3 &\le 16\\
2y_1 + 3y_2 + 3y_3 &\le 8\\
2y_1 + y_2 + y_3 &\le 4\\
y_1,\ y_2,\ y_3 &\ge 0
\end{aligned}
$$

We introduce slack variables x_1, x_2, and x_3 to obtain the initial system:

$$
\begin{aligned}
3y_1 + 4y_2 + 5y_3 + x_1 &= 16\\
2y_1 + 3y_2 + 3y_3\phantom{{}+x_1} + x_2 &= 8\\
2y_1 + y_2 + y_3\phantom{{}+x_1+x_2} + x_3 &= 4\\
-16y_1 - 14y_2 - 12y_3\phantom{{}+x_1+x_2+x_3} + P &= 0
\end{aligned}
$$

The simplex tableau for this problem is:

$$
\begin{array}{c}
\begin{array}{ccccccc}
\ y_1 & y_2 & y_3 & x_1 & x_2 & x_3 & P
\end{array}\\
\begin{array}{c}
x_1\\ x_2\\ x_3\\ \\ P
\end{array}
\left[
\begin{array}{ccccccc|c}
3 & 4 & 5 & 1 & 0 & 0 & 0 & 16\\
2 & 3 & 3 & 0 & 1 & 0 & 0 & 8\\
② & 1 & 1 & 0 & 0 & 1 & 0 & 4\\
\hdashline
-16 & -14 & -12 & 0 & 0 & 0 & 1 & 0
\end{array}
\right]
\begin{array}{l}
\frac{16}{3} = 5.33\\[4pt]
\frac{8}{2} = 4\\[4pt]
\frac{4}{2} = 2
\end{array}
\end{array}
$$

$$\tfrac{1}{2} R_3 \to R_3$$

$$\sim \begin{bmatrix} 3 & 4 & 5 & 1 & 0 & 0 & 0 & | & 16 \\ 2 & 3 & 3 & 0 & 1 & 0 & 0 & | & 8 \\ ① & \frac{1}{2} & \frac{1}{2} & 0 & 0 & \frac{1}{2} & 0 & | & 2 \\ \hline -16 & -14 & -12 & 0 & 0 & 0 & 1 & | & 0 \end{bmatrix}$$

$(-3)R_3 + R_1 \rightarrow R_1,\ (-2)R_3 + R_2 \rightarrow R_2,$ and $16R_3 + R_4 \rightarrow R_4$

$$\sim \begin{bmatrix} 0 & \frac{5}{2} & \frac{7}{2} & 1 & 0 & -\frac{3}{2} & 0 & | & 10 \\ 0 & ② & 2 & 0 & 1 & -1 & 0 & | & 4 \\ 1 & \frac{1}{2} & \frac{1}{2} & 0 & 0 & \frac{1}{2} & 0 & | & 2 \\ \hline 0 & -6 & -4 & 0 & 0 & 8 & 1 & | & 32 \end{bmatrix} \begin{matrix} \frac{10}{5/2} = 4 \\ \frac{4}{2} = 2 \\ \frac{2}{1/2} = 4 \end{matrix}$$

$\frac{1}{2}R_2 \rightarrow R_2$

$$\sim \begin{bmatrix} 0 & \frac{5}{2} & \frac{7}{2} & 1 & 0 & -\frac{3}{2} & 0 & | & 10 \\ 0 & ① & 1 & 0 & \frac{1}{2} & -\frac{1}{2} & 0 & | & 2 \\ 1 & \frac{1}{2} & \frac{1}{2} & 0 & 0 & \frac{1}{2} & 0 & | & 2 \\ \hline 0 & -6 & -4 & 0 & 0 & 8 & 1 & | & 32 \end{bmatrix}$$

$\left(-\frac{5}{2}\right)R_2 + R_1 \rightarrow R_1,\ \left(-\frac{1}{2}\right)R_2 + R_3 \rightarrow R_3,$
and $6R_2 + R_4 \rightarrow R_4$

$$\begin{array}{c} \\ y_3 \\ y_2 \\ y_1 \\ P \end{array} \sim \begin{array}{cccccccc} y_1 & y_2 & y_3 & x_1 & x_2 & x_3 & P \\ \left[\begin{matrix} 0 & 0 & 1 & 1 & -\frac{5}{4} & -\frac{1}{4} & 0 & | & 5 \\ 0 & 1 & 1 & 0 & \frac{1}{2} & -\frac{1}{2} & 0 & | & 2 \\ 1 & 0 & 0 & 0 & -\frac{1}{4} & \frac{3}{4} & 0 & | & 1 \\ \hline 0 & 0 & 2 & 0 & 3 & 5 & 1 & | & 44 \end{matrix}\right] \end{array}$$

Optimal solution: min $C = 44$
at $x_1 = 0,\ x_2 = 3,\ x_3 = 5$.

35. The first and second inequalities must be rewritten before forming the dual.

Minimize $C = 5x_1 + 4x_2 + 5x_3 + 6x_4$
Subject to:
$$\begin{aligned} -x_1 - x_2 \qquad\qquad &\geq -12 \\ -x_3 - x_4 &\geq -25 \\ x_1 \qquad + x_3 \qquad &\geq 20 \\ x_2 \qquad + x_4 &\geq 15 \\ x_1,\ x_2,\ x_3,\ x_4 &\geq 0 \end{aligned}$$

The matrices corresponding to the given problem and the dual problem are:

$$A = \begin{bmatrix} -1 & -1 & 0 & 0 & | & -12 \\ 0 & 0 & -1 & -1 & | & -25 \\ 1 & 0 & 1 & 0 & | & 20 \\ 0 & 1 & 0 & 1 & | & 15 \\ \hline 5 & 4 & 5 & 6 & | & 1 \end{bmatrix} \quad \text{and} \quad A^T = \begin{bmatrix} -1 & 0 & 1 & 0 & | & 5 \\ -1 & 0 & 0 & 1 & | & 4 \\ 0 & -1 & 1 & 0 & | & 5 \\ 0 & -1 & 0 & 1 & | & 6 \\ \hline -12 & -25 & 20 & 15 & | & 1 \end{bmatrix}$$

The dual problem is: Maximize $P = -12y_1 - 25y_2 + 20y_3 + 15y_4$
Subject to:
$$\begin{aligned} -y_1 \qquad\quad + y_3 \qquad &\leq 5 \\ -y_1 \qquad\qquad + y_4 &\leq 4 \\ -y_2 + y_3 \qquad &\leq 5 \\ -y_2 \qquad + y_4 &\leq 6 \\ y_1,\ y_2,\ y_3,\ y_4 &\geq 0 \end{aligned}$$

We introduce the slack variables x_1, x_2, x_3, and x_4 to obtain the initial system:

$$
\begin{aligned}
-y_1 & & + y_3 & & + x_1 & & & & & = 5 \\
-y_1 & & & + y_4 & & + x_2 & & & & = 4 \\
& -y_2 & + y_3 & & & & + x_3 & & & = 5 \\
& -y_2 & & + y_4 & & & & + x_4 & & = 6 \\
+12y_1 & + 25y_2 & - 20y_3 & - 15y_4 & & & & & + P & = 0
\end{aligned}
$$

The simplex tableau for this problem is:

$$
\begin{array}{c}
\begin{array}{ccccccccc}
y_1 & y_2 & y_3 & y_4 & x_1 & x_2 & x_3 & x_4 & P
\end{array} \\
\begin{array}{c}
x_1 \\ x_2 \\ x_3 \\ x_4 \\ P
\end{array}
\left[
\begin{array}{ccccccccc|c}
-1 & 0 & 1 & 0 & 1 & 0 & 0 & 0 & 0 & 5 \\
-1 & 0 & 0 & 1 & 0 & 1 & 0 & 0 & 0 & 4 \\
0 & -1 & \textcircled{1} & 0 & 0 & 0 & 1 & 0 & 0 & 5 \\
0 & -1 & 0 & 1 & 0 & 0 & 0 & 1 & 0 & 6 \\
\hline
12 & 25 & -20 & -15 & 0 & 0 & 0 & 0 & 1 & 0
\end{array}
\right]
\begin{array}{l}
\frac{5}{1} = 5 \\ \\ \frac{5}{1} = 5 \\ \\ \\
\end{array}
\end{array}
$$

$(-1)R_3 + R_1 \rightarrow R_1$ and $20R_3 + R_5 \rightarrow R_5$

$$
\sim
\left[
\begin{array}{ccccccccc|c}
-1 & 1 & 0 & 0 & 1 & 0 & -1 & 0 & 0 & 0 \\
-1 & 0 & 0 & \textcircled{1} & 0 & 1 & 0 & 0 & 0 & 4 \\
0 & -1 & 1 & 0 & 0 & 0 & 1 & 0 & 0 & 5 \\
0 & -1 & 0 & 1 & 0 & 0 & 0 & 1 & 0 & 6 \\
\hline
12 & 5 & 0 & -15 & 0 & 0 & 20 & 0 & 1 & 100
\end{array}
\right]
\begin{array}{l}
\\ \frac{4}{1} = 4 \\ \\ \frac{6}{1} = 6 \\ \\
\end{array}
$$

$(-1)R_2 + R_4 \rightarrow R_4$ and $15R_2 + R_5 \rightarrow R_5$

$$
\sim
\left[
\begin{array}{ccccccccc|c}
-1 & 1 & 0 & 0 & 1 & 0 & -1 & 0 & 0 & 0 \\
-1 & 0 & 0 & 1 & 0 & 1 & 0 & 0 & 0 & 4 \\
0 & -1 & 1 & 0 & 0 & 0 & 1 & 0 & 0 & 5 \\
\textcircled{1} & -1 & 0 & 0 & 0 & -1 & 0 & 1 & 0 & 2 \\
\hline
-3 & 5 & 0 & 0 & 0 & 15 & 20 & 0 & 1 & 160
\end{array}
\right]
$$

$3R_4 + R_5 \rightarrow R_5$, $R_4 + R_1 \rightarrow R_1$, and $R_4 + R_2 \rightarrow R_2$

$$
\begin{array}{c}
\begin{array}{ccccccccc}
y_1 & y_2 & y_3 & y_4 & x_1 & x_2 & x_3 & x_4 & P
\end{array} \\
\begin{array}{c}
x_1 \\ y_4 \\ y_3 \\ y_1 \\ P
\end{array}
\sim
\left[
\begin{array}{ccccccccc|c}
0 & 0 & 0 & 0 & 1 & -1 & -1 & 1 & 0 & 2 \\
0 & -1 & 0 & 1 & 0 & 0 & 0 & 1 & 0 & 6 \\
0 & -1 & 1 & 0 & 0 & 0 & 1 & 0 & 0 & 5 \\
1 & -1 & 0 & 0 & 0 & -1 & 0 & 1 & 0 & 2 \\
\hline
0 & 2 & 0 & 0 & 0 & 12 & 20 & 3 & 1 & 166
\end{array}
\right]
\end{array}
$$

Optimal solution:
min $C = 166$ at $x_1 = 0$,
$x_2 = 12$, $x_3 = 20$, $x_4 = 3$.

37. (A) Let x_1 = the number of hours the Cedarburg plant is operated,
x_2 = the number of hours the Grafton plant is operated,
and x_3 = the number of hours the West Bend plant is operated.

The mathematical model for this problem is:

Minimize $C = 70x_1 + 75x_2 + 90x_3$

Subject to: $20x_1 + 10x_2 + 20x_3 \geq 300$
$10x_1 + 20x_2 + 20x_3 \geq 200$
$x_1, \ x_2, \ x_3 \geq \ 0$

Divide each of the problem constraint inequalities by 10 to simplify the calculations. The matrices corresponding to the given problem and the dual problem are:

$$A = \left[\begin{array}{ccc|c} 2 & 1 & 2 & 30 \\ 1 & 2 & 2 & 20 \\ \hline 70 & 75 & 90 & 1 \end{array} \right] \quad \text{and} \quad A^T = \left[\begin{array}{cc|c} 2 & 1 & 70 \\ 1 & 2 & 75 \\ 2 & 2 & 90 \\ \hline 30 & 20 & 1 \end{array} \right] \quad \text{respectively.}$$

Thus, the dual problem is:

Maximize $P = 30y_1 + 20y_2$

Subject to: $2y_1 + \ y_2 \leq 70$
$y_1 + 2y_2 \leq 75$
$2y_1 + 2y_2 \leq 90$
$y_1, \ y_2 \geq \ 0$

We introduce slack variables x_1, x_2, and x_3 to obtain the initial system:

$$\begin{array}{rcl} 2y_1 + \ y_2 + x_1 \qquad\qquad\quad & = & 70 \\ y_1 + 2y_2 \qquad + x_2 \qquad\quad & = & 75 \\ 2y_1 + 2y_2 \qquad\qquad + x_3 \quad & = & 90 \\ -30y_1 - 20y_2 \qquad\qquad\qquad + P & = & 0 \end{array}$$

The simplex tableau for this problem is:

$$\begin{array}{c} \quad\; y_1 \; y_2 \; x_1 \; x_2 \; x_3 \; P \\ \begin{array}{c} x_1 \\ x_2 \\ x_3 \\ P \end{array} \left[\begin{array}{cccccc|c} ② & 1 & 1 & 0 & 0 & 0 & 70 \\ 1 & 2 & 0 & 1 & 0 & 0 & 75 \\ 2 & 2 & 0 & 0 & 1 & 0 & 90 \\ \hline -30 & -20 & 0 & 0 & 0 & 1 & 0 \end{array} \right] \end{array} \begin{array}{l} \frac{70}{2} = 35 \\ \frac{75}{1} = 75 \\ \frac{90}{2} = 45 \end{array} \quad \sim \quad \left[\begin{array}{cccccc|c} ① & \frac{1}{2} & \frac{1}{2} & 0 & 0 & 0 & 35 \\ 1 & 2 & 0 & 1 & 0 & 0 & 75 \\ 2 & 2 & 0 & 0 & 1 & 0 & 90 \\ \hline -30 & -20 & 0 & 0 & 0 & 1 & 0 \end{array} \right]$$

$$\frac{1}{2} R_1 \rightarrow R_1$$

$$(-1)R_1 + R_2 \rightarrow R_2, \ (-2)R_1 + R_3 \rightarrow R_3,$$
$$\text{and } 30R_1 + R_4 \rightarrow R_4$$

$$\sim \begin{bmatrix} 1 & \frac{1}{2} & \frac{1}{2} & 0 & 0 & 0 & 35 \\ 0 & \frac{3}{2} & -\frac{1}{2} & 1 & 0 & 0 & 40 \\ 0 & ① & -1 & 0 & 1 & 0 & 20 \\ \hdashline 0 & -5 & 15 & 0 & 0 & 1 & 1050 \end{bmatrix} \begin{matrix} \frac{35}{1/2} = 70 \\[4pt] \frac{40}{3/2} = \frac{80}{3} \approx 26.67 \\[4pt] \frac{20}{1} = 20 \end{matrix}$$

$$\left(-\tfrac{1}{2}\right)R_3 + R_1 \to R_1, \ \left(-\tfrac{3}{2}\right)R_3 + R_2 \to R_2, \text{ and } 5R_3 + R_4 \to R_4$$

$$\sim \begin{array}{c} y_1 \\ x_2 \\ y_2 \\ \\ P \end{array} \begin{matrix} y_1 & y_2 & x_1 & x_2 & x_3 & P \end{matrix} \\ \begin{bmatrix} 1 & 0 & 1 & 0 & -\frac{1}{2} & 0 & 25 \\ 0 & 0 & 1 & 1 & -\frac{3}{2} & 0 & 10 \\ 0 & 1 & -1 & 0 & 1 & 0 & 20 \\ \hdashline 0 & 0 & 10 & 0 & 5 & 1 & 1150 \end{bmatrix}$$

The minimal production cost is $1150 when the Cedarburg plant is operated 10 hours per day, the West Bend plant is operated 5 hours per day, and the Grafton plant is not used.

(B) If the demand for deluxe ice cream increases to 300 gallons per day and all other data remains the same, then the matrices for this problem and the dual problem are:

$$A = \begin{bmatrix} 2 & 1 & 2 & 30 \\ 1 & 2 & 2 & 30 \\ \hline 70 & 75 & 90 & 1 \end{bmatrix} \quad \text{and} \quad A^T = \begin{bmatrix} 2 & 1 & 70 \\ 1 & 2 & 75 \\ 2 & 2 & 90 \\ \hline 30 & 30 & 1 \end{bmatrix} \quad \text{respectively.}$$

Thus, the dual problem is:

Maximize $P = 30y_1 + 30y_2$

Subject to: $2y_1 + y_2 \le 70$
$y_1 + 2y_2 \le 75$
$2y_1 + 2y_2 \le 90$
$y_1, \ y_2 \ge 0$

We introduce slack variables x_1, x_2, and x_3 to obtain the initial system:

$$\begin{aligned} 2y_1 + y_2 + x_1 \qquad\qquad &= 70 \\ y_1 + 2y_2 \qquad + x_2 \qquad &= 75 \\ 2y_1 + 2y_2 \qquad\qquad + x_3 \quad &= 90 \\ -30y_1 - 30y_2 \qquad\qquad\qquad + P &= 0 \end{aligned}$$

The simplex tableau for this problem is:

$$\begin{array}{c} \\ x_1 \\ x_2 \\ x_3 \\ P \end{array} \begin{array}{cccccc} y_1 & y_2 & x_1 & x_2 & x_3 & P \\ \left[\begin{array}{cccccc|c} ② & 1 & 1 & 0 & 0 & 0 & 70 \\ 1 & 2 & 0 & 1 & 0 & 0 & 75 \\ 2 & 2 & 0 & 0 & 1 & 0 & 90 \\ \hline -30 & -30 & 0 & 0 & 0 & 1 & 0 \end{array}\right] \end{array} \begin{array}{l} \frac{70}{2} = 35 \\ \frac{75}{1} = 75 \\ \frac{90}{2} = 45 \end{array}$$

$$\frac{1}{2} R_1 \rightarrow R_1$$

Note: Either column 1 or column 2 can be used as the pivot column. We chose column 1.

$$\left[\begin{array}{cccccc|c} ① & \frac{1}{2} & \frac{1}{2} & 0 & 0 & 0 & 35 \\ 1 & 2 & 0 & 1 & 0 & 0 & 75 \\ 2 & 2 & 0 & 0 & 1 & 0 & 90 \\ \hline -30 & -30 & 0 & 0 & 0 & 1 & 0 \end{array}\right] \sim \left[\begin{array}{cccccc|c} 1 & \frac{1}{2} & \frac{1}{2} & 0 & 0 & 0 & 35 \\ 0 & \frac{3}{2} & -\frac{1}{2} & 1 & 0 & 0 & 40 \\ 0 & ① & -1 & 0 & 1 & 0 & 20 \\ \hline 0 & -15 & 15 & 0 & 0 & 1 & 1050 \end{array}\right] \begin{array}{l} \frac{35}{1/2} = 70 \\ \frac{40}{3/2} = \frac{80}{3} \approx 26.67 \\ \frac{20}{1} = 20 \end{array}$$

$$(-1)R_1 + R_2 \rightarrow R_2,$$
$$(-2)R_1 + R_3 \rightarrow R_3,$$
$$30R_1 + R_4 \rightarrow R_4$$

$$\left(-\frac{1}{2}\right) R_3 + R_1 \rightarrow R_1,$$
$$\left(-\frac{3}{2}\right) R_3 + R_2 \rightarrow R_2,$$
$$15R_3 + R_4 \rightarrow R_4$$

$$\begin{array}{c} \\ y_1 \\ x_2 \\ y_2 \\ P \end{array} \begin{array}{cccccc} y_1 & y_2 & x_1 & x_2 & x_3 & P \\ \left[\begin{array}{cccccc|c} 1 & 0 & 1 & 0 & -\frac{1}{2} & 0 & 25 \\ 0 & 0 & 1 & 1 & -\frac{3}{2} & 0 & 10 \\ 0 & 1 & -1 & 0 & 1 & 0 & 20 \\ \hline 0 & 0 & 0 & 0 & 15 & 1 & 1350 \end{array}\right] \end{array}$$

The minimal production cost is \$1350 when the West Bend plant is operated 15 hours per day, and the Cedarburg and Grafton plants are not used.

(C) If the demand for deluxe ice cream increases to 400 gallons per day and all other data remains the same, then the matrices for this problem and the dual are:

$$A = \left[\begin{array}{ccc|c} 2 & 1 & 2 & 30 \\ 1 & 2 & 2 & 40 \\ \hline 70 & 75 & 90 & 1 \end{array}\right] \quad \text{and} \quad A^T = \left[\begin{array}{cc|c} 2 & 1 & 70 \\ 1 & 2 & 75 \\ 2 & 2 & 90 \\ \hline 30 & 40 & 1 \end{array}\right] \quad \text{respectively.}$$

Thus, the dual problem is:

Maximize $P = 30y_1 + 40y_2$

Subject to: $2y_1 + y_2 \leq 70$
$$y_1 + 2y_2 \leq 75$$
$$2y_1 + 2y_2 \leq 90$$
$$y_1, \ y_2 \geq 0$$

We introduce slack variables x_1, x_2, and x_3 to obtain the initial system:

$$\begin{aligned}
2y_1 + \ \ y_2 + x_1 \ \ \ \ \ \ \ \ \ \ \ \ \ \ \ \ &= 70 \\
y_1 + 2y_2 \ \ \ \ \ \ \ + x_2 \ \ \ \ \ \ \ \ \ \ &= 75 \\
2y_1 + 2y_2 \ \ \ \ \ \ \ \ \ \ \ \ + x_3 \ \ \ \ &= 90 \\
-30y_1 - 40y_2 \ \ \ \ \ \ \ \ \ \ \ \ \ \ \ \ + P &= 0
\end{aligned}$$

The simplex tableau for this problem is:

$$\begin{array}{c}
\begin{array}{ccccccc}
 & y_1 & y_2 & x_1 & x_2 & x_3 & P
\end{array} \\
\begin{array}{c}
x_1 \\ x_2 \\ x_3 \\ \\ P
\end{array}
\left[\begin{array}{cccccc|c}
2 & 1 & 1 & 0 & 0 & 0 & 70 \\
1 & ② & 0 & 1 & 0 & 0 & 75 \\
2 & 2 & 0 & 0 & 1 & 0 & 90 \\
\hline
-30 & -40 & 0 & 0 & 0 & 1 & 0
\end{array}\right]
\end{array}$$

$\dfrac{70}{1} = 70$

$\dfrac{75}{2} = 37.5$

$\dfrac{90}{2} = 45$

$$\tfrac{1}{2} R_2 \to R_2$$

$$\sim \left[\begin{array}{cccccc|c}
2 & 1 & 1 & 0 & 0 & 0 & 70 \\
\tfrac{1}{2} & ① & 0 & \tfrac{1}{2} & 0 & 0 & \tfrac{75}{2} \\
2 & 2 & 0 & 0 & 1 & 0 & 90 \\
\hline
-30 & -40 & 0 & 0 & 0 & 1 & 0
\end{array}\right] \sim \left[\begin{array}{cccccc|c}
\tfrac{3}{2} & 0 & 1 & -\tfrac{1}{2} & 0 & 0 & \tfrac{65}{2} \\
\tfrac{1}{2} & 1 & 0 & \tfrac{1}{2} & 0 & 0 & \tfrac{75}{2} \\
① & 0 & 0 & -1 & 1 & 0 & 15 \\
\hline
-10 & 0 & 0 & 20 & 0 & 1 & 1500
\end{array}\right]$$

$\dfrac{65/2}{3/2} \approx 21.67$

$\dfrac{75/2}{1/2} = 75$

$\dfrac{15}{1} = 15$

$$(-1)R_2 + R_1 \to R_1, \qquad \left(-\tfrac{3}{2}\right) R_3 + R_1 \to R_1,$$

$$(-2)R_2 + R_3 \to R_3, \qquad \left(-\tfrac{1}{2}\right) R_3 + R_2 \to R_2,$$

$$40R_2 + R_4 \to R_4 \qquad \qquad 10R_3 + R_4 \to R_4$$

$$\begin{array}{c}
\begin{array}{ccccccc}
 & y_1 & y_2 & x_1 & x_2 & x_3 & P
\end{array} \\
\begin{array}{c}
x_1 \\ y_2 \\ y_1 \\ \\ P
\end{array}
\sim \left[\begin{array}{cccccc|c}
0 & 0 & 1 & 1 & -\tfrac{3}{2} & 0 & 10 \\
0 & 1 & 0 & 1 & -\tfrac{1}{2} & 0 & 30 \\
1 & 0 & 0 & -1 & 1 & 0 & 15 \\
\hline
0 & 0 & 0 & 10 & 10 & 1 & 1650
\end{array}\right]
\end{array}$$

The minimal production cost is \$1650 when the Grafton plant and West Bend plant are each operated 10 hours per day, and the Cedarburg plant is not used.

39. Let x_1 = the number of single-sided drives from Associated Electronics,
   x_2 = the number of double-sided drives from Associated Electronics,
   x_3 = the number of single-sided drives from Digital Drives,
and x_4 = the number of double-sided drives from Digital Drives.

The mathematical model is: Minimize $C = 250x_1 + 350x_2 + 290x_3 + 320x_4$

$$\begin{aligned}
\text{Subject to:} \quad x_1 + x_2 \ \ \ \ \ \ \ \ \ \ \ \ &\leq 1000 \\
x_3 + x_4 &\leq 2000 \\
x_1 \ \ \ \ \ \ + x_3 \ \ \ \ \ \ &\geq 1200 \\
x_2 \ \ \ \ \ \ + x_4 &\geq 1600 \\
x_1, \ x_2, \ x_3, \ x_4 &\geq 0
\end{aligned}$$

We multiply the first two problem constraints by -1 to obtain inequalities of the $\geq$ type. The model becomes:

Minimize $C = 250x_1 + 350x_2 + 290x_3 + 320x_4$

Subject to:
$$
\begin{aligned}
-x_1 - x_2 &\geq -1000 \\
-x_3 - x_4 &\geq -2000 \\
x_1 + x_3 &\geq 1200 \\
x_2 + x_4 &\geq 1600 \\
x_1, x_2, x_3, x_4 &\geq 0
\end{aligned}
$$

The matrices for this problem and the dual problem are:

$$
A = \left[\begin{array}{cccc|c}
-1 & -1 & 0 & 0 & -1000 \\
0 & 0 & -1 & -1 & -2000 \\
1 & 0 & 1 & 0 & 1200 \\
0 & 1 & 0 & 1 & 1600 \\
\hline
250 & 350 & 290 & 320 & 1
\end{array}\right]
\quad \text{and} \quad
A^T = \left[\begin{array}{cccc|c}
-1 & 0 & 1 & 0 & 250 \\
-1 & 0 & 0 & 1 & 350 \\
0 & -1 & 1 & 0 & 290 \\
0 & -1 & 0 & 1 & 320 \\
\hline
-1000 & -2000 & 1200 & 1600 & 1
\end{array}\right]
$$

Thus, the dual problem is:

Maximize $P = -1000y_1 - 2000y_2 + 1200y_3 + 1600y_4$

Subject to:
$$
\begin{aligned}
-y_1 + y_3 &\leq 250 \\
-y_1 + y_4 &\leq 350 \\
-y_2 + y_3 &\leq 290 \\
-y_2 + y_4 &\leq 320 \\
y_1, y_2, y_3, y_4 &\geq 0
\end{aligned}
$$

We introduce slack variables x_1, x_2, x_3, and x_4 to obtain the initial system:

$$
\begin{aligned}
-y_1 + y_3 + x_1 &= 250 \\
-y_1 + y_4 + x_2 &= 350 \\
-y_2 + y_3 + x_3 &= 290 \\
-y_2 + y_4 + x_4 &= 320 \\
1000y_1 + 2000y_2 - 1200y_3 - 1600y_4 + P &= 0
\end{aligned}
$$

The simplex tableau for this problem is:

	y_1	y_2	y_3	y_4	x_1	x_2	x_3	x_4	P	
x_1	-1	0	1	0	1	0	0	0	0	250
x_2	-1	0	0	1	0	1	0	0	0	350
x_3	0	-1	1	0	0	0	1	0	0	290
x_4	0	-1	0	①	0	0	0	1	0	320
P	1000	2000	-1200	-1600	0	0	0	0	1	0

$$\frac{350}{1} = 350 \qquad \frac{320}{1} = 320$$

$(-1)R_4 + R_2 \rightarrow R_2$ and $1600R_4 + R_5 \rightarrow R_5$

$$
\sim
\begin{bmatrix}
-1 & 0 & \textcircled{1} & 0 & 1 & 0 & 0 & 0 & 0 & | & 250 \\
-1 & 1 & 0 & 0 & 0 & 1 & 0 & -1 & 0 & | & 30 \\
0 & -1 & 1 & 0 & 0 & 0 & 1 & 0 & 0 & | & 290 \\
0 & -1 & 0 & 1 & 0 & 0 & 0 & 1 & 0 & | & 320 \\
\hline
1000 & 400 & -1200 & 0 & 0 & 0 & 0 & 1600 & 1 & | & 512{,}000
\end{bmatrix}
\begin{matrix}
\frac{250}{1} = 250 \\[6pt] \\
\frac{290}{1} = 290 \\ \\
\end{matrix}
$$

$$(-1)R_1 + R_3 \to R_3 \text{ and } 1200R_1 + R_5 \to R_5$$

$$
\sim
\begin{bmatrix}
-1 & 0 & 1 & 0 & 1 & 0 & 0 & 0 & 0 & | & 250 \\
-1 & 1 & 0 & 0 & 0 & 1 & 0 & -1 & 0 & | & 30 \\
\textcircled{1} & -1 & 0 & 0 & -1 & 0 & 1 & 0 & 0 & | & 40 \\
0 & -1 & 0 & 1 & 0 & 0 & 0 & 1 & 0 & | & 320 \\
\hline
-200 & 400 & 0 & 0 & 1200 & 0 & 0 & 1600 & 1 & | & 812{,}000
\end{bmatrix}
$$

$$R_3 + R_1 \to R_1,\quad R_3 + R_2 \to R_2,\quad \text{and } 200R_3 + R_5 \to R_5$$

	y_1	y_2	y_3	y_4	x_1	x_2	x_3	x_4	P	
y_3	0	-1	1	0	0	0	1	0	0	290
x_2	0	0	0	0	-1	1	1	-1	0	70
$\sim\ y_1$	1	-1	0	0	-1	0	1	0	0	40
y_4	0	-1	0	1	0	0	0	1	0	320
P	0	200	0	0	1000	0	200	1600	1	820,000

The minimal purchase cost is \$820,000 when 1000 single-sided and no double-sided drives are ordered from Associated Electronics, and 200 single-sided and 1600 double-sided drives are ordered from Digital Drives.

41. Let x_1 = the number of ounces of food L,
x_2 = the number of ounces of food M,
x_3 = the number of ounces of food N.

Mathematical model: Minimize $C = 20x_1 + 24x_2 + 18x_3$
Subject to: $20x_1 + 10x_2 + 10x_3 \geq 300$
$10x_1 + 10x_2 + 10x_3 \geq 200$
$10x_1 + 15x_2 + 10x_3 \geq 240$
$x_1,\ x_2,\ x_3 \geq 0$

Divide the first two problem constraints by 10 and the third by 5. This will simplify the calculations.

$$
A = \begin{bmatrix}
2 & 1 & 1 & | & 30 \\
1 & 1 & 1 & | & 20 \\
2 & 3 & 2 & | & 48 \\
\hline
20 & 24 & 18 & | & 1
\end{bmatrix}
\quad \text{and} \quad
A^T = \begin{bmatrix}
2 & 1 & 2 & | & 20 \\
1 & 1 & 3 & | & 24 \\
1 & 1 & 2 & | & 18 \\
\hline
30 & 20 & 48 & | & 1
\end{bmatrix}
$$

The dual problem is: Maximize $P = 30y_1 + 20y_2 + 48y_3$

Subject to: $2y_1 + y_2 + 2y_3 \leq 20$

$y_1 + y_2 + 3y_3 \leq 24$

$y_1 + y_2 + 2y_3 \leq 18$

$y_1, \; y_2, \; y_3 \geq 0$

We introduce slack variables x_1, x_2, and x_3 to obtain the initial system:

$$2y_1 + \;\; y_2 + 2y_3 + x_1 \qquad\qquad = 20$$
$$y_1 + \;\; y_2 + 3y_3 \qquad + x_2 \qquad = 24$$
$$y_1 + \;\; y_2 + 2y_3 \qquad\quad + x_3 \quad = 18$$
$$-30y_1 - 20y_2 - 24y_3 \qquad\qquad\quad + P = \;\; 0$$

The simplex tableau for this problem is:

$$
\begin{array}{c}
 \\
x_1 \\
x_2 \\
x_3 \\
P
\end{array}
\begin{array}{c}
\begin{array}{ccccccc}
y_1 & y_2 & y_3 & x_1 & x_2 & x_3 & P
\end{array} \\
\left[
\begin{array}{ccccccc|c}
2 & 1 & 2 & 1 & 0 & 0 & 0 & 20 \\
1 & 1 & ③ & 0 & 1 & 0 & 0 & 24 \\
1 & 1 & 2 & 0 & 0 & 1 & 0 & 18 \\
\hdashline
-30 & -20 & -48 & 0 & 0 & 0 & 1 & 0
\end{array}
\right]
\end{array}
\begin{array}{c}
\frac{20}{2} = 10 \\
\frac{24}{3} = 8 \\
\frac{18}{2} = 9 \\
\;
\end{array}
$$

$\frac{1}{3}R_2 \rightarrow R_2$

$$
\sim
\left[
\begin{array}{ccccccc|c}
2 & 1 & 2 & 1 & 0 & 0 & 0 & 20 \\
\frac{1}{3} & \frac{1}{3} & ① & 0 & \frac{1}{3} & 0 & 0 & 8 \\
1 & 1 & 2 & 0 & 0 & 1 & 0 & 18 \\
\hdashline
-30 & -20 & -48 & 0 & 0 & 0 & 1 & 0
\end{array}
\right]
$$

$(-2)R_2 + R_1 \rightarrow R_1$, $(-2)R_2 + R_3 \rightarrow R_3$, and $48R_2 + R_4 \rightarrow R_4$

$$
\sim
\left[
\begin{array}{ccccccc|c}
④/③ & \frac{1}{3} & 0 & 1 & -\frac{2}{3} & 0 & 0 & 4 \\
\frac{1}{3} & \frac{1}{3} & 1 & 0 & \frac{1}{3} & 0 & 0 & 8 \\
\frac{1}{3} & \frac{1}{3} & 0 & 0 & -\frac{2}{3} & 1 & 0 & 2 \\
\hdashline
-14 & -4 & 0 & 0 & 16 & 0 & 1 & 384
\end{array}
\right]
\begin{array}{c}
\frac{4}{4/3} = 3 \\
\frac{8}{1/3} = 24 \\
\frac{2}{1/3} = 6 \\
\;
\end{array}
$$

$\frac{3}{4}R_1 \rightarrow R_1$

$$\sim \begin{bmatrix} ① & \frac{1}{4} & 0 & \frac{3}{4} & -\frac{1}{2} & 0 & 0 & \Big| & 3 \\ \frac{1}{3} & \frac{1}{3} & 1 & 0 & \frac{1}{3} & 0 & 0 & \Big| & 8 \\ \frac{1}{3} & \frac{1}{3} & 0 & 0 & -\frac{2}{3} & 1 & 0 & \Big| & 2 \\ \hline -14 & -4 & 0 & 0 & 16 & 0 & 1 & \Big| & 384 \end{bmatrix}$$

$$\left(-\frac{1}{3}\right)R_1 + R_2 \rightarrow R_2, \quad \left(-\frac{1}{3}\right)R_1 + R_3 \rightarrow R_3, \text{ and } 14R_1 + R_4 \rightarrow R_4$$

$$\sim \begin{bmatrix} 1 & \frac{1}{4} & 0 & \frac{3}{4} & -\frac{1}{2} & 0 & 0 & \Big| & 3 \\ 0 & \frac{1}{4} & 1 & -\frac{1}{4} & \frac{1}{2} & 0 & 0 & \Big| & 7 \\ 0 & ⓵\hspace{-0.6em}\tfrac{1}{4} & 0 & -\frac{1}{4} & -\frac{1}{2} & 1 & 0 & \Big| & 1 \\ \hline 0 & -\frac{1}{2} & 0 & \frac{21}{2} & 9 & 0 & 1 & \Big| & 426 \end{bmatrix} \quad \begin{array}{l} \frac{3}{1/4} = 12 \\[4pt] \frac{7}{1/4} = 28 \\[4pt] \frac{1}{1/4} = 4 \end{array}$$

$$4R_3 \rightarrow R_3$$

$$\sim \begin{bmatrix} 1 & \frac{1}{4} & 0 & \frac{3}{4} & -\frac{1}{2} & 0 & 0 & \Big| & 3 \\ 0 & \frac{1}{4} & 1 & -\frac{1}{4} & \frac{1}{2} & 0 & 0 & \Big| & 7 \\ 0 & ① & 0 & -1 & -2 & 4 & 0 & \Big| & 4 \\ \hline 0 & -\frac{1}{2} & 0 & \frac{21}{2} & 9 & 0 & 1 & \Big| & 426 \end{bmatrix} \sim \begin{array}{c} \\ y_1 \\ y_3 \\ y_2 \\ \\ P \end{array} \begin{bmatrix} y_1 & y_2 & y_3 & x_1 & x_2 & x_3 & P & & \\ 1 & 0 & 0 & 1 & 0 & -1 & 0 & \Big| & 2 \\ 0 & 0 & 1 & 0 & 1 & -1 & 0 & \Big| & 6 \\ 0 & 1 & 0 & -1 & -2 & 4 & 0 & \Big| & 4 \\ \hline 0 & 0 & 0 & 10 & 8 & 2 & 1 & \Big| & 428 \end{bmatrix}$$

$$\left(-\frac{1}{4}\right)R_3 + R_1 \rightarrow R_1, \quad \left(-\frac{1}{4}\right)R_3 + R_2 \rightarrow R_2,$$

$$\text{and } \frac{1}{2}R_3 + R_4 \rightarrow R_4$$

The minimal cholesterol intake is 428 units when 10 ounces of food L, 8 ounces of food M, and 2 ounces of food N are used.

43. Let x_1 = the number of students bused from North Division to Central,
$\quad\;\; x_2$ = the number of students bused from North Division to Washington,
$\quad\;\; x_3$ = the number of students bused from South Division to Central,
and x_4 = the number of students bused from South Division to Washington.
The mathematical model for this problem is:

$$\begin{aligned} &\text{Minimize } C = 5x_1 + 2x_2 + 3x_3 + 4x_4 \\ &\text{Subject to: } \begin{aligned}[t] x_1 + x_2 \quad\quad\quad\;\;\; &\geq 300 \\ x_3 + x_4 &\geq 500 \\ x_1 \quad\;\; + x_3 \quad\;\; &\leq 400 \\ x_2 \quad\;\; + x_4 &\leq 500 \\ x_1,\; x_2,\; x_3,\; x_4 &\geq 0 \end{aligned} \end{aligned}$$

We multiply the last two problem constraints by -1 so that all the constraints are of the $\geq$ type. The model becomes:

$$\begin{aligned} &\text{Minimize } C = 5x_1 + 2x_2 + 3x_3 + 4x_4 \\ &\text{Subject to: } \begin{aligned}[t] x_1 + x_2 \quad\quad\quad\quad\; &\geq 300 \\ x_3 + x_4 &\geq 500 \\ -x_1 \quad\;\; - x_3 \quad\;\; &\geq -400 \\ -x_2 \quad\;\; - x_4 &\geq -500 \\ x_1,\; x_2,\; x_3,\; x_4 &\geq 0 \end{aligned} \end{aligned}$$

The matrices for this problem and the dual problem are:

$$A = \begin{bmatrix} 1 & 1 & 0 & 0 & | & 300 \\ 0 & 0 & 1 & 1 & | & 500 \\ -1 & 0 & -1 & 0 & | & -400 \\ 0 & -1 & 0 & -1 & | & -500 \\ 5 & 2 & 3 & 4 & | & 1 \end{bmatrix} \quad \text{and} \quad A^T = \begin{bmatrix} 1 & 0 & -1 & 0 & | & 5 \\ 1 & 0 & 0 & -1 & | & 2 \\ 0 & 1 & -1 & 0 & | & 3 \\ 0 & 1 & 0 & -1 & | & 4 \\ 300 & 500 & -400 & -500 & | & 1 \end{bmatrix}$$

The dual problem is: Maximize $P = 300y_1 + 500y_2 - 400y_3 - 500y_4$

$$\text{Subject to: } \begin{aligned} y_1 \quad\quad - y_3 \quad\quad &\le 5 \\ y_1 \quad\quad\quad\quad - y_4 &\le 2 \\ y_2 - y_3 \quad\quad &\le 3 \\ y_2 \quad\quad - y_4 &\le 4 \\ y_1,\ y_2,\ y_3,\ y_4 &\ge 0 \end{aligned}$$

We introduce slack variables x_1, x_2, x_3, and x_4 to obtain the initial system:

$$\begin{aligned} y_1 \quad\quad\quad - y_3 \quad\quad\quad + x_1 \quad\quad\quad\quad\quad\quad &= 5 \\ y_1 \quad\quad\quad\quad\quad - y_4 \quad\quad + x_2 \quad\quad\quad\quad &= 2 \\ y_2 \quad - y_3 \quad\quad\quad\quad\quad + x_3 \quad\quad &= 3 \\ y_2 \quad\quad\quad - y_4 \quad\quad\quad\quad\quad + x_4 &= 4 \\ -300y_1 - 500y_2 + 400y_3 + 500y_4 \quad\quad\quad\quad\quad\quad + P &= 0 \end{aligned}$$

The simplex tableau for this problem is:

	y_1	y_2	y_3	y_4	x_1	x_2	x_3	x_4	P		
x_1	1	0	-1	0	1	0	0	0	0	5	
x_2	1	0	0	-1	0	1	0	0	0	2	
x_3	0	①	-1	0	0	0	1	0	0	3	$\frac{3}{1} = 3$
x_4	0	1	0	-1	0	0	0	1	0	4	$\frac{4}{1} = 4$
P	-300	-500	400	500	0	0	0	0	1	0	

$$(-1)R_3 + R_4 \rightarrow R_4 \text{ and } 500R_4 + R_5 \rightarrow R_5$$

$$\sim \begin{bmatrix} 1 & 0 & -1 & 0 & 1 & 0 & 0 & 0 & 0 & | & 5 \\ ① & 0 & 0 & -1 & 0 & 1 & 0 & 0 & 0 & | & 2 \\ 0 & 1 & -1 & 0 & 0 & 0 & 1 & 0 & 0 & | & 3 \\ 0 & 0 & 1 & -1 & 0 & 0 & -1 & 1 & 0 & | & 1 \\ \hline -300 & 0 & -100 & 500 & 0 & 0 & 500 & 0 & 1 & | & 1500 \end{bmatrix} \begin{matrix} \frac{5}{1} = 5 \\ \frac{2}{1} = 2 \\ \\ \\ \end{matrix}$$

$$(-1)R_2 + R_1 \rightarrow R_1 \text{ and } 300R_2 + R_5 \rightarrow R_5$$

$$\sim \begin{bmatrix} 0 & 0 & -1 & 1 & 1 & -1 & 0 & 0 & 0 & | & 3 \\ 1 & 0 & 0 & -1 & 0 & 1 & 0 & 0 & 0 & | & 2 \\ 0 & 1 & -1 & 0 & 0 & 0 & 1 & 0 & 0 & | & 3 \\ 0 & 0 & \textcircled{1} & -1 & 0 & 0 & -1 & 1 & 0 & | & 1 \\ \hline 0 & 0 & -100 & 200 & 0 & 300 & 500 & 0 & 1 & | & 2100 \end{bmatrix}$$

$$R_4 + R_1 \rightarrow R_1, \quad R_4 + R_3 \rightarrow R_3, \quad \text{and} \quad 100R_4 + R_5 \rightarrow R_5$$

$$\sim \begin{array}{c} \\ x_1 \\ y_1 \\ y_2 \\ y_3 \\ P \end{array} \begin{array}{c} \begin{array}{ccccccccc} y_1 & y_2 & y_3 & y_4 & x_1 & x_2 & x_3 & x_4 & P \end{array} \\ \begin{bmatrix} 0 & 0 & 0 & 0 & 1 & -1 & -1 & 1 & 0 & | & 4 \\ 1 & 0 & 0 & -1 & 0 & 1 & 0 & 0 & 0 & | & 2 \\ 0 & 1 & 0 & -1 & 0 & 0 & 0 & 1 & 0 & | & 4 \\ 0 & 0 & 1 & -1 & 0 & 0 & -1 & 1 & 0 & | & 1 \\ \hline 0 & 0 & 0 & 100 & 0 & 300 & 400 & 100 & 1 & | & 2200 \end{bmatrix} \end{array}$$

The minimal cost is $2200 when 300 students are bused from North Division to Washington, 400 students are bused from South Division to Central, and 100 students are bused from South Division to Washington. No students are bused from North Division to Central.

EXERCISE 5-6

Things to remember:

Given a linear programming problem with an objective function to be maximized and problem constraints that are a combination of ≥ and ≤ inequalities as well as equations. The solution method is called the BIG M method.

1. THE BIG M METHOD—INTRODUCING SLACK, SURPLUS AND ARTIFICIAL VARIABLES TO FORM THE MODIFIED PROBLEM

   STEP 1. If any problem constraints have negative constants on the right-hand side, multiply both sides by -1 to obtain a constraint with a nonnegative constant. [If the constraint is an inequality, this will reverse the direction of the inequality.]

   STEP 2. Introduce a SLACK VARIABLE in each ≤ constraint.

   STEP 3. Introduce a SURPLUS VARIABLE and an ARTIFICIAL VARIABLE in each ≥ constraint.

   STEP 4. Introduce an artificial variable in each = constraint.

   STEP 5. For each artificial variable a_i, add $-Ma_i$ to the objective function. Use the same constant M for all artificial variables.

2. THE BIG M METHOD—SOLVING THE PROBLEM

STEP 1. Form the preliminary simplex tableau for the modified problem.

STEP 2. Use row operations to eliminate the M's in the bottom row of the preliminary simplex tableau in the columns corresponding to the artificial variables. The resulting tableau is the initial simplex tableau.

STEP 3. Solve the modified problem by applying the simplex method to the initial simplex tableau found in Step 2.

STEP 4. Relate the solution of the modified problem to the original problem.

    (a) If the modified problem has no solution, then the original problem has no solution.

    (b) If all artificial variables are zero in the solution to the modified problem, then delete the artificial variables to find a solution to the original problem.

    (c) If any artificial variables are nonzero in the solution to the modified problem, then the original problem has no solution.

---

1. (A) We introduce a slack variable s_1 to convert the first inequality ($\leq$) into an equation, and we use a surplus variable s_2 and an artificial variable a_1 to convert the second inequality ($\geq$) into an equation. The modified problem is: Maximize $P = 5x_1 + 2x_2 - Ma_1$

$$\text{Subject to: } \begin{aligned} x_1 + 2x_2 + s_1 \qquad\qquad &= 12 \\ x_1 + x_2 \qquad - s_2 + a_1 &= 4 \\ x_1, \ x_2, \ s_1, \ s_2, \ a_1 &\geq 0 \end{aligned}$$

(B) The preliminary simplex tableau for the modified problem is:

$$
\begin{array}{cccccc}
x_1 & x_2 & s_1 & s_2 & a_1 & P
\end{array}
$$

$$
\left[\begin{array}{cccccc|c}
1 & 2 & 1 & 0 & 0 & 0 & 12 \\
1 & 1 & 0 & -1 & 1 & 0 & 4 \\
\hline
-5 & -2 & 0 & 0 & M & 1 & 0
\end{array}\right]
\sim
\left[\begin{array}{cccccc|c}
1 & 2 & 1 & 0 & 0 & 0 & 12 \\
1 & 1 & 0 & -1 & 1 & 0 & 4 \\
\hline
-M-5 & -M-2 & 0 & M & 0 & 1 & -4M
\end{array}\right]
$$

$(-M)R_2 + R_3 \rightarrow R_3$

Thus, the initial simplex tableau is:

$$
\begin{array}{cccccc}
x_1 & x_2 & s_1 & s_2 & a_1 & P
\end{array}
$$

$$
\left[\begin{array}{cccccc|c}
1 & 2 & 1 & 0 & 0 & 0 & 12 \\
1 & 1 & 0 & -1 & 1 & 0 & 4 \\
\hline
-M-5 & -M-2 & 0 & M & 0 & 1 & -4M
\end{array}\right]
$$

(C) We use the simplex method to solve the modified problem.

$$
\begin{array}{c}
\begin{array}{ccccccc} x_1 & x_2 & s_1 & s_2 & a_1 & & \end{array}\\
\begin{array}{c} s_1 \\ a_1 \\ \\ P \end{array}
\left[\begin{array}{cccccc|c}
1 & 2 & 1 & 0 & 0 & 0 & 12 \\
\textcircled{1} & 1 & 0 & -1 & 1 & 0 & 4 \\
\hline
-M-5 & -M-2 & 0 & M & 0 & 1 & -4M
\end{array}\right]
\begin{array}{l} \frac{12}{1}=12 \\[6pt] \frac{4}{1}=4 \end{array}
\end{array}
$$

$$(-1)R_2 + R_1 \rightarrow R_1 \text{ and } (M+5)R_2 + R_3 \rightarrow R_3$$

$$
\sim
\left[\begin{array}{cccccc|c}
0 & 1 & 1 & \textcircled{1} & -1 & 0 & 8 \\
1 & 1 & 0 & -1 & 1 & 0 & 4 \\
\hline
0 & 3 & 0 & -5 & M+5 & 1 & 20
\end{array}\right]
\begin{array}{c}
\begin{array}{cccccc} x_1 & x_2 & s_1 & s_2 & a_1 & P \end{array}\\
\begin{array}{c} s_2 \\ x_1 \\ \\ P \end{array}
\left[\begin{array}{cccccc|c}
0 & 1 & 1 & 1 & -1 & 0 & 8 \\
1 & 2 & 1 & 0 & 0 & 0 & 12 \\
\hline
0 & 8 & 5 & 0 & M & 1 & 60
\end{array}\right]
\end{array}
$$

$$R_1 + R_2 \rightarrow R_2 \text{ and } 5R_1 + R_3 \rightarrow R_3$$

Thus, the optimal solution of the modified problem is: max $P = 60$ at $x_1 = 12$, $x_2 = 0$, $s_1 = 0$, $s_2 = 8$, $a_1 = 0$.

(D) The optimal solution of the original problem is: max $P = 60$ at $x_1 = 12$, $x_2 = 0$.

3. (A) We introduce the slack variable s_1 and the artificial variable a_1 to obtain the modified problem: Maximize $P = 3x_1 + 5x_2 - Ma_1$

$$
\begin{aligned}
\text{Subject to: } 2x_1 + x_2 + s_1 &= 8 \\
x_1 + x_2 + a_1 &= 6 \\
x_1,\ x_2,\ s_1,\ a_1 &\geq 0
\end{aligned}
$$

(B) The preliminary simplex tableau for the modified problem is:

$$
\begin{array}{c}
\begin{array}{ccccc} x_1 & x_2 & s_1 & a_1 & P \end{array}\\
\left[\begin{array}{ccccc|c}
2 & 1 & 1 & 0 & 0 & 8 \\
1 & 1 & 0 & 1 & 0 & 6 \\
\hline
-3 & -5 & 0 & M & 1 & 0
\end{array}\right]
\sim
\left[\begin{array}{ccccc|c}
2 & 1 & 1 & 0 & 0 & 8 \\
1 & 1 & 0 & 1 & 0 & 6 \\
\hline
-M-3 & -M-5 & 0 & 0 & 1 & -6M
\end{array}\right]
\end{array}
$$

$$(-M)R_2 + R_3 \rightarrow R_3$$

Thus, the initial simplex tableau is:

$$
\begin{array}{c}
\begin{array}{ccccc} x_1 & x_2 & s_1 & a_1 & P \end{array}\\
\begin{array}{c} s_1 \\ a_1 \\ \\ P \end{array}
\left[\begin{array}{ccccc|c}
2 & 1 & 1 & 0 & 0 & 8 \\
1 & 1 & 0 & 1 & 0 & 6 \\
\hline
-M-3 & -M-5 & 0 & 0 & 1 & -6M
\end{array}\right]
\end{array}
$$

(C) We use the simplex method to solve the modified problem.

$$
\begin{array}{c}
\begin{array}{c}
s_1 \\
a_1 \\
P
\end{array}
\begin{bmatrix}
x_1 & x_2 & s_1 & a_1 & & \\
2 & 1 & 1 & 0 & 0 & 8 \\
1 & \boxed{1} & 0 & 1 & 0 & 6 \\
\hline
-M-3 & -M-5 & 0 & 0 & 1 & -6M
\end{bmatrix}
\begin{array}{l}
\frac{8}{1}=8 \\
\frac{6}{1}=6
\end{array}
\sim
\begin{array}{c}
s_1 \\
x_2 \\
P
\end{array}
\begin{bmatrix}
x_1 & x_2 & s_1 & a_1 & P & \\
1 & 0 & 1 & -1 & 0 & 2 \\
1 & 1 & 0 & 1 & 0 & 6 \\
\hline
2 & 0 & 0 & M+5 & 1 & 30
\end{bmatrix}
\end{array}
$$

$(-1)R_2 + R_1 \to R_1$ and $(M+5)R_2 + R_3 \to R_3$

Thus, the optimal solution of the modified problem is max $P = 30$ at $x_1 = 0$, $x_2 = 6$, $s_1 = 2$, $a_1 = 0$.

(D) The optimal solution of the original problem is: max $P = 30$ at $x_1 = 0$, $x_2 = 6$.

5. (A) We introduce slack, surplus, and artificial variables to obtain the modified problem: Maximize $P = 4x_1 + 3x_2 - Ma_1$

Subject to:
$$
\begin{array}{rrrrrl}
-x_1 & + 2x_2 & + s_1 & & & = 2 \\
x_1 & + x_2 & & - s_2 & + a_1 & = 4 \\
& & & & x_1, x_2, s_1, s_2, a_1 & \geq 0
\end{array}
$$

(B) The preliminary simplex tableau for the modified problem is:

$$
\begin{bmatrix}
x_1 & x_2 & s_1 & s_2 & a_1 & P & \\
-1 & 2 & 1 & 0 & 0 & 0 & 2 \\
1 & 1 & 0 & -1 & 1 & 0 & 4 \\
\hline
-4 & -3 & 0 & 0 & M & 1 & 0
\end{bmatrix}
\sim
\begin{bmatrix}
-1 & 2 & 1 & 0 & 0 & 0 & 2 \\
1 & 1 & 0 & -1 & 1 & 0 & 4 \\
\hline
-M-4 & -M-3 & 0 & M & 0 & 1 & -4M
\end{bmatrix}
$$

$(-M)R_2 + R_3 \to R_3$

Thus, the initial simplex tableau is:

$$
\begin{array}{c}
s_1 \\
a_1 \\
P
\end{array}
\begin{bmatrix}
x_1 & x_2 & s_1 & s_2 & a_1 & P & \\
-1 & 2 & 1 & 0 & 0 & 0 & 2 \\
1 & 1 & 0 & -1 & 1 & 0 & 4 \\
\hline
-M-4 & -M-3 & 0 & M & 0 & 1 & -4M
\end{bmatrix}
$$

(C) We use the simplex method to solve the modified problem:

$$
\begin{bmatrix}
x_1 & x_2 & s_1 & s_2 & a_1 & P & \\
-1 & 2 & 1 & 0 & 0 & 0 & 2 \\
\boxed{1} & 1 & 0 & -1 & 1 & 0 & 4 \\
\hline
-M-4 & -M-3 & 0 & M & 0 & 1 & -4M
\end{bmatrix}
\sim
\begin{bmatrix}
x_1 & x_2 & s_1 & s_2 & a_1 & P & \\
0 & 3 & 1 & -1 & 1 & 0 & 6 \\
1 & 1 & 0 & -1 & 1 & 0 & 4 \\
\hline
0 & 1 & 0 & -4 & M+4 & 1 & 16
\end{bmatrix}
$$

$R_2 + R_1 \to R_1$, $(M+4)R_2 + R_3 \to R_3$

No optimal solution exists because the elements in the pivot column (the s_2 column) above the dashed line are negative.

7. (A) We introduce slack, surplus, and artificial variables to obtain the modified problem: Maximize $P = 5x_1 + 10x_2 - Ma_1$

$$
\begin{aligned}
\text{Subject to:} \quad & x_1 + x_2 + s_1 && = 3 \\
& 2x_1 + 3x_2 && - s_2 + a_1 = 12 \\
& x_1, \; x_2, \; s_1, \; s_2, \; a_1 && \geq 0
\end{aligned}
$$

(B) The preliminary simplex tableau for the modified problem is:

	x_1	x_2	s_1	s_2	a_1	P	
s_1	1	1	1	0	0	0	3
a_1	2	3	0	-1	1	0	12
P	-5	-10	0	0	M	1	0

$$\begin{array}{ccccccc}
1 & 1 & 1 & 0 & 0 & 0 & 3 \\
2 & 3 & 0 & -1 & 1 & 0 & 12 \\
-2M-5 & -3M-10 & 0 & M & 0 & 1 & -12M
\end{array}$$

$$(-M)R_2 + R_3 \to R_3$$

Thus, the initial simplex tableau is:

	x_1	x_2	s_1	s_2	a_1	P	
s_1	1	1	1	0	0	0	3
a_1	2	3	0	-1	1	0	12
P	$-2M-5$	$-3M-10$	0	M	0	1	$-12M$

(C) Applying the simplex method to the initial tableau, we have:

$$\begin{array}{ccccccc}
1 & ① & 1 & 0 & 0 & 0 & 3 \\
2 & 3 & 0 & -1 & 1 & 0 & 12 \\
-2M-5 & -3M-10 & 0 & M & 0 & 1 & -12M
\end{array}$$

$$(-3)R_1 + R_2 \to R_2, \quad (3M+10)R_1 + R_3 \to R_3$$

	x_1	x_2	s_1	s_2	a_1	P	
x_2	1	1	1	0	0	0	3
~ a_1	-1	0	-3	-1	1	0	3
P	$M+5$	0	$3M+10$	M	0	1	$-3M+30$

The optimal solution of the modified problem is: max $P = -3M + 30$ at $x_1 = 0$, $x_2 = 3$, $s_1 = 0$, $s_2 = 0$, and $a_1 = 3$.

(D) The original problem does not have an optimal solution, since the artificial variable a_1 in the solution of the modified problem has a nonzero value.

9. To minimize $P = 2x_1 - x_2$, we maximize $T = -P = -2x_1 + x_2$. Introducing slack, surplus, and artificial variables, we obtain the modified problem:

Maximize $T = -2x_1 + x_2 - Ma_1$

Subject to:
$$x_1 + x_2 + s_1 \qquad\qquad = 8$$
$$5x_1 + 3x_2 \qquad - s_2 + a_1 = 30$$
$$x_1, \ x_2, \ s_1, \ s_2, \ a_1 \geq 0$$

The preliminary simplex tableau for this problem is:

$$
\begin{array}{cccccc}
x_1 & x_2 & s_1 & s_2 & a_1 & T \\
\end{array}
$$

$$
\left[
\begin{array}{cccccc|c}
1 & 1 & 1 & 0 & 0 & 0 & 8 \\
5 & 3 & 0 & -1 & 1 & 0 & 30 \\
\hline
2 & -1 & 0 & 0 & M & 1 & 0 \\
\end{array}
\right]
$$

$(-M)R_2 + R_3 \rightarrow R_3$

$$
\begin{array}{ccccccc}
 & x_1 & x_2 & s_1 & s_2 & a_1 & T \\
\end{array}
$$

$$
\sim
\begin{array}{c}
s_1 \\
a_1 \\
\\
T
\end{array}
\left[
\begin{array}{cccccc|c}
1 & 1 & 1 & 0 & 0 & 0 & 8 \\
\circled{5} & 3 & 0 & -1 & 1 & 0 & 30 \\
\hline
-5M+2 & -3M-1 & 0 & M & 0 & 1 & -30M \\
\end{array}
\right]
\begin{array}{l}
\frac{8}{1} = 8 \\
\frac{30}{5} = 6 \\
\end{array}
$$

(This is the initial simplex tableau.) $\frac{1}{5}R_2 \rightarrow R_2$

$$
\sim
\left[
\begin{array}{cccccc|c}
1 & 1 & 1 & 0 & 0 & 0 & 8 \\
\circled{1} & \frac{3}{5} & 0 & -\frac{1}{5} & \frac{1}{5} & 0 & 6 \\
\hline
-5M+2 & -3M-1 & 0 & M & 0 & 1 & -30M \\
\end{array}
\right]
$$

$(-1)R_2 + R_1 \rightarrow R_1, \quad (5M - 2)R_2 + R_3 \rightarrow R_3$

$$
\sim
\left[
\begin{array}{cccccc|c}
0 & \circled{\frac{2}{5}} & 1 & \frac{1}{5} & -\frac{1}{5} & 0 & 2 \\
1 & \frac{3}{5} & 0 & -\frac{1}{5} & \frac{1}{5} & 0 & 6 \\
\hline
0 & -\frac{11}{5} & 0 & \frac{2}{5} & M-\frac{2}{5} & 1 & -12 \\
\end{array}
\right]
\begin{array}{l}
\frac{2}{2/5} = 5 \\
\frac{6}{3/5} = 10 \\
\end{array}
$$

$\frac{5}{2}R_1 \rightarrow R_1$

$$
\sim
\left[
\begin{array}{cccccc|c}
0 & \circled{1} & \frac{5}{2} & \frac{1}{2} & -\frac{1}{2} & 0 & 5 \\
1 & \frac{3}{5} & 0 & -\frac{1}{5} & \frac{1}{5} & 0 & 6 \\
\hline
0 & -\frac{11}{5} & 0 & \frac{2}{5} & M-\frac{2}{5} & 1 & -12 \\
\end{array}
\right]
$$

$\sim$
$$
\begin{array}{c}
 \\
x_2 \\
x_1 \\
\\
T
\end{array}
\begin{array}{cccccc}
x_1 & x_2 & s_1 & s_2 & a_1 & T \\
\end{array}
\left[
\begin{array}{cccccc|c}
0 & 1 & \frac{5}{2} & \frac{1}{2} & -\frac{1}{2} & 0 & 5 \\
1 & 0 & -\frac{3}{2} & -\frac{1}{2} & \frac{1}{2} & 0 & 3 \\
\hline
0 & 0 & \frac{11}{2} & \frac{3}{2} & M-\frac{3}{2} & 1 & -1 \\
\end{array}
\right]
$$

$\left(-\frac{3}{5}\right)R_1 + R_2 \rightarrow R_2, \quad \left(\frac{11}{5}\right)R_1 + R_3 \rightarrow R_3$

Thus, the optimal solution is: max $T = -1$ at $x_1 = 3$, $x_2 = 5$, and min $P = -\max T = 1$.

The modified problem for maximizing $P = 2x_1 - x_2$ subject to the given constraints is: Maximize $P = 2x_1 - x_2 - Ma_1$

$$\text{Subject to: } \begin{aligned} x_1 + x_2 + s_1 \quad\quad\quad &= 8 \\ 5x_1 + 3x_2 \quad - s_2 + a_1 &= 30 \\ x_1, \ x_2, \ s_1, \ s_2, \ a_1 &\geq 0 \end{aligned}$$

The preliminary simplex tableau for the modified problem is:

$$
\begin{array}{cccccc}
x_1 & x_2 & s_1 & s_2 & a_1 & P \\
\end{array}
$$

$$
\left[
\begin{array}{cccccc|c}
1 & 1 & 1 & 0 & 0 & 0 & 8 \\
5 & 3 & 0 & -1 & 1 & 0 & 30 \\
\hline
-2 & 1 & 0 & 0 & M & 1 & 0
\end{array}
\right]
$$

$$(-M)R_2 + R_3 \rightarrow R_3$$

$$
\begin{array}{ccccccc}
 & x_1 & x_2 & s_1 & s_2 & a_1 & P \\
\end{array}
$$

$$
\begin{array}{c}
s_1 \\
\sim \ a_1 \\
\\
P
\end{array}
\left[
\begin{array}{cccccc|c}
1 & 1 & 1 & 0 & 0 & 0 & 8 \\
⑤ & 3 & 0 & -1 & 1 & 0 & 30 \\
\hline
-5M-2 & -3M+1 & 0 & M & 0 & 1 & -30M
\end{array}
\right]
\begin{array}{c}
\frac{8}{1} = 8 \\
\frac{30}{5} = 6 \\
\\
\end{array}
$$

(This is the initial simplex tableau.) $\frac{1}{5}R_2 \rightarrow R_2$

$$
\sim \left[
\begin{array}{cccccc|c}
1 & 1 & 1 & 0 & 0 & 0 & 8 \\
① & \frac{3}{5} & 0 & -\frac{1}{5} & \frac{1}{5} & 0 & 6 \\
\hline
-5M-2 & -3M+1 & 0 & M & 0 & 1 & -30M
\end{array}
\right]
\sim \left[
\begin{array}{cccccc|c}
0 & \frac{2}{5} & 1 & ①\!\!\frac{1}{5} & -\frac{1}{5} & 0 & 2 \\
1 & \frac{3}{5} & 0 & -\frac{1}{5} & \frac{1}{5} & 0 & 6 \\
\hline
0 & \frac{11}{5} & 0 & -\frac{2}{5} & M+\frac{2}{5} & 1 & 12
\end{array}
\right]
$$

$$(-1)R_2 + R_1 \rightarrow R_1, \ (5M+2)R_2 + R_3 \rightarrow R_3 \qquad\qquad 5R_1 \rightarrow R_1$$

$$
\sim \left[
\begin{array}{cccccc|c}
0 & 2 & 5 & 1 & -1 & 0 & 10 \\
1 & \frac{3}{5} & 0 & -\frac{1}{5} & \frac{1}{5} & 0 & 6 \\
\hline
0 & \frac{11}{5} & 0 & -\frac{2}{5} & M+\frac{2}{5} & 1 & 12
\end{array}
\right]
$$

$$
\begin{array}{ccccccc}
 & x_1 & x_2 & s_1 & s_2 & a_1 & P \\
\end{array}
$$

$$
\sim \begin{array}{c}
s_2 \\
x_2 \\
\\
P
\end{array}
\left[
\begin{array}{cccccc|c}
0 & 2 & 5 & 1 & -1 & 0 & 10 \\
1 & 1 & 1 & 0 & 0 & 0 & 8 \\
\hline
0 & 3 & 2 & 0 & M & 1 & 16
\end{array}
\right]
$$

$$\frac{1}{5}R_1 + R_2 \rightarrow R_2 \text{ and } \frac{2}{5}R_1 + R_3 \rightarrow R_3$$

Thus, the optimal solution is: max $P = 16$ at $x_1 = 8$, $x_2 = 0$.

11. We introduce slack, surplus, and artificial variables to obtain the modified problem: Maximize $P = 2x_1 + 5x_2 - Ma_1$

$$\text{Subject to: } \begin{aligned} x_1 + 2x_2 + s_1 \quad\quad\quad\quad &= 18 \\ 2x_1 + x_2 \quad + s_2 \quad\quad &= 21 \\ x_1 + x_2 \quad\quad - s_3 + a_1 &= 10 \\ x_1, \ x_2, \ s_1, \ s_2, \ s_3, \ a_1 &\geq 0 \end{aligned}$$

The preliminary simplex tableau for this problem is:

$$
\begin{array}{ccccccc}
x_1 & x_2 & s_1 & s_2 & s_3 & a_1 & P
\end{array}
$$

$$
\left[
\begin{array}{ccccccc|c}
1 & 2 & 1 & 0 & 0 & 0 & 0 & 18 \\
2 & 1 & 0 & 1 & 0 & 0 & 0 & 21 \\
1 & 1 & 0 & 0 & -1 & 1 & 0 & 10 \\
\hdashline
-2 & -5 & 0 & 0 & 0 & M & 1 & 0
\end{array}
\right]
$$

$$(-M)R_3 + R_4 \rightarrow R_4$$

$$
\begin{array}{cccccccc}
 & x_1 & x_2 & s_1 & s_2 & s_3 & a_1 & P
\end{array}
$$

$$
\sim
\begin{array}{c}
s_1 \\ s_2 \\ a_1 \\ P
\end{array}
\left[
\begin{array}{ccccccc|c}
1 & ② & 1 & 0 & 0 & 0 & 0 & 18 \\
2 & 1 & 0 & 1 & 0 & 0 & 0 & 21 \\
1 & 1 & 0 & 0 & -1 & 1 & 0 & 10 \\
\hdashline
-M-2 & -M-5 & 0 & 0 & M & 0 & 1 & -10M
\end{array}
\right]
\begin{array}{l}
\dfrac{18}{2} = 9 \\[4pt]
\dfrac{21}{1} = 21 \ \text{(This is the initial} \\
\hphantom{\dfrac{21}{1} = 21 }\ \text{simplex tableau.)} \\[4pt]
\dfrac{10}{1} = 10
\end{array}
$$

$$\tfrac{1}{2}R_1 \rightarrow R_1$$

$$
\sim
\left[
\begin{array}{ccccccc|c}
\frac{1}{2} & ① & \frac{1}{2} & 0 & 0 & 0 & 0 & 9 \\
2 & 1 & 0 & 1 & 0 & 0 & 0 & 21 \\
1 & 1 & 0 & 0 & -1 & 1 & 0 & 10 \\
\hdashline
-M-2 & -M-5 & 0 & 0 & M & 0 & 1 & -10M
\end{array}
\right]
$$

$$(-1)R_1 + R_2 \rightarrow R_2, \quad (-1)R_1 + R_3 \rightarrow R_3, \quad \text{and} \quad (M+5)R_1 + R_4 \rightarrow R_4$$

$$
\sim
\left[
\begin{array}{ccccccc|c}
\frac{1}{2} & 1 & \frac{1}{2} & 0 & 0 & 0 & 0 & 9 \\
\frac{3}{2} & 0 & -\frac{1}{2} & 1 & 0 & 0 & 0 & 12 \\
①\!\!\frac{1}{2} & 0 & -\frac{1}{2} & 0 & -1 & 1 & 0 & 1 \\
\hdashline
-\frac{1}{2}M+\frac{1}{2} & 0 & \frac{1}{2}M+\frac{5}{2} & 0 & M & 0 & 1 & -M+45
\end{array}
\right]
\begin{array}{l}
\dfrac{9}{1/2} = 18 \\[4pt]
\dfrac{12}{3/2} = 8 \\[4pt]
\dfrac{1}{1/2} = 2
\end{array}
$$

$$2R_3 \rightarrow R_3$$

$$
\sim
\left[
\begin{array}{ccccccc|c}
\frac{1}{2} & 1 & \frac{1}{2} & 0 & 0 & 0 & 0 & 9 \\
\frac{3}{2} & 0 & -\frac{1}{2} & 1 & 0 & 0 & 0 & 12 \\
① & 0 & -1 & 0 & -2 & 2 & 0 & 2 \\
\hdashline
-\frac{1}{2}M+\frac{1}{2} & 0 & \frac{1}{2}M+\frac{5}{2} & 0 & M & 0 & 1 & -M+45
\end{array}
\right]
$$

$$\left(-\tfrac{1}{2}\right)R_3 + R_1 \rightarrow R_1, \quad \left(-\tfrac{3}{2}\right)R_3 + R_2 \rightarrow R_2, \quad \text{and} \quad \left(\tfrac{1}{2}M-\tfrac{1}{2}\right)R_3 + R_4 \rightarrow R_4$$

$$
\begin{array}{c}
\quad\quad x_1 \quad x_2 \quad s_1 \quad s_2 \quad s_3 \quad a_1 \qquad P \\
\begin{array}{c} x_2 \\ \sim \quad s_2 \\ x_1 \\ \\ P \end{array}
\left[\begin{array}{ccccccc|c}
0 & 1 & 1 & 0 & 1 & -1 & 0 & 8 \\
0 & 0 & 1 & 1 & 3 & -3 & 0 & 9 \\
1 & 0 & -1 & 0 & -2 & 2 & 0 & 2 \\
\hline
0 & 0 & 3 & 0 & 1 & M-1 & 1 & 44
\end{array}\right]
\end{array}
$$

Optimal solution: max $P = 44$ at $x_1 = 2$, $x_2 = 8$.

13. We introduce surplus and artificial variables to obtain the modified problem: Maximize $P = 10x_1 + 12x_2 + 20x_3 - Ma_1 - Ma_2$

$$
\begin{aligned}
\text{Subject to: } 3x_1 + x_2 + 2x_3 - s_1 + a_1 &= 12 \\
x_1 - x_2 + 2x_3 + a_2 &= 6 \\
x_1, \ x_2, \ x_3, \ s_1, \ a_1, \ a_2 &\ge 0
\end{aligned}
$$

The preliminary simplex tableau for the modified problem is:

$$
\begin{array}{c}
x_1 \quad x_2 \quad x_3 \quad s_1 \quad a_1 \quad a_2 \quad P \\
\left[\begin{array}{ccccccc|c}
3 & 1 & 2 & -1 & 1 & 0 & 0 & 12 \\
1 & -1 & 2 & 0 & 0 & 1 & 0 & 6 \\
\hline
-10 & -12 & -20 & 0 & M & M & 1 & 0
\end{array}\right]
\end{array}
$$

$(-M)R_1 + R_3 \rightarrow R_3$

$$
\sim
\left[\begin{array}{ccccccc|c}
3 & 1 & 2 & -1 & 1 & 0 & 0 & 12 \\
1 & -1 & 2 & 0 & 0 & 1 & 0 & 6 \\
\hline
-3M-10 & -M-12 & -2M-20 & M & 0 & M & 1 & -12M
\end{array}\right]
$$

$(-M)R_2 + R_3 \rightarrow R_3$

$$
\begin{array}{c}
\quad\quad x_1 \quad\quad\quad x_2 \quad\quad\quad x_3 \quad\quad s_1 \quad a_1 \quad a_2 \quad P \\
\begin{array}{c} a_1 \\ \sim \quad a_2 \\ P \end{array}
\left[\begin{array}{ccccccc|c}
3 & 1 & 2 & -1 & 1 & 0 & 0 & 12 \\
1 & -1 & \textcircled{2} & 0 & 0 & 1 & 0 & 6 \\
\hline
-4M-10 & -12 & -4M-20 & M & 0 & 0 & 1 & -18M
\end{array}\right]
\begin{array}{l} \frac{12}{2} = 6 \\[4pt] \frac{6}{2} = 3 \end{array}
\end{array}
$$

$\frac{1}{2}R_2 \rightarrow R_2$

$$
\sim
\left[\begin{array}{ccccccc|c}
3 & 1 & 2 & -1 & 1 & 0 & 0 & 12 \\
\frac{1}{2} & -\frac{1}{2} & \textcircled{1} & 0 & 0 & \frac{1}{2} & 0 & 3 \\
\hline
-4M-10 & -12 & -4M-20 & M & 0 & 0 & 1 & -18M
\end{array}\right]
$$

$(-2)R_2 + R_1 \rightarrow R_1$ and $(4M+20)R_2 + R_3 \rightarrow R_3$

$$
\sim
\left[\begin{array}{ccccccc|c}
2 & \textcircled{2} & 0 & -1 & 1 & -1 & 0 & 6 \\
\frac{1}{2} & -\frac{1}{2} & 1 & 0 & 0 & \frac{1}{2} & 0 & 3 \\
\hline
-2M & -2M-22 & 0 & M & 0 & 2M+10 & 1 & -6M+60
\end{array}\right]
$$

$\frac{1}{2}R_1 \rightarrow R_1$

$$\sim \begin{bmatrix} 1 & \boxed{1} & 0 & -\frac{1}{2} & \frac{1}{2} & -\frac{1}{2} & 0 & 3 \\ \frac{1}{2} & -\frac{1}{2} & 1 & 0 & 0 & \frac{1}{2} & 0 & 3 \\ \hline -2M & -2M-22 & 0 & M & 0 & 2M+10 & 1 & -6M+60 \end{bmatrix}$$

$$\tfrac{1}{2}R_1 + R_2 \to R_2, \quad (2M+22)R_1 + R_3 \to R_3$$

$$\begin{array}{c} \\ x_2 \\ \sim \quad x_3 \\ P \end{array} \begin{array}{cccccccc} x_1 & x_2 & x_3 & s_1 & a_1 & & a_2 & P \\ \end{array}$$

$$\sim \begin{array}{c} x_2 \\ x_3 \\ P \end{array} \left[\begin{array}{ccccccc|c} 1 & 1 & 0 & -\frac{1}{2} & \frac{1}{2} & -\frac{1}{2} & 0 & 3 \\ 1 & 0 & 1 & -\frac{1}{4} & \frac{1}{4} & \frac{1}{4} & 0 & \frac{9}{2} \\ \hline 22 & 0 & 0 & -11 & M+11 & M-1 & 1 & 126 \end{array} \right]$$

No optimal solution exists because there are no positive numbers in the pivot column.

15. We will maximize $P = -C = 5x_1 + 12x_2 - 16x_3$ subject to the given constraints. Introduce slack, surplus, and artificial variables to obtain the modified problem:

Maximize $P = 5x_1 + 12x_2 - 16x_3 - Ma_1 - Ma_2$

$$\begin{array}{llll} \text{Subject to:} & x_1 + 2x_2 + x_3 + s_1 & & = 10 \\ & 2x_1 + 3x_2 + x_3 & - s_2 + a_1 & = 6 \\ & 2x_1 + x_2 - x_3 & + a_2 & = 1 \\ & x_1,\ x_2,\ x_3,\ s_1,\ s_2,\ a_1,\ a_2 \geq 0 \end{array}$$

The preliminary simplex tableau for the modified problem is:

$$\begin{array}{cccccccc} x_1 & x_2 & x_3 & s_1 & s_2 & a_1 & a_2 & P \\ \end{array}$$

$$\left[\begin{array}{cccccccc|c} 1 & 2 & 1 & 1 & 0 & 0 & 0 & 0 & 10 \\ 2 & 3 & 1 & 0 & -1 & 1 & 0 & 0 & 6 \\ 2 & 1 & -1 & 0 & 0 & 0 & 1 & 0 & 1 \\ \hline -5 & -12 & 16 & 0 & 0 & M & M & 1 & 0 \end{array} \right]$$

$$(-M)R_2 + R_4 \to R_4$$

$$\sim \left[\begin{array}{cccccccc|c} 1 & 2 & 1 & 1 & 0 & 0 & 0 & 0 & 10 \\ 2 & 3 & 1 & 0 & -1 & 1 & 0 & 0 & 6 \\ 2 & 1 & -1 & 0 & 0 & 0 & 1 & 0 & 1 \\ \hline -2M-5 & -3M-12 & -M+16 & 0 & M & 0 & M & 1 & -6M \end{array} \right]$$

$$(-M)R_3 + R_4 \to R_4$$

$$
\begin{array}{c}
\begin{array}{ccccccccc}
& x_1 & x_2 & x_3 & s_1 & s_2 & a_1 & a_2 & P
\end{array}
\end{array}
$$

	x_1	x_2	x_3	s_1	s_2	a_1	a_2	P		
s_1	1	2	1	1	0	0	0	0	10	$\frac{10}{2} = 5$
a_1	2	3	1	0	-1	1	0	0	6	$\frac{6}{3} = 2$
a_2	2	①	-1	0	0	0	1	0	1	$\frac{1}{1} = 1$
P	$-4M-5$	$-4M-12$	16	0	M	0	0	1	$-7M$	

$(-2)R_3 + R_1 \rightarrow R_1, \quad (-3)R_3 + R_2 \rightarrow R_2, \quad \text{and} \quad (4M+12)R_3 + R_4 \rightarrow R_4$

	-3	0	3	1	0	0	-2	0	8	$\frac{8}{3} \approx 2.67$
	-4	0	④	0	-1	1	-3	0	3	$\frac{3}{4} = .75$
	2	1	-1	0	0	0	1	0	1	
	$4M+19$	0	$-4M+4$	0	M	0	$4M+12$	1	$-3M+12$	

$\frac{1}{4}R_2 \rightarrow R_2$

	-3	0	3	1	0	0	-2	0	8
	-1	0	①	0	$-\frac{1}{4}$	$\frac{1}{4}$	$-\frac{3}{4}$	0	$\frac{3}{4}$
	2	1	-1	0	0	0	1	0	1
	$4M+19$	0	$-4M+4$	0	M	0	$4M+12$	1	$-3M+12$

$(-3)R_2 + R_1 \rightarrow R_1, \quad R_2 + R_3 \rightarrow R_3, \quad \text{and} \quad (4M-4)R_2 + R_4 \rightarrow R_4$

	x_1	x_2	x_3	s_1	s_2	a_1	a_2	P	
s_1	0	0	0	1	$\frac{3}{4}$	$-\frac{3}{4}$	$\frac{1}{4}$	0	$\frac{23}{4}$
x_3	-1	0	1	0	$-\frac{1}{4}$	$\frac{1}{4}$	$-\frac{3}{4}$	0	$\frac{3}{4}$
x_2	1	1	0	0	$-\frac{1}{4}$	$\frac{1}{4}$	$\frac{1}{4}$	0	$\frac{7}{4}$
P	23	0	0	0	1	$M-1$	$M+15$	1	9

Optimal solution:
min $C = -9$ at $x_1 = 0$, $x_2 = \frac{7}{4}$, and $x_3 = \frac{3}{4}$.

17. We introduce a slack and an artificial variable to obtain the modified problem: Maximize $P = 3x_1 + 5x_2 + 6x_3 - Ma_1$

$$
\begin{aligned}
\text{Subject to: } 2x_1 + x_2 + 2x_3 + s_1 &= 8 \\
2x_1 + x_2 - 2x_3 + a_1 &= 0 \\
x_1, x_2, x_3, s_1, a_1 &\geq 0
\end{aligned}
$$

The preliminary simplex tableau for the modified problem is:

	x_1	x_2	x_3	s_1	a_1	P	
s_1	2	1	2	1	0	0	8
a_1	2	1	-2	0	1	0	0
P	-3	-5	-6	0	M	1	0

	2	1	2	1	0	0	8
	②	1	-2	0	1	0	0
	$-3-2M$	$-5-M$	$-6+2M$	0	0	1	0

$(-M)R_2 + R_3 \rightarrow R_3 \qquad\qquad \frac{1}{2}R_2 \rightarrow R_2$

$$\sim \begin{bmatrix} 2 & 1 & 2 & 1 & 0 & 0 & | & 8 \\ ① & \frac{1}{2} & -1 & 0 & \frac{1}{2} & 0 & | & 0 \\ -3-2M & -5-M & -6+2M & 0 & 0 & 1 & | & 0 \end{bmatrix} \sim \begin{bmatrix} 0 & 0 & ④ & 1 & -1 & 0 & | & 8 \\ 1 & \frac{1}{2} & -1 & 0 & \frac{1}{2} & 0 & | & 0 \\ 0 & -\frac{7}{2} & -9 & 0 & \frac{3}{2}+M & 1 & | & 0 \end{bmatrix}$$

$(-2)R_2+R_1 \to R_1, \quad (3+2M)R_2+R_3 \to R_3 \qquad\qquad \frac{1}{4}R_1 \to R_1$

$$\sim \begin{bmatrix} 0 & 0 & ① & \frac{1}{4} & -\frac{1}{4} & 0 & | & 2 \\ 1 & \frac{1}{2} & -1 & 0 & \frac{1}{2} & 0 & | & 0 \\ 0 & -\frac{7}{2} & -9 & 0 & \frac{3}{2}+M & 1 & | & 0 \end{bmatrix} \sim \begin{bmatrix} 0 & 0 & 1 & \frac{1}{4} & -\frac{1}{4} & 0 & | & 2 \\ 1 & ②\!\!\frac{1}{2} & 0 & \frac{1}{4} & \frac{1}{4} & 0 & | & 2 \\ 0 & -\frac{7}{2} & 0 & \frac{9}{4} & -\frac{3}{4}+M & 1 & | & 18 \end{bmatrix}$$

$R_1+R_2 \to R_2, \quad 9R_1+R_3 \to R_3 \qquad\qquad 2R_2 \to R_2$

$$\sim \begin{bmatrix} 0 & 0 & 1 & \frac{1}{4} & -\frac{1}{4} & 0 & | & 2 \\ 2 & ① & 0 & \frac{1}{2} & \frac{1}{2} & 0 & | & 4 \\ 0 & -\frac{7}{2} & 0 & \frac{9}{4} & -\frac{3}{4}+M & 1 & | & 18 \end{bmatrix}$$

$$\begin{array}{c} \\ \\ \end{array}\qquad \begin{array}{ccccccc} x_1 & x_2 & x_3 & s_1 & & a_1 & P \end{array}$$

$$\sim \begin{array}{c} x_3 \\ x_2 \\ P \end{array} \begin{bmatrix} 0 & 0 & 1 & \frac{1}{4} & -\frac{1}{4} & 0 & | & 2 \\ 2 & 1 & 0 & \frac{1}{2} & \frac{1}{2} & 0 & | & 4 \\ 7 & 0 & 0 & 4 & 1+M & 1 & | & 32 \end{bmatrix}$$

$\frac{7}{2}R_2+R_3 \to R_3$

Optimal solution: max $P = 32$ at $x_1 = 0$, $x_2 = 4$, $x_3 = 2$.

19. We introduce slack, surplus, and artificial variables to obtain the modified problem: Maximize $P = 2x_1 + 3x_2 + 4x_3 - Ma_1$

$$\begin{aligned} \text{Subject to:} \quad x_1 + 2x_2 + x_3 + s_1 &= 25 \\ 2x_1 + x_2 + 2x_3 \qquad + s_2 &= 60 \\ x_1 + 2x_2 - x_3 \qquad\qquad - s_3 + a_1 &= 10 \\ x_1, \ x_2, \ x_3, \ s_1, \ s_2, \ s_3, \ a_1 &\geq 0 \end{aligned}$$

The preliminary simplex tableau for the modified problem is:

$$\begin{array}{cccccccc} x_1 & x_2 & x_3 & s_1 & s_2 & s_3 & a_1 & P \end{array}$$
$$\begin{bmatrix} 1 & 2 & 1 & 1 & 0 & 0 & 0 & 0 & | & 25 \\ 2 & 1 & 2 & 0 & 1 & 0 & 0 & 0 & | & 60 \\ 1 & 2 & -1 & 0 & 0 & -1 & 1 & 0 & | & 10 \\ -2 & -3 & -4 & 0 & 0 & 0 & M & 1 & | & 0 \end{bmatrix}$$

$(-M)R_3 + R_4 \to R_4$

$$\qquad\qquad \begin{array}{cccccccc} x_1 & & x_2 & & x_3 & s_1 & s_2 & s_3 & a_1 & P \end{array}$$

$$\sim \begin{array}{c} s_1 \\ s_2 \\ a_1 \\ P \end{array} \begin{bmatrix} 1 & 2 & 1 & 1 & 0 & 0 & 0 & 0 & | & 25 \\ 2 & 1 & 2 & 0 & 1 & 0 & 0 & 0 & | & 60 \\ 1 & ② & -1 & 0 & 0 & -1 & 1 & 0 & | & 10 \\ -2-M & -3-2M & -4+M & 0 & 0 & M & 0 & 1 & | & -10M \end{bmatrix}$$

$\frac{1}{2}R_3 \to R_3$

$$\sim \begin{bmatrix} 1 & 2 & 1 & 1 & 0 & 0 & 0 & 0 & 25 \\ 2 & 1 & 2 & 0 & 1 & 0 & 0 & 0 & 60 \\ \frac{1}{2} & ① & -\frac{1}{2} & 0 & 0 & -\frac{1}{2} & \frac{1}{2} & 0 & 5 \\ \hdashline -2-M & -3-2M & -4+M & 0 & 0 & M & 0 & 1 & -10M \end{bmatrix}$$

$$(-2)R_3 + R_1 \to R_1, \quad (-1)R_3 + R_2 \to R_2, \quad (3+2M)R_3 + R_4 \to R_4$$

$$\sim \begin{bmatrix} 0 & 0 & ② & 1 & 0 & 1 & -1 & 0 & 15 \\ \frac{3}{2} & 0 & \frac{5}{2} & 0 & 1 & \frac{1}{2} & -\frac{1}{2} & 0 & 55 \\ \frac{1}{2} & 1 & -\frac{1}{2} & 0 & 0 & -\frac{1}{2} & \frac{1}{2} & 0 & 5 \\ \hdashline -\frac{1}{2} & 0 & -\frac{11}{2} & 0 & 0 & -\frac{3}{2} & \frac{3}{2}+M & 1 & 15 \end{bmatrix}$$

$$\frac{1}{2}R_1 \to R_1$$

$$\sim \begin{bmatrix} 0 & 0 & ① & \frac{1}{2} & 0 & \frac{1}{2} & -\frac{1}{2} & 0 & \frac{15}{2} \\ \frac{3}{2} & 0 & \frac{5}{2} & 0 & 1 & \frac{1}{2} & -\frac{1}{2} & 0 & 55 \\ \frac{1}{2} & 1 & -\frac{1}{2} & 0 & 0 & -\frac{1}{2} & \frac{1}{2} & 0 & 5 \\ \hdashline -\frac{1}{2} & 0 & -\frac{11}{2} & 0 & 0 & -\frac{3}{2} & \frac{3}{2}+M & 1 & 15 \end{bmatrix}$$

$$\left(-\frac{5}{2}\right)R_1 + R_2 \to R_2, \quad \frac{1}{2}R_1 + R_3 \to R_3, \quad \frac{11}{2}R_1 + R_4 \to R_4$$

$$\sim \begin{bmatrix} 0 & 0 & 1 & \frac{1}{2} & 0 & \frac{1}{2} & -\frac{1}{2} & 0 & \frac{15}{2} \\ \frac{3}{2} & 0 & 0 & -\frac{5}{4} & 1 & -\frac{3}{4} & \frac{3}{4} & 0 & \frac{145}{4} \\ ⟨\frac{1}{2}⟩ & 1 & 0 & \frac{1}{4} & 0 & -\frac{1}{4} & \frac{1}{4} & 0 & \frac{35}{4} \\ \hdashline -\frac{1}{2} & 0 & 0 & \frac{11}{4} & 0 & \frac{5}{4} & -\frac{5}{4}+M & 1 & \frac{225}{4} \end{bmatrix}$$

$$2R_3 \to R_3$$

$$\sim \begin{bmatrix} 0 & 0 & 1 & \frac{1}{2} & 0 & \frac{1}{2} & -\frac{1}{2} & 0 & \frac{15}{2} \\ \frac{3}{2} & 0 & 0 & -\frac{5}{4} & 1 & -\frac{3}{4} & \frac{3}{4} & 0 & \frac{145}{4} \\ ① & 2 & 0 & \frac{1}{2} & 0 & -\frac{1}{2} & \frac{1}{2} & 0 & \frac{35}{2} \\ \hdashline -\frac{1}{2} & 0 & 0 & \frac{11}{4} & 0 & \frac{5}{4} & -\frac{5}{4}+M & 1 & \frac{225}{4} \end{bmatrix}$$

$$\left(-\frac{3}{2}\right)R_3 + R_2 \to R_2, \quad \frac{1}{2}R_3 + R_4 \to R_4$$

$$
\begin{array}{c}
\begin{array}{cccccccc} x_1 & x_2 & x_3 & s_1 & s_2 & s_3 & a_1 & P \end{array}\\
\sim
\begin{array}{c} x_3 \\ s_2 \\ x_1 \\ \\ P \end{array}
\left[
\begin{array}{cccccccc|c}
0 & 0 & 1 & \frac{1}{2} & 0 & \frac{1}{2} & -\frac{1}{2} & 0 & \frac{15}{2} \\
0 & -3 & 0 & -2 & 1 & 0 & 0 & 0 & 10 \\
1 & 2 & 0 & \frac{1}{2} & 0 & -\frac{1}{2} & \frac{1}{2} & 0 & \frac{35}{2} \\
\hdashline
0 & 1 & 0 & 3 & 0 & 1 & -1+M & 1 & 65
\end{array}
\right]
\end{array}
$$

Optimal solution: max $P = 65$ at $x_1 = \dfrac{35}{2}$, $x_2 = 0$, $x_3 = \dfrac{15}{2}$.

21. We introduce slack, surplus, and artificial variables to obtain the modified problem:

Maximize $P = x_1 + 2x_2 + 5x_3 - Ma_1 - Ma_2$

Subject to:
$$
\begin{aligned}
x_1 + 3x_2 + 2x_3 + s_1 \qquad\qquad\qquad\qquad &= 60 \\
2x_1 + 5x_2 + 2x_3 \qquad - s_2 \qquad + a_1 \qquad &= 50 \\
x_1 - 2x_2 + x_3 \qquad\qquad\quad - s_3 \qquad + a_2 &= 40 \\
x_1,\ x_2,\ x_3,\ s_1,\ s_2,\ s_3,\ a_1,\ a_2 &\geq 0
\end{aligned}
$$

The preliminary simplex tableau for the modified problem is:

$$
\begin{array}{c}
\begin{array}{ccccccccc} x_1 & x_2 & x_3 & s_1 & s_2 & a_1 & s_3 & a_2 & P \end{array}\\
\left[
\begin{array}{ccccccccc|c}
1 & 3 & 2 & 1 & 0 & 0 & 0 & 0 & 0 & 60 \\
2 & 5 & 2 & 0 & -1 & 1 & 0 & 0 & 0 & 50 \\
1 & -2 & 1 & 0 & 0 & 0 & -1 & 1 & 0 & 40 \\
\hdashline
-1 & -2 & -5 & 0 & 0 & M & 0 & M & 1 & 0
\end{array}
\right]
\end{array}
$$

$(-M)R_2 + R_4 \rightarrow R_4$

$$
\sim
\left[
\begin{array}{ccccccccc|c}
1 & 3 & 2 & 1 & 0 & 0 & 0 & 0 & 0 & 60 \\
2 & 5 & 2 & 0 & -1 & 1 & 0 & 0 & 0 & 50 \\
1 & -2 & 1 & 0 & 0 & 0 & -1 & 1 & 0 & 40 \\
\hdashline
-2M-1 & -5M-2 & -2M-5 & 0 & M & 0 & 0 & M & 1 & -50M
\end{array}
\right]
$$

$(-M)R_3 + R_4 \rightarrow R_4$

$$
\sim
\left[
\begin{array}{ccccccccc|c}
1 & 3 & 2 & 1 & 0 & 0 & 0 & 0 & 0 & 60 \\
2 & 5 & ② & 0 & -1 & 1 & 0 & 0 & 0 & 50 \\
1 & -2 & 1 & 0 & 0 & 0 & -1 & 1 & 0 & 40 \\
\hdashline
-3M-1 & -3M-2 & -3M-5 & 0 & M & 0 & M & 0 & 1 & -90M
\end{array}
\right]
$$

$\frac{1}{2}R_2 \rightarrow R_2$

$$
\sim
\left[
\begin{array}{ccccccccc|c}
1 & 3 & 2 & 1 & 0 & 0 & 0 & 0 & 0 & 60 \\
1 & \frac{5}{2} & ① & 0 & -\frac{1}{2} & \frac{1}{2} & 0 & 0 & 0 & 25 \\
1 & -2 & 1 & 0 & 0 & 0 & -1 & 1 & 0 & 40 \\
\hdashline
-3M-1 & -3M-2 & -3M-5 & 0 & M & 0 & M & 0 & 1 & -90M
\end{array}
\right]
$$

$(-2)R_2 + R_1 \rightarrow R_1,\quad (-1)R_2 + R_3 \rightarrow R_3,\quad (3M+5)R_2 + R_4 \rightarrow R_4$

$$\sim \begin{bmatrix} -1 & -2 & 0 & 1 & ① & -1 & 0 & 0 & 0 & | & 10 \\ 1 & \frac{5}{2} & 1 & 0 & -\frac{1}{2} & \frac{1}{2} & 0 & 0 & 0 & | & 25 \\ 0 & -\frac{9}{2} & 0 & 0 & \frac{1}{2} & -\frac{1}{2} & -1 & 1 & 0 & | & 15 \\ \hline 4 & \frac{9}{2}M + \frac{21}{2} & 0 & 0 & -\frac{1}{2}M - \frac{5}{2} & \frac{3}{2}M + \frac{5}{2} & M & 0 & 1 & | & -15M + 125 \end{bmatrix}$$

$$\frac{1}{2}R_1 + R_2 \to R_2, \quad \left(-\frac{1}{2}\right)R_1 + R_3 \to R_3, \quad \left(\frac{1}{2}M + \frac{5}{2}\right)R_1 + R_4 \to R_4$$

$$\sim \begin{bmatrix} -1 & -2 & 0 & 1 & 1 & -1 & 0 & 0 & 0 & | & 10 \\ \frac{1}{2} & \frac{3}{2} & 1 & \frac{1}{2} & 0 & 0 & 0 & 0 & 0 & | & 30 \\ ⟨\frac{1}{2}⟩ & -\frac{7}{2} & 0 & -\frac{1}{2} & 0 & 0 & -1 & 1 & 0 & | & 10 \\ \hline -\frac{1}{2}M + \frac{3}{2} & \frac{7}{2}M + \frac{11}{2} & 0 & \frac{1}{2}M + \frac{5}{2} & 0 & M & M & 0 & 1 & | & -10M + 150 \end{bmatrix}$$

$$2R_3 + R_1 \to R_1, \quad (-1)R_3 + R_2 \to R_2, \quad (M-3)R_3 + R_4 \to R_4, \quad 2R_3 \to R_3$$

$$\sim \begin{array}{c} \\ s_2 \\ x_3 \\ x_1 \\ \\ \end{array} \begin{array}{c} x_1 \;\; x_2 \;\; x_3 \;\; s_1 \;\; s_2 \;\; a_1 \;\; s_3 \;\; a_2 \;\;\;\;\; P \\ \begin{bmatrix} 0 & -9 & 0 & 0 & 1 & -1 & -2 & 2 & 0 & | & 30 \\ 0 & 5 & 1 & 1 & 0 & 0 & 1 & -1 & 0 & | & 20 \\ 1 & -7 & 0 & -1 & 0 & 0 & -2 & 2 & 0 & | & 20 \\ \hline 0 & 16 & 0 & 4 & 0 & M & 3 & M-3 & 1 & | & 120 \end{bmatrix} \end{array}$$

Optimal solution: max $P = 120$ at $x_1 = 20$, $x_2 = 0$, $x_3 = 20$.

23. (A) Refer to Problem 5.
   The graph of the feasible region is shown at the right. Since it is unbounded, $P = 4x_1 + 3x_2$ does not have a maximum value by Theorem 2(B) in Section 5.2.

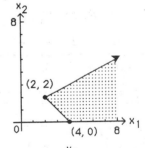

(B) Refer to Problem 7.
   The graph of the feasible region is empty. Therefore, $P = 5x_1 + 10x_2$ does not have a maximum value, by Theorem 2(C) in Section 5.2.

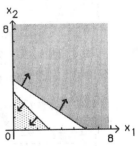

25. We will maximize $P = -C = -10x_1 + 40x_2 + 5x_3$
   Subject to: $x_1 + 3x_2 \qquad\quad + s_1 \qquad\quad = 6$
   $\qquad\qquad\qquad 4x_2 + x_3 \qquad\quad + s_2 = 3$
   $\qquad\qquad x_1, \; x_2, \; x_3, \; s_1, \; s_2 \geq 0$

where s_1, s_2 are slack variables. The simplex tableau for this problem is:

$$\begin{array}{c}
\begin{array}{cccccc} x_1 & x_2 & x_3 & s_1 & s_2 & P \end{array} \\
\begin{array}{c} s_1 \\ s_2 \\ P \end{array}
\left[\begin{array}{cccccc|c}
1 & 3 & 0 & 1 & 0 & 0 & 6 \\
0 & \boxed{4} & 1 & 0 & 1 & 0 & 3 \\
\hline
10 & -40 & -5 & 0 & 0 & 1 & 0
\end{array}\right]
\begin{array}{l} \frac{6}{3} = 2 \\ \frac{3}{4} = .75 \end{array}
\sim
\left[\begin{array}{cccccc|c}
1 & 3 & 0 & 1 & 0 & 0 & 6 \\
0 & \boxed{1} & \frac{1}{4} & 0 & \frac{1}{4} & 0 & \frac{3}{4} \\
\hline
10 & -40 & -5 & 0 & 0 & 1 & 0
\end{array}\right]
\end{array}$$

$$\tfrac{1}{4}R_2 \to R_2 \qquad\qquad (-3)R_2 + R_1 \to R_1 \text{ and } 40R_2 + R_3 \to R_3$$

$$\begin{array}{c}
\begin{array}{cccccc} x_1 & x_2 & x_3 & s_1 & s_2 & P \end{array} \\
\sim \begin{array}{c} x_1 \\ x_2 \\ P \end{array}
\left[\begin{array}{cccccc|c}
1 & 0 & -\frac{3}{4} & 1 & -\frac{3}{4} & 0 & \frac{15}{4} \\
0 & 1 & \frac{1}{4} & 0 & \frac{1}{4} & 0 & \frac{3}{4} \\
\hline
10 & 0 & 5 & 0 & 10 & 1 & 30
\end{array}\right]
\end{array}$$

Optimal solution: min $C = -30$ at $x_1 = 0$, $x_2 = \frac{3}{4}$, $x_3 = 0$.

27. Introduce slack, surplus, and artificial variables to obtain the modified problem: Maximize $P = -5x_1 + 10x_2 + 15x_3 - Ma_1$

$$\begin{array}{rl}
\text{Subject to: } & 2x_1 + 3x_2 + x_3 + s_1 = 24 \\
& x_1 - 2x_2 - 2x_3 - s_2 + a_1 = 1 \\
& x_1,\ x_2,\ x_3,\ s_1,\ s_2,\ a_1 \geq 0
\end{array}$$

The preliminary simplex tableau for the modified problem is:

$$\begin{array}{c}
\begin{array}{ccccccc} x_1 & x_2 & x_3 & s_1 & s_2 & a_1 & P \end{array} \\
\left[\begin{array}{ccccccc|c}
2 & 3 & 1 & 1 & 0 & 0 & 0 & 24 \\
1 & -2 & -2 & 0 & -1 & 1 & 0 & 1 \\
\hline
5 & -10 & -15 & 0 & 0 & M & 1 & 0
\end{array}\right]
\end{array}$$

$$(-M)R_2 + R_3 \to R_3$$

$$\begin{array}{c}
\begin{array}{ccccccc} x_1 & x_2 & x_3 & s_1 & s_2 & a_1 & P \end{array} \\
\sim \begin{array}{c} s_1 \\ a_1 \\ P \end{array}
\left[\begin{array}{ccccccc|c}
2 & 3 & 1 & 1 & 0 & 0 & 0 & 24 \\
\boxed{1} & -2 & -2 & 0 & -1 & 1 & 0 & 1 \\
\hline
-M+5 & 2M-10 & 2M-15 & 0 & M & 0 & 1 & -M
\end{array}\right]
\begin{array}{l} \frac{24}{2} = 12 \\ \frac{1}{1} = 1 \end{array}
\end{array}$$

$$(-2)R_2 + R_1 \to R_1 \text{ and } (M-5)R_2 + R_3 \to R_3$$

$$\sim
\left[\begin{array}{ccccccc|c}
0 & 7 & \boxed{5} & 1 & 2 & -2 & 0 & 22 \\
1 & -2 & -2 & 0 & -1 & 1 & 0 & 1 \\
\hline
0 & 0 & -5 & 0 & 5 & M-5 & 1 & -5
\end{array}\right]
\sim
\left[\begin{array}{ccccccc|c}
0 & \frac{7}{5} & \boxed{1} & \frac{1}{5} & \frac{2}{5} & -\frac{2}{5} & 0 & \frac{22}{5} \\
1 & -2 & -2 & 0 & -1 & 1 & 0 & 1 \\
\hline
0 & 0 & -5 & 0 & 5 & M-5 & 1 & -5
\end{array}\right]$$

$$\tfrac{1}{5}R_1 \to R_1 \qquad\qquad 2R_1 + R_2 \to R_2 \text{ and } 5R_1 + R_3 \to R_3$$

$$\begin{array}{c}
\\
x_3 \\
\sim \ x_1 \\
P
\end{array}
\begin{array}{c}
\begin{array}{ccccccc}
x_1 & x_2 & x_3 & s_1 & s_2 & a_1 & P
\end{array} \\
\left[\begin{array}{ccccccc|c}
0 & \frac{7}{5} & 1 & \frac{1}{5} & \frac{2}{5} & -\frac{2}{5} & 0 & \frac{22}{5} \\
1 & \frac{4}{5} & 0 & \frac{2}{5} & -\frac{1}{5} & \frac{1}{5} & 0 & \frac{49}{5} \\
\hline
0 & 7 & 0 & 1 & 7 & M-7 & 1 & 17
\end{array}\right]
\end{array}$$

Optimal solution: max $P = 17$

at $x_1 = \frac{49}{5}$, $x_2 = 0$, $x_3 = \frac{22}{5}$.

29. The matrices corresponding to the given problem and the dual problem are:

$$A = \left[\begin{array}{ccc|c}
1 & 3 & 0 & 6 \\
0 & 4 & 1 & 3 \\
\hline
10 & 40 & 5 & 1
\end{array}\right] \quad \text{and} \quad A^T = \left[\begin{array}{cc|c}
1 & 0 & 10 \\
3 & 4 & 40 \\
0 & 1 & 5 \\
\hline
6 & 3 & 1
\end{array}\right]$$

Thus, the dual problem is: Maximize $P = 6y_1 + 3y_2$

Subject to:
$$\begin{aligned}
y_1 & \leq 10 \\
3y_1 + 4y_2 & \leq 40 \\
y_2 & \leq 5 \\
y_1, \ y_2 & \geq 0
\end{aligned}$$

We introduce the slack variables x_1, x_2, and x_3 to obtain the initial system:

$$\begin{aligned}
y_1 \quad\quad + x_1 \quad\quad\quad &= 10 \\
3y_1 + 4y_2 \quad\quad + x_2 \quad\quad &= 40 \\
y_2 \quad\quad + x_3 \quad &= 5 \\
-6y_1 - 3y_2 \quad\quad\quad\quad + P &= 0
\end{aligned}$$

The simplex tableau for this problem is:

$$\begin{array}{c}
\\
x_1 \\
x_2 \\
x_3 \\
P
\end{array}
\begin{array}{c}
\begin{array}{cccccc}
y_1 & y_2 & x_1 & x_2 & x_3 & P
\end{array} \\
\left[\begin{array}{cccccc|c}
① & 0 & 1 & 0 & 0 & 0 & 10 \\
3 & 4 & 0 & 1 & 0 & 0 & 40 \\
0 & 1 & 0 & 0 & 1 & 0 & 5 \\
\hline
-6 & -3 & 0 & 0 & 0 & 1 & 0
\end{array}\right]
\end{array}
\begin{array}{l}
\frac{10}{1} = 10 \\
\frac{40}{3} \approx 13.33
\end{array}$$

$(-3)R_1 + R_2 \rightarrow R_2$ and $6R_1 + R_4 \rightarrow R_4$

$$\sim \left[\begin{array}{cccccc|c}
1 & 0 & 1 & 0 & 0 & 0 & 10 \\
0 & ④ & -3 & 1 & 0 & 0 & 10 \\
0 & 1 & 0 & 0 & 1 & 0 & 5 \\
\hline
0 & -3 & 6 & 0 & 0 & 1 & 60
\end{array}\right]
\begin{array}{l}
\frac{10}{4} = 2.5 \\
\frac{5}{1} = 5
\end{array}
\quad \sim \left[\begin{array}{cccccc|c}
1 & 0 & 1 & 0 & 0 & 0 & 10 \\
0 & ① & -\frac{3}{4} & \frac{1}{4} & 0 & 0 & \frac{5}{2} \\
0 & 1 & 0 & 0 & 1 & 0 & 5 \\
\hline
0 & -3 & 6 & 0 & 0 & 1 & 60
\end{array}\right]$$

$\frac{1}{4}R_2 \rightarrow R_2$

$(-1)R_2 + R_3 \rightarrow R_3$ and $3R_2 + R_4 \rightarrow R_4$

$$\sim \quad \begin{array}{c} y_1 \\ y_2 \\ x_3 \\ P \end{array} \begin{array}{cc} \begin{array}{cccccc} y_1 & y_2 & x_1 & x_2 & x_3 & P \end{array} & \\ \left[\begin{array}{cccccc|c} 1 & 0 & 1 & 0 & 0 & 0 & 10 \\ 0 & 1 & -\frac{3}{4} & \frac{1}{4} & 0 & 0 & \frac{5}{2} \\ 0 & 0 & \frac{3}{4} & -\frac{1}{4} & 1 & 0 & \frac{5}{2} \\ \hline 0 & 0 & \frac{15}{4} & \frac{3}{4} & 0 & 1 & \frac{135}{2} \end{array}\right] & \end{array}$$

Optimal solution: min $C = \dfrac{135}{2}$, $x_1 = \dfrac{15}{4}$, $x_2 = \dfrac{3}{4}$, $x_3 = 0$.

31. We introduce the slack variables s_1 and s_2 to obtain the initial system:

$$\begin{aligned} x_1 + 3x_2 + x_3 + s_1 \qquad\qquad &= 40 \\ 2x_1 + x_2 + 3x_3 \qquad + s_2 \quad\; &= 60 \\ -12x_1 - 9x_2 - 5x_3 \qquad\qquad + P &= 0 \end{aligned}$$

The simplex tableau for this problem is:

$$\begin{array}{c} s_1 \\ s_2 \\ {} \end{array} \begin{array}{cccccc} x_1 & x_2 & x_3 & s_1 & s_2 & P \end{array} \\ \left[\begin{array}{cccccc|c} 1 & 3 & 1 & 1 & 0 & 0 & 40 \\ ② & 1 & 3 & 0 & 1 & 0 & 60 \\ \hline -12 & -9 & -5 & 0 & 0 & 1 & 0 \end{array}\right] \begin{array}{l} \frac{40}{1} = 40 \\ \frac{60}{2} = 30 \end{array} \quad \sim \quad \left[\begin{array}{cccccc|c} 1 & 3 & 1 & 1 & 0 & 0 & 40 \\ ① & \frac{1}{2} & \frac{3}{2} & 0 & \frac{1}{2} & 0 & 30 \\ \hline -12 & -9 & -5 & 0 & 0 & 1 & 0 \end{array}\right]$$

$\frac{1}{2}R_2 \to R_2$ $\qquad\qquad\qquad\qquad\qquad\qquad\qquad\qquad (-1)R_2 + R_1 \to R_1$ and $12R_2 + R_3 \to R_3$

$$\sim \left[\begin{array}{cccccc|c} 0 & ⑤⁄₂ & -\frac{1}{2} & 1 & -\frac{1}{2} & 0 & 10 \\ 1 & \frac{1}{2} & \frac{3}{2} & 0 & \frac{1}{2} & 0 & 30 \\ \hline 0 & -3 & 13 & 0 & 6 & 1 & 360 \end{array}\right] \begin{array}{l} \frac{10}{5/2} = 4 \\ \frac{30}{1/2} = 60 \end{array} \quad \sim \left[\begin{array}{cccccc|c} 0 & ① & -\frac{1}{5} & \frac{2}{5} & -\frac{1}{5} & 0 & 4 \\ 1 & \frac{1}{2} & \frac{3}{2} & 0 & \frac{1}{2} & 0 & 30 \\ \hline 0 & -3 & 13 & 0 & 6 & 1 & 360 \end{array}\right]$$

$\frac{2}{5}R_1 \to R_1$ $\qquad\qquad\qquad\qquad\qquad\qquad\qquad\qquad \left(-\frac{1}{2}\right)R_1 + R_2 \to R_2$ and $3R_1 + R_3 \to R_3$

$$\sim \begin{array}{c} x_2 \\ x_1 \\ P \end{array} \begin{array}{cccccc} x_1 & x_2 & x_3 & s_1 & s_2 & P \end{array} \\ \left[\begin{array}{cccccc|c} 0 & 1 & -\frac{1}{5} & \frac{2}{5} & -\frac{1}{5} & 0 & 4 \\ 1 & 0 & \frac{8}{5} & -\frac{1}{5} & \frac{3}{5} & 0 & 28 \\ \hline 0 & 0 & \frac{62}{5} & \frac{6}{5} & \frac{27}{5} & 1 & 372 \end{array}\right]$$

Optimal solution: max $P = 372$ at $x_1 = 28$, $x_2 = 4$, $x_3 = 0$.

33. Let x_1 = the number of 16K modules
and x_2 = the number of 64K modules.
The mathematical model is: Maximize $P = 18x_1 + 30x_2$

$$\begin{aligned} \text{Subject to: } 10x_1 + 15x_2 &\le 2200 \\ 2x_1 + 4x_2 &\le 500 \\ x_1 \qquad &\ge 50 \\ x_1,\ x_2 &\ge 0 \end{aligned}$$

We introduce slack, surplus, and artificial variables to obtain the modified problem: Maximize $P = 18x_1 + 30x_2 - Ma_1$

$$\text{Subject to: } \begin{align} 10x_1 + 15x_2 + s_1 &= 2200 \\ 2x_1 + 4x_2 + s_2 &= 500 \\ x_1 - s_3 + a_1 &= 50 \\ x_1, \ x_2, \ s_1, \ s_2, \ s_3, \ a_1 &\geq 0 \end{align}$$

The preliminary simplex tableau for the modified problem is:

$$\begin{array}{ccccccc} x_1 & x_2 & s_1 & s_2 & s_3 & a_1 & P \end{array}$$

$$\left[\begin{array}{ccccccc|c} 10 & 15 & 1 & 0 & 0 & 0 & 0 & 2200 \\ 2 & 4 & 0 & 1 & 0 & 0 & 0 & 500 \\ 1 & 0 & 0 & 0 & -1 & 1 & 0 & 50 \\ \hline -18 & -30 & 0 & 0 & 0 & M & 1 & 0 \end{array} \right]$$

$(-M)R_3 + R_4 \rightarrow R_4$

$$\begin{array}{cccccccc} & x_1 & x_2 & s_1 & s_2 & s_3 & a_1 & P \end{array}$$

$$\sim \begin{array}{c} s_1 \\ s_2 \\ a_1 \\ P \end{array} \left[\begin{array}{ccccccc|c} 10 & 15 & 1 & 0 & 0 & 0 & 0 & 2200 \\ 2 & 4 & 0 & 1 & 0 & 0 & 0 & 500 \\ ① & 0 & 0 & 0 & -1 & 1 & 0 & 50 \\ \hline -M-18 & -30 & 0 & 0 & 0 & M & 1 & -50M \end{array} \right] \begin{array}{l} \frac{2200}{10} = 220 \\ \frac{500}{2} = 250 \\ \frac{50}{1} = 50 \end{array}$$

$(-10)R_3 + R_1 \rightarrow R_1$, $(-2)R_3 + R_2 \rightarrow R_2$, and $(M+18)R_3 + R_4 \rightarrow R_4$

$$\sim \left[\begin{array}{ccccccc|c} 0 & 15 & 1 & 0 & 10 & -10 & 0 & 1700 \\ 0 & ④ & 0 & 1 & 2 & -2 & 0 & 400 \\ 1 & 0 & 0 & 0 & -1 & 1 & 0 & 50 \\ \hline 0 & -30 & 0 & 0 & -18 & M+18 & 1 & 900 \end{array} \right] \begin{array}{l} \frac{1700}{15} = 113.33 \\ \frac{400}{4} = 100 \end{array}$$

$\frac{1}{4}R_2 \rightarrow R_2$

$$\sim \left[\begin{array}{ccccccc|c} 0 & 15 & 1 & 0 & 10 & -10 & 0 & 1700 \\ 0 & ① & 0 & \frac{1}{4} & \frac{1}{2} & -\frac{1}{2} & 0 & 100 \\ 1 & 0 & 0 & 0 & -1 & 1 & 0 & 50 \\ \hline 0 & -30 & 0 & 0 & -18 & M+18 & 1 & 900 \end{array} \right]$$

$(-15)R_2 + R_1 \rightarrow R_1$, $30R_2 + R_4 \rightarrow R_4$

$$\sim \left[\begin{array}{ccccccc|c} 0 & 0 & 1 & -\frac{15}{4} & ⑤⁄₂ & -\frac{5}{2} & 0 & 200 \\ 0 & 1 & 0 & \frac{1}{4} & \frac{1}{2} & -\frac{1}{2} & 0 & 100 \\ 1 & 0 & 0 & 0 & -1 & 1 & 0 & 50 \\ \hline 0 & 0 & 0 & \frac{15}{2} & -3 & M+3 & 1 & 3900 \end{array} \right] \begin{array}{l} \frac{200}{5/2} = 80 \\ \frac{100}{1/2} = 200 \end{array}$$

$\frac{2}{5}R_1 \rightarrow R_1$

$$\sim \begin{bmatrix} 0 & 0 & \frac{2}{5} & -\frac{3}{2} & \textcircled{1} & -1 & 0 & 80 \\ 0 & 1 & 0 & \frac{1}{4} & \frac{1}{2} & -\frac{1}{2} & 0 & 100 \\ 1 & 0 & 0 & 0 & -1 & 1 & 0 & 50 \\ \hline 0 & 0 & 0 & \frac{15}{2} & -3 & M+3 & 1 & 3900 \end{bmatrix}$$

$$\left(-\tfrac{1}{2}\right)R_1 + R_2 \to R_2, \quad R_1 + R_3 \to R_3, \quad 3R_1 + R_4 \to R_4$$

$$\sim \begin{array}{c} \\ s_3 \\ x_2 \\ x_1 \\ \\ \end{array} \begin{bmatrix} x_1 & x_2 & s_1 & s_2 & s_3 & a_1 & P & \\ 0 & 0 & \frac{2}{5} & -\frac{3}{2} & 1 & -1 & 0 & 80 \\ 0 & 1 & -\frac{1}{5} & 1 & 0 & 0 & 0 & 60 \\ 1 & 0 & \frac{2}{5} & -\frac{3}{2} & 0 & 0 & 0 & 130 \\ \hline 0 & 0 & \frac{6}{5} & 3 & 0 & M & 1 & 4140 \end{bmatrix}$$

The maximum profit is \$4140 when 130 16K modules and 60 64K modules are manufactured each day.

35. Let x_1 = the number of ads placed in the *Sentinel*,
    x_2 = the number of ads placed in the *Journal*,
and x_3 = the number of ads placed in the *Tribune*.

The mathematical model is: Minimize $C = 200x_1 + 200x_2 + 100x_3$
$$\begin{aligned} \text{Subject to:} \quad & x_1 + x_2 + x_3 \le 10 \\ & 2000x_1 + 500x_2 + 1500x_3 \ge 16{,}000 \\ & x_1, \; x_2, \; x_3 \ge 0 \end{aligned}$$

Divide the second constraint inequality by 100 to simplify the calculations, and introduce slack, surplus, and artificial variables to obtain the equivalent form:

Maximize $P = -C = -200x_1 - 200x_2 - 100x_3 - Ma_1$
$$\begin{aligned} \text{Subject to:} \quad & x_1 + x_2 + x_3 + s_1 = 10 \\ & 20x_1 + 5x_2 + 15x_3 - s_2 + a_1 = 160 \\ & x_1, \; x_2, \; x_3, \; s_1, \; s_2, \; a_1 \ge 0 \end{aligned}$$

The simplex tableau for the modified problem is:

$$\begin{array}{c} \\ \\ \\ \\ \end{array} \begin{bmatrix} x_1 & x_2 & x_3 & s_1 & s_2 & a_1 & P & \\ 1 & 1 & 1 & 1 & 0 & 0 & 0 & 10 \\ 20 & 5 & 15 & 0 & -1 & 1 & 0 & 160 \\ \hline 200 & 200 & 100 & 0 & 0 & M & 1 & 0 \end{bmatrix}$$

$$(-M)R_2 + R_3 \to R_3$$

$$
\begin{array}{c}
 \\
s_1 \\
\sim \quad a_1 \\
P
\end{array}
\begin{array}{cccccccc}
x_1 & x_2 & x_3 & s_1 & s_2 & a_1 & P & \\
\end{array}
$$

$$
\begin{array}{c}
s_1 \\
\sim \; a_1 \\
P
\end{array}
\left[
\begin{array}{ccccccc|c}
1 & 1 & 1 & 1 & 0 & 0 & 0 & 10 \\
\boxed{20} & 5 & 15 & 0 & -1 & 1 & 0 & 160 \\
\hline
-20M + 200 & -5M + 200 & -15M + 100 & 0 & M & 0 & 1 & -160M
\end{array}
\right]
\begin{array}{l}
\frac{10}{1} = 10 \\[4pt]
\frac{160}{20} = 8
\end{array}
$$

$$\frac{1}{20}R_2 \to R_2$$

$$
\sim
\left[
\begin{array}{ccccccc|c}
1 & 1 & 1 & 1 & 0 & 0 & 0 & 10 \\
\boxed{1} & \frac{1}{4} & \frac{3}{4} & 0 & -\frac{1}{20} & \frac{1}{20} & 0 & 8 \\
\hline
-20M + 200 & -5M + 200 & -15M + 100 & 0 & M & 0 & 1 & -160M
\end{array}
\right]
$$

$$(-1)R_2 + R_1 \to R_1 \quad \text{and} \quad (20M - 200)R_2 + R_3 \to R_3$$

$$
\sim
\left[
\begin{array}{ccccccc|c}
0 & \frac{3}{4} & \boxed{\frac{1}{4}} & 1 & \frac{1}{20} & -\frac{1}{20} & 0 & 2 \\
1 & \frac{1}{4} & \frac{3}{4} & 0 & -\frac{1}{20} & \frac{1}{20} & 0 & 8 \\
\hline
0 & 150 & -50 & 0 & 10 & M - 10 & 1 & -1600
\end{array}
\right]
\begin{array}{l}
\frac{2}{1/4} = 8 \\[4pt]
\frac{8}{3/4} = \frac{32}{3} \approx 10.67
\end{array}
$$

$$4R_1 \to R_1$$

$$
\sim
\left[
\begin{array}{ccccccc|c}
0 & 3 & \boxed{1} & 4 & \frac{1}{5} & -\frac{1}{5} & 0 & 8 \\
1 & \frac{1}{4} & \frac{3}{4} & 0 & -\frac{1}{20} & \frac{1}{20} & 0 & 8 \\
\hline
0 & 150 & -50 & 0 & 10 & M - 10 & 1 & -1600
\end{array}
\right]
$$

$$\left(-\frac{3}{4}\right)R_1 + R_2 \to R_2 \quad \text{and} \quad 50R_1 + R_3 \to R_3$$

$$
\begin{array}{c}
x_3 \\
\sim \; x_1 \\
P
\end{array}
\begin{array}{ccccccc}
x_1 & x_2 & x_3 & s_1 & s_2 & a_1 & P \\
\end{array}
$$

$$
\begin{array}{c}
x_3 \\
\sim \; x_1 \\
P
\end{array}
\left[
\begin{array}{ccccccc|c}
0 & 3 & 1 & 4 & \frac{1}{5} & -\frac{1}{5} & 0 & 8 \\
1 & -2 & 0 & -3 & -\frac{1}{5} & \frac{1}{5} & 0 & 2 \\
\hline
0 & 300 & 0 & 200 & 20 & M - 20 & 1 & -1200
\end{array}
\right]
$$

The minimal cost is \$1200 when two ads are placed in the *Sentinel*, no ads are placed in the *Journal*, and eight ads are placed in the *Tribune*.

37. Let x_1 = the number of bottles of brand A,
x_2 = the number of bottles of brand B,
and x_3 = the number of bottles of brand C.

The mathematical model is: Minimize $C = 0.6x_1 + 0.4x_2 + 0.9x_3$

Subject to: $10x_1 + 10x_2 + 20x_3 \geq 100$

$2x_1 + 3x_2 + 4x_3 \leq 24$

$x_1, x_2, x_3 \geq 0$

Divide the first inequality by 10, and introduce slack, surplus, and artificial variables to obtain the equivalent form:

Maximize $P = -10C = -6x_1 - 4x_2 - 9x_3 - Ma_1$

Subject to: $x_1 + x_2 + 2x_3 - s_1 + a_1 = 10$

$2x_1 + 3x_2 + 4x_3 + s_2 = 24$

$x_1, x_2, x_3, s_1, s_2, a_1 \geq 0$

The simplex tableau for the modified problem is:

$$
\begin{array}{ccccccc}
x_1 & x_2 & x_3 & s_1 & a_1 & s_2 & P \\
\end{array}
$$

$$
\left[\begin{array}{ccccccc|c}
1 & 1 & 2 & -1 & 1 & 0 & 0 & 10 \\
2 & 3 & 4 & 0 & 0 & 1 & 0 & 24 \\
\hline
6 & 4 & 9 & 0 & M & 0 & 1 & 0 \\
\end{array}\right]
$$

$(-M)R_1 + R_3 \rightarrow R_3$

$$
\begin{array}{c}
\\
a_1 \\
\sim s_2 \\
\\
\end{array}
\left[\begin{array}{ccccccc|c}
x_1 & x_2 & x_3 & s_1 & a_1 & s_2 & P & \\
1 & 1 & ② & -1 & 1 & 0 & 0 & 10 \\
2 & 3 & 4 & 0 & 0 & 1 & 0 & 24 \\
\hline
-M+6 & -M+4 & -2M+9 & M & 0 & 0 & 1 & -10M \\
\end{array}\right]
\begin{array}{c}
\frac{10}{2}=5 \\
\frac{24}{4}=6 \\
\end{array}
$$

$\frac{1}{2}R_1 \rightarrow R_1$

$$
\sim
\left[\begin{array}{ccccccc|c}
\frac{1}{2} & \frac{1}{2} & ① & -\frac{1}{2} & \frac{1}{2} & 0 & 0 & 5 \\
2 & 3 & 4 & 0 & 0 & 1 & 0 & 24 \\
\hline
-M+6 & -M+4 & -2M+9 & M & 0 & 0 & 1 & -10M \\
\end{array}\right]
$$

$(-4)R_1 + R_2 \rightarrow R_2$ and $(2M-9)R_1 + R_3 \rightarrow R_3$

$$
\sim
\left[\begin{array}{ccccccc|c}
\frac{1}{2} & \frac{1}{2} & 1 & -\frac{1}{2} & \frac{1}{2} & 0 & 0 & 5 \\
0 & ① & 0 & 2 & -2 & 1 & 0 & 4 \\
\hline
-\frac{3}{2} & -\frac{1}{2} & 0 & \frac{9}{2} & M-\frac{9}{2} & 0 & 1 & -45 \\
\end{array}\right]
\begin{array}{c}
\frac{5}{1/2}=10 \\
\frac{4}{1}=4 \\
\end{array}
$$

$\left(-\frac{1}{2}\right)R_2 + R_1 \rightarrow R_1$ and $\frac{1}{2}R_2 + R_3 \rightarrow R_3$

$$\begin{array}{c}\\x_3\\ \sim\ x_2 \\ P\end{array}
\begin{array}{c}x_1\ \ x_2\ \ x_3\ \ \ s_1\ \ \ \ \ a_1\ \ \ \ s_2\ \ P\\
\left[\begin{array}{ccccccc|c}
\frac{1}{2} & 0 & 1 & -\frac{3}{2} & \frac{3}{2} & -\frac{1}{2} & 0 & 3 \\
0 & 1 & 0 & 2 & -2 & 1 & 0 & 4 \\ \hline
\frac{3}{2} & 0 & 0 & \frac{11}{2} & M-\frac{11}{2} & \frac{1}{2} & 1 & -43
\end{array}\right]\end{array}$$

The minimal cost is $4.30 when 0 bottles of brand A, 4 bottles of brand B and 3 bottles of brand C are consumed.

39. Let x_1 = the number of cubic yards of mix A,
x_2 = the number of cubic yards of mix B,
and x_3 = the number of cubic yards of mix C.

The mathematical model is: Maximize $P = 12x_1 + 16x_2 + 8x_3$
Subject to: $16x_1 + 8x_2 + 16x_3 \geq 800$
$12x_1 + 8x_2 + 16x_3 \leq 700$
$x_1,\ x_2,\ x_3 \geq 0$

We simplify the inequalities, and introduce slack, surplus, and artificial variables to obtain the modified problem:

Maximize $P = 12x_1 + 16x_2 + 8x_3 - Ma_1$

Subject to: $4x_1 + 2x_2 + 4x_3 - s_1\ \ \ \ \ + a_1 = 200$
$3x_1 + 2x_2 + 4x_3\ \ \ \ \ \ \ + s_2\ \ \ \ = 175$
$x_1,\ x_2,\ x_3,\ s_1,\ s_2,\ a_1\ \ \geq 0$

The simplex tableau for the modified problem is:

$$\begin{array}{c}x_1\ \ \ \ x_2\ \ \ x_3\ \ \ s_1\ \ \ a_1\ \ s_2\ \ P\\
\left[\begin{array}{ccccccc|c}
4 & 2 & 4 & -1 & 1 & 0 & 0 & 200 \\
3 & 2 & 4 & 0 & 0 & 1 & 0 & 175 \\ \hline
-12 & -16 & -8 & 0 & M & 0 & 1 & 0
\end{array}\right]\end{array}$$

$(-M)R_1 + R_3 \rightarrow R_3$

$$\begin{array}{c}\\a_1\\ \sim\ s_2 \\ \end{array}
\begin{array}{c}x_1\ \ \ \ \ \ \ \ \ \ x_2\ \ \ \ \ \ \ \ x_3\ \ \ \ \ \ s_1\ \ \ a_1\ \ s_2\ \ P\\
\left[\begin{array}{ccccccc|c}
\textcircled{4} & 2 & 4 & -1 & 1 & 0 & 0 & 200 \\
3 & 2 & 4 & 0 & 0 & 1 & 0 & 175 \\ \hline
-4M-12 & -2M-16 & -4M-8 & M & 0 & 0 & 1 & -200M
\end{array}\right]\end{array}
\begin{array}{l}\frac{200}{4}=50\\[6pt]\frac{175}{3}\approx 58.33\end{array}$$

$\frac{1}{4}R_1 \rightarrow R_1$

$$\sim\left[\begin{array}{ccccccc|c}
\textcircled{1} & \frac{1}{2} & 1 & -\frac{1}{4} & \frac{1}{4} & 0 & 0 & 50 \\
3 & 2 & 4 & 0 & 0 & 1 & 0 & 175 \\ \hline
-4M-12 & -2M-16 & -4M-8 & M & 0 & 0 & 1 & -200M
\end{array}\right]$$

$(-3)R_1 + R_2 \rightarrow R_2$ and $(4M + 12)R_1 + R_3 \rightarrow R_3$

$$\sim \begin{bmatrix} 1 & \frac{1}{2} & 1 & -\frac{1}{4} & \frac{1}{4} & 0 & 0 & 50 \\ 0 & \textcircled{\tfrac{1}{2}} & 1 & \frac{3}{4} & -\frac{3}{4} & 1 & 0 & 25 \\ \hline 0 & -10 & 4 & -3 & M+3 & 0 & 1 & 600 \end{bmatrix} \begin{matrix} \frac{50}{1/2}=100 \\ \frac{25}{1/2}=50 \end{matrix}$$

$$2R_2 \to R_2$$

$$\sim \begin{bmatrix} 1 & \frac{1}{2} & 1 & -\frac{1}{4} & \frac{1}{4} & 0 & 0 & 50 \\ 0 & \textcircled{1} & 2 & \frac{3}{2} & -\frac{3}{2} & 2 & 0 & 50 \\ \hline 0 & -10 & 4 & -3 & M+3 & 0 & 1 & 600 \end{bmatrix}$$

$$\left(-\frac{1}{2}\right)R_2 + R_1 \to R_1 \text{ and } 10R_2 + R_3 \to R_3$$

The maximum amount of nitrogen is 1100 pounds when 25 cubic yards of mix A, 50 cubic yards of mix B, and 0 cubic yards of mix C are used.

$$\sim \begin{array}{c} x_1 \\ x_2 \\ {} \end{array} \begin{bmatrix} x_1 & x_2 & x_3 & s_1 & a_1 & s_2 & P & \\ 1 & 0 & 0 & -1 & 1 & -1 & 0 & 25 \\ 0 & 1 & 2 & \frac{3}{2} & -\frac{3}{2} & 2 & 0 & 50 \\ \hline 0 & 0 & 24 & 12 & M-12 & 20 & 1 & 1100 \end{bmatrix}$$

41. Let x_1 = the number of car frames produced in Milwaukee,
x_2 = the number of truck frames produced in Milwaukee,
x_3 = the number of car frames produced in Racine,
and x_4 = the number of truck frames produced in Racine.

The mathematical model for this problem is:
Maximize $P = 50x_1 + 70x_2 + 50x_3 + 70x_4$

Subject to:
$$\begin{aligned}
x_1 + x_3 &\leq 250 \\
x_2 + x_4 &\leq 350 \\
x_1 + x_2 &\leq 300 \\
x_3 + x_4 &\leq 200 \\
150x_1 + 200x_2 &\leq 50{,}000 \\
135x_3 + 180x_4 &\leq 35{,}000 \\
x_1, \ x_2, \ x_3, \ x_4 &\geq 0
\end{aligned}$$

43. Let x_1 = the number of barrels of A used in regular gasoline,
x_2 = the number of barrels of A used in premium gasoline,
x_3 = the number of barrels of B used in regular gasoline,
x_4 = the number of barrels of B used in premium gasoline,
x_5 = the number of barrels of C used in regular gasoline,
and x_6 = the number of barrels of C used in premium gasoline.

Cost $C = 28(x_1 + x_2) + 30(x_3 + x_4) + 34(x_5 + x_6)$
Revenue $R = 38(x_1 + x_3 + x_5) + 46(x_2 + x_4 + x_6)$
Profit $P = R - C = 10x_1 + 18x_2 + 8x_3 + 16x_4 + 4x_5 + 12x_6$

Thus, the mathematical model for this problem is:

Maximize $P = 10x_1 + 18x_2 + 8x_3 + 16x_4 + 4x_5 + 12x_6$

Subject to:
$$
\begin{aligned}
x_1 + x_2 \qquad\qquad\qquad\qquad\qquad &\le 40{,}000 \\
x_3 + x_4 \qquad\qquad\qquad &\le 25{,}000 \\
x_5 + x_6 &\le 15{,}000 \\
x_1 \qquad\quad + x_3 \qquad\quad + x_5 \qquad\quad &\ge 30{,}000 \\
x_2 \qquad\quad + x_4 \qquad\quad + x_6 &\ge 25{,}000 \\
-5x_1 \qquad\quad + 5x_3 \qquad\quad + 15x_5 \qquad\quad &\ge 0 \\
-15x_2 \qquad\quad - 5x_4 \qquad\quad + 5x_6 &\ge 0 \\
x_1, \; x_2, \; x_3, \; x_4, \; x_5, \; x_6 &\ge 0
\end{aligned}
$$

45. Let x_1 = the number of ounces of food L,
x_2 = the number of ounces of food M,
and x_3 = the number of ounces of food N.

The mathematical model for this problem is:

Minimize $C = 0.4x_1 + 0.6x_2 + 0.8x_3$

Subject to:
$$
\begin{aligned}
30x_1 + 10x_2 + 30x_3 &\ge 400 \\
10x_1 + 10x_2 + 10x_3 &\ge 200 \\
10x_1 + 30x_2 + 20x_3 &\ge 300 \\
8x_1 + 4x_2 + 6x_3 &\le 150 \\
60x_1 + 40x_2 + 50x_3 &\le 900 \\
x_1, \; x_2, \; x_3 &\ge 0
\end{aligned}
$$

47. Let x_1 = the number of students from A enrolled in school I,
x_2 = the number of students from A enrolled in school II,
x_3 = the number of students from B enrolled in school I,
x_4 = the number of students from B enrolled in school II,
x_5 = the number of students from C enrolled in school I,
and x_6 = the number of students from C enrolled in school II.

The mathematical model for this problem is:

Minimize $C = 4x_1 + 8x_2 + 6x_3 + 4x_4 + 3x_5 + 9x_6$

Subject to:
$$
\begin{aligned}
x_1 + x_2 \qquad\qquad\qquad\qquad\qquad &= 500 \\
x_3 + x_4 \qquad\qquad\qquad &= 1200 \\
x_5 + x_6 &= 1800 \\
x_1 \qquad\quad + x_3 \qquad\quad + x_5 \qquad\quad &\ge 1400 \\
x_2 \qquad\quad + x_4 \qquad\quad + x_6 &\ge 1400 \\
x_1 \qquad\quad + x_3 \qquad\quad + x_5 \qquad\quad &\le 2000 \\
x_2 \qquad\quad + x_4 \qquad\quad + x_6 &\le 2000 \\
x_1 \qquad\qquad\qquad\qquad\qquad &\le 300 \\
x_2 \qquad\qquad\qquad &\le 300 \\
x_3 \qquad\qquad\qquad &\le 720 \\
x_4 \qquad\qquad\qquad &\le 720 \\
x_5 \qquad\quad &\le 1080 \\
x_6 &\le 1080 \\
x_1, \; x_2, \; x_3, \; x_4, \; x_5, \; x_6 &\ge 0
\end{aligned}
$$

1. $2x_1 + x_2 \leq 8$
   $3x_1 + 9x_2 \leq 27$
   $x_1, x_2 \geq 0$

The graphs of the inequalities are shown at the right. The solution region is shaded; it is *bounded*.

The corner points are:
$(0, 0)$, $(0, 3)$, $(3, 2)$, $(4, 0)$

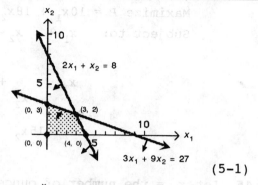

(5-1)

2. $3x_1 + x_2 \geq 9$
   $2x_1 + 4x_2 \geq 16$
   $x_1, x_2 \geq 0$

The graphs of the inequalities are shown at the right. The solution region is shaded; it is *unbounded*.

The corner points are:
$(0, 9)$, $(2, 3)$, $(8, 0)$

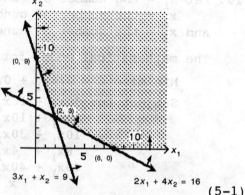

(5-1)

3. The feasible region is the solution set of the given inequalities, and is indicated by the shaded region in the graph at the right.

The corner points are $(0, 0)$, $(0, 5)$, $(2, 4)$, and $(4, 0)$.

The value of P at each corner point is:

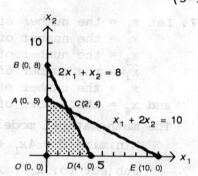

Corner Point	$P = 6x_1 + 2x_2$
$(0, 0)$	$P = 6(0) + 2(0) = 0$
$(0, 5)$	$P = 6(0) + 2(5) = 10$
$(2, 4)$	$P = 6(2) + 2(4) = 20$
$(4, 0)$	$P = 6(4) + 2(0) = 24$

Thus, the maximum occurs at $x_1 = 4$, $x_2 = 0$, and the maximum value is $P = 24$.

(5-2)

4. We introduce the slack variables s_1 and s_2 to obtain the system of equations:
   $2x_1 + x_2 + s_1 \qquad = 8$
   $x_1 + 2x_2 \qquad + s_2 = 10$

(5-3)

5. There are 2 basic and 2 nonbasic variables.

(5-3)

6. The basic solutions are given in the following table.

x_1	x_2	s_1	s_2	Intersection Point	Feasible?
0	0	8	10	O	Yes
0	8	0	-6	B	No
0	5	3	0	A	Yes
4	0	0	6	D	Yes
10	0	-12	0	E	No
2	4	0	0	C	Yes

(5-3)

7. The simplex tableau for Problem 3 is: Enter

$$
\begin{array}{c}
\text{Exit } s_1 \\
s_2 \\
P
\end{array}
\left[
\begin{array}{ccccc|c}
x_1 & x_2 & s_1 & s_2 & P & \\
\textcircled{2} & 1 & 1 & 0 & 0 & 8 \\
1 & 2 & 0 & 1 & 0 & 10 \\
\hline
-6 & -2 & 0 & 0 & 1 & 0
\end{array}
\right]
\begin{array}{l}
\frac{8}{2} = 4 \\
\frac{10}{1} = 10 \\
\end{array}
$$

(5-4)

8.

$$
\begin{array}{c}
s_1 \\
\sim \; s_2 \\
P
\end{array}
\left[
\begin{array}{ccccc|c}
x_1 & x_2 & s_1 & s_2 & P & \\
\textcircled{2} & 1 & 1 & 0 & 0 & 8 \\
1 & 2 & 0 & 1 & 0 & 10 \\
\hline
-6 & -2 & 0 & 0 & 1 & 0
\end{array}
\right]
\sim
\left[
\begin{array}{ccccc|c}
\textcircled{1} & \frac{1}{2} & \frac{1}{2} & 0 & 0 & 4 \\
1 & 2 & 0 & 1 & 0 & 10 \\
\hline
-6 & -2 & 0 & 0 & 1 & 0
\end{array}
\right]
$$

$$\tfrac{1}{2}R_1 \to R_1 \qquad\qquad (-1)R_1 + R_2 \to R_2 \text{ and } 6R_1 + R_3 \to R_3$$

$$
\begin{array}{c}
x_1 \\
\sim \; s_2 \\
P
\end{array}
\left[
\begin{array}{ccccc|c}
x_1 & x_2 & s_1 & s_2 & P & \\
1 & \frac{1}{2} & \frac{1}{2} & 0 & 0 & 4 \\
0 & \frac{3}{2} & -\frac{1}{2} & 1 & 0 & 6 \\
\hline
0 & 1 & 3 & 0 & 1 & 24
\end{array}
\right]
$$

Optimal solution: max $P = 24$ at $x_1 = 4$, $x_2 = 0$.

(5-4)

9. Enter

$$
\begin{array}{c}
x_2 \\
s_2 \\
\text{Exit } s_3 \\
P
\end{array}
\left[
\begin{array}{ccccccc|c}
x_1 & x_2 & x_3 & s_1 & s_2 & s_3 & P & \\
2 & 1 & 3 & -1 & 0 & 0 & 0 & 20 \\
3 & 0 & 4 & 1 & 1 & 0 & 0 & 30 \\
\textcircled{2} & 0 & 5 & 2 & 0 & 1 & 0 & 10 \\
\hline
-8 & 0 & -5 & 3 & 0 & 0 & 1 & 50
\end{array}
\right]
\begin{array}{l}
\frac{20}{2} = 10 \\
\frac{30}{3} = 10 \\
\frac{10}{2} = 5 \\
\end{array}
$$

The basic variables are x_2, s_2, s_3, and P, and the nonbasic variables are x_1, x_3, and s_1.

The first column is the pivot column and the third row is the pivot row. The pivot element is circled.

$$\sim \begin{bmatrix} 2 & 1 & 3 & -1 & 0 & 0 & 0 & | & 20 \\ 3 & 0 & 4 & 1 & 1 & 0 & 0 & | & 30 \\ ① & 0 & \frac{5}{2} & 1 & 0 & \frac{1}{2} & 0 & | & 5 \\ \hdashline -8 & 0 & -5 & 3 & 0 & 0 & 1 & | & 50 \end{bmatrix}$$

$$(-2)R_3 + R_1 \rightarrow R_1, \quad (-3)R_3 + R_2 \rightarrow R_2, \quad 8R_3 + R_4 \rightarrow R_4$$

$$\sim \begin{array}{c} x_2 \\ s_2 \\ x_1 \\ P \end{array} \begin{matrix} x_1 & x_2 & x_3 & s_1 & s_2 & s_3 & P \\ \begin{bmatrix} 0 & 1 & -2 & -3 & 0 & -1 & 0 & | & 10 \\ 0 & 0 & -\frac{7}{2} & -2 & 1 & -\frac{3}{2} & 0 & | & 15 \\ 1 & 0 & \frac{5}{2} & 1 & 0 & \frac{1}{2} & 0 & | & 5 \\ \hdashline 0 & 0 & 15 & 11 & 0 & 4 & 1 & | & 90 \end{bmatrix} \end{matrix}$$

(5-4)

10. (A) The basic feasible solution is: $x_1 = 0$, $x_2 = 2$, $s_1 = 0$, $s_2 = 5$, $P = 12$. Additional pivoting is required because the last row contains a negative indicator.

(B) The basic feasible solution is: $x_1 = 0$, $x_2 = 0$, $s_1 = 0$, $s_2 = 7$, $P = 22$. There is no optimal solution because there are no positive elements above the dashed line in the pivot column, column 1.

(C) The basic feasible solution is: $x_1 = 6$, $x_2 = 0$, $s_1 = 15$, $s_2 = 0$, $P = 10$. This is the optimal solution.

(5-4)

11. The feasible region is the solution set of the given inequalities and is indicated by the shaded region in the graph at the right.

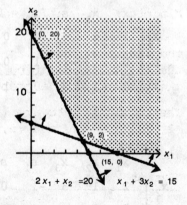

The corner points are $(0, 20)$, $(9, 2)$, and $(15, 0)$.

The value of C at each corner point is:

Corner Point	$C = 5x_1 + 2x_2$
$(0, 20)$	$C = 5(0) + 2(20) = 40$
$(9, 2)$	$C = 5(9) + 2(2) = 49$
$(15, 0)$	$C = 5(15) + 2(0) = 75$

The minimum occurs at $x_1 = 0$, $x_2 = 20$, and the minimum value is $C = 40$.

(5-2)

12. The matrices corresponding to the given problem and the dual problem are:

$$A = \begin{bmatrix} 1 & 3 & | & 15 \\ 2 & 1 & | & 20 \\ 5 & 2 & | & 1 \end{bmatrix} \quad \text{and} \quad A^T = \begin{bmatrix} 1 & 2 & | & 5 \\ 3 & 1 & | & 2 \\ 15 & 20 & | & 1 \end{bmatrix}$$

Thus, the dual problem is: Maximize $P = 15y_1 + 20y_2$
Subject to: $y_1 + 2y_2 \le 5$
$3y_1 + y_2 \le 2$
$y_1,\ y_2 \ge 0$ (5-5)

13. Introduce the slack variables x_1 and x_2 to obtain the initial system:

$$y_1 + 2y_2 + x_1 \qquad\qquad = 5$$
$$3y_1 + y_2 \qquad + x_2 \qquad = 2$$
$$-15y_1 - 20y_2 \qquad\qquad + P = 0 \qquad\qquad (5\text{-}5)$$

14. The first simplex tableau for the dual problem, Problem 12, is:

$$
\begin{array}{c}
\\
x_1 \\
x_2 \\
P
\end{array}
\begin{array}{c}
\begin{array}{cccccc}
y_1 & y_2 & x_1 & x_2 & P & \\
\end{array} \\
\left[
\begin{array}{ccccc|c}
1 & 2 & 1 & 0 & 0 & 5 \\
3 & 1 & 0 & 1 & 0 & 2 \\
\hline
-15 & -20 & 0 & 0 & 1 & 0
\end{array}
\right]
\end{array}
$$

             (5-5)

15. Using the simplex method, we have:

$$
\left[
\begin{array}{ccccc|c}
1 & 2 & 1 & 0 & 0 & 5 \\
3 & ① & 0 & 1 & 0 & 2 \\
\hline
-15 & -20 & 0 & 0 & 1 & 0
\end{array}
\right]
\begin{array}{l}
\frac{5}{2} = 2.5 \\
\\
\frac{2}{1} = 2
\end{array}
$$

$$
\begin{array}{c}
\\
x_1 \\
\sim\ y_2 \\
P
\end{array}
\begin{array}{c}
\begin{array}{cccccc}
y_1 & y_2 & x_1 & x_2 & P & \\
\end{array} \\
\left[
\begin{array}{ccccc|c}
-5 & 0 & 1 & -2 & 0 & 1 \\
3 & 1 & 0 & 1 & 0 & 2 \\
\hline
45 & 0 & 0 & 20 & 1 & 40
\end{array}
\right]
\end{array}
$$

$(-2)R_2 + R_1 \to R_1,\ 20R_2 + R_3 \to R_3$ Optimal solution: max $P = 40$ at
                                              $y_1 = 0$ and $y_2 = 2$.   (5-4)

16. Minimum $C = 40$ at $x_1 = 0$ and $x_2 = 20$.   (5-5)

17. The feasible region is the solution set of the given inequalities and is indicated by the shading in the graph at the right.

The corner points are $(0, 0)$, $(0, 6)$, $(2, 5)$, $(3, 4)$, and $(5, 0)$.

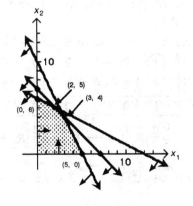

The value of P at each corner point is:

Corner Point	$P = 3x_1 + 4x_2$
$(0, 0)$	$P = 3(0) + 4(0) = 0$
$(0, 6)$	$P = 3(0) + 4(6) = 24$
$(2, 5)$	$P = 3(2) + 4(5) = 26$
$(3, 4)$	$P = 3(3) + 4(4) = 25$
$(5, 0)$	$P = 3(5) + 4(0) = 15$

Thus, the maximum occurs at $x_1 = 2$, $x_2 = 5$, and the maximum value is $P = 26$.

             (5-2)

18. We simplify the inequalities and introduce the slack variables s_1, s_2, and s_3 to obtain the equivalent form:

$$
\begin{aligned}
x_1 + 2x_2 + s_1 &= 12 \\
x_1 + x_2 \quad\; + s_2 &= 7 \\
2x_1 + x_2 \qquad\quad + s_3 &= 10 \\
-3x_1 - 4x_2 \qquad\qquad\quad\; + P &= 0
\end{aligned}
$$

The simplex tableau for this problem is:

$$
\begin{array}{c}
\text{Exit } s_1 \\
s_2 \\
s_3 \\
P
\end{array}
\left[
\begin{array}{cccccc|c}
x_1 & x_2 & s_1 & s_2 & s_3 & P & \\
1 & ② & 1 & 0 & 0 & 0 & 12 \\
1 & 1 & 0 & 1 & 0 & 0 & 7 \\
2 & 1 & 0 & 0 & 1 & 0 & 10 \\
\hline
-3 & -4 & 0 & 0 & 0 & 1 & 0
\end{array}
\right]
\begin{array}{l}
\frac{12}{2} = 6 \\
\frac{7}{1} = 7 \\
\frac{10}{1} = 10 \\
\end{array}
$$

(with **Enter** above x_2)

$$\tfrac{1}{2}R_1 \to R_1$$

$$
\sim
\left[
\begin{array}{cccccc|c}
\frac{1}{2} & ① & \frac{1}{2} & 0 & 0 & 0 & 6 \\
1 & 1 & 0 & 1 & 0 & 0 & 7 \\
2 & 1 & 0 & 0 & 1 & 0 & 10 \\
\hline
-3 & -4 & 0 & 0 & 0 & 1 & 0
\end{array}
\right]
\sim
\left[
\begin{array}{cccccc|c}
\frac{1}{2} & 1 & \frac{1}{2} & 0 & 0 & 0 & 6 \\
② & 0 & -\frac{1}{2} & 1 & 0 & 0 & 1 \\
\frac{3}{2} & 0 & -\frac{1}{2} & 0 & 1 & 0 & 4 \\
\hline
-1 & 0 & 2 & 0 & 0 & 1 & 24
\end{array}
\right]
\begin{array}{l}
\frac{6}{1/2} = 12 \\
\frac{1}{1/2} = 2 \\
\frac{4}{3/2} \approx 2.67
\end{array}
$$

(with circled $\tfrac{1}{2}$ in second row first column)

$$(-1)R_1 + R_2 \to R_2,\; (-1)R_1 + R_3 \to R_3, \qquad 2R_2 \to R_2$$

$$\text{and } 4R_1 + R_4 \to R_4$$

$$
\sim
\left[
\begin{array}{cccccc|c}
\frac{1}{2} & 1 & \frac{1}{2} & 0 & 0 & 0 & 6 \\
① & 0 & -1 & 2 & 0 & 0 & 2 \\
\frac{3}{2} & 0 & -\frac{1}{2} & 0 & 1 & 0 & 4 \\
\hline
-1 & 0 & 2 & 0 & 0 & 1 & 24
\end{array}
\right]
\sim
\begin{array}{c}
x_2 \\
x_1 \\
s_3 \\
P
\end{array}
\left[
\begin{array}{cccccc|c}
x_1 & x_2 & s_1 & s_2 & s_3 & P & \\
0 & 1 & 1 & -1 & 0 & 0 & 5 \\
1 & 0 & -1 & 2 & 0 & 0 & 2 \\
0 & 0 & 1 & -3 & 1 & 0 & 1 \\
\hline
0 & 0 & 1 & 2 & 0 & 1 & 26
\end{array}
\right]
$$

$$\left(-\tfrac{1}{2}\right)R_2 + R_1 \to R_1, \left(-\tfrac{3}{2}\right)R_2 + R_3 \to R_3,$$

$$\text{and } R_2 + R_4 \to R_4$$

Optimal solution: max $P = 26$ at $x_1 = 2$, $x_2 = 5$.

(5-4)

19. The feasible region is the solution set of the given inequalities and is indicated by the shaded region in the graph at the right.

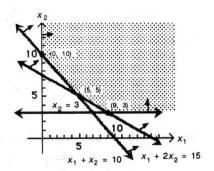

The corner points are $(0, 10)$, $(5, 5)$, and $(9, 3)$.

The value of C at each corner point is:

Corner Point	$C = 3x_1 + 8x_2$
$(0, 10)$	$C = 3(0) + 8(10) = 80$
$(5, 5)$	$C = 3(5) + 8(5) = 55$
$(9, 3)$	$C = 3(9) + 8(3) = 51$

Thus, the minimum occurs at $x_1 = 9$, $x_2 = 3$, and the minimum value is $C = 51$.

(5-2)

20. The matrices corresponding to the given problem and the dual problem are:

$$A = \begin{bmatrix} 1 & 1 & | & 10 \\ 1 & 2 & | & 15 \\ 0 & 1 & | & 3 \\ 3 & 8 & | & 1 \end{bmatrix} \quad \text{and} \quad A^T = \begin{bmatrix} 1 & 1 & 0 & | & 3 \\ 1 & 2 & 1 & | & 8 \\ 10 & 15 & 3 & | & 1 \end{bmatrix}$$

Thus, the dual problem is: Maximize $P = 10y_1 + 15y_2 + 3y_3$

$$\text{Subject to: } y_1 + y_2 \qquad \le 3$$
$$y_1 + 2y_2 + y_3 \le 8$$
$$y_1, \; y_2, \; y_3 \ge 0 \qquad \text{(5-5)}$$

21. Introduce the slack variables x_1 and x_2 to obtain the initial system:

$$y_1 + y_2 \qquad + x_1 \qquad = 3$$
$$y_1 + 2y_2 + y_3 \qquad + x_2 \qquad = 8$$
$$-10y_1 - 15y_2 - 3y_3 \qquad + P = 0$$

The simplex tableau for this problem is:

$$
\begin{array}{c}
\quad\; y_1 \quad y_2 \quad y_3 \quad x_1 \quad x_2 \quad\; P \\
\begin{array}{c} x_1 \\ x_2 \\ P \end{array}
\left[
\begin{array}{cccccc|c}
1 & ① & 0 & 1 & 0 & 0 & 3 \\
1 & 2 & 1 & 0 & 1 & 0 & 8 \\
\hline
-10 & -15 & -3 & 0 & 0 & 1 & 0
\end{array}
\right]
\begin{array}{l} \frac{3}{1} = 3 \\ \frac{8}{2} = 4 \end{array} \sim
\left[
\begin{array}{cccccc|c}
1 & 1 & 0 & 1 & 0 & 0 & 3 \\
-1 & 0 & ① & -2 & 1 & 0 & 2 \\
\hline
5 & 0 & -3 & 15 & 0 & 1 & 45
\end{array}
\right]
\end{array}
$$

$(-2)R_1 + R_2 \to R_2$ and $15R_1 + R_3 \to R_3$ $\qquad$ $3R_2 + R_3 \to R_3$

$$\begin{array}{c} \\ y_2 \\ \sim y_3 \\ \\ \end{array}\begin{array}{cccccc} y_1 & y_2 & y_3 & x_1 & x_2 & P \\ \end{array}\left[\begin{array}{cccccc|c} 1 & 1 & 0 & 1 & 0 & 0 & 3 \\ -1 & 0 & 1 & -2 & 1 & 0 & 2 \\ \hline 2 & 0 & 0 & 9 & 3 & 1 & 51 \end{array}\right]$$

Optimal solution: min $C = 51$ at $x_1 = 9$, $x_2 = 3$.

(5-5)

22. Introduce slack variables s_1 and s_2 to obtain the equivalent form:

$$\begin{aligned} x_1 - x_2 - 2x_3 + s_1 \quad\quad\quad &= 3 \\ 2x_1 + 2x_2 - 5x_3 \quad\quad + s_2 \quad &= 10 \\ -5x_1 - 3x_2 + 3x_3 \quad\quad\quad + P &= 0 \end{aligned}$$

The simplex tableau for this problem is:

Enter

$$\begin{array}{c} \\ \text{Exit } s_1 \\ s_2 \\ P \\ \end{array}\begin{array}{cccccc} x_1 & x_2 & x_3 & s_1 & s_2 & P \\ \end{array}\left[\begin{array}{cccccc|c} ① & -1 & -2 & 1 & 0 & 0 & 3 \\ 2 & 2 & -5 & 0 & 1 & 0 & 10 \\ \hline -5 & -3 & 3 & 0 & 0 & 1 & 0 \end{array}\right]\begin{array}{l} \frac{3}{1}=3 \\ \frac{10}{2}=5 \\ \\ \end{array}\sim\left[\begin{array}{cccccc|c} 1 & -1 & -2 & 1 & 0 & 0 & 3 \\ 0 & ④ & -1 & -2 & 1 & 0 & 4 \\ \hline 0 & -8 & -7 & 5 & 0 & 1 & 15 \end{array}\right]$$

$$(-2)R_1 + R_2 \to R_2 \text{ and } 5R_1 + R_3 \to R_3 \qquad\qquad \tfrac{1}{4}R_2 \to R_2$$

$$\begin{array}{cccccc} & & & & & & \\ x_1 & x_2 & x_3 & s_1 & s_2 & P \\ \end{array}$$

$$\sim\left[\begin{array}{cccccc|c} 1 & -1 & -2 & 1 & 0 & 0 & 3 \\ 0 & ① & -\frac{1}{4} & -\frac{1}{2} & \frac{1}{4} & 0 & 1 \\ \hline 0 & -8 & -7 & 5 & 0 & 1 & 15 \end{array}\right]\sim\left[\begin{array}{cccccc|c} 1 & 0 & -\frac{9}{4} & \frac{1}{2} & \frac{1}{4} & 0 & 4 \\ 0 & 1 & -\frac{1}{4} & -\frac{1}{2} & \frac{1}{4} & 0 & 1 \\ \hline 0 & 0 & -9 & 1 & 2 & 1 & 23 \end{array}\right]$$

$$R_2 + R_1 \to R_1 \text{ and } 8R_2 + R_3 \to R_3$$

No optimal solution exists; the elements in the pivot column (the x_3 column) above the dashed line are negative.

(5-4)

23. Introduce slack variables s_1 and s_2 to obtain the equivalent form:

$$\begin{aligned} x_1 - x_2 - 2x_3 + s_1 \quad\quad\quad &= 3 \\ x_1 + x_2 \quad\quad\quad\quad + s_2 \quad &= 5 \\ -5x_1 - 3x_2 + 3x_3 \quad\quad\quad + P &= 0 \end{aligned}$$

The simplex tableau for this problem is:

Enter

$$\begin{array}{c} \\ \text{Exit } s_1 \\ s_2 \\ P \\ \end{array}\begin{array}{cccccc} x_1 & x_2 & x_3 & s_1 & s_2 & P \\ \end{array}\left[\begin{array}{cccccc|c} ① & -1 & -2 & 1 & 0 & 0 & 3 \\ 1 & 1 & 0 & 0 & 1 & 0 & 5 \\ \hline -5 & -3 & 3 & 0 & 0 & 1 & 0 \end{array}\right]\begin{array}{l} \frac{3}{1}=3 \\ \frac{5}{1}=5 \\ \\ \end{array}\sim\left[\begin{array}{cccccc|c} 1 & -1 & -2 & 1 & 0 & 0 & 3 \\ 0 & ② & 2 & -1 & 1 & 0 & 2 \\ \hline 0 & -8 & -7 & 5 & 0 & 1 & 15 \end{array}\right]$$

$$(-1)R_1 + R_2 \to R_2 \text{ and } 5R_1 + R_3 \to R_3 \qquad\qquad \tfrac{1}{2}R_2 \to R_2$$

$$\sim \begin{bmatrix} 1 & -1 & -2 & 1 & 0 & 0 & | & 3 \\ 0 & \circled{1} & 1 & -\frac{1}{2} & \frac{1}{2} & 0 & | & 1 \\ \hline 0 & -8 & -7 & 5 & 0 & 1 & | & 15 \end{bmatrix}$$

$$\begin{array}{c} \\ x_1 \\ \sim x_2 \\ P \end{array} \begin{bmatrix} x_1 & x_2 & x_3 & s_1 & s_2 & P \\ 1 & 0 & -1 & \frac{1}{2} & \frac{1}{2} & 0 & | & 4 \\ 0 & 1 & 1 & -\frac{1}{2} & \frac{1}{2} & 0 & | & 1 \\ \hline 0 & 0 & 1 & 1 & 4 & 1 & | & 23 \end{bmatrix}$$

$R_2 + R_1 \rightarrow R_1$ and $8R_2 + R_3 \rightarrow R_3$ Optimal solution: max $P = 23$ at
$$x_1 = 4, \; x_2 = 1, \; x_3 = 0. \qquad (5\text{-}4)$$

24. (A) We introduce a surplus variable s_1 and an artificial variable a_1 to convert the first inequality ($\geq$) into an equation; we introduce a slack variable s_2 to convert the second inequality ($\leq$) into an equation.

The modified problem is: Maximize $P = x_1 + 3x_2 - Ma_1$

$$\begin{array}{lrl} \text{Subject to:} & x_1 + \; x_2 - s_1 + a_1 & = 6 \\ & x_1 + 2x_2 \qquad\qquad\quad + s_2 & = 8 \\ & x_1, \; x_2, \; s_1, \; s_2, \; a_1 & \geq 0 \end{array}$$

(B) The preliminary simplex tableau is:

$$\begin{array}{c} x_1 & x_2 & s_1 & a_1 & s_2 & P \\ \begin{bmatrix} 1 & 1 & -1 & 1 & 0 & 0 & | & 6 \\ 1 & 2 & 0 & 0 & 1 & 0 & | & 8 \\ \hline -1 & -3 & 0 & M & 0 & 1 & | & 0 \end{bmatrix} \end{array}$$

Now

$$\begin{array}{c} \\ a_1 \\ s_2 \\ P \end{array} \begin{array}{c} x_1 & x_2 & s_1 & a_1 & s_2 & P \\ \begin{bmatrix} 1 & 1 & -1 & 1 & 0 & 0 & | & 6 \\ 1 & 2 & 0 & 0 & 1 & 0 & | & 8 \\ \hline -1 & -3 & 0 & M & 0 & 1 & | & 0 \end{bmatrix} \end{array} \sim \begin{bmatrix} 1 & 1 & -1 & 1 & 0 & 0 & | & 6 \\ 1 & 2 & 0 & 0 & 1 & 0 & | & 8 \\ \hline -M-1 & -M-3 & M & 0 & 0 & 1 & | & -6M \end{bmatrix}$$

$$(-M)R_1 + R_3 \rightarrow R_3$$

Thus, the initial simplex tableau is:

$$\begin{array}{c} \\ a_1 \\ \sim s_1 \\ P \end{array} \begin{array}{c} x_1 & \qquad x_2 & \quad s_1 & a_1 & s_2 & P \\ \begin{bmatrix} 1 & 1 & -1 & 1 & 0 & 0 & | & 6 \\ 1 & 2 & 0 & 0 & 1 & 0 & | & 8 \\ \hline -M-1 & -M-3 & M & 0 & 0 & 1 & | & -6M \end{bmatrix} \end{array}$$

(C)

$$\begin{array}{c} \\ a_1 \\ \sim s_2 \\ P \end{array} \begin{array}{c} x_1 & \qquad x_2 & \quad s_1 & a_1 & s_2 & P \\ \begin{bmatrix} 1 & 1 & -1 & 1 & 0 & 0 & | & 6 \\ 1 & \circled{2} & 0 & 0 & 1 & 0 & | & 8 \\ \hline -M-1 & -M-3 & M & 0 & 0 & 1 & | & -6M \end{bmatrix} \begin{array}{l} \frac{6}{1} = 6 \\ \frac{8}{2} = 4 \end{array} \end{array}$$

$$\tfrac{1}{2}R_2 \rightarrow R_2$$

$$\sim \begin{bmatrix} 1 & 1 & -1 & 1 & 0 & 0 & | & 6 \\ \frac{1}{2} & ① & 0 & 0 & \frac{1}{2} & 0 & | & 4 \\ \hline -M-1 & -M-3 & M & 0 & 0 & 1 & | & -6M \end{bmatrix}$$

$$(-1)R_2 + R_1 \rightarrow R_1, \quad (M+3)R_2 + R_3 \rightarrow R_3$$

$$\sim \begin{bmatrix} ⓵ & 0 & -1 & 1 & -\frac{1}{2} & 0 & | & 2 \\ \frac{1}{2} & 1 & 0 & 0 & \frac{1}{2} & 0 & | & 4 \\ \hline -\frac{1}{2}M+\frac{1}{2} & 0 & M & 0 & \frac{1}{2}M+\frac{3}{2} & 1 & | & -2M+12 \end{bmatrix} \begin{array}{l} \frac{2}{1/2} = 4 \\ \frac{4}{1/2} = 8 \end{array}$$

$$2R_1 \rightarrow R_1$$

$$\sim \begin{bmatrix} ① & 0 & -2 & 2 & -1 & 0 & | & 4 \\ \frac{1}{2} & 1 & 0 & 0 & \frac{1}{2} & 0 & | & 4 \\ \hline -\frac{1}{2}M+\frac{1}{2} & 0 & M & 0 & \frac{1}{2}M+\frac{3}{2} & 1 & | & -2M+12 \end{bmatrix}$$

$$\left(-\frac{1}{2}\right)R_1 + R_2 \rightarrow R_2, \quad \left(\frac{1}{2}M - \frac{1}{2}\right)R_1 + R_3 \rightarrow R_3$$

$$\sim \begin{array}{c} \\ x_1 \\ x_2 \\ P \end{array} \begin{array}{c} \begin{matrix} x_1 & x_2 & s_1 & a_1 & s_2 & P \end{matrix} \\ \begin{bmatrix} 1 & 0 & -2 & 2 & -1 & 0 & | & 4 \\ 0 & 1 & 1 & -1 & 1 & 0 & | & 2 \\ \hline 0 & 0 & 1 & M-1 & 2 & 1 & | & 10 \end{bmatrix} \end{array}$$

The optimal solution to the modified problem is: Maximum $P = 10$ at $x_1 = 4$, $x_2 = 2$, $s_1 = 0$, $a_1 = 0$, $s_2 = 0$.

(D) Since $a_1 = 0$, the solution to the original problem is:
Maximum $P = 10$ at $x_1 = 4$, $x_2 = 2$. (5-4)

25. (A) We introduce a surplus variable s_1 and an artificial variable a_1 to convert the first inequality ($\geq$) into an equation; we introduce a slack variable s_2 to convert the second inequality ($\leq$) into an equation.

The modified problem is: Maximize $P = x_1 + x_2 - Ma_1$
Subject to: $x_1 + x_2 - s_1 + a_1 = 5$
$x_1 + 2x_2 + s_2 = 4$
$x_1, x_2, s_1, s_2, a_1 \geq 0$

(B) The preliminary simplex tableau is:

$$
\begin{array}{cccccc}
\ x_1 & x_2 & s_1 & a_1 & s_2 & P \\
\end{array}
$$

$$
\left[
\begin{array}{cccccc|c}
1 & 1 & -1 & 1 & 0 & 0 & 5 \\
1 & 2 & 0 & 0 & 1 & 0 & 4 \\
\hline
-1 & -1 & 0 & M & 0 & 1 & 0
\end{array}
\right]
$$

Now

$$
\left[
\begin{array}{cccccc|c}
1 & 1 & -1 & 1 & 0 & 0 & 5 \\
1 & 2 & 0 & 0 & 1 & 0 & 4 \\
\hline
-1 & -1 & 0 & M & 0 & 1 & 0
\end{array}
\right]
\sim
\begin{array}{c}
a_1 \\
s_2 \\
\\
P
\end{array}
\left[
\begin{array}{cccccc|c}
1 & 1 & -1 & 1 & 0 & 0 & 5 \\
1 & 2 & 0 & 0 & 1 & 0 & 4 \\
\hline
-M-1 & -M-1 & M & 0 & 0 & 1 & -5M
\end{array}
\right]
$$

$$(-M)R_1 + R_3 \rightarrow R_3 \qquad\qquad \text{Initial simplex tableau}$$

(C)

$$
\left[
\begin{array}{cccccc|c}
1 & 1 & -1 & 1 & 0 & 0 & 5 \\
①& 2 & 0 & 0 & 1 & 0 & 4 \\
\hline
-M-1 & -M-1 & M & 0 & 0 & 1 & -5M
\end{array}
\right]
$$

$$(-1)R_2 + R_1 \rightarrow R_1, \quad (M+1)R_2 + R_3 \rightarrow R_3$$

$$
\begin{array}{cccccc}
\ x_1 & x_2 & s_1 & a_1 & s_2 & P \\
\end{array}
$$

$$
\begin{array}{c}
a_1 \\
\sim\ x_1 \\
\\
P
\end{array}
\left[
\begin{array}{cccccc|c}
0 & -1 & -1 & 1 & -1 & 0 & 1 \\
1 & 2 & 0 & 0 & 1 & 0 & 4 \\
\hline
0 & M+1 & M & 0 & M+1 & 1 & -M+4
\end{array}
\right]
$$

The optimal solution to the modified problem is: $x_1 = 4$, $x_2 = 0$,
$s_1 = 0$, $a_1 = 1$, $s_2 = 0$, $P = -M + 4$. (5-6)

(D) Since $a_1 \neq 0$, the original problem does not have a solution.

26. Multiply the second inequality by -1 to obtain a positive number on the right-hand side. This yields the problem:

Maximize $P = 2x_1 + 3x_2 + x_3$

Subject to:
$$\begin{aligned}
x_1 - 3x_2 + x_3 &\le 7 \\
x_1 + x_2 - 2x_3 &\ge 2 \quad (\underline{\text{Note}}: \text{Direction of inequality is reversed.}) \\
3x_1 + 2x_2 - x_3 &= 4 \\
x_1, \ x_2, \ x_3 &\ge 0
\end{aligned}$$

Now, introduce a slack variable s_1 to convert the first inequality ($\le$) into an equation; introduce a surplus variable s_2 and an artificial variable a_1 to convert the second inequality (≥ 0) into an equation; introduce an artificial variable a_2 into the equation.

The modified problem is:

Maximize $P = 2x_1 + 3x_2 + x_3 - Ma_1 - Ma_2$

Subject to:
$$\begin{aligned}
x_1 - 3x_2 + x_3 + s_1 \qquad\qquad\qquad &= 7 \\
x_1 + x_2 - 2x_3 \qquad - s_2 + a_1 \qquad &= 2 \\
3x_1 + 2x_2 - x_3 \qquad\qquad\quad + a_2 &= 4 \\
x_1, \ x_2, \ x_3, \ s_1, \ s_2, \ a_1, \ a_2 &\ge 0
\end{aligned}$$
(5-6)

27. The geometric method solves maximization and minimization problems involving two decision variables. If the feasible region is bounded, there are no restrictions on the coefficients or constants. If the feasible region is unbounded and the coefficients of the objective function are positive, then a minimization problem has a solution, but a maximization does not. (5-2)

28. The basic simplex method with slack variables solves standard maximization problems involving $\leq$ constraints with nonnegative constants on the right side. (5-4)

29. The dual method solves minimization problems with positive coefficients in the objective function. (5-5)

30. The big M method solves any linear programming problem. (5-6)

31. Introduce slack variables s_1, s_2, s_3, and s_4 to obtain:
Maximize $P = 2x_1 + 3x_2$
Subject to:
$$
\begin{aligned}
x_1 + 2x_2 + s_1 &= 22 \\
3x_1 + x_2 + s_2 &= 26 \\
x_1 + s_3 &= 8 \\
x_2 + s_4 &= 10 \\
x_1,\ x_2,\ s_1,\ s_2,\ s_3,\ s_4 &\geq 0
\end{aligned}
$$

This system can be written in the initial form:
$$
\begin{aligned}
x_1 + 2x_2 + s_1 &= 22 \\
3x_1 + x_2 + s_2 &= 26 \\
x_1 + s_3 &= 8 \\
x_2 + s_4 &= 10 \\
-2x_1 - 3x_2 + P &= 0
\end{aligned}
$$

The simplex tableau for this problem is:

	x_1	x_2	s_1	s_2	s_3	s_4	P		
s_1	1	2	1	0	0	0	0	22	$\frac{22}{2} = 11$
s_2	3	1	0	1	0	0	0	26	$\frac{26}{1} = 26$
s_3	1	0	0	0	1	0	0	8	
s_4	0	①	0	0	0	1	0	10	$\frac{10}{1} = 10$
P	-2	-3	0	0	0	0	1	0	

$(-2)R_4 + R_1 \rightarrow R_1$, $\quad (-1)R_4 + R_2 \rightarrow R_2$,
$\quad 3R_4 + R_5 \rightarrow R_5$

Basic Solution	Corner Point
x_1 x_2 s_1 s_2 s_3 s_4	$(0, 0)$
0 0 22 26 8 10	

$$\begin{array}{c} \\ s_1 \\ s_2 \\ \sim \quad s_3 \\ x_2 \\ {} \end{array} \begin{array}{cccccccc} x_1 & x_2 & s_1 & s_2 & s_3 & s_4 & P & \\ \left[\begin{array}{ccccccc|c} \textcircled{1} & 0 & 1 & 0 & 0 & -2 & 0 & 2 \\ 3 & 0 & 0 & 1 & 0 & -1 & 0 & 16 \\ 1 & 0 & 0 & 0 & 1 & 0 & 0 & 8 \\ 0 & 1 & 0 & 0 & 0 & 1 & 0 & 10 \\ \hdashline -2 & 0 & 0 & 0 & 0 & 3 & 1 & 30 \end{array}\right] \end{array} \begin{array}{l} \frac{2}{1} = 2 \\ \frac{16}{3} = 5.33 \\ \frac{8}{1} = 8 \\ {} \\ {} \end{array}$$

$(-3)R_1 + R_2 \to R_2$, $\quad (-1)R_1 + R_3 \to R_3$,

$\quad\quad 2R_1 + R_5 \to R_5$

	Basic Solution						Corner Point
x_1	x_2	s_1	s_2	s_3	s_4		(0, 10)
0	10	2	16	8	0		

$$\begin{array}{c} \\ x_1 \\ s_2 \\ \sim \quad s_3 \\ x_2 \\ P \end{array} \begin{array}{cccccccc} x_1 & x_2 & s_1 & s_2 & s_3 & s_4 & P & \\ \left[\begin{array}{ccccccc|c} 1 & 0 & 1 & 0 & 0 & -2 & 0 & 2 \\ 0 & 0 & -3 & 1 & 0 & \textcircled{5} & 0 & 10 \\ 0 & 0 & -1 & 0 & 1 & 2 & 0 & 6 \\ 0 & 1 & 0 & 0 & 0 & 1 & 0 & 10 \\ \hdashline 0 & 0 & 2 & 0 & 0 & -1 & 1 & 34 \end{array}\right] \end{array} \begin{array}{l} {} \\ \frac{10}{5} = 2 \\ \frac{6}{2} = 3 \\ \frac{10}{1} = 10 \\ {} \end{array}$$

$\quad\quad \frac{1}{5}R_2 \to R_2$

	Basic Solution						Corner Point
x_1	x_2	s_1	s_2	s_3	s_4		(2, 10)
2	10	0	10	6	0		

$$\sim \left[\begin{array}{ccccccc|c} 1 & 0 & 1 & 0 & 0 & -2 & 0 & 2 \\ 0 & 0 & -\frac{3}{5} & \frac{1}{5} & 0 & \textcircled{1} & 0 & 2 \\ 0 & 0 & -1 & 0 & 1 & 2 & 0 & 6 \\ 0 & 1 & 0 & 0 & 0 & 1 & 0 & 10 \\ \hdashline 0 & 0 & 2 & 0 & 0 & -1 & 1 & 34 \end{array}\right]$$

$2R_2 + R_1 \to R_1$, $\quad (-2)R_2 + R_3 \to R_3$, $\quad (-1)R_2 + R_4 \to R_4$, $\quad R_2 + R_5 \to R_5$

$$\begin{array}{c} \\ x_1 \\ s_4 \\ \sim \; s_3 \\ x_2 \\ P \end{array} \begin{array}{cccccccc} x_1 & x_2 & s_1 & s_2 & s_3 & s_4 & P & \\ \left[\begin{array}{ccccccc|c} 1 & 0 & -\frac{1}{5} & \frac{2}{5} & 0 & 0 & 0 & 6 \\ 0 & 0 & -\frac{3}{5} & \frac{1}{5} & 0 & 1 & 0 & 2 \\ 0 & 0 & \frac{1}{5} & -\frac{2}{5} & 1 & 0 & 0 & 2 \\ 0 & 1 & \frac{3}{5} & -\frac{1}{5} & 0 & 0 & 0 & 8 \\ \hdashline 0 & 0 & \frac{7}{5} & \frac{1}{5} & 0 & 0 & 1 & 36 \end{array}\right] \end{array}$$

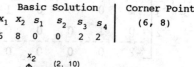

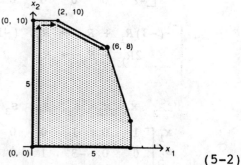

Basic Solution						Corner Point
x_1	x_2	s_1	s_2	s_3	s_4	(6, 8)
6	8	0	0	2	2	

Optimal solution:
    max $P = 36$ at $x_1 = 6$, $x_2 = 8$.

The graph of the feasible region and the path to the optimal solution is shown at the right.

(5-2)

32. Multiply the first constraint inequality by -1 to transform it into a $\geq$ inequality. Now the problem now is: Minimize $C = 3x_1 + 2x_2$

$$\begin{aligned} \text{Subject to: } -2x_1 - x_2 &\geq -20 \\ 2x_1 + x_2 &\geq 9 \\ x_1 + x_2 &\geq 6 \\ x_1,\; x_2 &\geq 0 \end{aligned}$$

The matrices corresponding to this problem and its dual are, respectively:

$$A = \left[\begin{array}{cc|c} -2 & -1 & -20 \\ 2 & 1 & 9 \\ 1 & 1 & 6 \\ \hline 3 & 2 & 1 \end{array}\right] \quad \text{and} \quad A^T = \left[\begin{array}{ccc|c} -2 & 2 & 1 & 3 \\ -1 & 1 & 1 & 2 \\ \hline -20 & 9 & 6 & 1 \end{array}\right]$$

Thus, the dual problem is: Maximize $P = -20y_1 + 9y_2 + 6y_3$

$$\begin{aligned} \text{Subject to: } -2y_1 + 2y_2 + y_3 &\leq 3 \\ -y_1 + y_2 + y_3 &\leq 2 \\ y_1,\; y_2,\; y_3 &\geq 0 \end{aligned}$$

We introduce slack variables x_1 and x_2 to obtain the initial system for the dual problem:

$$\begin{aligned} -2y_1 + 2y_2 + y_3 + x_1 \qquad\qquad &= 3 \\ -y_1 + y_2 + y_3 \qquad + x_2 \qquad &= 2 \\ 20y_1 - 9y_2 - 6y_3 \qquad\qquad + P &= 0 \end{aligned}$$

The simplex tableau for this problem is:

$$
\begin{array}{c}
\quad\quad y_1 \quad y_2 \quad y_3 \quad x_1 \quad x_2 \quad P \\
\begin{array}{c} x_1 \\ x_2 \\ \hline P \end{array}
\left[
\begin{array}{cccccc|c}
-2 & ② & 1 & 1 & 0 & 0 & 3 \\
-1 & 1 & 1 & 0 & 1 & 0 & 2 \\
\hline
20 & -9 & -6 & 0 & 0 & 1 & 0
\end{array}
\right]
\begin{array}{l} \frac{3}{2} = 1.5 \\ \frac{2}{1} = 2 \end{array}
\end{array}
$$

$$\sim
\left[
\begin{array}{cccccc|c}
-1 & ① & \frac{1}{2} & \frac{1}{2} & 0 & 0 & \frac{3}{2} \\
-1 & 1 & 1 & 0 & 1 & 0 & 2 \\
\hline
20 & -9 & -6 & 0 & 0 & 1 & 0
\end{array}
\right]
$$

$$\frac{1}{2} R_1 \rightarrow R_1 \qquad\qquad\qquad (-1)R_1 + R_2 \rightarrow R_2, \quad 9R_1 + R_3 \rightarrow R_3$$

$$\sim
\left[
\begin{array}{cccccc|c}
-1 & 1 & \frac{1}{2} & \frac{1}{2} & 0 & 0 & \frac{3}{2} \\
0 & 0 & ⓵⁄₂ & -\frac{1}{2} & 1 & 0 & \frac{1}{2} \\
\hline
11 & 0 & -\frac{3}{2} & \frac{9}{2} & 0 & 1 & \frac{27}{2}
\end{array}
\right]
\begin{array}{l} \frac{3/2}{1/2} = 3 \\ \frac{1/2}{1/2} = 1 \end{array}
$$

$$\sim
\left[
\begin{array}{cccccc|c}
-1 & 1 & \frac{1}{2} & \frac{1}{2} & 0 & 0 & \frac{3}{2} \\
0 & 0 & ① & -1 & 2 & 0 & 1 \\
\hline
11 & 0 & -\frac{3}{2} & \frac{9}{2} & 0 & 1 & \frac{27}{2}
\end{array}
\right]
$$

$$2R_2 \rightarrow R_2 \qquad\qquad\qquad \left(-\frac{1}{2}\right)R_2 + R_1 \rightarrow R_1, \quad \frac{3}{2}R_2 + R_3 \rightarrow R_3$$

$$
\begin{array}{c}
\quad\quad y_1 \quad y_2 \quad y_3 \quad x_1 \quad x_2 \quad P \\
\begin{array}{c} y_2 \\ y_3 \\ \hline P \end{array}
\sim
\left[
\begin{array}{cccccc|c}
-1 & 1 & 0 & 1 & -1 & 0 & 1 \\
0 & 0 & 1 & -1 & 2 & 0 & 1 \\
\hline
11 & 0 & 0 & 3 & 3 & 1 & 15
\end{array}
\right]
\end{array}
$$

Optimal solution: min $C = 15$ at $x_1 = 3$ and $x_2 = 3$. \hfill (5-5)

33. First convert the problem to a maximization problem by seeking the maximum of $P = -C = -3x_1 - 2x_2$. Next, introduce a slack variable s_1 into the first inequality to convert it into an equation; introduce surplus variables s_2 and s_3 and artificial variables a_1 and a_2 into the second and third inequalities to convert them into equations.

The modified problem is: Maximize $P = -3x_1 - 2x_2 - Ma_1 - Ma_2$

$$
\begin{aligned}
\text{Subject to: } 2x_1 + x_2 + s_1 \quad\quad\quad\quad\quad &= 20 \\
2x_1 + x_2 \quad\quad - s_2 \quad\quad + a_1 \quad\quad &= 9 \\
x_1 + x_2 \quad\quad\quad\quad - s_3 \quad + a_2 &= 6 \\
x_1, \, x_2, \, s_1, \, s_2, \, s_3, \, a_1, \, a_2 \geq \, &0
\end{aligned}
$$

The preliminary simplex tableau is:

$$
\begin{array}{c}
x_1 \quad x_2 \quad s_1 \quad s_2 \quad a_1 \quad s_3 \quad a_2 \quad P \\
\left[
\begin{array}{cccccccc|c}
2 & 1 & 1 & 0 & 0 & 0 & 0 & 0 & 20 \\
2 & 1 & 0 & -1 & 1 & 0 & 0 & 0 & 9 \\
1 & 1 & 0 & 0 & 0 & -1 & 1 & 0 & 6 \\
\hline
3 & 2 & 0 & 0 & M & 0 & M & 1 & 0
\end{array}
\right]
\end{array}
$$

$$(-M)R_2 + R_4 \rightarrow R_4$$

$$\sim \begin{bmatrix} 2 & 1 & 1 & 0 & 0 & 0 & 0 & 0 & \Big| & 20 \\ 2 & 1 & 0 & -1 & 1 & 0 & 0 & 0 & \Big| & 9 \\ 1 & 1 & 0 & 0 & 0 & -1 & 1 & 0 & \Big| & 6 \\ \hline -2M+3 & -M+2 & 0 & M & 0 & 0 & M & 1 & \Big| & -9M \end{bmatrix}$$

$$(-M)R_3 + R_4 \to R_4$$

$$\sim \begin{array}{c} s_1 \\ a_1 \\ a_2 \\ P \end{array} \begin{bmatrix} x_1 & x_2 & s_1 & s_2 & a_1 & s_3 & a_2 & P & & \\ 2 & 1 & 1 & 0 & 0 & 0 & 0 & 0 & \Big| & 20 \\ \boxed{2} & 1 & 0 & -1 & 1 & 0 & 0 & 0 & \Big| & 9 \\ 1 & 1 & 0 & 0 & 0 & -1 & 1 & 0 & \Big| & 6 \\ \hline -3M+3 & -2M+2 & 0 & M & 0 & M & 0 & 1 & \Big| & -15M \end{bmatrix} \begin{array}{l} \frac{20}{2}=10 \\ \frac{9}{2}=4.5 \\ \frac{6}{1}=6 \end{array}$$

$$\tfrac{1}{2}R_2 \to R_2$$

$$\sim \begin{bmatrix} 2 & 1 & 1 & 0 & 0 & 0 & 0 & 0 & \Big| & 20 \\ \boxed{1} & \frac{1}{2} & 0 & -\frac{1}{2} & \frac{1}{2} & 0 & 0 & 0 & \Big| & \frac{9}{2} \\ 1 & 1 & 0 & 0 & 0 & -1 & 1 & 0 & \Big| & 6 \\ \hline -3M+3 & -2M+2 & 0 & M & 0 & M & 0 & 1 & \Big| & -15M \end{bmatrix}$$

$$(-2)R_2 + R_1 \to R_1, \quad (-1)R_2 + R_3 \to R_3, \quad (3M-3)R_2 + R_4 \to R_4$$

$$\sim \begin{bmatrix} 0 & 0 & 1 & 1 & -1 & 0 & 0 & 0 & \Big| & 11 \\ 1 & \frac{1}{2} & 0 & -\frac{1}{2} & \frac{1}{2} & 0 & 0 & 0 & \Big| & \frac{9}{2} \\ 0 & \boxed{\frac{1}{2}} & 0 & \frac{1}{2} & -\frac{1}{2} & -1 & 1 & 0 & \Big| & \frac{3}{2} \\ \hline 0 & -\frac{1}{2}M+\frac{1}{2} & 0 & -\frac{1}{2}M+\frac{3}{2} & \frac{3}{2}M-\frac{3}{2} & M & 0 & 1 & \Big| & -\frac{3}{2}M-\frac{27}{2} \end{bmatrix} \begin{array}{l} \frac{9}{1/2}=18 \\ \frac{3/2}{1/2}=3 \end{array}$$

$$2R_3 \to R_3$$

$$\sim \begin{bmatrix} 0 & 0 & 1 & 1 & -1 & 0 & 0 & 0 & \Big| & 11 \\ 1 & \frac{1}{2} & 0 & -\frac{1}{2} & \frac{1}{2} & 0 & 0 & 0 & \Big| & \frac{9}{2} \\ 0 & \boxed{1} & 0 & 1 & -1 & -2 & 2 & 0 & \Big| & 3 \\ \hline 0 & -\frac{1}{2}M+\frac{1}{2} & 0 & -\frac{1}{2}M+\frac{3}{2} & \frac{3}{2}M-\frac{3}{2} & M & 0 & 1 & \Big| & -\frac{3}{2}M-\frac{27}{2} \end{bmatrix}$$

$$\left(-\tfrac{1}{2}\right)R_3 + R_2 \to R_2, \quad \left(\tfrac{1}{2}M - \tfrac{1}{2}\right)R_3 + R_4 \to R_4$$

$$\sim \begin{array}{c} s_1 \\ x_1 \\ x_2 \\ P \end{array} \begin{bmatrix} x_1 & x_2 & s_1 & s_2 & a_1 & s_3 & a_2 & P & & \\ 0 & 0 & 1 & 1 & -1 & 0 & 0 & 0 & \Big| & 11 \\ 1 & 0 & 0 & -1 & 1 & 1 & -1 & 0 & \Big| & 3 \\ 0 & 1 & 0 & 1 & -1 & -2 & 2 & 0 & \Big| & 3 \\ \hline 0 & 0 & 0 & 1 & M-1 & 1 & M-1 & 1 & \Big| & -15 \end{bmatrix}$$

Optimal solution: Max $P = -15$ at $x_1 = 3$, $x_2 = 3$. Thus, the optimal solution of the original problem is: Min $C = 15$ at $x_1 = 3$, $x_2 = 3$. (5-6)

34. Multiply the first two constraint inequalities by -1 to transform them into ≥ inequalitites. The problem now is:

Minimize $C = 15x_1 + 12x_2 + 15x_3 + 18x_4$

Subject to:
$$-x_1 - x_2 \qquad\qquad \geq -240$$
$$\qquad\quad - x_3 - x_4 \geq -500$$
$$x_1 \qquad + x_3 \qquad \geq 400$$
$$\qquad x_2 \qquad + x_4 \geq 300$$
$$x_1,\ x_2,\ x_3,\ x_4 \geq 0$$

The matrices corresponding to this problem and its dual are, respectively:

$$A = \begin{bmatrix} -1 & -1 & 0 & 0 & | & -240 \\ 0 & 0 & -1 & -1 & | & -500 \\ 1 & 0 & 1 & 0 & | & 400 \\ 0 & 1 & 0 & 1 & | & 300 \\ \hline 15 & 12 & 15 & 18 & | & 1 \end{bmatrix} \quad \text{and} \quad A^T = \begin{bmatrix} -1 & 0 & 1 & 0 & | & 15 \\ -1 & 0 & 0 & 1 & | & 12 \\ 0 & -1 & 1 & 0 & | & 15 \\ 0 & -1 & 0 & 1 & | & 18 \\ \hline -240 & -500 & 400 & 300 & | & 1 \end{bmatrix}$$

Thus, the dual problem is: Maximize $P = -240y_1 - 500y_2 + 400y_3 + 300y_4$

Subject to:
$$-y_1 \qquad\quad + y_3 \qquad\quad \leq 15$$
$$-y_1 \qquad\qquad\quad + y_4 \leq 12$$
$$\quad - y_2 + y_3 \qquad\quad \leq 15$$
$$\quad - y_2 \qquad + y_4 \leq 18$$
$$y_1,\ y_2,\ y_3,\ y_4 \geq 0$$

We introduce the slack variables x_1, x_2, x_3, and x_4 to obtain the initial system for the dual problem:

$$-y_1 \qquad + y_3 \qquad + x_1 \qquad\qquad\qquad = 15$$
$$-y_1 \qquad\qquad + y_4 \qquad + x_2 \qquad\qquad = 12$$
$$\quad - y_2 + y_3 \qquad\qquad\qquad + x_3 \qquad = 15$$
$$\quad - y_2 \qquad + y_4 \qquad\qquad\qquad + x_4 = 18$$
$$240y_1 + 500y_2 - 400y_3 - 300y_4 \qquad\qquad\qquad + P = 0$$

The simplex tableau for this problem is:

	y_1	y_2	y_3	y_4	x_1	x_2	x_3	x_4	P		
x_1	-1	0	①	0	1	0	0	0	0	15	$\frac{15}{1} = 15$
x_2	-1	0	0	1	0	1	0	0	0	12	
x_3	0	-1	1	0	0	0	1	0	0	15	$\frac{15}{1} = 15$
x_4	0	-1	0	1	0	0	0	1	0	18	
P	240	500	-400	-300	0	0	0	0	1	0	

[Note: Either element can be chosen as the pivot; we choose the element in the first row.]

$$(-1)R_1 + R_3 \rightarrow R_3, \quad 400R_1 + R_5 \rightarrow R_5$$

$$\sim \begin{bmatrix} -1 & 0 & 1 & 0 & 1 & 0 & 0 & 0 & 0 & | & 15 \\ -1 & 0 & 0 & ① & 0 & 1 & 0 & 0 & 0 & | & 12 \\ 1 & -1 & 0 & 0 & -1 & 0 & 1 & 0 & 0 & | & 0 \\ 0 & -1 & 0 & 1 & 0 & 0 & 0 & 1 & 0 & | & 18 \\ \hdashline -160 & 500 & 0 & -300 & 400 & 0 & 0 & 0 & 1 & | & 6000 \end{bmatrix}$$

$(-1)R_2 + R_4 \to R_4, \ 300R_2 + R_5 \to R_5$

$$\sim \begin{bmatrix} -1 & 0 & 1 & 0 & 1 & 0 & 0 & 0 & 0 & | & 15 \\ -1 & 0 & 0 & 1 & 0 & 1 & 0 & 0 & 0 & | & 12 \\ ① & -1 & 0 & 0 & -1 & 0 & 1 & 0 & 0 & | & 0 \\ 1 & -1 & 0 & 0 & 0 & -1 & 0 & 1 & 0 & | & 6 \\ \hdashline -460 & 500 & 0 & 0 & 400 & 300 & 0 & 0 & 1 & | & 9600 \end{bmatrix}$$

$R_3 + R_1 \to R_1, \ R_3 + R_2 \to R_2, \ (-1)R_3 + R_4 \to R_4, \ 460R_3 + R_5 \to R_5$

$$\sim \begin{bmatrix} 0 & -1 & 1 & 0 & 0 & 0 & 1 & 0 & 0 & | & 15 \\ 0 & -1 & 0 & 1 & -1 & 1 & 1 & 0 & 0 & | & 12 \\ 1 & -1 & 0 & 0 & -1 & 0 & 1 & 0 & 0 & | & 0 \\ 0 & 0 & 0 & 0 & ① & -1 & -1 & 1 & 0 & | & 6 \\ \hdashline 0 & 40 & 0 & 0 & -60 & 300 & 460 & 0 & 1 & | & 9600 \end{bmatrix}$$

$60R_4 + R_5 \to R_5$

$$\sim \begin{array}{c} \\ y_3 \\ y_4 \\ y_1 \\ x_4 \\ P \end{array} \begin{bmatrix} y_1 & y_2 & y_3 & y_4 & x_1 & x_2 & x_3 & x_4 & P & \\ 0 & -1 & 1 & 0 & 0 & 0 & 1 & 0 & 0 & | & 15 \\ 0 & -1 & 0 & 1 & -1 & 1 & 1 & 0 & 0 & | & 12 \\ 1 & -1 & 0 & 0 & -1 & 0 & 1 & 0 & 0 & | & 0 \\ 0 & 0 & 0 & 0 & 1 & -1 & -1 & 1 & 0 & | & 6 \\ \hdashline 0 & 40 & 0 & 0 & 0 & 240 & 400 & 60 & 1 & | & 9960 \end{bmatrix}$$

Optimal solution: min $C = 9960$ at $x_1 = 0$, $x_2 = 240$, $x_3 = 400$, $x_4 = 60$.

(5-5)

35. (A) Let x_1 = the number of regular sails
and x_2 = the number of competition sails.
The mathematical model for this problem is:

Maximize $P = 100x_1 + 200x_2$

Subject to: $2x_1 + \ 3x_2 \le 150$
$4x_1 + 10x_2 \le 380$
$x_1, \ x_2 \ge \ \ 0$

The feasible region is indicated by the shading in the graph below.

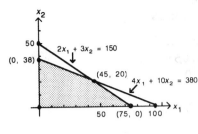

The corner points are $(0, 0)$, $(0, 38)$, $(45, 20)$, $(75, 0)$.

The value P at each corner point is:

Corner point	$P = 100x_1 + 200x_2$
$(0, 0)$	$P = 100(0) + 200(0) = 0$
$(0, 38)$	$P = 100(0) + 200(38) = 7,600$
$(45, 20)$	$P = 100(45) + 200(20) = 8,500$
$(75, 0)$	$P = 100(75) + 200(0) = 7,500$

Optimal solution: max $P = \$8,500$ when 45 regular and 20 competition sails are produced.

(B) The mathematical model for this problem is:

$$\text{Maximize } P = 100x_1 + 260x_2$$
$$\text{Subject to: } 2x_1 + 3x_2 \leq 150$$
$$4x_1 + 10x_2 \leq 380$$
$$x_1,\ x_2 \geq 0$$

The feasible region and the corner points are the same as in part (A). The value of P at each corner point is:

Corner point	$P = 100x_1 + 260x_2$
$(0, 0)$	$P = 100(0) + 260(0) = 0$
$(0, 38)$	$P = 100(0) + 260(38) = 9,880$
$(45, 20)$	$P = 100(45) + 260(20) = 9,700$
$(75, 0)$	$P = 100(75) + 260(0) = 7,500$

The maximum profit increases to $\$9,880$ when 38 competition and 0 regular sails are produced.

(C) The mathematical model for this problem is:

$$\text{Maximize } P = 100x_1 + 140x_2$$
$$\text{Subject to: } 2x_1 + 3x_2 \leq 150$$
$$4x_1 + 10x_2 \leq 380$$
$$x_1,\ x_2 \geq 0$$

The feasible region and the corner points are the same as in parts (A) and (B). The value of P at each corner point is:

Corner point	$P = 100x_1 + 140x_2$
$(0, 0)$	$P = 100(0) + 140(0) = 0$
$(0, 38)$	$P = 100(0) + 140(38) = 5,320$
$(45, 20)$	$P = 100(45) + 140(20) = 7,300$
$(75, 0)$	$P = 100(75) + 140(0) = 7,500$

The maximum profit decreases to $\$7,500$ when 0 competition and 75 regular sails are produced.

(5-2)

36. (A) Let x_1 = amount invested in oil stock
x_2 = amount invested in steel stock
x_3 = amount invested in government bonds

The mathematical model for this problem is:
Maximize $P = 0.12x_1 + 0.09x_2 + 0.05x_3$
Subject to: $x_1 + x_2 + x_3 \le 150{,}000$
$$x_1 \le 50{,}000$$
$$x_1 + x_2 - x_3 \le 25{,}000$$
$$x_1,\ x_2,\ x_3 \ge 0$$

Introduce slack variables s_1, s_2, s_3 to obtain the initial system:
$$x_1 + x_2 + x_3 + s_1 = 150{,}000$$
$$x_1 + s_2 = 50{,}000$$
$$x_1 + x_2 - x_3 + s_3 = 25{,}000$$
$$-0.12x_1 - 0.09x_2 - 0.05x_3 + P = 0$$
$$x_1,\ x_2,\ x_3,\ s_1,\ s_2,\ s_3 \ge 0$$

The simplex tableau for this problem is:

	x_1	x_2	x_3	s_1	s_2	s_3	P	
s_1	1	1	1	1	0	0	0	150,000
s_2	1	0	0	0	1	0	0	50,000
s_3	①	1	-1	0	0	1	0	25,000
P	-0.12	-0.09	-0.05	0	0	0	1	0

$(-1)R_3 + R_1 \to R_1,\ (-1)R_3 + R_2 \to R_1,\ 0.12R_3 + R_4 \to R_4$

0	0	2	1	0	-1	0	125,000	$\dfrac{125{,}000}{2} = 62{,}500$
0	-1	①	0	1	-1	0	25,000	$\dfrac{25{,}000}{1} = 25{,}000$
1	1	-1	0	0	1	0	25,000	
0	0.03	-0.17	0	0	0.12	1	3,000	

$(-2)R_2 + R_1 \to R_1,\ R_2 + R_3 \to R_3,\ (0.17)R_2 + R_4 \to R_4$

0	②	0	1	-2	1	0	75,000
0	-1	1	0	1	-1	0	25,000
1	0	0	0	1	0	0	50,000
0	-0.14	0	0	0.17	-0.05	1	7,250

$\dfrac{1}{2}R_1 \to R_1$

0	①	0	$\frac{1}{2}$	-1	$\frac{1}{2}$	0	37,500
0	-1	1	0	1	-1	0	25,000
1	0	0	0	1	0	0	50,000
0	-0.14	0	0	0.17	-0.05	1	7,250

$R_1 + R_2 \to R_2,\ 0.14R_1 + R_4 \to R_4$

	x_1	x_2	x_3	s_1	s_2	s_3	P	
x_2	0	1	0	$\frac{1}{2}$	-1	$\frac{1}{2}$	0	37,500
x_3	0	0	1	$\frac{1}{2}$	0	$-\frac{1}{2}$	0	62,500
x_1	1	0	0	0	1	0	0	50,000
P	0	0	0	0.07	0.03	0.02	1	12,500

The maximum return is \$12,500 when \$50,000 is invested in oil stock, \$37,500 in steel stock, and \$62,500 in government bonds.

(B) The mathematical model for this problem is:

$$\text{Maximize } P = 0.09x_1 + 0.12x_2 + 0.05x_3$$
$$\text{Subject to: } x_1 + x_2 + x_3 \le 150,000$$
$$x_1 \le 50,000$$
$$x_1 + x_2 - x_3 \le 25,000$$
$$x_1, \ x_2, \ x_3 \ge 0$$

Introduce slack variables s_1, s_2, s_3 to obtain the initial system:

$$x_1 + x_2 + x_3 + s_1 = 150,000$$
$$x_1 + s_2 = 50,000$$
$$x_1 + x_2 - x_3 + s_3 = 25,000$$
$$-0.09x_1 - 0.12x_2 - 0.05x_3 + P = 0$$

The simplex tableau for this problem is:

	x_1	x_2	x_3	s_1	s_2	s_3	P	
s_1	1	1	1	1	0	0	0	150,000
s_2	1	0	0	0	1	0	0	50,000
s_3	1	①	-1	0	0	1	0	25,000
P	-0.09	-0.12	-0.05	0	0	0	1	0

$(-1)R_3 + R_1 \to R_1, \ 0.12R_3 + R_4 \to R_4$

0	0	②	1	0	-1	0	125,000
1	0	0	0	1	0	0	50,000
1	1	-1	0	0	1	0	25,000
0.03	0	-0.17	0	0	0.12	1	3,000

$\frac{1}{2}R_1 \to R_1$

0	0	①	$\frac{1}{2}$	0	$-\frac{1}{2}$	0	62,500
1	0	0	0	1	0	0	50,000
1	1	-1	0	0	1	0	25,000
0.03	0	-0.17	0	0	0.12	1	3,000

$R_1 + R_3 \to R_3, \ 0.17R_1 + R_4 \to R_4$

$$\sim \begin{array}{c} x_3 \\ s_2 \\ x_2 \\ P \end{array} \left[\begin{array}{ccccccc|c} x_1 & x_2 & x_3 & s_1 & s_2 & s_3 & P & \\ 0 & 0 & 1 & \frac{1}{2} & 0 & -\frac{1}{2} & 0 & 62{,}500 \\ 1 & 0 & 0 & 0 & 1 & 0 & 0 & 50{,}000 \\ 1 & 1 & 0 & \frac{1}{2} & 0 & \frac{1}{2} & 0 & 87{,}500 \\ \hline 0.03 & 0 & 0 & 0.085 & 0 & 0.35 & 1 & 13{,}625 \end{array} \right]$$

The maximum return is \$13,625 when \$0 is invested in oil stock, \$87,500 is invested in steel stock, and \$62,500 in invested in government bonds.

(5-4)

37. Let x_1 = the number of motors from A to X,
x_2 = the number of motors from A to Y,
x_3 = the number of motors from B to X,
x_4 = the number of motors from B to Y.

The mathematical model for this problem is:
Minimize $C = 5x_1 + 8x_2 + 9x_3 + 7x_4$
Subject to:
$$\begin{aligned} x_1 + x_2 \quad\quad &\leq 1{,}500 \\ x_3 + x_4 &\leq 1{,}000 \\ x_1 \quad\quad + x_3 \quad\quad &\geq 900 \\ x_2 \quad\quad + x_4 &\geq 1{,}200 \end{aligned}$$

Multiply the first two inequalities by -1 to obtain $\geq$ inequalities. The model then becomes:
Minimize $C = 5x_1 + 8x_2 + 9x_3 + 7x_4$
Subject to:
$$\begin{aligned} -x_1 - x_2 \quad\quad &\geq -1{,}500 \\ -x_3 - x_4 &\geq -1{,}000 \\ x_1 \quad\quad + x_3 \quad\quad &\geq 900 \\ x_2 \quad\quad + x_4 &\geq 1{,}200 \\ x_1, \ x_2, \ x_3, \ x_4 &\geq 0 \end{aligned}$$

The matrices for this problem and the dual problem are:

$$A = \left[\begin{array}{cccc|c} -1 & -1 & 0 & 0 & -1{,}500 \\ 0 & 0 & -1 & -1 & -1{,}000 \\ 1 & 0 & 1 & 0 & 900 \\ 0 & 1 & 0 & 1 & 1{,}200 \\ \hline 5 & 8 & 9 & 7 & 1 \end{array} \right] \text{ and } A^T = \left[\begin{array}{cccc|c} -1 & 0 & 1 & 0 & 5 \\ -1 & 0 & 0 & 1 & 8 \\ 0 & -1 & 1 & 0 & 9 \\ 0 & -1 & 0 & 1 & 7 \\ \hline -1{,}500 & -1{,}000 & 900 & 1{,}200 & 1 \end{array} \right]$$

The dual problem is:
Maximize $P = -1{,}500y_1 - 1{,}000y_2 + 900y_3 + 1{,}200y_4$
Subject to:
$$\begin{aligned} -y_1 \quad\quad + y_3 \quad\quad &\leq 5 \\ -y_1 \quad\quad\quad + y_4 &\leq 8 \\ -y_2 + y_3 \quad\quad &\leq 9 \\ -y_2 \quad\quad + y_4 &\leq 7 \\ y_1, \ y_2, \ y_3, \ y_4 &\geq 0 \end{aligned}$$

Introduce slack variables x_1, x_2, x_3, x_4 to obtain the initial system:

$$-y_1 \quad\quad + y_3 \quad\quad + x_1 \quad\quad\quad\quad\quad = 5$$
$$-y_1 \quad\quad\quad\quad + y_4 \quad\quad + x_2 \quad\quad\quad\quad = 8$$
$$-y_2 + y_3 \quad\quad\quad\quad\quad + x_3 \quad\quad = 9$$
$$-y_2 \quad\quad + y_4 \quad\quad\quad\quad\quad + x_4 = 7$$
$$1{,}500y_1 + 1{,}000y_2 - 900y_3 - 1{,}200y_4 \quad + P = 0$$

The simplex tableau for this problem is:

	y_1	y_2	y_3	y_4	x_1	x_2	x_3	x_4	P	
x_1	-1	0	1	0	1	0	0	0	0	5
x_2	-1	0	0	1	0	1	0	0	0	8
x_3	0	-1	1	0	0	0	1	0	0	9
x_4	0	-1	0	①	0	0	0	1	0	7
P	1,500	1,000	-900	$-1{,}200$	0	0	0	0	1	0

$\frac{8}{1} = 8$ $\frac{7}{1} = 7$

$$(-1)R_4 + R_2 \rightarrow R_2, \quad 1{,}200R_4 + R_5 \rightarrow R_5$$

$$\sim
\begin{bmatrix}
-1 & 0 & ① & 0 & 1 & 0 & 0 & 0 & 0 & 5 \\
-1 & 1 & 0 & 0 & 0 & 1 & 0 & -1 & 0 & 1 \\
0 & -1 & 1 & 0 & 0 & 0 & 1 & 0 & 0 & 9 \\
0 & -1 & 0 & 1 & 0 & 0 & 0 & 1 & 0 & 7 \\
\hline
1{,}500 & -200 & -900 & 0 & 0 & 0 & 0 & 1{,}200 & 1 & 8{,}400
\end{bmatrix}$$

$\frac{5}{1} = 5$ $\frac{9}{1} = 9$

$$(-1)R_1 + R_3 \rightarrow R_3, \quad 900R_1 + R_5 \rightarrow R_5$$

$$\sim
\begin{bmatrix}
-1 & 0 & 1 & 0 & 1 & 0 & 0 & 0 & 0 & 5 \\
-1 & ① & 0 & 0 & 0 & 1 & 0 & -1 & 0 & 1 \\
1 & -1 & 0 & 0 & -1 & 0 & 1 & 0 & 0 & 4 \\
0 & -1 & 0 & 1 & 0 & 0 & 0 & 1 & 0 & 7 \\
\hline
600 & -200 & 0 & 0 & 900 & 0 & 0 & 1{,}200 & 1 & 12{,}900
\end{bmatrix}$$

$$R_2 + R_3 \rightarrow R_3, \quad R_2 + R_4 \rightarrow R_4, \quad 200R_2 + R_5 \rightarrow R_5$$

	y_1	y_2	y_3	y_4	x_1	x_2	x_3	x_4	P	
y_3	-1	0	1	0	1	0	0	0	0	5
y_2	-1	1	0	0	0	1	0	-1	0	1
x_3	0	0	0	0	-1	1	1	-1	0	5
y_4	-1	0	0	1	0	1	0	0	0	8
P	400	0	0	0	900	200	0	1,000	1	13,100

(with $\sim$ to the left)

Optimal soltuion: min $C = \$13{,}100$ when 900 motors are shipped from factory A to plant X, 200 motors are shipped from factory A to plant Y, 0 motors are shipped from factory B to plant X, and 1,000 motors are shipped from factory B to plant Y.   (5-5)

38. Let x_1 = number of pounds of long grain rice in Brand A
Let x_2 = number of pounds of long grain rice in Brand B
Let x_3 = number of pounds of wild rice in Brand A
Let x_4 = number of pounds of wild rice in Brand B

The mathematical model for this problem is:

Maximize $P = 0.8x_1 + 0.5x_2 - 1.9x_3 - 2.2x_4$

Subject to:
$$x_3 \geq 0.1(x_1 + x_3)$$
$$x_4 \geq 0.05(x_2 + x_4)$$
$$x_1 + x_2 \leq 8{,}000$$
$$x_3 + x_4 \leq 500$$
$$x_1, x_2, x_3, x_4 \geq 0$$

which is the same as:

Maximize $P = 0.8x_1 + 0.5x_2 - 1.9x_3 - 2.2x_4$

Subject to:
$$0.1x_1 \qquad - 0.9x_3 \qquad\qquad \leq 0$$
$$0.05x_2 \qquad\qquad - 0.95x_4 \leq 0$$
$$x_1 + x_2 \qquad\qquad\qquad \leq 8{,}000$$
$$x_3 + x_4 \leq 500$$
$$x_1, x_2, x_3, x_4 \geq 0$$

Introduce slack variables s_1, s_2, s_3, and s_4 to obtain the initial system:

	x_1	x_2	x_3	x_4	s_1	s_2	s_3	s_4	P		
s_1	(0.1)	0	-0.9	0	1	0	0	0	0	0	$\frac{0}{0.1} = 0$
s_2	0	0.05	0	-0.95	0	1	0	0	0	0	
s_3	1	1	0	0	0	0	1	0	0	8,000	$\frac{8000}{1} = 8{,}000$
s_4	0	0	1	1	0	0	0	1	0	500	
P	-0.8	-0.5	1.9	2.2	0	0	0	0	1	0	

$10R_1 \to R_1$

$$\sim \begin{bmatrix} ① & 0 & -9 & 0 & 10 & 0 & 0 & 0 & 0 & | & 0 \\ 0 & 0.05 & 0 & -0.95 & 0 & 1 & 0 & 0 & 0 & | & 0 \\ 1 & 1 & 0 & 0 & 0 & 0 & 1 & 0 & 0 & | & 8{,}000 \\ 0 & 0 & 1 & 1 & 0 & 0 & 0 & 1 & 0 & | & 500 \\ -0.8 & -0.5 & 1.9 & 2.2 & 0 & 0 & 0 & 0 & 1 & | & 0 \end{bmatrix}$$

$(-1)R_1 + R_3 \to R_3, \quad 0.8R_1 + R_5 \to R_5$

1	0	-9	0	10	0	0	0	0	0	
0	0.05	0	-0.95	0	1	0	0	0	0	
~ 0	1	9	0	-10	0	1	0	0	8,000	$\frac{8000}{9} \approx 888.9$
0	0	①	1	0	0	0	1	0	500	$\frac{500}{1} = 500$
0	-0.5	-5.3	2.2	8	0	0	0	1	0	

$9R_4 + R_1 \to R_1, \quad (-9)R_4 + R_3 \to R_3, \quad 5.3R_4 + R_5 \to R_5$

$$\sim \begin{bmatrix} 1 & 0 & 0 & 9 & 10 & 0 & 0 & 9 & 0 & | & 4,500 \\ 0 & (0.05) & 0 & -0.95 & 0 & 1 & 0 & 0 & 0 & | & 0 \\ 0 & 1 & 0 & -9 & -10 & 0 & 1 & -9 & 0 & | & 3,500 \\ 0 & 0 & 1 & 1 & 0 & 0 & 0 & 1 & 0 & | & 500 \\ 0 & -0.5 & 0 & 7.5 & 8 & 0 & 0 & 5.3 & 1 & | & 2,650 \end{bmatrix} \quad \begin{array}{l} \dfrac{0}{0.05} = 0 \\[2mm] \dfrac{3,500}{1} = 3,500 \end{array}$$

$$20R_2 \rightarrow R_2$$

$$\sim \begin{bmatrix} 1 & 0 & 0 & 9 & 10 & 0 & 0 & 9 & 0 & | & 4,500 \\ 0 & \textcircled{1} & 0 & -19 & 0 & 20 & 0 & 0 & 0 & | & 0 \\ 0 & 1 & 0 & -9 & -10 & 0 & 1 & -9 & 0 & | & 3,500 \\ 0 & 0 & 1 & 1 & 0 & 0 & 0 & 1 & 0 & | & 500 \\ 0 & -0.5 & 0 & 7.5 & 8 & 0 & 0 & 5.3 & 1 & | & 2,650 \end{bmatrix}$$

$$(-1)R_2 + R_3 \rightarrow R_3, \quad (0.5)R_2 + R_5 \rightarrow R_5$$

$$\sim \begin{bmatrix} 1 & 0 & 0 & 9 & 10 & 0 & 0 & 9 & 0 & | & 4,500 \\ 0 & 1 & 0 & -19 & 0 & 20 & 0 & 0 & 0 & | & 0 \\ 0 & 0 & 0 & \textcircled{10} & -10 & -20 & 1 & -9 & 0 & | & 3,500 \\ 0 & 0 & 1 & 1 & 0 & 0 & 0 & 1 & 0 & | & 500 \\ 0 & 0 & 0 & -2 & 8 & 10 & 0 & 5.3 & 1 & | & 2,650 \end{bmatrix} \quad \begin{array}{l} \dfrac{4,500}{9} = 500 \\[2mm] \dfrac{3,500}{10} = 350 \\[2mm] \dfrac{500}{1} = 500 \end{array}$$

$$\frac{1}{10}R_3 \rightarrow R_3$$

$$\sim \begin{bmatrix} 1 & 0 & 0 & 9 & 10 & 0 & 0 & 9 & 0 & | & 4,500 \\ 0 & 1 & 0 & -19 & 0 & 20 & 0 & 0 & 0 & | & 0 \\ 0 & 0 & 0 & \textcircled{1} & -1 & -2 & \frac{1}{10} & -\frac{9}{10} & 0 & | & 350 \\ 0 & 0 & 1 & 1 & 0 & 0 & 0 & 1 & 0 & | & 500 \\ 0 & 0 & 0 & -2 & 8 & 10 & 0 & 5.3 & 1 & | & 2,650 \end{bmatrix}$$

$$(-9)R_3 + R_1 \rightarrow R_1, \quad 19R_3 + R_2 \rightarrow R_2, \quad (-1)R_3 + R_4 \rightarrow R_4, \quad 2R_3 + R_5 \rightarrow R_5$$

$$\sim \begin{array}{c} \\ x_1 \\ x_2 \\ x_4 \\ x_3 \\ P \end{array} \begin{array}{cccccccccc} x_1 & x_2 & x_3 & x_4 & s_1 & s_2 & s_{31} & s_4 & P & \\ \left[\begin{array}{ccccccccc|c} 1 & 0 & 0 & 0 & 19 & 18 & -\frac{9}{10} & \frac{171}{10} & 0 & 1,350 \\ 0 & 1 & 0 & 0 & -19 & -18 & \frac{19}{10} & -\frac{171}{10} & 0 & 6,650 \\ 0 & 0 & 0 & 1 & -1 & -2 & \frac{1}{10} & -\frac{9}{10} & 0 & 350 \\ 0 & 0 & 1 & 0 & 1 & 2 & -\frac{1}{10} & \frac{19}{10} & 0 & 150 \\ \hline 0 & 0 & 0 & 0 & 6 & 6 & \frac{1}{5} & \frac{7}{2} & 1 & 3,350 \end{array}\right] \end{array}$$

The maximum profit is \$3,350 when 1,350 pounds of long grain rice and 150 pounds of wild rice are used to produce 1,500 pounds of brand A, and 6,650 pounds of long grain rice and 350 pounds of wild rice are used to produce 7,000 pounds of brand B.   (5-5)

39. Let x_1 = number of grams of mix A
x_2 = number of grams of mix B

The constraints are:
vitamins: $2x_1 + 5x_2 \geq 850$
$2x_1 + 4x_2 \geq 800$
$4x_1 + 5x_2 \geq 1,150$
$x_1, x_2 \geq 0$

The feasible region is indicated by
the shading in the graph at the
right. The corner points are:
(0, 230), (100, 150), (300, 50), (425, 0).

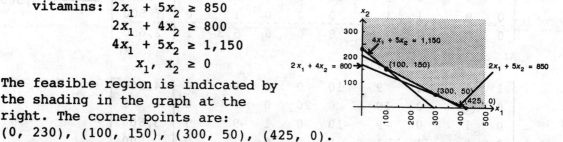

(A) The mathematical model for this problem is:
minimize $C = 0.04x_1 + 0.09x_2$ subject to the contraints given above.

The value of C at each corner point is:

Corner Point	$C = 0.04x_1 + 0.09x_2$
(0, 230)	$C = 0.04(0) + 0.09(230) = 20.70$
(100, 150)	$C = 0.04(100) + 0.09(150) = 17.50$
(300, 50)	$C = 0.04(300) + 0.09(50) = 16.50$
(425, 0)	$C = 0.04(425) + 0.09(0) = 17.00$

The minimum cost is \$16.50 when 300 grams of mix A and 50 grams of
mix B are used.

(B) The mathematical model for this problem is:
minimize $C = 0.04x_1 + 0.06x_2$ subject to the constraints given above.

The value of C at each corner point is:

Corner Point	$C = 0.04x_1 + 0.06x_2$
(0, 230)	$C = 0.04(0) + 0.06(230) = 13.80$
(100, 150)	$C = 0.04(100) + 0.06(150) = 13.00$
(300, 50)	$C = 0.04(300) + 0.06(50) = 15.00$
(425, 0)	$C = 0.04(425) + 0.06(0) = 17.00$

The minimum cost decreases to \$13.00 when 100 grams of mix A and 150
grams of mix B are used.

(C) The mathematical model for this problem is:
minimize $C = 0.04x_1 + 0.12x_2$ subject to the constraints given above.

The value of C at each corner point is:

Corner Point	$C = 0.04x_1 + 0.12x_2$
(0, 230)	$C = 0.04(0) + 0.12(230) = 27.60$
(100, 150)	$C = 0.04(100) + 0.12(150) = 22.00$
(300, 50)	$C = 0.04(300) + 0.12(50) = 18.00$
(425, 0)	$C = 0.04(425) + 0.12(0) = 17.00$

The minimum cost increases to \$17.00 when 425 grams of mix A and 0
grams of mix B are used.

(5-2)

6 PROBABILITY

Things to remember:

<u>1</u>. Let A be a set with finitely many elements. Then $n(A)$ denotes the number of elements in A.

<u>2</u>. ADDITION PRINCIPLE (for counting)
   For any two sets A and B,
   $$n(A \cup B) = n(A) + n(B) - n(A \cap B)$$
   If A and B are disjoint, i.e., if $A \cap B = \varnothing$, then
   $$n(A \cup B) = n(A) + n(B)$$

<u>3</u>. MULTIPLICATION PRINCIPLE (for counting)
   (a) If two operations O_1 and O_2 are performed in order, with N_1 possible outcomes for the first operation and N_2 possible outcomes for the second operation, then there are
   $$N_1 \cdot N_2$$
   possible combined outcomes of the first operation followed by the second.
   (b) In general, if n operations O_1, O_2, ..., O_n are performed in order with possible number of outcomes N_1, N_2, ..., N_n, respectively, then there are
   $$N_1 \cdot N_2 \cdot ... \cdot N_n$$
   possible combined outcomes of the operations performed in the given order.

---

1. $n(A) = 75 + 40 = 115$

3. $n(A \cup B) = 75 + 40 + 95 = 210$
   $n[(A \cup B)'] = 90$
   Thus, $n(U) = n(A \cup B) + n[(A \cup B)'] = 210 + 90 = 300$
   [Note: $U = (A \cup B) \cup (A \cup B)'$ and $(A \cup B) \cap (A \cup B)' = \varnothing$]

5. $B' = A \cap B' \cup (A \cup B)'$ and $(A \cap B') \cap (A \cup B)' = \varnothing$.
   Thus, $n(B') = n(A \cap B') + n[(A \cup B)']$
   $\qquad\qquad = 75 + 90 = 165$

7. $n(A \cup B) = n(A) + n(B) - n(A \cap B)$
$$= 115 + 135 - 40 = 210$$

Note, also, that $A \cup B = (A \cap B') \cup (A \cap B) \cup (A' \cap B)$, and
$(A \cap B') \cap (A \cap B) = \emptyset$, $(A \cap B') \cap (A' \cap B) = \emptyset$,
$(A \cap B) \cap (A' \cap B) = \emptyset$.

So, $n(A \cup B) = n(A \cap B') + n(A \cap B) + n(A' \cap B)$
$$= 75 + 40 + 95 = 210$$

9. $n(A' \cap B) = 95$

11. $(A \cap B) \cup (A \cap B)' = U$ and $(A \cap B) \cap (A \cap B)' = \emptyset$

Thus, $n(U) = n(A \cap B) + n[(A \cap B)']$

or $n[(A \cap B)'] = n(U) - n(A \cap B) = 300 - 40 = 260$

13. (A) Tree Diagram

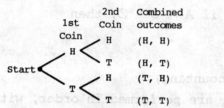

	2nd	Combined
1st	Coin	outcomes
Coin	H	(H, H)
	T	(H, T)
	H	(T, H)
	T	(T, T)

Thus, there are 4 ways.

(B) Multiplication Principle
O_1: 1st coin
N_1: 2 ways
O_2: 2nd coin
N_2: 2 ways

Thus, there are
$N_1 \cdot N_2 = 2 \cdot 2 = 4$ ways.

15. (A) Tree Diagram

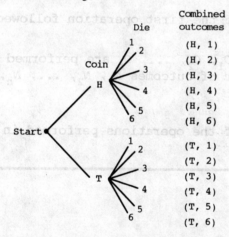

Die	Combined outcomes
1	(H, 1)
2	(H, 2)
3	(H, 3)
4	(H, 4)
5	(H, 5)
6	(H, 6)
1	(T, 1)
2	(T, 2)
3	(T, 3)
4	(T, 4)
5	(T, 5)
6	(T, 6)

Thus, there are 12 combined outcomes.

(B) Multiplication Principle
O_1: Coin
N_1: 2 outcomes
O_2: Die
N_2: 6 outcomes

Thus, there are
$N_1 \cdot N_2 = 2 \cdot 6 = 12$ combined outcomes

17. $A = (A \cap B') \cup (A \cap B)$ and $B = (A' \cap B) \cup (A \cap B)$

Thus, $n(A) = n(A \cap B') + n(A \cap B)$, so
$$n(A \cap B') = n(A) - n(A \cap B) = 80 - 20 = 60$$
$$n(A \cap B) = 20.$$
$$n(B) = n(A' \cap B) + n(A \cap B) \text{ so}$$
$$n(A' \cap B) = n(B) - n(A \cap B) = 50 - 20 = 30$$

Also, $A \cup B = (A' \cap B) \cup (A \cap B) \cup (A \cap B')$ and

$U = (A \cup B) \cup (A' \cap B')$ where $(A \cup B) \cap (A' \cap B') = \emptyset$

Thus, $n(U) = n(A \cup B) + n(A' \cap B')$ so

$n(A' \cap B') = n(U) - n(A \cup B)$

$= 200 - (60 + 20 + 30) = 200 - 110 = 90$

19. $n(A \cup B) = n(A) + n(B) - n(A \cap B)$

So $n(A \cap B) = n(A) + n(B) - n(A \cup B)$

$= 25 + 55 - 60 = 20$

$n(A \cap B') = n(A) - n(A \cap B) = 25 - 20 = 5$

$n(A' \cap B) = n(B) - n(A \cap B) = 55 - 20 = 35$

and $n(A' \cap B') = n(U) - n(A \cup B)$

$= 100 - 60 = 40$

21. $n(A \cap B) = 30$

$n(A \cap B') = n(A) - n(A \cap B) = 70 - 30 = 40$

$n(A' \cap B) = n(B) - n(A \cap B) = 90 - 30 = 60$

$n(A' \cap B') = n(U) - [n(A \cap B') + n(A \cap B) + n(A' \cap B)]$

$= 200 - [40 + 30 + 60] = 200 - 130 = 70$

Therefore,

	A	A'	Totals
B	30	60	90
B'	40	70	110
Totals	70	130	200

23. $n(A \cap B) = n(A) + n(B) - n(A \cup B) = 45 + 55 - 80 = 20$

$n(A \cap B') = n(A) - n(A \cap B) = 45 - 20 = 25$

$n(A' \cap B) = n(B) - n(A \cap B) = 55 - 20 = 35$

$n(A' \cap B') = n(U) - n(A \cup B) = 100 - 80 = 20$

Therefore,

	A	A'	Totals
B	20	35	55
B'	25	20	45
Totals	45	55	100

25. Using the Multiplication Principle:

O_1: Choose the color $\quad\quad$ O_3: Choose the interior
N_1: 5 ways $\quad\quad\quad\quad\quad\quad\quad$ N_3: 4 ways

O_2: Choose the transmission $\quad$ O_4: Choose the engine
N_2: 3 ways $\quad\quad\quad\quad\quad\quad\quad$ N_4: 2 ways

Thus, there are

$N_1 \cdot N_2 \cdot N_3 \cdot N_4 = 5 \cdot 3 \cdot 4 \cdot 2 = 120$ different variations of this model car.

27. (A) Number of four-letter code words, no letter repeated.

O_1: Selecting the first letter
N_1: 6 ways

O_2: Selecting the second letter
N_2: 5 ways

O_3: Selecting the third letter
N_3: 4 ways

O_4: Selecting the fourth letter
N_4: 3 ways

Thus, there are

$N_1 \cdot N_2 \cdot N_3 \cdot N_4 = 6 \cdot 5 \cdot 4 \cdot 3 = 360$

possible code words. Note that this is the number of permutations of 6 objects taken 4 at a time:

$$P_{6,4} = \frac{6!}{(6-4)!} = \frac{6 \cdot 5 \cdot 4 \cdot 3 \cdot 2!}{2!} = 360$$

(B) Number of four-letter code words, allowing repetition.

O_1: Selecting the first letter
N_1: 6 ways

O_2: Selecting the second letter
N_2: 6 ways

O_3: Selecting the third letter
N_3: 6 ways

O_4: Selecting the fourth letter
N_4: 6 ways

Thus, there are

$N_1 \cdot N_2 \cdot N_3 \cdot N_4 = 6 \cdot 6 \cdot 6 \cdot 6 = 6^4 = 1296$

possible code words.

(C) Number of four-letter code words, adjacent letters different.

O_1: Selecting the first letter
N_1: 6 ways

O_2: Selecting the second letter
N_2: 5 ways

O_3: Selecting the third letter
N_3: 5 ways

O_4: Selecting the fourth letter
N_4: 5 ways

Thus, there are

$N_1 \cdot N_2 \cdot N_3 \cdot N_4 = 6 \cdot 5 \cdot 5 \cdot 5 = 6 \cdot 5^3 = 750$

possible code words.

29. (A) Number of five-digit combinations, no digit repeated.

O_1: Selecting the first digit
N_1: 10 ways

O_2: Selecting the second digit
N_2: 9 ways

O_3: Selecting the third digit
N_3: 8 ways

O_4: Selecting the fourth digit
N_4: 7 ways

O_5: Selecting the fifth digit
N_5: 6 ways

Thus, there are

$N_1 \cdot N_2 \cdot N_3 \cdot N_4 \cdot N_5 = 10 \cdot 9 \cdot 8 \cdot 7 \cdot 6 = 30,240$

possible combinations

(B) Number of five-digit combinations, allowing repetition.

O_1: Selecting the first digit
N_1: 10 ways

O_2: Selecting the second digit
N_2: 10 ways

O_3: Selecting the third digit
N_3: 10 ways

O_4: Selecting the fourth digit
N_4: 10 ways

O_5: Selecting the fifth digit
N_5: 10 ways

Thus, there are

$N_1 \cdot N_2 \cdot N_3 \cdot N_4 \cdot N_5 = 10 \cdot 10 \cdot 10 \cdot 10 \cdot 10 = 10^5 = 100,000$
possible combinations

(C) Number of five digit combinations, if successive digits must be different.

O_1: Selecting the first digit
N_1: 10 ways

O_2: Selecting the second digit
N_2: 9 ways

O_3: Selecting the third digit
N_3: 9 ways

O_4: Selecting the fourth digit
N_4: 9 ways

O_5: Selecting the fifth digit
N_5: 9 ways

Thus, there are

$N_1 \cdot N_2 \cdot N_3 \cdot N_4 \cdot N_5 = 10 \cdot 9 \cdot 9 \cdot 9 \cdot 9 = 10 \cdot 9^4 = 65,610$ possible combinations

31. (A) Letters and/or digits may be repeated.

O_1: Selecting the first letter
N_1: 26 ways

O_2: Selecting the second letter
N_2: 26 ways

O_3: Selecting the third letter
N_3: 26 ways

O_4: Selecting the first digit
N_4: 10 ways

O_5: Selecting the second digit
N_5: 10 ways

O_6: Selecting the third digit
N_6: 10 ways

Thus, there are

$N_1 \cdot N_2 \cdot N_3 \cdot N_4 \cdot N_5 \cdot N_6 = 26 \cdot 26 \cdot 26 \cdot 10 \cdot 10 \cdot 10 = 17,576,000$
different license plates.

(B) No repeated letters and no repeated digits are allowed.

O_1: Select the three letters, no letter repeated
N_1: $26 \cdot 25 \cdot 24 = 15,600$ ways

O_2: Select the three numbers, no number repeated
N_2: $10 \cdot 9 \cdot 8 = 720$ ways

Thus, there are

$N_1 \cdot N_2 = 15,600 \cdot 720 = 11,232,000$

different license plates with no letter or digit repeated.

33.

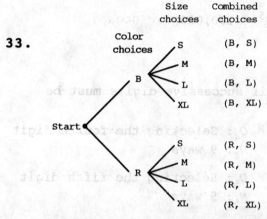

There are 8 combined choices; the color can be chosen in 2 ways and the size can be chosen in 4 ways. The number of possible combined choices is $2 \cdot 4 = 8$ just as in Example 3.

35. Let T = the people who play tennis, and
  G = the people who play golf.

Then $n(T) = 32$, $n(G) = 37$, $n(T \cap G) = 8$ and $n(U) = 75$.

Thus, $n(T \cup G) = n(T) + n(G) - n(T \cap G) = 32 + 37 - 8 = 61$

The set of people who play neither tennis nor golf is represented by $T' \cap G'$. Since $U = (T \cup G) \cup (T' \cap G')$ and $(T \cup G) \cap (T' \cap G') = \emptyset$, it follows that $n(T' \cap G') = n(U) - n(T \cup G) = 75 - 61 = 14$.

There are 14 people who play neither tennis nor golf.

37. Let F = the people who speak French, and
  G = the people who speak German.

Then $n(F) = 42$, $n(G) = 55$, $n(F' \cap G') = 17$ and $n(U) = 100$. Since $U = (F \cup G) \cup (F' \cap G')$ and $(F \cup G) \cap (F' \cap G') = \emptyset$, it follows that

$n(F \cup G) = n(U) - n(F' \cap G') = 100 - 17 = 83$

Now $n(F \cup G) = n(F) + n(G) - n(F \cap G)$, so

$n(F \cap G) = n(F) + n(G) - n(F \cup G)$

$= 42 + 55 - 83 = 14$

There are 14 people who speak both French and German.

39. (A)

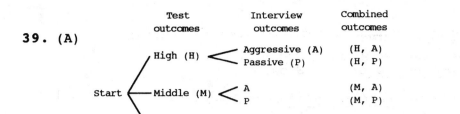

(B) Operation 1: Test scores can be classified into three groups, high, middle, or low:

$$N_1 = 3$$

Operation 2: Interviews can be classified into two groups, aggressive or passive:

$$N_2 = 2$$

The total possible combined classifications is:

$$N_1 \cdot N_2 = 3 \cdot 2 = 6$$

41. O_1: Travel from home to airport and back $\qquad$ O_3: Fly to second city
N_1: 2 ways $\qquad$ N_3: 2 ways

O_2: Fly to first city $\qquad$ O_4: Fly to third city
N_2: 3 ways $\qquad$ N_4: 1 way

Thus, there are
$$N_1 \cdot N_2 \cdot N_3 \cdot N_4 = 2 \cdot 3 \cdot 2 \cdot 1 = 12$$
different travel plans.

43. Let U = the group of people surveyed
M = people who own a microwave oven, and
V = people who own a VCR.

Then $n(U) = 1200$, $n(M) = 850$, $n(V) = 740$ and $n(M \cap V) = 580$. Now draw a Venn diagram.

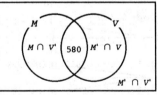

From this diagram, we see that
$$n(M \cap V') = n(M) - n(M \cap V) = 850 - 580 = 270$$
$$n(M' \cap V) = n(V) - n(M \cap V) = 740 - 580 = 160$$
$$n(M \cup V) = n(M \cap V') + n(M \cap V) + n(M' \cap V)$$
$$= 580 + 270 + 160 = 1010$$
and $n(M' \cap V') = n(U) - n(M \cup V) = 1200 - 1010 = 190$

Thus,
(A) $n(M \cup V) = 1010$ $\qquad$ (B) $n(M' \cap V') = 190$ $\qquad$ (C) $n(M \cap V') = 270$

45. Let U = group of people surveyed
H = group of people who receive HBO
S = group of people who receive Showtime.

Then, $n(U) = 8,000$, $n(H) = 2,450$, $n(S) = 1,940$ and $n(H' \cap S') = 5,180$

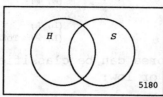

Now, $n(H \cup S) = n(U) - n(H' \cap S') = 8,000 - 5,180 = 2,820$

Since $n(H \cup S) = n(H) + n(S) - n(H \cap S)$, we have

$$n(H \cap S) = n(H) + n(S) - n(H \cup S) = 2,450 + 1,940 - 2,820 = 1,570$$

Thus, 1,570 subscribers receive both channels.

47. From the table:

(A) The number of males aged 20-24 *and* below minimum wage is: 102 (the element in the (2, 2) position in the body of the table.)

(B) The number of females aged 20 or older *and* at minimum wage is: $186 + 503 = 689$ (the sum of the elements in the (3, 2) and (3, 3) positions.)

(C) The number of workers who are *either* aged 16-19 *or* are males at minimum wage is:
$$343 + 118 + 367 + 251 + 154 + 237 = 1,470$$

(D) The number of workers below minimum wage is: $379 + 993 = 1,372$.

49. (A)

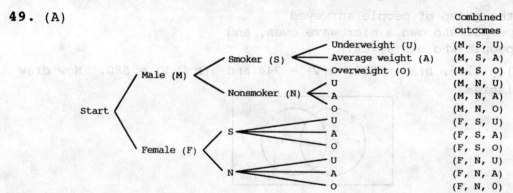

(B) Operation 1: Two classifications, male and female; $N_1 = 2$.

Operation 2: Two classifications, smoker and nonsmoker; $N_2 = 2$.

Operation 3: Three classifications, underweight, average weight, and overweight; $N_3 = 3$.

Thus the total possible combined classifications
$= N_1 \cdot N_2 \cdot N_3 = 2 \cdot 2 \cdot 3 = 12$

51. F = number of individuals who contributed to the first campaign
S = number of individuals who contributed to the second campaign.

Then $n(F) = 1,475$, $n(S) = 2,350$ and $n(F \cap S) = 920$.

Now, $n(F \cup S) = n(F) + n(S) - n(F \cap S)$
$$= 1,475 + 2,350 - 920$$
$$= 2,905$$

Thus, 2,905 individuals contributed to either the first campaign or the second campaign.

EXERCISE 6-2

Things to remember:

1. FACTORIAL

   For n a natural number,
   $$n! = n(n - 1)(n - 2) \cdot \ldots \cdot 3 \cdot 2 \cdot 1$$
   $$0! = 1$$
   $$n! = n(n - 1)!$$

   [NOTE: Many calculators have an $\boxed{n!}$ key or its equivalent.]

2. PERMUTATIONS

   A PERMUTATION of a set of distinct objects is an arrangement of the objects in a specific order, without repetitions. The number of permutations of n distinct objects without repetition, denoted by $P_{n,n}$, is:
   $$P_{n,n} = n(n - 1) \cdot \ldots \cdot 3 \cdot 2 \cdot 1 = n! \quad (n \text{ factors})$$

3. PERMUTATIONS OF n OBJECTS TAKEN r AT A TIME

   A permutation of a set of n distinct objects taken r at a time without repetition is an arrangement of the r objects in a specific order. The number of permutations of n objects taken r at a time, denoted by $P_{n,r}$, is given by:
   $$P_{n,r} = n(n - 1)(n - 2) \cdot \ldots \cdot (n - r + 1)$$
   $$(r \text{ factors})$$
   or $P_{n,r} = \dfrac{n!}{(n - r)!} \quad 0 \leq r \leq n$

   [Note: $P_{n,n} = \dfrac{n!}{(n - n)!} = \dfrac{n!}{0!} = n!$, the number of permutations of n objects taken n at a time. Remember, by definition, $0! = 1$.]

<u>4</u>. COMBINATIONS OF *n* OBJECTS TAKEN *r* AT A TIME

A combination of a set of *n* distinct objects taken *r* at a time without repetition is an *r*-element subset of the set of *n* objects. (The arrangement of the elements in the subset does not matter.) The number of combinations of *n* objects taken *r* at a time, denoted by $C_{n,r}$ or by $\binom{n}{r}$ is given by:

$$C_{n,r} = \binom{n}{r} = \frac{P_{n,r}}{r!} = \frac{n!}{r!(n-r)!} \qquad 0 \le r \le n$$

<u>5</u>. NOTE: In a permutation, the ORDER of the objects counts. In a combination, order does not count.

---

1. $4! = 4 \cdot 3 \cdot 2 \cdot 1 = 24$

3. $\dfrac{9!}{8!} = \dfrac{9 \cdot 8!}{8!} = 9$

5. $\dfrac{11!}{8!} = \dfrac{11 \cdot 10 \cdot 9 \cdot 8!}{8!} = 990$

7. $\dfrac{5!}{2!3!} = \dfrac{5 \cdot 4 \cdot 3!}{2 \cdot 1 \cdot 3!} = 10$

9. $\dfrac{7!}{4!(7-4)!} = \dfrac{7!}{4!3!} = \dfrac{7 \cdot 6 \cdot 5 \cdot 4!}{4! \cdot 3 \cdot 2 \cdot 1} = 35$

11. $\dfrac{7!}{7!(7-7)!} = \dfrac{7!}{7!0!} = \dfrac{1}{1} = 1$

13. $P_{5,3} = \dfrac{5!}{(5-3)!} = \dfrac{5!}{2!} = \dfrac{5 \cdot 4 \cdot 3 \cdot 2!}{2!} = 60$

15. $P_{52,4} = \dfrac{52!}{(52-4)!} = \dfrac{52!}{48!} = \dfrac{52 \cdot 51 \cdot 50 \cdot 49 \cdot 48!}{48!} = 6,497,400$

17. $C_{5,3} = \dfrac{5!}{3!(5-3)!} = \dfrac{5!}{3!2!} = \dfrac{5 \cdot 4 \cdot 3!}{3! \cdot 2 \cdot 1} = 10$

19. $C_{52,4} = \dfrac{52!}{4!(52-4)!} = \dfrac{52!}{4!48!} = \dfrac{52 \cdot 51 \cdot 50 \cdot 49 \cdot 48!}{4 \cdot 3 \cdot 2 \cdot 1 \cdot 48!} = 270,725$

21. The number of different finishes (win, place, show) for the ten horses is the number of permutations of 10 objects 3 at a time. This is:

$$P_{10,3} = \frac{10!}{(10-3)!} = \frac{10!}{7!} = \frac{10 \cdot 9 \cdot 8 \cdot 7!}{7!} = 720$$

23. (A) The number of ways that a three-person subcommittee can be selected from a seven-member committee is the number of combinations (since order *is not* important in selecting a subcommittee) of 7 objects 3 at a time. This is:

$$C_{7,3} = \frac{7!}{3!(7-3)!} = \frac{7!}{3!4!} = \frac{7 \cdot 6 \cdot 5 \cdot 4!}{3 \cdot 2 \cdot 1 \cdot 4!} = 35$$

(B) The number of ways a president, vice-president, and secretary can be chosen from a committee of 7 people is the number of permutations (since order *is* important in choosing 3 people for the positions) of 7 objects 3 at a time. This is:

$$P_{7,3} = \frac{7!}{(7-3)!} = \frac{7!}{4!} = \frac{7 \cdot 6 \cdot 5 \cdot 4!}{4!} = 7 \cdot 6 \cdot 5 = 210$$

25. Calculate $x!$, 3^x, x^3 for $x = 1, 2, 3, \ldots$
For $x \geq 7$, $x! > 3^x$; for $x \geq 1$, $3^x > x^3$ (except for $x = 3$). In general, $x!$ grows much faster than 3^x and 3^x grows much faster than x^3.

27. This is a "combinations" problem. We want the number of ways to select 5 objects from 13 objects with order not counting. This is:

$$C_{13,5} = \frac{13!}{5!(13-5)!} = \frac{13!}{5!8!} = \frac{13 \cdot 12 \cdot 11 \cdot 10 \cdot 9 \cdot 8!}{5 \cdot 4 \cdot 3 \cdot 2 \cdot 1 \cdot 8!} = 1287$$

29. The five spades can be selected in $C_{13,5}$ ways and the two hearts can be selected in $C_{13,2}$ ways. Applying the Multiplication Principle, we have:

$$\text{Total number of hands} = C_{13,5} \cdot C_{13,2} = \frac{13!}{5!(13-5)!} \cdot \frac{13!}{2!(13-2)!}$$

$$= \frac{13!}{5!8!} \cdot \frac{13!}{2!11!} = 100,386$$

31. The three appetizers can be selected in $C_{8,3}$ ways. The four main courses can be selected in $C_{10,4}$ ways. The two desserts can be selected in $C_{7,2}$ ways. Now, applying the Multiplication Principle, the total number of ways in which the above can be selected is given by:

$$C_{8,3} \cdot C_{10,4} \cdot C_{7,2} = \frac{8!}{3!(8-3)!} \cdot \frac{10!}{4!(10-4)!} \cdot \frac{7!}{2!(7-2)!} = 246,960$$

33. (A) A chord joins two distinct points. Thus, the total number of chords is given by:

$$C_{8,2} = \frac{8!}{2!(8-2)!} = \frac{8!}{2!6!} = \frac{8 \cdot 7 \cdot 6!}{2 \cdot 1 \cdot 6!} = 28$$

(B) Each triangle requires three distinct points. Thus, there are

$$C_{8,3} = \frac{8!}{3!(8-3)!} = \frac{8!}{3!5!} = \frac{8 \cdot 7 \cdot 6 \cdot 5!}{3 \cdot 2 \cdot 1 \cdot 5!} = 56 \text{ triangles.}$$

(C) Each quadrilateral requires four distinct points. Thus, there are

$$C_{8,4} = \frac{8!}{4!(8-4)!} = \frac{8!}{4!4!} = \frac{8 \cdot 7 \cdot 6 \cdot 5 \cdot 4!}{4 \cdot 3 \cdot 2 \cdot 1 \cdot 4!} = 70 \text{ quadrilaterals.}$$

35. (A) Two people.

O_1: First person selects a chair O_2: Second person selects a chair
N_1: 5 ways N_2: 4 ways

Thus, there are
$N_1 \cdot N_2 = 5 \cdot 4 = 20$
ways to seat two people in a row of 5 chairs. Note that this is $P_{5,2}$.

(B) Three people. There will be $P_{5,3}$ ways to seat 3 people in a row of 5 chairs:

$$P_{5,3} = \frac{5!}{(5-3)!} = \frac{5!}{2!} = \frac{5 \cdot 4 \cdot 3 \cdot 2!}{2!} = 60$$

(C) Four people. The number of ways to seat 4 people in a row of 5 chairs is given by:

$$P_{5,4} = \frac{5!}{(5-4)!} = \frac{5!}{1!} = 5 \cdot 4 \cdot 3 \cdot 2 = 120$$

(D) Five people. The number of ways to seat 5 people in a row of 5 chairs is given by:

$$P_{5,5} = \frac{5!}{(5-1)!} = \frac{5!}{0!} = 5! = 120$$

37. (A) The distinct positions are taken into consideration. The number of starting teams is given by:

$$P_{8,5} = \frac{8!}{(8-5)!} = \frac{8!}{3!} = \frac{8 \cdot 7 \cdot 6 \cdot 5 \cdot 4 \cdot 3!}{3!} = 6720$$

(B) The distinct positions are not taken into consideration. The number of starting teams is given by:

$$C_{8,5} = \frac{8!}{5!(8-5)!} = \frac{8!}{5!3!} = \frac{8 \cdot 7 \cdot 6 \cdot 5!}{5! \cdot 3 \cdot 2 \cdot 1} = 56$$

(C) Either Mike or Ken, but not both, must start; distinct positions are not taken into consideration.

O_1: Select either Mike or Ken

N_1: 2 ways

O_2: Select 4 players from the remaining 6

N_2: $C_{6,4}$

Thus, the number of starting teams is given by:

$$N_1 \cdot N_2 = 2 \cdot C_{6,4} = 2 \cdot \frac{6!}{4!(6-4)!} = 2 \cdot \frac{6 \cdot 5 \cdot 4!}{4! \cdot 2 \cdot 1} = 30$$

39. For many calculators, $k = 69$, but your calculator may be different. Note that $k!$ may also be calculated as $P_{k,k}$. On a TI-85, the largest integer k for which $k!$ can be calculated using $P_{k,k}$ is 449.

41. (A) Three printers are to be selected for the display. The *order* of selection does not count. Thus, the number of ways to select the 3 printers from 24 is:

$$C_{24,3} = \frac{24!}{3!(24-3)!} = \frac{24 \cdot 23 \cdot 22(21!)}{3 \cdot 2 \cdot 1(21!)} = 2,024$$

(B) Nineteen of the 24 printers are not defective. Thus, the number of ways to select 3 non-defective printers is:

$$C_{19,3} = \frac{19!}{3!(19-3)!} = \frac{19 \cdot 18 \cdot 17(16!)}{3 \cdot 2 \cdot 1(16!)} = 969$$

43. (A) There are 8 + 12 + 10 = 30 stores in all. The jewelry store chain will select 10 of these stores to close. Since order does not count here, the total number of ways to select the 10 stores to close is:

$$C_{30,10} = \frac{30!}{10!(30 - 10)!} = \frac{30 \cdot 29 \cdot 28 \cdot 27 \cdot 26 \cdot 25 \cdot 24 \cdot 23 \cdot 22 \cdot 21(20!)}{10 \cdot 9 \cdot 8 \cdot 7 \cdot 6 \cdot 5 \cdot 4 \cdot 3 \cdot 2 \cdot 1(20!)}$$
$$= 30,045,015$$

(B) The number of ways to close 2 stores in Georgia is: $C_{8,2}$
The number of ways to close 5 stores in Florida is: $C_{12,5}$
The number of ways to close 3 stores in Alabama is: $C_{10,3}$

By the multiplication principle, the total number of ways to select the 10 stores for closing is:

$$C_{8,2} \cdot C_{12,5} \cdot C_{10,3} = \frac{8!}{2!(8 - 2)!} \cdot \frac{12!}{5!(12 - 5)!} \cdot \frac{10!}{3!(10 - 3)!}$$
$$= \frac{8 \cdot 7 \cdot 6!}{2 \cdot 1 \cdot 6!} \cdot \frac{12 \cdot 11 \cdot 10 \cdot 9 \cdot 8(7!)}{5 \cdot 4 \cdot 3 \cdot 2 \cdot 1(7!)} \cdot \frac{10 \cdot 9 \cdot 8(7!)}{3 \cdot 2 \cdot 1(7!)}$$
$$= 28 \cdot 792 \cdot 120 = 2,661,120$$

45. (A) Three females can be selected in $C_{6,3}$ ways. Two males can be selected in $C_{5,2}$ ways. Applying the Multiplication Principle, we have:

Total number of ways $= C_{6,3} \cdot C_{5,2} = \dfrac{6!}{3!(6 - 3)!} \cdot \dfrac{5!}{2!(5 - 2)!} = 200$

(B) Four females and one male can be selected in $C_{6,4} \cdot C_{5,1}$ ways. Thus,

$$C_{6,4} \cdot C_{5,1} = \frac{6!}{4!(6 - 4)!} \cdot \frac{5!}{1!(5 - 1)!} = 75$$

(C) Number of ways in which 5 females can be selected is:

$$C_{6,5} = \frac{6!}{5!(6 - 5)!} = 6$$

(D) Number of ways in which 5 people can be selected is:

$$C_{6+5,5} = C_{11,5} = \frac{11!}{5!(11 - 5)!} = 462$$

(E) At least four females includes four females and five females. Four females and one male can be selected in 75 ways [see part (B)]. Five females can be selected in 6 ways [see part (C)]. Thus,

Total number of ways $= C_{6,4} \cdot C_{5,1} + C_{6,5} = 75 + 6 = 81$

47. (A) Select 3 samples from 8 blood types, no two samples having the same type. This is a permutation problem. The number of different examinations is:

$$P_{8,3} = \frac{8!}{(8 - 3)!} = \frac{8!}{5!} = \frac{8 \cdot 7 \cdot 6 \cdot 5!}{5!} = 336$$

(B) Select 3 samples from 8 blood types, repetition is allowed.

O_1: Select the first sample O_3: Select the third sample
N_1: 8 ways N_3: 8 ways

O_2: Select the second sample
N_2: 8 ways

Thus, the number of different examinations in this case is:

$N_1 \cdot N_2 \cdot N_3 = 8 \cdot 8 \cdot 8 = 8^3 = 512$

49. This is a permutations problem. The number of buttons is given by:

$$P_{4,2} = \frac{4!}{(4-2)!} = \frac{4!}{2!} = \frac{4 \cdot 3 \cdot 2!}{2!} = 12$$

EXERCISE 6-3

Things to remember:

<u>1.</u>   SAMPLE SPACE

A set S is a SAMPLE SPACE for an experiment if:

(a) Each element of S is an outcome of the experiment.

(b) Each outcome of the experiment corresponds to one and only one element of S.

Each element in the sample space is called a SIMPLE OUTCOME or SIMPLE EVENT.

<u>2.</u>   EVENT

Given a sample space S. An EVENT E is any subset of S (including the empty set $\varnothing$ and the sample space S). An event with only one element is called a SIMPLE EVENT; an event with more than one element is a COMPOUND EVENT. An event E *occurs* if the result of performing the experiment is one of the simple events in E.

<u>3.</u>   There is no one correct sample space for a given experiment. When specifying a sample space for an experiment, include as much detail as necessary to answer all questions of interest regarding the outcomes of the experiment. When in doubt, choose a sample space with more elements rather than fewer.

4. PROBABILITIES FOR SIMPLE EVENTS

Given a sample space
$$S = \{e_1, e_2, \ldots, e_n\}.$$
To each simple event e_i assign a real number denoted by $P(e_i)$, called the PROBABILITY OF THE EVENT e_i. These numbers can be assigned in an arbitrary manner provided the following two conditions are satisfied:

(a) $0 \leq P(e_i) \leq 1$

   (The probability of a simple event is a number between 0 and 1, inclusive.)

(b) $P(e_1) + P(e_2) + \ldots + P(e_n) = 1$

   (The sum of the probabilities of all simple events in the sample space is 1.)

Any probability assignment that meets these two conditions is called an ACCEPTABLE PROBABILITY ASSIGNMENT.

5. PROBABILITY OF AN EVENT E

Given an acceptable probability assignment for the simple events in a sample space S, the probability of an arbitrary event E, denoted $P(E)$, is defined as follows:

(a) $P(E) = 0$ if E is the empty set.

(b) If E is a simple event, then $P(E)$ has already been assigned.

(c) If E is a compound event, then $P(E)$ is the sum of the probabilities of all the simple events in E.

(d) If $E = S$, then $P(E) = P(S) = 1$ [this follows from 4(b)].

6. STEPS FOR FINDING THE PROBABILITY OF AN EVENT E

(a) Set up an appropriate sample space S for the experiment.

(b) Assign acceptable probabilities to the simple events in S.

(c) To obtain the probability of an arbitrary event E, add the probabilities of the simple events in E.

7. EMPIRICAL PROBABILITY

If an experiment is conducted n times and event E occurs with FREQUENCY $f(E)$, then the ratio $f(E)/n$ is called the RELATIVE FREQUENCY of the occurrence of event E in n trials. The EMPIRICAL PROBABILITY of E, denoted by $P(E)$, is given by the number (if it exists) that the relative frequency $f(E)/n$ approaches as n gets larger and larger. For any particular n, the relative frequency $f(E)/n$ is also called the APPROXIMATE EMPIRICAL PROBABILITY of event E:

$$P(E) \approx \frac{\text{Frequency of occurrence of } E}{\text{Total number of trials}} = \frac{f(E)}{n}$$

(The larger n is, the better the approximation.)

<u>8.</u>   PROBABILITIES UNDER AN EQUALLY LIKELY ASSUMPTION

If, in a sample space

$$S = \{e_1, e_2, \ldots, e_n\},$$

each simple event is as likely to occur as any other, then $P(e_i) = \frac{1}{n}$, for $i = 1, 2, \ldots, n$, i.e. assign the same probability, $1/n$, to each simple event. The probability of an arbitrary event E in this case is:

$$P(E) = \frac{\text{Number of elements in } E}{\text{Number of elements in } S} = \frac{n(E)}{n(S)}$$

---

1. $P(E) = 1$ means that the occurrence of E is certain.

3. Let B = boy and G = girl. Then

$$S = \{(B, B), (B, G), (G, B), (G, G)\}$$

where (B, B) means both children are boys, (B, G) means the first child is a boy, the second is a girl, and so on. The event E corresponding to having two children of opposite sex is $E = \{(B, G), (G, B)\}$. Since the simple events are equally likely,

$$P(E) = \frac{n(E)}{n(S)} = \frac{2}{4} = \frac{1}{2}.$$

5. (A) Reject; $P(G) = -0.35$ — no probability can be negative.

(B) Reject; $P(J) + P(G) + P(P) + P(S) = 0.32 + 0.28 + 0.24 + 0.30$
$$= 1.14 \neq 1$$

(C) Acceptable; each probability is between 0 and 1 (inclusive), and the sum of the probabilities is 1.

7. $P(\{J, P\}) = P(J) + P(P) = 0.26 + 0.30 = 0.56.$

9. $S = \{(B, B, B), (B, B, G), (B, G, B), (B, G, G), (G, B, B), (G, B, G),$
$$\qquad (G, G, B), (G, G, G)\}$$
$$E = \{(B, B, G)\}$$
Since the events are equally likely and $n(S) = 8$, $P(E) = \frac{1}{8}.$

11. The number of three-digit sequences with no digit repeated is $P_{10,3}$.
Since the possible opening combinations are equally likely, the probability of guessing the right combination is:

$$\frac{1}{P_{10,3}} = \frac{1}{10 \cdot 9 \cdot 8} = \frac{1}{720} \approx 0.0014$$

13. Let S = the set of five-card hands. Then $n(S) = C_{52,5}$.
Let A = "five black cards." Then $n(A) = C_{26,5}$.
Since individual hands are equally likely to occur:

$$P(A) = \frac{n(A)}{n(S)} = \frac{C_{26,5}}{C_{52,5}} = \frac{\dfrac{26!}{5!21!}}{\dfrac{52!}{5!47!}} = \frac{26 \cdot 25 \cdot 24 \cdot 23 \cdot 22}{52 \cdot 51 \cdot 50 \cdot 49 \cdot 48} \approx 0.025$$

15. S = set of five-card hands; $n(S) = C_{52,5}$.

F = "five face cards"; $n(F) = C_{12,5}$.

Since individual hands are equally likely to occur:

$$P(F) = \frac{n(F)}{n(S)} = \frac{C_{12,5}}{C_{52,5}} = \frac{\frac{12!}{5!7!}}{\frac{52!}{5!47!}} = \frac{12\cdot 11\cdot 10\cdot 9\cdot 8}{52\cdot 51\cdot 50\cdot 49\cdot 48} \approx 0.000305$$

17. S = {all the days in a year} (assume 365 days; exclude leap year); Equivalently, number the days of the year, beginning with January 1. Then $S = \{1, 2, 3, \ldots, 365\}$.

Assume that each day is as likely as any other day for a person to be born. Then the probability of each simple event is: $\frac{1}{365}$.

19. $n(S) = P_{5,5} = 5! = 120$

Let A = all notes inserted into the correct envelopes. Then $n(A) = 1$ and

$$P(A) = \frac{n(A)}{n(S)} = \frac{1}{120} \approx 0.00833$$

21. Using the sample space shown in Figure 3, we have

$n(S) = 36$, $n(A) = 1$,

where Event A = "Sum being 2":

$$P(A) = \frac{n(A)}{n(S)} = \frac{1}{36}$$

23. Let E = "Sum being 6." Then $n(E) = 5$. Thus, $P(E) = \frac{n(E)}{n(S)} = \frac{5}{36}$.

25. Let E = "Sum being less than 5." Then $n(E) = 6$. Thus,

$$P(E) = \frac{n(E)}{n(S)} = \frac{6}{36} = \frac{1}{6}.$$

27. Let E = "Sum not 7 or 11." Then $n(E) = 28$ and $P(E) = \frac{n(E)}{n(S)} = \frac{28}{36} = \frac{7}{9}$.

29. E = "Sum being 1" is not possible. Thus, $P(E) = 0$.

31. Let E = "Sum is divisible by 3" = "Sum is 3, 6, 9, or 12." Then

$n(E) = 12$ and $P(E) = \frac{n(E)}{n(S)} = \frac{12}{36} = \frac{1}{3}$.

33. Let E = "Sum is 7 or 11." Then $n(E) = 8$. Thus, $P(E) = \frac{n(E)}{n(S)} = \frac{8}{36} = \frac{2}{9}$.

35. Let E = "Sum is divisible by 2 or 3" = "Sum is 2, 3, 4, 6, 8, 9, 10, 12." Then $n(E) = 24$, and $P(E) = \frac{n(E)}{n(S)} = \frac{24}{36} = \frac{2}{3}$.

For Problems 37—41, the sample space S is given by:

$S = \{(H, H, H), (H, H, T), (H, T, H), (H, T, T)\}$

The outcomes are equally likely and $n(S) = 4$.

37. Let E = "1 head." Then $n(E) = 1$ and $P(E) = \dfrac{n(E)}{n(S)} = \dfrac{1}{4}$.

39. Let E = "3 heads." Then $n(E) = 1$ and $P(E) = \dfrac{n(E)}{n(S)} = \dfrac{1}{4}$.

41. Let E = "More than 1 head." Then $n(E) = 3$ and $P(E) = \dfrac{n(E)}{n(S)} = \dfrac{3}{4}$.

For Problems 43—49, the sample space S is given by:

$$S = \begin{Bmatrix} (1,\ 1),\ (1,\ 2),\ (1,\ 3) \\ (2,\ 1),\ (2,\ 2),\ (2,\ 3) \\ (3,\ 1),\ (3,\ 2),\ (3,\ 3) \end{Bmatrix}$$

The outcomes are equally likely and n(S) = 9.

43. Let E = "Sum is 2." Then $n(E) = 1$ and $P(E) = \dfrac{n(E)}{n(S)} = \dfrac{1}{9}$.

45. Let E = "Sum is 4." Then $n(E) = 3$ and $P(E) = \dfrac{n(E)}{n(S)} = \dfrac{3}{9} = \dfrac{1}{3}$.

47. Let E = "Sum is 6." Then $n(E) = 1$ and $P(E) = \dfrac{n(E)}{n(S)} = \dfrac{1}{9}$.

49. Let E = "Sum is odd" = "Sum is 3 or 5." Then $n(E) = 4$ and
$P(E) = \dfrac{n(E)}{n(S)} = \dfrac{4}{9}$.

For Problems 51—57, the sample space S is the set of all 5-card hands.
Then $n(S) = C_{52,5}$. The outcomes are equally likely.

51. Let E = "5 cards, jacks through aces." Then $n(E) = C_{16,5}$. Thus,

$$P(E) = \frac{C_{16,5}}{C_{52,5}} = \frac{\dfrac{16!}{5!11!}}{\dfrac{52!}{5!47!}} = \frac{16 \cdot 15 \cdot 14 \cdot 13 \cdot 12}{52 \cdot 51 \cdot 50 \cdot 49 \cdot 48} \approx 0.00168.$$

53. Let E = "4 aces." Then $n(E) = 48$ (the remaining card can be any one of the 48 cards which are not aces). Thus,

$$P(E) = \frac{48}{C_{52,5}} = \frac{48}{\dfrac{52!}{5!47!}} = \frac{48 \cdot 5!}{52 \cdot 51 \cdot 50 \cdot 49 \cdot 48} = \frac{5 \cdot 4 \cdot 3 \cdot 2}{52 \cdot 51 \cdot 50 \cdot 49} \approx 0.0000185$$

55. Let E = "Straight flush, ace high." Then $n(E) = 4$ (one such hand in each suit). Thus,

$$P(E) = \frac{4}{C_{52,5}} = \frac{4 \cdot 5!}{52 \cdot 51 \cdot 50 \cdot 49 \cdot 48} = \frac{480}{52 \cdot 51 \cdot 50 \cdot 49 \cdot 48} \approx 0.0000015$$

57. Let E = "2 aces and 3 queens." The number of ways to get 2 aces is $C_{4,2}$ and the number of ways to get 3 queens is $C_{4,3}$.

Thus, $n(E) = C_{4,2} \cdot C_{4,3} = \dfrac{4!}{2!2!} \cdot \dfrac{4!}{3!1!} = \dfrac{4 \cdot 3}{2} \cdot \dfrac{4}{1} = 24$

and

$P(E) = \dfrac{n(E)}{n(S)} = \dfrac{24}{C_{52,5}} = \dfrac{24 \cdot 5!}{52 \cdot 51 \cdot 50 \cdot 59 \cdot 48} \approx 0.000009.$

59. (A) The sample space S is the set of all possible permutations of the 12 brands taken 4 at a time, and $n(S) = P_{12,4}$. Thus, the probability of selecting 4 brands and identifying them correctly, with no answer repeated, is:

$P(E) = \dfrac{1}{P_{12,4}} = \dfrac{1}{\dfrac{12!}{(12-4)!}} = \dfrac{1}{12 \cdot 11 \cdot 10 \cdot 9} \approx 0.000084$

(B) Allowing repetition, $n(S) = 12^4$ and the probability of identifying them correctly is:

$P(F) = \dfrac{1}{12^4} \approx 0.000048$

61. (A) Total number of applicants = 6 + 5 = 11.

$n(S) = C_{11,5} = \dfrac{11!}{5!(11-5)!} = 462$

The number of ways that three females and two males can be selected is:

$C_{6,3} \cdot C_{5,2} = \dfrac{6!}{3!(6-3)!} \cdot \dfrac{5!}{2!(5-2)!} = 20 \cdot 10 = 200$

Thus, $P(A) = \dfrac{C_{6,3} \cdot C_{5,2}}{C_{11,5}} = \dfrac{200}{462} = 0.433$

(B) $P(\text{4 females and 1 male}) = \dfrac{C_{6,4} \cdot C_{5,1}}{C_{11,5}} = 0.162$

(C) $P(\text{5 females}) = \dfrac{C_{6,5}}{C_{11,5}} = 0.013$

(D) $P(\text{at least four females}) = P(\text{4 females and 1 male}) + P(\text{5 females})$

$= \dfrac{C_{6,4} \cdot C_{5,1}}{C_{11,5}} + \dfrac{C_{6,5}}{C_{11,5}}$

$= 0.162 + 0.013 \text{ [refer to parts (B) and (C)]}$

$= 0.175$

63. (A) The sample space S consists of the number of permutations of the 8 blood types chosen 3 at a time. Thus, $n(S) = P_{8,3}$ and the probability of guessing the three types in a sample correctly is:

$P(E) = \dfrac{1}{P_{8,3}} = \dfrac{1}{\dfrac{8!}{(8-3)!}} = \dfrac{1}{8 \cdot 7 \cdot 6} \approx 0.0030$

(B) Allowing repetition, $n(S) = 8^3$ and the probabilty of guessing the three types in a sample correctly is:

$$P(E) = \frac{1}{8^3} \approx 0.0020$$

65. (A) The total number of ways of selecting a president and a vice-president from the 11 members of the council is:

$P_{11,2}$, i.e., $n(S) = P_{11,2}$.

The total number of ways of selecting the president and the vice-president from the 6 democrats is $P_{6,2}$. Thus, if E is the event "The president and vice-president are both Democrats," then

$$P(E) = \frac{P_{6,2}}{P_{11,2}} = \frac{\dfrac{6!}{(6-2)!}}{\dfrac{11!}{(11-2)!}} = \frac{6 \cdot 5}{11 \cdot 10} = \frac{30}{110} \approx 0.273.$$

(B) The total number of ways of selecting a committee of 3 from the 11 members of the council is:

$C_{11,3}$, i.e., $n(S) = C_{11,3} = \dfrac{11!}{3!(11-3)!} = \dfrac{11 \cdot 10 \cdot 9 \cdot 8!}{3 \cdot 2 \cdot 1 \cdot 8!} = 165$

If we let F be the event "The majority are Republicans," which is the same as having either 2 Republicans and 1 Democrat or all 3 Republicans, then

$$n(F) = C_{5,2} \cdot C_{6,1} + C_{5,3} = \frac{5!}{2!(5-2)!} \cdot \frac{6!}{1!(6-1)!} + \frac{5!}{3!(5-3)!}$$

$$= 10 \cdot 6 + 10 = 70.$$

Thus,

$$P(F) = \frac{n(F)}{n(S)} = \frac{70}{165} \approx 0.424.$$

EXERCISE 6-4

Things to remember:

1. UNION AND INTERSECTION OF EVENTS

If A and B are two events in a sample space S, then the UNION of A and B, denoted by $A \cup B$, and the INTERSECTION of A and B, denoted by $A \cap B$, are defined as follows:

$A \cup B = \{e \in S \mid e \in A \text{ OR } e \in B\}$ $A \cap B = \{e \in S \mid e \in A \text{ AND } e \in B\}$

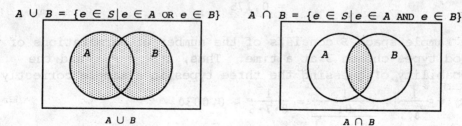

$A \cup B$ $A \cap B$

Furthermore, we define:

The **event A or B** to be $A \cup B$.

The **event A and B** to be $A \cap B$.

2. PROBABILITY OF A UNION OF TWO EVENTS

For any events A and B,

(a) $P(A \cup B) = P(A) + P(B) - P(A \cap B)$.

If A and B are MUTUALLY EXCLUSIVE $(A \cap B = \varnothing)$, then

(b) $P(A \cup B) = P(A) + P(B)$.

3. PROBABILITY OF COMPLEMENTS

For any event E, $E \cup E' = S$ and $E \cap E' = \varnothing$. Thus,

$$P(E) = 1 - P(E')$$

$$P(E') = 1 - P(E)$$

4. PROBABILITY TO ODDS

If $P(E)$ is the probability of the event E, then:

(a) **Odds for** $E = \dfrac{P(E)}{1 - P(E)} = \dfrac{P(E)}{P(E')}$ $[P(E) \neq 1]$

(b) **Odds against** $E = \dfrac{P(E')}{P(E)}$ $[P(E) \neq 0]$

[NOTE: When possible, odds are expressed as ratios of whole numbers.]

5. ODDS TO PROBABILTY

If the odds for an event E are $\dfrac{a}{b}$, then the probability of E is:

$$P(E) = \frac{a}{a + b}$$

---

1. Let E be the event "failing within 90 days."
   Then E' = "not failing within 90 days."

   $P(E') = 1 - P(E)$ (using 2)

   $\quad\quad = 1 - .003$

   $\quad\quad = .997$

3. Let Event A = "a number less than 3" = {1, 2}.
   Let Event B = "a number greater than 7" = {8, 9, 10}.

   Since $A \cap B = \varnothing$, A and B are mutually exclusive. So, using 1(b),

   $P(A \cup B) = P(A) + P(B) = \dfrac{n(A)}{n(S)} + \dfrac{n(B)}{n(S)} = \dfrac{2}{10} + \dfrac{3}{10} = \dfrac{1}{2}$

5. Let Event A = "an even number" = {2, 4, 6, 8, 10}.
   Let Event B = "a number divisible by 3" = {3, 6, 9}.
   Since $A \cap B = \{6\} \neq \varnothing$, A and B are not mutually exclusive. So, using 1(a),

   $P(A \cup B) = P(A) + P(B) - P(A \cap B) = \dfrac{5}{10} + \dfrac{3}{10} - \dfrac{1}{10} = \dfrac{7}{10}$

7. $P(A) = \dfrac{35 + 5}{35 + 5 + 20 + 40}$

$= \dfrac{40}{100} = .4$

9. $P(B) = \dfrac{5 + 20}{35 + 5 + 20 + 40}$

$= \dfrac{25}{100} = .25$

11. $P(A \cap B) = \dfrac{5}{35 + 5 + 20 + 40}$

$= \dfrac{5}{100} = .05$

13. $P(A' \cap B) = \dfrac{20}{35 + 5 + 20 + 40}$

$= \dfrac{20}{100} = .2$

15. $P(A \cup B) = \dfrac{35 + 5 + 20}{35 + 5 + 20 + 40}$

$= \dfrac{60}{100} = .6$

17. $P(A' \cup B) = \dfrac{20 + 40 + 5}{35 + 5 + 20 + 40}$

$= \dfrac{65}{100} = .65$

19. $P(\text{sum of 5 or 6}) = P(\text{sum of 5}) + P(\text{sum of 6})$ [using $\underline{1}$(b)]

$= \dfrac{4}{36} + \dfrac{5}{36} = \dfrac{9}{36} = \dfrac{1}{4}$ or $.25$

21. $P(\text{1 on first die or 1 on second die})$ [using $\underline{1}$(a)]

$= P(\text{1 on first die}) + P(\text{1 on second die}) - P(\text{1 on both dice})$

$= \dfrac{6}{36} + \dfrac{6}{36} - \dfrac{1}{36} = \dfrac{11}{36}$

23. Use $\underline{3}$ to find the odds for Event E.

(A) $P(E) = \dfrac{3}{8}$, $P(E') = 1 - P(E) = \dfrac{5}{8}$

    Odds for $E = \dfrac{P(E)}{P(E')}$

$= \dfrac{3/8}{5/8} = \dfrac{3}{5}$ (3 to 5)

    Odds against $E = \dfrac{P(E')}{P(E)}$

$= \dfrac{5/8}{3/8} = \dfrac{5}{3}$ (5 to 3)

(B) $P(E) = \dfrac{1}{4}$, $P(E') = 1 - P(E) = \dfrac{3}{4}$

    Odds for $E = \dfrac{P(E)}{P(E')}$

$= \dfrac{1/4}{3/4} = \dfrac{1}{3}$ (1 to 3)

    Odds against $E = \dfrac{P(E')}{P(E)}$

$= \dfrac{3/4}{1/4} = \dfrac{3}{1}$ (3 to 1)

(C) $P(E) = .4$, $P(E') = 1 - P(E) = .6$

    Odds for $E = \dfrac{P(E)}{P(E')}$

$= \dfrac{.4}{.6} = \dfrac{2}{3}$ (2 to 3)

    Odds against $E = \dfrac{P(E')}{P(E)}$

$= \dfrac{.6}{.4} = \dfrac{3}{2}$ (3 to 2)

(D) $P(E) = .55$, $P(E') = 1 - P(E) - .45$

    Odds for $E = \dfrac{P(E)}{P(E')}$

$= \dfrac{.55}{.45} = \dfrac{11}{9}$ (11 to 9)

    Odds against $E = \dfrac{P(E')}{P(E)}$

$= \dfrac{.45}{.55} = \dfrac{9}{11}$ (9 to 11)

25. Use $\underline{4}$ to find the probabilty of event E.

(A) Odds for $E = \dfrac{3}{8}$

$P(E) = \dfrac{3}{3 + 8} = \dfrac{3}{11}$

(B) Odds for $E = \dfrac{11}{7}$

$P(E) = \dfrac{11}{11 + 7} = \dfrac{11}{18}$

(C) Odds for $E = \dfrac{4}{1}$

$$P(E) = \dfrac{4}{4+1} = \dfrac{4}{5} = .8$$

(D) Odds for $E = \dfrac{49}{51}$

$$P(E) = \dfrac{49}{49+51} = \dfrac{49}{100} = .49$$

27. Odds for $E = \dfrac{P(E)}{P(E')} = \dfrac{1/2}{1/2} = 1.$

The odds in favor of getting a head in a single toss of a coin are 1 to 1.

29. The sample space for this problem is:

$S = \{HHH, HHT, THH, HTH, TTH, HTT, THT, TTT\}$

Let Event E = "getting at least 1 head."
Let Event E' = "getting no heads."

Thus, $\dfrac{P(E)}{P(E')} = \dfrac{7/8}{1/8} = \dfrac{7}{1}$

The odds in favor of getting at least 1 head are 7 to 1.

31. Let Event E = "getting a number greater than 4."
Let Event E' = "not getting a number greater than 4."

Thus, $\dfrac{P(E')}{P(E)} = \dfrac{4/6}{2/6} = \dfrac{2}{1}$

The odds against getting a number greater than 4 in a single roll of a die are 2 to 1.

33. Let Event E = "getting 3 or an even number" = $\{2, 3, 4, 6\}$.
Let Event E' = "not getting 3 or an even number" = $\{1, 5\}$.

Thus, $\dfrac{P(E')}{P(E)} = \dfrac{2/6}{4/6} = \dfrac{1}{2}$

The odds against getting 3 or an even number are 1 to 2.

35. Let E = "rolling a five." Then $P(E) = \dfrac{n(E)}{n(S)} = \dfrac{4}{36} = \dfrac{1}{9}$ and $P(E') = \dfrac{8}{9}$.

(A) Odds for $E = \dfrac{1/9}{8/9} = \dfrac{1}{8}$ (1 to 8)

(B) The house should pay \$8 for the game to be fair (see Example 6).

37. (A) Let E = "sum is less than 4 or greater than 9." Then

$$P(E) = \dfrac{10 + 30 + 120 + 80 + 70}{1000} = \dfrac{310}{1000} = \dfrac{31}{100} = .31 \text{ and } P(E') = \dfrac{69}{100}.$$

Thus,

Odds for $E = \dfrac{31/100}{69/100} = \dfrac{31}{69}$

(B) Let F = "sum is even or divisible by 5." Then

$$P(F) = \dfrac{10 + 50 + 110 + 170 + 120 + 70 + 70}{1000} = \dfrac{600}{1000} = \dfrac{6}{10} = .6$$

and $P(F') = \dfrac{4}{10}$. Thus,

Odds for $F = \dfrac{6/10}{4/10} = \dfrac{6}{4} = \dfrac{3}{2}$

39. Let A = "drawing a face card" (Jack, Queen, King)
and B = "drawing a club."

Then $P(A \cup B) = P(A) + P(B) - P(A \cap B) = \frac{12}{52} + \frac{13}{52} - \frac{3}{52} = \frac{22}{52} = \frac{11}{26}$

$P[(A \cup B)'] = \frac{15}{26}$

Odds for $A \cup B = \frac{11/26}{15/26} = \frac{11}{15}$

41. Let A = "drawing a black card"
and B = "drawing an ace."

$P(A \cup B) = P(A) + P(B) - P(A \cap B) = \frac{26}{52} + \frac{4}{52} - \frac{2}{52} = \frac{28}{52} = \frac{7}{13}$

$P[(A \cup B)'] = \frac{6}{13}$

Odds for $A \cup B = \frac{7/13}{6/13} = \frac{7}{6}$

43. The sample space S is the set of all 5-card hands and $n(S) = C_{52,5}$

Let E = "getting at least one diamond."
Then E' = "no diamonds" and $n(E) = C_{39,5}$.

Thus, $P(E') = \dfrac{C_{39,5}}{C_{52,5}}$, and

$P(E) = 1 - \dfrac{C_{39,5}}{C_{52,5}} = 1 - \dfrac{\frac{39!}{5!34!}}{\frac{52!}{5!47!}} = 1 - \dfrac{39 \cdot 38 \cdot 37 \cdot 36 \cdot 35}{52 \cdot 51 \cdot 50 \cdot 49 \cdot 48} \approx 1 - .22 = .78.$

45. The number of numbers less than or equal to 1000 which are divisible by 6 is the largest integer in $\frac{1000}{6}$ or 166.

The number of numbers less than or equal to 1000 which are divisible by 8 is the largest integer in $\frac{1000}{8}$ or 125.

The number of numbers less than or equal to 1000 which are divisible by both 6 and 8 is the same as the number of numbers which are divisible by 24. This is the largest integer in $\frac{1000}{24}$ or 41.

Thus, if A is the event "selecting a number which is divisible by either 6 or 8," then

$n(A) = 166 + 125 - 41 = 250$ and $P(A) = \frac{250}{1000} = .25.$

47. In general, for three events A, B, and C,

$$P(A \cup B \cup C) = P(A) + P(B) + P(C) - P(A \cap B) - P(A \cap C) - P(B \cap C)$$
$$+ P(A \cap B \cap C)$$

Therefore,

$(*)$ $P(A \cup B \cup C) = P(A) + P(B) + P(C) - P(A \cap B)$

will hold if A and C, and B and C are mutually exclusive. Note that, if either $A \cap C = \emptyset$, or $B \cap C = \emptyset$, then $A \cap B \cap C = \emptyset$.

Equation (*) will also hold if A, B, and C are mutually exclusive, in which case
$$P(A \cup B \cup C) = P(A) + P(B) + P(C)$$

49. From Example 5,
$$P(E) = 1 - \frac{365!}{365^n(365 - n)!} = 1 - \frac{1}{365^n} \cdot \frac{365!}{(365 - n)!}$$
$$= 1 - \frac{1}{(365)^n} \cdot P_{365,n}$$
$$= 1 - \frac{P_{365,n}}{(365)^n}$$

For calculators with a $P_{n,r}$ key, this form involves fewer calculator steps. Also, 365! produces an overflow error on many calculators, while $P_{365,n}$ does not produce an overflow error for many values of n.

51. S = set of all lists of n birth months, $n \leq 12$. Then
$n(S) = 12 \cdot 12 \cdot \ldots \cdot 12$ (n times) $= 12^n$.

Let E = "at least two people have the same birth month."
Then E' = "no two people have the same birth month."

$n(E') = 12 \cdot 11 \cdot 10 \cdot \ldots \cdot [12 - (n - 1)]$
$$= \frac{12 \cdot 11 \cdot 10 \cdot \ldots \cdot [12 - (n - 1)](12 - n)[12 - (n + 1)] \cdot \ldots \cdot 3 \cdot 2 \cdot 1}{(12 - n)[12 - (n + 1)] \cdot \ldots \cdot 3 \cdot 2 \cdot 1}$$
$$= \frac{12!}{(12 - n)!}$$

Thus, $P(E') = \dfrac{\dfrac{12!}{(12 - n)!}}{12^n} = \dfrac{12!}{12^n(12 - n)!}$ and $P(E) = 1 - \dfrac{12!}{12^n(12 - n)!}$.

53. Odds for $E = \dfrac{P(E)}{P(E')} = \dfrac{P(E)}{1 - P(E)} = \dfrac{a}{b}$. Therefore,
$bP(E) = a[1 - P(E)] = a - aP(E)$.
Thus, $aP(E) + bP(E) = a$
$(a + b)P(E) = a$
$$P(E) = \frac{a}{a + b}$$

55. Venn diagram:

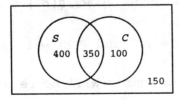

Let S be the event that the student owns a stereo and C be the event that the student owns a car.

The table corresponding to the given data is as follows:

	C	C'	Total
S	350	400	750
S'	100	150	250
Total	450	550	1000

The corresponding probabilities are:

	C	C'	Total
S	.35	.40	.75
S'	.10	.15	.25
Total	.45	.55	1.00

From the above table:

(A) $P(C \text{ or } S) = P(C \cup S) = P(C) + P(S) - P(C \cap S)$
$$= .45 + .75 - .35$$
$$= .85$$

(B) $P(C' \cap S') = .15$

57. (A) Using the table, we have:

$P(M_1 \text{ or } A) = P(M_1 \cup A) = P(M_1) + P(A) - P(M_1 \cap A)$
$$= .2 + .3 - .05$$
$$= .45$$

(B) $P[(M_2 \cap A') \cup (M_3 \cap A')] = P(M_2 \cap A') + P(M_3 \cap A')$
$$= .2 + .35 \text{ (from the table)}$$
$$= .55$$

59. Let K = "defective keyboard"
and D = "defective disk drive."

Then $K \cup D$ = "either a defective keyboard or a defective disk drive"
and $(K \cup D)'$ = "neither the keyboard nor the disk drive is defective"
$$= K' \cap D'$$

$P(K \cup D) = P(K) + P(D) - P(K \cap D) = .06 + .05 - .01 = .1$

Thus, $P(K' \cap D') = 1 - .1 = .9$.

61. The sample space S is the set of all possible 10-element samples from the 60 watches, and $n(S) = C_{60,10}$. Let E be the event that a sample contains at least one defective watch. Then E' is the event that a sample contains no defective watches. Now, $n(E') = C_{51,10}$.

Thus, $P(E') = \dfrac{C_{51,10}}{C_{60,10}} = \dfrac{\frac{51!}{10!41!}}{\frac{60!}{10!50!}} \approx .17$ and $P(E) \approx 1 - .17 = .83$.

Therefore, the probabilty that a sample will be returned is .83.

63. The given information is displayed in the Venn diagram:

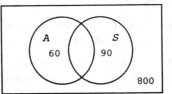

A = suffers from loss of appetite

S = suffers from loss of sleep

Thus, we can conclude that $n(A \cap S) = 1000 - (60 + 90 + 800) = 50.$

$P(A \cap S) = \dfrac{50}{1000} = .05$

65. (A) "Unaffiliated or no preference" $= U \cup N.$

$P(U \cup N) = P(U) + P(N) - P(U \cap N)$

$= \dfrac{150}{1000} + \dfrac{85}{1000} - \dfrac{15}{1000} = \dfrac{220}{1000} = \dfrac{11}{50} = .22$

Therefore, $P[(U \cup N)'] = 1 - \dfrac{11}{50} = \dfrac{39}{50}$ and

Odds for $U \cup N = \dfrac{11/50}{39/50} = \dfrac{11}{39}$

(B) "Affliated with a party and prefers candidate A" $= (D \cup R) \cap A.$

$P[(D \cup R) \cap A] = \dfrac{300}{1000} = \dfrac{3}{10} = .3$

The odds against this event are:

$\dfrac{1 - 3/10}{3/10} = \dfrac{7/10}{3/10} = \dfrac{7}{3}$

67. Let S = the set of all three-person groups from the total group.
Let E = the set of all three-person groups with at least one black.
Let E' = the set of all three-person groups with no blacks.
First, find $P(E')$, then use $P(E) = 1 - P(E')$ to find $P(E).$

$P(E') = \dfrac{n(E')}{n(E)} = \dfrac{C_{15,3}}{C_{20,3}} \approx .4$

$P(E) = 1 - P(E') \approx .6$

EXERCISE 6-5

Things to remember:

1. CONDITIONAL PROBABILITY

For events A and B in a sample space S, the CONDITIONAL PROBABILITY of A given B, denoted $P(A \mid B)$, is defined by

$$P(A \mid B) = \dfrac{P(A \cap B)}{P(B)}, \quad P(B) \neq 0$$

2. PRODUCT RULE

For events A and B, $P(A) \neq 0$, $P(B) \neq 0$, in a sample space S,

$$P(A \cap B) = P(A) \cdot P(B \mid A) = P(B) \cdot P(A \mid B).$$

[Note: We can use either $P(A) \cdot P(B \mid A)$ or $P(B) \cdot P(A \mid B)$ to compute $P(A \cap B).$]

<u>3</u>. PROBABILITY TREES

Given a sequence of probability experiments. To compute the probabilities of combined outcomes:

Step 1. Draw a tree diagram corresponding to all combined outcomes of the sequence of experiments.

Step 2. Assign a probability to each tree branch. (This is the probability of the occurrence of the event on the right end of the branch subject to the occurrence of all events on the path leading to the event on the right end of the branch. The probability of the occurrence of a combined outcome that corresponds to a path through the tree is the product of all branch probabilities on the path.*)

Step 3. Use the results in steps 1 and 2 to answer various questions related to the sequence of experiments as a whole.

If *A* and *B* are independent events with nonzero probabilities in a sample space *S*, then

(*) $P(A \mid B) = P(A)$ and $P(B \mid A) = P(B)$

If either equation in (*) holds, then *A* and *B* are independent.

<u>4</u>. INDEPENDENCE

Let *A* and *B* be any events in a sample space *S*. Then *A* and *B* are INDEPENDENT if and only if

$$P(A \cap B) = P(A) \cdot P(B).$$

Otherwise, *A* and *B* are DEPENDENT.

<u>5</u>. INDEPENDENT SET OF EVENTS

A set of events is said to be INDEPENDENT if for each finite subset $\{E_1, E_2, ..., E_k\}$

$$P(E_1 \cap E_2 \cap ... \cap E_k) = P(E_1) P(E_2) ... P(E_k)$$

---

1. $P(A) = .50$
See the given table.

3. $P(D) = .20$
See the given table.

5. $P(A \cap D) = .10$
See the given table for occurrences of both *A* and *D*.

7. $P(C \cap D) =$ probability of occurrences of both *C* and *D* = .06.

9. $P(A \mid D) = \dfrac{P(A \cap D)}{P(D)} = \dfrac{0.10}{0.20} = .50$ **11.** $P(C \mid D) = \dfrac{P(C \cap D)}{P(D)} = \dfrac{0.06}{0.20} = .30$

13. Events A and D are independent if $P(A \cap D) = P(A) \cdot P(D)$:

$P(A \cap D) = .10$

$P(A) \cdot P(D) = (.50)(.20) = .10$

Thus, A and D are independent.

15. $P(C \cap D) = .06$

$P(C) \cdot P(D) = (.20)(.20) = .04$

Since $P(C \cap D) \neq P(C) \cdot P(D)$, C and D are dependent.

17. (A) Let $H_8 =$ "a head on the eighth toss." Since each toss is independent of the other tosses, $P(H_8) = \frac{1}{2}$.

(B) Let $H_i =$ "a head on the ith toss." Since the tosses are independent,

$$P(H_1 \cap H_2 \cap \cdots \cap H_8) = P(H_1)P(H_2) \cdots P(H_8) = \left(\frac{1}{2}\right)^8 = \frac{1}{2^8} = \frac{1}{256}.$$

Similarly, if $T_i =$ "a tail on the ith toss," then

$$P(T_1 \cap T_2 \cap \cdots \cap T_8) = P(T_1)P(T_2) \cdots P(T_8) = \frac{1}{2^8} = \frac{1}{256}. \quad \text{Finally, if}$$

$H =$ "all heads" and $T =$ "all tails," then $H \cap T = \varnothing$ and

$$P(H \cup T) = P(H) + P(T) = \frac{1}{256} + \frac{1}{256} = \frac{2}{256} = \frac{1}{128} \approx .00781.$$

19. Given the table:

e_i	1	2	3	4	5
P_i	.3	.1	.2	.3	.1

$E =$ "pointer lands on an even number" $= \{2, 4\}$.

$F =$ "pointer lands on a number less than 4" $= \{1, 2, 3\}$.

(A) $P(F \mid E) = \dfrac{P(F \cap E)}{P(E)} = \dfrac{P(2)}{P(2) + P(4)} = \dfrac{.1}{.1 + .3} = \dfrac{.1}{.4} = \dfrac{1}{4}$

(B) $P(E \cap F) = P(2) = .1$

$P(E) = .4$,

$P(F) = P(1) + P(2) + P(3) = .3 + .1 + .2 = .6,$

and

$P(E)P(F) = (.4)(.6) = .24 \neq P(E \cap F).$

Thus, E and F are dependent.

21. From the probability tree,

(A) $P(M \cap S) = (.3)(.6) = .18$

(B) $P(R) = P(N \cap R) + P(M \cap R) = (.7)(.2) + (.3)(.4)$

$= .14 + .12$

$= .26$

23. $E_1 = \{HH, HT\}$ and $P(E_1) = \frac{1}{2}$

$E_2 = \{TH, TT\}$ and $P(E_2) = \frac{1}{2}$

$E_4 = \{HH, TH\}$ and $P(E_4) = \frac{1}{2}$

(A) Since $E_1 \cap E_4 = \{HH\} \neq \emptyset$, E_1 and E_4 **are not** mutually exclusive.

Since $P(E_1 \cap E_4) = P(HH) = \frac{1}{4} = P(E_1) \cdot P(E_4)$, E_1 and E_4 are independent.

(B) Since $E_1 \cap E_2 = \emptyset$, E_1 and E_2 **are** mutually exclusive.

Since $P(E_1 \cap E_2) = 0$ and $P(E_1) \cdot P(E_2) = \frac{1}{4}$, $P(E_1 \cap E_2) \neq P(E_1) \cdot P(E_2)$.
Therefore, E_1 and E_2 are dependent.

25. Let E_i = "even number on the ith throw," $i = 1, 2$,
and O_i = "odd number on the ith throw," $i = 1, 2$.

Then $P(E_i) = \frac{1}{2}$ and $P(O_i) = \frac{1}{2}$, $i = 1, 2$.

The probability tree for this experiment is shown at the right.

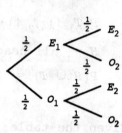

$P(E_1 \cap E_2) = \left(\frac{1}{2}\right)\left(\frac{1}{2}\right) = \frac{1}{4}$

$P(E_1 \cup E_2) = P(E_1) + P(E_2) - P(E_1 \cap E_2) = \frac{1}{2} + \frac{1}{2} - \frac{1}{4} = \frac{3}{4}$.

27. Let C = "first card is a club,"
and H = "second card is a heart."

(A) Without replacement, the probability tree is as shown at the right.

Thus, $P(C \cap H) = \left(\frac{1}{4}\right)\left(\frac{13}{51}\right) \approx .0637$.

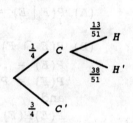

(B) With replacement, the draws are independent and

$P(C \cap H) = \left(\frac{1}{4}\right)\left(\frac{1}{4}\right) = \frac{1}{16} = .0625$.

29. G = "the card is black" = {spade or club} and $P(G) = \frac{1}{2}$.

H = "the card is divisible by 3" = {3, 6, or 9}. $P(H) = \frac{12}{52} = \frac{3}{13}$

$P(H \cap G)$ = {3, 6, or 9 of clubs or spades} = $\frac{6}{52} = \frac{3}{26}$

(A) $P(H \mid G) = \dfrac{P(H \cap G)}{P(G)} = \dfrac{3/26}{1/2} = \dfrac{6}{26} = \dfrac{3}{13}$

(B) $P(H \cap G) = \dfrac{3}{26} = P(H) \cdot P(G)$

Thus, H and G are independent.

31. (A) $S = \{BB,\ BG,\ GB,\ GG\}$

$A = \{BB,\ GG\}$ and $P(A) = \dfrac{2}{4} = \dfrac{1}{2}$

$B = \{BG,\ GB,\ GG\}$ and $P(B) = \dfrac{3}{4}$

$A \cap B = \{GG\}$.

$P(A \cap B) = \dfrac{1}{4}$ and $P(A) \cdot P(B) = \dfrac{1}{2} \cdot \dfrac{3}{4} = \dfrac{3}{8}$

Thus, $P(A \cap B) \neq P(A) \cdot P(B)$ and the events are dependent.

(B) $S = \{BBB,\ BBG,\ BGB,\ BGG,\ GBB,\ GBG,\ GGB,\ GGG\}$

$A = \{BBB,\ GGG\}$

$B = \{BGG,\ GBG,\ GGB,\ GGG\}$

$A \cap B = \{GGG\}$

$P(A) = \dfrac{2}{8} = \dfrac{1}{4}$, $P(B) = \dfrac{4}{8} = \dfrac{1}{2}$, and $P(A \cap B) = \dfrac{1}{8}$

Since $P(A \cap B) = \dfrac{1}{8} = P(A) \cdot P(B)$, A and B are independent.

33. (A) The probability tree with replacement is as follows:

(B) The probability tree without replacement is as follows:

35. Let E = At least one ball was red = $\{R_1 \cap R_2,\ R_1 \cap W_2,\ W_1 \cap R_2\}$.

(A) With replacement [see the probability tree in Problem 33(A)]:

$P(E) = P(R_1 \cap R_2) + P(R_1 \cap W_2) + P(W_1 \cap R_2)$

$ = \dfrac{4}{49} + \dfrac{10}{49} + \dfrac{10}{49} = \dfrac{24}{49}$

(B) Without replacement [see the probability tree in Problem 33(B)]:

$P(E) = P(R_1 \cap R_2) + P(R_1 \cap W_2) + P(W_1 \cap R_2) = \dfrac{1}{21} + \dfrac{5}{21} + \dfrac{5}{21} = \dfrac{11}{21}$

37. $n(S) = C_{9,2} = \dfrac{9!}{2!(9-2)!}$ (total number of balls $= 2 + 3 + 4 = 9$)

$$= \frac{9 \cdot 8 \cdot 7!}{2 \cdot 1 \cdot 7!} = 36$$

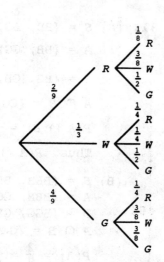

Let A = Both balls are the same color.

$n(A)$ = (No. of ways 2 red balls are selected)
        + (No. of ways 2 white balls are selected)
        + (No. of ways 2 green balls are selected)

$$= C_{2,2} + C_{3,2} + C_{4,2}$$

$$= \frac{2!}{2!(2-2)!} + \frac{3!}{2!(3-2)!} + \frac{4!}{2!(4-2)!}$$

$$= 1 + 3 + 6 = 10$$

$$P(A) = \frac{n(A)}{n(S)} = \frac{10}{36} = \frac{5}{18}$$

Alternatively, the probability tree for this experiment is shown at the right.

And $P(RR,\ WW,\ \text{or}\ GG) = P(RR) + P(WW) + P(GG)$

$$= \left(\frac{2}{9}\right)\left(\frac{1}{8}\right) + \left(\frac{1}{3}\right)\left(\frac{1}{4}\right) + \left(\frac{4}{9}\right)\left(\frac{3}{8}\right) = \frac{2}{72} + \frac{1}{12} + \frac{12}{72} = \frac{20}{72} = \frac{5}{18}$$

39. The probability tree for this experiment is:

(A) $P(\$16) = \left(\dfrac{1}{4}\right)\left(\dfrac{2}{3}\right)\left(\dfrac{1}{2}\right) + \left(\dfrac{1}{2}\right)\left(\dfrac{1}{3}\right)\left(\dfrac{1}{2}\right)$

$$= \frac{1}{12} + \frac{1}{12} = \frac{1}{6} \approx .167$$

(B) $P(\$17) = \left(\dfrac{1}{2}\right)\left(\dfrac{1}{3}\right)\left(\dfrac{1}{2}\right) + \left(\dfrac{1}{2}\right)\left(\dfrac{1}{3}\right)\left(\dfrac{1}{2}\right) + \left(\dfrac{1}{4}\right)\left(\dfrac{2}{3}\right)\left(\dfrac{1}{2}\right)$

$$= \frac{1}{12} + \frac{1}{12} + \frac{1}{12} = \frac{1}{4} = .25$$

(C) Let A = "$10 on second draw." Then

$$P(A) = \left(\frac{1}{4}\right)\left(\frac{1}{3}\right) + \left(\frac{1}{2}\right)\left(\frac{1}{3}\right) = \frac{1}{12} + \frac{1}{6} = \frac{1}{4} = .25$$

41. Assume that A and B are independent events with $P(A) \neq 0$, $P(B) \neq 0$. Then, by definition

(*) $P(A \cap B) = P(A) \cdot P(B)$.

Now,

$$P(A \mid B) = \frac{P(A \cap B)}{P(B)} \quad \text{(definition of conditional probability)}$$

$$= \frac{P(A) \cdot P(B)}{P(B)} \quad \text{(by *)}$$

$$= P(A)$$

Also,

$$P(B \mid A) = \frac{P(B \cap A)}{P(A)} = \frac{P(A \cap B)}{P(A)}$$
$$= \frac{P(A) \cdot P(B)}{P(A)}$$
$$= P(B)$$

43. Assume $P(A) \neq 0$. Then $P(A \mid A) = \dfrac{P(A \cap A)}{P(A)} = \dfrac{P(A)}{P(A)} = 1$.

45. If A and B are mutually exclusive, then $A \cap B = \varnothing$ and $P(A \cap B) = P(\varnothing) = 0$. Also, if $P(A) \neq 0$ and $P(B) \neq 0$, then $P(A) \cdot P(B) \neq 0$. Therefore, $P(A \cap B) = 0 \neq P(A) \cdot P(B)$, and events A and B are dependent.

47. (A)

To strike	Hourly H	Salary S	Salary + bonus B	Total
Yes (Y)	.400	.180	.020	.600
No (N)	.150	.120	.130	.400
Total	.550	.300	.150	1.000

[**Note**: The probability table above was derived from the table given in the problem by dividing each entry by 1000.]

Referring to the table in part (A):

(B) $P(Y \mid H) = \dfrac{P(Y \cap H)}{P(H)} = \dfrac{.400}{.55} \approx .727$

(C) $P(Y \mid B) = \dfrac{P(Y \cap B)}{P(B)} = \dfrac{.02}{.15} \approx .133$

(D) $P(S) = .300$

$P(S \mid Y) = \dfrac{P(S \cap Y)}{P(Y)} = \dfrac{.180}{.60} = .300$

(E) $P(H) = .550$

$P(H \mid Y) = \dfrac{P(H \cap Y)}{P(Y)} = \dfrac{.400}{.600} \approx .667$

(F) $P(B \cap N) = .130$

(G) S and Y are independent since $P(S \mid Y) = P(S) = .300$

(H) H and Y are dependent since $P(H \mid Y) \approx .667$ is not equal to $P(H) = .550$.

(I) $P(B \mid N) = \dfrac{P(B \cap N)}{P(N)}$
$= \dfrac{.130}{.400}$ (from table)
$= .325$

and $P(B) = .150$. Since $P(B \mid N) \neq P(B)$, B and N are dependent.

49. The probability tree for this experiment is:

(A) $P(\$26,000) = \left(\frac{1}{2}\right)\left(\frac{1}{3}\right)\left(\frac{1}{2}\right) + \left(\frac{1}{4}\right)\left(\frac{2}{3}\right)\left(\frac{1}{2}\right)$

$= \frac{1}{12} + \frac{1}{12} = \frac{1}{6} \approx .167$

(B) $P(\$31,000) = \left(\frac{1}{2}\right)\left(\frac{1}{3}\right)\left(\frac{1}{2}\right) + \left(\frac{1}{2}\right)\left(\frac{1}{3}\right)\left(\frac{1}{2}\right) + \left(\frac{1}{4}\right)\left(\frac{2}{3}\right)\left(\frac{1}{2}\right)$

$= \frac{3}{12} = \frac{1}{4} = .25$

(C) Let A = "$20 on third draw." Then
$P(A) = \left(\frac{1}{4}\right)\left(\frac{2}{3}\right)\left(\frac{1}{2}\right) + \left(\frac{1}{2}\right)\left(\frac{1}{3}\right)\left(\frac{1}{2}\right) + \left(\frac{1}{2}\right)\left(\frac{1}{3}\right)\left(\frac{1}{2}\right)$

$= \frac{3}{12} = \frac{1}{4} = .25$

51. (A)

	C	C'	Totals
R	0.06	0.44	0.50
R'	0.02	0.48	0.50
Total	0.08	0.92	1.00

(B) $P(C) = 0.08$, $P(R) = 0.50$, $P(R \cap C) = 0.06$

Since $0.06 = P(R \cap C) \neq P(R) \cdot P(C) = 0.04$, R and C are **dependent**.

(C) $P(C \mid R) = \dfrac{P(C \cap R)}{P(R)} = \dfrac{0.06}{0.50} = 0.12$ and $P(C) = 0.08$

Since $P(C \mid R) > P(C)$, cancer is more likely to be developed if the red die is used. The FDA should ban the use of the red die.

(D) The new probability table is

	C	C'	Totals
R	0.02	0.48	0.50
R'	0.06	0.44	0.50
Total	0.08	0.92	1.00

Now $P(C \mid R) = \dfrac{P(C \cap R)}{P(R)} = \dfrac{0.02}{0.5} = 0.04$ and $P(C) = 0.08$

Since $P(C \mid R) < P(C)$ it appears that the red die reduces the development of cancer. Therefore, the use of the die should not be banned.

53. (A)

	Below 90 A	90—120 B	Above 120 C	Total
Female (F)	.130	.286	.104	.520
Male (F')	.120	.264	.096	.480
Total	.250	.550	.200	1.000

[Note: The probability table above was derived from the table given in the problem by dividing each entry by 1000.]

Referring to the table in part (A):

(B) $P(A \mid F) = \dfrac{P(A \cap F)}{P(F)} = \dfrac{.130}{.520} \approx .250$

(C) $P(C \mid F) = \dfrac{P(C \cap F)}{P(F)} = \dfrac{.104}{.520} \approx .200$

$P(A \mid F') = \dfrac{P(A \cap F')}{P(F')} = \dfrac{.120}{.480} = .250$

$P(C \mid F') = \dfrac{P(C \cap F')}{P(F')} = \dfrac{.096}{.480} = .200$

(D) $P(A) = .25$

(E) $P(B) = .55$

$P(A \mid F) = \dfrac{P(A \cap F)}{P(F)} = \dfrac{.130}{.520} = .250$

$P(B \mid F') = \dfrac{P(B \cap F')}{P(F')} = \dfrac{.264}{.480} = .550$

(F) $P(F \cap C) = .104$

(G) No, the results in parts (B), (C), (D), and (E) imply that A, B, and C are independent of F and F'.

EXERCISE 6-6

Things to remember:

1. BAYES' FORMULA

Let U_1, U_2, ..., U_n be n mutually exclusive events whose union is the sample space S. Let E be an arbitrary event in S such that $P(E) \neq 0$. Then

$$P(U_1 \mid E) = \frac{P(U_1 \cap E)}{P(E)}$$

$$= \frac{P(U_1 \cap E)}{P(U_1 \cap E) + P(U_2 \cap E) + \cdots + P(U_n \cap E)}$$

$$= \frac{P(E \mid U_1)P(U_1)}{P(E \mid U_1)P(U_1) + \cdots + P(E \mid U_n)P(U_n)}$$

Similar results hold for U_2, U_3, ..., U_n.

2. BAYES' FORMULA AND PROBABILITY TREES

$$P(U_1 \mid E) = \frac{\text{Product of branch probabilities leading to } E \text{ through } U_1}{\text{Sum of all branch probabilities leading to } E}$$

Similar results hold for U_2, U_3, ..., U_n.

---

1. $P(M \cap A) = P(M) \cdot P(A \mid M) = (.6)(.8) = .48$

3. $P(A) = P(M \cap A) + P(N \cap A) = P(M)P(A \mid M) + P(N)P(A \mid N)$
$$= (.6)(.8) + (.4)(.3) = .60$$

5. $P(M \mid A) = \dfrac{P(M \cap A)}{P(M \cap A) + P(N \cap A)} = \dfrac{.48}{.60}$ (see Problems 1 and 3)
$$= .80$$

7. Referring to the Venn diagram:
$$P(U_1 \mid R) = \frac{P(U_1 \cap R)}{P(R)} = \frac{\dfrac{25}{100}}{\dfrac{60}{100}} = \frac{25}{60} = \frac{5}{12} \approx .417$$

Using Bayes' formula:
$$P(U_1 \mid R) = \frac{P(U_1 \cap R)}{P(U_1 \cap R) + P(U_2 \cap R)} = \frac{P(U_1)P(R \mid U_1)}{P(U_1)P(R \mid U_1) + P(U_2)P(R \mid U_2)}$$

$$= \frac{\left(\dfrac{40}{100}\right)\left(\dfrac{25}{40}\right)}{\left(\dfrac{40}{100}\right)\left(\dfrac{25}{40}\right) + \left(\dfrac{60}{100}\right)\left(\dfrac{35}{60}\right)} = \frac{.25}{.25 + .35} = \frac{.25}{.60} = \frac{5}{12} \approx .417$$

9. $P(U_1 \mid R') = \dfrac{P(U_1 \cap R')}{P(R')} = \dfrac{\dfrac{15}{100}}{1 - P(R)}$ (from the Venn diagram)

$$= \frac{\dfrac{15}{100}}{1 - \dfrac{60}{100}} = \frac{\dfrac{15}{100}}{\dfrac{40}{100}} = \frac{3}{8} = .375$$

Using Bayes' formula:
$$P(U_1 \mid R') = \frac{P(U_1 \cap R')}{P(R')} = \frac{P(U_1)P(R' \mid U_1)}{P(U_1 \cap R') + P(U_2 \cap R')}$$

$$= \frac{P(U_1)P(R' \mid U_1)}{P(U_1)P(R' \mid U_1) + P(U_2)P(R' \mid U_2)} = \frac{\left(\dfrac{40}{100}\right)\left(\dfrac{15}{40}\right)}{\left(\dfrac{40}{100}\right)\left(\dfrac{15}{40}\right) + \left(\dfrac{60}{100}\right)\left(\dfrac{25}{60}\right)}$$

$$= \frac{.15}{.15 + .25} = \frac{15}{40} = \frac{3}{8} = .375$$

11. $P(U \mid C) = \dfrac{P(U \cap C)}{P(C)} = \dfrac{P(U \cap C)}{P(U \cap C) + P(V \cap C) + P(W \cap C)}$

$$= \frac{(.2)(.4)}{(.2)(.4) + (.5)(.2) + (.3)(.6)} \qquad [\underline{\text{Note}}: \text{Recall } P(A \cap B) = P(A) \cdot P(B \mid A).]$$

$$= \frac{.08}{.36} \approx .222$$

13. $P(W \mid C) = \dfrac{P(W \cap C)}{P(C)} = \dfrac{P(W \cap C)}{P(W \cap C) + P(V \cap C) + P(U \cap C)}$

$$= \frac{(.3)(.6)}{(.3)(.6) + (.5)(.2) + (.2)(.4)} \qquad (\text{see Problem 11})$$

$$= \frac{.18}{.36} = .5$$

15. $P(V \mid C) = \dfrac{P(V \cap C)}{P(C)} = \dfrac{P(V \cap C)}{P(V \cap C) + P(W \cap C) + P(U \cap C)}$

$$= \frac{(.5)(.2)}{(.5)(.2) + (.3)(.6) + (.2)(.4)} = \frac{.1}{.36} = .278$$

17. From the Venn diagram,

$$P(U_1 \mid R) = \frac{5}{5 + 15 + 20} = \frac{5}{40} = \frac{1}{8} = .125$$

or

$$= \frac{P(U_1 \cap R)}{P(R)} = \frac{\frac{5}{100}}{\frac{40}{100}} = .125$$

Using Bayes' formula:

$$P(U_1 \mid R) = \frac{P(U_1 \cap R)}{P(U_1 \cap R) + P(U_2 \cap R) + P(U_3 \cap R)} = \frac{\frac{5}{100}}{\frac{5}{100} + \frac{15}{100} + \frac{20}{100}}$$

$$= \frac{.05}{.05 + .15 + .2} = \frac{.05}{.40} = .125$$

19. From the Venn diagram,

$$P(U_3 \mid R) = \frac{20}{5 + 15 + 20} = \frac{20}{40} = .5$$

Using Bayes' formula:

$$P(U_3 \mid R) = \frac{P(U_3 \cap R)}{P(U_1 \cap R) + P(U_2 \cap R) + P(U_3 \cap R)} = \frac{\frac{20}{100}}{\frac{5}{100} + \frac{15}{100} + \frac{20}{100}}$$

$$= \frac{.2}{.05 + .15 + .2} = \frac{.2}{.4} = .5$$

21. From the Venn diagram,

$$P(U_2 \mid R) = \frac{15}{5 + 15 + 20} = \frac{15}{40} = .375$$

Using Bayes' formula:

$$P(U_2 \mid R) = \frac{P(U_2 \cap R)}{P(U_1 \cap R) + P(U_2 \cap R) + P(U_3 \cap R)} = \frac{\dfrac{15}{100}}{\dfrac{5}{100} + \dfrac{15}{100} + \dfrac{20}{100}}$$

$$= \frac{.15}{.05 + .15 + .2} = \frac{.15}{.40}$$

$$= \frac{3}{8} = .375$$

23. From the given tree diagram, we have:

$$P(A) = \frac{1}{4} \qquad\qquad P(A') = \frac{3}{4}$$

$$P(B \mid A) = \frac{1}{5} \qquad\qquad P(B \mid A') = \frac{3}{5}$$

$$P(B' \mid A) = \frac{4}{5} \qquad\qquad P(B' \mid A') = \frac{2}{5}$$

We want to find the following:

$$P(B) = P(B \cap A) + P(B \cap A') = P(A)P(B \mid A) + P(A')P(B \mid A')$$

$$= \left(\frac{1}{4}\right)\left(\frac{1}{5}\right) + \left(\frac{3}{4}\right)\left(\frac{3}{5}\right) = \frac{1}{20} + \frac{9}{20} = \frac{10}{20} = \frac{1}{2}$$

$$P(B') = 1 - P(B) = 1 - \frac{1}{2} = \frac{1}{2}$$

$$P(A \mid B) = \frac{P(A \cap B)}{P(B)} = \frac{P(A)P(B \mid A)}{P(B)} = \frac{\left(\frac{1}{4}\right)\left(\frac{1}{5}\right)}{\frac{1}{2}} = \frac{\frac{1}{20}}{\frac{1}{2}} = \frac{1}{10}$$

Thus, $P(A' \mid B) = 1 - P(A \mid B) = 1 - \frac{1}{10} = \frac{9}{10}.$

$$P(A \mid B') = \frac{P(A \cap B')}{P(B')} = \frac{P(A)P(B' \mid A)}{P(B')} = \frac{\left(\frac{1}{4}\right)\left(\frac{4}{5}\right)}{\frac{1}{2}} = \frac{\frac{4}{20}}{\frac{1}{2}} = \frac{2}{5}$$

Thus, $P(A' \mid B') = 1 - P(A \mid B') = 1 - \frac{2}{5} = \frac{3}{5}.$

Therefore, the tree diagram for this problem is as shown at the right.

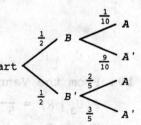

The following tree diagram is to be used for Problems 25 and 27.

$$\text{Start} \begin{cases} .5 \quad U_1 \text{ (urn 1)} \begin{cases} \frac{1}{5} = .2 \quad W \text{ (white)} \\ \frac{4}{5} = .8 \quad R \text{ (red)} \end{cases} \\ .5 \quad U_2 \text{ (urn 2)} \begin{cases} \frac{3}{5} = .6 \quad W \\ \frac{2}{5} = .4 \quad R \end{cases} \end{cases}$$

25. $P(U_1 \mid W) = \dfrac{P(U_1 \cap W)}{P(W)} = \dfrac{P(U_1 \cap W)}{P(U_1 \cap W) + P(U_2 \cap W)}$

$\qquad = \dfrac{P(U_1)P(W \mid U_1)}{P(U_1)P(W \mid U_1) + P(U_2)P(W \mid U_2)} = \dfrac{(.5)(.2)}{(.5)(.2) + (.5)(.6)}$

$\qquad\qquad\qquad\qquad\qquad\qquad\qquad = \dfrac{.1}{.4} = .25$

27. $P(U_2 \mid R) = \dfrac{P(U_2 \cap R)}{P(R)} = \dfrac{P(U_2 \cap R)}{P(U_2 \cap R) + P(U_1 \cap R)}$

$\qquad = \dfrac{P(U_2)P(R \mid U_2)}{P(U_2)P(R \mid U_2) + P(U_1)P(R \mid U_1)} = \dfrac{(.5)(.4)}{(.5)(.4) + (.5)(.8)}$

$\qquad\qquad\qquad\qquad\qquad\qquad\qquad = \dfrac{.4}{1.2} = \dfrac{1}{3} \approx .333$

29. $P(W_1 \mid W_2) = \dfrac{P(W_1 \cap W_2)}{P(W_2)} = \dfrac{P(W_1)P(W_2 \mid W_1)}{P(R_1 \cap W_2) + P(W_1 \cap W_2)}$

$\qquad = \dfrac{P(W_1)P(W_2 \mid W_1)}{P(R_1)P(W_2 \mid R_1) + P(W_1)P(W_2 \mid W_1)} = \dfrac{\left(\frac{5}{9}\right)\left(\frac{4}{8}\right)}{\left(\frac{4}{9}\right)\left(\frac{5}{8}\right) + \left(\frac{5}{9}\right)\left(\frac{4}{8}\right)} = \dfrac{\frac{20}{72}}{\frac{20}{72} + \frac{20}{72}}$

$\qquad\qquad\qquad\qquad\qquad\qquad\qquad = \dfrac{20}{40} = \dfrac{1}{2} \text{ or } .5$

31. $P(U_{R_1} \mid U_{R_2}) = \dfrac{P(U_{R_1} \cap U_{R_2})}{P(U_{R_2})} = \dfrac{P(U_{R_1})P(U_{R_2} \mid U_{R_1})}{P(U_{W_1} \cap U_{R_2}) + P(U_{R_1} \cap U_{R_2})}$

$\qquad = \dfrac{P(U_{R_1})P(U_{R_2} \mid U_{R_1})}{P(U_{W_1})P(U_{R_2} \mid U_{W_1}) + P(U_{R_1})P(U_{R_2} \mid U_{R_1})}$

$\qquad = \dfrac{\left(\frac{7}{10}\right)\left(\frac{5}{10}\right)}{\left(\frac{3}{10}\right)\left(\frac{4}{10}\right) + \left(\frac{7}{10}\right)\left(\frac{5}{10}\right)} = \dfrac{.35}{.12 + .35}$

$$= \frac{.35}{.47} = \frac{35}{47} \approx .745$$

The tree diagram follows:

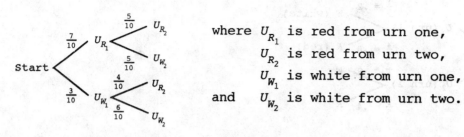

where U_{R_1} is red from urn one,
U_{R_2} is red from urn two,
U_{W_1} is white from urn one,
and U_{W_2} is white from urn two.

33. Suppose $c = e$. then
$$P(M) = ac + be = ac + bc = c(a + b) = c \quad (a + b = 1)$$
and
$$P(M \mid U) = \frac{P(M \cap U)}{P(U)} = \frac{ac}{a} = c$$
Therefore, M and U are independent.
Alternatively, note that
$$P(M) = c, \ P(U) = a \quad \text{and}$$
$$P(M \cap U) = ac = P(M) \cdot P(U),$$
which implies that M and U are independent.

35.

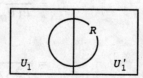

Here the tree diagram (not shown as separate image): Start branches $\frac{13}{52}$ to H_1 (heart) and $\frac{39}{52}$ to $\overline{H}_1$ (not a heart); H_1 branches $\frac{12}{51}$ to H_2 (heart) and $\frac{39}{51}$ to $\overline{H}_2$ (not a heart); $\overline{H}_1$ branches $\frac{13}{51}$ to H_2 and $\frac{38}{51}$ to $\overline{H}_2$.

$$P(H_1 \mid H_2) = \frac{P(H_1 \cap H_2)}{P(H_2)} = \frac{P(H_1 \cap H_2)}{P(H_1 \cap H_2) + P(\overline{H}_1 \cap H_2)}$$

$$= \frac{P(H_1)P(H_2 \mid H_1)}{P(H_1)P(H_2 \mid H_1) + P(\overline{H}_1)P(H_2 \mid \overline{H}_1)} = \frac{\frac{13}{52} \cdot \frac{12}{51}}{\frac{13}{52} \cdot \frac{12}{51} + \frac{39}{52} \cdot \frac{13}{51}}$$

$$= \frac{13(12)}{13(12) + 39(13)} = \frac{12}{51} \approx .235$$

37. Consider the following Venn diagram:

$$P(U_1 \mid R) = \frac{P(U_1 \cap R)}{P(U_1 \cap R) + P(U_1' \cap R)} \quad \text{and} \quad P(U_1' \mid R) = \frac{P(U_1' \cap R)}{P(U_1 \cap R) + P(U_1' \cap R)}$$

Adding these two equations, we obtain:

$$P(U_1 \mid R) + P(U_1' \mid R) = \frac{P(U_1 \cap R)}{P(U_1 \cap R) + P(U_1' \cap R)} + \frac{P(U_1' \cap R)}{P(U_1 \cap R) + P(U_1' \cap R)}$$

$$= \frac{P(U_1 \cap R) + P(U_1' \cap R)}{P(U_1 \cap R) + P(U_1' \cap R)} = 1$$

39. Consider the following tree diagram:

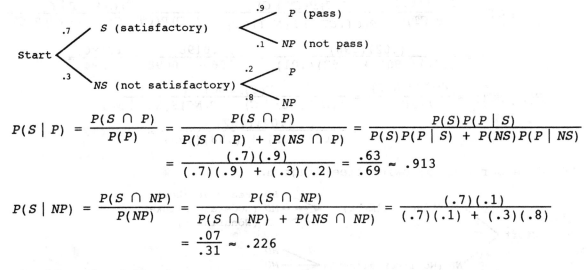

$$P(S \mid P) = \frac{P(S \cap P)}{P(P)} = \frac{P(S \cap P)}{P(S \cap P) + P(NS \cap P)} = \frac{P(S)P(P \mid S)}{P(S)P(P \mid S) + P(NS)P(P \mid NS)}$$

$$= \frac{(.7)(.9)}{(.7)(.9) + (.3)(.2)} = \frac{.63}{.69} \approx .913$$

$$P(S \mid NP) = \frac{P(S \cap NP)}{P(NP)} = \frac{P(S \cap NP)}{P(S \cap NP) + P(NS \cap NP)} = \frac{(.7)(.1)}{(.7)(.1) + (.3)(.8)}$$

$$= \frac{.07}{.31} \approx .226$$

41. Consider the following tree diagram:

$$P(A \mid D) = \frac{P(A \cap D)}{P(D)}, \text{ where}$$

$$P(D) = P(A \cap D) + P(B \cap D) + P(C \cap D)$$

$$= P(A)P(D \mid A) + P(B)P(D \mid B) + P(C)P(D \mid C)$$

$$= (.2)(.01) + (.40)(.03) + (.40)(.02)$$

$$= .002 + .012 + .008$$

$$= .022$$

Thus, $P(A \mid D) = \dfrac{P(A \cap D)}{P(D)} = \dfrac{P(A)P(D \mid A)}{P(D)} = \dfrac{(.20)(.01)}{.022} = \dfrac{.002}{.022} = \dfrac{2}{22}$ or .091

Similarly,

$$P(B \mid D) = \frac{P(B \cap D)}{P(D)} = \frac{P(B)P(D \mid B)}{P(D)} = \frac{(.40)(.03)}{.022} = \frac{.012}{.022} = \frac{6}{11} \text{ or } .545,$$

and $P(C \mid D) = \dfrac{P(C \cap D)}{P(D)} = \dfrac{P(C)P(D \mid C)}{P(D)} = \dfrac{(.40)(.02)}{.022} = \dfrac{.008}{.022} = \dfrac{4}{11}$ or .364.

43. Consider the following tree diagram:

$$P(C \mid CT) = \frac{P(C \cap CT)}{P(CT)} = \frac{P(C)P(CT \mid C)}{P(C \cap CT) + P(NC \cap CT)} = \frac{P(C)P(CT \mid C)}{P(C)P(CT \mid C) + P(NC)P(CT \mid NC)}$$

$$= \frac{(.02)(.98)}{(.02)(.98) + (.98)(.01)} = \frac{.0196}{.0196 + .0098} = \frac{.0196}{.0294} = .6667$$

$$P(C \mid NCT) = \frac{P(C \cap NCT)}{P(NCT)} = \frac{P(C)P(NCT \mid C)}{P(C)P(NCT \mid C) + P(NC)P(NCT \mid NC)}$$

$$= \frac{(.02)(.02)}{(.02)(.02) + (.98)(.99)} \approx .000412$$

45. Consider the following tree diagram.

$$P(L \mid HD) = \frac{P(L \cap HD)}{P(HD)} = \frac{P(L)P(HD \mid L)}{P(L \cap HD) + P(NL \cap HD)} = \frac{P(L)P(HD \mid L)}{P(L)P(HD \mid L) + P(NL)P(HD \mid NL)}$$

$$= \frac{(.07)(.4)}{(.07)(.4) + (.93)(.1)} \quad \text{(from the tree diagram)}$$

$$= \frac{.028}{.028 + .093} = \frac{.028}{.121} = \frac{28}{121} \approx .231$$

$$P(L \mid ND) = \frac{P(L \cap ND)}{P(ND)} = \frac{P(L)P(ND \mid L)}{P(L \cap ND) + P(NL \cap ND)} = \frac{P(L)P(ND \mid L)}{P(L)P(ND \mid L) + P(NL)P(ND \mid NL)}$$

$$= \frac{(.07)(.1)}{(.07)(.1) + (.93)(.2)} \quad \text{(from the tree diagram)}$$

$$= \frac{.007}{.007 + .186} = \frac{.007}{.194} = \frac{7}{194} \approx .036$$

47. Consider the following tree diagram.

$$P(L \mid LT) = \frac{P(L \cap LT)}{P(LT)} = \frac{P(L \cap LT)}{P(L \cap LT) + P(\overline{L} \cap LT)} = \frac{(.5)(.8)}{(.5)(.8) + (.5)(.05)}$$

$$= \frac{.4}{.425} \approx .941$$ If the test indicates that the subject was lying, then he was lying with a probability of 0.941.

$$P(\overline{L} \mid LT) = \frac{P(\overline{L} \cap LT)}{P(LT)} = \frac{(.5)(.05)}{(.5)(.8) + (.5)(.05)}$$

$$= \frac{.05}{.85} \approx .0588$$ If the test indicates that the subject was lying, there is still a probability of 0.0588 that he was not lying.

EXERCISE 6-7

Things to remember:

1. **RANDOM VARIABLE**

   A random variable is a function that assigns a numerical value to each simple event in a sample space S.

2. **PROBABILITY DISTRIBUTION OF A RANDOM VARIABLE X**

   A probability function $P(X = x) = p(x)$ is a PROBABILITY DISTRIBUTION OF THE RANDOM VARIABLE X if

   (a) $0 \le p(x) \le 1$, $x \in \{x_1, x_2, \ldots, x_n\}$,

   (b) $p(x_1) + p(x_2) + \cdots + p(x_n) = 1$,

   where $\{x_1, x_2, \ldots, x_n\}$ are values of X.

3. **EXPECTED VALUE OF A RANDOM VARIABLE X**

   Given the probability distribution for the random variable X:

   $$\left. \begin{array}{l} x_i: x_1, x_2, \ldots, x_m \\ p_i: p_1, p_2, \ldots, p_m \end{array} \right\} p_i = p(x_i)$$

   The expected value of X, denoted by $E(X)$, is given by the formula:

   $$E(X) = x_1 p_1 + x_2 p_2 + \cdots + x_m p_m$$

4. Steps for computing the expected value of a random variable X.

   (a) Form the probability distribution for the random variable X.

   (b) Multiply each image value of X, x_i, by its corresponding probability of occurrence, p_i, then add the results.

---

1. Expected value of X:

$$E(X) = -3(.3) + 0(.5) + 4(.2) = -0.1$$

3. Assign the number 0 to the event of observing zero heads, the number 1 to the event of observing one head, and the number 2 to the event of observing two heads. The probability distribution for X, then, is:

x_i	0	1	2
p_i	$\frac{1}{4}$	$\frac{1}{2}$	$\frac{1}{4}$

[Note: One head can occur two ways out of a total of four different ways (HT, TH).]

Hence, $E(X) = 0 \cdot \frac{1}{4} + 1 \cdot \frac{1}{2} + 2 \cdot \frac{1}{4} = 1$.

5. Assign a payoff of $1 to the event of observing a head and -$1 to the event of observing a tail. Thus, the probability distribution for X is:

x_i	1	-1
p_i	$\frac{1}{2}$	$\frac{1}{2}$

Hence, $E(X) = 1 \cdot \frac{1}{2} + (-1) \cdot \frac{1}{2} = 0$. The game is fair.

7. The table shows a payoff or probability distribution for the game.

Net gain	x_i	-3	-2	-1	0	1	2
	p_i	$\frac{1}{6}$	$\frac{1}{6}$	$\frac{1}{6}$	$\frac{1}{6}$	$\frac{1}{6}$	$\frac{1}{6}$

[Note: A payoff valued at -$3 is assigned to the event of observing a "1" on the die, resulting in a net gain of -$3, and so on.]

Hence, $E(X) = -3 \cdot \frac{1}{6} - 2 \cdot \frac{1}{6} - 1 \cdot \frac{1}{6} + 0 \cdot \frac{1}{6} + 1 \cdot \frac{1}{6} + 2 \cdot \frac{1}{6} = -\frac{1}{2}$ or -$0.50.

The game is not fair.

9. The probability distribution is:

Number of Heads	Gain, x_i	Probability, p_i
0	2	$\frac{1}{4}$
1	-3	$\frac{1}{2}$
2	2	$\frac{1}{4}$

The expected value is:

$$E(X) = 2 \cdot \frac{1}{4} + (-3) \cdot \frac{1}{2} + 2 \cdot \frac{1}{4} = 1 - \frac{3}{2} = -\frac{1}{2} \text{ or } -\$0.50.$$

11. In 4 rolls of a die, the total number of possible outcomes is $6 \cdot 6 \cdot 6 \cdot 6 = 6^4$. Thus, $n(S) = 6^4 = 1296$. The total number of outcomes that contain no 6's is $5 \cdot 5 \cdot 5 \cdot 5 = 5^4$. Thus, if E is the event "At least one 6," then $n(E) = 6^4 - 5^4 = 671$ and

$$P(E) = \frac{n(E)}{n(S)} = \frac{671}{1296} \approx 0.5178.$$

First, we compute the expected value to you.
The payoff table is:

x_i	-\$1	\$1
P_i	0.5178	0.4822

The expected value to you is:
$$E(X) = (-1)(0.5178) + 1(0.4822) = -0.0356 \text{ or } -\$0.036$$

The expected value to her is:
$$E(X) = 1(0.5178) + (-1)(0.4822) = 0.0356 \text{ or } \$0.036$$

13. Let x = amount you should lose if a 6 turns up.
The payoff table is:

	1	2	3	4	5	6
x_i	\$5	\$5	\$10	\$10	\$10	\$x
P_i	$\frac{1}{6}$	$\frac{1}{6}$	$\frac{1}{6}$	$\frac{1}{6}$	$\frac{1}{6}$	$\frac{1}{6}$

Now $E(X) = 5\left(\frac{1}{6}\right) + 5\left(\frac{1}{6}\right) + 10\left(\frac{1}{6}\right) + 10\left(\frac{1}{6}\right) + 10\left(\frac{1}{6}\right) + x\left(\frac{1}{6}\right) = \frac{40}{6} + \frac{x}{6}$

The game is fair if and only if $E(X) = 0$:

solving $\frac{40}{6} + \frac{x}{6} = 0$

gives $x = -40$

Thus, you should **lose** \$40 for the game to be fair.

15. $P(\text{sum} = 7) = \frac{6}{36} = \frac{1}{6}$

$P(\text{sum} = 11 \text{ or } 12) = P(\text{sum} = 11) + P(\text{sum} = 12) = \frac{2}{36} + \frac{1}{36} = \frac{3}{36} = \frac{1}{12}$

$P(\text{sum other than } 7, 11, \text{ or } 12) = 1 - P(\text{sum} = 7, 11, \text{ or } 12)$
$$= 1 - \frac{9}{36} = \frac{27}{36} = \frac{3}{4}$$

Let x_1 = sum is 7, x_2 = sum is 11 or 12, x_3 = sum is not 7, 11, or 12, and let t denote the amount you "win" if x_3 occurs. Then the payoff table is:

x_i	-\$10	\$11	t
P_i	$\frac{1}{6}$	$\frac{1}{12}$	$\frac{3}{4}$

The expected value is:
$$E(X) = -10\left(\frac{1}{6}\right) + 11\left(\frac{1}{12}\right) + t\left(\frac{3}{4}\right) = \frac{-10}{6} + \frac{11}{12} + \frac{3t}{4}$$

The game is fair if $E(X) = 0$, i.e., if
$$\frac{-10}{6} + \frac{11}{12} + \frac{3}{4}t = 0 \quad \text{or} \quad \frac{3}{4}t = \frac{10}{6} - \frac{11}{12} = \frac{20}{12} - \frac{11}{12} = \frac{9}{12} = \frac{3}{4}$$

Therefore, $t = \$1$.

17. Course A_1: $E(X) = (-200)(.1) + 100(.2) + 400(.4) + 100(.3)$
$= -20 + 20 + 160 + 30$
$= \$190$

Course A_2: $E(X) = (-100)(.1) + 200(.2) + 300(.4) + 200(.3)$
$= -10 + 40 + 120 + 60$
$= \$210$

A_2 will produce the largest expected value, and that value is $210.

19. The probability of winning $35 is $\frac{1}{38}$ and the probability of losing $1 is $\frac{37}{38}$. Thus, the payoff table is:

x_i	$35	-$1
P_i	$\frac{1}{38}$	$\frac{37}{38}$

The expected value of the game is:
$$E(X) = 35\left(\frac{1}{38}\right) + (-1)\left(\frac{37}{38}\right) = \frac{35 - 37}{38} = \frac{-1}{19} \approx -0.0526 \text{ or } E(X) = -5.26\cent \text{ or } -5\cent$$

21. Let p = probability of winning. Then $1 - p$ is the probability of losing and the payoff table is:

	W	L
x_i	99,900	-100
p_i	p	$1 - p$

Since $E(X) = 100$, we have
$$99,900(p) - 100(1 - p) = 100$$
$$99,900p - 100 + 100p = 100$$
$$100,000p = 200$$
$$p = 0.002$$

The probability of winning is 0.002. Since the expected value is positive, you should play the game. In the long run, you will win $100 per game.

23.

p_i		x_i
$\frac{1}{5000}$	chance of winning	$499
$\frac{3}{5000}$	chance of winning	$99
$\frac{5}{5000}$	chance of winning	$19
$\frac{20}{5000}$	chance of winning	$4
$\frac{4971}{5000}$	chance of losing	$1 [Note: $5000 - (1 + 3 + 5 + 20) = 4971$.]

The payoff table is:

x_i	$499	$99	$19	$4	−$1
P_i	0.0002	0.0006	0.001	0.004	0.9942

Thus,

$E(X) = 499(0.0002) + 99(0.0006) + 19(0.001) + 4(0.004) - 1(0.9942)$
$= -0.80$

or

$E(X) = -\$0.80$ or $-80¢$

25. (A) Total number of simple events $= n(S) = C_{10,2} = \dfrac{10!}{2!(10-2)!}$

$$= \dfrac{10!}{2!8!} = \dfrac{10 \cdot 9}{2} = 45$$

$P(\text{zero defective}) = P(0) = \dfrac{C_{7,2}}{45}$ [Note: None defective means 2 selected fom 7 nondefective.]

$$= \dfrac{\frac{7!}{2!5!}}{45} = \dfrac{21}{45} = \dfrac{7}{15}$$

$P(\text{one defective}) = P(1) = \dfrac{C_{3,1} \cdot C_{7,1}}{45} = \dfrac{21}{45} = \dfrac{7}{15}$

$P(\text{two defective}) = P(2) = \dfrac{C_{3,2}}{45}$ [Note: Two defectives selected from 3 defectives.]

$$= \dfrac{3}{45} = \dfrac{1}{15}$$

The probability distribution is as follows:

x_i	0	1	2
P_i	$\frac{7}{15}$	$\frac{7}{15}$	$\frac{1}{15}$

(B) $E(X) = 0\left(\dfrac{7}{15}\right) + 1\left(\dfrac{7}{15}\right) + 2\left(\dfrac{1}{15}\right) = \dfrac{9}{15} = \dfrac{3}{5} = 0.6$

27. (A) The total number of simple events $= n(S) = C_{1000,5}$.

$P(0 \text{ winning tickets}) = P(0) = \dfrac{C_{997,5}}{C_{1000,5}} = \dfrac{997 \cdot 996 \cdot 995 \cdot 994 \cdot 993}{1000 \cdot 999 \cdot 998 \cdot 997 \cdot 996} \approx 0.985$

$P(1 \text{ winning ticket}) = P(1) = \dfrac{C_{3,1} \cdot C_{997,4}}{C_{1000,5}} = \dfrac{3 \cdot \frac{997!}{4!(993)!}}{\frac{1000!}{5!(995)!}} \approx 0.0149$

$$P(2 \text{ winning tickets}) = P(2) = \frac{C_{3,2} \cdot C_{997,3}}{C_{1000,5}} = \frac{3 \cdot \frac{997!}{3!(994)!}}{\frac{1000!}{5!(995)!}} \approx 0.0000599$$

$$P(3 \text{ winning tickets}) = P(3) = \frac{C_{3,3} \cdot C_{997,2}}{C_{1000,5}} = \frac{1 \cdot \frac{997!}{2!(995)!}}{\frac{1000!}{5!(995)!}} \approx 0.00000006$$

The payoff table is as follows:

x_i	−$5	$195	$395	$595
P_i	0.985	0.0149	0.0000599	0.00000006

(B) The expected value to you is:
$$E(X) = (-5)(0.985) + 195(0.0149) + 395(0.0000599) + 595(0.00000006)$$
$$\approx -\$2.00$$

29. The payoff table is as follows:

Gain	x_i	$4850	−$150
	p_i	0.01	0.99

[<u>Note</u>: 5000 − 150 = 4850, the gain with a probability of 0.01 if stolen.]

Hence, $E(X) = 4850(0.01) - 150(0.99) = -\$100$

31. The payoff table for site A is as follows:

x_i	30 million	−3 million
p_i	0.2	0.8

Hence $E(X) = 30(0.2) - 3(0.8)$
$= 6 - 2.4$
$= \$3.6$ million

The payoff table for site B is as follows:

x_i	70 million	−4 million
p_i	0.1	0.9

Hence, $E(X) = 70(0.1) + (-4)(0.9)$
$= 7 - 3.6$
$= \$3.4$ million

The company should choose site A with $E(X) = \$3.6$ million.

33. Using $\underline{4}$,
$$E(X) = 0(0.12) + 1(0.36) + 2(0.38) + 3(0.14) = 1.54$$

35. Action A_1: $E(X) = 10(0.3) + 5(0.2) + 0(0.5) = \$4.00$
Action A_2: $E(X) = 15(0.3) + 3(0.1) + 0(0.6) = \$4.80$
Action A_2 is the better choice.

1. (A) We construct the following tree diagram for the experiment:

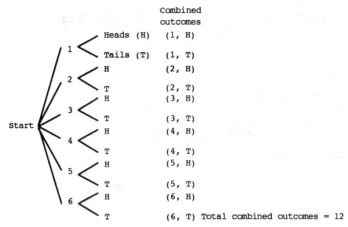

(B) Operation 1: Six possible outcomes, 1, 2, 3, 4, 5, or 6; $N_1 = 6$.
Operation 2: Two possible outcomes, heads (H) or tails (T); $N_2 = 2$.

Using the Multiplication Principle, the total combined outcomes = $N_1 \cdot N_2 = 6 \cdot 2 = 12$. (6-1)

2. (A) $n(A) = 30 + 35 = 65$ (B) $n(B) = 35 + 40 = 75$
(C) $n(A \cap B) = 35$ (D) $n(A \cup B) = 65 + 75 - 35 = 105$
      or $n(A \cup B) = 30 + 35 + 40 = 105$

(E) $n(U) = 30 + 35 + 40 + 45 = 150$ (F) $n(A') = n(U) - n(A) = 150 - 65 = 85$
(G) $n([A \cap B]') = n(U) - n(A \cap B)$ (H) $n([A \cup B]') = n(U) - n(A \cup B)$
$= 150 - 35 = 115$ $= 150 - 105 = 45$
(6-1)

3. $C_{6,2} = \dfrac{6!}{2!(6-2)!} = \dfrac{6!}{2!4!}$ $P_{6,2} = \dfrac{6!}{(6-2)!} = \dfrac{6!}{4!}$

$= \dfrac{6 \cdot 5 \cdot 4!}{2 \cdot 1 \cdot 4!} = 15$ $= \dfrac{6 \cdot 5 \cdot 4!}{4!} = 30$ (6-2)

4. Operation 1: First person can choose the seat in 6 different ways; $N_1 = 6$.
Operation 2: Second person can choose the seat in 5 different ways; $N_2 = 5$.
Operation 3: Third person can choose the seat in 4 different ways; $N_3 = 4$.
Operation 4: Fourth person can choose the seat in 3 different ways; $N_4 = 3$.
Operation 5: Fifth person can choose the seat in 2 different ways; $N_5 = 2$.
Operation 6: Sixth person can choose the seat in 1 way; $N_6 = 1$.
Using the Multiplication Principle, the total number of different
arrangements that can be made is $6 \cdot 5 \cdot 4 \cdot 3 \cdot 2 \cdot 1 = 720$. (6-1)

5. This is a permutations problem. The permutations of 6 objects taken 6
at a time is:
$$P_{6,6} = \frac{6!}{(6-6)!} = 6! = 720$$ (6-2)

6. First, we calculate the number of 5-card combinations that can be dealt from 52 cards:

$$n(S) = C_{52,5} = \frac{52!}{51! \cdot 47!} = 2,598,960$$

We then calculate the number of 5-club combinations that can be obtained from 13 clubs:

$$n(E) = C_{13,5} = \frac{13!}{5! \cdot 8!} = 1287$$

Thus,

$$P(5 \text{ clubs}) = P(E) = \frac{n(E)}{n(S)} = \frac{1287}{2,598,960} \approx 0.0005 \tag{6-3}$$

7. $n(S)$ is computed by using the permutation formula:

$$n(S) = P_{15,2} = \frac{15!}{(15-2)!} = 15 \cdot 14 = 210$$

Thus, the probability that Betty will be president and Bill will be treasurer is:

$$\frac{n(E)}{n(S)} = \frac{1}{210} \approx 0.0048 \tag{6-3}$$

8. (A) The total number of ways of drawing 3 cards from 10 with order taken into account is given by:

$$P_{10,3} = \frac{10!}{(10-3)!} = \frac{10 \cdot 9 \cdot 8 \cdot 7!}{7!} = 720$$

Thus, the probability of drawing the code word "dig" is:

$$P(\text{"dig"}) = \frac{1}{720} \approx 0.0014$$

(B) The total number of ways of drawing 3 cards from 10 without regard to order is given by:

$$C_{10,3} = \frac{10!}{3!(10-3)!} = \frac{10 \cdot 9 \cdot 8 \cdot 7!}{3! 7!} = 120$$

Thus, the probability of drawing the 3 cards "d," "i," and "g" (in so e order) is:

$$P(\text{"d," "i," "g"}) = \frac{1}{120} \approx 0.0083 \tag{6-3}$$

9. $P(\text{person having side effects}) = \frac{f(E)}{n} = \frac{50}{1000} = 0.05 \tag{6-3}$

10. The payoff table is as follows:

x_i	-$2	-$1	$0	$1	$2
p_i	$\frac{1}{5}$	$\frac{1}{5}$	$\frac{1}{5}$	$\frac{1}{5}$	$\frac{1}{5}$

Hence,

$$E(X) = (-2) \cdot \frac{1}{5} + (-1) \cdot \frac{1}{5} + 0 \cdot \frac{1}{5} + 1 \cdot \frac{1}{5} + 2 \cdot \frac{1}{5} = 0$$

The game is fair. $\tag{6-7}$

11. $P(A) = .3$, $P(B) = .4$, $P(A \cap B) = .1$

(A) $P(A') = 1 - P(A) = 1 - .3 = .7$

(B) $P(A \cup B) = P(A) + P(B) - P(A \cap B) = .3 + .4 - .1 = .6$ (6-4)

12. Since the spinner cannot land on R and G simultaneously, $R \cap G = \emptyset$. Thus,

$P(R \cup G) = P(R) + P(G) = .3 + .5 = .8$

The odds for an event E are: $\dfrac{P(E)}{P(E')}$

Thus, the odds for landing on either R or G are: $\dfrac{P(R \cup G)}{P[(R \cup G)']} = \dfrac{.8}{.2} = \dfrac{8}{2}$

or the odds are 8 to 2.

       (6-4)

13. If the odds for an event E are a to b, then $P(E) = \dfrac{a}{a+b}$. Thus, the

probability of rolling an 8 before rolling a 7 is: $\dfrac{5}{11} \approx .455$.   (6-4)

14. $P(T) = .27$ (6-5)   15. $P(Z) = .20$ (6-5)

16. $P(T \cap Z) = .02$ (6-5)   17. $P(R \cap Z) = .03$ (6-5)

18. $P(R \mid Z) = \dfrac{P(R \cap Z)}{P(Z)} = \dfrac{.03}{.20} = .15$ (6-5)

19. $P(Z \mid R) = \dfrac{P(Z \cap R)}{P(R)} = \dfrac{.03}{.23} \approx .1304$ (6-5)

20. $P(T \mid Z) = \dfrac{P(T \cap Z)}{P(Z)} = \dfrac{.02}{.20} = .10$ (6-5)

21. No, because $P(T \cap Z) = .02 \neq P(T) \cdot P(Z) = (.27)(.20) = .054$.   (6-5)

22. Yes, because $P(S \cap X) = .10 = P(S) \cdot P(X) = (.5)(.2)$.   (6-5)

23. $P(A) = .4$ from the tree diagram.   (6-5)

24. $P(B \mid A) = .2$ from the tree diagram.   (6-5)

25. $P(B \mid A') = .3$ from the tree diagram.   (6-5)

26. $P(A \cap B) = P(A)P(B \mid A) = (.4)(.2) = .08$ (6-5)

27. $P(A' \cap B) = P(A')P(B \mid A') = (.6)(.3) = .18$ (6-5)

28. $P(B) = P(A \cap B) + P(A' \cap B)$

$\qquad = P(A)P(B \mid A) + P(A')P(B \mid A')$

$\qquad = (.4)(.2) + (.6)(.3)$

$\qquad = .08 + .18$

$\qquad = .26$ (6-5)

29. $P(A \mid B) = \dfrac{P(A \cap B)}{P(B)} = \dfrac{P(A)P(B \mid A)}{P(A \cap B) + P(A' \cap B)} = \dfrac{P(A)P(B \mid A)}{P(A)P(B \mid A) + P(A')P(B \mid A')}$

$$= \dfrac{(.4)(.2)}{(.4)(.2) + (.6)(.3)} \quad \text{(from the tree diagram)}$$

$$= \dfrac{.08}{.26} = \dfrac{8}{26} \text{ or } .307 \approx .31 \qquad (6\text{-}6)$$

30. $P(A \mid B') = \dfrac{P(A \cap B')}{P(B')} = \dfrac{P(A)P(B' \mid A)}{1 - P(B)} = \dfrac{(.4)(.8)}{1 - .26}$ $[P(B) = .26, \text{ see Problem 28.}]$

$$= \dfrac{.32}{.74} = \dfrac{16}{37} \text{ or } .432 \qquad (6\text{-}6)$$

31. Let E = "born in June, July or August."

(A) Empirical Probability:
$$P(E) = \dfrac{f(E)}{n} = \dfrac{10}{32} = \dfrac{5}{16}$$

(B) Theoretical Probability:
$$P(E) = \dfrac{n(E)}{n(S)} = \dfrac{3}{12} = \dfrac{1}{4}$$

(C) As the sample size in part (A) increases, the approximate empirical probability of event E approaches the theoretical probability of event E. $\qquad (6\text{-}3)$

32. S = {HH, HT, TH, TT}.

The probabilities for 2 "heads," 1 "head," and 0 "heads" are, respectively, $\dfrac{1}{4}$, $\dfrac{1}{2}$, and $\dfrac{1}{4}$. Thus, the payoff table is:

x_i	$5	–$4	$2
P_i	0.25	0.5	0.25

$E(X) = 0.25(5) + 0.5(-4) + 0.25(2) = -0.25$ or $-\$0.25$

The game is not fair. $\qquad (6\text{-}7)$

33. S = {(1,1), (2,2), (3,3), (1,2), (2,1), (1,3), (3,1), (2,3), (3,2)}
$n(S) = 3 \cdot 3 = 9$

(A) $P(A) = \dfrac{n(A)}{n(S)} = \dfrac{3}{9} = \dfrac{1}{3}$ $\quad [A = \{(1,1), (2,2), (3,3)\}]$

(B) $P(B) = \dfrac{n(B)}{n(S)} = \dfrac{2}{9}$ $\qquad [B = \{(2,3), (3,2)\}]$ $\qquad (6\text{-}5)$

34. (A) $P(\text{jack or queen}) = P(\text{jack}) + P(\text{queen}) = \dfrac{4}{52} + \dfrac{4}{52} = \dfrac{8}{52} = \dfrac{2}{13}$

[<u>Note</u>: jack $\cap$ queen = $\varnothing$.]

The odds for drawing a jack or queen are 2 to 11.

(B) $P(\text{jack or spade}) = P(\text{jack}) + P(\text{spade}) - P(\text{jack and spade})$
$$= \dfrac{4}{52} + \dfrac{13}{52} - \dfrac{1}{52} = \dfrac{16}{52} = \dfrac{4}{13}$$

The odds for drawing a jack or a spade are 4 to 9.

(C) $P(\text{ace}) = \frac{4}{52} = \frac{1}{13}$. Thus,

$P(\text{card other than an ace}) = 1 - P(\text{ace}) = 1 - \frac{1}{13} = \frac{12}{13}$ (6-4)

35. (A) The probability of rolling a 5 is $\frac{4}{36} = \frac{1}{9}$.

Thus, the odds for rolling a five are 1 to 8.

(B) Let x = amount house should pay (and return the $1 bet). Then, for the game to be fair,

$$E(X) = x\left(\frac{1}{9}\right) + (-1)\left(\frac{8}{9}\right) = 0$$

$$\frac{x}{9} - \frac{8}{9} = 0$$

$$x = 8$$

Thus, the house should pay $8. (6-4)

36. Event E_1 = 2 heads; $f(E_1)$ = 210.
Event E_2 = 1 head; $f(E_2)$ = 480.
Event E_3 = 0 heads; $f(E_3)$ = 310.
Total number of trials = 1000.

(A) The empirical probabilities for the events above are as follows:

$$P(E_1) = \frac{210}{1000} = 0.21$$

$$P(E_2) = \frac{480}{1000} = 0.48$$

$$P(E_3) = \frac{310}{1000} = 0.31$$

(B) Sample space S = {HH, HT, TH, TT}.

$$P(2 \text{ heads}) = \frac{1}{4} = 0.25$$

$$P(1 \text{ head}) = \frac{2}{4} = 0.5$$

$$P(0 \text{ heads}) = \frac{1}{4} = 0.25$$

(C) Using part (B), the expected frequencies for each outcome are as follows:

2 heads = $1000 \cdot \frac{1}{4}$ = 250

1 head = $1000 \cdot \frac{2}{4}$ = 500

0 heads = $1000 \cdot \frac{1}{4}$ = 250 (6-3, 6-7)

37. Using the multiplication principle, the man has 5 children, $5 \cdot 3 = 15$ grandchildren, and $5 \cdot 3 \cdot 2 = 30$ greatgrandchildren, for a total of $5 + 15 + 30 = 50$ descendents. (6-1)

38. The individual tosses of a coin are independent events (the coin has no memory). Therefore, $P(H) = \frac{1}{2}$. (6-5)

39. (A) The sample space S is given by:

$$S = \{(1,1), (1,2), (1,3), (1,4), (1,5), (1,6),$$
$$(2,1), (2,2), (2,3), (2,4), (2,5), (2,6),$$
$$(3,1), (3,2), (3,3), (3,4), (3,5), (3,6),$$
$$(4,1), (4,2), (4,3), (4,4), (4,5), (4,6),$$
$$(5,1), (5,2), (5,3), (5,4), (5,5), (5,6),$$
$$(6,1), (6,2), (6,3), (6,4), (6,5), (6,6)\}$$

Sum 2, Sum 3, Sum 4, Sum 5

[Note: Event (2,3) means 2 on the the first die and 3 on the second die.]

The probability distribution corresponding to this sample space is:

Sum x_i	2	3	4	5	6	7	8	9	10	11	12
Probability p_i	$\frac{1}{36}$	$\frac{2}{36}$	$\frac{3}{36}$	$\frac{4}{36}$	$\frac{5}{36}$	$\frac{6}{36}$	$\frac{5}{36}$	$\frac{4}{36}$	$\frac{3}{36}$	$\frac{2}{36}$	$\frac{1}{36}$

(B) $E(X) = 2\left(\frac{1}{36}\right) + 3\left(\frac{2}{36}\right) + 4\left(\frac{3}{36}\right) + 5\left(\frac{4}{36}\right) + 6\left(\frac{5}{36}\right) + 7\left(\frac{6}{36}\right) + 8\left(\frac{5}{36}\right)$

$+ 9\left(\frac{4}{36}\right) + 10\left(\frac{3}{36}\right) + 11\left(\frac{2}{36}\right) + 12\left(\frac{1}{36}\right) = 7$ (6-7)

40. The event A that corresponds to the sum being divisible by 4 includes sums 4, 8, and 12. This set is:

$A = \{(1, 3), (2, 2), (3, 1), (2, 6), (3, 5), (4, 4), (5, 3), (6, 2), (6, 6)\}$

The event B that corresponds to the sum being divisible by 6 includes sums 6 and 12. This set is:

$B = \{(1, 5), (2, 4), (3, 3), (4, 2), (5, 1), (6, 6)\}$

$P(A) = \frac{n(A)}{n(S)} = \frac{9}{36} = \frac{1}{4}$

$P(B) = \frac{n(B)}{n(S)} = \frac{6}{36} = \frac{1}{6}$

$P(A \cap B) = \frac{1}{36}$ [Note: $A \cap B = \{(6, 6)\}$]

$P(A \cup B) = \frac{14}{36}$ or $\frac{7}{18}$ [Note: $A \cup B = \{(1, 3), (2, 2), (3, 1), (2, 6),$
$(3, 5), (4, 4), (5, 3), (6, 2), (6, 6),$
$(1, 5), (2, 4), (3, 3), (4, 2), (5, 1)\}$] (6-4)

41. The function P cannot be a probability function because:
(a) P cannot be negative. [Note: $P(e_2) = -0.2$.]
(b) P cannot have a value greater than 1. [Note: $P(e_4) = 2$.]
(c) The sum of the values of P must equal 1. [Note: $P(e_1) + P(e_2)$
$+ P(e_3) + P(e_4) = 0.1 + (-0.2) + 0.6 + 2 = 2.5 \neq 1$.] (6-3)

42. Since $n(A \cup B) = n(A) + n(B) - n(A \cap B)$, we have

$80 = 50 + 45 - n(A \cap B)$

and $n(A \cap B) = 15$

Now, $n(B') = n(U) - n(B) = 100 - 45 = 55$

$n(A') = n(U) - n(A) = 100 - 50 = 50$

$n(A \cap B') = 50 - 15 = 35$

$n(B \cap A') = 45 - 15 = 30$

$n(A' \cap B') = 55 - 35 = 20$

Thus,

	A	A'	Totals
B	15	30	45
B'	35	20	55
Totals	50	50	100

(6-4)

43. (A) $P(\text{odd number}) = P(1) + P(3) + P(5) = .2 + .3 + .1 = .6$

(B) Let $E =$ "number less than 4,"
and $F =$ "odd number."
Now, $E \cap F = \{1, 3\}$, $F = \{1, 3, 5\}$.

$$P(E \mid F) = \frac{P(E \cap F)}{P(F)} = \frac{.2 + .3}{.6} = \frac{5}{6}$$

(6-5)

44. Let $E =$ "card is red" and $F =$ "card is an ace." Then $F \cap E =$ "card is a red ace."

(A) $P(F \mid E) = \dfrac{P(F \cap E)}{P(E)} = \dfrac{2/52}{26/52} = \dfrac{1}{13}$

(B) $P(F \cap E) = \dfrac{1}{26}$, and $P(E) = \dfrac{1}{2}$, $P(F) = \dfrac{1}{13}$. Thus,

$P(F \cap E) = P(E) \cdot P(F)$, and E and F are independent.

(6-5)

45.

	Number of ways of completing operation under condition:		
Operation	No letter repeated	Letters can be repeated	Adjacent letters not alike
O_1	8	8	8
O_2	7	8	7
O_3	6	8	7

Total outcomes, without repeating letters $= 8 \cdot 7 \cdot 6 = 336$.
Total outcomes, with repeating letters $= 8 \cdot 8 \cdot 8 = 512$.
Total outcomes, with adjacent letters not alike $= 8 \cdot 7 \cdot 7 = 392$.

(6-1)

46. (A) This is a permutations problem.

$$P_{6,3} = \frac{6!}{(6-3)!} = \frac{6 \cdot 5 \cdot 4 \cdot 3!}{3!} = 120$$

(B) This is a combinations problem.

$$C_{5,2} = \frac{5!}{2!(5-2)!} = \frac{5 \cdot 4 \cdot 3!}{2 \cdot 1 \cdot 3!} = 10$$

(6-2)

47. (A) The tree diagram with replacement is:

(B) The tree diagram without replacement is:

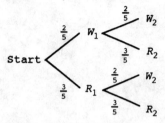

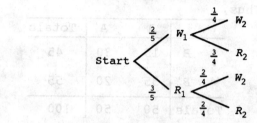

$$P(W_1 \cap R_2) = P(W_1)P(R_2 \mid W_1)$$
$$= \frac{2}{5} \cdot \frac{3}{5} = \frac{6}{25} \approx .24$$

$$P(W_1 \cap R_2) = P(W_1)P(R_2 \mid W_1)$$
$$= \frac{2}{5} \cdot \frac{3}{4} = \frac{6}{20} = .3$$

(6-5)

48. Part (B) involves dependent events because

$$P(R_2 \mid W_1) = \frac{3}{4}$$

$$P(R_2) = P(W_1 \cap R_2) + P(R_1 \cap R_2) = \frac{6}{20} + \frac{6}{20} = \frac{12}{20} = \frac{3}{5}$$

and $\quad P(R_2 \mid W_1) \neq P(R_2)$

The events in part (A) are independent.

(6-5)

49. (A) Using the tree diagram in Problem 47(A), we have:

$$P(\text{zero red balls}) = P(W_1 \cap W_2) = P(W_1)P(W_2) = \frac{2}{5} \cdot \frac{2}{5} = \frac{4}{25} = .16$$

$$P(\text{one red ball}) = P(W_1 \cap R_2) + P(R_1 \cap W_2)$$
$$= P(W_1)P(R_2) + P(R_1)P(W_2)$$
$$= \frac{2}{5} \cdot \frac{3}{5} + \frac{3}{5} \cdot \frac{2}{5} = \frac{12}{25} = .48$$

$$P(\text{two red balls}) = P(R_1 \cap R_2) = P(R_1)P(R_2) = \frac{3}{5} \cdot \frac{3}{5} = \frac{9}{25} = .36$$

Thus, the probability distribution is:

Number of red balls x_i	Probability p_i
0	.16
1	.48
2	.36

The expected number of red balls is:

$$E(X) = 0(.16) + 1(.48) + 2(.36) = .48 + .72 = 1.2$$

(B) Using the tree diagram in Problem 47(B), we have:

$$P(\text{zero red balls}) = P(W_1 \cap W_2) = P(W_1)P(W_2 \mid W_1) = \frac{2}{5} \cdot \frac{1}{4} = \frac{1}{10} = .1$$

$$\begin{aligned} P(\text{one red ball}) &= P(W_1 \cap R_2) + P(R_1 \cap W_2) \\ &= P(W_1)P(R_2 \mid W_1) + P(R_1)P(W_2 \mid R_1) \\ &= \frac{2}{5} \cdot \frac{3}{4} + \frac{3}{5} \cdot \frac{2}{4} = \frac{12}{20} = \frac{3}{5} = .6 \end{aligned}$$

$$P(\text{two red balls}) = P(R_1 \cap R_2) = P(R_1)P(R_2 \mid R_1) = \frac{3}{5} \cdot \frac{2}{4} = \frac{6}{20} = .3$$

Thus, the probability distribution is:

Number of red balls x_i	Probability p_i
0	.1
1	.6
2	.3

The expected number of red balls is:

$$E(X) = 0(.1) + 1(.6) + 2(.3) = 1.2 \qquad\qquad (6\text{--}3)$$

50. The tree diagram for this problem is as follows:

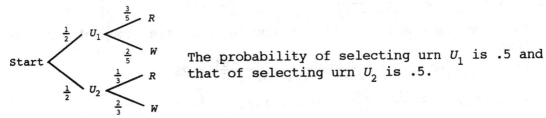

The probability of selecting urn U_1 is .5 and that of selecting urn U_2 is .5.

(A) $P(R \mid U_1) = \dfrac{3}{5}$ (B) $P(R \mid U_2) = \dfrac{1}{3}$

(C) $\begin{aligned} P(R) &= P(R \cap U_1) + P(R \cap U_2) \\ &= P(U_1)P(R \mid U_1) + P(U_2)P(R \mid U_2) \\ &= \frac{1}{2} \cdot \frac{3}{5} + \frac{1}{2} \cdot \frac{1}{3} = \frac{28}{60} = \frac{7}{15} \approx .4667 \end{aligned}$

(D) $\begin{aligned} P(U_1 \mid R) &= \frac{P(U_1 \cap R)}{P(R)} = \frac{P(U_1)P(R \mid U_1)}{P(U_1)P(R \mid U_1) + P(U_2)P(R \mid U_2)} \\ &= \frac{\frac{1}{2} \cdot \frac{3}{5}}{\frac{1}{2} \cdot \frac{3}{5} + \frac{1}{2} \cdot \frac{1}{3}} = \frac{\frac{3}{10}}{\frac{7}{15}} = \frac{9}{14} \approx .6429 \end{aligned}$

(E) $P(U_2 \mid W) = \dfrac{P(U_2 \cap W)}{P(W)} = \dfrac{P(U_2)P(W \mid U_2)}{P(U_2)P(W \mid U_2) + P(U_1)P(W \mid U_1)}$

$= \dfrac{\frac{1}{2} \cdot \frac{2}{3}}{\frac{1}{2} \cdot \frac{2}{3} + \frac{1}{2} \cdot \frac{2}{5}} = \dfrac{\frac{2}{3}}{\frac{16}{15}} = \dfrac{5}{8} = .625$

(F) $P(U_1 \cap R) = P(U_1)P(R \mid U_1) = \frac{1}{2} \cdot \frac{3}{5} = .3$

[Note: In parts (A)–(F), we derived the values of the probabilities from the tree diagram.]   (6-5, 6-6)

51. No, because $P(R \mid U_1) \neq P(R)$. (See Problem 50.)   (6-5)

52. $n(S) = C_{52,5}$

(A) Let A be the event "all diamonds." Then $n(A) = C_{13,5}$. Thus,

$P(A) = \dfrac{n(A)}{n(S)} = \dfrac{C_{13,5}}{C_{52,5}}.$

(B) Let B be the event "3 diamonds and 2 spades." Then $n(B) = C_{13,3} \cdot C_{13,2}$. Thus,

$P(B) = \dfrac{n(B)}{n(S)} = \dfrac{C_{13,3} \cdot C_{13,2}}{C_{52,5}}.$ (6-3)

53. $n(S) = C_{10,4} = \dfrac{10!}{4!(10-4)!} = \dfrac{10 \cdot 9 \cdot 8 \cdot 7 \cdot 6!}{4 \cdot 3 \cdot 2 \cdot 1 \cdot 6!} = 210$

Let A be the event "The married couple is in the group of 4 people." Then

$n(A) = C_{2,2} \cdot C_{8,2} = 1 \cdot \dfrac{8!}{2!(8-2)!} = \dfrac{8 \cdot 7 \cdot 6!}{2 \cdot 1 \cdot 6!} = 28.$

Thus, $P(A) = \dfrac{n(A)}{n(S)} = \dfrac{28}{210} = \dfrac{2}{15} \approx 0.1333.$ (6-3)

54. By the multiplication principle, there are
$N_1 \cdot N_2 \cdot N_3$
branches in the tree diagram.   (6-1)

55. Events S and H are mutually exclusive. Hence, $P(S \cap H) = 0$, while $P(S) \neq 0$ and $P(H) \neq 0$. Therefore,
$P(S \cap H) \neq P(S) \cdot P(H)$
which implies that S and F are dependent.   (6-5)

56. Let E_2 be the event "2 heads."

(A) From the table, $f(E_2) = 350$. Thus, the approximate empirical probability of obtaining 2 heads is:

$P(E_2) \sim \dfrac{f(E_2)}{n} = \dfrac{350}{1000} = 0.350$

(B) S = {HHH, HHT, HTH, HTT, THH, THT, TTH, TTT}

The theoretical probability of obtaining 2 heads is:

$$P(E_2) = \frac{n(E_2)}{n(S)} = \frac{3}{8} = 0.375$$

(C) The expected frequency of obtaining 2 heads in 1000 tosses of 3 fair coins is:

$$f(E_2) = 1000(0.375) = 375 \qquad\qquad (6\text{-}3,\ 6\text{-}7)$$

57. On one roll of the dice, the probability of getting a double six is $\frac{1}{36}$ and the probability of not getting a double six is $\frac{35}{36}$.

On two rolls of the dice there are $(36)^2$ possible outcomes. There are 71 ways to get at least one double six, namely a double six on the first roll and any one of the 35 other outcomes on the second roll, or a double six on the second roll and any one of the 35 other outcomes on the first roll, or a double six on both rolls. Thus, the probability of at least one double six on two rolls is $\frac{71}{(36)^2}$ and the probability of no double sixes is:

$$1 - \frac{71}{(36)^2} = \frac{(36)^2 - 2\cdot 36 + 1}{(36)^2} = \frac{(36-1)^2}{(36)^2} = \left(\frac{35}{36}\right)^2$$

Let E be the event "At least one double six." Then E' is the event "No double sixes." Continuing with the reasoning above, we conclude that, in 24 rolls of the die,

$$P(E') = \left(\frac{35}{36}\right)^{24} \approx 0.5086$$

Therefore, $P(E) = 1 - 0.5086 = 0.4914$.

The payoff table is:

x_i	1	−1
P_i	0.4914	0.5086

and $E(X)$ = 1(0.4914) + (−1)(0.5086)

$\qquad\qquad$ = 0.4914 − 0.5086

$\qquad\qquad$ = −0.0172

Thus, your expectation is −$0.0172.
Your friend's expectation is $0.0172.
The game is not fair. $\qquad\qquad$ (6-7)

58. (A) This is a permutations problem.

$$P_{10,3} = \frac{10!}{(10-3)!} = \frac{10!}{7!} = 10\cdot 9\cdot 8 = 720$$

(B) The number of ways in which women are selected for all three positions is given by:

$$P_{6,3} = \frac{6!}{(6-3)!} = \frac{6!}{3!} = 6 \cdot 5 \cdot 4 = 120$$

Thus, $P(\text{three women are selected}) = \frac{P_{6,3}}{P_{10,3}} = \frac{120}{720} = \frac{1}{6}$

(C) This is a combinations problem.

$$C_{10,3} = \frac{10!}{3!(10-3)!} = \frac{10 \cdot 9 \cdot 8 \cdot 7!}{3 \cdot 2 \cdot 1 \cdot 7!} = 120$$

(D) Let Event D = majority of team members will be women. Then

$$n(D) = \text{team has 3 women} + \text{team has 2 women}$$
$$= C_{6,3} + C_{6,2} \cdot C_{4,1}$$
$$= \frac{6!}{3!(6-3)!} + \frac{6!}{2!(6-2)!} \cdot \frac{4!}{1!(4-1)!} = 20 + 15 \cdot 4 = 80$$

Thus,

$$P(D) = \frac{n(D)}{n(S)} = \frac{C_{6,3} + C_{6,2} \cdot C_{4,1}}{C_{10,3}} = \frac{80}{120} = \frac{2}{3} \qquad \text{(6-1, 6-2, 6-3)}$$

59. Draw a Venn diagram with: A = Chess players, B = Checker players.

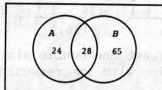

Now, $n(A \cap B) = 28$, $n(A \cap B') = n(A) - n(A \cap B) = 52 - 28 = 24$
$n(B \cap A') = n(B) - n(A \cap B) = 93 - 28 = 65$

Since there are 150 people in all,

$$n(A' \cap B') = n(U) - [n(A \cap B') + n(A \cap B) + n(B \cap A')]$$
$$= 150 - (24 + 28 + 65) = 150 - 117 = 33 \qquad \text{(6-1)}$$

60. The total number of ways that 3 people can be selected from a group of 10 is:

$$C_{10,3} = \frac{10!}{3!(10-3)!} = \frac{10 \cdot 9 \cdot 8 \cdot 7!}{3 \cdot 2 \cdot 1 \cdot 7!} = 120$$

The number of ways of selecting *no* women is:

$$C_{7,3} = \frac{7!}{3!(7-3)!} = \frac{7 \cdot 6 \cdot 5 \cdot 4!}{3 \cdot 2 \cdot 1 \cdot 4!} = 35$$

Thus, the number of samples of 3 people that contain at least one woman is $120 - 35 = 85$.

Therefore, if event A is "At least one woman is selected," then

$$P(A) = \frac{n(A)}{n(S)} = \frac{85}{120} = \frac{17}{24} \approx 0.708. \qquad \text{(6-3)}$$

61. $P(\text{second heart} \mid \text{first heart}) = P(H_2 \mid H_1) = \frac{12}{51} \approx .235$

[<u>Note</u>: One can see that $P(H_2 \mid H_1) = \frac{12}{51}$ directly.] $\qquad$ (6-5)

62. $P(\text{first heart} \mid \text{second heart}) = P(H_1 \mid H_2)$

$$= \frac{P(H_1 \cap H_2)}{P(H_2)} = \frac{P(H_1)P(H_2 \mid H_1)}{P(H_2)}$$

$$= \frac{P(H_1)P(H_2 \mid H_1)}{P(H_1 \cap H_2) + P(H_1' \cap H_2)}$$

$$= \frac{P(H_1)P(H_2 \mid H_1)}{P(H_1)P(H_2 \mid H_1) + P(H_1')P(H_2 \mid H_1')}$$

$$= \frac{\dfrac{13}{52} \cdot \dfrac{12}{51}}{\dfrac{13}{52} \cdot \dfrac{12}{51} + \dfrac{39}{52} \cdot \dfrac{13}{51}} = \frac{12}{51} \approx .235 \qquad (6\text{-}5)$$

63. Since each die has 6 faces, there are $6 \cdot 6 = 36$ possible pairs for the two up faces.

A sum of 2 corresponds to having $(1, 1)$ as the up faces. This sum can be obtained in $3 \cdot 3 = 9$ ways (3 faces on the first die, 3 faces on the second). Thus,

$P(2) = \dfrac{9}{36} = \dfrac{1}{4}.$

A sum of 3 corresponds to the two pairs $(2, 1)$ and $(1, 2)$. The number of such pairs is $2 \cdot 3 + 3 \cdot 2 = 12$. Thus,

$P(3) = \dfrac{12}{36} = \dfrac{1}{3}.$

A sum of 4 corresponds to the pairs $(3, 1)$, $(2, 2)$, $(1, 3)$. There are $1 \cdot 3 + 2 \cdot 2 + 3 \cdot 1 = 10$ such pairs. Thus,

$P(4) = \dfrac{10}{36}.$

A sum of 5 corresponds to the pairs $(2, 3)$ and $(3, 2)$. There are $2 \cdot 1 + 1 \cdot 2 = 4$ such pairs. Thus,

$P(5) = \dfrac{4}{36} = \dfrac{1}{9}.$

A sum of 6 corresponds to the pair $(3, 3)$ and there is one such pair. Thus,

$P(6) = \dfrac{1}{36}.$

(A) The probability distribution for X is:

x_i	2	3	4	5	6
P_i	$\dfrac{9}{36}$	$\dfrac{12}{36}$	$\dfrac{10}{36}$	$\dfrac{4}{36}$	$\dfrac{1}{36}$

(B) The expected value is:

$$E(X) = 2\left(\frac{9}{36}\right) + 3\left(\frac{12}{36}\right) + 4\left(\frac{10}{36}\right) + 5\left(\frac{4}{36}\right) + 6\left(\frac{1}{36}\right) = \frac{120}{36} = \frac{10}{3} \qquad (6\text{-}7)$$

64. The payoff table is:

x_i	-\$1.50	-\$0.50	\$0.50	\$1.50	\$2.50
P_i	$\frac{9}{36}$	$\frac{12}{36}$	$\frac{10}{36}$	$\frac{4}{36}$	$\frac{1}{36}$

and $E(X) = \frac{9}{36}(-1.50) + \frac{12}{36}(-0.50) + \frac{10}{36}(0.50) + \frac{4}{36}(1.50) + \frac{1}{36}(2.50)$

$\qquad = -0.375 - 0.167 + 0.139 + 0.167 + 0.069$

$\qquad = -0.167$ or $-\$0.167 \approx \$0.17$

The game is not fair. The game would be fair if you paid \$3.50 - \$0.17 = \$3.33 to play.

$\hfill$ (6-7)

65. Operation 1: Two possible outcomes, boy or girl, $N_1 = 2$.
Operation 2: Two possible outcomes, boy or girl, $N_2 = 2$.
Operation 3: Two possible outcomes, boy or girl, $N_3 = 2$.
Operation 4: Two possible outcomes, boy or girl, $N_4 = 2$.
Operation 5: Two possible outcomes, boy or girl, $N_5 = 2$.

Using the Multiplication Principle, the total combined outcomes is:
$N_1 \cdot N_2 \cdot N_3 \cdot N_4 \cdot N_5 = 2 \cdot 2 \cdot 2 \cdot 2 \cdot 2 = 32$.

If order pattern is not taken into account, there would be only 6 possible outcomes: families with 0, 1, 2, 3, 4, or 5 boys.

$\hfill$ (6-1)

66. The tree diagram for this experiment is:

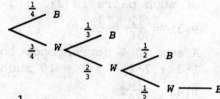

(A) $P(\text{black on the fourth draw}) = \frac{3}{4} \cdot \frac{2}{3} \cdot \frac{1}{2} = \frac{1}{4}$

The odds for black on the fourth draw are 1 to 3.

(B) Let x = amount house should pay (and return the \$1 bet). Then, for the game to be fair:

$E(X) = x\left(\frac{1}{4}\right) + (-1)\left(\frac{3}{4}\right) = 0$

$\qquad\qquad \frac{x}{4} - \frac{3}{4} = 0$

$\qquad\qquad\qquad x = 3$

Thus, the house should pay \$3.

$\hfill$ (6-4, 6-6)

67. $n(S) = 10 \cdot 10 \cdot 10 \cdot 10 \cdot 10 = 10^5$

Let event A = "at least two people identify the same book." Then
A' = "each person identifies a different book," and

$$n(A') = 10 \cdot 9 \cdot 8 \cdot 7 \cdot 6 = \frac{10!}{5!}$$

Thus, $P(A') = \dfrac{\frac{10!}{5!}}{10^5} = \dfrac{10!}{5!10^5}$ and $P(A) = 1 - \dfrac{10!}{5!10^5} \approx 1 - .3 = .7.$ (6-4)

68. The number of routes starting from A and visiting each of the 5 stores exactly once is the number of permutations of 5 objects taken 5 at a time, i.e.,

$$P_{5,5} = \frac{5!}{(5-5)!} = 120.$$ (6-1)

69. Draw a Venn diagram with:

S = people who have invested in stocks, and
B = people who have invested in bonds.

Then $n(U) = 1000$, $n(S) = 340$, $n(B) = 480$ and $n(S \cap B) = 210$

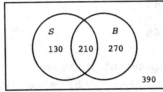

$n(S \cap B') = 340 - 210 = 130 \qquad n(B \cap S') = 480 - 210 = 270$
$n(S' \cap B') = 1000 - (130 + 210 + 270) = 1000 - 610 = 390$

(A) $n(S \cup B) = n(S) + n(B) - n(S \cap B) = 340 + 480 - 210 = 610.$

(B) $n(S' \cap B') = 390$

(C) $n(B \cap S') = 270$ (6-1)

70.

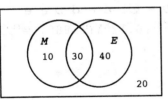

Event M = Reads the morning paper.
Event E = Reads the evening paper.

(A) $P(\text{reads a daily paper}) = P(M \text{ or } E) = P(M \cup E)$

$$= P(M) + P(E) - P(M \cap E)$$

$$= \frac{40}{100} + \frac{70}{100} - \frac{30}{100} = .8$$

(B) $P(\text{does not read a daily paper}) = \dfrac{20}{100}$ (from the Venn diagram)

$$= .2$$

$$\text{or} = 1 - P(M \cup E) \quad [\text{i.e., } P((M \cup E)')]$$

$$= 1 - .8 = .2$$

(C) P(reads exactly one daily paper) $= \dfrac{10 + 40}{100}$ (from the Venn diagram)

$$= .5$$

or $P((M \cap E') \text{ or } (M' \cap E))$

$$= P(M \cap E') + P(M' \cap E)$$

$$= \dfrac{10}{100} + \dfrac{40}{100} = .5 \qquad (6\text{-}1, \ 6\text{-}4)$$

71. Let A be the event that a person has seen the advertising and P be the event that the person purchased the product. Given:

$P(A) = .4$ and $P(P \mid A) = .85$

We want to find:

$P(A \cap P) = P(A)P(P \mid A) = (.4)(.85) = .34$

$\qquad\qquad\qquad (6\text{-}5)$

72. (A) $\qquad P(A) = \dfrac{290}{1000} = 0.290$

$\qquad\qquad P(B) = \dfrac{290}{1000} = 0.290$

$\qquad P(A \cap B) = \dfrac{100}{1000} = 0.100$

$\qquad P(A \mid B) = \dfrac{100}{290} = 0.345$

$\qquad P(B \mid A) = \dfrac{100}{290} = 0.345$

(B) A and B are **not** independent because

$\qquad 0.100 = P(A \cap B) \neq P(A) \cdot P(B) = (0.290)(0.290) = 0.084$

(C) $\qquad P(C) = \dfrac{880}{1000} = 0.880$

$\qquad\qquad P(D) = \dfrac{120}{1000} = 0.120$

$\qquad P(C \cap D) = 0$

$\qquad P(C \mid D) = P(D \mid C) = 0$

(D) C and D are mutually exclusive since $C \cap D \neq \varnothing$. C and D are

$\qquad$ dependent since $0 = P(C \cap D) \neq P(C) \cdot P(D) = (0.120)(0.880) = 0.106$

$\qquad\qquad\qquad (6\text{-}5)$

73. The payoff table for plan A is:

x_i	10 million	-2 million
P_i	0.8	0.2

Hence, $E(X) = 10(0.8) - 2(0.2) = 8 - 0.4 = \$7.6$ million.

The payoff table for plan B is:

x_i	12 million	-2 million
P_i	0.7	0.3

Hence, $E(X) = 12(0.7) - 2(0.3) = 8.4 - 0.6 = \$7.8$ million.
Plan B should be chosen.

$\qquad\qquad\qquad (6\text{-}7)$

74. The payoff table is:

Gain	x_i	\$270	−\$30
	P_i	0.08	0.92

[<u>Note</u>: 300 − 30 = 270 is the "gain" if the bicycle is stolen.]

Hence, $E(X) = 270(0.08) - 30(0.92) = 21.6 - 27.6 = -\$6.$ (6–7)

75. $n(S) = C_{12,4} = \dfrac{12!}{4!(12-4)!} = \dfrac{12 \cdot 11 \cdot 10 \cdot 9 \cdot 8!}{4 \cdot 3 \cdot 2 \cdot 1 \cdot 8!} = 495$

The number of samples that contain *no* substandard parts is:

$C_{10,4} = \dfrac{10!}{4!(10-4)!} = \dfrac{10 \cdot 9 \cdot 8 \cdot 7 \cdot 6!}{4 \cdot 3 \cdot 2 \cdot 1 \cdot 6!} = 210$

Thus, the number of samples that have at least one defective part is 495 − 210 = 285. If E is the event "The shipment is returned," then

$P(E) = \dfrac{n(E)}{n(S)} = \dfrac{285}{495} \approx 0.576.$ (6–2, 6–4)

76. $n(S) = C_{12,3} = \dfrac{12!}{3!(12-3)!} = \dfrac{12 \cdot 11 \cdot 10 \cdot 9!}{3 \cdot 2 \cdot 1 \cdot 9!} = 220$

A sample will either have 0, 1, or 2 defective circuit boards.

$P(0) = \dfrac{C_{10,3}}{C_{12,3}} = \dfrac{\frac{10!}{3!(10-3)!}}{220} = \dfrac{\frac{10 \cdot 9 \cdot 8 \cdot 7!}{3 \cdot 2 \cdot 1 \cdot 7!}}{220} = \dfrac{120}{220} = \dfrac{12}{22}$

$P(1) = \dfrac{C_{2,1} \cdot C_{10,2}}{C_{12,3}} = \dfrac{2 \cdot \frac{10!}{2!(10-2)!}}{220} = \dfrac{90}{220} = \dfrac{9}{22}$

$P(2) = \dfrac{C_{2,2} \cdot C_{10,1}}{220} = \dfrac{10}{220} = \dfrac{1}{22}$

(A) The probability distribution of X is:

x_i	0	1	2
P_i	$\frac{12}{22}$	$\frac{9}{22}$	$\frac{1}{22}$

(B) $E(X) = 0\left(\dfrac{12}{22}\right) + 1\left(\dfrac{9}{22}\right) + 2\left(\dfrac{1}{22}\right) = \dfrac{11}{22} = \dfrac{1}{2}$ (6–7)

77. Let Event NH = individual with normal heart,
        Event MH = individual with minor heart problem,
        Event SH = individual with severe heart problem,
and Event P = individual passes the cardiogram test.

Then, using the notation given above, we have:

$P(NH) = .82$
$P(MH) = .11$
$P(SH) = .07$
$P(P \mid NH) = .95$

$$P(P \mid MH) = .30$$
$$P(P \mid SH) = .05$$

We want to find $P(NH \mid P) = \dfrac{P(NH \cap P)}{P(P)} = \dfrac{P(NH)P(P \mid NH)}{P(NH \cap P) + P(MH \cap P) + P(SH \cap P)}$

$$= \dfrac{P(NH)P(P \mid NH)}{P(NH)P(P \mid NH) + P(MH)P(P \mid MH) + P(SH)P(P \mid SH)}$$

$$= \dfrac{(.82)(.95)}{(.82)(.95) + (.11)(.30) + (.07)(.05)} = .955$$

(6-6)

78. The tree diagram for this problem is as follows:

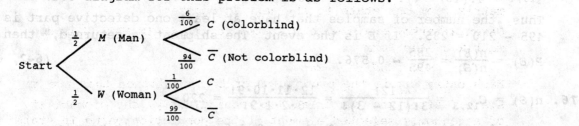

We now compute

$$P(M \mid C) = \dfrac{P(M \cap C)}{P(C)} = \dfrac{P(M \cap C)}{P(M \cap C) + P(W \cap C)} = \dfrac{P(M)P(C \mid M)}{P(M)P(C \mid M) + P(W)P(C \mid W)}$$

$$= \dfrac{\dfrac{1}{2} \cdot \dfrac{6}{100}}{\dfrac{1}{2} \cdot \dfrac{6}{100} + \dfrac{1}{2} \cdot \dfrac{1}{100}} = \dfrac{6}{7} \approx .857$$

(6-6)

79. According to the empirical probabilities, candidate A should have won the election. Since candidate B won the election one week later, either some of the students changed their minds during the week, or the 30 students in the math class were not representative of the entire student body, or both.

(6-3)

7 MARKOV CHAINS

Things to remember:

1. MARKOV CHAINS

   A MARKOV CHAIN, or PROCESS, is a sequence of experiments, trials, or observations such that the transition probability matrix from one state to the next is constant.

   Given a Markov chain with n states, a kth STATE MATRIX is a matrix of the form
   $$S_k = [s_{k1} \quad s_{k2} \quad \cdots \quad s_{kn}]$$

   Each entry s_{ki} is the proportion of the population that are in state i after the kth trial, or, equivalently, the probability of a randomly selected element of the population being in state i after the kth trial. The sum of all the entries in the kth state matrix S_k must be 1.

   A TRANSITION MATRIX is a constant square matrix P of order n such that the entry in the ith row and jth column indicates the probability of the system moving from the ith state to the jth state on the next observation or trial. The sum of the entries in each row must be 1.

2. COMPUTING STATE MATRICES FOR A MARKOV CHAIN

   If S_0 is the initial state matrix and P is the transition matrix for a Markov chain, then the subsequent state matrices are given by:

   $$\begin{aligned}
   S_1 &= S_0 P \qquad \text{First-state matrix} \\
   S_2 &= S_1 P \qquad \text{Second-state matrix} \\
   S_3 &= S_2 P \qquad \text{Third-state matrix} \\
   &\;\;\vdots \\
   S_k &= S_{k-1} P \quad k\text{th-state matrix}
   \end{aligned}$$

3. POWERS OF A TRANSITION MATRIX

   If P is the transition matrix and S_0 is an initial state matrix for a Markov chain, then the kth state matrix is given by
   $$S_k = S_0 P^k$$

   The entry in the ith row and jth column of P^k indicates the probability of the system moving from the ith state to the jth state in k observations or trials. The sum of the entries in each row of P^k is 1.

1. $S_1 = S_0 P = \begin{bmatrix} 1 & 0 \end{bmatrix} \begin{bmatrix} .8 & .2 \\ .4 & .6 \end{bmatrix} = \begin{matrix} A & B \\ [.8 & .2] \end{matrix}$

A: $(1)(.8) + (0)(.4) = .8$

B: $(1)(.2) + (0)(.6) = .2$

3. $S_1 = S_0 P = \begin{bmatrix} .5 & .5 \end{bmatrix} \begin{bmatrix} .8 & .2 \\ .4 & .6 \end{bmatrix} = \begin{matrix} A & B \\ [.6 & .4] \end{matrix}$

A: $(.5)(.8) + (.5)(.4) = .6$

B: $(.5)(.2) + (.5)(.6) = .4$

5. $S_2 = S_1 P = \begin{bmatrix} .8 & .2 \end{bmatrix} \begin{bmatrix} .8 & .2 \\ .4 & .6 \end{bmatrix}$ (from Problem 1)

$$= \begin{matrix} A & B \\ [.72 & .28] \end{matrix}$$

The probability of being in state A after two trials is .72; the probability of being in state B after two trials is .28.

7. $S_2 = S_1 P = \begin{bmatrix} .6 & .4 \end{bmatrix} \begin{bmatrix} .8 & .2 \\ .4 & .6 \end{bmatrix}$ (from Problem 3)

$$= \begin{matrix} A & B \\ [.64 & .36] \end{matrix}$$

The probability of being in state A after two trials is .64; the probability of being in state B after two trials is .36.

9.

$$\begin{matrix} & A & B \\ A & \begin{bmatrix} .4 & .6 \\ B & .7 & .3 \end{bmatrix} \end{matrix}$$

11.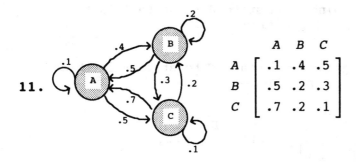

$$\begin{array}{c} \\ A \\ B \\ C \end{array} \begin{array}{ccc} A & B & C \\ \left[\begin{array}{ccc} .1 & .4 & .5 \\ .5 & .2 & .3 \\ .7 & .2 & .1 \end{array} \right] \end{array}$$

13.
$$\begin{array}{ll} 0 + .5 + a = 1 & \text{implies } a = .5 \\ b + 0 + .4 = 1 & \text{implies } b = .6 \\ .2 + c + .1 = 1 & \text{implies } c = .7 \end{array}$$

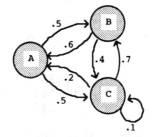

15.
$$\begin{array}{ll} 0 + a + .3 = 1 & \text{implies } a = .7 \\ 0 + b + 0 = 1 & \text{implies } b = 1 \\ c + .8 + 0 = 1 & \text{implies } c = .2 \end{array}$$

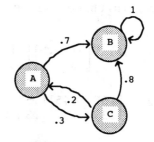

17. The probability of staying in state A is .3.
The probability of staying in state B is .1.

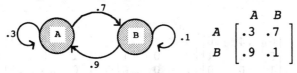

$$\begin{array}{c} \\ A \\ B \end{array} \begin{array}{cc} A & B \\ \left[\begin{array}{cc} .3 & .7 \\ .9 & .1 \end{array} \right] \end{array}$$

19. The probability of staying in state A is .6.
The probability of staying in state B is .3.
Since the probability of staying in state C is 1, the probability of going from state C to state A is 0 and the probability of going from state C to state B is 0.

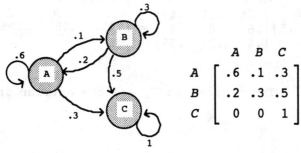

$$\begin{array}{c} \\ A \\ B \\ C \end{array} \begin{array}{ccc} A & B & C \\ \left[\begin{array}{ccc} .6 & .1 & .3 \\ .2 & .3 & .5 \\ 0 & 0 & 1 \end{array} \right] \end{array}$$

21. Using P^2, the probability of going from state A to state B in two trials is the element in the $(1,2)$ position—$.35$.

23. Using P^3, the probability of going from state C to state A in three trials is the number in the $(3,1)$ position—$.212$.

25. $S_2 = S_0 P^2 = \begin{bmatrix} 1 & 0 & 0 \end{bmatrix} \begin{bmatrix} .43 & .35 & .22 \\ .25 & .37 & .38 \\ .17 & .27 & .56 \end{bmatrix} = \begin{matrix} A & B & C \\ [.43 & .35 & .22] \end{matrix}$

These are the probabilities of going from state A to states A, B, and C, respectively, in two trials.

27. $S_3 = S_0 P^3 = \begin{bmatrix} 0 & 0 & 1 \end{bmatrix} \begin{bmatrix} .35 & .348 & .302 \\ .262 & .336 & .402 \\ .212 & .298 & .49 \end{bmatrix} = \begin{matrix} A & B & C \\ [.212 & .298 & .49] \end{matrix}$

These are the probabilities of going from state C to states A, B, and C, respectively, in three trials.

29. $P = \begin{bmatrix} .1 & .9 \\ .6 & .4 \end{bmatrix}$, $P^2 = \begin{bmatrix} .1 & .9 \\ .6 & .4 \end{bmatrix} \begin{bmatrix} .1 & .9 \\ .6 & .4 \end{bmatrix} = \begin{bmatrix} .55 & .45 \\ .3 & .7 \end{bmatrix}$;

$P^4 = P^2 \cdot P^2 = \begin{bmatrix} .55 & .45 \\ .3 & .7 \end{bmatrix} \begin{bmatrix} .55 & .45 \\ .3 & .7 \end{bmatrix} = \begin{matrix} & A & B \\ A & \begin{bmatrix} .4375 & .5625 \\ B & .375 & .625 \end{bmatrix} \end{matrix}$

$S_4 = S_0 P^4 = \begin{bmatrix} .8 & .2 \end{bmatrix} \begin{bmatrix} .4375 & .5625 \\ .375 & .625 \end{bmatrix} = \begin{matrix} A & B \\ [.425 & .575] \end{matrix}$

31. $P = \begin{bmatrix} 0 & .4 & .6 \\ 0 & 0 & 1 \\ 1 & 0 & 0 \end{bmatrix}$, $P^2 = \begin{bmatrix} 0 & .4 & .6 \\ 0 & 0 & 1 \\ 1 & 0 & 0 \end{bmatrix} \begin{bmatrix} 0 & .4 & .6 \\ 0 & 0 & 1 \\ 1 & 0 & 0 \end{bmatrix} = \begin{bmatrix} .6 & 0 & .4 \\ 1 & 0 & 0 \\ 0 & .4 & .6 \end{bmatrix}$;

$P^4 = P^2 \cdot P^2 = \begin{bmatrix} .6 & 0 & .4 \\ 1 & 0 & 0 \\ 0 & .4 & .6 \end{bmatrix} \begin{bmatrix} .6 & 0 & .4 \\ 1 & 0 & 0 \\ 0 & .4 & .6 \end{bmatrix} = \begin{matrix} & A & B & C \\ A & \begin{bmatrix} .36 & .16 & .48 \\ B & .6 & 0 & .4 \\ C & .4 & .24 & .36 \end{bmatrix} \end{matrix}$

$S_4 = S_0 P^4 = \begin{bmatrix} .2 & .3 & .5 \end{bmatrix} \begin{bmatrix} .36 & .16 & .48 \\ .6 & 0 & .4 \\ .4 & .24 & .36 \end{bmatrix} = \begin{matrix} A & B & C \\ [.452 & .152 & .396] \end{matrix}$

33. $S_k = S_0 P^k = \begin{bmatrix} 1 & 0 \end{bmatrix} P^k$; The entries in S_k are the entries in the first row of P^k.

35. (A) $P^2 = \begin{bmatrix} .2 & .2 & .3 & .3 \\ 0 & 1 & 0 & 0 \\ .2 & .2 & .1 & .5 \\ 0 & 0 & 0 & 1 \end{bmatrix}\begin{bmatrix} .2 & .2 & .3 & .3 \\ 0 & 1 & 0 & 0 \\ .2 & .2 & .1 & .5 \\ 0 & 0 & 0 & 1 \end{bmatrix} = \begin{bmatrix} .1 & .3 & .09 & .51 \\ 0 & 1 & 0 & 0 \\ .06 & .26 & .07 & .61 \\ 0 & 0 & 0 & 1 \end{bmatrix}$

$P^4 = \begin{bmatrix} .1 & .3 & .09 & .51 \\ 0 & 1 & 0 & 0 \\ .06 & .26 & .07 & .61 \\ 0 & 0 & 0 & 1 \end{bmatrix}\begin{bmatrix} .1 & .3 & .09 & .51 \\ 0 & 1 & 0 & 0 \\ .06 & .26 & .07 & .61 \\ 0 & 0 & 0 & 1 \end{bmatrix}$

$= \begin{array}{c} A \\ B \\ C \\ D \end{array}\begin{array}{cccc} A & B & C & D \end{array}$
$= \begin{array}{c} A \\ B \\ C \\ D \end{array}\begin{bmatrix} .0154 & .3534 & .0153 & .6159 \\ 0 & 1 & 0 & 0 \\ .0102 & .2962 & .0103 & .6833 \\ 0 & 0 & 0 & 1 \end{bmatrix}$

(B) The probability of going from state A to state D in 4 trials is the element in the (1,4) position: .6159.

(C) The element in the (3,2) position: .2962.

(D) The element in the (2,1) position: 0.

37. If $P = \begin{bmatrix} a & 1-a \\ 1-b & b \end{bmatrix}$ is a probability matrix

then $0 \le a \le 1$, $0 \le b \le 1$

$P^2 = \begin{bmatrix} a & 1-a \\ 1-b & b \end{bmatrix}\begin{bmatrix} a & 1-a \\ 1-b & b \end{bmatrix}$

$= \begin{bmatrix} a^2 + (1-a)(1-b) & a(1-a) + (1-a)b \\ (1-b)a + b(1-b) & (1-b)(1-a) + b^2 \end{bmatrix}$

Now, $a^2 + (1-a)(1-b) \ge 0$

and $a(1-a) + (1-a)b = (1-a)(a+b) \ge 0$

since $0 \le a \le 1$ and $0 \le b \le 1$.

Also,

$a^2 + (1-a)(1-b) + (1-a)(a+b) = a^2 + (1-a)[1-b+a+b]$
$= a^2 + (1-a)(1+a)$
$= a^2 + 1 - a^2$
$= 1$

Therefore, the elements in the first row of P^2 are nonnegative and their sum is 1. The same arguments apply to the elements in the second row of P^2. Thus, P^2 is a probability matrix.

39. $P = \begin{bmatrix} .4 & .6 \\ .2 & .8 \end{bmatrix}$

(A) Let $S_0 = [0 \quad 10]$. Then

$S_2 = S_0 P^2 = [.24 \quad .76]$

$S_4 = S_0 P^4 = [.2496 \quad .7504]$

$S_8 = S_0 P^8 = [.24999936 \quad .7500006]$

S_k is approaching $[.25 \quad .75]$

(B) Let $S_0 = [1 \quad 0]$. Then

$$S_2 = S_0 P^2 = [.28 \quad .72]$$
$$S_4 = S_0 P^4 = [.2512 \quad .7488]$$
$$S_8 = S_0 P^8 = [.25000192 \quad .7499980]$$

S_k is approaching $[.25 \quad .75]$

(C) Let $S_0 = [.5 \quad .5]$. Then

$$S_2 = S_0 P^2 = [.26 \quad .74]$$
$$S_4 = S_0 P^4 = [.2504 \quad .7496]$$
$$S_8 = S_0 P^8 = [.25000064 \quad .74999936]$$

S_k is approaching $[.25 \quad .75]$

(D) $[.25 \quad .75] \begin{bmatrix} .4 & .6 \\ .2 & .8 \end{bmatrix} = [.25 \quad .75]$

(E) The state matrices S_k appear to approach the same matrix $[.25 \quad .75]$, regardless of the values in the initial state matrix S_0.

41. $P^2 = \begin{bmatrix} .28 & .72 \\ .24 & .76 \end{bmatrix}$, $P^4 = \begin{bmatrix} .2512 & .7488 \\ .2496 & .7504 \end{bmatrix}$

$P^8 = \begin{bmatrix} .25000192 & .7499980 \\ .24999936 & .7500006 \end{bmatrix} \cdots$

The matrices P^k are approaching $Q = \begin{bmatrix} .25 & .75 \\ .25 & .75 \end{bmatrix}$; the rows of Q are the same as the matrix $S = [.25 \quad .75]$ in Problem 39.

43. Let R denote "rain" and R' "not rain".

(A)

R = Rain
R' = No rain

(B)
$$\begin{array}{c} R \\ R' \end{array} \begin{bmatrix} R & R' \\ .4 & .6 \\ .06 & .94 \end{bmatrix}$$

(C) Rain on Saturday: $P^2 = \begin{array}{c} R \\ R' \end{array} \begin{bmatrix} R & R' \\ .196 & .804 \\ .0804 & .9196 \end{bmatrix}$

The probability that it will rain on Saturday is .196.

Rain on Sunday: $P^3 = \begin{array}{c} R \\ R' \end{array} \begin{bmatrix} R & R' \\ .12664 & .87336 \\ .087336 & .912664 \end{bmatrix}$

The probability that it will rain on Sunday is .12664.

45. (A)

.8 ⟲ (X) ⇄ (X') ⟳ .8 with .2 between them

(B)

$$P = \begin{array}{c} \\ X \\ X' \end{array} \begin{array}{c} \begin{array}{cc} X & X' \end{array} \\ \begin{bmatrix} .8 & .2 \\ .2 & .8 \end{bmatrix} \end{array}$$

(C) $\begin{array}{cc} & \begin{array}{cc} X & \quad X' \end{array} \\ S = [.2 & .8] \end{array}$

$$S_1 = SP = [.2 \quad .8]\begin{bmatrix} .8 & .2 \\ .2 & .8 \end{bmatrix} = [.32 \quad .68]$$

32% will be using brand X one week later.

$$S_2 = SP^2 = [.2 \quad .8]\begin{bmatrix} .68 & .32 \\ .32 & .68 \end{bmatrix} = [.392 \quad .608]$$

39.2% will be using brand X two weeks later.

47. (A)

N = National Property
U = United Family
O = Other companies

(B)

$$P = \begin{array}{c} \\ N \\ U \\ O \end{array} \begin{array}{c} \begin{array}{ccc} N & U & O \end{array} \\ \begin{bmatrix} .65 & .25 & .10 \\ .10 & .85 & .05 \\ .15 & .35 & .50 \end{bmatrix} \end{array}$$

(C) $\begin{array}{ccc} & N & \;\; U \;\;\;\; O \end{array}$
$S = [.50 \quad .30 \quad .20]$

$\qquad\qquad\qquad\qquad N \qquad U \qquad O$
After one year: $SP = [.385 \quad .45 \quad .165]$
38.5% will be insured by National Property after one year.

$\qquad\qquad\qquad\qquad N \qquad\quad U \qquad\quad O$
After two years: $SP^2 = [.32 \quad .5365 \quad .1435]$
32% will be insured by National Property after two years.

(D) 45% of the homes will be insured by United Family after one year; 53.65% will be insured by United Family after two years.

49. (A)

B = Beginning agent
I = Intermediate agent
T = Terminated agent
Q = Qualified agent

(B)

$$P = \begin{array}{c} \\ B \\ I \\ T \\ Q \end{array} \begin{array}{c} \begin{array}{cccc} B & I & T & Q \end{array} \\ \begin{bmatrix} .5 & .4 & .1 & 0 \\ 0 & .6 & .1 & .3 \\ 0 & 0 & 1 & 0 \\ 0 & 0 & 0 & 1 \end{bmatrix} \end{array}$$

$$\begin{array}{cccc} & B & I & T & Q \end{array}$$

(C) $S = [1 \quad 0 \quad 0 \quad 0]$

$$\begin{array}{ccccc} & B & I & T & Q \end{array}$$

After one year: $SP^2 = [.25 \quad .44 \quad .19 \quad .12]$

The probability that a beginning agent will be promoted to qualified agent within one year (i.e., after 2 reviews) is: .12.

$$\begin{array}{ccccc} & B & I & T & Q \end{array}$$

After two years: $SP^4 = [.0625 \quad .2684 \quad .3079 \quad .3612]$

The probability that a beginning agent will be promoted to qualified agent within two years (i.e., after 4 reviews) is: .3612.

51. (A) $P = \begin{array}{c} \text{HMO} \\ \text{PPO} \\ \text{FFS} \end{array} \begin{bmatrix} \begin{array}{ccc} \text{HMO} & \text{PPO} & \text{FFS} \\ .80 & .15 & .05 \\ .20 & .70 & .10 \\ .25 & .30 & .45 \end{array} \end{bmatrix}$

$$\begin{array}{ccc} \text{HMO} & \text{PPO} & \text{FFS} \end{array}$$

(B) $S = [.20 \quad .25 \quad .55]$

$$\begin{array}{ccc} \text{HMO} & \text{PPO} & \text{FFS} \end{array}$$

$SP = [.3475 \quad .37 \quad .2825]$

34.75% were enrolled in the HMO; 37% were enrolled in the PPO; 28.25% were enrolled in the FFS.

$$\begin{array}{ccc} \text{HMO} & \text{PPO} & \text{FFS} \end{array}$$

(C) $SP^2 = [.422625 \quad .395875 \quad .1815]$

42.2625% will be enrolled in the HMO; 39.5875% will be enrolled in the PPO; 18.15% will be enrolled in the FFS.

53. (A) $P = \begin{array}{c} H \\ R \end{array} \begin{bmatrix} \begin{array}{cc} H & R \\ .88 & .12 \\ .05 & .95 \end{array} \end{bmatrix} \begin{array}{l} H = \text{Homeowner} \\ R = \text{Renter} \end{array}$

$$\begin{array}{cc} H & R \end{array}$$

(B) $S = [.451 \quad .549]$

1990: $SP = [.451 \quad .549] \begin{bmatrix} .88 & .12 \\ .05 & .95 \end{bmatrix} = [.42433 \quad .57567]$

42.433% were homeowners in 1990.

$$\begin{array}{cc} H & R \end{array}$$

2000: $SP^2 \approx [.40219 \quad .59781]$

(C) 40.219% will be homeowners in 2000.

Things to remember:

1. STATIONARY MATRIX FOR A MARKOV CHAIN

   The state matrix $S = [s_1 \quad s_2 \quad ... \quad s_n]$ is a STATIONARY MATRIX for a Markov chain with transition matrix P if

   $\quad\quad SP = S$

   where $s_i \geq 0$, $i = 1, ..., n$ and $s_1 + s_2 + ... + s_n = 1$.

2. REGULAR MARKOV CHAINS

   A transition matrix P is REGULAR if some power of P has only positive entries. A Markov chain is a REGULAR MARKOV CHAIN if its transition matrix is regular.

3. PROPERTIES OF REGULAR MARKOV CHAINS

   Let P be the transition matrix for a regular Markov chain.

   (A) There is a unique stationary matrix S which can be found by solving the equation

   $\quad\quad SP = S$

   (B) Given any initial state matrix S_0, the state matrices S_k approach the stationary matrix S.

   (C) The matrices P^k approach a limiting matrix $\overline{P}$ where each row of $\overline{P}$ is equal to the stationary matrix S.

---

1. $P = \begin{bmatrix} .1 & .9 \\ .7 & .3 \end{bmatrix}$ is regular; all entries are positive

3. $P = \begin{bmatrix} .2 & .8 \\ 1 & 0 \end{bmatrix}$, $P^2 = \begin{bmatrix} .84 & .16 \\ .2 & .8 \end{bmatrix}$

P is regular since P^2 has only positive entries.

5. $P = \begin{bmatrix} 1 & 0 \\ .6 & .4 \end{bmatrix}$, $P^2 = \begin{bmatrix} 1 & 0 \\ .84 & .16 \end{bmatrix}$, $P^3 = \begin{bmatrix} 1 & 0 \\ .936 & .064 \end{bmatrix}$, $P^4 = \begin{bmatrix} 1 & 0 \\ .9744 & .0256 \end{bmatrix}$, ...

It appears that P^k will have the form $\begin{bmatrix} 1 & 0 \\ a & b \end{bmatrix}$ for all positive integers k. Thus P is **not** regular.

7. $P = \begin{bmatrix} .3 & .4 & .3 \\ 0 & 0 & 1 \\ .4 & .2 & .4 \end{bmatrix}$, $P^2 = \begin{bmatrix} .21 & .18 & .61 \\ .4 & .2 & .4 \\ .28 & .24 & .48 \end{bmatrix}$

P is regular since P^2 has only positive entries.

9. $P = \begin{bmatrix} 0 & 1 & 0 \\ .4 & 0 & .6 \\ 0 & 1 & 0 \end{bmatrix}$, $P^2 = \begin{bmatrix} .4 & 0 & .6 \\ 0 & 1 & 0 \\ .4 & 0 & .6 \end{bmatrix}$, $P^3 = \begin{bmatrix} 0 & 1 & 0 \\ .4 & 0 & .6 \\ 0 & 1 & 0 \end{bmatrix}$, ...

In general, $P^{2k-1} = P$, $k = 1, 2, ...$

and $\quad P^{2k} = P^2$, $k = 1, 2, ...$

Clearly, P is not regular.

11. $P = \begin{bmatrix} 0 & 0 & 1 \\ .8 & .1 & .1 \\ 0 & 1 & 0 \end{bmatrix}$, $P^2 = \begin{bmatrix} 0 & 1 & 0 \\ .08 & .11 & .81 \\ .8 & .1 & .1 \end{bmatrix}$, $P^3 = \begin{bmatrix} .8 & .1 & .1 \\ .088 & .821 & .091 \\ .08 & .11 & .81 \end{bmatrix}$

P is regular since P^3 has only positive entries.

13. Let $S = [s_1, \ s_2]$, and solve the system:

$[s_1 \quad s_2] \begin{bmatrix} .1 & .9 \\ .6 & .4 \end{bmatrix} = [s_1 \quad s_2]$, $s_1 + s_2 = 1$

which is equivalent to

$$\begin{array}{lll} .1s_1 + .6s_2 = s_1 & & -.9s_1 + .6s_2 = 0 \\ .9s_1 + .4s_2 = s_2 & \text{or} & .9s_1 - .6s_2 = 0 \\ s_1 + s_2 = 1 & & s_1 + s_2 = 1 \end{array}$$

The solution is: $s_1 = .4$, $s_2 = .6$

The stationary matrix $S = [.4 \quad .6]$; the limiting matrix $\overline{P} = \begin{bmatrix} .4 & .6 \\ .4 & .6 \end{bmatrix}$.

15. Let $S = [s_1, \ s_2]$, and solve the system:

$[s_1 \quad s_2] \begin{bmatrix} .5 & .5 \\ .3 & .7 \end{bmatrix} = [s_1 \quad s_2]$, $s_1 + s_2 = 1$

which is equivalent to

$$\begin{array}{lll} .5s_1 + .3s_2 = s_1 & & -.5s_1 + .3s_2 = 0 \\ .5s_1 + .7s_2 = s_2 & \text{or} & .5s_1 - .3s_2 = 0 \\ s_1 + s_2 = 1 & & s_1 + s_2 = 1 \end{array}$$

The solution is: $s_1 = \dfrac{3}{8} = .375$, $s_2 = \dfrac{5}{8} = .625$

The stationary matrix $S = [.375 \quad .625]$; the limiting matrix $\overline{P} = \begin{bmatrix} .375 & .625 \\ .375 & .625 \end{bmatrix}$.

17. Let $S = [s_1 \quad s_2 \quad s_3]$, and solve the system:

$[s_1 \quad s_2 \quad s_3] \begin{bmatrix} .5 & .1 & .4 \\ .3 & .7 & 0 \\ 0 & .6 & .4 \end{bmatrix} = [s_1 \quad s_2 \quad s_3]$, $s_1 + s_2 + s_3 = 1$

which is equivalent to

$$\begin{array}{lll} .5s_1 + .3s_2 \qquad = s_1 & & -.5s_1 + .3s_2 \qquad = 0 \\ .1s_1 + .7s_2 + .6s_3 = s_2 & \text{or} & .1s_1 - .3s_2 + .6s_3 = 0 \\ .4s_1 \qquad + .4s_3 = s_3 & & .4s_1 \qquad - .6s_3 = 0 \\ s_1 + s_2 + s_3 = 1 & & s_1 + s_2 + s_3 = 1 \end{array}$$

From the first and third equations, we have $s_2 = \dfrac{5}{3} s_1$, and $s_3 = \dfrac{2}{3} s_1$.

Substituting these values into the fourth equation, we get:

$s_1 + \dfrac{5}{3} s_1 + \dfrac{2}{3} s_1 = 1$ or $\dfrac{10}{3} s_1 = 1$

Therefore, $s_1 = .3$, $s_2 = .5$, $s_3 = .2$.

The stationary matrix $S = [.3 \quad .5 \quad .2]$;

the limiting matrix $\overline{P} = \begin{bmatrix} .3 & .5 & .2 \\ .3 & .5 & .2 \\ .3 & .5 & .2 \end{bmatrix}$.

19. Let $S = [s_1 \quad s_2 \quad s_3]$, and solve the system:

$$[s_1 \quad s_2 \quad s_3] \begin{bmatrix} .8 & .2 & 0 \\ .5 & .1 & .4 \\ 0 & .6 & .4 \end{bmatrix} = [s_1 \quad s_2 \quad s_3], \ s_1 + s_2 + s_3 = 1$$

which is equivalent to

$$
\begin{array}{ll}
.8s_1 + .5s_2 \qquad\quad = s_1 & \qquad -.2s_1 + .5s_2 \qquad\qquad = 0 \\
.2s_1 + .1s_2 + .6s_3 = s_2 \quad \text{or} & \qquad .2s_1 - .9s_2 + .6s_3 = 0 \\
\qquad\quad .4s_2 + .4s_3 = s_3 & \qquad\qquad\quad .4s_2 - .6s_3 = 0 \\
s_1 + \quad s_2 + \quad s_3 = 1 & \qquad s_1 + \quad s_2 + \quad s_3 = 1
\end{array}
$$

From the first and third equations, we have $s_1 = \frac{5}{2}s_2$, and $s_3 = \frac{2}{3}s_2$.
Substituting these values into the fourth equation, we get:

$$\frac{5}{2}s_2 + s_2 + \frac{2}{3}s_2 = 1 \ \text{ or } \ \frac{25}{6}s_2 = 1$$

Therefore, $s_2 = \frac{6}{25} = .24$, $s_1 = .6$, $s_3 = .16$.
The stationary matrix $S = [.6 \quad .24 \quad .16]$;

the limiting matrix $\overline{P} = \begin{bmatrix} .6 & .24 & .16 \\ .6 & .24 & .16 \\ .6 & .24 & .16 \end{bmatrix}$.

21. $P = \begin{bmatrix} .51 & .49 \\ .27 & .73 \end{bmatrix}$, $P^2 = \begin{bmatrix} .3924 & .6076 \\ .3348 & .6652 \end{bmatrix}$, $P^4 \sim \begin{bmatrix} .3574 & .6426 \\ .3541 & .6459 \end{bmatrix}$,

$P^8 \sim \begin{bmatrix} .3553 & .6447 \\ .3553 & .6447 \end{bmatrix}$, $P^{16} \sim \begin{bmatrix} .3553 & .6447 \\ .3553 & .6447 \end{bmatrix}$

Therefore, $S \sim [.3553 \quad .6447]$.

23. $P = \begin{bmatrix} .5 & .5 & 0 \\ 0 & .5 & .5 \\ .8 & .1 & .1 \end{bmatrix}$, $P^2 = \begin{bmatrix} .25 & .5 & .25 \\ .4 & .3 & .3 \\ .48 & .46 & .06 \end{bmatrix}$, $P^4 = \begin{bmatrix} .3825 & .39 & .2275 \\ .364 & .428 & .208 \\ .3328 & .4056 & .2616 \end{bmatrix}$,

$P^8 \sim \begin{bmatrix} .3640 & .4084 & .2277 \\ .3642 & .4095 & .2262 \\ .3620 & .4095 & .2285 \end{bmatrix}$, $P^{16} \sim \begin{bmatrix} .3636 & .4091 & .2273 \\ .3636 & .4091 & .2273 \\ .3636 & .4091 & .2273 \end{bmatrix}$

Therefore, $S \sim [.3636 \quad .4091 \quad .2273]$.

25. (A)

(B) $P = \begin{array}{c} \\ \text{Red} \\ \text{Blue} \end{array} \begin{array}{cc} \text{Red} & \text{Blue} \\ \begin{bmatrix} .4 & .6 \\ .2 & .8 \end{bmatrix} \end{array}$

(C) Let $S = [s_1 \quad s_2]$ and solve the system:

$$[s_1 \quad s_2] \begin{bmatrix} .4 & .6 \\ .2 & .8 \end{bmatrix} = [s_1 \quad s_2], \ s_1 + s_2 = 1$$

which is equivalent to

$$
\begin{array}{ll}
.4s_1 + .2s_2 = s_1 & \qquad -.6s_1 + .2s_2 = 0 \\
.6s_1 + .8s_2 = s_2 \quad \text{or} & \qquad .6s_1 - .2s_2 = 0 \\
s_1 + \quad s_2 = 1 & \qquad s_1 + \quad s_2 = 1
\end{array}
$$

The solution is: $s_1 = .25$, $s_2 = .75$.

Thus, the stationary matrix $S = [.25 \quad .75]$. In the long run, the red urn will be selected 25% of the time and the blue urn 75% of the time.

27. (A) $s_1 = [.2 \quad .8] \begin{bmatrix} 0 & 1 \\ 1 & 0 \end{bmatrix} = [.8 \quad .2]$

$s_2 = s_1 P = [.8 \quad .2] \begin{bmatrix} 0 & 1 \\ 1 & 0 \end{bmatrix} = [.2 \quad .8]$

$s_3 = s_2 P = [.8 \quad .2]$

and so on.

The state matrices alternate between $[.2 \quad .8]$ and $[.8 \quad .2]$; they do not approach a "limiting" matrix.

(B) $s_1 = [.5 \quad .5] \begin{bmatrix} 0 & 1 \\ 1 & 0 \end{bmatrix} = [.5 \quad .5]$

$s_2 = s_1 P = [.5 \quad .5] \begin{bmatrix} 0 & 1 \\ 1 & 0 \end{bmatrix} = [.5 \quad .5]$

and so on.

Thus, $s_1 = s_2 = s_3 = \dots = [.5 \quad .5] = s_0$; s_0 is a stationary matrix.

(C) $P = \begin{bmatrix} 0 & 1 \\ 1 & 0 \end{bmatrix}$, $P^2 = \begin{bmatrix} 1 & 0 \\ 0 & 1 \end{bmatrix}$, $P^3 = \begin{bmatrix} 0 & 1 \\ 1 & 0 \end{bmatrix}$, $\dots$

The powers of P alternate between P and the identity, I; they do not approach a limiting matrix.

(D) Parts (B) and (C) are not valid for this matrix. Since P is not regular, this does not contradict Theorem 1.

29. (A) $RP = [1 \quad 0 \quad 0] \begin{bmatrix} 1 & 0 & 0 \\ .2 & .2 & .6 \\ 0 & 0 & 1 \end{bmatrix} = [1 \quad 0 \quad 0]$

Therefore R is a stationary matrix for P.

$SP = [0 \quad 0 \quad 1] \begin{bmatrix} 1 & 0 & 0 \\ .2 & .2 & .6 \\ 0 & 0 & 1 \end{bmatrix} = [0 \quad 0 \quad 1]$

Therefore S is a stationary matrix for P.

The powers of P have the form $\begin{bmatrix} 1 & 0 & 0 \\ a & b & c \\ 0 & 0 & 1 \end{bmatrix}$

Therefore P is not regular. As a result, P may have more than one stationary matrix.

(B) Following the hint, let
$$T = a[1 \quad 0 \quad 0] + (1 - a)[0 \quad 0 \quad 1], \quad 0 < a < 1.$$
$$= [a \quad 0 \quad 1 - a]$$

Now, $TP = [a \quad 0 \quad 1 - a]\begin{bmatrix} 1 & 0 & 0 \\ .2 & .2 & .6 \\ 0 & 0 & 1 \end{bmatrix} = [a \quad 0 \quad 1 - a] = T$

Thus, $[a \quad 0 \quad 1 - a]$ is a stationary matrix for P for every a with $0 < a < 1$. Note that if $a = 1$, then $T = R$, and if $a = 0$, $T = S$. If we let $a = .5$, then $T = [.5 \quad 0 \quad .5]$ is a stationary matrix.

(C) P has infinitely many stationary matrices.

31. $\overline{P} = \begin{bmatrix} 1 & 0 & 0 \\ .25 & 0 & .75 \\ 0 & 0 & 1 \end{bmatrix}$

Each row of $\overline{P}$ is a stationary matrix for P. As we saw in Problem 29, part (B),
$$T = [a \quad 0 \quad 1 - a]$$
is a stationary matrix for P for each a where $0 \le a \le 1$; $a = 1$ gives the first row of $\overline{P}$, $a = .25$ gives the second row; $a = 0$ gives the third row.

33. The transition matrix is

$$\begin{array}{cc} & H \qquad N \\ P = \begin{matrix} H \\ N \end{matrix} & \begin{bmatrix} .89 & .11 \\ .29 & .71 \end{bmatrix} \end{array} \quad \begin{array}{l} H = \text{home trackage} \\ N = \text{national pool} \end{array}$$

Calculating powers of P, we have
$$P^2 = \begin{bmatrix} .824 & .176 \\ .464 & .536 \end{bmatrix}, \quad P^4 \sim \begin{bmatrix} .7606 & .2394 \\ .6310 & .3690 \end{bmatrix}, \quad P^8 \sim \begin{bmatrix} .7296 & .2704 \\ .7128 & .2872 \end{bmatrix},$$
$$P^{16} \sim \begin{bmatrix} .7251 & .2749 \\ .7248 & .2752 \end{bmatrix}$$

In the long run, 72.5% of the company's box cars will be on its home trackage.

35. (A) $S_1 = [.433 \quad .567]\begin{bmatrix} .93 & .07 \\ .2 & .8 \end{bmatrix} = [.51609 \quad .48391] \sim [.516 \quad .484]$

$S_2 = S_0P^2 = [.433 \quad .567]\begin{bmatrix} .8789 & .1211 \\ .346 & .654 \end{bmatrix} = [.5767457 \quad .4232543]$

$$\sim [.577 \quad .423]$$

(B)

Year	Data %	Model %
1970	43.3	43.3
1980	51.5	51.6
1990	57.5	57.7

(C) $P^4 \sim \begin{bmatrix} .8143 & .1856 \\ .5303 & .4696 \end{bmatrix}$, $P^8 \sim \begin{bmatrix} .7616 & .2384 \\ .6810 & .3190 \end{bmatrix}$, $P^{16} \sim \begin{bmatrix} .7424 & .2576 \\ .7359 & .2641 \end{bmatrix}$,

$P^{32} \sim \begin{bmatrix} .7408 & .2592 \\ .7407 & .2593 \end{bmatrix}$

In the long run, 74.1% of the female population will be in the labor force.

37. The transition matrix for this problem is:

	GTT	NCJ	Dash
GTT	.75	.05	.20
NCJ	.15	.75	.1
Dash	.05	.10	.85

To find the steady-state matrix, we solve the system

$$[s_1 \quad s_2 \quad s_3] \begin{bmatrix} .75 & .05 & .20 \\ .15 & .75 & .10 \\ .05 & .10 & .85 \end{bmatrix} = [s_1 \quad s_2 \quad s_3], \ s_1 + s_2 + s_3 = 1$$

which is equivalent to the system of equations
$$.75s_1 + .15s_2 + .05s_3 = s_1$$
$$.05s_1 + .75s_2 + .1s_3 = s_2$$
$$.20s_1 + .10s_2 + .85s_3 = s_3$$
$$s_1 + s_2 + s_3 = 1$$

The solution of this system is $s_1 = .25$, $s_2 = .25$, $s_3 = .5$.

Thus, the expected market share of each company is:
GTT – 25%; NCJ – 25%; and Dash – 50%.

39. The transition matrix for this problem is:

	Poor	Satisfactory	Preferred
Poor	.60	.40	0
Satisfactory	.20	.60	.20
Preferred	0	.20	.80

To find the steady-state matrix, we solve the system

$$[s_1 \quad s_2 \quad s_3] \begin{bmatrix} .60 & .40 & 0 \\ .20 & .60 & .20 \\ 0 & .20 & .80 \end{bmatrix} = [s_1 \quad s_2 \quad s_3], \ s_1 + s_2 + s_3 = 1$$

which is equivalent to the system of equations:
$$.6s_1 + .2s_2 = s_1$$
$$.4s_1 + .6s_2 + .2s_3 = s_2$$
$$.2s_2 + .8s_3 = s_3$$
$$s_1 + s_2 + s_3 = 1$$

The solution of this system is $s_1 = .20$, $s_2 = .40$, and $s_3 = .40$.

Thus, the expected percentage in each category is:
poor – 20%; satisfactory – 40%; and preferred – 40%.

41. The transition matrix is:

$$P = \begin{bmatrix} .4 & .1 & .3 & .2 \\ .3 & .2 & .2 & .3 \\ .1 & .2 & .2 & .5 \\ .3 & .3 & .1 & .3 \end{bmatrix}$$

$S_0 P = [.3 \quad .3 \quad .4 \quad 0]P = [.25 \quad .17 \quad .23 \quad .35] = S_1$

$S_1 P = [.25 \quad .17 \quad .23 \quad .35]P = [.28 \quad .21 \quad .19 \quad .32] = S_2$

$S_2 P = [.28 \quad .21 \quad .19 \quad .32]P = [.29 \quad .20 \quad .20 \quad .31] = S_3$

$S_3 P = [.29 \quad .20 \quad .20 \quad .31]P = [.29 \quad .20 \quad .20 \quad .31] = S_4$

Thus, $S = [.29 \quad .20 \quad .20 \quad .31]$ is the steady-state matrix. The expected market share for the two Acme soaps is $.20 + .31 = .51$ or 51%.

43. To find the stationary solution, we solve the system

$$[s_1 \quad s_2 \quad s_3]\begin{bmatrix} .5 & .5 & 0 \\ .25 & .5 & .25 \\ 0 & .5 & .5 \end{bmatrix} = [s_1 \quad s_2 \quad s_3], \quad s_1 + s_2 + s_3 = 1,$$

which is equivalent to:

$$\begin{array}{ll} .5s_1 + .25s_2 \qquad\quad = s_1 & \qquad -.5s_1 + .25s_2 \qquad\quad = 0 \\ .5s_1 + .5s_2 + .5s_3 = s_2 \quad \text{or} & \qquad .5s_1 - .5s_2 + .5s_3 = 0 \\ \qquad\quad .25s_2 + .5s_3 = s_3 & \qquad\qquad\quad .25s_2 - .5s_3 = 0 \\ s_1 + s_2 + s_3 = 1 & \qquad s_1 + s_2 + s_3 = 1 \end{array}$$

The solution of this system is $s_1 = .25$, $s_2 = .5$, $s_3 = .25$.
Thus, the stationary matrix is $S = [.25 \quad .5 \quad .25]$.

45. (A)

	Rapid transit	Auto

Initial-state matrix = $[.25 \qquad .75]$

(B) Second-state matrix = $[.25 \quad .75]\begin{bmatrix} .8 & .2 \\ .3 & .7 \end{bmatrix} = [.425 \quad .575]$

Thus, 42.5% will be using the new system after one month.

Third-state matrix = $[.425 \quad .575]\begin{bmatrix} .8 & .2 \\ .3 & .7 \end{bmatrix} = [.5125 \quad .4875]$

Thus, 51.25% will be using the new system after two months.

(C) To find the stationary solution, we solve the system

$$[s_1 \quad s_2]\begin{bmatrix} .8 & .2 \\ .3 & .7 \end{bmatrix} = [s_1 \quad s_2], \quad s_1 + s_2 = 1,$$

which is equivalent to:
$$\begin{array}{ll} .8s_1 + .3s_2 = s_1 & \qquad -.2s_1 + .3s_2 = 0 \\ .2s_1 + .7s_2 = s_2 \quad \text{or} & \qquad .2s_1 - .3s_2 = 0 \\ s_1 + s_2 = 1 & \qquad s_1 + s_2 = 1 \end{array}$$

The solution of this system of linear equations is $s_1 = .6$ and $s_2 = .4$. Thus, the stationary solution is $S = [.6 \quad .4]$, which means that 60% of the commuters will use rapid transit and 40% will travel by automobile after the system has been in service for a long time.

47. (A) $s_1 = [.309 \quad .691]\begin{bmatrix} .61 & .39 \\ .21 & .79 \end{bmatrix} = [.3336 \quad .6664] \sim [.334 \quad .666]$

$s_2 = s_0 P^2 = [.309 \quad .691]\begin{bmatrix} .454 & .546 \\ .294 & .706 \end{bmatrix} = [.34344 \quad .65656]$

$\sim [.343 \quad .657]$

(B)

Year	Data %	Model %
1970	30.9	30.9
1980	33.3	33.4
1990	34.4	34.3

(C) $P^4 \sim \begin{bmatrix} .367 & .633 \\ .341 & .659 \end{bmatrix}$, $P^8 \sim \begin{bmatrix} .350 & .650 \\ .350 & .650 \end{bmatrix}$, $P^{16} \sim \begin{bmatrix} .350 & .650 \\ .350 & .650 \end{bmatrix}$

In the long run, 35% of the population will live in the south region.

EXERCISE 7-3

Things to remember:

1. **ABSORBING STATES AND TRANSIENT STATES**

   A state in a Markov chain is an ABSORBING STATE if once the state is entered, it is impossible to leave. A nonabsorbing state is called a TRANSIENT STATE.

2. **ABSORBING STATES AND TRANSITION MATRICES**

   A state in a Markov chain is ABSORBING if and only if the row of the transition matrix correspondig to the state has a 1 on the main diagonal and zeros elsewhere.

3. **ABSORBING MARKOV CHAINS**

   A Markov chain is an ABSORBING CHAIN if

   (A) There is at least one absorbing state.

   (B) It is possible to go from each nonabsorbing state to at least one absorbing state in a finite number of steps.

<u>4</u>. STANDARD FORMS FOR ABSORBING MARKOV CHAINS

A transition matrix for an absorbing Markov chain is a STANDARD FORM if the rows and columns are labeled so that all the absorbing states precede all the nonabsorbing states. (There may be more than one standard form.) Any standard form can always be partitioned into four submatrices:

$$\begin{array}{cc} & A \quad T \\ \begin{array}{c} A \\ T \end{array} & \left[\begin{array}{c:c} I & 0 \\ \hdashline R & Q \end{array}\right] \end{array} \left[\begin{array}{l} A = \text{All absorbing states} \\ T = \text{All nonabsorbing states} \end{array}\right]$$

where I is an identity matrix and 0 is a zero matrix.

<u>5</u>. LIMITING MATRICES FOR ABSORBING MARKOV CHAINS

If a standard form P for an absorbing Markov chain is partitioned as

$$P = \left[\begin{array}{c:c} I & 0 \\ \hdashline R & Q \end{array}\right]$$

then P^k approaches a matrix $\overline{P}$ as k increases, where

$$\overline{P} = P = \left[\begin{array}{c:c} I & 0 \\ \hdashline FR & 0 \end{array}\right]$$

The matrix F is given by $F = (I - Q)^{-1}$ and is called the FUNDAMENTAL MATRIX for P.

The identity matrix used to form the fundamental matrix F must be the same size as the matrix Q.

<u>6</u>. PROPERTIES OF THE LIMITING MATRIX $\overline{P}$

If P is a standard form transition matrix for an absorbing Markov chain, F is the fundamental matrix, and $\overline{P}$ is the limiting matrix, then

(A) The entry in row i and column j of $\overline{P}$ is the long run probability of going from state i to state j. For the nonabsorbing states, these probabilities are also the entries in the matrix FR used to form $\overline{P}$.

(B) The sum of the entries in each row of the fundamental matrix F is the average number of trials it will take to go from each nonabsorbing state to some absorbing state.

[Note that the rows of both F and FR correspond to the nonabsorbing states in the order given in the standard form P.]

---

1. By <u>2</u>, states B and C are absorbing states.

3. By <u>2</u>, states A and D are absorbing states.

5. B is an absorbing state; the diagram represents an absorbing Markov chain since it is possible to go from states A and C to state B in a finite number of steps.

7. *C* is an absorbing state; the diagram does not represent an absorbing Markov chain since it is not possible to go from either states *A* or *D* to state *C*.

9. The transition diagram is represented by the matrix:

$$
\begin{array}{c}
 \\ A \\ B \\ C
\end{array}
\begin{array}{ccc}
A & B & C \\
\end{array}
\left[\begin{array}{ccc}
.2 & .5 & .3 \\
0 & 1 & 0 \\
.5 & .1 & .4
\end{array}\right]
$$

A standard form for this matrix is:

$$
\begin{array}{c}
 \\ B \\ A \\ C
\end{array}
\begin{array}{ccc}
B & A & C \\
\end{array}
\left[\begin{array}{ccc}
1 & 0 & 0 \\
.5 & .2 & .3 \\
.1 & .5 & .4
\end{array}\right]
$$

11. The transition diagram is represented by the matrix

$$
\begin{array}{c}
 \\ A \\ B \\ C \\ D
\end{array}
\begin{array}{cccc}
A & B & C & D \\
\end{array}
\left[\begin{array}{cccc}
.3 & .4 & .2 & .1 \\
0 & 1 & 0 & 0 \\
0 & .4 & .3 & .3 \\
0 & 0 & 0 & 1
\end{array}\right]
$$

A standard form for this matrix

$$
\begin{array}{c}
 \\ B \\ D \\ A \\ C
\end{array}
\begin{array}{cccc}
B & D & A & C \\
\end{array}
\left[\begin{array}{cccc}
1 & 0 & 0 & 0 \\
0 & 1 & 0 & 0 \\
.4 & .1 & .3 & .2 \\
.4 & .3 & 0 & .3
\end{array}\right]
$$

13. A standard form for

$$
P = \begin{array}{c}
 \\ A \\ B \\ C
\end{array}
\begin{array}{ccc}
A & B & C \\
\end{array}
\left[\begin{array}{ccc}
.2 & .3 & .5 \\
1 & 0 & 0 \\
0 & 0 & 1
\end{array}\right]
$$

is:

$$
\begin{array}{c}
 \\ C \\ A \\ B
\end{array}
\begin{array}{ccc}
C & A & B \\
\end{array}
\left[\begin{array}{ccc}
1 & 0 & 0 \\
.5 & .2 & .3 \\
0 & 1 & 0
\end{array}\right]
$$

15. A standard form for

$$
P = \begin{array}{c}
 \\ A \\ B \\ C \\ D
\end{array}
\begin{array}{cccc}
A & B & C & D \\
\end{array}
\left[\begin{array}{cccc}
.1 & .2 & .3 & .4 \\
0 & 1 & 0 & 0 \\
.5 & .2 & .2 & .1 \\
0 & 0 & 0 & 1
\end{array}\right]
$$

is:

$$
\begin{array}{c}
 \\ B \\ D \\ A \\ C
\end{array}
\begin{array}{cccc}
B & D & A & C \\
\end{array}
\left[\begin{array}{cccc}
1 & 0 & 0 & 0 \\
0 & 1 & 0 & 0 \\
.2 & .4 & .1 & .3 \\
.2 & .1 & .5 & .2
\end{array}\right]
$$

17. For

$$
P = \begin{array}{c}
 \\ A \\ B \\ C
\end{array}
\begin{array}{ccc}
A & B & C \\
\end{array}
\left[\begin{array}{ccc}
1 & 0 & 0 \\
0 & 1 & 0 \\
.1 & .4 & .5
\end{array}\right]
$$

we have $R = [.1 \quad .4]$ and $Q = [.5]$.

The limiting matrix $\overline{P}$ has the form

$$
\overline{P} = \left[\begin{array}{cc|c}
1 & 0 & 0 \\
0 & 1 & 0 \\
\hline
F & R & 0
\end{array}\right]
$$

where $F = (I - Q)^{-1} = ([1] - [.5])^{-1} = [.5]^{-1} = [2]$
and $FR = [2][.1 \quad .4] = [.2 \quad .8]$.
Thus,

$$
\overline{P} = \begin{array}{c}
 \\ A \\ B \\ C
\end{array}
\begin{array}{ccc}
A & B & C \\
\end{array}
\left[\begin{array}{ccc}
1 & 0 & 0 \\
0 & 1 & 0 \\
.2 & .8 & 0
\end{array}\right]
$$

Let $P(i$ to $j)$ denote the probability of going from state i to state j.
Then $P(C$ to $A) = .2$, $P(C$ to $B) = .8$
Since $F = [2]$, it will take an average of 2 trials to go from C to either A or B.

19. For

$$P = \begin{array}{c} \\ A \\ B \\ C \end{array} \begin{array}{ccc} A & B & C \\ \begin{bmatrix} 1 & 0 & 0 \\ .2 & .6 & .2 \\ .4 & .2 & .4 \end{bmatrix} \end{array}$$

we have $R = \begin{bmatrix} .2 \\ .4 \end{bmatrix}$ and $Q = \begin{bmatrix} .6 & .2 \\ .2 & .4 \end{bmatrix}$

The limiting matrix $\overline{P}$ has the form

$$\overline{P} = \left[\begin{array}{c|c} 1 & 0 \\ \hline F \quad R & 0 \end{array}\right] \text{ where } F = (I - Q)^{-1} = \left(\begin{bmatrix} 1 & 0 \\ 0 & 1 \end{bmatrix} - \begin{bmatrix} .6 & .2 \\ .2 & .4 \end{bmatrix}\right)^{-1}$$

$$= \begin{bmatrix} .4 & -.2 \\ -.2 & .6 \end{bmatrix}^{-1} = \begin{bmatrix} \frac{2}{5} & -\frac{1}{5} \\ -\frac{1}{5} & \frac{3}{5} \end{bmatrix}^{-1}$$

We use row operations to find the inverse:

$$\begin{bmatrix} \frac{2}{5} & -\frac{1}{5} & 1 & 0 \\ -\frac{1}{5} & \frac{3}{5} & 0 & 1 \end{bmatrix} \sim \begin{bmatrix} 1 & -\frac{1}{2} & \frac{5}{2} & 0 \\ -\frac{1}{5} & \frac{3}{5} & 0 & 1 \end{bmatrix} \sim \begin{bmatrix} 1 & -\frac{1}{2} & \frac{5}{2} & 0 \\ 0 & \frac{1}{2} & \frac{1}{2} & 1 \end{bmatrix} \sim \begin{bmatrix} 1 & -\frac{1}{2} & \frac{5}{2} & 0 \\ 0 & 1 & 1 & 2 \end{bmatrix}$$

$$\left(\tfrac{5}{2}\right)R_1 \to R_1 \qquad \left(\tfrac{1}{5}\right)R_1 + R_2 \to R_2 \qquad 2R_2 \to R_2 \qquad \left(\tfrac{1}{2}\right)R_2 + R_1 \to R_1$$

$$\sim \begin{bmatrix} 1 & 0 & 3 & 1 \\ 0 & 1 & 1 & 2 \end{bmatrix}$$

Thus, $F = \begin{bmatrix} 3 & 1 \\ 1 & 2 \end{bmatrix}$ and $FR = \begin{bmatrix} 3 & 1 \\ 1 & 2 \end{bmatrix}\begin{bmatrix} .2 \\ .4 \end{bmatrix} = \begin{bmatrix} 1 \\ 1 \end{bmatrix}$

Now

$$\overline{P} = \begin{array}{c} \\ A \\ B \\ C \end{array} \begin{array}{ccc} A & B & C \\ \begin{bmatrix} 1 & 0 & 0 \\ 1 & 0 & 0 \\ 1 & 0 & 0 \end{bmatrix} \end{array}$$

$P(B \text{ to } A) = 1$, $P(C \text{ to } A) = 1$

It will take an average of 4 trials to go from B to A; it will take an average of 3 trials to go from C to A.

21. For

$$P = \begin{array}{c} \\ A \\ B \\ C \\ D \end{array} \begin{array}{cccc} A & B & C & D \\ \begin{bmatrix} 1 & 0 & 0 & 0 \\ 0 & 1 & 0 & 0 \\ .1 & .2 & .6 & .1 \\ .2 & .2 & .3 & .3 \end{bmatrix} \end{array}$$

we have $R = \begin{bmatrix} .1 & .2 \\ .2 & .2 \end{bmatrix}$ and $Q = \begin{bmatrix} .6 & .1 \\ .3 & .3 \end{bmatrix}$

The limiting matrix $\overline{P}$ has the form

$$\overline{P} = \left[\begin{array}{c|cc} I & 0 & 0 \\ & 0 & 0 \\ \hline FR & & 0 \end{array}\right]$$

where $F = (I - Q)^{-1} = \left(\begin{bmatrix} 1 & 0 \\ 0 & 1 \end{bmatrix} - \begin{bmatrix} .6 & .1 \\ .3 & .3 \end{bmatrix}\right)^{-1}$

$$= \begin{bmatrix} .4 & -.1 \\ -.3 & .7 \end{bmatrix}^{-1} = \begin{bmatrix} \frac{2}{5} & -\frac{1}{10} \\ -\frac{3}{10} & \frac{7}{10} \end{bmatrix}^{-1}$$

We use row operations to find the inverse:

$$\begin{bmatrix} \frac{2}{5} & -\frac{1}{10} & | & 1 & 0 \\ -\frac{3}{10} & \frac{7}{10} & | & 0 & 1 \end{bmatrix} \sim \begin{bmatrix} 1 & -\frac{1}{4} & | & \frac{5}{2} & 0 \\ -\frac{3}{10} & \frac{7}{10} & | & 0 & 1 \end{bmatrix} \sim \begin{bmatrix} 1 & -\frac{1}{4} & | & \frac{5}{2} & 0 \\ 0 & \frac{5}{8} & | & \frac{3}{4} & 1 \end{bmatrix} \sim \begin{bmatrix} 1 & -\frac{1}{4} & | & \frac{5}{2} & 0 \\ 0 & 1 & | & \frac{6}{5} & \frac{8}{5} \end{bmatrix}$$

$$\left(\tfrac{5}{2}\right)R_1 \to R_1 \qquad \left(\tfrac{3}{10}\right)R_1 + R_2 \to R_2 \qquad \left(\tfrac{8}{5}\right)R_2 \to R_2 \qquad \left(\tfrac{1}{4}\right)R_2 + R_1 \to R_1$$

$$\sim \begin{bmatrix} 1 & 0 & | & \frac{14}{5} & \frac{2}{5} \\ 0 & 1 & | & \frac{6}{5} & \frac{8}{5} \end{bmatrix}$$

Thus, $F = \begin{bmatrix} \frac{14}{5} & \frac{2}{5} \\ \frac{6}{5} & \frac{8}{5} \end{bmatrix} = \begin{bmatrix} 2.8 & .4 \\ 1.2 & 1.6 \end{bmatrix}$,

$$FR = \begin{bmatrix} 2.8 & .4 \\ 1.2 & 1.6 \end{bmatrix}\begin{bmatrix} .1 & .2 \\ .2 & .2 \end{bmatrix} = \begin{bmatrix} .36 & .64 \\ .44 & .56 \end{bmatrix}$$

and

$$\overline{P} = \begin{array}{c} A \\ B \\ C \\ D \end{array}\begin{bmatrix} \begin{array}{cccc} A & B & C & D \end{array} \\ 1 & 0 & 0 & 0 \\ 0 & 1 & 0 & 0 \\ .36 & .64 & 0 & 0 \\ .44 & .56 & 0 & 0 \end{bmatrix}$$

$P(C \text{ to } A) = .36, \ P(C \text{ to } B) = .64,$
$P(D \text{ to } A) = .44, \ P(D \text{ to } B) = .56$

It will take an average of 3.2 trials to go from C to either A or B; it will take an average of 2.8 trials to go from D to either A or B.

23. (A) $S_0\overline{P} = \begin{bmatrix} 0 & 0 & 1 \end{bmatrix}\begin{bmatrix} 1 & 0 & 0 \\ 0 & 1 & 0 \\ .2 & .8 & 0 \end{bmatrix} = \begin{bmatrix} .2 & .8 & 0 \end{bmatrix}$

(B) $S_0\overline{P} = \begin{bmatrix} .2 & .5 & .3 \end{bmatrix}\begin{bmatrix} 1 & 0 & 0 \\ 0 & 1 & 0 \\ .2 & .8 & 0 \end{bmatrix} = \begin{bmatrix} .26 & .74 & 0 \end{bmatrix}$

25. (A) $S_0\overline{P} = \begin{bmatrix} 0 & 0 & 1 \end{bmatrix}\begin{bmatrix} 1 & 0 & 0 \\ 1 & 0 & 0 \\ 1 & 0 & 0 \end{bmatrix} = \begin{bmatrix} 1 & 0 & 0 \end{bmatrix}$

(B) $S_0\overline{P} = \begin{bmatrix} .2 & .5 & .3 \end{bmatrix}\begin{bmatrix} 1 & 0 & 0 \\ 1 & 0 & 0 \\ 1 & 0 & 0 \end{bmatrix} = \begin{bmatrix} 1 & 0 & 0 \end{bmatrix}$

27. (A) $S_0\overline{P} = \begin{bmatrix} 0 & 0 & 0 & 1 \end{bmatrix}\begin{bmatrix} 1 & 0 & 0 & 0 \\ 0 & 1 & 0 & 0 \\ .36 & .64 & 0 & 0 \\ .44 & .56 & 0 & 0 \end{bmatrix} = \begin{bmatrix} .44 & .56 & 0 & 0 \end{bmatrix}$

(B) $S_0\overline{P} = \begin{bmatrix} 0 & 0 & 1 & 0 \end{bmatrix}\begin{bmatrix} 1 & 0 & 0 & 0 \\ 0 & 1 & 0 & 0 \\ .36 & .64 & 0 & 0 \\ .44 & .56 & 0 & 0 \end{bmatrix} = \begin{bmatrix} .36 & .64 & 0 & 0 \end{bmatrix}$

(C) $S_0\overline{P} = [0 \quad 0 \quad .4 \quad .6] \begin{bmatrix} 1 & 0 & 0 & 0 \\ 0 & 1 & 0 & 0 \\ .36 & .64 & 0 & 0 \\ .44 & .56 & 0 & 0 \end{bmatrix} = [.408 \quad .592 \quad 0 \quad 0]$

(D) $S_0\overline{P} = [.1 \quad .2 \quad .3 \quad .4] \begin{bmatrix} 1 & 0 & 0 & 0 \\ 0 & 1 & 0 & 0 \\ .36 & .64 & 0 & 0 \\ .44 & .56 & 0 & 0 \end{bmatrix} = [.384 \quad .616 \quad 0 \quad 0]$

29. By Theorem 2, P has a limiting matrix:

$$P^4 \sim \begin{bmatrix} 1 & 0 & 0 & 0 \\ 0 & 1 & 0 & 0 \\ .6364 & .362 & 0 & 0 \\ .7364 & .262 & 0 & 0 \end{bmatrix}, \quad P^8 \sim \begin{bmatrix} 1 & 0 & 0 & 0 \\ 0 & 1 & 0 & 0 \\ .6375 & .3625 & 0 & 0 \\ .7375 & .2625 & 0 & 0 \end{bmatrix}$$

$$P^{16} \sim \begin{bmatrix} 1 & 0 & 0 & 0 \\ 0 & 1 & 0 & 0 \\ .6375 & .3625 & 0 & 0 \\ .7375 & .2625 & 0 & 0 \end{bmatrix}; \quad \overline{P} = \begin{bmatrix} 1 & 0 & 0 & 0 \\ 0 & 1 & 0 & 0 \\ .6375 & .3625 & 0 & 0 \\ .7375 & .2625 & 0 & 0 \end{bmatrix}$$

31. By Theorem 2, P has a limiting matrix:

$$P^4 \sim \begin{bmatrix} 1 & 0 & 0 & 0 & 0 \\ 0 & 1 & 0 & 0 & 0 \\ .0724 & .8368 & .0625 & .011 & .0173 \\ .174 & .7792 & 0 & .0279 & .0189 \\ .4312 & .5472 & 0 & .0126 & .009 \end{bmatrix},$$

$$P^{16} \sim \begin{bmatrix} 1 & 0 & 0 & 0 & 0 \\ 0 & 1 & 0 & 0 & 0 \\ .0875 & .9125 & 0 & 0 & 0 \\ .1875 & .8125 & 0 & 0 & 0 \\ .4375 & .5625 & 0 & 0 & 0 \end{bmatrix}, \quad P^{32} \sim \begin{bmatrix} 1 & 0 & 0 & 0 & 0 \\ 0 & 1 & 0 & 0 & 0 \\ .0875 & .9125 & 0 & 0 & 0 \\ .1875 & .8125 & 0 & 0 & 0 \\ .4375 & .5625 & 0 & 0 & 0 \end{bmatrix};$$

$$\overline{P} = \begin{bmatrix} 1 & 0 & 0 & 0 & 0 \\ 0 & 1 & 0 & 0 & 0 \\ .0875 & .9125 & 0 & 0 & 0 \\ .1875 & .8125 & 0 & 0 & 0 \\ .4375 & .5625 & 0 & 0 & 0 \end{bmatrix}$$

33. _Step 1_. Transition diagram:

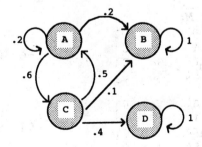

Standard form:

$$M = \begin{array}{c} \\ B \\ D \\ A \\ C \end{array}\begin{array}{c} \begin{array}{cccc} B & D & A & C \end{array} \\ \left[\begin{array}{cccc} 1 & 0 & 0 & 0 \\ 0 & 1 & 0 & 0 \\ .2 & 0 & .2 & .6 \\ .1 & .4 & .5 & 0 \end{array}\right] \end{array}$$

Step 2. Limiting matrix:

For M, we have $R = \begin{bmatrix} .2 & 0 \\ .1 & .4 \end{bmatrix}$ and $Q = \begin{bmatrix} .2 & .6 \\ .5 & 0 \end{bmatrix}$

The limiting matrix $\overline{M}$ has the form:

$$\overline{M} = \left[\begin{array}{c|c} I & 0 \\ \hline FR & 0 \end{array}\right]$$

where $F = (I - Q)^{-1} = \left(\begin{bmatrix} 1 & 0 \\ 0 & 1 \end{bmatrix} - \begin{bmatrix} .2 & .6 \\ .5 & 0 \end{bmatrix} \right)^{-1} = \begin{bmatrix} .8 & -.6 \\ -.5 & 1 \end{bmatrix}^{-1}$

$$= \begin{bmatrix} \frac{4}{5} & -\frac{3}{5} \\ -\frac{1}{2} & 1 \end{bmatrix}^{-1}$$

We use row operations to find the inverse:

$$\left[\begin{array}{cc|cc} \frac{4}{5} & -\frac{3}{5} & 1 & 0 \\ -\frac{1}{2} & 1 & 0 & 1 \end{array}\right] \sim \left[\begin{array}{cc|cc} 1 & -\frac{3}{4} & \frac{5}{4} & 0 \\ -\frac{1}{2} & 1 & 0 & 1 \end{array}\right] \sim \left[\begin{array}{cc|cc} 1 & -\frac{3}{4} & \frac{5}{4} & 0 \\ 0 & \frac{5}{8} & \frac{5}{8} & 1 \end{array}\right]$$

$\left(\frac{5}{4}\right)R_1 \rightarrow R_1$ $\qquad \left(\frac{1}{2}\right)R_1 + R_2 \rightarrow R_2$ $\quad \left(\frac{8}{5}\right)R_2 \rightarrow R_2$

$$\sim \left[\begin{array}{cc|cc} 1 & -\frac{3}{4} & \frac{5}{4} & 0 \\ 0 & 1 & 1 & \frac{8}{5} \end{array}\right] \sim \left[\begin{array}{cc|cc} 1 & 0 & 2 & \frac{6}{5} \\ 0 & 1 & 1 & \frac{8}{5} \end{array}\right]$$

$\left(\frac{3}{4}\right)R_2 + R_1 \rightarrow R_1$

Thus, $F = \begin{bmatrix} 2 & 1.2 \\ 1 & 1.6 \end{bmatrix}$ and $FR = \begin{bmatrix} 2 & 1.2 \\ 1 & 1.6 \end{bmatrix}\begin{bmatrix} .2 & 0 \\ .1 & .4 \end{bmatrix} = \begin{bmatrix} .52 & .48 \\ .36 & .64 \end{bmatrix}$

Therefore, $\overline{M} = \begin{array}{c} \\ B \\ D \\ A \\ C \end{array}\begin{array}{c} \begin{array}{cccc} B & D & A & C \end{array} \\ \left[\begin{array}{cccc} 1 & 0 & 0 & 0 \\ 0 & 1 & 0 & 0 \\ .52 & .48 & 0 & 0 \\ .36 & .64 & 0 & 0 \end{array}\right] \end{array}$

Step 3. Transition diagram for $\overline{M}$:

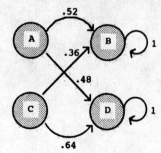

Limiting matrix for P:

$$\overline{P} = \begin{array}{c} \\ A \\ B \\ C \\ D \end{array} \begin{array}{cccc} A & B & C & D \\ \left[\begin{array}{cccc} 0 & .52 & 0 & .48 \\ 0 & 1 & 0 & 0 \\ 0 & .36 & 0 & .64 \\ 0 & 0 & 0 & 1 \end{array}\right] \end{array}$$

35. $P^4 \sim \begin{bmatrix} .1276 & .426 & .0768 & .3696 \\ 0 & 1 & 0 & 0 \\ .064 & .29 & .102 & .544 \\ 0 & 0 & 0 & 1 \end{bmatrix}$, $P^8 \sim \begin{bmatrix} .0212 & .5026 & .0176 & .4585 \\ 0 & 1 & 0 & 0 \\ .0147 & .3468 & .0153 & .6231 \\ 0 & 0 & 0 & 0 \end{bmatrix}$

$P^{32} \sim \begin{bmatrix} 0 & .52 & 0 & .48 \\ 0 & 1 & 0 & 0 \\ 0 & .36 & 0 & .64 \\ 0 & 0 & 0 & 1 \end{bmatrix}$

37. Let $S = [x \quad 1 - x \quad 0]$, $0 \le x \le 1$. Then

$$SP = [x \quad 1 - x \quad 0] \begin{bmatrix} 1 & 0 & 0 \\ 0 & 1 & 0 \\ .1 & .5 & .4 \end{bmatrix} = [x \quad 1 - x \quad 0]$$

Thus, S is a stationary matrix for P.

A stationary matrix for an absorbing Markov chain with two absorbing states and one nonabsorbing state will have one of the forms

$$[x \quad 1 - x \quad 0], \quad [x \quad 0 \quad 1 - x], \quad [0 \quad x \quad 1 - x]$$

39. A transition matrix for this problem is:

$$P = \begin{array}{c} \\ F \\ G \\ A \\ B \end{array} \begin{array}{cccc} F & G & A & B \\ \left[\begin{array}{cccc} 1 & 0 & 0 & 0 \\ .1 & .8 & .1 & 0 \\ .1 & .4 & .4 & .1 \\ 0 & 0 & 0 & 1 \end{array}\right] \end{array}$$

A standard form for this matrix is:

$$M = \begin{array}{c} \\ F \\ B \\ G \\ A \end{array} \begin{array}{cccc} F & B & G & A \\ \left[\begin{array}{cccc} 1 & 0 & 0 & 0 \\ 0 & 1 & 0 & 0 \\ .1 & 0 & .8 & .1 \\ .1 & .1 & .4 & .4 \end{array}\right] \end{array}$$

For this matrix, we have:

$$R = \begin{bmatrix} .1 & 0 \\ .1 & .1 \end{bmatrix} \quad \text{and} \quad Q = \begin{bmatrix} .8 & .1 \\ .4 & .4 \end{bmatrix}$$

The limiting matrix for M has the form:

$$\overline{M} = \left[\begin{array}{c|c} I & 0 \\ \hline FR & 0 \end{array}\right]$$

where $F = (I - Q)^{-1} = \left(\begin{bmatrix} 1 & 0 \\ 0 & 1 \end{bmatrix} - \begin{bmatrix} .8 & .1 \\ .4 & .4 \end{bmatrix} \right)^{-1} = \begin{bmatrix} .2 & -.1 \\ -.4 & .6 \end{bmatrix}^{-1}$

$$= \begin{bmatrix} \frac{1}{5} & -\frac{1}{10} \\ -\frac{2}{5} & \frac{3}{5} \end{bmatrix}^{-1}$$

We use row operations to find the inverse:

$$\begin{bmatrix} \frac{1}{5} & -\frac{1}{10} & \Big| & 1 & 0 \\ -\frac{2}{5} & \frac{3}{5} & \Big| & 0 & 1 \end{bmatrix} \sim \begin{bmatrix} 1 & -\frac{1}{2} & \Big| & 5 & 0 \\ -\frac{2}{5} & \frac{3}{5} & \Big| & 0 & 1 \end{bmatrix} \sim \begin{bmatrix} 1 & -\frac{1}{2} & \Big| & 5 & 0 \\ 0 & \frac{2}{5} & \Big| & 2 & 1 \end{bmatrix} \sim \begin{bmatrix} 1 & -\frac{1}{2} & \Big| & 5 & 0 \\ 0 & 1 & \Big| & 5 & \frac{5}{2} \end{bmatrix}$$

$$5R_1 \to R_1 \qquad \left(\tfrac{2}{5}\right)R_1 + R_2 \to R_2 \qquad \left(\tfrac{5}{2}\right)R_2 \to R_2 \qquad \left(\tfrac{1}{2}\right)R_2 + R_1 \to R_1$$

$$\sim \begin{bmatrix} 1 & 0 & \Big| & \frac{15}{2} & \frac{5}{4} \\ 0 & 1 & \Big| & 5 & \frac{5}{2} \end{bmatrix}$$

Thus, $F = \begin{bmatrix} 7.5 & 1.25 \\ 5 & 2.5 \end{bmatrix}$ and $FR = \begin{bmatrix} .875 & .125 \\ .75 & .25 \end{bmatrix}$

Therefore,

$$\bar{M} = \begin{array}{c} F \\ B \\ G \\ A \end{array} \begin{bmatrix} 1 & 0 & 0 & 0 \\ 0 & 1 & 0 & 0 \\ .875 & .125 & 0 & 0 \\ .75 & .25 & 0 & 0 \end{bmatrix} \begin{array}{c} \\ \end{array}$$

with columns labeled $F\quad B\quad G\quad A$

(A) In the long run, 75% of the accounts in arrears will pay in full.

(B) In the long run, 12.5% of the accounts in good standing will become bad debts.

(C) The average number of months that an account in arrears will either be paid in full or classified as a bad debt is:
$$5 + 2.5 = 7.5 \text{ months}$$

41. A transition matrix in standard form for this problem is:

$$P = \begin{array}{c} A \\ B \\ C \\ N \end{array} \begin{bmatrix} 1 & 0 & 0 & 0 \\ 0 & 1 & 0 & 0 \\ 0 & 0 & 1 & 0 \\ .6 & .3 & .11 & .8 \end{bmatrix}$$

with columns labeled $A\quad B\quad C\quad N$

For this matrix, we have $R = [.6 \quad .3 \quad .11]$ and $Q = [.8]$.

The limiting matrix for P has the form:
$$\bar{P} = \begin{bmatrix} I & 0 \\ \hline FR & 0 \end{bmatrix}$$

where $F = (I - Q)^{-1} = ([1] - [.8])^{-1} = [.2]^{-1} = 5$

Now, $FR = [5][.6 \quad .3 \quad .11] = [.3 \quad .15 \quad .55]$

and

$$\bar{P} = \begin{array}{c} A \\ B \\ C \\ N \end{array} \begin{bmatrix} 1 & 0 & 0 & 0 \\ 0 & 1 & 0 & 0 \\ 0 & 0 & 1 & 0 \\ .3 & .15 & .55 & 0 \end{bmatrix}$$

with columns labeled $A\quad B\quad C\quad N$

(A) In the long run, the market share of each company is:
Company A— 30%; Company B— 15%; and Company C— 55%.

(B) On the average, it will take 5 years for a department to decide to use a calculator from one of these companies in their courses.

43. Let I denote ICU, C denote CCW, D denote "died", and R denote "released". A transition matrix in standard form for this problem is:

$$P = \begin{array}{c} \\ D \\ R \\ I \\ C \end{array} \overset{\begin{array}{cccc} D & R & I & C \end{array}}{\begin{bmatrix} 1 & 0 & 0 & 0 \\ 0 & 1 & 0 & 0 \\ .02 & 0 & .46 & .52 \\ .01 & .22 & .04 & .73 \end{bmatrix}}$$

For this matrix, we have

$$R = \begin{bmatrix} .02 & 0 \\ .01 & .22 \end{bmatrix} \quad \text{and} \quad Q = \begin{bmatrix} .46 & .52 \\ .04 & .73 \end{bmatrix}$$

The limiting matrix for P has the form:

$$\overline{P} = \left[\begin{array}{c|c} I & 0 \\ \hline FR & 0 \end{array} \right]$$

where $F = (I - Q)^{-1} = \left(\begin{bmatrix} 1 & 0 \\ 0 & 1 \end{bmatrix} - \begin{bmatrix} .46 & .52 \\ .04 & .73 \end{bmatrix} \right)^{-1}$

$$= \begin{bmatrix} .54 & -.52 \\ -.04 & .27 \end{bmatrix}^{-1} = \begin{bmatrix} 2.16 & 4.16 \\ .32 & 4.32 \end{bmatrix}$$

Now, $FR = \begin{bmatrix} 2.16 & 4.16 \\ .32 & 4.32 \end{bmatrix} \begin{bmatrix} .02 & 0 \\ .01 & .22 \end{bmatrix} = \begin{bmatrix} .0848 & .9152 \\ .0496 & .9504 \end{bmatrix}$

and

$$\overline{P} = \begin{array}{c} \\ D \\ R \\ I \\ C \end{array} \overset{\begin{array}{cccc} D & R & I & C \end{array}}{\begin{bmatrix} 1 & 0 & 0 & 0 \\ 0 & 1 & 0 & 0 \\ .0848 & .9152 & 0 & 0 \\ .0496 & .9504 & 0 & 0 \end{bmatrix}}$$

(A) In the long run, 91.52% of the patients are released from the hospital.

(B) In the long run, 4.96% of the patients in the CCW die without being released from the hospital.

(C) The average number of days a patient in the ICU will stay in the hospital is:
$$2.16 + 4.16 = 6.32 \text{ days}$$

45. A transition matrix in standard form for this problem is:

$$P = \begin{array}{c} \\ L \\ R \\ F \\ B \end{array} \overset{\begin{array}{cccc} L & R & F & B \end{array}}{\begin{bmatrix} 1 & 0 & 0 & 0 \\ 0 & 1 & 0 & 0 \\ \frac{1}{4} & \frac{1}{4} & 0 & \frac{1}{2} \\ \frac{2}{5} & \frac{1}{5} & \frac{2}{5} & 0 \end{bmatrix}}$$

For this matrix we have:

$$R = \begin{bmatrix} \frac{1}{4} & \frac{1}{4} \\ \frac{2}{5} & \frac{1}{5} \end{bmatrix} \quad \text{and} \quad Q = \begin{bmatrix} 0 & \frac{1}{2} \\ \frac{2}{5} & 0 \end{bmatrix}$$

The limiting matrix for P has the form:

$$\overline{P} = \left[\begin{array}{c|c} I & 0 \\ \hline FR & 0 \end{array}\right]$$

where $F = (I - Q)^{-1} = \left(\begin{bmatrix} 1 & 0 \\ 0 & 1 \end{bmatrix} - \begin{bmatrix} 0 & \frac{1}{2} \\ \frac{2}{5} & 0 \end{bmatrix}\right)^{-1} = \begin{bmatrix} 1 & -\frac{1}{2} \\ -\frac{2}{5} & 1 \end{bmatrix}^{-1}$

We use row operations to find the inverse:

$$\begin{bmatrix} 1 & -\frac{1}{2} & 1 & 0 \\ -\frac{2}{5} & 1 & 0 & 1 \end{bmatrix} \sim \begin{bmatrix} 1 & -\frac{1}{2} & 1 & 0 \\ 0 & \frac{4}{5} & \frac{2}{5} & 1 \end{bmatrix} \sim \begin{bmatrix} 1 & -\frac{1}{2} & 1 & 0 \\ 0 & 1 & \frac{1}{2} & \frac{5}{4} \end{bmatrix} \sim \begin{bmatrix} 1 & 0 & \frac{5}{4} & \frac{5}{8} \\ 0 & 1 & \frac{1}{2} & \frac{5}{4} \end{bmatrix}$$

$$\left(\frac{2}{5}\right)R_1 + R_2 \rightarrow R_2 \qquad \left(\frac{5}{4}\right)R_2 \rightarrow R_2 \qquad \left(\frac{1}{2}\right)R_2 + R_1 \rightarrow R_1$$

Thus, $F = \begin{bmatrix} \frac{5}{4} & \frac{5}{8} \\ \frac{1}{2} & \frac{5}{4} \end{bmatrix}$ and $FR = \begin{bmatrix} \frac{5}{4} & \frac{5}{8} \\ \frac{1}{2} & \frac{5}{4} \end{bmatrix}\begin{bmatrix} \frac{1}{4} & \frac{1}{4} \\ \frac{2}{5} & \frac{1}{5} \end{bmatrix} = \begin{bmatrix} \frac{9}{16} & \frac{7}{16} \\ \frac{5}{8} & \frac{3}{8} \end{bmatrix}$.

Now,

$$\begin{array}{c} \\ L \\ \overline{P} = \begin{array}{c} L \\ R \\ F \\ B \end{array} \end{array} \begin{array}{cccc} L & R & F & B \end{array}$$

$$\overline{P} = \begin{array}{c} L \\ R \\ F \\ B \end{array}\begin{bmatrix} 1 & 0 & 0 & 0 \\ 0 & 1 & 0 & 0 \\ \frac{9}{16} & \frac{7}{16} & 0 & 0 \\ \frac{5}{8} & \frac{3}{8} & 0 & 0 \end{bmatrix}$$

(A) The long run probability that a rat placed in room B will end up in room R is $\frac{3}{8} = .375$.

(B) The average number of exits that a rat placed in room B will choose until it finds food is:
$$\frac{1}{2} + \frac{5}{4} = \frac{7}{4} = 1.75$$

CHAPTER 7 REVIEW

1. $S_1 = S_0 P = [.3 \quad .7]\begin{bmatrix} .6 & .4 \\ .2 & .8 \end{bmatrix} = \overset{A \qquad B}{[.32 \quad .68]}$

$S_2 = S_1 P = [.32 \quad .68]\begin{bmatrix} .6 & .4 \\ .2 & .8 \end{bmatrix} = \overset{A \qquad B}{[.328 \quad .672]}$

The probability of being in state A after one trial is .32; after two trials .328. The probability of being in state B after one trial is .68; after two trials .672.

(7-1)

2. A is an absorbing state; the chain is absorbing since it is possible to go from state B to state A.

(7-2, 7-3)

3. There are no absorbing states since there are no 1's on the main diagonal.

P is regular since $P^2 = \begin{bmatrix} .7 & .3 \\ .21 & .79 \end{bmatrix}$ has only positive entries. (7-2, 7-3)

4. $P = \begin{bmatrix} 0 & 1 \\ 1 & 0 \end{bmatrix}$ has no absorbing states. Since P^k, $k = 1, 2, 3, \ldots$,

alternates between $\begin{bmatrix} 0 & 1 \\ 1 & 0 \end{bmatrix}$ and $\begin{bmatrix} 1 & 0 \\ 0 & 1 \end{bmatrix}$, P is not regular. (7-2, 7-3)

5. $P = \begin{array}{c} \\ A \\ B \\ C \end{array} \begin{array}{c} A \quad B \quad C \\ \begin{bmatrix} 0 & 1 & 0 \\ .1 & 0 & .9 \\ 0 & 1 & 0 \end{bmatrix} \end{array}$

There are no absorbing states.

P^k, $k = 1, 2, 3, \ldots$, alternates between $\begin{bmatrix} 0 & 1 & 0 \\ .1 & 0 & .9 \\ 0 & 1 & 0 \end{bmatrix}$ and $\begin{bmatrix} .1 & 0 & .9 \\ 0 & 1 & 0 \\ .1 & 0 & .9 \end{bmatrix}$.

Thus, P is not regular. (7-1, 7-2, 7-3)

6. $P = \begin{array}{c} \\ A \\ B \\ C \end{array} \begin{array}{c} A \quad B \quad C \\ \begin{bmatrix} 0 & 1 & 0 \\ .1 & .2 & .7 \\ 0 & 0 & 1 \end{bmatrix} \end{array}$

C is an absorbing state. The chain is absorbing since it is possible to go from state A to state C (via B) and from state B to state C. (7-1, 7-2, 7-3)

7. $P = \begin{array}{c} \\ A \\ B \\ C \end{array} \begin{array}{c} A \quad B \quad C \\ \begin{bmatrix} 0 & 0 & 1 \\ .1 & .2 & .7 \\ 0 & 1 & 0 \end{bmatrix} \end{array}$

There are no absorbing states since there are no 1's on the main diagonal.

P is regular since $P^3 = \begin{bmatrix} .1 & .2 & .7 \\ .074 & .388 & .538 \\ .02 & .74 & .24 \end{bmatrix}$ has only positive entries.

(7-1, 7-2, 7-3)

8. $P = \begin{array}{c} \\ A \\ B \\ C \\ D \end{array} \begin{array}{c} A \quad B \quad C \quad D \\ \begin{bmatrix} .3 & .2 & 0 & .5 \\ 0 & 1 & 0 & 0 \\ 0 & 0 & .2 & .8 \\ 0 & 0 & .3 & .7 \end{bmatrix} \end{array}$

B is an absorbing state. The chain is not absorbing since it is not possible to go from state C to B, nor is it possible to go from state D to state B. (7-1, 7-2, 7-3)

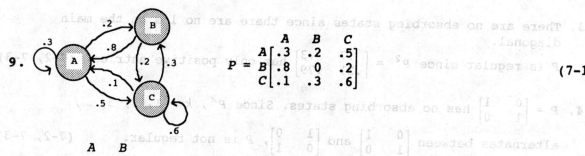

9. $P = \begin{matrix} & A & B & C \\ A & \begin{bmatrix} .3 & .2 & .5 \\ B & .8 & 0 & .2 \\ C & .1 & .3 & .6 \end{bmatrix} \end{matrix}$ (7-1)

10. $P = \begin{matrix} & A & B \\ A & \begin{bmatrix} .4 & .6 \\ B & .9 & .1 \end{bmatrix} \end{matrix}$

(A) $P^2 = \begin{matrix} & A & B \\ A & \begin{bmatrix} .7 & .3 \\ B & .45 & .55 \end{bmatrix} \end{matrix}$

The probability of going from state A to state B in two trials is .3.

(B) $P^3 = \begin{matrix} & A & B \\ A & \begin{bmatrix} .55 & .45 \\ B & .675 & .325 \end{bmatrix} \end{matrix}$

The probability of going from state B to state A in three trials is .675. (7-1)

11. Let $S = [s_1 \quad s_2]$ and solve the system:

$$[s_1 \quad s_2] \begin{bmatrix} .4 & .6 \\ .2 & .8 \end{bmatrix} = [s_1 \quad s_2], \quad s_1 + s_2 = 1$$

which is equivalent to

$$\begin{array}{ll} .4s_1 + .2s_2 = s_1 & \\ .6s_1 + .8s_2 = s_2 & \text{or} \\ s_1 + s_2 = 1 & \end{array} \qquad \begin{array}{l} -.6s_1 + .2s_2 = 0 \\ .6s_1 - .2s_2 = 0 \\ s_1 + s_2 = 1 \end{array}$$

The solution is: $s_1 = .25$, $s_2 = .75$

The stationary matrix $S = \begin{matrix} A & B \\ [.25 & .75] \end{matrix}$

The limiting matrix $\overline{P} = \begin{matrix} & A & B \\ A & \begin{bmatrix} .25 & .75 \\ B & .25 & .75 \end{bmatrix} \end{matrix}$ (7-2)

12. Let $S = [s_1 \quad s_2 \quad s_3]$ and solve the system:

$$[s_1 \quad s_2 \quad s_3] \begin{bmatrix} .4 & .6 & 0 \\ .5 & .3 & .2 \\ 0 & .8 & .2 \end{bmatrix} = [s_1 \quad s_2 \quad s_3], \quad s_1 + s_2 + s_3 = 1$$

which is equivalent to

$$\begin{array}{ll} .4s_1 + .5s_2 = s_1 & \\ .6s_1 + .3s_2 + .8s_3 = s_2 & \text{or} \\ .2s_2 + .2s_3 = s_3 & \\ s_1 + s_2 + s_3 = 1 & \end{array} \qquad \begin{array}{l} -.6s_1 + .5s_2 = 0 \\ .6s_1 - .7s_2 + .8s_3 = 0 \\ .2s_2 - .8s_3 = 0 \\ s_1 + s_2 + s_3 = 1 \end{array}$$

From the first and third equations, we have $s_1 = \frac{5}{6}s_2$ and $s_3 = \frac{1}{4}s_2$.

Substituting these values into the fourth equation, we get

$$\frac{5}{6}s_2 + s_2 + \frac{1}{4}s_2 = 1 \quad \text{or} \quad \frac{25}{12}s_2 = 1 \quad \text{and} \quad s_2 = .48$$

Therefore, $s_1 = .4$, $s_2 = .48$, $s_3 = .12$.

The stationary matrix $S = \begin{array}{ccc} A & B & C \\ [.4 & .48 & .12] \end{array}$.

The limiting matrix $\overline{P} = \begin{array}{c} \\ A \\ B \\ C \end{array}\begin{array}{ccc} A & B & C \\ \left[.4\right. & .48 & .12 \\ .4 & .48 & .12 \\ \left..4\right. & .48 & .12 \end{array}$ (7-2)

13. For $P = \begin{array}{c} \\ A \\ B \\ C \end{array}\begin{array}{ccc} A & B & C \\ \left[1\right. & 0 & 0 \\ 0 & 1 & 0 \\ \left..3\right. & .1 & .6 \end{array}$

we have $R = [.3 \quad .1]$ and $Q = [.6]$. The limiting matrix $\overline{P}$ has the form

$$\overline{P} = \left[\begin{array}{cc|c} 1 & 0 & 0 \\ 0 & 1 & 0 \\ \hline F & R & 0 \end{array}\right]$$

where $F = (I - Q)^{-1} = ([1] - [.6])^{-1} = [.4]^{-1} = \left[\frac{5}{2}\right]$

and $FR = \left[\frac{5}{2}\right][.3 \quad .1] = [.75 \quad .25]$.

Thus, $\overline{P} = \begin{array}{c} \\ A \\ B \\ C \end{array}\begin{array}{ccc} A & B & C \\ \left[1\right. & 0 & 0 \\ 0 & 1 & 0 \\ \left..75\right. & .25 & 0 \end{array}$

$P(C \text{ to } A) = .75$, $P(C \text{ to } B) = .25$. Since $F = \left[\frac{5}{2}\right]$ it will take an average of 2.5 trials to go from C to either A or B.   (7-3)

14. For $P = \begin{array}{c} \\ A \\ B \\ C \\ D \end{array}\begin{array}{cccc} A & B & C & D \\ \left[1\right. & 0 & 0 & 0 \\ 0 & 1 & 0 & 0 \\ .1 & .5 & .2 & .2 \\ \left..1\right. & .1 & .4 & .4 \end{array}$

we have $R = \begin{bmatrix} .1 & .5 \\ .1 & .1 \end{bmatrix}$ and $Q = \begin{bmatrix} .2 & .2 \\ .4 & .4 \end{bmatrix}$.

The limiting matrix $\overline{P}$ has the form

$$\overline{P} = \left[\begin{array}{c|c} I & 0 \\ \hline FR & 0 \end{array}\right]$$

where $F = (I - Q)^{-1} = \left(\begin{bmatrix} 1 & 0 \\ 0 & 1 \end{bmatrix} - \begin{bmatrix} .2 & .2 \\ .4 & .4 \end{bmatrix}\right)^{-1} = \begin{bmatrix} .8 & -.2 \\ -.4 & .6 \end{bmatrix}^{-1} = \begin{bmatrix} \frac{4}{5} & -\frac{1}{5} \\ -\frac{2}{5} & \frac{3}{5} \end{bmatrix}^{-1}$

We use row operations to find the inverse:

$$\begin{bmatrix} \frac{4}{5} & -\frac{1}{5} & 1 & 0 \\ -\frac{2}{5} & \frac{3}{5} & 0 & 1 \end{bmatrix} \sim \begin{bmatrix} 1 & -\frac{1}{4} & \frac{5}{4} & 0 \\ -\frac{2}{5} & \frac{3}{5} & 0 & 1 \end{bmatrix} \sim \begin{bmatrix} 1 & -\frac{1}{4} & \frac{5}{4} & 0 \\ 0 & \frac{1}{2} & \frac{1}{2} & 1 \end{bmatrix} \sim \begin{bmatrix} 1 & -\frac{1}{4} & \frac{5}{4} & 0 \\ 0 & 1 & 1 & 2 \end{bmatrix}$$

$$\left(\frac{5}{4}\right)R_1 \rightarrow R_1 \qquad \left(\frac{2}{5}\right)R_1 + R_2 \rightarrow R_2 \qquad 2R_2 \rightarrow R_2 \qquad \left(\frac{1}{4}\right)R_2 + R_1 \rightarrow R_1$$

$$\sim \begin{bmatrix} 1 & 0 & \frac{3}{2} & \frac{1}{2} \\ 0 & 1 & 1 & 2 \end{bmatrix}$$

Thus, $F = \begin{bmatrix} \frac{3}{2} & \frac{1}{2} \\ 1 & 2 \end{bmatrix} = \begin{bmatrix} 1.5 & .5 \\ 1 & 2 \end{bmatrix}$, $FR = \begin{bmatrix} 1.5 & .5 \\ 1 & 2 \end{bmatrix}\begin{bmatrix} .1 & .5 \\ .1 & .1 \end{bmatrix} = \begin{bmatrix} .2 & .8 \\ .3 & .7 \end{bmatrix}$.

and $\overline{P} = \begin{array}{c} \\ A \\ B \\ C \\ D \end{array} \begin{array}{cccc} A & B & C & D \\ \begin{bmatrix} 1 & 0 & 0 & 0 \\ 0 & 1 & 0 & 0 \\ .2 & .8 & 0 & 0 \\ .3 & .7 & 0 & 0 \end{bmatrix} \end{array}$

$P(C \text{ to } A) = .2$, $P(C \text{ to } B) = .8$, $P(D \text{ to } A) = .3$, $P(D \text{ to } B) = .7$.
It takes an average of 2 trials to go from C to either A or B; it takes
an average of three trials to go from D to A or B. (7–3)

15. $P = \begin{array}{c} A \\ B \end{array}\begin{array}{cc} A & B \\ \begin{bmatrix} .4 & .6 \\ .2 & .8 \end{bmatrix}\end{array}$, $P^4 \approx \begin{bmatrix} .2512 & .7488 \\ .2496 & .7504 \end{bmatrix}$, $P^8 \approx \begin{bmatrix} .2500 & .7499 \\ .2499 & .7500 \end{bmatrix}$; $\overline{P} = \begin{array}{c} A \\ B \end{array}\begin{array}{cc} A & B \\ \begin{bmatrix} .25 & .75 \\ .25 & .75 \end{bmatrix}\end{array}$

(7–3)

16. $P = \begin{array}{c} A \\ B \\ C \end{array}\begin{array}{ccc} A & B & C \\ \begin{bmatrix} .4 & .6 & 0 \\ .5 & .3 & .2 \\ 0 & .8 & .2 \end{bmatrix}\end{array}$, $P^4 \approx \begin{bmatrix} .4066 & .4722 & .1212 \\ .3935 & .4895 & .117 \\ .404 & .468 & .128 \end{bmatrix}$,

$P^8 \approx \begin{bmatrix} .4001 & .4799 & .1200 \\ .3999 & .4802 & .1199 \\ .4001 & .4798 & .1201 \end{bmatrix}$; $\overline{P} = \begin{array}{c} A \\ B \\ C \end{array}\begin{array}{ccc} A & B & C \\ \begin{bmatrix} .4 & .48 & .12 \\ .4 & .48 & .12 \\ .4 & .48 & .12 \end{bmatrix}\end{array}$

(7–3)

17. $P = \begin{array}{c} A \\ B \\ C \end{array}\begin{array}{ccc} A & B & C \\ \begin{bmatrix} 1 & 0 & 0 \\ 0 & 1 & 0 \\ .3 & .1 & .6 \end{bmatrix}\end{array}$, $P^4 \approx \begin{bmatrix} 1 & 0 & 0 \\ 0 & 1 & 0 \\ .6528 & .2176 & .1296 \end{bmatrix}$,

$P^8 \approx \begin{bmatrix} 1 & 0 & 0 \\ 0 & 1 & 0 \\ .7374 & .2458 & .01680 \end{bmatrix}$, $P^{16} \approx \begin{bmatrix} 1 & 0 & 0 \\ 0 & 1 & 0 \\ .7498 & .2499 & 0 \end{bmatrix}$;

$\overline{P} = \begin{array}{c} A \\ B \\ C \end{array}\begin{array}{ccc} A & B & C \\ \begin{bmatrix} 1 & 0 & 0 \\ 0 & 1 & 0 \\ .75 & .25 & 0 \end{bmatrix}\end{array}$

(7–3)

18. $P = \begin{array}{c} \\ A \\ B \\ C \\ D \end{array} \begin{array}{cccc} A & B & C & D \\ \end{array}$
$$P = \begin{array}{c} A \\ B \\ C \\ D \end{array} \begin{bmatrix} 1 & 0 & 0 & 0 \\ 0 & 1 & 0 & 0 \\ .1 & .5 & .2 & .2 \\ .1 & .1 & .4 & .4 \end{bmatrix}, \quad P^4 \sim \begin{bmatrix} 1 & 0 & 0 & 0 \\ 0 & 1 & 0 & 0 \\ .1784 & .7352 & .0432 & .0432 \\ .2568 & .5704 & .0864 & .0864 \end{bmatrix},$$

$$P^8 \sim \begin{bmatrix} 1 & 0 & 0 & 0 \\ 0 & 1 & 0 & 0 \\ .1972 & .7916 & .0056 & .0056 \\ .2944 & .6832 & .0112 & .0112 \end{bmatrix}, \quad P^{16} \sim \begin{bmatrix} 1 & 0 & 0 & 0 \\ 0 & 1 & 0 & 0 \\ .2000 & .7999 & 0 & 0 \\ .2999 & .6997 & 0 & 0 \end{bmatrix},$$

$$\overline{P} = \begin{array}{c} A \\ B \\ C \\ D \end{array} \begin{array}{cccc} A & B & C & D \\ \end{array} \begin{bmatrix} 1 & 0 & 0 & 0 \\ 0 & 1 & 0 & 0 \\ .2 & .8 & 0 & 0 \\ .3 & .7 & 0 & 0 \end{bmatrix}$$ (7-3)

19. A standard form for the given matrix is:

$$P = \begin{array}{c} B \\ D \\ A \\ C \end{array} \begin{array}{cccc} B & D & A & C \\ \end{array} \begin{bmatrix} 1 & 0 & 0 & 0 \\ 0 & 1 & 0 & 0 \\ .1 & .1 & .6 & .2 \\ .2 & .2 & .3 & .3 \end{bmatrix}$$ (7-3)

20. We will determine the limiting matrix of:

$$P = \begin{array}{c} A \\ B \\ C \end{array} \begin{array}{ccc} A & B & C \\ \end{array} \begin{bmatrix} 0 & 1 & 0 \\ 0 & 0 & 1 \\ .2 & .6 & .2 \end{bmatrix}$$

by solving

$$[s_1 \quad s_2 \quad s_3] \begin{bmatrix} 0 & 1 & 0 \\ 0 & 0 & 1 \\ .2 & .6 & .2 \end{bmatrix} = [s_1 \quad s_2 \quad s_3], \quad s_1 + s_2 + s_3 = 1.$$

The corresponding system of equations is:

$$\begin{array}{rl} .2s_3 = s_1 & \qquad s_1 \quad\quad - .2s_3 = 0 \\ s_1 \quad + .6s_3 = s_2 & \text{or} \quad s_1 - s_2 + .6s_3 = 0 \\ s_2 + .2s_3 = s_3 & \qquad s_2 - .8s_3 = 0 \\ s_1 + s_2 + s_3 = 1 & \qquad s_1 + s_2 + s_3 = 0 \end{array}$$

From the first and third equations, we have $s_1 = .2s_3$ and $s_2 = .8s_3$.
Substituting these values into the fourth equation gives

$$.2s_3 + .8s_3 + s_3 = 1 \quad \text{and} \quad s_3 = .5$$

It now follows that $s_1 = .1$ and $s_2 = .4$. Thus, $s = [.1 \quad .4 \quad .5]$ and

$$\overline{P} = \begin{array}{c} A \\ B \\ C \end{array} \begin{array}{ccc} A & B & C \\ \end{array} \begin{bmatrix} .1 & .4 & .5 \\ .1 & .4 & .5 \\ .1 & .4 & .5 \end{bmatrix}$$

(A) $\begin{bmatrix} 0 & 0 & 1 \end{bmatrix} \begin{bmatrix} .1 & .4 & .5 \\ .1 & .4 & .5 \\ .1 & .4 & .5 \end{bmatrix} = \begin{array}{ccc} A & B & C \\ [.1 & .4 & .5] \end{array}$

(B) $\begin{bmatrix} .5 & .3 & .2 \end{bmatrix} \begin{bmatrix} .1 & .4 & .5 \\ .1 & .4 & .5 \\ .1 & .4 & .5 \end{bmatrix} = \begin{array}{ccc} A & B & C \\ [.1 & .4 & .5] \end{array}$ (7-3)

21. The transition matrix:

$$P = \begin{array}{c} A \\ B \\ C \end{array} \begin{array}{ccc} A & B & C \\ \begin{bmatrix} 1 & 0 & 0 \\ 0 & 1 & 0 \\ .2 & .6 & .2 \end{bmatrix} \end{array}$$

is the standard form for an absorbing Markov chain with two absorbing and one nonabsorbing states. For this matrix, we have:

$R = [.2 \quad .6]$ and $Q = [.2]$.

The limiting matrix has the form

$$\overline{P} = \left[\begin{array}{c|c} I & 0 \\ \hline FR & 0 \end{array} \right]$$

where $F = (I - Q)^{-1} = ([1] - [.2])^{-1} = [.8]^{-1} = [1.25]$

Thus, $FR = [1.25][.2 \quad .6] = [.25 \quad .75]$ and

$$\overline{P} = \begin{array}{c} A \\ B \\ C \end{array} \begin{array}{ccc} A & B & C \\ \begin{bmatrix} 1 & 0 & 0 \\ 0 & 1 & 0 \\ .25 & .75 & 0 \end{bmatrix} \end{array}$$

(A) $\begin{bmatrix} 0 & 0 & 1 \end{bmatrix} \begin{bmatrix} 1 & 0 & 0 \\ 0 & 1 & 0 \\ .25 & .75 & 0 \end{bmatrix} = \begin{array}{ccc} A & B & C \\ [.25 & .75 & 0] \end{array}$

(B) $\begin{bmatrix} .5 & .3 & .2 \end{bmatrix} \begin{bmatrix} 1 & 0 & 0 \\ 0 & 1 & 0 \\ .25 & .75 & 0 \end{bmatrix} = \begin{array}{ccc} A & B & C \\ [.55 & .45 & 0] \end{array}$ (7-3)

22. (A)

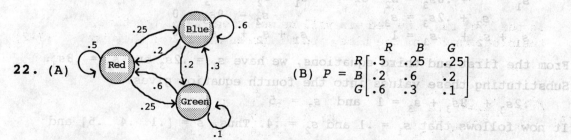

(B) $P = \begin{array}{c} R \\ B \\ G \end{array} \begin{array}{ccc} R & B & G \\ \begin{bmatrix} .5 & .25 & .25 \\ .2 & .6 & .2 \\ .6 & .3 & .1 \end{bmatrix} \end{array}$

(C) The chain is regular since it has only positive entries.

(D) Let $S = [s_1 \quad s_2 \quad s_3]$ and solve the system:

$$[s_1 \quad s_2 \quad s_3] \begin{bmatrix} .5 & .25 & .25 \\ .2 & .6 & .2 \\ .6 & .3 & .1 \end{bmatrix} = [s_1 \quad s_2 \quad s_3], \ s_1 + s_2 + s_3 = 1$$

which is equivalent to:

$$\begin{aligned}
s_1 + s_2 + s_3 &= 1 \\
.5s_1 + .2s_2 + .6s_3 &= s_1 \\
.25s_1 + .6s_2 + .3s_3 &= s_2 \\
.25s_1 + .2s_2 + .1s_3 &= s_3
\end{aligned}
\qquad \text{or} \qquad
\begin{aligned}
s_1 + s_2 + s_3 &= 1 \\
-.5s_1 + .2s_2 + .6s_3 &= 0 \\
.25s_1 - .4s_2 + .3s_3 &= 0 \\
.25s_1 + .2s_2 - .9s_3 &= 0
\end{aligned}$$

We use row operations to solve this system; but first multiply the second, third and fourth equations by 10 to simplify the calculations.

$$\begin{bmatrix} 1 & 1 & 1 & | & 1 \\ -5 & 2 & 6 & | & 0 \\ \frac{5}{2} & -4 & 3 & | & 0 \\ \frac{5}{2} & 2 & -9 & | & 0 \end{bmatrix} \sim \begin{bmatrix} 1 & 1 & 1 & | & 1 \\ 0 & 7 & 11 & | & 5 \\ 0 & -\frac{13}{2} & \frac{1}{2} & | & -\frac{5}{2} \\ 0 & -\frac{1}{2} & -\frac{23}{2} & | & -\frac{5}{2} \end{bmatrix} \sim \begin{bmatrix} 1 & 1 & 1 & | & 1 \\ 0 & 1 & 23 & | & 5 \\ 0 & -\frac{13}{2} & \frac{1}{2} & | & -\frac{5}{2} \\ 0 & 7 & 11 & | & 5 \end{bmatrix}$$

$$5R_1 + R_2 \to R_2 \qquad\qquad -2R_4 \to R_4 \qquad\qquad (-1)R_2 + R_1 \to R_1$$

$$\left(-\frac{5}{2}\right)R_1 + R_3 \to R_3 \qquad\qquad R_2 \leftrightarrow R_4 \qquad\qquad \left(\frac{13}{2}\right)R_2 + R_3 \to R_3$$

$$\left(-\frac{5}{2}\right)R_1 + R_4 \to R_4 \qquad\qquad\qquad\qquad\qquad (-7)R_2 + R_4 \to R_4$$

$$\sim \begin{bmatrix} 1 & 0 & -22 & | & -4 \\ 0 & 1 & 23 & | & 5 \\ 0 & 0 & 150 & | & 30 \\ 0 & 0 & -150 & | & -30 \end{bmatrix} \sim \begin{bmatrix} 1 & 0 & -22 & | & -4 \\ 0 & 1 & 23 & | & 5 \\ 0 & 0 & 1 & | & \frac{1}{5} \\ 0 & 0 & -150 & | & -30 \end{bmatrix} \sim \begin{bmatrix} 1 & 0 & 0 & | & \frac{2}{5} \\ 0 & 1 & 0 & | & \frac{2}{5} \\ 0 & 0 & 1 & | & \frac{1}{5} \\ 0 & 0 & 0 & | & 0 \end{bmatrix}$$

$$\frac{1}{150}R_3 \to R_3 \qquad\qquad 22R_3 + R_1 \to R_1$$

$$(-23)R_3 + R_2 \to R_2$$

$$150R_3 + R_4 \to R_4$$

The solution is $s_1 = 0.4$, $s_2 = 0.4$, $s_3 = 0.2$ and

$$\bar{P} = \begin{array}{c} \\ R \\ B \\ G \end{array}\begin{array}{c} \begin{array}{ccc} R & B & G \end{array} \\ \begin{bmatrix} .4 & .4 & .2 \\ .4 & .4 & .2 \\ .4 & .4 & .2 \end{bmatrix} \end{array}$$

In the long run, the red urn will be selected 40% of the time, the blue urn 40% of the time, and the green urn 20% of the time. (7-2)

23. (A)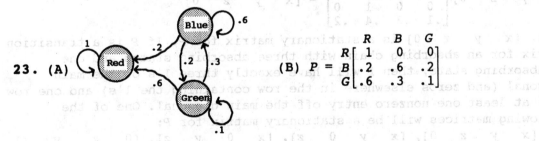

(B) $P = \begin{array}{c} \\ R \\ B \\ G \end{array}\begin{array}{c} \begin{array}{ccc} R & B & G \end{array} \\ \begin{bmatrix} 1 & 0 & 0 \\ .2 & .6 & .2 \\ .6 & .3 & .1 \end{bmatrix} \end{array}$

(C) State R is an absorbing state. The chain is absorbing since it is possible to go from states B and G to state R in a finite number (namely 1) steps.

(D) For $P = \begin{bmatrix} 1 & 0 & 0 \\ .2 & .6 & .2 \\ .6 & .3 & .1 \end{bmatrix}$ we have $R = \begin{bmatrix} .2 \\ .6 \end{bmatrix}$ and $Q = \begin{bmatrix} .6 & .2 \\ .3 & .1 \end{bmatrix}$.

The limiting matrix $\overline{P}$ has the form:

$$\overline{P} = \left[\begin{array}{c|c} I & 0 \\ \hline FR & 0 \end{array} \right]$$

where $F = (I - Q)^{-1} = \left(\begin{bmatrix} 1 & 0 \\ 0 & 1 \end{bmatrix} - \begin{bmatrix} .6 & .2 \\ .3 & .1 \end{bmatrix} \right)^{-1} = \begin{bmatrix} .4 & -.2 \\ -.3 & .9 \end{bmatrix}^{-1}$

$$= \begin{bmatrix} \frac{2}{5} & -\frac{1}{5} \\ -\frac{3}{10} & \frac{9}{10} \end{bmatrix}^{-1}$$

We use row operations to find the inverse:

$$\left[\begin{array}{cc|cc} \frac{2}{5} & -\frac{1}{5} & 1 & 0 \\ -\frac{3}{10} & \frac{9}{10} & 0 & 1 \end{array} \right] \sim \left[\begin{array}{cc|cc} 1 & -\frac{1}{2} & \frac{5}{2} & 0 \\ -\frac{3}{10} & \frac{9}{10} & 0 & 1 \end{array} \right] \sim \left[\begin{array}{cc|cc} 1 & -\frac{1}{2} & \frac{5}{2} & 0 \\ 0 & \frac{3}{4} & \frac{3}{4} & 1 \end{array} \right] \sim \left[\begin{array}{cc|cc} 1 & -\frac{1}{2} & \frac{5}{2} & 0 \\ 0 & 1 & 1 & \frac{4}{3} \end{array} \right]$$

$$\left(\frac{5}{2} \right) R_1 \to R_1 \qquad \left(\frac{3}{10} \right) R_1 + R_2 \to R_2 \qquad \frac{4}{3} R_2 \to R_2 \qquad \left(\frac{1}{2} \right) R_2 + R_1 \to R_1$$

$$\left[\begin{array}{cc|cc} 1 & 0 & 3 & \frac{2}{3} \\ 0 & 1 & 1 & \frac{4}{3} \end{array} \right]$$

Thus, $F = \begin{bmatrix} 3 & \frac{2}{3} \\ 1 & \frac{4}{3} \end{bmatrix}$ and $FR = \begin{bmatrix} 3 & \frac{2}{3} \\ 1 & \frac{4}{3} \end{bmatrix} \begin{bmatrix} \frac{1}{5} \\ \frac{3}{5} \end{bmatrix} = \begin{bmatrix} 1 \\ 1 \end{bmatrix}$.

Now, $\overline{P} = \begin{array}{c} \\ R \\ B \\ G \end{array} \begin{array}{c} \begin{array}{ccc} R & B & G \end{array} \\ \begin{bmatrix} 1 & 0 & 0 \\ 1 & 0 & 0 \\ 1 & 0 & 0 \end{bmatrix} \end{array}$

Once the red urn is selected, the blue and green urns will never be selected again. It will take an average of 3.67 trials to reach the red urn from the blue urn and an average of 2.33 trials to reach the red urn from the green urn. (7-3)

24. $[x \quad y \quad z \quad 0] \begin{bmatrix} 1 & 0 & 0 & 0 \\ 0 & 1 & 0 & 0 \\ 0 & 0 & 1 & 0 \\ .1 & .3 & .4 & .2 \end{bmatrix} = [x \quad y \quad z \quad 0]$

Thus, $[x \quad y \quad z \quad 0]$ is a stationary matrix for P. If P is a transition matrix for an absorbing chain with three absorbing states and one nonabsorbing state, then P will have exactly three 1's on the main diagonal (and zeros elsewhere in the row containing the 1's) and one row with at least one nonzero entry off the main diagonal. One of the following matrices will be a stationary matrix for P:

$[x \quad y \quad z \quad 0]$, $[x \quad y \quad 0 \quad z]$, $[x \quad 0 \quad y \quad z]$, $[0 \quad x \quad y \quad z]$

where $x + y + z = 1$. The position of the zero corresponds to the one row of P which has a nonzero entry off the main diagonal. (7-2, 7-3)

25. No such chain exists; if the chain has an absorbing state, then a corresponding transition matrix P will have a row containing a 1 on the main diagonal and zeros elsewhere. All powers of P will have the same row, and so the chain cannot be regular. (7-2, 7-3)

26. No such chain exists. The reasoning in Problem 25 applies here as well. (7-2, 7-3)

27. No such chain exists. By Theorem 1, Section 7.2 a regular Markov chain has a unique stationary matrix. (7-2)

28. $S = [1 \quad 0 \quad 0]$ and $S' = [0 \quad 1 \quad 0]$ are both stationary matrices for

$$P = \begin{array}{c} \\ A \\ B \\ C \end{array}\begin{array}{c} \begin{array}{ccc} A & B & C \end{array} \\ \begin{bmatrix} 1 & 0 & 0 \\ 0 & 1 & 0 \\ .6 & .3 & .1 \end{bmatrix} \end{array} \tag{7-3}$$

29. $P = \begin{array}{c} \\ A \\ B \end{array}\begin{array}{c} \begin{array}{cc} A & B \end{array} \\ \begin{bmatrix} 0 & 1 \\ 1 & 0 \end{bmatrix} \end{array}$ has no limiting matrix; $P^{2k} = \begin{bmatrix} 1 & 0 \\ 0 & 1 \end{bmatrix}$ and $P^{2k+1} = \begin{bmatrix} 0 & 1 \\ 1 & 0 \end{bmatrix}$
for all positive integers k. (7-2, 7-3)

30. No such chain exists. By Theorem 1, Section 7.2, a regular Markov chain has a unique limiting matrix. (7-2)

31. No such chain exists. By Theorem 2, Section 7.3, an absorbing Markov chain has a limiting matrix. (7-3)

32. (A)

(B) $P = \begin{array}{c} \\ x \\ x' \end{array}\begin{array}{c} \begin{array}{cc} x & x' \end{array} \\ \begin{bmatrix} .7 & .3 \\ .5 & .5 \end{bmatrix} \end{array}$ (C) $S = [.2 \quad .8]$

(D) $S_1 = SP = [.2 \quad .8]\begin{bmatrix} .7 & .3 \\ .5 & .5 \end{bmatrix} = [.54 \quad .46]$

54% of the consumers will use brand x on the next purchase.

(E) To find the stationary matrix $S = [s_1 \quad s_2]$, we need to solve:

$$[s_1 \quad s_2]\begin{bmatrix} .7 & .3 \\ .5 & .5 \end{bmatrix} = [s_1 \quad s_2], \; s_1 + s_2 = 1$$

This yields the system of equations:

$$\begin{array}{lll} .7s_1 + .5s_2 = s_1 & & -.3s_1 + .5s_2 = 0 \\ .3s_1 + .5s_2 = s_2 & \text{or} & .3s_1 - .5s_2 = 0 \\ s_1 + s_2 = 1 & & s_1 + s_2 = 1 \end{array}$$

The solution is $s_1 = .625$, $s_2 = .375$. Thus, $S = [.625 \quad .375]$.

(F) Brand X will have 62.5% of the market in the long run. (7-2)

33. A transition matrix in standard form for this problem is:

$$P = \begin{array}{c} \\ A \\ B \\ C \\ M \end{array} \begin{array}{cccc} A & B & C & M \\ \left[\begin{array}{cccc} 1 & 0 & 0 & 0 \\ 0 & 1 & 0 & 0 \\ 0 & 0 & 1 & 0 \\ .06 & .08 & .11 & .75 \end{array}\right] \end{array}$$

For this matrix, $R = [.06 \quad .08 \quad .11]$ and $Q = [.75]$.
The limiting matrix for P has the form:

$$\overline{P} = \left[\begin{array}{c|c} I & 0 \\ \hline FR & 0 \end{array}\right]$$

where $F = (I - Q)^{-1} = ([1] - [.75])^{-1} = [.25]^{-1} = [4]$.
Thus, $FR = [4][.06 \quad .08 \quad .11] = [.24 \quad .32 \quad .44]$

$$\text{and } \overline{P} = \begin{array}{c} \\ A \\ B \\ C \\ M \end{array} \begin{array}{cccc} A & B & C & M \\ \left[\begin{array}{cccc} 1 & 0 & 0 & 0 \\ 0 & 1 & 0 & 0 \\ 0 & 0 & 1 & 0 \\ .24 & .32 & .44 & 0 \end{array}\right] \end{array}$$

(A) In the long run, brand A will have 24% of the market, brand B will have 32% and brand C will have 44%.

(B) A company will wait an average of 4 years before converting to one of the new milling machines.

(7-3)

34. (A) $S_1 = [.106 \quad .894]\begin{bmatrix} .853 & .147 \\ .553 & .447 \end{bmatrix} = [.5848 \quad .4152]$

$S_2 = [.5848 \quad .4152]\begin{bmatrix} .853 & .147 \\ .553 & .447 \end{bmatrix} = [.72844 \quad .27156]$

Rounding to three decimal places, we have
$S_1 = [.585 \quad .415]$ and $S_2 = [.728 \quad .272]$

(B)

Year	Data %	Model %
1984	10.6	10.6
1988	58.0	58.5
1992	72.5	72.8

(C) To find the long term behavior, we will calculate the stationary matrix and the limiting matrix:

$$[s_1 \quad s_2]\begin{bmatrix} .853 & .147 \\ .553 & .447 \end{bmatrix} = [s_1 \quad s_2], \ s_1 + s_2 = 1$$

$$\begin{array}{ll} .853s_1 + .553s_2 = s_1 & -.147s_1 + .553s_2 = 0 \\ .147s_1 + .447s_2 = s_2 \quad \text{or} & .147s_1 - .553s_2 = 0 \\ s_1 + s_2 = 1 & s_1 + s_2 = 1 \end{array}$$

The solution is: $s_1 = .79$, $s_2 = .21$.

Thus, $S = [.79 \quad .21]$, and the limiting matrix is:

$$\bar{P} = \begin{matrix} v \\ v' \end{matrix} \begin{bmatrix} \overset{v}{.79} & \overset{v'}{.21} \\ .79 & .21 \end{bmatrix}$$

79% of the households will own a VCR in the long run. $\qquad$ (7-2)

35. A transition matrix in standard form for this problem is:

$$P = \begin{matrix} F \\ L \\ T \\ A \end{matrix} \begin{bmatrix} \overset{F}{1} & \overset{L}{0} & \overset{T}{0} & \overset{A}{0} \\ 0 & 1 & 0 & 0 \\ 0 & .05 & .8 & .15 \\ .17 & .03 & 0 & .8 \end{bmatrix}$$

where F = "Fellow", A = "Associate", T = "Trainee", L = leaves

For this matrix, $R = \begin{bmatrix} 0 & .05 \\ .17 & .03 \end{bmatrix}$ and $Q = \begin{bmatrix} .8 & .15 \\ 0 & .8 \end{bmatrix}$.

The limiting matrix for P has the form

$$\bar{P} = \left[\begin{array}{c|c} I & 0 \\ \hline FR & 0 \end{array} \right]$$

where $F = (I - Q)^{-1} = \left(\begin{bmatrix} 1 & 0 \\ 0 & 1 \end{bmatrix} - \begin{bmatrix} .8 & .15 \\ 0 & .8 \end{bmatrix} \right)^{-1} = \begin{bmatrix} .2 & -.15 \\ 0 & .2 \end{bmatrix}^{-1}$

We use row operations to calculate the inverse:

$$\begin{bmatrix} .2 & -.15 & | & 1 & 0 \\ 0 & .2 & | & 0 & 1 \end{bmatrix} \sim \begin{bmatrix} 1 & -.75 & | & 5 & 0 \\ 0 & 1 & | & 0 & 5 \end{bmatrix} \sim \begin{bmatrix} 1 & 0 & | & 5 & 3.75 \\ 0 & 1 & | & 0 & 5 \end{bmatrix}$$

$$5R_1 \to R_1 \qquad (.75)R_2 + R_1 \to R_1$$
$$5R_2 \to R_2$$

Thus, $F = \begin{bmatrix} 5 & 3.75 \\ 0 & 5 \end{bmatrix}$ and $FR = \begin{bmatrix} 5 & 3.75 \\ 0 & 5 \end{bmatrix} \begin{bmatrix} 0 & .05 \\ .17 & .03 \end{bmatrix} = \begin{bmatrix} .6375 & .3625 \\ .85 & .15 \end{bmatrix}$

The limiting matrix is:

$$\bar{P} = \begin{matrix} F \\ L \\ T \\ A \end{matrix} \begin{bmatrix} \overset{F}{1} & \overset{L}{0} & \overset{T}{0} & \overset{A}{0} \\ 0 & 1 & 0 & 0 \\ .6375 & .3625 & 0 & 0 \\ .85 & .15 & 0 & 0 \end{bmatrix}$$

(A) In the long run, 63.75% of the trainees will become Fellows.

(B) In the long run, 15% of the Associates will leave the company.

(C) A trainee remains in the program an average of $5 + 3.75 = 8.75$ years. $\qquad$ (7-3)

36. We shall find the limiting matrix for:

$$P = \begin{matrix} R \\ P \\ W \end{matrix} \begin{bmatrix} \overset{R}{1} & \overset{P}{0} & \overset{W}{0} \\ .5 & .5 & 0 \\ 0 & 1 & 0 \end{bmatrix}$$

We have $R = \begin{bmatrix} .5 \\ 0 \end{bmatrix}$ and $Q = \begin{bmatrix} .5 & 0 \\ 1 & 0 \end{bmatrix}$.

The limiting matrix for P will have the form:

$$\bar{P} = \left[\begin{array}{c|c} I & 0 \\ \hline FR & 0 \end{array}\right]$$

where $F = (I - Q)^{-1} = \left(\begin{bmatrix} 1 & 0 \\ 0 & 1 \end{bmatrix} - \begin{bmatrix} .5 & 0 \\ 1 & 0 \end{bmatrix}\right)^{-1} = \begin{bmatrix} .5 & 0 \\ -1 & 1 \end{bmatrix}^{-1} = \begin{bmatrix} \frac{1}{2} & 0 \\ -1 & 1 \end{bmatrix}^{-1}$

We use row operations to find the inverse:

$$\begin{bmatrix} \frac{1}{2} & 0 & | & 1 & 0 \\ -1 & 1 & | & 0 & 1 \end{bmatrix} \sim \begin{bmatrix} 1 & 0 & | & 2 & 0 \\ -1 & 1 & | & 0 & 1 \end{bmatrix} \sim \begin{bmatrix} 1 & 0 & | & 2 & 0 \\ 0 & 1 & | & 2 & 1 \end{bmatrix}$$

$\qquad 2R_1 \rightarrow R_1 \qquad\qquad R_1 + R_2 \rightarrow R_2$

Thus, $F = \begin{bmatrix} 2 & 0 \\ 2 & 1 \end{bmatrix}$ and $FR = \begin{bmatrix} 2 & 0 \\ 2 & 1 \end{bmatrix}\begin{bmatrix} .5 \\ 0 \end{bmatrix} = \begin{bmatrix} 1 \\ 1 \end{bmatrix}$

The limiting matrix $\bar{P}$ is:

$$\bar{P} = \begin{array}{c} \\ R \\ P \\ W \end{array}\begin{array}{c} \begin{array}{ccc} R & P & W \end{array} \\ \begin{bmatrix} 1 & 0 & 0 \\ 1 & 0 & 0 \\ 1 & 0 & 0 \end{bmatrix} \end{array}$$

From this matrix, we conclude that eventually all of the flowers will be red.
$\hfill$ (7-3)

37. (A) $S_1 = [.028 \quad .972]\begin{bmatrix} .751 & .249 \\ .001 & .999 \end{bmatrix} = [.022 \quad .978]$

$\qquad S_2 = [.022 \quad .978]\begin{bmatrix} .751 & .249 \\ .001 & .999 \end{bmatrix} = [.0175 \quad .9825]$

(B)

Year	Data %	Model %
1980	2.8	2.8
1985	2.3	2.2
1990	1.9	1.75

(C) To find the long term behavior, we calculate the stationary matrix and the limiting matrix.

$$[s_1 \quad s_2]\begin{bmatrix} .751 & .249 \\ .001 & .999 \end{bmatrix} = [s_1 \quad s_2], \quad s_1 + s_2 = 1$$

$\qquad .751s_1 + .001s_2 = s_1 \qquad\qquad -.249s_1 + .001s_2 = 0$
$\qquad .249s_1 + .999s_2 = s_2 \quad \text{or} \quad .249s_1 - .001s_2 = 0$
$\qquad\quad s_1 + \quad s_2 = 1 \qquad\qquad\qquad s_1 + \quad s_2 = 1$

The solution is: $s_1 = .004$, $s_2 = .996$.

The stationary matrix $S = [.004 \quad .996]$ and the limiting matrix:

$$P = \begin{bmatrix} .004 & .996 \\ .004 & .996 \end{bmatrix}$$

In the long run, only 0.4% of the population will live on a farm.
$\hfill$ (7-2)

8 THE DERIVATIVE

EXERCISE 8-1
───

Things to remember:

1. **AVERAGE RATE OF CHANGE**

   For $y = f(x)$, the AVERAGE RATE OF CHANGE FROM $x = a$ TO $x = a + h$ is

   $$\frac{f(a + h) - f(a)}{(a + h) - a} = \frac{f(a + h) - f(a)}{h} \qquad h \neq 0$$

   The expression $\frac{f(a + h) - f(a)}{h}$ is called the DIFFERENCE QUOTIENT.

2. **INSTANTANEOUS RATE OF CHANGE**

   For $y = f(x)$, the INSTANTANEOUS RATE OF CHANGE AT $x = a$ is

   $$\lim_{h \to 0} \frac{f(a + h) - f(a)}{h}$$

   if the limit exists.

3. **SECANT LINE**

   A line through two points on the graph of a function is called a SECANT LINE. If $(a, f(a))$ and $((a + h), f(a + h))$ are two points on the graph of $y = f(x)$, then

   $$\text{Slope of secant line} = \frac{f(a + h) - f(a)}{h} \qquad \text{[Difference quotient]}$$

4. **SLOPE OF A GRAPH**

   For $y = f(x)$, the SLOPE OF THE GRAPH at the point $(a, f(a))$ is given by

   $$\lim_{h \to 0} \frac{f(a + h) - f(a)}{h}$$

   provided the limit exists. The slope of the graph is also the SLOPE OF THE TANGENT LINE at the point $(a, f(a))$.

───

1. $f(4) - f(1) = 3(4)^2 - 3(1)^2 = 48 - 3 = 45$

3. Average rate of change $= \dfrac{f(4) - f(1)}{4 - 1} = \dfrac{3(4)^2 - 3(1)^2}{3} = \dfrac{45}{3} = 15$

5. Slope of secant line $= \dfrac{f(4) - f(1)}{4 - 1} = \dfrac{3(4)^2 - 3(1)^2}{3} = \dfrac{45}{3} = 15$

7.

h	-0.1	-0.01	-0.001	$\to 0 \leftarrow$	0.001	0.01	0.1
$\dfrac{f(1 + h) - f(1)}{h}$	5.7	5.97	5.997	$\to 6 \leftarrow$	6.003	6.03	6.3

9. By Problem 7, instantaneous rate of change = 6.

11. By Problem 7, instantaneous velocity = 6 ft/sec.

13. By Problem 7, slope of the graph = 6.

15. Average velocity $= \dfrac{f(6) - f(4)}{6 - 4} = \dfrac{10(6)^2 - 10(4)^2}{2} = \dfrac{360 - 160}{2}$

$\qquad\qquad\qquad = 100$ ft/sec

17.

h	-0.1	-0.01	-0.001	$\to 0 \leftarrow$	0.001	0.01	0.1
$\dfrac{f(4 + h) - f(4)}{h}$	79	79.9	79.99	$\to 80 \leftarrow$	80.01	80.1	81

Instantaneous velocity at $x = 4$: 80 ft/sec

19. (A) Slope of secant line through $(2, f(2))$, $(4, f(4))$:

$\dfrac{f(4) - f(2)}{4 - 2} = \dfrac{[4^2 - 2(4) - 4] - [2^2 - 2(2) - 4]}{2} = \dfrac{4 - (-4)}{2} = 4$

(B) Slope of secant line through $(2, f(2))$, $(3, f(3))$:

$\dfrac{f(3) - f(2)}{3 - 2} = \dfrac{[3^2 - 2(3) - 4] - [2^2 - 2(2) - 4]}{1} = -1 - (-4) = 3$

(C)

h	-0.1	-0.01	-0.001	$\to 0 \leftarrow$	0.001	0.01	0.1
$\dfrac{f(2 + h) - f(2)}{h}$	1.9	1.99	1.999	$\to 2 \leftarrow$	2.001	2.01	2.1

(D)

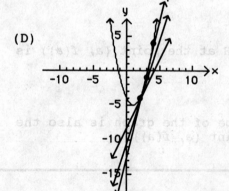

21. At $x = -1$, slope = 1; at $x = 3$, slope = -2

23. At $x = -3$, slope = -5; at $x = -1$, slope = 0; at $x = 1$, slope = -1; at $x = 3$, slope = 4

25.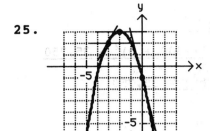

27.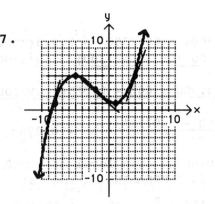

29. 8

31. 0.25 or $\frac{1}{4}$

33.

x	-3	-2	-1	0	1	2	3
slope	-3	-2	-1	0	1	2	3

slope function: $y = x$

35. The slope of the line is m. The slope of the graph of $f(x) = mx + b$ at any point on the graph is also m.

37. $\dfrac{f(1 + h) - f(1)}{h} = \dfrac{3(1 + h)^2 - 3(1)^2}{h} = \dfrac{3(1 + 2h + h^2) - 3}{h}$

$\quad\quad = \dfrac{3 + 6h + 3h^2 - 3}{h} = \dfrac{h(6 + 3h)}{h} = 6 + 3h;$

As $h \to 0$, $6 + 3h \to 6$.
The slope of the graph at $x = 1$ is 6.

39.

h	-0.1	-0.01	-0.001	$\to$	0	$\leftarrow$	0.001	0.01	0.1
$\dfrac{f(0 + h) - f(0)}{h}$	-1	-1	-1	$\to$ -1	$\neq$	1 $\leftarrow$	1	1	1

The slope of the graph is not defined at (0, 0).

41. (A) Average rate of change, 1975—1985: $\dfrac{34.9 - 36.1}{10} = -0.12$ hrs/yr

  (B) Average rate of change, 1975—1990: $\dfrac{10.01 - 4.53}{15} = \$0.37$ per yr

43. (A) $R(1000) = 60(1000) - 0.025(1000)^2 = \$35,000$

$\dfrac{R(1000 + h) - R(1000)}{h} = \dfrac{60(1000 + h) - 0.025(1000 + h)^2 - 35,000}{h}$

$= \dfrac{60,000 + 60h - 0.025(1,000,000 + 2000h + h^2) - 35,000}{h}$

$= \dfrac{10h - 0.025h^2}{h} = 10 - 0.025h \to 10$ as $h \to 0$

At a production level of 1,000 car seats, the revenue is \$35,000 and is INCREASING at the rate of \$10 per seat.

(B) $R(1300) = 60(1300) - 0.025(1300)^2 = \$35,750$

$$\frac{R(1300 + h) - R(1300)}{h} = \frac{60(1300 + h) - 0.025(1300 + h)^2 - 35,750}{h}$$

$$= \frac{78,000 + 60h - 0.025(1,690,000 + 2600h + h^2) - 35,750}{h}$$

$$= \frac{-5h - 0.025h^2}{h} = -5 - 0.025h \rightarrow -5 \text{ as } h \rightarrow 0$$

At a production level of 1,300 car seats, the revenue is \$35,750 and is DECREASING at the rate of \$5 per seat.

45. $f(3) = -150(3)^2 + 770(3) + 10,400 = 11,360.$

$$\frac{f(3 + h) - f(3)}{h} = \frac{-150(3 + h)^2 + 770(3 + h) + 10,400 - 11,360}{h}$$

$$= \frac{-150(9 + 6h + h^2) + 2310 + 770h + 10,400 - 11,360}{h}$$

$$= \frac{-130h - 150h^2}{h} = -130 - 150h \rightarrow -130 \text{ as } h \rightarrow 0.$$

In 1990, the annual production was 11,360,000 metric tons and was DECREASING at the rate of 130,000 metric tons per year.

47. (A) Average rate of change, 1988—1991: $\frac{728.6 - 526.2}{3} \approx \$67.5 \text{ billion/yr}$

(B) Average rate of change, 1987—1989: $\frac{20.7 - 17.3}{2} = \$1.7 \text{ billion/yr}$

49. $f(30) = 0.008(30)^2 - 0.9(30) + 29.6 = 9.8$

$$\frac{f(30 + h) - f(30)}{h} = \frac{0.008(30 + h)^2 - 0.9(30 + h) + 29.6 - 9.8}{h}$$

$$= \frac{0.008(900 + 60h + h^2) - 27 - 0.9h + 29.6 - 9.8}{h}$$

$$= \frac{-0.42h + 0.008h^2}{h} = -0.42 + 0.008h$$

In 1990, the number of male infant deaths per 100,000 births was 9.8 and was DECREASING at the rate of 0.42 deaths per 100,000 births per year.

EXERCISE 8-2

Things to remember:

1. LIMIT

We write

$$\lim_{x \rightarrow c} f(x) = L \text{ or } f(x) \rightarrow L \text{ as } x \rightarrow c$$

if the functional value $f(x)$ is close to the single real number L whenever x is close to but not equal to c (on either side of c).

[Note: The existence of a limit at c has nothing to do with the value of the function at c. In fact, c may not even be in the domain of f (see Examples 2 and 3). However, the function must be defined on both sides of c.]

2. ONE-SIDED LIMITS

We write $\lim\limits_{x \to c^-} f(x) = K$ [$x \to c^-$ is read "x approaches c from the left" and means $x \to c$ and $x < c$] and call K the LIMIT FROM THE LEFT or LEFT-HAND LIMIT if $f(x)$ is close to K whenever x is close to c, but to the left of c on the real number line.

We write $\lim\limits_{x \to c^+} f(x) = L$ [$x \to c^+$ is read "x approaches c from the right" and means $x \to c$ and $x > c$] and call L the LIMIT FROM THE RIGHT or RIGHT-HAND LIMIT if $f(x)$ is close to L whenever x is close to c, but to the right of c on the real number line.

3. EXISTENCE OF A LIMIT

In order for a limit to exist, the limit from the left and the limit from the right must both exist, and must be equal.

4. PROPERTIES OF LIMITS

Let f and g be two functions and assume that

$$\lim_{x \to c} f(x) = L \qquad \lim_{x \to c} g(x) = M$$

where L and M are real numbers (both limits exist). Then,

(a) $\lim\limits_{x \to c}[f(x) + g(x)] = \lim\limits_{x \to c} f(x) + \lim\limits_{x \to c} g(x) = L + M,$

(b) $\lim\limits_{x \to c}[f(x) - g(x)] = \lim\limits_{x \to c} f(x) - \lim\limits_{x \to c} g(x) = L - M,$

(c) $\lim\limits_{x \to c} kf(x) = k \lim\limits_{x \to c} f(x) = kL$ for any constant $k,$

(d) $\lim\limits_{x \to c}[f(x)g(x)] = \left(\lim\limits_{x \to c} f(x)\right)\left(\lim\limits_{x \to c} g(x)\right) = LM,$

(e) $\lim\limits_{x \to c} \dfrac{f(x)}{g(x)} = \dfrac{\lim\limits_{x \to c} f(x)}{\lim\limits_{x \to c} g(x)} = \dfrac{L}{M}$ if $M \neq 0,$

(f) $\lim\limits_{x \to c} \sqrt[n]{f(x)} = \sqrt[n]{\lim\limits_{x \to c} f(x)} = \sqrt[n]{L}$ ($L \geq 0$ for n even).

5. LIMIT OF A POLYNOMIAL FUNCTION

If $f(x)$ is a polynomial function and c is any real number, then

$$\lim_{x \to c} f(x) = f(c)$$

1.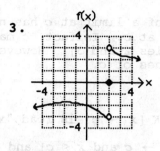

3.

5.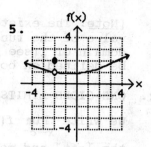

7. (A) $\lim\limits_{x \to 0^-} f(x) = 2$ (B) $\lim\limits_{x \to 0^+} f(x) = 2$ (C) $\lim\limits_{x \to 0} f(x) = 2$ (D) $f(0) = 2$

9. (A) $\lim\limits_{x \to 2^-} f(x) = 1$ (B) $\lim\limits_{x \to 2^+} f(x) = 2$ (C) $\lim\limits_{x \to 2} f(x)$ does not exist

   (D) $f(2) = 2$

11. (A) $\lim\limits_{x \to 1^-} g(x) = 1$ (B) $\lim\limits_{x \to 1^+} g(x) = 2$ (C) $\lim\limits_{x \to 1} g(x) =$ does not exist

   (D) $g(1)$ does not exist

13. (A) $\lim\limits_{x \to 3^-} g(x) = 1$ (B) $\lim\limits_{x \to 3^+} g(x) = 1$ (C) $\lim\limits_{x \to 3} g(x) = 1$ (D) $g(3) = 3$

15. $\lim\limits_{x \to 3}[f(x) - g(x)] = \lim\limits_{x \to 3} f(x) - \lim\limits_{x \to 3} g(x)$ [Property $\underline{4}$(b)]
$$= 5 - 9 = -4$$

17. $\lim\limits_{x \to 3} 4g(x) = 4 \lim\limits_{x \to 3} g(x)$ [Property $\underline{4}$(c)]
$$= 4 \cdot 9 = 36$$

19. $\lim\limits_{x \to 3} \dfrac{f(x)}{g(x)} = \dfrac{\lim\limits_{x \to 3} f(x)}{\lim\limits_{x \to 3} g(x)}$ [since $\lim\limits_{x \to 3} g(x) \neq 0$, Property $\underline{4}$(e)]
$$= \frac{5}{9}$$

21. $\lim\limits_{x \to 3} \sqrt{f(x)} = \sqrt{\lim\limits_{x \to 3} f(x)} = \sqrt{5}$ [Property $\underline{4}$(f)]

23. $\lim\limits_{x \to 3} \dfrac{f(x) + g(x)}{2f(x)} = \dfrac{\lim\limits_{x \to 3}[f(x) + g(x)]}{\lim\limits_{x \to 3} 2f(x)}$ [since $\lim\limits_{x \to 3} 2f(x) \neq 0$]
$$= \frac{\lim\limits_{x \to 3} f(x) + \lim\limits_{x \to 3} g(x)}{2 \cdot \lim\limits_{x \to 3} f(x)} = \frac{5 + 9}{2 \cdot 5} = \frac{14}{10} = \frac{7}{5} = 1.4$$

25. $\lim\limits_{x \to 5}(2x^2 - 3) = 2(5)^2 - 3$ [since $f(x) = 2x^2 - 3$ is continuous at $x = 5$]
$$= 47$$

27. $\lim\limits_{x \to 2} \dfrac{5x}{2 + x^2} = \dfrac{5(2)}{2 + (2)^2} \quad \left[f(x) = \dfrac{5x}{2 + x^2} \text{ is continuous at } x = 2 \right]$

$\qquad\qquad = \dfrac{10}{6} = \dfrac{5}{3}$

29. $\lim\limits_{x \to 2} (x + 1)^3 (2x - 1)^2 = \lim\limits_{x \to 2} (x + 1)^3 \cdot \lim\limits_{x \to 2} (2x - 1)^2$

$\qquad\qquad\qquad = (2 + 1)^3 \cdot (2 \cdot 2 - 1)^2 = 3^3 \cdot 3^2 = 3^5 = 243$

31. $\lim\limits_{x \to -1} \sqrt{5 - 4x} = \sqrt{\lim\limits_{x \to -1} (5 - 4x)} = \sqrt{5 - 4(-1)} = \sqrt{9} = 3$

33. Since $\dfrac{x^2 - 9}{x + 3} = \dfrac{(x + 3)(x - 3)}{x + 3} = x - 3, \; x \neq -3,$

$\qquad \lim\limits_{x \to -3} \dfrac{x^2 - 9}{x + 3} = \lim\limits_{x \to -3} (x - 3) = -6.$

35. For $x > 1$, $|x - 1| = x - 1$. Thus, $\dfrac{|x - 1|}{x - 1} = \dfrac{x - 1}{x - 1} = 1$ for $x > 1$;

$\qquad$ and $\lim\limits_{x \to 1^+} \dfrac{|x - 1|}{x - 1} = 1.$

37. For $x < 1$, $|x - 1| = -(x - 1)$. Thus, $\dfrac{|x - 1|}{x - 1} = \dfrac{-(x - 1)}{x - 1} = -1$

$\qquad$ for $x < 1$, and $\lim\limits_{x \to 1^-} \dfrac{|x - 1|}{x - 1} = -1.$

39. It follows from Problems 35 and 37 that $\lim\limits_{x \to 1} \dfrac{|x - 1|}{x - 1}$ does not exist.

41. $\lim\limits_{x \to 1} \dfrac{x - 2}{x^2 - 2x} = \dfrac{1 - 2}{1 - 2} = 1 \quad \left[\underline{\text{Note}}: f(x) = \dfrac{x - 2}{x^2 - 2x} \text{ is continuous at } x = 1. \right]$

43. $\lim\limits_{x \to 2} \dfrac{x - 2}{x^2 - 2x}$ is a 0/0 indeterminate form.

$\qquad$ Thus, we try to manipulate the expression algebraically.

$\qquad \dfrac{x - 2}{x^2 - 2x} = \dfrac{x - 2}{x(x - 2)} = \dfrac{1}{x}, \; x \neq 2$

$\qquad$ Now, $\lim\limits_{x \to 2} \dfrac{x - 2}{x^2 - 2x} = \lim\limits_{x \to 2} \dfrac{1}{x} = \dfrac{1}{2}.$

45. $\lim\limits_{x \to 2} \dfrac{x^2 - x - 6}{x + 2} = \dfrac{2^2 - 2 - 6}{2 + 2} = -1$

47. $\lim\limits_{x \to -2} \dfrac{x^2 - x - 6}{x + 2}$ is a 0/0 indeterminate form.

$\qquad \dfrac{x^2 - x - 6}{x + 2} = \dfrac{(x - 3)(x + 2)}{x + 2} = x - 3, \; x \neq -2$

$\qquad$ Thus, $\lim\limits_{x \to -2} \dfrac{x^2 - x - 6}{x + 2} = \lim\limits_{x \to -2} (x - 3) = -5.$

49. $\lim\limits_{x \to 3}\left(\dfrac{x}{x+3} + \dfrac{x-3}{x^2-9}\right) = \lim\limits_{x \to 3}\dfrac{x}{x+3} + \lim\limits_{x \to 3}\dfrac{x-3}{x^2-9} = \dfrac{3}{6} + \lim\limits_{x \to 3}\dfrac{x-3}{x^2-9}$

Now, $\dfrac{x-3}{x^2-9} = \dfrac{x-3}{(x-3)(x+3)} = \dfrac{1}{x+3}$, $x \neq 3$.

Thus, $\lim\limits_{x \to 3}\left(\dfrac{x}{x+3} + \dfrac{x-3}{x^2-9}\right) = \dfrac{1}{2} + \lim\limits_{x \to 3}\dfrac{1}{x+3} = \dfrac{1}{2} + \dfrac{1}{6} = \dfrac{2}{3}$.

51. $\lim\limits_{x \to 0}\left(\sqrt{x^2+9} - \dfrac{x^2+3x}{x}\right) = \lim\limits_{x \to 0}\sqrt{x^2+9} - \lim\limits_{x \to 0}\dfrac{x(x+3)}{x}$

$\qquad\qquad = \sqrt{\lim\limits_{x \to 0}(x^2+9)} - \lim\limits_{x \to 0}(x+3) = \sqrt{9} - 3 = 0$

53. $f(x) = 3x + 1$

$\lim\limits_{h \to 0}\dfrac{f(2+h)-f(2)}{h} = \lim\limits_{h \to 0}\dfrac{3(2+h)+1-(3\cdot 2+1)}{h}$

$\qquad\qquad = \lim\limits_{h \to 0}\dfrac{6+3h+1-7}{h} = \lim\limits_{h \to 0}\dfrac{3h}{h} = \lim\limits_{h \to 0}3 = 3$

55. $f(x) = x^2 + 1$

$\lim\limits_{h \to 0}\dfrac{f(2+h)-f(2)}{h} = \lim\limits_{h \to 0}\dfrac{(2+h)^2+1-(2^2+1)}{h}$

$\qquad\qquad = \lim\limits_{h \to 0}\dfrac{4+4h+h^2+1-5}{h} = \lim\limits_{h \to 0}\dfrac{4h+h^2}{h}$

$\qquad\qquad = \lim\limits_{h \to 0}(4+h) = 4$

57. $f(x) = 5$

$\lim\limits_{h \to 0}\dfrac{f(2+h)-f(2)}{h} = \lim\limits_{h \to 0}\dfrac{5-5}{h} = \lim\limits_{h \to 0}0 = 0$

59. $f(x) = \sqrt{x} - 2$

$\lim\limits_{h \to 0}\dfrac{f(2+h)-f(2)}{h} = \lim\limits_{h \to 0}\dfrac{\sqrt{2+h}-2-(\sqrt{2}-2)}{h} = \lim\limits_{h \to 0}\dfrac{\sqrt{2+h}-\sqrt{2}}{h}$

$\qquad\qquad = \lim\limits_{h \to 0}\dfrac{\sqrt{2+h}-\sqrt{2}}{h} \cdot \dfrac{\sqrt{2+h}+\sqrt{2}}{\sqrt{2+h}+\sqrt{2}} = \lim\limits_{h \to 0}\dfrac{2+h-2}{h(\sqrt{2+h}+\sqrt{2})}$

$\qquad\qquad = \lim\limits_{h \to 0}\dfrac{h}{h(\sqrt{2+h}+\sqrt{2})} = \lim\limits_{h \to 0}\dfrac{1}{\sqrt{2+h}+\sqrt{2}} = \dfrac{1}{2\sqrt{2}}$

61. $f(x) = |x-2| - 3$

$\lim\limits_{h \to 0}\dfrac{f(2+h)-f(2)}{h} = \lim\limits_{h \to 0}\dfrac{|(2+h)-2|-3-(|2-2|-3)}{h}$

$\qquad\qquad = \lim\limits_{h \to 0}\dfrac{|h|-3+3}{h} = \lim\limits_{h \to 0}\dfrac{|h|}{h}$ does not exist.

63. Slope of the graph of $y = x^2 - 3$ at $(2, 1)$:

$\lim\limits_{h \to 0}\dfrac{f(2+h)-f(2)}{h} = \lim\limits_{h \to 0}\dfrac{(2+h)^2-3-1}{h} = \lim\limits_{h \to 0}\dfrac{4+4h+h^2-4}{h}$

$\qquad\qquad\qquad\qquad\qquad = \lim\limits_{h \to 0}4+h = 4$

65. Slope of the graph of $y = \sqrt{x}$ at $(4, 2)$:

$$\lim_{h \to 0} \frac{\sqrt{4 + h} - \sqrt{4}}{h} = \lim_{h \to 0} \frac{\sqrt{4 + h} - 2}{h} \cdot \frac{\sqrt{4 + h} + 2}{\sqrt{4 + h} + 2} = \lim_{h \to 0} \frac{4 + h - 4}{h[\sqrt{4 + h} + 2]}$$

$$= \lim_{h \to 0} \frac{1}{\sqrt{4 + h} + 2} = \frac{1}{4}$$

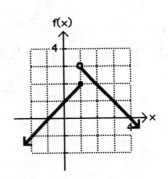

67. Instantaneous velocity at $x = 4$:

$$\lim_{h \to 0} \frac{f(4 + h) - f(4)}{h} = \lim_{h \to 0} \frac{10(4 + h)^2 - 160}{h} = \lim_{h \to 0} \frac{10(16 + 8h + h^2) - 160}{h}$$

$$= \lim_{h \to 0} \frac{80h + 10h^2}{h} = \lim_{h \to 0} 80 + 10h = 80 \text{ ft/sec}$$

69. At $x = -3$, slope $= -1$; at $x = -1$, slope $= 0$; at $x = 3$, slope $= 2$.

71. (A)

x	-0.1	-0.01	-0.001	-0.0001	$\to 0$
$\frac{1}{x}$	-10	-100	-1000	$-10,000$	

The limit does not exist; the values of $\frac{1}{x}$ are increasingly large negative numbers as x approaches 0 from the left.

(B)

x	$0 \leftarrow$	0.0001	0.001	0.01	0.1
$\frac{1}{x}$		$10,000$	1000	100	10

The limit does not exist; the values of $\frac{1}{x}$ are increasingly large positive numbers as x approaches 0 from the right.

73. (A) $\lim\limits_{x \to 1^-} f(x) = \lim\limits_{x \to 1^-} (1 + x) = 2$

$\lim\limits_{x \to 1^+} f(x) = \lim\limits_{x \to 1^+} (4 - x) = 3$

(B) $\displaystyle\lim_{x\to 1^-} f(x) = \lim_{x\to 1^-}(1 + 2x) = 3$

$\displaystyle\lim_{x\to 1^+} f(x) = \lim_{x\to 1^+}(4 - 2x) = 2$

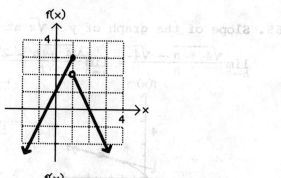

(C) $\displaystyle\lim_{x\to 1^-} f(x) = \lim_{x\to 1^-}(1 + mx) = 1 + m$

$\displaystyle\lim_{x\to 1^+} f(x) = \lim_{x\to 1^+}(4 - mx) = 4 - m$

$1 + m = 4 - m$

$2m = 3$

$m = \dfrac{3}{2}$

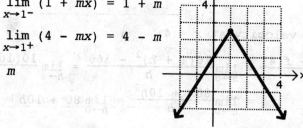

(D) The graph in (A) is broken at $x = 1$; it jumps up from $(1, 2)$ to $(1, 3)$.

The graph in (B) is also broken at $x = 1$; it jumps down from $(1, 3)$ to $(1, 2)$.

The graph in (C) is not broken; the two pieces meet at $\left(1, \dfrac{5}{2}\right)$.

75. $\displaystyle\lim_{h\to 0}\frac{(a + h)^2 - a^2}{h} = \lim_{h\to 0}\frac{a^2 + 2ah + h^2 - a^2}{h}$

$\qquad = \displaystyle\lim_{h\to 0}\frac{2ah + h^2}{h} = \lim_{h\to 0}(2a + h) = 2a$

77. $\displaystyle\lim_{h\to 0}\frac{\sqrt{a + h} - \sqrt{a}}{h} = \lim_{h\to 0}\frac{\sqrt{a + h} - \sqrt{a}}{h}\cdot\frac{\sqrt{a + h} + \sqrt{a}}{\sqrt{a + h} + \sqrt{a}} = \lim_{h\to 0}\frac{(a + h) - a}{h(\sqrt{a + h} + \sqrt{a})}$

$\qquad\qquad = \displaystyle\lim_{h\to 0}\frac{1}{\sqrt{a + h} + \sqrt{a}} = \frac{1}{2\sqrt{a}}$

79. (A) At $a = 1$,

$\displaystyle\lim_{h\to 0}\frac{f(1 + h) - f(1)}{h} = \lim_{h\to 0}\frac{(1 + h)^2 - 3(1 + h) + 1 - (-1)}{h}$

$\qquad\qquad = \displaystyle\lim_{h\to 0}\frac{1 + 2h + h^2 - 3 - 3h + 2}{h}$

$\qquad\qquad = \displaystyle\lim_{h\to 0}\frac{-h + h^2}{h} = \lim_{h\to 0}(-1 + h) = -1$

At $a = 2$,

$$\lim_{h \to 0} \frac{f(2 + h) - f(2)}{h} = \lim_{h \to 0} \frac{(2 + h)^2 - 3(2 + h) + 1 - (-1)}{h}$$

$$= \lim_{h \to 0} \frac{4 + 4h + h^2 - 6 - 3h + 2}{h}$$

$$= \lim_{h \to 0} \frac{h + h^2}{h} = \lim_{h \to 0} (1 + h) = 1$$

At $a = 3$,

$$\lim_{h \to 0} \frac{f(3 + h) - f(3)}{h} = \lim_{h \to 0} \frac{(3 + h)^2 - 3(3 + h) + 1 - (1)}{h}$$

$$= \lim_{h \to 0} \frac{9 + 6h + h^2 - 9 - 3h}{h}$$

$$= \lim_{h \to 0} \frac{3h + h^2}{h} = \lim_{h \to 0} (3 + h) = 3$$

(B) $\lim_{h \to 0} \dfrac{f(a + h) - f(a)}{h} = \lim_{h \to 0} \dfrac{(a + h)^2 - 3(a + h) + 1 - [a^2 - 3a + 1]}{h}$

$$= \lim_{h \to 0} \frac{a^2 + 2ah + h^2 - 3a - 3h - a^2 + 3a}{h}$$

$$= \lim_{h \to 0} \frac{(2a - 3)h + h^2}{h}$$

$$= \lim_{h \to 0} (2a - 3 + h) = 2a - 3$$

(C) At $a = 1$, $2(1) - 3 = -1$; at $a = 2$, $2(2) - 3 = 1$;
at $a = 3$, $2(3) - 3 = 3$

The slopes are the same as in part (A). In part (A), each value
required a new limit operation. In part (B) a single limit operation
produces the slope for all values of a.

81.

x	0.9	0.99	0.999	→ 1 ←	1.001	1.01	1.1
$\dfrac{x^{10} - 1}{x - 1}$	6.513	9.562	9.955	→ 10 ←	10.045	10.462	15.937

$$\lim_{x \to 1} \frac{x^{10} - 1}{x - 1} = 10$$

83.

x	−0.1	−0.01	−0.001	→ 0	← 0.001	0.01	0.1
$\dfrac{2^x - 1}{x}$	0.670	0.691	0.693	→ 0.693	← 0.693	0.696	0.718

$$\lim_{x \to 0} \frac{2^x - 1}{x} \approx 0.693$$

85.

x	-0.1	-0.01	-0.001	→ 0	← 0.001	0.01	0.1
$(1 + x)^{1/x}$	2.868	2.732	2.7196	→ 2.718	← 2.7169	2.705	2.594

$$\lim_{x \to 0} (1 + x)^{1/x} \approx 2.718$$

87. Typical values of n are 95 on a TI-81, 94 on a TI-82 and 126 on a TI-85.

89. $\lim_{x \to -2^-} f(x) = \lim_{x \to -2^+} f(x) = 2$

$\lim_{x \to 2^-} f(x) = \lim_{x \to 2^+} f(x) = 2$

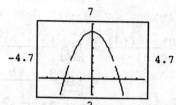

91. $\lim_{x \to 2^-} f(x) = -4$, $\lim_{x \to 2^+} f(x) = 4$ **93.** $\lim_{x \to -3^-} f(x) = -3$, $\lim_{x \to -3^+} f(x) = 3$,

$\lim_{x \to 3^-} f(x) = -3$, $\lim_{x \to 3^+} f(x) = 3$

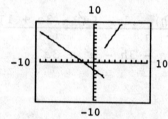

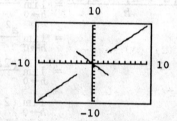

95. (A) $R(900) = 200(900) - 0.1(900)^2 = \$99,000$

$$\lim_{h \to 0} \frac{R(900 + h) - R(900)}{h} = \lim_{h \to 0} \frac{200(900 + h) - 0.1(900 + h)^2 - 99,000}{h}$$

$$= \lim_{h \to 0} \frac{180,000 + 200h - 0.1(810,000 + 1800h + h^2) - 99,000}{h}$$

$$= \lim_{h \to 0} \frac{20h - 0.1h^2}{h} = \lim_{h \to 0} (20 - 0.1h) = 20$$

At a production level of 900 units, the revenue is $99,000 and is INCREASING at the rate of $20 per jigsaw.

(B) $R(1,200) = 200(1,200) - 0.1(1,200)^2 = \$96,000$

$$\lim_{h \to 0} \frac{R(1,200 + h) - R(1,200)}{h}$$

$$= \lim_{h \to 0} \frac{200(1,200 + h) - 0.1(1,200 + h)^2 - 96,000}{h}$$

$$= \lim_{h \to 0} \frac{-40h - 0.1h^2}{h} = \lim_{h \to 0} (-40 - 0.1h) = -40$$

At a production level of 1,200 units, the revenue is $96,000 and is DECREASING at the rate of $40 per jigsaw.

97. $f(20) = 0.62(20)^2 - 20 + 5.1 = \$233.1$ billion

$$\lim_{h \to 0} \frac{f(20 + h) - f(20)}{h} = \lim_{h \to 0} \frac{0.62(20 + h)^2 - (20 + h) + 5.1 - 233.1}{h}$$

$$= \lim_{h \to 0} \frac{0.62(400 + 40h + h^2) - 20 - h + 5.1 - 233.1}{h}$$

$$= \lim_{h \to 0} \frac{23.8h + 0.62h^2}{h} = \lim_{h \to 0} (23.8 + 0.62h) = 23.8$$

In 1990, the revolving credit debt was \$233.1 bilion and was INCREASING at the rate of \$23.8 billion per year.

99. At $x = 9$(thousand) $y = 0.04(9)^2 - 3.66(9) + 100 = 70.3\%$.
Let $f(x) = 0.04x^2 - 3.66x + 100$. Then

$$\lim_{h \to 0} \frac{f(9 + h) - f(9)}{h} = \lim_{h \to 0} \frac{0.04(9 + h)^2 - 3.66(9 + h) + 100 - 70.3}{h}$$

$$= \lim_{h \to 0} \frac{0.04(81 + 18h + h^2) - 32.94 - 3.66h + 29.7}{h}$$

$$= \lim_{h \to 0} \frac{-2.94h + 0.04h^2}{h} = \lim_{h \to 0} (-2.94 + 0.04h)$$

$$= -2.94$$

The aveolar pressure at 9,000 feet is 70.3% of that at sea level and is DECREASING at the rate of 2.94% per 1,000 feet.

101. Let $y = f(t) = -0.03t^2 + 1.5t + 32$.

(A) $f(20) = -0.03(20)^2 + 1.5(20) + 32 = 50$ million

$$\lim_{h \to 0} \frac{f(20 + h) - f(20)}{h} = \lim_{h \to 0} \frac{-0.03(20 + h)^2 + 1.5(20 + h) + 32 - 50}{h}$$

$$= \lim_{h \to 0} \frac{-0.03(400 + 40h + h^2) + 30 + 1.5h - 18}{h}$$

$$= \lim_{h \to 0} \frac{0.3h - 0.03h^2}{h} = \lim_{h \to 0} (0.3 - 0.03h)$$

$$= 0.3$$

The school age population in 1970 was 50 million and was INCREASING at the rate of 0.3 million per year.

(B) $f(40) = -0.03(40)^2 + 1.5(40) + 32 = 44$ million

$$\lim_{h \to 0} \frac{f(40 + h) - f(40)}{h} = \lim_{h \to 0} \frac{-0.03(40 + h)^2 + 1.5(40 + h) + 32 - 44}{h}$$

$$= \lim_{h \to 0} \frac{-0.03(1,600 + 80h + h^2) + 60 + 1.5h - 12}{h}$$

$$= \lim_{h \to 0} \frac{-0.9h - 0.03h^2}{h} = \lim_{h \to 0} (-0.9 - 0.03)h$$

$$= -0.9$$

The school age population in 1990 was 44 million and was DECREASING at the rate of 0.9 million per year.

Things to remember:

1. THE DERIVATIVE

    For $y = f(x)$, we define THE DERIVATIVE OF f AT x, denoted by $f'(x)$, to be

    $$f'(x) = \lim_{h \to 0} \frac{f(x + h) - f(x)}{h} \quad \text{if the limit exists.}$$

    If $f'(x)$ exists for each x in the open interval (a, b), then f is said to be DIFFERENTIABLE OVER (a, b).

2. INTERPRETATIONS OF THE DERIVATIVE

    The derivative of a function f is a new function f'. The domain of f' is a subset of the domain of f. Interpretations of the derivative are:

    a. Slope of the tangent line. For each x in the domain of f', $f'(x)$ is the slope of the line tangent to the graph of f at the point $(x, f(x))$.

    b. Instantaneous rate of change. For each x in the domain of f', $f'(x)$ is the instantaneous rate of change of $y = f(x)$ with respect to x.

---

1. (A) $\dfrac{f(1) - f(0)}{1 - 0} = \dfrac{0 - (-1)}{1} = 1$; slope of secant line through $(0, f(0))$ and $(1, f(1))$.

    (B) $\dfrac{f(1 + h) - f(1)}{h} = \dfrac{(1 + h)^2 - 1 - 0}{h} = \dfrac{1 + 2h + h^2 - 1}{h} = \dfrac{h(2 + h)}{h}$
    $= 2 + h$, $h \neq 0$, slope of secant line through $(1 + h, f(1 + h))$ and $(1, f(1))$.

    (C) $\lim\limits_{h \to 0} \dfrac{f(1 + h) - f(1)}{h} = \lim\limits_{h \to 0} \dfrac{(1 + h)^2 - 1 - 0}{h} = \lim\limits_{h \to 0} (2 + h) = 2$;
    slope of tangent line at $(1, f(1))$.

3. $\lim\limits_{h \to 0} \dfrac{f(x + h) - f(x)}{h} = \lim\limits_{h \to 0} \dfrac{4hx - 3h + 2h^2}{h} = \lim\limits_{h \to 0} \dfrac{h(4x - 3 + 2h)}{h}$
    $$= \lim\limits_{h \to 0} (4x - 3 + 2h) = 4x - 3$$

5. $\lim\limits_{h \to 0} \dfrac{f(x + h) - f(x)}{h} = \lim\limits_{h \to 0} \dfrac{3hx^2 - 2xh + 3h^2x - h^2 + h^3}{h}$
    $$= \lim\limits_{h \to 0} \dfrac{h(3x^2 - 2x + 3hx - h + h^2)}{h}$$
    $$= \lim\limits_{h \to 0} (3x^2 - 2x + 3hx - h + h^2) = 3x^2 - 2x$$

7. $f(x) = 2x - 3$

   Step 1. Simplify $\dfrac{f(x + h) - f(x)}{h}$.

$$\frac{f(x + h) - f(x)}{h} = \frac{2(x + h) - 3 - (2x - 3)}{h}$$

$$= \frac{2x + 2h - 3 - 2x + 3}{h} = \frac{2h}{h} = 2$$

   Step 2. Evaluate $\lim\limits_{h \to 0} \dfrac{f(x + h) - f(x)}{h}$.

$$\lim_{h \to 0} \frac{f(x + h) - f(x)}{h} = \lim_{h \to 0} 2 = 2$$

   Thus, $f'(x) = 2$. Now $f'(1) = 2$, $f'(2) = 2$, $f'(3) = 2$.

9. $f(x) = 2 - x^2$

   Step 1. Simplify $\dfrac{f(x + h) - f(x)}{h}$.

$$\frac{f(x + h) - f(x)}{h} = \frac{2 - (x + h)^2 - (2 - x^2)}{h}$$

$$= \frac{2 - (x^2 + 2xh + h^2) - 2 + x^2}{h}$$

$$= \frac{-2xh - h^2}{h} = -2x - h$$

   Step 2. Evaluate $\lim\limits_{h \to 0} \dfrac{f(x + h) - f(x)}{h}$.

$$\lim_{h \to 0} \frac{f(x + h) - f(x)}{h} = \lim_{h \to 0} (-2x - h) = -2x$$

   Thus, $f'(x) = -2x$. Now $f'(1) = -2$, $f'(2) = -4$, $f'(3) = -6$.

11. At $x = -3$, $f'(-3) = -1$; at $x = 3$, $f'(3) = 2$; $f'(-1) = 0$

13. At $x = -5$, $f'(-5) = -3$; at $x = 1$, $f'(1) = 1$; at $x = 7$, $f'(7) = -3$;
   $f'(-2) = 0$, $f'(4) = 0$

15. $y = f(x) = x^2 + x$

   (A) $f(1) = 1^2 + 1 = 2$, $f(3) = 3^2 + 3 = 12$

   Slope of secant line: $\dfrac{f(3) - f(1)}{3 - 1} = \dfrac{12 - 2}{2} = 5$

   (B) $f(1) = 2$, $f(1 + h) = (1 + h)^2 + (1 + h) = 1 + 2h + h^2 + 1 + h$
   $= 2 + 3h + h^2$

   Slope of secant line: $\dfrac{f(1 + h) - f(1)}{h} = \dfrac{2 + 3h + h^2 - 2}{h} = 3 + h$

   (C) Slope of tangent line at $(1, f(1))$:

$$\lim_{h \to 0} \frac{f(1 + h) - f(1)}{h} = \lim_{h \to 0} (3 + h) = 3$$

   (D) Equation of tangent line at $(1, f(1))$:
   $y - f(1) = f'(1)(x - 1)$ or $y - 2 = 3(x - 1)$ and $y = 3x - 1$.

17. $f(x) = x^2 + x$

(A) Average velocity: $\dfrac{f(3) - f(1)}{3 - 1} = \dfrac{3^2 + 3 - (1^2 + 1)}{2} = \dfrac{12 - 2}{2}$

$$= 5 \text{ meters/sec.}$$

(B) Average velocity: $\dfrac{f(1 + h) - f(1)}{h} = \dfrac{(1 + h)^2 + (1 + h) - (1^2 + 1)}{h}$

$$= \dfrac{1 + 2h + h^2 + 1 + h - 2}{h}$$

$$= \dfrac{3h + h^2}{h} = 3 + h \text{ meters/sec.}$$

(C) Instantaneous velocity: $\displaystyle\lim_{h \to 0} \dfrac{f(1 + h) - f(1)}{h}$

$$= \lim_{h \to 0}(3 + h) = 3 \text{ meters/sec.}$$

19. $f(x) = 6x - x^2$

<u>Step 1</u>. Simplify $\dfrac{f(x + h) - f(x)}{h}$.

$$\dfrac{f(x + h) - f(x)}{h} = \dfrac{6(x + h) - (x + h)^2 - (6x - x^2)}{h}$$

$$= \dfrac{6x + 6h - (x^2 + 2xh + h^2) - 6x + x^2}{h}$$

$$= \dfrac{6h - 2xh - h^2}{h} = 6 - 2x - h$$

<u>Step 2</u>. Evaluate $\displaystyle\lim_{h \to 0} \dfrac{f(x + h) - f(x)}{h}$.

$$\lim_{h \to 0} \dfrac{f(x + h) - f(x)}{h} = \lim_{h \to 0}(6 - 2x - h) = 6 - 2x$$

Therefore, $f'(x) = 6 - 2x$. $f'(1) = 6 - 2(1) = 4$, $f'(2) = 6 - 2(2) = 2$, $f'(3) = 6 - 2(3) = 0$.

21. $f(x) = \sqrt{x} - 3$

<u>Step 1</u>. Simplify $\dfrac{f(x + h) - f(x)}{h}$

$$\dfrac{f(x + h) - f(x)}{h} = \dfrac{\sqrt{x + h} - 3 - (\sqrt{x} - 3)}{h}$$

$$= \dfrac{\sqrt{x + h} - \sqrt{x}}{h} \cdot \dfrac{\sqrt{x + h} + \sqrt{x}}{\sqrt{x + h} + \sqrt{x}}$$

$$= \dfrac{x + h - x}{h(\sqrt{x + h} + \sqrt{x})} = \dfrac{1}{\sqrt{x + h} + \sqrt{x}}$$

<u>Step 2</u>. Evaluate $\displaystyle\lim_{h \to 0} \dfrac{f(x + h) - f(x)}{h}$.

$$\lim_{h \to 0} \dfrac{f(x + h) - f(x)}{h} = \lim_{h \to 0} \dfrac{1}{\sqrt{x + h} + \sqrt{x}} = \dfrac{1}{2\sqrt{x}}$$

Therefore, $f'(x) = \dfrac{1}{2\sqrt{x}}$. $f'(1) = \dfrac{1}{2\sqrt{1}} = \dfrac{1}{2}$, $f'(2) = \dfrac{1}{2\sqrt{2}}$, $f'(3) = \dfrac{1}{2\sqrt{3}}$.

23. $f(x) = -\dfrac{1}{x}$

  Step 1. Simplify $\dfrac{f(x + h) - f(x)}{h}$.

$$\frac{f(x + h) - f(x)}{h} = \frac{-\dfrac{1}{x + h} - \left(-\dfrac{1}{x}\right)}{h} = \frac{-\dfrac{1}{x + h} + \dfrac{1}{x}}{h}$$

$$= \frac{\dfrac{-x + x + h}{x(x + h)}}{h} = \frac{1}{x(x + h)}$$

  Step 2. Evaluate $\displaystyle\lim_{h \to 0} \dfrac{f(x + h) - f(x)}{h}$.

$$\lim_{h \to 0} \frac{f(x + h) - f(x)}{h} = \lim_{h \to 0} \frac{1}{x(x + h)} = \frac{1}{x^2}$$

  Therefore, $f'(x) = \dfrac{1}{x^2}$. $f'(1) = \dfrac{1}{1^2} = 1$, $f'(2) = \dfrac{1}{2^2} = \dfrac{1}{4}$, $f'(3) = \dfrac{1}{3^2} = \dfrac{1}{9}$.

25. $F'(x)$ does exist at $x = a$.

27. $F'(x)$ does not exist at $x = c$; the graph has a vertical tangent line at $(c, F(c))$.

29. $F'(x)$ does not exist at $x = e$; F is not defined at $x = e$.

31. $F'(x)$ does exist at $x = g$.

33. $f(x) = x^2 - 4x$

  (A) Step 1. Simplify $\dfrac{f(x + h) - f(x)}{h}$.

$$\frac{f(x + h) - f(x)}{h} = \frac{(x + h)^2 - 4(x + h) - (x^2 - 4x)}{h}$$

$$= \frac{x^2 + 2xh + h^2 - 4x - 4h - x^2 + 4x}{h}$$

$$= \frac{2xh + h^2 - 4h}{h} = 2x - 4 + h$$

  Step 2. Evaluate $\displaystyle\lim_{h \to 0} \dfrac{f(x + h) - f(x)}{h}$.

$$\lim_{h \to 0} \frac{f(x + h) - f(x)}{h} = \lim_{h \to 0}(2x - 4 + h) = 2x - 4$$

  Therefore, $f'(x) = 2x - 4$.

  (B) $f'(0) = -4$, $f'(2) = 0$, $f'(4) = 4$

  (C) Since f is a quadratic function, the graph of f is a parabola.
  y intercept: $y = 0$
  x intercepts: $x = 0$, $x = 4$
  Vertex: $(2, -4)$

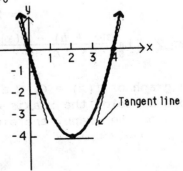

Tangent line

35. To find $v = f'(x)$, use the two-step process for the given distance function, $f(x) = 4x^2 - 2x$.

Step 1. $f(x + h) = 4(x + h)^2 - 2(x + h)$
$\qquad\qquad\quad = 4(x^2 + 2xh + h^2) - 2(x + h)$
$\qquad\qquad\quad = 4x^2 + 8xh + 4h^2 - 2x - 2h$

$f(x + h) - f(x) = (4x^2 + 8xh + 4h^2 - 2x - 2h) - (4x^2 - 2x)$
$\qquad\qquad\qquad\quad = 4x^2 + 8xh + 4h^2 - 2x - 2h - 4x^2 + 2x$
$\qquad\qquad\qquad\quad = 8xh + 4h^2 - 2h$
$\qquad\qquad\qquad\quad = h(8x + 4h - 2)$

$\dfrac{f(x + h) - f(x)}{h} = \dfrac{h(8x + 4h - 2)}{h}$
$\qquad\qquad\qquad\quad = 8x + 4h - 2, \; h \neq 0$

Step 2. $\lim\limits_{h \to 0} \dfrac{f(x + h) - f(x)}{h} = \lim\limits_{h \to 0}(8x + 4h - 2) = 8x - 2$

Thus, the velocity, $v = f'(x) = 8x - 2$
$\qquad\qquad\qquad\quad f'(1) = 8 \cdot 1 - 2 = 6$ feet per second
$\qquad\qquad\qquad\quad f'(3) = 8 \cdot 3 - 2 = 22$ feet per second
$\qquad\qquad\qquad\quad f'(5) = 8 \cdot 5 - 2 = 38$ feet per second

37.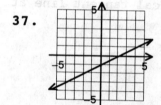

39.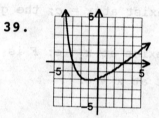

41. $f(x) = 2^x$; $f'(0) \approx 0.69$

43. $f(x) = \sqrt{2 + 2x - x^2}$; $f'(0) \approx 0.71$

45. (A) The graphs of g and h are vertical translations of the graph of f. All three functions should have the same derivative.

(B) Step 1. $\dfrac{m(x + h) - m(x)}{h} = \dfrac{(x + h)^2 + C - (x^2 + C)}{h}$

$\qquad\qquad\qquad\qquad = \dfrac{x^2 + 2xh + h^2 + C - x^2 - C}{h}$

$\qquad\qquad\qquad\qquad = \dfrac{2xh + h^2}{h} = \dfrac{h(2x + h)}{h}$

$\qquad\qquad\qquad\qquad = 2x + h \quad h \neq 0$

Step 2. $\lim\limits_{h \to 0} \dfrac{m(x + h) - m(x)}{h} = \lim\limits_{h \to 0}(2x + h) = 2x$; $m'(x) = 2x$

47. (A) The graph of $f(x) = C$, C a constant, is a horizontal line C units above or below the x axis depending on the sign of C. At any given point on the graph, the slope of the tangent line is 0.

(B) <u>Step 1</u>. $\dfrac{f(x + h) - f(x)}{h} = \dfrac{C - C}{h} = 0$

      <u>Step 2</u>. $\displaystyle\lim_{h \to 0} \dfrac{f(x + h) - f(x)}{h} = \lim_{h \to 0} 0 = 0$

49. The graph of $f(x) = \begin{cases} 2x, & x < 1 \\ 2, & x \geq 1 \end{cases}$ is:

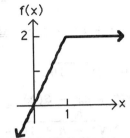

f is not differentiable at $x = 1$ because the graph of f has a sharp corner at this point.

51. $f(x) = |x|$

$\displaystyle\lim_{h \to 0} \dfrac{f(0 + h) - f(0)}{h} = \lim_{h \to 0} \dfrac{|0 + h| - |0|}{h} = \lim_{h \to 0} \dfrac{|h|}{h}$

The limit does not exist. Thus, f is not differentiable at $x = 0$.

53. $f(x) = \sqrt[3]{x} = x^{1/3}$

$\displaystyle\lim_{h \to 0} \dfrac{f(0 + h) - f(0)}{h} = \lim_{h \to 0} \dfrac{(0 + h)^{1/3} - 0^{1/3}}{h} = \lim_{h \to 0} \dfrac{h^{1/3}}{h} = \lim_{h \to 0} \dfrac{1}{h^{2/3}}$

The limit does not exist. Thus, f is not differentiable at $x = 0$.

55. $f(x) = 2x - x^2,\ 0 \leq x \leq 2$

    (A) For $0 < x < 2$:

$\displaystyle\lim_{h \to 0} \dfrac{f(x + h) - f(x)}{h} = \lim_{h \to 0} \dfrac{2(x + h) - (x + h)^2 - (2x - x^2)}{h}$

$\displaystyle \qquad\qquad = \lim_{h \to 0} \dfrac{2x + 2h - x^2 - 2xh - h^2 - 2x + x^2}{h}$

$\displaystyle \qquad\qquad = \lim_{h \to 0} \dfrac{2h - 2xh - h^2}{h} = \lim_{h \to 0} (2 - 2x - h) = 2 - 2x$

       Thus, $f'(x) = 2 - 2x,\ 0 < x < 2$.

    (B) For $x = 0$:

$\displaystyle\lim_{h \to 0^+} \dfrac{f(0 + h) - f(0)}{h} = \lim_{h \to 0^+} \dfrac{2(0 + h) - (0 + h)^2 - 0}{h}$

$\displaystyle \qquad\qquad = \lim_{h \to 0^+} \dfrac{2h - h^2}{h} = 2$

(C) For $x = 2$:

$$\lim_{h \to 0^-} \frac{f(2 + h) - f(2)}{h} = \lim_{h \to 0^-} \frac{2(2 + h) - (2 + h)^2 - (2 \cdot 2 - 2^2)}{h}$$

$$= \lim_{h \to 0^-} \frac{4 + 2h - 4 - 4h - h^2 - 0}{h}$$

$$= \lim_{h \to 0^-} -\frac{2h + h^2}{h} = -2$$

57. (A) $S(t) = 2\sqrt{t + 10}$

Step 1. $\dfrac{S(t + h) - S(t)}{h} = \dfrac{2\sqrt{t + h + 10} - 2\sqrt{t + 10}}{h}$

$$= \frac{2[\sqrt{t + h + 10} - \sqrt{t + 10}]}{h} \cdot \frac{\sqrt{t + h + 10} + \sqrt{t + 10}}{\sqrt{t + h + 10} + \sqrt{t + 10}}$$

$$= \frac{2[t + h + 10 - (t + 10)]}{h[\sqrt{t + h + 10} + \sqrt{t + 10}]} = \frac{2h}{h[\sqrt{t + h + 10} + \sqrt{t + 10}]}$$

$$= \frac{2}{\sqrt{t + h + 10} + \sqrt{t + 10}}$$

Step 2. $\displaystyle\lim_{h \to 0} \frac{S(t + h) - S(t)}{h} = \lim_{h \to 0} \frac{2}{\sqrt{t + h + 10} + \sqrt{t + 10}}$

$$= \frac{2}{2\sqrt{t + 10}} = \frac{1}{\sqrt{t + 10}};$$

$$S'(t) = \frac{1}{\sqrt{t + 10}}$$

(B) $S(15) = 2\sqrt{15 + 10} = 2\sqrt{25} = 10;$

$S'(15) = \dfrac{1}{\sqrt{15 + 10}} = \dfrac{1}{\sqrt{25}} = \dfrac{1}{5} = 0.2$

After 15 months, the total sales are $10 million and are INCREASING
at the rate of $0.2 million = $200,000 per month.

(C) The estimated total sales are $10.2 million after 16 months and
$10.4 million after 17 months.

59. (A) $A(t) = 100(1.06)^t$; $A(5) = 100(1.06)^5 = 133.82$

h	-0.1	-0.01	-0.001 → 0	← 0.001	0.01	0.1
$\dfrac{A(5 + h) - A(5)}{h}$	7.78	7.80	7.80 → 7.80 ← 7.80		7.80	7.82

$A'(5) \approx 7.80$

(B) After 5 years, the original $100 investment has grown to $133.82 and
is continuing to grow at the rate of $7.80 per year.

61. (A) In March, the price of the stock was $80 and was INCREASING at the
rate of $5 per month.

(B) The stock reached its highest price in April. The rate of change at
that point in time was 0.

63. (A) $P(t) = 80 + 12t - t^2$

Step 1. $\dfrac{P(t + h) - P(t)}{h} = \dfrac{80 + 12(t + h) - (t + h)^2 - [80 + 12t - t^2]}{h}$

$= \dfrac{80 + 12t + 12h - (t^2 + 2th + h^2) - 80 - 12t + t^2}{h}$

$= \dfrac{12h - 2th - h^2}{h} = \dfrac{h(12 - 2t - h)}{h} = 12 - 2t - h$

Step 2. $\lim\limits_{h \to 0} \dfrac{P(t + h) - P(t)}{h} = \lim\limits_{h \to 0} (12 - 2t - h) = 12 - 2t;$

$P'(t) = 12 - 2t$

(B) $P(3) = 80 + 12(3) - (3)^2 = 107; \quad P'(3) = 12 - 2(3) = 6$

After 3 hours, the ozone level is 107 ppb and is INCREASING at the rate of 6 ppb per hour.

65. (A) $P(t) = 26(1.02)^t$
$P(30) = 26(1.02)^{30} = 47.1$

h	-0.1	-0.01	$-0.001 \to 0$	$\leftarrow 0.001$	0.01	0.1
$\dfrac{P(30 + h) - P(30)}{h}$	0.932	0.933	$0.933 \to 0.933$	$\leftarrow 0.933$	0.933	0.934

Therefore, $P'(30) \approx 0.9$

(B) In 2010 there will be 47.1 million people aged 65 or older, and the number of people in this group will be INCREASING at the rate of 900,000 per year.

(C) The estimated population is 48 million in 2011 and 48.9 million in 2012.

EXERCISE 8-4

Things to remember:

1. DERIVATIVE NOTATION

   Given $y = f(x)$, then

   $$f'(x), \quad y', \quad \frac{dy}{dx}$$

   all represent the derivative of f at x.

2. DERIVATIVE OF A CONSTANT FUNCTION RULE

   If $f(x) = C$, C a constant, then $f'(x) = 0$. Also

   $$y' = 0 \text{ and } \frac{dy}{dx} = 0.$$

<u>3</u>. POWER RULE

If $f(x) = x^n$, n any real number, then

$\quad f'(x) = nx^{n-1}$.

Also, $y' = nx^{n-1}$ and $\dfrac{dy}{dx} = nx^{n-1}$

<u>4</u>. CONSTANT TIMES A FUNCTION RULE

If $y = f(x) = ku(x)$, where k is a constant, then

$\quad f'(x) = ku'(x)$.

Also,

$\quad y' = ku'$ and $\dfrac{dy}{dx} = k\dfrac{du}{dx}$.

<u>5</u>. SUM AND DIFFERENCE RULE

If $y = f(x) = u(x) \pm v(x)$, then

$\quad f'(x) = u'(x) \pm v'(x)$.

Also,

$\quad y' = u' \pm v'$ and $\dfrac{dy}{dx} = \dfrac{du}{dx} \pm \dfrac{dv}{dx}$

[<u>Note</u>: This rule generalizes to the sum and difference of any given number of functions.]

<u>6</u>. THE MARGINAL COST FUNCTION

If $C(x)$ is the total cost of producing x items, then the marginal cost function $C'(x)$ approximates the cost of producing one more item at a production level of x items.

---

1. $f(x) = 12$

$f'(x) = (12)' = 0$ (using <u>2</u>)

3. $\dfrac{d}{dx}(23) = 0$ (23 is a constant)

5. $y = x^{12}$

$\dfrac{dy}{dx} = 12x^{12-1}$ (using <u>3</u>)

$\quad = 12x^{11}$

7. $f(x) = x$

$f'(x) = 1x^{1-1}$ (using <u>3</u>)

$\quad = x^0$ ($x^0 = 1$)

$\quad = 1$

9. $y = x^{-7}$

$y' = -7x^{-7-1}$

$\quad = -7x^{-8}$

11. $y = x^{5/2}$

$\dfrac{dy}{dx} = \dfrac{5}{2}x^{5/2-1} = \dfrac{5}{2}x^{5/2-2/2} = \dfrac{5}{2}x^{3/2}$

13. $f(x) = \dfrac{1}{x^5} = x^{-5}$

$\dfrac{d}{dx}(x^{-5}) = -5x^{-5-1} = -5x^{-6}$

15. $f(x) = 2x^4$

$f'(x) = 2 \cdot 4x^3$ (using <u>4</u>)

$\quad = 8x^3$

17. $\frac{d}{dx}\left(\frac{1}{3}x^6\right) = \frac{1}{3}\frac{d}{dx}(x^6)$ (using $\underline{4}$)

$\qquad = \frac{1}{3} \cdot 6x^5 = 2x^5$

19. $y = \frac{x^5}{15} = \frac{1}{15}x^5$

$\qquad \frac{dy}{dx} = \frac{1}{15} \cdot 5x^4 = \frac{x^4}{3}$

21. $h(x) = 4f(x); \ h'(2) = 4 \cdot f'(2) = 4(3) = 12$

23. $h(x) = f(x) + g(x); \ h'(2) = f'(2) + g'(2) = 3 + (-1) = 2$

25. $h(x) = 2f(x) - 3g(x) + 7; \ h'(2) = 2f'(2) - 3g'(2)$

$\qquad\qquad\qquad\qquad\qquad\qquad\quad = 2(3) - 3(-1) = 9$

27. $\frac{d}{dx}(2x^{-5}) = 2\frac{d}{dx}(x^{-5}) = 2(-5)x^{-6} = -10x^{-6}$

29. $f(x) = \frac{4}{x^4} = 4x^{-4}$

$\qquad f'(x) = 4(-4)x^{-5} = -16x^{-5}$

31. $\frac{d}{dx}\left(\frac{-1}{2x^2}\right) = \frac{d}{dx}\left(-\frac{1}{2}x^{-2}\right) = -\frac{1}{2}\frac{d}{dx}(x^{-2})$

$\qquad\qquad\qquad = -\frac{1}{2}(-2)x^{-3} = x^{-3}$

33. $f(x) = -3x^{1/3}$

$\qquad f'(x) = -3\left(\frac{1}{3}\right)x^{1/3-1} = -x^{-2/3}$

35. $\frac{d}{dx}(2x^2 - 3x + 4) = \frac{d}{dx}(2x^2) - \frac{d}{dx}(3x) + \frac{d}{dx}(4)$ (using $\underline{5}$)

$\qquad\qquad\qquad\qquad = 2\frac{d}{dx}(x^2) - 3\frac{d}{dx}(x) + \frac{d}{dx}(4)$

$\qquad\qquad\qquad\qquad = 2 \cdot 2x - 3 \cdot 1 = 4x - 3$

37. $y = 3x^5 - 2x^3 + 5$

$\qquad \frac{dy}{dx} = (3x^5)' - (2x^3)' + (5)' = 15x^4 - 6x^2$

39. $\frac{d}{dx}(3x^{-4} + 2x^{-2}) = \frac{d}{dx}(3x^{-4}) + \frac{d}{dx}(2x^{-2}) = -12x^{-5} - 4x^{-3}$

41. $y = \frac{1}{2x} - \frac{2}{3x^3} = \frac{1}{2}x^{-1} - \frac{2}{3}x^{-3}$

$\qquad \frac{dy}{dx} = -\frac{1}{2}x^{-2} - \frac{2}{3}(-3)x^{-4} = -\frac{1}{2}x^{-2} + 2x^{-4}$

43. $\frac{d}{dx}(3x^{2/3} - 5x^{1/3}) = \frac{d}{dx}(3x^{2/3}) - \frac{d}{dx}(5x^{1/3})$

$\qquad\qquad\qquad = 3\left(\frac{2}{3}\right)x^{-1/3} - 5\left(\frac{1}{3}\right)x^{-2/3} = 2x^{-1/3} - \frac{5}{3}x^{-2/3}$

45. $\frac{d}{dx}\left(\frac{3}{x^{3/5}} - \frac{6}{x^{1/2}}\right) = \frac{d}{dx}(3x^{-3/5} - 6x^{-1/2}) = \frac{d}{dx}(3x^{-3/5}) - \frac{d}{dx}(6x^{-1/2})$

$\qquad\qquad\qquad = 3\left(\frac{-3}{5}\right)x^{-8/5} - 6\left(-\frac{1}{2}\right)x^{-3/2} = \frac{-9}{5}x^{-8/5} + 3x^{-3/2}$

47. $\dfrac{d}{dx}\dfrac{1}{\sqrt[3]{x}} = \dfrac{d}{dx}(x^{-1/3}) = -\dfrac{1}{3}x^{-4/3}$

49. $y = \dfrac{12}{\sqrt{x}} - 3x^{-2} + x = 12x^{-1/2} - 3x^{-2} + x$

$\dfrac{dy}{dx} = 12\left(-\dfrac{1}{2}\right)x^{-3/2} - 3(-2)x^{-3} + 1 = -6x^{-3/2} + 6x^{-3} + 1$

51. $f(x) = 6x - x^2$

(A) $f'(x) = 6 - 2x$

(B) Slope of the graph of f at $x = 2$: $f'(2) = 6 - 2(2) = 2$
Slope of the graph of f at $x = 4$: $f'(4) = 6 - 2(4) = -2$

(C) Tangent line at $x = 2$: $y - y_1 = m(x - x_1)$
$x_1 = 2$
$y_1 = f(2) = 6(2) - 2^2 = 8$
$m = f'(2) = 2$
Thus, $y - 8 = 2(x - 2)$ or $y = 2x + 4$.
Tangent line at $x = 4$: $y - y_1 = m(x - x_1)$
$x_1 = 4$
$y_1 = f(4) = 6(4) - 4^2 = 8$
$m = f'(4) = -2$
Thus, $y - 8 = -2(x - 4)$ or $y = -2x + 16$

(D) The tangent line is horizontal at the values $x = c$ such that
$f'(c) = 0$. Thus, we must solve the following:
$f'(x) = 6 - 2x = 0$
$2x = 6$
$x = 3$

53. $f(x) = 3x^4 - 6x^2 - 7$

(A) $f'(x) = 12x^3 - 12x$

(B) Slope of the graph of $x = 2$: $f'(2) = 12(2)^3 - 12(2) = 72$
Slope of the graph of $x = 4$: $f'(4) = 12(4)^3 - 12(4) = 720$

(C) Tangent line at $x = 2$: $y - y_1 = m(x - x_1)$, where $x_1 = 2$,
$y_1 = f(2) = 3(2)^4 - 6(2)^2 - 7 = 17$, $m = 72$.
$y - 17 = 72(x - 2)$ or $y = 72x - 127$

Tangent line at $x = 4$: $y - y_1 = m(x - x_1)$, where $x_1 = 4$,
$y_1 = f(4) = 3(4)^4 - 6(4)^2 - 7 = 665$, $m = 720$.
$y - 665 = 720(x - 4)$ or $y = 720x - 2215$

(D) Solve $f'(x) = 0$ for x:
$$12x^3 - 12x = 0$$
$$12x(x^2 - 1) = 0$$
$$12x(x - 1)(x + 1) = 0$$
$$x = -1, \; x = 0, \; x = 1$$

55. $f(x) = 176x - 16x^2$

(A) $v = f'(x) = 176 - 32x$

(B) $v\big|_{x=0} = f'(0) = 176$ feet/sec.

$v\big|_{x=3} = f'(3) = 176 - 32(3) = 80$ feet/sec.

(C) Solve $v = f'(x) = 0$ for x:
$$176 - 32x = 0$$
$$32x = 176$$
$$x = 5.5 \text{ seconds}$$

57. $f(x) = x^3 - 9x^2 + 15x$

(A) $v = f'(x) = 3x^2 - 18x + 15$

(B) $v\big|_{x=0} = f'(0) = 15$ feet/sec.

$v\big|_{x=3} = f'(3) = 3(3)^2 - 18(3) + 15 = -12$ feet/sec.

(C) Solve $v = f'(x) = 0$ for x:
$$3x^2 - 18x + 15 = 0$$
$$3(x^2 - 6x + 5) = 0$$
$$3(x - 5)(x - 1) = 0$$
$$x = 1, \; x = 5$$

59. $f(x) = x^2 - 3x - 4\sqrt{x} = x^2 - 3x - 4x^{1/2}$
$f'(x) = 2x - 3 - 2x^{-1/2}$
The graph of f has a horizontal tangent line at the value(s) of x where $f'(x) = 0$. Thus, we need to solve the equation
$$2x - 3 - 2x^{-1/2} = 0$$
By graphing the function $y = 2x - 3 - 2x^{-1/2}$, we see that there is one zero. To two decimal places, it is $x = 2.18$.

61. $f(x) = 3\sqrt[3]{x^4} - 1.5x^2 - 3x = 3x^{4/3} - 1.5x^2 - 3x$
$f'(x) = 4x^{1/3} - 3x - 3$
The graph of f has a horizontal tangent line at the value(s) of x where $f'(x) = 0$. Thus, we need to solve the equation
$$4x^{1/3} - 3x - 3 = 0$$
Graphing the function $y = 4x^{1/3} - 3x - 3$, we see that there is one zero. To two decimal places, it is $x = -2.90$.

63. $f(x) = 0.05x^4 - 0.1x^3 - 1.5x^2 - 1.6x + 3$

$f'(x) = 0.2x^3 + 0.3x^2 - 3x - 1.6$

The graph of f has a horizontal tangent line at the value(s) of x where $f'(x) = 0$. Thus, we need to solve the equation

$$0.2x^3 + 0.3x^2 - 3x - 1.6 = 0$$

By graphing the function $y = 0.2x^3 + 0.3x^2 - 3x - 1.6$ we see that there are three zeros. To two decimal places, they are

$$x_1 = -4.46, \quad x_2 = -0.52, \quad x_3 = 3.48$$

65. $f(x) = 0.2x^4 - 3.12x^3 + 16.25x^2 - 28.25x + 7.5$

$f'(x) = 0.8x^3 - 9.36x^2 + 32.5x - 28.25$

The graph of f has a horizontal tangent line at the value(s) of x where $f'(x) = 0$. Thus, we need to solve the equation

$$0.8x^3 - 9.36x^2 + 32.5x - 28.25 = 0$$

Graphing the function $y = 0.8x^3 - 9.36x^2 + 32.5x - 28.25$, we see that there is one zero. To two decimal places, it is $x = 1.30$.

67. $f(x) = ax^2 + bx + c$; $f'(x) = 2ax + b$.

The derivative is 0 at the vertex of the parabola:

$$2ax + b = 0$$

$$x = -\frac{b}{2a}$$

69. (A) $f(x) = x^3 + x$ (B) $f(x) = x^3$ (C) $f(x) = x^3 - x$

71. $f(x) = \dfrac{10x + 20}{x} = 10 + \dfrac{20}{x} = 10 + 20x^{-1}$

$f'(x) = -20x^{-2}$

73. $\dfrac{d}{dx}\left(\dfrac{x^4 - 3x^3 + 5}{x^2}\right) = \dfrac{d}{dx}\left(x^2 - 3x + \dfrac{5}{x^2}\right)$

$$= \frac{d}{dx}(x^2) - \frac{d}{dx}(3x) + \frac{d}{dx}(5x^{-2}) = 2x - 3 - 10x^{-3}$$

75. Let $f(x) = x^3$

<u>Step 1</u>. Simplify $\dfrac{f(x + h) - f(x)}{h}$.

$$\frac{f(x + h) - f(x)}{h} = \frac{(x + h)^3 - x^3}{h} = \frac{x^3 + 3x^2h + 3xh^2 + h^3 - x^3}{h}$$

$$= \frac{3x^2h + 3xh^2 + h^3}{h} = \frac{h(3x^2 + 3xh + h^2)}{h} = 3x^2 + 3xh + h^2 \quad (h \neq 0)$$

<u>Step 2</u>. Evaluate $\lim\limits_{h \to 0} \dfrac{f(x + h) - f(x)}{h}$.

$$\lim_{h \to 0} \frac{f(x + h) - f(x)}{h} = \lim_{h \to 0}(3x^2 + 3xh + h^2) = 3x^2$$

Therefore, $\dfrac{d}{dx}x^3 = 3x^2$.

77. $f(x) = x^{1/3}$; $f'(x) = \frac{1}{3}x^{-2/3} = \frac{1}{3x^{2/3}}$

The domain of f' is the set of all real numbers except $x = 0$. The graph of f is smooth, but it has a vertical tangent line at $(0, 0)$.

79. $C(x) = 800 + 60x - \frac{x^2}{4}$, $0 \le x \le 120$

(A) Marginal cost $= C'(x) = 60 - \frac{1}{2}x$

(B) $C'(60) = 60 - \frac{1}{2}(60) = 30$ or $30 per racket

Interpretation: At a production level of 60 rackets, the rate of change of total cost with respect to production is $30 per racket. Thus, the cost of producing one more racket at this level of production is approximately $30.

(C) $C(61) - C(60) = 800 + 60(61) - \frac{(61)^2}{4} - \left[800 + 60(60) - \frac{(60)^2}{4}\right]$
$$= 3529.75 - 3500 = 29.75$$
The actual cost of producing the 61st racket is $29.75.

(D) $C'(80) = 60 - \frac{1}{2}(80) = 20$ or $20 per racket

Interpretation: At a production level of 80 rackets, the rate of change of total cost with respect to production is $20 per racket. Thus, the cost of producing the 81st racket is approximately $20.

81. The approximate cost of producing the 101st oven is greater than the approximate cost of producing the 401st oven. Since the marginal costs are decreasing, the manufacturing process is becoming more efficient.

83. (A) $N(x) = 1,000 - \frac{3,780}{x} = 1,000 - 3,780x^{-1}$

$N'(x) = 3,780x^{-2} = \frac{3,780}{x^2}$

(B) $N'(10) = \frac{3,780}{(10)^2} = 37.8$

At the $10,000 level of advertising, sales are INCREASING at the rate of 37.8 boats per $1000 spent on advertising.

$N'(20) = \frac{3,780}{(20)^2} = 9.45$

At the $20,000 level of advertising, sales are INCREASING at the rate of 9.45 boats per $1000 spent on advertising.

85. $y = 590x^{-1/2}$, $30 \le x \le 75$

First, find $\frac{dy}{dx} = \frac{d}{dx}590x^{-1/2} = -295x^{-3/2} = \frac{-295}{x^{3/2}}$, the instantaneous rate of change of pulse when a person is x inches tall.

(A) The instantaneous rate of change of pulse rate at $x = 36$ is:
$$\frac{-295}{(36)^{3/2}} = \frac{-295}{216} = -1.37 \ (1.37 \text{ decrease in pulse rate})$$

(B) The instantaneous rate of change of pulse rate at $x = 64$ is:
$$\frac{-295}{(64)^{3/2}} = \frac{-295}{512} = -0.58 \ (0.58 \text{ decrease in pulse rate})$$

87. $y = 50\sqrt{x}, \ 0 \le x \le 9$

First, find $y' = (50\sqrt{x})' = (50x^{1/2})' = 25x^{-1/2}$
$$= \frac{25}{\sqrt{x}}, \text{ the rate of learning at the end of } x \text{ hours.}$$

(A) Rate of learning at the end of 1 hour:
$$\frac{25}{\sqrt{1}} = 25 \text{ items per hour}$$

(B) Rate of learning at the end of 9 hours:
$$\frac{25}{\sqrt{9}} = \frac{25}{3} = 8.33 \text{ items per hour}$$

EXERCISE 8-5

Things to remember:

1. **PRODUCT RULE**

   If
   $$y = f(x) = F(x)S(x)$$
   and if $F'(x)$ and $S'(x)$ exist, then
   $$f'(x) = F(x)S'(x) + S(x)F'(x).$$
   Also,
   $$y' = FS' + SF';$$
   $$\frac{dy}{dx} = F\frac{dS}{dx} + S\frac{dF}{dx}.$$

2. **QUOTIENT RULE**

   If
   $$y = f(x) = \frac{T(x)}{B(x)}$$
   and if $T'(x)$ and $B'(x)$ exist, then
   $$f'(x) = \frac{B(x)T'(x) - T(x)B'(x)}{[B(x)]^2}.$$
   Also,
   $$y' = \frac{BT' - TB'}{B^2};$$
   $$\frac{dy}{dx} = \frac{B\left(\frac{dT}{dx}\right) - T\left(\frac{dB}{dx}\right)}{B^2}.$$

1. $f(x) = 2x^3(x^2 - 2)$
   $f'(x) = 2x^3(x^2 - 2)' + (x^2 - 2)(2x^3)'$ [using $\underline{1}$ with $F(x) = 2x^3$,
   $\quad\quad = 2x^3(2x) + (x^2 - 2)(6x^2)$ $\quad\quad\quad\quad\quad\quad$ $S(x) = x^2 - 2$]
   $\quad\quad = 4x^4 + 6x^4 - 12x^2$
   $\quad\quad = 10x^4 - 12x^2$

3. $f(x) = (x - 3)(2x - 1)$
   $f'(x) = (x - 3)(2x - 1)' + (2x - 1)(x - 3)'$ (using $\underline{1}$)
   $\quad\quad = (x - 3)(2) + (2x - 1)(1)$
   $\quad\quad = 2x - 6 + 2x - 1$
   $\quad\quad = 4x - 7$

5. $f(x) = \dfrac{x}{x - 3}$
   $f'(x) = \dfrac{(x - 3)(x)' - x(x - 3)'}{(x - 3)^2}$ [using $\underline{2}$ with $T(x) = x$, $B(x) = x - 3$]
   $\quad\quad = \dfrac{(x - 3)(1) - x(1)}{(x - 3)^2} = \dfrac{-3}{(x - 3)^2}$

7. $f(x) = \dfrac{2x + 3}{x - 2}$
   $f'(x) = \dfrac{(x - 2)(2x + 3)' - (2x + 3)(x - 2)'}{(x - 2)^2}$ (using $\underline{2}$)
   $\quad\quad = \dfrac{(x - 2)(2) - (2x + 3)(1)}{(x - 2)^2} = \dfrac{2x - 4 - 2x - 3}{(x - 2)^2} = \dfrac{-7}{(x - 2)^2}$

9. $f(x) = (x^2 + 1)(2x - 3)$
   $f'(x) = (x^2 + 1)(2x - 3)' + (2x - 3)(x^2 + 1)'$ (using $\underline{1}$)
   $\quad\quad = (x^2 + 1)(2) + (2x - 3)(2x)$
   $\quad\quad = 2x^2 + 2 + 4x^2 - 6x$
   $\quad\quad = 6x^2 - 6x + 2$

11. $f(x) = \dfrac{x^2 + 1}{2x - 3}$
    $f'(x) = \dfrac{(2x - 3)(x^2 + 1)' - (x^2 + 1)(2x - 3)'}{(2x - 3)^2}$ (using $\underline{2}$)
    $\quad\quad = \dfrac{(2x - 3)(2x) - (x^2 + 1)(2)}{(2x - 3)^2}$
    $\quad\quad = \dfrac{4x^2 - 6x - 2x^2 - 2}{(2x - 3)^2} = \dfrac{2x^2 - 6x - 2}{(2x - 3)^2}$

13. $f(x) = (x^2 + 2)(x^2 - 3)$
    $f'(x) = (x^2 + 2)(x^2 - 3)' + (x^2 - 3)(x^2 + 2)'$
    $\quad\quad = (x^2 + 2)(2x) + (x^2 - 3)(2x)$
    $\quad\quad = 2x^3 + 4x + 2x^3 - 6x$
    $\quad\quad = 4x^3 - 2x$

15. $f(x) = \dfrac{x^2 + 2}{x^2 - 3}$

$$f'(x) = \frac{(x^2 - 3)(x^2 + 2)' - (x^2 + 2)(x^2 - 3)'}{(x^2 - 3)^2}$$

$$= \frac{(x^2 - 3)(2x) - (x^2 + 2)(2x)}{(x^2 - 3)^2} = \frac{2x^3 - 6x - 2x^3 - 4x}{(x^2 - 3)^2} = \frac{-10x}{(x^2 - 3)^2}$$

17. $h(x) = f(x)g(x); \quad h'(1) = f(1)g'(1) + g(1)f'(1)$

$$= 4(3) + 2(-2) = 12 - 4 = 8$$

19. $h(x) = \dfrac{g(x)}{f(x)}; \quad h'(1) = \dfrac{f(1)g'(1) - g(1)f'(1)}{[f(1)]^2}$

$$= \frac{4(3) - 2(-2)}{4^2} = \frac{16}{16} = 1$$

21. $h(x) = \dfrac{1}{f(x)}; \quad h'(1) = \dfrac{f(1)(0) - f'(1)}{[f(1)]^2} = \dfrac{-(-2)}{4^2} = \dfrac{1}{8}$

23. $f(x) = (2x + 1)(x^2 - 3x)$

$$f'(x) = (2x + 1)(x^2 - 3x)' + (x^2 - 3x)(2x + 1)'$$

$$= (2x + 1)(2x - 3) + (x^2 - 3x)(2)$$

$$= 6x^2 - 10x - 3$$

25. $y = (2x - x^2)(5x + 2)$

$$\frac{dy}{dx} = (2x - x^2)\frac{d}{dx}(5x + 2) + (5x + 2)\frac{d}{dx}(2x - x^2)$$

$$= (2x - x^2)(5) + (5x + 2)(2 - 2x)$$

$$= -15x^2 + 16x + 4$$

27. $y = \dfrac{5x - 3}{x^2 + 2x}$

$$y' = \frac{(x^2 + 2x)(5x - 3)' - (5x - 3)(x^2 + 2x)'}{(x^2 + 2x)^2}$$

$$= \frac{(x^2 + 2x)(5) - (5x - 3)(2x + 2)}{(x^2 + 2x)^2} = \frac{-5x^2 + 6x + 6}{(x^2 + 2x)^2}$$

29. $\dfrac{d}{dx}\left[\dfrac{x^2 - 3x + 1}{x^2 - 1}\right] = \dfrac{(x^2 - 1)\dfrac{d}{dx}(x^2 - 3x + 1) - (x^2 - 3x + 1)\dfrac{d}{dx}(x^2 - 1)}{(x^2 - 1)^2}$

$$= \frac{(x^2 - 1)(2x - 3) - (x^2 - 3x + 1)(2x)}{(x^2 - 1)^2}$$

$$= \frac{3x^2 - 4x + 3}{(x^2 - 1)^2}$$

31. $f(x) = (1 + 3x)(5 - 2x)$

First find $f'(x)$:

$$\begin{aligned} f'(x) &= (1 + 3x)(5 - 2x)' + (5 - 2x)(1 + 3x)' \\ &= (1 + 3x)(-2) + (5 - 2x)(3) \\ &= -2 - 6x + 15 - 6x \\ &= 13 - 12x \end{aligned}$$

An equation for the tangent line at $x = 2$ is:

$$y - y_1 = m(x - x_1)$$

where $x_1 = 2$, $y_1 = f(x_1) = f(2) = 7$, and $m = f'(x_1) = f'(2) = -11$.

Thus, we have:

$$y - 7 = -11(x - 2) \quad \text{or} \quad y = -11x + 29$$

33. $f(x) = \dfrac{x - 8}{3x - 4}$

First find $f'(x)$:

$$\begin{aligned} f'(x) &= \frac{(3x - 4)(x - 8)' - (x - 8)(3x - 4)'}{(3x - 4)^2} \\ &= \frac{(3x - 4)(1) - (x - 8)(3)}{(3x - 4)^2} = \frac{20}{(3x - 4)^2} \end{aligned}$$

An equation for the tangent line at $x = 2$ is:

$$y - y_1 = m(x - x_1)$$

where $x_1 = 2$, $y_1 = f(x_1) = f(2) = -3$, and $m = f'(x_1) = f'(2) = 5$.

Thus, we have: $y - (-3) = 5(x - 2)$ or $y = 5x - 13$

35. $f(x) = (2x - 15)(x^2 + 18)$

$$\begin{aligned} f'(x) &= (2x - 15)(x^2 + 18)' + (x^2 + 18)(2x - 15)' \\ &= (2x - 15)(2x) + (x^2 + 18)(2) \\ &= 6x^2 - 30x + 36 \end{aligned}$$

To find the values of x where $f'(x) = 0$, set: $f'(x) = 6x^2 - 30x + 36 = 0$

$$\begin{aligned} \text{or} \qquad x^2 - 5x + 6 &= 0 \\ (x - 2)(x - 3) &= 0 \end{aligned}$$

Thus, $x = 2$, $x = 3$.

37. $f(x) = \dfrac{x}{x^2 + 1}$

$$f'(x) = \frac{(x^2 + 1)(x)' - x(x^2 + 1)'}{(x^2 + 1)^2} = \frac{(x^2 + 1)(1) - x(2x)}{(x^2 + 1)^2} = \frac{1 - x^2}{(x^2 + 1)^2}$$

Now, set $f'(x) = \dfrac{1 - x^2}{(x^2 + 1)^2} = 0$

$$\begin{aligned} \text{or} \qquad 1 - x^2 &= 0 \\ (1 - x)(1 + x) &= 0 \end{aligned}$$

Thus, $x = 1$, $x = -1$.

39. $f(x) = x^3(x^4 - 1)$

First, we use the product rule:

$$\begin{aligned} f'(x) &= x^3(x^4 - 1)' + (x^4 - 1)(x^3)' \\ &= x^3(4x^3) + (x^4 - 1)(3x^2) \\ &= 7x^6 - 3x^2 \end{aligned}$$

Next, simplifying $f(x)$, we have $f(x) = x^7 - x^3$. Thus, $f'(x) = 7x^6 - 3x^2$.

41. $f(x) = \dfrac{x^3 + 9}{x^3}$

First, we use the quotient rule:

$$f'(x) = \frac{x^3(x^3 + 9)' - (x^3 + 9)(x^3)'}{(x^3)^2} = \frac{x^3(3x^2) - (x^3 + 9)(3x^2)}{x^6}$$

$$= \frac{-27x^2}{x^6} = \frac{-27}{x^4}$$

Next, simplifying $f(x)$, we have $f(x) = \dfrac{x^3 + 9}{x^3} = 1 + \dfrac{9}{x^3} = 1 + 9x^{-3}$

Thus, $f'(x) = -27x^{-4} = -\dfrac{27}{x^4}$.

43. $f(x) = (2x^4 - 3x^3 + x)(x^2 - x + 5)$

$f'(x) = (2x^4 - 3x^3 + x)(x^2 - x + 5)' + (x^2 - x + 5)(2x^4 - 3x^3 + x)'$

$\quad = (2x^4 - 3x^3 + x)(2x - 1) + (x^2 - x + 5)(8x^3 - 9x^2 + 1)$

$\quad = 4x^5 - 6x^4 + 2x^2 - 2x^4 + 3x^3 - x + 8x^5 - 8x^4 + 40x^3 - 9x^4 + 9x^3$

$\qquad\qquad\qquad\qquad\qquad\qquad\qquad\qquad - 45x^2 + x^2 - x + 5$

$\quad = 12x^5 - 25x^4 + 52x^3 - 42x^2 - 2x + 5$

45. $\dfrac{d}{dx} \dfrac{3x^2 - 2x + 3}{4x^2 + 5x - 1}$

$$= \frac{(4x^2 + 5x - 1)\dfrac{d}{dx}(3x^2 - 2x + 3) - (3x^2 - 2x + 3)\dfrac{d}{dx}(4x^2 + 5x - 1)}{(4x^2 + 5x - 1)^2}$$

$$= \frac{(4x^2 + 5x - 1)(6x - 2) - (3x^2 - 2x + 3)(8x + 5)}{(4x^2 + 5x - 1)^2}$$

$$= \frac{24x^3 + 30x^2 - 6x - 8x^2 - 10x + 2 - 24x^3 + 16x^2 - 24x - 15x^2 + 10x - 15}{(4x^2 + 5x - 1)^2}$$

$$= \frac{23x^2 - 30x - 13}{(4x^2 + 5x - 1)^2}$$

47. $y = 9x^{1/3}(x^3 + 5)$

$\dfrac{dy}{dx} = 9x^{1/3}\dfrac{d}{dx}(x^3 + 5) + (x^3 + 5)\dfrac{d}{dx}(9x^{1/3})$

$\quad = 9x^{1/3}(3x^2) + (x^3 + 5)\left(9 \cdot \dfrac{1}{3}x^{-2/3}\right) = 27x^{7/3} + (x^3 + 5)(3x^{-2/3})$

$\quad = 27x^{7/3} + \dfrac{3x^3 + 15}{x^{2/3}} = \dfrac{30x^3 + 15}{x^{2/3}}$

49. $f(x) = \dfrac{6\sqrt[3]{x}}{x^2 - 3} = \dfrac{6x^{1/3}}{x^2 - 3}$

$f'(x) = \dfrac{(x^2 - 3)(6x^{1/3})' - 6x^{1/3}(x^2 - 3)'}{(x^2 - 3)^2}$

$= \dfrac{(x^2 - 3)\left(6 \cdot \frac{1}{3}x^{-2/3}\right) - 6x^{1/3}(2x)}{(x^2 - 3)^2} = \dfrac{(x^2 - 3)(2x^{-2/3}) - 12x^{4/3}}{(x^2 - 3)^2}$

$= \dfrac{\frac{2(x^2 - 3)}{x^{2/3}} - 12x^{4/3}}{(x^2 - 3)^2} = \dfrac{2x^2 - 6 - 12x^2}{(x^2 - 3)^2 x^{2/3}} = \dfrac{-10x^2 - 6}{(x^2 - 3)^2 x^{2/3}}$

51. $\dfrac{d}{dx} \dfrac{x^3 - 2x^2}{\sqrt[3]{x^2}} = \dfrac{d}{dx} \dfrac{x^3 - 2x^2}{x^{2/3}}$

$= \dfrac{x^{2/3}\frac{d}{dx}(x^3 - 2x^2) - (x^3 - 2x^2)\frac{d}{dx}(x^{2/3})}{(x^{2/3})^2}$

$= \dfrac{x^{2/3}(3x^2 - 4x) - (x^3 - 2x^2)\left(\frac{2}{3}x^{-1/3}\right)}{x^{4/3}}$

$= x^{-2/3}(3x^2 - 4x) - \frac{2}{3}x^{-5/3}(x^3 - 2x^2)$

$= 3x^{4/3} - 4x^{1/3} - \frac{2}{3}x^{4/3} + \frac{4}{3}x^{1/3}$

$= -\frac{8}{3}x^{1/3} + \frac{7}{3}x^{4/3}$

53. $f(x) = \dfrac{(2x^2 - 1)(x^2 + 3)}{x^2 + 1}$

$f'(x) = \dfrac{(x^2 + 1)[(2x^2 - 1)(x^2 + 3)]' - (2x^2 - 1)(x^2 + 3)(x^2 + 1)'}{(x^2 + 1)^2}$

$= \dfrac{(x^2 + 1)[(2x^2 - 1)(x^2 + 3)' + (x^2 + 3)(2x^2 - 1)'] - (2x^2 - 1)(x^2 + 3)(2x)}{(x^2 + 1)^2}$

$= \dfrac{(x^2 + 1)[(2x^2 - 1)(2x) + (x^2 + 3)(4x)] - (2x^2 - 1)(x^2 + 3)(2x)}{(x^2 + 1)^2}$

$= \dfrac{(x^2 + 1)[4x^3 - 2x + 4x^3 + 12x] - [2x^4 + 5x^2 - 3](2x)}{(x^2 + 1)^2}$

$= \dfrac{(x^2 + 1)(8x^3 + 10x) - 4x^5 - 10x^3 + 6x}{(x^2 + 1)^2}$

$= \dfrac{8x^5 + 10x^3 + 8x^3 + 10x - 4x^5 - 10x^3 + 6x}{(x^2 + 1)^2}$

$= \dfrac{4x^5 + 8x^3 + 16x}{(x^2 + 1)^2}$

55. $f(x) = [u(x)]^2 = u(x) \cdot u(x)$
$f'(x) = u(x) \cdot u'(x) + u(x)u'(x) = 2u(x)u'(x)$

57. $f(x) = [u(x)]^n$
$f'(x) = n[u(x)]^{n-1} u'(x)$

59. $f(x) = \dfrac{5x - x^4}{x^2 + 1}$

Horizontal tangent line
at $x \approx 0.75$.

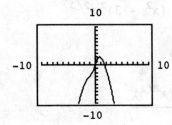

61. $f(x) = \dfrac{8 - x^3}{2 + x^2}$

Horizontal tangent lines
at $x \approx -1.76$, $x = 0$.

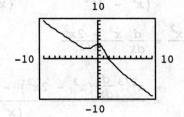

63. $f(x) = \dfrac{10x^2 + 9x}{x^4 + 2}$

Horizontal tangent lines
at $x \approx -1.67$, $x \approx -0.43$,
$x \approx 1.06$.

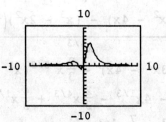

65. $S(t) = \dfrac{90t^2}{t^2 + 50}$

(A) $S'(t) = \dfrac{(t^2 + 50)(180t) - 90t^2(2t)}{(t^2 + 50)^2} = \dfrac{9000t}{(t^2 + 50)^2}$

(B) $S(10) = \dfrac{90(10)^2}{(10)^2 + 50} = \dfrac{9000}{150} = 60$;

$S'(10) = \dfrac{9000(10)}{[(10)^2 + 50]^2} = \dfrac{90,000}{22,500} = 4$

After 10 months, the total sales are 60,000 albums and the sales are INCREASING at the rate of 4,000 albums per month.

(C) The total sales after 11 months will be approximately 64,000 albums.

67. $x = \dfrac{4,000}{0.1p + 1}$, $10 \le p \le 70$

(A) $\dfrac{dx}{dp} = \dfrac{(0.1p + 1)(0) - 4,000(0.1)}{(0.1p + 1)^2} = \dfrac{-400}{(0.1p + 1)^2}$

(B) $x(40) = \dfrac{4{,}000}{0.1(40) + 1} = \dfrac{4{,}000}{5} = 800;$

$\dfrac{dx}{dp} = \dfrac{-400}{[0.1(40) + 1]^2} = \dfrac{-400}{25} = -16$

At a price level of \$40, the demand is 800 CD players and the demand is DECREASING at the rate of 16 CD players per dollar.

(C) At a price of \$41, the demand will be approximately 784 CD players.

69. $C(t) = \dfrac{0.14t}{t^2 + 1}$

(A) $C'(t) = \dfrac{(t^2 + 1)(0.14t)' - (0.14t)(t^2 + 1)'}{(t^2 + 1)^2}$

$= \dfrac{(t^2 + 1)(0.14) - (0.14t)(2t)}{(t^2 + 1)^2} = \dfrac{0.14 - 0.14t^2}{(t^2 + 1)^2} = \dfrac{0.14(1 - t^2)}{(t^2 + 1)^2}$

(B) $C'(0.5) = \dfrac{0.14(1 - [0.5]^2)}{([0.5]^2 + 1)^2} = \dfrac{0.14(1 - 0.25)}{(1.25)^2} = 0.0672$

Interpretation: At $t = 0.5$ hours, the concentration is increasing at the rate of 0.0672 units per hour.

$C'(3) = \dfrac{0.14(1 - 3^2)}{(3^2 + 1)^2} = \dfrac{0.14(-8)}{100} = -0.0112$

Interpretation: At $t = 3$ hours, the concentration is decreasing at the rate of 0.0112 units per hour.

71. $N(x) = \dfrac{100x + 200}{x + 32}$

(A) $N'(x) = \dfrac{(x + 32)(100x + 200)' - (100x + 200)(x + 32)'}{(x + 32)^2}$

$= \dfrac{(x + 32)(100) - (100x + 200)(1)}{(x + 32)^2}$

$= \dfrac{100x + 3200 - 100x - 200}{(x + 32)^2} = \dfrac{3000}{(x + 32)^2}$

(B) $N'(4) = \dfrac{3000}{(36)^2} = \dfrac{3000}{1296} \approx 2.31;$ $N'(68) = \dfrac{3000}{(100)^2} = \dfrac{3000}{10{,}000} = \dfrac{3}{10} = 0.30$

Things to remember:

1. GENERAL POWER RULE

    If $u(x)$ is a differentiable function, n is any real number, and

    $$y = f(x) = [u(x)]^n$$

    then

    $$f'(x) = n[u(x)]^{n-1}u'(x)$$

    This rule is often written more compactly as

    $$y' = nu^{n-1}u' \quad \text{or} \quad \frac{d}{dx}u^n = nu^{n-1}\frac{du}{dx}, \quad u = u(x)$$

1. $3; \dfrac{d}{dx}(3x + 4)^4 = 4(3x + 4)^3(3) = 12(3x + 4)^3$

3. $-4x; \dfrac{d}{dx}(4 - 2x^2)^3 = 3(4 - 2x^2)^2(-4x) = -12x(4 - 2x^2)^2$

5. $2 + 6x; \dfrac{d}{dx}(1 + 2x + 3x^2)^7 = 7(1 + 2x + 3x^2)^6(2 + 6x)$

    $$= 7(2 + 6x)(1 + 2x + 3x^2)^6$$

7. $f(x) = (2x + 5)^3$
    $f'(x) = 3(2x + 5)^2(2x + 5)'$
    $\quad\quad = 3(2x + 5)^2(2)$
    $\quad\quad = 6(2x + 5)^2$

9. $f(x) = (5 - 2x)^4$
    $f'(x) = 4(5 - 2x)^3(5 - 2x)'$
    $\quad\quad = 4(5 - 2x)^3(-2)$
    $\quad\quad = -8(5 - 2x)^3$

11. $f(x) = (3x^2 + 5)^5$
    $f'(x) = 5(3x^2 + 5)^4(3x^2 + 5)'$
    $\quad\quad = 5(3x^2 + 5)^4(6x)$
    $\quad\quad = 30x(3x^2 + 5)^4$

13. $f(x) = (x^3 - 2x^2 + 2)^8$
    $f'(x) = 8(x^3 - 2x^2 + 2)^7(x^3 - 2x^2 + 2)'$
    $\quad\quad = 8(x^3 - 2x^2 + 2)^7(3x^2 - 4x)$

15. $f(x) = (2x - 5)^{1/2}$
    $f'(x) = \dfrac{1}{2}(2x - 5)^{-1/2}(2x - 5)'$
    $\quad\quad = \dfrac{1}{2}(2x - 5)^{-1/2}(2) = \dfrac{1}{(2x - 5)^{1/2}}$

17. $f(x) = (x^4 + 1)^{-2}$
    $f'(x) = -2(x^4 + 1)^{-3}(x^4 + 1)'$
    $\quad\quad = -2(x^4 + 1)^{-3}(4x^3)$
    $\quad\quad = -8x^3(x^4 + 1)^{-3} = \dfrac{-8x^3}{(x^4 + 1)^3}$

19. $f(x) = (2x - 1)^3$

$f'(x) = 3(2x - 1)^2(2) = 6(2x - 1)^2$

Tangent line at $x = 1$: $y - y_1 = m(x - x_1)$ where $x_1 = 1$, $y_1 = f(1) = (2(1) - 1)^3 = 1$, $m = f'(1) = 6[2(1) - 1]^2 = 6$. Thus, $y - 1 = 6(x - 1)$ or $y = 6x - 5$.

The tangent line is horizontal at the value(s) of x such that $f'(x) = 0$:

$6(2x - 1)^2 = 0$

$2x - 1 = 0$

$x = \dfrac{1}{2}$

21. $f(x) = (4x - 3)^{1/2}$

$f'(x) = \dfrac{1}{2}(4x - 3)^{-1/2}(4) = \dfrac{2}{(4x - 3)^{1/2}}$

Tangent line at $x = 3$: $y - y_1 = m(x - x_1)$ where $x_1 = 3$, $y_1 = f(3) = (4 \cdot 3 - 3)^{1/2} = 3$, $f'(3) = \dfrac{2}{(4 \cdot 3 - 3)^{1/2}} = \dfrac{2}{3}$. Thus, $y - 3 = \dfrac{2}{3}(x - 3)$ or $y = \dfrac{2}{3}x + 1$.

The tangent line is horizontal at the value(s) of x such that $f'(x) = 0$.

Since $\dfrac{2}{(4x - 3)^{1/2}} \neq 0$ for all x $\left(x \neq \dfrac{3}{4}\right)$, there are no values of x where the tangent line is horizontal.

23. $y = 3(x^2 - 2)^4$

$\dfrac{dy}{dx} = 3 \cdot 4(x^2 - 2)^3(2x) = 24x(x^2 - 2)^3$

25. $y = 2(x^2 + 3x)^{-3}$

$\dfrac{dy}{dx} = 2 \cdot (-3)(x^2 + 3x)^{-4}(2x + 3) = -6(x^2 + 3x)^{-4}(2x + 3) = \dfrac{-6(2x + 3)}{(x^2 + 3x)^4}$

27. $y = \sqrt{x^2 + 8} = (x^2 + 8)^{1/2}$

$\dfrac{dy}{dx} = \dfrac{1}{2}(x^2 + 8)^{-1/2}(2x) = \dfrac{x}{(x^2 + 8)^{1/2}} = \dfrac{x}{\sqrt{x^2 + 8}}$

29. $y = \sqrt[3]{3x + 4} = (3x + 4)^{1/3}$

$\dfrac{dy}{dx} = \dfrac{1}{3}(3x + 4)^{-2/3}(3) = (3x + 4)^{-2/3} = \dfrac{1}{(3x + 4)^{2/3}} = \dfrac{1}{\sqrt[3]{(3x + 4)^2}}$

31. $y = (x^2 - 4x + 2)^{1/2}$

$\dfrac{dy}{dx} = \dfrac{1}{2}(x^2 - 4x + 2)^{-1/2}(2x - 4)$

$= \dfrac{2(x - 2)}{2(x^2 - 4x + 2)^{1/2}} = \dfrac{x - 2}{(x^2 - 4x + 2)^{1/2}}$

33. $y = \dfrac{1}{2x + 4} = (2x + 4)^{-1}$

$\dfrac{dy}{dx} = -1(2x + 4)^{-2}(2) = \dfrac{-2}{(2x + 4)^2}$

35. $y = \dfrac{1}{(x^3 + 4)^5} = (x^3 + 4)^{-5}$

$\dfrac{dy}{dx} = -5(x^3 + 4)^{-6}(3x^2) = -15x^2(x^3 + 4)^{-6} = \dfrac{-15x^2}{(x^3 + 4)^6}$

37. $y = \dfrac{1}{4x^2 - 4x + 1} = (4x^2 - 4x + 1)^{-1}$

$\dfrac{dy}{dx} = -1(4x^2 - 4x + 1)^{-2}(8x - 4) = \dfrac{-4(2x - 1)}{(4x^2 - 4x + 1)^2}$

$\qquad = \dfrac{-4(2x - 1)}{(2x - 1)^4} = \dfrac{-4}{(2x - 1)^3}$

39. $y = \dfrac{4}{\sqrt{x^2 - 3x}} = \dfrac{4}{(x^2 - 3x)^{1/2}} = 4(x^2 - 3x)^{-1/2}$

$\dfrac{dy}{dx} = 4\left[-\dfrac{1}{2}(x^2 - 3x)^{-3/2}\right](2x - 3) = \dfrac{-2(2x - 3)}{(x^2 - 3x)^{3/2}}$

41. $f(x) = x(4 - x)^3$

$f'(x) = x[(4 - x)^3]' + (4 - x)^3(x)'$

$\qquad = x(3)(4 - x)^2(-1) + (4 - x)^3(1)$

$\qquad = (4 - x)^3 - 3x(4 - x)^2 = (4 - x)^2[4 - x - 3x] = 4(4 - x)^2(1 - x)$

An equation for the tangent line to the graph of f at $x = 2$ is:

$y - y_1 = m(x - x_1)$ where $x_1 = 2$, $y_1 = f(x_1) = f(2) = 16$, and

$m = f'(x_1) = f'(2) = -16$. Thus, $y - 16 = -16(x - 2)$ or $y = -16x + 48$.

43. $f(x) = \dfrac{x}{(2x - 5)^3}$

$f'(x) = \dfrac{(2x - 5)^3(1) - x(3)(2x - 5)^2(2)}{[(2x - 5)^3]^2}$

$\qquad = \dfrac{(2x - 5)^3 - 6x(2x - 5)^2}{(2x - 5)^6} = \dfrac{(2x - 5) - 6x}{(2x - 5)^4} = \dfrac{-4x - 5}{(2x - 5)^4}$

An equation for the tangent line to the graph of f at $x = 3$ is:

$y - y_1 = m(x - x_1)$ where $x_1 = 3$, $y_1 = f(x_1) = f(3) = 3$, and

$m = f'(x_1) = f'(3) = -17$. Thus, $y - 3 = -17(x - 3)$ or $y = -17x + 54$.

45. $f(x) = x\sqrt{2x + 2} = x(2x + 2)^{1/2}$

$f'(x) = x[(2x + 2)^{1/2}]' + (2x + 2)^{1/2}(x)'$

$\qquad = x\left(\dfrac{1}{2}\right)(2x + 2)^{-1/2}(2) + (2x + 2)^{1/2}(1) = \dfrac{x}{(2x + 2)^{1/2}} + (2x + 2)^{1/2}$

$\qquad\qquad\qquad\qquad\qquad\qquad\qquad\qquad = \dfrac{3x + 2}{(2x + 2)^{1/2}}$

An equation for the tangent line to the graph of f at $x = 1$ is:

$y - y_1 = m(x - x_1)$ where $x_1 = 1$, $y_1 = f(x_1) = f(1) = 2$, and

$m = f'(x_1) = f'(1) = \dfrac{5}{2}$. Thus, $y - 2 = \dfrac{5}{2}(x - 1)$ or $y = \dfrac{5}{2}x - \dfrac{1}{2}$.

47. $f(x) = x^2(x - 5)^3$

$f'(x) = x^2[(x - 5)^3]' + (x - 5)^3(x^2)'$

$\quad = x^2(3)(x - 5)^2(1) + (x - 5)^3(2x)$

$\quad = 3x^2(x - 5)^2 + 2x(x - 5)^3 = 5x(x - 5)^2(x - 2)$

The tangent line to the graph of f is horizontal at the values of x such that $f'(x) = 0$. Thus, we set $5x(x - 5)^2(x - 2) = 0$ and $x = 0$, $x = 2$, $x = 5$.

49. $f(x) = \dfrac{x}{(2x + 5)^2}$

$f'(x) = \dfrac{(2x + 5)^2(x)' - x[(2x + 5)^2]'}{[(2x + 5)^2]^2}$

$\quad = \dfrac{(2x + 5)^2(1) - x(2)(2x + 5)(2)}{(2x + 5)^4} = \dfrac{2x + 5 - 4x}{(2x + 5)^3} = \dfrac{5 - 2x}{(2x + 5)^3}$

The tangent line to the graph of f is horizontal at the values of x such that $f'(x) = 0$. Thus, we set

$\dfrac{5 - 2x}{(2x + 5)^3} = 0$

$5 - 2x = 0$

and $x = \dfrac{5}{2}$.

51. $f(x) = \sqrt{x^2 - 8x + 20} = (x^2 - 8x + 20)^{1/2}$

$f'(x) = \dfrac{1}{2}(x^2 - 8x + 20)^{-1/2}(2x - 8)$

$\quad = \dfrac{x - 4}{(x^2 - 8x + 20)^{1/2}}$

The tangent line to the graph of f is horizontal at the values of x such that $f'(x) = 0$. Thus, we set

$\dfrac{x - 4}{(x^2 - 8x + 20)^{1/2}} = 0$

$x - 4 = 0$

and $x = 4$.

53. $f(x) = 4x - \sqrt{x^4 + 10}$

Horizontal tangent at $x \approx 2.32$.

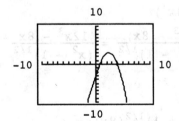

55. $f(x) = 5\sqrt{x^2 + 1} - \sqrt{x^4 + 1}$

Horizontal tangents at $x \approx -2.34, 0, 2.34$.

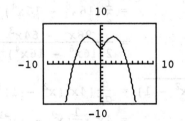

57. $f(x) = \sqrt{x^4 - 4x^3 + 6x + 10}$

Horizontal tangents
at $x \approx -0.64$, 0.83, 2.81.

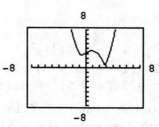

59. $\dfrac{d}{dx}[3x(x^2 + 1)^3] = 3x\dfrac{d}{dx}(x^2 + 1)^3 + (x^2 + 1)^3\dfrac{d}{dx}3x$

$\qquad\qquad = 3x\cdot 3(x^2 + 1)^2(2x) + (x^2 + 1)^3(3)$

$\qquad\qquad = 18x^2(x^2 + 1)^2 + 3(x^2 + 1)^3$

$\qquad\qquad = (x^2 + 1)^2[18x^2 + 3(x^2 + 1)]$

$\qquad\qquad = (x^2 + 1)^2(21x^2 + 3)$

$\qquad\qquad = 3(x^2 + 1)^2(7x^2 + 1)$

61. $\dfrac{d}{dx}\dfrac{(x^3 - 7)^4}{2x^3} = \dfrac{2x^3\dfrac{d}{dx}(x^3 - 7)^4 - (x^3 - 7)^4\dfrac{d}{dx}2x^3}{(2x^3)^2}$

$\qquad\qquad = \dfrac{2x^3\cdot 4(x^3 - 7)^3(3x^2) - (x^3 - 7)^4 6x^2}{4x^6}$

$\qquad\qquad = \dfrac{3(x^3 - 7)^3 x^2[8x^3 - 2(x^3 - 7)]}{4x^6}$

$\qquad\qquad = \dfrac{3(x^3 - 7)^3(6x^3 + 14)}{4x^4} = \dfrac{3(x^3 - 7)^3(3x^3 + 7)}{2x^4}$

63. $\dfrac{d}{dx}[(2x - 3)^2(2x^2 + 1)^3] = (2x - 3)^2\dfrac{d}{dx}(2x^2 + 1)^3 + (2x^2 + 1)^3\dfrac{d}{dx}(2x - 3)^2$

$\qquad\qquad = (2x - 3)^2 3(2x^2 + 1)^2(4x) + (2x^2 + 1)^3 2(2x - 3)(2)$

$\qquad\qquad = 12x(2x - 3)^2(2x^2 + 1)^2 + 4(2x^2 + 1)^3(2x - 3)$

$\qquad\qquad = 4(2x - 3)(2x^2 + 1)^2[3x(2x - 3) + (2x^2 + 1)]$

$\qquad\qquad = 4(2x - 3)(2x^2 + 1)^2(6x^2 - 9x + 2x^2 + 1)$

$\qquad\qquad = 4(2x - 3)(2x^2 + 1)^2(8x^2 - 9x + 1)$

65. $\dfrac{d}{dx}[4x^2\sqrt{x^2 - 1}] = \dfrac{d}{dx}[\sqrt{16x^4(x^2 - 1)}]$

$\qquad\qquad = \dfrac{d}{dx}[(16x^6 - 16x^4)^{1/2}]$

$\qquad\qquad = \dfrac{1}{2}(16x^6 - 16x^4)^{-1/2}(96x^5 - 64x^3)$

$\qquad\qquad = \dfrac{96x^5 - 64x^3}{2(16x^6 - 16x^4)^{1/2}} = \dfrac{8x^2(12x^3 - 8x)}{2\cdot 4x^2(x^2 - 1)^{1/2}} = \dfrac{12x^3 - 8x}{(x^2 - 1)^{1/2}}$

or
$\dfrac{d}{dx}[4x^2\sqrt{x^2 - 1}] = \dfrac{d}{dx}[4x^2(x^2 - 1)^{1/2}]$

$\qquad\qquad = 4x^2\cdot\dfrac{1}{2}(x^2 - 1)^{-1/2}(2x) + (x^2 - 1)^{1/2}(8x)$

$$= \frac{4x^3}{(x^2 - 1)^{1/2}} + 8x(x^2 - 1)^{1/2}$$

$$= \frac{4x^3 + 8x(x^2 - 1)}{(x^2 - 1)^{1/2}} = \frac{4x^3 + 8x^3 - 8x}{(x^2 - 1)^{1/2}} = \frac{12x^3 - 8x}{(x^2 - 1)^{1/2}}$$

67. $\dfrac{d}{dx} \dfrac{2x}{\sqrt{x - 3}} = \dfrac{(x - 3)^{1/2}(2) - 2x \cdot \frac{1}{2}(x - 3)^{-1/2}}{(x - 3)}$

$$= \frac{2(x - 3)^{1/2} - \dfrac{x}{(x - 3)^{1/2}}}{(x - 3)} = \frac{2(x - 3) - x}{(x - 3)(x - 3)^{1/2}}$$

$$= \frac{2x - 6 - x}{(x - 3)^{3/2}} = \frac{x - 6}{(x - 3)^{3/2}}$$

69. $\dfrac{d}{dx}\sqrt{(2x - 1)^3(x^2 + 3)^4} = \dfrac{d}{dx}[(2x - 1)^3(x^2 + 3)^4]^{1/2}$

$$= \frac{d}{dx}(2x - 1)^{3/2}(x^2 + 3)^2$$

$$= (2x - 1)^{3/2}\frac{d}{dx}(x^2 + 3)^2 + (x^2 + 3)^2\frac{d}{dx}(2x - 1)^{3/2}$$

$$= (2x - 1)^{3/2}2(x^2 + 3)(2x) + (x^2 + 3)^2 \cdot \frac{3}{2}(2x - 1)^{1/2}(2)$$

$$= (2x - 1)^{1/2}(x^2 + 3)[4x(2x - 1) + 3(x^2 + 3)]$$

$$= (2x - 1)^{1/2}(x^2 + 3)(8x^2 - 4x + 3x^2 + 9)$$

$$= (2x - 1)^{1/2}(x^2 + 3)(11x^2 - 4x + 9)$$

71. $C(x) = 10 + \sqrt{2x + 16} = 10 + (2x + 16)^{1/2}$, $0 \le x \le 50$

(A) $C'(x) = \dfrac{1}{2}(2x + 16)^{-1/2}(2) = \dfrac{1}{(2x + 16)^{1/2}}$

(B) $C'(24) = \dfrac{1}{[2(24) + 16]^{1/2}} = \dfrac{1}{(64)^{1/2}} = \dfrac{1}{8}$ or $12.50; at a production level of 24 calculators, total costs are INCREASING at the rate of $12.50 per calculator; also, the cost of producing the 25th calculator is approximately $12.50.

$C'(42) = \dfrac{1}{[2(42) + 16]^{1/2}} = \dfrac{1}{(100)^{1/2}} = \dfrac{1}{10}$ or $10.00; at a production level of 42 calculators, total costs are INCREASING at the rate of $10.00 per calculator; also the cost of producing the 43rd calculator is approximatley $10.00.

73. $x = 80\sqrt{p + 25} - 400 = 80(p + 25)^{1/2} - 400, \quad 20 \leq p \leq 100$

(A) $\dfrac{dx}{dp} = 80\left(\dfrac{1}{2}\right)(p + 25)^{-1/2}(1) = \dfrac{40}{(p + 25)^{1/2}}$

(B) At $p = 75$, $x = 80\sqrt{75 + 25} - 400 = 400$ and

$\dfrac{dx}{dp} = \dfrac{40}{(75 + 25)^{1/2}} = \dfrac{40}{(100)^{1/2}} = 4.$

At a price of \$75, the supply is 400 speakers, and the supply is INCREASING at a rate of 4 speakers per dollar.

75. $A = 1000\left(1 + \dfrac{1}{12}r\right)^{48}$

$\dfrac{dA}{dr} = 1000(48)\left(1 + \dfrac{1}{12}r\right)^{47}\left(\dfrac{1}{12}\right) = 4000\left(1 + \dfrac{1}{12}r\right)^{47}$

77. $y = (3 \times 10^6)\left[1 - \dfrac{1}{\sqrt[3]{(x^2 - 1)^2}}\right] = (3 \times 10^6)[1 - (x^2 - 1)^{-2/3}]$

$\dfrac{dy}{dx} = -(3 \times 10^6)\left(-\dfrac{2}{3}\right)(x^2 - 1)^{-5/3}(2x) = \dfrac{(4 \times 10^6)x}{(x^2 - 1)^{5/3}}$

79. $T = f(n) = 2n\sqrt{n - 2} = 2n(n - 2)^{1/2}$

(A) $f'(n) = 2n[(n - 2)^{1/2}]' + (n - 2)^{1/2}(2n)'$

$= 2n\left(\dfrac{1}{2}\right)(n - 2)^{-1/2}(1) + (n - 2)^{1/2}(2)$

$= \dfrac{n}{(n - 2)^{1/2}} + 2(n - 2)^{1/2}$

$= \dfrac{n + 2(n - 2)}{(n - 2)^{1/2}} = \dfrac{3n - 4}{(n - 2)^{1/2}}$

(B) $f'(11) = \dfrac{29}{3} = 9.67$; when the list contains 11 items, the learning time is increasing at the rate of 9.67 minutes per item;

$f'(27) = \dfrac{77}{5} = 15.4$; when the list contains 27 items, the learning time is increasing at the rate of 15.4 minutes per item.

Things to remember:

1. MARGINAL COST, REVENUE, AND PROFIT

   If x is the number of units of a product produced in some time interval, then:

   Total Cost = $C(x)$
   Marginal Cost = $C'(x)$
   Total Revenue = $R(x)$
   Marginal Revenue = $R'(x)$
   Total Profit = $P(x) = R(x) - C(x)$
   Marginal Profit = $P'(x) = R'(x) - C'(x)$
            = (Marginal Revenue) − (Marginal Cost)

   Marginal cost (or revenue or profit) is the instantaneous rate of change of cost (or revenue or profit) relative to production at a given production level.

2. MARGINAL COST AND EXACT COST

   If $C(x)$ is the cost of producing x items, then the marginal cost function approximates the exact cost of producing the $(x + 1)$st item:

        Marginal Cost   Exact Cost
          $C'(x)$ $\approx$ $C(x + 1) - C(x)$

   Similar interpretations can be made for total revenue and total profit functions.

3. BREAK-EVEN POINTS

   The BREAK-EVEN POINTS are the points where total revenue equals total cost.

4. MARGINAL AVERAGE COST, REVENUE, AND PROFIT

   If x is the number of units of a product produced in some time interval, then:

   Average Cost = $\overline{C}(x) = \dfrac{C(x)}{x}$ Cost per unit

   Marginal Average Cost = $\overline{C}'(x)$

   Average Revenue = $\overline{R}(x) = \dfrac{R(x)}{x}$ Revenue per unit

   Marginal Average Revenue = $\overline{R}'(x)$

   Average Profit = $\overline{P}(x) = \dfrac{P(x)}{x}$ Profit per unit

   Marginal Average Profit = $\overline{P}'(x)$

1. $C(x) = 2000 + 50x - 0.5x^2$

   (A) The exact cost of producing the 21st food processor is:

   $$C(21) - C(20) = 2000 + 50(21) - \frac{(21)^2}{2} - \left[2000 + 50(20) - \frac{(20)^2}{2}\right]$$
   $$= 2829.50 - 2800$$
   $$= 29.50 \text{ or } \$29.50$$

   (B) $C'(x) = 50 - x$
   $C'(20) = 50 - 20 = 30$ or $30

3. $C(x) = 60,000 + 300x$

   (A) $\overline{C}(x) = \frac{60,000 + 300x}{x} = \frac{60,000}{x} + 300 = 60,000x^{-1} + 300$

   $$\overline{C}(500) = \frac{60,000 + 300(500)}{500} = \frac{210,000}{500} = 420 \text{ or } \$420$$

   (B) $\overline{C}'(x) = -60,000x^{-2} = \frac{-60,000}{x^2}$

   $$\overline{C}'(500) = \frac{-60,000}{(500)^2} = -0.24 \text{ or } \$0.24$$

   Interpretation: At a production level of 500 frames, average cost is decreasing at the rate of 24¢ per frame.

   (C) The average cost per frame if 501 frames are produced is approximately $420 - $0.24 = $419.76.

5. $R(x) = 100x - 0.025x^2 = 100x - \frac{x^2}{40}$

   $R'(x) = 100 - \frac{x}{20}$

   (A) $R'(1600) = 100 - \frac{1600}{20} = 100 - 80 = 20$ or $20
   Interpretation: At a production level of 1600 radios, revenue is increasing at the rate of $20 per radio.

   (B) $R'(2500) = 100 - \frac{2500}{20} = -25$ or $-25
   Interpretation: At a production level of 2500 radios, revenue is decreasing at the rate of $25 per radio.

7. $P(x) = 30x - 0.5x^2 - 250$

   (A) The exact profit from the sale of the 26th skateboard is:

   $$P(26) - P(25) = 30(26) - \frac{(26)^2}{2} - 250 - \left[30(25) - \frac{(25)^2}{2} - 250\right]$$
   $$= 192 - 187.50$$
   $$= 4.50 \text{ or } \$4.50$$

   (B) $P'(x) = 30 - x$
   $P'(25) = 30 - 25 = 5$ or $5.00

9. $P(x) = 5x - \dfrac{x^2}{200} - 450$

$P'(x) = 5 - \dfrac{x}{100}$

(A) $P'(450) = 5 - \dfrac{450}{100} = 0.5$ or $0.50

Interpretation: At a production level of 450 cassettes, profit is increasing at the rate of 50¢ per cassette.

(B) $P'(750) = 5 - \dfrac{750}{100} = -2.5$ or $-$2.50

Interpretation: At a production level of 750 cassettes, profit is decreasing at the rate of $2.50 per cassette.

11. $P(x) = 30x - 0.03x^2 - 750$

Average profit: $\overline{P}(x) = \dfrac{P(x)}{x} = 30 - 0.03x - \dfrac{750}{x} = 30 - 0.03x - 750x^{-1}$

(A) At $x = 50$, $\overline{P}(50) = 30 - (0.03)50 - \dfrac{750}{50} = 13.50$ or $13.50.

(B) $\overline{P}'(x) = -0.03 + 750x^{-2} = -0.03 + \dfrac{750}{x^2}$

$\overline{P}'(50) = -0.03 + \dfrac{750}{(50)^2} = -0.03 + 0.3 = 0.27$ or $0.27; at a production level of 50 mowers, the average profit per mower is INCREASING at the rate of $0.27 per mower.

(C) The average profit per mower if 51 mowers are produced is approximately $13.50 + $0.27 = $13.77.

13. $p = 200 - \dfrac{x}{30}$, $C(x) = 72{,}000 + 60x$

(A) $C'(x) = 60$ or $60

(B) Revenue: $R(x) = xp = 200x - \dfrac{x^2}{30}$

(C) $R'(x) = 200 - \dfrac{x}{15}$

(D) $R'(1{,}500) = 200 - \dfrac{1{,}500}{15} = 100$; at a production level of 1,500 saws, revenue is INCREASING at the rate of $100 per saw.

$R'(4{,}500) = 200 - \dfrac{4{,}500}{15} = -100$; at a production level of 4,500 saws, revenue is DECREASING at the rate of $100 per saw.

(E) The graphs of $C(x)$ and $R(x)$ are shown below.

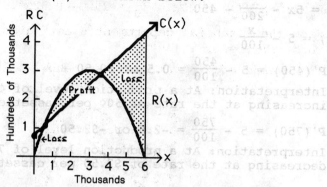

To find the break-even points, set $C(x) = R(x)$:

$$72,000 + 60x = 200x - \frac{x^2}{30}$$
$$x^2 - 4,200x + 2,160,000 = 0$$
$$(x - 600)(x - 3,600) = 0$$
$$x = 600 \quad \text{or} \quad x = 3,600$$

Now, $\quad C(600) = 72,000 + 60(600) \quad = 108,000;$
$\quad\quad\quad C(3,600) = 72,000 + 60(3,600) = 288,000.$
Thus, the break-even points are: $(600, 108,000)$, $(3,600, 288,000)$.

(F) $P(x) = R(x) - C(x) = 200x - \frac{x^2}{30} - [72,000 + 60x]$

$$= -\frac{x^2}{30} + 140x - 72,000$$

(G) $P'(x) = -\frac{x}{15} + 140$

(H) $P'(1,500) = -\frac{1,500}{15} + 140 = 40;$ at a production level of 1,500 saws, the profit is INCREASING at the rate of $40 per saw.

$P'(3,000) = -\frac{3,000}{15} + 140 = -60;$ at a production level of 3,000 saws, the profit is DECREASING at the rate of $60 per saw.

15. (A) We are given $p = 16$ when $x = 200$ and $p = 14$ when $x = 300$. Thus, we have the pair of equations:
$16 = 200m + b$
$14 = 300m + b$
Subtracting the second equation from the first, we get $-100m = 2$. Thus,
$m = -\frac{1}{50}.$
Substituting this into either equation yields $b = 20$. Therefore,
$p = -\frac{x}{50} + 20$ or $p = 20 - \frac{x}{50} = 20 - 0.02x.$

(B) $R(x) = x \cdot p(x) = 20x - \dfrac{x^2}{50} = 20x - 0.02x^2.$

(C) From the financial department's estimates, $m = 4$ and $b = 1400$.
Thus, $C(x) = 4x + 1400$.

(D) The graphs of $R(x)$ and $C(x)$ are shown below:

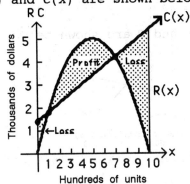

To find the break-even points, set $C(x) = R(x)$.
$$4x + 1400 = 20x - \dfrac{x^2}{50}$$
$$\dfrac{x^2}{50} - 16x + 1400 = 0$$
$$x^2 - 800x + 70,000 = 0$$
$$(x - 100)(x - 700) = 0$$
$$x = 100 \text{ or } x = 700$$
Now, $C(100) = 1800$ and $C(700) = 4200$. Thus, the break-even points
are $(100, 1800)$ and $(700, 4200)$.

(E) $P(x) = R(x) - C(x) = 20x - 0.02x^2 - (4x + 1400) = 16x - 0.02x^2 - 1400$

(F) $P(x) = 16x - 0.02x^2 - 1400$
$P'(x) = 16 - 0.04x$
$P'(250) = 16 - \dfrac{250}{25} = 6 \text{ or } \$6$

Interpretation: At a production level of 250 toasters, profit is
increasing at the rate of $6 per toaster.

$P'(475) = 16 - \dfrac{475}{25} = -3 \text{ or } -\$3$

Interpretation: At a production level of 475 toasters, profit is
decreasing at the rate of $3 per toaster.

17. Total cost: $C(x) = 24x + 21,900$
Total revenue: $R(x) = 200x - 0.2x^2, \ 0 \le x \le 1,000$

(A) $R'(x) = 200 - 0.4x$
The graph of R has a horizontal tangent line at the value(s) of
x where $R'(x) = 0$, i.e.,
$$200 - 0.4x = 0$$
$$\text{or } x = 500$$

(B) $P(x) = R(x) - C(x) = 200x - 0.2x^2 - (24x + 21,900)$
$$= 176x - 0.2x^2 - 21,900$$

(C) $P'(x) = 176 - 0.4x$. Setting $P'(x) = 0$, we have
$$176x - 0.4x = 0$$
or $x = 440$

(D) The graphs of C, R and P are shown below.

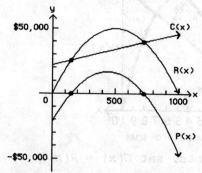

Break-even points: $R(x) = C(x)$
$$200x - 0.2x^2 = 24x + 21,900$$
$$0.2x^2 - 176x + 21,900 = 0$$
$$x = \frac{176 \pm \sqrt{(176)^2 - (4)(0.2)(21,900)}}{2(0.2)} \quad \text{(quadratic formula)}$$
$$= \frac{176 \pm \sqrt{30,976 - 17,520}}{0.4}$$
$$= \frac{176 \pm \sqrt{13,456}}{0.4} = \frac{176 \pm 116}{0.4} = 730, \ 150$$

Thus, the break-even points are: (730, 39,420) and (150, 25,500).

x-intercepts for P: $-0.2x^2 + 17.6x - 21,900 = 0$
$$\text{or } 0.2x^2 - 176x + 21,900 = 0$$
which is the same as the equation above.

Thus, $x = 150$ and $x = 730$.

19. Demand equation: $p = 20 - \sqrt{x} = 20 - x^{1/2}$
Cost equation: $C(x) = 500 + 2x$

(A) Revenue $R(x) = xp = x(20 - x^{1/2})$
$$\text{or } R(x) = 20x - x^{3/2}$$

(B) The graphs for R and C for $0 \le x \le 400$
are shown at the right.

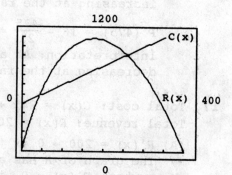

Break-even points (44, 588)
and (258, 1,016).

1. (A) $f(3) - f(1) = 2(3)^2 + 5 - [2(1)^2 + 5] = 16$

   (B) Average rate of change: $\dfrac{f(3) - f(1)}{3 - 1} = \dfrac{16}{2} = 8$

   (C) Slope of secant line: $\dfrac{f(3) - f(1)}{3 - 1} = \dfrac{16}{2} = 8$

   (D) Instantaneous rate of change at $x = 1$:

   Step 1. $\dfrac{f(1 + h) - f(1)}{h} = \dfrac{2(1 + h)^2 + 5 - [2(1)^2 + 5]}{h}$

   $= \dfrac{2(1 + 2h + h^2) + 5 - 7}{h} = \dfrac{4h + 2h^2}{h} = 4 + 2h$

   Step 2. $\lim\limits_{h \to 0} \dfrac{f(1 + h) - f(1)}{h} = \lim\limits_{h \to 0}(4 + 2h) = 4$

   (E) Slope of the tangent line at $x = 1$: 4

   (F) $f'(1) = 4$ (8-1, 8-3, 8-4)

2. $f'(-1) = -2$; $f'(1) = 1$ (8-1, (8-3)

3. (A) $\lim\limits_{x \to 1}(5f(x) + 3g(x)) = 5\lim\limits_{x \to 1}f(x) + 3\lim\limits_{x \to 1}g(x) = 5 \cdot 2 + 3 \cdot 4 = 22$

   (B) $\lim\limits_{x \to 1}[f(x)g(x)] = [\lim\limits_{x \to 1}f(x)][\lim\limits_{x \to 1}g(x)] = 2 \cdot 4 = 8$

   (C) $\lim\limits_{x \to 1}\dfrac{g(x)}{f(x)} = \dfrac{\lim\limits_{x \to 1}g(x)}{\lim\limits_{x \to 1}f(x)} = \dfrac{4}{2} = 2$ (8-2)

4. (A) $\lim\limits_{x \to 1^-}f(x) = 1$ (B) $\lim\limits_{x \to 1^+}f(x) = 1$ (C) $\lim\limits_{x \to 1}f(x) = 1$
   (D) $f(1) = 1$ (8-2)

5. (A) $\lim\limits_{x \to 2^-}f(x) = 2$ (B) $\lim\limits_{x \to 2^+}f(x) = 3$ (C) $\lim\limits_{x \to 2}f(x)$ does not exist
   (D) $f(2) = 3$ (8-2)

6. (A) $\lim\limits_{x \to 3^-}f(x) = 4$ (B) $\lim\limits_{x \to 3^+}f(x) = 4$ (C) $\lim\limits_{x \to 3}f(x) = 4$
   (D) $f(3)$ does not exist   (8-2)

7. $f'(x) = \lim\limits_{h \to 0}\dfrac{f(x + h) - f(x)}{h} = \lim\limits_{h \to 0}\dfrac{3x^2h + 3xh^2 + h^3 + 2xh + h^2}{h}$

   $= \lim\limits_{h \to 0}\dfrac{h(3x^2 + 3xh + h^2 + 2x + h)}{h}$

   $= \lim\limits_{h \to 0}(3x^2 + 3xh + h^2 + 2x + h)$

   $= 3x^2 + 2x$ (8-3)

8. (A) $h(x) = 2f(x) + 3g(x)$; $h(5) = 2f'(5) + 3g'(5) = 2(-1) + 3(-3) = -11$

(B) $h(x) = f(x)g(x)$; $h'(5) = f(5)g'(5) + g(5)f'(5) = 4(-3) + 2(-1) = -14$

(C) $h(x) = \dfrac{f(x)}{g(x)}$; $h'(5) = \dfrac{g(5)f'(5) - f(5)g'(5)}{[g(5)]^2} = \dfrac{2(-1) - 4(-3)}{2^2} = \dfrac{10}{4} = \dfrac{5}{2}$

(D) $h(x) = [f(x)]^2$; $h'(5) = 2f(5)f'(5) = 2(4)(-1) = -8$ (8-4, 8,5, 8-6)

9. $6x + 4$; $\dfrac{d}{dx}(3x^2 + 4x + 1)^5 = 5(3x^2 + 4x + 1)^4(6x + 4)$ (8-6)

10. $f(x) = 3x^4 - 2x^2 + 1$
$f'(x) = 12x^3 - 4x$ (8-4)

11. $f(x) = 2x^{1/2} - 3x$
$f'(x) = 2 \cdot \dfrac{1}{2}x^{-1/2} - 3 = \dfrac{1}{x^{1/2}} - 3$ (8-4)

12. $f(x) = 5$
$f'(x) = 0$
(8-4)

13. $f(x) = \dfrac{1}{2x^2} + \dfrac{x^2}{2}$

$= \dfrac{1}{2}x^{-2} + \dfrac{x^2}{2}$

$f'(x) = -x^{-3} + x$

$= -\dfrac{1}{x^3} + x$

(8-4)

14. $f(x) = (2x - 1)(3x + 2)$
$f'(x) = (2x - 1)(3) + (3x + 2)(2)$
$= 6x - 3 + 6x + 4$
$= 12x + 1$ (8-5)

15. $f(x) = (x^2 - 1)(x^3 - 3)$
$f'(x) = (x^2 - 1)(3x^2) + (x^3 - 3)(2x) = 3x^4 - 3x^2 + 2x^4 - 6x = 5x^4 - 3x^2 - 6x$
(8-5)

16. $f(x) = \dfrac{2x}{x^2 + 2}$

$f'(x) = \dfrac{(x^2 + 2)(2) - 2x(2x)}{(x^2 + 2)^2} = \dfrac{2x^2 + 4 - 4x^2}{(x^2 + 2)^2} = \dfrac{4 - 2x^2}{(x^2 + 2)^2}$ (8-5)

17. $f(x) = \dfrac{1}{3x + 2} = (3x + 2)^{-1}$

$f'(x) = -1(3x + 2)^{-2}(3) = \dfrac{-3}{(3x + 2)^2}$ (8-6)

18. $f(x) = (2x - 3)^3$
$f'(x) = 3(2x - 3)^2(2) = 6(2x - 3)^2$ (8-6)

19. $f(x) = (x^2 + 2)^{-2}$
$f'(x) = -2(x^2 + 2)^{-3}(2x) = \dfrac{-4x}{(x^2 + 2)^3}$ (8-6)

20. $f(x) = 0.5x^2 - 5$

(A) $\dfrac{f(4) - f(2)}{4 - 2} = \dfrac{0.5(4)^2 - 5 - [0.5(2)^2 - 5]}{2} = \dfrac{8 - 2}{2} = 3$

(B) $\dfrac{f(2 + h) - f(2)}{h} = \dfrac{0.5(2 + h)^2 - 5 - [0.5(2)^2 - 5]}{h}$

$= \dfrac{0.5(4 + 4h + h^2) - 5 + 3}{h}$

$= \dfrac{2h + 0.5h^2}{h} = \dfrac{h(2 + 0.5h)}{h} = 2 + 0.5h$

(C) $\displaystyle\lim_{h \to 0} \dfrac{f(2 + h) - f(2)}{h} = \lim_{h \to 0} (2 + 0.5h) = 2$ (8-1, 8-3)

21.

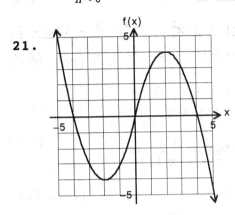

(8-1)

22. $y = 3x^4 - 2x^{-3} + 5$

$\dfrac{dy}{dx} = 12x^3 + 6x^{-4}$ (8-4)

23. $y = (2x^2 - 3x + 2)(x^2 + 2x - 1)$

$y' = (2x^2 - 3x + 2)(2x + 2) + (x^2 + 2x - 1)(4x - 3)$

$= 4x^3 - 6x^2 + 4x + 4x^2 - 6x + 4 + 4x^3 + 8x^2 - 4x - 3x^2 - 6x + 3$

$= 8x^3 + 3x^2 - 12x + 7$ (8-5)

24. $f(x) = \dfrac{2x - 3}{(x - 1)^2}$

$f'(x) = \dfrac{(x - 1)^2 2 - (2x - 3)2(x - 1)}{(x - 1)^4} = \dfrac{(x - 1)[2(x - 1) - 4x + 6]}{(x - 1)^4}$

$= \dfrac{(2x - 2 - 4x + 6)}{(x - 1)^3} = \dfrac{4 - 2x}{(x - 1)^3}$ (8-5)

25. $y = 2\sqrt{x} + \dfrac{4}{\sqrt{x}} = 2x^{1/2} + 4x^{-1/2}$

$y' = 2 \cdot \dfrac{1}{2} x^{-1/2} + 4\left(-\dfrac{1}{2}\right)x^{-3/2} = \dfrac{1}{x^{1/2}} - \dfrac{2}{x^{3/2}}$ or $\dfrac{1}{\sqrt{x}} - \dfrac{2}{\sqrt{x^3}}$ (8-4)

26. $\dfrac{d}{dx}[(x^2 - 1)(2x + 1)^2] = (x^2 - 1)\dfrac{d}{dx}(2x + 1)^2 + (2x + 1)^2\dfrac{d}{dx}(x^2 - 1)$

$= (x^2 - 1)[2(2x + 1)(2)] + (2x + 1)^2(2x)$

$= 2(2x + 1)[2(x^2 - 1) + x(2x + 1)]$

$= 2(2x + 1)(2x^2 - 2 + 2x^2 + x)$

$= 2(2x + 1)(4x^2 + x - 2)$ (8-5, 8-6)

27. $\dfrac{d}{dx}(x^3 - 5)^{1/3} = \dfrac{1}{3}(x^3 - 5)^{-2/3}(3x^2) = \dfrac{x^2}{(x^3 - 5)^{2/3}}$ (8-6)

28. $y = \dfrac{3x^2 + 4}{x^2}$

$\dfrac{dy}{dx} = \dfrac{x^2(6x) - (3x^2 + 4)2x}{[x^2]^2} = \dfrac{-8x}{x^4} = \dfrac{-8}{x^3}$

(8-4)

29. $\dfrac{d}{dx} \dfrac{(x^2 + 2)^4}{2x - 3} = \dfrac{(2x - 3)4(x^2 + 2)^3(2x) - (x^2 + 2)^4(2)}{(2x - 3)^2}$

$= \dfrac{2(x^2 + 2)^3[4x(2x - 3) - (x^2 + 2)]}{(2x - 3)^2}$

$= \dfrac{2(x^2 + 2)^3(8x^2 - 12x - x^2 - 2)}{(2x - 3)^2} = \dfrac{2(x^2 + 2)^3(7x^2 - 12x - 2)}{(2x - 3)^2}$

(8-5, 8-6)

30. $f(x) = x^2 + 4$
$f'(x) = 2x$

(A) The slope of the graph at $x = 1$ is $m = f'(1) = 2$.

(B) $f(1) = 1^2 + 4 = 5$
The tangent line at $(1, 5)$, where the slope $m = 2$, is:
$(y - 5) = 2(x - 1)$ [Note: $(y - y_1) = m(x - x_1)$.]
$\qquad y = 5 + 2x - 2$
$\qquad y = 2x + 3$

(8-3, 8-4)

31. $f(x) = x^3(x + 1)^2$
$f'(x) = x^3(2)(x + 1)(1) + (x + 1)^2(3x^2)$
$\qquad = 2x^3(x + 1) + 3x^2(x + 1)^2$

(A) The slope of the graph of f at $x = 1$ is:
$f'(1) = 2 \cdot 1^3(1 + 1) + 3 \cdot 1^2(1 + 1)^2 = 16$

(B) $f(1) = 1^3(1 + 1)^2 = 4$
An equation for the tangent line to the graph of f at $x = 1$ is
$y - 4 = 16(x - 1)$ or $y = 16x - 12$.

(8-3, 8-5)

32. $f(x) = 10x - x^2$
$f'(x) = 10 - 2x$
The tangent line is horizontal at the values of x such that $f'(x) = 0$:
$10 - 2x = 0$
$\qquad x = 5$

(8-4)

33. $f(x) = (x + 3)(x^2 - 45)$
$f'(x) = (x + 3)(2x) + (x^2 - 45)(1) = 3x^2 + 6x - 45$
Set $f'(x) = 0$:
$3x^2 + 6x - 45 = 0$
$x^2 + 2x - 15 = 0$
$(x - 3)(x + 5) = 0$
$\qquad x = 3, \; x = -5$

(8-5)

34. $f(x) = \dfrac{x}{x^2 + 4}$

$f'(x) = \dfrac{(x^2 + 4)(1) - x(2x)}{(x^2 + 4)^2} = \dfrac{4 - x^2}{(x^2 + 4)^2}$

Set $f'(x) = 0$:

$$\dfrac{4 - x^2}{(x^2 + 4)^2} = 0$$

$$4 - x^2 = 0$$

$$(2 - x)(2 + x) = 0$$

$$x = 2, \ x = -2 \qquad\qquad (8\text{-}5)$$

35. $f(x) = x^2(2x - 15)^3$

$f'(x) = x^2(3)(2x - 15)^2(2) + (2x - 15)^3(2x)$

$\quad = (2x - 15)^2[6x^2 + 4x^2 - 30x]$

$\quad = (2x - 15)^2 10x(x - 3)$

Set $f'(x) = 0$:

$10x(x - 3)(2x - 15)^2 = 0$

$$x = 0, \ x = 3, \ x = \dfrac{15}{2} \qquad\qquad (8\text{-}5)$$

36.

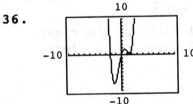

sHorizontal tangent lines at
$x \approx -1.37, \ 0.60, \ 1.52$ $(8\text{-}4)$

37.

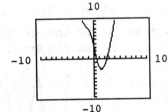

Horizontal tangent line
at $x \approx 1.43$ $(8\text{-}5)$

38.

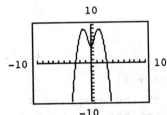

Horizontal tangent line at $x \approx -1.41, \ 1.41$ $(8\text{-}6)$

39. $y = f(x) = 16x^2 - 4x$

(A) Velocity function: $v(x) = f'(x) = 32x - 4$

(B) $v(3) = 32(3) - 4 = 92$ ft/sec $(8\text{-}4)$

40. $y = f(x) = 96x - 16x^2$

(A) Velocity function: $v(x) = f'(x) = 96 - 32x$

(B) $v(x) = 0$ when $96 - 32x = 0$

$$32x = 96$$

$$x = 3 \text{ sec} \qquad\qquad (8\text{-}4)$$

41.

h	-0.1	-0.01	-0.001	$\to 0$	$\leftarrow 0.001$	0.01	0.1
$\dfrac{f(h) - f(0)}{h}$	1.49	1.60	1.61	$\to 1.61$	$\leftarrow 1.61$	1.62	1.75

$f'(0) \approx 1.61$

(8-1, 8-3)

42. $f'(0) \approx -1.39$

(8-1, 8-3)

43. (A) $f(x) = x^3$, $g(x) = (x - 4)^3$, $h(x) = (x + 3)^3$

The graph of g is the graph of f shifted 4 units to the right; the graph of h is the graph of f shifted 3 units to the left.

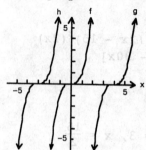

(B) $f'(x) = 3x^2$, $g'(x) = 3(x - 4)^2$, $h'(x) = 3(x + 3)^2$

The graph of g' is the graph of f' shifted 4 units to the right; the graph of h' is the graph of f' shifted 3 units to the left.

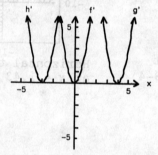

(1-2, 8-6)

44. (A) $g(x) = f(x + k)$; $g'(x) = f'(x + k)(1) = f'(x + k)$

The graph of g is a horizontal translation of the graph of f.
The graph of g' is a horizontal translation of the graph of f'.

(B) $g(x) = f(x) + k$, $g'(x) = f'(x)$

The graph of g is a vertical translation of the graph of f (up k units if $k > 0$, down k units if $k < 0$). The graph of g' is the same as the graph of f'.

(1-2, 8-6)

45. $\lim\limits_{x \to 3} \dfrac{2x - 3}{x + 5} = \dfrac{2(3) - 3}{3 + 5} = \dfrac{6 - 3}{8} = \dfrac{3}{8}$

(8-2)

46. $\lim\limits_{x \to 3}(2x^2 - x + 1) = 2 \cdot 3^2 - 3 + 1 = 16$

(8-2)

47. $\lim\limits_{x \to 0} \dfrac{2x}{3x^2 - 2x} = \lim\limits_{x \to 0} \dfrac{2x}{x(3x - 2)} = \lim\limits_{x \to 0} \dfrac{2}{3x - 2} = \dfrac{2}{-2} = -1$ (8-2)

48. $\lim\limits_{x \to 3} \dfrac{x - 3}{x^2 - 9} = \lim\limits_{x \to 3} \dfrac{\cancel{x - 3}}{(x + 3)\cancel{(x - 3)}}$

$\qquad\qquad = \dfrac{1}{3 + 3} = \dfrac{1}{6}$ (8-2)

49. For $x < 4$, $|x - 4| = -(x - 4) = 4 - x$

$\lim\limits_{x \to 4^-} \dfrac{|x - 4|}{x - 4} = \lim\limits_{x \to 4^-} \dfrac{-(x - 4)}{x - 4} = \lim\limits_{x \to 4^-} (-1) = -1$ (8-2)

50. For $x \geq 4$, $|x - 4| = x - 4$

$\lim\limits_{x \to 4^+} \dfrac{|x - 4|}{x - 4} = \lim\limits_{x \to 4^+} \dfrac{x - 4}{x - 4} = \lim\limits_{x \to 4^+} 1 = 1$ (8-2)

51. It follows from Exercises 49 and 50 that

$\lim\limits_{x \to 4} \dfrac{|x - 4|}{x - 4}$ does not exist.   (8-2)

52. $\lim\limits_{h \to 0} \dfrac{[(2 + h)^2 - 1] - [2^2 - 1]}{h} = \lim\limits_{h \to 0} \dfrac{4 + 4h + h^2 - 1 - 3}{h}$

$\qquad\qquad\qquad\qquad = \lim\limits_{h \to 0} \dfrac{4h + h^2}{h} = \lim\limits_{h \to 0} (4 + h) = 4$ (8-2)

53. $f(x) = x^2 + 4$

$\lim\limits_{h \to 0} \dfrac{f(2 + h) - f(2)}{h} = \lim\limits_{h \to 0} \dfrac{[(2 + h)^2 + 4] - [2^2 + 4]}{h}$

$\qquad\qquad\qquad\qquad = \lim\limits_{h \to 0} \dfrac{4 + 4h + h^2 + 4 - 8}{h} = \lim\limits_{h \to 0} \dfrac{4h + h^2}{h}$

$\qquad\qquad\qquad\qquad = \lim\limits_{h \to 0} (4 + h) = 4$ (8-2)

54. Let $f(x) = \dfrac{1}{x + 2}$

$\lim\limits_{h \to 0} \dfrac{f(x + h) - f(x)}{h} = \lim\limits_{h \to 0} \dfrac{\dfrac{1}{(x + h) + 2} - \dfrac{1}{x + 2}}{h}$

$\qquad\qquad\qquad\qquad = \lim\limits_{h \to 0} \dfrac{x + 2 - (x + h + 2)}{h(x + h + 2)(x + 2)}$

$\qquad\qquad\qquad\qquad = \lim\limits_{h \to 0} \dfrac{-h}{h(x + h + 2)(x + 2)}$

$\qquad\qquad\qquad\qquad = \lim\limits_{h \to 0} \dfrac{-1}{(x + h + 2)(x - 2)} = \dfrac{-1}{(x + 2)^2}$ (8-2)

55. (A) $\lim\limits_{x \to -2^-} f(x) = -6$, $\lim\limits_{x \to -2^+} f(x) = 6$; $\lim\limits_{x \to -2} f(x)$ does not exist

    (B) $\lim\limits_{x \to 0} f(x) = 4$

(C) $\lim\limits_{x \to 2^-} f(x) = 2$, $\lim\limits_{x \to 2^+} f(x) = -2$; $\lim\limits_{x \to 2} f(x)$ does not exist

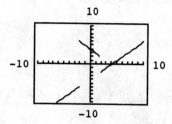

(8-2)

56. $f(x) = x^2 - x$

<u>Step 1</u>. Simplify $\dfrac{f(x + h) - f(x)}{h}$.

$$\frac{f(x + h) - f(x)}{h} = \frac{[(x + h)^2 - (x + h)] - (x^2 - x)}{h}$$

$$= \frac{x^2 + 2xh + h^2 - x - h - x^2 + x}{h}$$

$$= \frac{2xh + h^2 - h}{h} = 2x + h - 1$$

<u>Step 2</u>. Evaluate $\lim\limits_{h \to 0} \dfrac{f(x + h) - f(x)}{h}$.

$$\lim\limits_{h \to 0} \frac{f(x + h) - f(x)}{h} = \lim\limits_{h \to 0}(2x + h - 1) = 2x - 1$$

Thus, $f'(x) = 2x - 1$.

(8-3)

57. $f(x) = \sqrt{x} - 3$

<u>Step 1</u>. Simplify $\dfrac{f(x + h) - f(x)}{h}$.

$$\frac{f(x + h) - f(x)}{h} = \frac{[\sqrt{x + h} - 3] - (\sqrt{x} - 3)}{h}$$

$$= \frac{\sqrt{x + h} - \sqrt{x}}{h} \cdot \frac{\sqrt{x + h} + \sqrt{x}}{\sqrt{x + h} + \sqrt{x}} = \frac{x + h - x}{h[\sqrt{x + h} + \sqrt{x}]}$$

$$= \frac{1}{\sqrt{x + h} + \sqrt{x}}$$

<u>Step 2</u>. Evaluate $\lim\limits_{h \to 0} \dfrac{f(x + h) - f(x)}{h}$.

$$\lim\limits_{h \to 0} \frac{f(x + h) - f(x)}{h} = \lim\limits_{h \to 0} \frac{1}{\sqrt{x + h} + \sqrt{x}} = \frac{1}{2\sqrt{x}}$$

(8-3)

58. f is not differentiable at $x = 0$, since f is not continuous at 0. (8-3)

59. f is not differentiable at $x = 1$; the curve has a vertical tangent line at this point.

(8-3)

60. f is not differentiable at $x = 2$; the curve has a "corner" at this point. (8-3)

61. f is differentiable at $x = 3$. In fact, $f'(3) = 0$. (8-3)

62. $f(x) = (x - 4)^4(x + 3)^3$
$$f'(x) = (x - 4)^4(3)(x + 3)^2(1) + (x + 3)^3(4)(x - 4)^3(1)$$
$$= (x - 4)^3(x + 3)^2[3(x - 4) + 4(x + 3)]$$
$$= 7x(x - 4)^3(x + 3)^2 \qquad (8-5, \ 8-6)$$

63. $f(x) = \dfrac{x^5}{(2x + 1)^4}$
$$f'(x) = \frac{(2x + 1)^4(5x^4) - x^5(4)(2x + 1)^3(2)}{[(2x + 1)^4]^2}$$
$$= \frac{(2x + 1)(5x^4) - 8x^5}{(2x + 1)^5} = \frac{2x^5 + 5x^4}{(2x + 1)^5} = \frac{x^4(2x + 5)}{(2x + 1)^5} \qquad (8-5, \ 8-6)$$

64. $f(x) = \dfrac{\sqrt{x^2 - 1}}{x} = \dfrac{(x^2 - 1)^{1/2}}{x}$
$$f'(x) = \frac{x\left(\frac{1}{2}\right)(x^2 - 1)^{-1/2}(2x) - (x^2 - 1)^{1/2}(1)}{x^2} = \frac{\dfrac{x^2}{(x^2 - 1)^{1/2}} - (x^2 - 1)^{1/2}}{x^2}$$
$$= \frac{1}{x^2(x^2 - 1)^{1/2}} = \frac{1}{x^2\sqrt{x^2 - 1}} \qquad (8-5, \ 8-6)$$

65. $f(x) = \dfrac{x}{\sqrt{x^2 + 4}} = \dfrac{x}{(x^2 + 4)^{1/2}}$
$$f'(x) = \frac{(x^2 + 4)^{1/2}(1) - x\left(\frac{1}{2}\right)(x^2 + 4)^{-1/2}(2x)}{[(x^2 + 4)^{1/2}]^2}$$
$$= \frac{(x^2 + 4)^{1/2} - \dfrac{x^2}{(x^2 + 4)^{1/2}}}{(x^2 + 4)} = \frac{4}{(x^2 + 4)^{3/2}} \qquad (8-5, \ 8-6)$$

66. $f(x) = x^{1/5}$; $f'(x) = \dfrac{1}{5}x^{-4/5} = \dfrac{1}{5x^{4/5}}$

The domain of f' is all real numbers except $x = 0$. At $x = 0$, the graph of f is smooth, but the tangent line to the graph at $(0, 0)$ is vertical. (8-3)

67. $f(x) = \begin{cases} x^2 - m & \text{if } x \le 1 \\ -x^2 + m & \text{if } x > 1 \end{cases}$

(A)

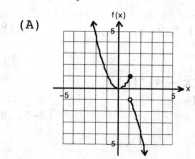

$\lim\limits_{x \to 1^-} f(x) = 1, \quad \lim\limits_{x \to 1^+} f(x) = -1$

(B)

$\lim\limits_{x \to 1^-} f(x) = -1, \quad \lim\limits_{x \to 1^+} f(x) = 1$

(C) $\lim\limits_{x \to 1^-} f(x) = 1 - m, \quad \lim\limits_{x \to 1^+} f(x) = -1 + m$

We want $1 - m = -1 + m$ which implies $m = 1$.

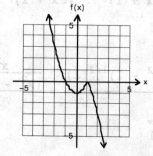

(D) The graphs in (A) and (B) have jumps at $x = 1$; the graph in (C) does not.

(8-2)

68. $f(x) = 1 - |x - 1|, \; 0 \le x \le 2$

(A) $\lim\limits_{h \to 0^-} \dfrac{f(1 + h) - f(1)}{h} = \lim\limits_{h \to 0^-} \dfrac{1 - |1 + h - 1| - 1}{h} = \lim\limits_{h \to 0^-} \dfrac{-|h|}{h}$

$= \lim\limits_{h \to 0^-} \dfrac{h}{h} = 1 \quad (|h| = -h \text{ if } h < 0)$

(B) $\lim\limits_{h \to 0^+} \dfrac{f(1 + h) - f(1)}{h} = \lim\limits_{h \to 0^+} \dfrac{1 - |1 + h - 1| - 1}{h} = \lim\limits_{h \to 0^+} \dfrac{-|h|}{h}$

$= \lim\limits_{h \to 0^+} \dfrac{-h}{h} = -1 \quad (|h| = h \text{ if } h > 0)$

(C) $\lim\limits_{h \to 0} \dfrac{f(1 + h) - f(1)}{h}$ does not exist, since the left limit and the right limit are not equal.

(D) $f'(1)$ does not exist.

(8-3)

69. $C(x) = 10,000 + 200x - 0.1x^2$

(A) $C(101) - C(100) = 10,000 + 200(101) - 0.1(101)^2$
$$- [10,000 + 200(100) - 0.1(100)^2]$$
$$= 29,179.90 - 29,000$$
$$= \$179.90$$

(B) $C'(x) = 200 - 0.2x$
$C'(100) = 200 - 0.2(100)$
$$= 200 - 20$$
$$= \$180 \qquad\qquad\qquad (8\text{-}7)$$

70. $C(x) = 5,000 + 40x + 0.05x^2$

(A) Cost of producing 100 bicycles:
$C(100) = 5,000 + 40(100) + 0.05(100)^2$
$$= 9000 + 500 = 9500$$
Marginal cost:
$C'(x) = 40 + 0.1x$
$C'(100) = 40 + 0.1(100) = 40 + 10 = 50$

Interpretation: At a production level of 100 bicycles, the total cost is $9,500 and is increasing at the rate of $50 per additional bicycle.

(B) Average cost: $\overline{C}(x) = \dfrac{C(x)}{x} = \dfrac{5000}{x} + 40 + 0.05x$

$$\overline{C}(100) = \dfrac{5000}{100} + 40 + 0.05(100) = 50 + 40 + 5 = 95$$

Marginal average cost: $\overline{C}'(x) = -\dfrac{5000}{x^2} + 0.05$

$$\text{and } \overline{C}'(100) = -\dfrac{5000}{(100)^2} + 0.05 = -0.5 + 0.05 = -0.45$$

Interpretation: At a production level of 100 bicycles, the average cost is $95 and the marginal average cost is decreasing at a rate of $0.45 per additional bicycle. $\qquad (8\text{-}7)$

71. The approximate cost of producing the 201st printer is greater than that of producing the 601st printer (the slope of the tangent line at $x = 200$ is greater than the slope of the tangent line at $x = 600$). Since the marginal costs are decreasing, the manufacturing process is becoming more efficient. $\qquad (8\text{-}7)$

72. $p = 25 - 0.1x$, $C(x) = 2x + 9,000$

(A) Marginal cost: $C'(x) = 2$

Average cost: $\overline{C}(x) = \dfrac{C(x)}{x} = 2 + \dfrac{9,000}{x}$

Marginal cost: $\overline{C}'(x) = -\dfrac{9,000}{x^2}$

(B) Revenue: $R(x) = xp = 25x - 0.01x^2$
Marginal revenue: $R'(x) = 25 - 0.02x$
Average revenue: $\bar{R}(x) = \dfrac{R(x)}{x} = 25 - 0.01x$

Marginal average revenue: $\bar{R}'(x) = -0.01$

(C) Profit: $P(x) = R(x) - C(x) = 25x - 0.01x^2 - (2x + 9,000)$
$$= 23x - 0.01x^2 - 9,000$$
Marginal profit: $P'(x) = 23 - 0.02x$

Average profit: $\bar{P}(x) = \dfrac{P(x)}{x} = 23 - 0.01x - \dfrac{9,000}{x}$

Marginal average profit: $\bar{P}'(x) = -0.01 + \dfrac{9,000}{x^2}$

(D) Break-even points: $R(x) = C(x)$
$$25x - 0.01x^2 = 2x + 9,000$$
$$0.01x^2 - 23x + 9,000 = 0$$
$$x^2 - 2,300x + 900,000 = 0$$
$$(x - 500)(x - 1,800) = 0$$

Thus, the break-even points are: $x = 500$, $x = 1,800$.

(E) $P'(1,000) = 23 - 0.02(1000) = 3$; profit is increasing at the rate of \$3 per umbrella.

$P'(1,150) = 23 - 0.02(1,150) = 0$; profit is flat.

$P'(1,400) = 23 - 0.02(1,400) = -5$; profit is decreasing at the rate of \$5 per umbrella.

(F)

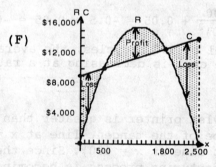

(8-7)

73. $N(t) = \dfrac{40t}{t + 2}$

(A) Average rate of change from $t = 3$ to $t = 6$:

$$\frac{N(6) - N(3)}{6 - 3} = \frac{\dfrac{40 \cdot 6}{6 + 2} - \dfrac{40 \cdot 3}{3 + 2}}{3} = \frac{30 - 24}{3} = 2 \text{ components per day}$$

(B) $N'(t) = \dfrac{(t + 2)(40) - 40t(1)}{(t + 2)^2} = \dfrac{80}{(t + 2)^2}$

$N'(3) = \dfrac{80}{25} = 3.2$ components per day

(8-5)

74. $N(t) = t\sqrt{4 + t} = t(4 + t)^{1/2}$

$N'(t) = t\left(\dfrac{1}{2}\right)(4 + t)^{-1/2} + (4 + t)^{1/2} = \dfrac{t}{2(4 + t)^{1/2}} + \dfrac{(4 + t)^{1/2}}{1}$

$\qquad\qquad = \dfrac{t + 2(4 + t)}{2(4 + t)^{1/2}} = \dfrac{8 + 3t}{2(4 + t)^{1/2}}$

$N(5) = 5\sqrt{4 + 5} = 15; \quad N'(t) = \dfrac{8 + 3(5)}{2(4 + 5)^{1/2}} = \dfrac{23}{6} = 3.833;$

After 5 months, the total sales are 15,000 pools and sales are INCREASING at the rate of 3,833 pools per month. (8-6)

75. (A) $A(t) = 5,000(1.07)^t; \quad A(10) = 5,000(1.07)^{10} \approx \$9836$

h	-0.1	-0.01	$-0.001 \rightarrow$	$0 \leftarrow$	0.001	0.01	0.1
$\dfrac{A(10 + h) - A(10)}{h}$	663.23	665.25	665.45 $\rightarrow$	665 $\leftarrow$	665.50	665.70	667.73

$\qquad A'(10) \approx 665$

(B) After 10 years, the amount in the account is \$9,836 and is GROWING at the rate of \$665 per year. (8-3)

76. $C(x) = 500(x + 1)^{-2}$

The instantaneous rate of change of concentration at x meters is:

$C'(x) = 500(-2)(x + 1)^{-3}$

$\qquad = -1000(x + 1)^{-3} = \dfrac{-1000}{(x + 1)^3}$

The rate of change of concentration at 9 meters is:

$C'(9) = \dfrac{-1000}{(9 + 1)^3} = \dfrac{-1000}{10^3} = -1$ part per million per meter

The rate of change of concentration at 99 meters is:

$C'(99) = \dfrac{-1000}{(99 + 1)^3} = \dfrac{-1000}{100^3} = \dfrac{-10^3}{10^6} = -10^{-3}$ or $= -\dfrac{1}{1000}$

$\qquad\qquad\qquad\qquad\qquad = -0.001$ parts per million per meter (8-4)

77. $F(t) = 98 + \dfrac{4}{\sqrt{t + 1}} = 98 + 4(t + 1)^{-1/2},$

$F'(t) = 4\left(-\dfrac{1}{2}\right)(t + 1)^{-3/2} = \dfrac{-2}{(t + 1)^{3/2}}$

$F(3) = 98 + \dfrac{4}{\sqrt{3 + 1}} = 100; \quad F'(3) = \dfrac{-2}{(3 + 1)^{3/2}} = -\dfrac{1}{4} = -0.25$

After 3 hours, the body temperature of the patient is $100°$ and is DECREASING at the rate of $0.25°$ per hour. (8-6)

78. $N(t) = 20\sqrt{t} = 20t^{1/2}$

The rate of learning is $N'(t) = 20\left(\dfrac{1}{2}\right)t^{-1/2} = 10t^{-1/2} = \dfrac{10}{\sqrt{t}}.$

(A) The rate of learning after one hour is $N'(1) = \dfrac{10}{\sqrt{1}}$

$\qquad\qquad\qquad\qquad\qquad\qquad\qquad = 10$ items per hour.

(B) The rate of learning after four hours is $N'(4) = \dfrac{10}{\sqrt{4}} = \dfrac{10}{2}$

$$= 5 \text{ items per hour.}$$

(8-4)

79. $M(t) = 40.7(1.01)^t$

(A) $M(50) = 40.7(1.01)^{50} \approx 66.9$

h	−0.1	−0.01	−0.001	→ 0	← 0.001	0.01	0.1
$\dfrac{M(50 + h) - M(50)}{h}$	0.666	0.666	0.666	→ 0.666	← 0.666	0.666	0.666

$M'(50) \approx 0.7$

(B) In 2010, there will be 66.9 million married couples and this number will be GROWING at the rate of 0.7 million = 700,000 couples per year.

(C) In 2011, there will be approximately 66.9 + 0.7 = 67.6 million married couples.

In 2012, there will be approximately 67.6 + 0.7 = 68.3 million married couples.

(8-3)

9 GRAPHING AND OPTIMIZATION

Things to remember:

1. **CONTINUITY**

   A function f is CONTINUOUS AT THE POINT $x = c$ if:

   (a) $\lim\limits_{x \to c} f(x)$ exists;

   (b) $f(c)$ exists;

   (c) $\lim\limits_{x \to c} f(x) = f(c)$

   If one or more of the three conditions fails, then f is DISCONTINUOUS at $x = c$.

   A function is CONTINUOUS ON THE OPEN INTERVAL (a, b) if it is continuous at each point on the interval.

2. **ONE-SIDED CONTINUITY**

   A function f is CONTINUOUS ON THE LEFT AT $x = c$ if $\lim\limits_{x \to c^-} f(x) = f(c)$; f is CONTINUOUS ON THE RIGHT AT $x = c$ if $\lim\limits_{x \to c^+} f(x) = f(c)$.

   The function f is continuous on the closed interval $[a, b]$ if it is continuous on the open interval (a, b), and is continuous on the right at a and continuous on the left at b.

3. **CONTINUITY PROPERTIES OF SOME SPECIFIC FUNCTIONS**

   (a) A constant function, $f(x) = k$, is continuous for all x.

   (b) For n a positive integer, $f(x) = x^n$ is continuous for all x.

   (c) A polynomial function
   $$P(x) = a_n x^n + a_{n-1} x^{n-1} + \ldots + a_1 x + a_0$$
   is continuous for all x.

   (d) A rational function
   $$R(x) = \frac{P(x)}{Q(x)},$$
   P and Q polynomial functions, is continuous for all x except those numbers $x = c$ such that $Q(c) = 0$.

   (e) For n an odd positive integer, $n > 1$, $\sqrt[n]{f(x)}$ is continuous wherever f is continuous.

(f) For n an even positive integer, $\sqrt[n]{f(x)}$ is continuous wherever f is continuous and non-negative.

4. VERTICAL ASYMPTOTES

Suppose that the limit of a function f fails to exist as x approaches c from the left because the values of $f(x)$ are becoming very large positive numbers (or very large negative numbers). This is denoted

$$\lim_{x \to c^-} f(x) = \infty \quad (\text{or } -\infty)$$

If this happens as x approaches c from the right, then

$$\lim_{x \to c^+} f(x) = \infty \quad (\text{or } -\infty)$$

If both one-sided limits exhibit the same behavior, then

$$\lim_{x \to c} f(x) = \infty \quad (\text{or } -\infty)$$

If any of the above hold, the line $x = c$ is a VERTICAL ASYMPTOTE for the graph of $y = f(x)$.

5. SIGN PROPERTIES ON AN INTERVAL (a, b)

If f is continuous or (a, b) and $f(x) \neq 0$ for all x in (a, b), then either $f(x) > 0$ for all x in (a, b) or $f(x) < 0$ for all x in (a, b).

6. CONSTRUCTING SIGN CHARTS

Given a function f:

Step 1. Find all partition numbers. That is:

(A) Find all numbers where f is discontinuous. (Rational functions are discontinuous for values of x that make a denominator 0.)

(B) Find all numbers where $f(x) = 0$. (For a rational function, this occurs where the numerator is 0 and the denominator is not 0.)

Step 2. Plot the numbers found in step 1 on a real number line, dividing the number line into intervals.

Step 3. Select a test number in each open interval determined in step 2, and evaluate $f(x)$ at each test number to determine whether $f(x)$ is positive (+) or negative (-) in each interval.

Step 4. Construct a sign chart using the real number line in step 2. This will show the sign of $f(x)$ on each open interval.

[*Note*: From the sign chart, it is easy to find the solution for the inequality $f(x) < 0$ or $f(x) > 0$.]

1. f is continuous at $x = 1$, since $\lim\limits_{x \to 1} f(x) = f(1) = 2$

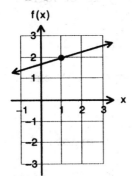

3. f is discontinuous at $x = 1$, since $\lim\limits_{x \to 1} f(x) \neq f(1)$

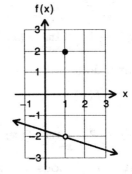

5. $\lim\limits_{x \to 1^-} f(x) = -2 = \lim\limits_{x \to 1^+} f(x)$ implies $\lim\limits_{x \to 1} f(x) = -2$;

f is continuous at $x = 1$, since $\lim\limits_{x \to 1} f(x) = f(1) = -2$

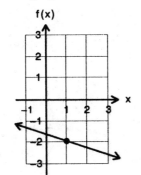

7. $\lim\limits_{x \to 1^-} f(x) = 2$, $\lim\limits_{x \to 1^+} f(x) = -2$ implies $\lim\limits_{x \to 1} f(x)$ does not exist;

f is discontinuous at $x = 1$, since $\lim\limits_{x \to 1} f(x)$ does not exist

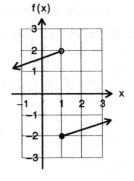

9.

11.

13. $f(x) = 2x - 3$ is a polynomial function. Therefore, f is continuous for all x [$\underline{3}$(c)].

15. $h(x) = \dfrac{2}{x - 5}$ is a rational function and the denominator $x - 5$ is 0 when $x = 5$. Thus, h is continuous for all x except $x = 5$ [$\underline{3}$(d)].

17. $g(x) = \dfrac{x - 5}{(x - 3)(x + 2)}$ is a rational function and the denominator

$(x - 3)(x + 2)$ is 0 when $x = 3$ or $x = -2$. Thus, g is continuous for all x except $x = 3$, $x = -2$.

19. (A) $\lim\limits_{x \to 0^-} f(x) = 1$, $\lim\limits_{x \to 0^+} f(x) = 1$, $\lim\limits_{x \to 0} f(x) = 1$, $f(1) = 1$.

(B) f **is** continuous at $x = 0$ since $\lim\limits_{x \to 0} f(x)$ exists, $f(0)$ exists and $\lim\limits_{x \to 0} f(x) = f(0)$.

21. (A) $\lim\limits_{x \to 1^-} f(x) = 2$, $\lim\limits_{x \to 1^+} f(x) = 1$, $\lim\limits_{x \to 1} f(x)$ does not exist, $f(1) = 1$.

(B) f **is not** continuous at $x = 1$ since $\lim\limits_{x \to 0} f(x)$ does not exist.

23. (A) $\lim\limits_{x \to -2^-} f(x) = 1$, $\lim\limits_{x \to -2^+} f(x) = 1$, $\lim\limits_{x \to -2} f(x) = 1$, $f(-2) = 3$.

(B) f **is not** continuous at $x = -2$ since $\lim\limits_{x \to -2} f(x) \neq f(-2)$.

25. $f(x) = \begin{cases} 2 \text{ if } x \text{ is an integer} \\ 1 \text{ if } x \text{ is not an integer} \end{cases}$

(A) The graph of f is:

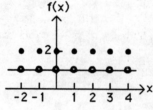

(B) $\lim\limits_{x \to 2} f(x) = 1$ (C) $f(2) = 2$

(D) f is not continuous at $x = 2$ since $\lim\limits_{x \to 2} f(x) \neq f(2)$.

(E) f is discontinuous at $x = n$ for all integers n.

27. $f(x) = \dfrac{1}{x + 3}$

f is discontinuous at $x = -3$; $\lim\limits_{x \to -3^-} f(x) = -\infty$, $\lim\limits_{x \to -3^+} f(x) = \infty$;

the line $x = -3$ is a vertical asymptote.

29. $h(x) = \dfrac{x^2 + 4}{x^2 - 4} = \dfrac{x^2 + 4}{(x - 2)(x + 2)}$

h is discontinuous at $x = -2$; $\lim\limits_{x \to -2^-} h(x) = \infty$ and $\lim\limits_{x \to -2^+} h(x) = -\infty$;

the line $x = -2$ is a vertical asymptote; h is discontinuous at $x = 2$; $\lim\limits_{x \to 2^-} h(x) = -\infty$ and $\lim\limits_{x \to 2^+} h(x) = \infty$; the line $x = 2$ is a vertical asymptote.

31. $F(x) = \dfrac{x^2 - 4}{x^2 + 4}$

$x^2 + 4 \neq 0$ for all x ($x^2 + 4 \geq 4$). Therefore, F is continuous for all real numbers x; there are no vertical asymptotes.

33. $H(x) = \dfrac{x^2 - 2x - 3}{x^2 - 4x + 3} = \dfrac{(x - 3)(x + 1)}{(x - 3)(x - 1)} = \dfrac{x + 1}{x - 1}$, $x \neq 3$

H is discontinuous at $x = 1$; $\lim\limits_{x \to 1^-} H(x) = -\infty$ and $\lim\limits_{x \to 1^+} H(x) = \infty$; the line $x = 1$ is a vertical asymptote; H is discontinuous at $x = 3$ (H is not defined at $x = 3$); $\lim\limits_{x \to 3} H(x) = 2$; there is not a vertical asymptote at $x = 3$.

35. $T(x) = \dfrac{8x - 16}{x^4 - 8x^3 + 16x^2} = \dfrac{8(x - 2)}{x^2(x - 4)^2}$

T is discontinuous at $x = 0$; $\lim\limits_{x \to 0^-} T(x) = -\infty$ and $\lim\limits_{x \to 0^+} T(x) = -\infty$. Therefore $\lim\limits_{x \to 0} T(x) = -\infty$; the line $x = 0$ (the y axis) is a vertical asymptote. T is discontinuous at $x = 4$; $\lim\limits_{x \to 4^-} T(x) = \infty$ and $\lim\limits_{x \to 4^+} T(x) = \infty$. Therefore, $\lim\limits_{x \to 4} T(x) = \infty$; the line $x = 4$ is a vertical asymptote.

37. $x^2 - x - 12 < 0$

Let $f(x) = x^2 - x - 12 = (x - 4)(x + 3)$. Then f is continuous for all x and $f(-3) = f(4) = 0$. Thus, $x = -3$ and $x = 4$ are partition numbers.

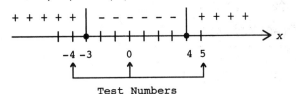

Test Numbers	
x	$f(x)$
-4	8 $(+)$
0	-12 $(-)$
5	8 $(+)$

Thus, $x^2 - x - 12 < 0$ for:

$-3 < x < 4$ (inequality notation)

$(-3, 4)$ (interval notation)

39. $x^2 + 21 > 10x$ or $x^2 - 10x + 21 > 0$

Let $f(x) = x^2 - 10x + 21 = (x - 7)(x - 3)$. Then f is continuous for all x and $f(3) = f(7) = 0$. Thus, $x = 3$ and $x = 7$ are partition numbers.

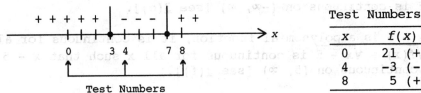

Test Numbers	
x	$f(x)$
0	21 $(+)$
4	-3 $(-)$
8	5 $(+)$

Thus, $x^2 - 10x + 21 > 0$ for:

$x < 3$ or $x > 7$ (inequality notation)

$(-\infty, 3) \cup (7, \infty)$ (interval notation)

41. $\dfrac{x^2 + 5x}{x - 3} > 0$

Let $f(x) = \dfrac{x^2 + 5x}{x - 3} = \dfrac{x(x + 5)}{x - 3}$. Then f is discontinuous at $x = 3$ and $f(0) = f(-5) = 0$. Thus, $x = -5$, $x = 0$, and $x = 3$ are partition numbers.

Test Numbers

x	$f(x)$
-6	$-\frac{2}{3}$ (−)
-1	1 (+)
1	-3 (−)
4	36 (+)

Thus, $\dfrac{x^2 + 5x}{x - 3} > 0$ for: $-5 < x < 0$ or $x > 3$ (inequality notation)

$(-5, 0) \cup (3, \infty)$ (interval notation)

43. $f(x) = x^3 - 3x^2 - 2x + 5$.
Partition numbers: $x_1 \approx -1.33$, $x_2 \approx 1.20$, $x_3 \approx 3.13$

(A) $f(x) > 0$ on $(-1.33, 1.20) \cup (3.13, \infty)$

(B) $f(x) < 0$ on $(-\infty, -1.33) \cup (1.20, 3.13)$

45. $f(x) = x^4 - 6x^2 + 3x + 5$.
Partition numbers: $x_1 \approx -2.53$, $x_2 \approx -0.72$

(A) $f(x) > 0$ on $(-\infty, -2.53) \cup (-0.72, \infty)$

(B) $f(x) < 0$ on $(-2.53, -0.72)$

47. $f(x) = \dfrac{x^3 + x + 6}{-x^3 - 2x + 5}$
Partition numbers: $x_1 \approx -1.63$, $x_2 \approx 1.33$

(A) $f(x) > 0$ on $(-1.63, 1.33)$

(B) $f(x) < 0$ on $(-\infty, -1.63) \cup (1.33, \infty)$

49. $F(x) = 2x^8 - 3x^4 + 5$ is a polynomial function. Thus F is continuous for all x, i.e., F is continuous on $(-\infty, \infty)$ [see $\underline{3}$(c)].

51. Since $f(x) = x - 5$ is a polynomial function, it is continuous for all x. Thus, $g(x) = \sqrt{f(x)} = \sqrt{x - 5}$ is continuous for all x such that $x - 5 \geq 0$, i.e., g is continuous on $[5, \infty)$ [see $\underline{3}$(f)].

53. Since $f(x) = x - 5$ is continuous for all x, $K(x) = \sqrt[3]{x - 5}$ is continuous for all x, i.e., K is continuous on $(-\infty, \infty)$ [see $\underline{3}$(e)].

55. $f(x) = \dfrac{x^2 - 1}{x^2 - 3x + 2} = \dfrac{x^2 - 1}{(x - 1)(x - 2)}$

Since f is a rational function, it is continuous for all x except the numbers $x = c$ at which the denominator is 0. Thus, f is continuous for all x except $x = 1$ and $x = 2$, i.e., f is continuous on $(-\infty, 1)$, $(1, 2)$, and $(2, \infty)$.

57. The graph of f is shown at the right. This function is discontinuous at $x = 1$. [$\lim\limits_{x \to 1} f(x)$ does not exist.]

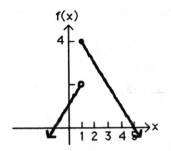

59. The graph of f is:

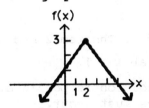

This function is continuous for all x. [<u>Note</u>: $\lim\limits_{x \to 2} f(x) = f(2) = 3$.]

61. The graph of f is:

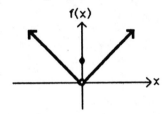

This function is discontinuous at $x = 0$. [<u>Note</u>: $\lim\limits_{x \to 0} f(x) = 0 \neq f(0) = 1$.]

63. (A) Yes; g is continuous on $(-1, 2)$.

(B) Since $\lim\limits_{x \to -1^+} g(x) = -1 = g(-1)$, g is continuous from the right at $x = -1$.

(C) Since $\lim\limits_{x \to 2^-} g(x) = 2 = g(2)$, g is continuous from the left at $x = 2$.

(D) Yes; g is continuous on the closed interval $[-1, 2]$.

65. (A) Since $\lim\limits_{x \to 0^+} f(x) = f(0) = 0$, f is continuous from the right at $x = 0$.

(B) Since $\lim\limits_{x \to 0^-} f(x) = -1 \neq f(0) = 0$, f is not continuous from the left at $x = 0$.

(C) f is continuous on the open interval $(0, 1)$.

(D) f is *not* continuous on the closed interval [0, 1] since
$\lim\limits_{x \to 1^-} f(x) = 0 \neq f(1) = 1$, i.e., f is not continuous from the left at
$x = 1$.

(E) f is continuous on the half-closed interval [0, 1).

67. x intercepts: $x = -5, 2$

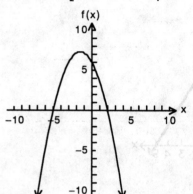

69. x intercepts: $x = -6, -1, 4$

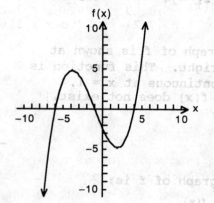

71. $f(x) = \dfrac{2}{1 - x} \neq 0$ for all x. This does not contradict Theorem 2 because f
is not continuous on $(-1, 3)$; f is discontinuous at $x = 1$.

73. The following sketches illustrate that either condition is possible.
Theorem 2 implies that one of these two conditions must occur.

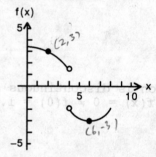

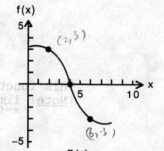

75. (A) $P(x) = \begin{cases} 0.32 & 0 < x \le 1 \\ 0.55 & 1 < x \le 2 \\ 0.78 & 2 < x \le 3 \\ 1.01 & 3 < x \le 4 \\ 1.24 & 4 < x \le 5 \end{cases}$

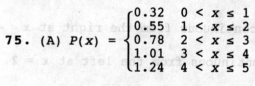

The graph of P is:

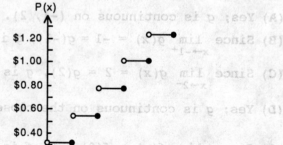

(B) $\lim\limits_{x \to 4.5} P(x) = \$1.24$; $P(4.5) = \$1.24$

(C) $\lim\limits_{x \to 4} P(x)$ does not exist; $P(4) = \$1.01$

(D) P is continuous at $x = 4.5$;
P is not continuous at $x = 4$.

77. (A) $p(x) = \begin{cases} 0.49 & 150 \le x < 250 \\ 0.39 & 250 \le x < 500 \\ 0.29 & 500 \le x < 1000 \\ 0.24 & 1000 \le x \le 1500 \end{cases}$

The graph of p is:

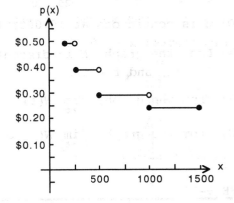

(B) p is discontinuous at $x = 250$, $x = 500$, $x = 1,000$.
In each case, the limit from the left
is greater than the limit from the
right reflecting the corresponding
drop in price at these order
quantities.

(C) $C(x) = \begin{cases} 0.49x & 150 \le x < 250 \\ 0.39x & 250 \le x < 500 \\ 0.29x & 500 \le x < 1000 \\ 0.24x & 1000 \le x \le 1500 \end{cases}$

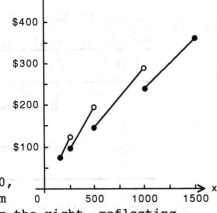

The graph of C is shown at the right.

(D) C is discontinuous at $x = 150$, 250, 500,
and $1,000$. In each case, the limit from
the left is greater than the limit from the right, reflecting
savings to the customer due to the corresponding drop in price at
these order quantities.

79. (A) $E(s) = \begin{cases} 1000, & 0 \le s \le 10,000 \\ 1000 + 0.05(s - 10,000), & 10,000 < s < 20,000 \\ 1500 + 0.05(s - 10,000), & s \ge 20,000 \end{cases}$

The graph of E is:

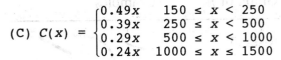

(B) From the graph, $\lim\limits_{s \to 10,000} E(s) = \$1000$ and $E(10,000) = \$1000$.

(C) From the graph, $\lim\limits_{s \to 20,000} E(s)$ does not exist. $E(20,000) = \$2000$.

(D) E is continuous at $10,000$; E is not continuous at $20,000$.

81. (A) From the graph, N is discontinuous at $t = t_2$, $t = t_3$, $t = t_4$, $t = t_6$, and $t = t_7$.

(B) From the graph, $\lim\limits_{t \to t_5} N(t) = 7$ and $N(t_5) = 7$.

(C) From the graph, $\lim\limits_{t \to t_3} N(t)$ does not exist; $N(t_3) = 4$.

EXERCISE 9-2

Things to remember:

1. INCREASING AND DECREASING FUNCTIONS

   For the interval (a, b):

$f'(x)$	$f(x)$	Graph of f	Examples
+	Increases ↗	Rises ↗	
–	Decreases ↘	Falls ↘	

2. CRITICAL VALUES

   The values of x in the domain of f where $f'(x) = 0$ or $f'(x)$ does not exist are called the CRITICAL VALUES of f. The critical values of f are always partition numbers for f', but f' may have partition numbers that are not critical values.

3. LOCAL EXTREMA

   Given a function f. The value $f(c)$ is a LOCAL MAXIMUM of f if there is an interval (m, n) containing c such that $f(x) \leq f(c)$ for all x in (m, n). The value $f(e)$ is a LOCAL MINIMUM of f if there is an interval (p, q) containing e such that $f(x) \geq f(e)$ for all x in (p, q). Local maxima and local minima are called LOCAL EXTREMA.

   A point on the graph where a local extremum occurs is also called a TURNING POINT.

<u>**4.**</u> **EXISTENCE OF LOCAL EXTREMA: CRITICAL VALUES**

If f is continuous on the interval $(a,\ b)$ and $f(c)$ is a local extremum, then either $f'(c) = 0$ or $f'(c)$ does not exist (is not defined).

<u>**5.**</u> **FIRST DERIVATIVE TEST FOR LOCAL EXTREMA**

Let c be a critical value of f [$f(c)$ is defined and either $f'(c) = 0$ or $f'(c)$ is not defined.]

Construct a sign chart for $f'(x)$ close to and on either side of c.

Sign Chart	$f(c)$
$f'(x)$ $\xrightarrow[\substack{m \quad c \quad n}]{(\underset{}{\quad}\ \ ---\,\vdots\,+++\ \ )}x$ $f(x)$ Decreasing $\vdots$ Increasing	$f(c)$ is a local minimum. If $f'(x)$ changes from negative to positive at c, then $f(c)$ is a local minimum.
$f'(x)$ $\xrightarrow[\substack{m \quad c \quad n}]{(\ \ +++\,\vdots\,---\ \ )}x$ $f(x)$ Increasing $\vdots$ Decreasing	$f(c)$ is a local maximum. If $f'(x)$ changes from positive to negative at c, then $f(c)$ is a local maximum.
$f'(x)$ $\xrightarrow[\substack{m \quad c \quad n}]{(\ \ ---\,\vdots\,---\ \ )}x$ $f(x)$ Decreasing $\vdots$ Decreasing	$f(c)$ is not a local extremum. If $f'(x)$ does not change sign at c, then $f(c)$ is neither a local maximum nor a local minimum.
$f'(x)$ $\xrightarrow[\substack{m \quad c \quad n}]{(\ \ +++\,\vdots\,+++\ \ )}x$ $f(x)$ Increasing $\vdots$ Increasing	$f(c)$ is not a local extremum. If $f'(x)$ does not change sign at c, then $f(c)$ is neither a local maximum nor a local minimum.

<u>**6.**</u> **INTERCEPTS AND LOCAL EXTREMA FOR POLYNOMIAL FUNCTIONS**

If $f(x) = a_n x^n + a_{n-1} x^{n-1} + \ldots + a_1 x + a_0$, $a_n \neq 0$ is an nth degree polynomial then f has at most n x intercepts and at most $n-1$ local extrema.

---

1. $(a,\ b),\ (d,\ f),\ (g,\ h)$ **3.** $(b,\ c),\ (c,\ d),\ (f,\ g)$

5. $x = c,\ d,\ f$ **7.** $x = b,\ f$

9. f has a local maximum at $x = a$, and a local minimum at $x = c$; f does not have a local extremum at $x = b$ or at $x = d$.

11.

x	$f'(x)$	$f(x)$	GRAPH OF f
$(-\infty, -1)$	+	Increasing	Rising
$x = -1$	0	Neither local maximum nor local minimum	Horizontal tangent
$(-1, 1)$	+	Increasing	Rising
$x = 1$	0	Local maximum	Horizontal tangent
$(1, \infty)$	−	Decreasing	Falling

Using this information together with the points $(-2, -1)$, $(-1, 1)$, $(0, 2)$, $(1, 3)$, $(2, 1)$ on the graph, we have

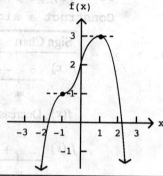

13.

x	$f'(x)$	$f(x)$	GRAPH OF $f(x)$
$(-\infty, -1)$	−	Decreasing	Falling
$x = -1$	0	Local minimum	Horizontal tangent
$(-1, 0)$	+	Increasing	Rising
$x = 0$	Not defined	Local maximum	Vertical tangent line
$(0, 2)$	−	Decreasing	Falling
$x = 2$	0	Neither local maximum nor local minimum	Horizontal tangent
$(2, \infty)$	−	Decreasing	Falling

Using this information together with the points $(-2, 2)$, $(-1, 1)$, $(0, 2)$, $(2, 1)$, $(4, 0)$ on the graph, we have

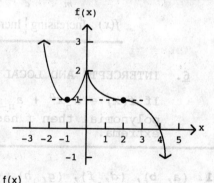

15.

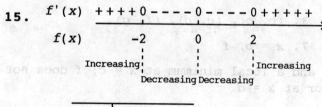

$f'(x)$ $++++0----0----0+++++$

$f(x)$ −2   0   2

Increasing | Decreasing | Decreasing | Increasing

x	−2	0	2
$f(x)$	4	0	−4

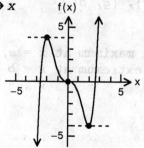

17.

$f'(x)$ $+ + + + 0 - - - \text{ND} - - - 0 + + + + +$

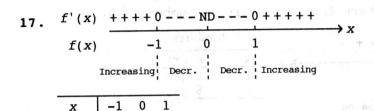

$f(x)$ -1 0 1

Increasing ¦ Decr. ¦ Decr. ¦ Increasing

x	-1	0	1
$f(x)$	2	0	2

19. $f_1' = g_4$ **21.** $f_3' = g_6$ **23.** $f_5' = g_2$

25. $f(x) = x^2 - 16x + 12$
$f'(x) = 2x - 16$
f' is continuous for all x and
$f'(x) = 2x - 16 = 0$
            $x = 8$
Thus, $x = 8$ is a partition number for f'.
Next, we construct a sign chart for f' (partition number is 8).

$f'(x)$ $- - - - - - - - - \ \ \ | + + + +$

                                    → x

         0 $8\ 9$

$f(x)$ Decreasing ¦ Increasing

Test Numbers

x	$f'(x)$
0	$-16\ (-)$
9	$2\ (+)$

Therefore, f is decreasing on $(-\infty, 8)$ and increasing on $(8, \infty)$; f has a local minimum at $x = 8$.

27. $f(x) = 4 + 10x - x^2$
$f'(x) = 10 - 2x$
f' is continuous for all x and
$f'(x) = 10 - 2x = 0$
            $x = 5$
Thus, $x = 5$ is a partition number for f'.
Next, we construct a sign chart for f' (partition number is 5).

$f'(x)$ $+ + + + + \ | - - - -$

                                    → x

         0 $5\ 6$

$f(x)$ Increasing ¦ Decreasing

Test Numbers

x	$f'(x)$
0	$10\ (+)$
6	$-2\ (-)$

Therefore, f is increasing on $(-\infty, 5)$ and decreasing on $(5, \infty)$; f has a local maximum at $x = 5$.

29. $f(x) = 2x^3 + 4$
$f'(x) = 6x^2$
f' is continuous for all x and
$f'(x) = 6x^2 = 0$
          $x = 0$
Thus, $x = 0$ is a partition number for f'.

Next, we construct a sign chart for f' (partition number is 0).

$f'(x)$

```
              + + + + ┊ + + + +
         ────┼─┼─┼─●─┼─┼────────→ x
              -1 0 1
```

$f(x)$ Increasing ┊ Increasing

Test Numbers	
x	$f'(x)$
-1	6 (+)
1	6 (+)

Therefore, f is increasing for all x; i.e., on $(-\infty, \infty)$; f has no local extrema.

31. $f(x) = 2 - 6x - 2x^3$
$f'(x) = -6 - 6x^2$
f' is continuous for all x and
$f'(x) = -6 - 6x^2 = 0$
$\qquad -6(1 + x^2) = 0$
There are no real numbers that satisfy this equation.
The sign chart for f' is:

$f'(x)$

```
              - - - - - - - - -
         ───────────────●──────────→ x
                        0
```

$f(x)$ Decreasing

Test Numbers	
x	$f'(x)$
0	-6 (-)

Thus, f is decreasing for all x; i.e., on $(-\infty, \infty)$; f has no local extrema.

33. $f(x) = x^3 - 12x + 8$
$f'(x) = 3x^2 - 12$
f' is continuous for all x and
$f'(x) = 3x^2 - 12 = 0$
$\qquad 3(x^2 - 4) = 0$
$3(x - 2)(x + 2) = 0$
Thus, $x = -2$ and $x = 2$ are partition numbers for f'.
Next, we construct the sign chart for f' (partition numbers are -2 and 2).

$f'(x)$

```
        ┊                ┊
   + + + + 0 - - - - - - - 0 + + + +
  ──┼───●───┼───────●───┼──→ x
    -3 -2       0      2 3
```

$f(x)$ Increasing ┊ Decreasing ┊ Increasing

Test Numbers	
x	$f'(x)$
-3	15 (+)
0	-12 (-)
3	15 (+)

Therefore, f is increasing on $(-\infty, -2)$ and on $(2, \infty)$, f is decreasing on $(-2, 2)$; f has a local maximum at $x = -2$ and a local minimum at $x = 2$.

35. $f(x) = x^3 - 3x^2 - 24x + 7$
$f'(x) = 3x^2 - 6x - 24$
f' is continuous for all x and
$f'(x) = 3x^2 - 6x - 24 = 0$
$\qquad 3(x^2 - 2x - 8) = 0$
$\qquad 3(x + 2)(x - 4) = 0$
Thus, $x = -2$ and $x = 4$ are partition numbers for f'.
The sign chart for f' is:

f'(x) + + + + 0 − − − − − − 0 + + + +

$f(x)$ Increasing   Decreasing   Increasing

Test Numbers	
x	$f'(x)$
-3	21 (+)
0	-24 (−)
5	21 (+)

Therefore, f is increasing on $(-\infty, -2)$ and on $(4, \infty)$, f is decreasing on $(-2, 4)$; f has a local maximum at $x = -2$ and a local minimum at $x = 4$.

37. $f(x) = 2x^2 - x^4$
$f'(x) = 4x - 4x^3$
f' is continuous for all x and
$f'(x) = 4x - 4x^3 = 0$
$\qquad 4x(1 - x^2) = 0$
$4x(1 - x)(1 + x) = 0$
Thus, $x = -1$, $x = 0$, and $x = 1$ are partition numbers for f'.

The sign chart for f' is:

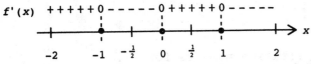

f'(x) + + + + + 0 − − − − − 0 + + + + + 0 − − − − −

$f(x)$ Increasing   Decreasing   Increasing   Decreasing

Test Numbers	
x	$f'(x)$
-2	24 (+)
$-\frac{1}{2}$	$-\frac{3}{2}$ (−)
$\frac{1}{2}$	$\frac{3}{2}$ (+)
2	-24 (−)

Therefore, f is increasing on $(-\infty, -1)$ and on $(0, 1)$, f is decreasing on $(-1, 0)$ and on $(1, \infty)$; f has local maxima at $x = -1$ and $x = 1$ and a local minimum at $x = 0$.

39. $f(x) = 4 + 8x - x^2$
$f'(x) = 8 - 2x$
f' is continuous for all x and
$f'(x) = 8 - 2x = 0$
$\qquad\qquad x = 4$
Thus, $x = 4$ is a partition number for f'.

The sign chart for f' is:

$f'(x)$ + + + + 0 - - - -

$f(x)$ Increasing | Decreasing

Test Numbers	
x	$f'(x)$
0	8 (+)
5	-2 (-)

Therefore, f is increasing on $(-\infty, 4)$ and decreasing on $(4, \infty)$; f has a local maximum at $x = 4$.

x	$f'(x)$	f	GRAPH OF f
$(-\infty, 4)$	+	Increasing	Rising
$x = 4$	0	Local maximum	Horizontal tangent
$(4, \infty)$	–	Decreasing	Falling

x	$f(x)$
0	4
4	20

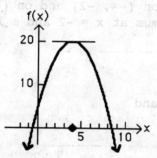

41. $f(x) = x^3 - 3x + 1$
$f'(x) = 3x^2 - 3$
f' is continuous for all x and
$f'(x) = 3x^2 - 3 = 0$
$ 3(x^2 - 1) = 0$
$3(x + 1)(x - 1) = 0$
Thus, $x = -1$ and $x = 1$ are partition numbers for f'.
The sign chart for f' is:

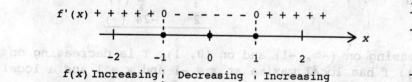

$f'(x)$ + + + + + 0 - - - - - - 0 + + + + +

$f(x)$ Increasing | Decreasing | Increasing

Test Numbers	
x	$f'(x)$
-2	9 (+)
0	-3 (-)
2	9 (+)

Therefore, f is increasing on $(-\infty, -1)$ and on $(1, \infty)$, f is decreasing on $(-1, 1)$; f has a local maximum at $x = -1$ and a local minimum at $x = 1$.

x	$f'(x)$	f	GRAPH OF f
$(-\infty, -1)$	+	Increasing	Rising
$x = -1$	0	Local maximum	Horizontal tangent
$(-1, 1)$	–	Decreasing	Falling
$x = 1$	0	Local minimum	Horizontal tangent
$(1, \infty)$	+	Increasing	Rising

x	$f(x)$
-1	3
0	1
1	-1

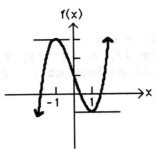

43. $f(x) = 10 - 12x + 6x^2 - x^3$
$f'(x) = -12 + 12x - 3x^2$
f' is continuous for all x and
$f'(x) = -12 + 12x - 3x^2 = 0$
$\qquad -3(x^2 - 4x + 4) = 0$
$\qquad\qquad -3(x - 2)^2 = 0$
Thus, $x = 2$ is a partition number for f'.
The sign chart for f' is:

$f'(x)$ $- - - - - 0 - - - - -$ $\rightarrow x$

      0   1   2   3

$f(x)$ Decreasing │ Decreasing

Test Numbers

x	$f'(x)$
0	-12 (−)
3	-3 (−)

Therefore, f is decreasing for all x, i.e., on $(-\infty, \infty)$, and there is a horizontal tangent line at $x = 2$.

x	$f'(x)$	f	GRAPH OF f
$(-\infty, 2)$	–	Decreasing	Falling
$x = 2$	0		Horizontal tangent
$x > 2$	–	Decreasing	Falling

x	$f(x)$
0	10
2	2

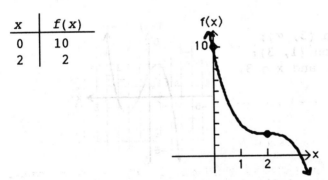

45. Critical values: $x = -1.26$; increasing on $(-1.26, \infty)$; decreasing on $(-\infty, -1.26)$; local minimum at $x = -1.26$

47. Critical values: $x = -0.43$, $x = 0.54$, $x = 2.14$; increasing on $(-0.43, 0.54)$ and $(2.14, \infty)$; decreasing on $(-\infty, -0.43)$ and $(0.54, 2.14)$; local maximum at $x = 0.54$, local minima at $x = -0.43$ and $x = 2.14$

49. Increasing on $(-1, 2)$ $[f'(x) > 0]$;
decreasing on $(-\infty, -1)$ and on $(2, \infty)$ $[f'(x) < 0]$;
local minimum at $x = -1$; local maximum at $x = 2$.

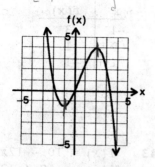

51. Increasing on $(-1, 2)$ and on $(2, \infty)$ $[f'(x) > 0]$;
decreasing on $(-\infty, -1)$ $[f'(x) < 0]$;
local minimum at $x = -1$.

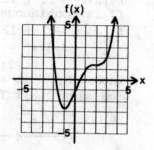

53. $f'(x) > 0$ on $(-\infty, -1)$ and on $(3, \infty)$;
$f'(x) < 0$ on $(-1, 3)$; $f'(x) = 0$ at $x = -1$ and $x = 3$.

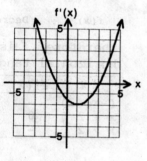

55. $f'(x) > 0$ on $(-2, 1)$ and on $(3, \infty)$;
$f'(x) < 0$ on $(-\infty, -2)$ and on $(1, 3)$:
$f'(x) = 0$ at $x = -2$, $x = 1$, and $x = 3$.

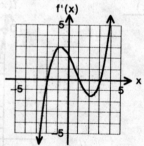

57. $f(x) = \dfrac{x - 1}{x + 2}$ [<u>Note</u>: f is not defined at $x = -2$.]

$f'(x) = \dfrac{(x + 2)(1) - (x - 1)(1)}{(x + 2)^2} = \dfrac{3}{(x + 2)^2}$

Critical values: $f'(x) \neq 0$ for all x, and f' is defined at all points *in the domain of f* (i.e., -2 is not a critical value since -2 is not in the domain of f). Thus, f does not have any critical values; $x = -2$ is a partition number for f'.

The sign chart for f' is:

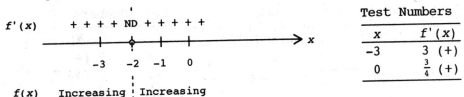

Test Numbers	
x	$f'(x)$
-3	3 (+)
0	$\frac{3}{4}$ (+)

$f(x)$ Increasing ┊ Increasing

Therefore, f is increasing on $(-\infty, -2)$ and on $(-2, \infty)$; f has no local extrema.

59. $f(x) = x + \frac{4}{x}$ [<u>Note</u>: f is not defined at $x = 0$.]

$f'(x) = 1 - \frac{4}{x^2}$

Critical values: $x = 0$ is *not* a critical value of f since 0 is not in the domain of f, but $x = 0$ is a partition number for f'.

$f'(x) = 1 - \frac{4}{x^2} = 0$

$x^2 - 4 = 0$

$(x + 2)(x - 2) = 0$

Thus, the critical values are $x = -2$ and $x = 2$; $x = -2$ and $x = 2$ are also partition numbers for f'.

The sign chart for f' is:

$f'(x)$ + + + + 0 - - ND - - 0 + + + +

```
  ┊     ┊     ┊
──┼──●──┼──○──┼──●──┼──→ x
 -3 -2 -1  0  1  2  3
  ┊     ┊     ┊
```

$f(x)$ Increasing ┊ Decreasing ┊ Increasing

Test Numbers	
x	$f'(x)$
-3	$\frac{5}{9}$ (+)
-1	-3 (−)
1	-3 (−)
3	$\frac{5}{9}$ (+)

Therefore, f is increasing on $(-\infty, -2)$ and on $(2, \infty)$, f is decreasing on $(-2, 0)$ and on $(0, 2)$; f has a local maximum at $x = -2$ and a local minimum at $x = 2$.

61. $f(x) = 1 + \frac{1}{x} + \frac{1}{x^2}$ [<u>Note</u>: f is not defined at $x = 0$.]

$f'(x) = -\frac{1}{x^2} - \frac{2}{x^3}$

Critical values: $x = 0$ is not a critical value of f since 0 is not in the domain of f; $x = 0$ is a partition number for f'.

$f'(x) = -\frac{1}{x^2} - \frac{2}{x^3} = 0$

$-x - 2 = 0$

$x = -2$

Thus, the critical value is $x = -2$; -2 is also a partition number for f'.

The sign chart for f' is:

$f'(x)$ - - - - - 0 + + + + + ND - - - - -

$$\xrightarrow{\quad\quad\quad\quad\quad\quad\quad\quad\quad\quad} x$$

$$-3 \quad -2 \quad -1 \quad 0 \quad 1$$

$f(x)$ Decreasing | Increasing | Decreasing

Test Numbers	
x	$f'(x)$
-3	$-\frac{1}{27}$ (−)
-1	1 (+)
1	-3 (−)

Therefore, f is increasing on $(-2, 0)$ and f is decreasing on $(-\infty, -2)$ and on $(0, \infty)$; f has a local minimum at $x = -2$.

63. $f(x) = \dfrac{x^2}{x - 2}$ [Note: f is not defined at $x = 2$.]

$f'(x) = \dfrac{(x - 2)(2x) - x^2(1)}{(x - 2)^2} = \dfrac{x^2 - 4x}{(x - 2)^2}$

Critical values: $x = 2$ is *not* a critical value of f since 2 is not in the domain of f; $x = 2$ is a partition number for f'.

$f'(x) = \dfrac{x^2 - 4x}{(x - 2)^2} = 0$

$x^2 - 4x = 0$

$x(x - 4) = 0$

Thus, the critical values are $x = 0$ and $x = 4$; 0 and 4 are also partition numbers for f'.

The sign chart for f' is:

$f'(x)$ + + + + 0 − − ND − − 0 + + + +

$$\xrightarrow{\quad\quad\quad\quad\quad\quad\quad\quad\quad\quad} x$$

$$-1 \quad 0 \quad 1 \quad 2 \quad 3 \quad 4 \quad 5$$

$f(x)$ Increasing | Decreasing | Increasing

Test Numbers	
x	$f'(x)$
-1	$\frac{5}{9}$ (+)
1	-3 (−)
3	-3 (−)
5	$\frac{5}{9}$ (+)

Therefore, f is increasing on $(-\infty, 0)$ and on $(4, \infty)$, f is decreasing on $(0, 2)$ and on $(2, 4)$; f has a local maximum at $x = 0$ and a local minimum at $x = 4$.

65. $f(x) = x^4(x - 6)^2$

$f'(x) = x^4(2)(x - 6)(1) + (x - 6)^2(4x^3)$

$\quad\quad = 2x^3(x - 6)[x + 2(x - 6)]$

$\quad\quad = 2x^3(x - 6)(3x - 12)$

$\quad\quad = 6x^3(x - 4)(x - 6)$

Thus, the critical values of f are $x = 0$, $x = 4$, and $x = 6$.

Now we construct the sign chart for f' ($x = 0$, $x = 4$, $x = 6$ are partition numbers).

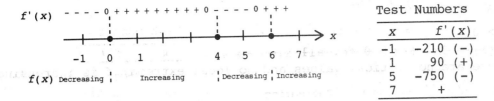

	Test Numbers	
x	f'(x)	
-1	-210	(-)
1	90	(+)
5	-750	(-)
7		+

Therefore, f is increasing on $(0, 4)$ and on $(6, \infty)$, f is decreasing on $(-\infty, 0)$ and on $(4, 6)$; f has a local maximum at $x = 4$ and local minima at $x = 0$ and $x = 6$.

67. $f(x) = 3(x - 2)^{2/3} + 4$

$f'(x) = 3\left(\dfrac{2}{3}\right)(x - 2)^{-1/3} = \dfrac{2}{(x - 2)^{1/3}}$

Critical values: f' is not defined at $x = 2$. [Note: $f(2)$ is defined, $f(2) = 4$.] $f'(x) \neq 0$ for all x. Thus, the critical value for f is $x = 2$; $x = 2$ is also a partition number for f'.

	Test Numbers	
x	f'(x)	
1	-2	(-)
3	2	(+)

Therefore, f is increasing on $(2, \infty)$ and decreasing on $(-\infty, 2)$; f has a local minimum at $x = 2$.

69. $f(x) = 2\sqrt{x} - x = 2x^{1/2} - x, \quad x > 0$

$f'(x) = x^{-1/2} - 1 = \dfrac{1}{x^{1/2}} - 1 = \dfrac{1 - \sqrt{x}}{\sqrt{x}}, \quad x > 0$

Critical values: $f'(x) = \dfrac{1 - \sqrt{x}}{\sqrt{x}} = 0, \quad x > 0$

$1 - \sqrt{x} = 0$
$\sqrt{x} = 1$
$x = 1$

Thus, the critical value for f is $x = 1$; $x = 1$ is also a partition number for f'.

The sign chart for f' is:

	Test Numbers	
x	f'(x)	
$\frac{1}{4}$	1	(+)
4	$-\frac{1}{2}$	(-)

Therefore, f is increasing on $(0, 1)$ and decreasing on $(1, \infty)$; f has a local maximum at $x = 1$.

71. Let $f(x) = x^3 + kx$

(A) $k > 0$

$f'(x) = 3x^2 + k > 0$ for all x.
There are no critical values and no local extrema; f is increasing on $(-\infty, \infty)$.

(B) $k < 0$

$f'(x) = 3x^2 + k;\ 3x^2 + k = 0$

$$x^2 = -\frac{k}{3}$$

$$x = \pm\sqrt{-\frac{k}{3}}$$

Critical vlaues: $x = -\sqrt{-\frac{k}{3}},\ x = \sqrt{-\frac{k}{3}}$;

$f'(x)$ $+++++0--------0+++++$

$-\sqrt{-\frac{k}{3}}$ 0 $\sqrt{-\frac{k}{3}}$

f is increasing on $\left(-\infty,\ -\sqrt{-\frac{k}{3}}\right)$ and on $\left(\sqrt{-\frac{k}{3}},\ \infty\right)$; f is decreasing on $\left(-\sqrt{-\frac{k}{3}},\ \sqrt{-\frac{k}{3}}\right)$; f has a local maximum at $x = -\sqrt{-\frac{k}{3}}$ and a local minimum at $x = \sqrt{-\frac{k}{3}}$.

(C) The only critical value is $x = 0$. There are no extrema, the function is increasing for all x.

73. (A) The marginal profit function, P', is positive on $(0, 600)$, zero at $x = 600$, and negative on $(600, 1,000)$.

(B)

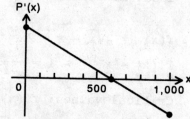

75. (A) The price function, $B(t)$, decreases for the first 15 months to a local minimum, increases for the next 40 months to a local maximum, and then decreases for the remaining 15 months.

(B)

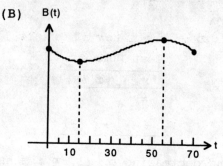

77. $C(x) = \dfrac{x^2}{20} + 20x + 320$

(A) $\overline{C}(x) = \dfrac{C(x)}{x} = \dfrac{x}{20} + 20 + \dfrac{320}{x}$

(B) Critical values:

$$\overline{C}'(x) = \frac{1}{20} - \frac{320}{x^2} = 0$$
$$x^2 - 320(20) = 0$$
$$x^2 - 6400 = 0$$
$$(x - 80)(x + 80) = 0$$

Thus, the critical value of $\overline{C}$ on the interval $(0, 150)$ is $x = 80$. Next, construct the sign chart for $\overline{C}'$ ($x = 80$ is a partition number for $\overline{C}'$).

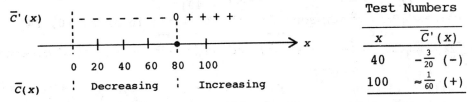

Test Numbers

x	$\overline{C}'(x)$
40	$-\frac{3}{20}$ (−)
100	$\approx \frac{1}{60}$ (+)

Therefore, $\overline{C}$ is increasing for $80 < x < 150$ and decreasing for $0 < x < 80$; $\overline{C}$ has a local minimum at $x = 80$.

79. $P(x) = R(x) - C(x)$
$P'(x) = R'(x) - C'(x)$
Thus, if $R'(x) > C'(x)$ on the interval (a, b), then $P'(x) = R'(x) - C'(x) > 0$ on this interval and P is increasing.

81. $C(t) = \dfrac{0.14t}{t^2 + 1}$, $0 < t < 24$

$C'(t) = \dfrac{(t^2 + 1)(0.14) - 0.14t(2t)}{(t^2 + 1)^2} = \dfrac{0.14(1 - t^2)}{(t^2 + 1)^2}$

Critical values: C' is continuous for all t on the interval $(0, 24)$:

$$C'(t) = \frac{0.14(1 - t^2)}{(t^2 + 1)^2} = 0$$
$$1 - t^2 = 0$$
$$(1 - t)(1 + t) = 0$$

Thus, the critical value of C on the interval $(0, 24)$ is $t = 1$. The sign chart for C' ($t = 1$ is a partition number for C') is:

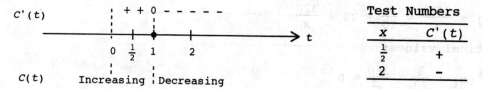

Test Numbers	
x	$C'(t)$
$\frac{1}{2}$	+
2	−

Therefore, C is increasing for $0 < t < 1$ and decreasing for $1 < t < 24$; C has a local maximum at $t = 1$.

83. $P(t) = \dfrac{8.4t}{t^2 + 49} + 0.1, \quad 0 < t < 24$

$P'(t) = \dfrac{(t^2 + 49)(8.4) - 8.4t(2t)}{(t^2 + 49)^2} = \dfrac{8.4(49 - t^2)}{(t^2 + 49)^2}$

Critical values: P is continuous for all t on the interval $(0, 24)$:

$P'(t) = \dfrac{8.4(49 - t^2)}{(t^2 + 49)^2} = 0$

$49 - t^2 = 0$

$(7 - t)(7 + t) = 0$

Thus, the critical value of P on $(0, 24)$ is $t = 7$.

The sign chart for P' ($t = 7$ is a partition number for P') is:

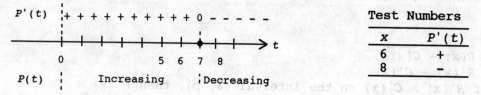

Test Numbers	
x	$P'(t)$
6	+
8	−

Therefore, P is increasing for $0 < t < 7$ and decreasing for $7 < t < 24$; P has a local maximum at $t = 7$.

EXERCISE 9-3

Things to remember:

1. SECOND DERIVATIVE

   For $y = f(x)$, the SECOND DERIVATIVE of f, provided it exists, is:

   $f''(x) = \dfrac{d}{dx} f'(x)$

   Other notations for $f''(x)$ are:

   $\dfrac{d^2 y}{dx^2}$ and y''.

<u>2</u>. CONCAVITY

For the interval (a, b):

$f''(x)$	$f'(x)$	Graph of $y = f(x)$	Example
+	Increasing	Concave upward	
−	Decreasing	Concave downward	

<u>3</u>. INFLECTION POINT

An INFLECTION POINT is a point on the graph of a function where the concavity changes (from upward to downward, or from downward to upward).

If $y = f(x)$ is continuous on (a, b) and has an inflection point at $x = c$, then either $f''(c) = 0$ or $f''(c)$ does not exist.

<u>4</u>. SECOND-DERIVATIVE TEST FOR LOCAL MAXIMA AND MINIMA

Let c be a critical value for $f(x)$.

$f'(c)$	$f''(c)$	Graph of f is	$f(c)$	Example
0	+	Concave upward	Local minimum	
0	−	Concave downward	Local maximum	
0	0	?	Test fails	

The first-derivative test must be used whenever $f''(c) = 0$ [or $f''(c)$ does not exist].

---

1. From the graph, f is concave upward on (a, c), (c, d), and (e, g).

3. $f''(x) < 0$ on (d, e) and (g, h)

5. $f'(x)$ is increasing on (a, c), (c, d), and (e, g).

7. From the graph, f has inflection points at $x = d$, e, and g.

9. $f'(x) > 0$, $f''(x) > 0$; (c)

11. $f'(x) < 0$, $f''(x) > 0$; (d)

13. f has a local minimum at $x = 2$.

15. Unable to determine from the given information ($f'(-3) = f''(-3) = 0$).

17. Neither a local maximum nor a local minimum at $x = 6$; $x = 6$ is not a critical value of f.

19.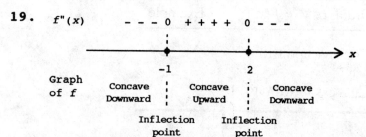

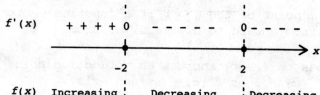

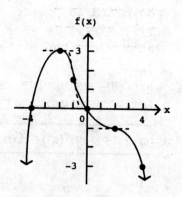

Using this information together with the points $(-4, 0)$, $(-2, 3)$, $(-1, 1.5)$, $(0, 0)$, $(2, -1)$, $(4, -3)$ on the graph, we have

21.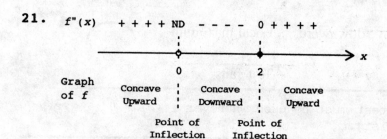

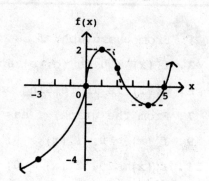

Using this information together with the points $(-3, -4)$, $(0, 0)$, $(1, 2)$, $(2, 1)$, $(4, -1)$, $(5, 0)$ on the graph, we have

23.

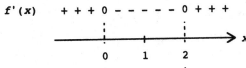

$f''(x)$ $-----0 + + + + +$

Graph of f Concave Downward   Concave Upward

$f'(x)$ $+ + + 0 - - - - - - 0 + + +$

$f(x)$ Increasing   Decreasing   Increasing

Local maximum   Local minimum

x	0	1	2
$f(x)$	2	0	-2

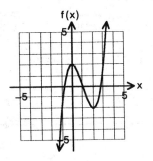

25.

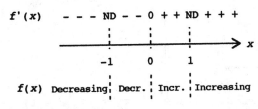

$f''(x)$ $- - - ND + + + + ND - - -$

Graph of f Concave downward   Concave upward   Concave downward

$f'(x)$ $- - - ND - - 0 + + ND + + +$

$f(x)$ Decreasing   Decr.   Incr.   Increasing

x	-1	0	1
$f(x)$	0	-2	0

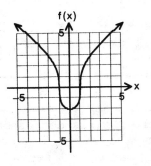

27. $f(x) = x^3 - 2x^2 - 1$

$f'(x) = 3x^2 - 4x$

$f''(x) = 6x - 4$

29. $y = 2x^5 - 3$

$\dfrac{dy}{dx} = 10x^4$

$\dfrac{d^2y}{dx^2} = 40x^3$

31. $y = 1 - 2x + x^3$

$y' = -2 + 3x^2$

$y'' = 6x$

33. $y = (x^2 - 1)^3$

$y' = 3(x^2 - 1)^2(2x) = 6x(x^2 - 1)^2$

$y'' = 6x(2)(x^2 - 1)(2x) + (x^2 - 1)^2(6)$

$\quad = 24x^2(x^2 - 1) + 6(x^2 - 1)^2$

$\quad = 6(x^2 - 1)[4x^2 + x^2 - 1]$

$\quad = 6(x^2 - 1)(5x^2 - 1)$

35. $f(x) = 3x^{-1} + 2x^{-2} + 5$

$f'(x) = -3x^{-2} - 4x^{-3}$

$f''(x) = 6x^{-3} + 12x^{-4}$

37. $f(x) = 2x^2 - 8x + 6$

$f'(x) = 4x - 8 = 4(x - 2)$

$f''(x) = 4$

Critical value: $x = 2$

Now, $f''(2) = 4 > 0$. Therefore, $f(2) = 2 \cdot 2^2 - 8 \cdot 2 + 6 = -2$ is a local minimum.

39. $f(x) = 2x^3 - 3x^2 - 12x - 5$

$f'(x) = 6x^2 - 6x - 12 = 6(x^2 - x - 2) = 6(x - 2)(x + 1)$

$f''(x) = 12x - 6 = 6(2x - 1)$

Critical values: $x = 2, -1$.

Now, $f''(2) = 6(2 \cdot 2 - 1) = 18 > 0$.

Therefore, $f(2) = 2 \cdot 2^3 - 3 \cdot 2^2 - 12 \cdot 2 - 5 = -25$ is a local minimum.

$f''(-1) = 6[2(-1) - 1] = -18 < 0$.

Therefore, $f(-1) = 2(-1)^3 - 3(-1)^2 - 12(-1) - 5 = 2$ is a local maximum.

41. $f(x) = 3 - x^3 + 3x^2 - 3x$

$f'(x) = -3x^2 + 6x - 3 = -3(x^2 - 2x + 1) = -3(x - 1)^2$

$f''(x) = -6(x - 1)$

Critical value: $x = 1$

Now, $f''(1) = -6(1 - 1) = 0$.

Thus, the second-derivative test fails. Since $f'(x) = -3(x - 1)^2 < 0$ for all $x \neq 1$, $f(x)$ is decreasing on $(-\infty, \infty)$. Therefore, $f(x)$ has no local extrema.

43. $f(x) = x^4 - 8x^2 + 10$

$f'(x) = 4x^3 - 16x = 4x(x^2 - 4) = 4x(x + 2)(x - 2)$

$f''(x) = 12x^2 - 16$

Critical values: $x = 0, -2, 2$

Now $f''(0) = 0 - 16 = -16 < 0$. Therefore, $f(0) = 10$ is a local maximum. $f''(-2) = 12(-2)^2 - 16 = 32 > 0$. Therefore, $f(-2) = -6$ is a local minimum. $f''(2) = 12 \cdot 2^2 - 16 = 32 > 0$. Therefore, $f(2) = -6$ is a local minimum.

45. $f(x) = x^6 + 3x^4 + 2$

$f'(x) = 6x^5 + 12x^3 = 6x^3(x^2 + 2)$

$f''(x) = 30x^4 + 36x^2 = 6x^2(5x^2 + 6)$

Critical value: $x = 0$ [Note: $x^2 + 2 \neq 0$ for all x.]

Now, $f''(0) = 0$. Thus, the second-derivative test fails, so the first-derivative test must be used.

The sign chart for f' (partition number is 0) is:

f'(x)   - - - - 0 + + + +

```
              +-------●-------+--------→ x
             -1       0       1
```

f(x)   Decreasing | Increasing

Test Numbers	
x	f'(x)
-1	-18 (−)
1	18 (+)

Therefore, $f(0) = 2$ is a local minimum of f.

47. $f(x) = x + \dfrac{16}{x}$

$f'(x) = 1 - \dfrac{16}{x^2} = \dfrac{x^2 - 16}{x^2} = \dfrac{(x + 4)(x - 4)}{x^2}$

$f''(x) = \dfrac{32}{x^3}$

Critical values: $x = -4, 4$ [Note: $x = 0$ is not a critical value, since $f(0)$ is not defined.]

$f''(-4) = \dfrac{32}{(-4)^3} = -\dfrac{1}{2} < 0$

Therefore, $f(-4) = -8$ is a local maximum.

$f''(4) = \dfrac{32}{(4)^3} = \dfrac{1}{2} > 0$

Therefore, $f(4) = 8$ is a local minimum.

49. $f(x) = x^2 - 4x + 5$

$f'(x) = 2x - 4$

$f''(x) = 2 > 0$ Thus, the graph of f is concave upward for all x; there are no inflection points.

51. $f(x) = x^3 - 18x^2 + 10x - 11$

$f'(x) = 3x^2 - 36x + 10$

$f''(x) = 6x - 36$

Now, $f''(x) = 6x - 36 = 0$

$x - 6 = 0$

$x = 6$

The sign chart for f'' (partition number is 6) is:

f''(x)   - - - - - - 0 + + + +

```
          +++++++●++--------→ x
         0      5 6 7
```

Graph   Concave | Concave
of f   Downward | Upward

Test Numbers	
x	f''(x)
5	-6 (−)
7	6 (+)

Therefore, the graph of f is concave upward on $(6, \infty)$ and concave downward on $(-\infty, 6)$; there is an inflection point at $x = 6$.

53. $f(x) = x^4 - 24x^2 + 10x - 5$
$f'(x) = 4x^3 - 48x + 10$
$f''(x) = 12x^2 - 48$
Now, $f''(x) = 12x^2 - 48 = 0$
$12(x^2 - 4) = 0$
$12(x + 2)(x - 2) = 0$
$x = -2, 2$
The sign chart for f'' (partition numbers are -2 and 2) is:

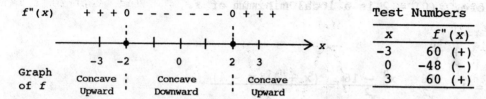

x	f"(x)
-3	60 (+)
0	-48 (−)
3	60 (+)

Test Numbers

Thus, the graph of f is concave upward on $(-\infty, -2)$ and on $(2, \infty)$, the graph is concave downward on $(-2, 2)$; the graph has inflection points at $x = -2$ and at $x = 2$.

55. $f(x) = -x^4 + 4x^3 + 3x + 7$
$f'(x) = -4x^3 + 12x^2 + 3$
$f''(x) = -12x^2 + 24x$
Now, $f''(x) = -12x^2 + 24x = 0$
$12x(-x + 2) = 0$
or $-12x(x - 2) = 0$
$x = 0, 2$
The sign chart for f'' (partition numbers are 0 and 2) is:

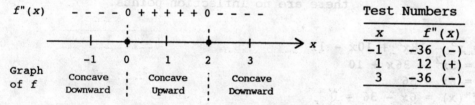

x	f"(x)
-1	-36 (−)
1	12 (+)
3	-36 (−)

Test Numbers

Thus, the graph of f is concave downward on $(-\infty, 0)$ and on $(2, \infty)$. The graph is concave upward on $(0, 2)$; the graph has inflection points at $x = 0$ and at $x = 2$.

57. $f(x) = x^3 - 6x^2 + 16$
$f'(x) = 3x^2 - 12x = 3x(x - 4)$
$f''(x) = 6x - 12 = 6(x - 2)$
Critical values: $x = 0, 4$
$f''(0) = -12 < 0$. Therefore, f has a local maximum at $x = 0$. $f''(4) = 6(4 - 2) = 12 > 0$. Therefore, f has a local minimum at $x = 4$.

The sign chart for $f''(x)$ (partition number is 2) is:

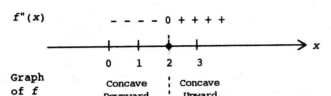

Test Numbers	
x	$f''(x)$
0	-12 $(-)$
3	6 $(+)$

The graph of f has an inflection point at $x = 2$. The graph of f is:

x	$f(x)$
0	16
2	0
4	-16

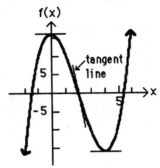

59. $f(x) = x^3 + x + 2$
$f'(x) = 3x^2 + 1$
$f''(x) = 6x$
Since $f'(x) = 3x^2 + 1 > 0$ for all x, f does not have any critical
values. Now, $f''(x) = 6x = 0$
$$x = 0$$
The sign chart for f'' (partition number is 0) is:

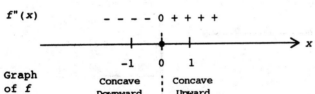

Test Numbers	
x	$f''(x)$
-1	-6 $(-)$
1	6 $(+)$

The graph of f has an
inflection point at $x = 0$.
The graph of f is:

x	$f(x)$
-1	0
0	2
1	4

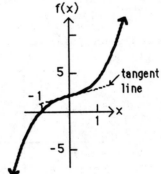

61. $f(x) = (2 - x)^3 + 1$
$f'(x) = 3(2 - x)^2(-1) = -3(2 - x)^2$
$f''(x) = -6(2 - x)(-1) = 6(2 - x)$
Critical value: $x = 2$
$f''(2) = 0$. Thus, the second derivative test fails. Note that $f'(x) = -3(2 - x)^2 < 0$ for all $x \neq 2$. Therefore, f is decreasing on $(-\infty, 2)$ and on $(2, \infty)$, and f does not have any local extrema. The sign chart for f'' (partition number is 2) is:

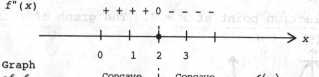

f"(x)

$+ + + + 0 - - - -$

0   1   2   3

Graph of f Concave Upward | Concave Downward

Test Numbers	
x	$f''(x)$
0	12 (+)
3	-6 (-)

The graph of f has an inflection point at $x = 2$.
The graph of f is:

x	$f(x)$
0	9
2	1
3	0

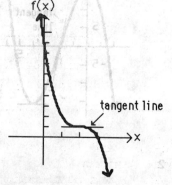

f(x)

tangent line

63. $f(x) = x^3 - 12x$
$f'(x) = 3x^2 - 12 = 3(x^2 - 4) = 3(x + 2)(x - 2)$
$f''(x) = 6x$
Critical values: $x = -2$, $x = 2$
$f''(-2) = 6(-2) = -12 < 0$. Therefore, f has a local maximum at $x = -2$.
$f''(2) = 6(2) = 12 > 0$. Therefore, f has a local minimum at $x = 2$. The sign chart for $f''(x) = 6x$ (partition number is 0) is:

f"(x)

$- - - - 0 + + + +$

-1   0   1

Graph of f Concave Downward | Concave Upward

Test Numbers	
x	$f''(x)$
-1	-6 (-)
1	6 (+)

The graph of f has an inflection point at $x = 0$. The graph of f is:

x	$f(x)$
-2	+16
0	0
2	-16

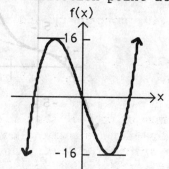

f(x)

16

-16

65.

x	$f'(x)$	$f(x)$
$-\infty < x < -1$	Positive and decreasing	Increasing and concave downward
$x = -1$	x intercept	Local maximum
$-1 < x < 0$	Negative and decreasing	Decreasing and concave downward
$x = 0$	Local minimum	Inflection point
$0 < x < 2$	Negative and increasing	Decreasing and concave upward
$x = 2$	Local maximum	Inflection point
$2 < x < \infty$	Negative and decreasing	Decreasing and concave downward

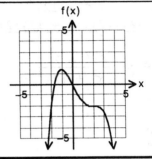

67.

x	$f'(x)$	$f(x)$
$-\infty < x < -2$	Negative and increasing	Decreasing and concave upward
$x = -2$	Local maximum	Inflection point
$-2 < x < 0$	Negative and decreasing	Decreasing and concave downward
$x = 0$	Local minimum	Inflection point
$0 < x < 2$	Negative and increasing	Decreasing and concave upward
$x = 2$	Local maximum	Inflection point
$2 < x < \infty$	Negative and decreasing	Decreasing and concave downward

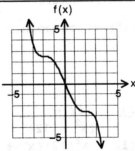

69. Inflection point at $x = -1.40$; concave upward on $(-1.40, \infty)$; concave downward on $(-\infty, -1.40)$

71. Inflection points at $x = -0.61$, $x = 0.66$, and $x = 1.74$; concave upward on $(-0.61, 0.66)$ and $(1.74, \infty)$; concave downward on $(-\infty, -0.61)$ and $(0.66, 1.74)$

73. If $f'(x)$ has a local extremum at $x = c$, then $f'(x)$ must change from increasing to decreasing, or from decreasing to increasing at $x = c$. It follows from this that the graph of f must change its concavity at $x = c$ and so there must be an inflection point at $x = c$.

75. If there is an inflection point on the graph of f at $x = c$, then the graph changes its concavity at $x = c$. Consequently, f' must change from increasing to decreasing, or from decreasing to increasing at $x = c$ and so $x = c$ is a local extremum of f'.

77. $f(x) = \dfrac{1}{x^2 + 12}$

$f'(x) = \dfrac{(x^2 + 12)(0) - 1(2x)}{(x^2 + 12)^2} = \dfrac{-2x}{(x^2 + 12)^2}$

$f''(x) = \dfrac{(x^2 + 12)^2(-2) - (-2x)(2)(x^2 + 12)(2x)}{(x^2 + 12)^4}$

$ = \dfrac{(x^2 + 12)(-2) + 8x^2}{(x^2 + 12)^3} = \dfrac{6x^2 - 24}{(x^2 + 12)^3}$

Now $f''(x) = \dfrac{6x^2 - 24}{(x^2 + 12)^3} = 0$

$ 6x^2 - 24 = 0$

$ 6(x^2 - 4) = 0$

$ 6(x + 2)(x - 2) = 0$

$ x = -2,\ 2$

The sign chart for f'' (partition numbers are -2 and 2) is:

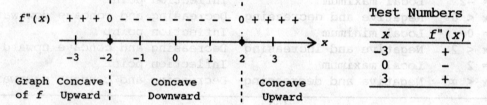

Test Numbers	
x	$f''(x)$
-3	$+$
0	$-$
3	$+$

Thus, the graph of f has inflection points at $x = -2$ and $x = 2$.

79. $f(x) = \dfrac{x}{x^2 + 12}$

$f'(x) = \dfrac{(x^2 + 12)(1) - x(2x)}{(x^2 + 12)^2} = \dfrac{12 - x^2}{(x^2 + 12)^2}$

$f''(x) = \dfrac{(x^2 + 12)^2(-2x) - (12 - x^2)(2)(x^2 + 12)(2x)}{(x^2 + 12)^4}$

$ = \dfrac{(x^2 + 12)(-2x) - 4x(12 - x^2)}{(x^2 + 12)^3} = \dfrac{2x^3 - 72x}{(x^2 + 12)^3} = \dfrac{2x(x^2 - 36)}{(x^2 + 12)^3}$

Now $f''(x) = \dfrac{2x(x^2 - 36)}{(x^2 + 12)^3} = 0$

$ 2x(x + 6)(x - 6) = 0$

$ x = 0,\ -6,\ 6$

The sign chart for f'' (partition numbers are 0, -6, 6) is:

$f''(x)$ - - - 0 + + + + + 0 - - - - - 0 + + +

Test Numbers	
x	$f''(x)$
-7	−
-1	+
1	−
7	+

Thus, the graph of f has inflection points at $x = -6$, $x = 0$, and $x = 6$.

81. The graph of the CPI is concave up.

83. The graph of C is increasing and concave down. Therefore, the graph of C' is positive and decreasing. Since the marginal costs are decreasing, the production process is becoming more efficient.

85. $R(x) = xp = 1296x - 0.12x^3$, $0 < x < 80$
$R'(x) = 1296 - 0.36x^2$
Critical values: $R'(x) = 1296 - 0.36x^2 = 0$
$$x^2 = \frac{1296}{0.36} = 3600$$
$$x = \pm 60$$
Thus, $x = 60$ is the only critical value on the interval $(0, 80)$.
$R''(x) = -0.72x$
$R''(60) = -43.2 < 0$
(A) R has a local maximum at $x = 60$.
(B) Since $R''(x) = -0.72x < 0$ for $0 < x < 80$, R is concave downward on this interval.

87. $N(x) = -3x^3 + 225x^2 - 3600x + 17,000$, $10 \le x \le 40$
$N'(x) = -9x^2 + 450x - 3600$
$N''(x) = -18x + 450$
(A) To determine when N' is increasing or decreasing, we must solve the inequalities $N''(x) > 0$ and $N''(x) < 0$, respectively. Now
$N''(x) = -18x + 450 = 0$
$x = 25$

The sign chart for N'' (partition number is 25) is:

$N''(x)$ + + + + + 0 - - - - -

Test Numbers	
x	$N''(x)$
10	270 (+)
30	-90 (−)

$N'(x)$ Increasing | Decreasing

Thus N' is increasing on $(10, 25)$ and decreasing on $(25, 40)$.

(B) Using the results in (A), the graph of N has an inflection point at $x = 25$.

(C)

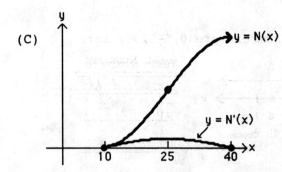

(D) Using the results in (A),
N' has a local maximum at
$x = 25$:
$N'(25) = 2025$

89. $N(t) = 1000 + 30t^2 - t^3$, $0 \le t \le 20$
$N'(t) = 60t - 3t^2$
$N''(t) = 60 - 6t$

(A) To determine when N' is increasing or decreasing, we must solve the
inequalities $N''(t) > 0$ and $N''(t) < 0$, respectively. Now
$N''(t) = 60 - 6t = 0$
$t = 10$

The sign chart for N'' (partition number is 10) is:

$N''(t)$ $+ + + + 0 - - - -$

$N'(t)$ Increasing | Decreasing

Test Numbers	
t	$N''(t)$
0	60 (+)
20	−60 (−)

Thus, N' is increasing on $(0, 10)$ and decreasing on $(10, 20)$.

(B) From the results in (A), the graph of N has an inflection point at
$t = 10$.

(C)

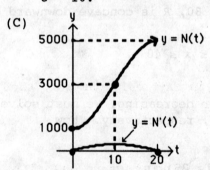

(D) Using the results in (A), N' has a local maximum at $t = 10$:
$N'(10) = 300$

91. $T(n) = 0.08n^3 - 1.2n^2 + 6n$, $n \ge 0$
$T'(n) = 0.24n^2 - 2.4n + 6$, $n \ge 0$
$T''(n) = 0.48n - 2.4$

(A) To determine when the rate of change of T, i.e., T', is increasing
or decreasing, we must solve the inequalities $T''(n) > 0$ and
$T''(n) < 0$, respectively. Now
$T''(n) = 0.48n - 2.4 = 0$
$n = 5$

The sign chart for T'' (partition number is 5) is:

$T''(n)$ $- - - - 0 + + + +$

         1   5   10 n

$T'(n)$ Decreasing | Increasing

Test Numbers	
t	$N''(t)$
1	-1.92 $(-)$
10	2.4 $(+)$

Thus, T' is increasing on $(5, \infty)$ and decreasing on $(0, 5)$.

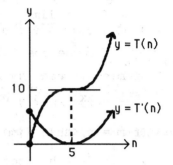

(B) Using the results in (A), the graph of T has an inflection point at $n = 5$. The graphs of T and T' are shown at the right.

(C) Using the results in (A), T' has a local minimum at $n = 5$:
$$T'(5) = 0.24(5)^2 - 2.4(5) + 6 = 0$$

EXERCISE 9-4

Things to remember:

1. **HORIZONTAL ASYMPTOTES**

   A line $y = b$ is a HORIZONTAL ASYMPTOTE for the graph of $y = f(x)$ if $f(x)$ approaches b as either x increases without bound or decreases without bound. That is,
   $$\lim_{x \to \infty} f(x) = b \quad \text{or} \quad \lim_{x \to -\infty} f(x) = b$$

2. **LIMITS AT INFINITY FOR POWER FUNCTIONS**
   If p is a positive real number and k is any real constant, then

   a. $\lim_{x \to -\infty} \dfrac{k}{x^p} = 0$ b. $\lim_{x \to \infty} \dfrac{k}{x^p} = 0$

   c. $\lim_{x \to -\infty} kx^p = \pm\infty$ d. $\lim_{x \to \infty} kx^p = \pm\infty$

   provided that x^p names a real number for negative values of x. The limits in 3 and 4 will be either $-\infty$ or ∞, depending on k and p.

3. **LIMITS AT INFINITY FOR POLYNOMIAL FUNCTIONS**
   If
   $$p(x) = a_n x^n + a_{n-1} x^{n-1} + \dots + a_1 x + a_0, \ a_n \neq 0, \ n \geq 1$$
   then
   $$\lim_{x \to \infty} p(x) = \lim_{x \to \infty} a_n x^n = \pm\infty$$
   and
   $$\lim_{x \to -\infty} p(x) = \lim_{x \to -\infty} a_n x^n = \pm\infty$$
   Each limit will be either $-\infty$ or ∞, depending on a_n and n.

<u>4</u>. LIMITS AT INFINITY AND HORIZONTAL ASYMPTOTES FOR RATIONAL FUNCTIONS

If
$$f(x) = \frac{a_m x^m + a_{m-1} x^{m-1} + \ldots + a_1 x + a_0}{b_n x^n + b_{n-1} x^{n-1} + \ldots + b_1 x + b_0}, \quad a_m \neq 0, \; b_n \neq 0$$

then
$$\lim_{x \to \infty} f(x) = \lim_{x \to \infty} \frac{a_m x^m}{b_n x^n} \quad \text{and} \quad \lim_{x \to -\infty} f(x) = \lim_{x \to -\infty} \frac{a_m x^m}{b_n x^n}$$

There are three possible cases for these limits:

a. If $m < n$, then $\lim\limits_{x \to \infty} f(x) = \lim\limits_{x \to -\infty} f(x) = 0$ and the line $y = 0$ (the x axis) is a horizontal asymptote for $f(x)$.

b. If $m = n$, then $\lim\limits_{x \to \infty} f(x) = \lim\limits_{x \to -\infty} f(x) = \dfrac{a_m}{b_n}$ and the line $y = \dfrac{a_m}{b_n}$ is a horizontal asymptote for $f(x)$.

c. If $m > n$, then each limit will be ∞ or $-\infty$, depending on m, n, a_m, and b_n, and $f(x)$ does not have a horizontal asymptote.

<u>5</u>. LOCATING VERTICAL ASYMPTOTES

Let $f(x) = \dfrac{n(x)}{d(x)}$, where both n and d are continuous at $x = c$.

If at $x = c$ the denominator $d(x)$ is 0 and the numerator $n(x)$ is not 0, then the line $x = c$ is a vertical asymptote for the graph of f.

[*Note*: Since a rational function is a ratio of two polynomial functions and polynomial functions are continuous for all real numbers, this theorem includes rational functions as a special case.]

<u>6</u>. A GRAPHING STRATEGY FOR $y = f(x)$

Omit any of the following steps if procedures involved appear to be too difficult or impossible (what may seem too difficult now, will become less so with a little practice).

<u>Step 1. Analyze $f(x)$</u>:

(A) Find the domain of f. [The domain of f is the set of all real numbers x that produce real values for $f(x)$.]

(B) Find intercepts. [The y intercept is $f(0)$, if it exists; the x intercepts are the solutions to $f(x) = 0$, if they exist.]

(C) Find asymptotes. [Use Theorems 3 and 4, if they apply, otherwise calculate limits at points of discontinuity and as x increases and decreases without bound.]

Step 2. Analyze $f'(x)$: Find any critical values for $f(x)$ and any partition numbers for $f'(x)$. [Remember, every critical value for $f(x)$ is also a partition number for $f'(x)$, but some partition numbers for $f'(x)$ may not be critical values for $f(x)$.] Construct a sign chart for $f'(x)$, determine the intervals where $f(x)$ is increasing and decreasing, and find local maxima and minima.

Step 3. Analyze $f''(x)$: Construct a sign chart for $f''(x)$, determine where the graph of f is concave upward and concave downward, and find any inflection points.

Step 4. Sketch the graph of f: Draw asymptotes and locate intercepts, local maxima and minima, and inflection points. Sketch in what you know from steps 1—3. In regions of uncertainty, use point-by-point plotting to complete the graph.

---

1. From the graph, $f'(x) < 0$ on $(-\infty, b)$, $(0, e)$, (e, g).

3. From the graph, $f(x)$ is increasing on (b, d), $(d, 0)$, (g, ∞).

5. From the graph, $f(x)$ has a local maximum at $x = 0$.

7. From the graph, $f''(x) < 0$ on $(-\infty, a)$, (d, e), (h, ∞).

9. The graph of f is concave upward on (a, d), (e, h).

11. From the graph, f has inflection points at $x = a$ and $x = h$.

13. From the graph, the lines $x = d$ and $x = e$ are vertical asymptotes.

15. Step 1. Analyze $f(x)$:

(A) Domain: All real numbers
(B) Intercepts: y-intercept: 0
               x-intercepts: $-4, 0, 4$

Step 2. Analyze $f'(x)$:

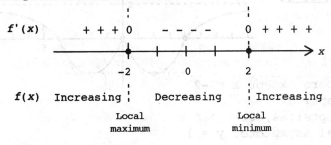

Step 3. Analyze $f''(x)$:

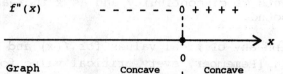

Step 4. Sketch the graph of f:

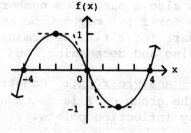

17. Step 1. Analyze $f(x)$:

(A) Domain: All real numbers

(B) Intercepts: y-intercept: 0
x-intercepts: -4, 0, 4

(C) Asymptotes: Horizontal asymptote: $y = 2$

Step 2. Analyze $f'(x)$:

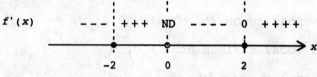

Step 3. Analyze $f''(x)$:

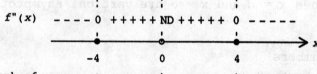

Step 4. Sketch the graph of f:

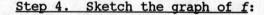

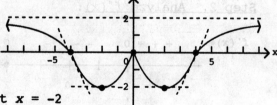

19. Step 1. Analyze $f(x)$:

(A) Domain: All real numbers except $x = -2$

(B) Intercepts: y-intercept: 0
x-intercepts: -4, 0

(C) Asymptotes: Horizontal asymptote: $y = 1$
Vertical asymptote: $x = -2$

Step 2. Analyze $f'(x)$:

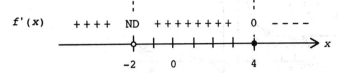

$f'(x)$ $++++$ ND $+++++++$ 0 $----$

−2   0   4

$f(x)$ Increasing   Increasing   Decreasing

Local maximum

Step 3. Analyze $f''(x)$:

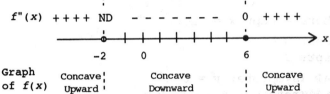

$f''(x)$ $++++$ ND $-------$ 0 $++++$

−2   0   6

Graph   Concave   Concave   Concave
of $f(x)$ Upward   Downward   Upward

Step 4. Sketch the graph of f:

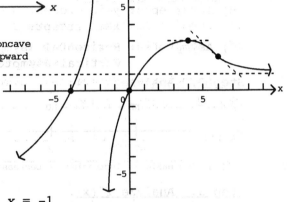

21. Step 1. Analyze $f(x)$:

(A) Domain: All real numbers except $x = -1$

(B) Intercepts: y-intercept: −1
              x-intercept: 1

(C) Asymptotes: Horizontal asymptote: $y = 1$
               Vertical asymptote: $x = -1$

Step 2. Analyze $f'(x)$:

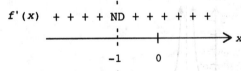

$f'(x)$ $++++$ ND $++++++$

−1   0

$f(x)$ Increasing   Increasing

Step 3. Analyze $f''(x)$:

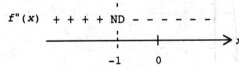

$f''(x)$ $++++$ ND $------$

−1   0

Graph   Concave   Concave
of f Upward   Downward

Step 4. Sketch the graph of f:

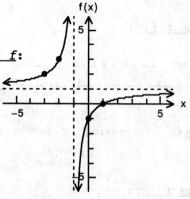

23. Step 1. Analyze $f(x)$:

(A) Domain: All real numbers except $x = -2$, $x = 2$

(B) Intercepts: y-intercept: 0
                  x-intercept: 0

(C) Asymptotes: Horizontal asymptote: $y = 0$
               Vertical asymptotes: $x = -2$, $x = 2$

Step 2. Analyze $f'(x)$:

$f'(x)$ $- - - -$ ND $+ + + + + +$ ND $- - - -$

                   -2 0   2

$f(x)$ Decreasing ┆ Increasing ┆ Decreasing

Step 3. Analyze $f''(x)$:

$f''(x)$ $- - - -$ ND $- -$ 0 $+ +$ ND $+ + + +$

                   -2 0   2

Graph   Concave ┆Concave┆Concave┆ Concave
of f Downward ┆Downward┆Upward┆ Upward

                        Inflection
                          Point

Step 4. Sketch the graph of f:

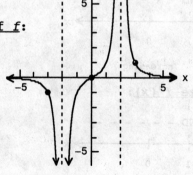

25. $\lim\limits_{x \to \infty} (4x^3 + 2x - 9) = \lim\limits_{x \to \infty} 4x^3 = \infty$ (by $\underline{3}$)

27. $\lim\limits_{x \to -\infty} (-3x^6 + 9x^5 + 4) = \lim\limits_{x \to -\infty} (-3x^6) = -\infty$ (by $\underline{3}$)

29. $\lim\limits_{x \to \infty} \dfrac{4x + 7}{5x - 9} = \dfrac{4}{5}$ (by $\underline{4}$(b))

31. $\lim\limits_{x \to \infty} \dfrac{5x^2 + 11}{7x - 2} = \infty$ (by 4(c))

33. $\lim\limits_{x \to -\infty} \dfrac{7x^4 - 11x^2}{6x^5 + 3} = 0$ (by $\underline{4}$(a))

35. $f(x) = \dfrac{2x}{x + 2}$; f is a rational function

<u>Horizontal asymptotes</u>: $\dfrac{a_n x^n}{b_m x^m} = \dfrac{2x}{x} = 2$ and the line $y = 2$ is a horizontal asymptote.

<u>Vertical asymptotes</u>: Using $\underline{5}$, $D(-2) = 0$ and $N(-2) = -4 \neq 0$. Thus, the line $x = -2$ is a vertical asymptote.

37. $f(x) = \dfrac{x^2 + 1}{x^2 - 1} = \dfrac{x^2 + 1}{(x - 1)(x + 1)}$; f is a rational function

<u>Horizontal asymptotes</u>: $\dfrac{a_n x^n}{b_m x^m} = \dfrac{x^2}{x^2} = 1$ and the line $y = 1$ is a horizontal asymptote.

<u>Vertical asymptotes</u>: Using $\underline{5}$, $D(-1) = 0$ and $N(-1) = 2$, so the line $x = -1$ is a vertical asymptote; $D(1) = 0$, $N(1) = 2$, so the line $x = 1$ is a vertical asymptote.

39. $f(x) = \dfrac{x^3}{x^2 + 6}$; f is a rational function

<u>Horizontal asymptotes</u>: $\dfrac{a_n x^n}{b_m x^m} = \dfrac{x^3}{x^2} = x$; f does not have a horizontal asymptote.

<u>Vertical asymptotes</u>: Using $\underline{5}$, $D(x) = x^2 + 6 \neq 0$ for all x; f has no vertical asymptotes.

41. $f(x) = \dfrac{x}{x^2 + 4}$; f is a rational function

<u>Horizontal asymptotes</u>: $\dfrac{a_n x^n}{b_m x^m} = \dfrac{x}{x^2} = \dfrac{1}{x}$; the line $y = 0$ (the x-axis) is a horizontal asymptote.

<u>Vertical asymptotes</u>: Using $\underline{5}$, $D(x) = x^2 + 4 \neq 0$ for all x; f has no vertical asymptotes.

43. $f(x) = \dfrac{x^2}{x - 3}$; f is a rational function

<u>Horizontal asymptotes</u>: $\dfrac{a_n x^n}{b_m x^m} = \dfrac{x^2}{x} = x$; f does not have a horizontal asymptote.

<u>Vertical asymptotes</u>: Using $\underline{5}$, $D(3) = 0$, $N(3) = 9$, so the line $x = 3$ is a vertical asymptote.

45. $f(x) = \dfrac{2x^2 + 3x - 2}{x^2 - x - 2} = \dfrac{(2x - 1)(x + 2)}{(x - 2)(x + 1)}$; f is a rational function

<u>Horizontal asymptotes</u>: $\dfrac{a_n x^n}{b_m x^m} = \dfrac{2x^2}{x^2} = 2$; the line $y = 2$ is a horizontal asymptote.

<u>Vertical asymptotes</u>: Using 5, $D(2) = 0$, $N(2) = 12$, so the line $x = 2$ is a vertical asymptote; $D(-1) = 0$, $N(-1) = -3$, so the line $x = -1$ is a vertical asymptote.

47. $f(x) = \dfrac{2x^2 - 5x + 2}{x^2 - x - 2} = \dfrac{(2x - 1)(x - 2)}{(x - 2)(x + 1)} = \dfrac{2x - 1}{x + 1}$, $x \neq 2$

<u>Horizontal asymptotes</u>: $\dfrac{a_n x^n}{b_m x^m} = \dfrac{2x^2}{x^2} = 2$; the line $y = 2$ is a horizontal asymptote.

<u>Vertical asymptotes</u>: Using 5, $D(-1) = 0$, $N(-1) = -3$, so the line $x = -1$ is a vertical asymptote. Since $\lim\limits_{x \to 2} f(x) = \lim\limits_{x \to 2} \dfrac{2x - 1}{x + 1} = 1$, f does not have a vertical asymptote at $x = 2$.

49. $f(x) = x^2 - 6x + 5$

<u>Step 1. Analyze $f(x)$</u>:
(A) Domain: All real numbers, $(-\infty, \infty)$.
(B) Intercepts: y intercept: $f(0) = 5$
$\qquad\qquad\quad x$ intercepts: $x^2 - 6x + 5 = 0$
$\qquad\qquad\qquad\qquad\quad (x - 5)(x - 1) = 0$
$\qquad\qquad\qquad\qquad\qquad\qquad x = 1, 5$
(C) Asymptotes: Since f is a polynomial, there are no horizontal or vertical asymptotes.

<u>Step 2. Analyze $f'(x)$</u>:
$f'(x) = 2x - 6 = 2(x - 3)$
Critical value: $x = 3$
Partition number: $x = 3$
Sign chart for f':

Test Numbers	
x	$f'(x)$
0	-6 $(-)$
4	2 $(+)$

$f'(x)$ $- - - - - - \; | \; + + + + +$ $\longrightarrow x$

$\qquad\qquad$ 0 $\;$ 1 $\;$ 2 $\;$ 3 $\;$ 4

$f(x)$ $\qquad$ Decreasing $\;|\;$ Increasing

Thus, f is decreasing on $(-\infty, 3)$ and increasing on $(3, \infty)$; f has a local minimum at $x = 3$.

Step 3. Analyze $f''(x)$:
$f''(x) = 2 > 0$ for all x.
Thus, the graph of f is concave upward on $(-\infty, \infty)$.
Step 4. Sketch the graph of f:

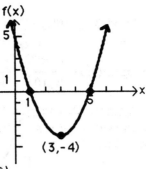

(3, -4)

51. $f(x) = x^3 - 6x^2$
Step 1. Analyze $f(x)$:
(A) Domain: All real numbers, $(-\infty, \infty)$.

(B) Intercepts: y intercept: $f(0) = 0$
$\qquad\qquad\quad x$ intercepts: $x^3 - 6x^2 = 0$
$\qquad\qquad\qquad\qquad\qquad\quad x^2(x - 6) = 0$
$\qquad\qquad\qquad\qquad\qquad\qquad\quad x = 0, 6$

(C) Asymptotes: Since f is a polynomial, there are no horizontal or vertical asymptotes.

Step 2. Analyze $f'(x)$:
$f'(x) = 3x^2 - 12x = 3x(x - 4)$
Critical values: $x = 0$, $x = 4$
Partition numbers: $x = 0$, $x = 4$
The sign chart for f' is:

$f'(x)$ $+ + + + 0 - - - - - 0 + + + +$

```
        +---+---+---+---+---+---+---+---> x
         -1  0           3   4   5
```

$f(x)$ Increasing ┊ Decreasing ┊ Increasing
$\qquad\qquad\qquad$ Local $\qquad$ Local
$\qquad\qquad\qquad$ Maximum $\quad$ Minimum

Test Numbers	
x	$f'(x)$
-1	15 (+)
3	-9 (−)
5	15 (+)

Thus, f is increasing on $(-\infty, 0)$ and on $(4, \infty)$; f is decreasing on $(0, 4)$; f has a local maximum at $x = 0$ and a local minimum at $x = 4$.

Step 3. Analyze $f''(x)$:
$f''(x) = 6x - 12 = 6(x - 2)$
Partition numbers for f'': $x = 2$

Sign chart for f'':

$f''(x)$ $\qquad\qquad - - - - 0 + + + +$

```
        +---+---+---+---> x
         0   1   2   3
```

Graph $\qquad$ Concave ┊ Concave
of f $\qquad\quad$ Downward ┊ Upward
$\qquad\qquad\qquad\quad$ Inflection
$\qquad\qquad\qquad\qquad$ Point

Test Numbers	
x	$f''(x)$
1	-6 (−)
3	6 (+)

Thus, the graph of f is concave downward on $(-\infty, 2)$ and concave upward on $(2, \infty)$; there is an inflection point at $x = 2$.

Step 4. Sketch the graph of f:

x	$f(x)$
0	0
2	-16
4	-32
6	0

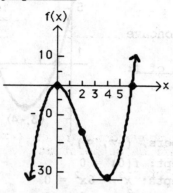

53. $f(x) = (x + 4)(x - 2)^2$

Step 1. Analyze $f(x)$:

(A) Domain: All real numbers, $(-\infty, \infty)$.

(B) Intercepts: y intercept: $f(0) = 4(-2)^2 = 16$

x intercepts: $(x + 4)(x - 2)^2 = 0$

$$x = -4, 2$$

(C) Asymptotes: Since f is a polynomial, there are no horizontal or vertical asymptotes.

Step 2. Analyze $f'(x)$:

$f'(x) = (x + 4)2(x - 2)(1) + (x - 2)^2(1)$
$= (x - 2)[2(x + 4) + (x - 2)]$
$= (x - 2)(3x + 6)$
$= 3(x - 2)(x + 2)$

Critical values: $x = -2$, $x = 2$
Partition numbers: $x = -2$, $x = 2$
Sign chart for f':

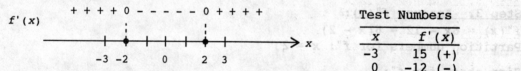

Test Numbers	
x	$f'(x)$
-3	15 (+)
0	-12 (-)
3	15 (+)

Thus, f is increasing on $(-\infty, -2)$ and on $(2, \infty)$; f is decreasing on $(-2, 2)$; f has a local maximum at $x = -2$ and a local minimum at $x = 2$.

Step 3. Analyze $f''(x)$:

$f''(x) = 3(x + 2)(1) + 3(x - 2)(1) = 6x$
Partition number for f'': $x = 0$

Thus, the graph of f is concave downward on $(-\infty, 2)$ and concave upward on $(2, \infty)$; there is an inflection point at $x = 2$.

Sign chart for f'':

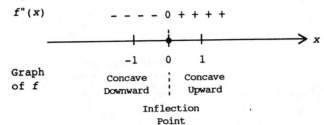

f"(x)   − − − − 0 + + + +

          −1   0   1

Graph
of f   Concave   Concave
          Downward   Upward

          Inflection
          Point

Test Numbers	
x	f"(x)
1	−6 (−)
1	6 (+)

Thus, the graph of f is concave downward on $(-\infty, 0)$ and concave upward on $(0, \infty)$; there is an inflection point at $x = 0$.

Step 4. Sketch the graph of f:

x	f(x)
−2	32
0	16
2	0

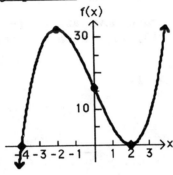

55. $f(x) = 8x^3 - 2x^4$

Step 1. Analyze $f(x)$:

(A) Domain: All real numbers, $(-\infty, \infty)$.

(B) Intercepts: y intercept: $f(0) = 0$
x intercepts: $8x^3 - 2x^4 = 0$
$2x^3(4 - x) = 0$
$x = 0, 4$

(C) Asymptotes: No horizontal or vertical asymptotes.

Step 2. Analyze $f'(x)$:

$f'(x) = 24x^2 - 8x^3 = 8x^2(3 - x)$
Critical values: $x = 0, x = 3$
Partition numbers: $x = 0, x = 3$
Sign chart for f':

          + + + + 0 + + + + + 0 − − − −

f'(x)

          −1   0   1   3   4

f(x)   Increasing ┊ Increasing ┊ Decreasing

          Local
          Maximum

Test Numbers	
x	f'(x)
−1	32 (+)
1	16 (+)
4	−128 (−)

Thus, f is increasing on $(-\infty, 3)$ and decreasing on $(3, \infty)$; f has a local maximum at $x = 3$.

Step 3. Analyze $f''(x)$:

$f''(x) = 48x - 24x^2 = 24x(2 - x)$

Partition numbers for f'': $x = 0$, $x = 2$

Sign chart for f'':

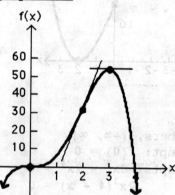

Test Numbers	
x	$f''(x)$
-1	-72 (-)
1	24 (+)
3	-72 (-)

Thus, the graph of f is concave downward on $(-\infty, 0)$ and on $(2, \infty)$; the graph is concave upward on $(0, 2)$; there are inflection points at $x = 0$ and $x = 2$.

Step 4. Sketch the graph of f:

x	$f(x)$
0	0
2	32
3	54

57. $f(x) = \dfrac{x + 3}{x - 3}$

Step 1. Analyze $f(x)$:

(A) Domain: All real numbers except $x = 3$.

(B) Intercepts: y intercept: $f(0) = \dfrac{3}{-3} = -1$

x intercepts: $\dfrac{x + 3}{x - 3} = 0$

$x + 3 = 0$

$x = -3$

(C) Asymptotes:

Horizontal asymptote: $\lim\limits_{x \to \infty} \dfrac{x + 3}{x - 3} = \lim\limits_{x \to \infty} \dfrac{x\left(1 + \dfrac{3}{x}\right)}{x\left(1 - \dfrac{3}{x}\right)} = 1.$

Thus, $y = 1$ is a horizontal asymptote.

Vertical asymptote: The denominator is 0 at $x = 3$ and the numerator is not 0 at $x = 3$. Thus, $x = 3$ is a vertical asymptote.

Step 2. Analyze f'(x):

$$f'(x) = \frac{(x - 3)(1) - (x + 3)(1)}{(x - 3)^2} = \frac{-6}{(x - 3)^2} = -6(x - 3)^{-2}$$

Critical values: None
Partition number: $x = 3$
Sign chart for f':

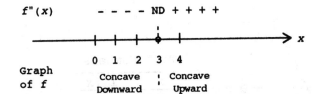

f'(x)   - - - - - ND - - - -

                      0 1 2 3 4   x

f(x)   Decreasing ¦ Decreasing

Test Numbers

x	f'(x)
2	-6 (-)
4	-6 (-)

Thus, f is decreasing on $(-\infty, 3)$ and on $(3, \infty)$; there are no local extrema.

Step 3. Analyze f"(x):

$$f''(x) = 12(x - 3)^{-3} = \frac{12}{(x - 3)^3}$$

Partition number for f'': $x = 3$
Sign chart for f'':

f"(x)   - - - - - ND + + + +

                      0 1 2 3 4   x

Graph   Concave ¦ Concave
of f   Downward ¦ Upward

Test Numbers

x	f"(x)
2	-12 (-)
4	12 (+)

Thus, the graph of f is concave downward on $(-\infty, 3)$ and concave upward on $(3, \infty)$.

Step 4. Sketch the graph of f:

x	f(x)
-3	0
0	-1
5	4

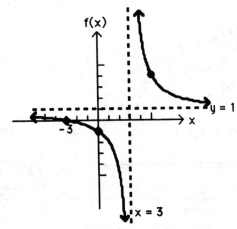

59. $f(x) = \dfrac{x}{x - 2}$

Step 1. Analyze $f(x)$:

(A) Domain: All real numbers except $x = 2$.

(B) Intercepts: y intercept: $f(0) = \dfrac{0}{-2} = 0$

x intercepts: $\dfrac{x}{x - 2} = 0$

$x = 0$

(C) Asymptotes:

<u>Horizontal asymptote</u>: $\displaystyle\lim_{x \to \infty} \dfrac{x}{x - 2} = \lim_{x \to \infty} \dfrac{x}{x\left(1 - \dfrac{2}{x}\right)} = 1$.

Thus, $y = 1$ is a horizontal asymptote.

<u>Vertical asymptote</u>: The denominator is 0 at $x = 2$ and the numerator is not 0 at $x = 2$. Thus, $x = 2$ is a vertical asymptote.

Step 2. Analyze $f'(x)$:

$f'(x) = \dfrac{(x - 2)(1) - x(1)}{(x - 2)^2} = \dfrac{-2}{(x - 2)^2} = -2(x - 2)^{-2}$

Critical values: None
Partition number: $x = 2$
Sign chart for f':

	Test Numbers	
	x	$f'(x)$
	0	$-\frac{1}{2}$ (−)
	3	−2 (−)

$f'(x)$ — − − − − ND − − − −

$f(x)$ Decreasing ¦ Decreasing

(number line marked at 0 1 2 3)

Thus, f is decreasing on $(-\infty, 2)$ and on $(2, \infty)$; there are no local extrema.

Step 3. Analyze $f''(x)$:

$f''(x) = 4(x - 2)^{-3} = \dfrac{4}{(x - 2)^3}$

Partition number for f'': $x = 2$
Sign chart for f'':

	Test Numbers	
	x	$f''(x)$
	0	$-\frac{1}{2}$ (−)
	3	4 (+)

$f''(x)$ — − − − − ND + + + +

Graph
of f Concave ¦ Concave
         Downward ¦ Upward

(number line marked at 0 1 2 3)

Thus, the graph of f is concave downward on $(-\infty, 2)$ and concave upward on $(2, \infty)$.

Step 4. Sketch the graph of f:

x	$f(x)$
0	0
4	2

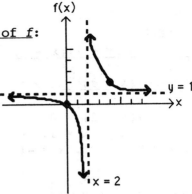

61. For any $n \geq 1$, $\lim\limits_{n \to \infty} (a_n x^n + a_{n-1} x^{n-1} + \dots + a_1 x + a_0)$

$$= \lim\limits_{x \to \infty} a_n x^n = \begin{cases} \infty & \text{if } a_n > 0 \\ -\infty & \text{if } a_n < 0 \end{cases}$$

63. $p(x) = x^3 - 2x^2$; $p'(x) = 3x^2 - 4x$; $p''(x) = 6x - 4$

(A) $\lim\limits_{x \to \infty} p'(x) = \lim\limits_{x \to \infty} 3x^2 = \infty$

$\lim\limits_{x \to \infty} p''(x) = \lim\limits_{x \to \infty} 6x = \infty$

The graph of f is increasing and concave upward for large positive values of x.

(B) $\lim\limits_{x \to -\infty} p'(x) = \lim\limits_{x \to -\infty} 3x^2 = \infty$

$\lim\limits_{x \to -\infty} p'(x) = \lim\limits_{x \to -\infty} 6x = -\infty$

The graph of f is increasing and concave downward for large negative values of x.

65. For $|x|$ very large, $f(x) = x + \dfrac{1}{x} \approx x$. Thus, the line $y = x$ is an oblique asymptote.

Step 1. Analyze $f(x)$:

(A) Domain: All real numbers except $x = 0$.

(B) Intercepts: y intercept: There is no y intercept since f is not defined at $x = 0$.

x intercepts: $x + \dfrac{1}{x} = 0$

$x^2 + 1 = 0$

Thus, there are no x intercepts.

(C) Asymptotes:

Oblique asymptote: $y = x$

Vertical asymptote: $f(x) = x + \dfrac{1}{x} = \dfrac{x^2 + 1}{x}$

The denominator is 0 at $x = 0$ and the numerator is not 0 at $x = 0$. Thus, $x = 0$ is a vertical asymptote.

<u>Step 2.</u> <u>Analyze $f'(x)$</u>:

$$f'(x) = 1 - \frac{1}{x^2} = \frac{x^2 - 1}{x^2} = \frac{(x + 1)(x - 1)}{x^2}$$

Critical values: $\dfrac{(x - 1)(x + 1)}{x^2} = 0$

$$(x - 1)(x + 1) = 0$$
$$x = 1, -1$$

Partition numbers: $x = 0$, $x = 1$, $x = -1$
Sign chart for f':

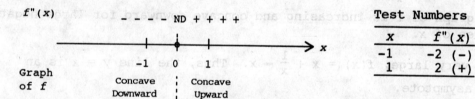

Test Numbers	
x	$f'(x)$
-2	$\frac{3}{4}$ (+)
$-\frac{1}{2}$	-3 (−)
$\frac{1}{2}$	-3 (−)
2	$\frac{3}{4}$ (+)

Thus, f is increasing on $(-\infty, -1)$ and on $(1, \infty)$; f is decreasing on $(-1, 0)$ and on $(0, 1)$; f has a local maximum at $x = -1$ and a local minimum at $x = 1$.

<u>Step 3.</u> <u>Analyze $f''(x)$</u>:

$$f''(x) = 2x^{-3} = \frac{2}{x^3}$$

Partition number for f'': $x = 0$
Sign chart for f'':

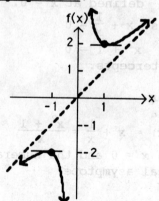

Test Numbers	
x	$f''(x)$
-1	-2 (−)
1	2 (+)

Thus, the graph of f is concave downward on $(-\infty, 0)$ and concave upward on $(0, \infty)$.

<u>Step 4.</u> <u>Sketch the graph of f</u>:

x	$f(x)$
-1	-2
1	2

67. $f(x) = x^3 - x$

<u>Step 1. Analyze $f(x)$:</u>

(A) Domain: All real numbers, $(-\infty, \infty)$.

(B) Intercepts: y intercept: $f(0) = 0^3 - 0 = 0$

x intercepts: $x^3 - x = 0$

$$x(x^2 - 1) = 0$$
$$x(x - 1)(x + 1) = 0$$
$$x = 0, 1, -1$$

(C) Asymptotes: There are no asymptotes.

<u>Step 2. Analyze $f'(x)$:</u>

$f'(x) = 3x^2 - 1 = (\sqrt{3}x + 1)(\sqrt{3}x - 1)$

Critical values: $x = -\dfrac{\sqrt{3}}{3}$, $x = \dfrac{\sqrt{3}}{3}$

Partition numbers: $x = -\dfrac{\sqrt{3}}{3}$, $x = \dfrac{\sqrt{3}}{3}$

Sign chart for f':

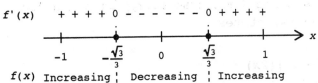

Test Numbers	
x	$f'(x)$
-1	2 (+)
0	-1 (−)
1	2 (+)

Thus, f is increasing on $\left(-\infty, -\dfrac{\sqrt{3}}{3}\right)$ and on $\left(\dfrac{\sqrt{3}}{3}, \infty\right)$; f is decreasing on $\left(-\dfrac{\sqrt{3}}{3}, \dfrac{\sqrt{3}}{3}\right)$; f has a local maximum at $x = -\dfrac{\sqrt{3}}{3}$ and a local minimum at $x = \dfrac{\sqrt{3}}{3}$.

<u>Step 3. Analyze $f''(x)$:</u>

$f''(x) = 6x$

Partition number for f'': $x = 0$

Sign chart for f'':

```
f"(x)        - - - - 0 + + + +
      ────────────────┼────────────────► x
                      0

Graph          Concave  │  Concave
of f           Downward │  Upward
                        │
                   Inflection
                     Point
```

Thus, the graph of f is concave downward on $(-\infty, 0)$ and concave upward on $(0, \infty)$. There is an inflection point at $x = 0$.

Step 4. Sketch the graph of f:

x	$f(x)$
-1	0
$-\frac{\sqrt{3}}{3}$	≈ 0.4
0	0
$\frac{\sqrt{3}}{3}$	≈ -0.4

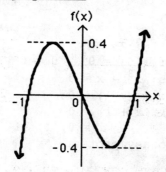

69. $f(x) = (x^2 + 3)(9 - x^2)$

Step 1. Analyze $f(x)$:

(A) Domain: All real numbers, $(-\infty, \infty)$.

(B) Intercepts: y intercept: $f(0) = 3(9) = 27$

x intercepts: $(x^2 + 3)(9 - x^2) = 0$

$(3 - x)(3 + x) = 0$

$x = 3, -3$

(C) Asymptotes: There are no asymptotes.

Step 2. Analyze $f'(x)$:

$f'(x) = (x^2 + 3)(-2x) + (9 - x^2)(2x)$

$= 2x[9 - x^2 - (x^2 + 3)]$

$= 2x(6 - 2x^2)$

$= 4x(\sqrt{3} + x)(\sqrt{3} - x)$

Critical values: $x = 0$, $x = -\sqrt{3}$, $x = \sqrt{3}$

Partition numbers: $x = 0$, $x = -\sqrt{3}$, $x = \sqrt{3}$

Sign chart for f':

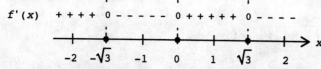

Test Numbers	
x	$f'(x)$
-2	8 $(+)$
-1	-8 $(-)$
1	8 $(+)$
2	-8 $(-)$

Thus, f is increasing on $(-\infty, -\sqrt{3})$ and on $(0, \sqrt{3})$; f is decreasing on $(-\sqrt{3}, 0)$ and on $(\sqrt{3}, \infty)$; f has local maxima at $x = -\sqrt{3}$ and $x = \sqrt{3}$ and a local minimum at $x = 0$.

Step 3. Analyze $f''(x)$:

$f''(x) = 2x(-4x) + (6 - 2x^2)(2) = 12 - 12x^2 = -12(x - 1)(x + 1)$

Partition numbers for f'': $x = 1$, $x = -1$

Sign chart for f'':

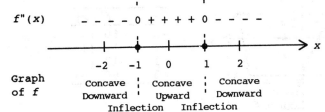

Test Numbers	
x	$f''(x)$
-2	-36 (-)
0	12 (+)
2	-36 (-)

Thus, the graph of f is concave downward on $(-\infty, -1)$ and on $(1, \infty)$; the graph of f is concave upward on $(-1, 1)$; the graph has inflection points at $x = -1$ and $x = 1$.

Step 4. Sketch the graph of f:

x	$f(x)$
$-\sqrt{3}$	36
-1	32
0	27
1	32
$\sqrt{3}$	36

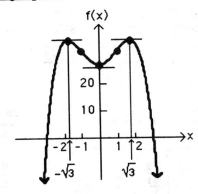

71. $f(x) = (x^2 - 4)^2$

Step 1. Analyze $f(x)$:

(A) Domain: All real numbers, $(-\infty, \infty)$.

(B) Intercepts: y intercept: $f(0) = (-4)^2 = 16$

$\qquad\qquad\quad x$ intercepts: $(x^2 - 4)^2 = 0$

$\qquad\qquad\qquad\qquad [(x - 2)(x + 2)]^2 = 0$

$\qquad\qquad\qquad\qquad (x - 2)^2(x + 2)^2 = 0$

$\qquad\qquad\qquad\qquad\qquad\qquad x = 2, -2$

(C) Asymptotes: There are no asymptotes.

Step 2. Analyze $f'(x)$:

$f'(x) = 2(x^2 - 4)(2x) = 4x(x - 2)(x + 2)$

Critical values: $x = 0$, $x = 2$, $x = -2$

Partition numbers: Same as critical values.

Sign chart for f':

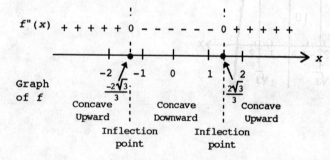

Test Numbers	
x	$f'(x)$
-3	-60 (−)
-1	12 (+)
1	-12 (−)
3	60 (+)

Thus, f is decreasing on $(-\infty, -2)$ and on $(0, 2)$; f is increasing on $(-2, 0)$ and on $(2, \infty)$; f has local minima at $x = -2$ and $x = 2$ and a local maximum at $x = 0$.

Step 3. Analyze $f''(x)$:

$$f''(x) = 4x(2x) + (x^2 - 4)(4) = 12x^2 - 16 = 12\left(x^2 - \frac{4}{3}\right)$$

$$= 12\left(x - \frac{2\sqrt{3}}{3}\right)\left(x + \frac{2\sqrt{3}}{3}\right)$$

Partition numbers for f'': $x = \frac{2\sqrt{3}}{3}$, $x = \frac{-2\sqrt{3}}{3}$

Sign chart for f'':

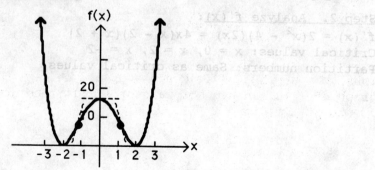

Test Numbers	
x	$f''(x)$
-2	32 (+)
0	-16 (−)
2	32 (+)

Thus, the graph of f is concave upward on $\left(-\infty, \frac{-2\sqrt{3}}{3}\right)$ and on $\left(\frac{2\sqrt{3}}{3}, \infty\right)$; the graph of f is concave downward on $\left(\frac{-2\sqrt{3}}{3}, \frac{2\sqrt{3}}{3}\right)$; the graph has inflection points at $x = \frac{-2\sqrt{3}}{3}$ and $x = \frac{2\sqrt{3}}{3}$.

Step 4. Sketch the graph of f:

x	$f(x)$
-2	0
$-\frac{2\sqrt{3}}{3}$	$\frac{64}{9}$
0	16
$\frac{2\sqrt{3}}{3}$	$\frac{64}{9}$
2	0

73. $f(x) = 2x^6 - 3x^5$

Step 1. **Analyze $f(x)$:**
(A) Domain: All real numbers, $(-\infty, \infty)$.
(B) Intercepts: y intercept: $f(0) = 2 \cdot 0^6 - 3 \cdot 0^5 = 0$
  x intercepts: $2x^6 - 3x^5 = 0$
  $x^5(2x - 3) = 0$
  $x = 0, \frac{3}{2}$

(C) Asymptotes: There are no asymptotes.

Step 2. **Analyze $f'(x)$:**
$$f'(x) = 12x^5 - 15x^4 = 12x^4\left(x - \frac{5}{4}\right)$$

Critical values: $x = 0, \ x = \frac{5}{4}$

Partition numbers: Same as critical values.
Sign chart for f':

	Test Numbers	
x	$f'(x)$	
-1	-27	$(-)$
1	-3	$(-)$
2	144	$(+)$

$f'(x)$ $- - - - \ 0 \ - - - - - - \ 0 \ + + + +$

$f(x)$ Decreasing ¦ Decreasing ¦ Increasing
Local minimum

Thus, f is decreasing on $(-\infty, 0)$ and $\left(0, \frac{5}{4}\right)$; f is increasing on $\left(\frac{5}{4}, \infty\right)$;

f has a local minimum at $x = \frac{5}{4}$.

Step 3. **Analyze $f''(x)$:**
$f''(x) = 60x^4 - 60x^3 = 60x^3(x - 1)$
Partition numbers for f'': $x = 0, \ x = 1$
Sign chart for f'':

	Test Numbers	
x	$f''(x)$	
-1	120	$(+)$
$\frac{1}{2}$	$-\frac{15}{4}$	$(-)$
2	480	$(+)$

$f''(x)$ $+ + + + \ 0 \ - - - - \ 0 \ + + + +$

Graph of f Concave Upward ¦ Concave Downward ¦ Concave Upward
Inflection point   Inflection point

Thus, the graph of f is concave upward on $(-\infty, 0)$ and on $(1, \infty)$; the graph of f is concave downward on $(0, 1)$; the graph has inflection points at $x = 0$ and $x = 1$.

Step 4. Sketch the graph of f:

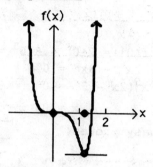

x	f(x)
0	0
1	-1
$\frac{5}{4}$	$\approx$ -1.5

75. $f(x) = \dfrac{x}{x^2 - 4} = \dfrac{x}{(x - 2)(x + 2)}$

Step 1. Analyze $f(x)$:

(A) Domain: All real numbers except $x = 2$, $x = -2$.

(B) Intercepts: y intercept: $f(0) = \dfrac{0}{-4} = 0$

x intercept: $\dfrac{x}{x^2 - 4} = 0$

$x = 0$

(C) Asymptotes:
Horizontal asymptote:

$$\lim_{x \to \infty} \frac{x}{x^2 - 4} = \lim_{x \to \infty} \frac{x}{x^2\left(1 - \dfrac{4}{x^2}\right)} = \lim_{x \to \infty} \frac{1}{x}\left(\frac{1}{1 - \dfrac{4}{x^2}}\right) = 0$$

Thus, $y = 0$ (the x axis) is a horizontal asymptote.

Vertical asymptotes: The denominator is 0 at $x = 2$ and $x = -2$. The numerator is nonzero at each of these points. Thus, $x = 2$ and $x = -2$ are vertical asymptotes.

Step 2. Analyze $f'(x)$:

$f'(x) = \dfrac{(x^2 - 4)(1) - x(2x)}{(x^2 - 4)^2} = \dfrac{-(x^2 + 4)}{(x^2 - 4)^2}$

Critical values: None ($x^2 + 4 \neq 0$ for all x)
Partition numbers: $x = 2$, $x = -2$
Sign chart for f':

```
f'(x)   - - - - - ND - - - - - - - - ND - - - - -
       ┼───────●───┼───┼───┼───●───┼──────→ x
          -2  -1   0   1   2

f(x) Decreasing ┆ Decreasing ┆ Decreasing
```

Thus, f is decreasing on $(-\infty, -2)$, on $(-2, 2)$, and on $(2, \infty)$; f has no local extrema.

Step 3. Analyze $f''(x)$:

$$f''(x) = \frac{(x^2 - 4)^2(-2x) - [-(x^2 + 4)](2)(x^2 - 4)(2x)}{(x^2 - 4)^4}$$

$$= \frac{(x^2 - 4)(-2x) + 4x(x^2 + 4)}{(x^2 - 4)^3} = \frac{2x^3 + 24x}{(x^2 - 4)^3} = \frac{2x(x^2 + 12)}{(x^2 - 4)^3}$$

Partition numbers for f'': $x = 0$, $x = 2$, $x = -2$

Sign chart for f'':

$f''(x)$ _ _ _ ND + + + + 0 - - - - ND + + +

Graph of f Concave Downward ¦ Concave Upward ¦ Concave Downward ¦ Concave Upward

Inflection point

Test Numbers

x	$f''(x)$
-3	$-\frac{126}{125}$ (−)
-1	$\frac{26}{27}$ (+)
1	$-\frac{26}{27}$ (−)
3	$\frac{126}{127}$ (+)

Thus, the graph of f is concave downward on $(-\infty, -2)$ and on $(0, 2)$; the graph of f is concave upward on $(-2, 0)$ and on $(2, \infty)$; the graph has an inflection point at $x = 0$.

Step 4. Sketch the graph of f:

x	$f(x)$
0	0
1	$-\frac{1}{3}$
-1	$\frac{1}{3}$
3	$\frac{3}{5}$
-3	$-\frac{3}{5}$

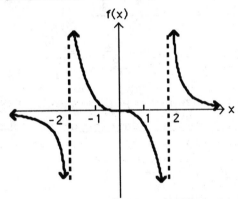

77. $f(x) = \dfrac{1}{1 + x^2}$

Step 1. Analyze $f(x)$:

(A) Domain: All real numbers ($1 + x^2 \neq 0$ for all x).

(B) Intercepts: y intercept: $f(0) = 1$

$\quad\quad\quad\quad\quad$ x intercept: $\dfrac{1}{1 + x^2} \neq 0$ for all x; no x intercepts

(C) Asymptotes:

$\quad$ Horizontal asymptote: $\lim\limits_{x \to \infty} \dfrac{1}{1 + x^2} = 0$

$\quad$ Thus, $y = 0$ (the x axis) is a horizontal asymptote.

$\quad$ Vertical asymptotes: Since $1 + x^2 \neq 0$ for all x, there are no vertical asymptotes.

<u>Step 2. Analyze $f'(x)$:</u>

$f'(x) = \dfrac{(1 + x^2)(0) - 1(2x)}{(1 + x^2)^2} = \dfrac{-2x}{(1 + x^2)^2}$

Critical values: $x = 0$

Partition numbers: $x = 0$

Sign chart for f':

Test Numbers	
x	$f'(x)$
-1	$\frac{1}{2}$ (+)
1	$-\frac{1}{2}$ (−)

Thus, f is increasing on $(-\infty, 0)$; f is decreasing on $(0, \infty)$; f has a local maximum at $x = 0$.

<u>Step 3. Analyze $f''(x)$:</u>

$f''(x) = \dfrac{(1 + x^2)^2(-2) - (-2x)(2)(1 + x^2)2x}{(1 + x^2)^4} = \dfrac{(-2)(1 + x^2) + 8x^2}{(1 + x^2)^3}$

$\qquad = \dfrac{6x^2 - 2}{(1 + x^2)^3} = \dfrac{6\left(x + \frac{\sqrt{3}}{3}\right)\left(x - \frac{\sqrt{3}}{3}\right)}{(1 + x^2)^3}$

Partition numbers for f'': $x = -\dfrac{\sqrt{3}}{3}$, $x = \dfrac{\sqrt{3}}{3}$

Sign chart for f'':

Test Numbers	
x	$f''(x)$
-1	$\frac{1}{2}$ (+)
0	-2 (−)
1	$\frac{1}{2}$ (+)

Thus, the graph of f is concave upward on $\left(-\infty, \dfrac{-\sqrt{3}}{3}\right)$ and on $\left(\dfrac{\sqrt{3}}{3}, \infty\right)$; the graph of f is concave downward on $\left(\dfrac{-\sqrt{3}}{3}, \dfrac{\sqrt{3}}{3}\right)$; the graph has inflection points at $x = \dfrac{-\sqrt{3}}{3}$ and $x = \dfrac{\sqrt{3}}{3}$.

Step 4. Sketch the graph of f:

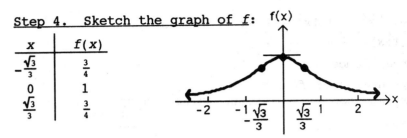

x	$f(x)$
$-\frac{\sqrt{3}}{3}$	$\frac{3}{4}$
0	1
$\frac{\sqrt{3}}{3}$	$\frac{3}{4}$

79. $f(x) = -x^4 - x^3 + 2x^2 - 2x + 3$

Step 1. Analyze $f(x)$:

(A) Domain: all real numbers

(B) Intercepts: y-intercept: $f(0) = 3$
     x-intercepts: $x \approx -2.40,\ 1.16$

(C) Asymptotes: None

Step 2. Analyze $f'(x)$: $f'(x) = -4x^3 - 3x^2 + 4x - 2$
Critical value: $x \approx -1.58$
f is increasing on $(-\infty, -1.58)$; f is decreasing on $(-1.58, \infty)$; f has a local maximum at $x = -1.58$

Step 3. Analyze $f''(x)$: $f''(x) = -12x^2 - 6x + 4$
The graph of f is concave downward on $(-\infty, -0.88)$ and $(0.38, \infty)$; the graph of f is concave upward on $(-0.88, 0.38)$; the graph has inflection points at $x = -0.88$ and $x = 0.38$.

81. $f(x) = x^4 - 5x^3 + 3x^2 + 8x - 5$

Step 1. Analyze $f(x)$:

(A) Domain: all real numbers

(B) Intercepts: y-intercept: $f(0) = -5$
     x-intercepts: $x \approx -1.18,\ 0.61,\ 1.87,\ 3.71$

(C) Asymptotes: None

Step 2. Analyze $f'(x)$: $f'(x) = 4x^3 - 15x^2 + 6x + 8$
Critical values: $x \approx -0.53,\ 1.24,\ 3.04$
f is decreasing on $(-\infty, -0.53)$ and $(1.24, 3.04)$; f is increasing on $(-0.53, 1.24)$ and $(3.04, \infty)$; f has local minima at $x = -0.53$ and 3.04; f has a local maximum at $x = 1.24$

Step 3. Analyze $f''(x)$: $f''(x) = 12x^2 - 30x + 6$
The graph of f is concave upward on $(-\infty, 0.22)$ and $(2.28, \infty)$; the graph of f is concave downward on $(0.22, 2.28)$; the graph has inflection points at $x = 0.22$ and 2.28.

83. $f(x) = 0.01x^5 + 0.03x^4 - 0.4x^3 - 0.5x^2 + 4x + 3$

Step 1. Analyze $f(x)$:

(A) Domain: all real numbers

(B) Intercepts: y-intercept: $f(0) = 3$
$\qquad\qquad\quad$ x-intercepts: $x \approx -6.68, -3.64, -0.72$

(C) Asymptotes: None

Step 2. Analyze $f'(x)$: $f'(x) = 0.05x^4 + 0.12x^3 - 1.2x^2 - x + 4$
Critical values: $x \approx -5.59, -2.27, 1.65, 3.82$
f is increasing on $(-\infty, -5.59)$, $(-2.27, 1.65)$, and $(3.82, \infty)$; f is decreasing on $(-5.59, -2.27)$ and $(1.65, 3.82)$; f has local minima at $x = -2.27$ and 3.82; f has local maxima at $x = -5.59$ and 1.65

Step 3. Analyze $f''(x)$: $f''(x) = 0.2x^3 + 0.36x^2 - 2.4x - 1$
The graph of f is concave downward on $(-\infty, -4.31)$ and $(-0.40, 2.91)$; the graph of f is concave upward on $(-4.31, -0.40)$ and $(2.91, \infty)$; the graph has inflection points at $x = -4.31, -0.40$ and 2.91.

85. $R(x) = xp = 1296x - 0.12x^3$, $0 < x < 80$

Step 1. Analyze $R(x)$:
(A) Domain: $0 < x < 80$ or $(0, 80)$.
(B) Intercepts: There are no intercepts on $(0, 80)$.
(C) Asymptotes: Since R is a polynomial, there are no asymptotes.

Step 2. Analyze $R'(x)$:
$R'(x) = 1296 - 0.36x^2$
$\qquad\;\; = -0.36(x^2 - 3600)$
$\qquad\;\; = -0.36(x - 60)(x + 60)$, $0 < x < 80$
Critical values: [on $(0, 80)$]: $x = 60$
Partition numbers: $x = 60$
Sign chart for R':

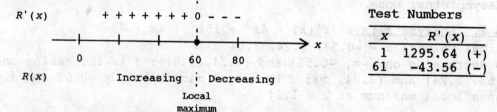

Test Numbers	
x	$R'(x)$
1	1295.64 (+)
61	−43.56 (−)

Thus, R is increasing on $(0, 60)$ and decreasing on $(60, 80)$, R has a local maximum at $x = 60$.

Step 3. Analyze $R''(x)$:
$R''(x) = -0.72x < 0$ for $0 < x < 80$
Thus, the graph of R is concave downward on $(0, 80)$.

<u>Step 4.</u> <u>Sketch the graph of R</u>:

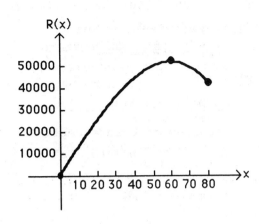

87. $P(x) = \dfrac{2x}{1 - x}$, $0 \le x < 1$

(A) $P'(x) = \dfrac{(1 - x)(2) - 2x(-1)}{(1 - x)^2} = \dfrac{2}{(1 - x)^2}$

$P'(x) > 0$ for $0 \le x < 1$. Thus, P is increasing on $(0, 1)$.

(B) From (A), $P'(x) = 2(1 - x)^{-2}$. Thus,

$P''(x) = -4(1 - x)^{-3}(-1) = \dfrac{4}{(1 - x)^3}$.

$P''(x) > 0$ for $0 \le x < 1$, and the graph of P is concave upward on $(0, 1)$.

(C) Since the domain of P is $[0, 1)$, there are no horizontal asymptotes. The denominator is 0 at $x = 1$ and the numerator is nonzero there. Thus, $x = 1$ is a vertical asymptote.

(D) $P(0) = \dfrac{2 \cdot 0}{1 - 0} = 0$.

Thus, the origin is both an x and a y intercept of the graph.

(E) The graph of P is:

x	$P(x)$
0	0
$\frac{1}{2}$	2
$\frac{3}{4}$	6

89. $C(n) = 3200 + 250n + 50n^2$, $0 < n < \infty$

(A) Average cost per year:

$$\overline{C}(n) = \frac{C(n)}{n} = \frac{3200}{n} + 250 + 50n, \ 0 < n < \infty$$

(B) Graph $\overline{C}(n)$:

Step 1. Analyze $\overline{C}(n)$:
Domain: $0 < n < \infty$
Intercepts: C intercept: None $(n > 0)$
n intercepts: $\frac{3200}{n} + 250 + 50n > 0$ on $(0, \infty)$;
there are no n intercepts.
Asymptotes: For large n, $\overline{C}(n) = \frac{3200}{n} + 250 + 50n \approx 250 + 50n$.

Thus, $y = 250 + 50n$ is an oblique asymptote. As $n \to 0$, $\overline{C}(n) \to \infty$;
thus, $n = 0$ is a vertical asymptote.

Step 2. Analyze $\overline{C}'(n)$:

$$\overline{C}'(n) = -\frac{3200}{n^2} + 50 = \frac{50n^2 - 3200}{n^2} = \frac{50(n^2 - 64)}{n^2}$$

$$= \frac{50(n - 8)(n + 8)}{n^2}, \ 0 < n < \infty$$

Critical value: $n = 8$

Sign chart for $\overline{C}'$:

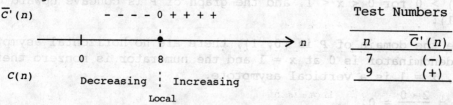

$\overline{C}'(n)$ $- - - - 0 + + + +$

n	$\overline{C}'(n)$
7	$(-)$
9	$(+)$

Test Numbers

$C(n)$ Decreasing $\vdots$ Increasing

Local
minimum

Thus, $\overline{C}$ is decreasing on $(0, 8)$ and increasing on $(8, \infty)$; $n = 8$ is
a local minimum.

Step 3: Analyze $\overline{C}''(n)$:

$$\overline{C}''(n) = \frac{6400}{n^3}, \ 0 < n < \infty$$

$\overline{C}''(n) > 0$ on $(0, \infty)$. Thus, the graph of $\overline{C}$ is concave upward
on $(0, \infty)$.

Step 4. Sketch the graph of $\overline{C}$:

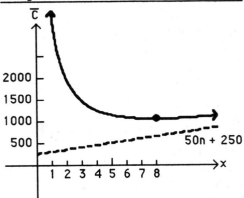

(C) The average cost per year is a minimum when $n = 8$ years.

91. $C(x) = 1000 + 5x + 0.1x^2$, $0 < x < \infty$.

(A) The average cost function is: $\overline{C}(x) = \dfrac{1000}{x} + 5 + 0.1x$.

Now, $\overline{C}'(x) = -\dfrac{1000}{x^2} + \dfrac{1}{10} = \dfrac{x^2 - 10,000}{10x^2} = \dfrac{(x + 100)(x - 100)}{10x^2}$

Sign chart for $\overline{C}'$:

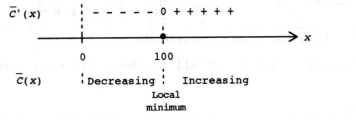

Test Numbers	
x	$\overline{C}'(x)$
1	≈ -1000 (−)
101	$\approx \frac{1}{500}$ (+)

Thus, $\overline{C}$ is decreasing on $(0, 100)$ and increasing on $(100, \infty)$; $\overline{C}$ has a minimum at $x = 100$.

Since $\overline{C}''(x) = \dfrac{2000}{x^3} > 0$ for $0 < x < \infty$, the graph of $\overline{C}$ is concave upward on $(0, \infty)$. The line $x = 0$ is a vertical asymptote and the line

$y = 5 + 0.1x$ is an oblique asymptote for the graph of $\overline{C}$.
The marginal cost function is $C'(x) = 5 + 0.2x$.

The graphs of $\overline{C}$ and C' are:

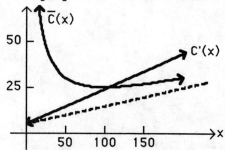

(B) The minimum average cost is:

$$\overline{C}(100) = \frac{1000}{100} + 5 + \frac{1}{10}(100) = 25$$

93. $C(t) = \dfrac{0.14t}{t^2 + 1}$

<u>Step 1. Analyze $C(t)$:</u>
Domain: $t \geq 0$, i.e., $[0, \infty)$
Intercepts: y intercept: $C(0) = 0$

$\qquad\qquad$ t intercepts: $\dfrac{0.14t}{t^2 + 1} = 0$

$\qquad\qquad\qquad\qquad\qquad t = 0$

Asymptotes:

<u>Horizontal asymptote</u>: $\displaystyle\lim_{t \to \infty} \frac{0.14t}{t^2 + 1} = \lim_{t \to \infty} \frac{0.14t}{t^2\left(1 + \frac{1}{t^2}\right)} = \lim_{t \to \infty} \frac{0.14}{t\left(1 + \frac{1}{t^2}\right)} = 0$

Thus, $y = 0$ (the t axis) is a horizontal asymptote.

<u>Vertical asymptotes</u>: Since $t^2 + 1 > 0$ for all t, there are no vertical asymptotes.

<u>Step 2. Analyze $C'(t)$:</u>

$$C'(t) = \frac{(t^2 + 1)(0.14) - 0.14t(2t)}{(t^2 + 1)^2} = \frac{0.14(1 - t^2)}{(t^2 + 1)^2} = \frac{0.14(1 - t)(1 + t)}{(t^2 + 1)^2}$$

Critical values on $[0, \infty)$: $t = 1$
Sign chart for C':

Test Numbers	
t	$C'(t)$
0	(+)
2	(−)

$C'(t)$ $\quad$ $+ + + + + \, 0 - - - - -$

$C(t)$ $\qquad$ Increasing $\vdots$ Decreasing
$\qquad\qquad\qquad$ Local
$\qquad\qquad\qquad$ maximum

Thus, C is increasing on $(0, 1)$ and decreasing on $(1, \infty)$; C has a maximum value at $t = 1$.

Step 3. Analyze $C''(t)$:

$$C''(t) = \frac{(t^2 + 1)^2(-0.28t) - 0.14(1 - t^2)(2)(t^2 + 1)(2t)}{(t^2 + 1)^4}$$

$$= \frac{(t^2 + 1)(-0.28t) - 0.56t(1 - t^2)}{(t^2 + 1)^3} = \frac{0.28t^3 - 0.84t}{(t^2 + 1)^3}$$

$$= \frac{0.28t(t^2 - 3)}{(t^2 + 1)^3} = \frac{0.28t(t - \sqrt{3})(t + \sqrt{3})}{(t^2 + 1)^3}, \quad 0 \le t < \infty$$

Partition numbers for C'' on $[0, \infty)$: $t = \sqrt{3}$

Sign chart for C'':

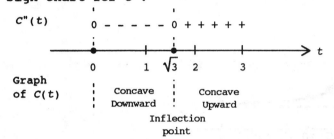

Test Numbers	
t	$C''(t)$
1	-0.07 $(-)$
2	≈ 0.005 $(+)$

Thus, the graph of C is concave downward on $(0, \sqrt{3})$ and concave upward on $(\sqrt{3}, \infty)$; the graph has an inflection point at $t = \sqrt{3}$.

Step 4. Sketch the graph of $C(t)$:

t	$C(t)$
0	0
1	0.07
$\sqrt{3}$	≈ 0.06

95. $N(t) = \dfrac{5t + 20}{t} = 5 + 20t^{-1}, \quad 1 \le t \le 30$

Step 1. Analyze $N(t)$:

Domain: $1 \le t \le 30$, or $[1, 30]$.
Intercepts: There are no t or N intercepts.
Asymptotes: Since N is defined only for $1 \le t \le 30$, there are no horizontal asymptotes. Also, since $t \ne 0$ on $[1, 30]$, there are no vertical asymptotes.

Step 2. Analyze $N'(t)$:

$N'(t) = -20t^{-2} = \dfrac{-20}{t^2}, \quad 1 \le t \le 30$

Since $N'(t) < 0$ for $1 \le t \le 30$, N is decreasing on $(1, 30)$; N has no local extrema.

Step 3. Analyze $N''(t)$:

$N''(t) = \dfrac{40}{t^3}$, $1 \le t \le 30$

Since $N''(t) > 0$ for $1 \le t \le 30$, the graph of N is concave upward on $(1, 30)$.

Step 4. Sketch the graph of N:

t	$N(t)$
1	25
5	9
10	7
30	5.67

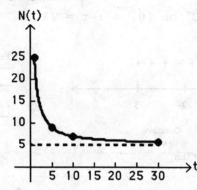

EXERCISE 9-5

Things to remember:

1. A function f continuous on a closed interval $[a, b]$ assumes both an absolute maximum and an absolute minimum on that interval. Absolute extrema (if they exist) must always occur at critical values or at endpoints.

2. STEPS FOR FINDING ABSOLUTE EXTREMA:

   To find the absolute maximum and absolute minimum of a function f on the closed interval $[a, b]$:

   (a) Verify that f is continuous on $[a, b]$.

   (b) Determine the critical values of f on the open interval (a, b).

   (c) Evaluate f at the endpoints a and b and at the critical values found in (b).

   (d) The absolute maximum $f(x)$ on $[a, b]$ is the largest of the values found in step (c).

   (e) The absolute minimum $f(x)$ on $[a, b]$ is the smallest of the values found in step (c).

<u>3</u>. SECOND–DERIVATIVE TEST FOR ABSOLUTE MAXIMUM AND MINIMUM

Suppose f is continuous on an interval I and has only one critical value C on I:

$f'(c)$	$f''(c)$	$f(c)$	Example
0	+	Absolute minimum	
0	−	Absolute maximum	
0	0	Test fails	

<u>4</u>. STRATEGY FOR SOLVING APPLIED OPTIMIZATION PROBLEMS

<u>Step 1</u>: Introduce variables and a function f, including the domain I of f, and then construct a mathematical model of the form:

Maximize (or minimize) f on the interval I

<u>Step 2</u>: Find the absolute maximum (or minimum) value of f on the interval I and the value(s) of x where this occurs.

<u>Step 3</u>: Use the solution to the mathematical model to answer the questions asked in the problem.

1. On [0, 10]; absolute minimum: $f(0) = 1$; absolute maximum: $f(10) = 9$

3. On [0, 8]; absolute minimum: $f(0) = 1$; absolute maximum: $f(3) = 8$

5. On [4, 6]; absolute minimum: $f(6) = 3$; absolute maximum: $f(4) = 7$

7. On [1, 5]; absolute minimum: $f(1) = f(5) = 5$; absolute maximum: $f(3) = 8$

9. $f(x) = x^2 - 4x + 5$, $I = (-\infty, \infty)$
$f'(x) = 2x - 4 = 2(x - 2)$
$x = 2$ is the *only critical value* on I, and $f(2) = 2^2 - 4 \cdot 2 + 5 = 1$.
$f''(x) = 2$
$f''(2) = 2 > 0$. Therefore, $f(2) = 1$ is the absolute minimum.
The function does not have an absolute maximum.

11. $f(x) = 10 + 8x - x^2$, $I = (-\infty, \infty)$
$f'(x) = 8 - 2x = 2(4 - x)$
$x = 4$ is the *only critical value* on I, and $f(4) = 10 + 32 - 16 = 26$.
$f''(x) = -2$
$f''(4) = -2 < 0$. Therefore, $f(4) = 26$ is the absolute maximum.
The function does not have an absolute minimum.

13. $f(x) = 1 - x^3$, $I = (-\infty, \infty)$

$f'(x) = -3x^2$

$x = 0$ is the *only critical value* on I, and $f(0) = 1 - 0^3 = 1$.

$f''(x) = -6x$

$f''(0) = 0$. Therefore, the test fails.

Since $f'(x) = -3x^2 < 0$ on $(-\infty, 0)$ and on $(0, \infty)$, f is decreasing on I and f does not have any absolute extrema.

15. $f(x) = 24 - 2x - \dfrac{8}{x}$, $x > 0$

$f'(x) = -2 + \dfrac{8}{x^2} = \dfrac{-2x^2 + 8}{x^2} = \dfrac{-2(x^2 - 4)}{x^2} = \dfrac{-2(x + 2)(x - 2)}{x^2}$

Critical values: $x = 2$ [<u>Note</u>: $x = -2$ is not a critical value, since the domain of f is $x > 0$.]

$f''(x) = -\dfrac{16}{x^3}$ and $f''(2) = -\dfrac{16}{8} = -2 < 0$.

Thus, the absolute maximum value of f is $f(2) = 24 - 2\cdot 2 - \dfrac{8}{2} = 16$.

17. $f(x) = 5 + 3x + \dfrac{12}{x^2}$, $x > 0$

$f'(x) = 3 - \dfrac{24}{x^3} = \dfrac{3x^3 - 24}{x^3} = \dfrac{3(x^3 - 8)}{x^3} = \dfrac{3(x - 2)(x^2 + 2x + 4)}{x^3}$

Critical value: $x = 2$

$f''(x) = \dfrac{72}{x^4}$ and $f''(2) = \dfrac{72}{2^4} = \dfrac{72}{16} > 0$.

Thus, the absolute minimum of f is: $f(2) = 5 + 3(2) + \dfrac{12}{2^2} = 14$.

19. $f(x) = x^3 - 6x^2 + 9x - 6$

$f'(x) = 3x^2 - 12x + 9 = 3(x^2 - 4x + 3) = 3(x - 3)(x - 1)$

Critical values: $x = 1, 3$

(A) On the interval $[-1, 5]$: $f(-1) = -1 - 6 - 9 - 6 = -22$

$\qquad\qquad\qquad\qquad\qquad f(1) = 1 - 6 + 9 - 6 = -2$

$\qquad\qquad\qquad\qquad\qquad f(3) = 27 - 54 + 27 - 6 = -6$

$\qquad\qquad\qquad\qquad\qquad f(5) = 125 - 150 + 45 - 6 = 14$

Thus, the absolute maximum of f is $f(5) = 14$, and the absolute minimum of f is $f(-1) = -22$.

(B) On the interval $[-1, 3]$: $f(-1) = -22$

$\qquad\qquad\qquad\qquad\qquad f(1) = -2$

$\qquad\qquad\qquad\qquad\qquad f(3) = -6$

Absolute maximum of f: $f(1) = -2$

Absolute minimum of f: $f(-1) = -22$

(C) On the interval $[2, 5]$: $f(2) = 8 - 24 + 18 - 6 = -4$

$\qquad\qquad\qquad\qquad\qquad f(3) = -6$

$\qquad\qquad\qquad\qquad\qquad f(5) = 14$

Absolute maximum of f: $f(5) = 14$

Absolute minimum of f: $f(3) = -6$

21. $f(x) = (x - 1)(x - 5)^3 + 1$

$f'(x) = (x - 1)3(x - 5)^2 + (x - 5)^3$

$\qquad = (x - 5)^2(3x - 3 + x - 5)$

$\qquad = (x - 5)^2(4x - 8)$

Critical values: $x = 2, 5$

(A) Interval $[0, 3]$: $f(0) = (-1)(-5)^3 + 1 = 126$

$\qquad\qquad\qquad f(2) = (2 - 1)(2 - 5)^3 + 1 = -26$

$\qquad\qquad\qquad f(3) = (3 - 1)(3 - 5)^3 + 1 = -15$

Absolute maximum of f: $f(0) = 126$
Absolute minimum of f: $f(2) = -26$

(B) Interval $[1, 7]$: $f(1) = 1$

$\qquad\qquad\qquad f(2) = -26$

$\qquad\qquad\qquad f(5) = 1$

$\qquad\qquad\qquad f(7) = (7 - 1)(7 - 5)^3 + 1 = 6 \cdot 8 + 1 = 49$

Absolute maximum of f: $f(7) = 49$
Absolute minimum of f: $f(2) = -26$

(C) Interval $[3, 6]$: $f(3) = (3 - 1)(3 - 5)^3 + 1 = -15$

$\qquad\qquad\qquad f(5) = 1$

$\qquad\qquad\qquad f(6) = (6 - 1)(6 - 5)^3 + 1 = 6$

Absolute maximum of f: $f(6) = 6$
Absolute minimum of f: $f(3) = -15$

23. Let one length = x and the other = $10 - x$.
Since neither length can be negative, we have $x \geq 0$ and $10 - x \geq 0$, or $x \leq 10$. We want the maximum value of the product $x(10 - x)$, where $0 \leq x \leq 10$.

Let $f(x) = x(10 - x) = 10x - x^2$; domain $I = [0, 10]$

$\quad f'(x) = 10 - 2x$; $x = 5$ is the only critical value

$\quad f''(x) = -2$

$\quad f''(5) = -2 < 0$

Thus, $f(5) = 25$ is the absolute maximum; divide the line in half.

25. Let one number = x. Then the other number = $x + 30$.

$f(x) = x(x + 30) = x^2 + 30x$; domain $I = (-\infty, \infty)$

$f'(x) = 2x + 30$; $x = -15$ is the only critical value

$f''(x) = 2$

$f''(-15) = 2 > 0$

Thus, the absolute minimum of f occurs at $x = -15$. The numbers, then, are -15 and $-15 + 30 = 15$.

27. Let x = the length of the rectangle
and y = the width of the rectangle.
Then, $2x + 2y = 100$

$\qquad\quad x + y = 50$

$\qquad\qquad y = 50 - x$

We want to find the maximum of the area:
$A(x) = x \cdot y = x(50 - x) = 50x - x^2$.

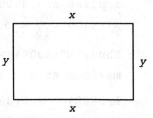

Since $x \geq 0$ and $y \geq 0$, we must have $0 \leq x \leq 50$. [Note: $A(0) = A(50) = 0$.]
$A'(x) = \dfrac{dA}{dx} = 50 - 2x$; $x = 25$ is the only critical value.

Now, $A'' = -2$ and $A''(25) = -2 < 0$. Thus, $A(25)$ is the absolute maximum.
The maximum area is $A(25) = 25(50 - 25) = 625$ cm^2, which means that the rectangle is actually a square with sides measuring 25 cm each.

29. Let the rectangle of fixed area A have dimensions x and y. Then $A = xy$ and $y = \dfrac{A}{x}$.

The cost of the fence is
$$C = 2Bx + 2By = 2Bx + \frac{2AB}{x}, \quad x > 0$$

Thus, we want to find the absolute minimum of
$$C(x) = 2Bx + \frac{2AB}{x}, \quad x > 0$$

Since $\lim\limits_{x \to 0} C(x) = \lim\limits_{x \to \infty} C(x) = \infty$, and $C(x) > 0$ for all $x > 0$, we can conclude that C has an absolute minimum on $(0, \infty)$. This agrees with our intuition that there should be a cheapest way to build the fence.

31. Let x and y be the dimensions of the rectangle and let C be the fixed amount which can be spent. Then
$$C = 2Bx + 2By \quad \text{and} \quad y = \frac{C - 2Bx}{2B}$$

The area enclosed by the fence is:
$$A = xy = x\left[\frac{C - 2Bx}{2B}\right]$$

Thus, we want to find the absolute maximum value of
$$A(x) = \frac{C}{2B}x - x^2, \quad 0 \leq x \leq \frac{C}{2B}$$

Since $A(x)$ is a continuous function on the closed interval $\left[0, \dfrac{C}{2B}\right]$, it has an absolute maximum value. This agrees with our intuition that there should be a largest rectangular area that can be enclosed with a fixed amount of fencing.

33. (A) Revenue $R(x) = x \cdot p(x) = x\left(200 - \dfrac{x}{30}\right) = 200x - \dfrac{x^2}{30}$, $0 \leq x \leq 6{,}000$

$R'(x) = 200 - \dfrac{2x}{30} = 200 - \dfrac{x}{15}$

Now $R'(x) = 200 - \dfrac{x}{15} = 0$
implies $x = 3000$.

$R''(x) = -\dfrac{1}{15} < 0$.

Thus, $R''(3000) = -\dfrac{1}{15} < 0$ and we conclude that R has an absolute maximum at $x = 3000$. The maximum revenue is

$R(3000) = 200(3000) - \dfrac{(3000)^2}{30} = \$300{,}000$

(B) Profit $P(x) = R(x) - C(x) = 200x - \dfrac{x^2}{30} - (72{,}000 + 60x)$

$$= 140x - \dfrac{x^2}{30} - 72{,}000$$

$P'(x) = 140 - \dfrac{x}{15}$

Now $140 - \dfrac{x}{15} = 0$ implies $x = 2{,}100$.

$P''(x) = -\dfrac{1}{15}$ and $P''(2{,}100) = -\dfrac{1}{15} < 0$. Thus, the maximum profit occurs when 2,100 television sets are produced. The maximum profit is

$$P(2{,}100) = 140(2{,}100) - \dfrac{(2{,}100)^2}{30} - 72{,}000 = \$75{,}000$$

the price that the company should charge is

$$p(2{,}100) = 200 - \dfrac{2{,}100}{30} = \$130 \text{ for each set.}$$

(C) If the government taxes the company \$5 for each set, then the profit $P(x)$ is given by

$$P(x) = 200x - \dfrac{x^2}{30} - (72{,}000 + 60x) - 5x$$

$$= 135x - \dfrac{x^2}{30} - 72{,}000.$$

$P'(x) = 135 - \dfrac{x}{15}$.

Now $135 - \dfrac{x}{15} = 0$ implies $x = 2{,}025$.

$P''(x) = -\dfrac{1}{15}$ and $P''(2{,}025) = -\dfrac{1}{15} < 0$. Thus, the maximum profit in this case occurs when 2,025 television sets are produced. The maximum profit is

$$P(2{,}025) = 135(2{,}025) - \dfrac{(2{,}025)^2}{30} - 72{,}000$$

$$= \$64{,}687.50$$

and the company should charge $p(2{,}025) = 200 - \dfrac{2{,}025}{30}$

$$= \$132.50 \text{ per set.}$$

35. Let x = number of dollar increases in the rate per day. Then $200 - 5x$ = total number of cars rented and $30 + x$ = rate per day.
Total income = (total number of cars rented)(rate)

$y(x) = (200 - 5x)(30 + x)$, $0 \le x \le 40$

$y'(x) = (200 - 5x)(1) + (30 + x)(-5)$

$\qquad = 200 - 5x - 150 - 5x$

$\qquad = 50 - 10x$

$\qquad = 10(5 - x)$

Thus, $x = 5$ is the only critical value and $y(5) = (200 - 25)(30 + 5) = 6125$.

$y''(x) = -10$

$y''(5) = -10 < 0$

Therefore, the absolute maximum income is $y(5) = \$6125$ when the rate is \$35 per day.

37. Let x = number of additional trees planted per acre. Then
$30 + x$ = total number of trees per acre and $50 - x$ = yield per tree.
Yield per acre = (total number of trees per acre)(yield per tree)
$$y(x) = (30 + x)(50 - x), \quad 0 \le x \le 20$$
$$y'(x) = (30 + x)(-1) + (50 - x)$$
$$= 20 - 2x$$
$$= 2(10 - x)$$
The only critical value is $x = 10$, $y(10) = 40(40) = 1600$ pounds per acre.
$y''(x) = -2$
$y''(10) = -2 < 0$
Therefore, the absolute maximum yield is $y(10) = 1600$ pounds per acre
when the number of trees per acre is 40.

39. Volume = $V(x) = (12 - 2x)(8 - 2x)x, \quad 0 \le x \le 4$
$$= 96x - 40x^2 + 4x^3$$
$$V'(x) = 96 - 80x + 12x^2$$
$$= 4(24 - 20x + 3x^2)$$

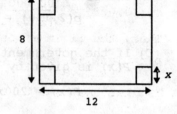

We solve $24 - 20x + 3x^2 = 0$ by using the
quadratic formula:
$$x = \frac{20 \pm \sqrt{400 - 4 \cdot 24 \cdot 3}}{6} = \frac{10 \pm 2\sqrt{7}}{3}$$

Thus, $x = \dfrac{10 - 2\sqrt{7}}{3} \approx 1.57$ is the only critical value on the interval
$[0, 4]$.
$V''(x) = -80 + 24x$
$V''(1.57) = -80 + 24(1.57) < 0$
Therefore, a square with a side of length $x = 1.57$ inches should be cut
from each corner to obtain the maximum volume.

41. Area = 800 square feet = xy (1)
Cost = $18x + 6(2y + x)$
From (1), we have $y = \dfrac{800}{x}$.

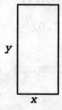

Hence, cost $C(x) = 18x + 6\left(\dfrac{1600}{x} + x\right)$, or

$$C(x) = 24x + \frac{9600}{x}, \quad x > 0,$$

$$C'(x) = 24 - \frac{9600}{x^2} = \frac{24(x^2 - 400)}{x^2} = \frac{24(x - 20)(x + 20)}{x^2}.$$

Therefore, $x = 20$ is the only critical value.

$$C''(x) = \frac{19,200}{x^3}$$

$C''(20) = \dfrac{19,200}{8000} > 0$. Therefore, $x = 20$ for the

minimum cost.
The dimensions of the fence are shown in the
diagram at the right.

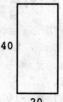

40

20
(Expensive
side)

43. Let x = number of books produced each printing. Then, the number of printings = $\frac{50,000}{x}$.

Cost = $C(x)$ = cost of storage + cost of printing

$$= \frac{x}{2} + \frac{50,000}{x}(1000), \; x > 0$$

[<u>Note</u>: $\frac{x}{2}$ is the average number in storage each day.]

$$C'(x) = \frac{1}{2} - \frac{50,000,000}{x^2} = \frac{x^2 - 100,000,000}{2x^2} = \frac{(x + 10,000)(x - 10,000)}{2x^2}$$

Critical value: $x = 10,000$

$$C''(x) = \frac{100,000,000}{x^3}$$

$$C''(10,000) = \frac{100,000,000}{(10,000)^3} > 0$$

Thus, the minimum cost occurs when $x = 10,000$ and the number of printings is $\frac{50,000}{10,000} = 5$.

45. (A) Let the cost to lay the pipe on the land be 1 unit; then the cost to lay the pipe in the lake is 1.4 units.

$$C(x) = \text{total cost} = (1.4)\sqrt{x^2 + 25} + (1)(10 - x), \; 0 \le x \le 10$$

$$= (1.4)(x^2 + 25)^{1/2} + 10 - x$$

$$C'(x) = (1.4)\frac{1}{2}(x^2 + 25)^{-1/2}(2x) - 1$$

$$= (1.4)x(x^2 + 25)^{-1/2} - 1$$

$$= \frac{1.4x - \sqrt{x^2 + 25}}{\sqrt{x^2 + 25}}$$

$C'(x) = 0$ when $1.4x - \sqrt{x^2 + 25} = 0$ or $1.96x^2 = x^2 + 25$

$$.96x^2 = 25$$

$$x^2 = \frac{25}{.96} = 26.04$$

$$x = \pm 5.1$$

Thus, the critical value is $x = 5.1$.

$$C''(x) = (1.4)(x^2 + 25)^{-1/2} + (1.4)x\left(-\frac{1}{2}\right)(x^2 + 25)^{-3/2}2x$$

$$= \frac{1.4}{(x^2 + 25)^{1/2}} - \frac{(1.4)x^2}{(x^2 + 25)^{3/2}} = \frac{35}{(x^2 + 25)^{3/2}}$$

$$C''(5.1) = \frac{35}{[(5.1)^2 + 25]^{3/2}} > 0$$

Thus, the cost will be a minimum when $x = 5.1$.

Note that:
$C(0) = (1.4)\sqrt{25} + 10 = 17$
$C(5.1) = (1.4)\sqrt{51.01} + (10 - 5.1) = 14.9$
$C(10) = (1.4)\sqrt{125} = 15.65$

Thus, the absolute minimum occurs when $x = 5.1$ miles.

(B) $C(x) = (1.1)\sqrt{x^2 + 25} + (1)(10 - x)$, $0 \le x \le 10$

$C'(x) = \dfrac{(1.1)x - \sqrt{x^2 + 25}}{\sqrt{x^2 + 25}}$

$C'(x) = 0$ when $1.1x - \sqrt{x^2 + 25} = 0$ or $(1.21)x^2 = x^2 + 25$

$$.21x^2 = 25$$
$$x^2 = \frac{25}{.21} = 119.05$$
$$x = \pm 10.91$$

Critical value: $x = 10.91 > 10$, i.e., there are no critical values on the interval $[0, 10]$. Now,

$C(0) = (1.1)\sqrt{25} + 10 = 15.5$,

$C(10) = (1.1)\sqrt{125} \approx 12.30$.

Therefore, the absolute minimum occurs when $x = 10$ miles.

47. $C(t) = 30t^2 - 240t + 500$, $0 \le t \le 8$
$C'(t) = 60t - 240$; $t = 4$ is the only critical value.
$C''(t) = 60$
$C''(4) = 60 > 0$
Now, $C(0) = 500$

$C(4) = 30(4)^2 - 240(4) + 500 = 20$,

$C(8) = 30(8)^2 - 240(8) + 500 = 500$.

Thus, 4 days after a treatment, the concentration will be minimum; the minimum concentration is 20 bacteria per cm^3.

49. Let x = the number of mice ordered in each order. Then the number of orders = $\dfrac{500}{x}$.

$C(x) = \text{Cost} = \dfrac{x}{2} \cdot 4 + \dfrac{500}{x}(10)$ [Note: Cost = cost of feeding + cost of order, $\dfrac{x}{2}$ is the average number of mice at any one time.]

$C(x) = 2x + \dfrac{5000}{x}$, $0 < x \le 500$

$C'(x) = 2 - \dfrac{5000}{x^2} = \dfrac{2x^2 - 5000}{x^2} = \dfrac{2(x^2 - 2500)}{x^2} = \dfrac{2(x + 50)(x - 50)}{x^2}$

Critical value: $x = 50$ (-50 is not a critical value, since the domain of C is $x > 0$.

$C''(x) = \dfrac{10,000}{x^3}$ and $C''(50) = \dfrac{10,000}{50^3} > 0$

Therefore, the minimum cost occurs when 50 mice are ordered each time. The total number of orders is $\dfrac{500}{50} = 10$.

51. $H(t) = 4t^{1/2} - 2t$, $0 \le t \le 2$
$H'(t) = 2t^{-1/2} - 2$
Thus, $t = 1$ is the only critical value.
Now, $H(0) = 4 \cdot 0^{1/2} - 2(0) = 0$,

$H(1) = 4 \cdot 1^{1/2} - 2(1) = 2$,

$H(2) = 4 \cdot 2^{1/2} - 4 \approx 1.66$.

Therefore, $H(1)$ is the absolute maximum, and after one month the maximum height will be 2 feet.

53. $N(t) = 30 + 12t^2 - t^3$, $0 \le t \le 8$

The rate of increase $= R(t) = N'(t) = 24t - 3t^2$, and

$R'(t) = N''(t) = 24 - 6t$.

Thus, $t = 4$ is the only critical value of $R(t)$.

Now, $R(0) = 0$,

$\qquad R(4) = 24 \cdot 4 - 3 \cdot 4^2 = 48$,

$\qquad R(8) = 24 \cdot 8 - 3 \cdot 8^2 = 0$.

Therefore, the absolute maximum value of R occurs when $t = 4$; the maximum rate of increase will occur four years from now.

CHAPTER 9 REVIEW

1. The function f is increasing on (a, c_1), (c_3, c_6). (9-2, 9-3)

2. $f'(x) < 0$ on (c_1, c_3), (c_6, b). (9-2, 9-3)

3. The graph of f is concave downward on (a, c_2), (c_4, c_5), (c_7, b).

(9-2, 9-3)

4. A local minimum occurs at $x = c_3$. (9-2)

5. The absolute maximum occurs at $x = c_6$. (9-5)

6. $f'(x)$ appears to be zero at $x = c_1$, c_3, c_5. (9-2)

7. $f'(x)$ does not exist at $x = c_6$. (9-2)

8. $x = c_2$, c_4, c_5, c_7 are inflection points. (9-3)

9. (A) From the graph, $\lim\limits_{x \to 1} f(x)$ does not exist since

$\qquad \lim\limits_{x \to 1^-} f(x) = 2 \ne \lim\limits_{x \to 1^+} f(x) = 3$.

(B) $f(1) = 3$

(C) f is NOT continuous at $x = 1$, since $\lim\limits_{x \to 1} f(x)$ does not exist. (9-1)

10. (A) $\lim\limits_{x \to 2} f(x) = 2$ (B) $f(2)$ is not defined

(C) f is NOT continous at $x = 2$ since $f(2)$ is not defined. (9-1)

11. (A) $\lim\limits_{x \to 3} f(x) = 1$ (B) $f(3) = 1$

(C) f is continous at $x = 3$ since $\lim\limits_{x \to 3} f(x) = f(3)$. (9-1)

12.

$f''(x) \quad ---- \ 0 \ +++ \ 0 \ ----- \ \text{ND} \ ----$

```
          ●         ●         ○              → x
         -1         0         2
```

Graph	Concave	Concave	Concave	Concave
of f	Downward	Upward	Downward	Downward

Point of   Point of

Inflection   Inflection

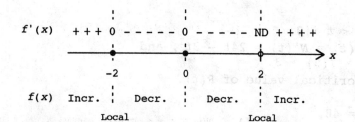

$f(x)$ Incr. Decr. Decr. Incr.

Local maximum

Local minimum

Using this information together with the points $(-3, 0)$, $(-2, 3)$, $(-1, 2)$, $(0, 0)$, $(2, -3)$, $(3, 0)$ on the graph, we have

(9-4)

13. Domain: all real numbers
  Intercepts: y intercept: $f(0) = 0$
         x-intercepts: $x = 0$
  Asymptotes: Horizontal asymptote: $y = 2$
         no vertical asymptotes
  Critical values: $x = 0$

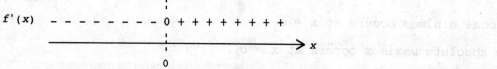

$f'(x)$ $- - - - - - - - - - 0 + + + + + + + +$

$f(x)$ Decreasing Increasing

Local minimum

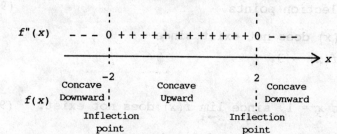

$f''(x)$ $- - - 0 + + + + + + + + + + + + 0 - - -$

$f(x)$ Concave Downward Concave Upward Concave Downward

Inflection point   Inflection point

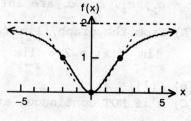

(9-4)

14. $f(x) = x^4 + 5x^3$

  $f'(x) = 4x^3 + 15x^2$

  $f''(x) = 12x^2 + 30x$ (9-3)

15. $y = 3x + \dfrac{4}{x}$

  $y' = 3 - \dfrac{4}{x^2}$

  $y'' = \dfrac{8}{x^3}$ (9-3)

16. From the graph:
  (A) $\lim\limits_{x \to 2^-} f(x) = 4$ (B) $\lim\limits_{x \to 2^+} f(x) = 6$

  (C) $\lim\limits_{x \to 2} f(x)$ does not exist since $\lim\limits_{x \to 2^-} f(x) \neq \lim\limits_{x \to 2^+} f(x)$

  (D) $f(2) = 6$ (E) No, since $\lim\limits_{x \to 2} f(x)$ does not exist.   (9-1)

17. From the graph:
(A) $\lim\limits_{x \to 5^-} f(x) = 3$ (B) $\lim\limits_{x \to 5^+} f(x) = 3$ (C) $\lim\limits_{x \to 5} f(x) = 3$ (D) $f(5) = 3$

(E) Yes, since $\lim\limits_{x \to 5} f(x) = f(5) = 3$. (9-1)

18. $x^2 - x < 12$ or $x^2 - x - 12 < 0$
Let $f(x) = x^2 - x - 12 = (x + 3)(x - 4)$. Then f is continuous for all x and $f(-3) = f(4) = 0$. Thus, $x = -3$ and $x = 4$ are partition numbers.

$f(x)$ $+ + + + +$ | $- - - - - -$ | $+ + + +$ → x

$-4\ -3 \qquad 0 \qquad\quad 4\ 5$

Test Numbers

x	$f(x)$
-4	8 (+)
0	-12 (-)
5	8 (+)

Thus, $x^2 - x < 12$ for: $-3 < x < 4$ or $(-3, 4)$. (9-1)

19. $\dfrac{x - 5}{x^2 + 3x} > 0$ or $\dfrac{x - 5}{x(x + 3)} > 0$
Let $f(x) = \dfrac{x - 5}{x(x + 3)}$. Then f is discontinuous at $x = 0$ and $x = -3$, and $f(5) = 0$. Thus, $x = -3$, $x = 0$, and $x = 5$ are partition numbers.

$f(x)$ $- - -$ | $+ +$ | $- - - -$ | $+ + +$ → x

$-4\ -3 \quad -1\ 0\ 1 \qquad 5\ 6$

Test Numbers

x	$f(x)$
-4	$-\frac{9}{4}$ (-)
-1	3 (+)
1	-1 (-)
6	$\frac{1}{54}$ (+)

Thus, $\dfrac{x - 5}{x^2 + 3x} > 0$ for $-3 < x < 0$ or $x > 5$
or $(-3, 0) \cup (5, \infty)$. (9-1)

20. $x^3 + x^2 - 4x - 2 > 0$
Let $f(x) = x^3 + x^2 - 4x - 2$. The f is continuous for all x and $f(x) = 0$ at $x \approx -2.34$, -0.47 and 1.81.

$f(x)$ $- - - 0 + + + 0 - - - - 0 + + +$

$-2.34 \quad -0.47\ 0 \quad 1.81$ → x

Thus, $x^3 + x^2 - 4x - 2 > 0$ for $-2.34 < x < -0.47$ or $1.81 < x < \infty$,
or $(-2.34, -0.47) \cup (1.81, \infty)$. (9-1)

21. $f(x) = x^3 - 18x^2 + 81x$

   (A) Domain: f is defined for all real numbers.

   (B) Intercepts: y intercept: $f(0) = 0^3 - 18(0)^2 + 81(0) = 0$

$$x \text{ intercepts: } x^3 - 18x^2 + 81x = 0$$
$$x(x^2 - 18x + 81) = 0$$
$$x(x - 9)^2 = 0$$
$$x = 0, \ 9$$

   (C) Since f is a polynomial, there are no horizontal or vertical asymptotes.   (9-4)

22. $f(x) = x^3 - 18x^2 + 81x$

   $f'(x) = 3x^2 - 36x + 81 = 3(x^2 - 12x + 27) = 3(x - 3)(x - 9)$

   (A) Critical values: $x = 3$, $x = 9$

   (B) Partition numbers: $x = 3$, $x = 9$

   (C) Sign chart for f':

	Test Numbers	
	x	$f'(x)$
	0	81 (+)
	5	-24 (−)
	10	21 (+)

$f'(x)$ + + + + 0 - - - - - 0 + + + +

$f(x)$ Increasing ¦ Decreasing ¦ Increasing

Local Maximum   Local Minimum

   Thus, f is increasing on $(-\infty, 3)$ and on $(9, \infty)$; f is decreasing on $(3, 9)$.

   (D) There is a local maximum at $x = 3$ and a local minimum at $x = 9$.   (9-4)

23. $f'(x) = 3x^2 - 36x + 81$

   $f''(x) = 6x - 36 = 6(x - 6)$

   Thus, $x = 6$ is a partition number for f''.

   (A) Sign chart for f'':

	Test Numbers	
	x	$f''(x)$
	0	-36 (−)
	7	6 (+)

$f''(x)$ - - - - - - 0 + + + + + +

Graph of f Concave Downward ¦ Concave Upward

Inflection Point

   Thus, the graph of f is concave downward on $(-\infty, 6)$ and concave upward on $(6, \infty)$.

   (B) The point $x = 6$ is an inflection point.   (9-4)

24. The graph of f:

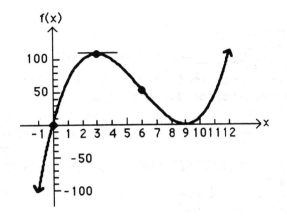

(9-4)

25. $f(x) = \dfrac{3x}{x + 2}$

(A) The domain of f is all real numbers except $x = -2$.

(B) Intercepts: y intercept: $f(0) = \dfrac{3(0)}{0 + 2} = 0$

x intercepts: $\dfrac{3x}{x + 2} = 0$

$3x = 0$

$x = 0$

(C) Asymptotes:

<u>Horizontal asymptotes</u>: $\dfrac{a_n x^n}{b_m x^m} = \dfrac{3x}{x} = 3$

Thus, the line $y = 3$ is a horizontal asymptote.

<u>Vertical asymptote(s)</u>: The denominator is 0 at $x = -2$ and the numerator is nonzero at $x = -2$. Thus, the line $x = -2$ is a vertical asymptote.

(9-4)

26. $f(x) = \dfrac{3x}{x + 2}$

$f'(x) = \dfrac{(x + 2)(3) - 3x(1)}{(x + 2)^2} = \dfrac{6}{(x + 2)^2}$

(A) Critical values: $f'(x) = \dfrac{6}{(x + 2)^2} \neq 0$ for all x ($x \neq -2$).

Thus, f does not have any critical values.

(B) Partition numbers: $x = -2$ is a partition number for f'.

(C) Sign chart for f':

	Test Numbers	
	x	$f'(x)$
	-3	6 (+)
	0	$\frac{3}{2}$ (+)

$f'(x)$ $\quad + + + \text{ND} + + + + + +$

$\xrightarrow{\qquad \qquad \qquad \qquad}$ x

$-3 \ -2 \ -1 \ 0 \ 1$

$f(x)$ $\quad$ Increasing $\vdots$ Increasing

Thus, f is increasing on $(-\infty, -2)$ and on $(-2, \infty)$.

(D) $f'(x) > 0$ for all x ($x \neq -2$). Thus, from (C), f does not have any local extrema.

(9-4)

27. $f'(x) = \dfrac{6}{(x + 2)^2} = 6(x + 2)^{-2}$

$f''(x) = -12(x + 2)^{-3} = \dfrac{-12}{(x + 2)^3}$

(A) Partition numbers for f'': $x = -2$
   Sign chart for f'':

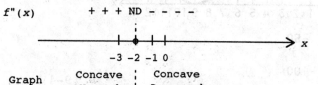

Test Numbers	
x	$f''(x)$
-3	12 (+)
0	$-\frac{3}{2}$ (−)

$f''(x)$ $+ + +$ ND $- - - -$

–3 –2 –1 0

Graph   Concave   Concave
of f Upward   Downward

The graph of f is concave upward on $(-\infty, -2)$ and concave downward on $(-2, \infty)$.

(B) The graph of f does not have any inflection points.   (9-4)

28. The graph of f is:

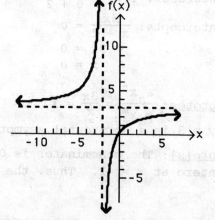

(9-4)

29.

x	$f'(x)$	$f(x)$
$-\infty < x < -2$	Negative and increasing	Decreasing and concave upward
$x = -2$	x intercept	Local minimum
$-2 < x < -1$	Positive and increasing	Increasing and concave upward
$x = -1$	Local maximum	Inflection point
$-1 < x < 1$	Positive and decreasing	Increasing and concave downward
$x = 1$	Local minimum	Inflection point
$1 < x < \infty$	Positive and increasing	Increasing and concave upward

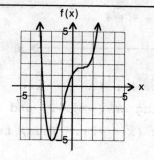

(9-3)

30. The graph in (C) could be the graph of $y = f''(x)$. (9-3)

31. $f(x) = x^3 - 6x^2 - 15x + 12$
$f'(x) = 3x^2 - 12x - 15$
$3x^2 - 12x - 15 = 0$
$3(x^2 - 4x - 5) = 0$
$3(x - 5)(x + 1) = 0$
Thus, $x = -1$ and $x = 5$ are critical values of f.
$f''(x) = 6x - 12$
Now, $f''(-1) = 6(-1) - 12 = -18 < 0$. Thus, f has a local maximum at $x = -1$.
Also, $f''(5) = 6(5) - 12 = 18 > 0$ and f has a local minimum at $x = 5$.
(9-3)

32. $y = f(x) = x^3 - 12x + 12$, $-3 \le x \le 5$
$f'(x) = 3x^2 - 12$
Critical values: f' is defined for all x:
$f'(x) = 3x^2 - 12 = 0$
$3(x^2 - 4) = 0$
$3(x - 2)(x + 2) = 0$
Thus, the critical values of f are: $x = -2$, $x = 2$.
$f(-3) = (-3)^3 - 12(-3) + 12 = 21$
$f(-2) = (-2)^3 - 12(-2) + 12 = 28$
$f(2) = 2^3 - 12(2) + 12 = -4$ Absolute minimum
$f(5) = 5^3 - 12(5) + 12 = 77$ Absolute maximum
(9-5)

33. $y = f(x) = x^2 + \dfrac{16}{x^2}$, $x > 0$

$f'(x) = 2x - \dfrac{32}{x^3} = \dfrac{2x^4 - 32}{x^3} = \dfrac{2(x^4 - 16)}{x^3} = \dfrac{2(x - 2)(x + 2)(x^2 + 4)}{x^3}$

$f''(x) = 2 + \dfrac{96}{x^4}$

The only critical value of f in the interval $(0, \infty)$ is $x = 2$.
Since

$f''(2) = 2 + \dfrac{96}{2^4} = 8 > 0$,

$f(2) = 8$ is the absolute minimum of f on $(0, \infty)$. (9-5)

34. $f(x) = 2x^2 - 3x + 1$. Since f is a polynomial function, f is continuous for all x, i.e., f is continuous on $(-\infty, \infty)$. (9-1)

35. $f(x) = \dfrac{1}{x + 5}$. Since f is a rational function, f is continuous for all x such that the denominator $x + 5 \ne 0$, i.e., for all x such that $x \ne -5$. Thus, f is continuous on $(-\infty, -5)$ and on $(-5, \infty)$. (9-1)

36. $f(x) = \dfrac{x - 3}{x^2 - x - 6}$

Since f is a rational function, f is continuous for all x such that the denominator $x^2 - x - 6 \neq 0$. Now $x^2 - x - 6 = (x - 3)(x + 2) = 0$ for $x = -2$ and $x = 3$. Thus, f is continuous on $(-\infty, -2)$, $(-2, 3)$, and $(3, \infty)$.

(9-1)

37. $f(x) = \sqrt{x - 3}$. f is continuous for all x such that $x - 3 \geq 0$, or $x \geq 3$. Thus, f is continuous on $[3, \infty)$.

(9-1)

38. $f(x) = \sqrt[3]{1 - x^2}$. f is continuous for all x such that $g(x) = 1 - x^2$ is continuous. Since g is a polynomial, g is continuous for all x. Thus, f is continuous on $(-\infty, \infty)$.

(9-1)

39. $\lim\limits_{x \to 2^-} \dfrac{x}{2 - x} = \infty$ (9-1)   **40.** $\lim\limits_{x \to 2^+} \dfrac{x}{2 - x} = -\infty$ (9-1)

41. $\lim\limits_{x \to 2} \dfrac{x}{2 - x}$ does not exist (9-1)   **42.** $\lim\limits_{x \to 2} \dfrac{x}{(2 - x)^2} = \infty$ (9-1)

43. $\lim\limits_{x \to \infty} (-3x^5) = -\infty$ (9-4) **44.** $\lim\limits_{x \to -\infty} (3x^2 - 4x^3) = \lim\limits_{x \to -\infty} (-4x^3) = \infty$ (9-4)

45. $\lim\limits_{x \to \infty} (4x^6 - 2x^5) = \lim\limits_{x \to \infty} 4x^6 = \infty$

(9-4)

46. $\lim\limits_{x \to \infty} \dfrac{6x^3 + 4x^2 + 5}{3x^3 + 2x + 7} = \lim\limits_{x \to \infty} \dfrac{6x^3}{3x^3} = \lim\limits_{x \to \infty} 2 = 2$

(9-4)

47. $\lim\limits_{x \to \infty} \dfrac{6x^4 + 4x^2 + 5}{3x^3 + 2x + 7} = \lim\limits_{x \to \infty} \dfrac{6x^4}{3x^3} = \lim\limits_{x \to \infty} 2x = \infty$

(9-4)

48. $\lim\limits_{x \to \infty} \dfrac{6x^3 + 4x^2 + 5}{3x^4 + 2x + 7} = \lim\limits_{x \to \infty} \dfrac{6x^3}{3x^4} = \lim\limits_{x \to \infty} \dfrac{2}{x} = 0$

(9-4)

49. $f(x) = \dfrac{x}{x^2 + 9}$

$\lim\limits_{x \to \infty} f(x) = \lim\limits_{x \to \infty} \dfrac{x}{x^2 + 9} = 0$

Thus, the line $y = 0$, or the x axis, is a horizontal asymptote. Since $x^2 + 9 \neq 0$ for all x, there are no vertical asymptotes.

(9-4)

50. $f(x) = \dfrac{x^3}{x^2 - 9}$

$\lim\limits_{x \to \infty} f(x) = \lim\limits_{x \to \infty} \dfrac{x^3}{x^2} = \lim\limits_{x \to \infty} x = \infty$

Thus, there are no horizontal asymptotes. Since the denominator $x^2 - 9 = (x + 3)(x - 3) = 0$ when $x = -3$ and when $x = 3$, and since the numerator $x^3 \neq 0$ at these values, the lines $x = -3$ and $x = 3$ are vertical asymptotes.

(9-4)

51. Yes. Consider f on the interval $[a, b]$. Since f is a polynomial, f is continuous on $[a, b]$. Therefore, f has an absolute maximum on $[a, b]$. Since f has a local minimum at $x = a$ and $x = b$, the absolute maximum of f on $[a, b]$ must occur at some point c in (a, b); f has a local maximum at $x = c$. (9-5)

52. No, increasing/decreasing properties are stated in terms of intervals in the domain of f. A correct statement is: $f(x)$ is decreasing on $(-\infty, 0)$ and $(0, \infty)$. (9-2)

53. A critical value for $f(x)$ is a partition number for $f'(x)$ that is also in the domain of f. However, $f'(x)$ may have partition numbers that are not in the domain of f and hence are not critical values for $f(x)$. For example, let $f(x) = \dfrac{1}{x}$. Then $f'(x) = -\dfrac{1}{x^2}$ and 0 is a partition number for $f'(x)$, but 0 is NOT a critical value for $f(x)$ since it is not in the domain of f. (9-2)

54. $f(x) = 6x^2 - x^3 + 8$, $0 \le x \le 4$
$f'(x) = 12x - 3x^2$
$f''(x) = 12 - 6x$

Now, $f''(x)$ is defined for all x and $f''(x) = 12 - 6x = 0$ implies $x = 2$. Thus, f' has a critical value at $x = 2$. Since this is the only critical value of f' and $(f'(x))'' = f'''(x) = -6$ so that $f'''(2) = -6 < 0$, it follows that $f'(2) = 12$ is the absolute maximum of f'. The graph is shown at the right.

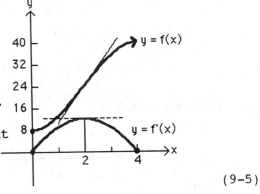

(9-5)

55. Let $x > 0$ be one of the numbers. Then $\dfrac{400}{x}$ is the other number. Now, we have:

$$S(x) = x + \frac{400}{x}, \quad x > 0,$$

$$S'(x) = 1 - \frac{400}{x^2} = \frac{x^2 - 400}{x^2} = \frac{(x - 20)(x + 20)}{x^2}$$

Thus, $x = 20$ is the only critical value of S on $(0, \infty)$.

$$S''(x) = \frac{800}{x^3} \quad \text{and} \quad S''(20) = \frac{800}{8000} = \frac{1}{10} > 0$$

Therefore, $S(20) = 20 + \dfrac{400}{20} = 40$ is the absolute minimum sum, and this occurs when each number is 20. (9-5)

56. $f(x) = (x - 1)^3(x + 3)$

<u>Step 1. Analyze $f(x)$</u>:
(A) Domain: All real numbers.
(B) Intercepts: y intercept: $f(0) = (-1)^3(3) = -3$
                x intercepts: $(x - 1)^3(x + 3) = 0$
                                      $x = 1, -3$

(C) Asymptotes: Since f is a polynomial (of degree 4), the graph of f has no asymptotes.

Step 2. Analyze $f'(x)$:

$$f'(x) = (x - 1)^3(1) + (x + 3)(3)(x - 1)^2(1)$$
$$= (x - 1)^2[(x - 1) + 3(x + 3)]$$
$$= 4(x - 1)^2(x + 2)$$

Critical values: $x = -2$, $x = 1$
Partition numbers: $x = -2$, $x = 1$
Sign chart for f':

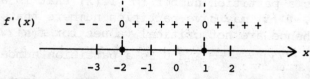

Test Numbers	
x	$f'(x)$
-3	-64 (−)
0	8 (+)
2	16 (+)

Thus, f is decreasing on $(-\infty, -2)$; f is increasing on $(-2, 1)$ and $(1, \infty)$; f has a local minimum at $x = -2$.

Step 3. Analyze $f''(x)$:

$$f''(x) = 4(x - 1)^2(1) + 4(x + 2)(2)(x - 1)(1)$$
$$= 4(x - 1)[(x - 1) + 2(x + 2)]$$
$$= 12(x - 1)(x + 1)$$

Partition numbers for f'': $x = -1$, $x = 1$.
Sign chart for f'':

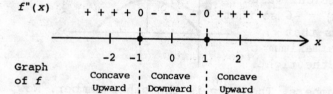

Test Numbers	
x	$f''(x)$
-2	36 (+)
0	-12 (−)
2	36 (+)

Thus, the graph of f is concave upward on $(-\infty, -1)$ and on $(1, \infty)$; the graph of f is concave downward on $(-1, 1)$; the graph has inflection points at $x = -1$ and at $x = 1$.

Step 4. Sketch the graph of f:

x	$f(x)$
-2	-27
0	-3
1	0

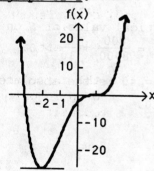

(9-4)

57. $f(x) = x^4 + x^3 + 4x^2 - 3x + 4$.

Step 1. Analyze $f(x)$:

(A) Domain: All real numbers (f is a polynomial function)

(B) Intercepts: y-intercept: $f(0) = 4$

x-intercepts: $x \approx 0.79, 1.64$

(C) Asymptotes: Since f is a polynomial function (of degree 4), the graph of f has no asymptotes.

Step 2. Analyze $f'(x)$:

$f'(x) = 4x^3 + 3x^2 - 8x - 3$

Critical values: $x \approx -1.68, -0.35, 1.28$; f is increasing on $(-1.68, -0.35)$ and $(1.28, \infty)$; f is decreasing on $(-\infty, -1.68)$ and $(-0.35, 1.28)$. f has local minima at $x = -1.68$ and $x = 1.28$. f has a local maximum at $x = -0.35$.

Step 3. Analyze $f''(x)$:

$f''(x) = 12x^2 + 6x - 8$

The graph of f is concave downward on $(-1.10, 0.60)$; the graph of f is concave upward on $(-\infty, -1.10)$ and $(0.60, \infty)$; the graph has inflection points at $x \approx -1.10$ and 0.60. (9-4)

58. (A)

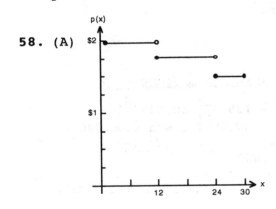

(B) p is discontinuous at $x = 12$ and $x = 24$. In each case, the limit from the left is greater then the limit from the right, reflecting the corresponding drop in price at these order quantities.

(C)

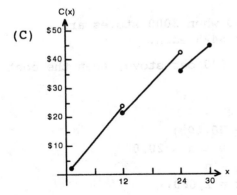

(D) C is discontinuous at $x = 12$ and $x = 24$. In each case, the limit from the left is greater than the limit from the right, reflecting savings to the customer due to the corresponding drop in price at these order quantities. (9-1)

59. (A) For the first 15 months, the price is increasing and concave down, with a local maximum at $t = 15$. For the next 15 months, the price is decreasing and concave down, with an inflection point at $t = 30$. For the next 15 months, the price is decreasing and concave up, with a local minimum at $t = 45$. For the remaining 15 months, the price is increasing and concave up.

(B)

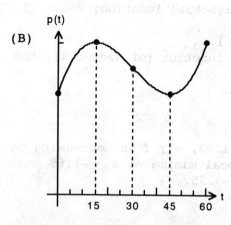

$p(t)$

15   30   45   60 t

$(9\text{-}1)$

60. (A) $R(x) = xp(x) = 500x - 0.025x^2$, $0 \le x \le 20{,}000$
$R'(x) = 500 - 0.05x$; $500 - 0.05x = 0$
$$x = 10{,}000$$
Thus, $x = 10{,}000$ is a critical value.

Now, $R(0) = 0$
    $R(10{,}000) = 2{,}500{,}000$
    $R(20{,}000) = 0$

Thus, $R(10{,}000) = \$2{,}500{,}000$ is the absolute maximum of R.

(B) $P(x) = R(x) - C(x) = 500x - 0.025x^2 - (350x + 50{,}000)$
      $= 150x - 0.025x^2 - 50{,}000$, $0 \le x \le 20{,}000$
$P'(x) = 150 - 0.05x$; $150 - 0.05x = 0$
$$x = 3{,}000$$

Now, $P(0) = -50{,}000$
    $P(3{,}000) = 175{,}000$
    $P(20{,}000) = -7{,}050{,}000$

Thus, the maximum profit is $\$175{,}000$ when 3000 stoves are manufactured and sold at $p(3{,}000) = \$425$ each.

(C) If the government taxes the company $\$20$ per stove, then the cost equation is:
    $C(x) = 370x + 50{,}000$
and
    $P(x) = 500x - 0.025x^2 - (370x + 50{,}000)$
        $= 130x - 0.025x^2 - 50{,}000$, $0 \le x \le 20{,}000$
  $P'(x) = 130 - 0.05x$; $130 - 0.05x = 0$
$$x = 2{,}600$$

The maximum profit is $P(2{,}600) = \$119{,}000$ when 2,600 stoves are produced and sold for $p(2{,}600) = \$435$ each. $(9\text{-}5)$

61.

$5/ft (top)
$5/ft (left) $15/ft (right)
$5/ft (bottom)

Let x be the length and y the width of the rectangle.

(A) $C(x, y) = 5x + 5x + 5y + 15y = 10x + 20y$

Also, Area $A = xy = 5000$, so $y = \dfrac{5000}{x}$

and $C(x) = 10x + \dfrac{100,000}{x}$, $x \geq 0$

Now, $C'(x) = 10 - \dfrac{100,000}{x^2}$ and

$10 - \dfrac{100,000}{x^2} = 0$ implies $10x^2 = 100,000$

$$x^2 = 10,000$$
$$x = \pm 100$$

Thus, $x = 100$ is the critical value.

Now, $C''(x) = \dfrac{200,000}{x^3}$ and $C''(100) = \dfrac{200,000}{1,000,000} = 0.2 > 0$

and the most economical (i.e. least cost) fence will have dimensions: length $x = 100$ feet and width $y = \dfrac{5000}{100} = 50$ feet.

(B) We want to maximize $A = xy$ subject to
$C(x, y) = 10x + 20y = 3000$ or $x = 300 - 2y$
Thus, $A = y(300 - 2y) = 300y - 2y^2$, $0 \leq y \leq 150$.
Now, $A'(y) = 300 - 4y$ and
$300 - 4y = 0$ implies $y = 75$.
Therefore, $y = 75$ is the critical value.

Now, $A''(y) = -4$ and $A''(75) = -4 < 0$. Thus, A has an absolute maximum when $y = 75$. Therefore the dimensions of the rectangle that will enclose maximum area are:
length $x = 300 - 2(75) = 150$ feet and width $y = 75$ feet.   (9-5)

62. $C(x) = 4000 + 10x + 0.1x^2$, $x > 0$

Average cost $= \overline{C}(x) = \dfrac{4000}{x} + 10 + 0.1x$

Marginal cost $= C'(x) = 10 + \dfrac{2}{10}x = 10 + 0.2x$

The graph of $C'(x)$ is a straight line with slope $\dfrac{1}{5}$ and y intercept 10.

$\overline{C}'(x) = \dfrac{-4000}{x^2} + \dfrac{1}{10} = \dfrac{-40,000 + x^2}{10x^2} = \dfrac{(x + 200)(x - 200)}{10x^2}$

Thus, $\overline{C}'(x) < 0$ on $(0, 200)$ and $\overline{C}'(x) > 0$ on $(200, \infty)$. Therefore, $\overline{C}(x)$ is decreasing on $(0, 200)$, increasing on $(200, \infty)$, and a minimum occurs at $x = 200$.

$$\text{Min } \overline{C}(x) = \overline{C}(200) = \frac{4000}{200} + 10 + \frac{1}{10}(200) = 50$$

$$\overline{C}''(x) = \frac{8000}{x^3} > 0 \text{ on } (0, \infty).$$

Therefore, the graph of $\overline{C}(x)$ is concave upward on $(0, \infty)$.

Using this information and point-by-point plotting (use a calculator), the graphs of $C'(x)$ and $\overline{C}(x)$ are as shown in the diagram at the right.
The line $y = 0.1x + 10$ is an oblique asymptote for $y = \overline{C}(x)$.

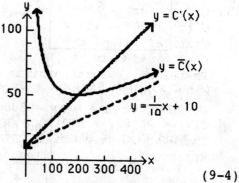

(9-4)

63. Let x = the number of dollars increase in the nightly rate, $x \geq 0$. Then $200 - 4x$ rooms will be rented at $(40 + x)$ dollars per room. [Note: Since $200 - 4x \geq 0$, $x \leq 50$.] The cost of service for $200 - 4x$ rooms at \$8 per room is $8(200 - 4x)$. Thus:

$$\begin{aligned}\text{Gross profit: } P(x) &= (200 - 4x)(40 + x) - 8(200 - 4x)\\ &= (200 - 4x)(32 + x)\\ &= 6400 + 72x - 4x^2, \quad 0 \leq x \leq 50\end{aligned}$$

$$P'(x) = 72 - 8x$$

Critical value: $72 - 8x = 0$

$$x = 9$$

Now, $P(0) = 6400$

$\quad\;\; P(9) = 6724$ Absolute maximum

$\quad\;\, P(50) = 0$

Thus, the maximum gross profit is \$6724 and this occurs at $x = 9$, i.e., the rooms should be rented at \$49 per night.

(9-5)

64. Let x = number of times the company should order. Then, the number of discs per order = $\dfrac{7200}{x}$. The average number of unsold discs is given by:

$$\frac{7200}{2x} = \frac{3600}{x}$$

Total cost: $C(x) = 5x + 0.2\left(\dfrac{3600}{x}\right), \quad x > 0$

$$C(x) = 5x + \frac{720}{x}$$

$$C'(x) = 5 - \frac{720}{x^2} = \frac{5x^2 - 720}{x^2} = \frac{5(x^2 - 144)}{x^2}$$

$$= \frac{5(x + 12)(x - 12)}{x^2}$$

Critical value: $x = 12$ [Note: $x > 0$, so $x = -12$ is not a critical value.]

$C''(x) = \dfrac{1440}{x^3}$ and $C''(12) = \dfrac{1440}{12^3} > 0$

Therefore, $C(x)$ is a minimum when $x = 12$.

(9-5)

65. $C(t) = 20t^2 - 120t + 800$, $0 \leq t \leq 9$
$C'(t) = 40t - 120 = 40(t - 3)$
Critical value: $t = 3$
$C''(t) = 40$ and $C''(3) = 40 > 0$
Therefore, a local minimum occurs at $t = 3$.
$C(3) = 20(3^2) - 120(3) + 800 = 620$ Absolute minimum
$C(0) = 800$
$C(9) = 20(81) - 120(9) + 800 = 1340$
Therefore, the bacteria count will be at a minimum three days after a treatment. (9-2)

66. $N = 10 + 6t^2 - t^3$, $0 \leq t \leq 5$
$\frac{dN}{dt} = 12t - 3t^2$
Now, find the critical values of the rate function $R(t)$:
$R(t) = \frac{dN}{dt} = 12t - 3t^2$

$R'(t) = \frac{dR}{dt} = \frac{d^2N}{dt^2} = 12 - 6t$

Critical value: $t = 2$
$R''(t) = -6$ and $R''(2) = -6 < 0$
$R(0) = 0$
$R(2) = 12$ Absolute maximum
$R(5) = -15$
Therefore, $R(t)$ has an absolute maximum at $t = 2$. The rate of increase will be a maximum after two years. (9-2)

10 ADDITIONAL DERIVATIVE TOPICS

EXERCISE 10-1

Things to remember:

1. THE NUMBER e

    The irrational number e is defined by
    $$e = \lim_{n \to \infty} \left(1 + \frac{1}{n}\right)^n$$
    or alternatively,
    $$e = \lim_{s \to 0} (1 + s)^{1/s}$$
    $$e = 2.7182818...$$

2. CONTINUOUS COMPOUND INTEREST

    $$A = Pe^{rt}$$
    where P = Principal
     r = Annual nominal interest rate compounded continuously
     t = Time in years
     A = Amount at time t

---

1. $A = \$1000e^{0.1t}$
    When $t = 2$, $A = \$1000e^{(0.1)2} = \$1000e^{0.2} = \$1221.40$.
    When $t = 5$, $A = \$1000e^{(0.1)5} = \$1000e^{0.5} = \$1628.72$.
    When $t = 8$, $A = \$1000e^{(0.1)8} = \$1000e^{0.8} = \$2225.54$

3. $2 = e^{0.06t}$
    Take the natural log of both sides of this equation
    $$\ln(e^{0.06t}) = \ln 2$$
    $$0.06t \ln e = \ln 2$$
    $$0.06t = \ln 2 \qquad (\ln e = 1)$$
    $$t = \frac{\ln 2}{0.06} \approx 11.55$$

5. $3 = e^{0.1t}$
    $$\ln(e^{0.1t}) = \ln 3$$
    $$0.1t = \ln 3$$
    $$t = \frac{\ln 3}{0.1} \approx 10.99$$

7. $2 = e^{5r}$
    $$\ln(e^{5r}) = \ln 2$$
    $$5r = \ln 2$$
    $$r = \frac{\ln 2}{5} \approx 0.14$$

9.

n	$\left(1 + \dfrac{1}{n}\right)^n$
10	2.59374
100	2.70481
1000	2.71692
10,000	2.71815
100,000	2.71827
1,000,000	2.71828
10,000,000	2.71828
$\downarrow$	$\downarrow$
∞	$e = 2.7182818\ldots$

11. The graphs of $y_1 = \left(1 + \dfrac{1}{n}\right)^n$,

$y_2 = 2.718281828 \approx e$, and

$y_3 = \left(1 + \dfrac{1}{n}\right)^{n+1}$ for $0 \le n \le 20$ are

given below:

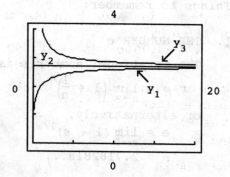

13. $A = Pe^{rt} = \$20,000e^{0.12(8.5)}$
$= \$20,000e^{1.02}$
$\approx \$55,463.90$

15. $A = Pe^{rt}$
$\$20,000 = Pe^{0.07(10)} = Pe^{0.7}$
Therefore,
$P = \dfrac{\$20,000}{e^{0.7}} = \$20,000e^{-0.7}$
$\approx \$9931.71$

17. $30,000 = 20,000e^{4r}$
$e^{4r} = 1.5$
$4r = \ln 1.5$
$r = \dfrac{\ln 1.5}{4} \approx 0.1014 \quad \text{or} \quad 10.14\%$

19. $P = 10,000e^{-0.08t}, \ 0 \le t \le 50$

(A)

t	0	10	20	30	40	50
P	10,000	4493.30	2019	907.18	407.62	183.16

The graph of P is shown below.

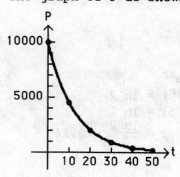

(B) $\lim\limits_{t \to \infty} 10,000e^{-0.08t} = 0$

21.
$$2P = Pe^{0.25t}$$
$$e^{0.25t} = 2$$
$$\ln(e^{0.25t}) = \ln 2$$
$$0.25t = \ln 2$$
$$t = \frac{\ln 2}{0.25} \approx 2.77 \text{ years}$$

23.
$$2P = Pe^{r(5)}$$
$$\ln(e^{5r}) = \ln 2$$
$$5r = \ln 2$$
$$r = \frac{\ln 2}{5} \approx .1386 \text{ or } = 13.86\%$$

25. The total investment in the two accounts is given by
$$A = 10,000e^{0.072t} + 10,000(1 + 0.084)^t$$
$$= 10,000[e^{0.072t} + (1.084)^t]$$

On a graphing utility, locate the intersection point of
$y_1 = 10,000[e^{0.072x} + (1.084)^x]$
and $y_2 = 35,000$.

The result is: $x = t \approx 7.3$ years.

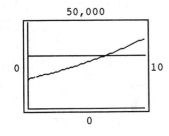

27. (A) $A = Pe^{rt}$; set $A = 2P$ (B)
$$2P = Pe^{rt}$$
$$e^{rt} = 2$$
$$rt = \ln 2$$
$$t = \frac{\ln 2}{r}$$

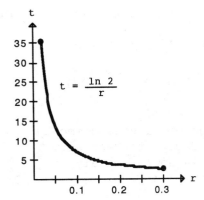

$$t = \frac{\ln 2}{r}$$

In theory, r could be any positive number. However, the restrictions on r are reasonable in the sense that most investments would be expected to earn between 2% and 30%.

(C) $r = 5\%$; $t = \dfrac{\ln 2}{0.05} \approx 13.86$ years

$r = 10\%$; $t = \dfrac{\ln 2}{0.10} \approx 6.93$ years

$r = 15\%$; $t = \dfrac{\ln 2}{0.15} \approx 4.62$ years

$r = 20\%$; $t = \dfrac{\ln 2}{0.20} \approx 3.47$ years

$r = 25\%$; $t = \dfrac{\ln 2}{0.25} \approx 2.77$ years

$r = 30\%$; $t = \dfrac{\ln 2}{0.30} \approx 2.31$ years

29.
$$Q = Q_0 e^{-0.0004332t}$$
$$\frac{1}{2}Q_0 = Q_0 e^{-0.0004332t}$$
$$e^{-0.0004332t} = \frac{1}{2}$$
$$\ln(e^{-0.0004332t}) = \ln\left(\frac{1}{2}\right) = \ln 1 - \ln 2$$
$$-0.0004332t = -\ln 2 \quad (\ln 1 = 0)$$
$$t = \frac{\ln 2}{0.0004332}$$
$$\approx \frac{0.6931}{0.0004332} \approx 1599.95$$

Thus, the half-life of radium is approximately 1600 years.

31.
$$Q = Q_0 e^{rt} \quad (r < 0)$$
$$\frac{1}{2}Q_0 = Q_0 e^{r(30)}$$
$$e^{30r} = \frac{1}{2}$$
$$\ln(e^{30r}) = \ln\left(\frac{1}{2}\right) = \ln 1 - \ln 2$$
$$30r = -\ln 2 \quad (\ln 1 = 0)$$
$$r = \frac{-\ln 2}{30} \approx \frac{-0.6931}{30}$$
$$\approx -0.0231$$

Thus, the continuous compound rate of decay of the cesium isotope is approximately −0.0231.

33.
$$2P_0 = P_0 e^{0.02t} \quad \text{or} \quad e^{0.02t} = 2$$
Thus, $\ln(e^{0.02t}) = \ln 2$
and $\qquad 0.02t = \ln 2$.
Therefore, $t = \dfrac{\ln 2}{0.02} \approx 34.66$ years.

35.
$$2P_0 = P_0 e^{r(20)}$$
$$e^{20r} = 2$$
$$\ln(e^{20r}) = \ln 2$$
$$20r = \ln 2$$
$$r = \frac{\ln 2}{20} \approx 0.0347$$
$$\text{or } 3.47\%$$

37.
$$A = Pe^{rt}$$
$$1.68 \times 10^{14} = 5 \times 10^9 e^{0.02t}$$
$$e^{0.02t} = \frac{1.68 \times 10^{14}}{5 \times 10^9} = 33,600$$
$$\ln(e^{0.02t}) = \ln(33,600)$$
$$0.02t = \ln 33,600$$
$$t = \frac{\ln 33,600}{0.02} \approx 521.11$$

Thus, there will be one square yard of land per person in approximately 521 years.

Things to remember:

1. DERIVATIVES OF THE NATURAL LOGARITHMIC AND EXPONENTIAL FUNCTIONS

$$\frac{d}{dx} \ln x = \frac{1}{x} \qquad \frac{d}{dx} e^x = e^x$$

---

1. $f(x) = 6e^x - 7 \ln x$

$f'(x) = 6 \frac{d}{dx} e^x - 7 \frac{d}{dx} \ln x = 6e^x - 7\left(\frac{1}{x}\right) = 6e^x - \frac{7}{x}$

3. $f(x) = 2x^e + 3e^x$

$f'(x) = 2 \frac{d}{dx} x^e + 3 \frac{d}{dx} e^x = 2ex^{e-1} + 3e^x$

[Note: $e \approx 2.71828$ is a constant and so we use the power rule on the first term.]

5. $f(x) = \ln x^5 = 5 \ln x$ (Property of logarithms)

$f'(x) = 5 \frac{d}{dx} \ln x = 5\left(\frac{1}{x}\right) = \frac{5}{x}$

7. $f(x) = (\ln x)^2$

$f'(x) = 2(\ln x) \frac{d}{dx} \ln x$ (Power rule for functions)

$\qquad = 2(\ln x)\frac{1}{x} = \frac{2 \ln x}{x}$

9. $f(x) = x^4 \ln x$

$f'(x) = x^4 \frac{d}{dx} \ln x + \ln x \frac{d}{dx} x^4$ (Product rule)

$\qquad = x^4\left(\frac{1}{x}\right) + (\ln x)4x^3 = x^3 + 4x^3 \ln x = x^3(1 + 4 \ln x)$

11. $f(x) = x^3 e^x$

$f'(x) = x^3 \frac{d}{dx} e^x + e^x \frac{d}{dx} x^3$ (Product rule)

$\qquad = x^3 e^x + e^x 3x^2 = x^2 e^x(x + 3)$

13. $f(x) = \dfrac{e^x}{x^2 + 9}$

$f'(x) = \dfrac{(x^2 + 9) \frac{d}{dx} e^x - e^x \frac{d}{dx}(x^2 + 9)}{(x^2 + 9)^2}$ (Quotient rule)

$\qquad = \dfrac{(x^2 + 9)e^x - e^x(2x)}{(x^2 + 9)^2} = \dfrac{e^x(x^2 - 2x + 9)}{(x^2 + 9)^2}$

15. $f(x) = \dfrac{\ln x}{x^4}$

$$f'(x) = \frac{x^4 \dfrac{d}{dx} \ln x - \ln x \dfrac{d}{dx} x^4}{(x^4)^2} \quad \text{(Quotient rule)}$$

$$= \frac{x^4\left(\dfrac{1}{x}\right) - (\ln x)4x^3}{x^8} = \frac{x^3 - 4x^3 \ln x}{x^8} = \frac{1 - 4 \ln x}{x^5}$$

17. $f(x) = (x + 2)^3 \ln x$

$$f'(x) = (x + 2)^3 \frac{d}{dx} \ln x + (\ln x) \frac{d}{dx}(x + 2)^3$$

$$= (x + 2)^3\left(\frac{1}{x}\right) + (\ln x)[3(x + 2)^2(1)]$$

$$= 3(x + 2)^2 \ln x + \frac{(x + 2)^3}{x} = (x + 2)^2\left[3 \ln x + \frac{x + 2}{x}\right]$$

19. $f(x) = (x + 1)^3 e^x$

$$f'(x) = (x + 1)^3 \frac{d}{dx} e^x + e^x \frac{d}{dx}(x + 1)^3$$

$$= (x + 1)^3 e^x + e^x(3)(x + 1)^2(1)$$

$$= (x + 1)^2 e^x[x + 1 + 3] = (x + 1)^2(x + 4)e^x$$

21. $f(x) = \dfrac{x^2 + 1}{e^x}$

$$f'(x) = \frac{e^x \dfrac{d}{dx}(x^2 + 1) - (x^2 + 1)\dfrac{d}{dx}e^x}{(e^x)^2} = \frac{e^x(2x) - (x^2 + 1)e^x}{e^{2x}} = \frac{2x - x^2 - 1}{e^x}$$

23. $f(x) = x(\ln x)^3$

$$f'(x) = x \frac{d}{dx}(\ln x)^3 + (\ln x)^3 \frac{d}{dx}x$$

$$= x(3)(\ln x)^2\left(\frac{1}{x}\right) + (\ln x)^3(1) = (\ln x)^2[3 + \ln x]$$

25. $f(x) = (4 - 5e^x)^3$

$$f'(x) = 3(4 - 5e^x)^2(-5e^x) = -15e^x(4 - 5e^x)^2$$

27. $f(x) = \sqrt{1 + \ln x} = (1 + \ln x)^{1/2}$

$$f'(x) = \frac{1}{2}(1 + \ln x)^{-1/2}\left(\frac{1}{x}\right)$$

$$= \frac{1}{2x(1 + \ln x)^{1/2}} = \frac{1}{2x\sqrt{1 + \ln x}}$$

29. $f(x) = xe^x - e^x$

$$f'(x) = x \frac{d}{dx} e^x + e^x \frac{d}{dx} x - \frac{d}{dx} e^x = xe^x + e^x - e^x = xe^x$$

31. $f(x) = 2x^2 \ln x - x^2$

$f'(x) = 2x^2 \dfrac{d}{dx} \ln x + \ln x \dfrac{d}{dx} 2x^2 - \dfrac{d}{dx} x^2 = 2x^2\left(\dfrac{1}{x}\right) + 4x \ln x - 2x$

$\qquad\qquad\qquad\qquad\qquad\qquad\qquad\qquad\qquad = 4x \ln x$

33. $f(x) = e^x$

$f'(x) = \dfrac{d}{dx} e^x = e^x$

The tangent line at $x = 1$ has an equation of the form
$y - y_1 = m(x - x_1)$
where $x_1 = 1$, $y_1 = f(1) = e$, and $m = f'(1) = e$. Thus, we have:
$y - e = e(x - 1)$ or $y = ex$

35. $f(x) = \ln x$

$f'(x) = \dfrac{d}{dx}(\ln x) = \dfrac{1}{x}$

The tangent line at $x = e$ has an equation of the form
$y - y_1 = m(x - x_1)$

where $x_1 = e$, $y_1 = f(e) = \ln e = 1$, and $m = f'(e) = \dfrac{1}{e}$. Thus, we have:

$y - 1 = \dfrac{1}{e}(x - e)$ or $y = \dfrac{1}{e}x$

37. An equation for the tangent line to the graph of $f(x) = e^x$ at the point $(3, f(3)) = (3, e^3)$ is:

$\qquad\qquad y - e^3 = e^3(x - 3)$
$\qquad$ or $\qquad y = xe^3 - 2e^3 = e^3(x - 2)$

Clearly, $y = 0$ when $x = 2$, that is the tangent line passes through the point $(2, 0)$.

In general, an equation for the tangent line to the graph of $f(x) = e^x$ at the point $(c, f(c)) = (c, e^c)$ is:

$\qquad\qquad y - e^c = e^c(x - c)$
$\qquad$ or $\qquad y = e^c(x - [c - 1])$

Thus, the tangent line at the point (c, e^c) passes through $(c - 1, 0)$; then tangent line at the point $(4, e^4)$ passes through $(3, 0)$.

39. $f(x) = 4x - x \ln x,\ x > 0$

$f'(x) = 4 - \left[x \dfrac{d}{dx} \ln x - (\ln x) \dfrac{d}{dx} x\right]$

$\qquad = 4 - \left[x\left(\dfrac{1}{x}\right) - (\ln x)(1)\right] = 3 - \ln x,\ x > 0$

Critical values: $f'(x) = 3 - \ln x = 0$
$\qquad\qquad\qquad\qquad\qquad\qquad \ln x = 3$
$\qquad\qquad\qquad\qquad\qquad\qquad\qquad x = e^3$

Thus, $x = e^3$ is the only critical value of f on $(0, \infty)$.

Now, $f''(x) = \frac{d}{dx}(3 - \ln x) = -\frac{1}{x}$

and $f''(e^3) = -\frac{1}{e^3} < 0$.

Therefore, f has a maximum value at $x = e^3$, and $f(e^3) = 4e^3 - e^3 \ln e^3$
$= 4e^3 - 3e^3 = e^3 \approx 20.086$ is the absolute maximum of f.

41. $f(x) = \frac{e^x}{x}$, $x > 0$

$f'(x) = \frac{x \frac{d}{dx} e^x - e^x \frac{d}{dx} x}{x^2} = \frac{xe^x - e^x(1)}{x^2} = \frac{e^x(x - 1)}{x^2}$, $x > 0$

Critical values: $f'(x) = \frac{e^x(x - 1)}{x^2} = 0$

$e^x(x - 1) = 0$

$x = 1$ [<u>Note</u>: $e^x \neq 0$ for all x.]

Thus, $x = 1$ is the only critical value of f on $(0, \infty)$.

Sign chart for f' [<u>Note</u>: This approach is a litle easier than calculating $f''(x)$]:

$f'(x)$	$- - - - - - 0 + + + + +$
	$\xrightarrow{\hspace{4cm}} x$
	$\quad 0 \qquad\qquad 1 \qquad\qquad 2$
$f(x)$	Decreasing $\vdots$ Increasing

Test Numbers

x	$f'(x)$
$\frac{1}{2}$	$-2\sqrt{e}$ (−)
2	$\frac{e^2}{4}$ (+)

By the first derivative test, f has a minimum value at $x = 1$;
$f(1) = \frac{e}{1} = e \approx 2.718$ is the absolute minimum of f.

43. $f(x) = \frac{1 + 2 \ln x}{x}$, $x > 0$

$f'(x) = \frac{x \frac{d}{dx}(1 + 2 \ln x) - (1 + 2 \ln x) \frac{d}{dx} x}{x^2} = \frac{x\left(\frac{2}{x}\right) - (1 + 2 \ln x)(1)}{x^2}$

$= \frac{1 - 2 \ln x}{x^2}$, $x > 0$

Critical values: $f'(x) = \frac{1 - 2 \ln x}{x^2} = 0$

$1 - 2 \ln x = 0$

$\ln x = \frac{1}{2}$

$x = e^{1/2} = \sqrt{e} \approx 1.65$

Sign chart for f':

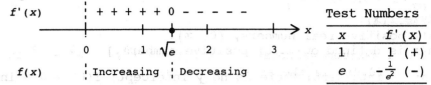

f'(x)   + + + + + 0 - - - - -

0   1 $\sqrt{e}$ 2   3   x

f(x)   Increasing   Decreasing

Test Numbers	
x	f'(x)
1	1 (+)
e	$-\frac{1}{e^2}$ (−)

By the first derivative test, f has a maximum value at $x = \sqrt{e}$;

$$f(\sqrt{e}) = \frac{1 + 2\ln\sqrt{e}}{\sqrt{e}} = \frac{1 + 2\left(\frac{1}{2}\right)}{\sqrt{e}} = \frac{2}{\sqrt{e}} \approx 1.213 \text{ is the absolute maximum of } f.$$

45. $f(x) = 1 - e^x$

Step 1. Analyze $f(x)$:
(A) Domain: All real numbers, $(-\infty, \infty)$.

(B) Intercepts: y intercept: $f(0) = 1 - e^0 = 0$
             x intercept: $1 - e^x = 0$
                       $e^x = 1$
                       $x = 0$

(C) Asymptotes:
Horizontal asymptote: $\lim\limits_{x \to -\infty} (1 - e^x) = 1 - \lim\limits_{x \to -\infty} e^x = 1 - 0 = 1$

$$[\lim\limits_{x \to \infty}(1 - e^x) = -\infty]$$

Thus, $y = 1$ is a horizontal asymptote.
Vertical asymptotes: There are no vertical asymptotes.

Step 2. Analyze $f'(x)$:
$f'(x) = -e^x$
Critical values: $f'(x) = -e^x$ is continuous and nonzero (negative) for all x; there are no critical values.
Partition numbers: There are no partition numbers. Since $f'(x) < 0$ for all x, f is decreasing on $(-\infty, \infty)$; f has no local extrema.

Step 3. Analyze $f''(x)$:
$f''(x) = -e^x < 0$ for all x
Thus, the graph of f is concave downward on $(-\infty, \infty)$.

Step 4. Sketch the graph of f:

x	f(x)
0	0
1	$1 - e \approx -1.718$
-1	$1 - \frac{1}{e}$

f(x)

1    $y = 1$

x

47. $f(x) = x - \ln x$

Step 1. Analyze $f(x)$:

(A) Domain: All positive real numbers, $(0, \infty)$.
   [Note: $\ln x$ is defined only for positive numbers.]

(B) Intercepts: y intercept: There is no y intercept; $f(0) = 0 - \ln(0)$
   is not defined.
   x intercept: $x - \ln x = 0$
   $\ln x = x$
   Since the graph of $y = \ln x$ is below the graph of $y = x$, there are
   no solutions to this equation; there are no x intercepts.

(C) Asymptotes:
   Horizontal asymptote: None
   Vertical asymptotes: Since $\lim\limits_{x \to 0^+} \ln x = -\infty$, $\lim\limits_{x \to 0^+} (x - \ln x) = \infty$.
   Thus, $x = 0$ is a vertical asymptote for
   $f(x) = x - \ln x$.

Step 2. Analyze $f'(x)$:

$f'(x) = 1 - \dfrac{1}{x} = \dfrac{x - 1}{x}$, $x > 0$

Critical values: $\dfrac{x - 1}{x} = 0$
   $x = 1$

Partition numbers: $x = 1$

Sign chart for $f'(x) = \dfrac{x - 1}{x}$:

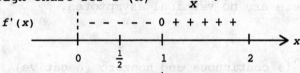

Test Numbers	
x	$f'(x)$
$\frac{1}{2}$	$-1 \ (-)$
2	$\frac{1}{2} \ (+)$

Thus, f is decreasing on $(0, 1)$ and increasing on $(1, \infty)$; f has a local minimum at $x = 1$.

Step 3. Analyze $f''(x)$:

$f''(x) = \dfrac{1}{x^2}$, $x > 0$

Thus, $f''(x) > 0$ and the graph of f is concave upward on $(0, \infty)$.

Step 4. Sketch the graph of f:

x	$f(x)$
0.1	≈ 2.4
1	1
10	≈ 7.7

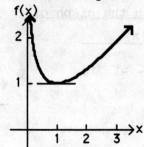

49. $f(x) = (3 - x)e^x$

Step 1. Analyze $f(x)$:

(A) Domain: All real numbers, $(-\infty, \infty)$.

(B) Intercepts: y intercept: $f(0) = (3 - 0)e^0 = 3$
x intercept: $(3 - x)e^x = 0$
$$3 - x = 0$$
$$x = 3$$

(C) Asymptotes:

   Horizontal asymptote: Consider the behavior of f as $x \to \infty$ and
   as $x \to -\infty$.
   Using the following tables,

x	-1	-10	-20
$f(x)$	1.47	0.00059	0.000000047

x	5	10
$f(x)$	-296.83	$-154,185.26$

   we conclude that $\lim\limits_{x \to -\infty} f(x) = 0$ and $\lim\limits_{x \to \infty} f(x)$ does not exist. Because
   of the first limit, $y = 0$ is a horizontal asymptote.

   Vertical asymptotes: There are no vertical asymptotes.

Step 2. Analyze $f'(x)$:

$f'(x) = (3 - x)e^x + e^x(-1) = (2 - x)e^x$
Critical values: $(2 - x)e^x = 0$
$$x = 2 \quad [\underline{\text{Note}}: e^x > 0]$$
Partition numbers: $x = 2$
Sign chart for f':

Test Numbers	
x	$f'(x)$
0	2 (+)
3	$-e^3$ (−)

Thus, f is increasing on $(-\infty, 2)$ and decreasing on $(2, \infty)$; f has a
local maximum at $x = 2$.

Step 3. Analyze $f''(x)$:

$f''(x) = (2 - x)e^x + e^x(-1) = (1 - x)e^x$
Partition number for f'': $x = 1$
Sign chart for f'':

Test Numbers	
x	$f''(x)$
0	1 (+)
2	$-e^2$ (−)

Thus, the graph of f is concave upward on $(-\infty, 1)$ and concave downward
on $(1, \infty)$; the graph has an inflection point at $x = 1$.

Step 4.   Sketch the graph of f:

x	$f(x)$
0	3
2	$e^2 \approx 7.4$
3	0

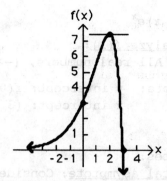

51. $f(x) = x^2 \ln x$.

Step 1.   Analyze $f(x)$:

(A) Domain: All positive numbers, $(0, \infty)$.

(B) Intercepts: y intercept: There is no y intercept.
                  x intercept: $x^2 \ln x = 0$
                  $\ln x = 0$
                  $x = 1$

(C) Asymptotes: Consider the behavior of f as $x \to \infty$ and as $x \to 0$.   It is
    clear that $\lim\limits_{x \to \infty} f(x)$ does not exist; f is unbounded as x approaches ∞.

    The following table indicates that f approaches 0 as x approaches 0.

x	1	0.1	0.01	0.001
$f(x)$	0	-0.023	-0.00046	-0.000007

    Thus, there are no vertical or horizontal asymptotes.

Step 2.   Analyze $f'(x)$:

$f'(x) = x^2\left(\dfrac{1}{x}\right) + (\ln x)(2x) = x(1 + 2 \ln x)$

Critical values: $x(1 + 2 \ln x) = 0$
                 $1 + 2 \ln x = 0$ [Note: $x > 0$]
                 $\ln x = -\dfrac{1}{2}$
                 $x = e^{-1/2} = \dfrac{1}{\sqrt{e}} \approx 0.6065$

Partition number: $x = \dfrac{1}{\sqrt{e}} \approx 0.6065$

Sign chart for f':

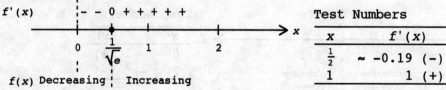

Test Numbers	
x	$f'(x)$
$\frac{1}{2}$	≈ -0.19 $(-)$
1	1 $(+)$

Thus, f is decreasing on $(0, e^{-1/2})$ and increasing on $(e^{-1/2}, \infty)$; f has
a local minimum at $x = e^{-1/2}$.

Step 3. Analyze $f''(x)$:

$$f''(x) = x\left(\frac{2}{x}\right) + (1 + 2 \ln x) = 3 + 2 \ln x$$

Partition number for f'': $3 + 2 \ln x = 0$

$$\ln x = -\frac{3}{2}$$

$$x = e^{-3/2} \approx 0.2231$$

Sign chart for f'':

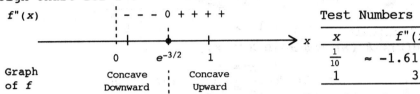

	Test Numbers	
	x	$f''(x)$
	$\frac{1}{10}$	≈ -1.61 (–)
	1	3 (+)

Thus, the graph of f is concave downward on $(0, e^{-3/2})$ and concave upward on $(e^{-3/2}, \infty)$; the graph has an inflection point at $x = e^{-3/2}$.

Step 4. Sketch the graph of f:

x	$f(x)$
$e^{-3/2}$	≈ -0.075
$e^{-1/2}$	≈ -0.18
1	0

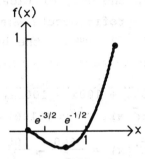

53. $f(x) = e^x - 2x^2 \qquad -\infty < x < \infty$

$f'(x) = e^x - 4x$

Critical values:

Solve $f'(x) = e^x - 4x = 0$

To two decimal places, $x = 0.36$ and $x = 2.15$

Increasing/Decreasing: $f(x)$ is increasing on $(-\infty, 0.36)$ and on $(2.15, \infty)$; $f(x)$ is decreasing on $(0.36, 2.15)$

Local extrema: $f(x)$ has a local maximum at $x = 0.36$ and a local minimum at $x = 2.15$

55. $f(x) = 20 \ln x - e^x \qquad 0 < x < \infty$

$f'(x) = \dfrac{20}{x} - e^x$

Critical values:

Solve $f'(x) = \dfrac{20}{x} - e^x = 0$

To two decimal places, $x = 2.21$

Increasing/Decreasing: $f(x)$ is increasing on $(0, 2.21)$ and decreasing on $(2.21, \infty)$

Local extrema: $f(x)$ has a local maximum at $x = 2.21$

57. On a graphing utility, graph $y_1 = e^x$ and $y_2 = x^4$. Rounded off to two decimal places, the points of intersection are: $(-0.82, 0.44)$, $(1.43, 4.18)$, $(8.61, 5503.66)$.

59. Demand: $p = 5 - \ln x$, $5 \leq x \leq 50$

Revenue: $R = xp = x(5 - \ln x) = 5x - x \ln x$

Cost: $C = x(1) = x$

Profit = Revenue - Cost: $P = 5x - x \ln x - x$

$$\text{or} \quad P(x) = 4x - x \ln x$$

$$P'(x) = 4 - x\left(\frac{1}{x}\right) - \ln x$$

$$= 3 - \ln x$$

Critical value(s): $P'(x) = 3 - \ln x = 0$

$$\ln x = 3$$

$$x = e^3$$

$P''(x) = -\dfrac{1}{x}$ and $P''(e^3) = -\dfrac{1}{e^3} < 0$.

Since $x = e^3$ is the only critical value and $P''(e^3) < 0$, the maximum weekly profit occurs when $x = e^3 \approx 20.09$ and the price $p = 5 - \ln(e^3) = 2$. Thus, the hot dogs should be sold at \$2.

61. Cost: $C(x) = 600 + 100x - 100 \ln x$, $x \geq 1$

Average cost: $\overline{C}(x) = \dfrac{600}{x} + 100 - \dfrac{100}{x} \ln x$

$$\overline{C}'(x) = \frac{-600}{x^2} - \frac{100}{x^2} + \frac{100 \ln x}{x^2} = \frac{-700 + 100 \ln x}{x^2}, \quad x \geq 1$$

Critical value(s): $\overline{C}'(x) = \dfrac{-700 + 100 \ln x}{x^2} = 0$

$$-700 + 100 \ln x = 0$$

$$\ln x = 7$$

$$x = e^7$$

$$\overline{C}''(x) = \frac{x^2 \dfrac{100}{x} - (-700 + 100 \ln x)(2x)}{x^4}$$

$$= \frac{100x + 1400x - 200x \ln x}{x^4} = \frac{1500 - 200 \ln x}{x^3}$$

$$\overline{C}''(e^7) = \frac{1500 - 200 \ln(e^7)}{e^{21}} = \frac{100}{e^{21}} > 0$$

Since $x = e^7$ is the only critical value and $\overline{C}''(e^7) > 0$, the **minimum average cost** is

$$\overline{C}(e^7) = \frac{600}{e^7} + 100 - \frac{100}{e^7} \ln(e^7) = \frac{600}{e^7} + 100 - \frac{700}{e^7} = 100 - \frac{100}{e^7} \approx 99.91$$

Thus, the minimal average cost is approximately \$99.91.

63. Demand: $p = 10e^{-x} = \dfrac{10}{e^x}$, $0 \le x \le 2$

Revenue: $R(x) = xp = 10xe^{-x} = \dfrac{10x}{e^x}$

(A) $R'(x) = \dfrac{e^x(10) - 10xe^x}{e^{2x}} = \dfrac{10 - 10x}{e^x}$, $0 \le x \le 2$

Critical value(s): $\dfrac{10 - 10x}{e^x} = 0$

$$10 - 10x = 0$$
$$x = 1$$

$R''(x) = \dfrac{e^x(-10) - (10 - 10x)e^x}{e^{2x}} = \dfrac{-20 + 10x}{e^x}$

$R''(1) = -\dfrac{10}{e} < 0$

Now, $R(0) = 0$

$\qquad R(1) = \dfrac{10}{e} \approx 3.68$ Absolute maximum

$\qquad R(2) = \dfrac{20}{e^2} \approx 2.71$

Thus, the maximum weekly revenue occurs at price $p = \dfrac{10}{e} \approx \$3.68$.

The maximum weekly revenue is $R(1) = 3.68$ thousand dollars, or $\$3680$.

(B) The sign chart for R' is:

$R'(x)$ $\quad$ + + + + + 0 - - - - -

$R(x)$ $\quad$ Increasing $\vdots$ Decreasing

Test Numbers

x	$f'(x)$
0	10 (+)
2	$-\dfrac{10}{e^2}$ (−)

Thus, R is increasing on $(0, 1)$ and decreasing on $(1, 2)$; the maximum value of R occurs at $x = 1$, as noted in (A).

$R''(x) = 10e^{-x}(x - 2) < 0$ on $(0, 2)$
Thus, the graph of R is concave downward on $(0, 2)$. The graph is shown at the right.

x	$R(x)$
0	0
1	3.68
2	2.71

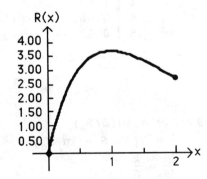

65. $P(x) = 17.5(1 + \ln x)$, $10 \le x \le 100$

$$P'(x) = \frac{17.5}{x}$$

$$P'(40) = \frac{17.5}{40} \approx 0.44$$

$$P'(90) = \frac{17.5}{90} \approx 0.19$$

Thus, at the 40 pound weight level, blood pressure would increase at the rate of 0.44 mm of mercury per pound of weight gain; at the 90 pound weight level, blood pressure would increase at the rate of 0.19 mm of mercury per pound of weight gain.

67. $C(t) = 4.35e^{-t} = \dfrac{4.35}{e^t}$, $0 \le t \le 5$

(A) $C'(t) = \dfrac{-4.35e^t}{e^{2t}} = \dfrac{-4.35}{e^t} = -4.35e^{-t}$

$C'(1) = -4.35e^{-1} \approx -1.60$

$C'(4) = -4.35e^{-4} \approx -0.08$

Thus, after one hour, the concentration is decreasing at the rate of 1.60 mg/ml per hour; after four hours, the concentration is decreasing at the rate of 0.08 mg/ml per hour.

(B) $C'(t) = -4.35e^{-t} < 0$ on $(0, 5)$

Thus, C is decreasing on $(0, 5)$; there are no local extrema.

$C''(t) = \dfrac{4.35e^t}{e^{2t}} = \dfrac{4.35}{e^t} = 4.35e^{-t} > 0$ on $(0, 5)$

Thus, the graph of C is concave upward on $(0, 5)$. The graph of C is shown at the right.

t	$C(t)$
0	4.35
1	1.60
4	0.08
5	0.03

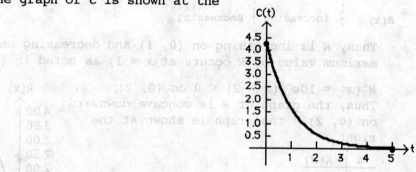

69. $R = k \ln(S/S_0)$

$\quad = k[\ln S - \ln S_0]$

$\dfrac{dR}{dS} = \dfrac{k}{S}$

Things to remember:

1. **COMPOSITE FUNCTIONS**

    A function m is the composite of functions f and g (in this order) if

    $$m(x) = f[g(x)].$$

    The domain of m is the set of all numbers x in the domain of g such that $g(x)$ is in the domain of f.

2. **CHAIN RULE**

    If $y = f(u)$ and $u = g(x)$, define the composite function
    $$y = m(x) = f[g(x)].$$
    Then
    $$\frac{dy}{dx} = \frac{dy}{du} \cdot \frac{du}{dx}$$
    provided $\frac{dy}{du}$ and $\frac{du}{dx}$ exist.

    Or, equivalently, $m'(x) = f'[g(x)]g'(x)$ provided $f'[g(x)]$ and $g'(x)$ exist.

3. **GENERAL DERIVATIVE RULES**

    (a) $\frac{d}{dx}[f(x)]^n = n[f(x)]^{n-1}f'(x)$

    (b) $\frac{d}{dx}\ln[f(x)] = \frac{1}{f(x)} \cdot f'(x)$

    (c) $\frac{d}{dx}e^{f(x)} = e^{f(x)} \cdot f'(x)$

4. **OTHER LOGARITHMIC AND EXPONENTIAL FUNCTIONS**

    $\frac{d}{dx}[\log_b x] = \frac{1}{\ln b} \cdot \frac{1}{x}, \ b \neq e$

    $\frac{d}{dx}[b^x] = b^x \ln b, \ b \neq e$

---

1. Let $u = g(x) = 2x + 5$ and $f(u) = u^3$. Then $y = f(u) = u^3$.

3. Let $u = g(x) = 2x^2 + 7$ and $f(u) = \ln u$. Then $y = f(u) = \ln u$.

5. Let $u = g(x) = x^2 - 2$ and $f(u) = e^u$. Then $y = f(u) = e^u$.

7. $y = u^2$, $u = 2 + e^x$. Thus, $y = (2 + e^x)^2$ and $\dfrac{dy}{dx} = \dfrac{dy}{du} \cdot \dfrac{du}{dx} = 2u(e^x)$

$$= 2e^x(2 + e^x).$$

9. $y = e^u$, $u = 2 - x^4$. Thus, $y = e^{2-x^4}$ and $\dfrac{dy}{dx} = \dfrac{dy}{du} \cdot \dfrac{du}{dx} = e^u(-4x^3)$

$$= -4x^3 e^{2-x^4}.$$

11. $y = \ln u$, $u = 4x^5 - 7$, so $y = \ln(4x^5 - 7)$ and

$$\dfrac{dy}{dx} = \dfrac{dy}{du} \cdot \dfrac{du}{dx} = \dfrac{1}{u}(20x^4) = \dfrac{1}{4x^5 - 7}(20x^4) = \dfrac{20x^4}{4x^5 - 7}$$

13. $\dfrac{d}{dx} \ln(x - 3) = \dfrac{1}{x - 3}(1)$ (using $\underline{3b}$)

$$= \dfrac{1}{x - 3}$$

15. $\dfrac{d}{dt} \ln(3 - 2t) = \dfrac{1}{3 - 2t}(-2)$ (using $\underline{3b}$)

$$= \dfrac{-2}{3 - 2t}$$

17. $\dfrac{d}{dx} 3e^{2x} = 3\dfrac{d}{dx} e^{2x} = 3e^{2x}(2)$ (using $\underline{3c}$)

$$= 6e^{2x}$$

19. $\dfrac{d}{dt} 2e^{-4t} = 3\dfrac{d}{dt} e^{-4t} = 2e^{-4t}(-4) = -8e^{-4t}$

21. $\dfrac{d}{dx} 100e^{-0.03x} = 100\dfrac{d}{dx} e^{-0.03x} = 100e^{-0.03x}(-0.03) = -3e^{-0.03x}$

23. $\dfrac{d}{dx} \ln(x + 1)^4 = \dfrac{d}{dx} 4\ln(x + 1) = 4\dfrac{d}{dx} \ln(x + 1) = 4\dfrac{1}{x + 1}(1) = \dfrac{4}{x + 1}$

25. $\dfrac{d}{dx}(2e^{2x} - 3e^x + 5) = 2\dfrac{d}{dx} e^{2x} - 3\dfrac{d}{dx} e^x + \dfrac{d}{dx} 5 = 2e^{2x}(2) - 3e^x = 4e^{2x} - 3e^x$

27. $\dfrac{d}{dx} e^{3x^2-2x} = e^{3x^2-2x}(6x - 2) = (6x - 2)e^{3x^2-2x}$

29. $\dfrac{d}{dt} \ln(t^2 + 3t) = \dfrac{1}{t^2 + 3t}(2t + 3) = \dfrac{2t + 3}{t^2 + 3t}$

31. $\dfrac{d}{dx} \ln(x^2 + 1)^{1/2} = \dfrac{d}{dx} \dfrac{1}{2} \ln(x^2 + 1) = \dfrac{1}{2}\dfrac{d}{dx} \ln(x^2 + 1)$

$$= \dfrac{1}{2}\left(\dfrac{1}{x^2 + 1}\right)(2x) = \dfrac{x}{x^2 + 1}$$

33. $\dfrac{d}{dt}[\ln(t^2 + 1)]^4 = 4[\ln(t^2 + 1)]^3 \dfrac{1}{t^2 + 1}(2t) = \dfrac{8t}{t^2 + 1}[\ln(t^2 + 1)]^3$

35. $\dfrac{d}{dx}(e^{2x} - 1)^4 = 4(e^{2x} - 1)^3[e^{2x}(2)] = 8e^{2x}(e^{2x} - 1)^3$

37. $\dfrac{d}{dx} \dfrac{e^{2x}}{x^2 + 1} = \dfrac{(x^2 + 1)\dfrac{d}{dx} e^{2x} - e^{2x}\dfrac{d}{dx}(x^2 + 1)}{(x^2 + 1)^2} = \dfrac{(x^2 + 1)e^{2x}(2) - e^{2x}(2x)}{(x^2 + 1)^2}$

$$= \dfrac{2e^{2x}(x^2 - x + 1)}{(x^2 + 1)^2}$$

39. $\frac{d}{dx}(x^2 + 1)e^{-x} = (x^2 + 1)\frac{d}{dx}e^{-x} + e^{-x}\frac{d}{dx}(x^2 + 1)$

$$= (x^2 + 1)e^{-x}(-1) + e^{-x}(2x) = e^{-x}(2x - x^2 - 1)$$

41. $\frac{d}{dx}e^{-x}\ln x = e^{-x}\frac{d}{dx}\ln x + \ln x \frac{d}{dx}e^{-x} = e^{-x}\left(\frac{1}{x}\right) + (\ln x)(e^{-x})(-1)$

$$= \frac{e^{-x}}{x} - e^{-x}\ln x = \frac{e^{-x}[1 - x\ln x]}{x}$$

43. $\frac{d}{dx}\frac{1}{\ln(1 + x^2)} = \frac{d}{dx}[\ln(1 + x^2)]^{-1} = -1[\ln(1 + x^2)]^{-2}\frac{d}{dx}\ln(1 + x^2)$

$$= -[\ln(1 + x^2)]^{-2}\frac{1}{1 + x^2}(2x) = \frac{-2x}{(1 + x^2)[\ln(1 + x^2)]^2}$$

45. $\frac{d}{dx}\sqrt[3]{\ln(1 - x^2)} = \frac{d}{dx}[\ln(1 - x^2)]^{1/3} = \frac{1}{3}[\ln(1 - x^2)]^{-2/3}\frac{d}{dx}\ln(1 - x^2)$

$$= \frac{1}{3}[\ln(1 - x^2)]^{-2/3}\frac{1}{1 - x^2}(-2x) = \frac{-2x}{3(1 - x^2)[\ln(1 - x^2)]^{2/3}}$$

47. $f(x) = 1 - e^{-x}$

> **Step 1. Analyze f(x):**
> (A) Domain: All real numbers, $(-\infty, \infty)$.
>
> (B) Intercepts: y intercept: $f(0) = 1 - e^{-0} = 0$
> x intercept: $1 - e^{-x} = 0$
> $e^{-x} = 1$
> $x = 0$
>
> (C) Asymptotes:
> <u>Horizontal asymptote</u>: $\lim\limits_{x \to \infty}(1 - e^{-x}) = \lim\limits_{x \to \infty}\left(1 - \frac{1}{e^x}\right) = 1$
> $\lim\limits_{x \to -\infty}(1 - e^{-x})$ does not exist.
> $y = 1$ is a horizontal asymptote.
> <u>Vertical asymptotes</u>: There are no vertical asymptotes.
>
> **Step 2. Analyze f'(x):**
> $f'(x) = -e^{-x}(-1) = e^{-x}$
> Since $e^{-x} > 0$ for all x, f is increasing on $(-\infty, \infty)$; there are no local extrema.
>
> **Step 3. Analyze f"(x):**
> $f''(x) = e^{-x}(-1) = -e^{-x}$
> Since $-e^{-x} < 0$ for all x, the graph of f is concave downward on $(-\infty, \infty)$.

Step 4. Sketch the graph of f:

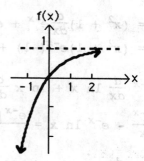

x	$f(x)$
0	0
-1	$\approx$ -1.72
1	$\approx$ 0.63

49. $f(x) = \ln(1 - x)$

Step 1. Analyze $f(x)$:

(A) Domain: All real numbers x such that $1 - x > 0$, i.e., $x < 1$ or $(-\infty, 1)$.

(B) Intercepts: y intercept: $f(0) = \ln(1 - 0) = \ln 1 = 0$

x intercepts: $\ln(1 - x) = 0$

$$1 - x = 1$$
$$x = 0$$

(C) Asymptotes:

Horizontal asymptote: $\lim\limits_{x \to -\infty} f(x) = \lim\limits_{x \to -\infty} \ln(1 - x)$ does not exist.

Thus, there are no horizontal asymptotes.

Vertical asymptote: From the table,

x	0.9	0.99	0.99999	0.9999999	$\to 1$
$f(x)$	-2.30	-4.61	-11.51	-16.12	$\to -\infty$

We conclude that $x = 1$ is a vertical asymptote.

Step 2. Analyze $f'(x)$:

$$f'(x) = \frac{1}{1 - x}(-1), \; x < 1$$

$$= \frac{1}{x - 1}$$

Now, $f'(x) = \dfrac{1}{x - 1} < 0$ on $(-\infty, 1)$.

Thus, f is decreasing on $(-\infty, 1)$; there are no critical values and no local extrema.

Step 3. Analyze $f''(x)$:

$$f'(x) = (x - 1)^{-1}$$

$$f''(x) = -1(x - 1)^{-2} = \frac{-1}{(x - 1)^2}$$

Since $f''(x) = \dfrac{-1}{(1 - x)^2} < 0$ on $(-\infty, 1)$, the graph of f is concave downward on $(-\infty, 1)$; there are no inflection points.

Step 4. Sketch the graph of f:

x	$f(x)$
0	0
-2	$\approx$ 1.10
.9	$\approx$ -2.30

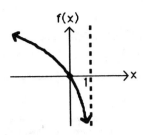

51. $f(x) = e^{-(1/2)x^2}$

Step 1. Analyze $f(x)$:

(A) Domain: All real numbers, $(-\infty, \infty)$.

(B) Intercepts: y intercept: $f(0) = e^{-(1/2)0} = e^0 = 1$

x intercepts: Since $e^{-(1/2)x^2} \neq 0$ for all x,
there are no x intercepts.

(C) Asymptotes: $\lim\limits_{x \to \infty} f(x) = \lim\limits_{x \to \infty} e^{-(1/2)x^2} = \lim\limits_{x \to \infty} \dfrac{1}{e^{(1/2)x^2}} = 0$

$\lim\limits_{x \to -\infty} f(x) = \lim\limits_{x \to -\infty} e^{-(1/2)x^2} = \lim\limits_{x \to -\infty} \dfrac{1}{e^{(1/2)x^2}} = 0$

Thus, $y = 0$ is a horizontal asymptote.

Since
$$f(x) = e^{-(1/2)x^2} = \frac{1}{e^{(1/2)x^2}} \quad \text{and} \quad g(x) = e^{(1/2)x^2} \neq 0 \text{ for all } x, \text{ there}$$
are no vertical asymptotes.

Step 2. Analyze $f'(x)$:

$f'(x) = e^{-(1/2)x^2}(-x) = -xe^{-(1/2)x^2}$
Critical values: $-xe^{-(1/2)x^2} = 0$
$$x = 0$$
Partition numbers: $x = 0$
Sign chart for f':

```
              + + + + + 0 - - - - -          Test Numbers
   f'(x)                                    ──────────────────
        ────┼────┼────●────────┼───→ x      x      f'(x)
           -1        0        1             ──────────────────
                                            -1    1/e^{1/2} (+)
                     ┊
   f(x)    Increasing ┊ Decreasing          1    -1/e^{1/2} (-)
                                            ──────────────────
```

Thus, f is increasing on $(-\infty, 0)$ and decreasing on $(0, \infty)$; f has a local maximum at $x = 0$.

Step 3. Analyze $f''(x)$:

$f''(x) = -xe^{-(1/2)x^2}(-x) - e^{-(1/2)x^2}$
$\qquad = e^{-(1/2)x^2}(x^2 - 1) = e^{-(1/2)x^2}(x - 1)(x + 1)$
Partition numbers for f'': $e^{-(1/2)x^2}(x - 1)(x + 1) = 0$
$$(x - 1)(x + 1) = 0$$
$$x = -1, \ 1$$

Sign chart for f'':

$f''(x)$ $+ + + + 0 - - - - 0 + + + +$

$$\xrightarrow{\hspace{8cm}} x$$

$\qquad\qquad$ -2 $\quad$ -1 $\quad$ 0 $\quad$ 1 $\quad$ 2

Graph
of f

$\quad$ Concave $\vdots$ Concave $\vdots$ Concave
$\quad$ Upward $\vdots$ Downward $\vdots$ Upward

Test Numbers

x	$f''(x)$
-2	$\frac{3}{e^2}$ $(+)$
0	-1 $(-)$
2	$\frac{3}{e^2}$ $(+)$

Thus, the graph of f is concave upward on $(-\infty, -1)$ and on $(1, \infty)$; the graph of f is concave downward on $(-1, 1)$; the graph has inflection points at $x = -1$ and at $x = 1$.

<u>Step 4.</u> $\quad$ <u>Sketch the graph of f</u>:

x	$f(x)$
0	1
-1	≈ 0.61
1	≈ 0.61

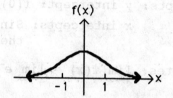

53. $y = 1 + w^2$, $w = \ln u$, $u = 2 + e^x$
Thus, $y = 1 + [\ln(2 + e^x)]^2$ and
$$\frac{dy}{dx} = 2[\ln(2 + e^x)]\left(\frac{1}{2 + e^x}\right)(e^x) = \frac{2e^x \ln(2 + e^x)}{2 + e^x}$$

55. $\dfrac{d}{dx} \log_2(3x^2 - 1) = \dfrac{1}{\ln 2} \cdot \dfrac{1}{3x^2 - 1} \cdot 6x = \dfrac{1}{\ln 2} \cdot \dfrac{6x}{3x^2 - 1}$

57. $\dfrac{d}{dx} 10^{x^2+x} = 10^{x^2+x}(\ln 10)(2x + 1) = (2x + 1)10^{x^2+x} \ln 10$

59. $\dfrac{d}{dx} \log_3(4x^3 + 5x + 7) = \dfrac{1}{\ln 3} \cdot \dfrac{1}{4x^3 + 5x + 7}(12x^2 + 5)$

$$= \frac{12x^2 + 5}{\ln 3(4x^3 + 5x + 7)}$$

61. $\dfrac{d}{dx} 2^{x^3-x^2+4x+1} = 2^{x^3-x^2+4x+1} \ln 2(3x^2 - 2x + 4)$

$$= \ln 2(3x^2 - 2x + 4)2^{x^3-x^2+4x+1}$$

63. Let $u = f(x)$ and $y = \ln u$. Then, by the chain rule,
$$\frac{dy}{dx} = \frac{dy}{du} \cdot \frac{du}{dx} = \frac{1}{u} \cdot f'(x) = \frac{1}{f(x)} \cdot f'(x).$$

65. (A) f is decreasing on $(-\infty, 0)$ but $g(x)$ is not negative on this interval.

(B) $f'(x) = \dfrac{1}{x^2 + 1}(2x) = \dfrac{2x}{x^2 + 1}$

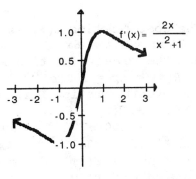

67. $f'(x) = \dfrac{1}{5(x^2 + 3)^4}[20(x^2 + 3)^3](2x) = \dfrac{8x}{x^2 + 3}$

$g'(x) = 4 \cdot \dfrac{1}{x^2 + 3}(2x) = \dfrac{8x}{x^2 + 3}$

For another way to see this, recall the properties of logarithms discussed in Section 2.3:

$f(x) = \ln[5(x^2 + 3)^4] = \ln 5 + \ln(x^2 + 3)^4 = \ln 5 + 4 \ln(x^2 + 3)$
$= \ln 5 + g(x)$

Now $\dfrac{d}{dx}f(x) = \dfrac{d}{dx}\ln 5 + \dfrac{d}{dx}g(x) = 0 + \dfrac{d}{dx}g(x) = \dfrac{d}{dx}g(x)$

Conclusion: $f'(x)$ and $g'(x)$ ARE the same function.

69. Price: $p = 100e^{-0.05x}$, $x \geq 0$
Revenue: $R(x) = xp = 100xe^{-0.05x}$
$\quad R'(x) = 100xe^{-0.05x}(-0.05) + 100e^{-0.05x}$
$\quad\quad = 100e^{-0.05x}(1 - 0.05x)$
Critical value(s): $R'(x) = 100e^{-0.05x}(1 - 0.05x) = 0$
$\quad\quad\quad\quad\quad\quad\quad\quad 1 - 0.05x = 0$
$\quad\quad\quad\quad\quad\quad\quad\quad\quad\quad\quad x = 20$
$R''(x) = 100e^{-0.05x}(-0.05) + (1 - 0.05x)100e^{-0.05x}(-0.05)$
$\quad\quad = 100e^{-0.05x}(0.0025x - 0.1)$
$R''(20) = -100e^{-1}(0.05) = \dfrac{-5}{e} < 0$

Since $x = 20$ is the only critical value and $R''(20) < 0$, the production level that maximizes the revenue is 20 units. The maximum revenue is $R(20) = 20(36.79) = 735.80$ or $\$735.80$, and the price is $p(20) = 36.79$ or $\$36.79$ each.

71. The cost function $C(x)$ is given by
$$C(x) = 400 + 6x$$
and the revenue function $R(x)$ is
$$R(x) = xp = 100xe^{-0.05x}$$
The profit function $P(x)$ is
$$P(x) = R(x) - C(x)$$
$$= 100xe^{-0.05x} - 400 - 6x$$
and $P'(x) = 100e^{-0.05x} - 5xe^{-0.05x} - 6$
We graph $y = P(x)$ and $y = P'(x)$ in the
viewing rectangle $0 \leq x \leq 50$,
$-400 \leq y \leq 300$

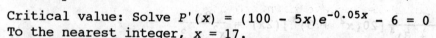

Critical value: Solve $P'(x) = (100 - 5x)e^{-0.05x} - 6 = 0$
To the nearest integer, $x = 17$.
$P(x)$ is increasing on $(0, 17)$ and decreasing on $(17, \infty)$; $P(x)$ has a
maximum at $x = 17$. Thus, the maximum profit $P(17) = \$224.61$ is realized
at a production level of 17 units at a price of $\$42.74$ per unit.

73. $S(t) = 300,000e^{-0.1t}$, $t \geq 0$
$S'(t) = 300,000e^{-0.1t}(-0.1) = -30,000e^{-0.1t}$
The rate of depreciation after one year is:
$S'(1) = -30,000e^{-0.1} \approx -\$27,145.12$ per year.
The rate of depreciation after five years is:
$S'(5) = -30,000e^{-0.5} \approx -\$18,195.92$ per year.
The rate of depreciation after ten years is:
$S'(10) = -30,000e^{-1} \approx -\$11,036.38$ per year.

75. Revenue: $R(t) = 200,000(1 - e^{-0.03t})$, $t \geq 0$
Cost: $C(t) = 4000 + 3000t$, $t \geq 0$
Profit: $P(t) = R(t) - C(t) = 200,000(1 - e^{-0.03t}) - (4000 + 3000t)$
$$= 200,000(1 - e^{-0.03t}) - 3000t - 4000$$
(A) $P'(t) = -200,000e^{-0.03t}(-0.03) - 3000 = 6000e^{-0.03t} - 3000$
   Critical value(s): $P'(t) = 6000e^{-0.03t} - 3000 = 0$
$$e^{-0.03t} = \frac{1}{2}$$
$$-0.03t = \ln\left(\frac{1}{2}\right) = -\ln 2$$
$$t = \frac{\ln 2}{0.03} \approx 23$$
$$P''(t) = 6000e^{-0.03t}(-0.03) = -180e^{-0.03t}$$
$$P''(23) = -180e^{-0.69} < 0$$
Since $t = 23$ is the only critical value and $P''(23) < 0$, 23 days of
TV promotion should be used to maximize profits. The maximum profit
is: $P(23) = 200,000(1 - e^{-0.03(23)}) - 3000(23) - 4000 \approx \$26,685$
The proportion of people buying the disk after t days is:
$p(t) = 1 - e^{-0.03t}$
Thus, $p(23) = 1 - e^{-0.03(23)} \approx 0.50$ or approximately 50%.

(B) From A, the sign chart for P' is:

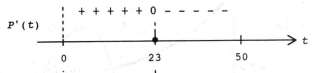

	Test Numbers	
t	$P'(t)$	
0	3000	(+)
50	−1661.22	(−)

Thus, P is increasing on $(0, 23)$ and decreasing on $(23, \infty)$; P has a maximum at $t = 23$.
Since $P''(t) = -180e^{-0.03t} < 0$ on $(0, \infty)$, the graph of P is concave downward on $(0, \infty)$; $P(0) = -4000$ and $P(50) \approx 0$.
The graph of P is shown at the right.

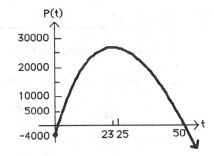

77. $P(x) = 40 + 25 \ln(x + 1)$ $0 \le x \le 65$

$$P'(x) = 25\left(\frac{1}{x+1}\right)(1) = \frac{25}{x+1}$$

$$P'(10) = \frac{25}{11} \approx 2.27$$

$$P'(30) = \frac{25}{31} \approx 0.81$$

$$P'(60) = \frac{25}{61} \approx 0.41$$

Thus, the rate of change of pressure at the end of 10 years is 2.27 millimeters of mercury per year; at the end of 30 years the rate of change is 0.81 millimeters of mercury per year; at the end of 60 years the rate of change is 0.41 millimeters of mercury per year.

79. $A(t) = 5000 \cdot 2^{2t}$
$A'(t) = 5000 \cdot 2^{2t}(2)(\ln 2) = 10{,}000 \cdot 2^{2t}(\ln 2)$
$A'(1) = 10{,}000 \cdot 2^{2}(\ln 2) = 40{,}000 \ln 2$
$\qquad \approx 27{,}726$ rate of change of bacteria at the end of the first hour.
$A'(5) = 10{,}000 \cdot 2^{2 \cdot 5}(\ln 2) = 10{,}000 \cdot 2^{10}(\ln 2)$
$\qquad \approx 7{,}097{,}827$ rate of change of bacteria at the end of the fifth hour.

81. $N(n) = 1{,}000{,}000e^{-0.09(n-1)}$, $1 \le n \le 20$
There are no asymptotes and no intercepts.
Using the first derivative:
$N'(n) = 1{,}000{,}000e^{-0.09(n-1)}(-0.09)$
$\qquad = -90{,}000e^{-0.09(n-1)} < 0$, $1 \le n \le 20$
Thus, N is decreasing on $(0, 20)$.
Using the second derivative:
$N''(n) = -90{,}000e^{-0.09(n-1)}(-0.09)$
$\qquad = 8100e^{-0.09(n-1)} > 0$, $1 \le n \le 20$
Thus, the graph of N is concave upward on $(0, 20)$.

The graph of N is:

n	N
1	1,000,000
20	$\approx$ 180,866

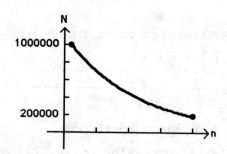

EXERCISE 10-4

Things to remember:

<u>1</u>. Let $y = y(x)$. Then

(a) $\dfrac{d}{dx} y^n = ny^{n-1}y'$ (General Power Rule)

(b) $\dfrac{d}{dx} \ln y = \dfrac{1}{y} \cdot y' = \dfrac{y'}{y}$

(c) $\dfrac{d}{dx} e^y = e^y \cdot y' = y'e^y$

---

1. $y - 3x^2 + 5 = 0$
Using implicit differentiation:
$$y' - \frac{d}{dx}(3x^2) + \frac{d}{dx}(5) = \frac{d}{dx}(0)$$
$$y' - 6x + 0 = 0$$
$$y' = 6x$$
$$y'\Big|_{(1, -2)} = 6 \cdot 1 = 6$$

3. $y^2 - 3x^2 + 8 = 0$
$$\frac{d}{dx}(y^2) - \frac{d}{dx}(3x^2) + \frac{d}{dx}(8) = \frac{d}{dx}(0)$$
$$2yy' - 6x + 0 = 0 \text{ (using } \underline{1})$$
$$2yy' = 6x$$
$$y' = \frac{3x}{y}$$
$$y'\Big|_{(2, 2)} = \frac{3 \cdot 2}{2} = 3$$

5. $y^2 + y - x = 0$
$$\frac{d}{dx}y^2 + \frac{d}{dx}y - \frac{d}{dx}x = \frac{d}{dx}(0)$$
$$2yy' + y' - 1 = 0$$
$$y'(2y + 1) = 1$$
$$y' = \frac{1}{2y + 1}$$
$$y' \text{ at } (2, 1) = \frac{1}{2 \cdot 1 + 1} = \frac{1}{3}$$

7. $xy - 6 = 0$
$$\frac{d}{dx}xy - \frac{d}{dx}6 = \frac{d}{dx}(0)$$
$$xy' + y - 0 = 0$$
$$xy' = -y$$
$$y' = -\frac{y}{x}$$
$$y' \text{ at } (2, 3) = -\frac{3}{2}$$

9. $2xy + y + 2 = 0$
$$2\frac{d}{dx}xy + \frac{d}{dx}y + \frac{d}{dx}2 = \frac{d}{dx}(0)$$
$$2xy' + 2y + y' + 0 = 0$$
$$y'(2x + 1) = -2y$$
$$y' = \frac{-2y}{2x + 1}$$
$$y' \text{ at } (-1, 2) = \frac{-2(2)}{2(-1) + 1} = 4$$

11. $x^2y - 3x^2 - 4 = 0$

$$\frac{d}{dx}x^2y - \frac{d}{dx}3x^2 - \frac{d}{dx}4 = \frac{d}{dx}(0)$$

$$x^2y' + y\frac{d}{dx}(x^2) - 6x - 0 = 0$$

$$x^2y' + y2x - 6x = 0$$

$$x^2y' = 6x - 2yx$$

$$y' = \frac{6x - 2yx}{x^2} \text{ or } \frac{6 - 2y}{x}$$

$$y'\Big|_{(2,\ 4)} = \frac{6\cdot2 - 2\cdot4\cdot2}{2^2} = \frac{12 - 16}{4} = -1$$

13. $e^y = x^2 + y^2$

$$\frac{d}{dx}e^y = \frac{d}{dx}x^2 + \frac{d}{dx}y^2$$

$$e^yy' = 2x + 2yy'$$

$$y'(e^y - 2y) = 2x$$

$$y' = \frac{2x}{e^y - 2y}$$

$$y'\Big|_{(1,\ 0)} = \frac{2\cdot1}{e^0 - 2\cdot0} = \frac{2}{1} = 2$$

15. $x^3 - y = \ln y$

$$\frac{d}{dx}x^3 - \frac{d}{dx}y = \frac{d}{dx}\ln y$$

$$3x^2 - y' = \frac{y'}{y}$$

$$3x^2 = \left(1 + \frac{1}{y}\right)y'$$

$$3x^2 = \frac{y + 1}{y}y'$$

$$y' = \frac{3x^2y}{y + 1}$$

$$y'\Big|_{(1,\ 1)} = \frac{3\cdot1^2\cdot1}{1 + 1} = \frac{3}{2}$$

17. $x \ln y + 2y = 2x^3$

$$\frac{d}{dx}[x \ln y] + \frac{d}{dx}2y = \frac{d}{dx}2x^3$$

$$\ln y\cdot\frac{d}{dx}x + x\frac{d}{dx}\ln y + 2y' = 6x^2$$

$$\ln y\cdot1 + x \cdot \frac{y'}{y} + 2y' = 6x^2$$

$$y'\left(\frac{x}{y} + 2\right) = 6x^2 - \ln y$$

$$y' = \frac{6x^2y - y \ln y}{x + 2y}$$

$$y'\Big|_{(1,\ 1)} = \frac{6\cdot1^2\cdot1 - 1\cdot\ln 1}{1 + 2\cdot1} = \frac{6}{3} = 2$$

19. $x^2 - t^2x + t^3 + 11 = 0$

$$\frac{d}{dt}x^2 - \frac{d}{dt}(t^2x) + \frac{d}{dt}t^3 + \frac{d}{dt}11 = \frac{d}{dt}0$$

$$2xx' - [t^2x' + x(2t)] + 3t^2 + 0 = 0$$

$$2xx' - t^2x' - 2tx + 3t^2 = 0$$

$$x'(2x - t^2) = 2tx - 3t^2$$

$$x' = \frac{2tx - 3t^2}{2x - t^2}$$

$$x'\Big|_{(-2,\ 1)} = \frac{2(-2)(1) - 3(-2)^2}{2(1) - (-2)^2}) = \frac{-4 - 12}{2 - 4} = \frac{-16}{-2} = 8$$

21. $xy - x - 4 = 0$

When $x = 2$, $2y - 2 - 4 = 0$, so $y = 3$. Thus, we want to find the equation of the tangent line at $(2, 3)$.

First, find y'.

$$\frac{d}{dx}xy - \frac{d}{dx}x - \frac{d}{dx}4 = \frac{d}{dx}0$$

$$xy' + y - 1 - 0 = 0$$

$$xy' = 1 - y$$

$$y' = \frac{1 - y}{x}$$

$$y'\Big|_{(2, 3)} = \frac{1 - 3}{2} = -1$$

Thus, the slope of the tangent line at $(2, 3)$ is $m = -1$. The equation of the line through $(2, 3)$ with slope $m = -1$ is:

$$(y - 3) = -1(x - 2)$$

$$y - 3 = -x + 2$$

$$y = -x + 5$$

23. $y^2 - xy - 6 = 0$

When $x = 1$,

$$y^2 - y - 6 = 0$$

$$(y - 3)(y + 2) = 0$$

$$y = 3 \quad \text{or} \quad -2.$$

Thus, we want to find the equations of the tangent lines at $(1, 3)$ and $(1, -2)$. First, find y'.

$$\frac{d}{dx}y^2 - \frac{d}{dx}xy - \frac{d}{dx}6 = \frac{d}{dx}0$$

$$2yy' - xy' - y - 0 = 0$$

$$y'(2y - x) = y$$

$$y' = \frac{y}{2y - x}$$

$$y'\Big|_{(1, 3)} = \frac{3}{2(3) - 1} = \frac{3}{5} \qquad \text{[Slope at (1, 3)]}$$

The equation of the tangent line at $(1, 3)$ with $m = \frac{3}{5}$ is:

$$(y - 3) = \frac{3}{5}(x - 1)$$

$$y - 3 = \frac{3}{5}x - \frac{3}{5}$$

$$y = \frac{3}{5}x + \frac{12}{5}$$

$$y'\Big|_{(1, -2)} = \frac{-2}{2(-2) - 1} = \frac{2}{5} \qquad \text{[Slope at (1, -2)]}$$

Thus, the equation of the tangent line at $(1, -2)$ with $m = \frac{2}{5}$ is:

$$(y + 2) = \frac{2}{5}(x - 1)$$

$$y + 2 = \frac{2}{5}x - \frac{2}{5}$$

$$y = \frac{2}{5}x - \frac{12}{5}$$

25. $xe^y = 1$

Implicit differentiation: $x \cdot \dfrac{d}{dx} e^y + e^y \dfrac{d}{dx} x = \dfrac{d}{dx} 1$

$$xe^y y' + e^y = 0$$

$$y' = -\frac{e^y}{xe^y} = -\frac{1}{x}$$

Solve for y: $e^y = \dfrac{1}{x}$

$$y = \ln\left(\frac{1}{x}\right) = -\ln x \quad \text{(see Section 2.3)}$$

$$y' = -\frac{1}{x}$$

In this case, solving for y first and then differentiating is a little easier than differentiating implicitly.

27. $(1 + y)^3 + y = x + 7$

$$\frac{d}{dx}(1 + y)^3 + \frac{d}{dx} y = \frac{d}{dx} x + \frac{d}{dx} 7$$

$$3(1 + y)^2 y' + y' = 1$$

$$y'[3(1 + y)^2 + 1] = 1$$

$$y' = \frac{1}{3(1 + y)^2 + 1}$$

$$y' \Big|_{(2,\ 1)} = \frac{1}{3(1 + 1)^2 + 1} = \frac{1}{13}$$

29. $(x - 2y)^3 = 2y^2 - 3$

$$\frac{d}{dx}(x - 2y)^3 = \frac{d}{dx}(2y^2) - \frac{d}{dx}(3)$$

$$3(x - 2y)^2(1 - 2y') = 4yy' - 0 \qquad \text{[\underline{Note}: The chain rule is applied to the left-hand side.]}$$

$$3(x - 2y)^2 - 6(x - 2y)^2 y' = 4yy'$$

$$-6(x - 2y)^2 y' - 4yy' = -3(x - 2y)^2$$

$$-y'[6(x - 2y)^2 + 4y] = -3(x - 2y)^2$$

$$y' = \frac{3(x - 2y)^2}{6(x - 2y)^2 + 4y}$$

$$y' \Big|_{(1,\ 1)} = \frac{3(1 - 2 \cdot 1)^2}{6(1 - 2)^2 + 4} = \frac{3}{10}$$

31. $\sqrt{7 + y^2} - x^3 + 4 = 0$ or $(7 + y^2)^{1/2} - x^3 + 4 = 0$

$$\frac{d}{dx}(7 + y^2)^{1/2} - \frac{d}{dx} x^3 + \frac{d}{dx} 4 = \frac{d}{dx} 0$$

$$\frac{1}{2}(7 + y^2)^{-1/2} \frac{d}{dx}(7 + y^2) - 3x^2 + 0 = 0$$

$$\frac{1}{2}(7 + y^2)^{-1/2} 2yy' - 3x^2 = 0$$

$$\frac{yy'}{(7 + y^2)^{1/2}} = 3x^2$$

$$y' = \frac{3x^2(7 + y^2)^{1/2}}{y}$$

$$y' \Big|_{(2,\ 3)} = \frac{3 \cdot 2^2(7 + 3^2)^{1/2}}{3} = \frac{12(16)^{1/2}}{3} = 16$$

33. $\ln(xy) = y^2 - 1$

$$\frac{d}{dx}[\ln(xy)] = \frac{d}{dx}y^2 - \frac{d}{dx}1$$

$$\frac{1}{xy} \cdot \frac{d}{dx}(xy) = 2yy'$$

$$\frac{1}{xy}(x \cdot y' + y) = 2yy'$$

$$\frac{1}{y} \cdot y' - 2yy' + \frac{1}{x} = 0$$

$$xy' - 2xy^2y' + y = 0$$

$$y'(x - 2xy^2) = -y$$

$$y' = \frac{-y}{x - 2xy^2} = \frac{y}{2xy^2 - x}$$

$$y'\Big|_{(1,\ 1)} = \frac{1}{2 \cdot 1 \cdot 1^2 - 1} = 1$$

35. First find point(s) on the graph of the equation with abscissa $x = 1$: Setting $x = 1$, we have

$$y^3 - y - 1 = 2 \quad \text{or} \quad y^3 - y - 3 = 0$$

Graphing this equation on a graphing utility, we get $y \approx 1.67$.

Now, differentiate implicitly to find the slope of the tangent line at the point $(1, 1.67)$: $\frac{d}{dx}y^3 + x\frac{d}{dx}y + y\frac{d}{dx}x - \frac{d}{dx}x^3 = \frac{d}{dx}2$

$$3y^2y' - xy' - y - 3x^2 = 0$$

$$(3y^2 - x)y' = 3x^2 + y$$

$$y' = \frac{3x^2 + y}{3y^2 - x};$$

$$y'\Big|_{(1,\ 1.67)} = \frac{3 + 1.67}{3(1.67)^2 - 1} = \frac{4.67}{7.37} \approx 0.63$$

Tangent line: $y - 1.67 = 0.63(x - 1)$ or $y = 0.63x + 1.04$

37. $x = p^2 - 2p + 1000$

$$\frac{d(x)}{dx} = \frac{d(p^2)}{dx} - \frac{d(2p)}{dx} + \frac{d(1000)}{dx}$$

$$1 = 2p\frac{dp}{dx} - 2\frac{dp}{dx} + 0$$

$$1 = (2p - 2)\frac{dp}{dx}$$

Thus, $\frac{dp}{dx} = p' = \frac{1}{2p - 2}$.

39. $x = \sqrt{10,000 - p^2} = (10,000 - p^2)^{1/2}$

$$\frac{d}{dx}x = \frac{d}{dx}(10,000 - p^2)^{1/2}$$

$$1 = \frac{1}{2}(10,000 - p^2)^{-1/2}\frac{d}{dx}[10,000 - p^2]$$

$$1 = \frac{1}{2(10,000 - p^2)^{1/2}} \cdot (-2pp')$$

$$1 = \frac{-pp'}{\sqrt{10,000 - p^2}}$$

$$p' = \frac{-\sqrt{10,000 - p^2}}{p}$$

41. $(L + m)(V + n) = k$

$$\frac{d(L + m)(V + n)}{dV} = \frac{d(k)}{dV}$$

$$(V + n)\frac{d(L + m)}{dV} + (L + m)\frac{d(V + n)}{dV} = 0$$

$$(V + n)\frac{dL}{dV} + (L + m) = 0$$

$$\frac{dL}{dV} = \frac{-(L + m)}{V + n}$$

find $\frac{dL}{dV}$ using implicit differentiation

$\frac{d}{dL}(L+m)(V+n) = \frac{d}{dL}(k)$

$(L+m)(V+n)'+(L+m)'(V+n) = 0$

EXERCISE 10-5

Things to remember:

1. SUGGESTIONS FOR SOLVING RELATED RATE PROBLEMS

   (a) Sketch a figure.

   (b) Identify all relevant variables, including those whose rates are given and those whose rates are to be found.

   (c) Express all given rates and rates to be found as derivatives.

   (d) Find an equation connecting the variables in (b).

   (e) Differentiate the equation implicitly, using the chain rule where appropriate, and substitute in all given values.

   (f) Solve for the derivative that will give the unknown rate.

---

1. $y = 2x^2 - 1$

Differentiating with respect to t:

$$\frac{dy}{dt} = 4x\frac{dx}{dt}$$

$$\frac{dy}{dt} = 4 \cdot 30 \cdot 2 \quad \left(x = 30, \frac{dx}{dt} = 2\right)$$

$$= 240$$

3. $x^2 + y^2 = 25$

Differentiating with respect to t:

$$2x\frac{dx}{dt} + 2y\frac{dy}{dt} = 0$$

$$2y\frac{dy}{dt} = -2x\frac{dx}{dt}$$

$$\frac{dy}{dt} = -\frac{x}{y}\frac{dx}{dt} \quad \left(\text{at } x = 3,\ y = 4,\ \text{and } \frac{dx}{dt} = -3\right)$$

$$= -\frac{3}{4}(-3) = \frac{9}{4}$$

5. $x^2 + xy + 2 = 0$

Differentiating with respect to t:

$$2x\frac{dx}{dt} + x\frac{dy}{dt} + y\frac{dx}{dt} = 0$$

$$x\frac{dy}{dt} = -2x\frac{dx}{dt} - y\frac{dx}{dt}$$

$$\frac{dy}{dt} = -2\frac{dx}{dt} - \frac{y}{x}\frac{dx}{dt}$$

Given: $\frac{dx}{dt} = -1$ when $x = 2$ and $y = -3$. Thus, at $(2, -3)$ and $\frac{dx}{dt} = -1$,

$\frac{dy}{dt} = -2(-1) - \frac{-3}{2}(-1) = 2 - \frac{3}{2} = \frac{1}{2}$.

7. $xy = 36$

Differentiate with respect to t:

$$\frac{d(xy)}{dt} = \frac{d(36)}{dt}$$

$$x\frac{dy}{dt} + y\frac{dx}{dt} = 0$$

Given: $\frac{dx}{dt} = 4$ when $x = 4$ and $y = 9$. Therefore,

$$4\frac{dy}{dt} + 9(4) = 0$$

$$4\frac{dy}{dt} = -36$$

and $\frac{dy}{dt} = -9$.

The y coordinate is decreasing at 9 units per second.

9.

From the triangle,
$x^2 + y^2 = z^2$
or $x^2 + 16 = z^2$, since $y = 4$.

Differentiate with respect to t:

$$2x\frac{dx}{dt} = 2z\frac{dz}{dt}$$

or $x\frac{dx}{dt} = z\frac{dz}{dt}$

Given: $\frac{dz}{dt} = -3$. Also, when $x = 30$, $900 + 16 = z^2$ or $z = \sqrt{916}$.

Therefore,

$$30\frac{dx}{dt} = \sqrt{916}(-3) \qquad \text{and} \qquad \frac{dx}{dt} = \frac{-3\sqrt{916}}{30} = \frac{-\sqrt{916}}{10} \approx \frac{-30.27}{10}$$

$$\approx -3.03 \text{ feet per second.}$$

[<u>Note</u>: The negative sign indicates that the distance between the boat and the dock is decreasing.]

11. Area: $A = \pi R^2$

$$\frac{dA}{dt} = \frac{d\pi R^2}{dt} = \pi \cdot 2R\frac{dR}{dt}$$

Given: $\quad \frac{dR}{dt} = 2 \text{ ft/sec}$

$$\frac{dA}{dt} = 2\pi R \cdot 2 = 4\pi R$$

$$\frac{dA}{dt}\Big|_{R = 10 \text{ ft}} = 4\pi(10) = 40\pi \text{ ft}^2/\text{sec}$$

$$\approx 126 \text{ ft}^2/\text{sec}$$

13. $V = \frac{4}{3}\pi R^3$

$$\frac{dV}{dt} = \frac{4}{3}\pi 3R^2\frac{dR}{dt} = 4\pi R^2\frac{dR}{dt}$$

Given: $\quad \frac{dR}{dt} = 3 \text{ cm/min}$

$$\frac{dV}{dt} = 4\pi R^2 3 = 12\pi R^2$$

$$\frac{dV}{dt}\Big|_{R = 10 \text{ cm}} = 12\pi(10)^2 = 1200\pi$$

$$\approx 3768 \text{ cm}^3/\text{min}$$

15. $\frac{P}{T} = k \qquad\qquad (1)$

$P = kT$

Differentiate with respect to t:

$$\frac{dP}{dt} = k\frac{dT}{dt}$$

Given: $\frac{dT}{dt} = 3$ degrees per hour, $T = 250°$, $P = 500$ pounds per square inch.

From (1), for $T = 250$ and $P = 500$,

$$k = \frac{500}{250} = 2.$$

Thus, we have

$$\frac{dP}{dt} = 2\frac{dT}{dt}$$

$$\frac{dP}{dt} = 2(3) = 6$$

Pressure increases at 6 pounds per square inch per hour.

17. By the Pythagorean theorem,

$$x^2 + y^2 = 10^2$$

or $\quad x^2 + y^2 = 100 \qquad\qquad (1)$

Differentiate with respect to t:

$$2x\frac{dx}{dt} + 2y\frac{dy}{dt} = 0$$

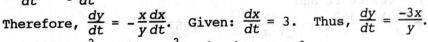

Therefore, $\frac{dy}{dt} = -\frac{x}{y}\frac{dx}{dt}$. Given: $\frac{dx}{dt} = 3$. Thus, $\frac{dy}{dt} = \frac{-3x}{y}$.

From (1), $y^2 = 100 - x^2$ and, when $x = 6$,

$$y^2 = 100 - 6^2$$
$$= 100 - 36 = 64.$$

Thus, $y = 8$ when $x = 6$, and

$$\frac{dy}{dt}\Big|_{(6,\ 8)} = \frac{-3(6)}{8} = \frac{-18}{8} = \frac{-9}{4} \text{ ft/sec.}$$

19. y = length of shadow

x = distance of man from light

z = distance of tip of shadow from light

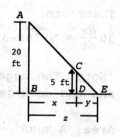

We want to compute $\frac{dz}{dt}$. Triangles ABE and CDE

are similar triangles; thus, the ratios of corresponding sides are equal.

Therefore, $\frac{z}{20} = \frac{y}{5} = \frac{z-x}{5}$ [Note: $y = z - x$.]

or $\frac{z}{20} = \frac{z-x}{5}$

$z = 4(z - x)$

$z = 4z - 4x$

$4x = 3z$

Differentiate with respect to t:

$4\frac{dx}{dt} = 3\frac{dz}{dt}$

$\frac{dz}{dt} = \frac{4}{3}\frac{dx}{dt}$

Given: $\frac{dx}{dt} = 5$. Thus, $\frac{dz}{dt} = \frac{4}{3}(5) = \frac{20}{3}$ ft/sec.

21. $V = \frac{4}{3}\pi r^3$ (1)

Differentiate with respect to t:

$\frac{dV}{dt} = 4\pi r^2 \frac{dr}{dt}$ and $\frac{dr}{dt} = \frac{1}{4\pi r^2} \cdot \frac{dV}{dt}$

Since $\frac{dV}{dt} = 4$ cu ft/sec,

$\frac{dr}{dt} = \frac{1}{4\pi r^2}(4) = \frac{1}{\pi r^2}$ ft/sec   (2)

At $t = 1$ minute $= 60$ seconds,

$V = 4(60) = 240$ cu ft and, from (1),

$r^3 = \frac{3V}{4\pi} = \frac{3(240)}{4\pi} = \frac{180}{\pi}$; $r = \left(\frac{180}{\pi}\right)^{1/3} \approx 3.855$.

From (2)

$\frac{dr}{dt} = \frac{1}{\pi(3.855)^2} \approx 0.0214$ ft/sec

At $t = 2$ minutes $= 120$ seconds,

$V = 4(120) = 480$ cu ft and

$r^3 = \frac{3V}{4\pi} = \frac{3(480)}{4\pi} = \frac{360}{\pi}$; $r = \left(\frac{360}{\pi}\right)^{1/3} \approx 4.857$

From (2),

$\frac{dr}{dt} = \frac{1}{\pi(4.857)^2} \approx 0.0135$ ft/sec

To find the time at which $\frac{dr}{dt} = 100$ ft/sec, solve

$$\frac{1}{\pi r^2} = 100$$

$$r^2 = \frac{1}{100\pi}$$

$$r = \frac{1}{\sqrt{100\pi}} = \frac{1}{10\sqrt{\pi}}$$

Now, when $r = \frac{1}{10\sqrt{\pi}}$,

$$V = \frac{4}{3}\pi\left(\frac{1}{10\sqrt{\pi}}\right)^3$$

$$= \frac{4}{3} \cdot \frac{1}{1000\sqrt{\pi}}$$

$$= \frac{1}{750\sqrt{\pi}}$$

Since the volume at time t is $4t$, we have

$$4t = \frac{1}{750\sqrt{\pi}} \quad \text{and}$$

$$t = \frac{1}{3000\sqrt{\pi}} \approx 0.00019 \text{ secs.}$$

23. $y = e^x + x + 1$; $\frac{dx}{dt} = 3$.

Differentiate with respect to t:

$$\frac{dy}{dt} = e^x \frac{dx}{dt} + \frac{dx}{dt} = e^x(3) + 3 = 3(e^x + 1)$$

To find where the point crosses the x axis, use a graphing utility to solve

$$e^x + x + 1 = 0$$

The result is $x \approx -1.278$.

Now, at $x = -1.278$,

$$\frac{dy}{dt} = 3(e^{-1.278} + 1) \approx 3.835 \text{ units/sec.}$$

25. $C = 90,000 + 30x$ (1)

$R = 300x - \dfrac{x^2}{30}$ (2)

$P = R - C$ (3)

(A) Differentiating (1) with respect to t:

$$\frac{dC}{dt} = \frac{d(90,000)}{dt} + \frac{d(30x)}{dt}$$

$$\frac{dC}{dt} = 30\frac{dx}{dt}$$

Thus, $\dfrac{dC}{dt} = 30(500)$ $\left(\dfrac{dx}{dt} = 500\right)$

$$= \$15,000 \text{ per week.}$$

Costs are increasing at $15,000 per week at this production level.

(B) Differentiating (2) with respect to t:

$$\frac{dR}{dt} = \frac{d(300x)}{dt} - \frac{d\frac{x^2}{30}}{dt}$$

$$= 300\frac{dx}{dt} - \frac{2x}{30}\frac{dx}{dt}$$

$$= \left(300 - \frac{x}{15}\right)\frac{dx}{dt}$$

Thus, $\frac{dR}{dt} = \left(300 - \frac{6000}{15}\right)(500)$ $\left(x = 6000, \frac{dx}{dt} = 500\right)$

$$= (-100)500$$

$$= -\$50,000 \text{ per week.}$$

Revenue is decreasing at \$50,000 per week at this production level.

(C) Differentiating (3) with respect to t:

$$\frac{dP}{dt} = \frac{dR}{dt} - \frac{dC}{dt}$$

Thus, from parts (A) and (B), we have:

$$\frac{dP}{dt} = -50,000 - 15,000 = -\$65,000$$

Profits are decreasing at \$65,000 per week at this production level.

27. $S = 60,000 - 40,000e^{-0.0005x}$

Differentiating implicitly with respect to t, we have

$$\frac{ds}{dt} = -40,000(-0.0005)e^{-0.0005x}\frac{dx}{dt} \text{ and } \frac{ds}{dt} = 20e^{-0.0005x}\frac{dx}{dt}$$

Now, for $x = 2000$ and $\frac{dx}{dt} = 300$, we have

$$\frac{ds}{dt} = 20(300)e^{-0.0005(2000)}$$

$$= 6000e^{-1} = 2,207$$

Thus, sales are increasing at the rate of \$2,207 per week.

29. Price p and demand x are related by the equation
$$2x^2 + 5xp + 50p^2 = 80,000 \qquad (1)$$
Differentiating implicitly with respect to t, we have
$$4x\frac{dx}{dt} + 5x\frac{dp}{dt} + 5p\frac{dx}{dt} + 100p\frac{dp}{dt} = 0 \qquad (2)$$

(A) From (2), $\frac{dx}{dt} = \dfrac{-(5x + 100p)\frac{dp}{dt}}{4x + 5p}$

Setting $p = 30$ in (1), we get
$$2x^2 + 150x + 45,000 = 80,000$$
or $x^2 + 75x - 17,500 = 0$

Thus, $x = \dfrac{-75 \pm \sqrt{(75)^2 + 70,000}}{2}$

$$= \frac{-75 \pm 275}{2} = 100, -175$$

Since $x \geq 0$, $x = 100$

Now, for $x = 100$, $p = 30$ and $\frac{dp}{dt} = 2$, we have

$$\frac{dx}{dt} = \frac{-[5(100) + 100(30)] \cdot 2}{4(100) + 5(30)} = -\frac{7000}{550} \text{ and } \frac{dx}{dt} = -12.73$$

The demand is decreasing at the rate of -12.73 units/month.

(B) From (2), $\dfrac{dp}{dt} = \dfrac{-(4x + 5p)\dfrac{dx}{dt}}{(5x + 100p)}$

Setting $x = 150$ in (1), we get

$$45,000 + 750p + 50p^2 = 80,000$$

or $\qquad p^2 + 15p - 700 = 0$

and $p = \dfrac{-15 \pm \sqrt{225 + 2800}}{2} = \dfrac{-15 \pm 55}{2} = -35, \ 20$

Since $p \geq 0$, $p = 20$.

Now, for $x = 150$, $p = 20$ and $\frac{dx}{dt} = -6$, we have

$$\frac{dp}{dt} = -\frac{[4(150) + 5(20)](-6)}{5(150) + 100(20)} = \frac{4200}{2750} \approx 1.53$$

Thus, the price is increasing at the rate of \$1.53 per month.

31. Volume $V = \pi R^2 h$, where $h =$ thickness of the circular oil slick. Since $h = 0.1 = \frac{1}{10}$, we have:

$$V = \frac{\pi}{10} R^2$$

Differentiating with respect to t:

$$\frac{dV}{dt} = \frac{d\left(\frac{\pi}{10} R^2\right)}{dt} = \frac{\pi}{10} 2R \frac{dR}{dt} = \frac{\pi}{5} R \frac{dR}{dt}$$

Given: $\frac{dR}{dt} = 0.32$ when $R = 500$. Therefore,

$$\frac{dV}{dt} = \frac{\pi}{5}(500)(0.32) = 100\pi(0.32) \approx 100.48 \text{ cubic feet per minute.}$$

CHAPTER 10 REVIEW

1. $A(t) = 2000e^{0.09t}$
$A(5) = 2000e^{0.09(5)} = 2000e^{0.45} \approx 3136.62 \quad \text{or} \quad \$3136.62$
$A(10) = 2000e^{0.09(10)} = 2000e^{0.9} \approx 4919.21 \quad \text{or} \quad \$4919.21$
$A(20) = 2000e^{0.09(20)} = 2000e^{1.8} \approx 12,099.29 \quad \text{or} \quad \$12,099.29 \qquad (10\text{-}1)$

2. $\dfrac{d}{dx}(2 \ln x + 3e^x) = 2\dfrac{d}{dx} \ln x + 3\dfrac{d}{dx} e^x = \dfrac{2}{x} + 3e^x \qquad (10\text{-}2)$

3. $\dfrac{d}{dx} e^{2x-3} = e^{2x-3} \dfrac{d}{dx}(2x - 3) \qquad \text{(by the chain rule)}$

$\qquad\qquad = 2e^{2x-3} \qquad\qquad\qquad\qquad\qquad\qquad\qquad\qquad (10\text{-}2)$

4. $y = \ln(2x + 7)$

$y' = \dfrac{1}{2x + 7}(2)$ (by the chain rule)

$\quad = \dfrac{2}{2x + 7}$ (10-2)

5. $y = \ln u$, where $u = 3 + e^x$.

(A) $y = \ln[3 + e^x]$

(B) $\dfrac{dy}{dx} = \dfrac{dy}{du} \cdot \dfrac{du}{dx} = \dfrac{1}{u}(e^x) = \dfrac{1}{3 + e^x}(e^x) = \dfrac{e^x}{3 + e^x}$ (10-3)

6. $\dfrac{d}{dx}2y^2 - \dfrac{d}{dx}3x^3 - \dfrac{d}{dx}5 = \dfrac{d}{dx}(0)$

$\quad 4yy' - 9x^2 - 0 = 0$

$\quad\quad\quad y' = \dfrac{9x^2}{4y}$

$\quad \dfrac{dy}{dx}\Big|_{(1,\ 2)} = \dfrac{9 \cdot 1^2}{4 \cdot 2} = \dfrac{9}{8}$ (10-4)

7. $y = 3x^2 - 5$

$\dfrac{dy}{dt} = \dfrac{d(3x^2)}{dt} - \dfrac{d(5)}{dt}$

$\dfrac{dy}{dt} = 6x\dfrac{dx}{dt}$

$x = 12; \quad \dfrac{dx}{dt} = 3$

$\dfrac{dy}{dt} = 6 \cdot 12 \cdot 3 = 216$ (10-4)

8. $y = 100e^{-0.1x}$

Step 1. Analyze $f(x)$:

(A) Domain: All real numbers, $(-\infty, \infty)$.

(B) Intercepts: y intercept: $f(0) = 100e^{-0.1(0)} = 100$

$\quad\quad\quad\quad\quad x$ intercept: Since $100e^{-0.1x} \neq 0$ for all x, there are no

$\quad\quad\quad\quad\quad\quad\quad\quad\quad\quad x$ intercepts.

(C) Asymptotes:

$\quad \lim\limits_{x \to \infty} 100e^{-0.1x} = \lim\limits_{x \to \infty} \dfrac{100}{e^{0.1x}} = 0$

$\quad \lim\limits_{x \to -\infty} 100e^{-0.1x}$ does not exist.

Thus, $y = 0$ is a horizontal asymptote. There are no vertical
asymptotes.

Step 2. Analyze $f'(x)$:

$y' = 100e^{-0.1x}(-0.1)$

$\quad = -10e^{-0.1x} < 0$ on $(-\infty, \infty)$

Thus, y is decreasing on $(-\infty, \infty)$; there are no local extrema.

Step 3. Analyze $f''(x)$:

$y'' = -10e^{-0.1x}(-0.1)$

$\quad = e^{-0.1x} > 0$ on $(-\infty, \infty)$

Thus, the graph of f is concave upward on $(-\infty, \infty)$; there are no
inflection points.

Step 4. Sketch the graph of f:

x	y
0	100
−1	~ 110
10	~ 37

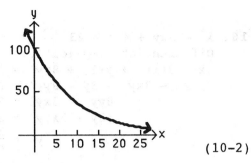

(10–2)

9. $\dfrac{d}{dz}[(\ln z)^7 + \ln z^7] = \dfrac{d}{dz}[\ln z]^7 + \dfrac{d}{dz}7\ln z$

$\qquad\qquad\qquad\qquad = 7[\ln z]^6 \dfrac{d}{dz}\ln z + 7\dfrac{d}{dz}\ln z$

$\qquad\qquad\qquad\qquad = 7[\ln z]^6 \dfrac{1}{z} + \dfrac{7}{z}$

$\qquad\qquad\qquad\qquad = \dfrac{7(\ln z)^6 + 7}{z} = \dfrac{7[(\ln z)^6 + 1]}{z}$ (10–3)

10. $\dfrac{d}{dx}x^6 \ln x = x^6 \dfrac{d}{dx}\ln x + (\ln x)\dfrac{d}{dx}x^6$

$\qquad\qquad\qquad = x^6\left(\dfrac{1}{x}\right) + (\ln x)6x^5 = x^5(1 + 6\ln x)$ (10–2)

11. $\dfrac{d}{dx}\left(\dfrac{e^x}{x^6}\right) = \dfrac{x^6 \dfrac{d}{dx}e^x - e^x \dfrac{d}{dx}x^6}{(x^6)^2} = \dfrac{x^6 e^x - 6x^5 e^x}{x^{12}} = \dfrac{xe^x - 6e^x}{x^7} = \dfrac{e^x(x-6)}{x^7}$ (10–2)

12. $y = \ln(2x^3 - 3x)$

$y' = \dfrac{1}{2x^3 - 3x}(6x^2 - 3) = \dfrac{6x^2 - 3}{2x^3 - 3x}$

(10–3)

13. $f(x) = e^{x^3 - x^2}$

$f'(x) = e^{x^3 - x^2}(3x^2 - 2x)$

$\qquad = (3x^2 - 2x)e^{x^3 - x^2}$ (10–3)

14. $y = e^{-2x}\ln 5x$

$\dfrac{dy}{dx} = e^{-2x}\left(\dfrac{1}{5x}\right)(5) + (\ln 5x)(e^{-2x})(-2)$

$\qquad = e^{-2x}\left(\dfrac{1}{x} - 2\ln 5x\right) = \dfrac{1 - 2x\ln 5x}{xe^{2x}}$ (10–3)

15. $f(x) = 1 + e^{-x}$

$f'(x) = e^{-x}(-1) = -e^{-x}$

An equation for the tangent line to the graph of f at $x = 0$ is:

$y - y_1 = m(x - x_1)$,

where $x_1 = 0$, $y_1 = f(0) = 1 + e^0 = 2$, and $m = f'(0) = -e^0 = -1$.

Thus, $y - 2 = -1(x - 0)$ or $y = -x + 2$.

An equation for the tangent line to the graph of f at $x = -1$ is

$y - y_1 = m(x - x_1)$,

where $x_1 = -1$, $y_1 = f(-1) = 1 + e$, and $m = f'(-1) = -e$. Thus,

$y - (1 + e) = -e[x - (-1)]$ or $y - 1 - e = -ex - e$ and $y = -ex + 1$.

(10–2)

16. $x^2 - 3xy + 4y^2 = 23$
Differentiate implicitly:
$$2x - 3(xy' + y \cdot 1) + 8yy' = 0$$
$$2x - 3xy' - 3y + 8yy' = 0$$
$$8yy' - 3xy' = 3y - 2x$$
$$(8y - 3x)y' = 3y - 2x$$
$$y' = \frac{3y - 2x}{8y - 3x}$$
$$y'\Big|_{(-1,\ 2)} = \frac{3 \cdot 2 - 2(-1)}{8 \cdot 2 - 3(-1)} = \frac{8}{19} \quad \text{[Slope at } (-1, 2)] \quad (10\text{-}4)$$

17.
$$x^3 - 2t^2 x + 8 = 0$$
$$3x^2 x' - (2t^2 x' + x \cdot 4t) + 0 = 0$$
$$3x^2 x' - 2t^2 x' - 4xt = 0$$
$$(3x^2 - 2t^2)x' = 4xt$$
$$x' = \frac{4xt}{3x^2 - 2t^2}$$
$$x'\Big|_{(-2,\ 2)} = \frac{4 \cdot 2 \cdot (-2)}{3(2^2) - 2(-2)^2} = \frac{-16}{12 - 8} = \frac{-16}{4} = -4 \quad (10\text{-}4)$$

18. $x - y^2 = e^y$
Differentiate implicitly:
$$1 - 2yy' = e^y y'$$
$$1 = e^y y' + 2yy'$$
$$1 = y'(e^y + 2y)$$
$$y' = \frac{1}{e^y + 2y}$$
$$y'\Big|_{(1,\ 0)} = \frac{1}{e^0 + 2 \cdot 0} = 1 \quad (10\text{-}4)$$

19. $\ln y = x^2 - y^2$
Differentiate implicitly:
$$\frac{y'}{y} = 2x - 2yy'$$
$$y'\left(\frac{1}{y} + 2y\right) = 2x$$
$$y'\left(\frac{1 + 2y^2}{y}\right) = 2x$$
$$y' = \frac{2xy}{1 + 2y^2}$$
$$y'\Big|_{(1,\ 1)} = \frac{2 \cdot 1 \cdot 1}{1 + 2(1)^2} = \frac{2}{3} \quad (10\text{-}4)$$

20. $y^2 - 4x^2 = 12$
Differentiate with respect to t:
$$2y\frac{dy}{dt} - 8x\frac{dx}{dt} = 0$$
Given: $\frac{dx}{dt} = -2$ when $x = 1$ and $y = 4$. Therefore,
$$2 \cdot 4\frac{dy}{dt} - 8 \cdot 1 \cdot (-2) = 0$$
$$8\frac{dy}{dt} + 16 = 0$$
$$\frac{dy}{dt} = -2.$$
The y coordinate is decreasing at 2 units per second. $\quad (10\text{-}5)$

21. From the figure, $x^2 + y^2 = 17^2$.

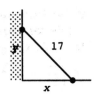

Differentiate with respect to t:

$$2x\frac{dx}{dt} + 2y\frac{dy}{dt} = 0 \quad \text{or} \quad x\frac{dx}{dt} + y\frac{dy}{dt} = 0$$

We are given $\frac{dx}{dt} = -0.5$ feet per second. Therefore,

$$x(-0.5) + y\frac{dy}{dt} = 0 \quad \text{or} \quad \frac{dy}{dt} = \frac{0.5x}{y} = \frac{x}{2y}$$

Now, when $x = 8$, we have: $\quad 8^2 + y^2 = 17^2$

$$y^2 = 289 - 64 = 225$$
$$y = 15$$

Therefore, $\left.\dfrac{dy}{dt}\right|_{(8,\ 15)} = \dfrac{8}{2(15)} = \dfrac{4}{15} \approx 0.27$ ft/sec.　　　　　(10-5)

22. $A = \pi R^2$. Given: $\dfrac{dA}{dt} = 24$ square inches per minute.

Differentiate with respect to t:

$$\frac{dA}{dt} = 2\pi R\frac{dR}{dt}$$
$$24 = 2\pi R\frac{dR}{dt}$$

Therefore, $\dfrac{dR}{dt} = \dfrac{24}{2\pi R} = \dfrac{12}{\pi R}$.

$\left.\dfrac{dR}{dt}\right|_{(R\ =\ 12)} = \dfrac{12}{\pi\cdot 12} = \dfrac{1}{\pi} \approx 0.318$ inches per minute　　　　　(10-5)

23. $f(x) = 11x - 2x \ln x,\ x > 0$

$$f'(x) = 11 - 2x\left(\frac{1}{x}\right) - (\ln x)(2)$$
$$= 11 - 2 - 2\ln x = 9 - 2\ln x,\ x > 0$$

Critical value(s): $f'(x) = 9 - 2\ln x = 0$

$$2\ln x = 9$$
$$\ln x = \frac{9}{2}$$
$$x = e^{9/2}$$

$$f''(x) = -\frac{2}{x} \quad \text{and} \quad f''(e^{9/2}) = -\frac{2}{e^{9/2}} < 0$$

Since $x = e^{9/2}$ is the only critical value, and $f''(e^{9/2}) < 0$, f has an absolute maximum at $x = e^{9/2}$. The absolute maximum is:

$$f(e^{9/2}) = 11e^{9/2} - 2e^{9/2}\ln(e^{9/2})$$
$$= 11e^{9/2} - 9e^{9/2}$$
$$= 2e^{9/2} \approx 180.03$$　　　　　(10-2)

24. $f(x) = 10xe^{-2x},\ x > 0$

$$f'(x) = 10xe^{-2x}(-2) + 10e^{-2x}(1) = 10e^{-2x}(1 - 2x),\ x > 0$$

Critical value(s): $f'(x) = 10e^{-2x}(1 - 2x) = 0$

$$1 - 2x = 0$$
$$x = \frac{1}{2}$$

$$f''(x) = 10e^{-2x}(-2) + 10(1 - 2x)e^{-2x}(-2)$$
$$= -20e^{-2x}(1 + 1 - 2x)$$
$$= -40e^{-2x}(1 - x)$$
$$f''\left(\frac{1}{2}\right) = -20e^{-1} < 0$$

Since $x = \frac{1}{2}$ is the only critical value, and $f''\left(\frac{1}{2}\right) = -20e^{-1} < 0$, f has an absolute maximum at $x = \frac{1}{2}$. The absolute maximum of f is:
$$f\left(\frac{1}{2}\right) = 10\left(\frac{1}{2}\right)e^{-2(1/2)}$$
$$= 5e^{-1} \approx 1.84 \tag{10-3}$$

25. $f(x) = 3x - x^2 + e^{-x}$, $x > 0$
$f'(x) = 3 - 2x - e^{-x}$, $x > 0$
Critical value(s): $f'(x) = 3 - 2x - e^{-x} = 0$
$$x \approx 1.373$$
$f''(x) = -2 + e^{-x}$ and $f''(1.373) = -2 + e^{-1.373} < 0$

Since $x \approx 1.373$ is the only critical value, and $f''(1.373) < 0$, f has an absolute maximum at $x = 1.373$. The absolute maximum of f is:
$$f(1.373) = 3(1.373) - (1.373)^2 + e^{-1.373}$$
$$\approx 2.487 \tag{10-2}$$

26. $f(x) = \dfrac{\ln x}{e^x}$, $x > 0$

$$f'(x) = \frac{e^x\left(\frac{1}{x}\right) - (\ln x)e^x}{(e^x)^2} = \frac{e^x\left(\frac{1}{x} - \ln x\right)}{e^{2x}}$$
$$= \frac{1 - x\ln x}{xe^x}, \quad x > 0$$

Critical value(s): $f'(x) = \dfrac{1 - x\ln x}{xe^x} = 0$
$$1 - x\ln x = 0$$
$$x\ln x = 1$$
$$x \approx 1.763$$

$$f''(x) = \frac{xe^x[-1 - \ln x] - (1 - x\ln x)(xe^x + e^x)}{x^2e^{2x}}$$
$$= \frac{-x(1 + \ln x) - (x + 1)(1 - x\ln x)}{x^2e^x};$$
$$f''(1.763) \approx \frac{-1.763(1.567) - (2.763)(0.000349)}{(1.763)^2e^{1.763}} < 0$$

Since $x = 1.763$ is the only critical value, and $f''(1.763) < 0$, f has an absolute maximum at $x = 1.763$. The absolute maximum of f is:
$$f(1.763) = \frac{\ln(1.763)}{e^{1.763}} \approx 0.097 \tag{10-2}$$

27. $f(x) = 5 - 5e^{-x}$

Step 1. Analyze $f(x)$:

(A) Domain: All real numbers, $(-\infty, \infty)$.

(B) Intercepts: y intercept: $f(0) = 5 - 5e^{-0} = 0$

x intercepts: $5 - 5e^{-x} = 0$

$$e^{-x} = 1$$

$$x = 0$$

(C) Asymptotes:

$$\lim_{x \to \infty} (5 - 5e^{-x}) = \lim_{x \to \infty} \left(5 - \frac{5}{e^x}\right) = 5$$

$$\lim_{x \to -\infty} (5 - 5e^{-x}) \text{ does not exist.}$$

Thus, $y = 5$ is a horizontal asymptote.

Since $f(x) = 5 - \dfrac{5}{e^x} = \dfrac{5e^x - 5}{e^x}$ and $e^x \neq 0$ for all x, there are no vertical asymptotes.

Step 2. Analyze $f'(x)$:

$f'(x) = -5e^{-x}(-1) = 5e^{-x} > 0$ on $(-\infty, \infty)$

Thus, f is increasing on $(-\infty, \infty)$; there are no local extrema.

Step 3. Analyze $f''(x)$:

$f''(x) = -5e^{-x} < 0$ on $(-\infty, \infty)$.

Thus, the graph of f is concave downward on $(-\infty, \infty)$; there are no inflection points.

Step 4. Sketch the graph of f:

x	$f(x)$
0	0
-1	-8.59
2	4.32

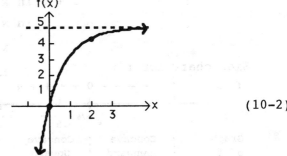

(10-2)

28. $f(x) = x^3 \ln x$

Step 1. Analyze $f(x)$:

(A) Domain: all positive real numbers, $(0, \infty)$.

(B) Intercepts: y intercept: Since $x = 0$ is not in the domain, there is no y intercept.

x intercepts: $x^3 \ln x = 0$

$$\ln x = 0$$

$$x = 1$$

(C) Asymptotes:

$$\lim_{x \to \infty} (x^3 \ln x) \text{ does not exist.}$$

It can be shown that $\lim_{x \to 0^+} (x^3 \ln x) = 0$. Thus, there are no horizontal or vertical asymptotes.

Step 2. Analyze $f'(x)$:

$$f'(x) = x^3\left(\frac{1}{x}\right) + (\ln x)3x^2$$

$$= x^2[1 + 3 \ln x], \quad x > 0$$

Critical values: $x^2[1 + 3 \ln x] = 0$

$$1 + 3 \ln x = 0 \quad (\text{since } x > 0)$$

$$\ln x = -\frac{1}{3}$$

$$x = e^{-1/3} \approx 0.72$$

Partition numbers: $x = e^{-1/3}$

Sign chart for f':

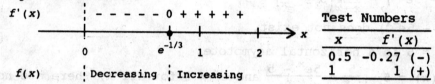

Test Numbers	
x	$f'(x)$
0.5	−0.27 (−)
1	1 (+)

Thus, f is decreasing on $(0, e^{-1/3})$ and increasing on $(e^{-1/3}, \infty)$; f has a local minimum at $x = e^{-1/3}$.

Step 3. Analyze $f''(x)$:

$$f''(x) = x^2\left(\frac{3}{x}\right) + (1 + 3 \ln x)2x$$

$$= x(5 + 6 \ln x), \quad x > 0$$

Partition numbers: $x(5 + 6 \ln x) = 0$

$$5 + 6 \ln x = 0$$

$$\ln x = -\frac{5}{6}$$

$$x = e^{-5/6} \approx 0.43$$

Sign chart for f'':

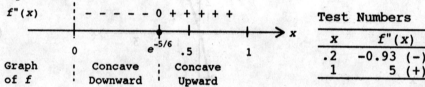

Test Numbers	
x	$f''(x)$
.2	−0.93 (−)
1	5 (+)

Thus, the graph of f is concave downward on $(0, e^{-5/6})$ and concave upward on $(e^{-5/6}, \infty)$; the graph has an inflection point at $x = e^{-5/6}$.

Step 4. Sketch the graph of f:

x	$f(x)$
$e^{-5/6}$	−0.07
$e^{-1/3}$	−0.12
1	0

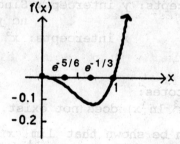

(10-2)

29. $y = w^3$, $w = \ln u$, $u = 4 - e^x$

(A) $y = [\ln(4 - e^x)]^3$

(B) $\dfrac{dy}{dx} = \dfrac{dy}{dw} \cdot \dfrac{dw}{du} \cdot \dfrac{du}{dx}$

$\qquad = 3w^2 \cdot \dfrac{1}{u} \cdot (-e^x) = 3[\ln(4 - e^x)]^2 \left(\dfrac{1}{4 - e^x}\right)(-e^x)$

$\qquad\qquad = \dfrac{-3e^x[\ln(4 - e^x)]^2}{4 - e^x}$ $\hfill$ (10-3)

30. $y = 5^{x^2-1}$

$y' = 5^{x^2-1}(\ln 5)(2x) = 2x5^{x^2-1}(\ln 5)$ $\hfill$ (10-3)

31. $\dfrac{d}{dx} \log_5(x^2 - x) = \dfrac{1}{x^2 - x} \cdot \dfrac{1}{\ln 5} \cdot \dfrac{d}{dx}(x^2 - x) = \dfrac{1}{\ln 5} \cdot \dfrac{2x - 1}{x^2 - x}$ $\hfill$ (10-3)

32. $\dfrac{d}{dx} \sqrt{\ln(x^2 + x)} = \dfrac{d}{dx}[\ln(x^2 + x)]^{1/2} = \dfrac{1}{2}[\ln(x^2 + x)]^{-1/2} \dfrac{d}{dx} \ln(x^2 + x)$

$\qquad = \dfrac{1}{2}[\ln(x^2 + x)]^{-1/2} \dfrac{1}{x^2 + x} \dfrac{d}{dx}(x^2 + x)$

$\qquad = \dfrac{1}{2}[\ln(x^2 + x)]^{-1/2} \cdot \dfrac{2x + 1}{x^2 + x} = \dfrac{2x + 1}{2(x^2 + x)[\ln(x^2 + x)]^{1/2}}$

$\hfill$ (10-3)

33. $e^{xy} = x^2 + y + 1$

Differentiate implicitly:

$\qquad \dfrac{d}{dx} e^{xy} = \dfrac{d}{dx} x^2 + \dfrac{d}{dx} y + \dfrac{d}{dx} 1$

$e^{xy}(xy' + y) = 2x + y'$

$xe^{xy}y' - y' = 2x - ye^{xy}$

$\qquad\quad y' = \dfrac{2x - ye^{xy}}{xe^{xy} - 1}$

$\qquad y' \Big|_{(0,\ 0)} = \dfrac{2 \cdot 0 - 0 \cdot e^0}{0 \cdot e^0 - 1} = 0$ $\hfill$ (10-4)

34. $A = \pi r^2$, $r \geq 0$

Differentiate with respect to t:

$\qquad \dfrac{dA}{dt} = 2\pi r \dfrac{dr}{dt} = 6\pi r$ since $\dfrac{dr}{dt} = 3$

The area increases at the rate $6\pi r$. This is smallest when $r = 0$; there is no largest value. $\hfill$ (10-5)

35. $y = x^3$

Differentiate with respect to t:

$\qquad \dfrac{dy}{dt} = 3x^2 \dfrac{dx}{dt}$

Solving for $\dfrac{dx}{dt}$, we get

$\qquad \dfrac{dx}{dt} = \dfrac{1}{3x^2} \cdot \dfrac{dy}{dt} = \dfrac{5}{3x^2}$ since $\dfrac{dy}{dt} = 5$

To find where $\frac{dx}{dt} > \frac{dy}{dt}$, solve the inequality

$$\frac{5}{3x^2} > 5$$

$$\frac{1}{3x^2} > 1$$

$$3x^2 < 1$$

$$-\frac{1}{\sqrt{3}} < x < \frac{1}{\sqrt{3}} \quad \text{or} \quad \frac{-\sqrt{3}}{3} < x < \frac{\sqrt{3}}{3} \tag{10-5}$$

36. (A) The compound interest formula is: $A = P(1 + r)^t$. Thus, the time for P to double when $r = 0.05$ and interest is compounded annually can be found by solving

$2P = P(1 + 0.05)^t$ or $2 = (1.05)^t$ for t.

$\ln(1.05)^t = \ln 2$

$t \ln(1.05) = \ln 2$

$$t = \frac{\ln 2}{\ln(1.05)} \approx 14.2 \text{ years}$$

(B) The continuous compound interest formula is: $A = Pe^{rt}$. Proceeding as above, we have

$2P = Pe^{0.05t}$ or $e^{0.05t} = 2$.

Therefore, $0.05t = \ln 2$ and

$$t = \frac{\ln 2}{.05} \approx 13.9 \text{ years} \tag{10-1}$$

37. $A(t) = 100e^{0.1t}$

$A'(t) = 100(0.1)e^{0.1t} = 10e^{0.1t}$

$A'(1) = 11.05$ or $11.05 per year

$A'(10) = 27.18$ or $27.18 per year $\tag{10-1}$

38. $R(x) = xp(x) = 1000xe^{-0.02x}$

$R'(x) = 1000[xD_xe^{-0.02x} + e^{-0.02x}D_xx]$

$\qquad = 1000[x(-0.02)e^{-0.02x} + e^{-0.02x}]$

$\qquad = (1000 - 20x)e^{-0.02x} \tag{10-3}$

39. From Problem 38,

$R'(x) = (1000 - 20x)e^{-0.02x}$

Critical value(s): $R'(x) = (1000 - 20x)e^{-0.02x} = 0$

$\qquad\qquad\qquad\qquad\qquad 1000 - 20x = 0$

$\qquad\qquad\qquad\qquad\qquad\qquad x = 50$

$R''(x) = (1000 - 20x)e^{-0.02x}(-0.02) + e^{-0.02x}(-20)$

$\qquad = e^{-0.02x}[0.4x - 20 - 20]$

$\qquad = e^{-0.02x}(0.4x - 40)$

$R''(50) = e^{-0.02(50)}[0.4(50) - 40] = -20e^{-1} < 0$

Since $x = 50$ is the only critical value and $R''(50) < 0$, R has an absolute maximum at a production level of 50 units. The maximum revenue is

$R(50) = 1000(50)e^{-0.02(50)} = 50,000e^{-1} \approx 18,394$ or \$18,394.

The price per unit at the production level of 50 units is

$p(50) = 1000e^{-0.02(50)} = 1000e^{-1} \approx 367.88$ or \$367.88. (10-3)

40. $R(x) = 1000xe^{-0.02x}$, $0 \le x \le 100$

Step 1. Analyze $R(x)$:

(A) Domain: $0 \le x \le 100$ or $[0, 100]$

(B) Intercepts: y intercept: $R(0) = 0$

                 x intercepts: $100xe^{-0.02x} = 0$

                                 $x = 0$

(C) Asymptotes: There are no horizontal or vertical asymptotes.

Step 2. Analyze $R'(x)$:

From Problems 47 and 48, $R'(x) = (1000 - 20x)e^{-0.02x}$ and $x = 50$ is a critical value.

Sign chart for R':

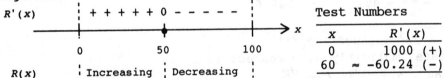

Test Numbers		
x	$R'(x)$	
0	1000	(+)
60	≈ -60.24	(−)

Thus, R is increasing on $(0, 50)$ and decreasing on $(50, 100)$; R has a maximum at $x = 50$.

Step 3. Analyze $R''(x)$:

$R''(x) = (0.4x - 40)e^{-0.02x} < 0$ on $(0, 100)$

Thus, the graph of R is concave downward on $(0, 100)$.

Step 4. Sketch the graph of R:

x	$R(x)$
0	0
50	18,394
100	13,533

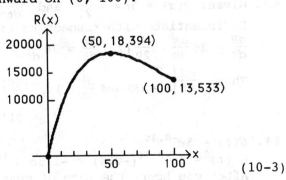

(10-3)

41. Cost: $C(x) = 200 + 50x - 50 \ln x$, $x \geq 1$

Average cost: $\overline{C} = \dfrac{C(x)}{x} = \dfrac{200}{x} + 50 - \dfrac{50}{x} \ln x$, $x \geq 1$

$\overline{C}'(x) = \dfrac{-200}{x^2} - \dfrac{50}{x}\left(\dfrac{1}{x}\right) + (\ln x)\dfrac{50}{x^2} = \dfrac{50(\ln x - 5)}{x^2}$, $x \geq 1$

Critical value(s): $\overline{C}'(x) = \dfrac{50(\ln x - 5)}{x^2} = 0$

$$\ln x = 5$$
$$x = e^5$$

Sign chart for $\overline{C}'$:

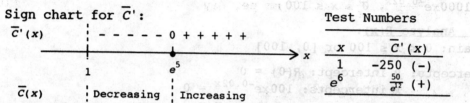

Test Numbers		
x	$\overline{C}'(x)$	
1	-250	$(-)$
e^6	$\frac{50}{e^{12}}$	$(+)$

By the first derivative test, $\overline{C}$ has a local minimum at $x = e^5$. Since this is the only critical value of $\overline{C}$, $\overline{C}$ has as absolute minimum at $x = e^5$. Thus, the minimal average cost is:

$\overline{C}(e^5) = \dfrac{200}{e^5} + 50 - \dfrac{50}{e^5} \ln(e^5)$

$\qquad = 50 - \dfrac{50}{e^5} \approx 49.66$ or $\$49.66$ \hfill (10-2)

42. $x = \sqrt{5000 - 2p^3} = (5000 - 2p^3)^{1/2}$

Differentiate implicitly with respect to x:

$1 = \dfrac{1}{2}(5000 - 2p^3)^{-1/2}(-6p^2)\dfrac{dp}{dx}$

$1 = \dfrac{-3p^2}{(5000 - 2p^3)^{1/2}}\dfrac{dp}{dx}$

$\dfrac{dp}{dx} = \dfrac{-(5000 - 2p^3)^{1/2}}{3p^2}$ \hfill (10-3)

43. Given: $R(x) = 36x - \dfrac{x^2}{20}$ and $\dfrac{dx}{dt} = 10$ when $x = 250$.

Differentiate with respect to t:

$\dfrac{dR}{dt} = 36\dfrac{dx}{dt} - \dfrac{1}{20}(2x)\dfrac{dx}{dt} = 36\dfrac{dx}{dt} - \dfrac{x}{10}\dfrac{dx}{dt}$

Thus, $\dfrac{dR}{dt}\Big|_{x = 250 \text{ and } \frac{dx}{dt} = 10} = 36(10) - \dfrac{250}{10}(10)$

$\qquad\qquad\qquad\qquad = \$110 \text{ per day}$ \hfill (10-5)

44. $C(t) = 5e^{-0.3t}$

$C'(t) = 5e^{-0.3t}(-0.3) = -1.5e^{-0.3t}$

After one hour, the rate of change of concentration is

$C'(1) = -1.5e^{-0.3(1)} = -1.5e^{-0.3} \approx -1.111$ mg/ml per hour.

After five hours, the rate of change of concentration is

$C'(5) = -1.5e^{-0.3(5)} = -1.5e^{-1.5} \approx -0.335$ mg/ml per hour. \hfill (10-3)

45. Given: $A = \pi R^2$ and $\frac{dA}{dt} = -45$ mm^2 per day (negative because the area is decreasing).

Differentiate with respect to t:
$$\frac{dA}{dt} = \pi 2R \frac{dR}{dt}$$
$$-45 = 2\pi R \frac{dR}{dt}$$
$$\frac{dR}{dt} = -\frac{45}{2\pi R}$$
$$\left.\frac{dR}{dt}\right|_{R\,=\,15} = \frac{-45}{2\pi \cdot 15} = \frac{-3}{2\pi} \approx -0.477 \text{ mm per day}$$
(10-5)

46. $N(t) = 10(1 - e^{-0.4t})$

(A) $N'(t) = -10e^{-0.4t}(-0.4) = 4e^{-0.4t}$
$N'(1) = 4e^{-0.4(1)} = 4e^{-0.4} \approx 2.68$.
Thus, learning is increasing at the rate of 2.68 units per day after 1 day.
$N'(5) = 4e^{-0.4(5)} = 4e^{-2} \approx 0.54$
Thus, learning is increasing at the rate of 0.54 units per day after 5 days.

(B) From (A), $N'(t) = 4e^{-0.4t} > 0$ on $(0, 10)$. Thus, N is increasing on $(0, 10)$.
$N''(t) = 4e^{-0.4t}(-0.4) = -1.6e^{-0.4t} < 0$ on $(0, 10)$.

Thus, the graph of N is concave downward on $(0, 10)$. The graph of N is:

t	$N(t)$
0	0
5	8.65
10	9.82

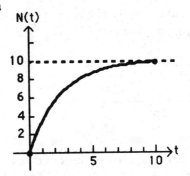

(10-3)

47. Given: $T = 2\left(1 + \frac{1}{x^{3/2}}\right) = 2 + 2x^{-3/2}$, and $\frac{dx}{dt} = 3$ when $x = 9$.

Differentiate with respect to t:
$$\frac{dT}{dt} = 0 + 2\left(-\frac{3}{2}x^{-5/2}\right)\frac{dx}{dt} = -3x^{-5/2}\frac{dx}{dt}$$

$$\left.\frac{dT}{dt}\right|_{x\,=\,9 \text{ and } \frac{dx}{dt}\,=\,3} = -3(9)^{-5/2}(3) = -3 \cdot 3^{-5} \cdot 3 = -3^{-3} = \frac{-1}{27}$$
$$\approx -0.037 \text{ minutes per operation per hour}$$
(10-5)

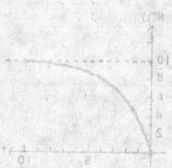

11 INTEGRATION

EXERCISE 11-1
_____

Things to remember:

1. A function $F(x)$ is an ANTIDERIVATIVE of $f(x)$ if $F'(x) = f(x)$.

2. THEOREM ON ANTIDERIVATIVES

   If the derivatives of two functions are equal on an open interval (a, b), then the functions can differ by at most a constant. Symbolically: If F and G are differentiable functions on the interval (a, b) and $F'(x) = G'(x)$ for all x in (a, b), then $F(x) = G(x) + k$ for some constant k.

3. The INDEFINITE INTEGRAL of $f(x)$, denoted

   $$\int f(x)\,dx,$$

   represents all antiderivatives of $f(x)$ and is given by

   $$\int f(x)\,dx = F(x) + C$$

   where $F(x)$ is any antiderivative of $f(x)$ and C is an arbitrary constant. The symbol $\int$ is called an INTEGRAL SIGN, the function $f(x)$ is called the INTEGRAND, and C is called the CONSTANT OF INTEGRATION.

4. Indefinite integration and differentiation are reverse operations (except for the addition of the constant of integration). This is expressed symbolically by:

   (a) $\dfrac{d}{dx}\left(\int f(x)\,dx\right) = f(x)$

   (b) $\int F'(x)\,dx = F(x) + C$

5. INDEFINITE INTEGRAL FORMULAS:

   (a) $\int k\,dx = kx + C$, k constant

   (b) $\int x^n dx = \dfrac{x^{n+1}}{n+1} + C$, $n \neq -1$

   (c) $\int e^x dx = e^x + C$

   (d) $\int \dfrac{dx}{x} = \ln|x| + C$, $x \neq 0$

<u>6.</u> INDEFINITE INTEGRATION PROPERTIES:

(a) $\displaystyle\int kf(x)\,dx = k\int f(x)\,dx$, k constant

(b) $\displaystyle\int [f(x) \pm g(x)]\,dx = \int f(x)\,dx \pm \int g(x)\,dx$

---

1. $\displaystyle\int 7\,dx = 7x + C$ [using <u>5</u>(a)]   <u>Check</u>: $(7x + C)' = 7$

3. $\displaystyle\int x^6 dx = \frac{x^{6+1}}{6+1} + C$ [using <u>5</u>(b)]

$\qquad = \dfrac{x^7}{7} + C$ <u>Check</u>: $\left(\dfrac{x^7}{7} + C\right)' = x^6$

5. $\displaystyle\int 8t^3 dt = 8\int t^3 dt$ [using <u>6</u>(a)]

$\qquad = 8\,\dfrac{t^{3+1}}{3+1} + C = 2t^4 + C$ <u>Check</u>: $(2t^4 + C)' = 8t^3$

7. $\displaystyle\int (2u + 1)\,du = \int 2u\,du + \int 1\,du$ [using <u>6</u>(b)]

$\qquad = 2\,\dfrac{u^2}{2} + u + C = u^2 + u + C$ <u>Check</u>: $(u^2 + u + C)' = 2u + 1$

9. $\displaystyle\int (3x^2 + 2x - 5)\,dx = \int 3x^2 dx + \int 2x\,dx - \int 5\,dx$

$\qquad = 3\int x^2\,dx + 2\int x\,dx - \int 5\,dx$

$\qquad = 3\,\dfrac{x^3}{3} + 2\,\dfrac{x^2}{2} - 5x + C = x^3 + x^2 - 5x + C$

<u>Check</u>: $(x^3 + x^2 - 5x + C)' = 3x^2 + 2x - 5$

11. $\displaystyle\int (s^4 - 8s^5)\,ds = \int s^4 ds - \int 8s^5 ds = \int s^4 ds - 8\int s^5 ds$

$\qquad = \dfrac{s^5}{5} - 8\,\dfrac{s^6}{6} + C = \dfrac{s^5}{5} - \dfrac{4s^6}{3} + C$

<u>Check</u>: $\left(\dfrac{s^5}{5} - \dfrac{4s^6}{3} + C\right)' = s^4 - 8s^5$

13. $\displaystyle\int 3e^t dt = 3\int e^t dt$ [using <u>5</u>(c)]

$\qquad = 3e^t + C$ <u>Check</u>: $(3e^t + C)' = 3e^t$

15. $\int 2z^{-1}dz = 2 \int \frac{1}{z} dz = 2 \ln|z| + C$ [using $\underline{5}$(d)]

Check: $(2 \ln|z| + C)' = \frac{2}{z}$

17.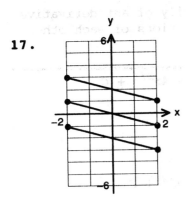

19.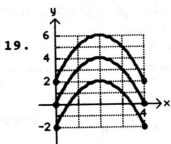

21. $\frac{dy}{dx} = 200x^4$

$y = \int 200x^4 dx = 200 \int x^4 dx = 200 \frac{x^5}{5} + C = 40x^5 + C$

23. $\frac{dP}{dx} = 24 - 6x$

$P = \int (24 - 6x) dx = \int 24 \ dx - \int 6x \ dx = \int 24 \ dx - 6 \int x \ dx$

$= 24x - \frac{6x^2}{2} + C = 24x - 3x^2 + C$

25. $\frac{dy}{du} = 2u^5 - 3u^2 - 1$

$y = \int (2u^5 - 3u^2 - 1) du = \int 2u^5 du - \int 3u^2 du - \int 1 \ du$

$= 2 \int u^5 du - 3 \int u^2 du - \int du$

$= \frac{2u^6}{6} - \frac{3u^3}{3} - u + C = \frac{u^6}{3} - u^3 - u + C$

27. $\frac{dy}{dx} = e^x + 3$

$y = \int (e^x + 3) dx = \int e^x dx + \int 3 \ dx = e^x + 3x + C$

29. $\frac{dx}{dt} = 5t^{-1} + 1$

$x = \int (5t^{-1} + 1) dt = \int 5t^{-1} dt + \int 1 \ dt = 5 \int \frac{1}{t} dt + \int dt$

$= 5 \ln|t| + t + C$

31. The graphs in this set ARE NOT graphs from a family of antiderivative functions since the graphs are not vertical translations of each other.

33. The graphs in this set could be graphs from a family of antiderivative functions since they appear to be vertical translations of each other.

35. $\displaystyle\int 6x^{1/2}dx = 6\int x^{1/2}dx = 6\,\frac{x^{(1/2)+1}}{\frac{1}{2}+1} + C = \frac{6x^{3/2}}{\frac{3}{2}} + C = 4x^{3/2} + C$

Check: $(4x^{3/2} + C)' = 4\left(\dfrac{3}{2}\right)x^{1/2} = 6x^{1/2}$

37. $\displaystyle\int 8x^{-3}dx = 8\int x^{-3}dx = 8\,\frac{x^{-3+1}}{-3+1} + C = \frac{8x^{-2}}{-2} + C = -4x^{-2} + C$

Check: $(-4x^{-2} + C)' = -4(-2)x^{-3} = 8x^{-3}$

39. $\displaystyle\int \frac{du}{\sqrt{u}} = \int \frac{du}{u^{1/2}} = \int u^{-1/2}du = \frac{u^{(-1/2)+1}}{-\frac{1}{2}+1} + C = \frac{u^{1/2}}{\frac{1}{2}} + C$

$\qquad\qquad = 2u^{1/2} + C \text{ or } 2\sqrt{u} + C$

Check: $(2u^{1/2} + C)' = 2\left(\dfrac{1}{2}\right)u^{-1/2} = \dfrac{1}{u^{1/2}} = \dfrac{1}{\sqrt{u}}$

41. $\displaystyle\int \frac{dx}{4x^3} = \frac{1}{4}\int x^{-3}dx = \frac{1}{4}\cdot\frac{x^{-2}}{-2} + C = \frac{-x^{-2}}{8} + C$

Check: $\left(\dfrac{-x^{-2}}{8} + C\right)' = \dfrac{1}{8}(-2)(-x^{-3}) = \dfrac{1}{4}x^{-3} = \dfrac{1}{4x^3}$

43. $\displaystyle\int \frac{du}{2u^5} = \frac{1}{2}\int u^{-5}du = \frac{1}{2}\cdot\frac{u^{-4}}{-4} + C = \frac{-u^{-4}}{8} + C$

Check: $\left(\dfrac{-u^{-4}}{8} + C\right)' = \dfrac{-1}{8}(-4)u^{-5} = \dfrac{u^{-5}}{2} = \dfrac{1}{2u^5}$

45. $\displaystyle\int \left(3x^2 - \frac{2}{x^2}\right)dx = \int 3x^2 dx - \int \frac{2}{x^2}dx$

$\qquad\qquad = 3\int x^2 dx - 2\int x^{-2}dx = 3\cdot\frac{x^3}{3} - \frac{2x^{-1}}{-1} + C = x^3 + 2x^{-1} + C$

Check: $(x^3 + 2x^{-1} + C)' = 3x^2 - 2x^{-2} = 3x^2 - \dfrac{2}{x^2}$

47. $\int \left(10x^4 - \dfrac{8}{x^5} - 2\right)dx = \int 10x^4 dx - \int 8x^{-5} dx - \int 2\ dx$

$$= 10\int x^4 dx - 8\int x^{-5} dx - \int 2\ dx$$

$$= \frac{10x^5}{5} - \frac{8x^{-4}}{-4} - 2x + C = 2x^5 + 2x^{-4} - 2x + C$$

Check: $(2x^5 + 2x^{-4} - 2x + C)' = 10x^4 - 8x^{-5} - 2 = 10x^4 - \dfrac{8}{x^5} - 2$

49. $\int \left(3\sqrt{x} + \dfrac{2}{\sqrt{x}}\right)dx = 3\int x^{1/2} dx + 2\int x^{-1/2} dx$

$$= \frac{3x^{3/2}}{\frac{3}{2}} + \frac{2x^{1/2}}{\frac{1}{2}} + C = 2x^{3/2} + 4x^{1/2} + C$$

Check: $(2x^{3/2} + 4x^{1/2} + C)' = 2\left(\frac{3}{2}\right)x^{1/2} + 4\left(\frac{1}{2}\right)x^{-1/2}$

$$= 3x^{1/2} + 2x^{-1/2} = 3\sqrt{x} + \frac{2}{\sqrt{x}}$$

51. $\int \left(\sqrt[3]{x^2} - \dfrac{4}{x^3}\right)dx = \int x^{2/3} dx - 4\int x^{-3} dx = \dfrac{x^{5/3}}{\frac{5}{3}} - \dfrac{4x^{-2}}{-2} + C$

$$= \frac{3x^{5/3}}{5} + 2x^{-2} + C$$

Check: $\left(\frac{3}{5}x^{5/3} + 2x^{-2} + C\right)' = \frac{3}{5}\left(\frac{5}{3}\right)x^{2/3} + 2(-2)x^{-3}$

$$= x^{2/3} - 4x^{-3} = \sqrt[3]{x^2} - \frac{4}{x^3}$$

53. $\int \dfrac{e^x - 3x}{4}dx = \int \left(\dfrac{e^x}{4} - \dfrac{3x}{4}\right)dx = \dfrac{1}{4}\int e^x dx - \dfrac{3}{4}\int x\ dx$

$$= \frac{1}{4}e^x - \frac{3}{4}\cdot\frac{x^2}{2} + C = \frac{1}{4}e^x - \frac{3x^2}{8} + C$$

Check: $\left(\frac{1}{4}e^x - \frac{3x^2}{8} + C\right)' = \frac{1}{4}e^x - \frac{6x}{8} = \frac{1}{4}e^x - \frac{3}{4}x$

55. $\int (2z^{-3} + z^{-2} + z^{-1})dz = 2\int z^{-3} dz + \int z^{-2} dz + \int \dfrac{1}{z} dz$

$$= \frac{2z^{-2}}{-2} + \frac{z^{-1}}{-1} + \ln|z| + C = -z^{-2} - z^{-1} + \ln|z| + C$$

Check: $(-z^{-2} - z^{-1} + \ln|z| + C)' = 2z^{-3} + z^{-2} + \dfrac{1}{z}$

57.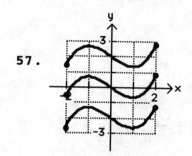

59.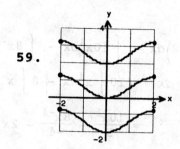

61. $\dfrac{dy}{dx} = 2x - 3$

$$y = \int (2x - 3)\, dx = 2 \int x\, dx - \int 3\, dx = \dfrac{2x^2}{2} - 3x + C = x^2 - 3x + C$$

Given $y(0) = 5$: $5 = 0^2 - 3(0) + C$. Hence, $C = 5$ and $y = x^2 - 3x + 5$.

63. $C'(x) = 6x^2 - 4x$

$$C(x) = \int (6x^2 - 4x)\, dx = 6 \int x^2 dx - 4 \int x\, dx = \dfrac{6x^3}{3} - \dfrac{4x^2}{2} + C = 2x^3 - 2x^2 + C$$

Given $C(0) = 3000$: $3000 = 2(0^3) - 2(0^2) + C$. Hence, $C = 3000$ and
$C(x) = 2x^3 - 2x^2 + 3000$.

65. $\dfrac{dx}{dt} = \dfrac{20}{\sqrt{t}}$

$$x = \int \dfrac{20}{\sqrt{t}}\, dt = 20 \int t^{-1/2} dt = 20 \dfrac{t^{1/2}}{\dfrac{1}{2}} + C = 40\sqrt{t} + C$$

Given $x(1) = 40$: $40 = 40\sqrt{1} + C$ or $40 = 40 + C$. Hence, $C = 0$ and
$x = 40\sqrt{t}$.

67. $\dfrac{dy}{dx} = 2x^{-2} + 3x^{-1} - 1$

$$y = \int (2x^{-2} + 3x^{-1} - 1)\, dx = 2 \int x^{-2} dx + 3 \int x^{-1} dx - \int dx$$

$$= \dfrac{2x^{-1}}{-1} + 3 \ln|x| - x + C = \dfrac{-2}{x} + 3 \ln|x| - x + C$$

Given $y(1) = 0$: $0 = -\dfrac{2}{1} + 3 \ln|1| - 1 + C$. Hence, $C = 3$ and
$y = -\dfrac{2}{x} + 3 \ln|x| - x + 3$.

69. $\dfrac{dx}{dt} = 4e^t - 2$

$$x = \int (4e^t - 2)\, dt = 4 \int e^t dt - \int 2\, dt = 4e^t - 2t + C$$

Given $x(0) = 1$: $1 = 4e^0 - 2(0) + C = 4 + C$. Hence, $C = -3$ and
$x = 4e^t - 2t - 3$.

71. $\dfrac{dy}{dx} = 4x - 3$

$$y = \int (4x - 3)\,dx = 4\int x\,dx - \int 3\,dx = \frac{4x^2}{2} - 3x + C = 2x^2 - 3x + C$$

Given $y(2) = 3$: $3 = 2 \cdot 2^2 - 3 \cdot 2 + C$. Hence, $C = 1$ and $y = 2x^2 - 3x + 1$.

73. $\displaystyle\int \frac{2x^4 - x}{x^3}\,dx = \int \left(\frac{2x^4}{x^3} - \frac{x}{x^3}\right)dx$

$$= 2\int x\,dx - \int x^{-2}\,dx = \frac{2x^2}{2} - \frac{x^{-1}}{-1} + C = x^2 + x^{-1} + C$$

75. $\displaystyle\int \frac{x^5 - 2x}{x^4}\,dx = \int \left(\frac{x^5}{x^4} - \frac{2x}{x^4}\right)dx$

$$= \int x\,dx - 2\int x^{-3}\,dx = \frac{x^2}{2} - \frac{2x^{-2}}{-2} + C = \frac{x^2}{2} + x^{-2} + C$$

77. $\displaystyle\int \frac{x^2 e^x - 2x}{x^2}\,dx = \int \left(\frac{x^2 e^x}{x^2} - \frac{2x}{x^2}\right)dx = \int e^x\,dx - 2\int x^{-1}\,dx = e^x - 2\ln|x| + C$

79. $\dfrac{dM}{dt} = \dfrac{t^2 - 1}{t^2}$

$$M = \int \frac{t^2 - 1}{t^2}\,dt = \int \left(\frac{t^2}{t^2} - \frac{1}{t^2}\right)dt = \int dt - \int t^{-2}\,dt = t - \frac{t^{-1}}{-1} + C = t + \frac{1}{t} + C$$

Given $M(4) = 5$: $5 = 4 + \frac{1}{4} + C$ or $C = 5 - \frac{17}{4} = \frac{3}{4}$. Hence, $M = t + \frac{1}{t} + \frac{3}{4}$.

81. $\dfrac{dy}{dx} = \dfrac{5x + 2}{\sqrt[3]{x}}$

$$y = \int \frac{5x + 2}{\sqrt[3]{x}}\,dx = \int \left(\frac{5x}{x^{1/3}} + \frac{2}{x^{1/3}}\right)dx = 5\int x^{2/3}\,dx + 2\int x^{-1/3}\,dx$$

$$= \frac{5x^{5/3}}{\frac{5}{3}} + \frac{2x^{2/3}}{\frac{2}{3}} + C = 3x^{5/3} + 3x^{2/3} + C$$

Given $y(1) = 0$: $0 = 3 \cdot 1^{5/3} + 3 \cdot 1^{2/3} + C$. Hence, $C = -6$ and
$y = 3x^{5/3} + 3x^{2/3} - 6$.

83. $p'(x) = -\dfrac{10}{x^2}$

$$p(x) = \int -\frac{10}{x^2}\,dx = -10\int x^{-2}\,dx = \frac{-10x^{-1}}{-1} + C = \frac{10}{x} + C$$

Given $p(1) = 20$: $20 = \frac{10}{1} + C = 10 + C$. Hence, $C = 10$ and

$$p(x) = \frac{10}{x} + 10.$$

85. $\overline{C}'(x) = -\dfrac{1,000}{x^2}$

$$\overline{C}(x) = \int \overline{C}'(x)\,dx = \int -\frac{1,000}{x^2}\,dx = -1,000\int x^{-2}\,dx$$
$$= -1,000\,\frac{x^{-1}}{-1} + C$$
$$= \frac{1,000}{x} + C$$

Given $\overline{C}(100) = 25$: $\quad \dfrac{1,000}{100} + C = 25$

$$C = 15$$

Thus, $\overline{C}(x) = \dfrac{1,000}{x} + 15$.

Cost function: $C(x) = x\overline{C}(x) = 15x + 1,000$
Fixed costs: $C(0) = \$1,000$

87. (A) The cost function increases from 0 to 8. The graph is concave downward from 0 to 4 and concave upward from 4 to 8. There is an inflection point at $x = 4$.

(B) $C(x) = \displaystyle\int C'(x)\,dx = \int (3x^2 - 24x + 53)\,dx$

$$= 3\int x^2\,dx - 24\int x\,dx + \int 53\,dx$$
$$= x^3 - 12x^2 + 53x + K$$

Since $C(0) = 30$, we have $K = 30$ and
$$C(x) = x^3 - 12x^2 + 53x + 30.$$

$C(4) = 4^3 - 12(4)^2 + 53(4) + 30 = \$114$ thousand
$C(8) = 8^3 - 12(8)^2 + 53(8) + 30 = \$198$ thousand

(C)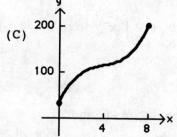

(D) Manufacturing plants are often inefficient at low and high levels of production.

89. $S'(t) = -25t^{2/3}$

$$S(t) = \int S'(t)\,dt = \int -25t^{2/3}\,dt = -25\int t^{2/3}\,dt = -25\,\frac{t^{5/3}}{5/3} + C = -15t^{5/3} + C$$

Given $S(0) = 2000$: $-15(0)^{5/3} + C = 2000$. Hence, $C = 2000$ and
$S(t) = -15t^{5/3} + 2000$. Now, we want to find t such that $S(t) = 800$,
that is: $-15t^{5/3} + 2000 = 800$
$$-15t^{5/3} = -1200$$
$$t^{5/3} = 80$$
and $\qquad\qquad\qquad t = 80^{3/5} \approx 14$

Thus, the company should manufacture the computer for 14 months.

91. $S'(t) = -25t^{2/3} - 70$

$$S(t) = \int S'(t)\,dt = \int (-25t^{2/3} - 70)\,dt$$
$$= -25 \int t^{2/3}\,dt - \int 70\,dt$$
$$= -25\,\frac{t^{5/3}}{5/3} - 70t + C$$
$$= -15t^{5/3} - 70t + C$$

Given $S(0) = 2{,}000$ implies $C = 2{,}000$ and
$$S(t) = 2{,}000 - 15t^{5/3} - 70t$$

Graphing $y_1 = 2{,}000 - 15t^{5/3} - 70t$, $y_2 = 800$ on $0 \le x \le 10$, $0 \le y \le 1000$, we see that the point of intersection is $x \approx 8.92066$, $y = 800$.
So we get $t \approx 8.92$ months.

93. $L'(x) = g(x) = 2400x^{-1/2}$

$$L(x) = \int g(x)\,dx = \int 2400x^{-1/2}\,dx = 2400 \int x^{-1/2}\,dx = 2400\,\frac{x^{1/2}}{1/2} + C$$
$$= 4800\,x^{1/2} + C$$

Given $L(16) = 19{,}200$: $19{,}200 = 4800(16)^{1/2} + C = 19{,}200 + C$. Hence, $C = 0$ and $L(x) = 4800x^{1/2}$.
$L(25) = 4800(25)^{1/2} = 4800(5) = 24{,}000$ labor hours.

95. $\dfrac{dW}{dh} = 0.0015h^2$

$$W = \int 0.0015h^2\,dh = 0.0015 \int h^2\,dh = 0.0015\,\frac{h^3}{3} + C = 0.0005h^3 + C$$

Given $W(60) = 108$: $108 = 0.0005(60)^3 + C$ or $108 = 108 + C$.
Hence, $C = 0$ and $W(h) = 0.0005h^3$. Now $5'10'' = 70''$ and
$W(70) = 0.0005(70)^3 = 171.5$ lb.

97. $\dfrac{dN}{dt} = 400 + 600\sqrt{t}$, $0 \le t \le 9$

$$N = \int (400 + 600\sqrt{t})\,dt = \int 400\,dt + 600 \int t^{1/2}\,dt$$
$$= 400t + 600\,\frac{t^{3/2}}{3/2} + C = 400t + 400t^{3/2} + C$$

Given $N(0) = 5000$: $5000 = 400(0) + 400(0)^{3/2} + C$. Hence, $C = 5000$ and
$N(t) = 400t + 400t^{3/2} + 5000$.
$N(9) = 400(9) + 400(9)^{3/2} + 5000 = 3600 + 10{,}800 + 5000 = 19{,}400$

Things to remember:

1. **REVERSING THE CHAIN RULE**

   The chain rule formula for differentiating a composite function:
   $$\frac{d}{dx} f[g(x)] = f'[g(x)]g'(x),$$
   yields the integral formula
   $$\int f'[g(x)]g'(x)\,dx = f[g(x)] + C$$

2. **GENERAL INDEFINITE INTEGRAL FORMULAS (Version 1)**

   (a) $\displaystyle\int [f(x)]^n f'(x)\,dx = \frac{[f(x)]^{n+1}}{n+1} + C, \; n \neq -1$

   (b) $\displaystyle\int e^{f(x)} f'(x)\,dx = e^{f(x)} + C$

   (c) $\displaystyle\int \frac{1}{f(x)} f'(x)\,dx = \ln|f(x)| + C$

3. **DIFFERENTIALS**

   If $y = f(x)$ defines a differentiable function, then:

   (a) The DIFFERENTIAL dx of the independent variable x is an arbitrary real number.

   (b) The DIFFERENTIAL dy of the dependent variable y is defined as the product of $f'(x)$ and dx; that is: $dy = f'(x)\,dx$.

4. **GENERAL INDEFINITE INTEGRAL FORMULAS (Version 2)**

   (a) $\displaystyle\int u^n\,du = \frac{u^{n+1}}{n+1} + C, \; n \neq -1$

   (b) $\displaystyle\int e^u\,du = e^u + C$

   (c) $\displaystyle\int \frac{1}{u}\,du = \ln|u| + C$

5. **INTEGRATION BY SUBSTITUTION**

   (a) Select a substitution that appears to simplify the integrand. In particular, try to select u so that du is a factor in the integrand.

   (b) Express the integrand entirely in terms of u and du, completely eliminating the original variable and its differential.

   (c) Evaluate the new integral, if possible.

   (d) Express the antiderivative found in Step (c) in terms of the original variable.

1. $\int (x^2 - 4)^5 2x\ dx$

Let $u = x^2 - 4$, then $du = 2x\ dx$ and

$$\int (x^2 - 4)^5 2x\ dx = \int u^5 du = \frac{u^6}{6} + C \quad \text{[using formula \underline{4}(a)]}$$

$$= \frac{(x^2 - 4)^6}{6} + C$$

Check: $\frac{d}{dx}\left[\frac{(x^2 - 4)^6}{6} + C\right] = \frac{1}{6}(6)(x^2 - 4)^5(2x) = (x^2 - 4)^5 2x$

3. $\int e^{4x} 4\ dx$

Let $u = 4x$, then $du = 4\ dx$ and

$$\int e^{4x} 4\ dx = \int e^u du = e^u + C \quad \text{[using formula \underline{4}(b)]}$$

$$= e^{4x} + C$$

Check: $\frac{d}{dx}[e^{4x} + C] = e^{4x}(4)$

5. $\int \frac{1}{2t + 3}\ 2\ dt$

Let $u = 2t + 3$, then $du = 2\ dt$ and

$$\int \frac{1}{2t + 3}\ 2\ dt = \int \frac{1}{u}\ du = \ln|u| + C \quad \text{[using formula \underline{4}(c)]}$$

$$= \ln|2t + 3| + C$$

Check: $\frac{d}{dt}[\ln|2t + 3| + C] = \frac{1}{2t + 3}(2)$

7. $\int (3x - 2)^7 dx$

Let $u = 3x - 2$, then $du = 3\ dx$ and

$$\int (3x - 2)^7 dx = \int (3x - 2)^7 \frac{3}{3} dx = \frac{1}{3}\int (3x - 2)^7 3\ dx = \frac{1}{3}\int u^7 du$$

$$= \frac{1}{3} \cdot \frac{u^8}{8} + C = \frac{(3x - 2)^8}{24} + C$$

Check: $\frac{d}{dx}\left[\frac{(3x - 2)^8}{24} + C\right] = \frac{1}{24}(8)(3x - 2)^7(3) = (3x - 2)^7$

9. Let $u = x^2 + 3$, then $du = 2x\ dx$.

$$\int (x^2 + 3)^7 x\ dx = \int (x^2 + 3)^7 \frac{2}{2} x\ dx = \frac{1}{2}\int (x^2 + 3)^7 2x\ dx$$

$$= \frac{1}{2}\int u^7 du = \frac{1}{2} \cdot \frac{u^8}{8} + C = \frac{u^8}{16} + C = \frac{(x^2 + 3)^8}{16} + C$$

Check: $\dfrac{d}{dx}\left[\dfrac{(x^2+3)^8}{16}+C\right] = \dfrac{1}{16}(8)(x^2+3)^7(2x) = (x^2+3)^7 x$

11. Let $u = -0.5t$, then $du = -0.5\ dt$.

$$\int 10e^{-0.5t}dt = 10\int e^{-0.5t}dt = 10\int e^{-0.5t}\ \dfrac{(-0.5)}{-0.5}\ dt$$

$$= \dfrac{10}{-0.5}\int e^u du = -20e^u + C = -20e^{-0.5t} + C$$

Check: $\dfrac{d}{dt}[-20e^{-0.5t}+C] = -20e^{-0.5t}(-0.5) = 10e^{-0.5t}$

13. Let $u = 10x + 7$, then $du = 10\ dx$.

$$\int \dfrac{1}{10x+7}dx = \int \dfrac{1}{10x+7}\cdot \dfrac{10}{10}dx = \dfrac{1}{10}\int \dfrac{1}{10x+7}10\ dx$$

$$= \dfrac{1}{10}\int \dfrac{1}{u}du = \dfrac{1}{10}\ln|u| + C = \dfrac{1}{10}\ln|10x+7| + C$$

Check: $\dfrac{d}{dx}\left[\dfrac{1}{10}\ln|10x+7| + C\right] = \left(\dfrac{1}{10}\right)\dfrac{1}{10x+7}(10) = \dfrac{1}{10x+7}$

15. Let $u = 2x^2$, then $du = 4x\ dx$.

$$\int xe^{2x^2}dx = \int e^{2x^2}x\cdot \dfrac{4}{4}dx = \dfrac{1}{4}\int e^{2x^2}(4x)dx$$

$$= \dfrac{1}{4}\int e^u du = \dfrac{1}{4}e^u + C = \dfrac{1}{4}e^{2x^2} + C$$

Check: $\dfrac{d}{dx}\left[\dfrac{1}{4}e^{2x^2} + C\right] = \dfrac{1}{4}e^{2x^2}(4x) = xe^{2x^2}$

17. Let $u = x^3 + 4$, then $du = 3x^2 dx$.

$$\int \dfrac{x^2}{x^3+4}dx = \int \dfrac{1}{x^3+4}\cdot \dfrac{3}{3}x^2 dx = \dfrac{1}{3}\int \dfrac{1}{x^3+4}(3x^2)dx$$

$$= \dfrac{1}{3}\int \dfrac{1}{u}du = \dfrac{1}{3}\ln|u| + C = \dfrac{1}{3}\ln|x^3+4| + C$$

Check: $\dfrac{d}{dx}\left[\dfrac{1}{3}\ln|x^3+4| + C\right] = \left(\dfrac{1}{3}\right)\dfrac{1}{x^3+4}(3x^2) = \dfrac{x^2}{x^3+4}$

19. Let $u = 3t^2 + 1$, then $du = 6t\,dt$.

$$\int \frac{t}{(3t^2 + 1)^4}\,dt = \int (3t^2 + 1)^{-4}t\,dt = \int (3t^2 + 1)^{-4}\frac{6}{6}t\,dt$$

$$= \frac{1}{6}\int (3t^2 + 1)^{-4}6t\,dt = \frac{1}{6}\int u^{-4}\,du$$

$$= \frac{1}{6}\cdot\frac{u^{-3}}{-3} + C = \frac{-1}{18}(3t^2 + 1)^{-3} + C$$

Check: $\dfrac{d}{dt}\left[\dfrac{-1}{18}(3t^2 + 1)^{-3} + C\right] = \left(\dfrac{-1}{18}\right)(-3)(3t^2 + 1)^{-4}(6t) = \dfrac{t}{(3t^2 + 1)^4}$

21. Let $u = 4 - x^3$, then $du = -3x^2\,dx$.

$$\int \frac{x^2}{(4 - x^3)^2}\,dx = \int (4 - x^3)^{-2}x^2\,dx = \int (4 - x^3)^{-2}\left(\frac{-3}{-3}\right)x^2\,dx$$

$$= \frac{-1}{3}\int (4 - x^3)^{-2}(-3x^2)\,dx = \frac{-1}{3}\int u^{-2}\,du = \frac{-1}{3}\cdot\frac{u^{-1}}{-1} + C$$

$$= \frac{1}{3}(4 - x^3)^{-1} + C$$

Check: $\dfrac{d}{dx}\left[\dfrac{1}{3}(4 - x^3)^{-1} + C\right] = \dfrac{1}{3}(-1)(4 - x^3)^{-2}(-3x^2) = \dfrac{x^2}{(4 - x^3)^2}$

23. $\displaystyle\int x\sqrt{x + 4}\,dx$

Let $u = x + 4$, then $du = dx$ and $x = u - 4$.

$$\int x\sqrt{x + 4}\,dx = \int (u - 4)u^{1/2}\,du = \int (u^{3/2} - 4u^{1/2})\,du$$

$$= \frac{u^{5/2}}{\frac{5}{2}} - \frac{4u^{3/2}}{\frac{3}{2}} + C = \frac{2}{5}u^{5/2} - \frac{8}{3}u^{3/2} + C$$

$$= \frac{2}{5}(x + 4)^{5/2} - \frac{8}{3}(x + 4)^{3/2} + C \quad \text{(since } u = x + 4)$$

Check: $\dfrac{d}{dx}\left[\dfrac{2}{5}(x + 4)^{5/2} - \dfrac{8}{3}(x + 4)^{3/2} + C\right]$

$$= \frac{2}{5}\left(\frac{5}{2}\right)(x + 4)^{3/2}(1) - \frac{8}{3}\left(\frac{3}{2}\right)(x + 4)^{1/2}(1)$$

$$= (x + 4)^{3/2} - 4(x + 4)^{1/2} = (x + 4)^{1/2}[(x + 4) - 4] = x\sqrt{x + 4}$$

25. $\displaystyle\int \frac{x}{\sqrt{x-3}}\,dx$

Let $u = x - 3$, then $du = dx$ and $x = u + 3$.

$$\int \frac{x}{\sqrt{x-3}}\,dx = \int \frac{u+3}{u^{1/2}}\,du = \int (u^{1/2} + 3u^{-1/2})\,du = \frac{u^{3/2}}{3/2} + \frac{3u^{1/2}}{1/2} + C$$

$$= \frac{2}{3}u^{3/2} + 6u^{1/2} + C = \frac{2}{3}(x-3)^{3/2} + 6(x-3)^{1/2} + C$$

(since $u = x - 3$)

Check: $\displaystyle\frac{d}{dx}\left[\frac{2}{3}(x-3)^{3/2} + 6(x-3)^{1/2} + C\right]$

$$= \frac{2}{3}\left(\frac{3}{2}\right)(x-3)^{1/2}(1) + 6\left(\frac{1}{2}\right)(x-3)^{-1/2}(1)$$

$$= (x-3)^{1/2} + \frac{3}{(x-3)^{1/2}} = \frac{x-3+3}{(x-3)^{1/2}} = \frac{x}{\sqrt{x-3}}$$

27. $\displaystyle\int x(x-4)^9\,dx$

Let $u = x - 4$, then $du = dx$ and $x = u + 4$.

$$\int x(x-4)^9\,dx = \int (u+4)u^9\,du = \int (u^{10} + 4u^9)\,du$$

$$= \frac{u^{11}}{11} + \frac{4u^{10}}{10} + C = \frac{(x-4)^{11}}{11} + \frac{2}{5}(x-4)^{10} + C$$

Check: $\displaystyle\frac{d}{dx}\left[\frac{(x-4)^{11}}{11} + \frac{2}{5}(x-4)^{10} + C\right]$

$$= \frac{1}{11}(11)(x-4)^{10}(1) + \frac{2}{5}(10)(x-4)^9(1)$$

$$= (x-4)^9[(x-4) + 4] = x(x-4)^9$$

29. Let $u = 1 + e^{2x}$, then $du = 2e^{2x}\,dx$.

$$\int e^{2x}(1 + e^{2x})^3\,dx = \int (1 + e^{2x})^3 \frac{2}{2}e^{2x}\,dx = \frac{1}{2}\int (1 + e^{2x})^3 2e^{2x}\,dx$$

$$= \frac{1}{2}\int u^3\,du = \frac{1}{2}\cdot\frac{u^4}{4} + C = \frac{1}{8}(1 + e^{2x})^4 + C$$

Check: $\displaystyle\frac{d}{dx}\left[\frac{1}{8}(1 + e^{2x})^4 + C\right] = \left(\frac{1}{8}\right)(4)(1 + e^{2x})^3 e^{2x}(2) = e^{2x}(1 + e^{2x})^3$

31. Let $u = 4 + 2x + x^2$, then $du = (2 + 2x)dx = 2(1 + x)dx$.

$$\int \frac{1 + x}{4 + 2x + x^2}\,dx = \int \frac{1}{4 + 2x + x^2} \cdot \frac{2(1 + x)}{2}\,dx = \frac{1}{2}\int \frac{1}{4 + 2x + x^2}2(1 + x)\,dx$$

$$= \frac{1}{2}\int \frac{1}{u}\,du = \frac{1}{2}\ln|u| + C = \frac{1}{2}\ln|4 + 2x + x^2| + C$$

Check: $\dfrac{d}{dx}\left[\dfrac{1}{2}\ln|4 + 2x + x^2| + C\right] = \left(\dfrac{1}{2}\right)\dfrac{1}{4 + 2x + x^2}(2 + 2x) = \dfrac{1 + x}{4 + 2x + x^2}$

33. Let $u = x^2 + x + 1$, then $du = (2x + 1)dx$.

$$\int (2x + 1)e^{x^2+x+1}dx = \int e^u du = e^u + C = e^{x^2+x+1} + C$$

Check: $\dfrac{d}{dx}[e^{x^2+x+1} + C] = e^{x^2+x+1}(2x + 1)$

35. Let $u = e^x - 2x$, then $du = (e^x - 2)dx$.

$$\int (e^x - 2x)^3(e^x - 2)\,dx = \int u^3 du = \frac{u^4}{4} + C = \frac{(e^x - 2x)^4}{4} + C$$

Check: $\dfrac{d}{dx}\left[\dfrac{(e^x - 2x)^4}{4} + C\right] = \dfrac{1}{4}(4)(e^x - 2x)^3(e^x - 2) = (e^x - 2x)^3(e^x - 2)$

37. Let $u = x^4 + 2x^2 + 1$, then $du = (4x^3 + 4x)dx = 4(x^3 + x)dx$.

$$\int \frac{x^3 + x}{(x^4 + 2x^2 + 1)^4}\,dx = \int (x^4 + 2x^2 + 1)^{-4}\frac{4}{4}(x^3 + x)\,dx$$

$$= \frac{1}{4}\int (x^4 + 2x^2 + 1)^{-4}4(x^3 + x)\,dx$$

$$= \frac{1}{4}\int u^{-4}du = \frac{1}{4}\cdot\frac{u^{-3}}{-3} + C = \frac{-u^{-3}}{12} + C$$

$$= \frac{-(x^4 + 2x^2 + 1)^{-3}}{12} + C$$

Check: $\dfrac{d}{dx}\left[-\dfrac{1}{12}(x^4 + 2x^2 + 1)^{-3} + C\right] = \left(-\dfrac{1}{12}\right)(-3)(x^4 + 2x^2 + 1)^{-4}(4x^3 + 4x)$

$$= (x^4 + 2x^2 + 1)^{-4}(x^3 + x)$$

39. (A) Differentiate $F(x) = \ln|2x - 3| + C$ to see if you get the integrand
$$f(x) = \frac{1}{2x - 3}$$

(B) Wrong; $\dfrac{d}{dx}[\ln|2x - 3| + C] = \dfrac{1}{2x - 3}(2) = \dfrac{2}{2x - 3} \neq \dfrac{1}{2x - 3}$

(C) Let $u = 2x - 3$, then $du = 2\,dx$

$$\int \frac{1}{2x-3}\,dx = \int \frac{1}{2x-3}\cdot\frac{2}{2}\,dx = \frac{1}{2}\int \frac{1}{2x-3}\,2\,dx$$

$$= \frac{1}{2}\int \frac{1}{u}\,du$$

$$= \frac{1}{2}\ln|u| + C$$

$$= \frac{1}{2}\ln|2x-3| + C$$

Check: $\dfrac{d}{dx}\left[\dfrac{1}{2}\ln|2x-3| + C\right] = \dfrac{1}{2}\cdot\dfrac{1}{2x-3}\cdot 2 = \dfrac{1}{2x-3}$

41. (A) Differentiate $F(x) = e^{x^4} + C$ to see if you get the integrand $f(x) = x^3 e^{x^4}$.

(B) Wrong; $\dfrac{d}{dx}[e^{x^4} + c] = e^{x^4}(4x^3) = 4x^3 e^{x^4} \neq x^3 e^{x^4}$

(C) Let $u = x^4$, then $du = 4x^3\,dx$

$$\int x^3 e^{x^4}\,dx = \int \frac{4}{4}x^3 e^{x^4}\,dx = \frac{1}{4}\int 4\,x^3 e^{x^4}\,dx$$

$$= \frac{1}{4}\int e^u\,du$$

$$= \frac{1}{4}e^u + C$$

$$= \frac{1}{4}e^{x^4} + C$$

Check: $\dfrac{d}{dx}\left[\dfrac{1}{4}e^{x^4} + C\right] = \dfrac{1}{4}e^{x^4}(4x^3) = x^3 e^{x^4}$

43. (A) Differentiate $F(x) = \dfrac{(x^2-2)^2}{3x} + C$ to see if you get the integrand $f(x) = 2(x^2-2)^2$

(B) Wrong; $\dfrac{d}{dx}\left[\dfrac{(x^2-2)^2}{3x} + C\right] = \dfrac{3x\cdot 2(x^2-2)(2x) - (x^2-2)^2\cdot 3}{9x^2}$

$$= \frac{(x^2-2)[9x^2 + 6]}{9x^2} = \frac{9x^4 - 12x^2 - 12}{9x^2}$$

$$= \frac{3x^4 - 4x^2 - 4}{3x^2}$$

$$\neq 2(x^2-2)^2$$

(C) $\displaystyle\int 2(x^2-2)^2\,dx = 2\int (x^4 - 4x^2 + 4)\,dx$

$$= 2\cdot\left[\frac{1}{5}x^5 - \frac{4}{3}x^3 + 4x\right] + C$$

$$= \frac{2}{5}x^5 - \frac{8}{3}x^3 + 8x + C$$

Check: $\dfrac{d}{dx}\left[\dfrac{2}{5}x^5 - \dfrac{8}{3}x^3 + 8x + C\right] = 2x^4 - 8x^2 + 8 = 2[x^4 - 4x^2 + 4]$

$$= 2(x^2 - 2)^2$$

45. Let $u = 3x^2 + 7$, then $du = 6x\ dx$.

$$\int x\sqrt{3x^2 + 7}\ dx = \int (3x^2 + 7)^{1/2}x\ dx = \int (3x^2 + 7)^{1/2}\dfrac{6}{6}x\ dx$$

$$= \dfrac{1}{6}\int u^{1/2}du = \dfrac{1}{6}\cdot\dfrac{u^{3/2}}{\frac{3}{2}} + C = \dfrac{1}{9}(3x^2 + 7)^{3/2} + C$$

Check: $\dfrac{d}{dx}\left[\dfrac{1}{9}(3x^2 + 7)^{3/2} + C\right] = \dfrac{1}{9}\left(\dfrac{3}{2}\right)(3x^2 + 7)^{1/2}(6x) = x(3x^2 + 7)^{1/2}$

47. $\displaystyle\int x(x^3 + 2)^2 dx = \int x(x^6 + 4x^3 + 4)\,dx = \int (x^7 + 4x^4 + 4x)\,dx$

$$= \dfrac{x^8}{8} + \dfrac{4}{5}x^5 + 2x^2 + C$$

Check: $\dfrac{d}{dx}\left[\dfrac{x^8}{8} + \dfrac{4}{5}x^5 + 2x^2 + C\right] = x^7 + 4x^4 + 4x$

$$= x(x^6 + 4x^3 + 4) = x(x^3 + 2)^2$$

49. $\displaystyle\int x^2(x^3 + 2)^2 dx$

Let $u = x^3 + 2$, then $du = 3x^2 dx$.

$$\int x^2(x^3 + 2)^2 dx = \int (x^3 + 2)^2\dfrac{3x^2}{3}\,dx = \dfrac{1}{3}\int (x^3 + 2)^2 3x^2 dx$$

$$= \dfrac{1}{3}\int u^2 du = \dfrac{1}{3}\cdot\dfrac{u^3}{3} + C = \dfrac{1}{9}u^3 + C = \dfrac{1}{9}(x^3 + 2)^3 + C$$

Check: $\dfrac{d}{dx}\left[\dfrac{1}{9}(x^3 + 2)^3 + C\right] = \dfrac{1}{9}(3)(x^3 + 2)^2(3x^2) = x^2(x^3 + 2)^2$

51. Let $u = 2x^4 + 3$, then $du = 8x^3 dx$.

$$\int \dfrac{x^3}{\sqrt{2x^4 + 3}}\,dx = \int (2x^4 + 3)^{-1/2}x^3 dx = \int (2x^4 + 3)^{-1/2}\dfrac{8}{8}x^3 dx$$

$$= \dfrac{1}{8}\int u^{-1/2}du = \dfrac{1}{8}\cdot\dfrac{u^{1/2}}{\frac{1}{2}} + C = \dfrac{1}{4}(2x^4 + 3)^{1/2} + C$$

Check: $\dfrac{d}{dx}\left[\dfrac{1}{4}(2x^4 + 3)^{1/2} + C\right] = \dfrac{1}{4}\left(\dfrac{1}{2}\right)(2x^4 + 3)^{-1/2}(8x^3) = \dfrac{x^3}{(2x^4 + 3)^{1/2}}$

53. Let $u = \ln x$, then $du = \dfrac{1}{x}dx$.

$$\int \frac{(\ln x)^3}{x}dx = \int u^3 du = \frac{u^4}{4} + C = \frac{(\ln x)^4}{4} + C$$

Check: $\dfrac{d}{dx}\left[\dfrac{(\ln x)^4}{4} + C\right] = \dfrac{1}{4}(4)(\ln x)^3 \cdot \dfrac{1}{x} = \dfrac{(\ln x)^3}{x}$

55. Let $u = \dfrac{-1}{x} = -x^{-1}$, then $du = \dfrac{1}{x^2}dx$.

$$\int \frac{1}{x^2}e^{-1/x}dx = \int e^u du = e^u + C = e^{-1/x} + C$$

Check: $\dfrac{d}{dx}[e^{-1/x} + C] = e^{-1/x}\left(\dfrac{1}{x^2}\right) = \dfrac{1}{x^2}e^{-1/x}$

57. $\dfrac{dx}{dt} = 7t^2(t^3 + 5)^6$

Let $u = t^3 + 5$, then $du = 3t^2 dt$.

$$x = \int 7t^2(t^3 + 5)^6 dt = 7\int t^2(t^3 + 5)^6 dt = 7\int (t^3 + 5)^6 \frac{3}{3}t^2 dt$$

$$= \frac{7}{3}\int u^6 du = \frac{7}{3}\cdot\frac{u^7}{7} + C = \frac{1}{3}(t^3 + 5)^7 + C$$

59. $\dfrac{dy}{dt} = \dfrac{3t}{\sqrt{t^2 - 4}}$

Let $u = t^2 - 4$, then $du = 2t\, dt$.

$$y = \int \frac{3t}{(t^2 - 4)^{1/2}}dt = 3\int (t^2 - 4)^{-1/2}t\, dt = 3\int (t^2 - 4)^{-1/2}\frac{2}{2}t\, dt$$

$$= \frac{3}{2}\int u^{-1/2}du = \frac{3}{2}\cdot\frac{u^{1/2}}{\frac{1}{2}} + C = 3(t^2 - 4)^{1/2} + C$$

61. $\dfrac{dp}{dx} = \dfrac{e^x + e^{-x}}{(e^x - e^{-x})^2}$

Let $u = e^x - e^{-x}$, then $du = (e^x + e^{-x})dx$.

$$p = \int \frac{e^x + e^{-x}}{(e^x - e^{-x})^2}dx = \int (e^x - e^{-x})^{-2}(e^x + e^{-x})dx = \int u^{-2}du$$

$$= \frac{u^{-1}}{-1} + C = -(e^x - e^{-x})^{-1} + C$$

63. Let $v = au$, then $dv = a\ du$.

$$\int e^{au}du = \int e^{au}\,\frac{a}{a}\,du = \frac{1}{a}\int e^{au}a\ du = \frac{1}{a}\int e^v dv = \frac{1}{a}e^v + C = \frac{1}{a}e^{au} + C$$

Check: $\dfrac{d}{du}\left[\dfrac{1}{a}e^{au} + C\right] + \dfrac{1}{a}e^{au}(a) = e^{au}$

65. $p'(x) = \dfrac{-6000}{(3x + 50)^2}$

Let $u = 3x + 50$, then $du = 3\ dx$.

$$p(x) = \int \frac{-6000}{(3x + 50)^2}\,dx = -6000\int (3x + 50)^{-2}dx = -6000\int (3x + 50)^{-2}\,\frac{3}{3}\,dx$$

$$= -2000\int u^{-2}du = -2000\cdot\frac{u^{-1}}{-1} + C = \frac{2000}{3x + 50} + C$$

Given $p(150) = 4$:

$$4 = \frac{2000}{(3\cdot 150 + 50)} + C$$

$$4 = \frac{2000}{500} + C$$

$$C = 0$$

Thus, $p(x) = \dfrac{2000}{3x + 50}$.

Now, $2.50 = \dfrac{2000}{3x + 50}$

$2.50(3x + 50) = 2000$

$7.5x + 125 = 2000$

$7.5x = 1875$

$x = 250$

Thus, the demand is 250 bottles when the price is \$2.50.

67. $C'(x) = 12 + \dfrac{500}{x + 1}$, $x > 0$

$$C(x) = \int\left(12 + \frac{500}{x + 1}\right)dx = \int 12\ dx + 500\int\frac{1}{x + 1}dx \quad (u = x + 1,\ du = dx)$$

$$= 12x + 500\ \ln(x + 1) + C$$

Now, $C(0) = 2000$. Thus, $C(x) = 12x + 500\ \ln(x + 1) + 2000$. The average cost is:

$$\overline{C}(x) = 12 + \frac{500}{x}\ln(x + 1) + \frac{2000}{x}$$

and

$$\overline{C}(1000) = 12 + \frac{500}{1000}\ln(1001) + \frac{2000}{1000} = 12 + \frac{1}{2}\ln(1001) + 2$$

$$\approx 17.45 \text{ or } \$17.45 \text{ per pair of shoes}$$

69. $S'(t) = 10 - 10e^{-0.1t}$, $0 \le t \le 24$

(A) $S(t) = \int (10 - 10e^{-0.1t})\,dt = \int 10\,dt - 10\int e^{-0.1t}\,dt$

$\qquad = 10t - \dfrac{10}{-0.1}e^{-0.1t} + C = 10t + 100e^{-0.1t} + C$

Given $S(0) = 0$: $\quad 0 + 100e^0 + C = 0$

$\qquad\qquad\qquad\qquad 100 + C = 0$

$\qquad\qquad\qquad\qquad\qquad C = -100$

Total sales at time t:

$\qquad S(t) = 10t + 100e^{-0.1t} - 100$

(B) $S(12) = 10(12) + 100e^{-0.1(12)} - 100$

$\qquad\quad = 20 + 100e^{-1.2} \approx 50$

Total estimated sales for the first twelve months: $50 million.

(C) On a graphing utility, solve

$\qquad\quad 10t + 100e^{-0.1t} - 100 = 100$

or $\qquad\quad 10t + 100e^{-0.1t} = 200$

The result is: $t \approx 18.41$ months.

71. $Q(t) = \int R(t)\,dt = \int \left(\dfrac{100}{t+1} + 5\right)dt = 100\int \dfrac{1}{t+1}\,dt + \int 5\,dt$

$\qquad\qquad\qquad\qquad\qquad\qquad = 100\ln(t+1) + 5t + C$

Given $Q(0) = 0$:

$0 = 100\ln(1) + 0 + C$

Thus, $C = 0$ and $Q(t) = 100\ln(t+1) + 5t$, $0 \le t \le 20$.

$Q(9) = 100\ln(9+1) + 5(9) = 100\ln 10 + 45 \approx 275$ thousand barrels.

73. $W(t) = \int W'(t)\,dt = \int 0.2e^{0.1t}\,dt = \dfrac{0.2}{0.1}\int e^{0.1t}(0.1)\,dt = 2e^{0.1t} + C$

Given $W(0) = 2$:

$2 = 2e^0 + C$.

Thus, $C = 0$ and $W(t) = 2e^{0.1t}$.

The weight of the culture after 8 hours is given by:

$W(8) = 2e^{0.1(8)} = 2e^{0.8} \approx 4.45$ grams.

75. $\dfrac{dN}{dt} = \dfrac{-2000t}{1+t^2}$, $0 \le t \le 10$

Let $u = 1 + t^2$, then $du = 2t\,dt$.

$N = \int \dfrac{-2000t}{1+t^2}\,dt = \dfrac{-2000}{2}\int \dfrac{2t}{1+t^2}\,dt = -1000\int \dfrac{1}{u}\,du$

$\qquad = -1000\ln|u| + C = -1000\ln(1+t^2) + C$

Given $N(0) = 5000$:
$5000 = -1000 \ln(1) + C$
Hence, $C = 5000$ and $N(t) = 5000 - 1000 \ln(1 + t^2)$, $0 \leq t \leq 10$.
Now, $N(10) = 5000 - 1000 \ln(1 + 10^2)$
$\qquad\qquad = 5000 - 1000 \ln(101)$
$\qquad\qquad \approx 385$ bacteria per milliliter.

77. $N'(t) = 6e^{-0.1t}$, $0 \leq t \leq 15$

$N(t) = \int N'(t)\,dt = \int 6e^{-0.1t}\,dt = 6\int e^{-0.1t}\,dt$

$\qquad\qquad = \dfrac{6}{-0.1}\int e^{-0.1t}(-0.1)\,dt = -60e^{-0.1t} + C$

Given $N(0) = 40$:
$40 = -60e^0 + C$
Hence, $C = 100$ and $N(t) = 100 - 60e^{-0.1t}$, $0 \leq t \leq 15$.
The number of words per minute after completing the course is:
$N(15) = 100 - 60e^{-0.1(15)} = 100 - 60e^{-1.5} \approx 87$ words per minute.

79. $\dfrac{dE}{dt} = 5000(t + 1)^{-3/2}$, $t \geq 0$

Let $u = t + 1$, then $du = dt$

$E = \int 5000(t + 1)^{-3/2}\,dt = 5000\int (t + 1)^{-3/2}\,dt = 5000\int u^{-3/2}\,du$

$\qquad\qquad = 5000\dfrac{u^{-1/2}}{-\dfrac{1}{2}} + C = -10{,}000(t + 1)^{-1/2} + C$

$\qquad\qquad = \dfrac{-10{,}000}{\sqrt{t + 1}} + C$

Given $E(0) = 2000$:
$2000 = \dfrac{-10{,}000}{\sqrt{1}} + C$

Hence, $C = 12{,}000$ and $E(t) = 12{,}000 - \dfrac{10{,}000}{\sqrt{t + 1}}$.
The projected enrollment 15 years from now is:
$E(15) = 12{,}000 - \dfrac{10{,}000}{\sqrt{15 + 1}} = 12{,}000 - \dfrac{10{,}000}{\sqrt{16}} = 12{,}000 - \dfrac{10{,}000}{4}$

$\qquad\qquad\qquad\qquad = 9500$ students

EXERCISE 11-3

Things to remember:

1. A DIFFERENTIAL EQUATION is an equation that involves an unknown function and one or more of its derivatives. The ORDER of a differential equation is the order of the highest derivative of the unknown function.

2. A SLOPE FIELD for a first-order differential equation is obtained by drawing tangent line segments determined by the equation at each point in a grid.

3. EXPONENTIAL GROWTH LAW

If $\dfrac{dQ}{dt} = rQ$ and $Q(0) = Q_0$, then $Q(t) = Q_0 e^{rt}$, where

Q_0 = Amount at $t = 0$

r = Continuous compound growth rate (expressed as a decimal)

t = Time

Q = Quantity at time t

4. COMPARISON OF EXPONENTIAL GROWTH PHENOMENA

DESCRIPTION	MODEL	SOLUTION	GRAPH	USES
Unlimited growth: Rate of growth is proportional to the amount present	$\dfrac{dy}{dt} = ky$ $k,\ t > 0$ $y(0) = c$	$y = ce^{kt}$		• Short-term population growth (people, bacteria, etc.) • Growth of money at continuous compound interest • Price-supply curves • Depletion of natural resources
Exponential decay: Rate of growth is proportional to the amount present	$\dfrac{dy}{dt} = -ky$ $k,\ t > 0$ $y(0) = c$	$y = ce^{-kt}$		• Radioactive decay • Light absorption in water • Price-demand curves • Atmospheric pressure (t is altitude)
Limited growth: Rate of growth is proportional to the difference between the amount present and a fixed limit	$\dfrac{dy}{dt} = k(M - y)$ $k,\ t > 0$ $y(0) = 0$	$y = M(1 - e^{-kt})$		• Sales fads (e.g., skateboards) • Depreciation of equipment • Company growth • Learning
Logistic growth: Rate of growth is proportional to the amount present and to the difference between the amount present and a fixed limit	$\dfrac{dy}{dt} = ky(M - y)$ $k,\ t > 0$ $y(0) = \dfrac{M}{1 + c}$	$y = \dfrac{M}{1 - ce^{-kMt}}$		• Long-term population growth • Epidemics • Sales of new products • Rumor spread • Company growth

1. $\dfrac{dy}{dx} = e^{0.5x}; \quad \displaystyle\int \dfrac{dy}{dx}\, dx = \int e^{0.5x}\, dx$

$$y = \dfrac{e^{0.5x}}{0.5} + C$$

General solution: $y = 2e^{0.5x} + C$

3. $\dfrac{dy}{dx} = x^2 - x;\ \ y(0) = 0$

$$\int \dfrac{dy}{dx}\,dx = \int (x^2 - x)\,dx$$

$$y = \dfrac{1}{3}x^3 - \dfrac{1}{2}x^2 + C$$

Given $y(0) = 0$: $\dfrac{1}{3}(0)^3 - \dfrac{1}{2}(0)^2 + C = 0$

$$C = 0$$

Particular solution: $y = \dfrac{1}{3}x^3 - \dfrac{1}{2}x^2$

5. $\dfrac{dy}{dx} = -2xe^{-x^2};\ \ y(0) = 3$

$$\int \dfrac{dy}{dx}\,dx = \int -2xe^{-x^2}\,dx$$

$$y = \int -2xe^{-x^2}\,dx$$

Let $u = -x^2$, then $du = -2x\,dx$ and

$$\int -2xe^{-x^2}\,dx = \int e^u\,du = e^u + c = e^{-x^2} + c$$

Thus, $\qquad y = e^{-x^2} + c$

Given $y(0) = 3$: $\ 3 = e^0 + c$

$$3 = 1 + c$$
$$c = 2$$

Particular solution: $y = e^{-x^2} + 2$

7. Figure (b). When $x = 1$, $\dfrac{dy}{dx} = 1 - 1 = 0$ for any y. When $x = 0$,
$\dfrac{dy}{dx} = 0 - 1 = -1$ for any y. When $x = 2$, $\dfrac{dy}{dx} = 2 - 1 = 1$ for any y; and so
on. These facts are consistent with the slope-field in Figure (b); they
are not consistent with the slope-field in Figure (a).

9. $\dfrac{dy}{dx} = x - 1$

$$\int \dfrac{dy}{dx}\,dx = \int (x - 1)\,dx$$

General solution: $y = \dfrac{1}{2}x^2 - x + c$

Given $y(0) = -2$: $\dfrac{1}{2}(0)^2 - 0 + c = -2$

$$c = -2$$

Particular solution: $y = \dfrac{1}{2}x^2 - x - 2$

11.

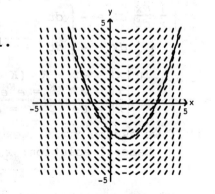

13. $\dfrac{dy}{dx} = -0.8y$

$$\int \frac{1}{y}\frac{dy}{dx}\,dx = \int -0.8\,dx$$
$$\int \frac{1}{y}\,dy = \int -0.8\,dx$$

$\ln|y| = -0.8x + K$ (K an arbitrary constant)
$\quad |y| = e^{-0.8x+K} = e^K e^{-0.8x}$
$\quad |y| = Ce^{-0.8x}$ where $C = e^K$

If we assume $y > 0$, we get:
    General solution: $y = Ce^{-0.8x}$

Note: The differential equation $\dfrac{dy}{dx} = -0.8x$ is the model for exponential decay with decay rate $r = 0.8$ (see $\underline{3}$). Thus,
$\quad y = Ce^{-0.8x}$

15. $\dfrac{dy}{dx} = 0.07y,\ y(0) = 1{,}000$

$$\int \frac{1}{y}\frac{dy}{dx}\,dx = \int 0.07\,dx$$
$$\int \frac{1}{y}\,dy = \int 0.07\,dx$$

$\ln|y| = 0.07x + K$ (K an arbitrary constant)
$\quad |y| = e^{0.07x+K} = e^K e^{0.07x}$
$\quad |y| = Ce^{0.07x}$ $(C = e^K)$

If we assume $y > 0$, we get
    General solution: $y = Ce^{0.07x}$

Given $y(0) = 1{,}000$: $1000 = Ce^0$
$$C = 1000$$
    Particular solution: $y = 1{,}000e^{0.07x}$

17. $\dfrac{dx}{dt} = -x$

$$\int \frac{1}{x}\frac{dx}{dt}\,dt = -\int dt$$
$$\int \frac{1}{x}\,dx = -\int dt$$

$\ln|x| = -t + K$ (K an arbitrary constant)
$\quad |x| = e^{-t+K} = e^K e^{-t}$
$\quad |x| = Ce^{-t},\ (C = e^K)$

If we assume $x > 0$, we get
    General solution: $x = Ce^{-t}$

19. Figure (c). When $y = 1$, $\frac{dy}{dx} = 1 - 1 = 0$ for any x.

When $y = 2$, $\frac{dy}{dx} = 1 - 2 = -1$ for any x; and so on. This is consistent with the slope-field in Figure (c); it is not consistent with the slope-field in Figure (d).

21. $y = 1 - Ce^{-x}$

$\frac{dy}{dx} = \frac{d}{dx}[1 - Ce^{-x}] = Ce^{-x}$

From the original equation,

$\quad Ce^{-x} = 1 - y$

Thus, we have

$\quad \frac{dy}{dx} = 1 - y$

and $y = 1 - Ce^{-x}$ is a solution of the differential equation for any number c.

Given $y(0) = 0$: $0 = 1 - Ce^{0} = 1 - c$

$\qquad\qquad\qquad c = 1$

Particular solution: $y = 1 - e^{-x}$

23.

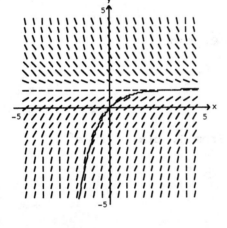

25.

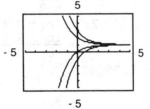

27. $y = 1{,}000e^{0.08t}$

$0 \le t \le 15$, $0 \le y \le 3{,}500$

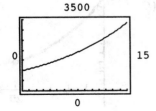

29. $p = 100e^{-0.05x}$

$0 \le x \le 30$, $0 \le p \le 100$

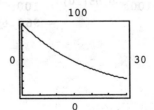

31. $N = 100(1 - e^{-0.05t})$
$0 \le t \le 100, \ 0 \le N \le 100$

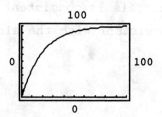

33. $N = \dfrac{1,000}{1 + 999e^{-0.4t}}$
$0 \le t \le 40, \ 0 \le N \le 1,000$

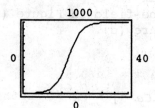

35. $\dfrac{dA}{dt} = 0.08A, \ A(0) = 1,000.$

This is an unlimited growth model. From $\underline{4}$, $A(t) = 1,000e^{0.08t}$.

37. $\dfrac{dA}{dt} = rA, \ A(0) = 8,000$

is an unlimited growth model. From $\underline{4}$, $A(t) = 8,000e^{rt}$.
Since $A(2) = 9,020$, we solve $8,000e^{2r} = 9,020$ for r.

$$8000e^{2r} = 9,020$$
$$e^{2r} = \frac{902}{800}$$
$$2r = \ln(902/800)$$
$$r = \frac{\ln(902/800)}{2} \approx 0.06$$

Thus, $A(t) = 8,000e^{0.06t}$.

39. (A) $\dfrac{dp}{dx} = rp, \ p(0) = 100$

This is an UNLIMITED GROWTH MODEL. From $\underline{4}$, $p(x) = 100e^{rx}$.
Since $p(5) = 77.88$, we have
$$77.88 = 100e^{5r}$$
$$e^{5r} = 0.7788$$
$$5r = \ln(0.7788)$$
$$r = \frac{\ln(0.7788)}{5} \approx -0.05$$

Thus, $p(x) = 100e^{-0.05x}$.

(B) $p(10) = 100e^{-0.05(10)} = 100e^{-0.5}$
$\approx \$60.65$ per unit

(C)

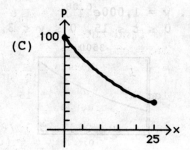

41. (A) $\dfrac{dN}{dt} = k(L - N)$; $N(0) = 0$

This is a LIMITED GROWTH MODEL. From $\underline{4}$, $N(t) = L(1 - e^{-kt})$.

Since $N(10) = 0.4L$, we have

$$0.4L = L(1 - e^{-10k})$$
$$1 - e^{-10k} = 0.4$$
$$e^{-10k} = 0.6$$
$$-10k = \ln(0.6)$$
$$k = \frac{\ln(0.6)}{-10} \approx 0.051$$

Thus, $N(t) = L(1 - e^{-0.051t})$.

(B) $N(5) = L[1 - e^{-0.051(5)}] = L[1 - e^{-0.255}] \approx 0.225L$

Approximately 22.5% of the possible viewers will have been exposed after 5 days.

(C) Solve $L(1 - e^{-0.051t}) = 0.8L$ for t:

$$1 - e^{-0.051t} = 0.8$$
$$e^{-0.051t} = 0.2$$
$$-0.051t = \ln(0.2)$$
$$t = \frac{\ln(0.2)}{-0.051} \approx 31.56$$

It will take 32 days for 80% of the possible viewers to be exposed.

(D)

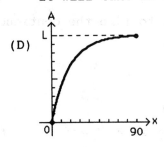

43. $\dfrac{dI}{dx} = -kI$, $I(0) = I_0$

This is an exponential decay model. From $\underline{4}$, $I(x) = I_0 e^{-kx}$ with $k = 0.00942$, we have

$$I(x) = I_0 e^{-0.00942x}$$

To find the depth at which the light is reduced to half of that at the surface, solve,

$$I_0 e^{-0.00942x} = \frac{1}{2} I_0$$

for x:

$$e^{-0.00942x} = 0.5$$
$$-0.00942x = \ln(0.5)$$
$$x = \frac{\ln(0.5)}{-0.00942} \approx 74 \text{ feet}$$

45. $\frac{dQ}{dt} = -0.04Q$, $Q(0) = Q_0$.

    (A) This is a model for exponential decay. From 4,
$$Q(t) = Q_0 e^{-0.04t}$$
    With $Q_0 = 3$, we have
$$Q(t) = 3e^{-0.04t}$$

    (B) $Q(10) = 3e^{-0.04(10)} = 3e^{-0.4} \approx 2.01$.
    There are approximately 2.01 milliliters in the body after 10 hours.

    (C)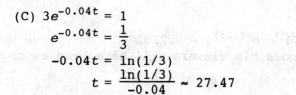

    (D)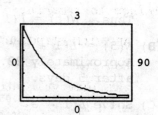

    It will take approximately 27.47 hours
    for Q to decrease to 1 milliliter.

47. Using the exponential decay model, we have $\frac{dy}{dt} = -ky$, $y(0) = 100$, $k > 0$
where $y = y(t)$ is the amount of cesium-137 present at time t. From 4,
$$y(t) = 100e^{-kt}$$
Since $y(3) = 93.3$, we solve $93.3 = 100e^{-3k}$ for k to find the continuous compound decay rate:
$$93.3 = 100e^{-3k}$$
$$e^{-3k} = 0.933$$
$$-3k = \ln(0.933)$$
$$k = \frac{\ln(0.933)}{-3} \approx 0.023117$$

49. From Example 3: $Q = Q_0 e^{-0.0001238t}$

Now, the amount of radioactive carbon-14 present is 5% of the original amount. Thus, $0.05Q_0 = Q_0 e^{-0.0001238t}$ or $e^{-0.0001238t} = 0.05$.

Therefore, $-0.0001238t = \ln(0.05) \approx -2.9957$ and $t \approx 24{,}200$ years.

51. $N(k) = 180e^{-0.11(k-1)}$, $1 \le k \le 10$
Thus, $N(6) = 180e^{-0.11(6-1)} = 180e^{-0.55} \approx 104$ times
and $N(10) = 180e^{-0.11(10-1)} = 180e^{-0.99} \approx 67$ times.

53. **(A)** $x(t) = \dfrac{400}{1 + 399e^{-0.4t}}$

        $x(5) = \dfrac{400}{1 + 399e^{(-0.4)5}} = \dfrac{400}{1 + 399e^{-2}} \approx \dfrac{400}{55} \approx 7$ people

        $x(20) = \dfrac{400}{1 + 399e^{(-0.4)20}} = \dfrac{400}{1 + 399e^{-8}} \approx 353$ people

    (B) $\lim\limits_{t \to \infty} x(t) = 400$.

(C)

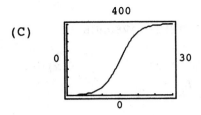

Things to remember:

1. APPROXIMATING AREA UNDER A GRAPH; LEFT SUMS AND RIGHT SUMS

   Let $f(x)$ be defined and positive on the interval $[a, b]$. Divide the interval into n equal subintervals of length $\Delta x = \dfrac{b - a}{n}$, with $x_1, x_2, x_3, \ldots, x_{n-1}$, the points of subdivision.

   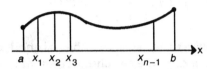

   Then

   $$L_n = f(a)\Delta x + f(x_1)\Delta x + f(x_2)\Delta x + \ldots + f(x_{n-1})\Delta x$$

   is called a LEFT SUM;

   $$R_n = f(x_1)\Delta x + f(x_2)\Delta x + f(x_3)\Delta x + \ldots + f(x_{n-1})\Delta x + f(x_n)\Delta x$$

   is called a RIGHT SUM;

   $$A_n = \frac{L_n + R_n}{2} \text{ is the average of } L_n \text{ and } R_n.$$

   Left and right sums and their averages are used to approximate the area under the graph of f. The exact area under the graph of $y = f(x)$ from $x = a$ to $x = b$ is denoted by the DEFINITE INTEGRAL SYMBOL

   $$\int_a^b f(x)\,dx = \left(\begin{array}{l}\text{area under the graph} \\ \text{from } x = a \text{ to } x = b\end{array}\right)$$

2. MONOTONE FUNCTIONS AND AREA UNDER THE GRAPH

   A function f is MONOTONE over an interval $[a, b]$ if it is either increasing over $[a, b]$ or decreasing over $[a, b]$.

   If f is a monotone function, then the area under the graph of f always lies between the left sum L_n and the right sum R_n for any integer n.

<u>3</u>. ERROR BOUNDS FOR LEFT AND RIGHT SUMS AND THEIR AVERAGE
(MONOTONE FUNCTIONS)

If $f(x)$ is monotonic on the interval $[a, b]$ and $I = \int_a^b f(x)\,dx$,

L_n = left sum, R_n = right sum, and $A_n = \dfrac{L_n + R_n}{2}$

then the following error bounds hold:

$$\left| I - L_n \right| \le \left| f(b) - f(a) \right| \frac{b - a}{n}$$

$$\left| I - R_n \right| \le \left| f(b) - f(a) \right| \frac{b - a}{n}$$

$$\left| I - A_n \right| \le \left| f(b) - f(a) \right| \frac{b - a}{2n}$$

<u>4</u>. DEFINITE INTEGRAL SYMBOL FOR FUNCTIONS WITH NEGATIVE VALUES

If $f(x)$ is positive for some values of x on $[a, b]$ and negative for
others, then the DEFINITE INTEGRAL SYMBOL

$$\int_a^b f(x)\,dx$$

represents the cumulative sum of the signed areas between the curve
$y = f(x)$ and the x axis where the areas above the x axis are counted
positively and the areas below the x axis are counted negatively
(see the figure where A and B are actual areas of the indicated
regions).

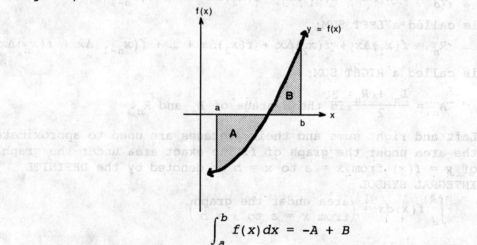

$$\int_a^b f(x)\,dx = -A + B$$

<u>5</u>. RATE, AREA, AND DISTANCE

If $r = r(t)$ is a positive rate function for an object moving on a
line, then

$$\int_a^b r(t)\,dt = \begin{pmatrix} \text{Net distance traveled} \\ \text{from } t = a \text{ to } t = b \end{pmatrix}$$

$$= \begin{pmatrix} \text{Total Change in Position} \\ \text{from } t = a \text{ to } t = b \end{pmatrix}$$

6. RATE, AREA, AND TOTAL CHANGE

If $y = F'(x)$ is a rate function (derivative), then the cumulated sum of the signed areas between the curve $y = F'(x)$ and the x axis from $x = a$ to $x = b$ represents the total net change in $F(x)$ from $x = a$ to $x = b$. Symbolically,

$$\int_a^b F'(x)\,dx = \begin{pmatrix} \text{Total net change in } F(x) \\ \text{from } x = a \text{ to } x = b \end{pmatrix}$$

1.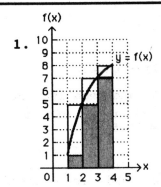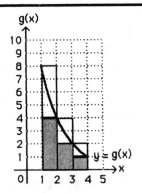

3. For Figure (a):

$$L_3 = f(1) \cdot 1 + f(2) \cdot 1 + f(3) \cdot 1$$
$$= 1 + 5 + 7 = 13$$

$$R_3 = f(2) \cdot 1 + f(3) \cdot 1 + f(4) \cdot 1$$
$$= 5 + 7 + 8 = 20$$

$$A_3 = \frac{L_3 + R_3}{2} = \frac{13 + 20}{2} = \frac{33}{2} = 16.5$$

For Figure (b):

$$L_3 = g(1) \cdot 1 + g(2) \cdot 1 + g(3) \cdot 1$$
$$= 8 + 4 + 2 = 14$$

$$R_3 = g(2) \cdot 1 + g(3) \cdot 1 + g(4) \cdot 1$$
$$= 4 + 2 + 1 = 7$$

$$A_3 = \frac{L_3 + R_3}{2} = \frac{14 + 7}{2} = 10.5$$

5. $L_3 \le \int_1^4 f(x)\,dx \le R_3$, $R_3 \le \int_1^4 g(x)\,dx \le L_3$; since f is increasing on $[1, 4]$, L_3 underestimates the area and R_3 overestimates the area; since g is decreasing on $[1, 4]$, L_3 overestimates the area and R_3 underestimates the area.

7. For Figure (a).
   Error bound for L_3 and R_3:

   $$\text{Error} \leq |f(4) - f(1)| \left(\frac{4-1}{3}\right) = |8 - 1| = 7$$

   Error bound for A_3:

   $$\text{Error} \leq \frac{7}{2} = 3.5$$

   For Figure (b).
   Error bound for L_3 and R_3:

   $$\text{Error} \leq |f(4) - f(1)| \left(\frac{4-1}{3}\right) = |1 - 8| = |-7| = 7$$

   Error bound for A_3:

   $$\text{Error} \leq \frac{7}{2} = 3.5$$

9. The exact area under the graph of $y = f(x)$ is within 3.5 units of the average of the left sum and right sum estimates, $A_3 = 16.5$.

11. $r(t)$ is a decreasing function.

    (A) $L_4 = r(0) \cdot 1 + r(1) \cdot 1 + r(2) \cdot 1 + r(3) \cdot 1$
        $= 128 + 96 + 64 + 32 = 320$

    $R_4 = r(1) \cdot 1 + r(2) \cdot 1 + r(3) \cdot 1 + r(4) \cdot 1$
        $= 96 + 64 + 32 + 0 = 192$

    $$A_4 = \frac{L_4 + R_4}{2} = \frac{320 + 192}{2} = 256$$

    Error bound for L_4 and R_4:

    $$\text{Error} \leq |f(4) - f(0)| \left(\frac{4-0}{4}\right) = 128$$

    Error bound for A_4:

    $$\text{Error} \leq \frac{128}{2} = 64$$

    (B) The height of each rectangle represents an instantaneous rate and the base of each rectangle is a time interval; rate *times* time *equals* distance.

    (C) We want to find n such that $|I - A_n| < 1$, that is:

    $$|f(4) - f(0)| \left(\frac{4-0}{2n}\right) < 1$$

    $$|0 - 128| \left(\frac{2}{n}\right) < 1$$

    $$\frac{256}{n} < 1 \text{ or } n > 256$$

13. $h(x)$ is an increasing function; $\Delta x = 100$

$$L_{10} = h(0)100 + h(100)100 + h(200)100 + \ldots + h(900)(100)$$
$$= [0 + 183 + 235 + 245 + 260 + 286 + 322 + 388 + 453 + 489]100$$
$$= (2,861)100 = 286,100 \text{ sq ft}$$

$$R_{10} = h(100)100 + h(200)100 + h(300)100 + \ldots + h(1,000)100$$
$$= [183 + 235 + 245 + 260 + 286 + 322 + 388 + 453 + 489 + 500]100$$
$$= (3,361)100 = 336,100 \text{ sq ft}$$

$$A_{10} = \frac{L_{10} + R_{10}}{2} = \frac{286,100 + 336,100}{2} = 311,100 \text{ sq ft}$$

Error bound for A_{10}:

$$\text{Error} \leq |h(1,000) - h(0)| \left(\frac{1000 - 0}{2 \cdot 10}\right) = 500(50) = 25,000 \text{ sq ft}$$

We want to find n such that $|I - A_n| \leq 2,500$:

$$|h(1000) - h(0)| \left(\frac{1000 - 0}{2n}\right) \leq 2,500$$

$$500 \left(\frac{500}{n}\right) \leq 2,500$$

$$250,000 \leq 2,500n$$

$$n \geq 100$$

15. $r(t)$ is a decreasing function; $\Delta t = 1$

(A) $L_7 = r(0)1 + r(1)1 + r(2)1 + r(3)1 + r(4)1 + r(5)1 + r(6)1$
$$= 110 + 85 + 63 + 45 + 29 + 16 + 5 = 353$$

$R_7 = r(1)1 + r(2)1 + r(3)1 + r(4)1 + r(5)1 + r(6)1 + r(7)1$
$$= 85 + 63 + 45 + 29 + 16 + 5 + 0 = 243$$

$$A_7 = \frac{L_7 + R_7}{2} = \frac{353 + 243}{2} = 298 \text{ ft}$$

Error bound for A_7:

$$\text{Error} \leq |r(7) - r(0)| \left(\frac{7 - 0}{2 \cdot 7}\right) = |0 - 110| \frac{1}{2} = 55 \text{ ft}$$

(B) We want to find n such that $|I - A_n| \leq 5$, that is:

$$|r(7) - r(0)| \left(\frac{7 - 0}{2n}\right) < 5$$

$$\frac{(110)7}{2n} < 5$$

$$10n > 770$$

$$n > 77$$

17. (A) $P = 2$

(B)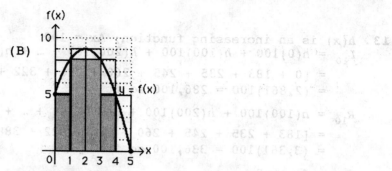

(C) To the left of $P = 2$, the left rectangles underestimate the true area and the right rectangles overestimate the true area. To the right of $P = 2$, the left rectangles overestimate the true area and the right rectangles underestimate the true area.

(D) N_1 = sum of areas of rectangles below the graph of f:

$$= 5 \cdot 1 + 8 \cdot 1 + 8 \cdot 1 + 5 \cdot 1 + 0 \cdot 1 = 26$$

N_2 = sum of areas of rectangles above the graph of f:

$$= 8 \cdot 1 + 9 \cdot 1 + 9 \cdot 1 + 8 \cdot 1 + 5 \cdot 1 = 39$$

Thus, $26 \le \int_0^5 f(x)\,dx \le 39$.

19. $f(x) = 0.25x^2 - 4$ on $[2, 5]$

$L_6 = f(2)\Delta x + f(2.5)\Delta x + f(3)\Delta x + f(3.5)\Delta x + f(4)\Delta x + f(4.5)\Delta x$

where $\Delta x = 0.5$

Thus,

$L_6 = [-3 - 2.44 - 1.75 - 0.94 + 0 + 1.06](0.5) = -3.53$

$R_6 = f(2.5)\Delta x + f(3)\Delta x + f(3.5)\Delta x + f(4)\Delta x + f(4.5)\Delta x + f(5)\Delta x$

where $\Delta x = 0.5$

Thus,

$R_6 = [-2.44 - 1.75 - 0.94 + 0 + 1.06 + 2.25](0.5) = -0.91$

$$A_6 = \frac{L_6 + R_6}{2} = \frac{-3.53 - 0.91}{2} = -2.22$$

Error bound for L_6 and R_6:

$$\text{Error} \le |f(5) - f(2)| \left(\frac{5 - 2}{6}\right) = |2.25 - (-3)|\,(0.5) = 2.63$$

Error bound for A_6:

$$\text{Error} \le \frac{2.63}{2} = 1.32$$

Geometrically, the definite integral over the interval $[2, 5]$ is the area of the region which lies above the x-axis minus the area of the region which lies below the x-axis. From the figure, if R_1 represents the region bounded by the graph of f and the x-axis for $2 \le x \le 4$ and R_2 represents the region bounded by the graph of f and the x-axis for $4 \le x \le 5$, then

$$\int_2^5 f(x)\,dx = \text{area}(R_2) - \text{area}(R_1)$$

21. $\int_1^2 x^x \, dx; \quad f(x) = x^x$

$$|I - R_n| \leq |f(2) - f(1)| \left(\frac{2-1}{n}\right) \leq 0.05$$

$$|4 - 1| \left(\frac{1}{n}\right) \leq 0.05$$

$$\frac{3}{n} \leq 0.05$$

$$n \geq \frac{3}{0.05} = 60$$

23. $\int_0^2 e^{-x^2} \, dx; \quad f(x) = e^{-x^2}$

$$|I - L_n| = |f(2) - f(0)| \left(\frac{2-0}{n}\right) \leq 0.005$$

$$|e^{-4} - e^0| \left(\frac{2}{n}\right) \leq 0.005$$

$$|0.018316 - 1| \left(\frac{2}{n}\right) \leq 0.005$$

$$(0.981684) \frac{2}{n} \leq 0.005$$

$$n \geq \frac{(0.981684)2}{0.005} \approx 393$$

25. Suppose f is monotonic on $[a, b]$. Let $I = \int_a^b f(x) \, dx$. Then

$$|I - L_n| \leq |f(b) - f(a)| \left(\frac{b-a}{n}\right)$$

Now, $\lim_{n \to \infty} |f(b) - f(a)| \left(\frac{b-a}{n}\right) = 0$

Thus, $\lim_{n \to \infty} |I - L_n| = 0$ which implies $I = \lim_{n \to \infty} L_n$

27. Let $a = 300$, $b = 900$, $n = 2$. Then $\Delta x = \frac{900 - 300}{2} = 300$

$$L_2 = f(300)(300) + f(600)(300)$$
$$= [400 + 300]300 = 210{,}000$$

$$R_2 = f(600)300 + f(900)(300)$$
$$= [300 + 200]300 = 150{,}000$$

$$A_2 = \frac{L_2 + R_2}{2} = \frac{210{,}000 + 150{,}000}{2} = \$180{,}000$$

Error bound for A_2:

$$\text{Error} \leq |f(900) - f(300)| \left(\frac{900 - 300}{2 \cdot 2}\right)$$

$$= |200 - 400| \, (150)$$
$$= 200(150) = \$30{,}000$$

29. Let $A'(t) = 800e^{0.08t}$, $a = 2$, $b = 6$, $n = 4$. Then $\Delta t = \dfrac{6 - 2}{4} = 1$.

$L_4 = A(2)1 + A(3)1 + A(4)1 + A(5)1$

$ = 939 + 1017 + 1102 + 1193 = \$4251$

$R_4 = A(3)1 + A(4)1 + A(5)1 + A(6)1$

$ = 1017 + 1102 + 1193 + 1293 = \$4605$

Since $A'(t)$ is increasing on $[2, 6]$,

$$\$4251 \leq \int_0^2 800e^{0.08t}\, dt \leq \$4605$$

31. First 60 days:

$L_3 = N(0)20 + N(20)20 + N(40)20$

$ = (10 + 51 + 68)20 = 2580$

$R_3 = N(20)20 + N(40)20 + N(60)20$

$ = (51 + 68 + 76)20 = 3900$

$A_3 = \dfrac{2580 + 3900}{2} = 3240$

Error bound for A_3:

$$\text{Error} \leq |N(60) - N(0)| \left(\dfrac{60 - 0}{2 \cdot 3} \right) = (76 - 10)(10) = 660 \text{ units}$$

Second 60 days:

$L_3 = N(60)20 + N(80)20 + N(100)20$

$ = (76 + 81 + 84)20 = 4820$

$R_3 = N(80)20 + N(100)20 + N(120)20$

$ = (81 + 84 + 86)20 = 5020$

$A_3 = \dfrac{4820 + 5020}{2} = 4920$

Error bound for A_3:

$$\text{Error} \leq |N(120) - N(60)| \left(\dfrac{120 - 60}{2 \cdot 3} \right) = (86 - 76)(10) = 100 \text{ units}$$

33. (A) Geometrically, $\displaystyle\int_{100}^{200} R'(x)\, dx$ is the area under the graph of $R'(x)$ on the interval $[100, 200]$. $\displaystyle\int_{100}^{200} R'(x)\, dx$ also represents the total change in revenue going from sales of 100 six packs per day to 200 six packs per day.

(B) $R'(x) = 8 - \dfrac{x}{25}$, $a = 100$, $b = 200$, $n = 4$; $\Delta x = \dfrac{200 - 100}{4} = 25$

$$L_4 = R'(100)25 + R'(125)25 + R'(150)25 + R'(175)25$$
$$= (4 + 3 + 2 + 1)25 = 250$$
$$R_4 = R'(125)25 + R'(150)25 + R'(175)25 + R'(200)25$$
$$= (3 + 2 + 1 + 0)25 = 150$$
$$A_4 = \frac{250 + 150}{2} = \$200$$

Error bound for A_4:

$$\text{Error} \leq |R'(200 - R'(100)| \left(\frac{200 - 100}{2 \cdot 4}\right)$$
$$= |0 - 4|(12.5) = 4(12.5) = \$50$$

(C) $R(200) - R(100) = 8(200) - \dfrac{(200)^2}{50} - \left[8(100) - \dfrac{(100)^2}{50}\right]$

$$= 1600 - 800 - (800 - 200)$$
$$= \$200$$

$R(200) - R(100)$ and $\displaystyle\int_{100}^{200} R'(x)\,dx$ each represents the total change in revenue going from sales of 100 six packs per day to 200 six packs per day. This suggests that

$$\int_{100}^{200} R'(x)\,dx = R(200) - R(100)$$

35. (A) $L_5 = A'(0)1 + A'(1)1 + A'(2)1 + A'(3)1 + A'(4)1$
$$= 0.90 + 0.81 + 0.74 + 0.67 + 0.60$$
$$= 3.72 \text{ sq cm}$$
$$R_5 = A'(1)1 + A'(2)1 + A'(3)1 + A'(4)1 + A'(5)1$$
$$= (0.81 + 0.74 + 0.67 + 0.60 + 0.55)$$
$$= 3.37 \text{ sq cm}$$

(B) Since $A'(t)$ is a decreasing function
$$3.37 \leq \int_0^5 A'(t)\,dt \leq 3.72$$

37. $L_3 = N'(6)2 + N'(8)2 + N'(10)2$
$$= (21 + 19 + 17)2 = 114$$
$$R_3 = N'(8)2 + N'(10)2 + N'(12)2$$
$$= (19 + 17 + 15)2 = 102$$
$$A_3 = \frac{L_3 + R_3}{2} = \frac{114 + 102}{2} = 108 \text{ code symbols}$$

Error bound for A_3:

$$\text{Error} \leq |N'(12) - N'(6)| \left(\frac{12 - 6}{2 \cdot 3}\right)$$
$$|15 - 21|(1) = 6 \text{ code symbols}$$

Things to remember:

1. **LEFT SUM, RIGHT SUM, MIDPOINT SUM**

   Let f be defined on the interval $[a, b]$. Partition $[a, b]$ into n equal subintervals of length $\Delta x = \dfrac{b - a}{n}$ with endpoints $a = x_0, x_1, x_2, \dots, x_{n-1}, x_n = b$.

   Then,

   (a) LEFT SUM $= L_n = \displaystyle\sum_{k=1}^{n} f(x_{k-1})\Delta x$

   $\qquad\qquad\qquad = f(x_0)\Delta x + f(x_1)\Delta x + \dots + f(x_{n-1})\Delta x$

   (b) RIGHT SUM $= R_n = \displaystyle\sum_{k=1}^{n} f(x_k)\Delta x$

   $\qquad\qquad\qquad = f(x_1)\Delta x + f(x_2)\Delta x + \dots + f(x_n)\Delta x$

   (c) MIDPOINT SUM $= M_n = \displaystyle\sum_{k=1}^{n} f\left(\dfrac{x_{k-1} + x_k}{2}\right)\Delta x$

   $\qquad\qquad\qquad = f\left(\dfrac{x_0 + x_1}{2}\right)\Delta x + f\left(\dfrac{x_1 + x_2}{2}\right)\Delta x + \dots$

   $\qquad\qquad\qquad\qquad\qquad\qquad\qquad\qquad + f\left(\dfrac{x_{n-1} + x_n}{2}\right)\Delta x$

2. **DEFINITION OF A DEFINITE INTEGRAL**

   Let f be a continuous function defined on the closed interval $[a, b]$, and let

   1. $a = x_0 < x_1 < \dots < x_{n-1} < x_n = b$

   2. $\Delta x_k = x_k - x_{k-1}$ for $k = 1, 2, \dots, n$

   3. $\Delta x_k \to 0$ as $n \to \infty$

   4. $x_{k-1} \le c_k \le x_k$ for $k = 1, 2, \dots, n$

   Then,

   $$\int_a^b f(x)\,dx = \lim_{n \to \infty} \sum_{k=1}^{n} f(c_k)\Delta x_k$$
   $$= \lim_{n \to \infty} [f(c_1)\Delta x_1 + f(c_2)\Delta x_2 + \dots + f(c_n)\Delta x_n]$$

   is called a DEFINITE INTEGRAL of f from a to b. The INTEGRAND is $f(x)$, the LOWER LIMIT is a and the UPPER LIMIT is b.

<u>3.</u> ERROR BOUNDS FOR L_n, R_n, M_n

Let
$$I = \int_a^b f(x)\,dx \qquad L_n = \text{Left Sum} \qquad R_n = \text{Right Sum}$$
$$M_n = \text{Midpoint Sum}$$

LEFT AND RIGHT SUM ERROR BOUND:

If $|f'(x)| \le B_1$ for all x on $[a,\ b]$, then
$$|I - L_n| \le \frac{B_1(b-a)^2}{2n}, \qquad |I - R_n| \le \frac{B_1(b-a)^2}{2n}$$

MIDPOINT ERROR BOUND:

If $|f''(x)| \le B_2$ for all x on $[a,\ b]$, then
$$|I - M_n| \le \frac{B_2(b-a)^3}{24n^2}$$

<u>4.</u> DEFINITE INTEGRAL PROPERTIES

(a) $\displaystyle\int_a^a f(x)\,dx = 0$

(b) $\displaystyle\int_a^b f(x)\,dx = -\int_b^a f(x)\,dx$

(c) $\displaystyle\int_a^b Kf(x)\,dx = K\int_a^b f(x)\,dx \qquad K$ is a constant

(d) $\displaystyle\int_a^b [f(x) \pm g(x)]\,dx = \int_a^b f(x)\,dx \pm \int_a^b g(x)\,dx$

(e) $\displaystyle\int_a^b f(x)\,dx = \int_a^c f(x)\,dx + \int_c^b f(x)\,dx$

<u>5.</u> FUNDAMENTAL THEOREM OF CALCULUS

If f is a continuous function on the closed interval $[a,\ b]$ and F is any antiderivative of f, then
$$\int_a^b f(x)\,dx = F(x)\ \Big|_a^b = F(b) - F(a);$$
$$F'(x) = f(x)$$

<u>6.</u> AVERAGE VALUE OF A CONTINUOUS FUNCTION OVER $[a,\ b]$

Let f be continuous on $[a,\ b]$. Then the AVERAGE VALUE of f over $[a,\ b]$ is:
$$\frac{1}{b-a}\int_a^b f(x)\,dx$$

1. $\int_a^0 f(x)\,dx = -2.33$ **3.** $\int_a^b f(x)\,dx = -2.33 + 10.67 = 8.34$

5. $\int_0^b \dfrac{f(x)}{10}\,dx = \dfrac{1}{10}\int_0^b f(x)\,dx = \dfrac{1}{10}(10.67) = 1.067$

7. $\int_2^3 2x\,dx = 2 \cdot \dfrac{x^2}{2}\Big|_2^3 = 3^2 - 2^2 = 5$ **9.** $\int_3^4 5\,dx = 5x\Big|_3^4 = 5\cdot 4 - 5\cdot 3 = 5$

11. $\int_1^3 (2x - 3)\,dx = (x^2 - 3x)\Big|_1^3 = (3^2 - 3\cdot 3) - (1^2 - 3\cdot 1) = 2$

13. $\int_0^4 (3x^2 - 4)\,dx = (x^3 - 4x)\Big|_0^4 = (4^3 - 4\cdot 4) - (0^3 - 4\cdot 0) = 48$

15. $\int_{-3}^4 (4 - x^2)\,dx = \left(4x - \dfrac{x^3}{3}\right)\Big|_{-3}^4 = \left(4\cdot 4 - \dfrac{4^3}{3}\right) - \left(4(-3) - \dfrac{(-3)^3}{3}\right)$

$$= 16 - \dfrac{64}{3} + 3 = -\dfrac{7}{3}$$

17. $\int_0^1 24x^{11}\,dx = 24\,\dfrac{x^{12}}{12}\Big|_0^1 = 2x^{12}\Big|_0^1 = 2\cdot 1^{12} - 2\cdot 0^{12} = 2$

19. $\int_0^1 e^{2x}\,dx = \dfrac{1}{2}e^{2x}\Big|_0^1 = \dfrac{1}{2}e^{2\cdot 1} - \dfrac{1}{2}e^{2\cdot 0} = \dfrac{1}{2}(e^2 - 1)$

21. $\int_1^{3.5} 2x^{-1}\,dx = 2\,\ln x\Big|_1^{3.5} = 2\,\ln 3.5 - 2\,\ln 1$

$$= 2\,\ln 3.5 \quad (\text{Recall: } \ln 1 = 0)$$

23. $\int_b^0 f(x)\,dx = -\int_0^b f(x)\,dx = -10.67$

25. $\int_c^0 f(x)\,dx = -\int_0^c f(x)\,dx = -\left[\int_0^b f(x)\,dx + \int_b^c f(x)\,dx\right]$

$$= -[10.67 - 5.63]$$
$$= -5.04$$

27. $\int_1^4 32t\,dt = 16t^2\Big|_1^4 = 16(4)^2 - 16(1)^2 = 256 - 16 = 240 \text{ ft}$

29. $\int_1^2 (2x^{-2} - 3)\,dx = (-2x^{-1} - 3x)\Big|_1^2 = \left(-\dfrac{2}{x} - 3x\right)\Big|_1^2$

$$= -\dfrac{2}{2} - 3\cdot 2 - \left(-\dfrac{2}{1} - 3\cdot 1\right) = -7 - (-5) = -2$$

31. $\int_1^4 3\sqrt{x}\,dx = 3\int_1^4 x^{1/2}\,dx = 3 \cdot \dfrac{2}{3}x^{3/2}\Big|_1^4 = 2x^{3/2}\Big|_1^4$

$$= 2\cdot 4^{3/2} - 2\cdot 1^{3/2} = 16 - 2 = 14$$

33. $\int_2^3 12(x^2 - 4)^5 x\, dx.$ Consider the indefinite integral $\int 12(x^2 - 4)^5 x\, dx.$

Let $u = x^2 - 4$, then $du = 2x\, dx.$

$$\int 12(x^2 - 4)^5 x\, dx = 6\int (x^2 - 4)^5 2x\, dx = 6\int u^5 du$$

$$= 6\frac{u^6}{6} + C = u^6 + C = (x^2 - 4)^6 + C$$

Thus,

$$\int_2^3 12(x^2 - 4)^5 x\, dx = (x^2 - 4)^6 \Big|_2^3 = (3^2 - 4)^6 - (2^2 - 4)^6 = 5^6 = 15,625.$$

35. $\int_3^9 \frac{1}{x - 1}\, dx$

Let $u = x - 1$.   Then $du = dx$ and $u = 8$ when $x = 9$, $u = 2$ when $x = 3$.
Thus,

$$\int_3^9 \frac{1}{x - 1}\, dx = \int_2^8 \frac{1}{u}\, du = \ln u \Big|_2^8 = \ln 8 - \ln 2 = \ln 4 \approx 1.386.$$

37. $\int_{-5}^{10} e^{-0.05x}\, dx$

Let $u = -0.05x$.   Then $du = -0.05\, dx$ and $u = -0.5$ when $x = 10$, $u = 0.25$
when $x = -5$.   Thus,

$$\int_{-5}^{10} e^{-0.05x}\, dx = -\frac{1}{0.05}\int_{-5}^{10} e^{-0.05x}(-0.05)\, dx = -\frac{1}{0.05}\int_{0.25}^{-0.5} e^{u}\, du$$

$$= -\frac{1}{0.05} e^u \Big|_{0.25}^{-0.5} = -\frac{1}{0.05}[e^{-0.5} - e^{0.25}]$$

$$= 20(e^{0.25} - e^{-0.5}) \approx 13.550$$

39. $\int_{-6}^{0} \sqrt{4 - 2x}\, dx$

Consider the indefinite integral $\int \sqrt{4 - 2x}\, dx = \int (4 - 2x)^{1/2}\, dx.$

Let $u = 4 - 2x$, then $du = -2\, dx.$

$$\int (4 - 2x)^{1/2}\, dx = -\frac{1}{2}\int (4 - 2x)^{1/2}(-2)\, dx = -\frac{1}{2}\int u^{1/2}\, du$$

$$= -\frac{1}{2} \cdot \frac{u^{3/2}}{\frac{3}{2}} + C = -\frac{1}{3}u^{3/2} + C = -\frac{1}{3}(4 - 2x)^{3/2} + C$$

Thus,

$$\int_{-6}^{0} (4 - 2x)^{1/2}\, dx = -\frac{1}{3}(4 - 2x)^{3/2} \Big|_{-6}^{0} = -\frac{1}{3}[4^{3/2} - 16^{3/2}]$$

$$= -\frac{1}{3}(8 - 64) = \frac{56}{3} \approx 18.667$$

41. $\int_{-1}^{7} \dfrac{x}{\sqrt{x + 2}}\,dx$

Consider the indefinite integral $\int \dfrac{x}{\sqrt{x + 2}}\,dx = \int x(x + 2)^{-1/2}\,dx$.

Let $u = x + 2$, then $du = dx$ and $x = u - 2$.

$\int x(x + 2)^{-1/2}\,dx = \int (u - 2)u^{-1/2}\,du = \int (u^{1/2} - 2u^{-1/2})\,du$

$\qquad = \dfrac{u^{3/2}}{\frac{3}{2}} - \dfrac{2u^{1/2}}{\frac{1}{2}} + C = \dfrac{2}{3}(x + 2)^{3/2} - 4(x + 2)^{1/2} + C$

Thus,

$\int_{-1}^{7} \dfrac{x}{\sqrt{x + 2}}\,dx = \left[\dfrac{2}{3}(x + 2)^{3/2} - 4(x + 2)^{1/2} \right]\Big|_{-1}^{7}$

$\qquad = \dfrac{2}{3}(9)^{3/2} - 4(9)^{1/2} - \left(\dfrac{2}{3}(1)^{3/2} - 4(1)^{1/2} \right)$

$\qquad = \dfrac{2}{3}(27) - 12 - \left(\dfrac{2}{3} - 4 \right) = 6 + \dfrac{10}{3} = \dfrac{28}{3} \approx 9.333.$

43. $\int_{0}^{1} (e^{2x} - 2x)^2 (e^{2x} - 1)\,dx = \dfrac{1}{2} \int_{0}^{1} (e^{2x} - 2x)^2 (2e^{2x} - 2)\,dx$

$\qquad = \dfrac{1}{2} \cdot \dfrac{(e^{2x} - 2x)^3}{3}\Big|_{0}^{1}$ $\quad$ [**Note** : The integrand

$\qquad = \dfrac{1}{6}(e^{2x} - 2x)^3\Big|_{0}^{1}$ $\qquad$ has the form $u^2\,du$; an

$\qquad = \dfrac{1}{6}[(e^2 - 2)^3 - 1]$ $\qquad$ antiderivative is

$\qquad \approx 25.918$ $\qquad\qquad\qquad \dfrac{u^3}{3} = \dfrac{(e^{2x} - 2x)^3}{3}.\Big]$

45. $\int_{-2}^{-1} (x^{-1} + 2x)\,dx = (\ln|x| + x^2)\Big|_{-2}^{-1}$

$\qquad = \ln|-1| + (-1)^2 - [\ln|-2| + (-2)^2]$

$\qquad = 1 - \ln 2 - 4$

$\qquad = -3 - \ln 2 \approx -3.693$

47. $f(x) = 500 - 50x$ on $[0, 10]$

(A) Ave $f(x) = \dfrac{1}{10 - 0} \int_{0}^{10} (500 - 50x)\,dx$ $\qquad$ (B)

$\qquad = \dfrac{1}{10}(500x - 25x^2)\Big|_{0}^{10}$

$\qquad = \dfrac{1}{10}[5{,}000 - 2{,}500] = 250$

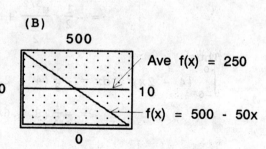

49. $f(t) = 3t^2 - 2t$ on $[-1, 2]$

(A) Ave $f(t) = \dfrac{1}{2 - (-1)} \displaystyle\int_{-1}^{2} (3t^2 - 2t)\,dt$ (B)

$$= \frac{1}{3}(t^3 - t^2)\Big|_{-1}^{2}$$

$$= \frac{1}{3}[4 - (-2)] = 2$$

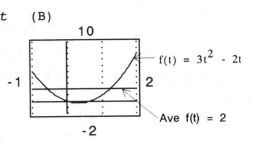

51. $f(x) = \sqrt[3]{x} = x^{1/3}$ on $[1, 8]$

(A) Ave $f(x) = \dfrac{1}{8 - 1} \displaystyle\int_{1}^{8} x^{1/3}\,dx$ (B)

$$= \frac{1}{7}\left(\frac{3}{4}x^{4/3}\right)\Big|_{1}^{8}$$

$$= \frac{3}{28}(16 - 1) = \frac{45}{28} \approx 1.61$$

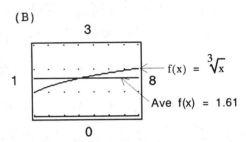

53. $f(x) = 4e^{-0.2x}$ on $[0, 10]$

(A) Ave $f(x) = \dfrac{1}{10 - 0} \displaystyle\int_{0}^{10} 4e^{-0.2x}\,dx$ (B)

$$= \frac{1}{10}(-20e^{-0.2x})\Big|_{0}^{10}$$

$$= \frac{1}{10}(20 - 20e^{-2}) \approx 1.73$$

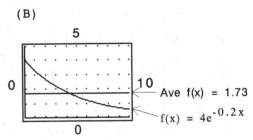

55. $f(x) = 0.25x^2 - 4$ on $[0, 8]$, $a = 0$, $b = 8$, $n = 4$;

$$\Delta x = \frac{8 - 0}{4} = 2$$

$$M_4 = f(1)\Delta x + f(3)\Delta x + f(5)\Delta x + f(7)\Delta x$$
$$= (-3.75 - 1.75 + 2.25 + 8.25)2 = 10$$

Thus, $I = \displaystyle\int_{0}^{8} (0.25x^2 - 4)\,dx \approx 10.$

Error bound:
$$f'(x) = 0.5x, \quad f''(x) = 0.5$$

From $\underline{3}$,

$$|I - M_4| \leq \frac{0.5(8 - 0)^3}{24(4)^2} = \frac{256}{384} = 0.67$$

Thus, $I = 10 \pm 0.67$

57. $I = \int_0^8 (0.25x^2 - 4)\,dx = \int_0^8 \left(\frac{1}{4}x^2 - 4\right)dx$

$$= \left(\frac{1}{12}x^3 - 4x\right)\Big|_0^8 = \frac{512}{12} - 32 = 10.67$$

$|I - M_4| = |10.67 - 10| = 0.67$

This error does lie within the error bound calculated in Problem 55.

59. $f''(x) = 0.5$ on $[0, 8]$

$$|I - M_n| \le \frac{0.5(8 - 0)^3}{24n^2} = \frac{256}{24n^2} = \frac{32}{3n^2};$$

$$\frac{32}{3n^2} \le 0.005$$

$$n^2 \ge \frac{32}{0.015} \approx 2133.33$$

$$n \ge 47$$

61. $\lim_{n \to \infty}[(1 - c_1^2)\Delta x + (1 - c_2^2)\Delta x + \ldots + (1 - c_n^2)\Delta x]$ where $\Delta x = \frac{5 - 2}{n}$ and

$c_k = 2 + k \cdot \frac{3}{n}$, $k = 1, 2, \ldots, n$ is a Riemann sum for $\int_2^5 (1 - x^2)\,dx$.

$$\int_2^5 (1 - x^2)\,dx = \left(x - \frac{1}{3}x^3\right)\Big|_2^5 = \left[5 - \frac{125}{3} - \left(2 - \frac{8}{3}\right)\right] = -36$$

63. $\lim_{n \to \infty}[(3c_1^2 - 2c_1 + 3)\Delta x + (3c_2^2 - 2c_2 + 3)\Delta x + \ldots + (3c_n^2 - 2c_n + 3)\Delta x]$,

where $\Delta x = \frac{12 - 2}{n}$ and $c_k = 2 + k \cdot \frac{10}{n}$, $k = 1, 2, \ldots, n$ is a Riemann sum

for $\int_2^{12} (3x^2 - 2x + 3)\,dx$.

$$\int_2^{12} (3x^2 - 2x + 3)\,dx = (x^3 - x^2 + 3x)\Big|_2^{12} = (12)^3 - (12)^2 + 36 - (10)$$

$$= 1{,}610$$

65. $\int_2^3 x\sqrt{2x^2 - 3}\,dx = \int_2^3 x(2x^2 - 3)^{1/2}\,dx$

$$= \frac{1}{4}\int_2^3 (2x^2 - 3)^{1/2}4x\,dx \qquad \left[\underline{\text{Note}} : \text{The integrand has the form } u^{1/2}du; \text{ the antiderivative}\right.$$

$$= \frac{1}{4}\left(\frac{2}{3}\right)(2x^2 - 3)^{3/2}\Big|_2^3 \qquad \left.\text{is } \frac{2}{3}u^{3/2} = \frac{2}{3}(2x^2 - 3)^{3/2}.\right]$$

$$= \frac{1}{6}[2(3)^2 - 3]^{3/2} - \frac{1}{6}[2(2)^2 - 3]^{3/2}$$

$$= \frac{1}{6}(15)^{3/2} - \frac{1}{6}(5)^{3/2} = \frac{1}{6}[15^{3/2} - 5^{3/2}] \approx 7.819$$

67. $\int_0^1 \dfrac{x - 1}{x^2 - 2x + 3}\,dx$

Consider the indefinite integral and let $u = x^2 - 2x + 3$.
Then $du = (2x - 2)\,dx = 2(x - 1)\,dx$.

$$\int \frac{x - 1}{x^2 - 2x + 3}\,dx = \frac{1}{2}\int \frac{2(x - 1)}{x^2 - 2x + 3}\,dx = \frac{1}{2}\int \frac{1}{u}\,du = \frac{1}{2}\ln|u| + C$$

Thus,

$$\int_0^1 \frac{x - 1}{x^2 - 2x + 3}\,dx = \frac{1}{2}\ln|x^2 - 2x + 3|\,\Big|_0^1$$

$$= \frac{1}{2}\ln 2 - \frac{1}{2}\ln 3 = \frac{1}{2}(\ln 2 - \ln 3) \approx -0.203$$

69. $\int_{-1}^1 \dfrac{e^{-x} - e^x}{(e^{-x} + e^x)^2}\,dx$

Consider the indefinite integral and let $u = e^{-x} + e^x$.
Then $du = (-e^{-x} + e^x)\,dx = -(e^{-x} - e^x)\,dx$.

$$\int \frac{e^{-x} - e^x}{(e^{-x} + e^x)^2}\,dx = -\int \frac{-(e^{-x} - e^x)}{(e^{-x} + e^x)^2}\,dx = -\int u^{-2}\,du = \frac{-u^{-1}}{-1} + C = \frac{1}{u} + C$$

Thus,

$$\int_{-1}^1 \frac{e^{-x} - e^x}{(e^{-x} + e^x)^2}\,dx = \frac{1}{e^{-x} + e^x}\,\Big|_{-1}^1 = \frac{1}{e^{-1} + e^1} - \frac{1}{e^{-(-1)} + e^{-1}}$$

$$= \frac{1}{e^{-1} + e} - \frac{1}{e^{-1} + e} = 0$$

71. $f(t) = \dfrac{1}{t}$ on $[1,\ 2]$

(A) $a = 1$, $b = 2$, $n = 5$, $\Delta x = \dfrac{2 - 1}{5} = 0.2$

partition $\{1,\ 1.2,\ 1.4,\ 1.6,\ 1.8,\ 2\}$
midpoints $\{1.1,\ 1.3,\ 1.5,\ 1.7,\ 1.9\}$

$$M_5 = f(1.1)\Delta x + f(1.3)\Delta x + f(1.5)\Delta x + f(1.7)\Delta x + f(1.9)\Delta x$$

$$\approx [0.9091 + 0.7692 + 0.6667 + 0.5882 + 0.5263]0.2$$

$$= (3.4595)0.2 = 0.6919$$

Error bound:

$$f'(t) = -\frac{1}{t^2}, \quad f''(t) = \frac{2}{t^3}$$

Max $|f''(t)|$ on $[1,\ 2]$ = max $\dfrac{2}{t^3}$ on $[1,\ 2]$ = 2

$$|I - M_5| \le \frac{2(2 - 1)^3}{24(5)^2} = \frac{1}{300} = 0.0033$$

Thus, $\ln 2 = 0.6919 \pm 0.0033$

(B) $\ln 2 \approx 0.6931$

(C) Error = $|\ln 2 - M_5| = |0.6931 - 0.6919| = 0.0012$

This error is within the bound determined in part (A).

73. $f(t) = \frac{1}{t}$ on $[1, 2]$; $f'(t) = -\frac{1}{t^2}$, $f''(t) = \frac{2}{t^3}$

Max $f''(t)$ on $[1, 2]$ = max $\frac{2}{t^3}$ on $[1, 2]$ = 2

$$|I - M_n| \le \frac{2(2-1)^3}{24n^2} = \frac{1}{12n^2};$$

$$\frac{1}{12n^2} \le 0.0005$$

$$n^2 \ge \frac{1}{12(0.0005)} \approx 166.67$$

$$n \ge 13$$

75. $C'(x) = 500 - \frac{x}{3}$ on $[300, 900]$

The increase in cost from a production level of 300 bikes per month to a production level of 900 bikes per month is given by:

$$\int_{300}^{900} \left(500 - \frac{x}{3}\right)dx = \left(500x - \frac{1}{6}x^2\right)\Big|_{300}^{900}$$

$$= 315{,}000 - (135{,}000)$$

$$= \$180{,}000$$

77. Total loss in value in the first 5 years:

$$V(5) - V(0) = \int_0^5 V'(t)dt = \int_0^5 500(t - 12)dt = 500\left(\frac{t^2}{2} - 12t\right)\Big|_0^5$$

$$= 500\left(\frac{25}{2} - 60\right) = -\$23{,}750$$

Total loss in value in the second 5 years:

$$V(10) - V(5) = \int_5^{10} V'(t)dt = \int_5^{10} 500(t - 12)dt = 500\left(\frac{t^2}{2} - 12t\right)\Big|_5^{10}$$

$$= 500\left[(50 - 120) - \left(\frac{25}{2} - 60\right)\right] = -\$11{,}250$$

79. $C(x) = 1 + 12x - x^2$, $0 \le x \le 12$

(A) For the average cash reserve for the first quarter, take $a = 0$, $b = 3$

$$\text{Ave } C(x) = \frac{1}{3 - 0}\int_0^3 (1 + 12x - x^2)dx$$

$$= \frac{1}{3}\left(x + 6x^2 - \frac{1}{3}x^3\right)\Big|_0^3$$

$$= \frac{1}{3}(48) = 16$$

Thus, Ave $C(x) = \$16{,}000$.

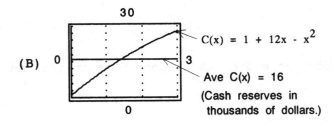

(B)   0 $C(x) = 1 + 12x - x^2$

   3

Ave $C(x) = 16$

(Cash reserves in thousands of dollars.)

81. (A) To find the useful life, set $C'(t) = R'(t)$ and solve for t.

$$\frac{1}{11}t = 5te^{-t^2}$$

$$e^{t^2} = 55$$

$$t^2 = \ln 55$$

$$t = \sqrt{\ln 55} \approx 2 \text{ years}$$

(B) The total profit accumulated during the useful life is:

$$P(2) - P(0) = \int_0^2 [R'(t) - C'(t)]dt = \int_0^2 \left(5te^{-t^2} - \frac{1}{11}t\right)dt$$

$$= \int_0^2 5te^{-t^2}dt - \int_0^2 \frac{1}{11}t\ dt$$

$$= -\frac{5}{2}\int_0^2 e^{-t^2}(-2t)dt - \frac{1}{11}\int_0^2 t\ dt$$

$$= -\frac{5}{2}e^{-t^2}\Big|_0^2 - \frac{1}{22}t^2\Big|_0^2$$

$$= -\frac{5}{2}e^{-4} + \frac{5}{2} - \frac{4}{22} = \frac{51}{22} - \frac{5}{2}e^{-4} \approx 2.272$$

[Note: In the first integral, the integrand has the form $e^u du$, where $u = -t^2$; an antiderivative is $e^u = e^{-t^2}$.]

Thus, the total profit is approximately \$2,272.

83. $C(x) = 60,000 + 300x$

(A) Average cost per unit:

$$\overline{C}(x) = \frac{C(x)}{x} = \frac{60,000}{x} + 300$$

$$\overline{C}(500) = \frac{60,000}{500} + 300 = \$420$$

(B) Ave $C(x) = \dfrac{1}{500}\displaystyle\int_0^{500}(60,000 + 300x)dx$

$$= \frac{1}{500}(60,000x + 150x^2)\Big|_0^{500}$$

$$= \frac{1}{500}(30,000,000 + 37,500,000) = \$135,000$$

(C) $\overline{C}(500)$ is the average cost per unit at a production level of 500 units; Ave $C(x)$ is the average value of the total cost as production increases from 0 units to 500 units.

85. $A'(t) = 800e^{0.08t}, \ t \geq 0$

The change in the account from the end of the second year to the end of the sixth year is given by:

$$\int_2^6 A'(t)\,dt = \int_2^6 800e^{0.08t}\,dt$$

$$= \left(\frac{800}{0.08}e^{0.08t}\right)\Big|_2^6$$

$$= 10{,}000(e^{0.48} - e^{0.16})$$

$$\approx \$4{,}425.64$$

87. Average price:

$$\text{Ave } S(x) = \frac{1}{30 - 20}\int_{20}^{30} 10(e^{0.02x} - 1)\,dx = \int_{20}^{30}(e^{0.02x} - 1)\,dx$$

$$= \int_{20}^{30} e^{0.02x}\,dx - \int_{20}^{30} dx$$

$$= \frac{1}{0.02}\int_{20}^{30} e^{0.02x}(0.02)\,dx - x\Big|_{20}^{30}$$

$$= 50e^{0.02x}\Big|_{20}^{30} - (30 - 20)$$

$$= 50e^{0.6} - 50e^{0.4} - 10$$

$$\approx 6.51 \text{ or } \$6.51$$

89. $g(x) = 2400x^{-1/2}$ and $L'(x) = g(x)$.

The number of labor hours to assemble the 17th through the 25th control units is:

$$L(25) - L(16) = \int_{16}^{25} g(x)\,dx = \int_{16}^{25} 2400x^{-1/2}\,dx = 2400(2)x^{1/2}\Big|_{16}^{25}$$

$$= 4800x^{1/2}\Big|_{16}^{25} = 4800[25^{1/2} - 16^{1/2}] = 4800 \text{ labor hours}$$

91. (A) The inventory function is obtained by finding the equation of the line joining $(0, 600)$ and $(3, 0)$.

Slope: $m = \frac{0 - 600}{3 - 0} = -200$, y intercept: $b = 600$

Thus, the equation of the line is: $I = -200t + 600$

(B) The average of I over $[0, 3]$ is given by:

$$\text{Ave } I(t) = \frac{1}{3 - 0}\int_0^3 I(t)\,dt = \frac{1}{3}\int_0^3 (-200t + 600)\,dt$$

$$= \frac{1}{3}(-100t^2 + 600t)\Big|_0^3$$

$$= \frac{1}{3}[-100(3^2) + 600(3) - 0]$$

$$= \frac{900}{3} = 300 \text{ units}$$

93. Rate of production: $R(t) = \dfrac{100}{t + 1} + 5$, $0 \le t \le 20$

Total production from year N to year M is given by:

$$P = \int_N^M R(t)\,dt = \int_N^M \left(\frac{100}{t + 1} + 5\right)dt = 100 \int_N^M \frac{1}{t + 1}\,dt + \int_N^M 5\,dt$$

$$= 100\,\ln|t + 1|\,\Big|_N^M + 5t\,\Big|_N^M$$

$$= 100\,\ln(M + 1) - 100\,\ln(N + 1) + 5(M - N)$$

Thus, for total production during the first 10 years, let $M = 10$ and $N = 0$.

$P = 100\,\ln 11 - 100\,\ln 1 + 5(10 - 0)$

$\quad = 100\,\ln 11 + 50 \approx 290$ thousand barrels

For the total production from the end of the 10th year to the end of the 20th year, let $M = 20$ and $N = 10$.

$P = 100\,\ln 21 - 100\,\ln 11 + 5(20 - 10)$

$\quad = 100\,\ln 21 - 100\,\ln 11 + 50 \approx 115$ thousand barrels

95. Let $P(t) = R(t) - C(t)$. Then the total accumulated profits over the five-year period are given by:

$$P(5) - P(0) = \int_0^5 P'(t)\,dt = \int_0^5 [R'(t) - C'(t)]\,dt$$

$$= \int_0^5 R'(t)\,dt - \int_0^5 C'(t)\,dt$$

Now, $C'(t) = 1{,}500$ (constant). Therefore,

$$\int_0^5 C'(t)\,dt = \int_0^5 1{,}500\,dt = 1{,}500t\,\Big|_0^5 = 7{,}500$$

Using a midpoint sum with $n = 5$ to approximate

$$\int_0^5 R'(t)\,dt,$$

we have $\Delta x = 1$ and

$$M_5 = R'\left(\frac{1}{2}\right)1 + R'\left(\frac{3}{2}\right)1 + R'\left(\frac{5}{2}\right)1 + R'\left(\frac{7}{2}\right)1 + R'\left(\frac{9}{2}\right)1$$

$$= 5{,}000 + 4{,}500 + 3{,}500 + 2{,}500 + 2{,}000$$

$$= 17{,}500$$

Therefore, the total accumulated profits are (approximately):

$$P(5) - P(0) = \int_0^5 R'(t)\,dt - \int_0^5 C'(t)\,dt \approx 17{,}500 - 7{,}500 = \$10{,}000$$

97.

x	300	900	1,500	2,100
$f(x)$	900	1,700	1,700	900

Using a midpoint sum with $n = 4$ and $\Delta x = 600$, we have

$$\int_0^{2,400} f(x)\,dx \approx M_4 = f(300)\Delta x + f(900)\Delta x + f(1{,}500)\Delta x + f(2{,}100)\Delta x$$

$$= [900 + 1{,}700 + 1{,}700 + 900]600$$

$$= (5{,}200)600 = 3{,}120{,}000 \text{ sq ft}$$

99. $W'(t) = 0.2e^{0.1t}$

The weight increase during the first eight hours is given by:

$$W(8) - W(0) = \int_0^8 W'(t)\,dt = \int_0^8 0.2e^{0.1t}\,dt = 0.2\int_0^8 e^{0.1t}\,dt$$

$$= \frac{0.2}{0.1}\int_0^8 e^{0.1t}(0.1)\,dt \qquad \text{(Let } u = 0.1t, \text{ then } du = 0.1dt.)$$

$$= 2e^{0.1t}\Big|_0^8 = 2e^{0.8} - 2 \approx 2.45 \text{ grams}$$

The weight increase during the second eight hours, i.e., from the 8th hour through the 16th hour, is given by:

$$W(16) - W(8) = \int_8^{16} W'(t)\,dt = \int_8^{16} 0.2e^{0.1t}\,dt = 2e^{0.1t}\Big|_8^{16}$$

$$= 2e^{1.6} - 2e^{0.8} \approx 5.45 \text{ grams}$$

101. Average temperature over time period $[0, 2]$ is given by:

$$\frac{1}{2-0}\int_0^2 C(t)\,dt = \frac{1}{2}\int_0^2 (t^3 - 2t + 10)\,dt = \frac{1}{2}\left(\frac{t^4}{4} - \frac{2t^2}{2} + 10t\right)\Big|_0^2$$

$$= \frac{1}{2}(4 - 4 + 20) = 10° \text{ Celsius}$$

103. Using a midpoint sum with $n = 3$ and $\Delta t = 1$, and estimating the values of $R(t)$ from the graph, we have

$$\int_0^3 R(t)\,dt \approx M_3 = R\left(\frac{1}{2}\right)1 + R\left(\frac{3}{2}\right)1 + R\left(\frac{5}{2}\right)1$$

$$= 0.3 + 0.5 + 0.3 = 1.1$$

Thus, the total volume of air inhaled is approximately 1.1 liters.

105. $P(t) = \dfrac{8.4t}{t^2 + 49} + 0.1,\ 0 \le t \le 24$

(A) Average fraction of people during the first seven months:

$$\frac{1}{7-0}\int_0^7 \left[\frac{8.4t}{t^2+49} + 0.1\right]dt = \frac{4.2}{7}\int_0^7 \frac{2t}{t^2+49}\,dt + \frac{1}{7}\int_0^7 0.1\,dt$$

$$= 0.6\ \ln(t^2+49)\Big|_0^7 + \frac{0.1}{7}t\Big|_0^7$$

$$= 0.6[\ln 98 - \ln 49] + 0.1$$

$$= 0.6\ \ln 2 + 0.1 \approx 0.516$$

(B) Average fraction of people during the first two years:

$$\frac{1}{24-0}\int_0^{24}\left[\frac{8.4t}{t^2+49} + 0.1\right]dt = \frac{4.2}{24}\int_0^{24}\frac{2t}{t^2+49}\,dt + \frac{1}{24}\int_0^{24}0.1\,dt$$

$$= 0.175\ \ln(t^2+49)\Big|_0^{24} + \frac{0.1}{24}t\Big|_0^{24}$$

$$= 0.175[\ln 625 - \ln 49] + 0.1 \approx 0.546$$

1. $\int (3t^2 - 2t)\,dt = 3\int t^2\,dt - 2\int t\,dt = 3 \cdot \dfrac{t^3}{3} - 2 \cdot \dfrac{t^2}{2} + C = t^3 - t^2 + C$ (11-1)

2. $\int_2^5 (2x - 3)\,dx = 2\int_2^5 x\,dx - 3\int_2^5 dx = x^2\Big|_2^5 - 3x\Big|_2^5$

$\qquad\qquad = (25 - 4) - (15 - 6) = 12$ (11-5)

3. $\int (3t^{-2} - 3)\,dt = 3\int t^{-2}\,dt - 3\int dt = 3 \cdot \dfrac{t^{-1}}{-1} - 3t + C = -3t^{-1} - 3t + C$ (11-1)

4. $\int_1^4 x\,dx = \dfrac{x^2}{2}\Big|_1^4 = \dfrac{16}{2} - \dfrac{1}{2} = \dfrac{15}{2}$ (11-5)

5. $\int e^{-0.5x}\,dx = \dfrac{e^{-0.5x}}{-0.5} + C = -2e^{-0.5x} + C$ (11-2)

6. $\int_1^5 \dfrac{2}{u}\,du = 2\int_1^5 \dfrac{du}{u} = 2\ln u\Big|_1^5 = 2\ln 5 - 2\ln 1 = 2\ln 5$ (11-5)

7. $\dfrac{dy}{dx} = 3x^2 - 2$

$\qquad y = f(x) = \int (3x^2 - 2)\,dx$

$\qquad f(x) = x^3 - 2x + C$
$\qquad f(0) = C = 4$
$\qquad f(x) = x^3 - 2x + 4$ (11-3)

8. The graph of an antiderivative function f is increasing on $(0, 2)$, decreasing on $(2, 4)$, concave down on $(0, 4)$; f has a local maximum at $x = 2$. The graphs of the antiderivative functions differ by a vertical translation. (11-1)

9.

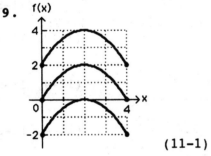

(11-1)

10. (A) $\int (8x^3 - 4x - 1)\,dx = 8\int x^3\,dx - 4\int x\,dx - \int dx$

$\qquad\qquad = 8 \cdot \dfrac{1}{4}x^4 - 4\,\dfrac{1}{2}x^2 - x + C$

$\qquad\qquad = 2x^4 - 2x^2 - x + C$ (11-1)

(B) $\int (e^t - 4t^{-1})\,dt = \int e^t - 4\int \dfrac{1}{t}\,dt$

$\qquad\qquad = e^t - 4\ln|t| + C$ (11-1)

11. $f(x) = x^2 + 1$, $a = 1$, $b = 5$, $n = 2$, $\Delta x = \dfrac{5 - 1}{2} = 2$;

$$M_2 = f(2)\Delta x + f(4)\Delta x$$
$$= 5 \cdot 2 + 17 \cdot 2 = 44$$

Error bound for M_2:

$$f'(x) = 2x, \quad f''(x) = 2$$
$$|I - M_2| \le \frac{2(5 - 1)^3}{24(2)^2} = \frac{128}{96} \approx 1.333$$

Thus, $\displaystyle\int_1^5 (x^2 + 1)\, dx = 44 \pm 1.333$

(11-5)

12. $\displaystyle\int_1^5 (x^2 + 1)\, dx = \left(\frac{1}{3}x^3 + x\right)\Big|_1^5 = \frac{125}{3} + 5 - \left(\frac{1}{3} + 1\right) \approx 45.333$

$|I - M_2| = 1.333$

(11-5)

13. Using the values of f in the table with $a = 1$, $b = 17$, $n = 4$,

$\Delta x = \dfrac{17 - 1}{4} = 4$;

$$M_4 = f(3)\Delta x + f(7)\Delta x + f(11)\Delta x + f(15)\Delta x$$
$$= [1.2 + 3.4 + 2.6 + 0.5]4 = 30.8$$

(11-5)

14. $f(x) = 6x^2 + 2x$ on $[-1, 2]$;

$$\text{Ave } f(x) = \frac{1}{2 - (-1)} \int_{-1}^2 (6x^2 + 2x)\, dx$$
$$= \frac{1}{3}(2x^3 + x^2)\Big|_{-1}^2 = \frac{1}{3}[20 - (-1)] = 7$$

(11-5)

15. width $= 2 - (-1) = 3$, height $= \text{Ave } f(x) = 7$

(11-5)

16. $\displaystyle\int_a^b 5f(x)\, dx = 5\int_a^b f(x)\, dx = 5(-2) = -10$

(11-4, 11-5)

17. $\displaystyle\int_b^c \frac{f(x)}{5}\, dx = \frac{1}{5}\int_b^c f(x)\, dx = \frac{1}{5}(2) = \frac{2}{5} = 0.4$

(11-4, 11-5)

18. $\displaystyle\int_b^d f(x)\, dx = \int_b^c f(x)\, dx + \int_c^d f(x)\, dx = 2 - 0.6 = 1.4$

(11-4, 11-5)

19. $\displaystyle\int_a^c f(x)\, dx = \int_a^b f(x)\, dx + \int_b^c f(x)\, dx = -2 + 2 = 0$

(11-4, 11-5)

20. $\displaystyle\int_0^d f(x)\, dx = \int_0^a f(x)\, dx + \int_a^b f(x)\, dx + \int_b^c f(x)\, dx + \int_c^d f(x)\, dx$

$$= 1 - 2 + 2 - 0.6 = 0.4$$

(11-4, 11-5)

21. $\displaystyle\int_b^a f(x)\, dx = -\int_a^b f(x)\, dx = -(-2) = 2$

(11-4, 11-5)

22. $\int_c^b f(x)\,dx = -\int_b^c f(x)\,dx = -2$ $\hspace{3cm}$ (11-4, 11-5)

23. $\int_d^0 f(x)\,dx = -\int_0^d f(x)\,dx = -0.4$ (from Problem 20) $\hspace{1cm}$ (11-4, 11-5)

24. Let f be an antiderivative function. Then:
f is increasing on $[0, 1]$ and $[3, 4]$ ($f'(x) > 0$);
f is decreasing on $[1, 3]$ ($f'(x) < 0$);
the graph of f is concave down on $[0, 2]$ (f' is decreasing); the graph
of f is concave up on $[2, 4]$ (f' is increasing); f has a local maximum
at $x = 1$; f has a local minimum at $x = 3$; there is an inflection point
at $x = 2$. The graphs of antiderivative functions differ by a vertical
translation. $\hspace{9cm}$ (11-1)

25.

(11-1)

26. (A) $\dfrac{dy}{dx} = \dfrac{2y}{x}$; $\left.\dfrac{dy}{dx}\right|_{(2,\,1)} = \dfrac{2(1)}{2} = 1$, $\left.\dfrac{dy}{dx}\right|_{(-2,\,-1)} = \dfrac{2(-1)}{-2} = 1$

$\hspace{0.9cm}$ (B) $\dfrac{dy}{dx} = \dfrac{2x}{y}$; $\left.\dfrac{dy}{dx}\right|_{(2,\,1)} = \dfrac{2(2)}{1} = 4$, $\left.\dfrac{dy}{dx}\right|_{(-2,\,-1)} = \dfrac{2(-2)}{-1} = 4$ $\hspace{0.8cm}$ (11-3)

27. $\dfrac{dy}{dx} = \dfrac{2y}{x}$; from the figure, the slopes at $(2, 1)$ and $(-2, -1)$ are
approximately equal to 1 as computed in Problem 26(A), not 4 as
computed in Problem 26(B). $\hspace{7cm}$ (11-3)

28. Let $y = Cx^2$. Then $\dfrac{dy}{dx} = 2Cx$. From the original equation, $C = \dfrac{y}{x^2}$ so

$\hspace{1cm}$ $\dfrac{dy}{dx} = 2x\left(\dfrac{y}{x^2}\right) = \dfrac{2y}{x}$ $\hspace{8cm}$ (11-3)

29. Letting $x = 2$ and $y = 1$ in $y = Cx^2$, we get
$\hspace{1.5cm}$ $1 = 4C$ so $C = \dfrac{1}{4}$ and $y = \dfrac{1}{4}x^2$
$\hspace{0.8cm}$ Letting $x = -2$ and $y = -1$ in $y = Cx^2$, we get
$\hspace{1.5cm}$ $-1 = 4C$ so $C = -\dfrac{1}{4}$ and $y = -\dfrac{1}{4}x^2$ $\hspace{4cm}$ (11-3)

30.

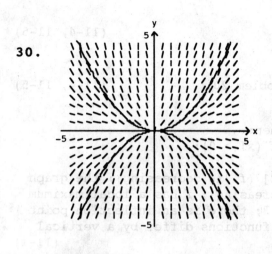

(11-3)

31.

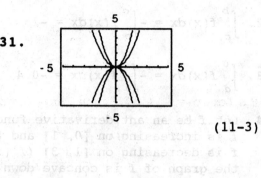

(11-3)

32. $\int \sqrt[3]{6x - 5}\, dx = \int (6x - 5)^{1/3} dx = \frac{1}{6}\int (6x - 5)^{1/3} 6\, dx$

$$= \frac{1}{6}\frac{(6x - 5)^{4/3}}{\frac{4}{3}} + C$$

$$= \frac{1}{8}(6x - 5)^{4/3} + C \qquad\qquad (11\text{-}1,\ 11\text{-}2)$$

33. $\int_0^1 10(2x - 1)^4 dx = 5\int_0^1 (2x - 1)^4 2\, dx = \frac{5(2x - 1)^5}{5}\Big|_0^1$

$$= (2x - 1)^5 \Big|_0^1 = 1 - (-1)^5 = 2 \qquad\qquad (11\text{-}5)$$

34. $\int \left(\frac{2}{x^2} - 2xe^{x^2}\right) dx = 2\int x^{-2} dx - \int 2xe^{x^2} dx = \frac{2x^{-1}}{-1} - e^{x^2} + C$

$$= -2x^{-1} - e^{x^2} + C \qquad\qquad (11\text{-}2)$$

35. $\int_0^4 \sqrt{x^2 + 4}\, x\, dx = \int_0^4 (x^2 + 4)^{1/2} x\, dx = \frac{1}{2}\int_0^4 (x^2 + 4)^{1/2} 2x\, dx$

$$= \frac{1}{2}\cdot\frac{(x^2 + 4)^{3/2}}{\frac{3}{2}}\Big|_0^4 = \frac{(x^2 + 4)^{3/2}}{3}\Big|_0^4 = \frac{(20)^{3/2} - 8}{3} \approx 27.148$$

(11-5)

36. $\int (e^{-2x} + x^{-1}) dx = \int e^{-2x} dx + \int \frac{1}{x} dx = -\frac{1}{2}\int e^{-2x}(-2)\, dx + \ln|x| + C$

$$= -\frac{1}{2}e^{-2x} + \ln|x| + C \qquad\qquad (11\text{-}2)$$

37. $\displaystyle\int_0^{10} 10e^{-0.02x}dx = 10\int_0^{10} e^{-0.02x}dx = \frac{10}{-0.02}\int_0^{10} e^{-0.02x}(-0.02)\,dx$

$$= -500e^{-0.02x}\Big|_0^{10} = -500e^{-0.2} + 500 \approx 90.635 \qquad (11\text{-}5)$$

38. Let $u = 1 + x^2$, then $du = 2x\,dx$.

$$\int_0^3 \frac{x}{1 + x^2}\,dx = \int_0^3 \frac{1}{1 + x^2}\frac{2}{2}x\,dx$$

$$= \frac{1}{2}\int_0^3 \frac{1}{1 + x^2}2x\,dx = \frac{1}{2}\ln(1 + x^2)\Big|_0^3$$

$$= \frac{1}{2}\ln 10 - \frac{1}{2}\ln 1 = \frac{1}{2}\ln 10 \approx 1.151 \qquad (11\text{-}5)$$

39. Let $u = 1 + x^2$, then $du = 2x\,dx$.

$$\int_0^3 \frac{x}{(1 + x^2)^2}\,dx = \int_0^3 (1 + x^2)^{-2}\frac{2}{2}x\,dx = \frac{1}{2}\int_0^3 (1 + x^2)^{-2}2x\,dx$$

$$= \frac{1}{2}\cdot\frac{(1 + x^2)^{-1}}{-1}\Big|_0^3 = \frac{-1}{2(1 + x^2)}\Big|_0^3 = -\frac{1}{20} + \frac{1}{2} = \frac{9}{20} = 0.45$$

$$(11\text{-}5)$$

40. Let $u = 2x^4 + 5$, then $du = 8x^3 dx$.

$$\int x^3(2x^4 + 5)^5 dx = \int (2x^4 + 5)^5 \frac{8}{8}x^3 dx = \frac{1}{8}\int u^5 du$$

$$= \frac{1}{8}\cdot\frac{u^6}{6} + C = \frac{(2x^4 + 5)^6}{48} + C \qquad (11\text{-}2)$$

41. Let $u = e^{-x} + 3$, then $du = -e^{-x}dx$.

$$\int \frac{e^{-x}}{e^{-x} + 3}\,dx = \int \frac{1}{e^{-x} + 3}\cdot\frac{(-1)}{(-1)}e^{-x}dx = -\int \frac{1}{u}\,du$$

$$= -\ln|u| + C = -\ln|e^{-x} + 3| + C = -\ln(e^{-x} + 3) + C$$

[<u>Note</u>: Absolute value not needed since $e^{-x} + 3 > 0$.] $\qquad (11\text{-}2)$

42. Let $u = e^x + 2$, then $du = e^x dx$.

$$\int \frac{e^x}{(e^x + 2)^2}\,dx = \int (e^x + 2)^{-2}e^x dx = \int u^{-2}du$$

$$= \frac{u^{-1}}{-1} + C = -(e^x + 2)^{-1} + C = \frac{-1}{(e^x + 2)} + C \qquad (11\text{-}2)$$

43. $\dfrac{dy}{dx} = 3x^{-1} - x^{-2}$

$$y = \int (3x^{-1} - x^{-2})\,dx = 3\int \frac{1}{x}\,dx - \int x^{-2}dx$$

$$= 3\ln|x| - \frac{x^{-1}}{-1} + C = 3\ln|x| + x^{-1} + C$$

Given $y(1) = 5$:
$5 = 3\ln 1 + 1 + C$ and $C = 4$
Thus, $y = 3\ln|x| + x^{-1} + 4$. $\qquad (11\text{-}2, 11\text{-}3)$

44. $\dfrac{dy}{dx} = 6x + 1$

$f(x) = y = \displaystyle\int (6x + 1)\,dx = \dfrac{6x^2}{2} + x + C = 3x^2 + x + C$

We have $y = 10$ when $x = 2$: $3(2)^2 + 2 + C = 10$

$\qquad\qquad\qquad\qquad\qquad\qquad\qquad C = 10 - 12 - 2 = -4$

Thus, the equation of the curve is $y = 3x^2 + x - 4$. $\hfill$ (11-3)

45. $r(t)$ given in the figure, $a = 0$, $b = 1$, $n = 5$, $\Delta t = \dfrac{5 - 0}{5} = 1$.

$L_5 = r(0)\Delta t + r(1)\Delta t + r(2)\Delta t + r(3)\Delta t + r(4)\Delta t$

$\quad = 160 + 128 + 96 + 64 + 32 = 480$ ft

$R_5 = r(1)\Delta t + r(2)\Delta t + r(3)\Delta t + r(4)\Delta t + r(5)\Delta t$

$\quad = 128 + 96 + 64 + 32 + 0 = 320$ ft

$A_5 = \dfrac{L_5 + R_5}{2} = \dfrac{480 + 320}{2} = 400$ ft

Error bound for L_5 and R_5.

Error $\le |r(5) - r(0)|\left(\dfrac{5 - 0}{5}\right) = |0 - 160| = 160$ ft

Error bound for A_5:

Error $\le |r(5) - r(0)|\left(\dfrac{5 - 0}{2 \cdot 5}\right) = |0 - 160|\dfrac{1}{2} = 80$ ft $\hfill$ (11-4)

46. The height of each rectangle represents an instantaneous rate and the base of each rectangle is a time interval; and rate *times* time *equals* distance. $\hfill$ (11-4)

47. We want to find n such that

$\qquad\qquad\qquad |I - A_n| < 1$:

$|r(5) - r(0)|\left(\dfrac{5 - 0}{2n}\right) < 1$

$\qquad |0 - 160|\dfrac{5}{2n} < 1$

$\qquad\qquad \dfrac{400}{n} < 1$

$\qquad\qquad\quad n > 400 \hfill$ (11-4)

48. The graph of r is a straight line.

Slope: $\dfrac{0 - 160}{5 - 0} = -32$; y-intercept: 160; equation: $r = -32t + 160$

Therefore, Height $= \displaystyle\int_0^5 (-32t + 160)\,dt$

$\qquad\qquad\qquad = (-16t^2 + 160t)\Big|_0^5$

$\qquad\qquad\qquad = 400$ ft $\hfill$ (11-4)

49. (A) $f(x) = 3\sqrt{x} = 3x^{1/2}$ on $[1, 9]$

Ave $f(x) = \dfrac{1}{9 - 1} \displaystyle\int_1^9 3x^{1/2}\,dx$

$= \dfrac{3}{8} \cdot \dfrac{x^{3/2}}{\frac{3}{2}}\bigg|_1^9 = \dfrac{1}{4}x^{3/2}\bigg|_1^9 = \dfrac{27}{4} - \dfrac{1}{4} = \dfrac{26}{4} = 6.5$

(B)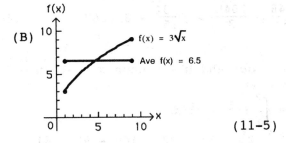

$(11\text{-}5)$

50. Let $u = \ln x$, then $du = \dfrac{1}{x}\,dx$.

$\displaystyle\int \dfrac{(\ln x)^2}{x}\,dx = \int (\ln x)^2 \dfrac{1}{x}\,dx = \int u^2\,du = \dfrac{u^3}{3} + C = \dfrac{(\ln x)^3}{3} + C$ $\qquad(11\text{-}2)$

51. $\displaystyle\int x(x^3 - 1)^2\,dx = \int x(x^6 - 2x^3 + 1)\,dx$ $\quad$ (square $x^3 - 1$)

$= \displaystyle\int (x^7 - 2x^4 + x)\,dx = \dfrac{x^8}{8} - \dfrac{2x^5}{5} + \dfrac{x^2}{2} + C$ $\qquad(11\text{-}2)$

52. Let $u = 6 - x$, then $x = 6 - u$ and $dx = -du$.

$\displaystyle\int \dfrac{x}{\sqrt{6 - x}}\,dx = -\int \dfrac{(6 - u)\,du}{u^{1/2}} = \int (u^{1/2} - 6u^{-1/2})\,du$

$= \dfrac{u^{3/2}}{\frac{3}{2}} - \dfrac{6u^{1/2}}{\frac{1}{2}} + C = \dfrac{2}{3}u^{3/2} - 12u^{1/2} + C$

$= \dfrac{2}{3}(6 - x)^{3/2} - 12(6 - x)^{1/2} + C$ $\qquad(11\text{-}2)$

53. $\displaystyle\int_0^7 x\sqrt{16 - x}\,dx.$ First consider the indefinite integral:

Let $u = 16 - x$, then $x = 16 - u$ and $dx = -du$.

$\displaystyle\int x\sqrt{16 - x}\,dx = -\int (16 - u)u^{1/2}\,du = \int (u^{3/2} - 16u^{1/2})\,du = \dfrac{u^{5/2}}{\frac{5}{2}} - \dfrac{16u^{3/2}}{\frac{3}{2}} + C$

$= \dfrac{2}{5}u^{5/2} - \dfrac{32}{3}u^{3/2} + C = \dfrac{2(16 - x)^{5/2}}{5} - \dfrac{32(16 - x)^{3/2}}{3} + C$

$$\int_0^7 x\sqrt{16 - x}\,dx = \left[\frac{2(16 - x)^{5/2}}{5} - \frac{32(16 - x)^{3/2}}{3}\right]\Bigg|_0^7$$

$$= \frac{2\cdot 9^{5/2}}{5} - \frac{32\cdot 9^{3/2}}{3} - \left(\frac{2\cdot 16^{5/2}}{5} - \frac{32\cdot 16^{3/2}}{3}\right)$$

$$= \frac{2\cdot 3^5}{5} - \frac{32\cdot 3^3}{3} - \left(\frac{2\cdot 4^5}{5} - \frac{32\cdot 4^3}{3}\right)$$

$$= \frac{486}{5} - 288 - \left(\frac{2048}{5} - \frac{2048}{3}\right) = \frac{1234}{15} \approx 82.267 \qquad (11\text{-}5)$$

54. Let $u = x + 1$, then $x = u - 1$, $dx = du$; and $u = 0$ when $x = -1$, $u = 2$ when $x = 1$.

$$\int_{-1}^{1} x(x + 1)^4\,dx = \int_0^2 (u - 1)u^4\,du = \int_0^2 (u^5 - u^4)\,du$$

$$= \left[\frac{u^6}{6} - \frac{u^5}{5}\right]\Bigg|_0^2 = \frac{2^6}{6} - \frac{2^5}{5} = \frac{32}{3} - \frac{32}{5} = \frac{160 - 96}{15} = \frac{64}{15} \approx 4.267$$

$$(11\text{-}5)$$

55. $\dfrac{dy}{dx} = 9x^2 e^{x^3}$, $f(0) = 2$

Let $u = x^3$, then $du = 3x^2\,dx$.

$$y = \int 9x^2 e^{x^3}\,dx = 3\int e^{x^3}\cdot 3x^2\,dx = 3\int e^u\,du = 3e^u + C = 3e^{x^3} + C$$

Given $f(0) = 2$:

$2 = 3e^0 + C = 3 + C$

Hence, $C = -1$ and $y = f(x) = 3e^{x^3} - 1$. $\qquad (11\text{-}3)$

56. $\dfrac{dN}{dt} = 0.06N$, $N(0) = 800$, $N > 0$

From the differential equation, $N(t) = Ce^{0.06t}$, where C is an arbitrary constant. Since $N(0) = 800$, we have

$800 = Ce^0 = C$.

Hence, $C = 800$ and $N(t) = 800e^{0.06t}$. $\qquad (11\text{-}3)$

57.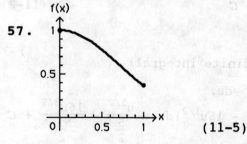

$(11\text{-}5)$

58. $f(x) = e^{-x^2}$ on $[0, 1]$; $a = 0$, $b = 1$, $n = 5$, $\Delta x = \dfrac{1 - 0}{5} = \dfrac{1}{5} = 0.2$

$$M_5 = f(0.1)\Delta x + f(0.3)\Delta x + f(0.5)\Delta x + f(0.7)\Delta x + f(0.9)\Delta x$$

$$= [0.9900 + 0.9139 + 0.7788 + 0.6126 + 0.4449]0.2$$

$$= 0.74805$$

$$(11\text{-}5)$$

59. $f(x) = e^{-x^2}$

$f'(x) = -2xe^{-x^2}$

$f''(x) = -2e^{-x^2} + 4x^2 e^{-x^2} = (4x^2 - 2)e^{-x^2}$

The graph of f'' is shown at the right.

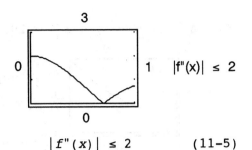

$|f''(x)| \leq 2$

(11-5)

60. From Problems 58 and 59, $M_5 \approx 0.74805$ and $|f''(x)| \leq 2$.

$$\text{Error} = |I - M_5| \leq \frac{B_2(b-a)^3}{24n^2} = \frac{2(1-0)^3}{24(5)^2} = \frac{1}{300} < 0.00334$$

Therefore,

$$I = M_5 \pm 0.00334 = 0.74805 \pm 0.00334$$

(11-5)

61. We want to find n such that

$$|I - M_n| \leq 0.0005:$$

$$|I - M_n| \leq \frac{2(1-0)^3}{24n^2} \leq 0.0005$$

$$\frac{1}{12n^2} \leq 0.0005$$

$$n^2 \geq \frac{1}{0.006} \approx 166.67$$

$$n \geq 12.9$$

Take $n \geq 13$.

(11-5)

62. $N = 50(1 - e^{-0.07t})$, $0 \leq t \leq 80$, $0 \leq N \leq 60$

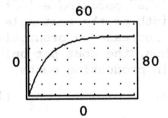

Limited growth

(11-3)

63. $p = 500e^{-0.03x}$, $0 \leq x \leq 100$, $0 \leq p \leq 500$

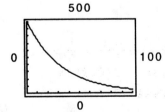

Exponential decay

(11-3)

64. $A = 200e^{0.08t}$, $0 \leq t \leq 20$, $0 \leq A \leq 1,000$

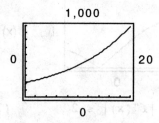

Unlimited growth

(11-3)

65. $N = \dfrac{100}{1 + 9e^{-0.3t}}$, $0 \leq t \leq 25$, $0 \leq N \leq 100$

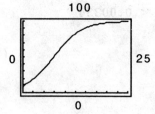

Logistic growth

(11-3)

66. $a = 200$, $b = 600$, $n = 2$, $\Delta x = \dfrac{600 - 200}{2} = 200$

$L_2 = C'(200)\Delta x + C'(400)\Delta x$

$\quad = [500 + 400]200 = \$180,000$

$R_2 = C'(400)\Delta x + C'(600)\Delta x$

$\quad = [400 + 300]200 = \$140,000$

$140,000 \leq \displaystyle\int_{200}^{600} C'(x)\,dx \leq 180,000$ (11-4)

67. The height of the rectangle, $C'(x)$, represents the marginal cost at a production level of x units; that is, $C'(x)$ is the approximate cost per unit at the production level of x units. The width of the rectangle represents the number of units involved in the increase in production. In the case of Problem 66, the width is 200. Thus, the cost per unit *times* the number of units *equals* the increase in production costs.

Note: The approximation improves as n increases. (11-4)

68. The graph of $C'(x)$ is a straight line with y-intercept = 600 and

$\quad$ slope $= \dfrac{300 - 600}{600 - 0} = -\dfrac{1}{2}$

Thus, $C'(x) = -\dfrac{1}{2}x + 600$

Increase in costs:

$\displaystyle\int_{200}^{600} \left(600 - \frac{1}{2}x\right) dx = \left(600x - \frac{1}{4}x^2\right)\Big|_{200}^{600}$

$\qquad\qquad\qquad\qquad = 270,000 - 110,000$

$\qquad\qquad\qquad\qquad = \$160,000$ (11-5)

69. The total change in profit for a production change from 10 units per week to 40 units per week is given by:

$$\int_{10}^{40} \left(150 - \frac{x}{10}\right) dx = \left(150x - \frac{x^2}{20}\right)\Big|_{10}^{40}$$

$$= \left(150(40) - \frac{40^2}{20}\right) - \left(150(10) - \frac{10^2}{20}\right)$$

$$= 5920 - 1495 = \$4425 \qquad (11\text{-}5)$$

70. $P'(x) = 100 - 0.02x$

$$P(x) = \int (100 - 0.02x)\, dx = 100x - 0.02\frac{x^2}{2} + C = 100x - 0.01x^2 + C$$

$P(0) = 0 - 0 + C = 0$

$\qquad\quad C = 0$

Thus, $P(x) = 100x - 0.01x^2$.

The profit on 10 units of production is given by:

$P(10) = 100(10) - 0.01(10)^2 = \$999 \qquad (11\text{-}3)$

71. The required definite integral is:

$$\int_{0}^{15} (60 - 4t)\, dt = (60t - 2t^2)\Big|_{0}^{15}$$

$$= 60(15) - 2(15)^2 = 450 \text{ or } 450{,}000 \text{ barrels}$$

The total production in 15 years is 450,000 barrels. $\qquad (11\text{-}5)$

72. Average inventory from $t = 3$ to $t = 6$:

$$\text{Ave } I(t) = \frac{1}{6 - 3}\int_{3}^{6} (10 + 36t - 3t^2)\, dt$$

$$= \frac{1}{3}[10t + 18t^2 - t^3]\Big|_{3}^{6}$$

$$= \frac{1}{3}[60 + 648 - 216 - (30 + 162 - 27)]$$

$$= 109 \text{ items} \qquad (11\text{-}5)$$

73. $S(x) = 8(e^{0.05x} - 1)$

Average price over the interval $[40, 50]$:

$$\text{Ave } S(x) = \frac{1}{50 - 40}\int_{40}^{50} 8(e^{0.05x} - 1)\, dx = \frac{8}{10}\int_{40}^{50} (e^{0.05x} - 1)\, dx$$

$$= \frac{4}{5}\left[\frac{e^{0.05x}}{0.05} - x\right]\Big|_{40}^{50}$$

$$= \frac{4}{5}[20e^{2.5} - 50 - (20e^2 - 40)]$$

$$= 16e^{2.5} - 16e^2 - 8 \approx \$68.70 \quad (11\text{-}5)$$

74. From the table, $a = 0$, $b = 50$, $n = 5$, $\Delta t = 10$.

$$L_5 = N(0)\Delta t + N(10)\Delta t + N(20)\Delta t + N(30)\Delta t + N(40)\Delta t$$
$$= [5 + 10 + 14 + 17 + 19]10 = 650$$

$$R_5 = N(10)\Delta t + N(20)\Delta t + N(30)\Delta t + N(40)\Delta t + N(50)\Delta t$$
$$= [10 + 14 + 17 + 19 + 20]10 = 800$$

$$A_5 = \frac{L_5 + R_5}{2} = \frac{650 + 800}{2} = 725 \text{ components}$$

Error bound for A_5:

$$\text{Error} \le |N(50) - N(0)| \frac{(50 - 0)}{2 \cdot 5} = |20 - 5|5 = 75 \tag{11-4}$$

75. To find the useful life, set $R'(t) = C'(t)$:

$$20e^{-0.1t} = 3$$
$$e^{-0.1t} = \frac{3}{20}$$
$$-0.1t = \ln\left(\frac{3}{20}\right) \approx -1.897$$
$$t = 18.97 \text{ or } 19 \text{ years}$$

$$\text{Total profit} = \int_0^{19} [R'(t) - C'(t)]dt = \int_0^{19} (20e^{-0.1t} - 3)dt$$

$$= 20\int_0^{19} e^{-0.1t}dt - \int_0^{19} 3\,dt = \frac{20}{-0.1}\int_0^{19} e^{-0.1t}(-0.1)\,dt - \int_0^{19} 3\,dt$$

$$= -200e^{-0.1t}\Big|_0^{19} - 3t\Big|_0^{19}$$

$$= -200e^{-1.9} + 200 - 57 \approx 113.086 \text{ or } \$113,086 \tag{11-5}$$

76. $S'(t) = 4e^{-0.08t}$, $0 \le t \le 24$. Therefore,

$$S(t) = \int 4e^{-0.08t}dt = \frac{4e^{-0.08t}}{-0.08} + C = -50e^{-0.08t} + C.$$

Now, $S(0) = 0$, so
$$0 = -50e^{-0.08(0)} + C = -50 + C.$$
Thus, $C = 50$, and $S(t) = 50(1 - e^{-0.08t})$ gives the total sales after t months.

Estimated sales after 12 months:
$$S(12) = 50(1 - e^{-0.08(12)}) = 50(1 - e^{-0.96}) \approx 31 \text{ or } \$31 \text{ million.}$$

To find the time to reach $40 million in sales, solve
$$40 = 50(1 - e^{-0.08t})$$
for t.
$$0.8 = 1 - e^{-0.08t}$$
$$e^{-0.08t} = 0.2$$
$$-0.08t = \ln(0.2)$$
$$t = \frac{\ln(0.2)}{-0.08} \approx 20 \text{ months} \tag{11-3}$$

77. Using a midpoint sum with $n = 6$, $\Delta t = \dfrac{12 - 0}{6} = 2$ and the values of C from the graph, we have:

$$\text{Ave } C(t) = \frac{1}{12 - 0}\int_0^{12} C(t)\,dt = \frac{1}{12}\int_0^{12} C(t)\,dt$$

$$\approx \frac{1}{12} M_6 = \frac{1}{12}[C(1)\Delta t + C(3)\Delta t + C(5)\Delta t + C(7)\Delta t + C(9)\Delta t + C(11)\Delta t]$$

$$= \frac{2}{12}[4 + 6 + 8 + 9 + 8 + 4] = \frac{39}{6} = 6.5 \text{ parts per million} \quad (11\text{-}5)$$

78. $\dfrac{dA}{dt} = -5t^{-2}$, $1 \le t \le 5$

$$A = \int -5t^{-2}\,dt = -5\int t^{-2}\,dt = -5 \cdot \frac{t^{-1}}{-1} + C = \frac{5}{t} + C$$

Now $A(1) = \dfrac{5}{1} + C = 5$. Therefore, $C = 0$ and

$$A(t) = \frac{5}{t}$$

$$A(5) = \frac{5}{5} = 1$$

The area of the wound after 5 days is 1 cm^2. $\qquad\qquad (11\text{-}3)$

79. The total amount of seepage during the first four years is given by:

$$T = \int_0^4 R(t)\,dt = \int_0^4 \frac{1000}{(1 + t)^2}\,dt = 1000\int_0^4 (1 + t)^{-2}\,dt = 1000\left.\frac{(1 + t)^{-1}}{-1}\right|_0^4$$

$$[\text{Let } u = 1 + t, \text{ then } du = dt.] = \left.\frac{-1000}{1 + t}\right|_0^4 = \frac{-1000}{5} + 1000 = 800 \text{ gallons}$$

$$(11\text{-}5)$$

80. (A) $A(t) = 770e^{0.01t}$, $t \ge 0$ (1995 is $t = 0$)

Population in year 2030: $t = 35$

$A(35) = 770e^{0.01(35)} = 770e^{0.35} \approx 1093$ million

(B) Time to double:

$$770e^{0.01t} = 1540$$
$$e^{0.01t} = 2$$
$$0.01t = \ln 2$$
$$t = \frac{\ln 2}{0.01} \approx 69.3$$

It will take approximately **70 years** for the population to double.

$$(11\text{-}3)$$

81. Let $Q = Q(t)$ be the amount of carbon-14 present in the bone at time t. Then,

$$\frac{dQ}{dt} = -0.0001238Q \quad \text{and} \quad Q(t) = Q_0 e^{-0.0001238t},$$

where Q_0 is the amount present originally (i.e., at the time the animal died). We want to find t such that $Q(t) = 0.04Q_0$.

$$0.04Q_0 = Q_0 e^{-0.0001238t}$$
$$e^{-0.0001238t} = 0.04$$
$$-0.0001238t = \ln 0.04$$
$$t = \frac{\ln 0.04}{-0.0001238} \approx 26,000 \text{ years} \tag{11-3}$$

82. $N'(t) = 7e^{-0.1t}$ and $N(0) = 25$.

$$N(t) = \int 7e^{-0.1t}dt = 7\int e^{-0.1t}dt = \frac{7}{-0.1}\int e^{-0.1t}(-0.1)dt$$

$$= -70e^{-0.1t} + C, \ 0 \le t \le 15$$

Given $N(0) = 25$: $25 = -70e^0 + C = -70 + C$

Hence, $C = 95$ and $N(t) = 95 - 70e^{-0.1t}$. The student would be expected to type $N(15) = 95 - 70e^{-0.1(15)} = 95 - 70e^{-1.5} \approx 79$ words per minute after completing the course. (11-3)

12 ADDITIONAL INTEGRATION TOPICS

Things to remember:

1. **AREA UNDER A CURVE**

    If f is continuous and $f(x) \geq 0$ over the interval $[a, b]$, then the area between $y = f(x)$ and the x-axis from $x = a$ to $x = b$ is given by the definite integral:

    $$A = \int_a^b f(x)\, dx$$

    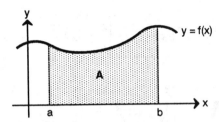

    If $f(x) \leq 0$ over the interval $[a, b]$, then the area between $y = f(x)$ and the x-axis from $x = a$ to $x = b$ is given by

    $$\int_a^b [-f(x)]\, dx.$$

    Finally, if $f(x)$ is positive for some values of x and negative for others, the area between the graph of f and the x axis can be obtained by dividing $[a, b]$ into subintervals on which f is always positive or always negative, finding the area over each subinterval, and then summing these areas.

2. **AREA BETWEEN TWO CURVES**

    If f and g are continuous and $f(x) \geq g(x)$ over the interval $[a, b]$, then the area bounded by $y = f(x)$ and $y = g(x)$, for $a \leq x \leq b$, is given exactly by:

    $$A = \int_a^b [f(x) - g(x)]\, dx.$$

    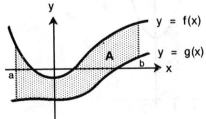

<u>3.</u>   INDEX OF INCOME CONCENTRATION

If $y = f(x)$ is the equation of a Lorenz curve, then the

Index of Income Concentration $= 2\int_0^1 [x - f(x)]\,dx.$

---

1. $A = \int_a^b g(x)\,dx$ **3.** $A = \int_a^b [-h(x)]\,dx$

5. Since the shaded region in Figure (c) is below the x-axis, $h(x) \leq 0$.
Thus, $\int_a^b h(x)\,dx$ represents the negative of the area of the region.

7. $A = \int_1^2 3x^2\,dx = \dfrac{3x^3}{3}\Big|_1^2 = x^3\Big|_1^2$
$= 2^3 - 1 = 7$

9. $A = \int_0^4 -[-2x - 1]\,dx = \int_0^4 [2x + 1]\,dx$
$= (x^2 + x)\Big|_0^4 = 20$

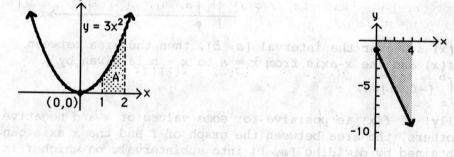

11. $A = \int_{-1}^0 (x^2 + 2)\,dx = \left(\dfrac{x^3}{3} + 2x\right)\Big|_{-1}^0$
$= 0 - \left(-\dfrac{1}{3} - 2\right) = \dfrac{7}{3} \approx 2.333$

13. $A = \int_{-1}^2 -[x^2 - 4]\,dx = \int_{-1}^2 [4 - x^2]\,dx$
$= \left(4x - \dfrac{x^3}{3}\right)\Big|_{-1}^2 = \left(8 - \dfrac{8}{3}\right) - \left(-4 + \dfrac{1}{3}\right)$
$= 9$

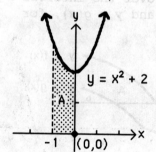

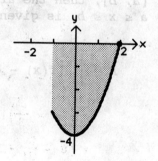

15. $A = \int_{-1}^{2} e^x dx = e^x \Big|_{-1}^{2}$

$\qquad = e^2 - e^{-1} \approx 7.021$

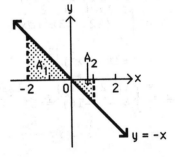

17. $A = \int_{0.5}^{1} -\left[-\frac{1}{t}\right] dt = \int_{0.5}^{1} \frac{1}{t} dt = \ln t \Big|_{0.5}^{1} = \ln 1 - \ln(0.5) \approx 0.693$

19. $A = \int_{a}^{b} [-f(x)] dx$ **21.** $a = \int_{b}^{c} f(x) dx + \int_{c}^{d} [-f(x)] dx$

23. $A = \int_{c}^{d} [f(x) - g(x)] dx$ **25.** $A = \int_{a}^{b} [f(x) - g(x)] dx + \int_{b}^{c} [g(x) - f(x)] dx$

27. Find the x-coordinates of the points of intersection of the two curves on $[a, d]$ by solving the equation $f(x) = g(x)$, $a \le x \le d$, to find $x = b$ and $x = c$. Then note that $f(x) \ge g(x)$ on $[a, b]$, $g(x) \ge f(x)$ on $[b, c]$ and $f(x) \ge g(x)$ on $[c, d]$.

Thus,

$$\text{Area} = \int_{a}^{b} [f(x) - g(x)] dx + \int_{b}^{c} [g(x) - f(x)] dx + \int_{c}^{d} [f(x) - g(x)] dx$$

29. $A = A_1 + A_2 = \int_{-2}^{0} -x \, dx + \int_{0}^{1} -(-x) dx$

$\qquad = -\int_{-2}^{0} x \, dx + \int_{0}^{1} x \, dx$

$\qquad = -\frac{x^2}{2}\Big|_{-2}^{0} + \frac{x^2}{2}\Big|_{0}^{1}$

$\qquad = -\left(0 - \frac{(-2)^2}{2}\right) + \left(\frac{1^2}{2} - 0\right)$

$\qquad = 2 + \frac{1}{2} = \frac{5}{2} = 2.5$

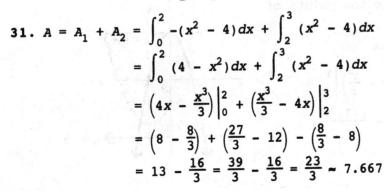

31. $A = A_1 + A_2 = \int_{0}^{2} -(x^2 - 4) dx + \int_{2}^{3} (x^2 - 4) dx$

$\qquad = \int_{0}^{2} (4 - x^2) dx + \int_{2}^{3} (x^2 - 4) dx$

$\qquad = \left(4x - \frac{x^3}{3}\right)\Big|_{0}^{2} + \left(\frac{x^3}{3} - 4x\right)\Big|_{2}^{3}$

$\qquad = \left(8 - \frac{8}{3}\right) + \left(\frac{27}{3} - 12\right) - \left(\frac{8}{3} - 8\right)$

$\qquad = 13 - \frac{16}{3} = \frac{39}{3} - \frac{16}{3} = \frac{23}{3} \approx 7.667$

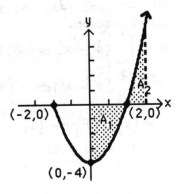

33. $A = A_1 + A_2 + A_3 = \int_{-3}^{-2} -[4 - x^2]\,dx + \int_{-2}^{2} (4 - x^2)\,dx + \int_{2}^{4} -[4 - x^2]\,dx$

$$= \int_{-3}^{-2} (x^2 - 4)\,dx + \int_{-2}^{2} (4 - x^2)\,dx + \int_{2}^{4} (x^2 - 4)\,dx$$

$$= \left(\frac{x^3}{3} - 4x\right)\Big|_{-3}^{-2} + \left(4x - \frac{x^3}{3}\right)\Big|_{-2}^{2} + \left(\frac{x^3}{3} - 4x\right)\Big|_{2}^{4}$$

$$= \left(-\frac{8}{3} + 8\right) - (-9 + 12) + \left(8 - \frac{8}{3}\right) - \left(-8 + \frac{8}{3}\right)$$

$$\qquad\qquad + \left(\frac{64}{3} - 16\right) - \left(\frac{8}{3} - 8\right)$$

$$= 13 + \frac{32}{3} \approx 23.667$$

35. $A = \int_{-1}^{2} [12 - (-2x + 8)]\,dx = \int_{-1}^{2} (2x + 4)\,dx$

$$= \left(\frac{2x^2}{2} + 4x\right)\Big|_{-1}^{2} = (x^2 + 4x)\Big|_{-1}^{2}$$

$$= (4 + 8) - (1 - 4)$$

$$= 12 + 3 = 15$$

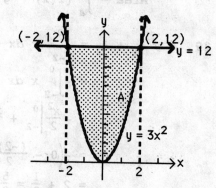

37. $A = \int_{-2}^{2} (12 - 3x^2)\,dx = \left(12x - \frac{3x^3}{3}\right)\Big|_{-2}^{2}$

$$= (12x - x^3)\Big|_{-2}^{2}$$

$$= (12\cdot2 - 2^3) - [12\cdot(-2) - (-2)^3]$$

$$= 16 - (-16) = 32$$

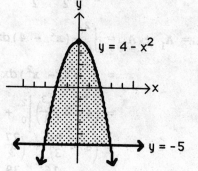

39. $(3, -5)$ and $(-3, -5)$ are the points of intersection.

$$A = \int_{-3}^{3} [4 - x^2 - (-5)]\,dx$$

$$= \int_{-3}^{3} (9 - x^2)\,dx = \left(9x - \frac{x^3}{3}\right)\Big|_{-3}^{3}$$

$$= \left(9\cdot3 - \frac{3^3}{3}\right) - \left(9(-3) - \frac{(-3)^3}{3}\right)$$

$$= 18 + 18 = 36$$

41. $A = \int_{-1}^{2} [(x^2 + 1) - (2x - 2)]\,dx$

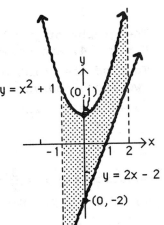

$= \int_{-1}^{2} (x^2 - 2x + 3)\,dx = \left(\dfrac{x^3}{3} - x^2 + 3x\right)\Big|_{-1}^{2}$

$= \left(\dfrac{8}{3} - 4 + 6\right) - \left(-\dfrac{1}{3} - 1 - 3\right)$

$= 3 - 4 + 6 + 1 + 3 = 9$

43. $A = \int_{1}^{2} \left[e^{0.5x} - \left(-\dfrac{1}{x}\right)\right]dx$

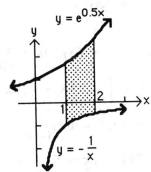

$= \int_{1}^{2} \left(e^{0.5x} + \dfrac{1}{x}\right)dx$

$= \left(\dfrac{e^{0.5x}}{0.5} + \ln|x|\right)\Big|_{1}^{2}$

$= 2e + \ln 2 - 2e^{0.5}$

≈ 2.832

45. The graphs of $y = 3 - 5x - 2x^2$ and $y = 2x^2 + 3x - 2$ are shown at the right. The x-coordinates of the points of intersection are: $x_1 = -2.5$, $x_2 = 0.5$.

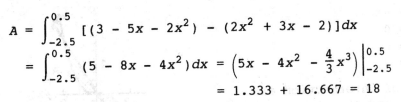

$A = \int_{-2.5}^{0.5} [(3 - 5x - 2x^2) - (2x^2 + 3x - 2)]\,dx$

$= \int_{-2.5}^{0.5} (5 - 8x - 4x^2)\,dx = \left(5x - 4x^2 - \dfrac{4}{3}x^3\right)\Big|_{-2.5}^{0.5}$

$= 1.333 + 16.667 = 18$

47. The graphs of $y = -0.5x + 2.25$ and $y = \dfrac{1}{x}$ are shown at the right. The x-coordinates of the points of intersection are: $x_1 = 0.5$, $x_2 = 4$.

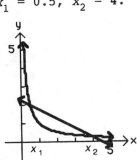

$A = \int_{0.5}^{4} \left[(-0.5x + 2.25) - \left(\dfrac{1}{x}\right)\right]dx$

$= \left(-\dfrac{1}{4}x^2 + \dfrac{9}{4}x - \ln x\right)\Big|_{0.5}^{4}$

$= [-4 + 9 - \ln 4] - [-0.0625 + 1.125 - \ln(0.5)]$

≈ 1.858

49. The graphs of $y = 10 - 2x$ and $y = 4 + 2x$, $0 \leq x \leq 4$, are shown at the right.

To find the point of intersection of the two lines, solve:

$$10 - 2x = 4 + 2x$$
$$-4x = -6$$
$$x = \frac{3}{2}$$

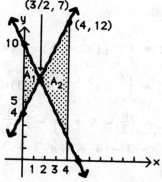

Substituting $x = \frac{3}{2}$ into either equation, we find $y = 7$. Now we have:

$$A = A_1 + A_2$$
$$= \int_0^{3/2} [(10 - 2x) - (4 + 2x)]\,dx + \int_{3/2}^4 [(4 + 2x) - (10 - 2x)]\,dx$$
$$= \int_0^{3/2} (6 - 4x)\,dx + \int_{3/2}^4 (4x - 6)\,dx$$
$$= (6x - 2x^2)\Big|_0^{3/2} + (2x^2 - 6x)\Big|_{3/2}^4$$
$$= 9 - \frac{9}{2} + (32 - 24) - \left(\frac{9}{2} - 9\right) = 17$$

51. The graphs are given at the right. To find the points of intersection, solve:

$$x^3 = 4x$$
$$x^3 - 4x = 0$$
$$x(x^2 - 4) = 0$$
$$x(x + 2)(x - 2) = 0$$

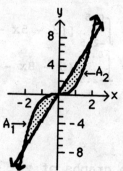

Thus, the points of intersection are $(-2, -8)$, $(0, 0)$, and $(2, 8)$.

$$A = A_1 + A_2 = \int_{-2}^0 (x^3 - 4x)\,dx + \int_0^2 (4x - x^3)\,dx$$

$$= \left(\frac{x^4}{4} - 2x^2\right)\Big|_{-2}^0 + \left(2x^2 - \frac{x^4}{4}\right)\Big|_0^2$$

$$= 0 - \left[\frac{(-2)^4}{4} - 2(-2)^2\right] + \left[2(2^2) - \frac{2^4}{4}\right] - 0$$

$$= -4 + 8 + 8 - 4 = 8$$

53. The graphs are given at the right.
To find the points of intersection, solve:

$$x^3 - 3x^2 - 9x + 12 = x + 12$$
$$x^3 - 3x^2 - 10x = 0$$
$$x(x^2 - 3x - 10) = 0$$
$$x(x - 5)(x + 2) = 0$$
$$x = -2, \ x = 0, \ x = 5$$

Thus, $(-2, 10)$, $(0, 12)$, and $(5, 17)$ are
the points of intersection.

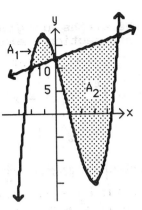

$$A = A_1 + A_2$$

$$= \int_{-2}^{0} [x^3 - 3x^2 - 9x + 12 - (x + 12)]\,dx$$

$$\qquad + \int_{0}^{5} [x + 12 - (x^3 - 3x^2 - 9x + 12)]\,dx$$

$$= \int_{-2}^{0} (x^3 - 3x^2 - 10x)\,dx + \int_{0}^{5} (-x^3 + 3x^2 + 10x)\,dx$$

$$= \left(\frac{x^4}{4} - x^3 - 5x^2\right)\Big|_{-2}^{0} + \left(-\frac{x^4}{4} + x^3 + 5x^2\right)\Big|_{0}^{5}$$

$$= -\left[\frac{(-2)^4}{4} - (-2)^3 - 5(-2)^2\right] + \left(\frac{-5^4}{4} + 5^3 + 5 \cdot 5^2\right)$$

$$= 8 + \frac{375}{4} = \frac{407}{4} = 101.75$$

55. The graphs are given at the right. To find the points of intersection,
solve:

$$x^4 - 4x^2 + 1 = x^2 - 3$$
$$x^4 - 5x^2 + 4 = 0$$
$$(x^2 - 4)(x^2 - 1) = 0$$
$$x = -2, \ -1, \ 1, \ 2$$

$$A = A_1 + A_2 + A_3$$

$$= \int_{-2}^{-1} [(x^2 - 3) - (x^4 - 4x^2 + 1)]\,dx + \int_{-1}^{1} [(x^4 - 4x^2 + 1) - (x^2 - 3)]\,dx$$

$$\qquad + \int_{1}^{2} [(x^2 - 3) - (x^4 - 4x^2 + 1)]\,dx$$

$$= \int_{-2}^{-1} (-x^4 + 5x^2 - 4)\,dx + \int_{-1}^{1} (x^4 - 5x^2 + 4)\,dx + \int_{1}^{2} (-x^4 + 5x^2 - 4)\,dx$$

$$= \left(-\frac{x^5}{5} + \frac{5}{3}x^3 - 4x\right)\Big|_{-2}^{-1} + \left(\frac{x^5}{5} - \frac{5}{3}x^3 + 4x\right)\Big|_{-1}^{1} + \left(-\frac{x^5}{5} + \frac{5}{3}x^3 - 4x\right)\Big|_{1}^{2}$$

$$= \left(\frac{1}{5} - \frac{5}{3} + 4\right) - \left(\frac{32}{5} - \frac{40}{3} + 8\right) + \left(\frac{1}{5} - \frac{5}{3} + 4\right) - \left(-\frac{1}{5} + \frac{5}{3} - 4\right)$$

$$\qquad + \left(-\frac{32}{5} + \frac{40}{3} - 8\right) - \left(-\frac{1}{5} + \frac{5}{3} - 4\right) = 8$$

57. The graphs are given below. The x-coordinates of the points of intersection are: $x_1 = -2$, $x_2 = 0.5$, $x_3 = 2$

$$A = A_1 + A_2$$

$$= \int_{-2}^{0.5} [(x^3 - x^2 + 2) - (-x^3 + 8x - 2)]dx$$

$$+ \int_{0.5}^{2} [(-x^3 + 8x - 2) - (x^3 - x^2 + 2)]dx$$

$$= \int_{-2}^{0.5} (2x^3 - x^2 - 8x + 4)dx + \int_{0.5}^{2} (-2x^3 + x^2 + 8x - 4)dx$$

$$= \left(\frac{1}{2}x^4 - \frac{1}{3}x^3 - 4x^2 + 4x\right)\Big|_{-2}^{0.5} + \left(-\frac{1}{2}x^4 + \frac{1}{3}x^3 + 4x^2 - 4x\right)\Big|_{0.5}^{2}$$

$$= \left(\frac{1}{32} - \frac{1}{24} - 1 + 2\right) - \left(8 + \frac{8}{3} - 16 - 8\right)$$

$$+ \left(-8 + \frac{8}{3} + 16 - 8\right) - \left(-\frac{1}{32} + \frac{1}{24} + 1 - 2\right)$$

$$= 18 + \frac{1}{16} - \frac{1}{12} \approx 17.979$$

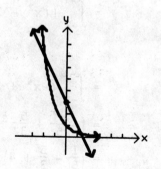

59. The graphs are given at the right. The x-coordinates of the points of intersection are: $x_1 \approx -1.924$, $x_2 \approx 1.373$

$$A = \int_{-1.924}^{1.373} [(3 - 2x) - e^{-x}]dx$$

$$= (3x - x^2 + e^{-x})\Big|_{-1.924}^{1.373}$$

$$\approx 2.487 - (-2.626) = 5.113$$

61. The graphs are given at the right. The x-coordinates of the points of intersection are: $x_1 \approx -2.247$, $x_2 \approx 0.264$, $x_3 \approx 1.439$

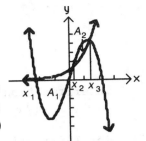

$$A = A_1 + A_2 = \int_{-2.247}^{0.264} [e^x - (5x - x^3)]dx$$

$$+ \int_{0.264}^{1.439} [5x - x^3 - e^x]dx$$

$$= \left(e^x - \frac{5}{2}x^2 + \frac{1}{4}x^4\right)\Big|_{-2.247}^{0.264} + \left(\frac{5}{2}x^2 - \frac{1}{4}x^4 - e^x\right)\Big|_{0.264}^{1.439}$$

$$\approx (1.129) - (-6.144) + (-0.112) - (-1.129) = 8.290$$

63. $\displaystyle\int_5^{10} R(t)\,dt = \int_5^{10}\left(\frac{100}{t+10} + 10\right)dt = 100\int_5^{10}\frac{1}{t+10}\,dt + \int_5^{10} 10\,dt$

$$= 100 \ln(t+10)\Big|_5^{10} + 10t\Big|_5^{10}$$

$$= 100 \ln 20 - 100 \ln 15 + 10(10 - 5)$$

$$= 100 \ln 20 - 100 \ln 15 + 50 \approx 79$$

The total production from the end of the fifth year to the end of the tenth year is approximately 79 thousand barrels.

65. To find the useful life, set $R'(t) = C'(t)$ and solve for t:

$$9e^{-0.3t} = 2$$

$$e^{-0.3t} = \frac{2}{9}$$

$$-0.3t = \ln\frac{2}{9}$$

$$-0.3t \approx -1.5$$

$$t \approx 5 \text{ years}$$

$$\int_0^5 [R'(t) - C'(t)]dt] = \int_0^5 [9e^{-0.3t} - 2]dt$$

$$= 9\int_0^5 e^{-0.3t}dt - \int_0^5 2\,dt = \frac{9}{-0.3}e^{-0.3t}\Big|_0^5 - 2t\Big|_0^5$$

$$= -30e^{-1.5} + 30 - 10$$

$$= 20 - 30e^{-1.5} \approx 13.306$$

The total profit over the useful life of the game is approximately $13,306.

67. For 1935: $f(x) = x^{2.4}$

Index of Income Concentration $= 2\displaystyle\int_0^1 [x - f(x)] = 2\int_0^1 (x - x^{2.4})\,dx$

$$= 2\left(\frac{x^2}{2} - \frac{x^{3.4}}{3.4}\right)\Big|_0^1$$

$$= 2\left(\frac{1}{2} - \frac{1}{3.4}\right) \approx 0.412$$

For 1947: $g(x) = x^{1.6}$

Index of Income Concentration $= 2\int_0^1 [x - g(x)]dx = 2\int_0^1 (x - x^{1.6})dx$

$$= 2\left(\frac{x^2}{2} - \frac{x^{2.6}}{2.6}\right)\Big|_0^1$$

$$= 2\left(\frac{1}{2} - \frac{1}{2.6}\right) \approx 0.231$$

Interpretation: Income was more equally distributed in 1947.

69. For 1963: $f(x) = x^{10}$

Index of Income Concentration $= 2\int_0^1 [x - f(x)]dx = 2\int_0^1 (x - x^{10})dx$

$$= 2\left(\frac{x^2}{2} - \frac{x^{11}}{11}\right)\Big|_0^1$$

$$= 2\left(\frac{1}{2} - \frac{1}{11}\right) \approx 0.818$$

For 1983: $g(x) = x^{12}$

Index of Income Concentration $= 2\int_0^1 [x - g(x)]dx = 2\int_0^1 (x - x^{12})dx$

$$= 2\left(\frac{x^2}{2} - \frac{x^{13}}{13}\right)\Big|_0^1$$

$$= 2\left(\frac{1}{2} - \frac{1}{13}\right) \approx 0.846$$

Interpretation: Total assets were less equally distributed in 1983.

71. $W(t) = \int_0^{10} W'(t)dt = \int_0^{10} 0.3e^{0.1t}dt = 0.3\int_0^{10} e^{0.1t}dt$

$$= \frac{0.3}{0.1}e^{0.1t}\Big|_0^{10} = 3e^{0.1t}\Big|_0^{10} = 3e - 3 \approx 5.15$$

Total weight gain during the first 10 hours is approximately 5.15 grams.

73. $V = \int_2^4 \frac{15}{t}dt = 15\int_2^4 \frac{1}{t}dt = 15 \ln t\Big|_2^4$

$$= 15 \ln 4 - 15 \ln 2 = 15 \ln\left(\frac{4}{2}\right) = 15 \ln 2 \approx 10$$

Average number of words learned during the second 2 hours is 10.

Things to remember:

1. PROBABILITY DENSITY FUNCTION

A function f which satisfies the following three conditions:

a. $f(x) \geq 0$ for all real x.

b. The area under the graph of f over the interval $(-\infty, \infty)$ is exactly 1.

c. If $[c, d]$ is a subinterval of $(-\infty, \infty)$, then the probability that the outcome x of an experiment will be in the interval $[c, d]$, denoted Probability $(c \leq x \leq d)$, is given by

$$\text{Probability } (c \leq x \leq d) = \int_c^d f(x)\,dx$$

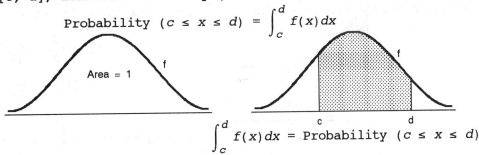

$$\int_c^d f(x)\,dx = \text{Probability } (c \leq x \leq d)$$

2. TOTAL INCOME FOR A CONTINUOUS INCOME STREAM

If $f(t)$ is the rate of flow of a continuous income stream, then the TOTAL INCOME produced during the time period from $t = a$ to $t = b$ is:

$$\text{Total income} = \int_a^b f(t)\,dt$$

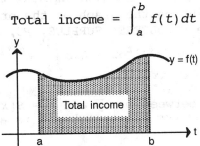

3. FUTURE VALUE OF A CONTINUOUS INCOME STREAM

If $f(t)$ is the rate of flow of a continuous income stream, $0 \leq t \leq T$, and if the income is continuously invested at a rate r, compounded continuously, then the FUTURE VALUE, FV, at the end of T years is given by:

$$FV = \int_0^T f(t)\,e^{r(T-t)}\,dt = e^{rT}\int_0^T f(t)\,e^{-rt}\,dt$$

The future value of a continuous income stream is the total value of all money produced by the continuous income stream (income and interest) at the end of T years.

<u>4</u>. CONSUMERS' SURPLUS

If $(\bar{x}, \bar{p})$ is a point on the graph of the price-demand equation $p = D(x)$ for a particular product, then the CONSUMERS' SURPLUS, CS, at a price level of $\bar{p}$ is

$$CS = \int_0^{\bar{x}} [D(x) - \bar{p}] dx$$

which is the area between $p = \bar{p}$ and $p = D(x)$ from $x = 0$ to $x = \bar{x}$.

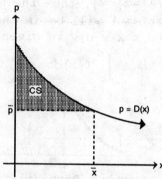

Consumer's surplus represents the total savings to consumers who are willing to pay more than $\bar{p}$ for the product but are still able to buy the product for $\bar{p}$.

<u>5</u>. PRODUCERS' SURPLUS

If $(\bar{x}, \bar{p})$ is a point on the graph of the price-supply equation $p = S(x)$, then the PRODUCERS' SURPLUS, PS, at a price level of $\bar{p}$ is

$$PS = \int_0^{\bar{x}} [\bar{p} - S(x)] dx$$

which is the area between $p = \bar{p}$ and $p = S(x)$ from $x = 0$ to $x = \bar{x}$.

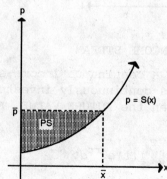

Producers' surplus represents the total gain to producers who are willing to supply units at a lower price than $\bar{p}$ but are still able to supply units at $\bar{p}$.

<u>6</u>. EQUILIBRIUM PRICE AND EQUILIBRIUM QUANTITY

If $p = D(x)$ and $p = S(x)$ are the price-demand and the price-supply equations, respectively, for a product and if $(\bar{x}, \bar{p})$ is the point of intersection of these equations, then $\bar{p}$ is called the EQUILIBRIUM PRICE and $\bar{x}$ is called the EQUILIBRIUM QUANTITY.

---

1. $f(x) = \begin{cases} \dfrac{2}{(x+2)^2}, & x \geq 0 \\ 0 & x < 0 \end{cases}$

(A) Probability $(0 \leq x \leq 6) = \displaystyle\int_0^6 f(x)\,dx = \int_0^6 \dfrac{2}{(x+2)^2}\,dx$

$$= 2\dfrac{(x+2)^{-1}}{-1}\Big|_0^6 = \dfrac{-2}{(x+2)}\Big|_0^6$$

$$= -\dfrac{1}{4} + 1 = \dfrac{3}{4} = 0.75$$

Thus, Probability $(0 \leq x \leq 6) = 0.75$

(B) Probability $(6 \leq x \leq 12) = \displaystyle\int_6^{12} f(x)\,dx = \int_6^{12} \dfrac{2}{(x+2)^2}\,dx$

$$= \dfrac{-2}{x+2}\Big|_6^{12} = -\dfrac{1}{7} + \dfrac{1}{4} = \dfrac{3}{28} \approx 0.11$$

(C)

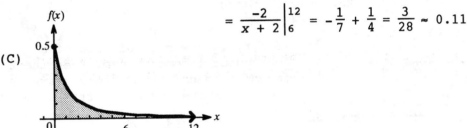

3. We want to find d such that

Probability $(0 \leq x \leq d) = \displaystyle\int_0^d f(x)\,dx = 0.8$:

$$\int_0^d f(x)\,dx = \int_0^d \dfrac{2}{(x+2)^2}\,dx = -\dfrac{2}{x+2}\Big|_0^d = \dfrac{-2}{d+2} + 1 = \dfrac{d}{d+2}$$

Now, $\dfrac{d}{d+2} = 0.8$

$$d = 0.8d + 1.6$$

$$0.2d = 1.6$$

$$d = 8 \text{ years}$$

5. $f(t) = \begin{cases} 0.01e^{-0.01t} & \text{if } t \geq 0 \\ 0 & \text{otherwise} \end{cases}$

(A) Since t is in months, the probability of failure during the warranty period of the first year is

$$\text{Probability } (0 \leq t \leq 12) = \int_0^{12} f(t)\,dt = \int_0^{12} 0.01e^{-0.01t}\,dt$$

$$= \frac{0.01}{-0.01} e^{-0.01t}\Big|_0^{12} = -1(e^{-0.12} - 1) \approx 0.11$$

(B) Probability $(12 \leq t \leq 24 = \int_{12}^{24} 0.01e^{-0.01t}\,dt = -1e^{-0.01t}\Big|_{12}^{24}$

$$= -1(e^{-0.24} - e^{-0.12}) \approx 0.10$$

7. Probability $(0 \leq t \leq \infty) = 1 = \int_0^\infty f(t)\,dt$

But, $\int_0^\infty f(t)\,dt = \int_0^{12} f(t)\,dt + \int_{12}^\infty f(t)\,dt$

Thus, Probability $(t \geq 12) = 1 - \text{Probability } (0 \leq t \leq 12)$
$$\approx 1 - 0.11 = 0.89$$

9. $f(t) = 2500$

Total income $= \int_0^5 2500\,dt = 2500t\Big|_0^5 = \$12,500$

11.

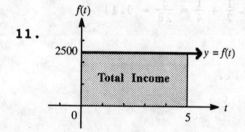

If $f(t)$ is the rate of flow of a continuous income stream, then the total income produced from 0 to 5 years is the area under the curve $y = f(t)$ from $t = 0$ to $t = 5$.

13. $f(t) = 400e^{0.05t}$

Total income $= \int_0^3 400e^{0.05t}\,dt = \frac{400}{0.05} e^{0.05t}\Big|_0^3 = 8000(e^{0.15} - 1) \approx \$1295$

15.

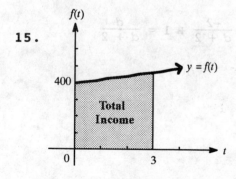

If $f(t)$ is the rate of flow of a continuous income stream, then the total income produced from 0 to 3 years is the area under the curve $y = f(t)$ from $t = 0$ to $t = 3$.

17. $f(t) = 2,000e^{0.05t}$
The amount in the account after 40 years is given by:
$$\int_0^{40} 2,000e^{0.05t}\,dt = 40,000e^{0.05t}\Big|_0^{40} = 295,562.24 - 40,000 \approx \$255,562$$

Since $\$2,000 \times 40 = \$80,000$ was deposited into the account, the interest earned is:
$$\$255,562 - \$80,000 = \$175,562$$

19. $f(t) = 1500e^{-0.02t}$, $r = 0.1$, $T = 4$
$$FV = e^{0.1(4)}\int_0^4 1500e^{-0.02t}e^{-0.1t}\,dt = 1500e^{0.4}\int_0^4 e^{-0.12t}\,dt$$

$$= \frac{1500e^{0.4}}{-0.12}e^{-0.12t}\Big|_0^4 = -12,500e^{0.4}(e^{-0.48} - 1)$$

$$= 12,500(e^{0.4} - e^{-0.08}) \approx \$7,109$$

21. Total Income $= \displaystyle\int_0^4 1,500e^{-0.02t}\,dt = -75,000e^{-0.02t}\Big|_0^4$
$$= 75,000 - 69,233.73 \approx \$5,766$$

From Problem 19,
$$\text{Interest Earned} = \$7,109 - \$5,766 = \$1,343$$

23. Clothing store: $f(t) = 12,000$, $r = 0.1$, $T = 5$.
$$FV = e^{0.1(5)}\int_0^5 12,000e^{-0.1t}\,dt = 12,000e^{0.5}\int_0^5 e^{-0.1t}\,dt$$

$$= \frac{12,000e^{0.5}}{-0.1}e^{-0.1t}\Big|_0^5 = -120,000e^{0.5}(e^{-0.5} - 1)$$

$$= 120,000(e^{0.5} - 1) \approx \$77,847$$

Computer store: $g(t) = 10,000e^{0.05t}$, $r = 0.1$, $T = 5$.
$$FV = e^{0.1(5)}\int_0^5 10,000e^{0.05t}e^{-0.1t}\,dt = 10,000e^{0.5}\int_0^5 e^{-0.05t}\,dt$$

$$= \frac{10,000e^{0.5}}{-0.05}e^{-0.05t}\Big|_0^5 = -200,000e^{0.5}(e^{-0.25} - 1)$$

$$= 200,000(e^{0.5} - e^{0.25}) \approx \$72,939$$
The clothing store is the better investment.

25. Bond: $P = \$10,000$, $r = 0.08$, $t = 5$.
$$FV = 10,000e^{0.08(5)} = 10,000e^{0.4} \approx \$14,918$$
Business: $f(t) = 2000$, $r = 0.08$, $T = 5$.
$$FV = e^{0.08(5)}\int_0^5 2000e^{-0.08t}\,dt = 2000e^{0.4}\int_0^5 e^{-0.08t}\,dt$$

$$= \frac{2000e^{0.4}}{-0.08}e^{-0.08t}\Big|_0^5 = -25,000e^{0.4}(e^{-0.4} - 1)$$

$$= 25,000(e^{0.4} - 1) \approx \$12,296$$
The bond is the better investment.

27. $f(t) = 9000$, $r = 0.12$, $T = 8$

$$FV = e^{0.12(8)} \int_0^8 9000e^{-0.12t}dt = 9000e^{0.96} \int_0^8 e^{-0.12t}dt$$

$$= \frac{9000e^{0.96}}{-0.12} e^{-0.12t} \Big|_0^8 = -75,000e^{0.96}(e^{-0.96} - 1)$$

$$= 75,000(e^{0.96} - 1) \approx \$120,877$$

The relationship between present value (*PV*) and future value (*FV*) at a continuously compounded interest rate r (expressed as a decimal) for t years is:

$FV = PVe^{rt}$ or $PV = FVe^{-rt}$

Thus, we have:

$PV = 120,877e^{-0.12(8)} = 120,877e^{-0.96} \approx 46,283$

Thus, the single deposit should be \$46,283.

29. $f(t) = k$, rate r (expressed as a decimal), years T:

$$FV = e^{rT} \int_0^T ke^{-rt}dt = ke^{rT} \int_0^T e^{-rt}dt = \frac{ke^{rT}}{-r} e^{-rt} \Big|_0^T$$

$$= -\frac{k}{r}e^{rT}(e^{-rT} - 1) = \frac{k}{r}(e^{rT} - 1)$$

31. $D(x) = 400 - \frac{1}{20}x$, $\overline{p} = 150$

First, find $\overline{x}$: $150 = 400 - \frac{1}{20}\overline{x}$

$$\overline{x} = 5000$$

$$CS = \int_0^{5000} \left[400 - \frac{1}{20}x - 150 \right]dx = \int_0^{5000} \left(250 - \frac{1}{20}x \right)dx$$

$$= \left(250x - \frac{1}{40}x^2 \right) \Big|_0^{5000} = \$625,000$$

33.

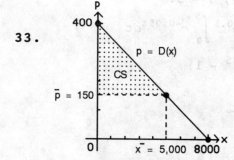

The shaded area is the consumers' surplus and represents the total savings to consumers who are willing to pay more than \$150 for a product but are still able to buy the product for \$150.

35. $p = S(x) = 10 + 0.1x + 0.0003x^2$, $\overline{p} = 67$.

First find $\overline{x}$: $67 = 10 + 0.1\overline{x} + 0.0003\overline{x}^2$

$$0.0003\overline{x}^2 + 0.1\overline{x} - 57 = 0$$

$$\overline{x} = \frac{-0.1 + \sqrt{0.01 + 0.0684}}{0.0006}$$

$$= \frac{-0.1 + 0.28}{0.0006} = 300$$

$$PS = \int_0^{300} [67 - (10 + 0.1x + 0.0003x^2)]\,dx$$

$$= \int_0^{300} (57 - 0.1x - 0.0003x^2)\,dx$$

$$= (57x - 0.05x^2 - 0.0001x^3) \Big|_0^{300} = \$9,900$$

37.

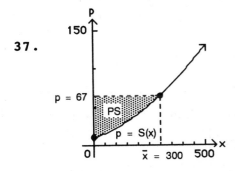

The area of the region PS is the producers' surplus and represents the total gain to producers who are willing to supply units at a lower price than \$67 but are still able to supply the product at \$67.

39. $p = D(x) = 50 - 0.1x$; $p = S(x) = 11 + 0.05x$

Equilibrium price: $D(x) = S(x)$
$$50 - 0.1x = 11 + 0.05x$$
$$39 = 0.15x$$
$$x = 260$$

Thus, $\overline{x} = 260$ and $\overline{p} = 50 - 0.1(260) = 24.$

$$CS = \int_0^{260} [(50 - 0.1x) - 24]\,dx = \int_0^{260} (26 - 0.1x)\,dx$$

$$= (26x - 0.05x^2) \Big|_0^{260}$$

$$= \$3,380$$

$$PS = \int_0^{260} [24 - (11 + 0.05x)]\,dx = \int_0^{260} [13 - 0.05x]\,dx$$

$$= (13x - 0.025x^2) \Big|_0^{260}$$

$$= \$1,690$$

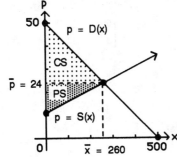

41. $D(x) = 80e^{-0.001x}$ and $S(x) = 30e^{0.001x}$

Equilibrium price: $D(x) = S(x)$

$$80e^{-0.001x} = 30e^{0.001x}$$

$$e^{0.002x} = \frac{8}{3}$$

$$0.002x = \ln\left(\frac{8}{3}\right)$$

$$\overline{x} = \frac{\ln\left(\frac{8}{3}\right)}{0.002} \approx 490$$

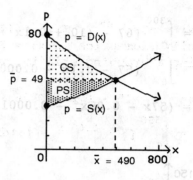

Thus, $\overline{p} = 30e^{0.001(490)} \approx 49$.

$$CS = \int_0^{490} [80e^{-0.001x} - 49]dx = \left(\frac{80e^{-0.001x}}{-0.001} - 49x\right)\Big|_0^{490}$$

$$= -80,000e^{-0.49} + 80,000 - 24,010 \approx \$6,980$$

$$PS = \int_0^{490} [49 - 30e^{0.001x}]dx = \left(49x - \frac{30e^{0.001x}}{0.001}\right)\Big|_0^{490}$$

$$= 24,010 - 30,000(e^{0.49} - 1) \approx \$5,041$$

43. $D(x) = 80 - 0.04x$; $S(x) = 30e^{0.001x}$

Equilibrium price: $D(x) = S(x)$

$$80 - 0.04x = 30e^{0.001x}$$

Using a graphing utility, we find that

$$\overline{x} \approx 614$$

Thus, $\overline{p} = 80 - (0.04)614 \approx 55$

$$CS = \int_0^{614} [80 - 0.04x - 55]dx = \int_0^{614} (25 - 0.04x)dx$$

$$= (25x - 0.02x^2)\Big|_0^{614}$$

$$\approx \$7,810$$

$$PS = \int_0^{614} [55 - (30e^{0.001x})]dx = \int_0^{614} (55 - 30e^{0.001x})dx$$

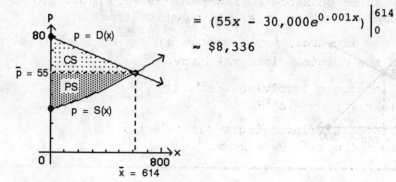

$$= (55x - 30,000e^{0.001x})\Big|_0^{614}$$

$$\approx \$8,336$$

45. $D(x) = 80e^{-0.001x}$; $S(x) = 15 + 0.0001x^2$

Equilibrium price: $D(x) = S(x)$

Using a graphing utility, we find that

$$\bar{x} \approx 556$$

Thus, $\bar{p} = 15 + 0.0001(556)^2 \approx 46$

$$CS = \int_0^{556} [80e^{-0.001x} - 46]dx = (-80,000e^{-0.001x} - 46x)\Big|_0^{556}$$

$$\approx \$8,544$$

$$PS = \int_0^{556} [46 - (15 + 0.0001x^2)]dx = \int_0^{556} (31 - 0.0001x^2)dx$$

$$= \left(31x - \frac{0.0001}{3}x^3\right)\Big|_0^{556}$$

$$\approx \$11,507$$

EXERCISE 12-3

Things to remember:

1. INTEGRATION-BY-PARTS FORMULA

$$\int u \, dv = uv - \int v \, du$$

2. INTEGRATION-BY-PARTS: SELECTION OF u AND dv

   (a) The product udv must equal the original integrand.

   (b) It must be possible to integrate dv (preferably by using standard formulas or simple substitutions.)

   (c) The new integral, $\int v \, du$, should not be any more involved than the original integral $\int u \, dv$.

   (d) For integrals involving $x^p e^{ax}$, try
   $$u = x^p; \quad dv = e^{ax}dx.$$

   (e) For integrals involving $x^p (\ln x)^q$, try
   $$u = (\ln x)^q; \quad dv = x^p dx.$$

---

1. $\int xe^{3x}dx$

Let $u = x$ and $dv = e^{3x}dx$. Then $du = dx$ and $v = \dfrac{e^{3x}}{3}$.

$\int xe^{3x}dx = \dfrac{xe^{3x}}{3} - \int \dfrac{e^{3x}}{3}dx = \dfrac{1}{3}xe^{3x} - \dfrac{1}{3}\int e^{3x}dx = \dfrac{1}{3}xe^{3x} - \dfrac{1}{9}e^{3x} + C$

3. $\int x^2 \ln x \, dx$

Let $u = \ln x$ and $dv = x^2 dx$. Then $du = \dfrac{dx}{x}$ and $v = \dfrac{x^3}{3}$.

$\int x^2 \ln x \, dx = (\ln x)\left(\dfrac{x^3}{3}\right) - \int \dfrac{x^3}{3}\cdot\dfrac{dx}{x} = \dfrac{1}{3}x^3 \ln x - \dfrac{1}{3}\int x^2 dx$

$= \dfrac{x^3 \ln x}{3} - \dfrac{1}{3}\cdot\dfrac{x^3}{3} + C = \dfrac{x^3 \ln x}{3} - \dfrac{x^3}{9} + C$

5. $\int xe^{-x}dx$

Let $u = x$ and $dv = e^{-x}dx$. Then $du = dx$ and $v = -e^{-x}$.

$\int xe^{-x}dx = x(-e^{-x}) - \int(-e^{-x})dx = -xe^{-x} + \int e^{-x}dx = -xe^{-x} - e^{-x} + C$

7. $\int xe^{x^2}dx = \int e^{x^2}\dfrac{2}{2}x \, dx = \dfrac{1}{2}\int e^{x^2}2x \, dx = \dfrac{1}{2}\int e^{u}du$

Let $u = x^2$, $\qquad\qquad = \dfrac{1}{2}e^{u} + C = \dfrac{1}{2}e^{x^2} + C$
then $du = 2x \, dx$.

9. $\int_0^1 (x - 3)e^x dx$

Let $u = (x - 3)$ and $dv = e^x dx$. Then $du = dx$ and $v = e^x$.

$\int (x - 3)e^x dx = (x - 3)e^x - \int e^x dx = (x - 3)e^x - e^x + C$

$\qquad\qquad = xe^x - 4e^x + C.$

Thus, $\int_0^1 (x - 3)e^x dx = (xe^x - 4e^x)\Big|_0^1 = (e - 4e) - (-4)$

$\qquad\qquad\qquad = -3e + 4 \approx -4.1548.$

11. $\int_1^3 \ln 2x \, dx$

Let $u = \ln 2x$ and $dv = dx$. Then $du = \dfrac{dx}{x}$ and $v = x$.

$\int \ln 2x \, dx = (\ln 2x)(x) - \int x\cdot\dfrac{dx}{x} = x \ln 2x - x + C$

Thus, $\int_1^3 \ln 2x \, dx = (x \ln 2x - x)\Big|_1^3 = (3 \ln 6 - 3) - (\ln 2 - 1) \approx 2.6821.$

13. $\int \dfrac{2x}{x^2 + 1}\, dx = \int \dfrac{1}{u}\, du = \ln|u| + C = \ln(x^2 + 1) + C$

Substitution: $u = x^2 + 1$
$\qquad\qquad\qquad du = 2x\, dx$

[Note: Absolute value not needed, since $x^2 + 1 \geq 0$.]

15. $\int \dfrac{\ln\, x}{x}\, dx = \int u\, du = \dfrac{u^2}{2} + C = \dfrac{(\ln\, x)^2}{2} + C$

Substitution: $u = \ln\, x$
$\qquad\qquad\qquad du = \dfrac{1}{x}\, dx$

17. $\int \sqrt{x}\, \ln\, x\, dx = \int x^{1/2}\, \ln\, x\, dx$

Let $u = \ln\, x$ and $dv = x^{1/2} dx$. Then $du = \dfrac{dx}{x}$ and $v = \dfrac{2}{3} x^{3/2}$.

$\int x^{1/2}\, \ln\, x\, dx = \dfrac{2}{3} x^{3/2}\, \ln\, x - \int \dfrac{2}{3} x^{3/2}\, \dfrac{dx}{x} = \dfrac{2}{3} x^{3/2}\, \ln\, x - \dfrac{2}{3} \int x^{1/2} dx$

$\qquad\qquad\qquad\qquad\qquad\qquad = \dfrac{2}{3} x^{3/2}\, \ln\, x - \dfrac{4}{9} x^{3/2} + C$

19.

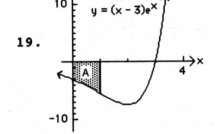

Since $f(x) = (x - 3)e^x < 0$ on $[0, 1]$, the integral represents the negative of the area between the graph of f and the x-axis from $x = 0$ to $x = 1$.

21.

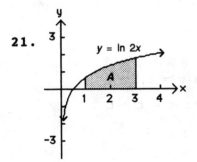

The integral represents the area between the curve $y = \ln\, 2x$ and the x-axis from $x = 1$ to $x = 3$.

23. $\int x^2 e^x dx$

Let $u = x^2$ and $dv = e^x dx$. Then $du = 2x\, dx$ and $v = e^x$.

$\int x^2 e^x dx = x^2 e^x - \int e^x(2x)\, dx = x^2 e^x - 2 \int x e^x dx$

$\int x e^x dx$ can be computed by using integration-by-parts again.

Let $u = x$ and $dv = e^x dx$. Then $du = dx$ and $v = e^x$.

$$\int xe^x dx = xe^x - \int e^x dx = xe^x - e^x + C$$

and

$$\int x^2 e^x dx = x^2 e^x - 2(xe^x - e^x) + C = x^2 e^x - 2xe^x + 2e^x + C$$

$$= (x^2 - 2x + 2)e^x + C$$

25. $\int xe^{ax} dx$

Let $u = x$ and $dv = e^{ax} dx$. Then $du = dx$ and $v = \dfrac{e^{ax}}{a}$.

$$\int xe^{ax} dx = x \cdot \frac{e^{ax}}{a} - \int \frac{e^{ax}}{a} dx = \frac{xe^{ax}}{a} - \frac{e^{ax}}{a^2} + C$$

27. $\int_1^e \dfrac{\ln x}{x^2} dx$

Let $u = \ln x$ and $dv = \dfrac{dx}{x^2}$. Then $du = \dfrac{dx}{x}$ and $v = \dfrac{-1}{x}$.

$$\int \frac{\ln x}{x^2} dx = (\ln x)\left(-\frac{1}{x}\right) - \int -\frac{1}{x} \cdot \frac{dx}{x} = -\frac{\ln x}{x} + \int \frac{dx}{x^2} = -\frac{\ln x}{x} - \frac{1}{x} + C$$

Thus, $\displaystyle\int_1^e \frac{\ln x}{x^2} dx = \left(-\frac{\ln x}{x} - \frac{1}{x}\right)\Bigg|_1^e = -\frac{\ln e}{e} - \frac{1}{e} - \left(-\frac{\ln 1}{1} - \frac{1}{1}\right)$

$$= -\frac{2}{e} + 1 \approx 0.2642.$$

[<u>Note</u>: $\ln e = 1$.]

29. $\displaystyle\int_0^2 \ln(x + 4) dx$

Let $t = x + 4$. Then $dt = dx$ and

$$\int \ln(x + 4) dx = \int \ln t\, dt.$$

Now, let $u = \ln t$ and $dv = dt$. Then $du = \dfrac{dt}{t}$ and $v = t$.

$$\int \ln t\, dt = t \ln t - \int t\left(\frac{1}{t}\right) dt = t \ln t - \int dt = t \ln t - t + C$$

Thus, $\displaystyle\int \ln(x + 4) dx = (x + 4) \ln(x + 4) - (x + 4) + C$

and

$$\int_0^2 \ln(x + 4) dx = [(x + 4) \ln(x + 4) - (x + 4)]\Big|_0^2$$

$$= 6 \ln 6 - 6 - (4 \ln 4 - 4) = 6 \ln 6 - 4 \ln 4 - 2 \approx 3.205.$$

31. $\int xe^{x-2} dx$

Let $u = x$ and $dv = e^{x-2} dx$. Then $du = dx$ and $v = e^{x-2}$.

$$\int xe^{x-2} dx = xe^{x-2} - \int e^{x-2} dx = xe^{x-2} - e^{x-2} + C$$

33. $\int x \ln(1 + x^2) dx$

Let $t = 1 + x^2$. Then $dt = 2x \, dx$ and

$$\int x \ln(1 + x^2) dx = \int \ln(1 + x^2) x \, dx = \int \ln t \frac{dt}{2} = \frac{1}{2} \int \ln t \, dt.$$

Now, for $\int \ln t \, dt$, let $u = \ln t$, $dv = dt$. Then $du = \frac{dt}{t}$ and $v = t$.

$$\int \ln t \, dt = t \ln t - \int t \left(\frac{1}{t}\right) dt = t \ln t - \int dt = t \ln t - t + C$$

Therefore,

$$\int x \ln(1 + x^2) dx = \frac{1}{2}(1 + x^2) \ln(1 + x^2) - \frac{1}{2}(1 + x^2) + C.$$

35. $\int e^x \ln(1 + e^x) dx$

Let $t = 1 + e^x$. Then $dt = e^x dx$ and

$$\int e^x \ln(1 + e^x) dx = \int \ln t \, dt.$$

Now, as shown in Problems 29 and 33,

$$\int \ln t \, dt = t \ln t - t + C.$$

Thus, $\int e^x \ln(1 + e^x) dx = (1 + e^x) \ln(1 + e^x) - (1 + e^x) + C.$

37. $\int (\ln x)^2 dx$

Let $u = (\ln x)^2$ and $dv = dx$. Then $du = \frac{2 \ln x}{x} dx$ and $v = x$.

$$\int (\ln x)^2 dx = x(\ln x)^2 - \int x \cdot \frac{2 \ln x}{x} dx = x(\ln x)^2 - 2 \int \ln x \, dx$$

$\int \ln x \, dx$ can be computed by using integration-by-parts again.

As shown in Problems 29 and 33,

$$\int \ln x \, dx = x \ln x - x + C.$$

Thus, $\int (\ln x)^2 dx = x(\ln x)^2 - 2(x \ln x - x) + C$

$$= x(\ln x)^2 - 2x \ln x + 2x + C.$$

39. $\int (\ln x)^3 dx$

Let $u = (\ln x)^3$ and $dv = dx$. Then $du = 3(\ln x)^2 \cdot \frac{1}{x} dx$ and $v = x$.

$$\int (\ln x)^3 dx = x(\ln x)^3 - \int x \cdot 3(\ln x)^2 \cdot \frac{1}{x} dx = x(\ln x)^3 - 3 \int (\ln x)^2 dx$$

Now, using Problem 37,

$$\int (\ln x)^2 dx = x(\ln x)^2 - 2x \ln x + 2x + C.$$

Therefore, $\int (\ln x)^3 dx = x(\ln x)^3 - 3[x(\ln x)^2 - 2x \ln x + 2x] + C$

$$= x(\ln x)^3 - 3x(\ln x)^2 + 6x \ln x - 6x + C.$$

41. $y = x - 2 - \ln x$, $1 \le x \le 4$

$y = 0$ at $x \approx 3.146$

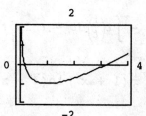

$$A = \int_1^{3.146} [-(x - 2 - \ln x)]dx + \int_{3.146}^4 (x - 2 - \ln x)dx$$

$$= \int_1^{3.146} (\ln x + 2 - x)dx + \int_{3.146}^4 (x - 2 - \ln x)dx$$

Now, $\int \ln x \, dx$ is found using integration-by-parts. Let $u = \ln x$ and

$dv = dx$. Then $du = \dfrac{1}{x} dx$ and $v = x$.

$$\int \ln x \, dx = x \ln x - \int x\left(\frac{1}{x}\right) dx = x \ln x - \int dx = x \ln x - x + C$$

Thus,

$$A = \left(x \ln x - x + 2x - \frac{1}{2}x^2\right)\Big|_1^{3.146} + \left(\frac{1}{2}x^2 - 2x - x \ln x + x\right)\Big|_{3.146}^4$$

$$= \left(x \ln x + x - \frac{1}{2}x^2\right)\Big|_1^{3.146} + \left(\frac{1}{2}x^2 - x - x \ln x\right)\Big|_{3.146}^4$$

$$\approx (1.803 - 0.5) + (-1.545 + 1.803) = 1.561$$

43. $y = 5 - xe^x$, $0 \le x \le 3$

$y = 0$ at $x \approx 1.327$

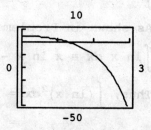

$$A = \int_0^{1.327} (5 - xe^x) dx + \int_{1.327}^3 [-(5 - xe^x)]dx$$

$$= \int_0^{1.327} (5 - xe^x) dx + \int_{1.327}^3 (xe^x - 5) dx$$

Now, $\int xe^x \, dx$ is found using integration-by-parts. Let $u = x$ and

$dv = e^x dx$. Then, $du = dx$ and $v = e^x$.

$$\int xe^x \, dx = xe^x - \int e^x dx = xe^x - e^x + C$$

Thus,

$$A = (5x - [xe^x - e^x]) \Big|_0^{1.327} + (xe^x - e^x - 5x) \Big|_{1.327}^3$$

$$\approx (5.402 - 1) + (25.171 - [-5.402]) \approx 34.98$$

45. Marginal profit: $P'(t) = 2t - te^{-t}$.
The total profit over the first 5 years is given by the definite integral:

$$\int_0^5 (2t - te^{-t})\,dt = \int_0^5 2t\;dt - \int_0^5 te^{-t}\;dt$$

We calculate the second integral using integration-by-parts. Let $u = t$ and $dv = e^{-t}\,dt$. Then $du = dt$ and $v = -e^{-t}$

$$\int te^{-t}\,dt = -te^{-t} - \int -e^{-t}\,dt = -te^{-t} - e^{-t} + C = -e^{-t}[t + 1] + C$$

Thus,

$$\text{Total profit} = t^2 \Big|_0^5 + (e^{-t}[t + 1]) \Big|_0^5$$

$$\approx 25 + (0.040 - 1) = 24.040$$

To the nearest million, the total profit is \$24 million.

47.

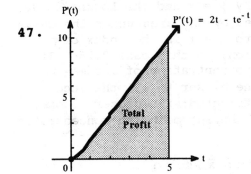

The total profit for the first five years (in millions of dollars) is the same as the area under the marginal profit function, $P'(t) = 2t - te^{-t}$, from $t = 0$ to $t = 5$.

49. Future Value $= e^{rT} \int_0^T f(t)e^{-rt}\,dt$. Now $r = 0.08$, $T = 5$,

$f(t) = 1000 - 200t$. Thus,

$$FV = e^{(0.08)5} \int_0^5 (1000 - 200t)e^{-0.08t}\,dt$$

$$= 1000e^{0.4} \int_0^5 e^{-0.08t}\,dt - 200e^{0.4} \int_0^5 te^{-0.08t}\,dt.$$

We calculate the second integral using integration-by-parts.

Let $u = t$, $dv = e^{-0.08t}\,dt$. Then $du = dt$ and $v = \dfrac{e^{-0.08t}}{-0.08}$.

$$\int te^{-0.08t}\,dt = \frac{te^{-0.08t}}{-0.08} - \int \frac{e^{-0.08t}}{-0.08}\,dt = -12.5te^{-0.08t} - \frac{e^{-0.08t}}{0.0064} + C$$

$$= -12.5te^{-0.08t} - 156.25e^{-0.08t} + C$$

Thus, we have:

$$FV = 1000e^{0.4}\frac{e^{-0.08t}}{-0.08}\Big|_0^5 - 200e^{0.4}[-12.5te^{-0.08t} - 156.25e^{-0.08t}]\Big|_0^5$$

$$= -12{,}500 + 12{,}500e^{0.4} - 200e^{0.4}[-62.5e^{-0.4} - 156.25e^{-0.4} + 156.25]$$

$$= -12{,}500 + 12{,}500e^{0.4} + 43{,}750 - 31{,}250e^{0.4}$$

$$= 31{,}250 - 18{,}750e^{0.4} \approx 3{,}278 \text{ or } \$3{,}278$$

51. Index of Income Concentration $= 2\int_0^1 (x - xe^{x-1})dx$

$$= 2\int_0^1 x\, dx - 2\int_0^1 xe^{x-1}dx$$

We calculate the second integral using integration-by-parts.
Let $u = x$, $dv = e^{x-1}dx$. Then $du = dx$, $v = e^{x-1}$.

$$\int xe^{x-1}dx = xe^{x-1} - \int e^{x-1}dx = xe^{x-1} - e^{x-1} + C$$

Therefore, $2\int_0^1 x\, dx - 2\int_0^1 xe^{x-1}dx = x^2\Big|_0^1 - 2[xe^{x-1} - e^{x-1}]\Big|_0^1$

$$= 1 - 2[1 - 1 + (e^{-1})]$$

$$= 1 - 2e^{-1} \approx 0.264.$$

53.

The area bounded by $y = x$ and the Lorenz curve $y = xe^{(x-1)}$ divided by the area under the curve $y = x$ from $x = 0$ to $x = 1$ is the index of income concentration, in this case 0.264. It is a measure of the concentration of income—the closer to zero, the closer to all the income being equally distributed; the closer to one, the closer to all the income being concentrated in a few hands.

55. $p = D(x) = 9 - \ln(x + 4)$; $\overline{p} = \$2.089$. To find $\overline{x}$, solve

$$9 - \ln(\overline{x} + 4) = 2.089$$

$$\ln(\overline{x} + 4) = 6.911$$

$$\overline{x} + 4 = e^{6.911} \quad \text{(take the exponential of both sides)}$$

$$\overline{x} \approx 1{,}000$$

Now,

$$CS = \int_0^{1{,}000} (D(x) - \overline{p})dx = \int_0^{1{,}000} [9 - \ln(x + 4) - 2.089]dx$$

$$= \int_0^{1{,}000} 6.911dx - \int_0^{1{,}000} \ln(x + 4)dx$$

To calculate the second integral, we first let $z = x + 4$ and $dz = dx$ to get

$$\int \ln(x + 4)dx = \int \ln z\, dz$$

Then we use integration-by-parts. Let $u = \ln z$ and $dv = dz$. Then $du = \frac{1}{z} dz$ and $v = z$.

$$\int \ln z \, dz = z \ln z - \int z \cdot \frac{1}{z} dz = z \ln z - z + C$$

Therefore,

$$\int \ln(x + 4) dx = (x + 4) \ln(x + 4) - (x + 4) + C$$

and

$$CS = 6.911x \Big|_0^{1,000} - [(x + 4) \ln(x + 4) - (x + 4)] \Big|_0^{1,000}$$

$$\approx 6911 - (5935.39 - 1.55) \approx \$977$$

57.

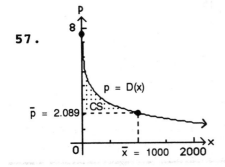

The area bounded by the price-demand equation, $p = 9 - \ln(x + 4)$, and the price equation, $y = \bar{p} = 2.089$, from $x = 0$ to $x = \bar{x} = 1,000$, represents the consumers' surplus. This is the amount saved by consumer's who are willing to pay more than \$2.089.

59. Average concentration: $= \dfrac{1}{5 - 0} \displaystyle\int_0^5 \dfrac{20 \ln(t + 1)}{(t + 1)^2} dt = 4 \displaystyle\int_0^5 \dfrac{\ln(t + 1)}{(t + 1)^2} dt$

$\displaystyle\int \dfrac{\ln(t + 1)}{(t + 1)^2} dt$ is found using integration-by-parts.

Let $u = \ln(t + 1)$ and $dv = (t + 1)^{-2} dt$.

Then $du = \dfrac{1}{t + 1} dt = (t + 1)^{-1} dt$ and $v = -(t + 1)^{-1}$.

$\displaystyle\int \dfrac{\ln(t + 1)}{(t + 1)^2} dt = -\dfrac{\ln(t + 1)}{t + 1} - \int - (t + 1)^{-1}(t + 1)^{-1} dt$

$\qquad = -\dfrac{\ln(t + 1)}{t + 1} + \displaystyle\int (t + 1)^{-2} dt = -\dfrac{\ln(t + 1)}{t + 1} - \dfrac{1}{t + 1} + C$

Therefore, the average concentration is:

$$\frac{1}{5} \int_0^5 \frac{20 \ln(t + 1)}{(t + 1)^2} dt = 4\left[-\frac{\ln(t + 1)}{t + 1} - \frac{1}{t + 1}\right] \Big|_0^5 = 4\left(-\frac{\ln 6}{6} - \frac{1}{6}\right) - 4(-\ln 1 - 1)$$

$$= 4 - \frac{2}{3} \ln 6 - \frac{2}{3} = \frac{1}{3}(10 - 2 \ln 6) \approx 2.1388 \text{ ppm}$$

61. Average number of voters $= \dfrac{1}{5} \displaystyle\int_0^5 (20 + 4t - 5te^{-0.1t}) dt$

$\qquad\qquad\qquad\qquad\qquad = \dfrac{1}{5} \displaystyle\int_0^5 (20 + 4t) dt - \displaystyle\int_0^5 te^{-0.1t} dt$

$\int te^{-0.1t}dt$ is found using integration-by-parts.

Let $u = t$ and $dv = e^{-0.1t}dt$. Then $du = dt$ and $v = \dfrac{e^{-0.1t}}{-0.1} = -10e^{-0.1t}$.

$\int te^{-0.1t}dt = -10te^{-0.1t} - \int -10e^{-0.1t}dt = -10te^{-0.1t} + 10\int e^{-0.1t}dt$

$\qquad\qquad = -10te^{-0.1t} + \dfrac{10e^{-0.1t}}{-0.1} + C = -10te^{-0.1t} - 100e^{-0.1t} + C$

Therefore, the average number of voters is:

$\dfrac{1}{5}\int_0^5 (20 + 4t)dt - \int_0^5 te^{-0.1t}dt$

$\qquad = \dfrac{1}{5}(20t + 2t^2)\Big|_0^5 - (-10te^{-0.1t} - 100e^{-0.1t})\Big|_0^5$

$\qquad = \dfrac{1}{5}(100 + 50) + (10te^{-0.1t} + 100e^{-0.1t})\Big|_0^5$

$\qquad = 30 + (50e^{-0.5} + 100e^{-0.5}) - 100$

$\qquad = 150e^{-0.5} - 70$

$\qquad \approx 20.98$ (thousands) or 20,980

EXERCISE 12-4

1. Use Formula 9 with $a = b = 1$.

$\int \dfrac{1}{x(1 + x)}dx = \dfrac{1}{1}\ln\left|\dfrac{x}{1 + x}\right| + C = \ln\left|\dfrac{x}{x + 1}\right| + C$

3. Use Formula 18 with $a = 3$, $b = 1$, $c = 5$, $d = 2$:

$\int \dfrac{1}{(3 + x)^2(5 + 2x)}dx = \dfrac{1}{3\cdot 2 - 5\cdot 1}\cdot\dfrac{1}{3 + x} + \dfrac{2}{(3\cdot 2 - 5\cdot 1)^2}\ln\left|\dfrac{5 + 2x}{3 + x}\right| + C$

$\qquad\qquad\qquad = \dfrac{1}{3 + x} + 2\ln\left|\dfrac{5 + 2x}{3 + x}\right| + C$

5. Use Formula 25 with $a = 16$ and $b = 1$:

$\int \dfrac{x}{\sqrt{16 + x}}dx = \dfrac{2(x - 2\cdot 16)}{3\cdot 1^2}\sqrt{16 + x} + C = \dfrac{2(x - 32)}{3}\sqrt{16 + x} + C$

7. Use Formula 37 with $a = 2$ $(a^2 = 4)$:

$\int \dfrac{1}{x\sqrt{x^2 + 4}}dx = \dfrac{1}{2}\ln\left|\dfrac{x}{2 + \sqrt{x^2 + 4}}\right| + C$

9. Use Formula 51 with $n = 2$:

$\int x^2\ln x\, dx = \dfrac{x^{2+1}}{2 + 1}\ln x - \dfrac{x^{2+1}}{(2 + 1)^2} + C = \dfrac{x^3}{3}\ln x - \dfrac{x^3}{9} + C$

11. First use Formula 5 with $a = 3$ and $b = 1$ to find the indefinite integral.

$$\int \frac{x^2}{3 + x} dx = \frac{(3 + x)^2}{2 \cdot 1^3} - \frac{2 \cdot 3(3 + x)}{1^3} + \frac{3^2}{1^3} \ln |3 + x| + C$$

$$= \frac{(3 + x)^2}{2} - 6(3 + x) + 9 \ln |3 + x| + C$$

Thus, $\int_1^3 \frac{x^2}{3 + x} dx = \left[\frac{(3 + x)^2}{2} - 6(3 + x) + 9 \ln |3 + x| \right]\Big|_1^3$

$$= \frac{(3 + 3)^2}{2} - 6(3 + 3) + 9 \ln |3 + 3|$$

$$- \left[\frac{(3 + 1)^2}{2} - 6(3 + 1) + 9 \ln |3 + 1| \right]$$

$$= 9 \ln \frac{3}{2} - 2 \approx 1.6492.$$

13. First use Formula 15 with $a = 3$, $b = c = d = 1$ to find the indefinite integral.

$$\int \frac{1}{(3 + x)(1 + x)} dx = \frac{1}{3 \cdot 1 - 1 \cdot 1} \ln \left| \frac{1 + x}{3 + x} \right| + C = \frac{1}{2} \ln \left| \frac{1 + x}{3 + x} \right| + C$$

Thus, $\int_0^7 \frac{1}{(3 + x)(1 + x)} dx = \frac{1}{2} \ln \left| \frac{1 + x}{3 + x} \right|\Big|_0^7 = \frac{1}{2} \ln \left| \frac{1 + 7}{3 + 7} \right| - \frac{1}{2} \ln \left| \frac{1}{3} \right|$

$$= \frac{1}{2} \ln \left| \frac{4}{5} \right| - \frac{1}{2} \ln \left| \frac{1}{3} \right| = \frac{1}{2} \ln \frac{12}{5} \approx 0.4377.$$

15. First use Formula 36 with $a = 3$ $(a^2 = 9)$ to find the indefinite integral:

$$\int \frac{1}{\sqrt{x^2 + 9}} dx = \ln \left| x + \sqrt{x^2 + 9} \right| + C$$

Thus, $\int_0^4 \frac{1}{\sqrt{x^2 + 9}} dx = \ln \left| x + \sqrt{x^2 + 9} \right|\Big|_0^4 = \ln \left| 4 + \sqrt{16 + 9} \right| - \ln \left| \sqrt{9} \right|$

$$= \ln 9 - \ln 3 = \ln 3 \approx 1.0986.$$

17. Consider Formula 35. Let $u = 2x$. Then $u^2 = 4x^2$, $x = \frac{u}{2}$, and $dx = \frac{du}{2}$.

$$\int \frac{\sqrt{4x^2 + 1}}{x^2} dx = \int \frac{\sqrt{u^2 + 1}}{\frac{u^2}{4}} \frac{du}{2} = 2 \int \frac{\sqrt{u^2 + 1}}{u^2} du$$

$$= 2 \left[-\frac{\sqrt{u^2 + 1}}{u} + \ln \left| u + \sqrt{u^2 + 1} \right| \right] + C$$

$$= 2 \left[-\frac{\sqrt{4x^2 + 1}}{2x} + \ln \left| 2x + \sqrt{4x^2 + 1} \right| \right] + C$$

$$= -\frac{\sqrt{4x^2 + 1}}{x} + 2 \ln \left| 2x + \sqrt{4x^2 + 1} \right| + C$$

19. Let $u = x^2$. Then $du = 2x\,dx$.

$$\int \frac{x}{\sqrt{x^4 - 16}}\,dx = \frac{1}{2}\int \frac{1}{\sqrt{u^2 - 16}}\,du$$

Now use Formula 43 with $a = 4$ $(a^2 = 16)$:

$$\frac{1}{2}\int \frac{1}{\sqrt{u^2 - 16}}\,du = \frac{1}{2}\ln\left| u + \sqrt{u^2 - 16}\right| + C = \frac{1}{2}\left|\ln x^2 + \sqrt{x^4 - 16}\right| + C$$

21. Let $u = x^3$. Then $du = 3x^2\,dx$.

$$\int x^2\sqrt{x^6 + 4}\,dx = \frac{1}{3}\int \sqrt{u^2 + 4}\,du$$

Now use Formula 32 with $a = 2$ $(a^2 = 4)$:

$$\frac{1}{3}\int \sqrt{u^2 + 4}\,du = \frac{1}{3}\cdot\frac{1}{2}\left[u\sqrt{u^2 + 4} + 4\ln\left| u + \sqrt{u^2 + 4}\right| \right] + C$$

$$= \frac{1}{6}\left[x^3\sqrt{x^6 + 4} + 4\ln\left| x^3 + \sqrt{x^6 + 4}\right| \right] + C$$

23. $$\int \frac{1}{x^3\sqrt{4 - x^4}}\,dx = \int \frac{x}{x^4\sqrt{4 - x^4}}\,dx$$

Let $u = x^2$. Then $du = 2x\,dx$.

$$\int \frac{x}{x^4\sqrt{4 - x^4}}\,dx = \frac{1}{2}\int \frac{1}{u^2\sqrt{4 - u^2}}\,du$$

Now use Formula 30 with $a = 2$ $(a^2 = 4)$:

$$\frac{1}{2}\int \frac{1}{u^2\sqrt{4 - u^2}}\,du = -\frac{1}{2}\cdot\frac{\sqrt{4 - u^2}}{4u} + C = \frac{-\sqrt{4 - x^4}}{8x^2} + C$$

25. $$\int \frac{e^x}{(2 + e^x)(3 + 4e^x)}\,dx = \int \frac{1}{(2 + u)(3 + 4u)}\,du$$

Substitution: $u = e^x$, $du = e^x dx$.

Now use Formula 15 with $a = 2$, $b = 1$, $c = 3$, $d = 4$:

$$\int \frac{1}{(2 + u)(3 + 4u)}\,du = \frac{1}{2\cdot4 - 3\cdot1}\ln\left|\frac{3 + 4u}{2 + u}\right| + C = \frac{1}{5}\ln\left|\frac{3 + 4e^x}{2 + e^x}\right| + C$$

27. $$\int \frac{\ln x}{x\sqrt{4 + \ln x}}\,dx = \int \frac{u}{\sqrt{4 + u}}\,du$$

Substitution: $u = \ln x$, $du = \frac{1}{x}dx$.

Use Formula 25 with $a = 4$, $b = 1$:

$$\int \frac{u}{\sqrt{4 + u}}\,du = \frac{2(u - 2\cdot4)}{3\cdot1^2}\sqrt{4 + u} + C = \frac{2(u - 8)}{3}\sqrt{4 + u} + C$$

$$= \frac{2(\ln x - 8)}{3}\sqrt{4 + \ln x} + C$$

29. Use Formula 47 with $n = 2$ and $a = 5$:

$$\int x^2 e^{5x}\,dx = \frac{x^2 e^{5x}}{5} - \frac{2}{5}\int xe^{5x}\,dx$$

To find $\int xe^{5x}dx$, use Formula 47 with $n = 1$, $a = 5$:

$$\int xe^{5x}dx = \frac{xe^{5x}}{5} - \frac{1}{5}\int e^{5x}dx = \frac{xe^{5x}}{5} - \frac{1}{5} \cdot \frac{e^{5x}}{5}$$

Thus, $\int x^2 e^{5x}dx = \frac{x^2 e^{5x}}{5} - \frac{2}{5}\left[\frac{xe^{5x}}{5} - \frac{1}{25}e^{5x}\right] + C = \frac{x^2 e^{5x}}{5} - \frac{2xe^{5x}}{25} + \frac{2e^{5x}}{125} + C.$

31. Use Formula 47 with $n = 3$ and $a = -1$.

$$\int x^3 e^{-x}dx = \frac{x^3 e^{-x}}{-1} - \frac{3}{-1}\int x^2 e^{-x}dx = -x^3 e^{-x} + 3\int x^2 e^{-x}dx$$

Now $\int x^2 e^{-x}dx = \frac{x^2 e^{-x}}{-1} - \frac{2}{-1}\int xe^{-x}dx = -x^2 e^{-x} + 2\int xe^{-x}dx$

and $\int xe^{-x}dx = \frac{xe^{-x}}{-1} - \frac{1}{-1}\int e^{-x}dx = -xe^{-x} - e^{-x}$, using Formula 47.

Thus, $\int x^3 e^{-x}dx = -x^3 e^{-x} + 3[-x^2 e^{-x} + 2(-xe^{-x} - e^{-x})] + C$

$$= -x^3 e^{-x} - 3x^2 e^{-x} - 6xe^{-x} - 6e^{-x} + C.$$

33. Use Formula 52 with $n = 3$:

$$\int (\ln x)^3 dx = x(\ln x)^3 - 3\int (\ln x)^2 dx$$

Now $\int (\ln x)^2 dx = x(\ln x)^2 - 2\int \ln x\, dx$ using Formula 52 again, and

$\int \ln x\, dx = x \ln x - x$ by Formula 49.

Thus, $\int (\ln x)^3 dx = x(\ln x)^3 - 3[x(\ln x)^2 - 2(x \ln x - x)] + C$
$$= x(\ln x)^3 - 3x(\ln x)^2 + 6x \ln x - 6x + C.$$

35. $\int_3^5 x\sqrt{x^2 - 9}\, dx$. First consider the indefinite integral.

Let $u = x^2 - 9$. Then $du = 2x\, dx$ or $x\, dx = \frac{1}{2}du$. Thus,

$$\int x\sqrt{x^2 - 9}\, dx = \frac{1}{2}\int u^{1/2}du = \frac{1}{2} \cdot \frac{u^{3/2}}{\frac{3}{2}} + C = \frac{1}{3}(x^2 - 9)^{3/2} + C.$$

Now, $\int_3^5 x\sqrt{x^2 - 9}\, dx = \frac{1}{3}(x^2 - 9)^{3/2}\Big|_3^5 = \frac{1}{3} \cdot 16^{3/2} = \frac{64}{3}.$

37. $\int_2^4 \frac{1}{x^2 - 1}dx$. Consider the indefinite integral:

$\int \frac{1}{x^2 - 1}dx = \frac{1}{2 \cdot 1}\ln\left|\frac{x - 1}{x + 1}\right| + C$, using Formula 13 with $a = 1$.

Thus, $\int_2^4 \frac{1}{x^2 - 1}dx = \frac{1}{2}\ln\left|\frac{x - 1}{x + 1}\right|\Big|_2^4 = \frac{1}{2}\ln\left|\frac{3}{5}\right| - \frac{1}{2}\ln\left|\frac{1}{3}\right| = \frac{1}{2}\ln\frac{9}{5} \approx 0.2939.$

39. $\displaystyle\int \frac{x+1}{x^2+2x}\,dx = \frac{1}{2}\int \frac{du}{u}$

Substitution: $u = x^2 + 2x$
$du = (2x+2)\,dx$
$= 2(x+1)\,dx$
$\frac{1}{2}\,du = (x+1)\,dx$

$\displaystyle = \frac{1}{2}\ln|u| + C$

$\displaystyle = \frac{1}{2}\ln|x^2+2x| + C$

41. $\displaystyle\int \frac{x+1}{x^2+3x}\,dx = \int \frac{x}{x(x+3)}\,dx + \int \frac{1}{x(x+3)}\,dx = \int \frac{1}{x+3}\,dx + \int \frac{1}{x(x+3)}\,dx$

Now $\displaystyle\int \frac{1}{x+3}\,dx = \ln|x+3|.$ Substitution: $u = x+3$, $du = dx$.

Use Formula 15 with $a = 0$, $b = 1$, $c = 3$, $d = 1$ on the second integral:

$\displaystyle\int \frac{1}{x(x+3)}\,dx = \frac{1}{0\cdot 1 - 1\cdot 3}\ln\left|\frac{x+3}{x}\right| = -\frac{1}{3}\ln\left|\frac{x+3}{x}\right|$

Thus, $\displaystyle\int \frac{x+1}{x^2+3x}\,dx = \ln|x+3| - \frac{1}{3}\ln\left|\frac{x+3}{x}\right| + C$

$\displaystyle = \ln|x+3| - \frac{1}{3}\ln|x+3| + \frac{1}{3}\ln|x| + C$

$\displaystyle = \frac{2}{3}\ln|x+3| + \frac{1}{3}\ln|x| + C$

43. $f(x) = \dfrac{10}{\sqrt{x^2+1}}$, $g(x) = x^2 + 3x$

The graphs of f and g are shown at the right.
The x-coordinates of the points of intersection
are: $x_1 \approx -3.70$, $x_2 \approx 1.36$

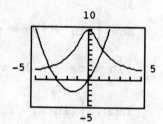

$A = \displaystyle\int_{-3.70}^{1.36}\left[\frac{10}{\sqrt{x^2+1}} - (x^2+3x)\right]dx$

$= 10\displaystyle\int_{-3.70}^{1.36}\frac{1}{\sqrt{x^2+1}}\,dx - \int_{-3.70}^{1.36}(x^2+3x)\,dx$

For the first integral, use Formula 36 with $a = 1$:

$A = \left(10\ln|x+\sqrt{x^2+1}|\right)\Big|_{-3.70}^{1.36} - \left(\frac{1}{3}x^3 + \frac{3}{2}x^2\right)\Big|_{-3.70}^{1.36}$

$\approx [11.15 - (-20.19)] - [3.61 - (3.65)] = 31.38$

45. $f(x) = x\sqrt{x+4}$, $g(x) = 1 + x$

The graphs of f and g are shown at the right.
The x-coordinates of the points of intersection
are: $x_1 \approx -3.49$, $x_2 \approx 0.83$

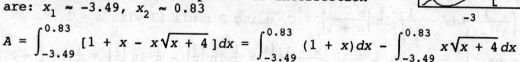

$A = \displaystyle\int_{-3.49}^{0.83}[1 + x - x\sqrt{x+4}]\,dx = \int_{-3.49}^{0.83}(1+x)\,dx - \int_{-3.49}^{0.83}x\sqrt{x+4}\,dx$

For the second integral, use Formula 22 with $a = 4$ and $b = 1$:

$$A = \left(x + \frac{1}{2}x^2\right)\Big|_{-3.49}^{0.83} - \left(\frac{2[3x - 8]}{15}\sqrt{(x + 4)^3}\right)\Big|_{-3.49}^{0.83}$$

$$\approx (1.17445 - 2.60005) - (-7.79850 + 0.89693) \approx 5.48$$

47. Find $\bar{x}$, the demand when the price $\bar{p} = 15$:

$$15 = \frac{7500 - 30\bar{x}}{300 - \bar{x}}$$

$$4500 - 15\bar{x} = 7500 - 30\bar{x}$$

$$15\bar{x} = 3000$$

$$\bar{x} = 200$$

Consumers' surplus:

$$CS = \int_0^{\bar{x}} [D(x) - \bar{p}]dx = \int_0^{200}\left[\frac{7500 - 30x}{300 - x} - 15\right]dx = \int_0^{200}\left[\frac{3000 - 15x}{300 - x}\right]dx$$

Use Formula 20 with $a = 3000$, $b = -15$, $c = 300$, $d = -1$:

$$CS = \left[\frac{-15x}{-1} + \frac{3000(-1) - (-15)(300)}{(-1)^2}\ln|300 - x|\right]\Big|_0^{200}$$

$$= [15x + 1500 \ln|300 - x|]\Big|_0^{200}$$

$$= 3000 + 1500 \ln(100) - 1500 \ln(300)$$

$$= 3000 + 1500 \ln\left(\frac{1}{3}\right) \approx 1352$$

Thus, the consumers' surplus is $1352.

49.

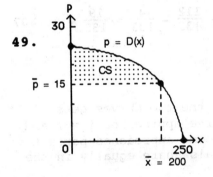

The shaded region represents the consumers' surplus.

51. $FV = e^{rT}\int_0^T f(t)e^{-rt}dt$

Now, $r = 0.1$, $T = 10$, $f(t) = 50t^2$.

$$FV = e^{(0.1)10}\int_0^{10} 50t^2e^{-0.1t}dt = 50e\int_0^{10} t^2e^{-0.1t}dt$$

To evaluate the integral, use Formula 47 with $n = 2$ and $a = -0.1$:

$$\int t^2e^{-0.1t}dt = \frac{t^2e^{-0.1t}}{-0.1} - \frac{2}{-0.1}\int te^{-0.1t}dt = -10t^2e^{-0.1t} + 20\int te^{-0.1t}dt$$

Now, using Formula 47 again:

$$\int te^{-0.1t}dt = \frac{te^{-0.1t}}{-0.1} - \frac{1}{-0.1}\int e^{-0.1t}dt = -10te^{-0.1t} + 10\frac{e^{-0.1t}}{-0.1}$$

$$= -10te^{-0.1t} - 100e^{-0.1t}$$

Thus, $\int t^2 e^{-0.1t}dt = -10t^2 e^{-0.1t} - 200te^{-0.1t} - 2000e^{-0.1t} + C.$

$$FV = 50e[-10t^2 e^{-0.1t} - 200te^{-0.1t} - 2000e^{-0.1t}] \Big|_0^{10}$$

$$= 50e[-1000e^{-1} - 2000e^{-1} - 2000e^{-1} + 2000] = 100,000e - 250,000$$

$$\approx 21,828 \text{ or } \$21,828$$

53. Index of Income Concentration:

$$2\int_0^1 [x - f(x)]dx = 2\int_0^1 \left[x - \frac{1}{2}x\sqrt{1 + 3x}\right]dx = \int_0^1 [2x - x\sqrt{1 + 3x}]dx$$

$$= \int_0^1 2x\,dx - \int_0^1 x\sqrt{1 + 3x}\,dx$$

For the second integral, use Formula 22 with $a = 1$ and $b = 3$:

$$= x^2 \Big|_0^1 - \frac{2(3\cdot 3x - 2\cdot 1)}{15(3)^2}\sqrt{(1 + 3x)^3} \Big|_0^1$$

$$= 1 - \frac{2(9x - 2)}{135}\sqrt{(1 + 3x)^3} \Big|_0^1$$

$$= 1 - \frac{14}{135}\sqrt{4^3} - \frac{4}{135}\sqrt{1^3} = 1 - \frac{112}{135} - \frac{4}{135} = \frac{19}{135} \approx 0.1407$$

55.

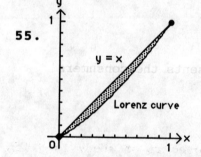

As the area bounded by the two curves gets smaller, the Lorenz curve approaches $y = x$ and the distribution of income approaches perfect equality—all individuals share equally in the income.

57. $S'(t) = \frac{t^2}{(1 + t)^2}; \quad S(t) = \int \frac{t^2}{(1 + t)^2}dt$

Use Formula 7 with $a = 1$ and $b = 1$:

$$S(t) = \frac{1 + t}{1^3} - \frac{1^2}{1^3(1 + t)} - \frac{2(1)}{1^3}\ln|1 + t| + C$$

$$= 1 + t - \frac{1}{1 + t} - 2\ln|1 + t| + C$$

Since $S(0) = 0$, we have $0 = 1 - 1 - 2\ln 1 + C$ and $C = 0$. Thus,

$$S(t) = 1 + t - \frac{1}{1 + t} - 2\ln|1 + t|.$$

Now, the total sales during the first two years (= 24 months) is given by:

$$S(24) = 1 + 24 - \frac{1}{1 + 24} - 2 \ln|1 + 24| = 24.96 - 2 \ln 25 \approx 18.5$$

Thus, total sales during the first two years is approximately $18.5 million.

59.

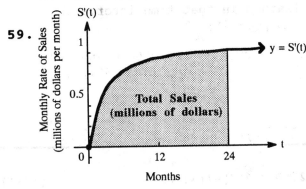

The total sales, in millions of dollars, over the first two years (24 months) is the area under the curve $y = S'(t)$ from $t = 0$ to $t = 24$.

61. $\frac{dR}{dt} = \frac{100}{\sqrt{t^2 + 9}}$. Therefore,

$$R = \int \frac{100}{\sqrt{t^2 + 9}}\, dt = 100 \int \frac{1}{\sqrt{t^2 + 9}}\, dt$$

Using Formula 36 with $a = 3$ ($a^2 = 9$), we have:

$R = 100 \ln\left| t + \sqrt{t^2 + 9} \right| + C$

Now $R(0) = 0$, so $0 = 100 \ln|3| + C$ or $C = -100 \ln 3$. Thus,

$R(t) = 100 \ln\left| t + \sqrt{t^2 + 9} \right| - 100 \ln 3$

and

$R(4) = 100 \ln(4 + \sqrt{4^2 + 9}) - 100 \ln 3$
$ = 100 \ln 9 - 100 \ln 3$
$ = 100 \ln 3 \approx 110 \text{ feet}$

63. $N'(t) = \frac{60}{\sqrt{t^2 + 25}}$

The number of items learned in the first twelve hours of study is given by:

$$N = \int_0^{12} \frac{60}{\sqrt{t^2 + 25}}\, dt = 60 \int_0^{12} \frac{1}{\sqrt{t^2 + 25}}\, dt$$

$$= 60 \left(\ln\left| t + \sqrt{t^2 + 25} \right| \right) \Big|_0^{12} , \text{ using Formula 36}$$

$$= 60 \left[\ln\left| 12 + \sqrt{12^2 + 25} \right| - \ln\sqrt{25} \right]$$

$$= 60(\ln 25 - \ln 5)$$

$$= 60 \ln 5 \approx 96.57 \text{ or 97 items}$$

65.

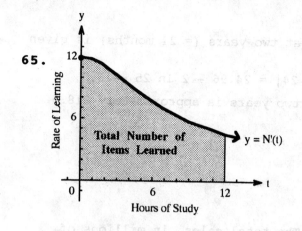

Rate of Learning

Total Number of Items Learned

$y = N'(t)$

Hours of Study

The area under the rate of learning curve, $y = N'(t)$, from $t = 0$ to $t = 12$ represents the total number of items learned in that time interval.

CHAPTER 12 REVIEW

1. $A = \int_a^b f(x)\, dx$ (12-1)   **2.** $A = \int_b^c [-f(x)]\, dx$ (12-1)

3. $A = \int_a^b f(x)\, dx + \int_b^c [-f(x)]\, dx$ (12-1)

4. $A = \int_{0.5}^1 [-\ln x]\, dx + \int_1^e \ln x\, dx$

We evaluate the integral using integration-by-parts. Let $u = \ln x$, $dv = dx$. Then $du = \frac{1}{x}\, dx$, $v = x$, and $\int \ln x\, dx =$

$x \ln x - \int x\left(\frac{1}{x}\right) dx = x \ln x - x + C$

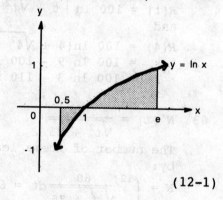

$y = \ln x$

Thus,

$A = -\int_{0.5}^1 \ln x\, dx + \int_1^e \ln x\, dx$

$= (-x \ln x + x)\Big|_{0.5}^1 + (x \ln x - x)\Big|_1^e$

$\approx (1 - 0.847) + (1) = 1.153$ (12-1)

5. $\int xe^{4x} dx$. Use integration-by-parts:

Let $u = x$ and $dv = e^{4x} dx$. Then $du = dx$ and $v = \frac{e^{4x}}{4}$.

$\int xe^{4x} dx = \frac{xe^{4x}}{4} - \int \frac{e^{4x}}{4} dx = \frac{xe^{4x}}{4} - \frac{e^{4x}}{16} + C$ (12-3, 12-4)

6. $\int x \ln x \, dx$. Use integration-by-parts:

Let $u = \ln x$ and $dv = x \, dx$. Then $du = \frac{1}{x} dx$ and $v = \frac{x^2}{2}$.

$$\int x \ln x \, dx = \frac{x^2 \ln x}{2} - \int \frac{1}{x} \cdot \frac{x^2}{2} dx = \frac{x^2 \ln x}{2} - \frac{1}{2} \int x \, dx = \frac{x^2 \ln x}{2} - \frac{x^2}{4} + C$$

(12-3, 12-4)

7. Use Formula 11 with $a = 1$ and $b = 1$.

$$\int \frac{1}{x(1 + x)^2} dx = \frac{1}{1(1 + x)} + \frac{1}{1^2} \ln \left| \frac{x}{1 + x} \right| + C = \frac{1}{1 + x} + \ln \left| \frac{x}{1 + x} \right| + C \quad (12-4)$$

8. Use Formula 28 with $a = 1$ and $b = 1$.

$$\int \frac{1}{x^2 \sqrt{1 + x}} dx = -\frac{\sqrt{1 + x}}{1 \cdot x} - \frac{1}{2 \cdot 1\sqrt{1}} \ln \left| \frac{\sqrt{1 + x} - \sqrt{1}}{\sqrt{1 + x} + \sqrt{1}} \right| + C$$

$$= -\frac{\sqrt{1 + x}}{x} - \frac{1}{2} \ln \left| \frac{\sqrt{1 + x} - 1}{\sqrt{1 + x} + 1} \right| + C \quad (12-4)$$

9. $A = \int_a^b [f(x) - g(x)] dx$ (12-1)   **10.** $A = \int_b^c [g(x) - f(x)] dx$ (12-1)

11. $A = \int_b^c [g(x) - f(x)] dx + \int_c^d [f(x) - g(x)] dx$ (12-1)

12. $A = \int_a^b [f(x) - g(x)] dx + \int_b^c [g(x) - f(x)] dx + \int_c^d [f(x) - g(x)] dx$ (12-1)

13. $A = \int_0^5 [(9 - x) - (x^2 - 6x + 9)] dx$

$= \int_0^5 (5x - x^2) \, dx$

$= \left(\frac{5}{2} x^2 - \frac{1}{3} x^3 \right) \Big|_0^5$

$= \frac{125}{2} - \frac{125}{3} = \frac{125}{6} \approx 20.833$

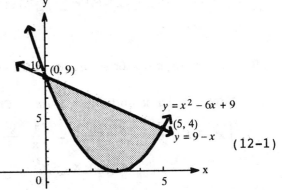

(12-1)

14. $\int_0^1 x e^x dx$. Use integration-by-parts.

Let $u = x$ and $dv = e^x dx$. Then $du = dx$ and $v = e^x$.

$$\int x e^x dx = x e^x - \int e^x dx = x e^x - e^x + C$$

Therefore, $\int_0^1 x e^x dx = (x e^x - e^x) \Big|_0^1 = 1 \cdot e - e - (0 \cdot 1 - 1)$

$= 1$ (12-3, 12-4)

15. Use Formula 38 with $a = 4$

$$\int_0^3 \frac{x^2}{\sqrt{x^2 + 16}}\, dx = \frac{1}{2}\left[\, x\sqrt{x^2 + 16} - 16\,\ln\left|\, x + \sqrt{x^2 + 16}\,\right|\,\right]\,\Big|_0^3$$

$$= \frac{1}{2}\left[\, 3\sqrt{25} - 16\,\ln(3 + \sqrt{25})\,\right] - \frac{1}{2}(-16\,\ln\sqrt{16})$$

$$= \frac{1}{2}[15 - 16\,\ln 8] + 8\,\ln 4$$

$$= \frac{15}{2} - 8\,\ln 8 + 8\,\ln 4 \approx 1.955 \tag{12-4}$$

16. Let $u = 3x$, then $du = 3\,dx$. Now, use Formula 40 with $a = 7$.

$$\int \sqrt{9x^2 - 49}\, dx = \frac{1}{3}\int \sqrt{u^2 - 49}\, du$$

$$= \frac{1}{3} \cdot \frac{1}{2}\left(\, u\sqrt{u^2 - 49} - 49\,\ln\left|\, u + \sqrt{u^2 - 49}\,\right|\,\right) + C$$

$$= \frac{1}{6}\left(\, 3x\sqrt{9x^2 - 49} - 49\,\ln\left|\, 3x + \sqrt{9x^2 - 49}\,\right|\,\right) + C \tag{12-4}$$

17. $\int te^{-0.5t}\, dt$. Use integration-by-parts.

Let $u = t$ and $dv = e^{-0.5t}\, dt$. Then $du = dt$ and $v = \dfrac{e^{-0.5t}}{-0.5}$.

$$\int te^{-0.5t}\, dt = \frac{-te^{-0.5t}}{0.5} + \int \frac{e^{-0.5t}}{0.5}\, dt = \frac{-te^{-0.5t}}{0.5} + \frac{e^{-0.5t}}{-0.25} + C$$

$$= -2te^{-0.5t} - 4e^{-0.5t} + C \tag{12-3, 12-4}$$

18. $\int x^2 \ln x\, dx$. Use integration-by-parts.

Let $u = \ln x$ and $dv = x^2 dx$. Then $du = \dfrac{1}{x}dx$ and $v = \dfrac{x^3}{3}$.

$$\int x^2 \ln x\, dx = \frac{x^3 \ln x}{3} - \int \frac{1}{x} \cdot \frac{x^3}{3}\, dx = \frac{x^3 \ln x}{3} - \frac{1}{3}\int x^2 dx$$

$$= \frac{x^3 \ln x}{3} - \frac{x^3}{9} + C \tag{12-3, 12-4}$$

19. Use Formula 48 with $a = 1$, $c = 1$, and $d = 2$.

$$\int \frac{1}{1 + 2e^x}\, dx = \frac{x}{1} - \frac{1}{1 \cdot 1}\,\ln\left|\, 1 + 2e^x\,\right| + C = x - \ln\left|\, 1 + 2e^x\,\right| + C \tag{12-4}$$

20. (A)

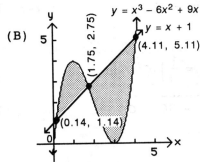

$$A = \int_0^2 [(x^3 - 6x^2 + 9x) - x]dx + \int_2^4 [x - (x^3 - 6x^2 + 9x)]dx$$

$$= \int_0^2 (x^3 - 6x^2 + 8x)dx + \int_2^4 (-x^3 + 6x^2 - 8x)dx$$

$$= \left(\frac{1}{4}x^4 - 2x^3 + 4x^2\right)\Big|_0^2 + \left(-\frac{1}{4}x^4 + 2x^3 - 4x^2\right)\Big|_2^4$$

$$= 4 + 4 = 8$$

(B)

The x-coordinates of the points of intersection are: $x_1 \approx 0.14$, $x_2 \approx 1.75$, $x_3 \approx 4.11$.

$$A = \int_{0.14}^{1.75} [(x^3 - 6x^2 + 9x) - (x + 1)]dx$$

$$+ \int_{1.75}^{4.11} [(x + 1) - (x^3 - 6x^2 + 9x)]dx$$

$$= \int_{0.14}^{1.75} (x^3 - 6x^2 + 8x - 1)dx + \int_{1.75}^{4.11} (1 - x^3 + 6x^2 - 8x)dx$$

$$= \left(\frac{1}{4}x^4 - 2x^3 + 4x^2 - x\right)\Big|_{0.14}^{1.75} + \left(x - \frac{1}{4}x^4 + 2x^3 - 4x^2\right)\Big|_{1.75}^{4.11}$$

$$= [2.126 - (-0.066)] + [4.059 - (-2.126)] \approx 8.38 \qquad (12-1)$$

21. $\int \dfrac{(\ln x)^2}{x} dx = \int u^2 du = \dfrac{u^3}{3} + C = \dfrac{(\ln x)^3}{3} + C$

Substitution: $u = \ln x$

$du = \dfrac{1}{x} dx$

$$(11-2)$$

22. $\int x(\ln x)^2 dx$. Use integration-by-parts.

Let $u = (\ln x)^2$ and $dv = x\, dx$. Then $du = 2(\ln x)\dfrac{1}{x}\, dx$ and $v = \dfrac{x^2}{2}$.

$\int x(\ln x)^2 dx = \dfrac{x^2(\ln x)^2}{2} - \int 2(\ln x)\dfrac{1}{x}\cdot\dfrac{x^2}{2} dx = \dfrac{x^2(\ln x)^2}{2} - \int x \ln x\, dx$

Let $u = \ln x$ and $dv = x\, dx$. Then $du = \dfrac{1}{x} dx$ and $v = \dfrac{x^2}{2}$.

$\int x \ln x\, dx = \dfrac{x^2 \ln x}{2} - \int \dfrac{x^2}{2}\cdot\dfrac{1}{x} dx = \dfrac{x^2 \ln x}{2} - \dfrac{1}{2}\int x\, dx = \dfrac{x^2 \ln x}{2} - \dfrac{x^2}{4} + C$

Thus, $\int x(\ln x)^2 dx = \dfrac{x^2(\ln x)^2}{2} - \left[\dfrac{x^2 \ln x}{2} - \dfrac{x^2}{4}\right] + C$

$= \dfrac{x^2(\ln x)^2}{2} - \dfrac{x^2 \ln x}{2} + \dfrac{x^2}{4} + C.$ $\qquad$ $(12-3,\ 12-4)$

23. Let $u = x^2 - 36$. Then $du = 2x\, dx$.

$\int \dfrac{x}{\sqrt{x^2 - 36}} dx = \int \dfrac{x}{(x^2 - 36)^{1/2}} dx = \dfrac{1}{2}\int \dfrac{1}{u^{1/2}} du = \dfrac{1}{2}\int u^{-1/2} du$

$= \dfrac{1}{2}\cdot\dfrac{u^{1/2}}{\frac{1}{2}} + C = u^{1/2} + C = \sqrt{x^2 - 36} + C$ $\qquad$ $(11-2)$

24. Let $u = x^2$, $du = 2x\, dx$.

Then use Formula 43 with $a = 6$.

$\int \dfrac{x}{\sqrt{x^4 - 36}} dx = \dfrac{1}{2}\int \dfrac{du}{\sqrt{u^2 - 36}} = \dfrac{1}{2} \ln\left| u + \sqrt{u^2 - 36}\right| + C$

$= \dfrac{1}{2} \ln\left| x^2 + \sqrt{x^4 - 36}\right| + C$ $\qquad$ $(12-4)$

25. $\int_0^4 x \ln(10 - x) dx$

Consider $\int x \ln(10 - x) dx = \int (10 - t)\ln t(-dt)$

Substitution: $t = 10 - x$

$dt = -dx$

$x = 10 - t$

$= \int t \ln t\, dt - 10\int \ln t\, dt.$

Now use integration-by-parts on the two integrals.

Let $u = \ln t$, $dv = t\, dt$. Then $du = \dfrac{1}{t} dt$, $v = \dfrac{t^2}{2}$.

$\int t \ln t\, dt = \dfrac{t^2}{2} \ln t - \int \dfrac{t^2}{2}\cdot\dfrac{1}{t} dt = \dfrac{t^2 \ln t}{2} - \dfrac{t^2}{4} + C$

Let $u = \ln t$, $dv = dt$. Then $du = \dfrac{1}{t} dt$, $v = t$.

$$\int \ln t\, dt = t \ln t - \int t \cdot \frac{1}{t} dt = t \ln t - t + C$$

Thus, $\displaystyle\int_0^4 x \ln(10 - x)\,dx = \left[\frac{(10 - x)^2 \ln(10 - x)}{2} - \frac{(10 - x)^2}{4}\right.$

$$\left. - 10(10 - x)\ln(10 - x) + 10(10 - x)\right]\Bigg|_0^4$$

$$= \frac{36 \ln 6}{2} - \frac{36}{4} - 10(6)\ln 6 + 10(6)$$

$$- \left[\frac{100 \ln 10}{2} - \frac{100}{4} - 10(10)\ln 10 + 10(10)\right]$$

$$= 18 \ln 6 - 9 - 60 \ln 6 + 60 - 50 \ln 10 + 25$$
$$+ 100 \ln 10 - 100$$

$$= 50 \ln 10 - 42 \ln 6 - 24 \approx 15.875. \quad (12\text{-}3, \ 12\text{-}4)$$

26. Use Formula 52 with $n = 2$.

$$\int (\ln x)^2 dx = x(\ln x)^2 - 2\int \ln x\, dx$$

Now use integration-by-parts to calculate $\displaystyle\int \ln x\, dx$.

Let $u = \ln x$, $dv = dx$. Then $du = \frac{1}{x} dx$, $v = x$.

$$\int \ln x\, dx = x \ln x - \int x \cdot \frac{1}{x} dx = x \ln x - x + C$$

Therefore, $\displaystyle\int (\ln x)^2 dx = x(\ln x)^2 - 2[x \ln x - x] + C$

$$= x(\ln x)^2 - 2x \ln x + 2x + C. \qquad (12\text{-}3, \ 12\text{-}4)$$

27. $\displaystyle\int x e^{-2x^2}\, dx$

Let $u = -2x^2$. Then $du = -4x\, dx$.

$$\int x e^{-2x^2}\, dx = -\frac{1}{4}\int e^u\, du = -\frac{1}{4} e^u + C$$

$$= -\frac{1}{4} e^{-2x^2} + C \qquad\qquad (11\text{-}2)$$

28. $\displaystyle\int x^2 e^{-2x}\, dx$. Use integration-by-parts. Let $u = x^2$ and $dv = e^{-2x}\, dx$.

Then $du = 2x\, dx$ and $v = -\frac{1}{2} e^{-2x}$.

$$\int x^2 e^{-2x}\, dx = -\frac{1}{2} x^2 e^{-2x} + \int x e^{-2x}\, dx$$

Now use integration-by-parts again. Let $u = x$ and $dv = e^{-2x}\, dx$.

Then $du = dx$ and $v = -\frac{1}{2} e^{-2x}$.

$$\int x e^{-2x}\, dx = -\frac{1}{2} x e^{-2x} + \frac{1}{2}\int e^{-2x}\, dx$$

$$= -\frac{1}{2} x e^{-2x} - \frac{1}{4} e^{-2x} + C$$

Thus,

$$\int x^2 e^{-2x}\, dx = -\frac{1}{2}x^2 e^{-2x} + \left[-\frac{1}{2}xe^{-2x} - \frac{1}{4}e^{-2x}\right] + C$$

$$= -\frac{1}{2}x^2 e^{-2x} - \frac{1}{2}xe^{-2x} - \frac{1}{4}e^{-2x} + C \qquad (12\text{-}3,\ 12\text{-}4)$$

29. (A) Probability $(0 \le t \le 1) = \displaystyle\int_0^1 0.21e^{-0.21t}\, dt$

$$= -e^{-0.21t}\Big|_0^1$$

$$= -e^{-0.21} + 1 \approx 0.189$$

(B) Probability $(1 \le t \le 2) = \displaystyle\int_1^2 0.21e^{-0.21t}\, dt$

$$= -e^{-0.21t}\Big|_1^2$$

$$= e^{-0.21} - e^{-0.42} \approx 0.154 \qquad (12\text{-}2)$$

30.

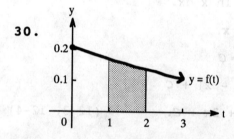

The probability that the product will fail during the second year of warranty is the area under the probability density function $y = f(t)$ from $t = 1$ to $t = 2$.

$(12\text{-}2)$

31. (A)

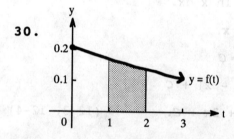

(B) Total income $= \displaystyle\int_1^4 2{,}500e^{0.05t}\, dt$

$$= 50{,}000e^{0.05t}\Big|_1^4$$

$$= 50{,}000[e^{0.2} - e^{0.05}] \approx \$8{,}507 \qquad (12\text{-}2)$$

32. $f(t) = 2{,}500e^{0.05t}$, $r = 0.15$, $T = 5$

(A) $FV = e^{(0.15)5} \int_0^5 2{,}500e^{0.05t} \, e^{-0.15t} \, dt = 2{,}500e^{0.75} \int_0^5 e^{-0.1t} \, dt$

$$= -25{,}000e^{0.75} \, e^{-0.1t} \Big|_0^5$$

$$= 25{,}000[e^{0.75} - e^{0.25}] \approx \$20{,}824$$

(B) Total income $= \int_0^5 2{,}500e^{0.05t} \, dt = 50{,}000e^{0.05t} \Big|_0^5$

$$= 50{,}000[e^{0.25} - 1]$$

$$\approx \$14{,}201$$

Interest $= FV -$ Total income $= \$20{,}824 - \$14{,}201 = \$6{,}623$ (12-2)

33. (A)

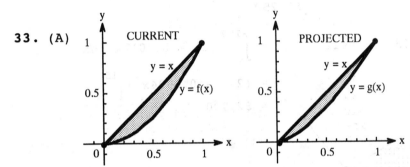

(B) The income will be more equally distributed 10 years from now since the area between $y = x$ and the projected Lorenz curve is less than the area between $y = x$ and the current Lorenz curve.

(C) Current:

Index of income concentration $= 2 \int_0^1 [x - (0.1x + 0.9x^2)] \, dx$

$$= 2 \int_0^1 (0.9x - 0.9x^2) \, dx = 2(0.45x^2 - 0.3x^3) \Big|_0^1 = 0.30$$

Projected:

Index of Income Concentration $= 2 \int_0^1 (x - x^{1.5}) \, dx$

$$= 2 \int_0^1 (x - x^{3/2}) \, dx = 2\left(\frac{1}{2}x^2 - \frac{2}{5}x^{5/2}\right) \Big|_0^1 = 2\left(\frac{1}{10}\right) = 0.2$$

Thus, income will be more equally distributed 10 years from now, as indicated in part (B).   (12-1)

34. (A) $p = D(x) = 70 - 0.2x$, $p = S(x) = 13 + 0.0012x^2$

Equilibrium price: $D(x) = S(x)$

$$70 - 0.2x = 13 + 0.0012x^2$$

$$0.0012x^2 + 0.2x - 57 = 0$$

$$x = \frac{-0.2 \pm \sqrt{0.04 + 0.2736}}{0.0024}$$

$$= \frac{-0.2 \pm 0.56}{0.0024}$$

Therefore, $\bar{x} = \dfrac{-0.2 + 0.56}{0.0024} = 150$, and $\bar{p} = 70 - 0.2(150) = 40$.

$$CS = \int_0^{150} (70 - 0.2x - 40)\,dx = \int_0^{150} (30 - 0.2x)\,dx$$

$$= (30x - 0.1x^2)\,\Big|_0^{150}$$

$$= \$2,250$$

$$PS = \int_0^{150} [40 - (13 + 0.0012x^2)]\,dx = \int_0^{150} (27 - 0.0012x^2)\,dx$$

$$= (27x - 0.0004x^3)\,\Big|_0^{150}$$

$$= \$2,700$$

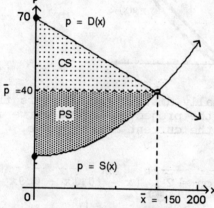

(B) $p = D(x) = 70 - 0.2x$, $p = S(x) = 13e^{0.006x}$

Equilibrium price: $D(x) = S(x)$

$$70 - 0.2x = 13e^{0.006x}$$

Using a graphing utility to solve for x, we get $\bar{x} \sim 170$ and $\bar{p} = 70 - 0.2(170) \approx 36$.

$$CS = \int_0^{170} (70 - 0.2x - 36)\,dx = \int_0^{170} (34 - 0.2x)\,dx$$

$$= (34x - 0.1x^2)\,\Big|_0^{170}$$

$$= \$2,890$$

$$PS = \int_0^{170} (36 - 13e^{0.006x})\, dx = (36x - 2{,}166.67e^{0.006x}) \Big|_0^{170}$$

$$\approx \$2{,}278$$

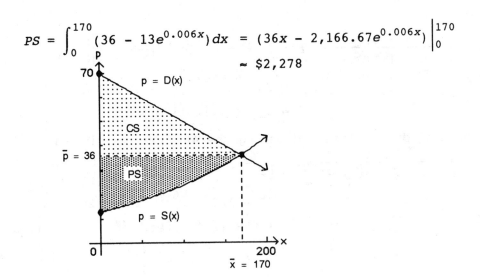

(12-2)

35. $R(t) = \dfrac{60t}{(t + 1)^2 (t + 2)}$

The amount of the drug eliminated during the first hour is given by

$$A = \int_0^1 \frac{60t}{(t + 1)^2 (t + 2)}\, dt$$

We will use the Table of Integration Formulas to calculate this integral. First, let $u = t + 2$. Then $t = u - 2$, $t + 1 = u - 1$, $du = dt$ and

$$\int \frac{60t}{(t + 1)^2 (t + 2)}\, dt = 60 \int \frac{u - 2}{(u - 1)^2 \cdot u}\, du$$

$$= 60 \int \frac{1}{(u - 1)^2}\, du - 120 \int \frac{1}{u(u - 1)^2}\, du$$

In the first integral, let $v = u - 1$, $dv = du$. Then

$$60 \int \frac{1}{(u - 1)^2}\, du = 60 \int v^{-2}\, dv = -60v^{-1} = \frac{-60}{u - 1}$$

For the second integral, use Formula 11 with $a = -1$, $b = 1$:

$$-120 \int \frac{1}{u(u - 1)^2}\, du = -120 \left[\frac{-1}{u - 1} + \ln \left| \frac{u}{u - 1} \right| \right]$$

Combining these results and replacing u by $t + 2$, we have:

$$\int \frac{60t}{(t + 1)^2 (t + 2)}\, dt = \frac{-60}{t + 1} + \frac{120}{t + 1} - 120 \ln \left| \frac{t + 2}{t + 1} \right| + C$$

$$= \frac{60}{t + 1} - 120 \ln \left| \frac{t + 2}{t + 1} \right| + C$$

Now,

$$A = \int_0^1 \frac{60t}{(t+1)^2(t+2)} \, dt = \left[\frac{60}{t+1} - 120 \ln\left(\frac{t+2}{t+1}\right) \right] \Big|_0^1$$

$$= 30 - 120 \ln\left(\frac{3}{2}\right) - 60 + 120 \ln 2$$

$$\approx 4.522 \text{ milliliters}$$

The amount of drug eliminated during the 4th hour is given by:

$$A = \int_3^4 \frac{60t}{(t+1)^2(t+2)} \, dt = \left[\frac{60}{t+1} - 120 \ln\left(\frac{t+2}{t+1}\right) \right] \Big|_3^4$$

$$= 12 - 120 \ln\left(\frac{6}{5}\right) - 15 + 120 \ln\left(\frac{5}{4}\right)$$

$$\approx 1.899 \text{ milliliters} \qquad\qquad (11\text{-}5, \ 12\text{-}4)$$

36.

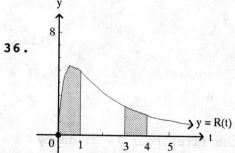

37. $f(t) = \begin{cases} \dfrac{4/3}{(t+1)^2} & 0 \le t \le 3 \\ 0 & \text{otherwise} \end{cases}$

(A) Probability $(0 \le t \le 1) = \int_0^1 \frac{4/3}{(t+1)^2} \, dt$

To calculate the integral, let $u = t + 1$, $du = dt$. Then,

$$\int \frac{4/3}{(t+1)^2} \, dt = \frac{4}{3} \int u^{-2} \, du = \frac{4}{3} \frac{u^{-1}}{-1} = -\frac{4}{3u} + C = \frac{-4}{3(t+1)} + C$$

Thus,

$$\int_0^1 \frac{4/3}{(t+1)^2} \, dt = \frac{-4}{3(t+1)} \Big|_0^1 = -\frac{2}{3} + \frac{4}{3} = \frac{2}{3} \approx 0.667$$

(B) Probability $(t \ge 1) = \int_1^3 \frac{4/3}{(t+1)^2} \, dt$

$$= \frac{-4}{3(t+1)} \Big|_1^3$$

$$= -\frac{1}{3} + \frac{2}{3} = \frac{1}{3} \approx 0.333 \qquad\qquad (12\text{-}2)$$

38.

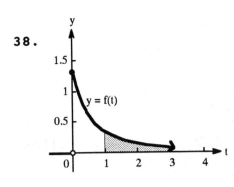

The probability that the doctor will spend more than an hour with a randomly selected patient is the area under the probability density function $y = f(t)$ from $t = 1$ to $t = 3$.

(12-2)

39. $N'(t) = \dfrac{100t}{(1 + t^2)^2}$. To find $N(t)$, we calculate

$$\int \frac{100t}{(1 + t^2)^2}\, dt$$

Let $u = 1 + t^2$. Then $du = 2t\, dt$, and

$$N(t) = \int \frac{100t}{(1 + t^2)^2}\, dt = 50 \int \frac{1}{u^2}\, du = 50 \int u^{-2}\, du$$

$$= -50\,\frac{1}{u} + C$$

$$= \frac{-50}{1 + t^2} + C$$

At $t = 0$, we have
$$N(0) = -50 + C$$

Therefore, $C = N(0) + 50$ and
$$N(t) = \frac{-50}{1 + t^2} + 50 + N(0)$$

Now,
$$N(3) = \frac{-5}{1 + 3^2} + 50 + N(0) = 45 + N(0)$$

Thus, the population will increase by 45 thousand during the next 3 years. (11-5, 12-1)

40. We want to find Probability $(t \geq 2) = \displaystyle\int_{2}^{\infty} f(t)\, dt$

Since

$$\int_{-\infty}^{\infty} f(t)\, dt = \int_{-\infty}^{2} f(t)\, dt + \int_{2}^{\infty} f(t)\, dt = 1,$$

$$\int_{2}^{\infty} f(t)\, dt = 1 - \int_{-\infty}^{2} f(t)\, dt = 1 - \int_{0}^{2} f(t)\, dt \quad \text{(since } f(t) = 0 \text{ for } t \leq 0\text{)}$$

$$= 1 - \text{Probability } (0 \leq t \leq 2)$$

Now, Probability $(0 \leq t \leq 2) = \int_0^2 0.5e^{-0.5t}\,dt$

$$= -e^{-0.5t} \Big|_0^2$$

$$= -e^{-1} + 1 \approx 0.632$$

Therefore, Probability $(t \geq 2) = 1 - 0.632 = 0.368$

(12-2)

13 MULTIVARIABLE CALCULUS

EXERCISE 13-1

Things to remember:

1. An equation of the form $z = f(x, y)$ describes a FUNCTION OF TWO INDEPENDENT VARIABLES if for each ordered pair (x, y) in the domain of f there is one and only one value of z determined by $f(x, y)$.

2. An equation of the form $w = f(x, y, z)$ describes a FUNCTION OF THREE INDEPENDENT VARIABLES if for each ordered triple (x, y, z) in the domain of f there is one and only one value of w determined by $f(x, y, z)$.

3. Functions of more than three independent variables are defined similarly.

---

1. $f(x, y) = 10 + 2x - 3y$
   $f(0, 0) = 10 + 2 \cdot 0 - 3 \cdot 0 = 10$
   $(x = 0$ and $y = 0)$

3. $f(x, y) = 10 + 2x - 3y$
   $f(-3, 1) = 10 + 2(-3) - 3(1) = 1$
   $(x = -3$ and $y = 1)$

5. $g(x, y) = x^2 - 3y^2$
   $g(0, 0) = 0^2 - 3 \cdot 0^2 = 0$
   $(x = 0$ and $y = 0)$

7. $g(x, y) = x^2 - 3y^2$
   $g(2, -1) = 2^2 - 3(-1)^2 = 1$
   $(x = 2$ and $y = -1)$

9. $A(x, y) = xy$

   $A(2, 3) = 2 \cdot 3 = 6$

   $(x = 2$ and $y = 3)$

11. $Q(M, C) = \dfrac{M}{C}(100)$

    $Q(12, 8) = \dfrac{12}{8}(100) = 150$

    $(M = 12$ and $C = 8)$

13. $V(r, h) = \pi r^2 h$
    $V(2, 4) = \pi \cdot 2^2 \cdot 4 = 16\pi$
    $(r = 2$ and $h = 4)$

15. $R(x, y) = -5x^2 + 6xy - 4y^2 + 200x + 300y$
    $R(1, 2) = -5(1)^2 + 6 \cdot 1 \cdot 2 - 4 \cdot 2^2 + 200 \cdot 1 + 300 \cdot 2$
    $\qquad = -5 + 12 - 16 + 200 + 600$
    $\qquad = 791$
    $(x = 1$ and $y = 2)$

17. $R(L, r) = .002\dfrac{L}{r^4}$

    $R(6, 0.5) = 0.002\dfrac{6}{(0.5)^4} = \dfrac{0.012}{0.0625} = 0.192$

    $(L = 6$ and $r = 0.5)$

19. $A(P, r, t) = P + Prt$
$A(100, 0.06, 3) = 100 + 100(0.06)3 = 118$
$(P = 100, r = 0.06,$ and $t = 3)$

21. $A(P, r, t) = Pe^{rt}$
$A(100, 0.08, 10) = 100e^{0.08 \cdot 10} = 100e^{0.8} \approx 222.55$
$(P = 100, r = 0.08,$ and $t = 10)$

23. $F(x, y) = x^2 + e^x y - y^2$; $F(x, 2) = x^2 + 2e^x - 4$.

We use a graphing utility to solve $F(x, 2) = 0$.
The graph of $u = F(x, 2)$ is shown at the right.

The solutions of $F(x, 2) = 0$ are: $x_1 \approx -1.926$,
$x_2 \approx 0.599$

25. $f(x, y) = x^2 + 2y^2$

$$\frac{f(x + h, y) - f(x, y)}{h} = \frac{(x + h)^2 + 2y^2 - (x^2 + 2y^2)}{h}$$

$$= \frac{x^2 + 2xh + h^2 + 2y^2 - x^2 - 2y^2}{h}$$

$$= \frac{2xh + h^2}{h} = \frac{h(2x + h)}{h} = 2x + h, \ h \neq 0$$

27. $f(x, y) = 2xy^2$

$$\frac{f(x + h, y) - f(x, y)}{h} = \frac{2(x + h)y^2 - 2xy^2}{h}$$

$$= \frac{2xy^2 + 2hy^2 - 2xy^2}{h} = \frac{2hy^2}{h} = 2y^2, \ h \neq 0$$

29. Coordinates of point $E = E(0, 0, 3)$.
Coordinates of point $F = F(2, 0, 3)$.

31. $f(x, y) = x^2$

(A) In the plane $y = c$, c any constant, the graph of $z = x^2$ is a parabola.

(B) Cross-section corresponding to $x = 0$: the y-axis

Cross-section corresponding to $x = 1$: the line passing through $(1, 0, 1)$ parallel to the y-axis.

Cross-section corresponding to $x = 2$: the line passing through $(2, 0, 4)$ parallel to the y-axis.

(C) The surface $z = x^2$ is a parabolic trough lying on the y-axis.

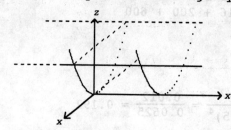

33. Monthly cost function = $C(x, y) = 2000 + 70x + 100y$

$$C(20, 10) = 2000 + 70 \cdot 20 + 100 \cdot 10 = \$4400$$
$$C(50, 5) = 2000 + 70 \cdot 50 + 100 \cdot 5 = \$6000$$
$$C(30, 30) = 2000 + 70 \cdot 30 + 100 \cdot 30 = \$7100$$

35. $R(p, q) = p \cdot x + q \cdot y = 200p - 5p^2 + 4pq + 300q - 4q^2 + 2pq$ or

$R(p, q) = -5p^2 + 6pq - 4q^2 + 200p + 300q$

$R(2, 3) = -5 \cdot 2^2 + 6 \cdot 2 \cdot 3 - 4 \cdot 3^2 + 200 \cdot 2 + 300 \cdot 3 = 1280$ or $1280

$R(3, 2) = -5 \cdot 3^2 + 6 \cdot 3 \cdot 2 - 4 \cdot 2^2 + 200 \cdot 3 + 300 \cdot 2 = 1175$ or $1175

37. $f(x, y) = 20x^{0.4}y^{0.6}$

$f(1250, 1700) = 20(1250)^{0.4}(1700)^{0.6}$

$\approx 20(17.3286)(86.7500) \approx 30,065$ units

39. $FV = F(P, i, n) = P\dfrac{(1 + i)^n - 1}{i}$

$F(2000, .09, 30) = 2000\dfrac{(1 + .09)^{30} - 1}{.09} \approx \$272,615.08$

41. $T(V, x) = \dfrac{33V}{x + 33}$

$T(70, 47) = \dfrac{33 \cdot 70}{47 + 33} = \dfrac{33 \cdot 70}{80} = 28.8 \approx 29$ minutes

$T(60, 27) = \dfrac{33 \cdot 60}{27 + 33} = 33$ minutes

43. $C(W, L) = 100\dfrac{W}{L}$

$C(6, 8) = 100\dfrac{6}{8} = 75$

$C(8.1, 9) = 100\dfrac{8.1}{9} = 90$

45. $Q(M, C) = \dfrac{M}{C}100$

$Q(12, 10) = \dfrac{12}{10}100 = 120$

$Q(10, 12) = \dfrac{10}{12}100 = 83.33 \approx 83$

EXERCISE 13-2

Things to remember:

<u>1</u>. Let $z = f(x, y)$ be a function of two independent variables. The PARTIAL DERIVATIVE OF z WITH RESPECT TO x, denoted by $\dfrac{\partial z}{\partial x}$, f_x, or $f_x(x, y)$, is given by

$$\frac{\partial z}{\partial x} = \lim_{h \to 0} \frac{f(x + h, y) - f(x, y)}{h}$$

provided this limit exists. Similarly, the PARTIAL DERIVATIVE OF z WITH RESPECT TO y, denoted by $\dfrac{\partial z}{\partial y}$, f_y, or $f_y(x, y)$, is given by

$$\frac{\partial z}{\partial y} = \lim_{k \to 0} \frac{f(x, y + k) - f(x, y)}{k}$$

provided this limit exists.

2. SECOND-ORDER PARTIAL DERIVATIVES

If $z = f(x, y)$, then:

$$\frac{\partial^2 z}{\partial x^2} = \frac{\partial\left(\frac{\partial z}{\partial x}\right)}{\partial x} = f_{xx}(x, y) = f_{xx}$$

$$\frac{\partial^2 z}{\partial x \partial y} = \frac{\partial\left(\frac{\partial z}{\partial y}\right)}{\partial x} = f_{yx}(x, y) = f_{yx}$$

$$\frac{\partial^2 z}{\partial y \partial x} = \frac{\partial\left(\frac{\partial z}{\partial x}\right)}{\partial y} = f_{xy}(x, y) = f_{xy}$$

$$\frac{\partial^2 z}{\partial y^2} = \frac{\partial\left(\frac{\partial z}{\partial y}\right)}{\partial y} = f_{yy}(x, y) = f_{yy}$$

Note: For the functions being considered in this text, the mixed partial derivatives f_{xy} and f_{yx} are equal, i.e.,

$$\frac{\partial^2 z}{\partial x \partial y} = \frac{\partial^2 z}{\partial y \partial x}.$$

---

1. $z = f(x, y) = 10 + 3x + 2y$
$\frac{\partial z}{\partial x} = 0 + 3 + 0 = 3$

3. $z = f(x, y) = 10 + 3x + 2y$
$f_y(x, y) = 0 + 0 + 2 = 2$
$f_y(1, 2) = 2$

5. $z = f(x, y) = 3x^2 - 2xy^2 + 1$
$\frac{\partial z}{\partial y} = 0 - 2x(2y) + 0 = -4xy$

7. $z = f(x, y) = 3x^2 - 2xy^2 + 1$
$f_x(x, y) = 6x - 2y^2 + 0 = 6x - 2y^2$
$f_x(2, 3) = 6 \cdot 2 - 2 \cdot 3^2 = -6$

9. $S(x, y) = 5x^2y^3$
$S_x(x, y) = 5y^3 2x = 10xy^3$

11. $S(x, y) = 5x^2y^3$
$S_y(x, y) = 5x^2 \cdot 3y^2 = 15x^2y^2$
$S_y(2, 1) = 15 \cdot 2^2 \cdot 1^2 = 60$

13. $C(x, y) = x^2 - 2xy + 2y^2 + 6x - 9y + 5$
$C_x(x, y) = 2x - 2y + 0 + 6 - 0 + 0 = 2x - 2y + 6$

15. $C_x(x, y) = 2x - 2y + 6$ (from Problem 13)
$C_x(2, 2) = 2 \cdot 2 - 2 \cdot 2 + 6 = 6$

17. $C_x(x, y) = 2x - 2y + 6$ (from Problem 13)
$C_{xy}(x, y) = 0 - 2 + 0 = -2$

19. $C_x(x, y) = 2x - 2y + 6$ (from Problem 13)
$C_{xx}(x, y) = 2 - 0 + 0 = 2$

21. $z = f(x, y) = e^{(2x+3y)}$

$\dfrac{\partial z}{\partial x} = e^{(2x+3y)} \dfrac{\partial}{\partial x}(2x + 3y) = e^{(2x+3y)}2 = 2e^{(2x+3y)}$

23. $z = f(x, y) = e^{(2x+3y)}$

$\dfrac{\partial z}{\partial y} = e^{(2x+3y)} \dfrac{\partial}{\partial y}(2x + 3y) = e^{(2x+3y)}3 = 3e^{(2x+3y)}$

$\dfrac{\partial^2 z}{\partial x \partial y} = \dfrac{\partial\left(\dfrac{\partial z}{\partial y}\right)}{\partial x} = \dfrac{\partial(3e^{(2x+3y)})}{\partial x} = 3e^{(2x+3y)}\dfrac{\partial}{\partial x}(2x + 3y) = 3e^{(2x+3y)}2 = 6e^{(2x+3y)}$

25. $z = f(x, y) = e^{(2x+3y)}$

$f_x = e^{(2x+3y)}\dfrac{\partial}{\partial x}(2x + 3y) = 2e^{(2x+3y)}$

$f_{xy} = \dfrac{\partial}{\partial y}\left(\dfrac{\partial z}{\partial x}\right) = 2e^{(2x+3y)}\dfrac{\partial}{\partial y}(2x + 3y) = 6e^{(2x+3y)}$

$f_{xy}(1, 0) = 6e^{(2\cdot 1+0)} = 6e^2$

27. $f_x = 2e^{(2x+3y)}$ (from Problem 21)

$f_{xx} = \dfrac{\partial}{\partial x}(2e^{(2x+3y)}) = 2e^{(2x+3y)}\dfrac{\partial}{\partial x}(2x + 3y) = 4e^{(2x+3y)}$

$f_{xx}(0, 1) = 4e^{(0+3\cdot 1)} = 4e^3$

29. $f(x, y) = (x^2 - y^3)^3$

$f_x(x, y) = 3(x^2 - y^3)^2\dfrac{\partial(x^2 - y^3)}{\partial x} = 3(x^2 - y^3)^2 2x = 6x(x^2 - y^3)^2$

$f_y(x, y) = 3(x^2 - y^3)^2\dfrac{\partial(x^2 - y^3)}{\partial y} = 3(x^2 - y^3)^2(-3y^2) = -9y^2(x^2 - y^3)^2$

31. $f(x, y) = (3x^2y - 1)^4$

$f_x(x, y) = 4(3x^2y - 1)^3\dfrac{\partial(3x^2y - 1)}{\partial x} = 4(3x^2y - 1)^3 6xy = 24xy(3x^2y - 1)^3$

$f_y(x, y) = 4(3x^2y - 1)^3\dfrac{\partial(3x^2y - 1)}{\partial y} = 4(3x^2y - 1)^3 3x^2 = 12x^2(3x^2y - 1)^3$

33. $f(x, y) = \ln(x^2 + y^2)$

$f_x(x, y) = \dfrac{1}{x^2 + y^2}\cdot\dfrac{\partial(x^2 + y^2)}{\partial x} = \dfrac{2x}{x^2 + y^2}$

$f_y(x, y) = \dfrac{1}{x^2 + y^2}\cdot\dfrac{\partial(x^2 + y^2)}{\partial y} = \dfrac{2y}{x^2 + y^2}$

35. $f(x, y) = y^2 e^{xy^2}$

$f_x(x, y) = y^2 e^{xy^2}\dfrac{\partial(xy^2)}{\partial x} = y^2 e^{xy^2}y^2 = y^4 e^{xy^2}$

$f_y(x, y) = y^2\dfrac{\partial(e^{xy^2})}{\partial y} + e^{xy^2}\dfrac{\partial(y^2)}{\partial y}$ (Product rule)

$\qquad = y^2 e^{xy^2}2yx + y^2 e^{xy^2}2y = 2xy^3 e^{xy^2} + 2ye^{xy^2}$

37. $f(x, y) = \dfrac{x^2 - y^2}{x^2 + y^2}$

Applying the quotient rule:

$$f_x(x, y) = \frac{(x^2 + y^2)\dfrac{\partial (x^2 - y^2)}{\partial x} - (x^2 - y^2)\dfrac{\partial (x^2 + y^2)}{\partial x}}{(x^2 + y^2)^2}$$

$$= \frac{(x^2 + y^2)(2x) - (x^2 - y^2)(2x)}{(x^2 + y^2)^2}$$

$$= \frac{2x^3 + 2y^2x - 2x^3 + 2y^2x}{(x^2 + y^2)^2} = \frac{4xy^2}{(x^2 + y^2)^2}$$

Again, applying the quotient rule:

$$f_y(x, y) = \frac{(x^2 + y^2)(-2y) - (x^2 - y^2)(2y)}{(x^2 + y^2)^2} = \frac{-4x^2 y}{(x^2 + y^2)^2}$$

39. (A) $f(x, y) = y^3 + 4y^2 - 5y + 3$

Since f is independent of x, $\dfrac{\partial f}{\partial x} = 0$

(B) If $g(x, y)$ depends on y only, that is, if $g(x, y) = G(y)$ is independent of x, then

$$\frac{\partial g}{\partial x} = 0$$

Clearly there are an infinite number of such functions.

41. $f(x, y) = x^2 y^2 + x^3 + y$

$f_x(x, y) = 2xy^2 + 3x^2$ $\qquad$ $f_y(x, y) = 2x^2 y + 1$

$f_{xx}(x, y) = 2y^2 + 6x$ $\qquad$ $f_{yx}(x, y) = 4xy$

$f_{xy}(x, y) = 4xy$ $\qquad$ $f_{yy}(x, y) = 2x^2$

43. $f(x, y) = \dfrac{x}{y} - \dfrac{y}{x}$

$f_x(x, y) = \dfrac{1}{y} + \dfrac{y}{x^2}$ $\qquad$ $f_y(x, y) = -\dfrac{x}{y^2} - \dfrac{1}{x}$

$f_{xx}(x, y) = -\dfrac{2y}{x^3}$ $\qquad$ $f_{yx}(x, y) = -\dfrac{1}{y^2} + \dfrac{1}{x^2}$

$f_{xy}(x, y) = -\dfrac{1}{y^2} + \dfrac{1}{x^2}$ $\qquad$ $f_{yy}(x, y) = \dfrac{2x}{y^3}$

45. $f(x, y) = xe^{xy}$

$f_x(x, y) = xye^{xy} + e^{xy}$ $\qquad$ $f_y(x, y) = x^2 e^{xy}$

$f_{xx}(x, y) = xy^2 e^{xy} + 2ye^{xy}$ $\qquad$ $f_{yx}(x, y) = x^2 ye^{xy} + 2xe^{xy}$

$f_{xy}(x, y) = x^2 ye^{xy} + 2xe^{xy}$ $\qquad$ $f_{yy}(x, y) = x^3 e^{xy}$

47. $P(x, y) = -x^2 + 2xy - 2y^2 - 4x + 12y - 5$

$P_x(x, y) = -2x + 2y - 0 - 4 + 0 - 0 = -2x + 2y - 4$

$P_y(x, y) = 0 + 2x - 4y - 0 + 12 - 0 = 2x - 4y + 12$

$P_x(x, y) = 0$ and $P_y(x, y) = 0$ when

$$-2x + 2y - 4 = 0 \qquad (1)$$
$$2x - 4y + 12 = 0 \qquad (2)$$

Add equations (1) and (2): $-2y + 8 = 0$
$$y = 4$$
Substitute $y = 4$ into (1): $-2x + 2 \cdot 4 - 4 = 0$
$$-2x + 4 = 0$$
$$x = 2$$
Thus, $P_x(x, y) = 0$ and $P_y(x, y) = 0$ when $x = 2$ and $y = 4$.

49. $F(x, y) = x^3 - 2x^2y^2 - 2x - 4y + 10$;
$F_x(x, y) = 3x^2 - 4xy^2 - 2$; $F_y(x, y) = -4x^2y - 4$.

Set $F_x(x, y) = 0$ and $F_y(x, y) = 0$ and solve simultaneously:

$$3x^2 - 4xy^2 - 2 = 0 \qquad (1)$$
$$-4x^2y - 4 = 0 \qquad (2)$$

From (2), $y = -\dfrac{1}{x^2}$. Substituting this into (1),

$$3x^2 - 4x\left(-\frac{1}{x^2}\right)^2 - 2 = 0$$

$$3x^2 - 4x\left(\frac{1}{x^4}\right) - 2 = 0$$

$$3x^5 - 2x^3 - 4 = 0$$

Using a graphing utility, we find that
$$x \approx 1.200$$
Then, $y \approx -0.694$.

51. $f(x, y) = \ln(x^2 + y^2)$ (see Problem 33)

$$f_x(x, y) = \frac{2x}{x^2 + y^2} \qquad\qquad f_y(x, y) = \frac{2y}{x^2 + y^2}$$

$$f_{xx}(x, y) = \frac{(x^2 + y^2)2 - 2x(2x)}{(x^2 + y^2)^2} \qquad f_{yy}(x, y) = \frac{(x^2 + y^2)2 - 2y(2y)}{(x^2 + y^2)^2}$$

$$= \frac{2(y^2 - x^2)}{(x^2 + y^2)^2} \qquad\qquad = \frac{2(x^2 - y^2)}{(x^2 + y^2)^2} = \frac{-2(y^2 - x^2)}{(x^2 + y^2)^2}$$

$$f_{xx}(x, y) + f_{yy}(x, y) = \frac{2(y^2 - x^2)}{(x^2 + y^2)^2} + \frac{-2(y^2 - x^2)}{(x^2 + y^2)^2} = 0$$

53. $f(x, y) = x^2 + 2y^2$

(A) $\displaystyle\lim_{h \to 0} \frac{f(x + h, y) - f(x, y)}{h} = \lim_{h \to 0} \frac{(x + h)^2 + 2y^2 - (x^2 + 2y^2)}{h}$

$$= \lim_{h \to 0} \frac{x^2 + 2xh + h^2 + 2y^2 - x^2 - 2y^2}{h}$$

$$= \lim_{h \to 0} \frac{h(2x + h)}{h} = \lim_{h \to 0} (2x + h)$$

$$= 2x$$

(B) $\lim_{k \to 0} \dfrac{f(x, y + k) - f(x, y)}{k} = \lim_{k \to 0} \dfrac{x^2 + 2(y + k)^2 - (x^2 + 2y^2)}{k}$

$$= \lim_{k \to 0} \dfrac{x^2 + 2(y^2 + 2yk + k^2) - x^2 - 2y^2}{k}$$

$$= \lim_{k \to 0} \dfrac{4yk + 2k^2}{k} = \lim_{k \to 0} (4y + 2k)$$

$$= 4y$$

55. $R(x, y) = 80x + 90y + 0.04xy - 0.05x^2 - 0.05y^2$
$C(x, y) = 8x + 6y + 20{,}000$
The profit $P(x, y)$ is given by:
$P(x, y) = R(x, y) - C(x, y)$

$$= 80x + 90y + 0.04xy - 0.05x^2 - 0.05y^2 - (8x + 6y + 20{,}000)$$

$$= 72x + 84y + 0.04xy - 0.05x^2 - 0.05y^2 - 20{,}000$$

Now
$P_x(x, y) = 72 + 0.04y - 0.1x$

and
$P_x(1200, 1800) = 72 + 0.04(1800) - 0.1(1200)$

$$= 72 + 72 - 120 = 24;$$

$P_y(x, y) = 84 + 0.04x - 0.1y$

and
$P_y(1200, 1800) = 84 + 0.04(1200) - 0.1(1800)$

$$= 84 + 48 - 180 = -48.$$

Thus, at the (1200, 1800) output level, profit will increase approximately \$24 per unit increase in production of type A calculators; and profit will decrease \$48 per unit increase in production of type B calculators.

57. $x = 200 - 5p + 4q$
$y = 300 - 4q + 2p$
$\dfrac{\partial x}{\partial p} = -5, \quad \dfrac{\partial y}{\partial p} = 2$

A \$1 increase in the price of brand A will decrease the demand for brand A by 5 pounds at any price level (p, q).

A \$1 increase in the price of brand A will increase the demand for brand B by 2 pounds at any price level (p, q).

59. $f(x, y) = 10x^{0.75}y^{0.25}$

(A) $f_x(x, y) = 10(0.75)x^{-0.25}y^{0.25} = 7.5x^{-0.25}y^{0.25}$

$\quad f_y(x, y) = 10(0.25)x^{0.75}y^{-0.75} = 2.5x^{0.75}y^{-0.75}$

(B) Marginal productivity of labor $= f_x(600, 100)$

$$= 7.5(600)^{-0.25}(100)^{0.25} \approx 4.79$$

Marginal productivity of capital $= f_y(600, 100)$

$$= 2.5(600)^{0.75}(100)^{-0.75} \approx 9.58$$

(C) The government should encourage the increased use of capital.

61. $x = f(p, q) = 8000 - 0.09p^2 + 0.08q^2$ (Butter)
    $y = g(p, q) = 15,000 + 0.04p^2 - 0.3q^2$ (Margarine)
    $f_q(p, q) = 0.08(2)q = 0.16q > 0$
    $g_p(p, q) = 0.04(2)p = 0.08p > 0$
    Thus, the products are competitive.

63. $x = f(p, q) = 800 - 0.004p^2 - 0.003q^2$ (Skis)
    $y = g(p, q) = 600 - 0.003p^2 - 0.002q^2$ (Ski boots)
    $f_q(p, q) = -0.003(2)q = -0.006q < 0$
    $g_p(p, q) = -0.003(2)p = -0.006p < 0$
    Thus, the products are complementary.

65. $A = f(w, h) = 15.64w^{0.425}h^{0.725}$
    (A) $f_w(w, h) = 15.64(0.425)w^{-0.575}h^{0.725} \approx 6.65w^{-0.575}h^{0.725}$
        $f_h(w, h) = 15.64(0.725)w^{0.425}h^{-0.275} \approx 11.34w^{0.425}h^{-0.275}$

    (B) $f_w(65, 57) = 6.65(65)^{-0.575}(57)^{0.725} \approx 11.31$
        For a 65 pound child 57 inches tall, the rate of change of surface area is approximately 11.31 square inches for a one-pound gain in weight, height held fixed.
        $f_h(65, 57) = 11.34(65)^{0.425}(57)^{-0.275} \approx 21.99$
        For a 65 pound child 57 inches tall, the rate of change of surface area is approximately 21.99 square inches for a one-inch gain in height, weight held fixed.

67. $C(W, L) = 100\dfrac{W}{L}$ $C_L(W, L) = -\dfrac{100W}{L^2}$

    $C_W(W, L) = \dfrac{100}{L}$ $C_L(6, 8) = -\dfrac{100 \times 6}{8^2}$

    $C_W(6, 8) = \dfrac{100}{8} = 12.5$ $= -\dfrac{600}{64} = -9.38$

The index increases 12.5 units per 1-inch increase in the width of the head (length held fixed) when $W = 6$ and $L = 8$.

The index decreases 9.38 units per 1-inch increase in length (width held fixed) when $W = 6$ and $L = 8$.

EXERCISE 13-3

Things to remember:

<u>1</u>. $f(a, b)$ is a LOCAL MAXIMUM if there exists a circular region in the domain of $f(x, y)$ with (a, b) as the center, such that $f(a, b) \geq f(x, y)$ for all (x, y) in the region. Similarly, $f(a, b)$ is a LOCAL MINIMUM if $f(a, b) \leq f(x, y)$ for all (x, y) in the region.

2. If $f(a, b)$ is either a local maximum of a local minimum for the function $f(x, y)$, and if $f_x(a, b)$ and $f_y(a, b)$ exist, then
$$f_x(a, b) = 0 \quad \text{and} \quad f_y(a, b) = 0.$$

3. SECOND-DERIVATIVE TEST FOR LOCAL EXTREMA FOR $z = f(x, y)$

Given:

(a) $f_x(a, b) = 0$ and $f_y(a, b) = 0$ [(a, b) is a critical point].

(b) All second-order partial derivatives of f exist in some circular region containing (a, b) as center.

(c) $A = f_{xx}(a, b)$, $B = f_{xy}(a, b)$, $C = f_{yy}(a, b)$.

Then:

i) If $AC - B^2 > 0$ and $A < 0$, then $f(a, b)$ is a local maximum.

ii) If $AC - B^2 > 0$ and $A > 0$, then $f(a, b)$ is a local minimum.

iii) If $AC - B^2 < 0$, then f has a saddle point at (a, b).

iv) If $AC - B^2 = 0$, then the test fails.

---

1. $f(x, y) = 6 - x^2 - 4x - y^2$
$f_x(x, y) = -2x - 4 = 0$
$$x = -2$$
$f_y(x, y) = -2y = 0$
$$y = 0$$
Thus, $(-2, 0)$ is a critical point.

$f_{xx} = -2$, $f_{xy} = 0$, $f_{yy} = -2$,
$f_{xx}(-2, 0) \cdot f_{yy}(-2, 0) - [f_{xy}(-2, 0)]^2 = (-2)(-2) - 0^2 = 4 > 0$
and $\qquad\qquad f_{xx}(-2, 0) = -2 < 0$.

Thus, $f(-2, 0) = 6 - (-2)^2 - 4(-2) - 0^2 = 10$ is a local maximum (using 3).

3. $f(x, y) = x^2 + y^2 + 2x - 6y + 14$
$f_x(x, y) = 2x + 2 = 0$
$$x = -1$$
$f_y(x, y) = 2y - 6 = 0$
$$y = 3$$
Thus, $(-1, 3)$ is a critical point.

$f_{xx} = 2$ $\qquad\qquad f_{xy} = 0$ $\qquad\qquad f_{yy} = 2$
$f_{xx}(-1, 3) = 2 > 0$ $\qquad f_{xy}(-1, 3) = 0$ $\qquad f_{yy}(-1, 3) = 2$
$f_{xx}(-1, 3) \cdot f_{yy}(-1, 3) - [f_{xy}(-1, 3)]^2 = 2 \cdot 2 - 0^2 = 4 > 0$
Thus, using 3, $f(-1, 3) = 4$ is a local minimum.

5. $f(x, y) = xy + 2x - 3y - 2$

$f_x = y + 2 = 0$

$\qquad y = -2$

$f_y = x - 3 = 0$

$\qquad x = 3$

Thus, $(3, -2)$ is a critical point.

$f_{xx} = 0 \qquad\qquad f_{xy} = 1 \qquad\qquad f_{yy} = 0$

$f_{xx}(3, -2) = 0 \qquad f_{xy}(3, -2) = 1 \qquad f_{yy}(3, -2) = 0$

$f_{xx}(3, -2) \cdot f_{yy}(3, -2) - [f_{xy}(3, -2)]^2 = 0 \cdot 0 - [1]^2 = -1 < 0$

Thus, using $\underline{3}$, f has a saddle point at $(3, -2)$.

7. $f(x, y) = -3x^2 + 2xy - 2y^2 + 14x + 2y + 10$

$f_x = -6x + 2y + 14 = 0 \qquad (1)$

$f_y = 2x - 4y + 2 = 0 \qquad (2)$

Solving (1) and (2) for x and y, we obtain $x = 3$ and $y = 2$. Thus, $(3, 2)$ is a critical point.

$f_{xx} = -6 \qquad\qquad f_{xy} = 2 \qquad\qquad f_{yy} = -4$

$f_{xx}(3, 2) = -6 < 0 \qquad f_{xy}(3, 2) = 2 \qquad f_{yy}(3, 2) = -4$

$f_{xx}(3, 2) \cdot f_{yy}(3, 2) - [f_{xy}(3, 2)]^2 = (-6)(-4) - 2^2 = 20 > 0$

Thus, using $\underline{3}$, $f(3, 2)$ is a local maximum and

$f(3, 2) = -3 \cdot 3^2 + 2 \cdot 3 \cdot 2 - 2 \cdot 2^2 + 14 \cdot 3 + 2 \cdot 2 + 10 = 33.$

9. $f(x, y) = 2x^2 - 2xy + 3y^2 - 4x - 8y + 20$

$f_x = 4x - 2y - 4 = 0 \qquad (1)$

$f_y = -2x + 6y - 8 = 0 \qquad (2)$

Solving (1) and (2) for x and y, we obtain $x = 2$ and $y = 2$. Thus, $(2, 2)$ is a critical point.

$f_{xx} = 4 \qquad\qquad f_{xy} = -2 \qquad\qquad f_{yy} = 6$

$f_{xx}(2, 2) = 4 > 0 \qquad f_{xy}(2, 2) = -2 \qquad f_{yy}(2, 2) = 6$

$f_{xx}(2, 2) \cdot f_{yy}(2, 2) - [f_{xy}(2, 2)]^2 = 4 \cdot 6 - [-2]^2 = 20 > 0$

Thus, using $\underline{3}$, $f(2, 2)$ is a local minimum and

$f(2, 2) = 2 \cdot 2^2 - 2 \cdot 2 \cdot 2 + 3 \cdot 2^2 - 4 \cdot 2 - 8 \cdot 2 + 20 = 8.$

11. $f(x, y) = e^{xy}$

$f_x = e^{xy} \dfrac{\partial (xy)}{\partial x} \qquad\qquad f_y = e^{xy} \dfrac{\partial (xy)}{\partial y}$

$\quad = e^{xy} y = 0 \qquad\qquad\qquad = e^{xy} x = 0$

$\qquad y = 0 \ (e^{xy} \neq 0) \qquad\qquad x = 0 \ (e^{xy} \neq 0)$

Thus, $(0, 0)$ is a critical point.

$f_{xx} = ye^{xy} \dfrac{\partial (xy)}{\partial x} \qquad f_{xy} = e^{xy} \cdot 1 + ye^{xy} x \qquad f_{yy} = xe^{xy} \dfrac{\partial (xy)}{\partial y}$

$\quad = ye^{xy} y \qquad\qquad\qquad = e^{xy} + xye^{xy} \qquad\qquad = x^2 e^{xy}$

$\quad = y^2 e^{xy}$

$f_{xx}(0, 0) = 0 \qquad\qquad f_{xy}(0, 0) = 1 + 0 = 1 \qquad f_{yy}(0, 0) = 0$

$f_{xx}(0, 0) \cdot f_{yy}(0, 0) - [f_{xy}(0, 0)]^2 = 0 - [1]^2 = -1 < 0$

Thus, using $\underline{3}$, $f(x, y)$ has a saddle point at $(0, 0)$.

13. $f(x, y) = x^3 + y^3 - 3xy$

$f_x = 3x^2 - 3y = 3(x^2 - y) = 0$

Thus, $y = x^2$.   (1)

$f_y = 3y^2 - 3x = 3(y^2 - x) = 0$

Thus, $y^2 = x$.   (2)

Combining (1) and (2), we obtain $x = x^4$ or $x(x^3 - 1) = 0$. Therefore, $x = 0$ or $x = 1$, and the critical points are $(0, 0)$ and $(1, 1)$.

$f_{xx} = 6x$ $f_{xy} = -3$ $f_{yy} = 6y$

For the critical point $(0, 0)$:

$f_{xx}(0, 0) = 0$ $f_{xy}(0, 0) = -3$ $f_{yy}(0, 0) = 0$

$f_{xx}(0, 0) \cdot f_{yy}(0, 0) - [f_{xy}(0, 0)]^2 = 0 - (-3)^2 = -9 < 0$

Thus, using $\underline{3}$, $f(x, y)$ has a saddle point at $(0, 0)$.

For the critical point $(1, 1)$:

$f_{xx}(1, 1) = 6$ $f_{xy}(1, 1) = -3$ $f_{yy}(1, 1) = 6$

$f_{xx}(1, 1) \cdot f_{yy}(1, 1) - [f_{xy}(1, 1)]^2 = 6 \cdot 6 - (-3)^2 = 27 > 0$

$f_{xx}(1, 1) > 0$

Thus, using $\underline{3}$, $f(1, 1)$ is a local minimum and

$f(1, 1) = 1^3 + 1^3 - 3 \cdot 1 \cdot 1 = 2 - 3 = -1$.

15. $f(x, y) = 2x^4 + y^2 - 12xy$

$f_x = 8x^3 - 12y = 0$

Thus, $y = \frac{2}{3}x^3$.

$f_y = 2y - 12x = 0$

Thus, $y = 6x$

Therefore, $6x = \frac{2}{3}x^3$

$x^3 - 9x = 0$

$x(x^2 - 9) = 0$

$x = 0, \; x = 3, \; x = -3$

Thus, the critical points are $(0, 0)$, $(3, 18)$, $(-3, -18)$. Now,

$f_{xx} = 24x^2$ $f_{xy} = -12$ $f_{yy} = 2$

For the critical point $(0, 0)$:

$f_{xx}(0, 0) = 0$ $f_{xy}(0, 0) = -12$ $f_{yy}(0, 0) = 2$

and

$f_{xx}(0, 0) \cdot f_{yy}(0, 0) - [f_{xy}(0, 0)]^2 = 0 \cdot 2 - (-12)^2 = -144$.

Thus, $f(x, y)$ has a saddle point at $(0, 0)$.

For the critical point $(3, 18)$:

$f_{xx}(3, 18) = 24 \cdot 3^2 = 216 > 0$ $f_{xy}(3, 18) = -12$ $f_{yy}(3, 18) = 2$

and

$f_{xx}(3, 18) \cdot f_{yy}(3, 18) - [f_{xy}(3, 18)]^2 = 216 \cdot 2 - (-12)^2 = 288 > 0$

Thus, $f(3, 18) = -162$ is a local minimum.

For the critical point $(-3, -18)$:
$$f_{xx}(-3, -18) = 216 > 0 \qquad f_{xy}(-3, -18) = -12 \qquad f_{yy}(-3, -18) = 2$$
and
$$f_{xx}(-3, -18) \cdot f_{yy}(-3, -18) - [f_{xy}(-3, -18)]^2 = 288 > 0$$
Thus, $f(-3, -18) = -162$ is a local minimum.

17. $f(x, y) = x^3 - 3xy^2 + 6y^2$
$f_x = 3x^2 - 3y^2 = 0$
Thus, $y^2 = x^2$ or $y = \pm x$.
$f_y = -6xy + 12y = 0$ or $-6y(x - 2) = 0$
Thus, $y = 0$ or $x = 2$.
Therefore, the critical points are $(0, 0)$, $(2, 2)$, and $(2, -2)$. Now,
$$f_{xx} = 6x \qquad f_{xy} = -6y \qquad f_{yy} = -6x + 12$$

For the critical point $(0, 0)$:
$$f_{xx}(0, 0) \cdot f_{yy}(0, 0) - [f_{xy}(0, 0]^2 = 0 \cdot 12 - 0^2 = 0$$
Thus, the second-derivative test fails.

For the critical point $(2, 2)$:
$$f_{xx}(2, 2) \cdot f_{yy}(2, 2) - [f_{xy}(2, 2)]^2 = 12 \cdot 0 - (-12)^2 = -144 < 0$$
Thus, $f(x, y)$ has a saddle point at $(2, 2)$.

For the critical point $(2, -2)$:
$$f_{xx}(2, -2) \cdot f_{yy}(2, -2) - [f_{xy}(2, -2)]^2 = 12 \cdot 0 - (12)^2 = -144 < 0$$
Thus, $f(x, y)$ has a saddle point at $(2, -2)$.

19. $f(x, y) = y^3 + 2x^2y^2 - 3x - 2y + 8$;
$f_x = 4xy^2 - 3$; $f_y = 3y^2 + 4x^2y - 2$
Set $f_x = 0$ and $f_y = 0$ to find the critical points:
$$4xy^2 - 3 = 0 \qquad (1)$$
$$3y^2 + 4x^2y - 2 = 0 \qquad (2)$$
From (1) $x = \dfrac{3}{4y^2}$. Substituting this into (2), we have
$$3y^2 + 4\left(\frac{3}{4y^2}\right)^2 y - 2 = 0$$
$$3y^2 + 4\left(\frac{9}{16y^4}\right)y - 2 = 0$$
$$12y^5 - 8y^3 + 9 = 0$$
Using a graphing utility, we find that $y \approx -1.105$ and $x \approx 0.614$.

Now, $f_{xx} = 4y^2$ and $f_{xx}(0.614, -1.105) \approx 4.884$
$\qquad f_{xy} = 8xy$ and $f_{xy}(0.614, -1.105) \approx -5.428$
$\qquad f_{yy} = 6y + 4x^2$ and $f_{yy}(0.614, -1.105) \approx -5.122$
$\qquad f_{xx}(0.614, -1.105) \cdot f_{yy}(0.614, -1.105) - [f_{xy}(0.614, -1.105)]^2$
$\qquad \approx -54.479 < 0$

Thus, $f(x, y)$ has a saddle point at $(0.614, -1.105)$.

21. $f(x, y) = x^2 \geq 0$ for all (x, y) and $f(x, y) = 0$ when $x = 0$. Thus, f has a local minimum at each point $(0, y, 0)$ on the y-axis.

23. $P(x, y) = R(x, y) - C(x, y)$
$$= 2x + 3y - (x^2 - 2xy + 2y^2 + 6x - 9y + 5)$$
$$= -x^2 + 2xy - 2y^2 - 4x + 12y - 5$$
$P_x = -2x + 2y - 4 = 0 \qquad (1)$
$P_y = 2x - 4y + 12 = 0 \qquad (2)$

Solving (1) and (2) for x and y, we obtain $x = 2$ and $y = 4$. Thus, $(2, 4)$ is a critical point.

$P_{xx} = -2$ and $P_{xx}(2, 4) = -2 < 0$
$P_{xy} = 2$ and $P_{xy}(2, 4) = 2$
$P_{yy} = -4$ and $P_{yy}(2, 4) = -4$

$P_{xx}(2, 4) \cdot P_{yy}(2, 4) - [P_{xy}(2, 4)]^2 = (-2)(-4) - [2]^2 = 4 > 0$

The maximum occurs when 2000 type A and 4000 type B calculators are produced. The maximum profit is given by $P(2, 4)$. Hence,

$\max P = P(2, 4) = -(2)^2 + 2 \cdot 2 \cdot 4 - 2 \cdot 4^2 - 4 \cdot 2 + 12 \cdot 4 - 5$
$$= -4 + 16 - 32 - 8 + 48 - 5 = \$15 \text{ million.}$$

25. $x = 116 - 30p + 20q$ (Brand A)
$y = 144 + 16p - 24q$ (Brand B)

(A)

p	q	x	y
10	12	56	16
11	11	6	56

(B) In terms of p and q, the cost function C is given by:
$C = 6x + 8y = 6(116 - 30p + 20q) + 8(144 + 16p - 24q)$
$$= 1848 - 52p - 72q$$

The revenue function R is given by:
$R = px + qy = p(116 - 30p + 20q) + q(144 + 16p - 24q)$
$$= 116p - 30p^2 + 20pq + 144q + 16pq - 24q^2$$
$$= -30p^2 + 36pq - 24q^2 + 116p + 144q$$

Thus, the profit $P = R - C$ is given by:
$P = -30p^2 + 36pq - 24q^2 + 116p + 144q - (1848 - 52p - 72q)$

$$= -30p^2 + 36pq - 24q^2 + 168p + 216q - 1848$$

Now, calculating P_p and P_q and setting these equal to 0, we have:
$P_p = -60p + 36q + 168 = 0 \qquad (1)$
$P_q = 36p - 48q + 216 = 0 \qquad (2)$

Solving (1) and (2) for p and q, we get $p = 10$ and $q = 12$. Thus, $(10, 12)$ is a critical point of the profit function P.

$P_{pp} = -60$ and $P_{pp}(10, 12) = -60$
$P_{pq} = 36$ and $P_{pq}(10, 12) = 36$
$P_{qq} = -48$ and $P_{qq}(10, 12) = -48$

$P_{pp} \cdot P_{qq} - [P_{pq}]^2 = (-60)(-48) - (36)^2 = 1584 > 0$

Since $P_{pp}(10, 12) = -60 < 0$, we conclude that the maximum profit occurs when $p = \$10$ and $q = \$12$. The maximum profit is:

$P(10, 12) = -30(10)^2 + 36(10)(12) - 24(12)^2 + 168(10) + 216(12) - 1848$
$= \$288$

27. The square of the distance from P to A is: $x^2 + y^2$
The square of the distance from P to B is:
$(x - 2)^2 + (y - 6)^2 = x^2 - 4x + y^2 - 12y + 40$
The square of the distance from P to C is:
$(x - 10)^2 + y^2 = x^2 - 20x + y^2 + 100$
Thus, we have:
$P(x, y) = 3x^2 - 24x + 3y^2 - 12y + 140$

$P_x = 6x - 24 = 0$ $\qquad\qquad P_y = 6y - 12 = 0$
$\qquad x = 4$ $\qquad\qquad\qquad\qquad y = 2$

Therefore, $(4, 2)$ is a critical point.

$P_{xx} = 6$ and $P_{xx}(4, 2) = 6 > 0$
$P_{xy} = 0$ and $P_{xy}(4, 2) = 0$
$P_{yy} = 6$ and $P_{yy}(4, 2) = 6$
$P_{xx} \cdot P_{yy} - [P_{xy}]^2 = 6 \cdot 6 - 0 = 36 > 0$
Therefore, P has a minimum at the point $(4, 2)$.

29. Let $x = $ length, $y = $ width, and $z = $ height. Then $V = xyz = 64$ or $z = \dfrac{64}{xy}$. The surface area of the box is:

$S = xy + 2xz + 4yz$ or $S(x, y) = xy + \dfrac{128}{y} + \dfrac{256}{x}$, $x > 0$, $y > 0$

$S_x = y - \dfrac{256}{x^2} = 0$ or $y = \dfrac{256}{x^2}$ (1)

$S_y = x - \dfrac{128}{y^2} = 0$ or $x = \dfrac{128}{y^2}$

Thus, $y = \dfrac{256}{\dfrac{(128)^2}{y^4}}$ or $y^4 - 64y = 0$

$\qquad\qquad\qquad\qquad y(y^3 - 64) = 0$ (Since $y > 0$, $y = 0$ does not
$\qquad\qquad\qquad\qquad$ and $\qquad y = 4$ yield a critical point.)

Setting $y = 4$ in (1), we find $x = 8$. Therefore, the critical point is $(8, 4)$.
Now we have:
$S_{xx} = \dfrac{512}{x^3}$ and $S_{xx}(8, 4) = 1 > 0$
$S_{xy} = 1$
$S_{yy} = \dfrac{256}{y^3}$ and $S_{yy}(8, 4) = 4$

$S_{xx}(8, 4) \cdot S_{yy}(8, 4) - [S_{xy}(8, 4)]^2 = 1 \cdot 4 - 1^2 = 3 > 0$

Thus, the dimensions that will require the least amount of material are:
Length $x = 8$ inches; Width $y = 4$ inches; Height $z = \dfrac{64}{8(4)} = 2$ inches.

31. Let x = length of the package, y = width, and z = height. Then
$x + 2y + 2z = 120$ (1)
Volume $= V = xyz$.

From (1), $z = \dfrac{120 - x - 2y}{2}$. Thus, we have:

$V(x, y) = xy\left(\dfrac{120 - x - 2y}{2}\right) = 60xy - \dfrac{x^2 y}{2} - xy^2$, $x > 0$, $y > 0$

$V_x = 60y - xy - y^2 = 0$

$\quad y(60 - x - y) = 0$

$\quad\quad 60 - x - y = 0$ (2)   (Since $y > 0$, $y = 0$ does not yield
$\quad\quad\quad\quad\quad\quad\quad\quad\quad\quad\quad\quad\quad$ a critical point.)

$V_y = 60x - \dfrac{x^2}{2} - 2xy = 0$

$\quad x\left(60 - \dfrac{x}{2} - 2y\right) = 0$

$\quad\quad 120 - x - 4y = 0$ (3)   (Since $x > 0$, $x = 0$ does not yield
$\quad\quad\quad\quad\quad\quad\quad\quad\quad\quad\quad\quad\quad$ a critical point.)

Solving (2) and (3) for x and y, we obtain $x = 40$ and $y = 20$. Thus, (40, 20) is the critical point.

$V_{xx} = -y$ $\quad\quad\quad\quad$ and $\quad V_{xx}(40, 20) = -20 < 0$
$V_{xy} = 60 - x - 2y$ and $\quad V_{xy}(40, 20) = 60 - 40 - 40 = -20$
$V_{yy} = -2x$ $\quad\quad\quad\quad$ and $\quad V_{yy}(40, 20) = -80$

$V_{xx}(40, 20) \cdot V_{yy}(40, 20) - [V_{xy}(40, 20)]^2 = (-20)(-80) - [-20]^2$

$\quad\quad\quad\quad\quad\quad\quad\quad\quad\quad\quad\quad\quad\quad\quad\quad\quad = 1600 - 400$
$\quad\quad\quad\quad\quad\quad\quad\quad\quad\quad\quad\quad\quad\quad\quad\quad\quad = 1200 > 0$

Thus, the maximum volume of the package is obtained when $x = 40$, $y = 20$, and $z = \dfrac{120 - 40 - 2 \cdot 20}{2} = 20$ inches. The package has dimensions:

Length $x = 40$ inches; Width $y = 20$ inches; Height $z = 20$ inches.

EXERCISE 13-4

Things to remember:

1. Any local maxima or minima of the function $z = f(x, y)$ subject to the constraint $g(x, y) = 0$ will be among those points (x_0, y_0) for which (x_0, y_0, λ_0) is a solution to the system:

$\quad\quad F_x(x, y, \lambda) = 0$
$\quad\quad F_y(x, y, \lambda) = 0$
$\quad\quad F_\lambda(x, y, \lambda) = 0$

where $F(x, y, \lambda) = f(x, y) + \lambda g(x, y)$, provided all the partial derivatives exist.

2. METHOD OF LAGRANGE MULTIPLIERS

[Note: Although stated for functions of two variables, the method also applies to functions of three or more variables.]

(a) Formulate the problem in the form:
Maximize (or Minimize) $z = f(x, y)$
Subject to: $g(x, y) = 0$

(b) Form the function F:
$$F(x, y, \lambda) = f(x, y) + \lambda g(x, y)$$

(c) Find the critical points (x_0, y_0, λ_0) for F, that is, solve the system:
$$F_x(x, y, \lambda) = 0$$
$$F_y(x, y, \lambda) = 0$$
$$F_\lambda(x, y, \lambda) = 0$$

(d) If (x_0, y_0, λ_0) is the only critical point of F, then assume that (x_0, y_0) is the solution to the problem. If F has more than one critical point, then evaluate $z = f(x, y)$ at (x_0, y_0) for each critical point (x_0, y_0, λ_0) of F. Assume that the largest of these values is the maximum value of $f(x, y)$ subject to the constraint $g(x, y) = 0$, and the smallest is the minimum value of $f(x, y)$ subject to the constraint $g(x, y) = 0$.

---

1. Step 1. Maximize $f(x, y) = 2xy$
Subject to: $g(x, y) = x + y - 6 = 0$

Step 2. $F(x, y, \lambda) = f(x, y) + \lambda g(x, y)$
$\qquad\qquad = 2xy + \lambda(x + y - 6)$

Step 3. $F_x = 2y + \lambda = 0 \qquad (1)$
$\quad\;\; F_y = 2x + \lambda = 0 \qquad (2)$
$\quad\;\; F_\lambda = x + y - 6 = 0 \quad (3)$

From (1) and (2), we obtain:
$$x = -\frac{\lambda}{2}, \; y = -\frac{\lambda}{2}$$
Substituting these into (3), we have:
$$-\frac{\lambda}{2} - \frac{\lambda}{2} - 6 = 0$$
$$\lambda = -6.$$
Thus, the critical point is $(3, 3, -6)$.

Step 4. Since $(3, 3, -6)$ is the only critical point for F, we conclude that max $f(x, y) = f(3, 3) = 2 \cdot 3 \cdot 3 = 18$.

3. <u>Step 1</u>. Minimize $f(x, y) = x^2 + y^2$
   Subject to: $g(x, y) = 3x + 4y - 25 = 0$

<u>Step 2</u>. $F(x, y, \lambda) = f(x, y) + \lambda g(x, y)$
   $$= x^2 + y^2 + \lambda(3x + 4y - 25)$$

<u>Step 3</u>. $F_x = 2x + 3\lambda = 0$ (1)

   $F_y = 2y + 4\lambda = 0$ (2)

   $F_\lambda = 3x + 4y - 25 = 0$ (3)

   From (1) and (2), we obtain:

   $$x = -\frac{3\lambda}{2}, \quad y = -2\lambda$$

   Substituting these into (3), we have:

   $$3\left(-\frac{3\lambda}{2}\right) + 4(-2\lambda) - 25 = 0$$

   $$\frac{25}{2}\lambda = -25$$

   $$\lambda = -2$$

   The critical point is $(3, 4, -2)$.

<u>Step 4</u>. Since $(3, 4, -2)$ is the only critical point for F, we conclude
   that min $f(x, y) = f(3, 4) = 3^2 + 4^2 = 25$.

5. <u>Step 1</u>. Maximize and minimize $f(x, y) = 2xy$
   Subject to: $g(x, y) = x^2 + y^2 - 18 = 0$

<u>Step 2</u>. $F(x, y, \lambda) = f(x, y) + \lambda g(x, y)$
   $$= 2xy + \lambda(x^2 + y^2 - 18)$$

<u>Step 3</u>. $F_x = 2y + 2\lambda x = 0$ (1)

   $F_y = 2x + 2\lambda y = 0$ (2)

   $F_\lambda = x^2 + y^2 - 18 = 0$ (3)

   From (1), (2), and (3), we obtain the critical points
   $(3, 3, -1)$, $(3, -3, 1)$, $(-3, 3, 1)$ and $(-3, -3, -1)$.

<u>Step 4</u>. $f(3, 3) = 2 \cdot 3 \cdot 3 = 18$

   $f(3, -3) = 2 \cdot 3(-3) = -18$

   $f(-3, 3) = 2(-3) \cdot 3 = -18$

   $f(-3, -3) = 2(-3)(-3) = 18$

   Thus, max $f(x, y) = f(3, 3) = f(-3, -3) = 18$;
   min $f(x, y) = f(3, -3) = f(-3, 3) = -18$.

7. Let x and y be the required numbers.

<u>Step 1</u>. Maximize $f(x, y) = xy$
   Subject to: $x + y = 10$ or $g(x, y) = x + y - 10 = 0$

<u>Step 2</u>. $F(x, y, \lambda) = xy + \lambda(x + y - 10)$

<u>Step 3</u>. $F_x = y + \lambda = 0$ (1)

   $F_y = x + \lambda = 0$ (2)

   $F_\lambda = x + y - 10 = 0$ (3)

From (1) and (2), we obtain:

$x = -\lambda, \ y = -\lambda$

Substituting these into (3), we have:

$\lambda = -5$

The critical point is $(5, 5, -5)$.

Step 4. Since $(5, 5, -5)$ is the only critical point for F, we conclude that max $f(x, y) = f(5, 5) = 5 \cdot 5 = 25$. Thus, the maximum product is 25 when $x = 5$ and $y = 5$.

9. Step 1. Minimize $f(x, y, z) = x^2 + y^2 + z^2$

Subject to: $g(x, y) = 2x - y + 3z + 28 = 0$

Step 2. $F(x, y, z, \lambda) = x^2 + y^2 + z^2 + \lambda(2x - y + 3z + 28)$

Step 3. $F_x = 2x + 2\lambda = 0 \qquad (1)$

$F_y = 2y - \lambda = 0 \qquad (2)$

$F_z = 2z + 3\lambda = 0 \qquad (3)$

$F_\lambda = 2x - y + 3z + 28 = 0 \quad (4)$

From (1), (2), and (3), we obtain:

$x = -\lambda, \ y = \dfrac{\lambda}{2}, \ z = -\dfrac{3}{2}\lambda$

Substituting these into (4), we have:

$$2(-\lambda) - \frac{\lambda}{2} + 3\left(-\frac{3}{2}\lambda\right) + 28 \ = 0$$

$$-\frac{14}{2}\lambda + 28 \ = 0$$

$$\lambda = 4$$

The critical point is $(-4, 2, -6, 4)$.

Step 4. Since $(-4, 2, -6, 4)$ is the only critical point for F, we conclude that min $f(x, y, z) = f(-4, 2, -6) = 56$.

11. Step 1. Maximize and minimize $f(x, y, z) = x + y + z$

Subject to: $g(x, y, z) = x^2 + y^2 + z^2 - 12 = 0$

Step 2. $F(x, y, z, \lambda) = f(x, y, z) + \lambda g(x, y, z)$

$= x + y + z + \lambda(x^2 + y^2 + z^2 - 12)$

Step 3. $F_x = 1 + 2x\lambda = 0 \qquad (1)$

$F_y = 1 + 2y\lambda = 0 \qquad (2)$

$F_z = 1 + 2z\lambda = 0 \qquad (3)$

$F_\lambda = x^2 + y^2 + z^2 - 12 = 0 \quad (4)$

From (1), (2), and (3), we obtain:

$x = -\dfrac{1}{2\lambda}, \ y = -\dfrac{1}{2\lambda}, \ z = -\dfrac{1}{2\lambda}$

Substituting these into (4), we have:

$$\left(-\frac{1}{2\lambda}\right)^2 + \left(-\frac{1}{2\lambda}\right)^2 + \left(-\frac{1}{2\lambda}\right)^2 - 12 = 0$$

$$\frac{3}{4\lambda^2} - 12 = 0$$

$$1 - 16\lambda^2 = 0$$

$$\lambda = \pm\frac{1}{4}$$

Thus, the critical points are $\left(2, 2, 2, -\frac{1}{4}\right)$ and $\left(-2, -2, -2, \frac{1}{4}\right)$.

Step 4. $f(2, 2, 2) = 2 + 2 + 2 = 6$
$f(-2, -2, -2) = -2 - 2 - 2 = -6$
Thus, max $f(x, y, z) = f(2, 2, 2) = 6$;
   min $f(x, y, z) = f(-2, -2, -2) = -6$.

13. Step 1. Maximize $f(x, y) = y + xy^2$
Subject to: $x + y^2 = 1$ or $g(x, y) = x + y^2 - 1 = 0$

Step 2. $F(x, y, \lambda) = y + xy^2 + \lambda(x + y^2 - 1)$

Step 3. $F_x = y^2 + \lambda = 0$ ⠀⠀⠀⠀(1)
$F_y = 1 + 2xy + 2y\lambda = 0$ ⠀⠀(2)
$F_\lambda = x + y^2 - 1 = 0$ ⠀⠀(3)

From (1), $\lambda = -y^2$ and from (3), $x = 1 - y^2$. Substituting these values into (2), we have

$$1 + 2(1 - y^2)y - 2y^3 = 0$$
or ⠀⠀$$4y^3 - 2y - 1 = 0$$

Using a graphing utility to solve this equation, we get
$y \approx 0.885$. Then $x \approx 0.217$ and max $f(x, y) = f(0.217, 0.885)$
≈ 1.055.

15. The constraint $g(x, y) = y - 5 = 0$ implies $y = 5$. Replacing y by 5 in
the function f, the problem reduces to maximizing the function $h(x) = f(x, 5)$, a function of one independent variable.

17. Step 1. Minimize cost function $C(x, y) = 6x^2 + 12y^2$
Subject to: $x + y = 90$ or $g(x, y) = x + y - 90 = 0$

Step 2. $F(x, y, \lambda) = 6x^2 + 12y^2 + \lambda(x + y - 90)$

Step 3. $F_x = 12x + \lambda = 0$ ⠀⠀⠀(1)
$F_y = 24y + \lambda = 0$ ⠀⠀⠀(2)
$F_\lambda = x + y - 90 = 0$ ⠀⠀(3)
From (1) and (2), we obtain
$$x = -\frac{\lambda}{12}, \; y = -\frac{\lambda}{24}$$

Substituting these into (3), we have:

$$-\frac{\lambda}{12} - \frac{\lambda}{24} - 90 = 0$$

$$\frac{3\lambda}{24} = -90$$

$$\lambda = -720$$

The critical point is $(60, 30, -720)$.

Step 4. Since $(60, 30, -720)$ is the only critical point for F, we conclude that:

$$\min C(x, y) = C(60, 30) = 6 \cdot 60^2 + 12 \cdot 30^2$$
$$= 21,600 + 10,800$$
$$= \$32,400$$

Thus, 60 of model A and 30 of model B will yield a minimum cost of $32,400 per week.

19. (A) Step 1. Maximize the production function $N(x, y) = 50x^{0.8}y^{0.2}$
Subject to the constraint: $C(x, y) = 40x + 80y = 400,000$
i.e., $g(x, y) = 40x + 80y - 400,000 = 0$

Step 2. $F(x, y, \lambda) = 50x^{0.8}y^{0.2} + \lambda(40x + 80y - 400,000)$

Step 3. $F_x = 40x^{-0.2}y^{0.2} + 40\lambda = 0$ (1)

$F_y = 10x^{0.8}y^{-0.8} + 80\lambda = 0$ (2)

$F_\lambda = 40x + 80y - 400,000 = 0$ (3)

From (1), $\lambda = -\dfrac{y^{0.2}}{x^{0.2}}$. From (2), $\lambda = -\dfrac{x^{0.8}}{8y^{0.8}}$.

Thus, we obtain

$$-\frac{y^{0.2}}{x^{0.2}} = -\frac{x^{0.8}}{8y^{0.8}} \quad \text{or} \quad x = 8y$$

Substituting into (3), we have:
$$320y + 80y - 400,000 = 0$$
$$y = 1000$$

Therefore, $x = 8000$, $\lambda \approx -0.6598$, and the critical point is $(8000, 1000, -0.6598)$. Thus, we conclude that:
$$\max N(x, y) = N(8000, 1000) = 50(8000)^{0.8}(1000)^{0.2}$$
$$\approx 263,902 \text{ units}$$

and production is maximized when 8000 labor units and 1000 capital units are used.

(B) The marginal productivity of money is $-\lambda \approx 0.6598$. The increase in production if an additional $50,000 is budgeted for production is:
$0.6598(50,000) = 32,990$ units

21. Let x = length, y = width, and z = height.

Step 1. Maximize volume $V = xyz$

Subject to: $S(x, y, z) = xy + 3xz + 3yz - 192 = 0$

Step 2. $F(x, y, z, \lambda) = xyz + \lambda(xy + 3xz + 3yz - 192)$

Step 3. $F_x = yz + \lambda(y + 3z) = 0$ (1)

$F_y = xz + \lambda(x + 3z) = 0$ (2)

$F_z = xy + \lambda(3x + 3y) = 0$ (3)

$F_\lambda = xy + 3xz + 3yz - 192 = 0$ (4)

Solving this system of equations, (1)-(4), simultaneously, yields:

$x = 8$, $y = 8$, $z = \dfrac{8}{3}$, $\lambda = -\dfrac{4}{3}$

Thus, the critical point is $\left(8,\ 8,\ \dfrac{8}{3},\ -\dfrac{4}{3}\right)$.

Step 4. Since $\left(8,\ 8,\ \dfrac{8}{3},\ -\dfrac{4}{3}\right)$ is the only critical point for F:

$\max V(x,\ y,\ z) = V\left(8,\ 8,\ \dfrac{8}{3}\right) = \dfrac{512}{3} \approx 170.67$

Thus, the dimensions that will maximize the volume of the box are: Length $x = 8$ inches; Width $y = 8$ inches; Height $z = \dfrac{8}{3}$ inches.

23. Step 1. Maximize $A = xy$

Subject to: $P(x, y) = y + 4x - 400 = 0$

Step 2. $F(x, y, \lambda) = xy + \lambda(y + 4x - 400)$

Step 3. $F_x = y + 4\lambda = 0$ (1)

$F_y = x + \lambda = 0$ (2)

$F_\lambda = y + 4x - 400 = 0$ (3)

From (1) and (2), we have:

$y = -4\lambda$ and $x = -\lambda$

Substituting these into (3), we obtain:

$-4\lambda - 4\lambda - 400 = 0$

Thus, $\lambda = -50$ and the critical point is $(50, 200, -50)$.

Step 4. Since $(50, 200, -50)$ is the only critical point for F,
$\max A(x,\ y) = A(50, 200) = 10,000$.
Therefore, $x = 50$ feet, $y = 200$ feet will produce the maximum area $A(50, 200) = 10,000$ square feet.

Things to remember:

<u>1</u>. For a set of n points (x_1, y_1), (x_2, y_2), ..., (x_n, y_n), the coefficients m and d of the least squares line
$$y = mx + d$$
are the solutions of the system of NORMAL EQUATIONS

$$\left(\sum_{k=1}^{n} x_k\right)m + nd = \sum_{k=1}^{n} y_k \tag{1}$$

$$\left(\sum_{k=1}^{n} x_k^2\right)m + \left(\sum_{k=1}^{n} x_k\right)d = \sum_{k=1}^{n} x_k y_k$$

and are given by the formulas

$$m = \frac{n\left(\sum_{k=1}^{n} x_k y_k\right) - \left(\sum_{k=1}^{n} x_k\right)\left(\sum_{k=1}^{n} y_k\right)}{n\left(\sum_{k=1}^{n} x_k^2\right) - \left(\sum_{k=1}^{n} x_k\right)^2} \tag{2}$$

and

$$d = \frac{\sum_{k=1}^{n} y_k - m\left(\sum_{k=1}^{n} x_k\right)}{n} \tag{3}$$

[<u>Note</u>: To find m and d, either solve system (1) directly, or use formulas (2) and (3). If the formulas are used, the value of m must be calculated first since it is used in formula (3).

1.

x_k	y_k	$x_k y_k$	x_k^2
1	1	1	1
2	3	6	4
3	4	12	9
4	3	12	16
Totals 10	11	31	30

Thus, $\sum_{k=1}^{4} x_k = 10$, $\sum_{k=1}^{4} y_k = 11$, $\sum_{k=1}^{4} x_k y_k = 31$, $\sum_{k=1}^{4} x_k^2 = 30.$

Substituting these values into formulas (2) and (3) for m and d, respectively, we have:

$$m = \frac{n\left(\sum_{k=1}^{n} x_k y_k\right) - \left(\sum_{k=1}^{n} x_k\right)\left(\sum_{k=1}^{n} y_k\right)}{n\left(\sum_{k=1}^{n} x_k^2\right) - \left(\sum_{k=1}^{n} x_k\right)^2} = \frac{4(31) - (10)(11)}{4(30) - (10)^2} = \frac{14}{20} = 0.7$$

$$d = \frac{\sum\limits_{k=1}^{n} y_k - m\left(\sum\limits_{k=1}^{n} x_k\right)}{n} = \frac{11 - 0.7(10)}{4} = 1$$

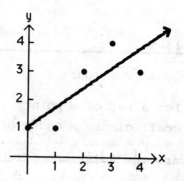

Thus, the least squares line is $y = mx + d = 0.7x + 1$. Refer to the graph at the right.

3.

	x_k	y_k	$x_k y_k$	x_k^2
	1	8	8	1
	2	5	10	4
	3	4	12	9
	4	0	0	16
Totals	10	17	30	30

Thus, $\sum\limits_{k=1}^{4} x_k = 10$, $\sum\limits_{k=1}^{4} y_k = 17$, $\sum\limits_{k=1}^{4} x_k y_k = 30$, $\sum\limits_{k=1}^{4} x_k^2 = 30$.

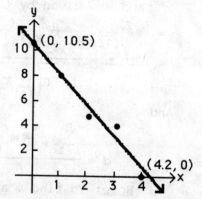

Substituting these values into system (1), we have:
$$10m + 4d = 17$$
$$30m + 10d = 30$$

The solution of this system is $m = -2.5$, $d = 10.5$. Thus, the least squares line is $y = mx + d = -2.5x + 10.5$. Refer to the graph at the right.

5.

	x_k	y_k	$x_k y_k$	x_k^2
	1	3	3	1
	2	4	8	4
	3	5	15	9
	4	6	24	16
Totals	10	18	50	30

Thus, $\sum\limits_{k=1}^{4} x_k = 10$, $\sum\limits_{k=1}^{4} y_k = 18$, $\sum\limits_{k=1}^{4} x_k y_k = 50$, $\sum\limits_{k=1}^{4} x_k^2 = 30$.

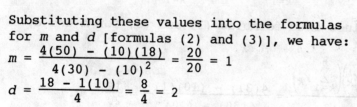

[Note: All points lie on the line.]

Substituting these values into the formulas for m and d [formulas (2) and (3)], we have:
$$m = \frac{4(50) - (10)(18)}{4(30) - (10)^2} = \frac{20}{20} = 1$$
$$d = \frac{18 - 1(10)}{4} = \frac{8}{4} = 2$$

Thus, the least squares line is $y = mx + d = x + 2$. Refer to the graph at the right.

7.

	x_k	y_k	$x_k y_k$	x_k^2
	0	10	0	0
	5	22	110	25
	10	31	310	100
	15	46	690	225
	20	51	1020	400
Totals	50	160	2130	750

Thus, $\sum_{k=1}^{5} x_k = 50$, $\sum_{k=1}^{5} y_k = 160$, $\sum_{k=1}^{5} x_k y_k = 2130$, $\sum_{k=1}^{5} x_k^2 = 750$.

Substituting these values into formulas (2) and (3) for m and d, respectively, we have:

$$m = \frac{5(2130) - (50)(160)}{5(750) - (50)^2} = \frac{2650}{1250} = 2.12$$

$$d = \frac{160 - 2.12(50)}{5} = \frac{54}{5} = 10.8$$

Thus, the least squares line is $y = 2.12x + 10.8$.
When $x = 25$, $y = 2.12(25) + 10.8 = 63.8$.

9.

	x_k	y_k	$x_k y_k$	x_k^2
	−1	14	−14	1
	1	12	12	1
	3	8	24	9
	5	6	30	25
	7	5	35	49
Totals	15	45	87	85

Thus, $\sum_{k=1}^{5} x_k = 15$, $\sum_{k=1}^{5} y_k = 45$, $\sum_{k=1}^{5} x_k y_k = 87$, $\sum_{k=1}^{5} x_k^2 = 85$.

Substituting these values into formulas (2) and (3) for m and d, respectively, we have:

$$m = \frac{5(87) - (15)(45)}{5(85) - (15)^2} = \frac{-240}{200} = -1.2$$

$$d = \frac{45 - (-1.2)(15)}{5} = 12.6$$

Thus, the least squares line is
$y = -1.2x + 12.6$.
When $x = 2$, $y = -1.2(2) + 12.6 = 10.2$.

11.

x_k	y_k	$x_k y_k$	x_k^2
0.5	25	12.5	0.25
2.0	22	44.0	4.00
3.5	21	73.5	12.25
5.0	21	105.0	25.00
6.5	18	117.0	42.25
9.5	12	114.0	90.25
11.0	11	121.0	121.00
12.5	8	100.0	156.25
14.0	5	70.0	196.00
15.5	1	15.5	240.25
Totals 80.0	144	772.5	887.50

Thus, $\displaystyle\sum_{k=1}^{10} x_k = 80$, $\displaystyle\sum_{k=1}^{10} y_k = 144$, $\displaystyle\sum_{k=1}^{10} x_k y_k = 772.5$, $\displaystyle\sum_{k=1}^{10} x_k^2 = 887.5$.

Substituting these values into formulas (2) and (3) for m and d, respectively, we have:

$$m = \frac{10(772.5) - (80)(144)}{10(887.5) - (80)^2} = \frac{-3795}{2475} \approx -1.53$$

$$d = \frac{144 - (-1.53)(80)}{10} = \frac{266.4}{10} = 26.64$$

Thus, the least squares line is
$y = -1.53x + 26.64$.
When $x = 8$, $y = -1.53(8) + 26.64 = 14.4$.

13. Minimize

$$F(a, b, c) = (a + b + c - 2)^2 + (4a + 2b + c - 1)^2$$
$$+ (9a + 3b + c - 1)^2 + (16a + 4b + c - 3)^2$$

$$F_a(a, b, c) = 2(a + b + c - 2) + 8(4a + 2b + c - 1)$$
$$+ 18(9a + 3b + c - 1) + 32(16a + 4b + c - 3)$$
$$= 708a + 200b + 60c - 126$$

$$F_b(a, b, c) = 2(a + b + c - 2) + 4(4a + 2b + c - 1)$$
$$+ 6(9a + 3b + c - 1) + 8(16a + 4b + c - 3)$$
$$= 200a + 60b + 20c - 38$$

$$F_c(a, b, c) = 2(a + b + c - 2) + 2(4a + 2b + c - 1)$$
$$+ 2(9a + 3b + c - 1) + 2(16a + 4b + c - 3)$$
$$= 60a + 20b + 8c - 14$$

The system is: $\quad F_a(a, b, c) = 0$
$$F_b(a, b, c) = 0$$
$$F_c(a, b, c) = 0$$

or: $\qquad 708a + 200b + 60c = 126$
$$200a + 60b + 20c = 38$$
$$60a + 20b + 8c = 14$$

The solution is $(a, b, c) = (0.75, -3.45, 4.75)$, which gives us the equation for the parabola shown at the right:
$y = ax^2 + bx + c$
or
$y = 0.75x^2 - 3.45x + 4.75$
The given points: $(1, 2)$, $(2, 1)$, $(3, 1)$, $(4, 3)$ also appear on the graph.

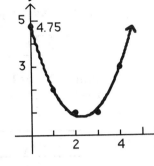

15. System (1) is:

$$\left(\sum_{k=1}^{n} x_k\right)m + nd = \sum_{k=1}^{n} y_k \qquad \text{(a)}$$

$$\left(\sum_{k=1}^{n} x_k^2\right)m + \left(\sum_{k=1}^{n} x_k\right)d = \sum_{k=1}^{n} x_k y_k \qquad \text{(b)}$$

Multiply equation (a) by $-\left(\sum_{k=1}^{n} x_k\right)$, equation (b) by n, and add the resulting equations. This will eliminate d from the system.

$$\left[-\left(\sum_{k=1}^{n} x_k\right)^2 + n\sum_{k=1}^{n} x_k^2\right]m = -\left(\sum_{k=1}^{n} x_k\right)\left(\sum_{k=1}^{n} y_k\right) + n\sum_{k=1}^{n} x_k y_k$$

Thus,

$$m = \frac{n\left(\sum_{k=1}^{n} x_k y_k\right) - \left(\sum_{k=1}^{n} x_k\right)\left(\sum_{k=1}^{n} y_k\right)}{n\left(\sum_{k=1}^{n} x_k^2\right) - \left(\sum_{k=1}^{n} x_k\right)^2}$$

which is equation (2). Solving equation (a) for d, we have

$$d = \frac{\sum_{k=1}^{n} y_k - m\left(\sum_{k=1}^{n} x_k\right)}{n}$$

which is equation (3).

17. (A) Suppose that $n = 5$ and $x_1 = -2$, $x_2 = -1$, $x_3 = 0$, $x_4 = 1$, $x_5 = 2$.

Then $\sum_{k=1}^{5} x_k = -2 - 1 + 0 + 1 + 2 = 0$. Therefore, from formula (2),

$$m = \frac{5\sum_{k=1}^{5} x_k y_k}{5\sum_{k=1}^{5} x_k^2} = \frac{\sum x_k y_k}{\sum x_k^2}$$

From formula (3), $d = \dfrac{\sum_{k=1}^{5} y_k}{5}$,

which is the average of y_1, y_2, y_3, y_4, and y_5.

(B) If the average of the x-coordinates is 0, then

$$\frac{\sum_{k=1}^{n} x_k}{n} = 0$$

Then all calculations will be the same as in part (A) with "n" instead of 5.

19. (A)

x_k	y_k	$x_k y_k$	x_k^2
0	23.8	0	0
1	16.5	16.5	1
2	19.0	38.0	4
3	29.0	87.0	9
4	37.9	151.6	16
5	51.2	256.0	25
6	61.1	366.6	36
Totals 21	238.5	915.7	91

Thus, $\sum_{k=1}^{7} x_k = 21$, $\sum_{k=1}^{7} y_k = 238.5$, $\sum_{k=1}^{7} x_k y_k = 915.7$, $\sum_{k=1}^{7} x_k^2 = 91$.

Substituting these values into the formulas for m and d, we have:

$$m = \frac{7(915.7) - 21(238.5)}{7(91) - (21)^2} = \frac{1401.4}{196} = 7.15$$

$$d = \frac{238.5 - 7.15(21)}{7} = \frac{88.35}{7} \approx 12.62$$

Thus, the least squares line is $y = 7.15x + 12.62$.

(B) 1998 corresponds to $x = 13$. The monthly production in the 13th year will be $y \approx 7.15(13) + 12.62 = 105.57$ or 105.57 thousand per month.

21. (A)

x_k	y_k	$x_k y_k$	x_k^2
5.0	2.0	10.0	25.00
5.5	1.8	9.9	30.25
6.0	1.4	8.4	36.00
6.5	1.2	7.8	42.25
7.0	1.1	7.7	49.00
Totals 30.0	7.5	43.8	182.50

Thus, $\sum_{k=1}^{5} x_k = 30$, $\sum_{k=1}^{5} y_k = 7.5$, $\sum_{k=1}^{5} x_k y_k = 43.8$, $\sum_{k=1}^{5} x_k^2 = 182.5$.

Substituting these values into the formulas for m and d, we have:

$$m = \frac{5(43.8) - (30)(7.5)}{5(182.5) - (30)^2} = \frac{-6}{12.5} = -0.48$$

$$d = \frac{7.5 - (-0.48)(30)}{5} = 4.38$$

Thus, a demand equation is $y = -0.48x + 4.38$.

(B) Cost: $C = 4y$

Revenue: $R = xy = -0.48x^2 + 4.38x$

Profit: $P = R - C = -0.48x^2 + 4.38x - 4(-0.48x + 4.38)$

or $P(x) = -0.48x^2 + 6.3x - 17.52$

Now, $P'(x) = -0.96x + 6.3$.

Critical value: $P'(x) = -0.96x + 6.3 = 0$

$$x = \frac{6.3}{0.96} \approx 6.56$$

$P''(x) = -0.96$ and $P''(6.56) = -0.96 < 0$

Thus, $P(x)$ has a maximum at $x = 6.56$; the price per bottle should be $6.56 to maximize the monthly profit.

23.

x_k	y_k	$x_k y_k$	x_k^2
50	15	750	2500
55	13	715	3025
60	10	600	3600
65	6	390	4225
70	2	140	4900
Totals 300	46	2595	18,250

Thus, $\sum\limits_{k=1}^{5} x_k = 300$, $\sum\limits_{k=1}^{5} y_k = 46$, $\sum\limits_{k=1}^{5} x_k y_k = 2595$, $\sum\limits_{k=1}^{5} x_k^2 = 18,250$.

Substitituting these values into the formulas for m and d, we have:

$m = \dfrac{5(2595) - (300)(46)}{5(18,250) - (300)^2} = \dfrac{-825}{1250} = -0.66$

$d = \dfrac{46 - (-0.66)300}{5} = 48.8$

(A) The least squares line for the data is $P = -0.66T + 48.8$.

(B) $P(57) = -0.66(57) + 48.8 = 11.18$ beats per minute.

25.

x_k	y_k	$x_k y_k$	x_k^2
1.7	51	86.7	2.89
2.1	49	102.9	4.41
2.3	53	121.9	5.29
2.4	36	86.4	5.76
3.6	65	234.0	12.96
3.7	35	129.5	13.69
4.7	29	136.3	22.09
6.2	40	248.0	38.44
7.1	34	241.4	50.41
7.4	29	214.6	54.76
8.7	20	174.0	75.69
11.9	23	273.7	141.61
Totals 61.8	464	2049.4	428.00

Thus, $\sum\limits_{k=1}^{12} x_k = 61.8$, $\sum\limits_{k=1}^{12} y_k = 464$, $\sum\limits_{k=1}^{12} x_k y_k = 2049.4$, $\sum\limits_{k=1}^{12} x_k^2 = 428$.

Substitituting these values into the formulas for m and d, we have:

$$m = \frac{12(2049.4) - (61.8)(464)}{12(428) - (61.8)^2} = \frac{-4082.4}{1316.76} = -3.1$$

$$d = \frac{464 - (-3.1)(61.8)}{12} = \frac{655.58}{12} = 54.6$$

(A) The least squares line for the data is $D = -3.1A + 54.6$.

(B) If $A = 3.0$, then $D = -3.1(3.0) + 54.6 \approx 45$ or 45%.

27. (A) Enter the data in a calculator or computer. (We used a TI-85.) The totals are:

$$n = 23, \qquad \sum x = 1098, \qquad \sum y = 343.61,$$
$$\sum x^2 = 73{,}860, \qquad \sum xy = 18{,}259.08$$

Now, the least squares line can be calculated either by using formulas (2) and (3), or by using the linear regression feature. We used the latter to get

$$m = 0.08653 \quad \text{and} \quad b = 10.81$$

Therefore, the least squares line is:
$$y = 0.08653x + 10.81$$

(B) Using the result in (A), an estimate for the winning height in the pole vault in the Olympic games of 2008 is:

$$y = 0.08653(112) + 10.81 \approx 20.50 \text{ feet}$$

EXERCISE 13-6

Things to remember:

GIVEN A FUNCTION $z = f(x, y)$:

1. $\int f(x, y)\,dx$ means antidifferentiate $f(x, y)$ with respect to x, holding y fixed.

   $\int f(x, y)\,dy$ means antidifferentiate $f(x, y)$ with respect to y, holding x fixed.

2. The DOUBLE INTEGRAL of $f(x, y)$ over the rectangle $R = \{(x, y) \mid a \le x \le b, c \le y \le d\}$ is:

$$\iint\limits_{R} f(x, y)\,dA = \int_{a}^{b}\left[\int_{c}^{d} f(x, y)\,dy\right]dx$$

$$= \int_{c}^{d}\left[\int_{a}^{b} f(x, y)\,dx\right]dy$$

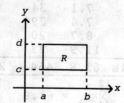

3. The AVERAGE VALUE of $f(x, y)$ over the rectangle
$R = \{(x, y) \mid a \leq x \leq b, c \leq y \leq d\}$ is:

$$\frac{1}{(b - a)(d - c)} \iint\limits_{R} f(x, y) \, dA$$

4. If $f(x, y) \geq 0$ over a rectangle
$R = \{(x, y) \mid a \leq x \leq b, c \leq y \leq d\}$,

then the VOLUME of the solid formed by graphing f over R is given by:

$$V = \iint\limits_{R} f(x, y) \, dA$$

---

1. (A) $\displaystyle\int 12x^2 y^3 \, dy = 12x^2 \int y^3 \, dy$ (x is treated as a constant.)

$\qquad\qquad = 12x^2 \dfrac{y^4}{4} + C(x)$ (The "constant" of integration is a function of x.)

$\qquad\qquad = 3x^2 y^4 + C(x)$

(B) $\displaystyle\int_0^1 12x^2 y^3 \, dy = 3x^2 y^4 \Big|_0^1 = 3x^2$

3. (A) $\displaystyle\int (4x + 6y + 5) \, dx$

$\qquad = \displaystyle\int 4x \, dx + \int (6y + 5) \, dx$ (y is treated as a constant.)

$\qquad = 2x^2 + (6y + 5)x + E(y)$ (The "constant" of integration is a function of y.)

$\qquad = 2x^2 + 6xy + 5x + E(y)$

(B) $\displaystyle\int_{-2}^3 (4x + 6y + 5) \, dx = (2x^2 + 6xy + 5x) \Big|_{-2}^3$

$\qquad\qquad = 2 \cdot 3^2 + 6 \cdot 3y + 5 \cdot 3 - [2(-2)^2 + 6(-2)y + 5(-2)]$
$\qquad\qquad = 30y + 35$

5. (A) $\displaystyle\int \frac{x}{\sqrt{y + x^2}} \, dx = \int (y + x^2)^{-1/2} x \, dx = \frac{1}{2} \int (y + x^2)^{-1/2} 2x \, dx$

Let $u = y + x^2$,
then $du = 2x \, dx$. $\qquad = \dfrac{1}{2} \displaystyle\int u^{-1/2} \, du$

$\qquad\qquad = u^{1/2} + E(y) = \sqrt{y + x^2} + E(y)$

(B) $\displaystyle\int_0^2 \frac{x}{\sqrt{y + x^2}} \, dx = \sqrt{y + x^2} \Big|_0^2 = \sqrt{y + 4} - \sqrt{y}$

7. $\displaystyle\int_{-1}^{2} \int_{0}^{1} 12x^2y^3dy \; dx = \int_{-1}^{2} \left[\int_{0}^{1} 12x^2y^3dy\right]dx = \int_{-1}^{2} 3x^2dx$ (see Problem 1)

$$= x^3 \Big|_{-1}^{2} = 8 + 1 = 9$$

9. $\displaystyle\int_{1}^{4} \int_{-2}^{3} (4x + 6y + 5)dx \; dy = \int_{1}^{4} \left[\int_{-2}^{3} (4x + 6y + 5)dx\right]dy$

$$= \int_{1}^{4} (30y + 35)dy \quad \text{(see Problem 3)}$$

$$= (15y^2 + 35y) \Big|_{1}^{4}$$

$$= 15 \cdot 4^2 + 35 \cdot 4 - (15 + 35) = 330$$

11. $\displaystyle\int_{1}^{5} \int_{0}^{2} \frac{x}{\sqrt{y + x^2}} dx \; dy = \int_{1}^{5} \left[\int_{0}^{2} \frac{x}{\sqrt{y + x^2}} dx\right]dy$

$$= \int_{1}^{5} (\sqrt{4 + y} - \sqrt{y})dy \quad \text{(see Problem 5)}$$

$$= \left[\frac{2}{3}(4 + y)^{3/2} - \frac{2}{3}y^{3/2}\right]\Big|_{1}^{5}$$

$$= \frac{2}{3}(9)^{3/2} - \frac{2}{3}(5)^{3/2} - \left(\frac{2}{3} \cdot 5^{3/2} - \frac{2}{3} \cdot 1^{3/2}\right)$$

$$= 18 - \frac{4}{3}(5)^{3/2} + \frac{2}{3} = \frac{56 - 20\sqrt{5}}{3}$$

13. $\displaystyle\iint\limits_{R} xy \; dA = \int_{0}^{2} \int_{0}^{4} xy \; dy \; dx = \int_{0}^{2} \left[\int_{0}^{4} xy \; dy\right]dx = \int_{0}^{2} \left[\frac{xy^2}{2}\Big|_{0}^{4}\right]dx$

$$= \int_{0}^{2} 8x \; dx = 4x^2 \Big|_{0}^{2} = 16$$

$\displaystyle\iint\limits_{R} xy \; dA = \int_{0}^{4} \int_{0}^{2} xy \; dy \; dx = \int_{0}^{4} \left[\int_{0}^{2} xy \; dx\right]dy = \int_{0}^{4} \left[\frac{x^2y}{2}\Big|_{0}^{2}\right]dy$

$$= \int_{0}^{4} 2y \; dy = y^2 \Big|_{0}^{4} = 16$$

15. $\displaystyle\iint\limits_{R} (x + y)^5dA = \int_{-1}^{1} \int_{1}^{2} (x + y)^5dy \; dx = \int_{-1}^{1} \left[\int_{1}^{2} (x + y)^5dy\right]dx$

$$= \int_{-1}^{1} \left[\frac{(x + y)^6}{6}\Big|_{1}^{2}\right]dx = \int_{-1}^{1} \left[\frac{(x + 2)^6}{6} - \frac{(x + 1)^6}{6}\right]dx$$

$$= \left[\frac{(x + 2)^7}{42} - \frac{(x + 1)^7}{42}\right]\Big|_{-1}^{1} = \frac{3^7}{42} - \frac{2^7}{42} - \frac{1}{42} = 49$$

$$\iint\limits_{R} (x + y)^5 dA = \int_1^2 \int_{-1}^1 (x + y)^5 dx \ dy = \int_1^2 \left[\int_{-1}^1 (x + y)^5 dx \right] dy$$

$$= \int_1^2 \left[\frac{(x + y)^6}{6} \Big|_{-1}^1 \right] dy = \int_1^2 \left[\frac{(y + 1)^6}{6} - \frac{(y - 1)^6}{6} \right] dy$$

$$= \left[\frac{(y + 1)^7}{42} - \frac{(y - 1)^7}{42} \right] \Big|_1^2 = \frac{3^7}{42} - \frac{1}{42} - \frac{2^7}{42} = 49$$

17. Average value $= \dfrac{1}{(5 - 1)[1 - (-1)]} \iint\limits_{R} (x + y)^2 dA$

$$= \frac{1}{8} \int_{-1}^1 \int_1^5 (x + y)^2 dx \ dy = \frac{1}{8} \int_{-1}^1 \left[\frac{(x + y)^3}{3} \Big|_1^5 \right] dy$$

$$= \frac{1}{8} \int_{-1}^1 \left[\frac{(5 + y)^3}{3} - \frac{(1 + y)^3}{3} \right] dy = \frac{1}{8} \left[\frac{(5 + y)^4}{12} - \frac{(1 + y)^4}{12} \right] \Big|_{-1}^1$$

$$= \frac{1}{96} [6^4 - 2^4 - 4^4] = \frac{32}{3}$$

19. Average value $= \dfrac{1}{(4 - 1)(7 - 2)} \iint\limits_{R} \dfrac{x}{y} dA = \dfrac{1}{15} \int_1^4 \int_2^7 \dfrac{x}{y} dy \ dx$

$$= \frac{1}{15} \int_1^4 \left[x \ln y \right]_2^7 dx = \frac{1}{15} \int_1^4 [x \ln 7 - x \ln 2] dx$$

$$= \frac{\ln 7 - \ln 2}{15} \int_1^4 x \ dx = \frac{\ln 7 - \ln 2}{15} \cdot \frac{x^2}{2} \Big|_1^4$$

$$= \frac{\ln 7 - \ln 2}{15} \left(\frac{4^2}{2} - \frac{1^2}{2} \right) = \frac{1}{2} (\ln 7 - \ln 2)$$

$$= \frac{1}{2} \ln \left(\frac{7}{2} \right) \approx 0.626$$

21. $V = \iint\limits_{R} (2 - x^2 - y^2) dA = \int_0^1 \int_0^1 (2 - x^2 - y^2) dy \ dx$

$$= \int_0^1 \left[\int_0^1 (2 - x^2 - y^2) dy \right] dx = \int_0^1 \left[\left(2y - x^2 y - \frac{y^3}{3} \right) \Big|_0^1 \right] dx$$

$$= \int_0^1 \left(2 - x^2 - \frac{1}{3} \right) dx = \int_0^1 \left(\frac{5}{3} - x^2 \right) dx = \left(\frac{5}{3} x - \frac{x^3}{3} \right) \Big|_0^1 = \frac{5}{3} - \frac{1}{3} = \frac{4}{3}$$

23. $V = \iint\limits_{R} (4 - y^2) dA = \int_0^2 \int_0^2 (4 - y^2) dx \ dy = \int_0^2 \left[\int_0^2 (4 - y^2) dx \right] dy$

$$= \int_0^2 \left[(4x - xy^2) \Big|_0^2 \right] dy = \int_0^2 (8 - 2y^2) dy = \left(8y - \frac{2}{3} y^3 \right) \Big|_0^2 = 16 - \frac{16}{3} = \frac{32}{3}$$

25. $\displaystyle\iint\limits_{R} xe^{xy}\,dA = \int_0^1 \int_1^2 xe^{xy}dy\,dx = \int_0^1\left[\int_1^2 xe^{xy}dy\right]dx$

$\displaystyle\qquad = \int_0^1\left[x\int_1^2 e^{xy}dy\right]dx = \int_0^1\left[x\cdot\frac{e^{xy}}{x}\Big|_1^2\right]dx = \int_0^1\left[e^{xy}\Big|_1^2\right]dx$

$\displaystyle\qquad = \int_0^1 (e^{2x} - e^x)\,dx = \left(\frac{e^{2x}}{2} - e^x\right)\Big|_0^1 = \frac{e^2}{2} - e - \left(\frac{1}{2} - 1\right)$

$\displaystyle\qquad = \frac{e^2}{2} - e + \frac{1}{2}$

27. $\displaystyle\iint\limits_{R}\frac{2y + 3xy^2}{1 + x^2}\,dA = \int_0^1\int_{-1}^1\frac{2y + 3xy^2}{1 + x^2}dy\,dx = \int_0^1\left[\int_{-1}^1\frac{2y + 3xy^2}{1 + x^2}dy\right]dx$

$\displaystyle\qquad = \int_0^1\left[\frac{1}{1 + x^2}(y^2 + xy^3)\Big|_{-1}^1\right]dx$

$\displaystyle\qquad = \int_0^1\left[\frac{1}{1 + x^2}(1 + x - [1 - x])\right]dx$

$\displaystyle\qquad = \int_0^1\frac{2x}{1 + x^2}dx = \ln(1 + x^2)\Big|_0^1 \qquad$ Substitution: $u = 1 + x^2$
$\displaystyle\qquad\qquad\qquad\qquad\qquad\qquad\qquad\qquad\qquad\qquad\qquad du = 2x\,dx$

$\displaystyle\qquad = \ln 2$

29. $\displaystyle\int_0^2\int_0^2 (1 - y)\,dx\,dy = \int_0^2\left[\int_0^2 (1 - y)\,dx\right]dy$

$\displaystyle\qquad = \int_0^2\left[(x - xy)\Big|_0^2\right]dy$

$\displaystyle\qquad = \int_0^2 (2 - 2y)\,dy$

$\displaystyle\qquad = (2y - y^2)\Big|_0^2 = 0$

Since $f(x, y) = 1 - y$ is NOT nonnegative over the rectangle $R = \{(x, y) \mid 0 \le x \le 2,\ 0 \le y \le 2\}$ the double integral does not represent the volume of solid.

31. $f(x, y) = x^3 + y^2 - e^{-x} - 1$ on $R = \{(x, y) \mid -2 \le x \le 2,\ -2 \le y \le 2\}$.

(A) Average value of f:

$\displaystyle\qquad \frac{1}{b - a}\cdot\frac{1}{d - c}\iint\limits_{R} f(x, y)\,dA$

$\displaystyle\qquad = \frac{1}{2 - (-2)}\cdot\frac{1}{2 - (-2)}\int_{-2}^2\int_{-2}^2 (x^3 + y^2 - e^{-x} - 1)\,dx\,dy$

$\displaystyle\qquad = \frac{1}{16}\int_{-2}^2\left[\left(\frac{1}{4}x^4 + xy^2 + e^{-x} - x\right)\Big|_{-2}^2\right]dy$

$$= \frac{1}{16} \int_{-2}^{2} [4y^2 + e^{-2} - e^2 - 4] dy$$

$$= \frac{1}{16} \left[\frac{4}{3} y^3 + e^{-2}y - e^2 y - 4y \right] \Big|_{-2}^{2}$$

$$= \frac{1}{16} \left[\frac{64}{3} + 4e^{-2} - 4e^2 - 16 \right] = \frac{1}{3} + \frac{1}{4} e^{-2} - \frac{1}{4} e^2$$

(B)

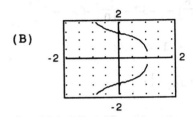

(C) $f(x, y) > 0$ at the points which lie to the right of the curve in part (B); $f(x, y) < 0$ at the points which lie to the left of the curve in part (B).

33. $S(x, y) = \frac{y}{1 - x}$, $0.6 \leq x \leq 0.8$, $5 \leq y \leq 7$.

The *average* total amount of spending is given by:

$$T = \frac{1}{(0.8 - 0.6)(7 - 5)} \iint_{R} \frac{y}{1 - x} dA = \frac{1}{0.4} \int_{0.6}^{0.8} \int_{5}^{7} \frac{y}{1 - x} dy \, dx$$

$$= \frac{1}{0.4} \int_{0.6}^{0.8} \left[\frac{1}{1 - x} \cdot \frac{y^2}{2} \Big|_{5}^{7} \right] dx = \frac{1}{0.4} \int_{0.6}^{0.8} \frac{1}{1 - x} \left(\frac{49}{2} - \frac{25}{2} \right) dx$$

$$= \frac{12}{0.4} \int_{0.6}^{0.8} \frac{1}{1 - x} dx = 30 \left[-\ln(1 - x) \right] \Big|_{0.6}^{0.8}$$

$$= 30[-\ln(0.2) + \ln(0.4)] = 30 \ln 2 \approx \$20.8 \text{ billion}$$

35. $N(x, y) = x^{0.75} y^{0.25}$, $10 \leq x \leq 20$, $1 \leq y \leq 2$

$$\text{Average value} = \frac{1}{(20 - 10)(2 - 1)} \int_{10}^{20} \int_{1}^{2} x^{0.75} y^{0.25} dy \, dx$$

$$= \frac{1}{10} \int_{10}^{20} \left[x^{0.75} \frac{y^{1.25}}{1.25} \Big|_{1}^{2} \right] dx = \frac{1}{10} \int_{10}^{20} \left[x^{0.75} \frac{2^{1.25} - 1}{1.25} \right] dx$$

$$= \frac{1}{12.5} (2^{1.25} - 1) \int_{10}^{20} x^{0.75} dx = \frac{1}{12.5} (2^{1.25} - 1) \frac{x^{1.75}}{1.75} \Big|_{10}^{20}$$

$$= \frac{1}{21.875} (2^{1.25} - 1)(20^{1.75} - 10^{1.75}) \approx 8.375 \text{ or } 8375 \text{ items}$$

37. $C = 10 - \frac{1}{10}d^2 = 10 - \frac{1}{10}(x^2 + y^2) = C(x, y)$, $-8 \le x \le 8$, $-6 \le y \le 6$

$$\text{Average concentration} = \frac{1}{16(12)}\int_{-8}^{8}\int_{-6}^{6}\left[10 - \frac{1}{10}(x^2 + y^2)\right]dy\ dx$$

$$= \frac{1}{192}\int_{-8}^{8}\left[10y - \frac{1}{10}\left(x^2y + \frac{y^3}{3}\right)\right]\Big|_{-6}^{6}dx$$

$$= \frac{1}{192}\int_{-8}^{8}\left\{60 - \frac{1}{10}\left(6x^2 + \frac{216}{3}\right) - \left[-60 - \frac{1}{10}\left(-6x^2 - \frac{216}{3}\right)\right]\right\}dx$$

$$= \frac{1}{192}\int_{-8}^{8}\left[120 - \frac{1}{10}(12x^2 + 144)\right]dx$$

$$= \frac{1}{192}\left[120x - \frac{1}{10}(4x^3 + 144x)\right]\Big|_{-8}^{8}$$

$$= \frac{1}{192}(1280) = \frac{20}{3} \approx 6.67 \text{ insects per square foot}$$

39. $C = 100 - 15d^2 = 100 - 15(x^2 + y^2) = C(x, y)$, $-2 \le x \le 2$, $-1 \le y \le 1$

$$\text{Average concentration} = \frac{1}{4(2)}\int_{-2}^{2}\int_{-1}^{1}[100 - 15(x^2 + y^2)]dy\ dx$$

$$= \frac{1}{8}\int_{-2}^{2}(100y - 15x^2y - 5y^3)\Big|_{-1}^{1}dx$$

$$= \frac{1}{8}\int_{-2}^{2}(190 - 30x^2)dx = \frac{1}{8}(190x - 10x^3)\Big|_{-2}^{2}$$

$$= \frac{1}{8}(600) = 75 \text{ parts per million}$$

41. $L = 0.0000133xy^2$, $2000 \le x \le 3000$, $50 \le y \le 60$

$$\text{Average length} = \frac{1}{10,000}\int_{2000}^{3000}\int_{50}^{60}0.0000133xy^2dy\ dx$$

$$= \frac{0.0000133}{10,000}\int_{2000}^{3000}\left[\frac{xy^3}{3}\Big|_{50}^{60}\right]dx$$

$$= \frac{0.0000133}{10,000}\int_{2000}^{3000}\frac{91,000}{3}x\ dx = \frac{1.2103}{30,000}\cdot\frac{x^2}{2}\Big|_{2000}^{3000}$$

$$= \frac{1.2103}{60,000}(5,000,000) \approx 100.86 \text{ feet}$$

43. $Q(x, y) = 100\left(\frac{x}{y}\right)$, $8 \le x \le 16$, $10 \le y \le 12$

$$\text{Average intelligence} = \frac{1}{16}\int_{8}^{16}\int_{10}^{12}100\left(\frac{x}{y}\right)dy\ dx = \frac{100}{16}\int_{8}^{16}\left[x\ln y\Big|_{10}^{12}\right]dx$$

$$= \frac{100}{16}\int_{8}^{16}x(\ln 12 - \ln 10)dx$$

$$= \frac{100(\ln 12 - \ln 10)}{16}\cdot\frac{x^2}{2}\Big|_{8}^{16}$$

$$= \frac{100(\ln 12 - \ln 10)}{32}(192)$$

$$= 600\ln(1.2) \approx 109.4$$

1. $f(x, y) = 2000 + 40x + 70y$
   $f(5, 10) = 2000 + 40 \cdot 5 + 70 \cdot 10 = 2900$
   $f_x(x, y) = 40$
   $f_y(x, y) = 70$ (13-1, 13-2)

2. $z = x^3 y^2$
   $\dfrac{\partial z}{\partial x} = 3x^2 y^2$

   $\dfrac{\partial^2 z}{\partial x^2} = \dfrac{\partial \left(\frac{\partial z}{\partial x}\right)}{\partial x} = \dfrac{\partial (3x^2 y^2)}{\partial x} = 6xy^2$

   $\dfrac{\partial z}{\partial y} = 2x^3 y$

   $\dfrac{\partial^2 z}{\partial x \partial y} = \dfrac{\partial \left(\frac{\partial z}{\partial y}\right)}{\partial x} = \dfrac{\partial (2x^3 y)}{\partial x} = 6x^2 y$ (13-2)

3. $\displaystyle\int (6xy^2 + 4y)\, dy = 6x \int y^2 dy + 4 \int y\, dy = 6x \cdot \dfrac{y^3}{3} + 4 \cdot \dfrac{y^2}{2} + C(x)$
   $$= 2xy^3 + 2y^2 + C(x) \qquad (13\text{-}6)$$

4. $\displaystyle\int (6xy^2 + 4y)\, dx = 6y^2 \int x\, dx + 4y \int dx = 6y^2 \cdot \dfrac{x^2}{2} + 4yx + E(y)$
   $$= 3x^2 y^2 + 4xy + E(y) \qquad (13\text{-}6)$$

5. $\displaystyle\int_0^1 \int_0^1 4xy\, dy\, dx = \int_0^1 \left[\int_0^1 4xy\, dy \right] dx = \int_0^1 \left[2xy^2 \Big|_0^1 \right] dx$
   $$= \int_0^1 2x\, dx = x^2 \Big|_0^1 = 1 \qquad (13\text{-}6)$$

6. $f(x, y) = 3x^2 - 2xy + y^2 - 2x + 3y - 7$
   $f(2, 3) = 3 \cdot 2^2 - 2 \cdot 2 \cdot 3 + 3^2 - 2 \cdot 2 + 3 \cdot 3 - 7 = 7$
   $f_y(x, y) = -2x + 2y + 3$
   $f_y(2, 3) = -2 \cdot 2 + 2 \cdot 3 + 3 = 5$ (13-1, 13-2)

7. $f(x, y) = -4x^2 + 4xy - 3y^2 + 4x + 10y + 81$
   $f_x(x, y) = -8x + 4y + 4 \qquad\qquad f_y(x, y) = 4x - 6y + 10$
   $f_{xx}(x, y) = -8 \qquad\qquad\qquad f_{yy}(x, y) = -6$
   $f_{xy}(x, y) = 4$
   Now, $f_{xx}(2, 3) \cdot f_{yy}(2, 3) - [f_{xy}(2, 3)]^2 = (-8)(-6) - 4^2 = 32.$ (13-2)

8. $f(x, y) = x + 3y$ and $g(x, y) = x^2 + y^2 - 10$.
   Let $F(x, y, \lambda) = f(x, y) + \lambda g(x, y) = x + 3y + \lambda(x^2 + y^2 - 10)$.
   Then, we have:
   $F_x = 1 + 2x\lambda$
   $F_y = 3 + 2y\lambda$
   $F_\lambda = x^2 + y^2 - 10$

Setting $F_x = F_y = F_\lambda = 0$, we obtain:

$$1 + 2x\lambda = 0 \qquad (1)$$
$$3 + 2y\lambda = 0 \qquad (2)$$
$$x^2 + y^2 - 10 = 0 \qquad (3)$$

From the first equation, $x = -\dfrac{1}{2\lambda}$; from the second equation, $y = -\dfrac{3}{2\lambda}$.

Substituting these into the third equation gives:

$$\frac{1}{4\lambda^2} + \frac{9}{4\lambda^2} - 10 = 0$$
$$40\lambda^2 = 10$$
$$\lambda^2 = \frac{1}{4}$$
$$\lambda = \pm\frac{1}{2}$$

Thus, the critical points are $\left(-1, -3, \dfrac{1}{2}\right)$ and $\left(1, 3, -\dfrac{1}{2}\right)$. $\qquad$ (13-4)

9.

	x_k	y_k	$x_k y_k$	x_k^2
	2	12	24	4
	4	10	40	16
	6	7	42	36
	8	3	24	64
Totals	20	32	130	120

Thus, $\displaystyle\sum_{k=1}^{4} x_k = 20$, $\displaystyle\sum_{k=1}^{4} y_k = 32$, $\displaystyle\sum_{k=1}^{4} x_k y_k = 130$, $\displaystyle\sum_{k=1}^{4} x_k^2 = 120$.

Substituting these values into the formulas for m and d, we have:

$$m = \frac{4\left(\displaystyle\sum_{k=1}^{4} x_k y_k\right) - \left(\displaystyle\sum_{k=1}^{4} x_k\right)\left(\displaystyle\sum_{k=1}^{4} y_k\right)}{4\left(\displaystyle\sum_{k=1}^{4} x_k^2\right) - \left(\displaystyle\sum_{k=1}^{4} x_k\right)^2} = \frac{4(130) - (20)(32)}{4(120) - (20)^2} = \frac{-120}{80} = -1.5$$

$$d = \frac{\displaystyle\sum_{k=1}^{4} y_k - (-1.5)\displaystyle\sum_{k=1}^{4} x_k}{4} = \frac{32 + (1.5)(20)}{4} = \frac{62}{4} = 15.5$$

Thus, the least squares line is:

$y = mx + d = -1.5x + 15.5$

When $x = 10$, $y = -1.5(10) + 15.5 = 0.5$. $\qquad$ (13-5)

10. $\displaystyle\iint\limits_{R} (4x + 6y)\,dA = \int_{-1}^{1}\int_{1}^{2} (4x + 6y)\,dy\ dx = \int_{-1}^{1}\left[\int_{1}^{2} (4x + 6y)\,dy\right] dx$

$\displaystyle\qquad = \int_{-1}^{1}\left[(4xy + 3y^2)\,\Big|_{1}^{2}\right] dx = \int_{-1}^{1} (8x + 12 - 4x - 3)\,dx$

$\displaystyle\qquad = \int_{-1}^{1} (4x + 9)\,dx = (2x^2 + 9x)\,\Big|_{-1}^{1} = 2 + 9 - (2 - 9) = 18$

$\displaystyle\iint\limits_{R} (4x + 6y)\,dA = \int_{1}^{2}\int_{-1}^{1} (4x + 6y)\,dx\ dy = \int_{1}^{2}\left[\int_{-1}^{1} (4x + 6y)\,dx\right] dy$

$\displaystyle\qquad = \int_{1}^{2}\left[(2x^2 + 6xy)\,\Big|_{-1}^{1}\right] dy = \int_{1}^{2} [2 + 6y - (2 - 6y)]\,dy$

$\displaystyle\qquad = \int_{1}^{2} 12y\ dy = 6y^2\,\Big|_{1}^{2} = 24 - 6 = 18$ $\qquad\qquad$ (13-6)

11. $f(x,\ y) = e^{x^2 + 2y}$

$f_x(x,\ y) = e^{x^2 + 2y}\cdot 2x = 2xe^{x^2 + 2y}$

$f_y(x,\ y) = e^{x^2 + 2y}\cdot 2 = 2e^{x^2 + 2y}$

$f_{xy}(x,\ y) = 2xe^{x^2 + 2y}\cdot 2 = 4xe^{x^2 + 2y}$ $\qquad\qquad$ (13-2)

12. $f(x,\ y) = (x^2 + y^2)^5$

$f_x(x,\ y) = 5(x^2 + y^2)^4\cdot 2x = 10x(x^2 + y^2)^4$

$f_{xy}(x,\ y) = 10x(4)(x^2 + y^2)^3\cdot 2y = 80xy(x^2 + y^2)^3$ $\qquad\qquad$ (13-2)

13. $f(x,\ y) = x^3 - 12x + y^2 - 6y$

$\qquad f_x(x,\ y) = 3x^2 - 12 \qquad\qquad\qquad f_y(x,\ y) = 2y - 6$

$\qquad 3x^2 - 12 = 0 \qquad\qquad\qquad\qquad\quad 2y - 6 = 0$

$\qquad\qquad x^2 = 4 \qquad\qquad\qquad\qquad\qquad\quad y = 3$

$\qquad\qquad x = \pm 2$

Thus, the critical points are (2, 3) an (-2, 3).

$f_{xx}(x,\ y) = 6x \qquad\quad f_{xy}(x,\ y) = 0 \qquad\quad f_{yy}(x,\ y) = 2$

For the critical point (2, 3):

$f_{xx}(2,\ 3) = 12 > 0$

$f_{xy}(2,\ 3) = 0$

$f_{yy}(2,\ 3) = 2$

$f_{xx}(2,\ 3)\cdot f_{yy}(2,\ 3) - [f_{xy}(2,\ 3)]^2 = 12\cdot 2 = 24 > 0$

Therefore, $f(2,\ 3) = 2^3 - 12\cdot 2 + 3^2 - 6\cdot 3 = -25$ is a local minimum.

For the critical point (-2, 3):

$f_{xx}(-2,\ 3) = -12 < 0$

$f_{xy}(-2,\ 3) = 0$

$f_{yy}(-2,\ 3) = 2$

$f_{xx}(-2,\ 3)\cdot f_{yy}(-2,\ 3) - [f_{xy}(-2,\ 3)]^2 = -12\cdot 2 - 0 = -24 < 0$

Thus, f has a saddle point at (-2, 3). $\qquad\qquad$ (13-3)

14. <u>Step 1.</u> Maximize $f(x, y) = xy$
Subject to: $g(x, y) = 2x + 3y - 24 = 0$

<u>Step 2.</u> $F(x, y, \lambda) = f(x, y) + \lambda g(x, y) = xy + \lambda(2x + 3y - 24)$

<u>Step 3.</u> $F_x = y + 2\lambda = 0$ (1)
$F_y = x + 3\lambda = 0$ (2)
$F_\lambda = 2x + 3y - 24 = 0$ (3)

From (1) and (2), we obtain:
$y = -2\lambda$ and $x = -3\lambda$

Substituting these into (3), we have:
$-6\lambda - 6\lambda - 24 = 0$
$\lambda = -2$

Thus, the critical point is $(6, 4, -2)$.

<u>Step 4.</u> Since $(6, 4, -2)$ is the only critical point for F, we conclude
that max $f(x, y) = f(6, 4) = 6 \cdot 4 = 24$.   (13-4)

15. <u>Step 1.</u> Minimize $f(x, y, z) = x^2 + y^2 + z^2$
Subject to: $2x + y + 2z = 9$ or $g(x, y, z) = 2x + y + 2z - 9 = 0$

<u>Step 2.</u> $F(x, y, z, \lambda) = x^2 + y^2 + z^2 + \lambda(2x + y + 2z - 9)$

<u>Step 3.</u> $F_x = 2x + 2\lambda = 0$ (1)
$F_y = 2y + \lambda = 0$ (2)
$F_z = 2z + 2\lambda = 0$ (3)
$F_\lambda = 2x + y + 2z - 9 = 0$ (4)

From equations (1), (2), and (3), we have:
$x = -\lambda$, $y = -\dfrac{\lambda}{2}$, and $z = -\lambda$

Substituting these into (4), we obtain:
$-2\lambda - \dfrac{\lambda}{2} - 2\lambda - 9 = 0$
$\dfrac{9}{2}\lambda = -9$
$\lambda = -2$

The critical point is: $(2, 1, 2, -2)$

<u>Step 4.</u> Since $(2, 1, 2, -2)$ is the only critical point for F, we
conclude that min $f(x, y, z) = f(2, 1, 2) = 2^2 + 1^2 + 2^2 = 9$.
        (13-4)

16.

	x_k	y_k	$x_k y_k$	x_k^2
	10	50	500	100
	20	45	900	400
	30	50	1,500	900
	40	55	2,200	1,600
	50	65	3,250	2,500
	60	80	4,800	3,600
	70	85	5,950	4,900
	80	90	7,200	6,400
	90	90	8,100	8,100
	100	110	11,000	10,000
Totals	550	720	45,400	38,500

Thus, $\displaystyle\sum_{k=1}^{10} x_k = 550$, $\displaystyle\sum_{k=1}^{10} y_k = 720$, $\displaystyle\sum_{k=1}^{10} x_k y_k = 45{,}400$, $\displaystyle\sum_{k=1}^{10} x_k^2 = 38{,}500$.

Substituting these values into the formulas for m and d, we have:

$$m = \frac{10(45{,}400) - (550)(720)}{10(38{,}500) - (550)^2} = \frac{58{,}000}{82{,}500} = \frac{116}{165}$$

$$d = \frac{720 - \left(\frac{116}{165}\right)550}{10} = \frac{100}{3}$$

Therefore, the least squares line is:

$$y = \frac{116}{165}x + \frac{100}{3} \approx 0.703x + 33.33 \tag{13-5}$$

17.

$$\frac{1}{(b-a)(d-c)} \iint\limits_{R} f(x, y)\,dA = \frac{1}{[8 - (-8)](27 - 0)} \int_{-8}^{8} \int_{0}^{27} x^{2/3} y^{1/3}\,dy\,dx$$

$$= \frac{1}{16 \cdot 27} \int_{-8}^{8} \left(\frac{3}{4} x^{2/3} y^{4/3} \Big|_{y=0}^{y=27}\right) dx$$

$$= \frac{1}{16 \cdot 27} \int_{-8}^{8} \frac{3^5}{4} x^{2/3}\,dx = \frac{9}{64} \int_{-8}^{8} x^{2/3}\,dx$$

$$= \frac{9}{64} \cdot \frac{3}{5} x^{5/3} \Big|_{-8}^{8} = \frac{9}{64} \cdot \frac{3}{5} [2^5 - (-2)^5]$$

$$= \frac{9}{64} \cdot \frac{3}{5} \cdot 2^6 = \frac{27}{5} \tag{13-6}$$

18. $\displaystyle V = \iint\limits_{R} (3x^2 + 3y^2)\,dA = \int_{0}^{1} \int_{-1}^{1} (3x^2 + 3y^2)\,dy\,dx = \int_{0}^{1} \left[\int_{-1}^{1} (3x^2 + 3y^2)\,dy\right] dx$

$$= \int_{0}^{1} \left[(3x^2 y + y^3) \Big|_{-1}^{1}\right] dx = \int_{0}^{1} [3x^2 + 1 - (-3x^2 - 1)]\,dx$$

$$= \int_{0}^{1} (6x^2 + 2)\,dx = (2x^3 + 2x) \Big|_{0}^{1} = 4 \text{ cubic units} \tag{13-6}$$

19. $f(x, y) = x + y;\ -10 \le x \le 10,\ -10 \le y \le 10$

Prediction: average value $= f(0, 0) = 0$.

Verification:

$$\text{average value} = \frac{1}{[10 - (-10)][10 - (-10)]} \int_{-10}^{10} \int_{-10}^{10} (x + y)\, dy\, dx$$

$$= \frac{1}{400} \int_{-10}^{10} \left[\left(xy + \frac{1}{2}y^2 \right) \Big|_{-10}^{10} \right] dy$$

$$= \frac{1}{400} \int_{-10}^{10} 20x\, dx$$

$$= \frac{1}{400}(10x^2) \Big|_{-10}^{10} = 0 \qquad\qquad (13\text{-}6)$$

20. $f(x, y) = \dfrac{e^x}{y + 10}$

(A) $S = \{x, y)\ |\ -a \le x \le a,\ -a \le y \le a\}$

The average value of f over S is given by:

$$\frac{1}{[a - (-a)][a - (-a)]} \int_{-a}^{a} \int_{-a}^{a} \frac{e^x}{y + 10}\, dx\, dy$$

$$= \frac{1}{4a^2} \int_{-a}^{a} \left[\frac{e^x}{y + 10} \Big|_{-a}^{a} \right] dy$$

$$= \frac{1}{4a^2} \int_{-a}^{a} \left(\frac{e^a}{y + 10} - \frac{e^{-a}}{y + 10} \right) dy$$

$$= \frac{e^a - e^{-a}}{4a^2} \int_{-a}^{a} \frac{1}{y + 10}\, dy$$

$$= \frac{e^a - e^{-a}}{4a^2} (\ln|y + 10|) \Big|_{-a}^{a}$$

$$= \frac{e^a - e^{-a}}{4a^2} [\ln(10 + a) - \ln(10 - a)]$$

$$= \frac{e^a - e^{-a}}{4a^2} \ln\!\left(\frac{10 + a}{10 - a} \right)$$

Now, $\dfrac{e^a - e^{-a}}{4a^2} \ln\!\left(\dfrac{10 + a}{10 - a} \right) = 5$

is equivalent to

$$(e^a - e^{-a}) \ln\!\left(\frac{10 + a}{10 - a} \right) - 20a^2 = 0.$$

Using a graphing utility, the graph of

$$f(x) = (e^x - e^{-x}) \ln\!\left(\frac{10 + x}{10 - x} \right) - 20x^2$$

is shown at the right and $f(x) = 0$ at $x \approx \pm 6.28$.

The dimensions of the square are: 12.56×12.56.

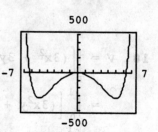

(B) To determine whether there is a square centered at $(0, 0)$ such that

$$\frac{e^a - e^{-a}}{4a^2} \ln\left(\frac{10 + a}{10 - a}\right) = 0.05,$$

graph,

$$f(x) = (e^x - e^{-x}) \ln\left(\frac{10 + x}{10 - x}\right) - 0.20x^2$$

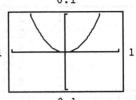

The result is shown at the right
and $f(x) = 0$ only at $x = 0$.

Thus, there does not exist a square centered
at $(0, 0)$ such that the average value of $f = 0.05$. (13-6)

21. $P(x, y) = -4x^2 + 4xy - 3y^2 + 4x + 10y + 81$

(A) $P_x(x, y) = -8x + 4y + 4$
$P_x(1, 3) = -8 \cdot 1 + 4 \cdot 3 + 4 = 8$

At the output level $(1, 3)$, profit will increase by \$8000 for 100
units increase in product A if the production of product B is held
fixed.

(B) $P_x = -8x + 4y + 4 = 0$ (1)
$P_y = 4x - 6y + 10 = 0$ (2)

Solving (1) and (2) for x and y, we obtain $x = 2$, $y = 3$.
Thus, $(2, 3)$ is a critical point.

$P_{xx} = -8$ $P_{yy} = -6$ $P_{xy} = 4$
$P_{xx}(2, 3) = -8 < 0$ $P_{yy}(2, 3) = -6$ $P_{xy}(2, 3) = 4$

$P_{xx}(2, 3) \cdot P_{yy}(2, 3) - [P_{xy}(2, 3)]^2 = (-8)(-6) - 4^2 = 32 > 0$

Thus, $P(2, 3)$ is a maximum and
$\max P(x, y) = P(2, 3) = -4 \cdot 2^2 + 4 \cdot 2 \cdot 3 - 3 \cdot 3^2 + 4 \cdot 2 + 10 \cdot 3 + 81$
$= -16 + 24 - 27 + 8 + 30 + 81$
$= 100.$

Thus, the maximum profit is \$100,000. This is obtained when 200
units of A and 300 units of B are produced per month.   (13-2, 13-3)

22. Minimize $S(x, y, z) = xy + 4yz + 3xz$
Subject to: $V(x, y, z) = xyz - 96 = 0$
Put $F(x, y, z, \lambda) = S(x, y, z) + \lambda V(x, y, z) = xy + 4yz + 3xz + \lambda(xyz - 96)$.
Then, we have:

$F_x = y + 3z + \lambda yz = 0$ (1)

$F_y = x + 4z + \lambda xz = 0$ (2)

$F_z = 4y + 3x + \lambda xy = 0$ (3)

$F_\lambda = xyz - 96 = 0$ (4)

Solving the system of equations, (1)–(4), simultaneously, yields $x = 8$,
$y = 6$, $z = 2$, and $\lambda = -1$. Thus, the critical point is $(8, 6, 2, -1)$ and
$S(8, 6, 2) = 8 \cdot 6 + 4 \cdot 6 \cdot 2 + 3 \cdot 8 \cdot 2 = 144$
is the minimum value of S subject to the constraint $V = xyz - 96 = 0$.

The dimensions of the box that will require the minimum amount of
material are:
Length $x = 8$ inches; Width $y = 6$ inches; Height $z = 2$ inches   (13-3)

23.

	x_k	y_k	$x_k y_k$	x_k^2
	1	2.0	2.0	1
	2	2.5	5.0	4
	3	3.1	9.3	9
	4	4.2	16.8	16
	5	4.3	21.5	25
Totals	15	16.1	54.6	55

Thus, $\sum\limits_{k=1}^{5} x_k = 15$, $\sum\limits_{k=1}^{5} y_k = 16.1$, $\sum\limits_{k=1}^{5} x_k y_k = 54.6$, $\sum\limits_{k=1}^{5} x_k^2 = 55$.

Substituting these values into the formulas for m and d, we have:

$m = \dfrac{5(54.6) - (15)(16.1)}{5(55) - (15)^2} = \dfrac{31.5}{50} \approx 0.63$

$d = \dfrac{16.1 - (0.63)(15)}{5} = 1.33$

Therefore, the least squares line is:
$y = 0.63x + 1.33$
When $x = 6$, $y = 0.63(6) + 1.33 = 5.11$, and the profit for the sixth year is estimated to be \$5.11 million. (13-4)

24. $N(x, y) = 10x^{0.8}y^{0.2}$

(A) $N_x(x, y) = 8x^{-0.2}y^{0.2}$

$N_x(40, 50) = 8(40)^{-0.2}(50)^{0.2} \approx 8.36$

$N_y(x, y) = 2x^{0.8}y^{-0.8}$

$N_y(40, 50) = 2(40)^{0.8}(50)^{-0.8} \approx 1.67$

Thus, at the level of 40 units of labor and 50 units of capital, the marginal productivity of labor is approximately 8.36 and the marginal productivity of capital is approximately 1.67. Management should encourage increased use of labor.

(B) <u>Step 1</u>. Maximize the production function $N(x, y) = 10x^{0.8}y^{0.2}$
Subject to the constraint: $C(x, y) = 100x + 50y = 10,000$
i.e., $g(x, y) = 100x + 50y - 10,000 = 0$

<u>Step 2</u>. $F(x, y, \lambda) = 10x^{0.8}y^{0.2} + \lambda(100x + 50y - 10,000)$

<u>Step 3</u>.
$F_x = 8x^{-0.2}y^{0.2} + 100\lambda = 0$ (1)
$F_y = 2x^{0.8}y^{-0.8} + 50\lambda = 0$ (2)
$F_\lambda = 100x + 50y - 10,000 = 0$ (3)

From equation (1), $\lambda = \dfrac{-0.08y^{0.2}}{x^{0.2}}$, and from (2),

$\lambda = \dfrac{-0.04x^{0.8}}{y^{0.8}}$. Thus, $\dfrac{0.08y^{0.2}}{x^{0.2}} = \dfrac{0.04x^{0.8}}{y^{0.8}}$ and $x = 2y$.

Substituting into (3) yields:

$$200y + 50y = 10,000$$
$$250y = 10,000$$
$$y = 40$$

Therefore, $x = 80$ and $\lambda \approx -0.0696$. The critical point is $(80, 40, -0.0696)$. Thus, we conclude that max $N(x, y) = N(80, 40) = 10(80)^{0.8}(40)^{0.2} \approx 696$ units.

Production is maximized when 80 units of labor and 40 units of capital are used.

The marginal productivity of money is $-\lambda \approx 0.0696$. The increase in production resulting from an increase of $2000 in the budget is:

$$0.0696(2000) \approx 139 \text{ units}$$

(C) Average number of units

$$= \frac{1}{(100 - 50)(40 - 20)} \int_{50}^{100} \int_{20}^{40} 10x^{0.8}y^{0.2}\, dy\, dx$$

$$= \frac{1}{(50)(20)} \int_{50}^{100} \left[\frac{10x^{0.8}y^{1.2}}{1.2} \Big|_{20}^{40} \right] dx = \frac{1}{1000} \int_{50}^{100} \frac{10}{1.2} x^{0.8}(40^{1.2} - 20^{1.2})\, dx$$

$$= \frac{40^{1.2} - 20^{1.2}}{120} \int_{50}^{100} x^{0.8}\, dx = \frac{40^{1.2} - 20^{1.2}}{120} \cdot \frac{x^{1.8}}{1.8} \Big|_{50}^{100}$$

$$= \frac{(40^{1.2} - 20^{1.2})(100^{1.8} - 50^{1.8})}{216} \approx \frac{(47.24)(2837.81)}{216} \approx 621$$

Thus, the average number of units produced is approximately 621.

(13-4)

25. $T(V, x) = \dfrac{33V}{x + 33} = 33V(x + 33)^{-1}$

$T_x(V, x) = -33V(x + 33)^{-2} = \dfrac{-33V}{(x + 33)^2}$

$T_x(70, 17) = \dfrac{-33(70)}{(17 + 33)^2} = \dfrac{-33(70)}{2500}$

$\qquad\qquad\quad = -0.924$ minutes per unit increase in depth when $V = 70$ cubic feet and $x = 17$ feet

(13-2)

26. $C = 100 - 24d^2 = 100 - 24(x^2 + y^2)$

$C(x, y) = 100 - 24(x^2 + y^2)$, $-2 \leq x \leq 2$, $-2 \leq y \leq 2$

Average concentration $= \dfrac{1}{4(4)} \int_{-2}^{2} \int_{-2}^{2} [100 - 24(x^2 + y^2)]\, dy\, dx$

$$= \frac{1}{16} \int_{-2}^{2} [100y - 24x^2y - 8y^3] \Big|_{-2}^{2} dx$$

$$= \frac{1}{16} \int_{-2}^{2} [400 - 96x^2 - 128]\, dx = \frac{1}{16} \int_{-2}^{2} (272 - 96x^2)\, dx$$

$$= \frac{1}{16} [272x - 32x^3] \Big|_{-2}^{2} = \frac{1}{16}(544 - 256) - \frac{1}{16}(-544 + 256)$$

$$= 18 + 18 = 36 \text{ parts per million}$$

(13-6)

27. $n(P_1, P_2, d) = 0.001 \dfrac{P_1 P_2}{d}$

$n(100{,}000, 50{,}000, 100) = 0.001 \dfrac{100{,}000 \times 50{,}000}{100} = 50{,}000$ (13-1)

28.

x_k	y_k	$x_k y_k$	x_k^2
30	60	1,800	900
50	75	3,750	2,500
60	80	4,800	3,600
70	85	5,950	4,900
90	90	8,100	8,100
Totals 300	390	24,400	20,000

Thus, $\displaystyle\sum_{k=1}^{5} x_k = 300$, $\displaystyle\sum_{k=1}^{5} y_k = 390$, $\displaystyle\sum_{k=1}^{5} x_k y_k = 24{,}400$, $\displaystyle\sum_{k=1}^{5} x_k^2 = 20{,}000$.

Substituting these values into the formulas for m and d, we have:

$m = \dfrac{5(24{,}400) - (300)(390)}{5(20{,}000) - (300)^2} = \dfrac{5000}{10{,}000} = 0.5$

$d = \dfrac{390 - 0.5(300)}{5} = \dfrac{240}{5} = 48$

Therefore, the least squares line is:
$y = 0.5x + 48$
When $x = 40$, $y = 0.5(40) + 48 = 68$. (13-5)

29. (A) Enter the data in a calculator or computer. (We used a TI-85.)
The totals are:
$n = 10$, $\sum x = 450$, $\sum y = 470.6$,
$\sum x^2 = 28{,}500$, $\sum xy = 25{,}062$

Now, the least squares line can be calculated either by using formulas (2) and (3) in Section 7-5, or by using the linear regression feature. We used the latter to get

$m = 0.4709$ and $b = 25.87$

Therefore, the least squares line is:
$y = 0.4709x + 25.87$

(B) Using the result in (A), an estimate for the population density in the year 2000 is:

$y = 0.4709(100) + 25.87 \approx 72.96$ persons/square mile (13-5)

30. (A) Enter the data in a calculator or computer. (We used a TI-85.)
The totals are:
$$n = 9, \quad \sum x = 589.44, \quad \sum y = 634.97,$$
$$\sum x^2 = 38,819.351, \quad \sum xy = 41,816.153$$

Now, the least squares line can be calculated either by using formulas (2) and (3) in Section 7-5, or by using the linear regression feature. We used the latter to get

$$m = 1.069 \quad \text{and} \quad b = 0.522$$

Therefore, the least squares line is:
$$y = 1.069x + 0.522$$

(B) Using the result in (A), an estimate for the life expectancy of a female corresponding to a life expectancy of a male of 60 years is:

$$y = 1.069(60) + 0.522 \approx 64.66 \text{ years} \hspace{3cm} (13-5)$$

EXERCISE 14-1

Things to remember:

1. A DIFFERENTIAL EQUATION is an equation involving an unknown function and one or more of its derivatives.

2. The ORDER of a differential equation is the order of the highest derivative of the unknown function present in the equation.

3. A SLOPE FIELD for a differential equation is obtained by drawing tangent line segments determined by the equation at each point in a grid. (Also, see Section 6.3.)

4. Remember the terms (a) general solution, (b) particular solution, and (c) initial condition.

---

1. Substitute $y = Cx^2$, $y' = 2Cx$ into the given differential equation:
$$x(2Cx) = 2(Cx^2)$$
$$2Cx^2 = 2Cx^2$$
Thus, $y = Cx^2$ is the general solution.

3. Substitute $y = \dfrac{C}{x}$, $y' = -\dfrac{C}{x^2}$ into the given differential equation:
$$x\left(-\frac{C}{x^2}\right) = -\frac{C}{x}$$
$$-\frac{C}{x} = -\frac{C}{x}$$
Thus, $y = \dfrac{C}{x}$ is the general solution.

5. (C)   **7.** (A)

9.   **11.**

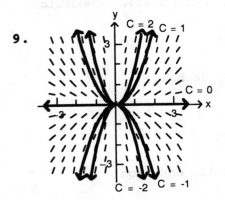

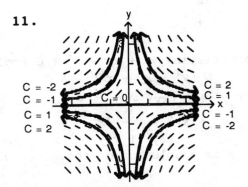

13. Substituting $y = Ce^x - 5x - 5$, $y' = Ce^x - 5$ in the differential equation, we have

$$Ce^x - 5 = (Ce^x - 5x - 5) + 5x \quad \text{or} \quad Ce^x - 5 = Ce^x - 5.$$

Thus, $y = Ce^x - 5x - 5$ is the general solution. Letting $x = 0$ and $y = 2$ in the general solution yields

$$2 = Ce^0 - 5(0) - 5 \quad \text{or} \quad 2 = C - 5 \quad \text{and} \quad C = 7.$$

Therefore, the particular solution satisfying the initial condition $y(0) = 2$ is

$$y = 7e^x - 5x - 5.$$

15. Substituting $y = e^x + Ce^{2x}$, $y' = e^x + 2Ce^{2x}$ in the differential equation, we have

$$e^x + 2Ce^{2x} = 2(e^x + Ce^{2x}) - e^x$$
$$e^x + 2Ce^{2x} = 2e^x + 2Ce^{2x} - e^x$$
$$e^x + 2Ce^{2x} = e^x + 2Ce^{2x}.$$

Thus, $y = e^x + Ce^{2x}$ is the general solution. Letting $x = 0$ and $y = -1$ in the general solution yields

$$-1 = e^0 + Ce^{2 \cdot 0} \quad \text{or} \quad -1 = 1 + C \quad \text{and} \quad C = -2.$$

Therefore, the particular solution satisfying the initial condition $y(0) = -1$ is

$$y = e^x - 2e^{2x}.$$

17. Substituting $y = x + \dfrac{C}{x}$, $y' = 1 - \dfrac{C}{x^2}$ in the differential equation, we have

$$x\left(1 - \frac{C}{x^2}\right) = 2x - \left(x + \frac{C}{x}\right)$$
$$x - \frac{C}{x} = 2x - x - \frac{C}{x}$$
$$x - \frac{C}{x} = x - \frac{C}{x}.$$

Thus, $y = x + \dfrac{C}{x}$ is the general solution. Letting $x = 2$ and $y = 3$ in the general solution yields

$$3 = 2 + \frac{C}{2} \quad \text{or} \quad 1 = \frac{C}{2} \quad \text{and} \quad C = 2.$$

Therefore, the particular solution satisfying the initial condition $y(2) = 3$ is

$$y = x + \frac{2}{x}.$$

19. Differentiating $y^3 + xy - x^3 = C$ implicitly, we have

$$D_x(y^3 + xy - x^3) = D_x C$$
$$D_x y^3 + D_x(xy) - D_x x^3 = 0$$
$$3y^2 y' + xy' + y - 3x^2 = 0$$
$$(3y^2 + x)y' = 3x^2 - y.$$

Thus, y is a solution of the given differential equation.

21. Differentiating $xy + e^{y^2} - x^2 = C$ implicitly, we have

$$D_x(xy + e^{y^2} - x^2) = D_xC$$

$$D_x(xy) + D_xe^{y^2} - D_xx^2 = 0$$

$$xy' + y + e^{y^2}D_xy^2 - 2x = 0$$

$$xy' + y + 2yy'e^{y^2} - 2x = 0$$

$$(x + 2ye^{y^2})y' = 2x - y.$$

Thus, y is a solution of the given differential equation.

23. Differentiating $y^2 + x^2 = C$ implicitly, we have

$$D_x(y^2 + x^2) = D_xC$$

$$D_xy^2 + D_xx^2 = 0$$

$$2yy' + 2x = 0$$

$$yy' = -x$$

Thus, y is a solution of the given differential equation. Substituting $x = 0$ and $y = 3$ in $y^2 + x^2 = C$ yields

$$3^2 + 0^2 = C \quad \text{or} \quad C = 9.$$

Therefore, the particular solution satisfying $y(0) = 3$ is a solution of the equation

$$y^2 + x^2 = 9 \quad \text{or} \quad y^2 = 9 - x^2.$$

This equation has two continuous solutions

$$y_1(x) = \sqrt{9 - x^2} \quad \text{and} \quad y_2(x) = -\sqrt{9 - x^2}.$$

Clearly, $y_1(x) = \sqrt{9 - x^2}$ is the solution that satisfies the initial condition $y(0) = 3$.

25. Differentiating $\ln(2 - y) = x + C$ implicitly, we have

$$D_x[\ln(2 - y)] = D_x(x + C)$$

$$\frac{1}{2 - y}(-y') = 1$$

$$-y' = 2 - y$$

$$y' = y - 2.$$

Thus, y is a solution of the given differential equation. Substituting $x = 0$ and $y = 1$ in $\ln(2 - y) = x + C$ yields

$$\ln(2 - 1) = 0 + C \quad \text{or} \quad \ln 1 = C \quad \text{and} \quad C = 0.$$

Therefore, the particular solution satisfying $y(0) = 1$ is a solution of the equation

$$\ln(2 - y) = x.$$

We can solve this equation for y by taking the exponential of both sides. We have

$$e^{\ln(2-y)} = e^x \quad \text{or} \quad 2 - y = e^x \quad \text{and} \quad y = 2 - e^x.$$

27. Given the general solution $y = 2 + Ce^{-x}$.

(A) Substituting $x = 0$ and $y = 1$ in the general solution yields

$$1 = 2 + Ce^0$$

$$C = -1.$$

Thus, the particular solution satisfying $y(0) = 1$ is $y_a = 2 - e^{-x}$.

(B) Substituting $x = 0$ and $y = 2$ in the general solution yields
$$2 = 2 + Ce^0$$
$$C = 0.$$
Thus, the particular solution satisfying $y(0) = 2$ is $y_b = 2$.

(C) Substituting $x = 0$ and $y = 3$ in the general solution yields
$$3 = 2 + Ce^0$$
$$C = 1.$$
Thus, the particular solution satisfying $y(0) = 3$ is $y_c = 2 + e^{-x}$.

The graphs of the particular solutions for $x \geq 0$ are shown at the right.

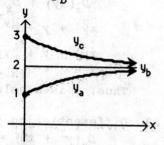

29. Given the general solution $y = 2 + Ce^x$.

(A) Substituting $x = 0$ and $y = 1$ in the general solution yields
$$1 = 2 + Ce^0$$
$$C = -1$$
Thus, the particular solution satisfying $y(0) = 1$ is $y_a = 2 - e^x$.

(B) Substituting $x = 0$ and $y = 2$ in the general solution yields
$$2 = 2 + Ce^0$$
$$C = 0.$$
Thus, the particular solution satisfying $y(0) = 2$ is $y_b = 2$.

(C) Substituting $x = 0$ and $y = 3$ in the general solution yields
$$3 = 2 + Ce^0$$
$$C = 1.$$
Thus, the particular solution satisfying $y(0) = 3$ is $y_c = 2 + e^x$.

The graphs of these solutions for $x \geq 0$ are shown at the right.

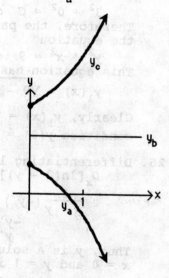

31. Given the general solution $y = \dfrac{10}{1 + Ce^{-x}}$.

(A) Substituting $x = 0$ and $y = 1$ in the general solution yields
$$1 = \frac{10}{1 + Ce^0} = \frac{10}{1 + C}.$$
Therefore, $1 + C = 10$ or $C = 9$.
Thus, the particular solution satisfying $y(0) = 1$ is $y_a = \dfrac{10}{1 + 9e^{-x}}$.

(B) Substituting $x = 0$ and $y = 10$ in the general solution yields
$$10 = \frac{10}{1 + Ce^0} = \frac{10}{1 + C}.$$
Therefore, $10 + 10C = 10$ and $C = 0$.
Thus, the particular solution satisfying $y(0) = 10$ is $y_b = 10$.

(C) Substituting $x = 0$ and $y = 20$ in the general solution yields

$$20 = \frac{10}{1 + Ce^0} = \frac{10}{1 + C}.$$

Therefore, $20 + 20C = 10$ and $C = -0.5$.
Thus, the particular solution satisfying
$y(0) = 20$ is

$$y_c = \frac{10}{1 - 0.5e^{-x}}.$$

The graphs of these solutions for $x \geq 0$ are
shown at the right.

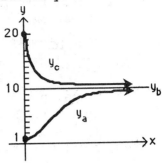

33. Given the general solution $y = Cx^3 + 2$.

(A) Substituting $x = 0$, $y = 2$ in the general solution yields

$$2 = C \cdot 0 + 2$$

This equation is satisfied for all values of C. Thus, $y = Cx^3 + 2$
satisfies the initial condition $y(0) = 2$ for any C.

(B) Substituting $x = 0$, $y = 0$ in the general solution yields

$$0 = C \cdot 0 + 2 \quad \text{or} \quad 0 = 2.$$

This equation is <u>not</u> satisfied for any value of C. There is <u>no</u>
particular solution of the differential equation which satisfies the
initial condition $y(0) = 0$.

(C) Substituting $x = 1$, $y = 1$ in the general solution yields

$$1 = C \cdot 1 + 2 \quad \text{and} \quad C = -1$$

Thus, $y = 2 - x^3$ is the particular solution of the differential
equation which satisfies the initial condition $y(1) = 1$.

35. (A) The graphs of $y = x + e^{-x}$, $y = x + 2e^{-x}$,
$y = x + 3e^{-x}$ and the graph of $y = x$ are
shown at the right.

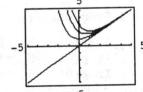

(B) Each of the graphs $y = x + Ce^{-x}$, $C = 1, 2, 3$,
lies above the line $y = x$, each decreases to
a local minimum and then increases,
approaching $y = x$ as x approaches ∞.

(C) The graphs of $y = x - e^{-x}$, $y = x - 2e^{-x}$,
$y = x - 3e^{-x}$ and the graph of $y = x$ are
shown at the right.

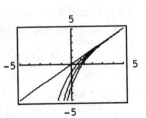

(D) Each of the graphs $y = x - Ce^{-x}$, $C = 1, 2, 3$,
is increasing and approaches $y = x$ as
x approaches ∞.

37. (A) Substitute $p = 5 - Ce^{-0.1t}$, $\dfrac{dp}{dt} = 0.1Ce^{-0.1t}$ into the differential equation:

$$0.1Ce^{-0.1t} = 0.5 - 0.1(5 - Ce^{-0.1t})$$
$$= 0.5 - 0.5 + 0.1Ce^{-0.1t}$$
$$= 0.1Ce^{-0.1t}$$

Thus, $p = 5 - Ce^{-0.1t}$ is the general solution and
$$\bar{p} = \lim_{t \to \infty} p(t) = \lim_{t \to \infty} (5 - Ce^{-0.1t}) = 5$$

(B) Setting $t = 0$ and $p = 1$ in the general solution yields:
$$1 = 5 - Ce^0 = 5 - C$$
and $\quad C = 4$

Thus, the particular solution satisfying the initial condition $p(0) = 1$ is:
$$p(t) = 5 - 4e^{-0.1t}$$

Setting $t = 0$ and $p = 10$ in the general solution yields:
$$10 = 5 - Ce^0 = 5 - C$$
and $\quad C = -5$

Thus, the particular solution satisfying the initial condition $p(0) = 10$ is:
$$p(t) = 5 + 5e^{-0.1t}$$

The graphs of these solutions are shown at the right.

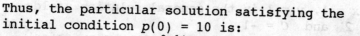

(C) From part (A), the equilibrium price $\bar{p} = 5$.

If $p(0) < 5$, then the price increases and approaches 5 as a limit.
If $p(0) > 5$, then the price decreases and approaches 5 as a limit.
If $p(0) = 5$, then $C = 0$ and $p(t) = 5$ for all t.

39. (A) Substitute $A(t) = Ce^{0.08t} - 2,500$, $\dfrac{dA}{dt} = 0.08Ce^{0.08t}$ into the differential equation:

$$0.08Ce^{0.08t} = 0.08(Ce^{0.08t} - 2,500) + 200$$
$$= 0.08Ce^{0.08t} - 200 + 200$$
$$= 0.08Ce^{0.08t}$$

Thus, $A(t) = Ce^{0.08t} - 2,500$ is the general solution.

(B) Setting $t = 0$ and $A = 0$ in the general solution yields:
$$0 = Ce^0 - 2{,}500 = C - 2{,}500$$
and $C = 2{,}500$
Thus, the particular solution satisfying $A(0) = 0$ is:
$$A_0(t) = 2{,}500e^{0.08t} - 2{,}500$$

Setting $t = 0$ and $A = 1{,}000$ in the general solution yields:
$$1000 = Ce^0 - 2{,}500 = C - 2{,}500$$
and $C = 3{,}500$
Thus, the particular solution satisfying $A(0) = 1{,}000$ is:
$$A_1(t) = 3{,}500e^{0.08t} - 2{,}500$$
The graphs of these solutions are at the right.

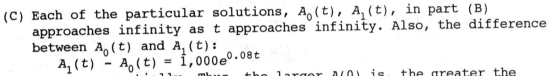

(C) Each of the particular solutions, $A_0(t)$, $A_1(t)$, in part (B) approaches infinity as t approaches infinity. Also, the difference between $A_0(t)$ and $A_1(t)$:
$$A_1(t) - A_0(t) = 1{,}000e^{0.08t}$$
grows exponentially. Thus, the larger $A(0)$ is, the greater the amount in the account.

41. (A) Substitute $N(t) = 200 - Ce^{-0.5t}$, $\dfrac{dN}{dt} = 0.5Ce^{-0.5t}$ into the differential equation:
$$\begin{aligned}0.5Ce^{-0.5t} &= 100 - 0.5(200 - Ce^{-0.5t})\\ &= 100 - 100 + 0.5Ce^{-0.5t}\\ &= 0.5Ce^{-0.5t}\end{aligned}$$

Thus, $N(T) = 200 - Ce^{-0.5t}$ is the general solution.
Now, $\overline{N} = \lim\limits_{t \to \infty} (200 - Ce^{-0.5t}) = 200 - C \lim\limits_{t \to \infty} e^{-0.5t} = 200$

The equilibrium size of the population is: $\overline{N} = 200$.

(B) Setting $t = 0$ and $N = 50$ in the general solution yields
$$50 = 200 - Ce^0 = 200 - C$$
and $C = 150$

Thus, the particular solution satisfying $N(0) = 50$ is:
$$N(t) = 200 - 150e^{-0.5t}$$

Setting $t = 0$ and $N = 300$ in the general solution yields:
$$300 = 200 - Ce^0 = 200 - C$$
and $C = -100$

Thus, the particular solution satisfying $N(0) = 300$ is:
$$N(t) = 200 + 100e^{-0.5t}$$
The graphs of these solutions are shown at the right.

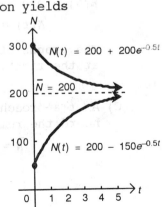

(C) If $N(0) < 200$, then number $N(t)$ of bacteria increases and approaches 200 as a limit; if $N(0) > 200$, then the number of bacteria decreases and approaches 200 as a limit. If $N(0) = 200$, then the number of bacteria remains constant for all t.

43. (A) Substitute $N(t) = Ce^{-2e^{-0.5t}}$, $\dfrac{dN}{dt} = Ce^{-2e^{-0.5t}}(-2e^{-0.5t})(-0.5) = Ce^{-0.5t}e^{-2e^{-0.5t}}$ into the differential equation:

$$Ce^{-0.5t}e^{-2e^{-0.5t}} = (Ce^{-2e^{-0.5t}})e^{-0.5t}$$
$$= Ce^{-0.5t}e^{-2e^{-0.5t}}$$

Thus, $N(t) = Ce^{-2e^{-0.5t}}$ is the general solution.

Now, $\overline{N} = \lim\limits_{t\to\infty} Ce^{-2e^{-0.5t}} = C \lim\limits_{t\to\infty} e^{-2e^{-0.5t}}$
$$= C \lim\limits_{t\to\infty} e^{-2/e^{0.5t}} = Ce^0 = C$$

(B) Setting $t = 0$ and $N = 100$ in the general solution yields:
$$100 = Ce^{-2e^0} = Ce^{-2}$$
and $\quad C = 100e^2$

Thus, the particular solution satisfying $N(0) = 100$ is:
$$N(t) = 100e^{2-2e^{-0.5t}}$$

Setting $t = 0$ and $N = 200$ in the general solution yields:
$$200 = Ce^{-2e^0} = Ce^{-2}$$
and $\quad C = 200e^2$

Thus, the particular solution satisfying $N(0) = 200$ is:
$$N(t) = 200e^{2-2e^{-0.5t}}$$

The graphs of these solutions are shown at the right.

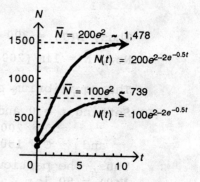

(C) As t approaches infinity, the number $N(t)$ of individuals who have heard the rumor approaches
$$\overline{N} = N(0)e^2$$

Things to remember:

1. The method of SEPARATION OF VARIABLES is applied to first-order differential equations of the form

    (A) $f(y)y' = g(x)$

    The GENERAL SOLUTION of equation (A) is given implicitly by the equation

    $$\int f(y)\,dy = \int g(x)\,dx$$

2. EXPONENTIAL GROWTH LAW

    If the rate of change with respect to time of a quantity y is proportional to the amount present, then $y = y(t)$ satisfies the differential equation

    $$\frac{dy}{dt} = ky$$

    where k is a constant.

3. LIMITED GROWTH LAW

    If the rate of change with respect to time of a quantity y is proportional to the difference between y and a limiting value M, then $y = y(t)$ satisfies the differential equation

    $$\frac{dy}{dt} = k(M - y)$$

    where k and M are constants.

4. LOGISTIC GROWTH LAW

    If the rate of change with respect to time of a quantity y is proportional to the product of y and the difference between y and a limiting value M, then $y = y(t)$ satisfies the differential equation

    $$\frac{dy}{dt} = ky(M - y)$$

    where k and M are constants.

---

1. $\dfrac{dy}{dt} = 100{,}000$

3. $\dfrac{dy}{dt} = k(10{,}000 - y)$, $k > 0$ constant

5. The annual sales of a company are $2 million initially, and are increasing at a rate (10%) proportional to the annual sales.

7. In a community of 5,000 people, one person started a rumor that is spreading at a rate proportional to the product of the number of people who have heard the rumor and the number who have not.

9. $y' = 1$; $y(0) = 2$

$$\int dy = \int dx$$

$y = x + C$.　　*General
solution*

Applying the initial condition
$y(0) = 2$, we have

　　$2 = 0 + C$　or　$C = 2$

Thus,

　　$y = x + 2$.　*Particular
solution*

13. $y' = y$; $y(0) = 10$.

$\dfrac{y'}{y} = 1$.　　Separate the variables

$$\int \frac{dy}{y} = \int dx$$

$\ln|y| = x + C$

$\quad y = e^{x+C}$

$\quad\quad = e^C e^x$ (put $K = e^C$)

$\quad\quad = Ke^x$.　　*General solution*

Applying the initial condition
$y(0) = 10$, we have

　　$10 = Ke^0$　or　$K = 10$.

Thus,

　　$y = 10e^x$.　*Particular
solution*

17. $y' = \dfrac{y}{x}$; $y(1) = 5$, $x > 0$.

$\dfrac{y'}{y} = \dfrac{1}{x}$.　　Separate the variables

$$\int \frac{dy}{y} = \int \frac{dx}{x}$$

$\ln|y| = \ln x + C$　(Note: $x > 0$,
$\quad y = e^{\ln x + C}$　　so $|x| = x$.)

$\quad\quad = e^C e^{\ln x}$

$\quad\quad = Kx$.　　(Note: $e^{\ln x} =$
$\quad\quad\quad\quad\quad x$, put $K = e^C$.)

Thus,

　　$y = Kx$.　*General solution*

Applying the initial condition
$y(1) = 5$, we have

　　$5 = K(1)$　or　$K = 5$.

Therefore,

　　$y = 5x$.　*Particular solution*

11. $y' = \dfrac{1}{\sqrt{x}} = \dfrac{1}{x^{1/2}}$; $y(1) = -2$

$$\int dy = \int \frac{dx}{x^{1/2}}$$

$y = 2x^{1/2} + C$　*General solution*

Applying the initial condition
$y(1) = -2$, we have

　　$-2 = 2(1)^{1/2} + C$　or　$C = -4$

Thus,

　　$y = 2x^{1/2} - 4$　*Particular
solution*

15. $y' = 25 - y$; $y(0) = 5$.

$\dfrac{y'}{25 - y} = 1$.　　Separate the variables

$$\int \frac{dy}{25 - y} = \int dx$$

$-\ln|25 - y| = x + C$

$\ln|25 - y| = -(x + C)$

$\quad 25 - y = e^{-x-C}$

$\quad 25 - y = e^{-C} e^{-x}$

$\quad 25 - y = Ke^{-x}$ (put $K = e^{-C}$)

$\quad\quad\quad y = 25 - Ke^{-x}$.　*General
solution*

Applying the initial condition
$y(0) = 5$, we have

　　$5 = 25 - Ke^0$　or　$K = 20$.

Thus,

　　$y = 25 - 20e^{-x}$.　*Particular
solution*

19. $y' = \dfrac{1}{y^2}$; $y(1) = 3$.

$y^2 y' = 1$

$$\int y^2 dy = \int dx$$

$\dfrac{y^3}{3} = x + C$

$\quad y^3 = 3x + 3C$

$\quad\quad = 3x + A$.　　(put $A = 3C$)

Thus,

　　$y = (3x + A)^{1/3}$. *General solution*

Applying the initial condition
$y(1) = 3$, we have

　　$3 = (3 + A)^{1/3}$

　　$27 = 3 + A$　and　$A = 24$.

Therefore,

　　$y = (3x + 24)^{1/3}$. *Particular
solution*

21. $y' = ye^x$; $y(0) = 3e$

$$\frac{y'}{y} = e^x$$

$$\int \frac{dy}{y} = \int e^x dx$$

$$\ln|y| = e^x + C$$

$$y = e^{e^x + C}$$

$$= e^C e^{e^x}$$

$$= Ke^{e^x}. \quad \textit{General solution}$$

Applying the initial condition $y(0) = 3e$, we have

$$3e = Ke^{e^0}$$

$$= Ke \quad \text{and} \quad K = 3.$$

Thus,

$$y = 3e^{e^x}. \quad \textit{Particular solution}$$

23. $y' = \dfrac{e^x}{e^y}$; $y(0) = \ln 2$.

$$e^y y' = e^x$$

$$\int e^y dy = \int e^x dx$$

$$e^y = e^x + C$$

$$y = \ln(e^x + C) \quad \textit{General solution}$$

Applying the initial condition $y(0) = \ln 2$, we have

$$\ln 2 = \ln(e^0 + C)$$

$$= \ln(1 + C).$$

Therefore,

$$1 + C = 2 \quad \text{and} \quad C = 1.$$

Thus,

$$y = \ln(e^x + 1). \quad \textit{Particular solution}$$

25. $y' = xy + x$; $y(0) = 2$

$$y' = x(y + 1)$$

$$\frac{y'}{y + 1} = x$$

$$\int \frac{dy}{y + 1} = \int x \, dx$$

$$\ln|y + 1| = \frac{x^2}{2} + C$$

$$y + 1 = e^{x^2/2 + C}$$

$$y + 1 = e^C e^{x^2/2}$$

$$y + 1 = Ke^{x^2/2}$$

$$y = Ke^{x^2/2} - 1. \quad \textit{General solution}$$

Applying the initial condition $y(0) = 2$, we have

$$2 = Ke^0 - 1 = K - 1 \quad \text{and} \quad K = 3.$$

Thus,

$$y = 3e^{x^2/2} - 1 \quad \textit{Particular solution}$$

27. $y' = (2 - y)^2 e^x$; $y(0) = 1$.

$$\frac{y'}{(2 - y)^2} = e^x$$

$$\int \frac{dy}{(2 - y)^2} = \int e^x dx$$

$$\frac{1}{2 - y} = e^x + C$$

$$2 - y = \frac{1}{e^x + C}$$

$$y = 2 - \frac{1}{e^x + C}. \quad \begin{array}{l}\textit{General}\\ \textit{solution}\end{array}$$

Applying the initial condition $y(0) = 1$, we have

$$1 = 2 - \frac{1}{e^0 + C}$$

$$\frac{1}{1 + C} = 1$$

$$1 + C = 1 \quad \text{and} \quad C = 0.$$

Thus,

$$y = 2 - \frac{1}{e^x} \quad \text{or} \quad y = 2 - e^{-x}.$$

$$\textit{Particular solution}$$

29. $y' = \dfrac{1 + x^2}{1 + y^2}$

$$(1 + y^2)y' = 1 + x^2 \quad \text{Separate the variables}$$

$$\int (1 + y^2) dy = \int (1 + x^2) dx$$

$$y + \frac{y^3}{3} = x + \frac{x^3}{3} + C. \quad \textit{General solution}$$

31. $xyy' = (1 + x^2)(1 + y^2)$

$\dfrac{yy'}{1 + y^2} = \dfrac{1 + x^2}{x}$ Separate the variables

$\displaystyle\int \dfrac{y\ dy}{1 + y^2} = \int\left(\dfrac{1}{x} + x\right)dx$

$\dfrac{1}{2}\displaystyle\int \dfrac{2y\ dy}{1 + y^2} = \ln|x| + \dfrac{x^2}{2} + C$

$\dfrac{1}{2}\ln(1 + y^2) = \ln|x| + \dfrac{x^2}{2} + C,$ *General solution*

or $\ln(1 + y^2) = \ln(x^2) + x^2 + C.$

33. $x^2 e^y y' = x^3 + x^3 e^y$

$x^2 e^y y' = x^3(1 + e^y)$

$\dfrac{e^y y'}{1 + e^y} = x$ Separate the variables

$\displaystyle\int \dfrac{e^y dy}{1 + e^y} = \int x\ dx$ (<u>Note</u>: Put $u = 1 + e^y$, $du = e^y dy$.)

$\ln(1 + e^y) = \dfrac{x^2}{2} + C.$ *General solution*

35. $xyy' = \ln x;\ y(1) = 1$

$yy' = \dfrac{\ln x}{x}$

$\displaystyle\int y\ dy = \int \dfrac{\ln x}{x}dx$ (<u>Note</u>: Put $u = \ln x$, $du = \dfrac{1}{x}dx$.)

$\dfrac{y^2}{2} = \dfrac{[\ln x]^2}{2} + C$

$y^2 = [\ln x]^2 + 2C$ or

$y^2 = [\ln x]^2 + A.$ *General solution.*

Solving this equation for y, we have

$y = \pm\sqrt{[\ln x]^2 + A}$

Applying the initial condition $y(1) = 1$, we choose the function

$y = \sqrt{[\ln x]^2 + A}$

and get

$1 = \sqrt{[\ln 1]^2 + A}$

$1 = \sqrt{0 + A}$ and $A = 1.$

Thus,

$y = \sqrt{[\ln x]^2 + 1}.$ *Particular solution*

37. $xy' = x\sqrt{y} + 2\sqrt{y};\ y(1) = 4$

$xy' = \sqrt{y}(x + 2)$

$\dfrac{y'}{\sqrt{y}} = \dfrac{x + 2}{x}$ Separate the variables

$\displaystyle\int \dfrac{dy}{\sqrt{y}} = \int\left(1 + \dfrac{2}{x}\right)dx$

$2y^{1/2} = x + 2\ln|x| + C$

$y^{1/2} = \dfrac{1}{2}(x + 2\ln|x| + C)$

$y = \dfrac{1}{4}(x + 2\ln|x| + C)^2$ or

$y = \dfrac{1}{4}[x + \ln(x^2) + C]^2.$

General solution

Applying the initial condition $y(1) = 4$, we have

$4 = \dfrac{1}{4}[1 + \ln(1^2) + C]^2$

$16 = (1 + C)^2$

$1 + C = 4$

$C = 3.$

Thus,

$y = \dfrac{1}{4}[x + \ln(x^2) + 3]^2.$

Particular solution

39. $yy' = xe^{-y^2};\ y(0) = 1$

$ye^{y^2}y' = x$

$\displaystyle\int ye^{y^2}\,dy = \int x\,dx$ (<u>Note</u>: Put $u = y^2$,
                                   $dy = 2y\,dy$.)

$\dfrac{1}{2}e^{y^2} = \dfrac{x^2}{2} + C$

$e^{y^2} = x^2 + 2C$

$e^{y^2} = x^2 + A$

$y^2 = \ln(x^2 + A)$. *General*
                        solution

Solving this equation for y, we have $y = \pm\sqrt{\ln(x^2 + A)}$.

We choose the function $y = \sqrt{\ln(x^2 + A)}$ and apply the initial condition $y(0) = 1$ to obtain

$1 = \sqrt{\ln(0 + A)}$

$\ln A = 1$

$A = e$

Thus, $y = \sqrt{\ln(x^2 + e)}$.   *Particular solution*

41. $y = M(1 - e^{-kt}),\ M > 0,\ k > 0$ constants.

   (A) At $t = 1,\ y = 3$:

         $3 = M(1 - e^{-k})$ so $M = \dfrac{3}{1 - e^{-k}}$

     At $t = 1.5,\ y = 4$

         $4 = M(1 - e^{1.5k})$ so $M = \dfrac{4}{1 - e^{-1.5k}}$

   (B) Using a graphing utility to graph the two equations in part (A), we find that $M \approx 7.3$ (and $k \approx 0.53$).

43. The model for this problem is $\dfrac{dA}{dt} = 0.12A;\ A(0) = 5000$.

       $\dfrac{dA}{A} = 0.12\,dt$

       $\displaystyle\int \dfrac{dA}{A} = \int 0.12\,dt$

       $\ln|A| = 0.12t + C$

           $A = e^{0.12t+C}$

           $A = e^{C}e^{0.12t}$

           $A = ke^{0.12t}$. *General solution*

Applying the initial condition $A(0) = 5000$, we have

      $5000 = ke^{0},\ k = 5000$.

Thus, $A = 5000e^{0.12t}$ is the particular solution.

When $t = 10$ years,

      $A = 5000e^{(0.12)10} = 5000e^{1.2} \approx \$16,600$.

45. (A) Let $s(t)$ denote the number of people who have heard about the new product. Then the model for this problem is

$$\frac{ds}{dt} = k(100,000 - s); \; s(0) = 0, \; s(7) = 20,000, \; k > 0.$$

$$\frac{ds}{(100,000 - s)} = k \, dt$$

$$\int \frac{ds}{100,000 - s} = \int k \, dt$$

$$-\ln|100,000 - s| = kt + C \qquad (0 < s < 100,000)$$

$$\ln|100,000 - s| = -kt - C$$

$$100,000 - s = e^{-kt-C}$$

$$100,000 - s = e^{-C}e^{-kt}$$

$$= Ae^{-kt}.$$

Thus,

$$s(t) = 100,000 - Ae^{-kt}. \quad \textit{(General solution)}$$

We use the conditions $s(0) = 0$, $s(7) = 20,000$ to determine the constants A and k.

$$s(0) = 0 = 100,000 - Ae^0, \; A = 100,000.$$

Thus, $s(t) = 100,000 - 100,000e^{-kt}.$

$$s(7) = 20,000 = 100,000 - 100,000e^{-7k}$$

$$100,000e^{-7k} = 80,000$$

$$e^{-7k} = 0.8$$

$$-7k = \ln(0.8)$$

$$k = \frac{-\ln(0.8)}{7}.$$

Therefore, the particular solution is

$$s(t) = 100,000 - 100,000e^{[\ln(0.8)/7]t}.$$

Finally, we want to find time t such that $s(t) = 50,000$:

$$50,000 = 100,000 - 100,000e^{[\ln(0.8)/7]t}$$

$$e^{[\ln(0.8)/7]t} = 0.5$$

$$\frac{\ln 0.8}{7}t = \ln(0.5).$$

Thus, $t = \dfrac{7 \ln(0.5)}{\ln(0.8)} \approx 22$ days.

(B) From part (A), we know that

$$s(t) = 100,000(1 - e^{-kt})$$

Since we want 50,000 people to be aware of the product after 14 days, we have

$$50,000 = 100,000(1 - e^{-14k})$$

$$1 - e^{-14k} = \frac{1}{2}$$

$$e^{-14k} = \frac{1}{2}$$

$$-14k = \ln(0.5)$$

$$k = -\frac{1}{14} \ln(0.5)$$

Thus, $s(t) = 100,000[1 - e^{(t/14) \ln(0.5)}]$

Now, $s(7) = 100,000[1 - e^{(7/14) \ln(0.5)}]$
$= 100,000[1 - e^{0.5 \ln(0.5)}] \approx 29,300$

Therefore, approximately 29,300 people must become aware of the product during the first 7 days in order to ensure that 50,000 will be aware of the product after 14 days.

47. Let $s(t)$ denote the percentage of the deodorizer that is present at time t. Then the model for this problem is

$$\frac{ds}{dt} = ks; \quad s(0) = 1, \quad s(30) = 0.5, \quad k < 0.$$

Separating the variables, we have

$$\frac{ds}{s} = k \, dt$$

$$\int \frac{ds}{s} = \int k \, dt$$

$$\ln|s| = kt + C$$
$$s = e^{kt+C}$$
$$s = Ae^{kt}. \quad \textit{General solution}$$

We use the conditions $s(0) = 1$, $s(30) = 0.5$, to evaluate the constants A and k.

$$s(0) = 1 = Ae^0, \quad A = 1.$$

Thus, $s(t) = e^{kt}$
$$s(30) = 0.5 = e^{30k}$$
$$30k = \ln(0.5)$$
$$k = \frac{\ln(0.5)}{30}.$$

Therefore, $s(t) = e^{[\ln(0.5)/30]t}$ *Particular solution*

Now, we want to find time t such that $s(t) = 0.1$:

$$0.1 = e^{[\ln(0.5)/30]t}$$
$$\frac{\ln(0.5)}{30} t = \ln(0.1)$$
$$t = \frac{30 \ln(0.1)}{\ln(0.5)} \approx 100 \text{ days.}$$

49. If $s(t)$ is the annual sales at time t, then the model for this problem is:

$$\frac{ds}{dt} = k(5 - s); \quad s(0) = 0, \quad s(4) = 1, \quad k > 0.$$

$$\frac{ds}{5 - s} = k \, dt$$

$$\int \frac{ds}{5 - s} = \int k \, dt$$

$$-\ln|5 - s| = kt + C$$
$$\ln|5 - s| = -kt - C$$
$$5 - s = e^{-kt-C}$$
$$5 - s = Ae^{-kt}$$
$$s = 5 - Ae^{-kt}.$$

We use the conditions $s(0) = 0$ and $s(4) = 1$ to evaluate the constants A and k.

$$s(0) = 0 = 5 - Ae^0, \quad A = 5.$$

Thus, $s(t) = 5 - 5e^{-kt}$

$$s(4) = 1 = 5 - 5e^{-4k}$$
$$-5e^{-4k} = -4$$
$$e^{-4k} = \frac{4}{5} = 0.8$$
$$-4k = \ln(0.8)$$
$$k = \frac{-\ln(0.8)}{4}.$$

Therefore, $s(t) = 5 - 5e^{[\ln(0.8)/4]t}$.

Now we want to find time t such that $s(t) = 4$:

$$4 = 5 - 5e^{[\ln(0.8)/4]t}$$
$$-5e^{[\ln(0.8)/4]t} = -1$$
$$e^{[\ln(0.8)/4]t} = \frac{1}{5} = 0.2$$
$$\frac{\ln(0.8)}{4} t = \ln(0.2).$$

Thus, $t = \frac{4 \ln(0.2)}{\ln(0.8)} \approx 29$ years.

51. $S(t) = M(1 - e^{-kt})$, $M > 0$, $k > 0$ constant

We have:
$$S(1) = M(1 - e^{-k}) = 2$$
and $\qquad S(3) = M(1 - e^{-3k}) = 5$

From the first equation, $M = \dfrac{2}{1 - e^{-k}}$. Substituting this in the second equation yields:

$$\frac{2}{(1 - e^{-k})} (1 - e^{-3k}) = 5$$
$$2 - 2e^{-3k} = 5 - 5e^{-k}$$
and $\qquad 2e^{-3k} - 5e^{-k} + 3 = 0 \qquad\qquad (1)$

Now let $t = e^{-k}$ ($k = -\ln t$). Then equation (1) becomes
$$2t^3 - 5t + 3 = 0$$

which factors into
$$(t - 1)(2t^2 + 2t - 3) = 0$$

The solutions are $t = 1$, $t \approx 0.82$ and $t \approx -1.82$.

Since $\ln(-1.82)$ is undefined and $\ln(1)$ implies $k = 0$, it follows that $k = -\ln(0.82) \approx 0.2$.

Now, $M = \dfrac{2}{(1 - e^{-0.2})} \approx 11$.

Thus, $k \approx 0.2$ and $M \approx \$11$ million.

53. Let $T(t)$ denote the temperature of the bar at time t. Then the model for this problem is

$$\frac{dT}{dt} = k(T - 800), \quad k < 0; \quad T(0) = 80, \quad T(2) = 200.$$

Separating the variables, we have

$$\frac{dT}{T - 800} = k\, dt$$

$$\int \frac{dT}{T - 800} = \int k\, dt$$

$$\ln|T - 800| = kt + C$$
$$T - 800 = e^{kt+C}$$
$$T = 800 + Ae^{kt} \quad \textit{General solution}$$

Using the conditions $T(0) = 80$ and $T(2) = 200$ to evaluate the constants A and k, we have

$$T(0) = 80 = 800 + Ae^0, \quad A = -720.$$

Thus, $T(t) = 800 - 720e^{kt}$.

$$T(2) = 200 = 800 - 720e^{2k}$$
$$720e^{2k} = 600$$
$$e^{2k} = \frac{5}{6}$$
$$2k = \ln\left(\frac{5}{6}\right)$$
$$k = \frac{1}{2} \ln\left(\frac{5}{6}\right)$$

Therefore, $T(t) = 800 - 720e^{1/2\ln(5/6)t}$.

Finally, we want to find t such that $T(t) = 500$. Thus,

$$500 = 800 - 720e^{1/2\ln(5/6)t}$$
$$720e^{1/2\ln(5/6)t} = 300$$
$$e^{1/2\ln(5/6)t} = \frac{5}{12}$$
$$\frac{1}{2} \ln\left(\frac{5}{6}\right) t = \ln\left(\frac{5}{12}\right)$$
$$t = \frac{2 \ln\left(\frac{5}{12}\right)}{\ln\left(\frac{5}{6}\right)} \approx 9.6 \text{ minutes.}$$

55. Let $T(t)$ denote the temperature of the pie at time t. Then the model for this problem is:

$$\frac{dT}{dt} = k(T - 25), \quad k < 0; \quad T(0) = 325, \quad T(1) = 225.$$

Separating the variables, we have $\dfrac{dT}{T - 25} = k\, dt$

$$\int \frac{dT}{T - 25} = \int k\, dt$$
$$\ln|T - 25| = kt + C$$
$$T - 25 = e^{kt+C}$$
$$T = 25 + Ae^{kt}$$

Using the initial conditions $T(0) = 325$ and $T(1) = 225$ to evaluate the constants A and k, we have:

$$T(0) = 325 = 25 + Ae^0, \quad A = 300$$

Thus, $T(t) = 25 + 300e^{kt}$.

$$T(1) = 225 = 25 + 300e^k$$
$$300e^k = 200$$
$$e^k = \frac{2}{3}$$
$$k = \ln\left(\frac{2}{3}\right)$$

Therefore, $T(t) = 25 + 300e^{t\ln(2/3)}$.

Finally, we want to determine $T(4)$: $\quad T(4) = 25 + 300e^{4\ln(2/3)}$

$$= 25 + 300e^{\ln(2/3)^4}$$
$$= 25 + 300\left(\frac{2}{3}\right)^4 \approx 84.26°\text{F}$$

57. If $P(t)$ is the number of bacteria present at time t, then the model for this problem is:

$$\frac{dP}{dt} = kP; \quad P(0) = 100, \quad P(1) = 140, \quad k > 0.$$

$$\frac{dP}{P} = k\, dt$$

$$\int \frac{dP}{P} = \int k\, dt$$

$$\ln|P| = kt + C$$
$$P = e^{kt+C}$$
$$P = Ae^{kt}. \quad \textit{General solution}$$

We use the conditions $P(0) = 100$ and $P(1) = 140$ to evaluate the constants A and k.

$$P(0) = 100 = Ae^0, \quad A = 100.$$

Thus,

$$P(t) = 100e^{kt}$$
$$P(1) = 140 = 100e^k$$
$$e^k = 1.4$$
$$k = \ln(1.4).$$

Therefore,

$$P(t) = 100e^{(\ln 1.4)t}. \quad \textit{Particular solution}$$

(A) When $t = 5$,
$$P(5) = 100e^{\ln(1.4)5}$$
$$\approx 538 \text{ bacteria.}$$

(B) When $P = 1000$,
$$1000 = 100e^{\ln(1.4)t}$$
$$e^{\ln(1.4)t} = 10$$
$$\ln(1.4)t = \ln 10$$
$$t = \frac{\ln 10}{\ln(1.4)} \approx 6.8 \text{ hours.}$$

59. If $P(t)$ is the number of people infected at time t, then the model for this problem is:

$$\frac{dP}{dt} = kP(50,000 - P); \quad P(0) = 100, \quad P(10) = 500.$$

$$\frac{dP}{P(50,000 - P)} = k\,dt$$

$$\int \frac{dP}{P(50,000 - P)} = \int k\,dt$$

$$\frac{1}{50,000} \int \left[\frac{1}{P} + \frac{1}{50,000 - P}\right] dP = kt + C$$

$$\frac{1}{50,000}[\ln P - \ln(50,000 - P)] = kt + C$$

$$\ln\left[\frac{P}{50,000 - P}\right] = 50,000(kt + C)$$

$$\frac{P}{50,000 - P} = e^{50,000kt + 50,000C}$$

$$= e^{50,000C}e^{50,000kt}$$

$$= Ae^{50,000kt}$$

Solving this equation for P, we obtain

$$P = (50,000 - P)Ae^{50,000kt}$$

$$P = \frac{50,000Ae^{50,000kt}}{1 + Ae^{50,000kt}},$$

which can be written

$$P(t) = \frac{50,000}{1 + Be^{-50,000kt}}, \quad B = \frac{1}{A}.$$

Using the conditions $P(0) = 100$ and $P(10) = 500$ to evaluate the constants B and k, we obtain

$$P(0) = 100 = \frac{50,000}{1 + Be^0}$$

$$100(1 + B) = 50,000$$

$$1 + B = 500$$

$$B = 499.$$

$$P(t) = \frac{50,000}{1 + 499e^{-50,000kt}}.$$

$$P(10) = 500 = \frac{50,000}{1 + 499e^{-500,000k}}$$

$$1 + 499e^{-500,000k} = 100$$

$$499e^{-500,000k} = 99$$

$$-500,000k = \ln\left(\frac{99}{499}\right)$$

$$k = -\frac{1}{500,000} \ln\left(\frac{99}{499}\right).$$

Therefore,

$$P(t) = \frac{50,000}{1 + 499e^{0.1\ln(99/499)t}}. \qquad \textit{Particular solution}$$

(A) When $t = 20$,
$$P(20) = \frac{50,000}{1 + 499e^{2\ln(99/499)}} \approx 2422 \text{ people.}$$

(B) When $P = 25,000$,
$$25,000 = \frac{50,000}{1 + 499e^{0.1\ln(99/499)t}}$$
$$1 + 499e^{0.1\ln(99/499)t} = 2$$
$$e^{0.1\ln(99/499)t} = \frac{1}{499}$$
$$0.1 \ln\left(\frac{99}{499}\right)t = \ln\left(\frac{1}{499}\right)$$
$$t = \frac{10 \ln\left(\frac{1}{499}\right)}{\ln\left(\frac{99}{499}\right)} \approx 38.4 \text{ days.}$$

61. $\dfrac{dI}{ds} = k\dfrac{I}{s}, \quad I > 0, \ s > 0.$

$$\frac{dI}{I} = \frac{k}{s}\,ds$$
$$\int \frac{dI}{I} = \int \frac{k}{s}\,ds$$
$$\ln I = k \ln s + C$$
$$\quad\ = \ln s^k + C \qquad (k \ln s = \ln s^k)$$
$$I = e^{\ln s^k + C}$$
$$\quad = e^C e^{\ln s^k}.$$
Therefore,
$$I = As^k. \qquad (\underline{\text{Note}}:\ e^{\ln s^k} = s^k.)$$

63. If $P(t)$ is the number of people who have heard the rumor at time t, then the model for this problem is:
$$\frac{dP}{dt} = kP(1000 - P); \quad P(0) = 5, \ P(1) = 10.$$

$$\frac{dP}{P(1000 - P)} = k\,dt$$
$$\int \frac{dP}{P(1000 - P)} = \int k\,dt$$
$$\frac{1}{1000} \int \left[\frac{1}{P} + \frac{1}{1000 - P}\right]dP = kt + C$$
$$\frac{1}{1000}[\ln P - \ln(1000 - P)] = kt + C$$
$$\ln\left(\frac{P}{1000 - P}\right) = 1000(kt + C)$$
$$\frac{P}{1000 - P} = e^{1000kt + 1000C}$$
$$\quad = e^{1000C}e^{1000kt}$$
$$\quad = Ae^{1000kt}$$

Solving this equation for P, we obtain

$$P = \frac{1000Ae^{1000kt}}{1 + Ae^{1000kt}} \quad \text{or} \quad P = \frac{1000}{1 + Be^{-1000kt}} \quad \left(B = \frac{1}{A}\right). \quad \textit{General solution}$$

Using the conditions $P(0) = 5$ and $P(1) = 10$ to evaluate the constants B and k, we have

$$P(0) = 5 = \frac{1000}{1 + Be^0}$$

$$5(1 + B) = 1000$$

$$1 + B = 200$$

$$B = 199.$$

Thus,

$$P(t) = \frac{1000}{1 + 199e^{-1000kt}}$$

$$P(1) = 10 = \frac{1000}{1 + 199e^{-1000k}}$$

$$1 + 199e^{-1000k} = 100$$

$$199e^{-1000k} = 99$$

$$e^{-1000k} = \frac{99}{199}$$

$$-1000k = \ln\left(\frac{99}{199}\right)$$

$$k = -\frac{1}{1000} \ln\left(\frac{99}{199}\right).$$

Therefore,

$$P(t) = \frac{1000}{1 + 199e^{\ln(99/199)t}}. \quad \textit{Particular solution}$$

(A) When $t = 7$,

$$P(7) = \frac{1000}{1 + 199e^{\ln(99/199)7}} \approx 400 \text{ people.}$$

(B) When $P = 850$,

$$850 = \frac{1000}{1 + 199e^{\ln(99/199)t}}$$

$$1 + 199e^{\ln(99/199)t} = \frac{1000}{850} = \frac{20}{17}$$

$$e^{\ln(99/199)t} = \frac{3}{3383}$$

$$\ln\left(\frac{99}{199}\right)t = \ln\left(\frac{3}{3383}\right)$$

$$t = \frac{\ln\left(\frac{3}{3383}\right)}{\ln\left(\frac{99}{199}\right)} \approx 10 \text{ days.}$$

Things to remember:

1. **SOLVING FIRST-ORDER LINEAR DIFFERENTIAL EQUATIONS**
   Step 1: Write the equation in the STANDARD FORM:
   (A) $y' + f(x)y = g(x)$
   Step 2: Compute the INTEGRATING FACTOR:
   $$I(x) = e^{\int f(x)\,dx}$$

   (Note: when evaluating $\int f(x)\,dx$, choose 0 for the
   constant of integration.)
   Step 3: Multiply both sides of (A) by the integrating factor.
   The left side will be in the form $[I(x)y]'$:
   $$[I(x)y]' = I(x)g(x)$$

   Step 4: Integrate both sides:
   $$I(x)y = \int I(x)g(x)\,dx$$

   (Note: when evaluating $\int I(x)g(x)\,dx$, include the

   arbitrary constant of integration.)

   Step 5: Solve for y to obtain the GENERAL SOLUTION:
   $$y = \frac{1}{I(x)}\int I(x)g(x)\,dx$$

2. **BASIC FORMULAS INVOLVING THE NATURAL LOGARITHM FUNCTION**

   If the domain of h is restricted so that $h(x) > 0$, then
   $$\int \frac{h'(x)}{h(x)}\,dx = \ln h(x) \quad \text{and} \quad e^{\ln h(x)} = h(x)$$

---

1. $\int 3x^2;\ \int (x^3 y' + 3x^2 y)\,dx = x^3 y$ 3. $-3e^{-3x};\ \int (e^{-3x}y' - 3e^{-3x}y)\,dx = e^{-3x}y$

5. $x^4;\ \int (x^4 y' + 4x^3 y)\,dx = x^4 y$

7. $e^{-0.5x};\ \int (e^{-0.5x}y' - 0.5e^{-0.5x}y)\,dx = e^{-0.5x}y$

9. $y' + 2y = 4$; $y(0) = 1$

Step 1: The equation is in standard form.

Step 2: Find the integrating factor.
$$f(x) = 2 \quad \text{and} \quad I(x) = e^{\int f(x)\,dx} = e^{\int 2\,dx} = e^{2x}$$

Step 3: Multiply both sides of the standard form by the integrating factor.
$$e^{2x}(y' + 2y) = e^{2x}(4)$$
$$e^{2x}y' + 2e^{2x}y = 4e^{2x}$$
$$[e^{2x}y]' = 4e^{2x}$$

Step 4: Integrate both sides.
$$\int [e^{2x}y]'\,dx = \int 4e^{2x}dx$$
$$e^{2x}y = 2e^{2x} + C$$

Step 5: Solve for y.
$$y = \frac{1}{e^{2x}}(2e^{2x} + C) = 2 + Ce^{-2x} \qquad \textit{General solution}$$

To find the particular solution satisfying the initial condition $y(0) = 1$, substitute $x = 0$, $y = 1$ in the general solution:
$$1 = 2 + Ce^0 = 2 + C.$$
Thus, $C = -1$ and the particular solution is
$$y = 2 - e^{-2x}$$

11. $y' + y = e^{-2x}$; $y(0) = 3$

Step 1: The equation is in standard form.

Step 2: Find the integrating factor.
$$f(x) = 1 \quad \text{and} \quad I(x) = e^{\int f(x)\,dx} = e^{\int 1\,dx} = e^{x}$$

Step 3: Multiply both sides of the standard form by the integrating factor.
$$e^{x}[y' + y] = e^{x} \cdot e^{-2x}$$
$$e^{x}y' + e^{x}y = e^{-x}$$
$$[e^{x}y]' = e^{-x}$$

Step 4: Integrate both sides.
$$\int [e^{x}y]'\,dx = \int e^{-x}dx$$
$$e^{x}y = \frac{e^{-x}}{-1} + C = -e^{-x} + C$$

Step 5: Solve for y.
$$y = \frac{1}{e^{x}}[-e^{-x} + C] = -e^{-2x} + Ce^{-x} \qquad \textit{General solution}$$

To find the particular solution satisfying the initial condition $y(0) = 3$, substitute $x = 0$, $y = 3$ in the general solution:
$$3 = -e^0 + Ce^0 = -1 + C$$
Thus, $C = 4$ and the particular solution is
$$y = -e^{-2x} + 4e^{-x}$$

13. $y' - y = 2e^x$; $y(0) = -4$

Step 1: The equation is in standard form.

Step 2: Find the integrating factor.
$$f(x) = -1 \quad \text{and} \quad I(x) = e^{\int f(x)\,dx} = e^{\int (-1)\,dx} = e^{-x}$$

Step 3: Multiply both sides of the standard form by the integrating factor.
$$e^{-x}[y' - y] = e^{-x}(2e^x)$$
$$e^{-x}y' - e^{-x}y = 2$$
$$[e^{-x}y]' = 2$$

Step 4: Integrate both sides.
$$\int [e^{-x}y]'\,dx = \int 2\,dx$$
$$e^{-x}y = 2x + C$$

Step 5: Solve for y.
$$y = \frac{1}{e^{-x}}[2x + C] = 2xe^x + Ce^x \quad \textit{General solution}$$

To find the particular solution satisfying the initial condition $y(0) = -4$, substitute $x = 0$, $y = -4$ in the general solution:
$$-4 = 2(0)e^0 + Ce^0 = C$$
Thus, $C = -4$ and the particular solution is
$$y = 2xe^x - 4e^x$$

15. $y' + y = 9x^2e^{-x}$; $y(0) = 2$

Step 1: The equation is in standard form.

Step 2: Find the integrating factor.
$$f(x) = 1 \quad \text{and} \quad I(x) = e^{\int f(x)\,dx} = e^{\int 1\,dx} = e^x$$

Step 3: Multiply both sides of the standard form by the integrating factor.
$$e^x[y' + y] = e^x(9x^2e^{-x})$$
$$e^xy' + e^xy = 9x^2$$
$$[e^xy]' = 9x^2$$

Step 4: Integrate both sides.
$$\int [e^xy]'\,dx = \int 9x^2\,dx$$
$$e^xy = 3x^3 + C$$

Step 5: Solve for y.
$$y = \frac{1}{e^x}(3x^3 + C) = 3x^3e^{-x} + Ce^{-x} \quad \textit{General solution}$$

To find the particular solution satisfying the initial condition $y(0) = 2$, substitute $x = 0$, $y = 2$ in the general solution:
$$2 = 3(0)^3e^0 + Ce^0 = C$$
Thus, $C = 2$ and the particular solution is
$$y = 3x^3e^{-x} + 2e^{-x}$$

17. $xy' + y = 2x$; $y(1) = 1$

    Step 1: Write the differential equation in standard form.

            Multiply both sides by $\frac{1}{x}$ to obtain:

$$y' + \frac{1}{x}y = 2$$

    Step 2: Find the integrating factor.

$$f(x) = \frac{1}{x} \quad \text{and} \quad I(x) = e^{\int f(x)\,dx} = e^{\int (1/x)\,dx} = e^{\ln x} = x$$

    Step 3: Multiply both sides of the standard form by the integrating factor.

$$x\left(y' + \frac{1}{x}y\right) = x(2)$$
$$xy' + y = 2x$$
$$[xy]' = 2x$$

    Step 4: Integrate both sides.

$$\int [xy]'\,dx = \int 2x\,dx$$
$$xy = x^2 + C$$

    Step 5: Solve for y.

$$y = \frac{1}{x}(x^2 + C) = x + \frac{C}{x} \quad \textit{General solution}$$

To find the particular solution satisfying the initial condition $y(1) = 1$, substitute $x = 1$, $y = 1$ in the general solution:

$$1 = 1 + \frac{C}{1} = 1 + C$$

Thus, $C = 0$ and the particular solution is
$$y = x$$

19. $xy' + 2y = 10x^3$; $y(2) = 8$

    Step 1: Write the differential equation in standard form.

            Multiply both sides by $\frac{1}{x}$ to obtain:

$$y' + \frac{2}{x}y = 10x^2$$

    Step 2: Find the integrating factor.

$$f(x) = \frac{2}{x} \quad \text{and} \quad I(x) = e^{\int f(x)\,dx} = e^{\int (2/x)\,dx} = e^{2\ln x} = e^{\ln x^2} = x^2$$

    Step 3: Multiply both sides of the standard form by the integrating factor.

$$x^2\left(y' + \frac{2}{x}y\right) = x^2(10x^2)$$
$$x^2 y' + 2xy = 10x^4$$
$$[x^2 y]' = 10x^4$$

<u>Step 4</u>: Integrate both sides.

$$\int [x^2 y]' \, dx = \int 10x^4 \, dx$$

$$x^2 y = 2x^5 + C$$

<u>Step 5</u>: Solve for y.

$$y = \frac{1}{x^2}(2x^5 + C) = 2x^3 + \frac{C}{x^2} \quad \text{General solution}$$

To find the particular solution satisfying the initial condition $y(2) = 8$, substitute $x = 2$, $y = 8$ in the general solution:

$$8 = 2(2)^3 + \frac{C}{2^2} = 16 + \frac{C}{4}$$

and $\quad -8 = \frac{C}{4}$.

Thus, $C = -32$ and the particular solution is

$$y = 2x^3 - \frac{32}{x^2}$$

21. $y' + xy = 5x$

$I(x) = e^{\int f(x) \, dx} = e^{\int x \, dx} = e^{x^2/2} \quad \text{Integrating factor}$

$y = \frac{1}{I(x)} \int I(x) g(x) \, dx$

$\quad = \frac{1}{e^{x^2/2}} \int e^{x^2/2} 5x \, dx$

$\quad = 5e^{-x^2/2} \int x e^{x^2/2} \, dx \qquad \left(u = \frac{x^2}{2}, \ du = x \, dx \right)$

$\quad = 5e^{-x^2/2}[e^{x^2/2} + C] = 5 + 5Ce^{-x^2/2}$

$y = 5 + Ae^{-x^2/2}. \quad \text{General solution}$

23. $y' - 2y = 4x$

$I(x) = e^{\int(-2) \, dx} = e^{-2x} \qquad \text{Integrating factor}$

$y = \frac{1}{I(x)} \int I(x) g(x) \, dx = \frac{1}{e^{-2x}} \int e^{-2x} 4x \, dx$

$\quad = 4e^{2x} \int x e^{-2x} \, dx \qquad \text{(Integrate by parts)}$

$\quad = 4e^{2x} \left[-\frac{1}{2} x e^{-2x} - \frac{1}{4} e^{-2x} + C \right]$

$\quad = -2x - 1 + 4Ce^{2x} \qquad (A = 4C)$

$y = -2x - 1 + Ae^{2x}. \quad \text{General solution}$

25. $xy' + y = xe^x$

First write the equation in standard form: $y' + \frac{1}{x}y = e^x$.

Then,

$I(x) = e^{\int f(x)\,dx} = e^{\int(1/x)\,dx} = e^{\ln x} = x$ *Integrating factor*

$y = \frac{1}{I(x)}\int I(x)\,g(x)\,dx$

$\quad = \frac{1}{x}\int xe^x\,dx$ (Integrate by parts)

$\quad = \frac{1}{x}[xe^x - e^x + C]$

$y = e^x - \frac{e^x}{x} + \frac{C}{x}.$ *General solution*

27. $xy' + y = x\ln x$

First write the equation in standard form: $y' + \frac{1}{x}y = \ln x$

Then,

$y' + \frac{1}{x}y = \ln x$

$I(x) = e^{\int f(x)\,dx} = e^{\int(1/x)\,dx} = e^{\ln x} = x$ *Integrating factor*

$y = \frac{1}{I(x)}\int I(x)\,g(x)\,dx$

$\quad = \frac{1}{x}\int x\ln x\,dx$ (Integrate by parts)

$\quad = \frac{1}{x}\left[\frac{x^2}{2}\ln x - \frac{1}{4}x^2 + C\right]$

$\quad = \frac{1}{2}x\ln x - \frac{1}{4}x + \frac{C}{x}.$ *General solution*

29. $2xy' + 3y = 20x$

First write the equation in standard form

$\quad\quad y' + \frac{3}{2x}y = 10$

Then,

$I(x) = e^{\int f(x)\,dx} = e^{\int(3/2x)\,dx} = e^{(3/2)\ln x}$

$\quad\quad\quad\quad\quad\quad\quad\quad = e^{\ln x^{3/2}} = x^{3/2}$ *Integrating factor*

$y = \frac{1}{I(x)}\int I(x)\,g(x)\,dx = \frac{1}{x^{3/2}}\int x^{3/2}\,10\,dx$

$\quad\quad\quad\quad\quad = 10x^{-3/2}\int x^{3/2}\,dx$

$\quad\quad\quad\quad\quad = 10x^{-3/2}\left[\frac{2}{5}x^{5/2} + C\right]$

$\quad\quad\quad\quad\quad = 4x + 10Cx^{-3/2}$ $(A = 10C)$

$\quad\quad\quad\quad\quad = 4x + Ax^{-3/2}$ *General solution*

31. (A) Substitute $y = \frac{1}{3}(x + 1)^3 + C$; $y' = (x + 1)^2$ into the differential equation to determine whether these substitutions reduce the equation to an identity.

(B) The solution is wrong:

$$\frac{1}{x}\int x(x + 1)^2\,dx \neq \int (x + 1)^2\,dx$$

(C) $y = \frac{1}{x}\int x(x + 1)^2\,dx = \frac{1}{x}\int (x^3 + 2x^2 + x)\,dx$

$$= \frac{1}{x}\left[\frac{1}{4}x^4 + \frac{2}{3}x^3 + \frac{1}{2}x^2 + C\right]$$

$$= \frac{1}{4}x^3 + \frac{2}{3}x^2 + \frac{1}{2}x + \frac{C}{x}$$

Thus, $y = \frac{1}{4}x^3 + \frac{2}{3}x^2 + \frac{1}{2}x + \frac{C}{x}$ *General solution*

Substituting y and $y' = \frac{3}{4}x^2 + \frac{4}{3}x + \frac{1}{2} - \frac{C}{x^2}$

into the differential equation yields:

$$\frac{3}{4}x^2 + \frac{4}{3}x + \frac{1}{2} - \frac{C}{x^2} + \frac{1}{x}\left(\frac{1}{4}x^3 + \frac{2}{3}x^2 + \frac{1}{2}x + \frac{C}{x}\right) = x^2 + 2x + 1$$

$$= (x + 1)^2$$

33. (A) Substitute $y = \frac{1}{2}e^{-x}$, $y' = -\frac{1}{2}e^{-x}$ into the differential equation to determine whether these substitutions reduce the equation to an identity.

(B) $y = \frac{1}{2}e^{-x}$ is **not** the general solution since the constant of integration has been omitted. However, $y = \frac{1}{2}e^{-x}$ is a particular solution.

(C) $y = \frac{1}{e^{3x}}\int e^{2x}\,dx = \frac{1}{e^{3x}}\left[\frac{1}{2}e^{2x} + C\right]$

$$= \frac{1}{2}e^{-x} + Ce^{-3x}$$

Thus, $y = \frac{1}{2}e^{-x} + Ce^{-3x}$ *General solution*

Substituting y and $y' = -\frac{1}{2}e^{-x} - 3Ce^{-3x}$ into the differential equation yields:

$$-\frac{1}{2}e^{-x} - 3Ce^{-3x} + 3\left(\frac{1}{2}e^{-x} + Ce^{-3x}\right) = e^{-x}$$

35. $y' = \dfrac{1 - y}{x}$

This equation can be rewritten as a first-order linear differential equation in the standard form.

(A) $y' + \dfrac{1}{x}y = \dfrac{1}{x}$.

Then $f(x) = \dfrac{1}{x}$ and the integrating factor is

$I(x) = e^{\int (1/x)\,dx} = e^{\ln x} = x.$

Thus, $y = \dfrac{1}{x}\displaystyle\int x \cdot \dfrac{1}{x}\,dx = \dfrac{1}{x}\displaystyle\int dx = \dfrac{1}{x}(x + C) = 1 + \dfrac{C}{x}.$

Using separation of variables on the original equation, we have:

$$\dfrac{y'}{1 - y} = \dfrac{1}{x}$$

$$\int \dfrac{dy}{1 - y} = \int \dfrac{1}{x}\,dx$$

$-\ln(1 - y) = \ln(x) + C$ (assuming $1 - y > 0$ and $x > 0$)

$\ln(1 - y) = -\ln(x) - C$

$1 - y = e^{-\ln x - C}$

$1 - y = e^{-C}e^{\ln x^{-1}}$

$1 - y = \dfrac{K}{x}$ $\left(e^{\ln x^{-1}} = \dfrac{1}{x}\right)$

$y = 1 + \dfrac{K}{x}$

37. $y' = \dfrac{2x + 2xy}{1 + x^2}$

This equation can be rewritten as a first-order linear differential equation in the standard form.

(A) $y' - \dfrac{2x}{1 + x^2}y = \dfrac{2x}{1 + x^2}$.

Then, $f(x) = \dfrac{-2x}{1 + x^2}$ and the integrating factor is:

$I(x) = e^{\int (-2x/1+x^2)\,dx} = e^{-\ln(1+x^2)} = \dfrac{1}{1 + x^2}$

Thus, $y = \dfrac{1}{\dfrac{1}{1 + x^2}}\displaystyle\int \dfrac{1}{1 + x^2} \cdot \dfrac{2x}{1 + x^2}\,dx$

$= (1 + x^2)\displaystyle\int \dfrac{2x}{(1 + x^2)^2}\,dx = (1 + x^2)\left[\dfrac{-1}{1 + x^2} + C\right]$

and $y = -1 + C(1 + x^2).$

Using separation of variables on the original equation, we have:

$$\frac{y'}{1 + y} = \frac{2x}{1 + x^2}$$

$$\int \frac{dy}{1 + y} = \int \frac{2x}{1 + x^2}\,dx$$

$$\ln|1 + y| = \ln(1 + x^2) + C$$

$$1 + y = e^{\ln(1+x^2)+C}$$

$$1 + y = e^C e^{\ln(1+x^2)}$$

$$1 + y = K(1 + x^2) \qquad (e^{\ln(1+x^2)} = 1 + x^2)$$

$$y = -1 + K(1 + x^2)$$

39. $y' = 2x(y + 1)$

This equation can be rewritten as a first-order linear differential equation in the standard form.

(A) $y' - 2xy = 2x$

Then, $f(x) = -2x$ and the integrating factor is:

$$I(x) = e^{\int -2x\,dx} = e^{-x^2}$$

Thus, $y = \dfrac{1}{e^{-x^2}} \displaystyle\int e^{-x^2} 2x\,dx = e^{x^2} \int 2x e^{-x^2}\,dx = e^{x^2}[-e^{-x^2} + C]$

and $\quad y = -1 + Ce^{x^2}$.

Using separation of variables on the original equation, we have:

$$\frac{y'}{y + 1} = 2x$$

$$\int \frac{dy}{y + 1} = \int 2x\,dx$$

$$\ln|y + 1| = x^2 + C$$

$$y + 1 = e^{x^2+C}$$

$$y + 1 = e^C e^{x^2}$$

$$= Ke^{x^2}$$

Thus, $y = -1 + Ke^{x^2}$.

41. $\dfrac{dy}{dt} = ky$.

This equation can be rewritten as

(A) $\dfrac{dy}{dt} - ky = 0$

Here $f(t) = -k$ and the integrating factor is:

$$I(t) = e^{\int -k\,dt} = e^{-kt}$$

Thus, $y = \dfrac{1}{e^{-kt}} \displaystyle\int e^{-kt}\cdot 0\,dt = e^{kt} \int 0\,dt = e^{kt}C$

and $\quad y = Ce^{kt}$.

43. The amount A in the account at any time t must satisfy
$$\frac{dA}{dt} - 0.04A = -4000.$$
Now $f(t) = -0.04$ and the integrating factor is:
$$I(t) = e^{\int(-0.04)dt} = e^{-0.04t}$$
Thus, $A = \dfrac{1}{e^{-0.04t}} \displaystyle\int -4000e^{-0.04t}\, dt$
$$= e^{0.04t}\left[\frac{-4000e^{-0.04t}}{-0.04} + C\right]$$
$A = 100,000 + Ce^{0.04t}$ *General solution*

Applying the initial condition $A(0) = 20,000$ yields:
$A(0) = 100,000 + Ce^0 = 20,000$ and $C = -80,000$
Thus, the amount in the account at any time t is:
$A(t) = 100,000 - 80,000e^{0.04t}$

To determine when the amount in the account is 0, we must solve $A(t) = 0$ for t:
$$100,000 - 80,000e^{0.04t} = 0$$
$$80,000e^{0.04t} = 100,000$$
$$e^{0.04t} = \frac{5}{4}$$
$$t = \frac{\ln\left(\frac{5}{4}\right)}{0.04} \approx 5.579$$

Thus, the account is depleted after 5.579 years. The total amount withdrawn from the account is:
$4000(5.579) = \$22,316.$

45. The amount in the account at any time t must satisfy
$$\frac{dA}{dt} - 0.05A = -1500.$$
Now $f(t) = -0.05$ and the integrating factor is:
$$I(t) = e^{\int(-0.05)dt} = e^{-0.05t}$$
Thus,
$$A = \frac{1}{e^{-0.05t}} \int -1500e^{-0.05t}dt = e^{0.05t}\left[\frac{-1500e^{-0.05t}}{-0.05} + C\right]$$
$$= 30,000 + Ce^{0.05t} \qquad \textit{General solution}$$

Applying the initial condition $A(0) = P$ yields:
$30,000 + C = P$
$\qquad C = P - 30,000$

Thus, the amount in the account at any time t is:
$$A(t) = 30,000 + (P - 30,000)e^{0.05t}$$

Since $A(10) = 0$, we have:
$$0 = 30,000 + (P - 30,000)e^{0.05(10)}$$
$$(P - 30,000)e^{0.5} = -30,000$$

Solving for the initial deposit P yields:
$$P = \frac{-30,000}{e^{0.5}} + 30,000 = 30,000(1 - e^{-0.5}) \approx 11,804$$
Thus, the initial deposit was \$11,804.

47. The amount in the account at any time t must satisfy
$$\frac{dA}{dt} - 0.08A = 2000.$$
Now $f(t) = -0.08$ and the integrating factor is
$$I(t) = e^{\int (-0.08)dt} = e^{-0.08t}$$
Thus,
$$A = \frac{1}{e^{-0.08t}} \int 2000e^{-0.08t}dt = e^{0.08t}\left[\frac{2000e^{-0.08t}}{-0.08} + C\right]$$
$$= -25,000 + Ce^{0.08t}$$

Applying the initial condition $A(0) = 7000$ yields:
$$7000 = -25,000 + C$$
and $\quad C = 32,000$

Thus, the amount in the account at any time t is:
$$A(t) = 32,000e^{0.08t} - 25,000$$

After 5 years, the amount in the account is
$$A(5) = 32,000e^{0.08(5)} - 25,000 = 32,000e^{0.4} - 25,000$$
$$\approx \$22,738.39$$

49. Let r (expressed as a decimal) be the interest rate. Then the amount in the account at any time t must satisfy
$$\frac{dA}{dt} - rA = 1,000$$
Now, $f(t) = -r$ and the integrating factor is:
$$I(t) = e^{\int -r\,dt} = e^{-rt}$$
Thus,
$$A(t) = \frac{1}{e^{-rt}} \int 1,000e^{-rt}\,dt$$
$$= \frac{1,000}{e^{-rt}}\left[-\frac{1}{r}e^{-rt} + C\right]$$
$$= Ce^{rt} - \frac{1,000}{r}$$

Since $A(0) = 10,000$, we have
$$10,000 = Ce^{0} - \frac{1,000}{r}$$
and $\quad C = 10,000 + \frac{1,000}{r}$

Therefore,
$$A(t) = 10{,}000e^{rt} + \frac{1{,}000}{r}(e^{rt} - 1)$$

Now, at $t = 10$, we have
$$35{,}000 = 10{,}000e^{10r} + \frac{1{,}000}{r}(e^{10r} - 1)$$

Solving this equation for r using a graphing utility, we find that
$$r \approx 0.0713$$
Expressed as a percentage, $r \approx 7.13\%$.

51. The equilibrium price at time t is the solution of the equation
$$95 - 5p(t) + 2p'(t) = 35 - 2p(t) + 3p'(t)$$
which satisfies the initial condition $p(0) = 30$.

The equation simplifies to
$$p'(t) + 3p(t) = 60,$$
a first-order linear equation. The integrating factor is:
$$I(t) = e^{\int 3\,dt} = e^{3t}$$
Thus,
$$p(t) = \frac{1}{e^{3t}}\int 60e^{3t}\,dt = e^{-3t}\left[\frac{60e^{3t}}{3} + C\right]$$
$$= 20 + Ce^{-3t} \qquad \textit{General solution}$$

Applying the initial condition yields
$$p(0) = 20 + C = 30$$
$$C = 10$$
Thus, the equilibrium price at time t is:
$$p(t) = 20 + 10e^{-3t}$$

The long-range equilibrium price is:
$$\bar{p} = \lim_{t\to\infty}(20 + 10e^{-3t}) = 20$$

53. Let $p(t)$ be the amount of pollutants in the tank at time t. The initial amount of pollutants in the tank is $p(0) = 2\cdot200 = 400$ pounds.

Pollutants are entering the tank at the constant rate of $3\cdot75 = 225$ pounds per hour.

The amount of water in the tank at time t is $200 + 25t$.

The amount of pollutants in each gallon of water at time t is $\dfrac{p(t)}{200 + 25t}$.

The rate at which pollutants are leaving the tank is
$$\frac{50p(t)}{200 + 25t} = \frac{2p(t)}{8 + t}.$$
Thus, the model for this problem is
$$p'(t) = 225 - \frac{2p(t)}{8 + t};\ p(0) = 400 \text{ or } p'(t) + \frac{2p(t)}{8 + t} = 225;\ p(0) = 400.$$

Now $f(t) = \dfrac{2}{8 + t}$ and the integrating factor is:
$$I(t) = e^{\int(2/(8+t))\,dt} = e^{2\ln(8+t)} = e^{\ln(8+t)^2} = (8 + t)^2$$

Thus,

$$p(t) = \frac{1}{(8 + t)^2} \int 225(8 + t)^2 dt = \frac{1}{(8 + t)^2} \left[225 \frac{(8 + t)^3}{3} + C \right]$$

$$p(t) = 75(8 + t) + \frac{C}{(8 + t)^2}. \quad \textit{General solution}$$

We use the initial condition $p(0) = 400$ to evaluate the constant C.

$$p(0) = 400 = 75(8) + \frac{C}{8^2}$$

$$\frac{C}{64} = -200$$

$$C = -12,800$$

$$p(t) = 75(8 + t) - \frac{12,800}{(8 + t)^2}. \quad \textit{Particular solution}$$

To find the total amount of pollutants in the tank after two hours, we evaluate $p(2)$:

$$p(2) = 75(10) - \frac{12,800}{(10)^2} = 750 - 128 = 622$$

After two hours, the tank contains 250 gallons of water. Thus, the rate at which pollutants are being released is

$$\frac{622}{250} \approx 2.5 \text{ pounds per gallon.}$$

55. Let $p(t)$ be the amount of pollutants in the tank at time t. The initial amount of pollutants in the tank is $p(0) = 2 \cdot 200 = 400$ pounds.

Pollutants are entering the tank at the constant rate of $3(50) = 150$ pounds per hour.

Since water is entering and leaving the tank at the same rate, the amount of water in the tank at all times t is 200 gallons.

The amount of pollutants in each gallon of water at time t is $\frac{p(t)}{200}$.

The rate at which pollutants are leaving the tank is $\frac{50p(t)}{200} = \frac{p(t)}{4}$.

Thus, the model for this problem is:

$$p'(t) = 150 - \frac{p(t)}{4}; \quad p(0) = 400 \quad \text{or} \quad p'(t) + \frac{1}{4}p(t) = 150; \quad p(0) = 400$$

Now $f(t) = \frac{1}{4}$ and the integrating factor is:

$$I(t) = e^{\int (1/4) dt} = e^{t/4}$$

Thus,

$$p(t) = \frac{1}{e^{t/4}} \int e^{t/4}(150) dt$$

$$= 150e^{-t/4} \int e^{t/4} dt = 150e^{-t/4}[4e^{t/4} + C]$$

$$= 600 + 150Ce^{-t/4} \quad (A = 150C)$$

$$p(t) = 600 + Ae^{-t/4}. \quad \textit{General solution}$$

We use the initial condition $p(0) = 400$ to evaluate the constant A.

$$p(0) = 400 = 600 + Ae^0, \quad A = -200$$

Therefore,

$$p(t) = 600 - 200e^{-t/4}. \qquad \textit{Particular solution}$$

To find the amount of pollutants in the tank after two hours, we evaluate $p(2)$:

$$p(2) = 600 - 200e^{-1/2} \approx 479 \text{ pounds}$$

The rate at which pollutants are being released after two hours is

$$\frac{600 - 200e^{-1/2}}{200} = 3 - e^{-1/2} \approx 2.4 \text{ pounds per gallon.}$$

57. From Problem 53, the amount of pollutants in the tank at time t is given by:

$$p(t) = 75(8 + t) - \frac{12,800}{(8 + t)^2}$$

To find the time t when the tank contains 1,000 pounds of pollutants, we solve the equation:

$$75(8 + t) - \frac{12,800}{(8 + t)^2} = 1,000$$

Using a graphing utility, we find that $t \approx 6.2$ hrs.

The graphs of $p(t)$ and $p = 1,000$ are shown at the right.

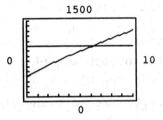

59. The model for this problem is:

$$\frac{dw}{dt} + 0.005w = \frac{2100}{3500} \quad \text{or} \quad \frac{dw}{dt} + 0.005w = \frac{3}{5}$$

Now $f(t) = 0.005$, and the integrating factor is:

$$I(t) = e^{\int 0.005dt} = e^{0.005t}$$

Thus,

$$w(t) = \frac{1}{e^{0.005t}} \int \frac{3}{5} e^{0.005t} \, dt$$

$$= \frac{3}{5} e^{-0.005t} \left[\frac{e^{0.005t}}{0.005} + k \right]$$

$$= 120 + \frac{3}{5} ke^{-0.005t}$$

$$= 120 + Ae^{-0.005t} \qquad \textit{General solution}$$

Applying the initial condition $w(0) = 160$, we have:

$$160 = 120 + Ae^0$$

$$A = 40$$

Thus,
$$w(t) = 120 + 40e^{-0.005t}.$$

How much will a person weigh after 30 days on the diet?
$$w(30) = 120 + 40e^{(-0.005)(30)} \approx 154 \text{ pounds}$$

Now, we want to find t such that $w(t) = 150$.
$$150 = 120 + 40e^{-0.005t}$$
$$40e^{-0.005t} = 30$$
$$e^{-0.005t} = \frac{3}{4}$$
$$-0.005t = \ln\left(\frac{3}{4}\right)$$
$$t = \frac{-\ln\left(\frac{3}{4}\right)}{0.005} \approx 58$$

Thus, it will take 58 days to lose 10 pounds.

Finally, $\lim\limits_{t \to \infty} w(t) = \lim\limits_{t \to \infty} (120 + 40e^{-0.005t}) = 120$, since $\lim\limits_{t \to \infty} e^{-0.005t} = 0$.
Therefore, the person's weight will approach 120 pounds if this diet is maintained for a long period.

61. The model for this problem is
$$\frac{dw}{dt} + 0.005w = \frac{1}{3500}C,$$
where C is to be determined.

Now, $f(t) = 0.005$, and the integrating factor is:
$$I(t) = e^{\int 0.005 dt} = e^{0.005t}$$

Thus,
$$w(t) = \frac{1}{e^{0.005t}} \int \frac{C}{3500} e^{0.005t} \, dt$$
$$= \frac{C}{3500} e^{-0.005t} \left[\frac{e^{0.005t}}{0.005} + k \right]$$
$$= \frac{C}{17.5} + \frac{Ck}{3500} e^{-0.005t}$$
$$= \frac{C}{17.5} + Ae^{-0.005t} \qquad \textit{General solution}$$

Applying the initial condition $w(0) = 130$, we have
$$130 = \frac{C}{17.5} + Ae^0$$
$$A = 130 - \frac{C}{17.5}$$
and
$$w(t) = \frac{C}{17.5} + \left(130 - \frac{C}{17.5} \right) e^{-0.005t}$$

Now, we want to determine C such that $w(30) = 125$.

$$125 = \frac{C}{17.5} + \left(130 - \frac{C}{17.5}\right)e^{-0.005(30)}$$

$$125 = \frac{C}{17.5} + \left(130 - \frac{C}{17.5}\right)e^{-0.15}$$

$$\frac{C}{17.5}(1 - e^{-0.15}) = 125 - 130e^{-0.15}$$

$$C = \frac{17.5(125 - 130e^{-0.15})}{1 - e^{-0.15}} \approx 1{,}647$$

Thus, the person should consume 1,647 calories per day.

63. The model for this problem is
$$\frac{dk}{dt} + \ell k = \lambda \ell,$$
where ℓ and λ are constants.

Now, $f(t) = \ell$, and the integrating factor is:
$$I(t) = e^{\int \ell \, dt} = e^{\ell t}$$
Thus,

$$k(t) = \frac{1}{e^{\ell t}}\int \ell \lambda e^{\ell t}\, dt = e^{-\ell t}\left[\ell \lambda \frac{e^{\ell t}}{\ell} + C\right] = \lambda + \frac{C}{\ell}e^{-\ell t}$$

$$= \lambda + Me^{-\ell t} \qquad \textit{General solution}$$

For Student A, $\ell = 0.8$ and $\lambda = 0.9$. Thus,
$$k(t) = 0.9 + Me^{-0.8t}.$$

Applying the initial condtion $k(0) = 0.1$ yields:
$$0.1 = 0.9 + Me^{0} \quad \text{or} \quad M = -0.8$$
and $k(t) = 0.9 - 0.8e^{-0.8t}$.

When $t = 6$, we have:
$$k(6) = 0.9 - 0.8e^{-0.8(6)} = 0.9 - 0.8e^{-4.8} \approx 0.8934 \text{ or } 89.34\%$$

For Student B, $\ell = 0.8$ and $\lambda = 0.7$. Thus,
$$k(t) = 0.7 + Me^{-0.8t}.$$

Applying the initial condition $k(0) = 0.4$ yields:
$$0.4 = 0.7 + Me^{0} \quad \text{or} \quad M = -0.3$$
and $k(t) = 0.7 - 0.3e^{-0.8t}$

When $t = 6$, we have:
$$k(6) = 0.7 - 0.3e^{-0.8(6)} = 0.7 - 0.3e^{-4.8} \approx 0.6975 \text{ or } 69.75\%$$

1. Substitute $y = C\sqrt{x}$, $y' = \dfrac{C}{2\sqrt{x}}$ into the given differential equation:

$$2x\left(\frac{C}{2\sqrt{x}}\right) = C\sqrt{x}$$

$$C\sqrt{x} = C\sqrt{x}$$

Thus, $y = C\sqrt{x}$ is the general solution. $\hspace{2cm}$ (14-1)

2. Substitute $y = 1 + Ce^{-x/3}$, $y' = -\dfrac{C}{3}e^{-x/3}$ into the given differential equation:

$$3\left(-\frac{C}{3}e^{-x/3}\right) + 1 + Ce^{-x/3} = 1$$

$$-Ce^{-x/3} + 1 + Ce^{-x/3} = 1$$

$$1 = 1$$

Thus, $y = 1 + Ce^{-x/3}$ is the general solution. $\hspace{1cm}$ (14-1)

3. (B) $\hspace{3cm}$ (14-1) $\hspace{2cm}$ 4. (A) $\hspace{3cm}$ (14-1)

5. 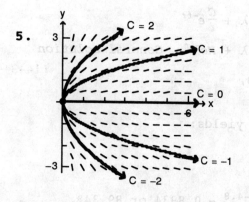 $\hspace{2cm}$ 6.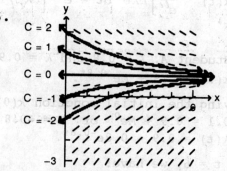

$\hspace{5cm}$ (14-1) $\hspace{4cm}$ (14-1)

7. $\dfrac{dy}{dt} = -k(y - 5)$, $k < 0$ $\hspace{1cm}$ (14-2) $\hspace{1cm}$ 8. $\dfrac{dy}{dt} = ky$, $k > 0$ $\hspace{1cm}$ (14-2)

9. A single person began the spread of a disease; the disease is spreading at a rate that is proportional to the product of the number of people who have the disease and the number who don't have it. (14-2)

10. There are 100 grams of a radioactive material at the instant a nuclear accident occurs, and the material is decaying at a rate proportional to the amount present. (14-2)

11. $y' = -\dfrac{4y}{x}$

$\dfrac{y'}{y} = \dfrac{-4}{x}$ Separate the variables

$\displaystyle\int \dfrac{dy}{y} = \int \dfrac{-4}{x}dx$

$\ln|y| = -4\ln|x| + C = \ln x^{-4} + C$

$\quad y = e^{\ln x^{-4}+C}$

$\quad\quad = e^{C}e^{\ln x^{-4}}$

$\quad y = Ax^{-4} = \dfrac{A}{x^4}$ *General solution*   (14-2)

12. $y' = \dfrac{-4y}{x} + x$

$y' + \dfrac{4}{x}y = x$ First-order linear equation

Integrating factor: $I(x) = e^{\int f(x)\,dx} = e^{\int (4/x)\,dx} = e^{4\ln x} = e^{\ln x^4} = x^4$

Therefore,

$y = \dfrac{1}{I(x)}\displaystyle\int I(x)g(x)\,dx, \quad g(x) = x$

$\quad = \dfrac{1}{x^4}\displaystyle\int x^4 x\,dx = \dfrac{1}{x^4}\int x^5\,dx = \dfrac{1}{x^4}\left[\dfrac{x^6}{6} + C\right]$

$y = \dfrac{x^2}{6} + \dfrac{C}{x^4}$ *General solution*   (14-3)

13. $y' = 3x^2y^2$

$\dfrac{y'}{y^2} = 3x^2$ Separate the variables

$\displaystyle\int \dfrac{dy}{y^2} = \int 3x^2\,dx$

$-\dfrac{1}{y} = x^3 + C$

$\quad y = \dfrac{-1}{x^3 + C}$ *General solution*   (14-2)

14. $y' = 2y - e^x$

$y' - 2y = -e^x$ First-order linear equation

Integrating factor: $I(x) = e^{\int (-2)\,dx} = e^{-2x}$

$y = \dfrac{1}{I(x)}\displaystyle\int I(x)g(x)\,dx, \quad g(x) = -e^x$

$\quad = \dfrac{1}{e^{-2x}}\displaystyle\int e^{-2x}(-e^x)\,dx = e^{2x}\int -e^{-x}\,dx = e^{2x}[e^{-x} + C]$

$y = e^x + Ce^{2x}$ *General solution*   (14-3)

15. $y' = \dfrac{5}{x}y + x^6$

$y' - \dfrac{5}{x}y = x^6$ First-order linear equation

Integrating factor: $I(x) = e^{\int(-5/x)dx} = e^{-5\ln x} = e^{\ln x^{-5}} = x^{-5}$

$y = \dfrac{1}{I(x)}\int I(x)g(x)\,dx, \quad g(x) = x^6$

$= \dfrac{1}{x^{-5}}\int x^{-5}x^6\,dx = x^5\int x\,dx = x^5\left[\dfrac{x^2}{2} + C\right]$

$y = \dfrac{x^7}{2} + Cx^5$ *General solution*   (14-3)

16. $y' = \dfrac{3 + y}{2 + x}$

$\dfrac{y'}{3 + y} = \dfrac{1}{2 + x}$ Separate the variables

$\int\dfrac{dy}{3 + y} = \int\dfrac{dx}{2 + x}$

$\ln|3 + y| = \ln(2 + x) + C$ (<u>Note</u>: $2 + x > 0$.)

$3 + y = e^{\ln(2+x)+C} = e^C(2 + x) = A(2 + x)$

$y = A(2 + x) - 3$ *General solution*   (14-2)

17. $y' = 10 - y; \; y(0) = 0$

$\dfrac{y'}{10 - y} = 1$ Separate the variables

$\int\dfrac{dy}{10 - y} = \int dx$

$-\ln|10 - y| = x + C$

$\ln|10 - y| = -x - C$

$10 - y = e^{-x-C}$

$\qquad = e^{-C}e^{-x} = Ae^{-x}$

$y = 10 - Ae^{-x}$. *General solution*

Applying the initial condition $y(0) = 0$, we have:

$0 = 10 - Ae^0, \; A = 10$

Thus, $y = 10 - 10e^{-x}$.   *Particular solution*   (14-2 or 14-3)

18. $y' + y = x; \; y(0) = 0$

Integrating factor: $I(x) = e^{\int 1\,dx} = e^x$

Thus, $y = \dfrac{1}{I(x)}\int I(x)g(x)\,dx, \quad g(x) = x$

$= \dfrac{1}{e^x}\int e^x x\,dx = e^{-x}\int xe^x\,dx = e^{-x}[xe^x - e^x + C]$

$y = x - 1 + Ce^{-x}$.   *General solution*

Applying the initial condition $y(0) = 0$, we have:

$0 = 0 - 1 + Ce^0, \; C = 1$

Therefore, $y = x - 1 + e^{-x}$. *Particular solution*   (14-3)

19. $y' = 2ye^{-x}$; $y(0) = 1$

$\dfrac{y'}{y} = 2e^{-x}$ Separate the variables

$\displaystyle\int \dfrac{dy}{y} = \int 2e^{-x}dx$

$\ln|y| = -2e^{-x} + C$

$\qquad y = e^{-2e^{-x}+C} = e^0 e^{-2e^{-x}}$

$\qquad y = Ae^{-2e^{-x}}$ *General solution*

Applying the initial condition $y(0) = 1$, we have:

$1 = Ae^{-2e^0} = Ae^{-2}$, $A = e^2$

Thus, $y = e^2 e^{-2e^{-x}}$.   *Particular solution*   (14-2)

20. $y' = \dfrac{2x - y}{x + 4}$; $y(0) = 1$

$y' + \dfrac{1}{x + 4}y = \dfrac{2x}{x + 4}$ First-order linear equation

Integrating factor: $I(x) = e^{\int (1/(x+4))dx} = e^{\ln(x+4)} = x + 4$

Thus, $y = \dfrac{1}{I(x)} \displaystyle\int I(x)g(x)dx$, $g(x) = \dfrac{2x}{x + 4}$

$\qquad = \dfrac{1}{x + 4}\displaystyle\int (x + 4)\dfrac{2x}{x + 4}dx = \dfrac{1}{x + 4}\int 2x\, dx = \dfrac{1}{x + 4}[x^2 + C]$

$\qquad y = \dfrac{x^2}{x + 4} + \dfrac{C}{x + 4}$.   *General solution*

Applying the initial condition $y(0) = 1$, we have:

$1 = 0 + \dfrac{C}{4}$, $C = 4$

Therefore, $y = \dfrac{x^2}{x + 4} + \dfrac{4}{x + 4} = \dfrac{x^2 + 4}{x + 4}$.   *Particular solution*   (14-3)

21. $y' = \dfrac{x}{y + 4}$; $y(0) = 0$

$(y + 4)y' = x$ Separate the variables

$\displaystyle\int (y + 4)dy = \int x\, dx$

$\qquad \dfrac{(y + 4)^2}{2} = \dfrac{x^2}{2} + C$

$\qquad (y + 4)^2 = x^2 + 2C$

or $(y + 4)^2 = x^2 + A$.   *General solution, implicit form*

Solving for y and applying the initial condition $y(0) = 0$, we have:

$y + 4 = \sqrt{x^2 + A}$

$0 + 4 = \sqrt{0 + A}$

$\qquad A = 16$

Therefore, $y = \sqrt{x^2 + 16} - 4$.   *Particular solution*   (14-2)

22. $y' + \dfrac{2}{x}y = \ln x; \; y(1) = 2$

Integrating factor: $I(x) = e^{\int (2/x)\,dx} = e^{2\ln x} = e^{\ln x^2} = x^2$

Thus, $y = \dfrac{1}{I(x)} \displaystyle\int I(x)g(x)\,dx, \quad g(x) = \ln x$

$\qquad = \dfrac{1}{x^2} \displaystyle\int x^2 \ln x \, dx \quad \text{(Integrate by parts)}$

$\qquad = \dfrac{1}{x^2}\left[\dfrac{1}{3}x^3 \ln x - \dfrac{x^3}{9} + C \right]$

$\quad y = \dfrac{1}{3}x \ln x - \dfrac{x}{9} + \dfrac{C}{x^2}. \qquad \text{General solution}$

Applying the initial condition $y(1) = 2$, we have:

$2 = \dfrac{1}{3}(1)\ln 1 - \dfrac{1}{9} + C, \; C = \dfrac{19}{9}$

Therefore, $y = \dfrac{1}{3}x \ln x - \dfrac{x}{9} + \dfrac{19}{9x^2}. \quad \textit{Particular solution} \qquad (14\text{-}3)$

23. $yy' = \dfrac{x(1 + y^2)}{1 + x^2}; \; y(0) = 1$

$\dfrac{yy'}{1 + y^2} = \dfrac{x}{1 + x^2} \qquad \text{Separate the variables}$

$\displaystyle\int \dfrac{y\,dy}{1 + y^2} = \int \dfrac{x\,dx}{1 + x^2}$

$\dfrac{1}{2}\ln(1 + y^2) = \dfrac{1}{2}\ln(1 + x^2) + C$

$\ln(1 + y^2) = \ln(1 + x^2) + 2C$

$1 + y^2 = e^{\ln(1+x^2)+2C}$

$\qquad = e^{2C}e^{\ln(1+x^2)}$

$\qquad = A(1 + x^2)$

$\quad y^2 = A(1 + x^2) - 1. \quad \textit{General solution, implicit form}$

Solving this equation for y and applying the initial condition $y(0) = 1$, we have:

$\quad y = \sqrt{A(1 + x^2) - 1}$

$\quad 1 = \sqrt{A - 1}$

$A - 1 = 1$

$\quad A = 2$

Thus, $y = \sqrt{2(1 + x^2) - 1} = \sqrt{1 + 2x^2}. \qquad \textit{Particular solution} \qquad (14\text{-}2)$

24. $y' + 2xy = 2e^{-x^2}$; $y(0) = 1$

Integrating factor: $I(x) = e^{\int 2x\,dx} = e^{x^2}$

Thus, $y = \dfrac{1}{I(x)} \int I(x) g(x)\,dx$, $g(x) = 2e^{-x^2}$

$$= \frac{1}{e^{x^2}} \int e^{x^2} 2e^{-x^2}\,dx = e^{-x^2} \int 2\,dx = e^{-x^2}[2x + C]$$

$$y = 2xe^{-x^2} + Ce^{-x^2}. \qquad\qquad\qquad\qquad \textit{General solution}$$

Applying the initial condition $y(0) = 1$, we have:

$1 = 2 \cdot 0 \cdot e^0 + Ce^0$, $C = 1$

Therefore, $y = 2xe^{-x^2} + e^{-x^2} = (2x + 1)e^{-x^2}.$ *Particular solution (14-3)*

25. $xy' - 4y = 8$

(A) <u>Solution using an integrating factor</u>:

<u>Step 1</u>: Write the differential equation in standard form:

Multiply both sides of the equation by $\dfrac{1}{x}$

$$y' - \frac{4}{x}y = \frac{8}{x}$$

<u>Step 2</u>: Find the integrating factor:

$$f(x) = -\frac{4}{x}, \quad I(x) = e^{\int f(x)\,dx} = e^{\int (-4/x)\,dx}$$

$$= e^{-4\ln x}$$
$$= e^{\ln(x^{-4})}$$
$$= x^{-4}$$

<u>Step 3</u>: Multiply both sides of the equation by the integrating factor:

$$x^{-4}(y' - 4x^{-1}y) = (8x^{-1})x^{-4}$$
$$x^{-4}y' - 4x^{-5}y = 8x^{-5}$$
$$[x^{-4}y]' = 8x^{-5}$$

<u>Step 4</u>: Integrate both sides:

$$\int [x^{-4}y]'\,dx = \int 8x^{-5}\,dx$$
$$x^{-4}y = -2x^{-4} + C$$

<u>Step 5</u>: Solve for y:

$$y = -2 + Cx^4 \qquad \textit{General solution}$$

(B) <u>Solution using separation of variables</u>

$$xy' - 4y = 8$$
$$xy' = 4y + 8$$
$$\frac{1}{4y + 8}y' = \frac{1}{x}$$
$$\int \frac{1}{4y + 8}\,dy = \int \frac{1}{x}\,dx$$
$$\frac{1}{4}\ln|4y + 8| = \ln|x| + K$$
$$\ln|4y + 8| = 4\ln|x| + L \qquad (L = 4K)$$
$$\ln|4y + 8| = \ln x^4 + L$$
$$|4y + 8| = e^{\ln x^4 + L} = e^L x^4$$
$$4y + 8 = Mx^4 \qquad (M = \pm e^L)$$
$$y = Cx^4 - 2 \qquad (C = M/4) \qquad\qquad (14\text{-}2,\ 14\text{-}3)$$

26. The equation $y' + y = x$ can be solved using an integrating factor; it cannot be solved by separating the variables.

Integrating factor:
$$f(x) = 1,\quad I(x) = e^{\int dx} = e^x$$
Multiply by e^x:
$$e^x y' + e^x y = xe^x$$
$$[e^x y]' = xe^x$$
Integrate:
$$e^x y = xe^x - e^x + C$$
Solve for y:
$$y = Ce^{-x} + x - 1 \qquad\qquad (14\text{-}3)$$

27. The equation $y' = xy^2$ can be solved by separating the variables; it cannot be solved using an integrating factor

$$y' = xy^2$$
$$\frac{1}{y^2}y' = x$$
$$\int \frac{1}{y^2}\,dy = \int x\,dx$$
$$-\frac{1}{y} = \frac{1}{2}x^2 + K$$
$$y = \frac{-1}{\frac{1}{2}x^2 + K} = \frac{-2}{x^2 + 2K}$$
$$y = \frac{-2}{x^2 + C} \qquad (C = 2K) \qquad\qquad (14\text{-}2)$$

28. $xy' - 5y = -10$

Step 1: Write the differential equation in standard form:
$$y' - \frac{5}{x}y = -\frac{10}{x}$$

Step 2: Find an integrating factor:
$$f(x) = -\frac{5}{x}; \quad I(x) = e^{\int (-5/x)\,dx} = e^{-5\ln x}$$
$$= e^{\ln x^{-5}} = x^{-5}$$

Step 3: Multiply by the integrating factor:
$$x^{-5}(y' - 5x^{-1}y) = x^{-5}(-10x^{-1})$$
$$x^{-5}y' - 5x^{-6}y = -10x^{-6}$$
$$(x^{-5}y)' = -10x^{-6}$$

Step 4: Integrate both sides:
$$\int (x^{-5}y)'\,dx = \int -10x^{-6}\,dx$$
$$x^{-5}y = 2x^{-5} + C$$

Step 5: Solve for y:
$$y = 2 + Cx^5 \qquad \textit{General solution}$$

(A) Applying the initial condition $y(0) = 2$, we have:
$$2 = 2 + 0 = 2$$
Thus, $y = 2 + Cx^5$ satisfies $y(0) = 2$ for all values of C.

(B) Applying the initial condition $y(0) = 0$, we have:
$$0 = 2 + 0 \quad \text{or} \quad 2 = 0$$
Thus, there is **no** particular solution that satisfies $y(0) = 0$.

(C) Applying the initial condition $y(1) = 1$, we have:
$$1 = 2 + C(1) \quad \text{and} \quad C = -1$$
Thus, $y = 2 - x^5$ is the particular solution that satisfies
$y(1) = 1$. (14-3)

29.
$$yy' = x$$
$$\int y\,dy = \int x\,dx$$
$$\frac{1}{2}y^2 = \frac{1}{2}x^2 + K$$
or $\quad y^2 = x^2 + C \quad (C = 2K) \qquad \textit{General solution}$

Applying the initial condition $y(0) = 4$, we have:
$$16 = 0^2 + C \quad \text{and} \quad C = 16$$
Thus, $y^2 = x^2 + 16 \quad \text{and} \quad y = \sqrt{x^2 + 16} \qquad \textit{Particular solution}$ (14-2)

30. (A)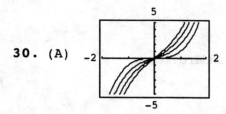

(B) The graphs are increasing and cross the x axis only at $x = 0$.

(C)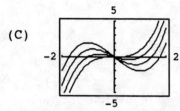

(D) The graphs have a local maximum, a local minimum, and cross the x axis three times. (14-1)

31. $y = M(1 - e^{-kt})$, $M > 0$, $k > 0$

(A) At $t = 2$, $4 = M(1 - e^{-2k})$ so $M = \dfrac{4}{1 - e^{-2k}}$

At $t = 5$, $7 = M(1 - e^{-5k})$ so $M = \dfrac{7}{1 - e^{-5k}}$

(B) Using a graphing utility to graph the two equations in part (A), we find that $M \approx 9.2$. (14-2)

32. Let $V(t)$ denote the value of the refrigerator at time t. Then the model for this problem is: $\dfrac{dV}{dt} = kV$; $V(0) = 500$, $V(20) = 25$

$$\frac{dV}{dt} - kV = 0$$

Integrating factor: $I(t) = e^{\int(-k)dt} = e^{-kt}$

Thus, $V = \dfrac{1}{I(t)} \int I(t)g(t)dt$, $g(t) = 0$

$\quad = \dfrac{1}{e^{-kt}} \int e^{-kt} \cdot 0 \ dt = e^{kt} \int 0 \ dt$

$\quad V = Ce^{kt}$. *General solution*

Applying the conditions $V(0) = 500$ and $V(20) = 25$, we have:

$\quad V(0) = 500 = Ce^0$, $C = 500$

Therefore, $V(t) = 500e^{kt}$.

$V(20) = 25 = 500e^{k(20)}$

$\quad e^{20k} = \dfrac{25}{500} = \dfrac{1}{20}$

$\quad 20k = \ln\left(\dfrac{1}{20}\right)$

$\quad k = \dfrac{1}{20} \ln\left(\dfrac{1}{20}\right)$

Therefore, $V(t) = 500e^{(1/20)\ln(1/20)t}$.

Finally, we want to calculate V when $t = 5$:

$V(5) = 500e^{(1/20)\ln(1/20)5} = 500e^{(1/4)\ln(1/20)} = 500e^{-0.25\ln 20} \approx \$236.44$

(14-2 or 14-3)

33. The model for this problem is:

$$\frac{ds}{dt} = k(200{,}000 - s); \quad s(0) = 0, \quad s(1) = 50{,}000, \quad k > 0$$

(A) $\dfrac{ds}{200{,}000 - s} = k\,dt \qquad$ Separate the variables

$$\int \frac{ds}{200{,}000 - s} = \int k\,dt$$

$$-\ln(200{,}000 - s) = kt + C \qquad (0 < s < 200{,}000)$$

$$200{,}000 - s = e^{-kt-C}$$

$$= e^{-C}e^{-kt}$$

$$= Ae^{-kt}$$

$$s = 200{,}000 - Ae^{-kt} \qquad \textit{General solution}$$

We use the conditions $s(0) = 0$ and $s(1) = 50{,}000$ to evaluate the constants A and k.

$$s(0) = 0 = 200{,}000 - Ae^{0}$$

$$A = 200{,}000$$

Thus, $s = 200{,}000 - 200{,}000e^{-kt}$.

$$s(1) = 50{,}000 = 200{,}000 - 200{,}000e^{-k}$$

$$200{,}000e^{-k} = 150{,}000$$

$$e^{-k} = \frac{150{,}000}{200{,}000} = \frac{3}{4}$$

$$-k = \ln\!\left(\frac{3}{4}\right)$$

Therefore, $s = 200{,}000 - 200{,}000e^{\ln(3/4)t}$.

Finally, we determine t such that $s(t) = 150{,}000$.

$$150{,}000 = 200{,}000 - 200{,}000e^{\ln(3/4)t}$$

$$200{,}000e^{\ln(3/4)t} = 50{,}000$$

$$e^{\ln(3/4)t} = \frac{50{,}000}{200{,}000} = \frac{1}{4}$$

$$\ln\!\left(\frac{3}{4}\right)t = \ln\!\left(\frac{1}{4}\right)$$

$$t = \frac{\ln(1/4)}{\ln(3/4)} \approx 5 \text{ years}$$

(B) From part (A), we know that

$$S(t) = 200{,}000 - 200{,}000e^{-kt}$$

Since we want the sales to be \$150,000 after 3 years, we have

$$150{,}000 = 200{,}000 - 200{,}000e^{-3k}$$

$$200{,}000e^{-3k} = 50{,}000$$

$$e^{-3k} = 0.25$$

$$-3k = \ln 0.25$$

$$k = -\frac{\ln(0.25)}{3}$$

Thus, $s(t) = 200{,}000 - 200{,}000e^{(t/3)\ln 0.25}$

Now, $s(1) = 200{,}000 - 200{,}000e^{(1/3)\ln 0.25} \approx 74{,}000$

Therefore, the sales in the first year should be \$74,000 to ensure that the sales after 3 years will be \$150,000. $\qquad$ (14-2)

34. (A) The equilibrium price $p(t)$ at time t satisfies
$S = D$; $p(0) = 75$. Thus, $100 + p + p' = 200 - p' - p$
$$2p' + 2p = 100$$
$$p' + p = 50$$

Integrating factor: $I(t) = e^{\int f(t)dt} = e^{\int 1 dt} = e^t$

Thus, $p = \dfrac{1}{I(t)} \displaystyle\int I(t)g(t)dt$, $\quad g(t) = 50$

$$= \frac{1}{e^t} \int e^t \cdot 50 \; dt = 50e^{-t}[e^t + C]$$
$$= 50 + 50Ce^{-t} \qquad (A = 50C)$$
$$p = 50 + Ae^{-t}. \qquad \textit{General solution}$$

(B) The equilibrium price $\bar{p}$ is given by
$$\bar{p} = \lim_{t \to \infty} (50 + 25e^{-t}) = 50 + 25 \lim_{t \to \infty} e^{-t} = 50.$$

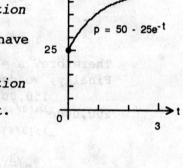

(C) Applying the initial condition $p(0) = 75$, we have:
$75 = 50 + Ae^0$ and $A = 25$
Therefore, $p_1 = 50 + 25e^{-t}$ *Particular solution*

Applying the initial condition $p(0) = 25$, we have
$25 = 50 + Ae^0$ and $A = -25$

Therefore, $p_2 = 50 - 25e^{-t}$ *Particular solution*

The graphs of p_1 and p_2 are shown at the right.

(D) For an initial price above the equilibrium
price $\bar{p} = 50$, the price of the commodity
decreases toward $\bar{p}$. For an initial price below the equilibrium
price $\bar{p}$, the price of the commodity increases toward $\bar{p}$. (14-1, 14-2)

35. The amount in the account at any time t must satisfy
$\dfrac{dA}{dt} - 0.05A = -5000.$
Now, $f(t) = -0.05$, and the integrating factor is:
$I(t) = e^{\int (-0.05)dt} = e^{-0.05t}$

Thus, $A = \dfrac{1}{e^{-0.05t}} \displaystyle\int -5000e^{-0.05t}dt = e^{0.05t}\left[-5000 \dfrac{e^{-0.05t}}{-0.05} + C\right]$

$A = 100,000 + Ce^{0.05t}$ *General solution*

Applying the initial condition $A(0) = 60,000$ yields:
$A(0) = 100,000 + Ce^0 = 60,000$
$$C = -40,000$$

Thus, the amount in the account at any time t is:
$A(t) = 100,000 - 40,000e^{0.05t}$
To determine when the amount in the account is 0, we must solve $A(t) = 0$ for t:
$$100,000 - 40,000e^{0.05t} = 0$$
$$e^{0.05t} = \frac{100,000}{40,000} = \frac{5}{2}$$
$$t = \frac{\ln\left(\frac{5}{2}\right)}{0.05} \approx 18.326$$
Thus, the account will be depleted after 18.326 years. The total amount withdrawn from the account is:
$5000(18.326) = \$91,630$ (14-3)

36. Let r be the continuous compound rate of interest (expressed in decimal form). Then the amount in the account at any time t is given by:
$$\frac{dA}{dt} = rA + 2,000$$
or $\quad \frac{dA}{dt} - rA = 2,000$

(Note: the method of separation of variables could also have been used to solve the equation.)

Now, $f(t) = -r$ and the integrating factor is:
$$I(t) = e^{\int -r\,dt} = e^{-rt}$$
Thus,
$$A = \frac{1}{e^{-rt}} \int 2,000e^{-rt}\,dt$$
$$= e^{rt}\left[-\frac{2,000}{r}e^{-rt} + C\right]$$
and $\quad A = Ce^{rt} - \frac{2,000}{r} \qquad$ *General solution*

Applying the initial condition $A(0) = 15,000$ yields:
$$A(0) = 15,000 = Ce^0 - \frac{2,000}{r}$$
$$C = 15,000 + \frac{2,000}{r}$$
Thus, the amount in the account at any time t is:
$$A(t) = \left(15,000 + \frac{2,000}{r}\right)e^{rt} - \frac{2,000}{r}$$

Now, at $t = 10$, $A(10) = 70,000$. Therefore, we have
$$70,000 = \left(15,000 + \frac{2,000}{r}\right)e^{10r} - \frac{2,000}{r}$$

Using a graphing utility to solve this equation for r, we find that $r = 0.0912$. Expressed as a percentage, $r = 9.12\%$. (14-2 or 14-3)

37. $\dfrac{dy}{dt} = 100 + e^{-t} - y; \ y(0) = 0$

$\dfrac{dy}{dt} + y = 100 + e^{-t}$

Integrating factor: $I(t) = e^{\int 1\,dt} = e^{t}$

Thus, $y = \dfrac{1}{I(t)} \displaystyle\int I(t)g(t)\,dt, \quad g(t) = 100 + e^{-t}$

$\qquad = \dfrac{1}{e^{t}} \displaystyle\int e^{t}(100 + e^{-t})\,dt = e^{-t}\int (100e^{t} + 1)\,dt = e^{-t}[100e^{t} + t + C]$

$\qquad y = 100 + te^{-t} + Ce^{-t}. \qquad$ *General solution*

Applying the initial condition, we have:
$0 = 100 + 0e^{0} + Ce^{0}, \ C = -100$
Therefore, $y = 100 + te^{-t} - 100e^{-t}. \quad$ *Particular solution* $\hfill (14\text{-}3)$

38. Let $p(t)$ be the amount of pollutants in the tank at time t. At $t = 0$, we have $p(0) = 0$.

Pollutants are entering the tank at the constant rate $2(75) = 150$ pounds per hour. The amount of water in the tank at time t is:
$\qquad 100 + 75t - 50t = 100 + 25t$

The amount of pollutants in each gallon of water at time t is:
$\qquad \dfrac{p(t)}{100 + 25t}$

The rate at which pollutants are leaving the tank at time t is:
$\qquad 50\left(\dfrac{p(t)}{100 + 25t}\right) = \dfrac{2p(t)}{4 + t}$

The mathematical model for this problem is:
$\qquad \dfrac{dp}{dt} = 150 - \dfrac{2p}{4 + t}; \ p(0) = 0$

(A) In standard form, the differential equation is:
$\qquad \dfrac{dp}{dt} + \dfrac{2}{4 + t}p = 150$

Now, $f(t) = \dfrac{2}{4 + t}$ and $I(t) = e^{\int [2/(4+t)]\,dt} = e^{2\ln(4+t)}$

$\qquad\qquad\qquad\qquad\qquad\qquad\qquad = e^{\ln(4+t)^{2}}$

$\qquad\qquad\qquad\qquad\qquad\qquad\qquad = (4 + t)^{2}$

Thus, $p(t) = \dfrac{1}{I(t)} \displaystyle\int I(t)g(t)\,dt, \ g(t) = 150$

$\qquad = \dfrac{1}{(4 + t)^{2}} \displaystyle\int (4 + t)^{2}\,150\ dt$

$\qquad = \dfrac{150}{(4 + t)^{2}}\left[\dfrac{(4 + t)^{3}}{3} + K\right]$

$\qquad = 50(4 + t) + \dfrac{C}{(4 + t)^{2}} \quad (C = 150K) \quad$ *General solution*

Applying the initial condition $p(0) = 0$, we have

$$0 = 50(4) + \frac{C}{(4)^2} \quad \text{and} \quad C = -3{,}200$$

Therefore, $p(t) = 50(4 + t) - \dfrac{3{,}200}{(4 + t)^2}$ *Particular solution*

Now, at $t = 2$,

$$p(2) = 50(6) - \frac{3{,}200}{6^2} \approx 211.1$$

There are approximately 211.1 pounds of pollutants in the tank after 2 hours.

(B) To find how long it will take for the tank to contain 700 pounds of pollutants, we solve the equation

$$50(4 + t) - \frac{3{,}200}{(4 + t)^2} = 700$$

for t using a graphing utility. The result is $t \approx 10.3$ hours. (14-3)

39. Let $p(t)$ be the bird population at time t. Then $\dfrac{dp}{dt} = k(p - 200)$, $k < 0$ constant.

(Note: this equation can be solved either by separating the variables or as a first order linear equation; we'll illustrate the latter.)

(A) $\dfrac{dp}{dt} = kp - 200k, \quad p(0) = 500$

In standard form, the differential equation is:

$$\frac{dp}{dt} - kp = -200k, \quad k < 0 \text{ constant}$$

Now, $f(t) = -k$ and $I(t) = e^{\int -k\,dt} = e^{-kt}$

Thus, $p(t) = \dfrac{1}{e^{-kt}} \displaystyle\int e^{-kt}(-200k)\,dt$

$$= e^{kt}[200e^{-kt} + C]$$

and $p(t) = 200 + Ce^{kt}$ *General solution*

Applying the initial condition $p(0) = 500$, yields

$$500 = 200 + C \quad \text{and} \quad C = 300$$

Therefore, $p(t) = 200 + 300e^{kt}$

Now, at $t = -5$, $p(-5) = 1{,}000$ and

$$1000 = 200 + 300e^{-5k}$$

$$e^{-5k} = \frac{8}{3}$$

$$-5k = \ln\!\left(\frac{8}{3}\right)$$

$$k = \frac{\ln(8/3)}{-5}$$

Thus, the bird population at any time t is:
$$p(t) = 200 + 300e^{-(t/5)\ln(8/3)}$$

The bird population 4 years from now will be:
$$p(4) = 200 + 300e^{-(4/5)\ln(8/3)} \approx 337 \text{ birds}$$

(B) From part (A), the general solution of
$$\frac{dp}{dt} = k(p - M), \quad k < 0 \text{ constant}$$
is $p(t) = M + Ce^{kt}$

Applying the initial condition $p(0) = 500$, we have
$$500 = M + C \quad \text{and} \quad C = 500 - M$$

Therefore,
$$p(t) = M + (500 - M)e^{kt}$$

Now, at $t = -5$, $p(-5) = 1{,}000$ and
$$1{,}000 = M + (500 - M)e^{-5k}$$
$$e^{-5k} = \frac{1{,}000 - M}{500 - M}$$
$$-5k = \ln\left(\frac{1{,}000 - M}{500 - M}\right)$$
$$k = -\frac{1}{5}\ln\left(\frac{1{,}000 - M}{500 - M}\right)$$

Thus, the bird population at any time t is
$$p(t) = M + (500 - M)e^{-(t/5)\ln[(1000-M)/(500-M)]}$$
$$= M + (500 - M)\left(\frac{1{,}000 - M}{500 - M}\right)^{-t/5}$$

Now, at $t = 4$, $p(4) = 400$. Using a graphing calculator to solve
$$400 = M + (500 - M)\left(\frac{1{,}000 - M}{500 - M}\right)^{-4/5}$$

we get $M \approx 357$ birds

(14-2 or 14-3)

40. Let $p(t)$ denote the number of people who have heard the rumor at time t. Then the model for this problem is:
$$\frac{dp}{dt} = k(200 - p); \quad p(0) = 1, \ p(2) = 10$$

$$\frac{dp}{200 - p} = k\, dt \qquad \text{Separating the variables}$$

$$\int \frac{dp}{200 - p} = \int k\, dt$$

$$-\ln(200 - p) = kt + C$$
$$\ln(200 - p) = -kt - C$$
$$200 - p = e^{-kt-C}$$
$$= e^{-C}e^{-kt}$$
$$= Ae^{-kt}$$
$$p = 200 - Ae^{-kt} \qquad \textit{General solution}$$

Apply the conditions $p(0) = 1$ and $p(2) = 10$ to evaluate the constants A and k:

$p(0) = 1 = 200 - Ae^0$, $A = 199$

Thus, $p = 200 - 199e^{-kt}$.

$p(2) = 10 = 200 - 199e^{-2k}$

$$199e^{-2k} = 190$$

$$e^{-2k} = \frac{190}{199}$$

$$-2k = \ln\left(\frac{190}{199}\right)$$

$$k = \frac{-1}{2}\ln\left(\frac{190}{199}\right)$$

Therefore, $p = 200 - 199e^{(1/2)\ln(190/199)t}$. *Particular solution*

(A) When $t = 5$,

$$p(5) = 200 - 199e^{(1/2)\ln(190/199)5}$$
$$= 200 - 199e^{(5/2)\ln(190/199)} \approx 23 \text{ people}$$

(B) Find t such that $p(t) = 100$.

$$100 = 200 - 199e^{(1/2)\ln(190/199)t}$$
$$199e^{(1/2)\ln(190/199)t} = 100$$
$$e^{(1/2)\ln(190/199)t} = \frac{100}{199}$$

$$\frac{1}{2}\ln\left(\frac{190}{199}\right)t = \ln\left(\frac{100}{199}\right)$$

$$t = \frac{2\ln\left(\frac{100}{199}\right)}{\ln\left(\frac{190}{199}\right)} \approx 30 \text{ days} \qquad (14\text{-}2)$$

15 PROBABILITY AND CALCULUS

EXERCISE 15-1
_____

Things to remember:

1. IMPROPER INTEGRALS

   Assume that f is continuous over the indicated interval.

   (a) $\int_a^\infty f(x)\,dx = \lim_{b\to\infty} \int_a^b f(x)\,dx$

   (b) $\int_{-\infty}^b f(x)\,dx = \lim_{a\to-\infty} \int_a^b f(x)\,dx$

   (c) $\int_{-\infty}^\infty f(x)\,dx = \int_{-\infty}^c f(x)\,dx + \int_c^\infty f(x)\,dx$

   where c is any point in $(-\infty, \infty)$

   If the indicated limit in (a) or (b) exists, or if both limits in (c) exist, then the corresponding improper integral is said to CONVERGE; otherwise, the improper integral is said to DIVERGE (and no value is assigned to it).

2. CAPITAL VALUE OF A PERPETUAL INCOME STREAM

   A continuous income stream is called PERPETUAL if it never stops producing income. The CAPITAL VALUE, CV, of a perpetual income stream $f(t)$ at a rate r compounded continuously is the present value over the time interval $[0, \infty)$. That is

   $$CV = \int_0^\infty f(t)e^{-rt}dt$$

---

1. $\int_1^\infty \dfrac{dx}{x^4} = \lim_{b\to\infty} \int_1^b \dfrac{dx}{x^4}$ (using 1(a))

$= \lim_{b\to\infty}\left(-\dfrac{1}{3x^3}\right)\Big|_1^b = -\lim_{b\to\infty}\dfrac{1}{3x^3}\Big|_1^b$

$= -\lim_{b\to\infty}\left(\dfrac{1}{3b^3} - \dfrac{1}{3}\right) = -\left(0 - \dfrac{1}{3}\right) = \dfrac{1}{3}$

Thus, the given improper integral *converges*.

3. $\int_0^\infty e^{-x/2}dx = \lim_{b \to \infty} \int_0^b e^{-x/2}dx = \lim_{b \to \infty} \frac{e^{-x/2}}{-\frac{1}{2}}\Big|_0^b = \lim_{b \to \infty} -2e^{-x/2}\Big|_0^b$

$\qquad = -2 \lim_{b \to \infty} e^{-x/2}\Big|_0^b = -2 \lim_{b \to \infty} (e^{-b/2} - 1)$

$\qquad = -2 \lim_{b \to \infty} \left(\frac{1}{e^{b/2}} - 1\right) = -2(0 - 1) = 2$

Thus, the given improper integral *converges*.

5. $\int_1^\infty \frac{dx}{\sqrt{x}} = \lim_{b \to \infty} \int_1^b \frac{dx}{\sqrt{x}}$ (using $\underline{1}$(a))

$\qquad = \lim_{b \to \infty} (2x^{1/2})\Big|_1^b = 2 \lim_{b \to \infty} (x^{1/2})\Big|_1^b = 2 \lim_{b \to \infty} (b^{1/2} - 1)$

Since $b^{1/2} \to \infty$ as $b \to \infty$, the limit does not exist; hence, the improper integral *diverges*.

7. $\int_0^\infty \frac{dx}{(x + 1)^2} = \lim_{b \to \infty} \int_0^b \frac{dx}{(x + 1)^2} = \lim_{b \to \infty} -\left(\frac{1}{(x + 1)}\right)\Big|_0^b$ [Note: If $u = x + 1$, then $du = dx$.]

$\qquad = -\lim_{b \to \infty} \left(\frac{1}{b + 1} - 1\right) = -(0 - 1) = 1$

Thus, the improper integral *converges*.

9. $\int_0^\infty \frac{dx}{(x + 1)^{2/3}} = \lim_{b \to \infty} \int_0^b \frac{dx}{(x + 1)^{2/3}}$

$\qquad = \lim_{b \to \infty} (3(x + 1)^{1/3})\Big|_0^b$ [Note: If $u = x + 1$, then $du = dx$.]

$\qquad = 3 \lim_{b \to \infty} (x + 1)^{1/3}\Big|_0^b = 3 \lim_{b \to \infty} ((b + 1)^{1/3} - 1)$

Since $(b + 1)^{1/3} \to \infty$ as $b \to \infty$, the limit does not exist; hence, the improper integral *diverges*.

11. $\int_1^\infty \frac{dx}{x^{0.99}} = \lim_{b \to \infty} \int_1^b \frac{dx}{x^{0.99}} = \lim_{b \to \infty} \frac{x^{0.01}}{0.01}\Big|_1^b$

$\qquad = 100 \lim_{b \to \infty} (x^{0.01})\Big|_1^b = 100 \lim_{b \to \infty} (b^{0.01} - 1)$

Since $b^{0.01} \to \infty$ as $b \to \infty$, the limit does not exist; hence, the improper integral *diverges*.

13. $0.3 \int_0^\infty e^{-0.3x}dx = 0.3 \lim_{b \to \infty} \int_0^b e^{-0.3x}dx = 0.3 \lim_{b \to \infty} \frac{e^{-0.3x}}{-0.3}\Big|_0^b$

$\qquad = -\lim_{b \to \infty} (e^{-0.3x})\Big|_0^b = -\lim_{b \to \infty} (e^{-0.3b} - 1)$

$\qquad = -\lim_{b \to \infty} \left(\frac{1}{e^{0.3b}} - 1\right)$

$\qquad = -(0 - 1)$ $\left[\underline{\text{Note}}: \frac{1}{e^{0.3b}} \to 0 \text{ as } b \to \infty.\right]$

$\qquad = 1$

Thus, the improper integral *converges*.

15. The graph of f is shown at the right. Since $f(x) = 0$ for $x < 0$ and $x > 2$,

$$\int_{-\infty}^{\infty} f(x)\,dx = \int_0^2 f(x)\,dx \qquad (f(x) = 0 \text{ for } x < 0 \text{ and } x > 2)$$

$$= \int_0^2 (1 + x^2)\,dx$$

$$= \left[\left(x + \frac{1}{3}x^3 \right) \Big|_0^2 \right]$$

$$= \frac{14}{3}$$

The integral *converges* and has value $\frac{14}{3}$.

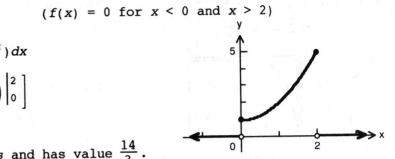

17. The graph of f is shown at the right. Since $f(x) = 0$ for $x < 0$,

$$\int_{-\infty}^{\infty} f(x)\,dx = \int_0^{\infty} f(x) \qquad (f(x) = 0 \text{ for } x < 0)$$

$$= \lim_{b \to \infty} \int_0^b e^{-0.1x}\,dx$$

$$= \lim_{b \to \infty} \left[-10e^{-0.1x} \Big|_0^b \right]$$

$$= \lim_{b \to \infty} [-10e^{-0.1b} + 10] = 10$$

The integral *converges* and has value 10.

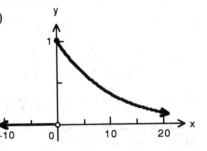

19. The graph of f is shown at the right. Since $f(x) = 0$ for $x < 2$,

$$\int_{-\infty}^{\infty} f(x)\,dx = \int_2^{\infty} f(x)\,dx \qquad (f(x) = 0 \text{ for } x < 2)$$

$$= \lim_{b \to \infty} \int_2^b \frac{4}{(x + 2)}\,dx$$

$$= \lim_{b \to \infty} \left[4 \ln(x + 2) \Big|_2^b \right]$$

$$= \lim_{b \to \infty} [4 \ln(b + 2) - 4 \ln 4]$$

Since $4 \ln(b + 2) \to \infty$ as $b \to \infty$, the integral *diverges*.

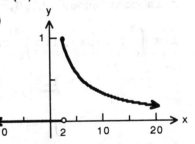

21. $F(b) = \displaystyle\int_1^b \frac{dx}{x^4} = \int_1^b x^{-4}\,dx$

$$= \frac{x^{-3}}{-3} \Big|_1^b = \frac{1}{3} - \frac{1}{3b^3}$$

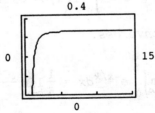

$$\lim_{b \to \infty} F(b) = \lim_{b \to \infty} \left(\frac{1}{3} - \frac{1}{3b^3} \right) = \frac{1}{3}$$

23. $F(b) = \int_0^b e^{-x/2}dx = -2e^{-x/2}\Big|_0^b$

$= 2 - 2e^{-b/2}$

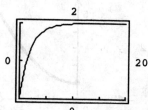

$\lim_{b \to \infty} F(b) = \lim_{b \to \infty}(2 - 2e^{-b/2}) = 2$

25. $F(b) = \int_1^b \frac{dx}{\sqrt{x}} = \int_1^b x^{-1/2}dx$

$= 2x^{1/2}\Big|_1^b = 2\sqrt{b} - 2$

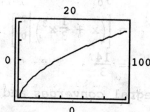

$\lim_{b \to \infty} F(b) = \lim_{b \to \infty}(2\sqrt{b} - 2) = \infty$

27. $F(b) = \int_0^b \frac{dx}{(x + 1)^2} = \int_1^{b+1} u^{-2}du$

$= -\frac{1}{u}\Big|_1^{b+1} = 1 - \frac{1}{b + 1}$

(Let $u = x + 1$. Then $du = dx$, $u = 1$ at $x = 0$, $u = b + 1$ at $x = b$)

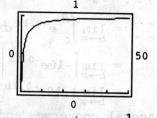

$\lim_{b \to \infty} F(b) = \lim_{b \to \infty}\left(1 - \frac{1}{b + 1}\right) = 1$

29. f is continuous on $[0, \infty)$ and $\int_0^\infty f(x)dx = L$ converges. Since

$L = \int_0^\infty f(x)dx = \lim_{b \to \infty} \int_0^b f(x)dx$

$= \lim_{b \to \infty}\left[\int_0^1 f(x)dx + \int_1^b f(x)dx\right]$

$= \int_0^1 f(x)dx + \lim_{b \to \infty}\int_1^b f(x)dx,$

it follows that

$\lim_{b \to \infty}\int_1^b f(x)dx = L - \int_0^1 f(x)dx$ exists.

Thus, $\int_1^\infty f(x)dx$ converges.

31. $\int_0^\infty \frac{1}{k}e^{-x/k}dx = \frac{1}{k}\lim_{b \to \infty}\int_0^b e^{-x/k}dx = \frac{1}{k}\lim_{b \to \infty}\frac{e^{-x/k}}{-\frac{1}{k}}\Big|_0^b = -\lim_{b \to \infty}e^{-x/k}\Big|_0^b$

$= -\lim_{b \to \infty}(e^{-b/k} - 1) = -\lim_{b \to \infty}\left(\frac{1}{e^{b/k}} - 1\right) = -(0 - 1) = 1$

Thus, the improper integral *converges*.

33. $\displaystyle\int_{-\infty}^{\infty} \frac{x}{1 + x^2}\,dx = \int_{-\infty}^{c} \frac{x}{1 + x^2}\,dx + \int_{c}^{\infty} \frac{x}{1 + x^2}\,dx$

where c is a real number, using 1(c). Let $c = 0$.

Consider $\displaystyle\int_{0}^{\infty} \frac{x}{1 + x^2}\,dx = \lim_{b\to\infty}\int_{0}^{b} \frac{x}{1 + x^2}\,dx$

$\displaystyle\int \frac{x}{1 + x^2}\,dx = \frac{1}{2}\int \frac{du}{u} = \frac{1}{2}\ln u + C = \frac{1}{2}\ln(1 + x^2) + C$

[Let $u = 1 + x^2$. Then $du = 2x\,dx$.]

Thus, $\displaystyle\lim_{b\to\infty}\int_{0}^{b} \frac{x}{1 + x^2}\,dx = \lim_{b\to\infty} \frac{1}{2}\ln(1 + x^2)\,\Big|_{0}^{b}$

$\displaystyle = \lim_{b\to\infty} \frac{1}{2}\ln(1 + b^2) - \frac{1}{2}\ln 1$

$\displaystyle = \frac{1}{2}\lim_{b\to\infty}\ln(1 + b^2)$

Since $\ln(1 + b^2) \to \infty$ as $b \to \infty$, this improper integral *diverges*.

Therefore, $\displaystyle\int_{-\infty}^{\infty} \frac{x}{1 + x^2}\,dx$ *diverges*.

35. $\displaystyle\int_{0}^{\infty} (e^{-x} - e^{-2x})\,dx = \lim_{b\to\infty}\int_{0}^{b} (e^{-x} - e^{-2x})\,dx = \lim_{b\to\infty}\left(-e^{-x} - \frac{e^{-2x}}{-2}\right)\Big|_{0}^{b}$

$\displaystyle = \lim_{b\to\infty}\left[-e^{-b} + \frac{e^{-2b}}{2} - \left(-1 + \frac{1}{2}\right)\right]$

$\displaystyle = \frac{1}{2} + \lim_{b\to\infty}\left(-e^{-b} + \frac{e^{-2b}}{2}\right) = \frac{1}{2} + 0 = \frac{1}{2}$

Thus, the improper integral *converges*.

37. $\displaystyle\int_{-\infty}^{0} \frac{dx}{\sqrt{1 - x}} = \lim_{a\to-\infty}\int_{a}^{0} \frac{dx}{\sqrt{1 - x}}$ (using 1(b))

$\displaystyle\int \frac{dx}{\sqrt{1 - x}} = \int -\frac{du}{u^{1/2}} = -\int u^{-1/2}\,du$

Substitution: $u = 1 - x$
$du = -dx$
$-du = dx$

$\displaystyle = -2u^{1/2} + C = -2\sqrt{1 - x} + C$

Therefore, $\displaystyle\lim_{a\to-\infty}\int_{a}^{0} \frac{dx}{\sqrt{1 - x}} = \lim_{a\to-\infty}\left(-2\sqrt{1 - x}\right)\Big|_{a}^{0} = \lim_{a\to-\infty}\left(-2 + 2\sqrt{1 - a}\right)$

Since $2\sqrt{1 - a} \to \infty$ as $a \to -\infty$, the limit does not exist; hence, the improper integral *diverges*.

39. $\displaystyle\int_{1}^{\infty} \frac{\ln x}{x}\,dx = \lim_{b\to\infty}\int_{1}^{b} \frac{\ln x}{x}\,dx = \lim_{b\to\infty}\int_{0}^{\ln b} u\,du$

Substitution: $u = \ln x$
$du = \frac{1}{x}\,dx$
Limits: $x = 1$ implies $u = 0$
$x = b$ implies $u = \ln b$

$\displaystyle = \lim_{b\to\infty}\frac{u^2}{2}\Big|_{0}^{\ln b} = \lim_{b\to\infty}\frac{(\ln b)^2}{2}$

Since $\ln b \to \infty$ as $b \to \infty$, the limit does not exist; hence, the improper integral *diverges*.

41. $CV = \int_0^\infty 6000e^{-0.12t} dt = \lim_{T\to\infty} \int_0^T 6000e^{-0.12t} dt$

$= \lim_{T\to\infty} \frac{6000}{-0.12} e^{-0.12t} \Big|_0^T = -50,000 \lim_{T\to\infty} (e^{-0.12T} - 1) = \$50,000$

43. $f(t) = 1500e^{0.04t}$

$CV = \int_0^\infty 1500e^{0.04t} e^{-0.09t} dt = \lim_{T\to\infty} \int_0^T 1500e^{-0.05t} dt$

$= \lim_{T\to\infty} \frac{1500}{-0.05} e^{-0.05t} \Big|_0^T = -30,000 \lim_{T\to\infty} (e^{-0.05T} - 1) = \$30,000$

45. At an interest rate of 15%,

$CV = \int_0^\infty 6000e^{-0.15t} dt = \lim_{T\to\infty} \int_0^T 6000e^{-0.15t} dt$

$= \lim_{T\to\infty} \left[-40,000e^{-0.15t} \Big|_0^T \right]$

$= \lim_{T\to\infty} [-40,000e^{-0.15T} + 40,000]$

$= \$40,000$

Increasing the interest rate to 15% decreases the capital value to $40,000.

At an interest rate of 10%,

$CV = \int_0^\infty 6000e^{-0.10t} dt = \lim_{T\to\infty} \int_0^T 6000e^{-0.1t} dt$

$= \lim_{T\to\infty} \left[-60,000e^{-0.1t} \Big|_0^T \right]$

$= \lim_{T\to\infty} [-60,000e^{-0.1T} + 60,000]$

$= \$60,000$

Decreasing the interest rate to 10% increases the capital value to $60,000.

In general, increasing the interest rate decreases the amount of capital required to establish the income stream.

47. $R(t) = 3e^{-0.2t} - 3e^{-0.4t}$

(A) Total production $= \int_0^\infty R(t) dt = \lim_{T\to\infty} \int_0^T R(t) dt$

$= \lim_{T\to\infty} \int_0^T (3e^{-0.2t} - 3e^{-0.4t}) dt$

$= \lim_{T\to\infty} \left[(-15e^{-0.2t} + 7.5e^{-0.4t}) \Big|_0^T \right]$

$= \lim_{T\to\infty} [-15e^{-0.2T} + 7.5e^{-0.4T} + 7.5]$

$= 7.5 \text{ billion ft}^3$

(B) 50% of 7.5 billion ft^3 is 3.75 billion ft^3. Thus, we need to solve

$$\int_0^T R(t)\,dt = \int_0^T (3e^{-0.2t} - 3e^{-0.4t})\,dt = 3.75$$

for T.

Now,

$$\int_0^T (3e^{-0.2t} - 3e^{-0.4t})\,dt = \left[(-15e^{-0.2t} + 7.5e^{-0.4t})\Big|_0^T\right]$$

$$= -15e^{-0.2T} + 7.5e^{-0.4T} + 7.5 = 3.75$$

and

$$-15e^{-0.2T} + 7.5e^{-0.4T} = -3.75$$

Using a graphing utility, we find that $T \approx 6.14$ years.

49. Total seepage $= \displaystyle\int_0^\infty \frac{500}{(1+t)^2}\,dt = \lim_{b\to\infty} 500 \int_0^b \frac{1}{(1+t)^2}\,dt$

$$= 500 \lim_{b\to\infty} \int_0^b \frac{1}{(1+t)^2}\,dt \qquad \begin{array}{l}\text{Substitution: } u = 1+t \\ \qquad\qquad\quad du = dt \end{array}$$

$$= 500 \lim_{b\to\infty} \int_1^{1+b} u^{-2}\,du \qquad \begin{array}{l}\text{Limits: } t = 0 \text{ implies } u = 1 \\ \qquad\quad t = b \text{ implies } u = 1 + b\end{array}$$

$$= 500 \lim_{b\to\infty} \left(-\frac{1}{u}\right)\Big|_1^{1+b}$$

$$= 500 \lim_{b\to\infty} \left(-\frac{1}{1+b} + 1\right) = 500 \text{ gallons}$$

51. Total $= \displaystyle\int_0^\infty R(t)\,dt = \lim_{T\to\infty} \int_0^T R(t)\,dt = \lim_{T\to\infty} \int_0^T \frac{400}{(5+t)^3}\,dt$

$$= \lim_{T\to\infty} \left[\frac{-200}{(5+t)^2}\Big|_0^T\right]$$

$$= \lim_{T\to\infty} \left[\frac{-200}{(5+T)^2} + 8\right] = 8$$

Thus, 8 million immigrants will enter the country under this policy.

EXERCISE 15-2

Things to remember:

1. CONTINUOUS RANDOM VARIBLE

A CONTINUOUS RANDOM VARIABLE X is a function that assigns a numeric value to each outcome of an experiment. The set of possible values of X is an interval of real numbers. This interval may be open or closed, and it may be bounded or unbounded.

2. PROBABILITY DENSITY FUNCTION

The function $f(x)$ is a PROBABILITY DENSITY FUNCTION for a continuous random variable X if:

(a) $f(x) \geq 0$ for all $x \in (-\infty, \infty)$

(b) $\displaystyle\int_{-\infty}^{\infty} f(x)\,dx = 1$

(c) The probability that x lies in the interval $[c, d]$ is given by
$$P(c \leq X \leq d) = \int_{c}^{d} f(x)\,dx.$$

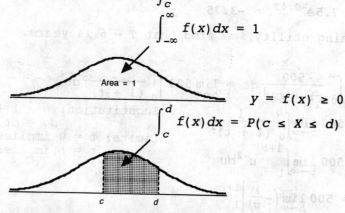

$$\int_{-\infty}^{\infty} f(x)\,dx = 1$$

Area = 1

$$y = f(x) \geq 0$$

$$\int_{c}^{d} f(x)\,dx = P(c \leq X \leq d)$$

Range of $X = (-\infty, \infty)$ = Domain of f.

3. CUMULATIVE DISTRIBUTION FUNCTION

If f is a probability density function, then the associated CUMULATIVE DISTRIBUTION FUNCTION F is defined by
$$F(x) = P(X \leq x) = \int_{-\infty}^{x} f(t)\,dt.$$

Furthermore,
$$P(c \leq X \leq d) = F(d) - F(c),$$

where F is an antiderivative of f, that is, $F' = f$.

4. PROPERTIES OF CUMULATIVE DISTRIBUTION FUNCTIONS

If f is a probability density function and
$$F(x) = \int_{-\infty}^{x} f(t)\,dt$$

is the associated cumulative distribution function, then:

(a) $F'(x) = f(x)$ wherever f is continuous.

(b) $0 \leq F(x) \leq 1$, $-\infty < x < \infty$.

(c) $F(x)$ is nondecreasing on $(-\infty, \infty)$.

1. The graph of $f(x)$ is shown at the right. From the graph we see that $f(x) \geq 0$ for $x \in (-\infty, \infty)$.

$$\int_{-\infty}^{\infty} f(x)\,dx = \int_0^4 \frac{1}{8}x\,dx$$

$$= \frac{1}{8} \cdot \frac{x^2}{2}\Big|_0^4$$

$$= \frac{1}{16}(16 - 0)$$

$$= 1$$

x	$f(x)$
0	0
1	$\frac{1}{8}$
4	$\frac{1}{2}$
$x > 4$	0

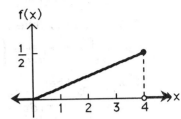

3. (A) Using 2(c),

$$P(1 < X < 3) = \int_1^3 \frac{1}{8}x\,dx = \frac{1}{8} \cdot \frac{x^2}{2}\Big|_1^3$$

$$= \frac{1}{16}(9 - 1) = \frac{8}{16} = \frac{1}{2}.$$

The graph is shown at the right.

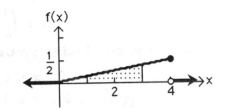

(B) $P(X \leq 2) = \int_{-\infty}^2 \frac{1}{8}x\,dx = \int_0^2 \frac{1}{8}x\,dx$ [since $f(x) = 0$ for $x \leq 0$]

$$= \frac{1}{8} \cdot \frac{x^2}{2}\Big|_0^2 = \frac{1}{16}(4) = \frac{1}{4}.$$

The graph is shown at the right.

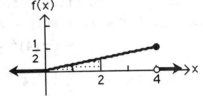

(C) $P(X > 3) = \int_3^\infty \frac{1}{8}x\,dx = \int_3^4 \frac{1}{8}x\,dx$ [since $f(x) = 0$ for $x > 4$]

$$= \frac{1}{8} \cdot \frac{x^2}{2}\Big|_3^4 = \frac{1}{16}(16 - 9) = \frac{7}{16}.$$

The graph is shown at the right.

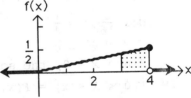

5. (A) $P(X = 1) = \int_1^1 f(x)\,dx = 0$

(B) $P(X > 5) = \int_5^\infty f(x)\,dx = \int_5^\infty 0\,dx = 0$ [$f(x) = 0$ for $x > 4$]

(C) $P(X < 5) = \int_{-\infty}^5 f(x)\,dx = \int_0^4 \frac{1}{8}x\,dx$ [$f(x) = 0$ when $x < 0$ and when $x > 4$]

$$= \frac{1}{8} \cdot \frac{x^2}{2}\Big|_0^4 = \frac{1}{16}(16) = 1.$$

7. If $x < 0$, then
$$F(x) = \int_{-\infty}^{x} f(t)\,dt = \int_{-\infty}^{x} 0\,dt = 0$$

If $0 \le x \le 4$, then
$$F(x) = \int_{-\infty}^{x} f(t)\,dt = \int_{-\infty}^{0} f(t)\,dt + \int_{0}^{x} f(t)\,dt = 0 + \int_{0}^{x} \frac{1}{8}t\,dt = \frac{1}{16}t^2\Big|_{0}^{x} = \frac{1}{16}x^2.$$

If $x > 4$, then
$$F(x) = \int_{-\infty}^{x} f(t)\,dt = \int_{-\infty}^{0} f(t)\,dt + \int_{0}^{4} f(t)\,dt + \int_{4}^{x} f(t)\,dt$$
$$= 0 + \int_{0}^{4} \frac{1}{8}t\,dt + 0 = \frac{1}{16}t^2\Big|_{0}^{4} = \frac{1}{16}(16 - 0) = 1.$$

Thus, the cumulative probability distribution function is:

$$F(x) = \begin{cases} 0 & x < 0 \\ \dfrac{1}{16}x^2 & 0 \le x \le 4 \\ 1 & x > 4 \end{cases}$$

The graph of $F(x)$ is shown at the right.

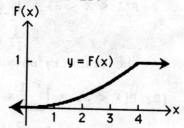

9. Using 3 and the cumulative probability distribution function F from Problem 7:

(A) $P(2 \le X \le 4) = F(4) - F(2)$
$$= \frac{1}{16}(4)^2 - \frac{1}{16}(2)^2$$
$$= 1 - \frac{1}{4} = \frac{3}{4}$$

(B) $P(0 < X < 2) = F(2) - F(0)$
$$= \frac{1}{16}(2)^2 - 0$$
$$= \frac{1}{4}$$

11. From Problem 7:

(A) $P(0 \le X \le x) = F(x) - F(0) = \frac{1}{16}x^2 - 0 = \frac{1}{16}x^2$

Now $\dfrac{1}{16}x^2 = \dfrac{1}{4}$

$x^2 = 4$

Since $x \ge 0$, the solution is $x = 2$.

(B) $P(0 \le X \le x) = F(x) - F(0) = \frac{1}{9}$ implies $\frac{1}{16}x^2 = \frac{1}{9}$

$$x^2 = \frac{16}{9}$$

Since $x \ge 0$, the solution is $x = \dfrac{4}{3}$.

13. From the graph of $f(x)$ shown below, we see that $f(x) \ge 0$ for $x \in (-\infty, \infty)$.

x	$f(x)$
0	2
2	$\frac{2}{27}$
4	$\frac{2}{125}$
$x < 0$	0

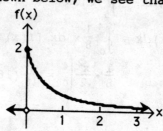

Also,

$$\int_{-\infty}^{\infty} f(x)\,dx = \int_{0}^{\infty} \frac{2}{(1+x)^3}\,dx = \lim_{R\to\infty} \int_{0}^{R} \frac{2}{(1+x)^3}\,dx$$

$$= \lim_{R\to\infty} \left(-\frac{1}{(1+x)^2}\right)\Big|_{0}^{R} = -\lim_{R\to\infty}\left(\frac{1}{(1+R)^2} - 1\right) = -(0-1) = 1.$$

15. (A) $P(1 \le X \le 4) = \int_{1}^{4} \frac{2}{(1+x)^3}\,dx = \frac{-1}{(1+x)^2}\Big|_{1}^{4} = \frac{-1}{(1+4)^2} - \frac{-1}{(1+1)^2}$

$$= \frac{-1}{25} + \frac{1}{4} = \frac{21}{100} = .21$$

(B) $P(X > 3) = \int_{3}^{\infty} \frac{2}{(1+x)^3}\,dx$

Now, $1 = \int_{-\infty}^{\infty} \frac{2}{(1+x)^3}\,dx = \int_{-\infty}^{3} \frac{2}{(1+x)^3}\,dx + \int_{3}^{\infty} \frac{2}{(1+x)^3}\,dx,$

so $\int_{3}^{\infty} \frac{2}{(1+x)^3}\,dx = 1 - \int_{-\infty}^{3} \frac{2}{(1+x)^3}\,dx = 1 - \int_{0}^{3} \frac{2}{(1+x)^3}\,dx,$

since $f(x) = 0$ for $x < 0$. Thus,

$$\int_{3}^{\infty} \frac{2}{(1+x)^3}\,dx = 1 - \left(\frac{-1}{(1+x)^2}\Big|_{0}^{3}\right) = 1 - \left(-\frac{1}{16} + 1\right) = \frac{1}{16}.$$

(C) $P(X \le 2) = \int_{-\infty}^{2} \frac{2}{(1+x)^3}\,dx = \int_{0}^{2} \frac{2}{(1+x)^3}\,dx = \frac{-1}{(1+x)^2}\Big|_{0}^{2} = -\frac{1}{9} + 1 = \frac{8}{9}$

17. If $x < 0$, then

$$F(x) = \int_{-\infty}^{x} f(t)\,dt = \int_{-\infty}^{x} 0\,dt = 0.$$

If $x \ge 0$, then

$$F(x) = \int_{-\infty}^{x} f(t)\,dt = \int_{-\infty}^{0} f(t)\,dt + \int_{0}^{x} f(t)\,dt = 0 + \int_{0}^{x} \frac{2}{(1+t)^3}\,dt$$

$$= \left(-\frac{1}{(1+t)^2}\right)\Big|_{0}^{x} = -\frac{1}{(1+x)^2} + 1 \quad \text{or} \quad 1 - \frac{1}{(1+x)^2}.$$

Thus, the cumulative probability distribution function is given by:

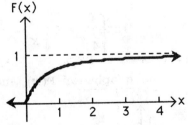

F(x)

$$F(x) = \begin{cases} 0 & x < 0 \\ 1 - \dfrac{1}{(1+x)^2} & x \ge 0 \end{cases}$$

The graph of $F(x)$ is shown at the right.

19. From Problem 17,

$$F(x) = \begin{cases} 0 & x < 0 \\ 1 - \dfrac{1}{(1+x)^2} & x \ge 0 \end{cases}$$

and $P(0 \le X \le x) = F(x) - F(0) = 1 - \dfrac{1}{(1+x)^2}.$

(A) Set $P(0 \le X \le x) = \dfrac{3}{4}$ and solve for x.

$$1 - \frac{1}{(1 + x)^2} = \frac{3}{4}$$

$$\frac{1}{(1 + x)^2} = \frac{1}{4}$$

$$(1 + x)^2 = 4$$

Since $x \geq 0$, we have $1 + x = 2$, so $x = 1$.

(B) $1 = P(X \leq x) + P(X > x)$. Therefore,

$$P(X > x) = 1 - P(X \leq x) = 1 - F(x) = 1 - \left(1 - \frac{1}{(1 + x)^2}\right) = \frac{1}{(1 + x)^2}.$$

Set $P(X > x) = \frac{1}{16}$ and solve for x.

$$\frac{1}{(1 + x)^2} = \frac{1}{16}$$

$$(1 + x)^2 = 16$$

$$1 + x = 4$$

$$x = 3$$

21. $f(x) = \begin{cases} \frac{3}{2}x - \frac{3}{4}x^2 & 0 \leq x \leq 2 \\ 0 & \text{otherwise} \end{cases}$

If $x < 0$, then

$$F(x) = \int_{-\infty}^{x} f(t)\,dt = \int_{-\infty}^{x} 0 \; dt = 0.$$

If $0 \leq x \leq 2$, then

$$F(x) = \int_{-\infty}^{x} f(t)\,dt = \int_{-\infty}^{0} f(t)\,dt + \int_{0}^{x} f(t)\,dt = 0 + \int_{0}^{x} \left(\frac{3}{2}t - \frac{3}{4}t^2\right)dt$$

$$= \left(\frac{3}{2} \cdot \frac{t^2}{2} - \frac{3}{4} \cdot \frac{t^3}{3}\right)\Big|_{0}^{x} = \left(\frac{3}{4}t^2 - \frac{1}{4}t^3\right)\Big|_{0}^{x} = \frac{3}{4}x^2 - \frac{1}{4}x^3.$$

If $x > 2$, then

$$F(x) = \int_{-\infty}^{x} f(t)\,dt = \int_{-\infty}^{0} f(t)\,dt + \int_{0}^{2} f(t)\,dt + \int_{2}^{x} f(t)\,dt$$

$$= 0 + \int_{0}^{2} \left(\frac{3}{2}t - \frac{3}{4}t^2\right)dt + 0 = \left(\frac{3}{4}t^2 - \frac{1}{4}t^3\right)\Big|_{0}^{2} = \frac{3}{4}(2)^2 - \frac{1}{4}(2)^3 = 1$$

Thus, the cumulative probability distribution function is:

$$F(x) = \begin{cases} 0 & x < 0 \\ \frac{3}{4}x^2 - \frac{1}{4}x^3 & 0 \leq x \leq 2 \\ 1 & x > 2 \end{cases}$$

The graphs of $F(x)$ and $f(x)$ are as follows:

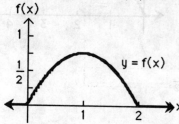

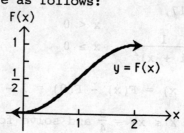

23. $f(x) = \begin{cases} \dfrac{1}{2} + \dfrac{1}{2}x^3 & -1 \le x \le 1 \\ 0 & \text{otherwise} \end{cases}$

If $x < -1$, then

$F(x) = \displaystyle\int_{-\infty}^{x} f(t)\,dt = \int_{-\infty}^{x} 0\,dt = 0$

If $-1 \le x \le 1$, then

$F(x) = \displaystyle\int_{-\infty}^{x} f(t)\,dt = \int_{-\infty}^{-1} f(t)\,dt + \int_{-1}^{x} f(t)\,dt = 0 + \int_{-1}^{x} \left(\dfrac{1}{2} + \dfrac{1}{2}t^3\right) dt$

$\qquad = \left(\dfrac{1}{2}t + \dfrac{1}{8}t^4\right)\Big|_{-1}^{x} = \dfrac{1}{2}x + \dfrac{1}{8}x^4 - \left(-\dfrac{1}{2} + \dfrac{1}{8}\right)$

$\qquad = \dfrac{3}{8} + \dfrac{1}{2}x + \dfrac{1}{8}x^4$

If $x > 1$, then

$F(x) = \displaystyle\int_{-\infty}^{x} f(t)\,dt = \int_{-\infty}^{-1} f(t)\,dt + \int_{-1}^{1} f(t)\,dt + \int_{1}^{x} f(t)\,dt$

$\qquad = 0 + \displaystyle\int_{-1}^{1} \left(\dfrac{1}{2} + \dfrac{1}{2}t^3\right) dt + 0$

$\qquad = \left(\dfrac{1}{2}t + \dfrac{1}{8}t^4\right)\Big|_{-1}^{1} = \left(\dfrac{1}{2} + \dfrac{1}{8}\right) - \left(-\dfrac{1}{2} + \dfrac{1}{8}\right)$

$\qquad = 1$

Thus, the cumulative probability distribution function is:

$F(x) = \begin{cases} 0 & x < -1 \\ \dfrac{3}{8} + \dfrac{1}{2}x + \dfrac{1}{8}x^4 & -1 \le x \le 1 \\ 1 & x > 1 \end{cases}$

The graphs of $f(x)$ and $F(x)$ are:

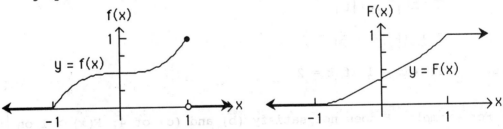

25. The graphs of

$F(x) = \begin{cases} 0 & x < 0 \\ \dfrac{3}{4}x^2 - \dfrac{1}{4}x^3 & 0 \le x \le 2 \\ 1 & x > 2 \end{cases}$ and $y = 0.2$ are shown below.

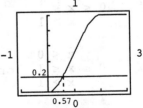

$F(x) = 0.2$ at $x \approx 0.57$

27. The graphs of
$$F(x) = \begin{cases} 0 & x < -1 \\ \dfrac{3}{8} + \dfrac{1}{2}x + \dfrac{1}{8}x^4 & -1 \le x \le 1 \\ 1 & x > 1 \end{cases} \quad \text{and} \quad y = 0.6 \text{ are shown below.}$$

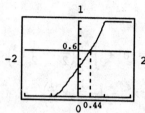

$F(x) = 0.6$ at $x \approx 0.44$

29. No such constant exists since
$$\int_{-\infty}^{\infty} f(x)\,dx = \int_{1}^{\infty} f(x)\,dx \qquad (f(x) = 0 \text{ for } x < 1)$$
$$= \int_{1}^{\infty} \frac{1}{x}\,dx$$
and the integral $\displaystyle\int_{1}^{\infty} \frac{1}{x}\,dx$ diverges.

31.
$$\int_{-\infty}^{\infty} f(x)\,dx = \int_{1}^{\infty} f(x)\,dx \qquad (f(x) = 0 \text{ for } x < 1)$$
$$= \int_{1}^{\infty} \frac{1}{x^3}\,dx = \lim_{b\to\infty} \int_{1}^{b} \frac{1}{x^3}\,dx$$
$$= \lim_{b\to\infty} \left[-\frac{1}{2x^2} \Big|_{1}^{b} \right]$$
$$= \lim_{b\to\infty} \left[\frac{-1}{2b^2} + \frac{1}{2} \right] = \frac{1}{2}$$
Thus, $\displaystyle\int_{-\infty}^{\infty} kf(x)\,dx = 1$ if $k = 2$.

33. No! For example, F does not satisfy (b) and (c) of $\underline{4}$; $F(x) > 1$ on $\left[0, \dfrac{1}{2}\right]$ and F is *decreasing* on $[0, 1]$.

35. The relationship between $f(x)$ and $F(x)$ is $F'(x) = f(x)$. Thus, if
$$F(x) = \begin{cases} 0 & x < 0 \\ x^2 & 0 \le x \le 1 \\ 1 & x > 1 \end{cases}$$
then
$$f(x) = \begin{cases} 0 & x < 0 \\ 2x & 0 \le x \le 1 \\ 0 & x > 1 \end{cases} \quad \text{or} \quad f(x) = \begin{cases} 2x & 0 \le x \le 1 \\ 0 & \text{otherwise} \end{cases}$$

37. $F(x) = \begin{cases} 0 & x < 0 \\ 6x^2 - 8x^3 + 3x^4 & 0 \le x \le 1 \\ 1 & x > 1 \end{cases}$

Thus,

$f(x) = \begin{cases} 0 & x < 0 \\ 12x - 24x^2 + 12x^3 & 0 \le x \le 1 \\ 0 & \text{otherwise} \end{cases}$ [<u>Note</u>: $f(x) = F'(x)$.]

or

$f(x) = \begin{cases} 12x(1 - 2x + x^2) & 0 \le x \le 1 \\ 0 & \text{otherwise} \end{cases}$

39. Given $f(x) = \begin{cases} x & 0 \le x \le 1 \\ 2 - x & 1 < x \le 2 \\ 0 & \text{otherwise} \end{cases}$

If $x < 0$, then

$$F(x) = \int_{-\infty}^{x} f(t)\,dt = \int_{-\infty}^{x} 0\,dt = 0.$$

If $0 \le x \le 1$, then

$$F(x) = \int_{-\infty}^{x} f(t)\,dt = \int_{-\infty}^{0} f(t)\,dt + \int_{0}^{x} f(t)\,dt = 0 + \int_{0}^{x} t\,dt = \frac{1}{2}t^2\Big|_{0}^{x} = \frac{1}{2}x^2.$$

If $1 < x \le 2$, then

$$F(x) = \int_{-\infty}^{x} f(t)\,dt = \int_{-\infty}^{0} f(t)\,dt + \int_{0}^{1} f(t)\,dt + \int_{1}^{x} f(t)\,dt$$

$$= 0 + \int_{0}^{1} t\,dt + \int_{1}^{x} (2 - t)\,dt = \frac{1}{2}t^2\Big|_{0}^{1} + \left(2t - \frac{1}{2}t^2\right)\Big|_{1}^{x}$$

$$= \frac{1}{2} + \left(2x - \frac{1}{2}x^2\right) - \left(2 - \frac{1}{2}\right) = \frac{1}{2} + 2x - \frac{1}{2}x^2 - \frac{3}{2} = 2x - \frac{1}{2}x^2 - 1.$$

If $x > 2$, then

$$F(x) = \int_{-\infty}^{x} f(t)\,dt = \int_{-\infty}^{0} f(t)\,dt + \int_{0}^{1} f(t)\,dt + \int_{1}^{2} f(t)\,dt + \int_{2}^{x} f(t)\,dt$$

$$= 0 + \int_{0}^{1} t\,dt + \int_{1}^{2} (2 - t)\,dt + 0 = \frac{1}{2}t^2\Big|_{0}^{1} + \left(2t - \frac{1}{2}t^2\right)\Big|_{1}^{2}$$

$$= \frac{1}{2} + \left[\left(4 - \frac{1}{2}\cdot 4\right) - \left(2 - \frac{1}{2}\right)\right] = 1.$$

Thus, $F(x)$ is given by:

$$F(x) = \begin{cases} 0 & x < 0 \\ \frac{1}{2}x^2 & 0 \le x < 1 \\ 2x - \frac{1}{2}x^2 - 1 & 1 \le x \le 2 \\ 1 & x > 2 \end{cases}$$

41. $f(x) = \begin{cases} .2 - .02x & 0 \le x \le 10 \\ 0 & \text{otherwise} \end{cases}$

(A) $\displaystyle\int_2^6 f(x)\,dx = \int_2^6 (0.2 - 0.02x)\,dx$

$$= \left[(0.2x - 0.01x^2) \Big|_2^6 \right] = 0.48$$

The probability that the daily demand for electricity is between 2 million and 6 million kilowatt hours is 0.48.

(B) $P(X \le 8) = \displaystyle\int_{-\infty}^8 f(x)\,dx = \int_0^8 f(x)\,dx = \int_0^8 (0.2 - 0.02x)\,dx$

$$= \left(0.2x - \frac{0.02x^2}{2} \right)\Big|_0^8 = 0.2(8) - 0.01(8)^2 = 1.6 - 0.64 = 0.96$$

(C) $P(X > 5) = \displaystyle\int_5^\infty f(x)\,dx = \int_5^{10} f(x)\,dx = \int_5^{10} (0.2 - 0.02x)\,dx$

$$= \left(0.2x - \frac{0.02x^2}{2} \right)\Big|_5^{10} = [0.2(10) - 0.01(10)^2] - [0.2(5) - 0.01(5)^2]$$

$$= 2 - 1 - (1 - 0.25) = 0.25$$

43. (A) $\displaystyle\int_5^{10} f(x)\,dx = \int_5^{10} \frac{1}{10} e^{-x/10}\,dx$

$$= \left[-e^{-x/10} \Big|_5^{10} \right] = e^{-1/2} - e^{-1} \approx 0.239$$

The probability that it takes the computer more than 5 seconds and less than 10 seconds to respond is 0.239.

(B) $P(0 \le X \le 1) = \displaystyle\int_0^1 f(x)\,dx = \int_0^1 \frac{1}{10} e^{-x/10}\,dx$

$$= \frac{1}{10} \int_0^1 e^{-x/10}\,dx = \frac{1}{10} \cdot \frac{e^{-x/10}}{-\frac{1}{10}} \Big|_0^1$$

$$= -(e^{-1/10} - 1) = 1 - e^{-1/10} \approx 0.0952$$

(C) $P(X > 4) = \displaystyle\int_4^\infty \frac{1}{10} e^{-x/10}\,dx = \lim_{R \to \infty} \frac{1}{10} \int_4^R e^{-x/10}\,dx = \lim_{R \to \infty} \left(-e^{-x/10} \Big|_4^R \right)$

$$= -\lim_{R \to \infty} (e^{-R/10} - e^{-4/10}) = -(0 - e^{-2/5}) = e^{-2/5} \approx 0.6703$$

45. (A) $P(X > 4) = \displaystyle\int_4^{10} f(x)\,dx$ [Note: 4 stands for 4000.]

$$= \int_4^{10} 0.003x\sqrt{100 - x^2}\,dx$$

$$= \frac{-0.003}{2} \int_4^{10} (100 - x^2)^{1/2}(-2x)\,dx$$ [Note: $d(100 - x^2) = -2x\,dx$.]

$$= -0.0015 \cdot \frac{2}{3} (100 - x^2)^{3/2} \Big|_4^{10} = -\frac{1}{1000}(100 - x^2)^{3/2} \Big|_4^{10}$$

$$= -\frac{1}{1000}(0 - (84)^{3/2})$$

$$= \frac{(84)^{3/2}}{1000} \approx 0.7699$$

(B) $P(0 \le x \le 8) = \displaystyle\int_0^8 f(x)\,dx = \int_0^8 0.003x\sqrt{100 - x^2}\,dx$

$$= -\frac{1}{1000}(100 - x^2)^{3/2} \Big|_0^8 \qquad \text{[refer to part (A)]}$$

$$= -\frac{1}{1000}\left[(100 - 64)^{3/2} - (100)^{3/2}\right]$$

$$= -\frac{1}{1000}(36^{3/2} - 100^{3/2}) = \frac{-1}{1000}(216 - 1000) = 0.784$$

(C) We must solve the following for x:

$$\int_0^x f(t)\,dt = 0.9$$

$$\int_0^x 0.003t\sqrt{100 - t^2}\,dt = 0.9$$

$$-\frac{1}{1000}(100 - t^2)^{3/2}\Big|_0^x = 0.9 \quad \text{[refer to part (A)]}$$

$$(100 - t^2)^{3/2}\Big|_0^x = -900$$

$$(100 - x^2)^{3/2} - (100)^{3/2} = -900$$

$$(100 - x^2)^{3/2} - 1000 = -900$$

$$(100 - x^2)^{3/2} = 100$$

$$100 - x^2 = 100^{2/3}$$

$$x^2 = 100 - 100^{2/3}$$

$$x = \sqrt{100 - 100^{2/3}} \approx 8.858 \quad \text{or} \quad 8858 \text{ pounds}$$

47. (A) $P(7 \le X) = \displaystyle\int_7^\infty f(x)\,dx = \int_7^{10} f(x)\,dx + \int_{10}^\infty f(x)\,dx$

$$= \int_7^{10} \frac{1}{5000}(10x^3 - x^4)\,dx + \int_{10}^\infty 0\,dx = \frac{1}{5000}\left(\frac{10}{4}x^4 - \frac{x^5}{5}\right)\Big|_7^{10} + 0$$

$$= \frac{1}{5000}\left(\left[\frac{5}{2}(10)^4 - \frac{1}{5}(10)^5\right] - \left[\frac{5}{2}(7)^4 - \frac{1}{5}(7)^5\right]\right) \approx 0.47178$$

(B) $P(X \le 5) = \displaystyle\int_{-\infty}^5 f(x)\,dx = \int_{-\infty}^0 f(x)\,dx + \int_0^5 f(x)\,dx$

$$= \int_{-\infty}^0 0\,dx + \int_0^5 \frac{1}{5000}(10x^3 - x^4)\,dx = 0 + \frac{1}{5000}\left(\frac{5}{2}x^4 - \frac{1}{5}x^5\right)\Big|_0^5$$

$$= \frac{1}{5000}\left[\frac{5}{2}(5)^4 - \frac{1}{5}(5)^5\right] = \frac{3}{16} = 0.1875$$

49. (A) $P(X \le 20) = \int_{-\infty}^{20} f(x)\,dx = \int_{0}^{20} \frac{800x}{(400 + x^2)^2}\,dx = 400 \int_{0}^{20} \frac{2x}{(400 + x^2)^2}\,dx$

$$= \frac{-400}{400 + x^2}\Big|_0^{20} = \frac{-400}{800} + \frac{400}{400} = 0.5$$

(B) $P(X > 15) = \int_{15}^{\infty} f(x)\,dx = 1 - \int_{-\infty}^{15} f(x)\,dx = 1 - \int_{0}^{15} \frac{800x}{(400 + x^2)^2}\,dx$

$$= 1 - \left(\frac{-400}{400 + x^2}\right)\Big|_0^{15} = 1 + \frac{400}{625} - 1 = 0.64$$

(C) We must solve the following for x:

$$\int_0^x f(t)\,dt = 0.8$$

$$\int_0^x \frac{800t}{(400 + t^2)^2}\,dt = 0.8$$

$$\frac{-400}{400 + t^2}\Big|_0^x = 0.8$$

$$\frac{-400}{400 + x^2} + 1 = 0.8$$

$$\frac{-400}{400 + x^2} = -0.2$$

$$-400 = -80 - 0.2x^2$$

$$0.2x^2 = 320$$

$$x^2 = 1600$$

$$x = 40 \text{ days}$$

51. (A) $P(X \ge 30) = \int_{30}^{\infty} f(x)\,dx = 1 - \int_{-\infty}^{30} f(x)\,dx = 1 - \int_0^{30} \frac{1}{20} e^{-x/20}\,dx$

$$= 1 + e^{-x/20}\Big|_0^{30} = 1 + e^{-30/20} - 1 = e^{-3/2} \approx 0.223$$

(B) $P(X \ge 80) = \int_{80}^{\infty} f(x)\,dx = 1 - \int_{-\infty}^{80} f(x)\,dx = 1 - \int_0^{80} \frac{1}{20} e^{-x/20}\,dx$

$$= 1 + e^{-x/20}\Big|_0^{80} = 1 + e^{-80/20} - 1 = e^{-4} \approx 0.018$$

EXERCISE 15-3

Things to remember:

<u>1.</u> EXPECTED VALUE AND STANDARD DEVIATION FOR A CONTINUOUS RANDOM VARIABLE

Let $f(x)$ be the probability density function for a continuous random variable X. The EXPECTED VALUE, or MEAN, of X is

$$\mu = E(X) = \int_{-\infty}^{\infty} xf(x)\,dx.$$

The VARIANCE is

$$V(X) = \int_{-\infty}^{\infty} (x - \mu)^2 f(x)\, dx,$$

and the STANDARD DEVIATION is

$$\sigma = \sqrt{V(X)}.$$

2. ALTERNATE FORMULA FOR VARIANCE:

$$V(X) = \int_{-\infty}^{\infty} x^2 f(x)\, dx - \mu^2$$

3. If m is the median, then it must satisfy

$$F(m) = P(X \le m) = \frac{1}{2}.$$

---

1. Using **1**,

$$\mu = E(X) = \int_{-\infty}^{\infty} xf(x)\, dx = \int_{0}^{2} x \cdot \frac{1}{2}x\, dx = \frac{1}{2} \cdot \frac{x^3}{3}\Big|_{0}^{2} = \frac{1}{6}(2^3) = \frac{8}{6} \text{ or } \frac{4}{3} \approx 1.333.$$

$$V(X) = \int_{-\infty}^{\infty} (x - \mu)^2 f(x)\, dx = \int_{0}^{2} \left(x - \frac{4}{3}\right)^2 \frac{1}{2}x\, dx = \frac{1}{2}\int_{0}^{2} \left(x^2 - \frac{8}{3}x + \frac{16}{9}\right) x\, dx$$

$$= \frac{1}{2}\int_{0}^{2} \left(x^3 - \frac{8}{3}x^2 + \frac{16}{9}x\right) dx = \frac{1}{2}\left(\frac{x^4}{4} - \frac{8}{9}x^3 + \frac{16}{18}x^2\right)\Big|_{0}^{2}$$

$$= \frac{1}{2}\left(\frac{16}{4} - \frac{64}{9} + \frac{32}{9}\right) = \frac{4}{18} = \frac{2}{9} \approx 0.222.$$

Thus, the standard deviation is $\sigma = \sqrt{V(X)} = \sqrt{\frac{2}{9}} = \frac{\sqrt{2}}{3} \approx 0.471.$

3. Using **1**,

$$\mu = E(X) = \int_{-\infty}^{\infty} xf(x)\, dx = \int_{2}^{5} \frac{1}{3}x\, dx = \frac{1}{6}x^2\Big|_{2}^{5} = \frac{25}{6} - \frac{4}{6} = \frac{7}{2} = 3.5$$

Using **2**,

$$V(X) = \int_{-\infty}^{\infty} x^2 f(x)\, dx - \mu^2 = \int_{2}^{5} \frac{1}{3}x^2\, dx - \frac{49}{4} = \frac{1}{9}x^3\Big|_{2}^{5} - \frac{49}{4}$$

$$= \frac{125}{9} - \frac{8}{9} - \frac{49}{4} = \frac{3}{4} = 0.75$$

Thus, $\sigma = \sqrt{V(X)} = \sqrt{\frac{3}{4}} = \frac{\sqrt{3}}{2} \approx 0.866.$

5. $\mu = E(X) = \int_{-\infty}^{\infty} xf(x)\, dx = \int_{1}^{2} x(4 - 2x)\, dx = \int_{1}^{2} (4x - 2x^2)\, dx = \left(2x^2 - \frac{2}{3}x^3\right)\Big|_{1}^{2}$

$$= 8 - \frac{16}{3} - \left(2 - \frac{2}{3}\right) = \frac{4}{3} \approx 1.333$$

Using 2,

$$V(X) = \int_{-\infty}^{\infty} x^2 f(x)\,dx - \mu^2 = \int_{1}^{2} x^2 (4 - 2x)\,dx - \frac{16}{9} = \int_{1}^{2} (4x^2 - 2x^3)\,dx - \frac{16}{9}$$

$$= \left(\frac{4}{3}x^3 - \frac{1}{2}x^4\right)\Big|_{1}^{2} - \frac{16}{9} = \frac{32}{3} - 8 - \left(\frac{4}{3} - \frac{1}{2}\right) - \frac{16}{9} = \frac{1}{18}$$

Thus, $\sigma = \sqrt{V(X)} = \sqrt{\frac{1}{18}} \approx 0.236$.

7. <u>Step 1</u>: Find the cumulative probability distribution function.

If $x < 0$, then $F(x) = 0$. If $0 \le x \le 1$, then

$$F(x) = \int_{-\infty}^{x} f(t)\,dt = \int_{0}^{x} 2t\,dt = t^2\Big|_{0}^{x} = x^2.$$

If $x > 1$, then

$$F(x) = \int_{-\infty}^{x} f(t)\,dt = \int_{-\infty}^{0} f(t)\,dt + \int_{0}^{1} f(t)\,dt + \int_{1}^{x} f(t)\,dt$$

$$= 0 + \int_{0}^{1} 2t\,dt + 0 = t^2\Big|_{0}^{1} = 1.$$

Thus, $F(x) = \begin{cases} 0 & \text{if } x < 0 \\ x^2 & \text{if } 0 \le x \le 1 \\ 1 & \text{if } x > 1 \end{cases}$

<u>Step 2</u>: Solve the equation $P(X \le m) = \frac{1}{2}$ for m, where m is the median.

$$F(m) = P(X \le m) = \frac{1}{2}$$

$$m^2 = \frac{1}{2}$$

$$m = \frac{1}{\sqrt{2}} \approx 0.707$$

9. <u>Step 1</u>: Find the cumulative probability distribution function.
If $x < 2$, then $F(x) = 0$. If $2 \le x \le 4$, then

$$F(x) = \int_{-\infty}^{x} f(t)\,dt = \int_{2}^{x} \frac{1}{6}t\,dt = \frac{1}{12}t^2\Big|_{2}^{x} = \frac{x^2}{12} - \frac{1}{3}.$$

If $x > 4$, then

$$F(x) = \int_{-\infty}^{x} f(t)\,dt = \int_{-\infty}^{2} f(t)\,dt + \int_{2}^{4} f(t)\,dt + \int_{4}^{x} f(t)\,dt$$

$$= 0 + \int_{2}^{4} \frac{1}{6}t\,dt + 0 = \frac{1}{12}t^2\Big|_{2}^{4} = 1.$$

Thus, $F(x) = \begin{cases} 0 & \text{if } x < 2 \\ \dfrac{x^2}{12} - \dfrac{1}{3} & \text{if } 2 \le x \le 4 \\ 1 & \text{if } x > 4 \end{cases}$

Solve the equation $P(X \leq m) = \frac{1}{2}$ for m.

$$F(m) = P(X \leq m) = \frac{1}{2}$$

$$\frac{m^2}{12} - \frac{1}{3} = \frac{1}{2}$$

$$\frac{m^2}{12} = \frac{5}{6}$$

$$m^2 = 10$$

$$m = \sqrt{10} \approx 3.162$$

11. <u>Step 1:</u> Find the cumulative probability distribution function.

If $x < 0$, then $F(x) = 0$. If $0 \leq x \leq 4$, then

$$F(x) = \int_{-\infty}^{x} f(t)\,dt = \int_{0}^{x} \left(\frac{1}{2} - \frac{1}{8}t\right)dt = \left(\frac{1}{2}t - \frac{1}{16}t^2\right)\Big|_{0}^{x} = \frac{1}{2}x - \frac{1}{16}x^2.$$

If $x > 4$, then

$$F(x) = \int_{-\infty}^{x} f(t)\,dt = \int_{-\infty}^{0} f(t)\,dt + \int_{0}^{4} f(t)\,dt + \int_{4}^{x} f(t)\,dt$$

$$= 0 + \int_{0}^{4} \left(\frac{1}{2} - \frac{1}{8}t\right)dt + 0$$

$$= \left(\frac{1}{2}t - \frac{1}{16}t^2\right)\Big|_{0}^{4} = 2 - 1 = 1$$

Thus, $F(x) = \begin{cases} 0 & \text{if } x < 0 \\ \frac{1}{2}x - \frac{1}{16}x^2 & \text{if } 0 \leq x \leq 4 \\ 1 & \text{if } x > 4 \end{cases}$

<u>Step 2:</u> Solve the equation $P(X \leq m) = \frac{1}{2}$ for m.

$$F(m) = P(X \leq m) = \frac{1}{2}$$

$$\frac{1}{2}m - \frac{1}{16}m^2 = \frac{1}{2}$$

$$m^2 - 8m + 8 = 0$$

Now, the roots of the quadratic equation are

$$\frac{8 \pm \sqrt{64 - 32}}{2} = 4 \pm 2\sqrt{2}.$$

Since m must lie in the interval $[0, 4]$, $m = 4 - 2\sqrt{2} \approx 1.172$.

13. $\mu = E(X) = \int_{-\infty}^{\infty} xf(x)\,dx = \int_{1}^{\infty} x \cdot \frac{4}{x^5}\,dx = \int_{1}^{\infty} \frac{4}{x^4}\,dx = \lim_{R \to \infty} \int_{1}^{R} \frac{4}{x^4}\,dx$

$$= \lim_{R \to \infty} \left(-\frac{4}{3x^3}\right) \Big|_{1}^{R} = \lim_{R \to \infty} \left(-\frac{4}{3R^3} + \frac{4}{3}\right) = \frac{4}{3}$$

$V(X) = \int_{-\infty}^{\infty} x^2 f(x)\,dx - \mu^2 = \int_{1}^{\infty} x^2 \cdot \frac{4}{x^5}\,dx - \frac{16}{9} = \int_{1}^{\infty} \frac{4}{x^3}\,dx - \frac{16}{9}$

$$= \lim_{R \to \infty} \int_{1}^{R} \frac{4}{x^3}\,dx - \frac{16}{9} = \lim_{R \to \infty} \left(-\frac{2}{x^2}\right) \Big|_{1}^{R} - \frac{16}{9} = \lim_{R \to \infty} \left(-\frac{2}{R^2} + 2\right) - \frac{16}{9} = \frac{2}{9}$$

$\sigma = \sqrt{V(X)} = \sqrt{\frac{2}{9}} = \frac{\sqrt{2}}{3} \approx 0.471$

15. $\mu = E(X) = \int_{-\infty}^{\infty} xf(x)\,dx = \int_{2}^{\infty} x \cdot \frac{64}{x^5}\,dx = \int_{2}^{\infty} \frac{64}{x^4}\,dx = \lim_{R \to \infty} \int_{2}^{R} \frac{64}{x^4}\,dx$

$$= \lim_{R \to \infty} \left(\frac{-64}{3x^3}\right) \Big|_{2}^{R} = \lim_{R \to \infty} \left(\frac{-64}{3R^3} + \frac{64}{24}\right) = \frac{8}{3} \approx 2.667$$

$V(X) = \int_{-\infty}^{\infty} x^2 f(x)\,dx - \mu^2 = \int_{2}^{\infty} x^2 \cdot \frac{64}{x^5}\,dx - \frac{64}{9} = \int_{2}^{\infty} \frac{64}{x^3}\,dx - \frac{64}{9}$

$$= \lim_{R \to \infty} \int_{2}^{R} \frac{64}{x^3}\,dx - \frac{64}{9} = \lim_{R \to \infty} \left(\frac{-32}{x^2}\right) \Big|_{2}^{R} - \frac{64}{9} = \lim_{R \to \infty} \left(\frac{-32}{R^2} + 8\right) - \frac{64}{9} = \frac{8}{9}$$

$\sigma = \sqrt{V(X)} = \sqrt{\frac{8}{9}} = \frac{2\sqrt{2}}{3} \approx 0.943$

17. <u>Step 1</u>: Find the cumulative probability distribution function.
If $x < 1$, then $F(x) = 0$. If $1 \leq x \leq e$, then

$$F(x) = \int_{-\infty}^{x} f(t)\,dt = \int_{-\infty}^{1} f(t)\,dt + \int_{1}^{x} f(t)\,dt$$

$$= 0 + \int_{1}^{x} \frac{1}{t}\,dt = \ln t \Big|_{1}^{x} = \ln x - \ln 1 = \ln x.$$

If $x > e$, then $F(x) = \int_{-\infty}^{x} f(t)\,dt = \int_{-\infty}^{1} f(t)\,dt + \int_{1}^{e} f(t)\,dt + \int_{e}^{x} f(t)\,dt$

$$= 0 + \int_{1}^{e} \frac{1}{t}\,dt + 0 = \ln t \Big|_{1}^{e} = \ln e - \ln 1 = 1.$$

Thus, $F(x) = \begin{cases} 0 & x < 1 \\ \ln x & 1 \leq x \leq e \\ 1 & x > e \end{cases}$

<u>Step 2</u>: Solve the equation $P(X \leq m) = \frac{1}{2}$ for m.

$$F(m) = P(X \leq m) = \frac{1}{2}$$

$$\ln m = \frac{1}{2} \quad (1 \leq x \leq e)$$

Thus, the median $m = e^{1/2} \approx 1.649$.

19. <u>Step 1</u>: Find the cumulative probabilty distribution function.
If $x < 0$, then $F(x) = 0$. If $0 \le x \le 2$, then

$$F(x) = \int_{-\infty}^{x} f(t)\,dt = \int_{0}^{x} \frac{4}{(2+t)^2}\,dt = \frac{-4}{2+t}\Big|_{0}^{x} = \frac{-4}{2+x} + 2.$$

If $x > 2$, then

$$F(x) = \int_{-\infty}^{x} f(t)\,dt = \int_{-\infty}^{0} f(t)\,dt + \int_{0}^{2} f(t)\,dt + \int_{2}^{x} f(t)\,dt$$

$$= 0 + \int_{0}^{2} \frac{4}{(2+t)^2}\,dt + 0 = \frac{-4}{2+t}\Big|_{0}^{2} = -1 + 2 = 1$$

Thus, $F(x) = \begin{cases} 0 & \text{if } x < 0 \\ \dfrac{-4}{2+x} + 2 & \text{if } 0 \le x \le 2 \\ 1 & \text{if } x > 2 \end{cases}$

<u>Step 2</u>: Solve the equation $P(X \le m) = \frac{1}{2}$ for m.

$$F(m) = P(X \le m) = \frac{1}{2}$$

$$\frac{-4}{2+m} + 2 = \frac{1}{2}$$

$$\frac{-4}{2+m} = \frac{-3}{2}$$

$$-6 - 3m = -8$$

$$m = \frac{2}{3}$$

21. <u>Step 1</u>: Find $F(x)$.
If $x < 0$, then $F(x) = 0$. If $x \ge 0$, then

$$F(x) = \int_{-\infty}^{x} f(t)\,dt = \int_{-\infty}^{0} f(t)\,dt + \int_{0}^{x} f(t)\,dt = 0 + \int_{0}^{x} \frac{1}{(1+t)^2}\,dt$$

$$= -\frac{1}{(1+t)}\Big|_{0}^{x} = -\left(\frac{1}{1+x} - 1\right) = 1 - \frac{1}{1+x} = \frac{x}{1+x}.$$

Thus,

$$F(x) = \begin{cases} 0 & x < 0 \\ \dfrac{x}{1+x} & x \ge 0 \end{cases}$$

<u>Step 2</u>: Solve $P(X \le m) = \frac{1}{2}$ for m.

$$F(m) = P(X \le m) = \frac{1}{2}$$

$$\frac{m}{1+m} = \frac{1}{2} \qquad (x \ge 0)$$

$$2m = 1 + m$$

Thus, the median $m = 1$.

23. Step 1: Find $F(x)$.

If $x < 0$, then $F(x) = 0$. If $x \geq 0$, then

$$F(x) = \int_{-\infty}^{x} f(t)\,dt = \int_{-\infty}^{0} f(t)\,dt + \int_{0}^{x} f(t)\,dt = 0 + \int_{0}^{x} 2e^{-2t}\,dt$$

$$= -e^{-2t}\Big|_{0}^{x} = -e^{-2x} + 1$$

Thus, $F(x) = \begin{cases} 0 & \text{if } x < 0 \\ 1 - e^{-2x} & \text{if } x \geq 0 \end{cases}$

Step 2: Solve $P(X \leq m) = \dfrac{1}{2}$ for m.

$$F(m) = P(X \leq m) = \frac{1}{2}$$
$$1 - e^{-2m} = \frac{1}{2}$$
$$e^{-2m} = \frac{1}{2}$$
$$-2m = \ln\frac{1}{2} = -\ln 2$$
$$m = \frac{\ln 2}{2} \approx 0.347$$

25. $F(x) = \dfrac{1}{2}$ for any x satisfying $0.5 \leq x \leq 2.5$. The median is *not* unique.

27. From the graph,

$$f(x) = \begin{cases} 0 & \text{if } x < 0 \\ 1 - x & \text{if } 0 \leq x \leq 1 \\ 0 & \text{if } 1 < x < 2 \\ x - 2 & \text{if } 2 \leq x \leq 3 \\ 0 & \text{if } x > 3 \end{cases}$$

Now $F(x) = \displaystyle\int_{-\infty}^{x} f(t)\,dt$ and:

for $-\infty < x < 0$, $F(x) = \displaystyle\int_{-\infty}^{x} f(t)\,dt = \int_{-\infty}^{x} 0\,dt = 0;$

for $0 \leq x < 1$,

$$F(x) = \int_{-\infty}^{x} f(t)\,dt = \int_{-\infty}^{0} 0\,dt + \int_{0}^{x} (1 - t)\,dt = \left[\left(t - \frac{1}{2}t^2\right)\Big|_{0}^{x}\right] = x - \frac{1}{2}x^2;$$

for $1 \leq x \leq 2$,

$$F(x) = \int_{-\infty}^{x} f(t)\,dt = \int_{-\infty}^{0} 0\,dt + \int_{0}^{1} (1 - t)\,dt + \int_{1}^{x} 0\,dt = \left[\left(t - \frac{1}{2}t^2\right)\Big|_{0}^{1}\right] = \frac{1}{2};$$

for $2 \leq x \leq 3$,

$$F(x) = \int_{-\infty}^{x} f(t)\,dt = \int_{-\infty}^{0} 0\,dt + \int_{0}^{1} (1 - t)\,dt + \int_{1}^{2} 0\,dt + \int_{2}^{x} (t - 2)\,dt$$

$$= \frac{1}{2} + \left[\left(\frac{1}{2}t^2 - 2t\right)\Big|_{2}^{x}\right]$$

$$= \frac{1}{2} + \frac{1}{2}x^2 - 2x + 2$$

$$= \frac{1}{2}x^2 - 2x + \frac{5}{2};$$

For $x > 3$

$$F(x) = \int_{-\infty}^{x} f(t)\,dt = \int_{-\infty}^{0} 0\,dt + \int_{0}^{1} (1-t)\,dt + \int_{1}^{2} 0\,dt + \int_{2}^{3} (t-2)\,dt + \int_{3}^{x} 0\,dt$$

$$= 0 + \frac{1}{2} + 0 + \frac{1}{2} = 1.$$

Thus,

$$F(x) = \begin{cases} 0 & \text{if } x < 0 \\ x - \frac{1}{2}x^2 & \text{if } 0 \le x < 1 \\ \frac{1}{2} & \text{if } 1 \le x \le 2 \\ \frac{1}{2}x^2 - 2x + \frac{5}{2} & \text{if } 2 < x \le 3 \\ 1 & \text{if } x > 3 \end{cases}$$

Now, $F(x) = \frac{1}{2}$ for any x satisfying $1 \le x \le 2$. The median is *not* unique.

29. Since f is a probability density function,

$$\int_{-\infty}^{\infty} f(x)\,dx = 1 \quad \text{and} \quad \int_{-\infty}^{\infty} xf(x)\,dx = \mu, \text{ the mean.}$$

Now,

$$\int_{-\infty}^{\infty} (ax + b)f(x)\,dx = \int_{-\infty}^{\infty} axf(x)\,dx + \int_{-\infty}^{\infty} bf(x)\,dx$$

$$= a\int_{-\infty}^{\infty} xf(x)\,dx + b\int_{-\infty}^{\infty} f(x)\,dx = a\mu + b.$$

31. Step 1: Find $F(x)$.
If $x < 0$, then $F(x) = 0$. If $0 \le x \le 2$, then

$$F(x) = \int_{-\infty}^{x} f(t)\,dt = \int_{-\infty}^{0} f(t)\,dt + \int_{0}^{x} f(t)\,dt$$

$$= 0 + \frac{1}{2}\int_{0}^{x} t\,dt = \frac{1}{4}t^2\Big|_{0}^{x} = \frac{1}{4}x^2.$$

If $x > 2$, then

$$F(x) = \int_{-\infty}^{x} f(t)\,dt = \int_{-\infty}^{0} f(t)\,dt + \int_{0}^{2} f(t)\,dt + \int_{2}^{x} f(t)\,dt$$

$$= 0 + \frac{1}{2}\int_{0}^{2} t\,dt + 0 = \frac{1}{4}t^2\Big|_{0}^{2} = 1.$$

Thus,

$$F(x) = \begin{cases} 0 & x < 0 \\ \frac{1}{4}x^2 & 0 \le x \le 2 \\ 1 & x > 2 \end{cases}$$

<u>Step 2</u>: In order to find the quartile point x_1, we solve the following for x_1:

$$F(x_1) = P(X \le x_1) = \frac{1}{4}$$
$$\frac{1}{4}x_1^2 = \frac{1}{4}$$
$$x_1^2 = 1$$
$$x_1 = 1$$

For the quartile point x_2 (or m), we solve the following for x_2:

$$F(x_2) = P(X \le x_2) = \frac{1}{2}$$
$$\frac{1}{4}x_2^2 = \frac{1}{2}$$
$$x_2^2 = 2$$
$$x_2 = \sqrt{2} \approx 1.414$$

For the quartile point x_3, we solve the following for x_3:

$$F(x_3) = P(X \le x_3) = \frac{3}{4}$$
$$\frac{1}{4}x_3^2 = \frac{3}{4}$$
$$x_3^2 = 3$$
$$x_3 = \sqrt{3} \approx 1.732$$

33. <u>Step 1</u>: Find $F(x)$.

If $x < 0$, then $F(x) = 0$. If $x \ge 0$, then

$$F(x) = \int_{-\infty}^{x} f(t)\,dt = \int_{-\infty}^{0} f(t)\,dt + \int_{0}^{x} f(t)\,dt = 0 + \int_{0}^{x} \frac{3}{(3+t)^2}\,dt$$

$$= \frac{-3}{3+t}\bigg|_0^x = \frac{-3}{3+x} + 1.$$

Thus, $F(x) = \begin{cases} 0 & \text{if } x < 0 \\ 1 - \dfrac{3}{3+x} & \text{if } x \ge 0 \end{cases}$

<u>Step 2</u>: For the quartile point x_1, we solve:

$$F(x_1) = P(X \le x_1) = \frac{1}{4}$$
$$1 - \frac{3}{3+x_1} = \frac{1}{4}$$
$$-\frac{3}{3+x_1} = -\frac{3}{4}$$
$$3 + x_1 = 4$$
$$x_1 = 1$$

For the quartile point x_2, we solve:

$$F(x_2) = P(X \le x_2) = \frac{1}{2}$$

$$1 - \frac{3}{3 + x_2} = \frac{1}{2}$$

$$-\frac{3}{3 + x_2} = -\frac{1}{2}$$

$$3 + x_2 = 6$$

$$x_2 = 3$$

For the quartile point x_3, we solve:

$$F(x_3) = P(X \le x_3) = \frac{3}{4}$$

$$1 - \frac{3}{3 + x_3} = \frac{3}{4}$$

$$-\frac{3}{3 + x_3} = -\frac{1}{4}$$

$$3 + x_3 = 12$$

$$x_3 = 9$$

35. Given $f(x) = \begin{cases} 4x - 4x^3 & 0 \le x \le 1 \\ 0 & \text{otherwise} \end{cases}$

Step 1. Find $F(x)$.

If $x < 0$, then $F(x) = 0$. If $0 \le x \le 1$, then

$$F(x) = \int_{-\infty}^{x} f(t)\,dt = \int_{-\infty}^{0} f(t)\,dt + \int_{0}^{x} f(t)\,dt$$

$$= 0 + \int_{0}^{x} (4t - 4t^3)\,dt = (2t^2 - t^4)\,\Big|_{0}^{x}$$

$$= 2x^2 - x^4$$

If $x > 1$, then

$$F(x) = \int_{-\infty}^{x} f(t)\,dt = \int_{-\infty}^{0} f(t)\,dt + \int_{0}^{1} f(t)\,dt + \int_{1}^{x} f(t)\,dt$$

$$= 0 + \int_{0}^{1} (4t - 4t^3)\,dt + 0 = (2t^2 - t^4)\,\Big|_{0}^{1} = 1$$

Thus, $F(x) = \begin{cases} 0 & x < 0 \\ 2x^2 - x^4 & 0 \le x \le 1 \\ 1 & x > 1 \end{cases}$

Step 2. Solve the equation $F(m) = \frac{1}{2}$ for m.

The graphs of F and $y = \frac{1}{2}$ are shown at the right.

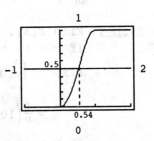

Thus, the median $m \approx 0.54$.

37. $f(x) = \begin{cases} \dfrac{1}{2x^2} + \dfrac{3}{2x^4} & \text{if } x \geq 1 \\ 0 & \text{if } x < 1 \end{cases}$

$\underline{\text{Step 1}}$: Find $F(x)$.

If $x < 1$, then $F(x) = 0$. If $x \geq 1$, then

$$F(x) = \int_{-\infty}^{x} f(t)\,dt = \int_{-\infty}^{1} 0\,dt + \int_{1}^{x} \left(\frac{1}{2t^2} + \frac{3}{2t^4}\right) dt$$

$$= \left[\left(-\frac{1}{2t} - \frac{3}{6t^3}\right)\Big|_{1}^{x}\right]$$

$$= 1 - \frac{1}{2x} - \frac{1}{2x^3}$$

Thus, $F(x) = \begin{cases} 0 & \text{if } x < 1 \\ 1 - \dfrac{1}{2x} - \dfrac{1}{2x^3} & \text{if } x \geq 1 \end{cases}$

$\underline{\text{Step 2}}$. Solve the equation $F(m) = \frac{1}{2}$ for m. The graphs of F and $y = \frac{1}{2}$ are shown below.

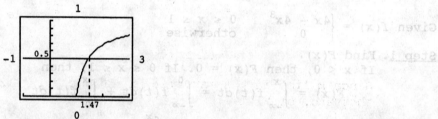

Thus, $m \approx 1.47$.

39. (A) The contractor's expected profit is given by:

$$E(X) = \int_{-\infty}^{\infty} xf(x)\,dx = \int_{6}^{10} x \cdot \frac{1}{8}(10 - x)\,dx = \frac{1}{8}\int_{6}^{10}(10x - x^2)\,dx$$

$$= \frac{1}{8}\left(\frac{10}{2}x^2 - \frac{x^3}{3}\right)\Big|_{6}^{10} = \frac{1}{8}\left[\left(5\cdot10^2 - \frac{1}{3}\cdot10^3\right) - \left(5\cdot6^2 - \frac{1}{3}\cdot6^3\right)\right]$$

$$= \frac{1}{8}\left(\frac{500}{3} - 108\right) = \frac{1}{8}\left(\frac{176}{3}\right) = \frac{22}{3} \approx 7.333 \quad \text{or} \quad \$7333$$

(B) $\underline{\text{Step 1}}$: Find $F(x)$.

If $x < 6$, then $F(x) = 0$. If $6 \leq x \leq 10$, then

$$F(x) = \int_{-\infty}^{x} f(t)\,dt = \int_{-\infty}^{6} f(t)\,dt + \int_{6}^{x} f(t)\,dt = 0 + \frac{1}{8}\int_{6}^{x}(10 - t)\,dt$$

$$= \frac{1}{8}\left(10t - \frac{t^2}{2}\right)\Big|_{6}^{x} = \frac{1}{8}\left[\left(10x - \frac{x^2}{2}\right) - \left(60 - \frac{36}{2}\right)\right] = \frac{1}{8}\left(10x - \frac{x^2}{2} - 42\right).$$

If $x > 10$, then

$$F(x) = \int_{-\infty}^{x} f(t)\,dt = \int_{-\infty}^{6} f(t)\,dt + \int_{6}^{10} f(t)\,dt + \int_{10}^{x} f(t)\,dt$$

$$= 0 + \frac{1}{8} \int_6^{10} (10 - t)\, dt + 0 = \frac{1}{8} \left(10t - \frac{t^2}{2} \right) \Big|_6^{10}$$

$$= \frac{1}{8} [(100 - 50) - (60 - 18)] = 1.$$

Thus,

$$F(x) = \begin{cases} 0 & x < 6 \\ \frac{1}{8} \left(10x - \frac{x^2}{2} - 42 \right) & 6 \le x \le 10 \\ 1 & x > 10 \end{cases}$$

<u>Step 2</u>: To find the median profit, m, we solve the following:

$$F(m) = P(X \le m) = \frac{1}{2}$$

$$\frac{1}{8} \left(10m - \frac{m^2}{2} - 42 \right) = \frac{1}{2}$$

$$10m - \frac{m^2}{2} - 42 = 4$$

$$m^2 - 20m + 92 = 0$$

$$m = \frac{20 \pm \sqrt{20^2 - 4(92)}}{2} \quad \text{(using the quadratic formula)}$$

$$= \frac{20 \pm \sqrt{32}}{2} = 10 \pm 2\sqrt{2}$$

Since m lies in the interval $[6, 10]$, $m = 10 - 2\sqrt{2} \approx \$7.172$ thousand, or $7,172.

41. <u>Step 1</u>: Find $F(x)$.

If $x < 0$, then $F(x) = 0$. If $x \ge 0$, then

$$F(x) = \int_{-\infty}^x f(t)\, dt = \int_{-\infty}^0 f(t)\, dt + \int_0^x f(t)\, dt = 0 + \int_0^x \frac{1}{3} e^{-t/3}\, dt$$

$$= -e^{-t/3} \Big|_0^x = -e^{-x/3} + 1.$$

Thus, $F(x) = \begin{cases} 0 & \text{if } x < 0 \\ 1 - e^{-x/3} & \text{if } x \ge 0 \end{cases}$

<u>Step 2</u>: Solve $P(X \le m) = \frac{1}{2}$ for m.

$$F(m) = P(X \le m) = \frac{1}{2}$$

$$1 - e^{-m/3} = \frac{1}{2}$$

$$e^{-m/3} = \frac{1}{2}$$

$$\frac{-m}{3} = \ln \frac{1}{2} = -\ln 2$$

$$m = 3 \ln 2 \approx 2.079 \text{ minutes}$$

43. The expected daily consumption is given by:

$$E(X) = \int_{-\infty}^{\infty} x f(x)\, dx = \int_{-\infty}^{\infty} \frac{x}{(1 + x^2)^{3/2}}\, dx$$

$$= \lim_{R \to \infty} \int_{0}^{R} \frac{x}{(1 + x^2)^{3/2}}\, dx \quad \left[\underline{\text{Note}}: \text{If } u = 1 + x^2, \text{ then } \frac{du}{2} = x\, dx \text{ and} \right.$$

$$\int \frac{x}{(1 + x^2)^{3/2}}\, dx = \frac{1}{2} \int \frac{du}{u^{3/2}} = \frac{1}{2} \cdot \frac{-2}{u^{1/2}}$$

$$\left. = -\frac{1}{(1 + x^2)^{1/2}}. \right]$$

$$= -\lim_{R \to \infty} \frac{1}{(1 + x^2)^{1/2}} \Big|_0^R = -\lim_{R \to \infty} \left(\frac{1}{(1 + R^2)^{1/2}} - \frac{1}{1} \right) = 1 \text{ or } 1 \text{ million gallons}$$

45. Mean life expectancy is given by:

$$E(X) = \mu = \int_{-\infty}^{\infty} x f(x)\, dx = \frac{1}{5000} \int_{0}^{10} x(10x^3 - x^4)\, dx = \frac{1}{5000} \left(10 \cdot \frac{x^5}{5} - \frac{x^6}{6} \right) \Big|_0^{10}$$

$$= \frac{1}{5000} \left(2 \cdot 10^5 - \frac{1}{6} \cdot 10^6 - 0 \right) = \frac{1}{5000} \left(\frac{1}{3} \cdot 10^5 \right) = \frac{20}{3} \approx 6.7 \text{ minutes}$$

47. <u>Step 1</u>: Find $F(x)$.

If $x < 0$, $F(x) = 0$. If $x \geq 0$, then

$$F(x) = \int_{-\infty}^{x} f(t)\, dt = \int_{-\infty}^{0} f(t)\, dt + \int_{0}^{x} f(t)\, dt = 0 + \int_{0}^{x} \frac{800t}{(400 + t^2)^2}\, dt$$

$$= \frac{-400}{400 + t^2} \Big|_0^x = \frac{-400}{400 + x^2} + 1.$$

Thus, $F(x) = \begin{cases} 0 & \text{if } x < 0 \\ 1 - \dfrac{400}{400 + x^2} & \text{if } x \geq 0 \end{cases}$

<u>Step 2</u>: Solve $P(X \leq m) = \frac{1}{2}$ for m.

$$1 - \frac{400}{400 + m^2} = \frac{1}{2}$$

$$\frac{-400}{400 + m^2} = \frac{-1}{2}$$

$$400 + m^2 = 800$$

$$m^2 = 400$$

$$m = 20 \text{ days}$$

49. The expected number of hours to learn the task is given by:

$$E(X) = \mu = \int_{-\infty}^{\infty} x f(x)\, dx = \int_{0}^{3} x\left(\frac{4}{9} x^2 - \frac{4}{27} x^3 \right) dx = \left(\frac{4}{9} \cdot \frac{x^4}{4} - \frac{4}{27} \cdot \frac{x^5}{5} \right) \Big|_0^3$$

$$= \frac{1}{9}(3^4) - \frac{4}{3^3(5)}(3^5) = 9 - \frac{36}{5} = \frac{9}{5} = 1.8 \text{ hours}$$

Things to remember:

1. UNIFORM PROBABILITY DENSITY FUNCTION

(a) $f(x) = \begin{cases} \dfrac{1}{b - a} & a \le x \le b \\ 0 & \text{otherwise} \end{cases}$

(b) $F(x) = \begin{cases} 0 & x < a \\ \dfrac{x - a}{b - a} & a \le x \le b \\ 1 & x > b \end{cases}$

(c) Mean: $\mu = \dfrac{1}{2}(a + b)$

(d) Median: $m = \dfrac{1}{2}(a + b)$

(e) Standard deviation: $\sigma = \dfrac{1}{\sqrt{12}}(b - a)$

2. EXPONENTIAL PROBABIILTY DENSITY FUNCTION

(a) $f(x) = \begin{cases} \dfrac{1}{\lambda} e^{-x/\lambda} & x \ge 0 \\ 0 & \text{otherwise} \end{cases}$

(b) $F(x) = \begin{cases} 1 - e^{-x/\lambda} & x \ge 0 \\ 0 & \text{otherwise} \end{cases}$

(c) Mean: $\mu = \lambda$

(d) Median: $m = \lambda \ln 2$

(e) Standard deviation: $\sigma = \lambda$

3. NORMAL PROBABILITY DENSITY FUNCTION

$f(x) = \dfrac{1}{\sigma\sqrt{2\pi}} e^{-(x-\mu)^2/2\sigma^2}, \quad \sigma > 0$

Mean: μ

Median: μ

Standard Deviation: σ

The graph of f is symmetric with respect to the line $x = \mu$.

1. Using 1(a) with $[a, b] = [0, 2]$, we have:

$$f(x) = \begin{cases} \dfrac{1}{2-0} & 0 \leq x \leq 2 \\ 0 & \text{otherwise} \end{cases} = \begin{cases} \dfrac{1}{2} & 0 \leq x \leq 2 \\ 0 & \text{otherwise} \end{cases}$$

Using 1(b) with $[a, b] = [0, 2]$, we have:

$$F(x) = \begin{cases} 0 & x < 0 \\ \dfrac{x-0}{2-0} & 0 \leq x \leq 2 \\ 1 & x > 2 \end{cases} = \begin{cases} 0 & x < 0 \\ \dfrac{x}{2} & 0 \leq x \leq 2 \\ 1 & x > 2 \end{cases}$$

3. Using 2(a) with $\lambda = \dfrac{1}{2}$, we have:

$$f(x) = \begin{cases} \dfrac{1}{1/2} e^{-x/(1/2)} & x \geq 0 \\ 0 & \text{otherwise} \end{cases} = \begin{cases} 2e^{-2x} & x \geq 0 \\ 0 & \text{otherwise} \end{cases}$$

Using 2(b) with $\lambda = \dfrac{1}{2}$, we have:

$$F(x) = \begin{cases} 1 - e^{-x/(1/2)} & x \geq 0 \\ 0 & \text{otherwise} \end{cases} = \begin{cases} 1 - e^{-2x} & x \geq 0 \\ 0 & \text{otherwise} \end{cases}$$

5. Using 1(c), (d), (e), with $[a, b] = [1, 5]$, we have:

Mean: $\mu = \dfrac{1}{2}(a + b) = \dfrac{1}{2}(1 + 5) = 3$

Median: $m = \dfrac{1}{2}(a + b) = \dfrac{1}{2}(1 + 6) = 3$

Standard deviation: $\sigma = \dfrac{1}{\sqrt{12}}(b - a) = \dfrac{1}{\sqrt{12}}(5 - 1) = \dfrac{4}{\sqrt{12}} = \dfrac{2}{\sqrt{3}} \approx 1.155$

7. Using 2(c), (d), (e), with $\lambda = 5$, we have:

Mean: $\mu = 5$

Median: $m = 5 \ln 2 \approx 3.466$

Standard deviation: $\sigma = 5$

9. From Table III in the text, the area under the standard normal curve between $z = 0$ and $z = 1$ is $A = 0.3413$.

11. The standard normal curve is symmetric with respect to the line $x = 0$. Thus, the area under the curve between $z = 0$ and $z = -3$ is the same as the area between $z = 0$ and $z = 3$. $A = 0.4987$.

13. From Table III, for $z = 0.9$, $A = 0.3159$.

15. From Table III, for $z = 2.47$, $A = 0.4932$.

17. $z = \dfrac{65 - 60}{10} = 1.5$

From Table III, we have the area corresponding to $z = 1.5$ is 0.4332.

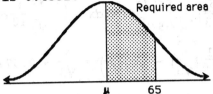

Required area

19. $z = \dfrac{83 - 50}{10} = 3.3$

From Table III, the area corresponding to $z = 3.3$ is 0.4995.

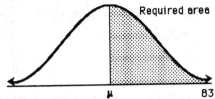

Required area

21. $z = \dfrac{45 - 50}{10} = -0.5$

From Table III, the area corresponding to $z = 0.5$ is 0.1915.

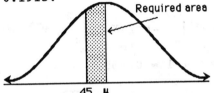

Required area

23. $z = \dfrac{42 - 50}{10} = -0.8$

From Table III, the area corresponding to $z = 0.8$ is 0.2881.

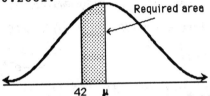

Required area

25. X is uniformly distributed on $[0, 4]$. The probability density function is

$$f(x) = \begin{cases} \dfrac{1}{4} & 0 \le x \le 4 \\ 0 & \text{otherwise} \end{cases}$$

and the mean is

$$\mu = \frac{1}{2}(0 + 4) = 2.$$

Thus, $P(X \le 2) = \displaystyle\int_0^2 f(x)\,dx = \int_0^2 \frac{1}{4}\,dx = \frac{1}{4}x\Big|_0^2 = \frac{1}{2}.$

27. X is an exponential random variable with $\mu = 1$. Then

$$f(x) = \begin{cases} e^{-x} & x \ge 0 \\ 0 & \text{otherwise} \end{cases}$$

and the mean is

$$\mu = 1.$$

Thus, $P(X \le 1) = \displaystyle\int_0^1 e^{-x}\,dx = -e^{-x}\Big|_0^1 = -e^{-1} + 1 = 1 - e^{-1} \approx 0.632.$

29. X is uniformly distributed on $[-5, 5]$. The mean is

$$\mu = \frac{1}{2}(-5 + 5) = 0$$

and the standard deviation is

$$\sigma = \frac{1}{\sqrt{12}}[5 - (-5)] = \frac{10}{\sqrt{12}} = \frac{5}{\sqrt{3}} \approx 2.887.$$

Also,

$$f(x) = \begin{cases} \dfrac{1}{10} & -5 \le x \le 5 \\ 0 & \text{otherwise} \end{cases}$$

Thus, $P(\mu - \sigma \le X \le \mu + \sigma) = P\left(\dfrac{-5}{\sqrt{3}} \le X \le \dfrac{5}{\sqrt{3}}\right) = \displaystyle\int_{-5/\sqrt{3}}^{5/\sqrt{3}} f(x)\,dx = \int_{-5/\sqrt{3}}^{5/\sqrt{3}} \dfrac{1}{10}\,dx$

$$= \dfrac{1}{10} x \Big|_{-5\sqrt{3}}^{5/\sqrt{3}} = \dfrac{1}{10}\left(\dfrac{5}{\sqrt{3}}\right) - \dfrac{1}{10}\left(\dfrac{-5}{\sqrt{3}}\right) = \dfrac{1}{\sqrt{3}} \approx 0.577.$$

31. X is an exponential random variable with median $m = 6 \ln 2$. Since $m = \lambda \ln 2$, we have $\lambda = 6$. Thus, mean $\mu = 6$ and standard deviation $\sigma = 6$. Also,

$$f(x) = \begin{cases} \dfrac{1}{6} e^{-x/6} & x \ge 0 \\ 0 & \text{otherwise} \end{cases}$$

Now, $P(\mu - \sigma \le X \le \mu + \sigma) = P(0 \le X \le 12) = \displaystyle\int_0^{12} f(x)\,dx = \int_0^{12} \dfrac{1}{6} e^{-x/6}\,dx$

$$= -e^{-x/6} \Big|_0^{12} = -e^{-2} + 1 \approx 0.865.$$

33. $\mu = 70$, $\sigma = 8$

z (for $x = 60$) $= \dfrac{60 - 70}{8} = -1.25$

z (for $x = 80$) $= \dfrac{80 - 70}{8} = 1.25$

Area $A_1 = 0.3944$. Area $A_2 = 0.3944$.
Total area $= A = A_1 + A_2 = 0.7888$.

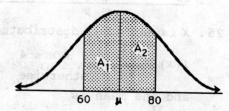

35. $\mu = 70$, $\sigma = 8$

z (for $x = 62$) $= \dfrac{62 - 70}{8} = -1.00$

z (for $x = 74$) $= \dfrac{74 - 70}{8} = 0.5$

Area $A_1 = 0.3413$. Area $A_2 = 0.1915$.
Total area $= A = A_1 + A_2 = 0.5328$.

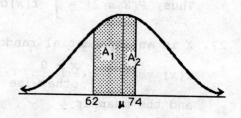

37. $\mu = 70$, $\sigma = 8$

z (for $x = 88$) $= \dfrac{88 - 70}{8} = 2.25$

Required area $= 0.5 - \begin{array}{l}\text{(area corresponding to}\\ \quad z = 2.25)\end{array}$

$$= 0.5 - 0.4878$$
$$= 0.0122$$

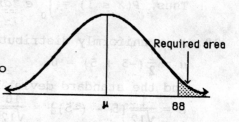

39. $\mu = 70$, $\sigma = 8$

z (for $x = 60$) $= \dfrac{60 - 70}{8} = -1.25$

Required area $= 0.5 - \begin{array}{c}\text{(area corresponding to}\\ z = 1.25)\end{array}$

$= 0.5 - 0.3944$

$= 0.1056$

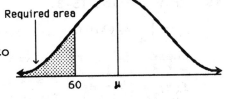

Required area

60   μ

41. Normal probability density function with $\mu = 0$:

$$f(x) = \frac{1}{\sigma\sqrt{2\pi}} e^{-x^2/2\sigma^2}, \quad \sigma > 0.$$

(A) $\sigma = 0.5$

$$P(-0.5 \le X \le 0.5) = \frac{1}{0.5\sqrt{2\pi}} \int_{-0.5}^{0.5} e^{-x^2/0.5} \, dx$$

Using a graphing utility with a numerical integration routine, we find that

$$P(-0.5 \le X \le 0.5) \approx 0.6827$$

(B) $\sigma = 1$

$$P(-1 \le X \le 1) = \frac{1}{\sqrt{2\pi}} \int_{-1}^{1} e^{-x^2/2} \, dx \approx 0.6827$$

(C) $\sigma = 2$

$$P(-2 \le X \le 2) = \frac{1}{2\sqrt{2\pi}} \int_{-2}^{2} e^{-x^2/8} \, dx \approx 0.6827$$

Interpretation: the probability of being within one standard deviation of the mean is always 0.6827.

43. Normal probability density function with $\mu = 0$:

$$f(x) = \frac{1}{\sigma\sqrt{2\pi}} e^{-x^2/2\sigma^2}, \quad \sigma > 0.$$

(A) $\sigma = 0.5$

$$P(-1.5 \le X \le 1.5) = \frac{1}{0.5\sqrt{2\pi}} \int_{-1.5}^{1.5} e^{-x^2/0.5} \, dx$$

Using a graphing utility with a numerical integration routine, we find that

$$P(-1.5 \le X \le 1.5) \approx 0.9973$$

(B) $\sigma = 1$

$$P(-3 \le X \le 3) = \frac{1}{\sqrt{2\pi}} \int_{-3}^{3} e^{-x^2/2} \, dx \approx 0.9973$$

(C) $\sigma = 2$

$$P(-6 \le X \le 6) = \frac{1}{2\sqrt{2\pi}} \int_{-6}^{6} e^{-x^2/8} \, dx \approx 0.9973$$

Interpretation: the probability of being within 3 standard deviations of the mean is always 0.9973.

45. From Section 15-3,

$$\mu = \int_{-\infty}^{\infty} x f(x)\, dx,$$

where

$$f(x) = \begin{cases} \dfrac{1}{b-a} & a \leq x \leq b \\ 0 & \text{otherwise} \end{cases}$$

Thus,

$$\mu = \int_{a}^{b} x \left(\frac{1}{b-a} \right) dx = \frac{1}{b-a} \int_{a}^{b} x\, dx = \frac{1}{b-a} \cdot \frac{x^2}{2}\Big|_{a}^{b} = \frac{1}{b-a}\left(\frac{b^2}{2} - \frac{a^2}{2} \right)$$

$$= \frac{1}{b-a} \cdot \frac{1}{2}(b-a)(b+a) = \frac{a+b}{2}.$$

47. $\displaystyle \int_{-\infty}^{\infty} x^2 f(x)\, dx = \int_{a}^{b} x^2 \left(\frac{1}{b-a} \right) dx = \frac{1}{b-a} \cdot \frac{x^3}{3}\Big|_{a}^{b} = \frac{1}{3} \cdot \frac{1}{b-a}(b^3 - a^3)$

$$= \frac{1}{3} \cdot \frac{1}{b-a}(b-a)(b^2 + ab + a^2) = \frac{1}{3}(b^2 + ab + a^2)$$

49. $f(x) = \begin{cases} \dfrac{p}{x^{p+1}} & \text{if } x \geq 1 \\ 0 & \text{otherwise} \end{cases} \quad p > 0$

(A) Clearly, $f(x) \geq 0$ for all x.

(B) $\displaystyle \int_{-\infty}^{\infty} f(x)\, dx = \int_{1}^{\infty} \frac{p}{x^{p+1}}\, dx = \lim_{b \to \infty} \int_{1}^{b} \frac{p}{x^{p+1}}\, dx$

$$= \lim_{b \to \infty} \left[-\frac{1}{x^p}\Big|_{1}^{b} \right]$$

$$= \lim_{b \to \infty} \left[1 - \frac{1}{b^p} \right] = 1$$

Thus, f is a probability density function.

The mean μ of f is given by:

$$\mu = \int_{-\infty}^{\infty} x f(x)\, dx = \int_{-\infty}^{\infty} x \left(\frac{p}{x^{p+1}} \right) dx$$

$$= \int_{1}^{\infty} \frac{p}{x^p}\, dx = \lim_{b \to \infty} \int_{1}^{b} \frac{p}{x^p}\, dx$$

$$= \lim_{b \to \infty} \left[\frac{p x^{-p+1}}{-p+1}\Big|_{1}^{b} \right], \quad p \neq 1$$

$$= \lim_{b \to \infty} \left[\frac{p b^{-p+1}}{1-p} - \frac{p}{1-p} \right]$$

$$= \frac{p}{p-1} \quad \text{provided } p > 1;$$

The limit does not exist if $p < 1$.

If $p = 1$, then

$$\int_{1}^{\infty} \frac{1}{x}\, dx = \lim_{b \to \infty} \int_{1}^{b} \frac{1}{x}\, dx$$

$$= \lim_{b \to \infty} [\ln b]$$

does not exist.

Thus,

$$\mu = \begin{cases} \dfrac{p}{p-1} & \text{if } p > 1 \\ \text{does not exist} & \text{if } 0 < p \leq 1 \end{cases}$$

51. The median m is given by

$$P(X \leq m) = \frac{1}{2}$$

Clearly, $m > 1$ since $f(x) = 0$ for $x < 1$.

Now,

$$P(X \leq m) = \int_{-\infty}^{m} f(x)\,dx = \int_{1}^{m} \frac{p}{x^{p+1}}\,dx$$

$$= \frac{-1}{x^p} \Big|_{1}^{m} = 1 - \frac{1}{m^p}$$

and

$$1 - \frac{1}{m^p} = \frac{1}{2}$$

$$\frac{1}{m^p} = \frac{1}{2}$$

$$m^p = 2$$

$$m = \sqrt[p]{2}$$

53. Using the uniform cumulative probability distribution function:

$$P(25 \leq X \leq 40) = F(40) - F(25)$$

$$= \frac{40 - 0}{40 - 0} - \frac{25 - 0}{40 - 0} \qquad [\underline{\text{Note}}: a = 0,\ b = 40.]$$

$$= 1 - \frac{25}{40} = \frac{15}{40} = \frac{3}{8} = 0.375$$

55. We are given that $\mu = 3$. Also $\lambda = \mu$. Thus, $\lambda = 3$, and we have:

$$P(0 \leq X \leq 2) = F(2) - F(0) = (1 - e^{-2/3}) - (1 - e^0)$$

$$= 1 - e^{-2/3} \approx 0.487$$

57. X is an exponential random variable and the median $m = 2$. Since $m = \lambda \ln 2$, we have $2 = \lambda \ln 2$ or $\lambda = \dfrac{2}{\ln 2}$. Now $P(X \leq 1) = F(1)$, where F is the cumulative probability distribution:

$$F(x) = \begin{cases} 1 - e^{-x/\lambda} & x \geq 0 \\ 0 & \text{otherwise} \end{cases}$$

Setting $\lambda = \dfrac{2}{\ln 2}$, we have

$$F(x) = \begin{cases} 1 - e^{-(x \ln 2)/2} & x \geq 0 \\ 0 & \text{otherwise} \end{cases}$$

Thus, $P(X \leq 1) = 1 - e^{-(\ln 2)/2} = 1 - e^{-(1/2)\ln 2}$

$$= 1 - e^{\ln (2)^{-1/2}}$$

$$= 1 - 2^{-1/2}$$
$$= 1 - \frac{1}{\sqrt{2}} \approx 0.293.$$

59. $\mu = 200{,}000$, $\sigma = 20{,}000$, $x \geq 240{,}000$

z (for $x = 240{,}000$) $= \dfrac{240{,}000 - 200{,}000}{20{,}000} = 2.0$

Fraction of the salesmen who would be expected to make annual sales of $240,000 or more

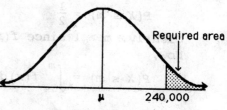

Required area

= Area A_1

= 0.5 - (area between μ and 240,000)

= 0.5 - 0.4772

= 0.0228

Thus, the percentage of salesmen expected to make annual sales of \$240,000 or more is 2.28%.

61. $x = 105$, $x = 95$, $\mu = 100$, $\sigma = 2$

z (for $x = 105$) $= \dfrac{105 - 100}{2} = 2.5$

z (for $x = 95$) $= \dfrac{95 - 100}{2} = -2.5$

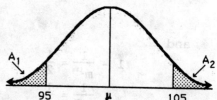

Fraction of parts to be rejected = Area $A_1 + A_2$

$= 1 - 2(\text{area corresponding to } z = 2.5)$

$= 1 - 2(0.4938)$

$= 0.0124$

Thus, the percentage of parts to be rejected is 1.24%.

63. (A) We are given that

$P(0 \leq X \leq 1) = 0.3 = F(1) - F(0) = 1 - e^{-1/\lambda} - (1 - e^0)$.

Thus,

$1 - e^{-1/\lambda} = 0.3$

$e^{-1/\lambda} = 0.7$

$-\dfrac{1}{\lambda} = \ln(0.7)$

$\lambda = -\dfrac{1}{\ln(0.7)}$

But $E(X) = \mu = \lambda = -\dfrac{1}{\ln(0.7)} \approx 2.8$ years.

(B) $P(X \geq 2.8) = \displaystyle\int_{2.8}^{\infty} f(x)\,dx = \lim_{R \to \infty} \int_{2.8}^{R} \frac{1}{\lambda} e^{-x/\lambda}\,dx$

$= \displaystyle\lim_{R \to \infty} \int_{2.8}^{R} \frac{1}{2.8} e^{-x/2.8}\,dx = \lim_{R \to \infty} \frac{1}{2.8} (-2.8 e^{-x/2.8}) \Big|_{2.8}^{R}$

$= \displaystyle\lim_{R \to \infty} (-e^{-R/2.8} + e^{-2.8/2.8})$

$= e^{-1} \approx 0.368$

65. $\mu = 240$, $\sigma = 20$

8 days = 192 hours = x

z (for $x = 192$) = $\dfrac{192 - 240}{20} = -2.4$

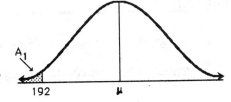

Fraction of people having this incision who would heal in 192 hours or less = Area A_1

= 0.5 - (area corresponding to $z = 2.4$)

= 0.5 - 0.4918

= 0.0082

Thus, the percentage of people who would heal in 8 days or less is 0.82%.

67. X is an exponential random variable with mean $\mu = 2$ (minutes).

Thus, $\lambda = 2$ and

$$P(X \geq 5) = \int_5^\infty f(x)\,dx = \int_5^\infty \frac{1}{2} e^{-x/2}\,dx = \lim_{R \to \infty} \int_5^R \frac{1}{2} e^{-x/2}\,dx = \lim_{R \to \infty} (-e^{-x/2}) \Big|_5^R$$

$$= \lim_{R \to \infty} (-e^{-R/5} + e^{-5/2}) = e^{-5/2} = e^{-2.5} \approx 0.082.$$

69. $\mu = 70$, $\sigma = 8$

We compute x_1, x_2, x_3, and x_4 corresponding to z_1, z_2, z_3, and z_4, respectively. The area between μ and x_3 is 0.2.

Hence, from the table, $z_3 = 0.52$ (approximately). Thus, we have:

$$0.52 = \frac{x_3 - 70}{8}$$

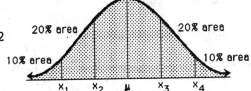

$x_3 - 70 = 4.16$ $\left[\underline{\text{Note}}: z = \dfrac{x - \mu}{\sigma}.\right] \approx 4.2$

and $x_3 = 74.2$.

Also, $x_2 = 70 - 4.2 = 65.8$

The area between μ and x_4 is 0.4. Hence, from the table, $z_4 = 1.28$ (approximately). Therefore:

$$1.28 = \frac{x_4 - 70}{8}$$

$x_4 - 70 = 10.24 \approx 10.2$

and $x_4 = 70 + 10.2 = 80.2$.

Also, $x_1 = 70 - 10.2 = 59.8$.

Thus, we have $x_1 = 59.8$, $x_2 = 65.8$, $x_3 = 74.2$, $x_4 = 80.2$. So, A's = 80.2 or greater, B's = 74.2 to 80.2, C's = 65.8 to 74.2, D's = 59.8 to 65.8, and F's = 59.8 or lower.

1. $\displaystyle\int_0^\infty e^{-2x}dx = \lim_{b\to\infty}\int_0^b e^{-2x}dx = \lim_{b\to\infty}\frac{e^{-2x}}{-2}\bigg|_0^b = \lim_{b\to\infty}\left(\frac{-e^{-2b}}{2} + \frac{1}{2}\right) = \frac{1}{2}$

Thus, the improper integral converges. (15-1)

2. $\displaystyle\int_0^\infty \frac{1}{x+1}dx = \lim_{b\to\infty}\int_0^b \frac{1}{x+1}dx = \lim_{b\to\infty}\int_1^{b+1}\frac{1}{u}du$

$\qquad\qquad\qquad = \lim_{b\to\infty}\ln u\bigg|_1^{b+1}$

$\qquad\qquad\qquad = \lim_{b\to\infty}\ln|b+1|$

$u = x + 1$
$du = dx$
$u = b + 1$ when $x = b$
$u = 1$ when $b = 0$

This limit does not exist. Thus, the improper integral diverges. (15-1)

3. $\displaystyle\int_1^\infty \frac{16\,dx}{x^3} = 16\lim_{b\to\infty}\int_1^b x^{-3}dx = 16\lim_{b\to\infty}\frac{x^{-2}}{-2}\bigg|_1^b = 16\lim_{b\to\infty}\left(\frac{-b^{-2}}{2} + \frac{1}{2}\right) = 16\left(\frac{1}{2}\right) = 8$

(15-1)

4. $\displaystyle P(0 \le X \le 1) = \int_0^1\left(1 - \frac{1}{2}x\right)dx$

$\qquad\qquad\quad = \left(x - \frac{1}{4}x^2\right)\bigg|_0^1$

$\qquad\qquad\quad = 1 - \frac{1}{4} = 0.75$

The graph of the function is given
at the right. (15-2)

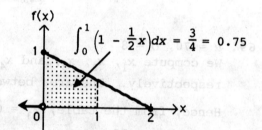

$\displaystyle\int_0^1\left(1 - \frac{1}{2}x\right)dx = \frac{3}{4} = 0.75$

5. $\displaystyle \mu = E(X) = \int_{-\infty}^\infty xf(x)\,dx = \int_0^2 x\left(1 - \frac{1}{2}x\right)dx = \int_0^2\left(x - \frac{1}{2}x^2\right)dx$

$\qquad\quad = \left(\frac{x^2}{2} - \frac{1}{6}x^3\right)\bigg|_0^2 = 2 - \frac{8}{6} = \frac{2}{3} \sim 0.6667$

$\displaystyle V(X) = \int_{-\infty}^\infty x^2 f(x)\,dx - \mu^2 = \int_0^2 x^2\left(1 - \frac{1}{2}x\right)dx - \left(\frac{2}{3}\right)^2$

$\qquad\quad = \int_0^2\left(x^2 - \frac{1}{2}x^3\right)dx - \frac{4}{9} = \left(\frac{x^3}{3} - \frac{1}{8}x^4\right)\bigg|_0^2 - \frac{4}{9}$

$\qquad\quad = \frac{8}{3} - \frac{16}{8} - \frac{4}{9} = \frac{16}{72} = \frac{2}{9} \sim 0.2222$

$\sigma = \sqrt{V(X)} = \sqrt{\frac{2}{9}} = \frac{\sqrt{2}}{3} \sim 0.4714$

(15-3)

6. When $x < 0$, then $F(x) = 0$. When $0 \le x \le 2$,

$$F(x) = \int_{-\infty}^{x} f(t)\,dt = \int_{-\infty}^{0} f(t)\,dt + \int_{0}^{x} f(t)\,dt = 0 + \int_{0}^{x}\left(1 - \frac{1}{2}t\right)dt$$

$$= \left(t - \frac{1}{4}t^2\right)\Big|_{0}^{x} = x - \frac{1}{4}x^2.$$

When $x > 2$,

$$F(x) = \int_{-\infty}^{x} f(t)\,dt = \int_{-\infty}^{0} f(t)\,dt + \int_{0}^{2} f(t)\,dt + \int_{2}^{x} f(t)\,dt$$

$$= 0 + \int_{0}^{2}\left(1 - \frac{1}{2}t\right)dt + 0 = \left(t - \frac{1}{4}t^2\right)\Big|_{0}^{2} = 2 - \frac{1}{4}\cdot 4 = 1.$$

Thus,

$$F(x) = \begin{cases} 0 & x < 0 \\ x - \frac{1}{4}x^2 & 0 \le x \le 2 \\ 1 & x > 2 \end{cases}$$

The graph of $F(x)$ is shown at the right.

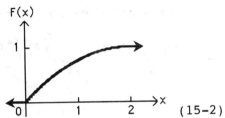
$F(x)$

(15-2)

7. We must solve the following for m:

$$F(m) = P(X \le m) = \frac{1}{2}$$

$$m - \frac{1}{4}m^2 = \frac{1}{2} \quad \text{(refer to Problem 6)}$$

$$m^2 - 4m + 2 = 0$$

$$m = \frac{4 \pm \sqrt{16 - 4(2)}}{2} = \frac{4 \pm \sqrt{8}}{2}$$

$$m = 2 \pm \sqrt{2}$$

Thus, the median is $x = 2 - \sqrt{2} \approx 0.5858$. (15-3)

8. $\mu = 100$, $\sigma = 10$, $x = 118$.

$$z = \frac{118 - 100}{10} = 1.8$$

$P(100 \le X \le 118) = P(0 \le Z \le 1.8) = $ area of region over the interval $[0, 1.8]$
$= 0.4641$ (from Table III). (15-4)

9. The uniform probability density function on $[5, 15]$ is given by:

$$f(x) = \begin{cases} \dfrac{1}{15 - 5} & \text{if } 5 \le x \le 15 \\ 0 & \text{otherwise} \end{cases}$$

Thus,

$$f(x) = \begin{cases} \dfrac{1}{10} & \text{if } 5 \le x \le 15 \\ 0 & \text{otherwise} \end{cases}$$

The cumulative distribution function is given by:

$$F(x) = \begin{cases} 0 & \text{if } x < 5 \\ \dfrac{1}{10}(x - 5) & \text{if } 5 \le x \le 15 \\ 1 & \text{if } x > 15 \end{cases}$$

(15-4)

10. The exponential probability density function with $\lambda = \frac{1}{5} = 0.2$ is given by:

$$f(x) = \begin{cases} \dfrac{1}{0.2}\, e^{-x/0.2} & \text{if } x \geq 0 \\ 0 & \text{otherwise} \end{cases}$$

Thus,

$$f(x) = \begin{cases} 5e^{-5x} & \text{if } x \geq 0 \\ 0 & \text{otherwise} \end{cases}$$

The cumulative distribution function is given by:

$$F(x) = \begin{cases} 0 & \text{if } x < 0 \\ 1 - e^{-5x} & \text{if } x \geq 0 \end{cases}$$

$$(15\text{-}4)$$

11.
$$P(1 \leq X \leq 4) = \int_1^4 \frac{5}{2} x^{-7/2}\, dx$$

$$= \frac{5}{2}\left(-\frac{2}{5}\right) x^{-5/2} \Big|_1^4$$

$$= -[(4)^{-5/2} - (1)^{-5/2}]$$

$$= -\left(\frac{1}{32} - 1\right) = \frac{31}{32}$$

$$\approx 0.9688$$

The graph is shown at the right.

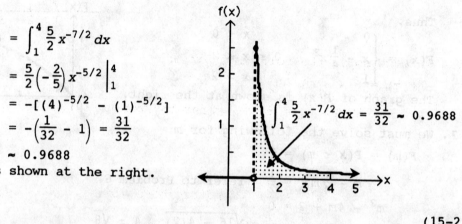

$$(15\text{-}2)$$

12.
$$\mu = E(X) = \int_{-\infty}^{\infty} x f(x)\, dx = \int_1^{\infty} x \cdot \frac{5}{2} x^{-7/2}\, dx$$

$$= \lim_{R \to \infty} \int_1^R \frac{5}{2} x^{-5/2}\, dx = \lim_{R \to \infty}\left(\frac{5}{2}\left(-\frac{2}{3}\right) x^{-3/2}\, \Big|_1^R\right)$$

$$= -\frac{5}{3} \lim_{R \to \infty}(R^{-3/2} - 1^{-3/2}) = -\frac{5}{3}(-1) = \frac{5}{3} \approx 1.667$$

$$V(X) = \int_{-\infty}^{\infty} x^2 f(x)\, dx - \mu^2 = \int_1^{\infty} x^2 \cdot \frac{5}{2} x^{-7/2}\, dx - \left(\frac{5}{3}\right)^2$$

$$= \lim_{R \to \infty} \int_1^R \frac{5}{2} x^{-3/2}\, dx - \frac{25}{9} = \lim_{R \to \infty}\left(\frac{5}{2}(-2) x^{-1/2}\, \Big|_1^R\right) - \frac{25}{9}$$

$$= -5 \lim_{R \to \infty}[R^{-1/2} - (1)^{-1/2}] - \frac{25}{9} = -5(-1) - \frac{25}{9} = 5 - \frac{25}{9} = \frac{20}{9} \approx 2.2222$$

$$\sigma = \sqrt{V(X)} = \sqrt{\frac{20}{9}} = \frac{2}{3}\sqrt{5} \approx 1.4907$$

$$(15\text{-}3)$$

13. When $x < 1$, $F(x) = 0$. When $x \geq 1$,

$$F(x) = \int_{-\infty}^{x} f(t)\,dt = \int_{-\infty}^{1} f(t)\,dt + \int_{1}^{x} f(t)\,dt = 0 + \int_{1}^{x} \frac{5}{2} t^{-7/2}\,dt$$

$$= \frac{5}{2}\left(-\frac{2}{5}\right) t^{-5/2}\Big|_{1}^{x} = -(x^{-5/2} - 1^{-5/2}) = 1 - x^{-5/2}.$$

Thus,

$$F(x) = \begin{cases} 1 - x^{-5/2} & x \geq 1 \\ 0 & \text{otherwise} \end{cases}$$

The graph of $F(x)$ is shown at the right.

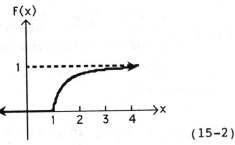

(15-2)

14. We must solve the following for m:

$$F(m) = P(X \leq m) = \frac{1}{2}$$

$$1 - m^{-5/2} = \frac{1}{2} \qquad \text{(refer to Problem 13)}$$

$$m^{-5/2} = \frac{1}{2}$$

$$\frac{1}{m^{5/2}} = \frac{1}{2}$$

$$m^{5/2} = 2 \qquad \text{(square both sides)}$$

$$m^5 = 4$$

$$m = (4)^{1/5}$$

$$m = 2^{2/5} \approx 1.32$$

(15-3)

15. $P(4 \leq X) = \displaystyle\int_{4}^{\infty} f(x)\,dx = e^{-2}$ (Solve for λ.)

$$= \int_{4}^{\infty} \frac{1}{\lambda} e^{-x/\lambda}\,dx = \lim_{R \to \infty} \frac{1}{\lambda} \int_{4}^{R} e^{-x/\lambda}\,dx = e^{-2}$$

$$= \lim_{R \to \infty} \frac{1}{\lambda} -\lambda e^{-x/\lambda}\Big|_{4}^{R} = -\lim_{R \to \infty} (e^{-R/\lambda} - e^{-4/\lambda}) = e^{-2}$$

$$= e^{-4/\lambda} = e^{-2}$$

Thus, we have $-\dfrac{4}{\lambda} = -2$

$$\lambda = 2.$$

The probability density function, with $\lambda = 2$, is:

$$f(x) = \begin{cases} \dfrac{1}{\lambda} e^{-x/\lambda} & x \geq 0 \\ 0 & \text{otherwise} \end{cases} = \begin{cases} \dfrac{1}{2} e^{-x/2} & x \geq 0 \\ 0 & \text{otherwise} \end{cases}$$

(15-4)

16. $P(0 \leq X \leq 2) = \int_0^2 f(x)\,dx = \int_0^2 \frac{1}{2} e^{-x/2}\,dx$ (refer to Problem 15)

$$= \frac{1}{2}(-2e^{-x/2})\Big|_0^2 = -(e^{-2/2} - e^0)$$

$$= -(e^{-1} - 1) \text{ or } (1 - e^{-1}) \approx 0.6321 \qquad (15\text{-}2)$$

17. $F(x) = \begin{cases} 1 - e^{-x/\lambda} & x \geq 0 \\ 0 & \text{otherwise} \end{cases} = \begin{cases} 1 - e^{-x/2} & x \geq 0 \\ 0 & \text{otherwise} \end{cases}$ [<u>Note</u>: $\lambda = 2$.]

$(15\text{-}4)$

18. $\mu = \lambda = 2$

$\sigma = \lambda = 2$

$m = \lambda \ln 2 = 2 \ln 2 \approx 1.3863 \qquad (15\text{-}4)$

19. (A) $\mu = 50$, $\sigma = 6$

z (for $x = 41$) $= \dfrac{41 - 50}{6} = -1.5$

z (for $x = 62$) $= \dfrac{62 - 50}{6} = 2.0$

Required area $= A_1 + A_2$

= (area corresponding to
    $z = 1.5$) + (area
    corresponding to $z = 2$)

= 0.4332 + 0.4772

= 0.9104

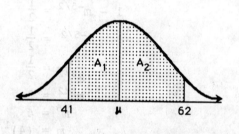

(B) z (for $x = 59$) $= \dfrac{59 - 50}{6} = 1.5$

Required area $= 0.5 - \begin{matrix}\text{(area corresponding to}\\ z = 1.5)\end{matrix}$

= 0.5 - 0.4332

= 0.0668

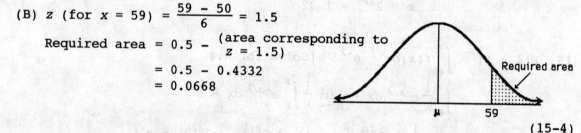

Required area

$(15\text{-}4)$

20. Given $\mu = 82$ and $\sigma = 8$.

(A) We first find the number of standard deviations that 84 and 94 are from the mean.

For $x = 84$: $z = \dfrac{84 - 82}{8} = \dfrac{2}{8} = 0.25$

For $x = 94$: $z = \dfrac{94 - 82}{8} = \dfrac{12}{8} = 1.5$

Now, $P(84 \leq X \leq 94) = P(0.25 \leq z \leq 1.5)$

$= 0.4332 - 0.0987$

$= 0.3345$

(B) For $x = 60$: $z = \dfrac{60 - 82}{8} = -\dfrac{22}{8} = -2.75$

$\qquad P(X \geq 60) = P(z \geq -2.75) = 0.4970 + 0.5000$
$\qquad\qquad\qquad\qquad\qquad\qquad = 0.9970$ $\hfill$ (15-4)

21. $\displaystyle\int_{-\infty}^{0} e^x dx = \lim_{a \to -\infty} \int_{a}^{0} e^x dx = \lim_{a \to -\infty} e^x \Big|_{a}^{0} = \lim_{a \to -\infty} (1 - e^a) = 1,$

since $e^a \to 0$ as $a \to -\infty$. $\hfill$ (15-1)

22. $\displaystyle\int_{0}^{\infty} \dfrac{1}{(x + 3)^2} dx = \lim_{b \to \infty} \int_{0}^{b} \dfrac{1}{(x + 3)^2} dx \qquad\qquad$ Substitution: $u = x + 3$
$\qquad\qquad\qquad\qquad\qquad\qquad\qquad\qquad\qquad\qquad\qquad\qquad\qquad\quad du = dx$

$\qquad\qquad = \lim_{b \to \infty} \dfrac{-1}{(x + 3)} \Big|_{0}^{b} = \lim_{b \to \infty} \left(\dfrac{-1}{b + 3} + \dfrac{1}{3} \right) = \dfrac{1}{3}$

Thus, the improper integral converges. $\hfill$ (15-1)

23. Yes. Since $\displaystyle\int_{-1}^{\infty} f(x)\,dx = \int_{-1}^{1} f(x)\,dx + \int_{1}^{\infty} f(x)\,dx$ and $\displaystyle\int_{-1}^{\infty} f(x)\,dx = L$ exists,

it follows that $\displaystyle\int_{1}^{\infty} f(x)\,dx = \int_{-1}^{\infty} f(x)\,dx - \int_{-1}^{1} f(x)\,dx = L - \int_{-1}^{1} f(x)\,dx$

exists. $\hfill$ (15-1)

24. $f(x) = \begin{cases} e^{-10x} & \text{if } x \geq 0 \\ 0 & \text{otherwise} \end{cases}$

$f(x) \geq 0$ on $(-\infty, \infty)$ and

$\qquad \displaystyle\int_{-\infty}^{\infty} f(x)\,dx = \int_{-\infty}^{0} f(x)\,dx + \int_{0}^{\infty} f(x)\,dx$

$\qquad\qquad = \displaystyle\int_{0}^{\infty} e^{-10x}\,dx = \lim_{b \to \infty} \int_{0}^{b} e^{-10x}\,dx$

$\qquad\qquad = \displaystyle\lim_{b \to \infty} \left[-\dfrac{1}{10} e^{-10x} \right]_{0}^{b}$

$\qquad\qquad = \displaystyle\lim_{b \to \infty} \left[\dfrac{1}{10} - \dfrac{1}{10} e^{-10b} \right] = \dfrac{1}{10}$

Therefore, let $k = 10$; $f(x) = 10e^{-10x}$ is a probability density
function. $\hfill$ (15-2)

25. $f(x) = \begin{cases} e^{10x} & \text{if } x \geq 0 \\ 0 & \text{otherwise} \end{cases}$

Since $\displaystyle\int_{-\infty}^{\infty} f(x)\,dx = \int_{-\infty}^{0} f(x)\,dx + \int_{0}^{\infty} f(x)\,dx$

$\qquad\qquad = \displaystyle\int_{0}^{\infty} e^{10x}\,dx = \lim_{b \to \infty} \int_{0}^{b} e^{10x}\,dx$

$\qquad\qquad = \displaystyle\lim_{b \to \infty} \left[\dfrac{1}{10} e^{10x} \right]_{0}^{b}$

$\qquad\qquad = \displaystyle\lim_{b \to \infty} \left[\dfrac{1}{10} e^{10b} - \dfrac{1}{10} \right] \text{ diverges,}$

no constant k exists. $\hfill$ (15-2)

26. X is an exponentially distributed random variable with median $m = 3 \ln 2$. It follows that the mean $\mu = 3$ and the probability density function is:

$$f(x) = \begin{cases} (1/3)e^{-x/3} & \text{if } x \geq 0 \\ 0 & \text{otherwise} \end{cases}$$

Now, $P(\underline{X}) \leq 3 = \displaystyle\int_{-\infty}^{3} f(x)\,dx = \int_{-\infty}^{0} f(x)\,dx + \int_{0}^{3} f(x)\,dx$

$$= \int_{0}^{3} \frac{1}{3} e^{-x/3}\,dx$$

$$= \frac{1}{3}\left[-3e^{-x/3} \right]_{0}^{3} = 1 - e^{-1} \approx 0.6321 \qquad (15\text{-}4)$$

27. $\mu = \displaystyle\int_{0}^{\infty} x f(x)\,dx = \int_{0}^{\infty} \frac{50x}{(x+5)^3}\,dx$

$$= \lim_{R \to \infty} \int_{0}^{R} \frac{50x}{(x+5)^3}\,dx = 50 \lim_{R \to \infty} \int_{0}^{R} \left[\frac{1}{(x+5)^2} - \frac{5}{(x+5)^3} \right] dx$$

$$= 50 \lim_{R \to \infty} \left(-\frac{1}{(x+5)} + \frac{5}{2} \cdot \frac{1}{(x+5)^2} \right) \Big|_{0}^{R}$$

$$= 50 \lim_{R \to \infty} \left[\left(-\frac{1}{(R+5)} + \frac{5}{2} \cdot \frac{1}{(R+5)^2} \right) - \left(-\frac{1}{5} + \frac{5}{2} \cdot \frac{1}{25} \right) \right]$$

$$= 50 \left(\frac{1}{5} - \frac{1}{10} \right) = 50 \left(\frac{1}{10} \right) = 5$$

Now find the cumulative probability distribution function. When $x < 0$, $F(x) = 0$. When $x \geq 0$, we have:

$$F(x) = \int_{-\infty}^{x} f(t)\,dt = \int_{-\infty}^{0} f(t)\,dt + \int_{0}^{x} f(t)\,dt = 0 + \int_{0}^{x} \frac{50}{(t+5)^3}\,dt$$

$$= -\frac{50}{2}\left(\frac{1}{(t+5)^2} \right) \Big|_{0}^{x} = -25 \left(\frac{1}{(x+5)^2} - \frac{1}{25} \right) = 1 - \frac{25}{(x+5)^2}$$

Thus, $F(x) = \begin{cases} 1 - \dfrac{25}{(x+5)^2} & x \geq 0 \\ 0 & \text{otherwise} \end{cases}$

Next, to find the median, m, we must solve the following for m:

$$F(m) = P(X \leq m) = \frac{1}{2}$$

$$1 - \frac{25}{(m+5)^2} = \frac{1}{2}$$

$$\frac{25}{(m+5)^2} = \frac{1}{2}$$

$$(m+5)^2 = 50$$

$$m + 5 = \sqrt{50}$$

$$m + 5 = 5\sqrt{2}$$

Therefore, the median, m, equals $5\sqrt{2} - 5 \approx 2.071$. $\qquad (15\text{-}3)$

28. $f(x) = \begin{cases} \dfrac{0.8}{x^2} + \dfrac{0.8}{x^5} & \text{if } x \geq 1 \\ 0 & \text{otherwise} \end{cases}$

The cumulative distribution function is:

$F(x) = \begin{cases} 0 & \text{if } x < 1 \\ 1 - \dfrac{0.8}{x} - \dfrac{0.2}{x^4} & \text{if } x \geq 1 \end{cases}$

To find the median m, we solve

$F(m) = P(X \leq m) = \dfrac{1}{2}$ for m:

$1 - \dfrac{0.8}{m} - \dfrac{0.2}{m^4} = \dfrac{1}{2}$

Using a graphing utility, we find that $m \sim 1.68$. (15-3)

29. Consider the integral $\displaystyle\int \dfrac{e^x}{(1 + e^x)^2}\, dx$.

If we let $u = 1 + e^x$, then $du = e^x\, dx$, and $\lim\limits_{x \to \infty} u = \infty$, $\lim\limits_{x \to -\infty} u = 1$. Thus,

$\displaystyle\int_{-\infty}^{\infty} \dfrac{e^x}{(1 + e^x)^2}\, dx = \int_{1}^{\infty} \dfrac{1}{u^2}\, du = \lim_{b \to \infty} \int_{1}^{b} \dfrac{1}{u^2}\, du$

$= \lim\limits_{b \to \infty} \left[-\dfrac{1}{u} \Big|_{1}^{b} \right]$

$= \lim\limits_{b \to \infty} \left[1 - \dfrac{1}{b} \right] = 1$ (15-1)

30. $\displaystyle\int_{-\infty}^{\infty} (ax^2 + bx + c)\, f(x) = a\int_{-\infty}^{\infty} x^2 f(x)\, dx + b\int_{-\infty}^{\infty} x f(x)\, dx + c\int_{-\infty}^{\infty} f(x)\, dx$

$= a(\sigma^2 + \mu^2) + b\mu + c,$

since $\displaystyle\int_{-\infty}^{\infty} x^2 f(x)\, dx = \sigma^2 + \mu^2$, $\displaystyle\int_{-\infty}^{\infty} x f(x)\, dx = \mu$, and $\displaystyle\int_{-\infty}^{\infty} f(x)\, dx = 1$. (15-3)

31. (A) The total production is given by:

$\displaystyle\int_{0}^{\infty} R(t)\, dt = \lim_{T \to \infty} \int_{0}^{T} R(t)\, dt$

$= \lim\limits_{T \to \infty} \int_{0}^{T} (12e^{-0.3t} - 12e^{-0.6t})\, dt$

$= \lim\limits_{T \to \infty} \left[(-40e^{-0.3t} + 20e^{-0.6t}) \Big|_{0}^{T} \right]$

$= \lim\limits_{T \to \infty} (-40e^{-0.3T} + 20e^{-0.6T} + 20) = 20$

Thus, the total production is 20 million barrels.

(B) To find when the well will reach 50% of the total production, we must solve

$\displaystyle\int_{0}^{T} R(t)\, dt = 10$

for T. Now,

$$\int_0^T R(t)\,dt = \int_0^T (12e^{-0.3t} - 12e^{-0.6t})\,dt$$

$$= \left[(-40e^{-0.3t} + 20e^{-0.6t}) \Big|_0^T \right]$$

$$= -40e^{-0.3T} + 20e^{-0.6T} + 20$$

Thus, we have
$$-40e^{-0.3T} + 20e^{-0.6T} + 20 = 10$$
or
$$-40e^{-0.3T} + 20e^{-0.6T} = -10$$

Using a graphing utility to solve this equation, we find that $T \approx 4.09$ years.　　　　　　　　　　　　　　　　　　(15-1)

32. $f(x) = \begin{cases} 0.02(1 - 0.01x) & \text{if } 0 \le x \le 100 \\ 0 & \text{otherwise} \end{cases}$

(A) $\displaystyle\int_{40}^{100} f(x)\,dx = \int_{40}^{100} 0.02(1 - 0.01x)\,dx$

$$= 0.02\left[(x - 0.005x^2) \Big|_{40}^{100} \right]$$

$$= 0.02[(100 - 50) - (40 - 8)] = 0.36$$

The probability that the weekly demand for popcorn is between 40 and 100 pounds is 0.36.

(B) $P(X \le 50) = \displaystyle\int_0^{50} f(x)\,dx = \frac{1}{50}\int_0^{50}(1 - 0.01x)\,dx = \frac{1}{50}(x - 0.005x^2) \Big|_0^{50}$

$$= \frac{1}{50}(50 - 0.005 \cdot 50^2) = 1 - 0.25 = 0.75$$

(C) Solve the following for x:

$$\int_0^x f(t)\,dt = 0.96 \quad (x = \text{number of pounds of popcorn})$$

$$\frac{1}{50}\int_0^x (1 - 0.01t)\,dt = 0.96$$

$$\frac{1}{50}(t - 0.005t^2) \Big|_0^x = 0.96$$

$$\frac{1}{50}(x - 0.005x^2) = 0.96$$

$$x - 0.005x^2 = 48$$

$$5x^2 - 1000x + 48{,}000 = 0$$

$$x^2 - 200x + 9600 = 0$$

$$(x - 80)(x - 120) = 0$$

$$x = 80 \quad \text{or} \quad x = 120$$

Thus, 80 pounds of popcorn must be on hand at the beginning of the week.　　　　　　　　　　　　　　　　　　(15-2)

Capital Value: $CV = \int_0^\infty 2400e^{-0.12t}dt = \lim_{T\to\infty}\int_0^T 2400e^{-0.12t}dt$

$$= \lim_{T\to\infty} 2400\left.\frac{e^{-0.12t}}{-0.12}\right|_0^T$$

$$= \lim_{T\to\infty} -20,000(e^{-0.12T} - 1)$$

$$= \$20,000 \tag{15-1}$$

4. $f(x) = \begin{cases} 6x(1-x) & \text{if } 0 \le x \le 1 \\ 0 & \text{otherwise} \end{cases}$

(A) $P(X \ge 0.2) = 1 - P(X < 0.2)$

$$= 1 - \int_{-\infty}^{0.2} f(x)\,dx$$

$$= 1 - \int_0^{0.2} f(x)\,dx$$

$$= 1 - \int_0^{0.2} 6x(1-x)\,dx$$

$$= 1 - \left[(3x^2 - 2x^3) \Big|_0^{0.2} \right]$$

$$= 1 - (0.12 - 0.016) = 0.896$$

(B) The expected value (mean) is given by:

$$\mu = \int_{-\infty}^\infty xf(x)\,dx = \int_0^1 (6x^2 - 6x^3)\,dx$$

$$= (2x^3 - \frac{3}{2}x^4)\Big|_0^1 = 2 - \frac{3}{2} = 0.5$$

The expected percentage is 50%.

(C) The cumulative distribution function is:

$$F(x) = \begin{cases} 0 & \text{if } x < 0 \\ 3x^2 - 2x^3 & \text{if } 0 \le x \le 1 \\ 1 & \text{if } x > 1 \end{cases}$$

To find the median m, we solve

$$F(m) = P(X \le m) = \frac{1}{2}$$

or $\qquad 3m^2 - 2m^3 = \frac{1}{2}$

for m. Using a graphing utility, we find that $m = 0.5$.
The median percentage is 50%. $\qquad$ (15-2, 15-3)

35. Mean failure time = $\mu = 4000$. As an exponential density function, it is expressed by $\lambda = \mu = 4000$. Thus,

$$f(x) = \begin{cases} \frac{1}{\lambda}e^{-x/\lambda} = \frac{1}{4000}e^{-x/4000} & x \ge 0 \\ 0 & \text{otherwise} \end{cases}$$

The cumulative distribution function is given by:

$$F(x) = \begin{cases} 1 - e^{-x/\lambda} \\ 0 \end{cases} = \begin{cases} 1 - e^{-x/4000} & x \ge 0 \\ 0 & \text{otherwise} \end{cases}$$

(A) $P(X \geq 4000) = \int_{4000}^{\infty} f(x)dx = 1 - F(4000)$

$= 1 - (1 - e^{-4000/4000}) = e^{-1} = 0.3679$

(B) $P(0 \leq X \leq 1000) = \int_{0}^{1000} f(x)dx = F(1000) - F(0)$

$= (1 - e^{-1000/4000}) - (1 - e^0) = 1 - e^{-.25} \approx 0.2212$

(15-4)

36. $\mu = 35,000$, $\sigma = 5,000$

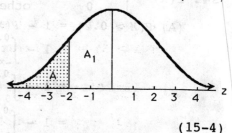

z (for $x = 25,000$) $= \dfrac{25,000 - 35,000}{5,000} = -2$

Required probability $=$ area A

$= 0.5 -$ area A_1

$= 0.5 - 0.4772$

$= 0.0228$

(15-4)

37. $\mu = 100$, $\sigma = 10$

(A) z (for $x = 91.5$) $= \dfrac{91.5 - 100}{10} = -0.85$

z (for $x = 108$) $= \dfrac{108.5 - 100}{10} = 0.85$

The probability of an applicant scoring between 92 and 108

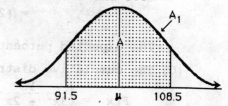

$\quad =$ area A

$\quad = 2 \cdot$ area A_1

$\quad = 2$(area corresponding to $z = 0.85$)

$\quad = 2(0.3023) = 0.6046$

Thus, the percentage of applicants
scoring between 92 and 108 is 60.46%.

(B) z (for $x = 114.5$) $= \dfrac{114.5 - 100}{10} = 1.45$

The probability of an applicant scoring 115 or higher

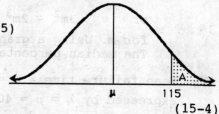

$\quad =$ area A

$\quad = 0.5 -$ (area corresponding to $z = 1.45$)

$\quad = 0.5 - 0.4265$

$\quad = 0.0735$

Thus, the percentage of applicants
scoring 115 or higher is 7.35%.

(15-4)

38. $f(x) = \begin{cases} \dfrac{10}{(x + 10)^2} & \text{if } x \geq 0 \\ 0 & \text{otherwise} \end{cases}$

(A) $\displaystyle\int_{2}^{8} f(x)dx = \int_{2}^{8} \dfrac{10}{(x + 10)^2} dx$

Let $u = x + 10$, then $du = dx$; $u = 12$ when $x = 2$, $u = 18$ when $x = 8$.

Thus,

$$\int_2^8 \frac{10}{(x+10)^2}\,dx = 10\int_{12}^{18} \frac{1}{u^2}\,du = 10\left[-\frac{1}{u}\Big|_{12}^{18}\right]$$

$$= 10\left(\frac{1}{12} - \frac{1}{18}\right) = \frac{5}{18} \approx 0.2778$$

The probability that the shelf-life of the drug is between 2 and 8 months is 0.2778.

(B) Probability that the drug is usable after five months is:

$$P(X > 5) = \int_5^\infty f(x)\,dx = \int_5^\infty \frac{10}{(x+10)^2}\,dx = \lim_{R\to\infty}\int_5^R \frac{10}{(x+10)^2}\,dx$$

$$= \lim_{R\to\infty} 10\left(-\frac{1}{x+10}\right)\Big|_5^R = -10\lim_{R\to\infty}\left(\frac{1}{R+10} - \frac{1}{5+10}\right)$$

$$= -10\left(-\frac{1}{15}\right) = \frac{2}{3} \approx 0.6667$$

(C) In order to find the median, m, we must solve the following for m:

$$P(X \le m) = \int_0^m f(x)\,dx = \frac{1}{2}$$

$$\int_0^m \frac{10}{(x+10)^2}\,dx = \frac{1}{2}$$

$$-\frac{10}{(x+10)}\Big|_0^m = \frac{1}{2}$$

$$-\left(\frac{10}{m+10} - 1\right) = \frac{1}{2}$$

$$-\frac{10}{m+10} + 1 = \frac{1}{2}$$

$$-\frac{10}{m+10} = -\frac{1}{2}$$

$$m + 10 = 20$$

$$m = 10 \text{ months} \hspace{3cm} (15\text{-}2,\ 15\text{-}3)$$

39. $f(x) = \begin{cases} \dfrac{1}{\lambda}e^{-x/\lambda} & x \ge 0 \\ 0 & \text{otherwise} \end{cases}$

$$P(X > 1) = \frac{1}{\lambda}\int_1^\infty e^{-x/\lambda}\,dx = e^{-2} \quad \text{(Given)}$$

Thus, $\quad \dfrac{1}{\lambda}\lim_{R\to\infty}\int_1^R e^{-x/\lambda}\,dx = e^{-2}$

$$\frac{1}{\lambda}(-\lambda)\lim_{R\to\infty}(e^{-x/\lambda})\Big|_1^R = e^{-2}$$

$$-1\cdot\lim_{R\to\infty}(e^{-R/\lambda} - e^{-1/\lambda}) = e^{-2}$$

$$e^{-1/\lambda} = e^{-2}$$

Thus,

$$-\frac{1}{\lambda} = -2$$

$$\lambda = \frac{1}{2}$$

Therefore, $f(x)$, with $\lambda = \frac{1}{2}$, is given by $f(x) = \begin{cases} 2e^{-2x} & x \geq 0 \\ 0 & \text{otherwise} \end{cases}$

(A) $P(X > 2) = \int_2^\infty f(x)\,dx = \int_2^\infty 2e^{-2x}\,dx = 2\lim_{R\to\infty} \int_2^R e^{-2x}\,dx$

$$= 2\lim_{R\to\infty}\left(-\frac{1}{2}e^{-2x}\Big|_2^R\right) = -\lim_{R\to\infty}(e^{-2R} - e^{-4}) = -(-e^{-4}) = e^{-4} \approx 0.0183$$

(B) Mean life expectancy:

$$\mu = \int_0^\infty xf(x)\,dx = \int_0^\infty x \cdot 2e^{-2x}\,dx$$

$$= 2\lim_{R\to\infty}\int_0^R xe^{-2x}\,dx \quad \text{(integration by parts; } u = x,\ dv = e^{-2x}\,dx\text{)}$$

$$= 2\lim_{R\to\infty}\left[x\left(-\frac{1}{2}e^{-2x}\right)\Big|_0^R + \frac{1}{2}\int_0^R e^{-2x}\,dx\right] = 2\lim_{R\to\infty}\left(-\frac{1}{2}xe^{-2x} - \frac{1}{4}e^{-2x}\right)\Big|_0^R$$

$$= -2\lim_{R\to\infty}\left[\left(\frac{1}{2}Re^{-2R} + \frac{1}{4}e^{-2R}\right) - \left(0 + \frac{1}{4}\right)\right]$$

$$= -2\left(-\frac{1}{4}\right) = \frac{1}{2} \text{ or } 0.5 \text{ month } or\ \mu = \lambda = \frac{1}{2} \qquad (15\text{-}2,\ 15\text{-}3)$$

40. $R(t) = 15e^{-0.2t} - 15e^{-0.3t}$

(A) The total amount of the drug that is eliminated by the body is given by:

$$\int_0^\infty R(t)\,dt = \lim_{T\to\infty}\int_0^T R(t)\,dt$$

$$= \lim_{T\to\infty}\int_0^T (15e^{-0.2t} - 15e^{-0.3t})\,dt$$

$$= \lim_{T\to\infty}\left[(-75e^{-0.2t} + 50e^{-0.3t})\Big|_0^T\right]$$

$$= \lim_{T\to\infty}[-75e^{-0.2T} + 50e^{-0.3T} + 25]$$

$$= 25 \text{ milliliters}$$

(B) To find how long it will take for 50% of the drug to be eliminated, we solve

$$\int_0^T R(t)\,dt = 0.5(25) = 12.5$$

for T:

$$\int_0^T (15e^{-0.2t} - 15e^{-0.3t})\,dt = 12.5$$

$$\left[(-75e^{-0.2t} + 50e^{-0.3t})\Big|_0^T\right] = 12.5$$

$$-75e^{-0.2T} + 50e^{-0.3T} + 25 = 12.5$$
$$-75e^{-0.2T} + 50e^{-0.3T} + 12.5 = 0$$

Using a graphing utility, we find that $T \approx 6.93$ hours.   (15-1)

41. $\mu = 108$, $\sigma = 12$

$$z = \frac{135 - 108}{12} = \frac{27}{12} = 2.25$$

$$P(X \geq 135) = P(Z \geq 2.25) = 0.5 - P(Z < 2.25)$$
$$= 0.5 - 0.4878$$
$$= 0.0122;$$

1.22% of the children can be expected to have IQ scores of 135 or more.
          (15-4)

42. $N'(t) = \dfrac{100t}{(1 + t^2)^2}$

Therefore, $N(t) = \displaystyle\int \dfrac{100t}{(1 + t^2)^2}\,dt = 50 \int \dfrac{1}{u^2}\,du$ Substitution: $u = 1 + t^2$
$$du = 2t\,dt$$

$$= \frac{50u^{-1}}{-1} + C = \frac{-50}{1 + t^2} + C$$

Now, $N(0) = \dfrac{-50}{1 + 0} + C$. Thus, $C = N(0) + 50$ and

$$N(t) = \frac{-50}{1 + t^2} + N(0) + 50$$

$$N(3) = \frac{-50}{1 + 3^2} + N(0) + 50 = N(0) + 45$$

Therefore, the voting population will increase by 45 thousand in 3 years. If the population grows indefinitely at this rate, then

$$\lim_{t \to \infty} N(t) = \lim_{t \to \infty} \left(\frac{-50}{1 + t^2} + N(0) + 50 \right) = N(0) + 50$$

Therefore, the total increase in voting population is 50 thousand.
          (15-1)